D0782614

Mobile DNA II

Mobile DNA II

EDITED BY

Nancy L. Craig

Howard Hughes Medical Institute and Department of Molecular Biology and Genetics
Johns Hopkins University School of Medicine
Baltimore, Maryland

Robert Craigie

Laboratory of Molecular Biology
National Institute of Diabetes and Digestive and Kidney Diseases
National Institutes of Health
Bethesda, Maryland

Martin Gellert

Laboratory of Molecular Biology
National Institute of Diabetes and Digestive and Kidney Diseases
National Institutes of Health
Bethesda, Maryland

Alan M. Lambowitz

Institute for Cellular and Molecular Biology,
Department of Chemistry and Biochemistry,
and Section of Molecular Genetics and Microbiology, School of Biological Sciences
The University of Texas
Austin, Texas

WASHINGTON, D.C.

Address editorial correspondence to ASM Press, 1752 N Street NW, Washington, DC 20036, USA

Send orders to ASM Press, P.O. Box 605, Herndon, VA 20172, USA
Phone: (800) 546-2416 or (703) 661-1593
Fax: (703) 661-1501
E-mail: books@asmusa.org
Online: www.asmpress.org

Library of Congress Cataloging-in-Publication Data

Mobile DNA II / edited by Nancy L. Craig . . . [et al.].
p. cm.
Includes bibliographical references and index.
ISBN 1-55581-209-0
1. Mobile genetic elements. I. Title: Mobile DNA 2. II. Title: Mobile DNA two. III. Craig, Nancy L.

QH452.3.M632 2001
572.8′69—dc21

2001045975

10 9 8 7 6 5 4 3 2 1

Cover illustration: Cre recombinase bound to the Holliday junction intermediate formed during site-specific recombination between 34-bp *loxP* sites, as observed in a crystal structure of the complex. The complex is similar to those described by Gopaul et al. (D. N. Gopaul, F. Guo, and G. D. Van Duyne, *EMBO J.* **17:** 4175–4187, 1998). Courtesy of Gregory D. Van Duyne.

CONTENTS

CONTRIBUTORS

R. Alazard
Laboratoire de Microbiologie et Génétique Moléculaires, CNRS, 118 Route de Narbonne, F-31062 Toulouse Cedex, France

Michael Aye
Department of Biological Chemistry, University of California, Irvine, Irvine, CA 92697-1700

Marco A. Azaro
Department of Molecular Biology, Cell Biology, and Biochemistry, Brown University, Providence, RI 02912

Tania A. Baker
Department of Biology and Howard Hughes Medical Institute, Massachusetts Institute of Technology, Cambridge, MA 02139

Alan Barbour
Departments of Microbiology and Molecular Genetics and of Medicine, University of California, Irvine, CA 92697-4025

François-Xavier Barre
Division of Molecular Genetics, Department of Biochemistry, University of Oxford, South Parks Rd., Oxford OX1 3QU, United Kingdom

Marlene Belfort
Molecular Genetics Program, Wadsworth Center, New York State Department of Health and School of Public Health, State University of New York at Albany, Albany, NY 12201-2002

Irantzu Bernales
Departamento de Biología Molecular, Universidad de Cantabria (Laboratorio Asociado al C.I.B., C.S.I.C.), C. Herrera Oria s/n, 39011 Santander, Spain

Jef D. Boeke
Department of Molecular Biology and Genetics, Johns Hopkins School of Medicine, 617 Hunterian Building, 725 N. Wolfe St., Baltimore, MD 21205-2185

Piet Borst
Division of Molecular Biology, The Netherlands Cancer Institute, Plesmanlaan 121, 1066 CX Amsterdam, The Netherlands

Kim Brügger
Institute of Molecular Biology, University of Copenhagen, Sølvgade 83H, DK-1307 Copenhagen K, Denmark

Alain Bucheton
Institut de Génétique Humaine, CNRS, 34396 Montpellier Cedex 5, France

Isabelle Busseau
Institut de Génétique Humaine, CNRS, 34396 Montpellier Cedex 5, France

Allan Campbell
Department of Biological Sciences, Stanford University, Stanford, CA 94305

George Chaconas
Department of Biochemistry, University of Western Ontario, London, Ontario N6A 5C1, Canada

Douglas L. Chalker
Division of Basic Sciences, Fred Hutchinson Cancer Research Center, Seattle, WA 98109

Michael Chandler
Laboratoire de Microbiologie et Génétique Moléculaires, CNRS, 118 Route de Narbonne, F-31062 Toulouse Cedex, France

Gordon Churchward
Department of Microbiology and Immunology, Emory University School of Medicine, Atlanta, GA 30322

Victor G. Corces
Department of Biology, The Johns Hopkins University, 3400 N. Charles St., Baltimore, MD 21218

Benoit Cousineau
Département de biochimie, Faculté de médecine, Université de Sherbrooke (CHUS), 3001, 12e Avenue Nord, Fleurimont, Québec J1H 5N4, Canada

Nancy L. Craig
Howard Hughes Medical Institute and Department of Molecular Biology and Genetics, 615 PCTB, Johns Hopkins School of Medicine, 725 N. Wolfe St., Baltimore, MD 21205

Robert Craigie
Laboratory of Molecular Biology, National Institute of Diabetes and Digestive and Kidney Diseases, National Institutes of Health, Bethesda, MD 20892

Brigid M. Davis
Howard Hughes Medical Institute and Division of Geographic Medicine and Infectious Diseases, New England Medical Center and Tufts University School of Medicine, Boston, MA 02111

P. G. DeBaryshe
Department of Biology, Massachusetts Institute of Technology, Cambridge, MA 02139

Prescott L. Deininger
Tulane Cancer Center and Department of Environmental Health Sciences, Tulane University Medical Center, 1430 Tulane Ave., New Orleans, LA 70112

Fernando de la Cruz
Departamento de Biología Molecular, Universidad de Cantabria (Laboratorio Asociado al C.I.B., C.S.I.C.), C. Herrera Oria s/n, 39011 Santander, Spain

Victoria Derbyshire
Molecular Genetics Program, Wadsworth Center, New York State Department of Health and School of Public Health, State University of New York at Albany, Albany, NY 12201-2002

Sandra Duharcourt
Division of Basic Sciences, Fred Hutchinson Cancer Research Center, Seattle, WA 98109

G. Duval-Valentin
Laboratoire de Microbiologie et Génétique Moléculaires, CNRS, 118 Route de Narbonne, F-31062 Toulouse Cedex, France

Thomas H. Eickbush
Department of Biology, University of Rochester, Rochester, NY 14627-0211

David Faguy
Department of Biology, University of New Mexico, Albuquerque, NM 87131-1091

Nina Fedoroff
Biology Department and Life Sciences Consortium, 519 Wartik Laboratory, The Pennsylvania State University, University Park, PA 16802

Cédric Feschotte
Departments of Botany and Genetics, University of Georgia, Athens, GA 30602

M. Pilar Garcillán-Barcia
Departamento de Biología Molecular, Universidad de Cantabria (Laboratorio Asociado al C.I.B., C.S.I.C.), C. Herrera Oria s/n, 39011 Santander, Spain

Roger A. Garrett
Institute of Molecular Biology, University of Copenhagen, Sølvgade 83H, DK-1307 Copenhagen K, Denmark

Martin Gellert
Laboratory of Molecular Biology, National Institute of Diabetes and Digestive and Kidney Diseases, National Institutes of Health, Bethesda, MD 20892-0540

Nicolas Gilbert
Departments of Human Genetics and Internal Medicine, The University of Michigan Medical School, Ann Arbor, MI 48109-0618

Ian Grainge
Section of Molecular Genetics and Microbiology and Institute of Cellular and Molecular Biology, University of Texas at Austin, Austin, TX 78712

Nigel D. F. Grindley
Department of Molecular Biophysics and Biochemistry, Yale University, New Haven, CT 06520-8114

James E. Haber
Rosenstiel Center, Brandeis University, Waltham, MA 02454-9110

David B. Haniford
Department of Biochemistry, University of Western Ontario, London, Ontario N6A 5C1, Canada

Rasika M. Harshey
Section of Molecular Genetics and Microbiology and Institute of Cellular and Molecular Biology, University of Texas at Austin, Austin, TX 78712

Masayori Inouye
Department of Biochemistry, Robert Wood Johnson Medical School, University of Medicine and Dentistry of New Jersey, 675 Hoes Ln., Piscataway, NJ 08854

Sumiko Inouye
Department of Biochemistry, Robert Wood Johnson Medical School, University of Medicine and Dentistry of New Jersey, 675 Hoes Ln., Piscataway, NJ 08854

Makkuni Jayaram
Section of Molecular Genetics and Microbiology and Institute of Cellular and Molecular Biology, University of Texas at Austin, Austin, TX 78712

Reid C. Johnson
Department of Biological Chemistry, UCLA School of Medicine, Los Angeles, CA 90095-1737

Margaret G. Kidwell
Department of Ecology and Evolutionary Biology, The University of Arizona, Tucson, AZ 85721

Robert M. Kotin
Laboratory of Biochemical Genetics, National Heart, Lung, and Blood Institute, Bethesda, MD 20892

Reinhard Kunze
Botany Institute II, University of Cologne, 50931 Cologne, Germany

Mariano Labrador
Department of Biology, The Johns Hopkins University, 3400 N. Charles St., Baltimore, MD 21218

Alan M. Lambowitz
Institute for Cellular and Molecular Biology, Department of Chemistry and Biochemistry, and Section of Molecular Genetics and Microbiology, School of Biological Sciences, University of Texas at Austin, Austin, TX 78712

Arthur Landy
Department of Molecular Biology, Cell Biology, and Biochemistry, Brown University, Providence, RI 02912

Henry L. Levin
Laboratory of Eukaryotic Gene Regulation, National Institute of Child Health and Human Development, National Institutes of Health, Bethesda, MD 20892

Damon R. Lisch
Department of Plant and Microbial Biology, University of California, Berkeley, Berkeley, CA 94720

C. Loot
Laboratoire de Microbiologie et Génétique Moléculaires, CNRS, 118 Route de Narbonne, F-31062 Toulouse Cedex, France

Jacques Mahillon
Laboratoire de Génétique Microbienne, Université catholique de Louvain, Place Croix du Sud, 2 Bte 12, B-1348 Louvain-la-Neuve, Belgium

Harmit S. Malik
Fred Hutchinson Cancer Research Center, Seattle, WA 98109

M. Victoria Mendiola
Departamento de Biología Molecular, Universidad de Cantabria (Laboratorio Asociado al C.I.B., C.S.I.C.), C. Herrera Oria s/n, 39011 Santander, Spain

Thomas Menees
Division of Cell Biology and Biophysics, University of Missouri—Kansas City, 2411 Holmes St., Kansas City, MO 64108

Kiyoshi Mizuuchi
Laboratory of Molecular Biology, National Institute of Diabetes and Digestive and Kidney Diseases, National Institutes of Health, Bethesda, MD 20892

John V. Moran
Departments of Human Genetics and Internal Medicine, The University of Michigan Medical School, Ann Arbor, MI 48109-0618

C. Normand
Laboratoire de Microbiologie et Génétique Moléculaires, CNRS, 118 Route de Narbonne, F-31062 Toulouse Cedex, France

Mary-Lou Pardue
Department of Biology, Massachusetts Institute of Technology, Cambridge, MA 02139

Monica M. Parker
Molecular Genetics Program, Wadsworth Center, New York State Department of Health and School of Public Health, State University of New York at Albany, Albany, NY 12201-2002

Ronald H. A. Plasterk
Hubrecht Laboratory, Uppsalalaan 8, 3584 CT Utrecht, The Netherlands

Gavin D. Recchia
Freehills Carter Smith Beadle, Sydney, New South Wales 2000, Australia

Peter Redder
Institute of Molecular Biology, University of Copenhagen, Sølvgade 83H, DK-1307 Copenhagen K, Denmark

William S. Reznikoff
Department of Biochemistry, University of Wisconsin, 433 Babcock Dr., Madison, WI 53706

Phoebe A. Rice
Department of Biochemistry and Molecular Biology, The University of Chicago, 920 E. 58th St., Chicago, IL 60637

Donald C. Rio
Department of Molecular and Cell Biology, University of California, Berkeley, Berkeley, CA 94720-3204

Hugh M. Robertson
Department of Entomology, University of Illinois, 320 Morrill Hall, MC118, 505 S. Goodwin, Urbana, IL 61801

P. Rousseau
Laboratoire de Microbiologie et Génétique Moléculaires, CNRS, 118 Route de Narbonne, F-31062 Toulouse Cedex, France

Astrid M. Roy-Engel
Tulane Cancer Center and Department of Environmental Health Sciences, Tulane University Medical Center, 1430 Tulane Ave., New Orleans, LA 70112

George N. Rudenko
Department of Biological Sciences, Stanford University, Stanford, CA 94305-5020

Suzanne B. Sandmeyer
Department of Biological Chemistry, University of California, Irvine, Irvine, CA 92697-1700

Brian Sauer
Developmental Biology Program, Oklahoma Medical Research Foundation, Oklahoma City, OK 73104

Qunxin She
Institute of Molecular Biology, University of Copenhagen, Sølvgade 83H, DK-1307 Copenhagen K, Denmark

David J. Sherratt
Division of Molecular Genetics, Department of Biochemistry, University of Oxford, South Parks Rd., Oxford OX1 3QU, United Kingdom

Tadashi Shimamoto
Faculty of Applied Biological Science, Hiroshima University, 1-4-4 Kagamiyama, Higashi-Hiroshima 739-8528, Japan

Richard H. Smith, Jr.
Laboratory of Biochemical Genetics, National Heart, Lung, and Blood Institute, Bethesda, MD 20892

Danielle Teninges
Institut de Génétique Humaine, CNRS, 34396 Montpellier Cedex 5, France

Gena Tribble
Section of Molecular Genetics and Microbiology and Institute of Cellular and Molecular Biology, University of Texas at Austin, Austin, TX 78712

C. Turlan
Laboratoire de Microbiologie et Génétique Moléculaires, CNRS, 118 Route de Narbonne, F-31062 Toulouse Cedex, France

Gregory D. Van Duyne
Department of Biochemistry and Biophysics and Howard Hughes Medical Institute, University of Pennsylvania School of Medicine, Philadelphia, PA 19104

Henri G. A. M. van Luenen
Division of Molecular Biology, The Netherlands Cancer Institute, Plesmanlaan 121, 1066 CX Amsterdam, The Netherlands

Daniel F. Voytas
Department of Zoology and Genetics, Iowa State University, Ames, IA 50011

Virginia Walbot
Department of Biological Sciences, Stanford University, Stanford, CA 94305-5020

Matthew K. Waldor
Howard Hughes Medical Institute and Division of Geographic Medicine and Infectious Diseases, New England Medical Center and Tufts University School of Medicine, Boston, MA 02111

Clifford F. Weil
Department of Agronomy, 1150 Lilly Hall, Purdue University, West Lafayette, IN 47907-1150

Susan R. Wessler
Departments of Botany and Genetics, University of Georgia, Athens, GA 30602

Kunitoshi Yamanaka
Department of Biochemistry, Robert Wood Johnson Medical School, University of Medicine and Dentistry of New Jersey, 675 Hoes Ln., Piscataway, NJ 08854

Meng-Chao Yao
Division of Basic Sciences, Fred Hutchinson Cancer Research Center, Seattle, WA 98109

Xiaoyu Zhang
Departments of Botany and Genetics, University of Georgia, Athens, GA 30602

PREFACE

When *Mobile DNA* (1) was published in 1989, it would have been hard to anticipate the speed and scope of the advances that would take place in the study of mobile DNAs. For example, we could not have guessed that by the time *Mobile DNA II* was published, transposable DNAs would have been used in human gene therapy. Moreover, we now have genomewide views of the propagation of these elements and their impact on genome evolution. The exciting and significant advances that have been made in a number of areas since the publication of *Mobile DNA* fully warrant the publication of another monograph as a completely new volume, with an even wider range of subjects to accommodate significant new work. We hope that *Mobile DNA II* will convey the excitement of those working in this field and facilitate the dissemination of information during this time when fundamental advances are being made. *Mobile DNA,* with its comprehensive information about virtually all mobile DNAs known at the time of its publication, will, of course, remain an important resource.

The collection of chapters in this new volume will be useful to those who study mobile DNAs. The book should also provide a valuable resource for those who work in areas to which the study of site-specific recombination has contributed: protein-nucleic acid interactions; DNA and chromosome structure; DNA replication, repair, and recombination; RNA splicing; and evolution. Also included are several chapters that explicitly focus on new uses of site-specific recombination for the manipulation of genes and chromosomes. We have also included an introduction that provides an overview of the many different types of recombination discussed in this volume as well as a chapter that focuses on the analysis of the chemical steps in a number of recombination reactions.

As revealed in many of the chapters, the analysis and dissection of a number of recombination systems have now moved far towards the mechanistic and structural level. To a degree unanticipated in *Mobile DNA,* work summarized here has revealed striking mechanistic similarities between transposition reactions involving elements that move exclusively by DNA intermediates and retrovirus-like elements for which the DNA substrate for recombination is generated via an RNA intermediate. The proteins that drive these related chemical reactions also share structural features. Indeed, the insights into human immunodeficiency virus DNA integration provided by these studies have led to new opportunities to identify inhibitors that may prove useful in blocking AIDS progression.

Another significant advance has come in understanding V(D)J recombination, a process that has become biochemically tractable only since *Mobile DNA* was published. A striking finding is that some of the chemical reactions that underlie V(D)J recombination are closely related in mechanism to the reactions that underlie certain transposition reactions. These findings raise interesting questions about the evolution of the immune system. It will be exceptionally interesting to know whether the RAG proteins that promote V(D)J recombination are also similar in structure to the transposases.

There have also been significant and striking advances in the dissection of elements that move via RNA intermediates, such as non-long terminal repeat elements and group II introns. The demonstration that group II introns can translocate by reverse splicing provides a mechanism to understand the spread of elements that were the likely progenitors of eukaryotic spliceosomal introns.

It has long been known that mobile elements are widespread and can influence chromosome structure and function. In addition to understanding how mobile elements can move, an important issue to understand is how their propagation can be limited. The full extent of the involvement of mobile DNAs in chromosome structure and function has only recently been revealed by genomic analysis. Indeed, we now appreciate that about 45% of the human genome is composed of mobile DNAs, most of which (luckily for us!) are relics incapable of further movement. The study of bacterial genomes in particular has revealed an amazing plasticity and dynamic structure that are often mediated by mobile DNA elements. There have also been significant advances in our understanding of the contributions of mobile DNAs to pathogenicity and to bacterial antibiotic resistance.

An indicator of the pace at which research on mobile DNAs is progressing is that several significant new superfamilies of transposons have been identified during the production of this book (see chapter 50). Miniature inverted-repeat transposable elements (MITEs), which are nonautonomous elements, are widespread in plants and animals (3). Although their transposition mechanism has not been directly established, a reasonable hypothesis is that they transpose via DNA-based breakage and joining reactions. Another significant class of mobile DNAs called helitrons is widespread in plant genomes (2); these DNAs likely move by a rolling-circle mechanism similar to that used by the bacterial IS*91* element (see chapter 37).

We are, of course, grateful to all the authors for their efforts in producing these outstanding chapters. It is the work that they describe that has made the publication of this volume appropriate and important. We are also very grateful to those who assisted in the production of *Mobile DNA II:* Patricia Eckhoff (Johns Hopkins University School of Medicine), who so ably managed the collection and distribution of the manuscripts; Gregory Payne (Senior Editor, ASM Press), who started this project; and Ken April (Production Editor, ASM Press), who brought the book to its final form.

Nancy L. Craig
Robert Craigie
Martin Gellert
Alan M. Lambowitz

REFERENCES

1. **Berg, D. E., and M. M. Howe (ed.).** 1989. *Mobile DNA*. American Society for Microbiology, Washington, D.C.
2. **Kapitonov, V. V., and J. Jurka.** 2001. Rolling-circle transposons in eukaryotes. *Proc. Natl. Acad. Sci. USA* 98:8714–8719.
3. **Zhang, X., C. Feschotte, Q. Zhang, N. Jiang, W. B. Eggleston, and S. R. Wessler.** 2001. P instability factor: an active maize transposon system associated with the amplification of Tourist-like MITEs and a new superfamily of transposases. *Proc. Natl. Acad. Sci. USA* 98:12572–12577.

PREFACE TO *MOBILE DNA* (1989)

This book documents the remarkable mobility of DNAs in procaryotic and eucaryotic genomes: the ability of various DNA segments to move to new sites, to invert, and to undergo deletion or amplification, generally without the extensive DNA sequence homology needed for classical recombination. Its chapters describe the variety of mechanisms by which these rearrangements occur, how they are regulated, their biological consequences, and ways in which transposable elements can be exploited as potent research tools. The many examples of mobile DNAs now known provide important exceptions to the general principle that gene and DNA sequence arrangements are stable and transmitted with great fidelity from parents to progeny. Some rearrangements, such as the movement of bacterial antibiotic resistance transposons, variation in surface antigen genes in certain pathogens, and gene amplification in mammalian cells, are typically quite rare, but are brought into prominence by selective pressure in large populations. At the other extreme, antibody gene rearrangements, DNA elimination and amplification in ciliated protozoans, integration of phage λ, and transposition of phage Mu can all be very frequent, but are regulated developmentally or restricted to specific cell types or nuclei.

The reader will be struck by the great diversity of mobile DNAs. *Escherichia coli* K-12 provides a particularly graphic example: the original strain contained one or more copies of at least nine different types of insertion (IS) sequences, a specific inversion system, prophage λ, and several other complete and cryptic prophages. Dozens of additional IS sequences and transposons found in other bacteria can also move in *E. coli* K-12. Spontaneous deletion events, probably involving several mechanisms, can eliminate various bacterial DNA segments including transposable elements. No less striking are the relationships among mobile DNAs in quite disparate organisms. In procaryotes there are many types of elements related to the ampicillin resistance transposon Tn*3*, both in gram-negative and gram-positive bacteria. In eucaryotes, the Ty and *copia* "retrotransposons" of *Saccharomyces cerevisiae* and *Drosophila melanogaster* are closely related to the retroviruses of vertebrates; LINE and SINE elements are found in diverse mammals and, like retrotransposons, use reverse transcriptases to proliferate, although they probably insert by other mechanisms. In addition, the I element of *D. melanogaster* resembles the mammalian LINEs in features of DNA sequence, but is similar to more conventional transposable elements in the regulation of its movement.

The reader will also find great diversity at the mechanistic level. Many events involve DNA breakage and joining, with little or no DNA synthesis, and are well understood, for example, integration and excision of phage λ; the resolution of cointegrates formed during gamma-delta and Tn*3* transposition; and specific DNA inversions that cause flagellar phase variation in *Salmonella typhimurium* and changes in the host range of phage Mu, and which also help drive replication of *S. cerevisiae* 2μm circle DNA. Nonreplicative cut-and-paste mechanisms are apparently involved in the transposition of Tn*5* and Tn*10* in bacteria and *Ac* in maize, and in the insertion of DNA copies of retroviruses and retrotransposons in diverse animal and plant species. Specific DNA breakage underlies macronuclear maturation in protozoa, development of the immune system, and some DNA rearrangements in fungal mitochondria. A paradigm for transposition events intimately linked with DNA replication is provided by the lytic growth of phage Mu. DNA synthesis also figures importantly in many spontaneous deletion and amplification events, mating type interconversion in yeasts, and perhaps changes in surface antigen genes in certain pathogens. The latter two phenomena also involve mechanisms similar to gene conversion seen in classical homologous recombination.

For us and many colleagues, the mobile DNA field is rooted in Barbara McClintock's seminal discovery of transposable controlling elements in maize. Using purely genetic analyses in the 1940s, she demonstrated that there are elements that can transpose to new chromosomal locations, alter the expression of nearby genes, and cause chromosome breakage, all in a developmentally regulated fashion. The prescience of her discoveries is all the more remarkable in that they were made without knowledge of DNA structure or the current recombinant DNA technologies. They preceded by several decades the discoveries of phage Mu, the bacterial insertion sequences and resistance transposons, and the mobile elements of *D. melanogaster* and other eucaryotes. Much of the initial excitement generated by discoveries of these other elements stemmed from the idea that they might be equivalent to McClintock's intriguing controlling elements. But the current interest in mobile DNA also comes from many other research areas: basic DNA structure, the nature of repetitive DNAs, the causes of mutation, the principles of genome evolution, the control of gene expression, the operation of developmental circuits, the virulence mechanisms of pathogenic microbes, the spread of antibiotic resistance in bacteria, and the emergence of drug resistance in human tumors.

This book is intended to serve scientists with diverse interests and a wide range of experience, from students just discovering the excitement and power of modern molecular genetics to teachers and senior researchers. The chapters detail phenomena from a great range of organisms: humans, their pathogens and parasites, bacteria and their viruses, yeasts and other fungi, protozoans, nematodes, fruit flies, frogs, mice, maize, slime molds, and snapdragons. The chapters differ in length and detail, reflecting variously the sophisticated understanding of intensively studied elements and the tantalizing early findings of emerging elements or systems. Each chapter provides insights into the current state of our knowledge and points to future research directions.

We are indebted to the authors for their informative, exciting and timely texts; to many colleagues worldwide for helpful critiques of individual chapters; and to members of the ASM Publications Department (Kirk Jensen, Dennis Burke, and Ellie Tupper) for their guidance and superb management of the editing and production process.

Douglas E. Berg
Martha M. Howe

I. INTRODUCTION TO MOBILE DNAs: WHAT THEY ARE, THEIR CELLULAR ROLES, HOW THEY MOVE, AND HOW TO EXPLOIT THEM

Mobile DNA II
Edited by N. L. Craig et al.

Chapter 1

Mobile DNA: an Introduction

NANCY L. CRAIG

INTRODUCTION

The once revolutionary idea that genomes contain segments of mobile DNA was actually first conceptualized before the discovery of the structure of DNA. Mobile DNA was first suggested by McClintock when she described "controlling elements" in maize (24–27). In the late 1940s, she noticed that there were genetic determinants in maize chromosomes that resulted in chromosomal breaks, hence their name "dissociater" (Ds), and found that these elements could move from place to place within the genome. She also discovered other specific determinants called Ac for Activator that could control the activity of Ds and that the Ac elements could also move from place to place. These studies actually provided two powerful ideas: (i) that genetic determinants could be mobile and (ii) that there were specific controllers of gene expression.

Other studies in the 1950s led to the realization that pieces of DNA that sometimes existed as plasmids or viruses could also insert into or excise from the bacterial chromosome. The transfer of bacterial markers between strains, i.e., bacterial sex, was determined to result from the integration of the fertility factor plasmid into the chromosome and subsequent transfer of chromosomal determinants to another cell (15, 22). It was also realized that bacteriophage lambda integrated into and excised from the bacterial chromosome (4). The transfer of certain kinds of unusual mutations in well-studied bacterial genes to bacterial viruses allowed the direct examination of their DNA by electron microscopy to show that these mutations resulted from the insertion of discrete mobile segments (16, 31), even in the days preceding molecular cloning (3).

Mobile DNA segments were also recognized by the discovery of antibiotic resistance determinants that could translocate between bacterial chromosomes and plasmids (7, 10, 11, 17). We now know that antibiotic resistance genes are often carried by transposable DNA segments (3).

Today, of course, we appreciate that a wide variety of different types of DNA move from place to place and that these recombination reactions underlie many different types of biological transactions. We also know that this mobility occurs by a variety of mechanisms (see chapter 2). The recombination reactions discussed in this volume focus on site-specific recombination reactions that involve the action of specialized DNA recombinases at specific sites in DNA. Such reactions are generally (but not always entirely) independent of the host's homologous recombination machinery. Mechanistic issues in all reactions include how the recombination sites are recognized by the recombination proteins, how the higher-order protein-DNA complexes that underlie many reactions are formed, how synapsis or juxtaposition of

Nancy L. Craig • Howard Hughes Medical Institute and Department of Molecular Biology and Genetics, 615 PCTB, Johns Hopkins School of Medicine, 725 N. Wolfe St., Baltimore, MD 21205.

recombination sites occurs, how strand exchange is executed, and how recombination is controlled.

These topics were discussed previously in *Mobile DNA* (1), which was written at a time when the mechanistic analysis of many site-specific recombination systems was just beginning. A notable feature of the work presented in this volume is that our understanding of many of these reactions has moved to the biochemical and structural level. However, it should be appreciated, especially as manifested in the chapters on genome structure and mobile DNA, that there are many more mobile DNAs to identify and understand.

CONSERVATIVE SITE-SPECIFIC RECOMBINATION

It is appropriate to begin a discussion of site-specific recombination reactions by considering conservative site-specific recombination. This is the kind of recombination reaction that underlies the integration/excision cycle of the bacteriophage lambda into and out of the *Escherichia coli* chromosome (see chapter 7). Genetic studies of these reactions in the 1950s provided one of the earliest examples of genome modification by a mobile DNA (4). Further extensive genetic analysis in the 1960s and 1970s identified and characterized the proteins and DNA sites involved in these reactions (12, 13). The lambda integration system was the first system for which an in vitro recombination reaction was established (28) and for which a purified system was developed (29). Having been localized to relatively small DNA segments by genetic means, the lambda recombination sites were actually among the first pieces of DNA to be sequenced (21). This sequencing spectacularly confirmed what was predicted from genetic studies, that strand exchange occurred in a short region of homology shared among the recombination sites and that each arm flanking a recombination site was distinct. Extensive genetic, biochemical, and structural analyses of the lambda system as well as other reactions in this category have provided deep insights into how these reactions occur.

We now know that conservative site-specific recombination reactions underlie a variety of biological transactions including various aspects of chromosome mechanics such as the integration/excision cycles of bacteriophages and other genetic elements (see chapters 9 and 10), the reduction of bacterial chromosome and plasmid dimers to monomers to facilitate the segregation of replicons to daughter cells (see chapters 6 and 8), the separation of different replicons (see chapter 14), and the control of gene expression (see chapter 13) (32) and DNA replication (see chapters 11 and 12) through the repositioning of controlling elements

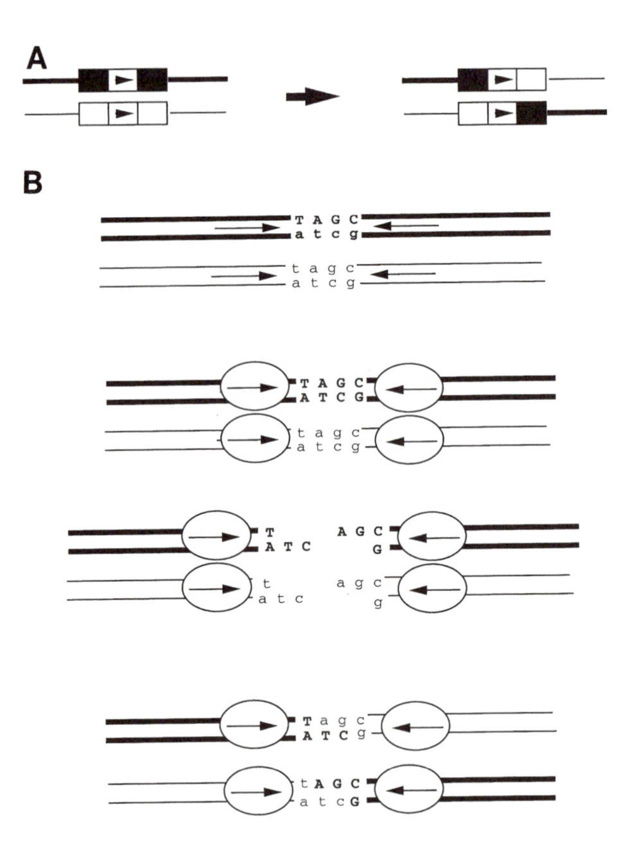

Figure 1. Mechanism of conservative site-specific recombination. (A) Conservative site-specific recombination between two sites that share a region of homology (arrow) and related flanking sequences. Recombination within the region of homology results in the reciprocal exchange of the segments that flank the region of homology. (B) Conservative site-specific recombination at higher resolution. There is a short region of homology between the two recombination sites that is flanked by recombinase binding sites (arrows) in inverted orientation at each edge of the core region. When the recombinase (ovals) binds to these sites, it is symmetrically positioned on each side of the core. The recombinase executes breakage, exchange, and rejoining reaction within the core to generate recombinants containing a region of heteroduplex resulting from the staggered positions of breakage and rejoining.

or gene segments and the assembly of active genes by the deletion of material between genes (2, 9, 20).

Conservative site-specific recombination occurs between two DNA sites that share a short (usually 6 to 8 bp) "core" region of homology (Fig. 1). Exchange occurs within these cores such that no sequence information is lost or gained and the products of these reactions are intact DNAs that lack breaks or other interruptions. Moreover, recombination occurs by a breakage and joining mechanism in simple salt solutions in the absence of ATP or DNA synthesis.

Strand exchange in conservative site-specific recombination occurs by the introduction of site-specific breaks at identical positions within the core of each partner duplex, followed by the rejoining of strands from partner duplexes. Thus, there is an exchange of DNA strands within the region of homology. In all known examples of this type of recombination, the positions of strand breakage and joining on each strand are staggered such that the recombinant products are heteroduplex within their core regions. This exchange of partner duplexes within the core results in the exchange of the arm sequences that flank the cores.

Conservative site-specific recombination is mediated by recombinases called integrases or, in some cases, resolvases or invertases, depending on the type of recombination executed. These recombinases bind to specific sequences positioned in inverted orientation at the edges of the central core regions of homology such that each recombinase is positioned precisely with respect to this core region. The recombinases make sequence-specific breaks within the core, cutting each recombination partner at the same position, thus allowing for reciprocal strand exchange within the region of core homology. These recombinases are sequence-specific topoisomerases; they make breaks in DNA that are accompanied by the formation of a high-energy phosphodiester bond between the integrase and one end of the broken DNA that preserves the energy of the DNA phosphodiester bond. The other DNA end is an exposed OH. Strand exchange occurs by the attack of the exposed OH on one partner DNA on the protein-DNA bond of the other partner DNA, regenerating an intact DNA strand formed of sequences from each partner, and releasing the protein. Because of this topoisomerase-like mechanism, this class of recombination reactions can occur in the absence of ATP and DNA replication.

In some cases, the integrase alone is sufficient to promote recombination (see chapters 6, 8, 11, 12, 14). In other cases, host proteins are involved (see chapters 7, 8, 10, 13) (14). These host proteins generally seem to be architectural elements that promote the assembly of the protein-DNA complexes in which recombination actually occurs by introducing DNA bends and/or flexibility.

The particular type of DNA reorganization that occurs by conservative site-specific recombination depends on the relative orientations of the recombination sites (Fig. 2). Integration/fusion occurs when the two recombination sites exist on two different DNAs. Excision/deletion/resolution occurs when the sites exist on the same substrate DNA in direct orientation as defined by the alignment of their core sequences. Inversion occurs between sites present on the same DNA in inverted orientation as defined by alignment of their core sequences. Not all systems are equally adept at all types of recombination (8). In general, each intramolecular system is specific for either resolution or inversion.

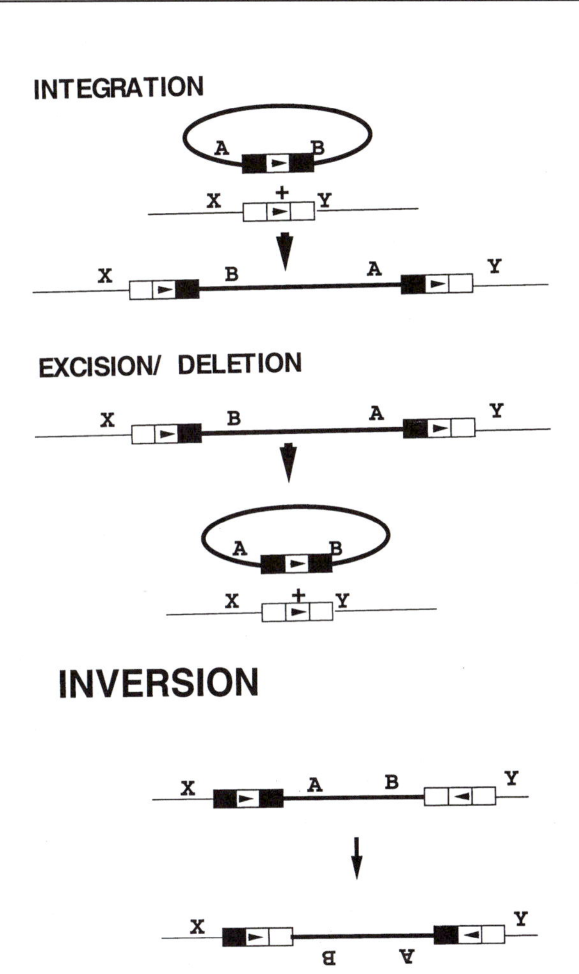

Figure 2. Different outcomes of conservative site-specific recombination are possible. Different relative orientations of the recombination sites in conservative site-specific recombination lead to different outcomes, although in all cases, the same breakage and joining events occur within the region of homology. When the recombination sites lie on different DNAs, recombination between them results in integration or fusion. When the two sites lie on the same DNA, two different outcomes are possible depending on the relative orientation of the sites. When the recombination sites are in direct orientation, excision/deletion/resolution of the material between the sites occurs to form an excised circle. When the sites are present in inverted orientation, the DNA segment between the two sites is inverted.

Many conservative site-specific recombination reactions have been studied to great depths and are described at high resolution including crystallographic analysis of recombinases and recombinase-DNA complexes (see chapters 6 to 8 and 12 to 14). This type of analysis has provided a wealth of information not only about recombination reactions but also about other features of nucleic acid-protein interactions that are applicable to many other reactions.

GENE ASSEMBLY BY DELETION: V(D)J RECOMBINATION

As described above, conservative site-specific recombination can be used to assemble genes from gene segments. Assembly of active genes by deletion is also a prominent feature of the immune system, although the recombination mechanism involved is very different from conservative site-specific recombination (see chapter 29).

The genes for antigen-binding receptors, the immunoglobulin heavy and light chains and the T-cell receptors, are actually assembled from multiple gene segments. These gene segments exist in large arrays or clusters at specific chromosomal regions. There are several different families of segments such that this combinatorial assembly pathway allows for the generation of numerous alternative antigen receptors. A reaction-specific recombinase formed by the RAG1 and RAG2 proteins mediates the excision of the DNA segment that separates the two gene coding segments that are to be joined. This intervening segment has at its boundaries specific binding sites for the RAG recombinase called recombination signal sequences (RSSs). The role of the RAG recombinase in this system is to execute the double-strand breaks in DNA that initiate recombination, whereas the actual joining step is mediated by other cellular functions (Fig. 3).

Recombination begins by the binding of the RAG recombinase to the RSSs of the DNA between the two gene segments to be joined. RAG then introduces specific nicks at the boundaries of the intervening segment, exposing the 3′-OH termini of the coding segments. These exposed 3′-OH ends then attack their complementary strands, generating hairpin ends on the termini of the gene coding segments. These hairpins are then opened. In some cases, the hairpin opening does not occur at the original site of hairpin formation so that hairpin opening results in overhanging sequences. The filling in of these overhangs can contribute additional diversity to the joints formed between the gene segments.

The gene segment DNAs are joined by the general cellular machinery that repairs double-strand breaks (19). The molecular mechanisms of these reactions are not yet known in detail.

Another prominent example of gene assembly through deletion occurs in ciliates (see chapter 30). Ciliates have two nuclei, one called the micronucleus, which is the germ-line nucleus, and the other called the macronucleus, which is the source of somatic gene expression. In the micronucleus, genes are interrupted by many intervening segments. The formation of functional genes in the somatic nucleus requires the excision of the intervening DNAs and the joining of the gene segments. The molecular mechanisms of these reactions remain to be established.

TRANSPOSITION

Transposition is the recombination reaction that mediates the movement of discrete DNA segments be-

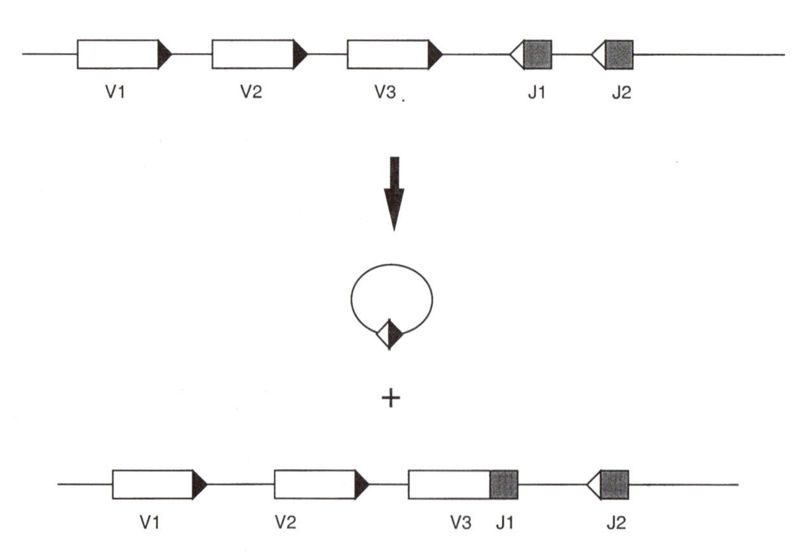

Figure 3. Assembly of genes from gene segments. Genes encoding proteins of the immunoglobulin superfamily are assembled by recombination. Multiple copies of different gene segments (V1 to V3, J1, and J2) exist. These segments are flanked by recombination signal sequences (open and filled triangles); recombination occurs only between a site of the open type and the filled type. The recombination reaction shown leads to the joining of the gene encoding segments, V3 and D1, to form a new hybrid gene.

tween many nonhomologous sites. These segments are variously called insertion sequences, transposons, and transposable elements. Such elements are widespread, having been found in virtually all organisms examined, and they account for considerable genome plasticity (see chapters 44, 46, and 47). Because these elements are mobile, they have the capacity to change host gene information, i.e., to make mutations, for example, by inserting into a gene and disrupting its coding sequence. Alternatively, insertion of a mobile element adjacent to a cellular gene may influence the expression of that gene by bringing a regulatory element on the mobile element near the host gene. In some organisms, a substantial fraction of spontaneous mutants derive from the movement of transposable elements. Indeed, it was changes in host gene expression resulting from the movement of a transposable element that led McClintock to her discovery of mobile DNA. Multiple copies of the element can also act as substrates for homologous recombination that can lead to alterations in genome structure and activity.

Transposable elements may also carry determinants such as antibiotic resistance genes (see chapters 14 and 18 to 20), catabolic genes, and virulence determinants. Other transposable elements contain viral genes, and insertion plays a key role in viral replication (see chapters 17 and 25 to 28).

There are two large classes of transposable elements: (i) those in which DNA is the actual substrate for recombination and (ii) those in which RNA is the actual substrate for recombination.

DNA-Mediated Recombination

Transposable elements are discrete pieces of DNA, bounded by terminal inverted repeats. These terminal inverted repeats are binding sites for the transposase that is usually encoded by the mobile element. The function of these terminal inverted repeats is to symmetrically position the recombinase at each end of the element. Transposition actually requires three DNA sites, one at each end of the transposon and another at the target site (Fig. 4). Binding of the transposase to the ends and to the target DNA mediates synapsis of the transposon ends and the target DNA. The transposase executes the DNA breakage reactions that disconnect the element from the flanking DNA in the donor site and the joining reactions that link the transposon to the new insertion site. In many reactions, the transposon is completely excised from the donor site in what is called "cut-and-paste" transposition and inserted into the target DNA to form a simple insertion.

The ends of the mobile element attack the target DNA at staggered positions such that the newly inserted transposon is flanked by short gaps. Host systems repair these gaps, resulting in the target sequence duplications that are characteristic of transposition. The length of this duplication is characteristic of each element, but typically, many different target sites are used and thus different target sequences duplicated.

Retroviruses such as HIV (see chapter 25) and retroviral-like elements, such as the yeast Ty elements, which resemble retroviruses except that they lack an extracellular phase (see chapters 26 to 28), also integrate into host genomes as DNA molecules to form simple insertions (Fig. 4). The RNA genomes of these elements are converted into these DNA recombination substrates by the action of an element-encoded reverse transcriptase. The termini of the resulting DNA copy of the element are recognized by its transposase which is usually called an integrase in these retroviral or retroviral-like elements. The integrase often removes a few nucleotides from the 3′ ends of the element to actually expose the reactive 3′ termini and joins those exposed ends to the target DNA. Retroviral-like transposable elements are also called long terminal repeat (LTR) elements because of the directly repeated segments at each end of the element. These LTRs play a critical role in the conversion of the element from its RNA to its DNA form.

In a few bacterial systems including bacteriophage Mu (see chapter 17) and likely Tn*3* (see chapter 14), single-strand nicks at the 3′ ends of the element occur rather than the double-strand breaks associated with other elements. The exposed 3′-OH ends of the transposon then attack the target DNA. These reactions result in a fusion product in which the transposon is covalently linked to both the donor backbone and the target DNA (6). Replication of the transposon segment and repair of the gaps result in the structure called a cointegrate, which contains two copies of the element and one copy each of the donor backbone and target DNA. In some systems, the mobile element also encodes a conservative site-specific recombinase and a cognate recombination site that promote the conversion of the cointegrate to a simple insertion, regenerating the donor DNA and a target DNA containing a copy of the element (see chapter 14).

From in vitro analysis of systems for both procaryotic and eucaryotic elements, we know that the transposition reactions described above occur by chemically related pathways (see chapter 2). The first step is the transposase-mediated hydrolysis of the phosphodiester bond that links the transposon to the flanking donor DNA, exposing the 3′-OH ends of the element. These exposed 3′-OH ends then directly attack the target DNA at staggered positions. Both of these steps are direct phosphoryl transfer reactions

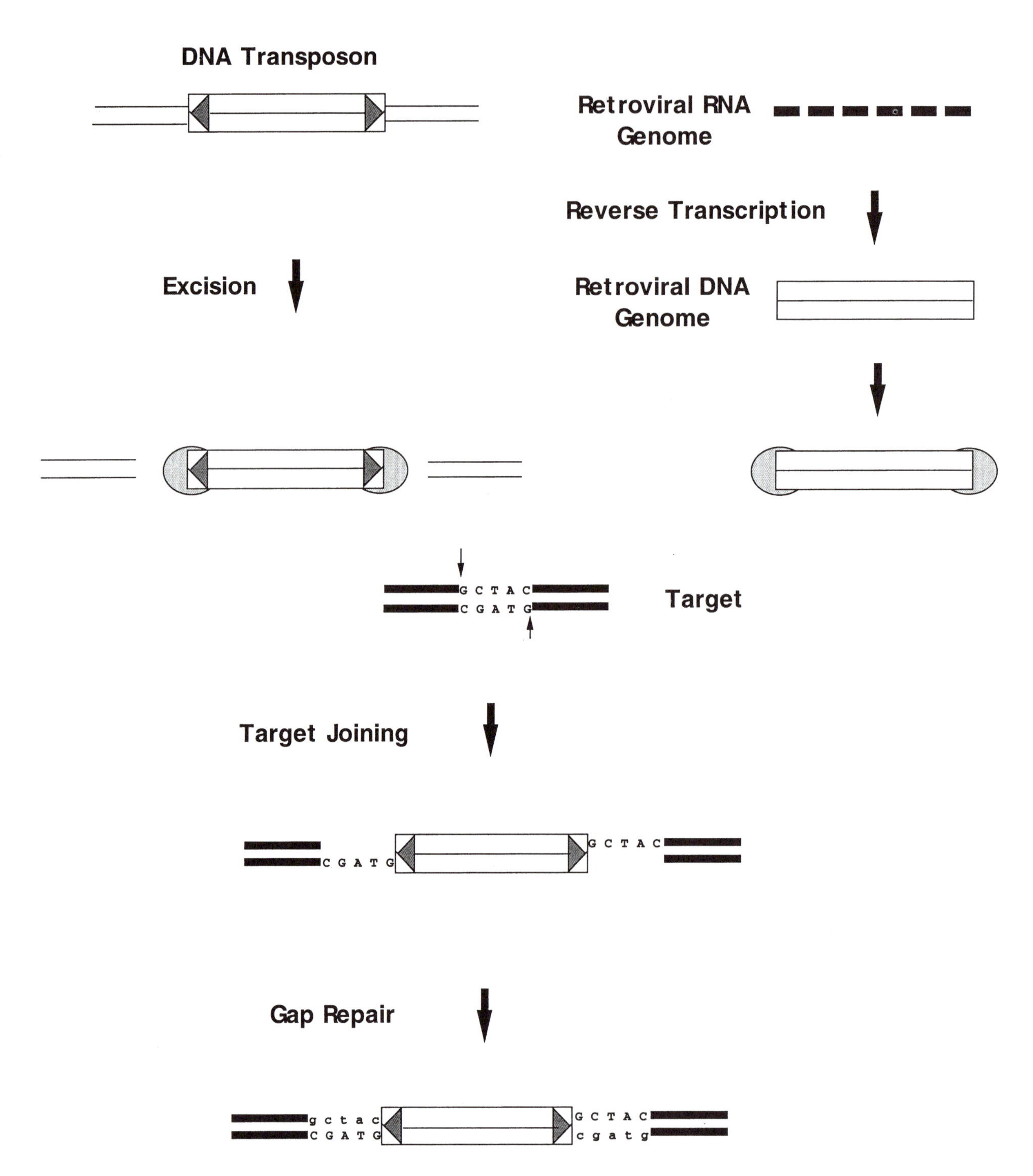

Figure 4. Transposition. (Left) The transposable element (white box) is a discrete DNA segment flanked by terminal inverted repeats (gray triangles) which are the binding sites for the transposase (gray circles). The transposase makes double-strand breaks at the termini of the element which separate it from the flanking donor DNA (thin lines). Each transposon 3′ end then attacks one strand of the target DNA (thick line) at staggered positions (arrows), generating a structure in which the transposon is flanked by short gaps. These gaps are then repaired by host functions, generating the target sequence duplications characteristic of transposon insertion. (Right) Transposition of a retrovirus-like element. The single-stranded RNA genome is converted to double-stranded DNA through the action of the element-encoded reverse transcriptase. The resulting DNA is functionally equivalent to the DNA intermediate that has been excised from the donor DNA during cut-and-paste transposition of elements whose life cycle involves only DNA. The transposase, called an integrase for these retrovirus-like elements, then joins the transposable element to the target DNA as described above. Repair of the resulting gaps again occurs by host functions.

that do not involve protein-DNA intermediates. Thus DNA replication (because of the staggered positions of target joining) and ligation are required to regenerate intact DNA. These transposition reactions are also performed by structurally related proteins (see chapters 17, 18, and 25).

RNA-Mediated Recombination

The recombination reactions described above are unified because the mobile element that joins to the target DNA is formed of DNA. There is another class of mobile elements called non-LTR transposons in which an RNA is the actual substrate of the recombination reaction (see chapters 33–35 and 49). An especially interesting aspect of some of these elements is that they are major constituents of many genomes, including the human genome. For example, a reasonable estimate is that perhaps 50% of the human genome is composed of non-LTR elements (see chapter 47). These elements have also been termed long interspersed nuclear elements (LINEs) and short interspersed nuclear elements (SINEs). Luckily for us, most of these elements are truncated and generally immobile, although a few cases of transposition of these elements which have resulted in human disease are known (18).

A full-length non-LTR element encodes two open reading frames, one of which encodes an RNA-binding protein and the other of which encodes a protein that has both endonuclease activity and reverse transcriptase activity. Although it has not yet been possible to perform a complete transposition reaction in vitro, biochemical analysis of a non-LTR element that inserts specifically into the rRNA genes of arthropods has established that recombination initiates by the introduction of a nick in the target DNA by the element-encoded endonuclease (23) (Fig. 5). The 3′-OH in the target DNA exposed by this reaction then serves as the primer for reverse transcriptase using the element RNA as a template to generate a DNA copy of the element linked to the target DNA. Because many of these non-LTR elements are not full length, it appears that DNA strand synthesis rarely extends to the 5′ end of the RNA copy of the element. It is not known how the next steps occur to complete element insertion, but they must involve joining the 5′ end of the newly synthesized DNA strand to the target DNA, removal of the RNA, and synthesis of the second strand of the element.

Another remarkable example of the use of an RNA recombination intermediate in transposition is found in the mobility of group II introns that are present in organellar DNAs and in some bacteria (see chapter 31). In these cases, the element inserts into a specific target site in a process called retrotransposition (5). Recombination is dependent on a multifunctional protein encoded by the intron itself. This protein has a maturase activity that promotes splicing, a reverse transcriptase activity, and a site-specific DNA endonuclease. Recombination initiates by the specific binding of the maturase to the intron in unspliced precursor RNA; the maturase then promotes splicing of the intron RNA. In the next step, the intron RNA reverse splices into one strand of the DNA target site in a reaction that is assisted by the intron-encoded protein. The endonuclease of the intron-encoded protein then makes a cut into the DNA strand that has not received the spliced intron. The 3′-OH generated by this nick serves as the primer for the reverse transcriptase of the intron-encoded protein to generate a DNA copy of the mobile element linked to its target site.

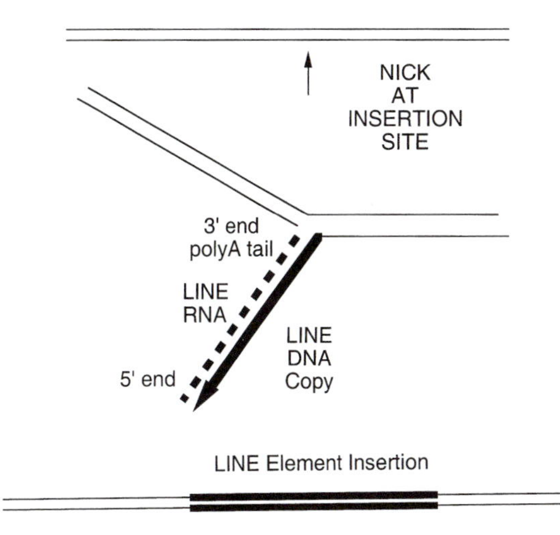

Figure 5. Transposition of a LINE element. Recombination is initiated by a single-stranded nick in the target DNA by an element-encoded endonuclease. The 3′-OH exposed by this nick acts as a primer for reverse transcription of the LINE element RNA by an element-encoded reverse transcriptase, resulting in one DNA strand of the LINE element joined to the target DNA. The next steps which join the 5′ end of the element to the target DNA and result in the synthesis of the second strand of the element into the target DNA are not yet understood.

GENE CONVERSION BY TARGETED HOMOLOGOUS RECOMBINATION

Several recombination systems mediate the alternate expression of multiple alleles of a gene by moving alternative gene copies or gene segments from a "silent" position to an active "expression site." This process can occur by homologous recombination which is targeted to particular sites via system-specific proteins

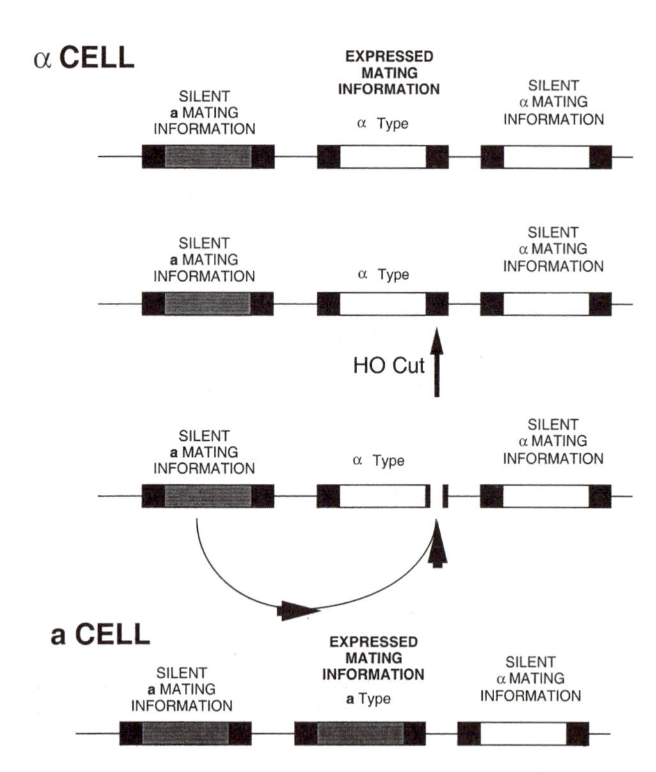

Figure 6. Yeast mating-type switching by directed gene conversion. Three different loci encode the transcription factors whose expression can determine the yeast mating type of **a** or α. Two of these loci are silent, i.e., the transcription factors are not expressed, whereas one site (center) is active and expresses the transcription factors. An α cell contains the information for the α cell transcription factors at the center expression site as well as at the silent right site (white boxes). Expression of the HO endonuclease leads to cleavage of the DNA at the expression site. The broken DNA invades homologous sequences bounding the mating-type information at the silent **a** site (gray box), allowing double-strand break repair to copy **a** information from the silent **a** site into the expression site. At each switch, the transfer of information is such that alternation between the left **a** and right α silent information occurs.

acting on a specific target DNA. The best understood example of this type of recombination occurs in the expression of alternate mating types by the yeast *Saccharomyces cerevisiae* (see chapter 39) (Fig. 6). Some introns also move by this DNA-mediated mechanism (see chapter 31), and some forms of antigenic variation also use this type of recombination (see chapters 40 and 41).

In *S. cerevisiae,* the mating type of a cell, **a** or α, is determined by the expression of alternative transcription factors from a locus called MAT. Also present in the genome are two other sites, one of which encodes the **a** cell information and the other of which encodes the α cell information. However, the mating-type information in these other locations is not expressed and they are said to be "silent." Yeast cells can switch between the **a** cell and α cell genotypes by a recombination reaction that transplants the sequence information from one of the silent loci into the expression MAT site. This switching process is initiated by the expression of a special endonuclease called HO that can specifically recognize and cut at the MAT expression locus. The introduction of the HO-specific break at MAT initiates a double-strand break repair reaction executed by the host's homologous recombination machinery using the information at the silent mating-type loci to replace the old information at the MAT site with information from the silent expression site. The expression of HO is highly regulated so that recombination occurs only under particular conditions.

Thus the system-specific protein, the HO endonuclease, acts to target homologous recombination and repair that will transfer DNA sequence information from one site to another, leading to the expression of alternative sets of genetic information.

GENOMES AND MOBILE DNA

Although we have been aware for some 50 years that there are mobile DNA segments, the recent descriptions of whole genomes have revealed a degree of genome plasticity that has been heretofore unappreciated. For example, the analysis of multiple bacterial genomes has revealed an extraordinary amount of gene transfer between bacteria; mobile DNA must play a key role in the acquisition of such DNAs (30) (see chapters 44 to 46). Moreover, we now know that considerable fractions of certain genomes—for example, the human genome—are actually made up of remnants of mobile DNAs (see chapter 47). No doubt, the continued study of genomes will lead us to more mobile DNAs to identify and understand and further investigation into mobile DNAs will help us to understand the structure and function of genomes.

REFERENCES

1. **Berg, D. E., and M. M. Howe (ed.).** 1989. *Mobile DNA.* American Society for Microbiology, Washington, D.C.
2. **Brusca, J. S., M. A. Hale, C. D. Carrasco, and J. W. Golden.** 1989. Excision of an 11-kilobase-pair DNA element from within the *nifD* gene in *Anabaena variabilis* heterocysts. *J. Bacteriol.* **171:**4138–4145.
3. **Bukari, A. I., J. A. Shapiro, and S. L. Adhya (ed.).** 1977. *DNA Insertion Elements, Plasmids and Episomes.* Cold Spring Harbor Laboratory, Cold Spring Harbor, N.Y.
4. **Campbell, A.** 1962. The episomes. *Adv. Genet.* **11:**101.
5. **Cousineau, B., S. Lawrence, D. Smith, and M. Belfort.** 2000. Retrotransposition of a bacterial group II intron. *Nature* **404:** 1018–1021.
6. **Craigie, R., and K. Mizuuchi.** 1985. Mechanism of transposi-

tion of bacteriophage Mu: structure of a transposition intermediate. *Cell* **41**:867–876.

7. **Egawa, R., and S. Mitsuhashi.** 1964. Newly isolated Hft lysate of an R factor. *Jpn. J. Bacteriol.* **19**:237.
8. **Gellert, M., and H. Nash.** 1987. Communication between segments of DNA during site-specific recombination. *Nature* **325**: 401–404.
9. **Golden, J. W., C. D. Carrasco, M. E. Mulligan, G. J. Schneider, and R. Haselkorn.** 1988. Deletion of a 55-kilobase-pair DNA element from the chromosome during heterocyst differentiation of *Anabaena* sp. strain PCC 7120. *J. Bacteriol.* **170**: 5034–5041.
10. **Harada, K., M. Kameda, M. Suzuki, and S. Mitsuhashi.** 1964. Drug resistance of enteric bacteria. III. Acquisition of transferability of nontransmissible R(TC) factor in cooperation with F factor and formation of FR(TC). *J. Bacteriol.* **88**:1257–1265.
11. **Harada, K., M. Kameda, M. Suzuki, S. Shigehara, and S. Mitsuhashi.** 1967. Drug resistance of enteric bacteria. VIII. Chromosomal location of nontransferable R factor in *Escherichia coli*. *J. Bacteriol.* **93**:1236–1241.
12. **Hershey, A. D. (ed.).** 1971. *The Bacteriophage Lambda*. Cold Spring Harbor Laboratory, Cold Spring Harbor, N.Y.
13. **Herskowitz, I.** 1973. Control of gene expression in bacteriophage lambda. *Annu. Rev. Genet.* **7**:289–324.
14. **Hodgman, T. C., H. Griffiths, and D. K. Summers.** 1998. Nucleoprotein architecture and ColE1 dimer resolution: a hypothesis. *Mol. Microbiol.* **29**:545–558.
15. **Jacob, F., and E. L. Wollman.** 1958. Les épisomes, éléments génétiques ajoutés. *C. R. Acad. Sci.* **247**:154–156.
16. **Jordan, E., H. Saedler, and P. Starlinger.** 1968. 0° and strong-polar mutations in the gal operon are insertions. *Mol. Gen. Genet.* **102**:353–365.
17. **Kameda, M., K. Harada, M. Suzuki, and S. Mitsuhashi.** 1965. Drug resistance of enteric bacteria. V. High frequency of transduction of R factors with bacteriophage epsilon. *J. Bacteriol.* **90**:1174.
18. **Kazazian, H. H., Jr., C. Wong, H. Youssoufian, A. F. Scott, D. G. Phillips, and S. E. Antonarakis.** 1988. Haemophilia A resulting from de novo insertion of L1 sequences represents a novel mechanism for mutation in man. *Nature* **332**:164–166.
19. **Khanna, K. K., and S. P. Jackson.** 2001. DNA double-strand breaks: signaling, repair and the cancer connection. *Nat. Genet.* **27**:247–253.
20. **Kunkel, B., R. Losick, and P. Stragier.** 1990. The *Bacillus subtilis* gene for the development transcription factor sigma K is generated by excision of a dispensable DNA element containing a sporulation recombinase gene. *Genes Dev.* **4**:525–535.
21. **Landy, A., and W. Ross.** 1977. Viral integration and excision: structure of the lambda att sites. *Science* **197**:1147–1160.
22. **Lederberg, J., and E. L. Tatum.** 1946. Gene recombination in *Escherichia coli*. *Nature* **158**:558.
23. **Luan, D. D., M. H. Korman, J. L. Jakubczak, and T. H. Eickbush.** 1993. Reverse transcription of R2Bm RNA is primed by a nick at the chromosomal target site: a mechanism for non-LTR retrotransposition. *Cell* **72**:595–605.
24. **McClintock, B.** 1952. Chromosome organization and genic expression. *Cold Spring Harbor Symp. Quant. Biol.* **16**:13–47.
25. **McClintock, B.** 1956. Controlling elements and the gene. *Cold Spring Harbor Symp. Quant. Biol.* **21**:197–216.
26. **McClintock, B.** 1948. Mutable loci in maize. *Carnegie Inst. Wash. Year Book* **47**:155–169.
27. **McClintock, B.** 1950. The origin and behavior of mutable loci in maize. *Proc. Natl. Acad. Sci. USA* **36**:344–349.
28. **Nash, H. A.** 1975. Integrative recombination of bacteriophage lambda DNA in vitro. *Proc. Natl. Acad. Sci. USA* **72**: 1072–1076.
29. **Nash, H. A., and C. A. Robertson.** 1981. Purification and properties of the *Escherichia coli* protein factor required for lambda integrative recombination. *J. Biol. Chem.* **256**: 9246–9253.
30. **Ochman, H., J. G. Lawrence, and E. A. Groisman.** 2000. Lateral gene transfer and the nature of bacterial innovation. *Nature* **405**:299–304.
31. **Shapiro, J. A.** 1969. Mutations caused by the insertion of genetic material into the galactose operon of *E. coli*. *J. Mol. Biol.* **40**:93–105.
32. **van de Putte, P., and N. Goosen.** 1992. DNA inversions in phages and bacteria. *Trends Genet.* **8**:457–462.

Mobile DNA II
Edited by N. L. Craig et al.

Chapter 2

Chemical Mechanisms for Mobilizing DNA

Kiyoshi Mizuuchi and Tania A. Baker

TWO CLASSES OF "MOBILIZING" RECOMBINASES

The DNA cutting and joining reactions necessary to move a segment of DNA are catalyzed by recombinase proteins that are polynucleotidyltransferases. Each recombinase is specific to a particular mobile DNA segment. Different types of elements, and their associated recombinases, continue to be discovered and characterized. Many such elements and their proteins are reviewed in individual chapters of this volume. Studies of two major classes of these recombinases have advanced to the point that detailed molecular mechanisms can be discussed. Here, we focus on the mechanisms of protein-catalyzed recombination by these two classes to provide examples of how the chemistry of DNA mobility is catalyzed and controlled.

The conservative site-specific recombinases are a well-characterized group of protein catalysts. Their mechanism of action is considered conservative because no high-energy cofactors are consumed, the total number of phosphodiester bonds remains unchanged by the reaction, and no DNA degradation or resynthesis is necessary. The integration of the bacteriophage λ DNA into the chromosome of its host by λ integrase protein is the prototypical example of this class. λ integrase, P1 Cre, and yeast Flp are related proteins, all members of the tyrosine recombinase family. Certain conjugative transposons are also mobilized by recombinases belonging to this class. The serine recombinases are a second subfamily of proteins that promote conservative site-specific recombination. The invertases (such as Hin and Gin) and the resolvases (such as Tn*3* and γδ resolvase) are members of this class.

The transposase/retroviral integrase family (referred to as transposases hereafter) is the second major class of recombinase. These proteins catalyze a set of mechanistically related reactions essential for DNA transposition mediated by bacterial transposons, insertion elements, and transposing bacteriophages, such as Mu. Translocation of P-elements and the Tc/mariner family of elements occurs by a similar mechanism, promoted by similar recombinases; several plant transposons are also thought to move by use of a similar process. Integration into the host chromosome of the DNA copies of the genomes of retroviruses and long terminal repeat (LTR) retrotransposons (such as yeast Ty elements) are also examples of this type of recombination. The RAG recombinase, which is involved in the initiation of V(D)J recombi-

Kiyoshi Mizuuchi • Laboratory of Molecular Biology, National Institute of Diabetes and Digestive and Kidney Diseases, National Institutes of Health, Bethesda, MD 20892. **Tania A. Baker** • Department of Biology and Howard Hughes Medical Institute, Massachusetts Institute of Technology, Cambridge, MA 02139.

nation to assemble gene segments during maturation of the vertebrate immune system, also seems to be a distantly related member of this recombinase family. There are, however, transposable elements that move by a distinct mechanism, such as the non-LTR retrotransposons; the proteins that catalyze these recombination reactions are distinct and are not discussed in detail here (see chapters 31 to 36).

Conservative Site-Specific Recombination

During conservative site-specific recombination, both strands of each recombination partner are cleaved at specific sites, exchanged, and religated to new partners. Four DNA strands must be cleaved and rejoined to complete recombination, and four recombinase subunits are believed to be the catalysts. The recombination pathways used by the serine and tyrosine recombinases are slightly different from each other (Color Plate 1 [see color insert]), although both involve the sequential formation and destruction of protein-DNA covalent intermediates that store the energy of the broken phosphodiester bonds. The serine recombinases cleave all four strands before strand exchange (Color Plate 1A). Double-strand cuts result in two nucleotide 3′-OH overhangs and, in the reaction intermediate, the 5′ ends are covalently attached to the protein via a phosphoserine linkage (hence, the name serine recombinase). In contrast, the tyrosine recombinases complete cleavage, exchange, and rejoining of one pair of DNA strands before initiating the same set of reactions on the other pair of strands (Color Plate 1B). In these reactions, the protein-bound intermediates are made through 3′-phosphotyrosine linkages, and the reaction proceeds through a Holliday junction intermediate rather than a pair of double-strand breaks. In both classes, each copy of the recombination site is organized as a pair of symmetrically positioned recombinase-binding sequences flanking a central short, often asymmetric, sequence known as the crossover region. DNA strand exchange takes place at the edges of this crossover region.

Transposition and Retroviral Integration

Transposases catalyze two types of chemical steps to mobilize a segment of DNA (Color Plate 2 [see color insert]). In the first of these steps, a pair of site-specific endonucleolytic cleavages separates the 3′-OH of the transposon DNA ends from the 5′-phosphoryl ends of the adjoining host DNA. This reaction, called donor DNA cleavage, occurs at each of the ends of the element DNA and results in a doubly nicked DNA molecule with the transposon terminating with free 3′-OH DNA ends. The second reaction catalyzed by the transposase is the direct attack of this 3′-OH DNA end on a new phosphodiester bond—the bond at the new insertion site. As a result of the direct attack, the transposon end is joined to the new DNA site and a nick is introduced into the target DNA. This transesterification reaction, called DNA strand transfer, occurs on the two ends of the element, with each end attacking a bond on one of the two strands of the DNA at the new insertion site (target site). The attack of the two transposon ends on the two strands of the target DNA is separated by 2 to 9 nucleotides with 5′ stagger. This distance is fixed for each element, and it results in short target sequence duplication that flanks the newly transposed copies of the element. The DNA intermediate generated by cleavage and strand transfer is converted to the final recombination product by cellular enzymes. This processing may involve extensive DNA synthesis or just a short DNA gap repair reaction.

The 3′ end cleavage and DNA strand transfer steps are shared features of the transposition mechanism. However, depending on the biology of an element, additional DNA cleavage and joining reactions may be required for recombination. When the two chemical steps described above are used by a chromosome-embedded transposon, the product is a branched DNA intermediate in which the two strands of the transposon are connected at each end to different segments of the host chromosome (Color Plate 2). Replication of the element DNA within this structure frequently causes chromosomal inversions and deletions that can be highly detrimental to the host cell. Many elements have developed ways to excise completely from their original DNA location before joining to a new DNA site and therefore avoid generating these major disruptions to the host genome.

Transposition via an excised transposon intermediate requires that both strands of the DNA at the junction between the element and its original flanking sequence be severed. Whereas cleavage of the "transferred strand" (which will later be directly joined to the target DNA) is always done by the transposase, different elements use different strategies to cleave the "nontransferred" (or "second") strand; three such mechanisms are well understood (Color Plate 3 [see color insert]) (reviewed in reference 65). For example, this strand can be cleaved by a protein catalyst other than the principal transposase. A restriction enzyme-like nuclease (the TnsA protein) carries out this reaction during Tn7 transposition (31, 60). The other two mechanisms of second-strand cleavage are promoted by the transposase itself, using an "intratransposon" attack of the initially cleaved 3′-OH end of the donor DNA on the second DNA strand. Attack can occur directly across the DNA duplex on a phosphodiester bond in the opposite strand. This reaction generates a "hairpin-structured" DNA end that must be opened by a second hydrolysis step before the element can recombine with a new target DNA site. IS*4* family elements, such as Tn*10* and Tn*5*, use this strategy (6, 37). This strand transfer reaction can also occur between the two ends of the mobile element. In this case, one cleaved 3′-OH end attacks the same DNA strand at the opposite DNA end. The resulting figure-eight-shaped intermediate is further processed to form a circularized copy of the transposon (see chapter 16 for further discussion). The ends of this "circularized" transposon are then exposed by additional hydrolysis steps to generate the excised transposon. The IS*3* family of transposons, such as IS*911*, uses this mechanism (65). In retroviruses and LTR retrotransposons, the direct precursor DNA for integration is generated from an RNA template by reverse transcription, essentially in the form of an excised transposon. Therefore, a specific mechanism to cleave the second strand is unnecessary for these elements (see chapter 25).

"Cross strand" strand transfer reactions are also found as part of the DNA cleavage mechanism in several related recombination reactions. A variation of this basic scheme is used by the RAG recombinases to cleave DNA for the initial steps of V(D)J recombination. However, in this case, the DNA hairpin is generated on the opposite side of the break than it is for the IS*4* transposons; that is, it is on the DNA segment that would be equivalent to the "flanking DNA" (the coding ends) rather than that equivalent to the transposon ends (the signal ends). This reflects the fact that the first strand cleavage for the V(D)J recombination takes place on the nontransferred strand instead of the transferred strand (see chapter 29). Some of the eukaryotic transposons, such as the snapdragon element Tam3, maize Ac, and *Drosophila* hobo, are expected to cleave DNA in this manner as well (4, 16, 67).

As with conservative site-specific recombination, the DNA cleavage and joining reactions promoted by transposases proceed without the need for external energy. The mechanistic basis of this energy conservation is distinct from the covalent protein-DNA intermediate strategy used by the site-specific recombinases. Transposases cleave more DNA phosphodiester bonds than they generate, so the recombination process, as a whole, is energetically downhill. Furthermore, during the DNA-joining step of strand transfer, a one-step transesterification reaction is used to make the new DNA phosphodiester. Because a phosphodiester bond is also cleaved in the process, this step is essentially isoenergic, and, as discussed later, is driven forward by product-binding energy. Note that, whereas the site-specific recombinases are very "tidy" protein catalysts that are able to complete the entire recombination reaction on their own, cellular DNA metabolic enzymes (i.e., polymerases and ligases) and their nucleotide substrates are essential to generate the final DNA products of transposition.

PRINCIPLES OF CATALYSIS

Phosphoryl Transfer in Recombination

Central to both types of DNA rearrangements are protein-catalyzed polynucleotidyl transfer steps. Polynucleotidyl transfer is a phosphoryl transfer in which the transferred group is a component of a polynucleotide chain. In these reactions, a scissile phosphate associated with a polynucleotide chain can be transferred to (i) water, resulting in hydrolysis of the phosphodiester bond; (ii) an amino acid side chain of a protein, to generate a covalent protein-nucleic acid linkage; or (iii) a second polynucleotide chain, to generate a recombinant strand. The necessary ingredients for catalysis of recombination are similar to those of other phosphoryl transferases such as nucleases, phosphatases, and kinases. Virtually all enzymatic phosphoryl transfer reactions proceed by an SN_2 in-line nucleophilic substitution mechanism (Color Plate 4A [see color insert]) (19, 39). The SN_2 mechanism predicts that the stereoconfiguration at the phosphorus center will be inverted with each reaction step, provided the scissile phosphate is made chiral by isotopic or thiosubstitution of one of the nonbridging oxygens. In every recombination reaction in which this has been tested, the observations have been consistent with this expectation (20, 23, 38, 48, 49, 66).

Enzymatic phosphoryl transfer reactions involving phosphodiesters are assumed here to proceed through a principally associative mechanism involving a pentacoordinated bipyramidal intermediate at the transition state as depicted in Color Plate 4A. In an associative mechanism, the incoming nucleophile's electron orbit establishes significant covalent character with the phosphorus center before the departure of the leaving group. In contrast, if the reaction path is closer to a dissociative mechanism, the phosphorus center partially loses its connectivity to the leaving oxyanion before establishing connection with the incoming group; thus, in a dissociative reaction, there is a metaphosphate (PO_3^-)-like transition state intermediate. Because of the difference in the transition state intermediate, the distinction between these two reaction paths is useful to consider when discussing likely catalytic mechanisms. Proof that the reaction paths used during enzymatic phosphoryl transfer are associative is still incomplete, however.

Components of the Active Center

What components are necessary constituents of an active center capable of catalyzing phosphoryl transfer at a DNA phosphodiester? One can consider three principal active-site components: (i) functional groups that interact with the incoming nucleophile; (ii) those that interact with the leaving group oxygen; and (iii) those that interact with one or more of the other three oxygens surrounding the phosphorus center, including the two nonbridging oxygens. A single functional group may play multiple roles within this scheme. Together, these components cooperate to stabilize the transition state(s), and assist in extracting a proton from the incoming group and donating a proton to the leaving group.

General acid/base catalysis probably is the most intuitively straightforward reaction mechanism, because a proton must be removed from the incoming group and donated to the leaving group during phosphoryl transfer. A general base can extract a proton from the incoming group to activate it as a nucleophile. On the other side of the reaction center, a general acid can donate a proton to the leaving group. A general base, after accepting a proton, can function as a general acid, and a general acid turns into a general base by losing a proton; thus, the functional capacities of these active-site components reverse as a consequence of the reaction.

A nice example of general acid/base catalysis is provided by RNase A, as illustrated in Color Plate 4B. (For details, see reference 59.) This enzyme uses the reversible nature of the general acid/base functional groups, His-12 and His-119, for the two-step hydrolysis of RNA phosphodiester. In the first step, His-12 works as a general base to extract proton from the 2′-oxygen, activating it as a nucleophile, while a general acid, the protonated His-119, donates a proton to the leaving 5′-oxygen. This step generates a 2′,3′-cyclic phosphate intermediate. After replacement of the leaving group 5′-OH with a water molecule, His-119 then functions as a general base to activate this bound water as a nucleophile, while His-12 donates the proton to the leaving 2′-oxygen to complete hydrolysis. Reaction mechanisms of this type often involve additional functional groups to stabilize the transition state; as discussed below, interaction between positively charged groups and the nonbridging oxygens of the scissile phosphate can provide this function.

Although attractive, general acid/base catalysis is not the only successful strategy used for phosphoryl transfer. If the transition state is sufficiently stabilized by the local electronic environment, the rate of the deprotonation of the nucleophile, or protonation of the leaving oxyanion need only be as fast as the overall rate of the chemistry. For example, a proton can be donated to the leaving group without involving a general acid catalyst—directly from the solvent or through a chain of water molecules—and proton-accepting side chains that are not functioning as primary catalysts. A similar scenario for deprotonation of the incoming nucleophile is also possible. Lewis acids, electron acceptors such as metal cations, or other electron-starved entities such as positively charged amino acid side chains, can then play the principal role in transition state stabilization. Divalent metal ions, such as Mg^{2+}, are particularly well suited as active site components for phosphoryl transfer, especially when water is the acceptor of the transfer (e.g., during hydrolysis). These ions can position a coordinated hydroxide anion for nucleophilic attack on the DNA. Metal ions greatly lower the pK_a values of water, such that a metal-bound hydroxyl ion is a potent nucleophile. For the same reason, metal-bound water can function as a general acid/base. Therefore, metal cations at the active site can in some cases function as a part of the general acid/base catalysis mechanism.

On the leaving group side of the active center, the developing negative charge on the leaving oxyanion can be stabilized again by an electron-starved entity, instead of by proton donation from a general acid. Thus, divalent metal ions can perform catalytic functions on both sides of the reaction center, as is seen in the two-metal catalysis mechanism proposed for the exonuclease active site in *Escherichia coli* DNA polymerase I and other phosphoryl transferases (5, 63) (Color Plate 4C). By simultaneously coordinating with several key components at the active site, the metal ions position themselves properly to assist substrate engagement and stabilize the proper active-site conformation. In the model, the two metal ions contact one of the nonbridging phosphate oxygen atoms. Here, again, they can act as Lewis acids, to polarize the P—O bond and make the phosphorus center a better target for nucleophilic attack, while positioning these oxygens to favor the equatorial planar configuration rather than the tetrahedral configuration.

The impact of polarization of the nonexchanging P—O bonds, or displacement of negative charge away from the phosphorus center, on the reaction rate would be different depending on the reaction mechanism. P—O bond polarization would disfavor a metaphosphate-like transition state intermediate (the expected intermediate for proteins that use a more dissociative mechanism) by making it difficult for the leaving group to dissociate first; these interactions would therefore be expected to slow the reaction rate. In contrast, a dissociative mechanism is thought to be used in transfer of phosphomonoesters (5a, 32, 39,

46) and P—O bond polarization should stabilize the pentacoordinated intermediates allied with this type of mechanism (5a). Thus, assuming that the chemical step of the recombinases discussed here proceeds via a relatively associative path, P—O bond polarization should contribute to the chemical reaction rate. This polarization role can be fulfilled by either metal cations or positively charged amino acid side chains. Conceptually, it is easy to understand how attractive forces between the substrate and the enzyme surface can be used to stabilize the transition state in the immediate vicinity of the reaction center. Thus, it is not a surprise that we frequently find positively charged functional groups and ions within the active centers of proteins that catalyze phosphoryl transfer.

In both the general acid/base catalysis and the two-metal catalysis mechanisms discussed above, the active-site architecture is essentially symmetric around the phosphorus center. Practically the same catalytic components play their roles on the two sides of the reaction center. After all, the SN_2 reaction is essentially symmetric. However, the active-site architecture need not be symmetric. For example, a metal ion may provide the catalytic function on one side of the reaction center, whereas the critical functional groups on the other side may be provided by a general acid/base, or by another electron-starved entity, such as a positively charged amino acid side chain, or a nonmetal cofactor. As described below, the conservative site-specific recombinases catalyze recombination without the use of metal ion cofactors, whereas the transposases seem to rely extensively on metal-mediated mechanisms.

Active-Site Plasticity

As discussed above, the active site for phosphoryl transfer reaction can be composed of different sets of catalytic components. Although proteins belonging to the same family may be expected to use similar sets of active-site components, one should be aware that this may not always be true. It may not be very difficult to switch the catalytic components of an existing enzyme. One dramatic example of artificial changes of an active site has been reported for *E. coli* RNase H, a protein structurally related to the transposases. Like other enzymes in the family, RNase H usually requires Mg^{2+} for catalysis. As in many nucleases that use this cofactor, the active-site carboxylate side chains and the phosphoryl oxygens at the reaction center are expected to directly coordinate this catalytic metal ion. Yet, the enzyme can function with hexaminecobalt in place of Mg^{2+} (35). This was unexpected because, unlike Mg^{2+}, the Co^{3+} is stably ligated, in this case to $(NH_3)_6$. Therefore, it is highly unlikely that Co^{3+} with its associated amine groups could occupy the same location and coordinate the active site groups and the scissile phosphate isosterically with Mg^{2+}. Furthermore, it is also unlikely that this cofactor could activate a hydroxide anion for nucleophilic attack in exactly the same way as Mg^{2+}. Therefore, there seems to be a significant change in the conformation of the active site to accommodate use of this cofactor in catalysis.

RNase H has also been successfully converted to a metal-independent nuclease by mutations. As expected, mutations that change the active-site carboxylate side chains (groups that usually coordinate Mg^{2+}) individually to nonacidic side chains, inactivate the enzyme. However, the simultaneous change of two of these acidic amino acids to arginines restores nuclease activity, and converts the enzyme into a metal-independent catalyst (11). Thus, it seems possible to modify the local active-site architecture, and thereby the catalytic strategy of this enzyme within an existing structural framework; it would be extraordinary if nature had not done this with some proteins as well.

ACTIVE SITES OF THE CONSERVATIVE SITE-SPECIFIC RECOMBINASES

The serine and tyrosine recombinases are unrelated enzymes that promote site-specific recombination using covalent protein–DNA intermediates and active sites devoid of metal ion cofactors. Of these two classes of proteins, the catalytic mechanisms used by the serine recombinases are less well understood; information about these proteins' active sites will be reviewed first, followed by a more detailed discussion of the tyrosine recombinases.

Sequence conservation among serine recombinase family members, biochemical studies of mutant enzymes, and structural studies of the γδ resolvase have identified several important active-site residues (references 33 and 71 and references therein). These include (residue numbers are those of γδ resolvase): Ser-10 that forms the covalent bond with the substrate DNA, Arg-8, Arg-68 and Arg-71, Gln-19, and Asp-67. Glu-124 and Arg-125 also may participate in catalysis, being donated to the active center from a neighboring subunit in the multimeric recombinase complex (71). Of these residues, Gln-19 and Asp-67 may function principally to position other residues (71). As discussed above, the participation of several positively charged arginines in metal-independent catalysis of phosphoryl transfer is understandable. However, because the active site in the only currently available structure of a recombinase-DNA complex seems not to be engaged with the scissile phosphate in an active configuration (71), the exact catalytic role of each residue remains unclear. In particular, whether any of the side chains participates as a general acid/base catalyst remains unknown.

Substantial recent progress in understanding the catalytic mechanism of the tyrosine family of site-specific recombinases has resulted from analysis of crystal structures of several family members, and a structurally related type 1B DNA topoisomerase. In particular, cocrystals of Cre recombinase bound to its DNA substrate (a pair of Lox sites) have been studied that capture the protein specifically at several different reaction stages (see chapter 6). This structural information has revealed the functions of individual active-site residues. In addition, parallel progress in understanding the catalytic mechanism of the type 1B topoisomerases, cousins of the recombinases, has aided elucidation of the catalytic strategies. Some of the details of the progress are reviewed in later chapters, and we focus here on the likely catalytic functions of several active-site residues.

Early sequence comparisons and characterization of mutant proteins identified an Arg-His-Arg triad and the tyrosine that forms the covalent bond with DNA as likely active-site residues for the tyrosine recombinases (3). More recent analysis identified two additional critical residues: a lysine located between the first arginine and the histidine of the triad, and a histidine (or a tryptophan) preceding the active-site tyrosine (10, 15, 69). The histidine residue of the triad is replaced by lysine in the type 1B topoisomerases whose structure has been determined.

In the type 1B topoisomerases, which are enzymes that function as monomers, all the amino acid residues that contribute to a functional active site are contributed by the same subunit. This is not necessarily the case for recombinases that function as tetramers. For example, experiments in which subunits of Flp recombinase mutated at different active-site residues were mixed together suggested that this protein has an active site constructed from residues donated by more than one subunit. The catalytic tyrosine of one subunit functions together with the Arg-His-Arg triad from a different subunit within the active, DNA-bound tetramer (references 12 and 44 and references therein). The generality of this feature among other members of the tyrosine recombinase family has been a subject of interest for some time. It is now clear from biochemical and structural studies that, as predicted from the subunit-mixing studies, the tyrosine nucleophile of Flp functions in *trans* to the remainder of the active site (13). In contrast, in all the other well-studied tyrosine recombinases the six key active-site residues are donated to the complex together from a single subunit (2, 7, 26, 30, 52). Thus, although the composite active site is an attractive means for coupling subunit assembly to enzyme activation, it is clearly not an essential means of regulation.

The roles of the active-site residues other than the tyrosine nucleophile are becoming clear. The Arg-His-Arg triad coordinates the scissile phosphate in the covalent intermediate structure (24, 26). In the precleavage complex of Cre, the His of the triad is located close enough to the active site tyrosine to make it a feasible candidate for a general base/acid catalyst (27). However, the corresponding lysine in vaccinia topoisomerase does not seem critical for enzyme activity (14), suggesting that this residue may not play the key role of a general base catalyst, at least for the topoisomerases. The two arginines of the triad almost certainly help polarizing the P—O bond and, at the same time, function to stabilize the pentacoordinated transition state. The second His (or Trp) residue also coordinates the scissile phosphate as judged by structural analysis (27, 55; also see reference 13). It has been suggested that this histidine might function as a general acid to donate a proton to the leaving 5′-oxyanion (25, 64). However, it is very unlikely that a tryptophan, present in the analogous position in some enzymes, could fulfill this role. Furthermore, substitution of this histidine in vaccinia topoisomerase to Asn or Gln, good hydrogen bond donors but poor general acids, does not seriously affect the enzyme activity (54), further suggesting that His-265 is not a general acid catalyst, unless the catalytic mechanism is changed for the mutant enzymes.

The importance of the conserved lysine between the Arg and His of the triad was first recognized by the mutational analysis of vaccinia virus topoisomerase (69) and recently confirmed for several tyrosine recombinases (10; G. van Duyne, personal communication). The role of this residue in catalysis was not immediately clear from the available structures. However, a relatively small movement of the position of the side chain in some of the DNA-bound structures could bring this side chain close to the 5′-OH leaving group. Thus, this lysine may play a role in protonating the leaving group (13, 26, 40). Consistent with this observation, strong evidence supports the conclusion that this residue is the general acid catalyst that donates a proton to the leaving 5′-oxyanion in the vaccinia virus topoisomerase reaction (40). The catalytic defect observed with the K167A mutant enzyme in the reaction where DNA is cleaved to form the covalent intermediate is largely reversed by a thiosubstitution of the leaving group oxygen. This is presumably because the lower pK_a of the —S(H) group makes it a better leaving group than the —O(H) that is usually present. With the wild-type enzyme, the same thiosubstitution has little impact on the cleavage rate, presumably because lysine is an efficient proton donor.

In addition to understanding the chemical roles of each functional group within the active site, it is important to understand how the conformational organization of the active site permits reaction selectivity and the properly ordered sequence of catalytic events. It has long been known that productive recombination requires a pair of DNA substrates with identical sequences in the crossover region. This crossover region (Color Plate 1) can essentially be any sequence, as long as it is the same on both recombining sites. The structure of the Cre-DNA complex in which the covalent intermediate has been trapped reveals an absence of extensive protein contacts involving the cut 5′ strand within the crossover region (26). Thus, the 5′-OH that will cleave the phosphotyrosine linkage of the covalent intermediate can be properly positioned for nucleophilic attack only if this strand can base pair with the new partner strand. Therefore, if the crossover region does not share a common sequence, the reaction will revert back exclusively to the substrate state. This structural organization therefore provides a powerful means for preventing recombination between mismatched sites. This physical mechanism for demanding sequence identity at the crossover region agrees well with an earlier model based on biochemical experiments (43, 53).

The tyrosine recombinases catalyze one pair of strand exchange reactions at a time: the first strand exchange forms a Holliday junction intermediate, and the second strand exchange resolves it (Color Plate 1B). This ordered catalytic sequence requires that, at any time, two of the four recombinase monomers within the catalytic tetramer adopt the active, catalytic conformation whereas the other two active sites must be forced into an inactive state. After the first strand exchange generates the Holliday intermediate, the two pairs of recombinase subunits must switch conformational states for the second strand exchange. This coordinated pairwise conformational change appears to be enforced by the architecture of the synaptic complex. The four recombinase subunts bound to the four "arms" of the substrate DNA are arranged around a pseudo-4-fold symmetry axis that coincides with a 2-fold axis (26). The structure of a recombinase subunit in the DNA-bound complex does not allow for all four of the DNA-bound subunits to adopt the same conformation. No acceptable conformation of the tetramer exists that would allow each of the four subunits to have the appropriate dimer interface in the active conformation at the same time; thus, the tetramer is always forced to contain two classes of dimers. These alternative conformations provide the structural basis for the pairwise coordinated switch between the engaged and disengaged active-site conformations that results in the sequential "one strand at a time" exchange mechanism (for more detail, see chapters 6 and 12).

ACTIVE SITES OF THE TRANSPOSASES AND RETROVIRAL INTEGRASES

Insight into the mechanism of transposition and retroviral integration is provided by biochemical and structural studies of multiple members of this protein family. The recent Tn*5* transposase-DNA complex also provides information regarding how the active sites engage their substrate DNA (17). Structures of domains from Mu transposase, several retroviral integrases, and the Tn*5* transposase reveal that proteins in this family have highly related catalytic domains (reviewed in references 21, 28, 58, and 72). Although these proteins represent three distant branches of the family tree (the integrases are close relatives), the structures surrounding the active centers of these proteins can nearly be superimposed. The structures also revealed a relationship, undetectable by primary sequence comparisons, to RNase H enzymes and the Holliday junction resolvase RuvC, placing the transposases as members of a large polynucleotidyltransferase superfamily (18, 57).

The organization of the catalytic domain (approximately 160 residues in its simplest form) is a mixed α/β-fold containing a five-stranded β-sheet flanked by α-helices. The domain contains the three invariant acidic amino acids (two Asps and a Glu), known as the DDE-motif residues, that mutagenesis in many family members has shown to be required specifically for catalysis of both the DNA cleavage and DNA strand transfer steps of recombination. Within the structures, these DDE-motif residues cluster together in the folded catalytic domain. These acidic residues have been proposed to coordinate the divalent metal ions, Mg^{2+} or Mn^{2+}, which are also required for catalysis of both DNA cleavage and DNA strand transfer.

Structural studies support the role of these essential acidic amino acids in coordinating metal ions in the active site. How these ions will be positioned in the catalytic center and the exact functions they will perform in the catalytic mechanism are not yet clear. In the Tn*5*-DNA cocrystal structure, one Mn^{2+} ion is bound to each active site, coordinated by two DDE-motif residues (Asp-97 and Glu-326) and the 3′-OH of the transferred DNA strand (17). This Mn^{2+} is attractively positioned for a role in activating this 3′-OH DNA end for nucleophilic attack on a new DNA phosphodiester. Although only one metal ion is observed in the current Tn*5* structure, as discussed above, many enzymes that catalyze phosphoryl transfer reactions use a two-metal-ion mechanism and it is currently unclear whether a one- or two-metal mechanism best describes catalysis by transposases. Evidence supporting a one-metal mechanism includes data indicating that the transposase cousin, RNase H, may require only a single catalytic metal ion (36). However, the RNase H active site differs significantly from that of transposases, and it may be misleading to consider this protein as a general model. Furthermore, mutagenesis studies of this enzyme (discussed above) reveal a surprising degree of plasticity in the active site, again highlighting the importance of caution in extrapolating general conclusions regarding mechanism from this one example.

Crystallographic studies that reveal metal ions bound to the retroviral integrase active sites are also ambiguous with respect to the number of divalent metal ions likely to be involved in catalysis. In the structures solved with Mg^{2+}, Mn^{2+}, or Ca^{2+}, a single metal ion is seen in the active site, but with Zn^{2+} or Cd^{2+}, two ions are observed (70). Furthermore, there is reason to suspect that none of these structures may accurately depict the fully functional active site. As seen with other phosphotransferases, the scissile phosphate of the substrate often contributes directly to metal ion coordination. None of the current RNase H or integrase structures have DNA engaged within their active sites, and the substrate DNA used in the Tn*5* structure lacks the terminal 5′-phosphate on the "nontransferred" strand (see above), which is expected to lie in the active site. Thus, additional experiments are needed to resolve how the metal ions are bound, and the roles they play in catalysis of the individual steps of transposition.

In addition to the DDE-motif residues, several amino acids have been implicated in catalysis and substrate docking to the active site. However, in contrast to the invariant nature of the DDE residues, these additional residues are not as highly conserved. Rather, specific amino acids appear to be conserved features of subgroups of transposases, but not common to the entire family. Biochemical experiments with HIV integrase help inform how DNA is interacting with this active site. Investigation of the regions of the protein responsible for different aspects of DNA substrate specificity established that the catalytic core domain is responsible for recognizing nucleotides immediately around the site of DNA cleavage (22), as well as the target DNA (62). Protein-DNA cross-linking experiments have indicated that a few residues (most notably Lys-159 and Gln-62) in this domain are in close contact with the cleavage site (34). These residues lie close to acidic residues in the active site and are well conserved among retroviral integrases. The synaptic complex structure of Tn*5* transposase reveals a direct role for a lysine in the equivalent position of HIV-1 Lys-159 (Lys-333 in the Tn*5* protein) in contacting the DNA in the active site (17). The nearby basic residue, Lys-330, makes an additional contact. As discussed earlier, lysines can play several different active-site functions; careful mutational analysis and additional structural information will be needed to pinpoint their catalytic role.

The IS*4* subfamily of transposases has additional conserved residues in their active sites that seem especially important for the DNA hairpin mechanism used by these proteins to cleave the nontransferred strand during transposon excision. This family of proteins has a motif known as the "YREK signature" (which conforms to the sequence pattern $YX_2RX_3EX_6K$), where E is the catalytic glutamate (56). The view of the protein's active site in the Tn*5* transposase synaptic complex structure provides a picture of how these residues function to assist in DNA hairpin formation. To form the DNA hairpin, the backbone of the nontransferred strand must be bent; the distortion of the DNA is extensive, with the thymine located adjacent to the last base of the DNA strand "flipped out" of its normal position; the 5′-phosphate of this residue is held in position by interactions with the tyrosine and arginine (Tyr-319 and Arg-322) of the YREK motif (17). The role for these residues in catalysis is supported by the deleterious effects of mutations at these positions in both the Tn*5* and Tn*10* transposases. Arg-322 is also poised to interact with the scissile phosphate and may play a catalytic role in reaction steps in addition to hairpin formation. A substitution mutation of the analogous arginine in Tn*10* transposase obliterates all of this protein's catalytic activities (8).

The key to understanding the catalytic mechanisms used by transposases is knowledge of how the different DNA segments are positioned and repositioned within the active sites as recombination progresses. It is now clear that a single active center promotes multiple distinct chemical steps. For Tn*5* and Tn*10* there are a total of four reaction steps: hydrolysis of the transferred strand, hairpinning, hydrolysis of the DNA hairpin, and DNA strand transfer to covalently join this cleaved DNA to a new target site. Although this strategy of using one catalytic center to catalyze multiple reactions is economical, it also requires that the active site have unusual flexibility in both the substrate and nucleophile binding sites. The active center, to bind and stabilize the intermediates during each of these reactions, must be occupied sequentially by three different phosphate groups: first one from the transferred DNA strand, then one from the nontransferred strand, and finally one from the target DNA. Moreover, when a particular phosphate group leaves the active site, it does not fully dissociate, but remains in the complex during the next reaction step. An interesting series of conformational changes must be required to rearrange the strands between the sequential recombination steps. Are these different scissile DNA strands always bound by the active site in the same fashion? Or are there multiple binding modes within the active-site pocket?

Recent experiments with Tn*10* transposase using chiral phosphorothioates at the scissile phosphate have shed light on how the "to be attacked" DNA phophodiesters may be positioned in the active center (38). Considered together with the structure of the Tn*5* complex, these biochemical data suggest that the scissile phosphate of the target DNA is not bound in exactly the same orientation as was the scissile phosphate of the earlier reaction steps, that is, the first strand cleavage, hairpin formation, or hairpin opening. Furthermore, the Tn*5* structure clearly suggests that the 3′ end of the transferred strand is likely to remain relatively fixed within the active site throughout recombination. If this is in fact the case, the various phosphates must then enter and exit the catalytic center as recombination proceeds. It will be essential to have structures of transposase family members poised for catalyzing all the reaction steps in a round of recombination to understand this process.

Even outside the direct neighborhood of the active site, the structural organization of transposases seems to influence catalytic activity. Transposases function as multimeric proteins. The IS4 family proteins work as dimers (17), Mu transposase is a tetramer (42), and HIV-1 integrase has been proposed to function as either a tetramer (73) or an octamer (29). A common feature of transposases is that they need to complete the assembly processes before becoming active catalysts. Both biochemical experiments and the Tn*5* synaptic complex structure provide insight into the mechanistic basis of the requirement for assembly. First, protein subunits in the synaptic complex contact both of the DNA ends of the tranposon; thus the subunits are not in the proper position for catalysis unless the complex is assembled and the DNA ends are paired. The Tn*5* transposase-DNA complex structure illustrates this feature dramatically; each of the subunits in the dimer makes extensive contacts with both of the Tn*5* DNA ends (17). This structure reveals, as has been shown biochemically for the Mu transposase (1, 51, 61, 68), that within the active synaptic complex, individual subunits recognize one DNA end via their sequence-specific DNA binding domain, whereas the same protein subunit donates its catalytic domain to the opposite DNA end. This interwoven subunit arrangement can explain why complex assembly is a prerequisite for catalytic activity. However, additional changes in the transposase proteins that occur during assembly and DNA binding are probably also involved. It is very likely that protein conformational changes that occur during assembly will prove to be essential for activation of the protein's catalytic function. Furthermore, the key feature of transposases is that once they are assembled, they promote the multiple steps of recombination without ever dissociating from the substrate DNA. This feature helps to orchestrate the chemical reactions on the two ends of the element DNA and must be kept in mind when one considers the catalytic strategies used by members of this protein family.

CATALYTIC STRATEGIES: RELATIONSHIP TO MECHANISMS OF REGULATION

The recombinases discussed here, although clearly biological "catalysts" because of their ability to enhance specific chemical reaction rates, are not conventional enzymes. They generally do not release their products and thus do not "turn over" and are required in stoichiometric quantities. This characteristic of recombinases is related to two important biological features that distinguish these proteins from traditional enzymes. First, the physiological demand on the reaction rate for phosphoryl transfer in recombination is usually very low, with many reactions occurring less than once every cell cycle. Second, the physiology also frequently demands that essentially all substrate molecules are converted to product, even when the "substrate" and the "product" are nearly isoenergetic.

The catalytic strategies used by the recombinases for chemical step catalysis therefore differ fundamentally from those for high-turnover enzymes that are true catalysts. True catalysts cannot influence the substrate/product equilibrium. If the reaction needs to proceed against an uphill free-energy difference, the catalytic mechanism must be coupled to the consumption of high-energy cofactors. Also, a high turnover rate demands that the activation energy barrier for the overall reaction be relatively small. Thus, excessively stabile enzyme-substrate and enzyme-product complexes must be avoided. In contrast, when high turnover rates are not needed, the protein-catalyst need not be constrained to release the product quickly. Thus, the reaction can take place within a stable protein-DNA complex that does not allow spontaneous product release, and product release can subsequently be coupled to additional cellular processes that involve an external energy source. Because recombination takes place within a stable protein-DNA complex, several physiologically beneficial control mechanisms can be built into the reaction. Many examples of these regulatory strategies can be found in the later chapters. We briefly summarize a few of these general mechanisms below.

Genome rearrangements are dangerous for the cell. Genome damage caused by recombination between an improper pair of sites or by a reaction that failed to go to completion can be especially deleterious. By demanding the assembly of a stable protein-DNA complex that includes all the proper substrate DNA sites and protein components before activation of the protein's DNA-cleaving activity, the danger of abortive partial reactions can be minimized. As described above, both site-specific recombinases and transposases have developed strategies to enforce complete complex assembly as a prerequisite to activation of the recombinase's catalytic function. These "assembly checkpoints" clearly contribute to avoidance of undesirable reactions.

The assembly of the recombination competent complex can also be the target of physiological control. For example, in many site-specific recombination reactions, the specific geometry of the DNA segments within the recombination complex appears to encourage productive synapses specifically between recombination sites that are in the proper orientation on the DNA. In prokaryotes, this "topological filter" effect is often augmented by the negative supercoiling of the substrate DNA. Examples of this type of control can be seen in many prokaryotic conservative site-specific recombination reactions and in Mu transposition (47). In addition, assembly of active recombinase complexes can be placed under the direct control of regulatory proteins, as illustrated by the regulation of Mu transposase assembly by the Mu transcriptional repressor (50). Efficient assembly of the Mu transposition complex requires participation of a transpositional enhancer element, a *cis*-acting DNA sequence near the left end of the Mu genome. To stimulate assembly, this enhancer sequence is transiently bound by a domain of the Mu transposase. This enhancer binding domain of transposase is homologous to the DNA binding domain of the transcriptional repressor; thus, the two proteins compete for binding to the same DNA sequence. As a result, when the repressor is produced, it directly blocks assembly of the recombination complex, even before repression of transcription takes effect.

Once the substrate-protein complex is assembled, recombination reactions appear to proceed toward the product-protein complex to an extent that goes beyond the equilibrium of a hypothetical uncatalyzed reaction. This is possible without the help of an external energy source such as ATP (at least during this reaction stage), because the product-protein complex can be more stable than the substrate-protein complex, and because dissociation of the product-protein complex is not required. Differential complex stability can be used not only to ensure the completion of the reaction, but also to switch the directionality of recombination, as seen in the λ integration/excision reaction (see chapter 7). Depending on the absence or presence of an accessory protein, Xis, different recombinase complexes are assembled; in the absence of Xis, the complexes formed clearly favor integrative recombination, whereas in the presence of Xis, excision is the predominant reaction.

The use of exceedingly stable protein-DNA complexes to drive recombination reactions to completion has the consequence that the stable product-protein complexes are unlikely to dissociate spontaneously within the time frame needed for continued cellular DNA metabolism. Thus, the disassembly of these complexes is likely to require an infusion of external energy. The best characterized example of this requirement is the use of the ClpX unfolding chaperone, a member of the Clp/Hsp100 ATPase family, to destabilize the Mu transposition complex at the junction between the recombination and replication stages of replicative transposition (41, 45). We anticipate that a similar requirement for catalysis of protein disassembly will be found for other recombinase complexes.

As illustrated in these examples, the structural organization of recombination complexes is intimately related to the mechanisms used for biological control. Perhaps one of the most satisfying aspects of the continued understanding of site-specific recombination and transposition is that, as the details of the protein-DNA complex structures and catalytic strategies are elucidated, they provide insight into the molecular basis of regulation. This field has been especially rich in fundamental insights, as a result of the parallel progress in analyzing related proteins that each function in a distinct biological niche. Despite substantial progress, numerous puzzles remain. Evidence suggests that the rewards of unlocking these puzzles will include basic insights into the chemical and physical bases of both biological catalysis and its control.

Acknowledgments. We thank Phoebe Rice, Ilana Goldhaber-Gordon, and Kenji Adzuma for helpful discussions and comments on the manuscript.

Work on transposases in T.A.B.'s lab is supported by the National Institutes of Health. Work in K.M.'s lab is supported in part by the NIH Intramural AIDS Targeted Antiviral Program. T.A.B. is an investigator of the Howard Hughes Medical Institute.

REFERENCES

1. **Aldaz, H., E. Schuster, and T. A. Baker.** 1996. The interwoven architecture of the Mu transposase couples DNA synapsis to catalysis. *Cell* **85**:257–269.
2. **Arciszewska, L. K., and D. J. Sherratt.** 1995. Xer site-specific recombination in vitro. *EMBO J.* **14**:2112–2120.
3. **Argos, P., A. Landy, K. Abremski, J. B. Egan, E. Haggard-Ljungquist, R. H. Hoess, M. L. Kahn, B. Kalionis, S. V. L. Narayana, L. S. Pierson III, N. Sternberg, and J. M. Leong.** 1986. The integrase family of site-specific recombinases: regional similarities and global diversity. *EMBO J.* **5**:433–440.
4. **Atkinson, P. W., W. D. Warren, and D. A. O'Brochta.** 1993. The hobo transposable element of Drosophila can be cross-mobilized in houseflies and excises like the Ac element of maize. *Proc. Natl. Acad. Sci. USA* **90**:9693–9697.
5. **Beese, L. S., and T. A. Steitz.** 1991. Structural basis for the 3′–5′ exonuclease activity of Echerichia coli DNA polymerase I: a two metal ion mechanism. *EMBO J.* **10**:25–33.

5a. **Benkovic, S. J., and K. J. Schray.** 1978. The mechanism of phosphoryl transfer, p. 493–527. *In* R. D. Gandour and R. L. Schowen (ed.), *Transition State of Biochemical Processes,* Plenum Press, New York, N.Y.

6. **Bhasin, A., I. Y. Goryshin, and W. S. Reznikoff.** 1999. Hairpin formation in Tn5 transposition. *J. Biol. Chem.* **274**: 37021–37029.
7. **Blakely G. W., A. O. Davidson, and D. J. Sherratt.** 1997. Binding and cleavage of nicked substrates by site-specific recombinases XerC and XerD. *J. Mol. Biol.* **265**:30–39.
8. **Bolland, S., and N. Kleckner.** 1996. The three chemical steps of Tn10/IS10 transposition involve repeated utilization of a single active site. *Cell* **84**:223–233.
9. **Brautigam, C. A., and T. A. Steitz.** 1998. Structural principles for the inhibition of the 3′–5′ exonuclease activity of Escherichia coli DNA polymerase I by phosphorothioates. *J. Mol. Biol.* **277**:363–377.
10. **Cao, Y., and F. Hayes.** 1999. A newly identified, essential catalytic residue in a critical secondary structure element in the integrase family of site-specific recombinases is conserved in a similar element in eucaryotic type IB topoisomerases. *J. Mol. Biol.* **289**:517–527.
11. **Casareno, R. L. B., D. Li, and J. A. Cowan.** 1995. Rational redesign of a metal-dependent nuclease. Engineering the active site of magnesium-dependent ribonuclease H to form an active "metal-independent" enzyme. *J. Am. Chem. Soc.* **117**: 11011–11012.
12. **Chen, J. W., B. R. Evans, S. H. Yang, H. Araki, Y. Oshima, and M. Jayaram.** 1992. Functional analysis of box I mutations in yeast site-specific recombinases Flp and R: pairwise complementation with recombinase variants lacking the active-site tyrosine. *Mol. Cell. Biol.* **12**:3757–3765.
13. **Chen, Y., U. Narendra, E. L. Iype, M. M. Cox, and P. A. Rice.** 2000. Crystal structure of a Flp recombinase-Holliday junction complex: assembly of an active oligomer by helix swapping. *Mol. Cell.* **6**:885–897.
14. **Cheng, C., L. K. Wang, J. Sekiguchi, and S. Shuman.** 1997. Mutational analysis of 39 residues of vaccinia DNA topoisomerase identifies Lys-220, Arg-223, and Asn-228 as important for covalent catalysis. *J. Biol. Chem.* **272**:8263–8269.
15. **Cheng, C., P. Kussie, N. Pavletich, and S. Shuman.** 1998. Conservation of structure and mechanism between eukaryotic topoisomerase I and site-specific recombinases. *Cell* **92**: 841–850.
16. **Coen, E. S., T. P. Robbins, J. Almeida, A. Hudson, and R. Carpenter.** 1989. Consequences and mechanism of transposition in Antirrhinum majus, p. 413–436. *In* D. E. Berg and M. M. Howe, (ed.), *Mobile DNA.* American Society for Microbiology, Washington, D.C.
17. **Davies, D. R., I. Y. Goryshin, W. S. Reznikoff, and I. Rayment.** 2000. Three-dimensional structure of the Tn5 synaptic complex transposition intermediate. *Science* **289**:77–85.
18. **Dyda, F., A. B. Hickman, T. M. Jenkins, A. Engelman, R. Craigie, and D. R. Davies.** 1994. Crystal structure of the catalytic domain of HIV-1 integrase: similarity to other polynucleotidyl transferases. *Science* **266**:1981–1986.
19. **Eckstein, F.** 1985. Nucleoside phosphorothioates. *Annu. Rev. Biochem.* **54**:367–402.
20. **Engelman, A., K. Mizuuchi, and R. Craigie.** 1991. HIV-1 DNA integration: mechanism of viral DNA cleavage and DNA strand transfer. *Cell* **67**:1211–1221.
21. **Esposito, D., and R. Craigie.** 1999. HIV integrase structure and function. *Adv. Virus Res.* **52**:319–333.
22. **Gerton, J. L., and P. O. Brown.** 1997. The core domain of HIV-1 integrase recognizes key features of its DNA substrates. *J. Biol. Chem.* **272**:25809–25815.
23. **Gerton, J. L., D. Herschlag, and P. O. Brown.** 1999. Stereospecificity of reactions catalyzed by HIV-1 integrase. *J. Biol. Chem.* **274**:33480–33487.
24. **Gopaul, D. N., F. Guo, and G. D. Van Duyne.** 1998. Structure of the Holliday junction intermediatein Cre-LoxP site-specific recombination. *EMBO J.* **17**:4175–4187.
25. **Grainge, I., and M. Jayaram.** 1999. The integrase family of recombinases: organization and function of the active site. *Mol. Microbiol.* **33**:449–456.
26. **Guo, F., D. N. Gopaul, and G. D. Van Duyne.** 1997. Structure of Cre recombinase complexed with DNA in a site-specific recombination synapse. *Nature* **389**:40–46.
27. **Guo, F., D. N. Gopaul, and G. D. Van Duyne.** 1999. Asymmetric DNA bending in the Cre-1oxP site-specific recombination synapse. *Proc. Natl. Acad. Sci. USA* **96**:7143–7148.
28. **Haren, L., B. Ton-Hoang, and M. Chandler.** 1999. Integrating DNA: transposases and retroviral integrases. *Annu. Rev. Microbiol.* **53**:245–281.
29. **Heuer, T. S., and P. O. Brown.** 1998. Photo-cross-linking studies suggest a model for the architecture of an active human immunodeficiency virus type 1 integrase-DNA complex. *Biochemistry* **37**:6667–6678.
30. **Hickman, A. B., S. Waninger, J. J. Scocca, and F. Dyda.** 1997. Molecular organization in site-specific recombination: the catalytic domain of bacteriophage HP1 integrase at 2.7 A resolution. *Cell* **89**:227–237.
31. **Hickman, A. B., Y. Li, S. V. Mathew, E. W. May, N. L. Craig, and F. Dyda.** 2000. Unexpected structural diversity in DNA recombination: the restriction endonuclease connection. *Mol. Cell* **5**:1025–1034.
32. **Hollfelder, F., and D. Herschlag.** 1995. The nature of the transition state for enzyme-catalyzed phosphoryl transfer. Hydrolysis of O-aryl phosphorothioates by alkaline phosphatase. *Biochemistry* **34**:12255–12264.
33. **Hughes, R. E., G. F. Hatfull, P. Rice, T. A. Steitz, and N. D. Grindley.** 1990. Cooperativity mutants of the gamma delta resolvase identify an essential interdimer interaction. *Cell* **63**: 1331–1338.
34. **Jenkins, T. M., D. Esposito, A. Engelman, and R. Craigie.** 1997. Critical contacts between HIV-1 integrase and viral DNA identified by structure-based analysis and photo-cross-linking. *EMBO J.* **16**:6849–6859.
35. **Jou, R., and J. A. Cowan.** 1991. Ribonuclease H activation by inert transition-metal complexes. Mechanistic probes for metallocofactors: insights on the metallobiochemistry of divalent magnesium ion. *J. Am. Chem. Soc.* **113**:6685–6686.
36. **Keck, J. L., E. R. Goedken, and S. Marqusee.** 1998. Activation/

attenuation model for RNase H. A one-metal mechanism with second-metal inhibition. *J. Biol. Chem.* **273**:34128–34133.

37. **Kennedy, A. K., A. Guhathakurta, N. Kleckner, and D. B. Haniford.** 1998. Tn10 transposition via a DNA hairpin intermediate. *Cell* **95**:125–134.
38. **Kennedy, A. K., D. B. Haniford, and K. Mizuuchi.** 2000. Single active site catalysis of the successive phosphoryl transfer steps by DNA transposases: insights from phosphorothioate stereoselectivity. *Cell* **101**:295–305.
39. **Knowles, J. R.** 1980. Enzyme-catalyzed phosphoryl transfer reactions. *Annu. Rev. Biochem.* **49**:877–919.
40. **Krogh, B. O., and S. Shuman.** 2000. Catalytic mechanism of DNA topoisomerase IB. *Mol. Cell* **5**:1035–1041.
41. **Kruklitis, R., D. J. Welty, and H. Nakai.** 1996. ClpX protein of Escherichia coli activates bacteriophage Mu transposase in the strand transfer complex for initiation of Mu DNA synthesis. *EMBO J.* **15**:935–944.
42. **Lavoie, B. D., B. S. Chan, R. G. Allison, and G. Chaconas.** 1991. Structural aspects of a higher order nucleoprotein complex: induction of an altered DNA structure at the Mu-host junction of the Mu type 1 transpososome. *EMBO J.* **10**: 3051–3059.
43. **Lee, J., and M. Jayaram.** 1995. Role of partner homology in DNA recombination. *J. Biol. Chem.* **270**:4042–4052.
44. **Lee, J., M. Jayaram, and I. Grainge.** 1999. Wild-type Flp recombinase cleaves DNA in trans. *EMBO J.* **18**:784–791.
45. **Levchenko, I., L. Luo, and T. A. Baker.** 1995. Disassembly of the Mu transposase tetramer by the ClpX chaperone. *Genes Dev.* **9**:2399–2408.
46. **Maegley, K. A., S. J. Admiraal, and D. Herschlag.** 1996. Ras-catalyzed hydrolysis of GTP: a new perspective from model studies. *Proc. Natl. Acad. Sci. USA* **93**:8160–8166.
47. **Mizuuchi, K.** 1992. Transpositional recombination: mechanistic insights from studies of Mu and other elements. *Annu. Rev. Biochem.* **61**:1011–1051.
48. **Mizuuchi, K., and K. Adzuma.** 1991. Inversion of the phosphate chirality at the target site of Mu DNA strand transfer: evidence for a one-step transesterification mechanism. *Cell* **66**: 129–140.
49. **Mizuuchi, K., T. J. Nobbs, S. E. Halford, K. Adzuma, and J. Qin.** 1999. A new method for determining the stereochemistry of DNA cleavage reactions: application to the SfiI and HpaII restriction endonucleases and to the MuA transposase. *Biochemistry* **38**:4640–4648.
50. **Mizuuchi, M., and K. Mizuuchi.** 1989. Efficient Mu transposition requires interaction of transposase with a DNA sequence at the Mu operator: implications for regulation. *Cell* **58**: 399–408.
51. **Namgoong, S. Y., and R. M. Harshey.** 1998. The same two monomers within a MuA tetramer provide the DDE domains for the strand cleavage and strand transfer steps of transposition. *EMBO J.* **17**:3775–3785.
52. **Nunes-Duby, S. E., R. S. Tirumalai, L. Dorgai, E. Yagil, R. A. Weisberg, and A. Landy.** 1994. Lambda integrase cleaves DNA in cis. *EMBO J.* **13**:4421–4430.
53. **Nunes-Duby, S. E., M. A. Azaro, and A. Landy.** 1995. Swapping DNA strands and sensing homology without branch migration in lambda site-specific recombination. *Curr. Biol.* **5**: 139–148.
54. **Petersen, B. O., and S. Shuman.** 1997. Histidine 265 is important for covalent catalysis by vaccinia topoisomerase and is conserved in all eukaryotic type I enzymes. *J. Biol. Chem.* **272**: 3891–3896.
55. **Redinbo, M. R., L. Stewart, P. Kuhn, J. J. Champoux, and W. G. Hol.** 1998. Crystal structure of human topoisomerase I in covalent and noncovalent complexes with DNA. *Science* **279**: 1504–1513.
56. **Rezsohazy, R., B. Hallet, J. Delcour, and J. Mahillon.** 1993. The IS4 family of insertion sequences: evidence for a conserved transposase motif. *Mol. Microbiol.* **9**:1283–1295.
57. **Rice, P., and K. Mizuuchi.** 1995. Structure of the bacteriophage Mu transposase core: a common structural motif for DNA transposition and retroviral integration. *Cell* **82**: 209–220.
58. **Rice, P., R. Craigie, and D. R. Davies.** 1996. Retroviral integrases and their cousins. *Curr. Opin. Struct. Biol.* **6**:76–83.
59. **Richards, F. M., and H. W. Wyckoff.** 1971. Bovine pancreatic ribonuclease, p. 647–806. *In* P. D. Boyer (ed.), *The Enzymes,* 3rd ed., vol. 4. Academic Press, New York, N.Y.
60. **Sarnovsky, R. J., E. W. May, and N. L. Craig.** 1996. The Tn7 transposase is a heteromeric complex in which DNA breakage and joining activities are distributed between different gene products. *EMBO J.* **15**:6348–6361.
61. **Savilahti, H., and K. Mizuuchi.** 1996. Mu transpositional recombination: donor DNA cleavage and strand transfer in trans by the Mu transposase. *Cell* **85**:271–280.
62. **Shibagaki, Y., and S. A. Chow.** 1997. Central core domain of retroviral integrase is responsible for target site selection. *J. Biol. Chem.* **272**:8361–8369.
63. **Steitz, T. A., and J. A. Steitz.** 1993. A general two-metal-ion mechanism for catalytic RNA. *Proc. Natl. Acad. Sci. USA* **90**: 6498–6502.
64. **Stewart, L., M. R. Redinbo, X. Qiu, W. G. J. Hol, and J. J. Champoux.** 1998. A model for the mechanism of human topoisomerase I. *Science* **279**:1534–1540.
65. **Turlan, C., and M. Chandler.** 2000. Playing second fiddle: second-strand processing and liberation of transposable elements from donor DNA. *Trends Microbiol.* **8**:268–274.
66. **van Gent, D. C., K. Mizuuchi, and M. Gellert.** 1996. Similarities between initiation of V(D)J recombination and retroviral integration. *Science* **271**:1592–1594.
67. **Weil, C. F., and R. Kunze.** 2000. Transposition of maize Ac/Ds transposable elements in the yeast Saccharomyces cerevisiae. *Nat. Genet.* **26**:187–190.
68. **Williams, T. L., E. L. Jackson, A. Carritte, and T. A. Baker.** 1999. Organization and dynamics of the Mu transpososome: recombination by communication between two active sites. *Genes Dev.* **13**:2725–2737.
69. **Wittschieben, J., and S. Shuman.** 1997. Mechanism of DNA transesterification by vaccinia topoisomerase: catalytic contributions of essential residues Arg-130, Gly-132, Tyr-136 and Lys-167. *Nucleic Acids Res.* **25**:3001–3008.
70. **Wlodawer, A.** 1999. Crystal structures of catalytic core domains of retroviral integrases and role of divalent cations in enzymatic activity. *Adv. Virus Res.* **52**:335–350.
71. **Yang, W., and T. A. Steitz.** 1995. Crystal structure of the site-specific recombinase gamma delta resolvase complexed with a 34 bp cleavage site. *Cell* **82**:193–207.
72. **Yang, W., and T. A. Steitz.** 1995. Recombining the structures of HIV integrase, RuvC and RNase H. *Structure* **3**:131–134.
73. **Zheng, R., T. M. Jenkins, and R. Craigie.** 1996. Zinc folds the N-terminal domain of HIV-1 integrase, promotes multimerization, and enhances catalytic activity. *Proc. Natl. Acad. Sci. USA* **93**:13659–13664.

Mobile DNA II
Edited by N. L. Craig et al.

Chapter 3

Putting Mobile DNA to Work: the Toolbox

Jef D. Boeke

INTRODUCTION

The past few years have seen the introduction of many new practical applications based on mobile nucleic acids and the reactions they effect. Many conservative site-specific reactions catalyzed by Cre-*lox,* yeast Flp, and lambda integrase systems have been put to work in a wide variety of host organisms and in vitro. In addition, many new methods based on DNA transposons, retrotransposons, and even mobile introns have recently become practical in vitro as well as in vivo, affording new levels of specificity and efficiency. Several biotechnology companies now prominently feature mobile DNA-based products in their catalogs. In this chapter, I review some mobile DNA tools currently being used in DNA sequencing for manipulating clones efficiently, and in functional analysis of more modest segments of DNA containing one or a few genes as well as entire genomes. I also cover the increasing use of transposons as genetic tags and as devices for the delivery and selective destruction of genes. Finally, I briefly review some recent applications of transposons as genetic markers.

TOOLS FOR DNA SEQUENCING AND FOR ANALYSIS OF GENE-SIZED SEGMENTS

The ability of transposons to serve as mobile priming sites useful for acquiring large amounts of sequencing with use of a universal primer was first pointed out by Ahmed (2). Ahmed's method exploited intramolecular deletions promoted by transposase to generate multiple sequenceable endpoints. Because it required the use of specialized, cumbersome, and large vectors, it never gained wide use. Moreover, each clone yielded only a single sequence of interest. Adachi et al. (1) reported using intermolecular transposition of a mini-Mu element into a target plasmid. Besides being much more flexible with regard to target plasmids, each insertion mutant clone yields two sequences, one primed from each end of the transposon. Subsequently, similar approaches based on Tn*5* (50) and γδ (Tn*1000* and its close relative, Tn*3*) (71) were reported.

More recently, transposition was performed in vitro to generate libraries of insertion mutants in a target plasmid of choice useful for gene sequencing and other purposes. The overall outline of this process is shown in Fig. 1. Mu transposition in vitro was first used to mutagenize Tn*7* (73). Ty1 was the first transposon to be commercialized as a sequencing tool (17, 19, 40). Subsequently, the Tn*552* element (25), variants of Tn*7* with reduced target specificity (6, 29, 69, 70), Tn*5* (8, 22, 23), Tn*10* (Y.-H. Sun, S. Bakshi, R. Chalmers, and C. M. Tang, submitted for publication), *mariner* (3, 42), and Mu (30, 31) were developed for similar purposes. All these systems have their

Jef D. Boeke • Department of Molecular Biology & Genetics, Johns Hopkins University School of Medicine, 617 Hunterian Building, 725 N. Wolfe St., Baltimore, MD 21205-2185.

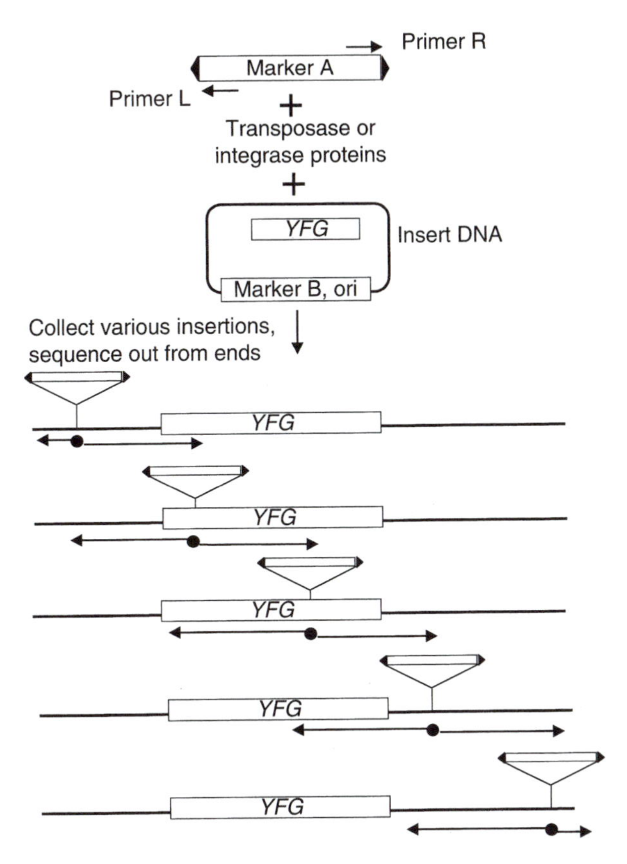

Figure 1. In vitro transposition as a method for sequencing DNA. A transposable element (box with black triangles; black triangles represent short terminal inverted repeats recognized by transposase or integrase) bearing a selectable marker and unique priming sites near each end (Primer L and Primer R, arrows) is mixed with the appropriate protein and a target plasmid bearing *YFG* or DNA segment of interest (bold line). After the transposition reaction occurs in vitro, bacteria expressing markers A and B are selected and DNAs are prepared. DNA from each individual transformant is sequenced with use of primers A and B, resulting in two divergent sequence reads (horizontal arrows) from each transposon which can be linked via the target-site duplication sequence (filled circle) and then assembled with the other sequence reads into a single contig.

own unique advantages and disadvantages, summarized in Table 1.

A key issue for many transposon-based tools is their degree of randomness. Ideally, there should be no target-site preference or specificity whatsoever, because this results in the largest diversity of sites of integration, which is what one is trying to achieve. In practice, this is never the case; presumably, at the very least, certain DNA/integration machinery interactions are disfavored because of local microvariations in DNA structure. For example, Ty1 may seem an odd choice as a tool in terms of randomness, for it is becoming clear that it is highly nonrandom in vivo, being preferentially inserted upstream of tRNA genes (18). Nevertheless, it is highly random in vitro, presumably because the in vivo randomness requires a specialized chromatin structure or other factors absent in vitro (7). A key advantage of the Ty1 system is that only 4 bp of Ty1 terminus are required for transposition; this means that the system offers the ultimate in flexibility in terms of engineering the "guts" of artificial transposons bearing a variety of useful bells and whistles. The major shortcoming of the Ty1 system currently is that the efficiency of the in vitro reaction is somewhat lower than those of some bacterial transposons. Still, it is straightforward to recover $>10^4$ Ty1 insertions per microgram of target plasmid. Tn*552* also appears to be quite random and, like Ty1, is able to insert into DNAs with considerable variations in base composition (25). Tn*7*, another transposon with a very high degree of target specificity in vivo, has been engineered mutationally to lose its target-site selectivity, making it another attractive choice for sequence analysis (6). A rigorous mathematical analysis of insertion randomness has not yet been applied to any of the popular in vitro transposition systems. A cross-transposon analysis of in vitro randomness into a wide variety of targets would be very useful. The major shortcoming of the Tn*7* system is that the minimal Tn*7* ends that function well in vitro are rather long, and this limits the amount of flanking nontransposon DNA that can be acquired. Its major strengths are its high in vitro efficiency and the fact that the system is commercially available. Another interesting advantage is that, despite its high efficiency in vitro, double insertions are largely avoided through the action of Tn*7* target immunity, which operates in vitro (see chapter 19).

Although in vitro transposition systems work well for plasmid and cosmid targets up to 50 kb long, their success rate with very large targets such as BACs and YACs is generally poor, although successes have been reported with transposition with Mu (H. Savilahti, personal communication) and Tn*10* (R. Chalmers, personal communication). This is probably because all these transposition reactions must be done in the presence of divalent cations like Mg^{2+} or Mn^{2+}, which are also essential cofactors for nucleases. The ability of BACs and YACs to transform their host will be destroyed by one double-strand cleavage; thus, they can be rather sensitive to this type of contamination, which could derive from the transposase preparation or even from the DNA preparation itself.

Related techniques were developed for the analysis of relatively small segments of DNA containing one or a few genes. In this chapter, I use the generic term "your favorite gene" (*YFG*) to refer to the target gene for analysis. Several types of applications fall into this category, and many of the same transposon

Table 1. In vitro transposition system features

Element	Efficiency	Randomness	Minimum end length (bp)	TSD length (bp)	Transposon length limit (kb)	Protein requirements	Target immunity	Commercial source	Reference(s)
Tn*5*	High	Demonstrated	256	9	~11 (intra-molecular)	1 pure protein	No	Epicentre Technologies	Chapter 18
Tn*7*	High	Demonstrated *tns*A mutant	Left, 150; right, 250	5	>2.5	3 pure proteins	Yes	New England BioLabs	Chapter 19
Tn*10*	Low	Demonstrated (*ats* mutant)	23	9	≥6	1 pure protein	No	None	R. Chalmers (personal communication)
Tn*552*	Medium	Demonstrated	48	6–9 (variable)	≥3	1 pure protein	No	None	25
Ty1	Low	Demonstrated	4	5	≥2.5	VLPs (complex)[a]	No	Applied BioSystems	17, 19
Mu	High	Demonstrated	47–50	5	≥6.8	1 pure protein	Possibly	Finnzymes	30, 31

[a]VLPs, virus-like particles.

players useful for DNA sequencing have been used for these purposes. The applications include insertional mutagenesis, formation of families of fusion proteins, and epitope tagging of *YFG*. Most of these maneuvers have been performed in a variety of bacterial and yeast species. The Ty1 system has also been used for a special application—the efficient generation of families of gene knockout and knockin constructs for mammals (76). When a knockout mutation is desired, one wants to introduce a mammalian selectable marker into a large segment of cloned DNA. Convenient restriction sites are rarely available, and the sequence of intronic regions may even be lacking. An in vitro transposition procedure can rapidly generate a family of potential knockout constructs, and the optimal one can be identified by sequencing out from the transposon end, finding those that inserted so the reporter is in-frame with an exon of the gene of interest in a desirable location (e.g., near or in the first coding exon) so that it will likely give a null phenotype.

For organisms like yeast, in which gene function can conveniently be assayed by complementation of a mutant phenotype by introducing a recombinant plasmid, transposons form a convenient insertional mutagen. One of the first such mutagens, Tn10-LUK, inserted both yeast and bacterial markers into recombinant plasmids, and was useful for identifying *YFG* within a large multigene insert. This family of transposons also used an ATG-less *lacZ* gene, called *'lacZ,* as a reporter of in-frame fusions (36). Although the original version of Tn10-LUK had limited usefulness because of its lack of randomness reflecting Tn*10* transposase's innate target-site specificity, later developments included an altered target specificity (*ats*) mutant that minimized this problem (5).

More recently, both in vivo Tn*3* elements (55) and in vitro Ty1 elements (46) have been designed for the functional analysis of yeast genes (or clusters of genes) cloned on plasmids; these elements bear yeast and bacterial selectable markers. The Tn*3* elements are transposed into the yeast shuttle plasmid bearing *YFG* inside bacterial cells. Then the plasmids bearing the Tn*3* hops are transformed into yeast cells. An example of a highly modified Tn*3* element, laden with bells and whistles allowing multiple types of gene or genome analysis, is shown in Color Plate 5 (see color insert). As the modified Tn*3* elements contain *'lacZ* or *GFP* reporters, the expression of the hybrid protein can be readily determined by fluorescence or indicator plates. Moreover, the reporter gene and selectable markers segment can be subsequently removed in vivo because it is flanked by *loxP* sites. This leaves behind a short segment bearing multiple epitope tags for the popular hemagglutinin antibody 12CA5 and consisting of an integral number of codons. Several insertion mutants generated by this procedure produced a readily detectable protein with wild-type or near-wild-type function, whereas other mutants had a conditional phenotype (55). An alternative strategy using Ty1 involved a transposon containing a single marker gene, *GFPsupF,* that could be selected in bacteria and maintained an open reading frame throughout the transposon. This element generates similar "tribrid" proteins detectable directly (by fluorescence microscopic screening) after the transformation of the recombinant plasmid (46).

FACILITATING SUBCLONING AND STRUCTURAL ANALYSIS

Two systems have recently been developed to attack the difficulties associated with the often complex subcloning steps required to ensure efficient recombinant protein expression. Often, multiple vector systems must be tested to identify one optimal for expression of a given protein. Care must be taken to ensure that translational start sites, epitope tags, protease cleavage sites, etc. are fused in the proper manner. These new methods bypass the need for restriction enzymes (not to mention appropriately placed restriction sites!) and DNA ligase, and numerous cloning reactions can be performed, in parallel, in a few hours. These methods can also be exploited to allow high-throughput computer-driven construction of large libraries of open reading frames (ORFs) into multiple expression vectors for, e.g., two-hybrid analyses, or protein expression for in vitro analyses and structural studies. Two systems designed to facilitate these maneuvers are based on bacterial site-specific recombination in vivo. One, the univector system (Fig. 2), uses the Cre-*lox* system, and the other, the Gateway system (Fig. 3), is based on the lambda Int/Xis (integrase/excision proteins).

The univector plasmid fusion system (44) uses Cre-*lox* to recombine two plasmids, a user-contructed univector containing the *YFG* coding sequence with a *lox* site adjacent to the ATG, and a host vector that places regulatory information (promoter, enhancer, epitope tags, etc.) upstream of the ATG. Thus, in a single step, a double-sized plasmid bearing an altered 5′ end is generated (Fig. 2). In a more complex procedure involving the sequential application of Cre-*lox* and a second site-specific recombination system, yeast

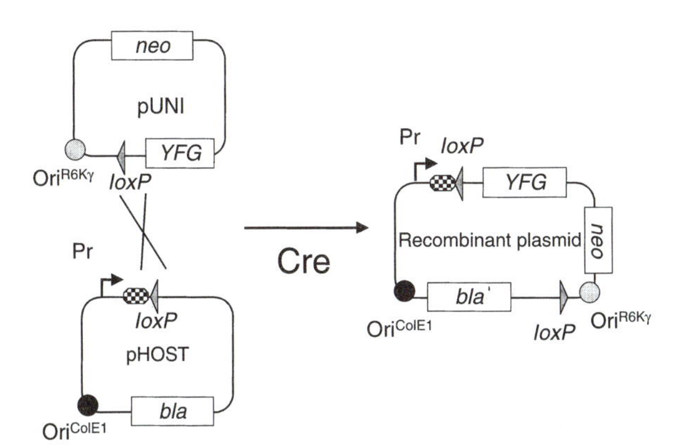

Figure 2. The univector system. A "univector" (pUNI) containing a *loxP* site (gray triangle) can be combined with one or more host vectors (pHOST) and can be efficiently and precisely recombined using Cre recombinase, generating a family of recombinant plasmids in which YFG is under the control of various promoters (hooked arrows), other 5′ control sequences, and epitope tags (checkerboard). The two plasmids contain compatible origins of replication R6K (gray dot) and Co1E1 (black dot). Adapted from reference 44.

RS-R, it is possible to manipulate the 3′ end of the cloned segment as well. The system is commercially available through Invitrogen as the Echo system.

The Gateway system (33, 74) consists of a user-designed entry vector in which the *YFG* coding sequence is flanked at both ends by two versions of the *attL* site. These two versions of the sequence are incompatible with each other because they contain different sets of point mutations. The entry vector itself can be constructed without the use of restriction enzymes and ligase from a PCR product bearing specially designed terminal *attB* sites that recombine with complementary *attP* sites flanking the counterselectable *ccdB* gene in the entry plasmid (Fig. 3C). Only correct recombinants (lacking *ccdB*) survive in wild-type *Escherichia coli* host strains. However, *ccdB* containing plasmids can be propagated in special strains bearing a *gyrA-462* mutation. This entry reaction requires an enzyme mix (BP Clonase) consisting of phage λ integrase (Int) and integration host factor (IHF) and generates (via two sequential conservative site-specific recombination reactions) an "entry" vector flanked by two different *attL* sites. The insert is then transferred to a variety of "destination" vectors, each containing *ccdB* flanked by two different *attR* sites corresponding to the *attL* sites used in the entry vector (Fig. 3B). This "destination reaction" is performed with "LR Clonase" (consisting of Int, Xis, and IHF). Recombinants can be selected at nearly 100% efficiency in this reaction as well. These destination vectors can contain altered regulatory regions at both the 5′ and 3′ ends. Hence it is possible to construct an extremely wide variety of expression or gene delivery constructs with this system, and the products generated are more compact than those made with the univector system. Moreover, most of the steps can be automated, allowing the design of high-throughput strategies to be used. The system is commercially available from Life Technologies.

Group I introns that encode homing endonucleases merit mention because these enzymes have begun to be more widely used as rare-cutters. Unlike typical restriction enzymes that recognize 4- to 8-bp target sequences which are often palindromic, these enzymes recognize nonpalindromic sequences of 12 to 40 bp.

However, although their target sequences are longer, one cannot assume that the homing endonucleases will have the same degree of base-for-base specificity as restriction endonucleases. Indeed, scanning through partially randomized cleavage site substrates allowed the identification of numerous variants that could be cut with varying efficiencies (4, 21). Thus, their observed sequence specificity is considerably lower than the length of their native target site might imply. Many interesting applications of

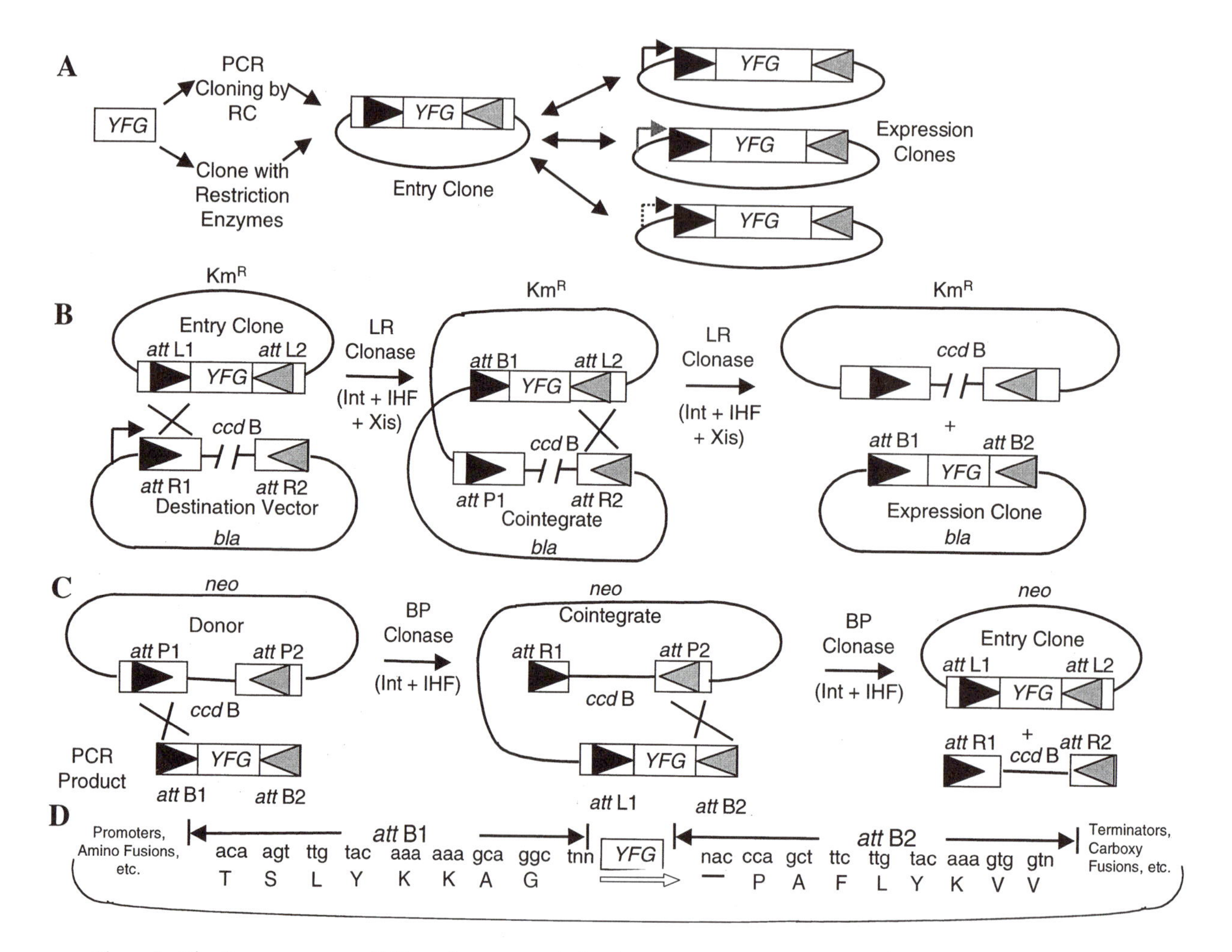

Figure 3. The Gateway system. PCR products or restriction fragments containing *YFG* can be manipulated by a series of steps based on the λ Int system to put its expression under the control of a wide variety of 5′ and/or 3′ regulatory sequences, epitope tags, protein fusions, etc. Boxed triangles represent λ *att* sites, labeled as to type; the triangles are black or gray to indicate type 1 and type 2 sites, respectively. The *ccdB* gene is a gene toxic to conventional strains of *E. coli* generally used for recombinant DNA work. (A) Overall flow of reactions. (B) Steps involved in generating a series of destination or expression clones from a single entry clone. (C) Generation of an entry clone from a *YFG* PCR product, into which terminal *attB* sites have been incorporated. Note that entry clones can also be generated by conventional restriction enzyme/ligase cloning steps. (D) Amino acid sequences that will be added to the N terminus (and, if desired, C terminus) of the Yfg protein. Adapted from reference 33.

these unusual enzymes are discussed in chapter 31. In addition, the impressively specific process of group II "retrointron" homing should be mentioned as a spectacular example of how a mobile element can be harnessed, in this case, as a disrupter of gene function. Because group II retrointron specificity is conferred by the retrointron RNA itself and not by a protein, these group II retrointrons can be re-engineered to designer specificities much more easily. This has recently been demonstrated quite impressively by the development of group II retrointrons specific for segments of HIV-1 and other DNAs (see chapter 31).

WHOLE-GENOME ANALYSES

In addition to their use as tools for analyzing gene-sized segments of DNA, transposons continue to acquire new roles in whole-genome analyses, especially as genome sequences are completed. The original uses of transposons as tools for genome analysis date back to Casadaban's pioneering work with Mu in *E. coli* (12, 13). Transposon tagging was developed for *Drosophila* in the early 1980s as a way to clone genes (57, 63, 67), and represents the first transposon-based, genome-wide approach to gene function. This

Table 2. Genome-wide mobile DNA-based functional genomics resources, "Security Council" organisms

Organism	Transposon	Resource type	Reference	URL
S. cerevisiae	Ty1	Gene function database	65	http://cmgm.stanford.edu/pbrown/
	Tn*3*	Gene function database, mutant strains	54	http://ygac.med.yale.edu/triples/triples.htm
D. melanogaster	Various	Catalog, strains	68	
A. thaliana	Ti plasmid	Localization information	16	
Mouse	Retrovirus	Mutant ES cells, flanking sequences	80	

technique culminated in a recent compendium of thousands of P insertions, including insertions of 25% of the essential *Drosophila* genes (68). A critical technical development along the way was the development of transposons bearing enhancer or promoter traps, which are transposons bearing enhancerless or promoterless reporter genes (15). These are especially useful tools in multicellular organisms, in which they permit the identification of genes by their expression patterns, dramatically revealed visually via the action of the attached reporter genes. Some genome-wide, transposon-based functional genomics and mutagenesis projects for widely studied organisms are summarized in Table 2.

An especially dramatic recent application of transposons to practical analysis of genomes of complex eukaryotes was described by the Hirochika laboratory, which isolated and harnessed an endogenous rice long terminal repeat (LTR) retrotransposon, Tos17. This element is highly active in tissue culture, in which it efficiently transposes into the coding regions of many host genes. In fact, this type of transposon may well be responsible for the very common mutations that arise spontaneously during tissue culture of many plant species. Tos17 is a low-copy-number element normally, but amplifies significantly during growth in tissue culture; individual plants regenerated from these callus cultures typically bear 5 to 30 new hops (34, 35). Thousands of mutant plant lines have been generated and screened phenotypically in this way, and most of the transposon insertions isolated indeed disrupt genes (H. Hirochika, personal communication).

Mammalian genomes have also been attacked with a variety of transposon-tagging strategies. An elegant strategy was developed by the group at Lexicon, in which mouse retroviruses were engineered to bear exon-trapping reporter genes (80). Mouse embryonic stem (ES) cells are transfected with a specially engineered retrovirus (Fig. 4) that contains a phosphoglycerate kinase promoter-driven reporter such as puromycin resistance. The latter gene lacks a polyadenylation signal but contains instead a splice donor site downstream of the Puro coding sequence; splicing to any cellular exon thus introduces a new polyadenylation signal, allowing Puro expression. A high-throughput method for amplifying the target gene exons flanking the retrovirus was developed. The flanking sequence information thus obtained is banked together with the mutant ES cells for later use in generating mice bearing the desired insertion mutation. In addition to the 3′ Puro reporter, the virus contains an upstream, internal polyadenylation signal designed to truncate incoming transcripts driven by the target gene's own promoter.

Are the resulting insertions null mutants? Indeed several mutations generated by this method result in a null phenotype. Nevertheless, it remains possible that low-level readthrough through the entire retrovirus genome could result in retention of partial gene function in cases where small amounts of gene product have biological activity. Alternatively, because two transcripts are generated from each insertion (in Fig. 4, exons 1 + 2 + 3 + reporter A and also reporter B, possibly in frame with exons 4 + 5), it is theoretically possible that intragenic complementation between the two resultant protein fragments could regenerate a functional protein. Nevertheless, these cases are probably the exception, not the rule. Once an investigator finds a homology between the Lexicon flanking exon database and the sequence of interest, it is possible (for a fee) to obtain the ES cells and, using these, to generate mice carrying the identified insertion mutation. This resource will become even more valuable once the full mouse genome sequence is available.

Since the completion of the sequencing of several microbial genomes and the yeast genome, a variety of innovative transposon-based approaches have been developed to determine gene function. One would like to know not only whether a given gene is essential for viability but also something about the function of each gene product. For the yeast genome, two

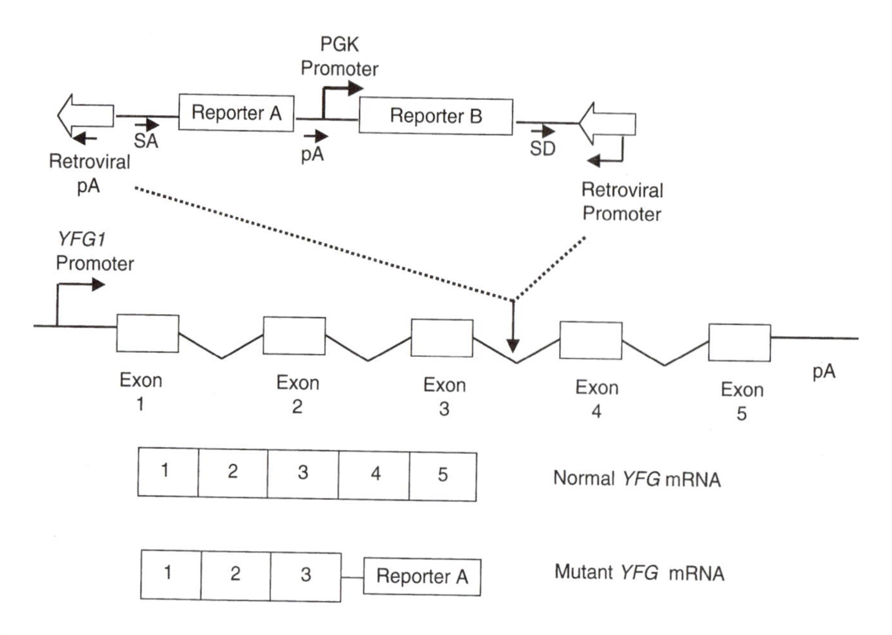

Figure 4. The Lexicon exon trap retrovirus. An exon trap retrovirus is engineered to contain two reporter genes A and B in orientations opposite to the retrovirus' natural transcription signals. Reporter B is inactive because it lacks a polyadenylation signal. Insertion of the retrovirus into a gene allows reporter B to acquire a polyadenylation signal by splicing (via the splice donor [SD]) signal engineered at the 3′ end of reporter B to adjacent cellular exons, resulting in expression of reporter B. At the same time, expression of *YFG* is disrupted because of the introduction of the polyadenylation signal upstream of the PGK promoter, which will both incorporate reporter A via the splice donor signal (SA) and truncate the *YFG* transcript via the polyadenylation signal (pA). Open arrows, LTRs.

transposon approaches have been used. The first is a direct extension of the modified tagged Tn*3* approach outlined above for individual genes, but extended to the entire yeast genome (9, 54). The modified Tn*3* elements are transposed in bacterial cells into yeast genomic DNA libraries (previously prepared in specialized plasmid vectors). These are then purified from *E. coli*, and the inserts are released via rare-cutting restriction enzymes flanking the inserts and used to transform yeast. Only yeasts expressing the truncated *'lacZ* reporter were picked (both mitotic and meiotic cells were examined), ensuring that only productive, in-frame fusions to target genes would be identified. Data on the localization of the resulting fusion protein (or epitope-tagged protein) were collected. The transposon flanks were also recovered and sequenced. Finally, the various insertion mutants were subjected to a battery of phenotypic tests. From this, a comprehensive database of yeast genes, potential functions, and localization of gene products emerges. One of the biggest surprises from this work was the number of fusions that expressed well but either did not correspond to an identified ORF or, despite high expression, appeared to be inserted out-of-frame relative to the previously identified host gene sequence.

In another variation of this theme, Tn*7* transposons modified by the addition of a yeast gene and a small plasmid allowing easy recovery of the transposon and its flanking DNA were inserted in vitro into large segments of sheared yeast genomic DNA. This DNA was then transformed directly into yeast, followed by selection for the yeast marker and screening for a phenotype of interest. This is a very efficient method for the screening of insertion mutations in yeast strains of any genetic background and for rapid identification of the disrupted gene (4a).

The second approach, genetic footprinting, exploits the endogenous yeast transposon Ty1 to characterize genes on the basis of function (Color Plate 6 [see color insert]). In this strategy, a Ty1 element tagged very close to one end with a unique oligonucleotide sequence (43) is allowed to transpose to very high levels throughout the genome. This generates a population of yeast strains, each bearing multiple insertions (65, 66). A PCR can then be conducted with a labeled primer corresponding to *YFG*, and an unlabeled primer specific for the tip of the marked Ty1 element. If transpositions occur throughout the gene and are not selected against (i.e., the gene is nonessential), numerous bands are seen when the PCR products are separated. The population can then be subjected to one or more genetic selections and, if a gene is required for a given selective condition, a "footprint" is seen on the gel pattern corresponding to the positions of insertions that inactivated the function of that target gene (Color Plate 6). This approach has now been used to analyze much of the yeast genome.

A variation on the footprinting theme for analyzing the genomes of naturally transformable bacteria referred to as GAMBIT (genomic analysis and mapping through in vitro transposition) has also been described (3). Large genomic segments are amplified by PCR. Marked Himar (*mariner* family) transposons are then inserted into these amplified segments in vitro, and the reaction mixture is transformed into the bacteria. The pool of selected transformants is then subjected to PCR with gene-specific and transposon end-specific primers (see genomic footprinting, above); the absence of bands from the selected pool implies that the disrupted region encodes an essential gene.

CONDITIONAL KNOCKOUTS AND OTHER CRE-*LOX* WIZARDRY

The Cre-*lox* system (see chapter 5 and reference 48a) and, to a considerable extent, the related FLP system (see chapter 11) have a distinguished history of practical applications in a wide variety of organisms. The simplicity of the system—a small (34-bp) directional binding site (*lox*) and a single protein (Cre) for which the expression problems have already been solved—has lent it to a wide variety of applications. When pairs of directly oriented *lox* sites flank a DNA sequence of interest, it is readily deleted when Cre is expressed (Color Plate 6). This simple excisive reaction has many uses. More complex reactions involving several sites placed on one or two different molecules can be used to effect more complex rearrangements such as insertion of a segment of DNA into a *lox*-containing DNA segment (such as the Univector strategy described above), exchanges of segments between two molecules, and even the formation of translocations. As a tool for manipulating the genomes of transgenic mice in vivo, the Cre-*lox* system is unparalleled.

The Cre-*lox* system was shown early on to be fully functional in yeast cells and has been used for a variety of purposes such as "recycling" selectable markers. For example, it is useful to be able to disrupt both copies of a gene in precisely the same way in a diploid organism. In both mammalian cells and yeast, recombination sites are introduced flanking the selectable marker (61). After expression of Cre, a screen (or selection) for cells that have lost the *YFG* transgene can be performed. This enables the second copy of *YFG* to be disrupted with use of the same strategy. When gene families are encountered in yeast, it is desirable to be able to inactivate sequentially all family members. This can be done by using different selectable markers for each gene family member or by intercrossing schemes, but both of these become very difficult when the number of genes in a family exceeds 5 or so. When genes are disrupted with selectable markers flanked by lox sites, they can be eliminated by expression of Cre; the same selectable marker can thus be used for multiple sequential rounds of gene disruption.

SITE-SPECIFIC INTERCHROMOSOME RECOMBINATION IN VIVO

A most powerful technology based on Cre is the activation and shutoff of *YFG* in transgenic animals. The ability to evaluate the phenotypic effects of gene activation and shutoff in mice has opened up all kinds of new possibilities. These techniques depend on the ability to perform introduction of foreign DNA via standard transgenesis techniques or by homology-dependent "knockin" recombination to place specific DNA segments within *YFG*. Two variants of these strategies are outlined in Fig. 5 and 6. In the first, an excisable "stop signal" (transcriptional terminator) is positioned just before *YFG*. Introduction of Cre (by introduction of a Cre expression plasmid or by intercrossing with a Cre transgenic mouse) results in re-

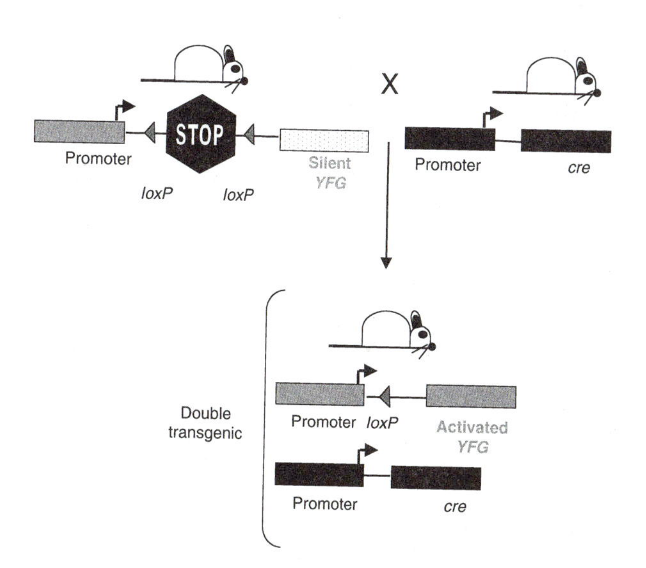

Figure 5. Use of Cre-*lox* to activate gene expression. First, a construct consisting of a synthetic "stop" signal (transcriptional terminator) flanked by *lox* sites is introduced next to *YFG* into ES cells by standard homology-dependent gene replacement techniques. Crossing the resultant knockin mouse by a Cre transgenic results in mice in which *YFG* is activated. Adapted from reference 60.

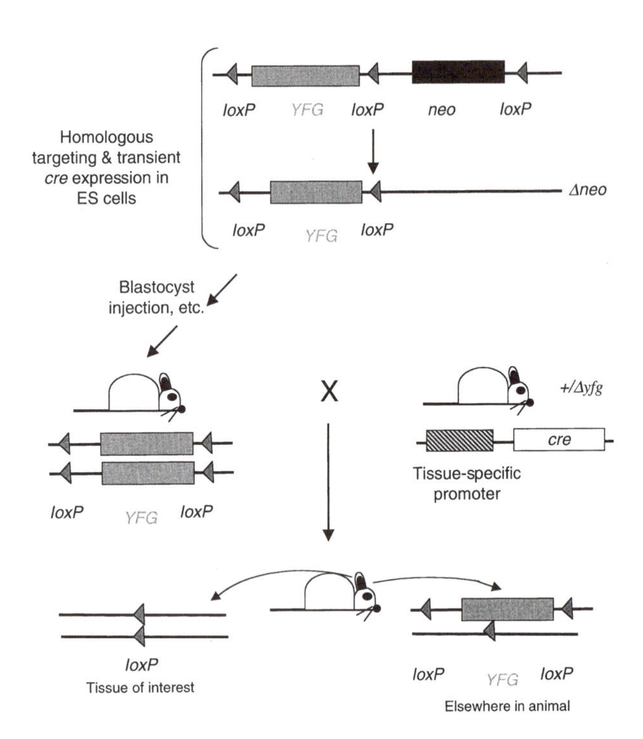

Figure 6. Using Cre-*lox* to inactivate gene expression; conditional gene disruption. First, a construct consisting of *YFG* flanked by *lox* sites and, on one side, by a *neo* gene and a third *lox* site is introduced into ES cells by standard homology-dependent gene replacement techniques. Neo-resistant cell lines are then transiently transfected with a Cre expression construct to eliminate the *neo* sequence. This is because the *neo* gene and associated enhancer sequences could have undesired effects on the expression of *YFG*. The mice are then bred to homozygosity; because the *lox* sites (hopefully) do not interfere with expression, a phenotypically normal mouse is expected. This mouse can then be bred by a strain heterozygous for Δ*YFG* and containing a *cre* transgene driven by a tissue-specific promoter. This leads to excision of the transgene specifically in the target tissue. Adapted from reference 60. A database of existing Cre-expressing mouse lines may be found at http://www.mshri.on.ca/nagy/Cre.html.

moval of the STOP signal and expression of *YFG*. This technique has many applications, including monitoring the phenotypic effects of expression of toxic genes such as oncogenes in vivo (41).

A major avenue to understanding the function of mammalian genes lies in the evaluation of essential genes. Deletion of these genes results in the inability to recover viable progeny, limiting the ability to infer function in different tissues or at different developmental stages. Although the analysis of the mutant embryos can provide valuable clues to function, one would ideally like to assess the effects of individually eliminating the function of *YFG* in each tissue or cell type. This can be accomplished by using the conditional knockout strategy (Fig. 5) in which *YFG* is flanked wholly or in part by *lox* sites (27, 72). The *lox* sites, being quite small, are expected to have little to no effect on *YFG* expression themselves. However, the inverted repeats within each Cre binding site can lead to inhibitory effects on translation when these sequences are located within 5′ untranslated regions of genes via formation of RNA hairpins. Thus, *lox* sites should be introduced to have little impact on native gene expression (e.g., outside the transcribed segment), which can be inferred from phenotypic normality. *Lox-YFG-lox* animals can then be crossed with *cre* transgenic animals previously shown to express Cre in only one or a small subset of tissues. The excisive reaction performed by Cre goes to >90% in vivo, meaning that most of the cells will lose expression of Yfg protein, provided Cre is actually expressed in every cell. With *cre* transgenic animals, this is not always the case, because position effects can lead to mosaic transgene expression, limiting the extent of *YFG* excision (60). Rather than screening *cre* transgenics for expression patterns of interest, it is preferable to "knock in" the Cre coding region to replace that of an endogenous gene expressed in the tissue, and at the developmental time of interest. A useful method for evaluating the expression pattern of a given *cre* transgenic or knockin construct is to backcross it to a *lox2-STOP-lacZ* mouse. LacZ expression will be limited to areas in which Cre is expressed.

TRANSPOSONS AS GENE DELIVERY VEHICLES

The discovery of the *Drosophila* P element as a gene delivery vehicle for *Drosophila* spurred excitement that it might prove to be a universal vector; however, this was not the case, even after many heroic attempts at re-engineering. Accordingly, much of the past 10 years has been spent searching for a broad host-range transposable element. A clear winner in this search has been the Tc1-mariner family of elements (see chapter 22). This class of elements has many attractive features that make it a useful generic eukaryotic gene delivery vehicle. First, there were many surprisingly close sequence relationships between mariners isolated from widely divergent host species (53). A very attractive aspect of these transposons is that a single protein is capable of performing the transposition reaction, and there is little evidence of host-specific factors. Indeed, the life cycle of these elements may depend on occasional horizontal transmission to distantly related organisms to escape "vertical inactivation" by mutations within a given lineage (32). Dependence on horizontal transmission would impose an innate requirement for a high degree of host independence. Indeed, by 1996 evidence was presented that the mariner-like *Hermes* element could transpose across genus boundaries, portending host factor independence (49). *Hermes* is from the housefly, and it was shown to transpose and deliver an eye color gene to the *Drosophila* germ line. In recent years, several groups have used mariner/Tc1 elements to perform mutagenesis and to deliver genes to various organisms. The transposon has many attractive features, but its main shortcoming is that it is quite specific for TA dinucleotides in every species examined.

A most dramatic example of the horizontal promiscuity of mariner-like elements was the harnessing of the *Drosophila mos1* element to develop an insertional mutagenesis system for the parasite *Leishmania* (28). Although there was essentially no classical genetics for this organism, transformation was developed and combined with phenotypic analysis to develop a system of cosmids used to isolate genes that complemented various mutant phenotypes. A method was required to mutagenize both the cosmids and the *Leishmania* genome, and the *mos1* element was shown to be useful for this purpose without much engineering other than addition of a trypanosomal splice acceptor signal (most trypanosomal genes are expressed via *trans*-splicing) to the transposase gene. Transposition was highly efficient, occurring in ~20% of transfected cells. Within the next few years, mariner-like elements, including *Drosophila mariner* itself, *mos1*, *Caenorhabditis elegans* Tc1, and the reconstructed hornfly *Himar* element, were harnessed for transposition in eubacteria (56), archaebacteria (81), mammalian cells (37), chicken eggs (64), mouse ES cells (45), fish (20, 51), human cells (62), and even adenovirus (3). The harnessing of the salmon "Sleeping Beauty" (SB) element represents an especially elegant case in which an apparently defunct vertebrate mariner element arose from the ashes through biotechnological ingenuity. Several mariner sequences had been cloned and sequenced from the salmon genome (as they had from various vertebrates) and most of these were defective as judged from the absence of a complete set of conserved structures and sequences in any individual element copy. A complete and functional element was then regenerated by reassembly of the wild-type sequence from bits and pieces of cloned element copies. The reconstructed element was not only active in salmon cells but, surprisingly, also highly active in human HeLa cells (37). Recent studies on the SB element indicate that it is active in some but not all vertebrate cell types; for example, it was active in mouse ES cells (45), opening up new avenues for insertional mutagenesis in the mouse.

Other promising gene delivery systems developed recently include homing group II introns (see chapter 31) and the plant Ac/Ds transposon system, the first DNA-based transposon shown to perform a complete cycle of transposition in yeast (75).

Another intriguing transposon-based approach for gene delivery involves the use of DNA-transposase complexes transfected directly into target cells. Although many attempts at this type of experiment have failed, recent success was reported using Tn*5* DNA protein complexes containing a linear Tn*5*-based transposition intermediate bearing a selectable marker and Tn*5* transposase (termed "transposomes") in several types of bacteria and even in yeast cells (22). It remains to be seen whether these Tn*5* transposomes can work in vertebrate cells and vertebrate chromatin, and if so, what their degree of randomness of insertion will be, but this is an extremely promising development. Although studies of the length of recombinant DNA that can be delivered with this technology have not yet been performed, this technology does not have the built-in packaging constraint imposed by viruses as gene delivery vehicles.

VIRUSES AS INTEGRATION MACHINES

Retroviruses are the classic gene-delivery vehicles; the fact that retroviruses could pick up cellular genes and use them to efficiently transduce host cells showed that this is an innate trait that could be exploited. Retroviruses have been in use for decades as gene delivery vectors in the laboratory and in the gene therapy industry. A very useful aspect of many retroviral vector systems is that packaging cell lines are available that express all necessary retroviral proteins to assemble particles. A retroviral vector, consisting of the retroviral LTRs, a viral RNA packaging signal, a selectable marker, and *YFG*, is then transfected into cells (Fig. 7). Lysates from the transfected cells can

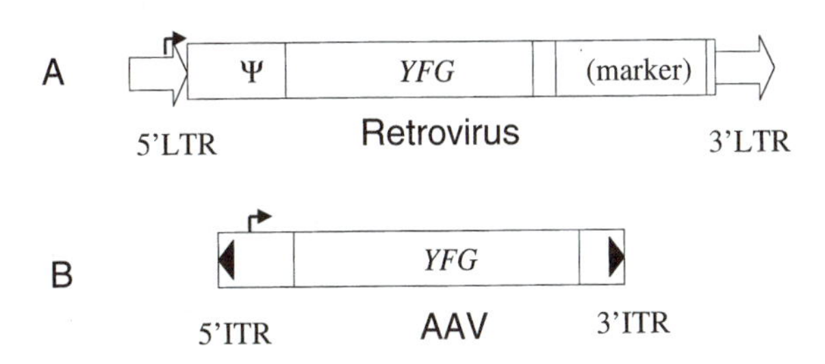

Figure 7. Viral vectors for gene therapy. (A) Retroviruses, Ψ, packaging signal; open arrows, LTRs. Retroviral vectors may or may not include a selectable marker gene. Usually the native LTR promoter is used to drive expression, although exogenous promoters can be used (Fig. 4). (B) AAV. ITR, inverted terminal repeat. Exogenous promoters are usually used to drive gene expression in these vectors; minimal promoters are often used to not exceed the packaging limit.

then be used as a source of "transducing particles" containing the vector RNA packaged into retroviral particles. Great effort has gone into designing packaging cells containing various expression cassettes representing the different viral genes that are required, so that the possibility of reconstructing a live "helper" retrovirus is greatly reduced. Another major technical improvement allowing the production of higher-titer stocks that can infect a broad host range has been to use retroviral vectors that carry the VSV-G protein instead of native viral envelope proteins. An excellent review (47) has been published on the many uses of retroviruses as vectors.

Retroviruses undoubtedly are highly effective vectors for the transduction of human and other animal cells; they are also potent agents of animal and human disease. In this section I highlight an aspect of recent work on retroviral vectors most relevant to this book, namely, attempts to reprogram the integration specificity (or rather, the lack thereof) of retroviruses.

A real safety issue exists with the application of retroviruses to gene therapy; for example, the relative randomness of retroviral integration is well known, as is the ability of retroviral insertion to activate oncogenes and inactivate tumor suppressor genes. This is an enormous practical disincentive to the widespread use of retroviruses as human gene therapy vectors. If only retroviral vectors could be encouraged to integrate at a preordained, safe site in the human genome! Retroviral integrase was indeed redirected in vitro to a specific target via a gene fusion approach in which retroviral integrase protein was fused to the DNA binding domain of lambda repressor (10). Similar approaches have been performed with the DNA binding proteins LexA (24, 38), and Zif268 (11). Although these recombinant proteins do show preference for integration into targets bearing the binding site of interest, there is still considerable integration into other sites. In addition, when the engineered integrase coding region is reintroduced into the viral genome, there are unexpected defects in multiplication of the engineered viruses for a variety of idiosyncratic reasons unrelated to integration. Nevertheless, the concept holds considerable promise.

In addition to retroviruses, the DNA parvoviruses are also capable of inserting into host DNA, in some cases with considerable innate site specificity. These viruses have many properties that make them excellent vehicles for gene delivery. Their natural life cycle is that they can be viewed as a kind of natural virus/transposon hybrid, although the detailed mechanism of their integration is still unclear. Structurally, they consist of a segment of DNA flanked by highly structured but relatively small terminal repeats (TRs) and two genes, *rep* and *cap,* which each produce multiple proteins based on differential splicing and polyadenylation (48). One heavily studied (and exploited commercially) is adeno-associated virus (AAV) (see chapter 38), so called because it is found in association with adenoviruses and is in fact incapable of multiplying by itself. It depends on a helper function provided by an adenovirus (or herpesviruses, or even by certain types of DNA damage) for multiplication. However, this requirement for a helper function is only manifest during the virion production phase. Once virions are made, AAV can infect a cell, replicate its single-stranded DNA genome, and, in some cell types, integrate that newly made DNA into the genome, all without the assistance of a helper virus. One most interesting feature of AAV is that it integrates position-specifically near the chromosome 19q telomere. Some of the advantages of AAV as a gene delivery vehicle include (i) the ability to generate high-titer stocks; (ii) stability of the virions; (iii) stable maintenance and, in some cases, stable integration of the virus; (iv) a broad range of cell types infected; (v) lack of association with any form of disease (AAV is generally found in humans); (vi) ability to infect nondividing cells; and (vii) site-specific integration of the wild-type virus (however, this is not a universal property [see below]). Some of the disadvantages include the packaging limit of AAV (less than 5 kb), the limitation in the size of genes that can be delivered, and the difficulty in growing the virus in the absence of potentially toxic helper viruses. The last problem has been overcome by transfection systems in which cells expressing just those adenovirus genes needed for virus production are cotransfected with AAV genes and with a recombinant AAV (rAAV) vector plasmid (26).

When the AAV genes *cap* and *rep* are replaced by YFG, the resulting rAAVs are still able to integrate in certain cell types, but lose their site specificity (58, 59). Rep is responsible for targeted integration, implying that Rep expression is needed in target cells for targeted integration. Indeed, expression of Rep alone in target cells can lead to the targeted integration of plasmids bearing viral inverted repeat structures (79). However, Rep expression is toxic to cells, so sustained Rep expression is undesirable. Studies recently identified the Rep domain capable of binding preferred targets and also raised the possibility of redirecting Rep to new targets via fusion approaches (14, 52).

However, a highly preferred tissue for AAV, especially from the viewpoint of efficient long-term production of therapeutic proteins, is muscle (39, 77). In these cells, conversion to double-stranded form, concatenation, and circle formation have occurred. Most of the DNA forms are episomal; integration probably occurs, but with a very low efficiency.

In another elegant recent twist on gene delivery in vertebrate cells, mariner elements were launched into mammalian genomes from an AAV "launching pad" (78). Although AAV vectors are single-stranded, they served as effective donors of transposition events, presumably because the AAV genome itself becomes double-stranded once the recombinant virus enters host cells. The success of this approach paves the way for new types of gene therapy vector that can integrate efficiently into a wide variety of mammalian cell types but confers the natural efficiency of gene transfer conferred by viruses.

This chapter focuses on the practical application of transposons as tangible tools for the manipulation of DNA sequences and of cellular phenotypes. However, mention should be made of another very practical use of endogenous transposon copies, even those that may have moved thousands or even millions of years ago, namely as tools for mapping and for understanding evolution of genomes, a topic beyond the scope of this chapter. A compelling example close to home for us (as human beings) is the use of human *Alu* sequences to deduce the origins and evolution of our species (see also chapter 47). Because these retrotransposons essentially never excise precisely, the transposonless "allele" of any given insertion can be considered ancestral. Indeed, relatively recently transposed *Alu* sequences (those absent from most other primates or found only in our closest relatives, the chimps and gorillas) can be quite polymorphic. Analysis of the patterns of these polymorphisms by clustering and principal coordinate analysis allowed the patterns of human migration to be observed, and allowed the conclusion that human populations evolved in Africa to be reached independently of the earlier evidence based on archaeology and studies of mitochondrial DNAs.

REFERENCES

1. **Adachi, T., M. Mizuuchi, E. A. Robinson, E. Appella, M. H. O'Dea, M. Gellert, and K. Mizuuchi.** 1987. DNA sequence of the E. coli gyrB gene: application of a new sequencing strategy. *Nucleic Acids Res.* **15:**771–784.
2. **Ahmed, A.** 1985. A rapid procedure for DNA sequencing using transposon-promoted deletions in Escherichia coli. *Gene* **39:** 305–310.
3. **Akerley, B. J., E. J. Rubin, A. Camilli, D. J. Lampe, H. M. Robertson, and J. J. Mekalanos.** 1998. Systematic identification of essential genes by in vitro mariner mutagenesis. *Proc. Natl. Acad. Sci. USA* **95:**8927–8932.
4. **Argast, G. M., K. M. Stephens, M. J. Emond, and R. J. Monnat, Jr.** 1998. I-PpoI and I-CreI homing site sequence degeneracy determined by random mutagenesis and sequential in vitro enrichment. *J. Mol. Biol.* **280:**345–353.

4a. **Bachman, N., M. Biery, J. D. Boeke, and M. L. Craig.** Tn7-mediated mutagenesis of *Saccharomyces cerevisiae* genomic DNA *in vitro*. *Methods Enzymol.*, in press.

5. **Bender, J., and N. Kleckner.** 1992. IS10 transposase mutations that specifically alter target site recognition. *EMBO J.* **11:** 741–750.
6. **Biery, M. C., F. J. Stewart, A. E. Stellwagen, E. A. Raleigh, and N. L. Craig.** 2000. A simple in vitro Tn7-based transposition system with low target site selectivity for genome and gene analysis. *Nucleic Acids Res.* **28:**1067–1077.
7. **Boeke, J. D., and S. E. Devine.** 1998. Yeast retrotransposons: finding a nice quiet neighborhood. *Cell* **93:**1087–1089.
8. **Braunstein, M., T. I. Griffin, J. I. Kriakov, S. T. Friedman, N. D. Grindley, and W. R. Jacobs, Jr.** 2000. Identification of genes encoding exported Mycobacterium tuberculosis proteins using a Tn552′phoA in vitro transposition system. *J. Bacteriol.* **182:** 2732–2740.
9. **Burns, N., B. Grimwade, P. B. Ross-Macdonald, E. Y. Choi, K. Finberg, G. S. Roeder, and M. Snyder.** 1994. Large-scale analysis of gene expression, protein localization, and gene disruption in Saccharomyces cerevisiae. *Genes Dev.* **8:** 1087–1105.
10. **Bushman, F. D.** 1994. Tethering human immunodeficiency virus 1 integrase to a DNA site directs integration to nearby sequences. *Proc. Natl. Acad. Sci. USA* **91:**9233–9237.
11. **Bushman, F. D., and M. D. Miller.** 1997. Tethering human immunodeficiency virus type 1 preintegration complexes to target DNA promotes integration at nearby sites. *J. Virol.* **71:** 458–464.
12. **Casadaban, M. J.** 1975. Fusion of the Escherichia coli lac genes to the ara promoter: a general technique using bacteriophage Mu-1 insertions. *Proc. Natl. Acad. Sci. USA* **72:**809–813.
13. **Casadaban, M. J.** 1976. Transposition and fusion of the lac genes to selected promoters in Escherichia coli using bacteriophage lambda and Mu. *J. Mol. Biol.* **104:**541–555.
14. **Cathomen, T., D. Collete, and M. D. Weitzman.** 2000. A chimeric protein containing the N terminus of the adeno-associated virus rep protein recognizes its target site in an in vivo assay. *J. Virol.* **74:**2372–2382.
15. **Cooley, L., R. Kelley, and A. Spradling.** 1988. Insertional mutagenesis of the Drosophila genome with single P elements. *Science* **239:**1121–1128.
16. **Cutler, S. R., D. W. Ehrhardt, J. S. Griffitts, and C. R. Somerville.** 2000. Random GFP::cDNA fusions enable visualization of subcellular structures in cells of Arabidopsis at a high frequency. *Proc. Natl. Acad. Sci. USA* **97:**3718–3723.
17. **Devine, S. E., and J. D. Boeke.** 1994. Efficient integration of artificial transposons into plasmid targets *in vitro*: a useful tool for DNA mapping, sequencing and genetic analysis. *Nucleic Acids Res.* **18:**3765–3772.
18. **Devine, S. E., and J. D. Boeke.** 1996. Integration of the yeast retrotransposon Ty1 is targeted to regions upstream of genes transcribed by RNA polymerase III. *Genes Dev.* **10:**620–633.
19. **Devine, S. E., S. L. Chissoe, Y. Eby, R. K. Wilson, and J. D. Boeke.** 1997. A transposon-based strategy for sequencing repetitive DNA in eukaryotic genomes. *Genome Res.* **7:** 551–563.
20. **Fadool, J. M., D. L. Hartl, and J. E. Dowling.** 1998. Transposition of the mariner element from Drosophila mauritiana in zebrafish. *Proc. Natl. Acad. Sci. USA* **95:**5182–5186.
21. **Gimble, F. S., and J. Wang.** 1996. Substrate recognition and induced DNA distortion by the PI-SceI endonuclease, an enzyme generated by protein splicing. *J. Mol. Biol.* **263:**163–180.
22. **Goryshin, I. Y., J. Jendrisak, L. M. Hoffman, R. Meis, and W. S. Reznikoff.** 2000. Insertional transposon mutagenesis by electroporation of released Tn5 transposition complexes. *Nat. Biotechnol.* **18:**97–100.
23. **Goryshin, I. Y., and W. S. Reznikoff.** 1998. Tn5 in vitro transposition. *J. Biol. Chem.* **273:**7367–7374.
24. **Goulaouic, H., and S. A. Chow.** 1996. Directed integration of viral DNA mediated by fusion proteins consisting of human

immunodeficiency virus type 1 integrase and *Escherichia coli* LexA protein. *J. Virol.* 70:37–46.

25. **Griffin, T. J. T., L. Parsons, A. E. Leschziner, J. DeVost, K. M. Derbyshire, and N. D. Grindley.** 1999. In vitro transposition of Tn552: a tool for DNA sequencing and mutagenesis. *Nucleic Acids Res.* **27:**3859–3865.
26. **Grimm, D., and J. A. Kleinschmidt.** 1999. Progress in adeno-associated virus type 2 vector production: promises and prospects for clinical use. *Hum. Gene Ther.* **10:**2445–2450.
27. **Gu, H., J. D. Marth, P. C. Orban, H. Mossmann, and K. Rajewsky.** 1994. Deletion of a DNA polymerase beta gene segment in T cells using cell type-specific gene targeting. *Science* **265:**103–106.
28. **Gueiros-Filho, F. J., and S. M. Beverley.** 1997. Trans-kingdom transposition of the Drosophila element mariner within the protozoan Leishmania. *Science* **276:**1716–1719.
29. **Gwinn, M. L., A. E. Stellwagen, N. L. Craig, J. F. Tomb, and H. O. Smith.** 1997. In vitro Tn7 mutagenesis of Haemophilus influenzae Rd and characterization of the role of atpA in transformation. *J. Bacteriol.* **179:**7315–7320.
30. **Haapa, S., S. Suomalainen, S. Eerikainen, M. Airaksinen, L. Paulin, and H. Savilahti.** 1999. An efficient DNA sequencing strategy based on the bacteriophage mu in vitro DNA transposition reaction. *Genome Res.* **9:**308–315.
31. **Haapa, S., S. Taira, E. Heikkinen, and H. Savilahti.** 1999. An efficient and accurate integration of mini-Mu transposons in vitro: a general methodology for functional genetic analysis and molecular biology applications. *Nucleic Acids Res.* **27:** 2777–2784.
32. **Hartl, D. L., A. R. Lohe, and E. R. Lozovskaya.** 1997. Modern thoughts on an ancient marinere: function, evolution, regulation. *Annu. Rev. Genet.* **31:**337–358.
33. **Hartley, J. L., G. F. Temple, and M. A. Brasch.** 2000. DNA cloning using in vitro site-specific recombination. *Genome Res.* **10:**1788–1795.
34. **Hirochika, H.** 1997. Retrotransposons of rice: their regulation and use for genome analysis. *Plant Mol. Biol.* **35:**231–240.
35. **Hirochika, H., K. Sugimoto, Y. Otsuki, H. Tsugawa, and M. Kanda.** 1996. Retrotransposons of rice involved in mutations induced by tissue culture. *Proc. Natl. Acad. Sci. USA* **93:** 7783–7788.
36. **Huisman, O., W. Raymond, K. U. Froehlich, P. Errada, N. Kleckner, D. Botstein, and A. Hoyt.** 1987. A Tn*10-lacZ-kan*^{r}-*URA3* gene fusion transposon for insertion mutagenesis and fusion analysis of yeast and bacterial genes. *Genetics* **116:** 191–199.
37. **Ivics, Z., P. B. Hackett, R. H. Plasterk, and Z. Izsvak.** 1997. Molecular reconstruction of Sleeping Beauty, a Tc1-like transposon from fish, and its transposition in human cells. *Cell* **91:**501–510.
38. **Katz, R. A., G. Merkel, and A. M. Skalka.** 1996. Targeting of retroviral integrase by fusion to a heterologous DNA binding domain: in vitro activities and incorporation of a fusion protein into viral particles. *Virology* **217:**178–190.
39. **Kessler, P. D., G. M. Podsakoff, X. Chen, S. A. McQuiston, P. C. Colosi, L. A. Matelis, G. J. Kurtzman, and B. J. Byrne.** 1996. Gene delivery to skeletal muscle results in sustained expression and systemic delivery of a therapeutic protein. *Proc. Natl. Acad. Sci. USA* **93:**14082–14087.
40. **Kimmel, B., M. J. Palazzolo, C. H. Martin, J. D. Boeke, and S. E. Devine.** 1997. Transposon-mediated DNA sequencing, p. 455–532. *In* B. Birren, E. Green, S. Klapholz, R. Myers, and J. Roskams (ed.), *Genome Analysis: a Laboratory Manual, Analyzing DNA*, vol. 1. Cold Spring Harbor Laboratory, Cold Spring Harbor, N.Y.
41. **Lakshmi, V. M., D. A. Bell, M. A. Watson, T. V. Zenser, and B. B. Davis.** 1995. N-Acetylbenzidine and N,N′-diacetylbenzidine formation by rat and human liver slices exposed to benzidine. *Carcinogenesis* **16:**1565–1571.
42. **Lampe, D. J., B. J. Akerley, E. J. Rubin, J. J. Mekalanos, and H. M. Robertson.** 1999. Hyperactive transposase mutants of the Himar1 mariner transposon. *Proc. Natl. Acad. Sci. USA* **96:**11428–11433.
43. **Lauermann, V., M. Hermankova, and J. D. Boeke.** 1997. Increased length of long terminal repeats inhibits Ty1 transposition and leads to the formation of tandem multimers. *Genetics* **145:**911–922.
44. **Liu, Q., M. Z. Li, D. Leibham, D. Cortez, and S. J. Elledge.** 1998. The univector plasmid-fusion system, a method for rapid construction of recombinant DNA without restriction enzymes. *Curr. Biol.* **8:**1300–1309.
45. **Luo, G., Z. Ivics, Z. Izsvak, and A. Bradley.** 1998. Chromosomal transposition of a Tc1/mariner-like element in mouse embryonic stem cells. *Proc. Natl. Acad. Sci. USA* **95:** 10769–10773.
46. **Merkulov, G. V., and J. D. Boeke.** 1998. Libraries of green fluorescent protein fusions generated by transposition in vitro. *Gene* **222:**213–222.
47. **Miller, A. D.** 1997. Development and applications of retroviral vectors, p. 437–475. *In* J. M. Coffin, S. H. Hughes, and H. E. Varmus (ed.), *Retroviruses*. Cold Spring Harbor Press, Cold Spring Harbor, N.Y.
48. **Muzyczka, N.** 1992. Use of adeno-associated virus as a general transduction vector for mammalian cells. *Curr. Top. Microb. Immunol.* **158:**97–129.

48a. **Nagy, A.** 2000. Cre recombinase: the universal reagent for genome tailing. *Genesis* **26:**99–109.

49. **O'Brochta, D. A., W. D. Warren, K. J. Saville, and P. W. Atkinson.** 1996. Hermes, a functional non-Drosophilid insect gene vector from Musca domestica. *Genetics* **142:**907–914.
50. **Phadnis, S. H., H. V. Huang, and D. E. Berg.** 1989. Tn5supF, a 264-base-pair transposon derived from Tn5 for insertion mutagenesis and sequencing DNAs cloned in phage lambda. *Proc. Natl. Acad. Sci. USA* **86:**5908–5912.
51. **Raz, E., H. G. van Luenen, B. Schaerringer, R. H. A. Plasterk, and W. Driever.** 1998. Transposition of the nematode Caenorhabditis elegans Tc3 element in the zebrafish Danio rerio. *Curr. Biol.* **8:**82–88.
52. **Rinaudo, D., S. Lamartina, G. Roscilli, G. Ciliberto, and C. Toniatti.** 2000. Conditional site-specific integration into human chromosome 19 by using a ligand-dependent chimeric adeno-associated virus/Rep protein. *J. Virol.* **74:**281–294.
53. **Robertson, H. M., and D. J. Lampe.** 1995. Recent horizontal transfer of a mariner transposable element among and between Diptera and Neuroptera. *Mol. Biol. Evol.* **12:**850–862.
54. **Ross-Macdonald, P., P. S. Coelho, T. Roemer, S. Agarwal, A. Kumar, R. Jansen, K. H. Cheung, A. Sheehan, D. Symoniatis, L. Umansky, M. Heidtman, F. K. Nelson, H. Iwasaki, K. Hager, M. Gerstein, P. Miller, G. S. Roeder, and M. Snyder.** 1999. Large-scale analysis of the yeast genome by transposon tagging and gene disruption. *Nature* **402:**413–418.
55. **Ross-Macdonald, P., A. Sheehan, G. S. Roeder, and M. Snyder.** 1997. A multipurpose transposon system for analyzing protein production, localization, and function in Saccharomyces cerevisiae. *Proc. Natl. Acad. Sci. USA* **94:**190–195.
56. **Rubin, E. J., B. J. Akerley, V. N. Novik, D. J. Lampe, R. N. Husson, and J. J. Mekalanos.** 1999. In vivo transposition of mariner-based elements in enteric bacteria and mycobacteria. *Proc. Natl. Acad. Sci. USA* **96:**1645–1650.
57. **Rubin, G. M., and A. C. Spradling.** 1982. Genetic transformation of Drosophila with transposable element vectors. *Science* **218:**348–353.

58. **Samulski, R. J.** 1993. Adeno-associated virus: integration at a specific chromosomal locus. *Curr. Opin. Genet. Dev.* **3:** 74–80.
59. **Samulski, R. J., X. Zhu, X. Xiao, J. D. Brook, D. E. Housman, N. Epstein, and L. A. Hunter.** 1991. Targeted integration of adeno-associated virus (AAV) into human chromosome 19. *EMBO J.* **10:**3941–3950.
60. **Sauer, B.** 1998. Inducible gene targeting in mice using the Cre/lox system. *Methods* **14:**381–392.
61. **Sauer, B.** 1994. Recycling selectable markers in yeast. *Biotechniques* **16:**1086–1088.
62. **Schouten, G. J., H. G. van Luenen, N. C. Verra, D. Valerio, and R. H. Plasterk.** 1998. Transposon Tc1 of the nematode Caenorhabditis elegans jumps in human cells. *Nucleic Acids Res.* **26:**3013–3017.
63. **Searles, L. L., R. S. Jokerst, P. M. Bingham, R. A. Voelker, and A. L. Greenleaf.** 1982. Molecular cloning of sequences from a Drosophila RNA polymerase II locus by P element transposon tagging. *Cell* **31:**585–592.
64. **Sherman, A., A. Dawson, C. Mather, H. Gilhooley, Y. Li, R. Mitchell, D. Finnegan, and H. Sang.** 1998. Transposition of the Drosophila element mariner into the chicken germ line. *Nat. Biotechnol.* **16:**1050–1053. (Erratum, **17:**81, 1999.)
65. **Smith, V., K. N. Chou, D. V. Lashkari, D. Botstein, and P. O. Brown.** 1996. Functional analysis of the genes of yeast chromosome V by genetic footprinting. *Science* **274:**2069–2074.
66. **Smith, V., D. Botstein, and P. O. Brown.** 1995. Genetic footprinting: a genomic strategy for determining a gene's function given its sequence. *Proc. Natl. Acad. Sci. USA* **92:**6479–6483.
67. **Spradling, A. C., and G. M. Rubin.** 1982. Transposition of cloned P elements into Drosophila germ line chromosomes. *Science* **218:**341–347.
68. **Spradling, A. C., D. Stern, A. Beaton, E. J. Rhem, T. Laverty, N. Mozden, S. Misra, and G. M. Rubin.** 1999. The Berkeley Drosophila Genome Project gene disruption project: single P-element insertions mutating 25% of vital Drosophila genes. *Genetics* **153:**135–177.
69. **Stellwagen, A. E., and N. L. Craig.** 1997. Avoiding self: two Tn7-encoded proteins mediate target immunity in Tn7 transposition. *EMBO J.* **16:**6823–6834.
70. **Stellwagen, A. E., and N. L. Craig.** 1997. Gain-of-function mutations in TnsC, an ATP-dependent transposition protein that activates the bacterial transposon Tn7. *Genetics* **145:** 573–585.
71. **Strathmann, M., B. A. Hamilton, C. A. Mayeda, M. I. Simon, E. M. Meyerowitz, and M. J. Palazzolo.** 1991. Transposon-facilitated DNA sequencing. *Proc. Natl. Acad. Sci. USA* **88:** 1247–1250.
72. **Tsien, J. Z., D. F. Chen, D. Gerber, C. Tom, E. H. Mercer, D. J. Anderson, M. Mayford, E. R. Kandel, and S. Tonegawa.** 1996. Subregion- and cell type-restricted gene knockout in mouse brain. *Cell* **87:**1317–1326.
73. **Waddell, C. S., and N. L. Craig.** 1988. Tn7 transposition: two transposition pathways directed by five Tn7-encoded genes. *Genes Dev.* **2:**137–149.
74. **Walhout, A. J., R. Sordella, X. Lu, J. L. Hartley, G. F. Temple, M. A. Brasch, N. Thierry-Mieg, and M. Vidal.** 2000. Protein interaction mapping in C. elegans using proteins involved in vulval development. *Science* **287:**116–122.
75. **Weil, C. F., and R. Kunze.** 2000. Transposition of maize Ac/Ds transposable elements in the yeast *Saccharomyces cerevisiae. Nat. Genet.* **26:**187–200.
76. **Westphal, C. H., and P. Leder.** 1997. Transposon-generated 'knock-out' and 'knock-in' gene-targeting constructs for use in mice. *Curr. Biol.* **7:**530–533.
77. **Xiao, X., J. Li, and R. J. Samulski.** 1996. Efficient long-term gene transfer into muscle tissue of immunocompetent mice by adeno-associated virus vector. *J. Virol.* **70:**8098–8108.
78. **Yant, S. R., L. Meuse, W. Chiu, Z. Ivics, Z. Izsvak, and M. A. Kay.** 2000. Somatic integration and long-term transgene expression in normal and haemophilic mice using a DNA transposon system. *Nat. Genet.* **25:**35–41.
79. **Young, S. M., Jr., D. M. McCarty, N. Degtyareva, and R. J. Samulski.** 2000. Roles of adeno-associated virus Rep protein and human chromosome 19 in site-specific recombination. *J. Virol.* **74:**3953–3966.
80. **Zambrowicz, B. P., G. A. Friedrich, E. C. Buxton, S. L. Lilleberg, C. Person, and A. T. Sands.** 1998. Disruption and sequence identification of 2,000 genes in mouse embryonic stem cells. *Nature* **392:**608–611.
81. **Zhang, J. K., M. A. Pritchett, D. J. Lampe, H. M. Robertson, and W. W. Metcalf.** 2000. In vivo transposon mutagenesis of the methanogenic archaeon methanosarcina acetivorans C2A using a modified version of the insect mariner-family transposable element himar1. *Proc. Natl. Acad. Sci. USA* **97:** 9665–9670.

Mobile DNA II
Edited by N. L. Craig et al.

Chapter 4

Chromosome Manipulation by Cre-*lox* Recombination

Brian Sauer

INTRODUCTION

Various naturally occurring biological systems use site-specific DNA recombination to precisely regulate gene expression in a temporal and spatial manner. Instances of recombination-mediated regulation are found in bacteria, ciliates, protozoa, and the immune system of vertebrates. With the biochemical understanding of site-specific recombination systems has come their adoption and reengineering to regulate gene expression in the laboratory, an early example being the use of the lambda Int system to control gene expression of recombinant proteins in *Escherichia coli* (14). The finding that the Cre DNA recombinase of bacteriophage P1 is functional not only in bacteria but also in eukaryotes (128) has led to a variety of new genetic strategies for chromosome engineering in higher cells (99, 118, 130). In particular, spatial and temporal regulation of Cre activity has led to the development of conditional somatic mutagenesis strategies that allow more precise determination of gene function in metazoans and plants. These strategies have been rapidly and widely adopted, particularly in the study of gene-modified and transgenic mice. In addition, Cre's modest biochemical requirements have led to the adoption of Cre for use in a variety of other DNA manipulation strategies both in vitro and in vivo.

BIOLOGICAL ROLES AND PROPERTIES

Cre is one of two site-specific DNA recombinases encoded by the genome of bacteriophage P1, and is a member of the Int family of DNA recombinases (107). See also chapters 6 to 11 and 14. Cre's efficient catalysis of DNA recombination at a unique site on the P1 genome neatly explains how the circularly permuted P1 genome comes to have a linear genetic map (141, 149, 165). The second recombinase, Cin, is a member of the resolvase family of DNA recombinases and catalyzes inversion of the C segment of the P1

Brian Sauer • Developmental Biology Program, Oklahoma Medical Research Foundation, Oklahoma City, OK 73104.

genome, thereby allowing the phage to alter its host range (71). Neither of the P1 recombinases interferes with recombination by the other.

Cre recombinase is the 38-kDa product of the P1 *cre* (*cy*clization *re*combination) gene (151) and catalyzes conservative site-specific DNA recombination at the 34-bp *loxP* (*lo*cus of *X*-over, *p*hage) on the P1 genome (67, 69). The *loxP* site is relatively small, consisting of two 13-bp inverted repeats flanking an asymmetric 8-bp core region that imparts overall directionality to the site. Recombination involves the sequential breaking and rejoining of DNA strands within the central core region of *loxP* by way of a transient phosphotyrosine protein-DNA linkage with the enzyme, and proceeds efficiently in vitro, requiring no accessory host protein or energy cofactors. Moreover, unlike many recombinases of the Int family, recombination is efficient on both linear and supercoiled DNA substrates. Cre-mediated recombination at two directly repeated *loxP* sites on the same DNA molecule, by convention called a $loxP^2$ configuration (2), results in precise excision of the intervening DNA segment as a covalently closed circle, whereas recombination at two sites in inverted orientation results in an inversion event (Fig. 1). Intermolecular recombination between *loxP* sites on separate DNA molecules can also occur, giving a reciprocal translocation if both DNAs are linear and an integration event if at least one is circular.

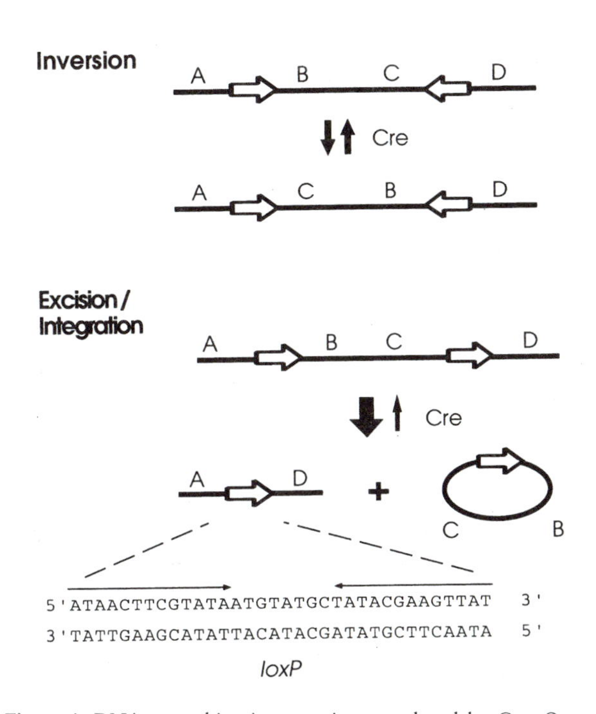

Figure 1. DNA recombination reactions catalyzed by Cre. Open arrows represent the 34-bp *loxP* site, and thin horizontal arrows indicate the 13-bp inverted repeat elements of the site. Rearrangement of the arbitrary genetic markers A, B, C, and D show the consequences of Cre-mediated DNA excision, integration, and inversion. The sequence of the spacer or core region (ATGTATGC) of *loxP* differs from that of the heterospecific *lox511* site (68) at only a single position (ATGTATAC).

Cre recombinase plays several roles in the biology of P1 (171), which provides insight into why Cre has become a useful laboratory tool for genomic manipulation. First, Cre ensures that the 100-kb linear DNA present in the P1 virion is cyclized after infection of *E. coli.* After infection of the host, cyclization of the circularly permuted and terminally redundant linear DNA of P1 virions primarily occurs by homologous recombination at the terminal redundancies after infection. However, should this fail (for example, after infection of a recombination-deficient host), Cre will efficiently cyclize the infecting DNA at *loxP* sites present in the terminal redundancies. The location of the *loxP* site just downstream of the P1 *pac* site on the P1 genome ensures that *loxP* sites will be present in the terminal redundancies in a good proportion of a P1 virion population. Because unidirectional headful packaging of P1 concatemeric DNA into the virion begins at *pac* and results in a greater-than-one-genome-size piece of DNA packaged into each capsid, the first packaged headful contains a *loxP* site in each terminal redundancy.

A second role for Cre is to help ensure maintenance of P1 DNA in lysogens (13). Because P1 does not usually integrate into the host's genome but instead resides in its host as a unit copy plasmid, formation of a dimer after DNA replication of the 90-kb plasmid genome by homologous recombination poses a special problem for faithful partitioning of P1 DNA into daughter cells (Fig. 2). Cre efficiently resolves these large P1 dimers and thus provides P1 a 40-fold

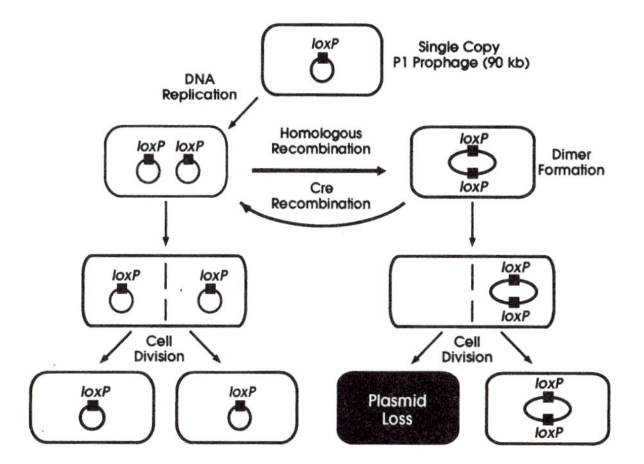

Figure 2. Stabilization of plasmid partitioning by Cre recombinase. The P1 genome in *E. coli* is represented by a circle, the *loxP* site by the small black box.

increase in plasmid stability by ensuring that two monomer P1 genomes are available for plasmid partitioning at bacterial cell division.

The modest biochemical requirements for Cre-mediated recombination (66), coupled with Cre's ability to efficiently catalyze DNA recombination at specific 34-bp sites, even when located at a considerable distance from each other, have led directly to Cre's adoption as a tool for genomic engineering in both prokaryotes and eukaryotes. Moreover, and unexpectedly, Cre also carries its own nuclear localization signal (50, 87) to provide efficient entry into the eukaryotic nucleus, although the recombinase is of prokaryotic origin, and this too has contributed to the widespread utility of Cre in eukaryotic cells. Uses span the range from simply removing unwanted DNA from the genome to orchestrating the generation of novel chromosome structures for the purpose of addressing questions of gene function and chromosome dynamics. Because Cre-mediated genome manipulation in bacteria and yeast not only has been extremely useful in its own right, but also has provided a glimpse of what can be accomplished in multicellular eukaryotes, this review will first briefly highlight several uses of Cre in microbial systems before turning its attention to uses of Cre in higher eukaryotes and, in particular, in genetically modified mice.

MANIPULATION OF PLASMIDS, PHAGES, AND VIRUSES

Bacteriophage lambda has been an extremely useful vector for constructing both genomic and cDNA libraries. Still, characterization of cloned DNA in these vectors is hampered by the large amounts of lambda vector DNA present, so inserts are generally subcloned into smaller, more manageable plasmid vectors. This process can be time-consuming and often relies on the presence of unique restriction sites that flank the insert but are not present in the cloned DNA. Because the DNA inserted between the two *loxP* sites in lambda *loxP*^{2} phages is removed efficiently upon infection of Cre$^+$ *E. coli* (2), lambda cloning vectors were designed in which inserts can be "automatically" subcloned by infection of Cre-expressing bacteria (22, 38, 112). These vectors carry two directly repeated *loxP* sites between which lie both a polylinker for DNA insertion and plasmid sequences for replication and selection in *E. coli.* Cre-mediated excision thus releases the genomic or cDNA insert from the phage to give a replicating plasmid carrying the cloned DNA and a single *loxP* site.

Excisive recombination by Cre in *E. coli* has also been useful in a mini-Tn*3* strategy of insertion mutagenesis (64, 65). By use of a mini-Tn*3* transposon into which *loxP* sites have been incorporated at the ends, insertion events can first be identified by fusion of the target gene with a *lacZ* reporter gene; these can then be converted to have only a small in-frame *loxP* site insertion by Cre-mediated removal of *lacZ* and other intervening sequences by recombination in bacteria. This procedure is thus useful for identifying regions within a protein that can tolerate the addition of 10 or so amino acid residues.

Similarly, linearized *loxP*^{2} plasmid DNA is efficiently cyclized by Cre in *E. coli* to give a replicating plasmid (135). Linear phage-packaged cointegrates derived from miniMu insertion into the bacterial chromosome (25) are also efficiently cyclized to give replicating plasmids upon infection of Cre-expressing bacteria (135). This cyclization strategy was incorporated into the design of both the phage P1 (148) and P1-derived PACmid (73) large DNA cloning vectors to allow efficient recovery of circular large insert-containing replicons in *E. coli* during library construction.

Conversely, integrative recombination by Cre in vitro allows the facile insertion of genetically manipulated DNA into large viral DNAs. Herpesviruses, for example, are sufficiently large (>130 kb) that cloning steps using restriction enzymes in vitro are relatively inefficient. As an alternative to using homologous recombination in cultured cells to introduce a transgene into such large viral genomes, a *loxP* site was introduced into several different herpesviral genomes (49, 138) so that a *loxP*- and transgene-containing plasmid can be efficiently integrated by Cre-mediated recombination in vitro, typically at a frequency of 10 to 50%. Transfection of the infectious recombinant herpesviral DNA into cultured cells then results in recombinant virus. Subsequent studies have adopted this strategy to facilitate generation of both recombinant baculoviral (113) and adenoviral (8) genomes. In a like manner, intermolecular recombination between two *loxP*-containing replicons in bacteria by Cre to give plasmid cointegrates has been exploited to facilitate engineering in vivo of the large Ti plasmid of *Agrobacterium* used for plant transformation (105).

BEGINNINGS AND USES IN YEAST

Marker Excision and Recycling of Selectable Markers

The molecular genetics of the yeast *Saccharomyces cerevisiae* is extraordinarily well developed, particularly in comparison with other eukaryotes. Even so, only a limited number of selectable markers for genetic manipulation exist in yeast. Sequential genetic manipulations can exhaust this repertoire, especially because many yeast-selectable markers function by complementing existing recessive auxotrophies of a host strain, so their utility is restricted by the particular auxotrophies present in the strain being used. This has led to the development of strategies to reuse or recycle selectable marker genes, for example, by flanking them with directly repeated DNA homology so that they can later be removed by homologous recombination (5).

An alternative strategy using Cre recombinase avoids reliance on homologous recombination for marker excision, thereby somewhat simplifying construction of the gene disruption vector. When expressed in yeast, the prokaryotic Cre recombinase efficiently catalyzes DNA recombination between directly repeated *loxP* sites engineered to flank a *LEU2* selectable marker gene, thus excising the marker gene from the genome (128). Placing *cre* under the control of the GAL1 promoter on a separate genetic element provides control of excision. There is no *cre* expression and no recombination occurs in glucose-grown cells, whereas a shift to galactose-containing media results in rapid and precise marker excision in essentially all cells in the culture.

Recyclable marker cassettes (marker genes flanked by directly repeated *loxP* sites) encompass not only the "classical" auxotrophic markers (132), but also the dominant *neo* selectable marker gene (59). In principle, these cassettes allow repeated gene disruption in yeast because, after Cre-mediated excision, the same selectable marker gene can be reused for an additional round of gene disruption by homologous targeting. One limitation, however, is that after several rounds of gene targeting and marker removal a single *loxP* site still resides at each targeted locus, and Cre-mediated DNA recombination between these *loxP*-tagged loci could generate unwelcome deletions or translocations (see below).

Similarly, homologous recombination in yeast can be used to allow epitope tagging of an endogenous gene by way of either a PCR strategy or a transposon-tagging strategy, with the selectable marker flanked by *loxR* sites so that it can be excised by Cre-mediated recombination (123). Moreover, *lox* sites have been designed to generate an in-frame fusion of the target gene with a reporter gene, either *lacZ* or green fluorescent protein. Subsequent Cre-mediated excision in this case not only removes the reporter gene (and selectable marker), but in so doing fuses the target gene to a downstream HA epitope tag. In this latter work, the selectable marker was excised from a cassette in which it was flanked by one *loxP* site and a modified *lox* site derived from the poorly recombinogenic variant *loxR* site (69). Although *loxP* × *loxR* recombination is not as efficient as *loxP* × *loxP* recombination, it nevertheless was driven to completion in virtually all cells by high-level production of Cre from the *GAL1* promoter.

Targeted DNA Integration and Interchromosomal Recombination

Cre catalyzes not only intramolecular but also intermolecular DNA recombination, and hence should be able to target circular DNA to a *loxP* site previously placed in the yeast genome, much like phage lambda integrates into the *E. coli* chromosome. Cre catalyzes the occasional but infrequent integration of the entire P1 genome into a specific *loxP*-like site, *loxB*, on the *E. coli* chromosome (28, 69). So, when an authentic *loxP* site was placed into the yeast genome, Cre was able to direct a *loxP*-equipped plasmid to integrate there with high specificity (137). Because integration at a *loxP* site can be reversed by a second round of Cre-mediated recombination (because *loxP* × *loxP* integration generates a product having *loxP* sites flanking the inserted DNA), it is critical to provide only a burst of Cre activity so that the incoming DNA will be trapped at the insertion site as Cre activity diminishes in the cell. Cre-mediated targeting of circular *loxP*-containing DNA to a chromosomal *loxP* site occurs 20- to 50-fold more frequently than homologous recombination targeting of the same circular plasmid to a homologous chromosomal site. Still, this frequency is ~3-fold less than can be achieved by homologous recombination using linearized DNA to stimulate double-strand break repair. One drawback of targeting with a plasmid carrying a single *loxP* site is that Cre necessarily integrates not only the desired transgene, but also the accompanying, and possibly unwanted, vector sequences. Integration of vector sequences can be circumvented by using a targeting plasmid carrying two *loxP* sites in direct orientation flanking the "unwanted" region. This allows Cre to excise vector sequences from the plasmid, thus permitting subsequent targeted integration of the desired daughter circular DNA (137).

The size of the *loxP* site, 34 bp, would seem to preclude the chance occurrence of a bonafide *lox* site in the yeast, or any eukaryotic, genome. However, not all bases are required in each of the two 13-bp repeats for Cre-mediated DNA recombination. In addition, the central spacer region tolerates some but not all base substitutions, although recombination will only occur efficiently between *lox* sites having homologous central spacer regions (68). This suggests that cryptic sites might exist in the genomes of yeast and other eukaryotes. A genetic selection strategy to detect sequences in yeast that could recombine with *loxP* identified at least 10 such sites, all of which recombine with *loxP* in yeast to give chromosome translocations and inversions, but at a frequency of less than 1×10^{-4} (129). All of the cryptic sites resemble *loxP* in having two inverted repeats, each with sequences similar to those of *loxP*, separated by an 8- or 9-bp spacer region.

Cryptic *lox* sites can be efficient targets for Cre-mediated recombination, and so may have utility in manipulating the genomes of higher eukaryotes. For example, a cryptic *lox* site adjacent to the *FAS1* gene of yeast becomes a hotspot for mitotic crossover events in the presence of Cre (131). Moreover, the *FAS1* site is recruitable for Cre-mediated DNA integration, but only if a "compatible" *lox* vector is used. Although Cre does not target *loxP*-containing vectors to the *FAS1* site, efficient targeting to the *FAS1* locus is obtained with vectors containing a mock *loxFAS1* site, a synthetic site having the arms or inverted repeats of *loxP* but the spacer region of the *FAS1* site.

One technical application of Cre's ability to catalyze DNA recombination between *loxP* sites that are far from each other is to cyclize large linear DNAs in yeast, such as YACs or portions of endogenous chromosomes (121). Cyclization of YAC DNA facilitates its purification and handling so that it can be used, for example, in constructing YAC transgenic mice. An intriguing use of Cre-mediated recombination in yeast is as a chromosome proximity sensor (23). By placing appropriately tagged *loxP* sites at various chromosome positions and measuring recombination between variously positioned pairs, several types of organization, including centromere clustering and homolog pairing, have been detected that govern the probability of interchromosomal collisions. This genetic approach to mapping the spatial organization of the genome in the nucleus promises to provide important insights into the role of chromosome disposition in DNA repair and gene expression.

GENE MANIPULATION IN MICE AND HIGHER EUKARYOTES

Marker Removal

Stable incorporation of exogenous DNA into the mammalian genome is a relatively rare event, especially for DNA targeted by homologous recombination. Thus, gene targeting in cultured cells necessitates that a selectable marker gene be included in the targeting vector. However, the presence in the genome of a selectable marker gene such as *neo* can result in unplanned disturbances to expression of adjacent genes (12, 43, 120). This problem is particularly acute for gene targeting in murine embryonic stem (ES) cells. Because ES cells are totipotent and can colonize the mouse germ line after injection into preimplantation blastocysts, gene targeting in ES cells allows the generation of precisely designed gene-modified mice. Not only can simple insertional mutations be generated by gene targeting, but also point mutations through use of homologous targeting and selection for a linked selectable marker gene. Thus, to avoid possible misinterpretation of phenotype in a gene-modified mouse because of a wayward *neo* gene, it has become particularly desirable to eliminate the selectable marker after gene targeting.

Because Cre is active in cultured mammalian cells (134, 136), transgenic mice (86, 110), plants (32, 108), and *Drosophila* (144), Cre can also be used for removal of a *loxP*-flanked selectable marker gene in higher eukaryotes. Not only does Cre-mediated excision allow the elimination of unpredictable effects due to the presence of a selectable marker gene, it also allows for a marker-recycling strategy, just as in yeast, so that the same marker can be used for a second round of stable transformation (3). The ability to reuse a selectable marker would help alleviate constraints on gene manipulation that might otherwise be imposed by the limited number of selectable marker genes available for molecular genetic manipulation in cultured cells. In addition, the ability to easily remove *loxP*-flanked DNA may become important in the generation of genetically modified food by allowing the removal of any objectionable antibiotic resistance gene used as a selectable marker gene in plants (31). Recombination-mediated marker elimination may thus become a valuable tool for preventing the "escape" of herbicide-resistant genes (126) from an engineered crop plant into a weed plant through cross-fertilization.

In cultured cells *loxP*-flanked DNA can be removed efficiently from the genome by transient expression of *cre* from a transfected DNA vector (58, 134) or from an infecting viral vector such as an adenovirus *cre* expression vector (7, 166). Transfection of DNA into cultured cells often results in only a small percentage of cells that actually take up DNA, however, and this is especially so for ES cells. Various strategies have therefore been enacted to enrich for Cre-mediated deletions after transfection. Incorporation of the thymidine kinase (TK) gene of herpes simplex virus into the *loxP*-flanked interval allows negative selection for ganciclovir-resistant colonies having the Cre-mediated deletion. Alternatively, ES cells productively transfected, and thus committed to Cre-mediated DNA recombination, can be rapidly identified by transfection with DNA vectors expressing a protein fusion of Cre with the green fluorescent protein (GFP) of *Aequorea victoria* and recovered by fluorescence-activated cell sorting (FACS) (50, 88). In a similar manner, enriched recovery of cells committed to Cre-mediated recombination has been obtained by selection for cells transiently resistant to puromycin after cotransfection with both a *cre* expression vector and a *pac* (puromycin acetylase) resistance marker (154).

Several different strategies have been used to excise a *loxP*-flanked genomic marker gene from all cells of a gene-modified mouse. A direct approach is to prepare fertilized zygotes from several donor females of the *loxP*-containing mouse strain and then to microinject them with either a *cre* expression DNA vector (9) or *cre* mRNA (34) to give a burst of Cre-mediated DNA recombination. Nearly all pups born from injected zygotes that have been reimplanted into recipient mothers show excision of the *loxP*-flanked DNA after these procedures. Alternatively, in vitro zygote manipulation and the attendant mouse surgeries can be bypassed by mating the *loxP*-tagged mice with a second transgenic or gene-modified mouse expressing Cre recombinase. A variety of promoters have been useful for driving *cre* expression in transgenic mice to allow marker excision: the cytomegalovirus (CMV) major immediate promoter (86, 139) that gives widespread expression, the adenovirus EIIa promoter (85) that directs expression specifically to fertilized zygotes, the mouse zona pellucida 3 (Zp3) promoter (91) that gives expression in growing oocytes, and the synaptonemal complex protein 1 (Sycp1) promoter (164) that directs expression specifically to an early stage of male meiosis. In all of these cases progeny mice born from crossing the *loxP*-tagged mouse with the *cre* transgenic mouse exhibited the desired deletion. To avoid generating "deleted" progeny with a mixed genetic background, the *loxP*-tagged mice should share the same mouse genetic background as the *cre* transgenic mice.

Cre-mediated excision of a transgene also allows for a simple strategy to reverse a desired transgene phenotype. For example, to reversibly immortalize mammalian cells, primary human foreskin fibroblasts, bovine aortic smooth muscle cells, and rabbit kidney cells were transduced using a retroviral vector carrying the SV40 T-Ag oncogene flanked by directly repeated *loxP* sites (168). Immortalization of these cells was then reversed at will by transfection with *cre* DNA or by infection with a *cre* retrovirus. Similarly, human myogenic cells have been reversibly immortalized, a strategy that may prove useful in ex vivo gene therapy of primary human myopathies (16). This approach may be useful in a clinical setting because reversibly immortalized primary adult human hepatocytes can be transplanted into the spleens of immunosuppressed rats and can save these rats from acute liver failure after 90% hepatectomy (79). In a somewhat similar strategy, random insertion of a transactivated promoter into transcribed sequences of the 3T3 cell genome led to the isolation of an insertion having a transformed phenotype due to production of an antisense transcript of the tagged gene (92). This interpretation was confirmed by Cre-mediated excision of the transactivator gene, which resulted in reversal of the transformed phenotype by elimination of antisense production.

Several other members of the Int family of recombinases also promote site-specific recombination when expressed in higher eukaryotes, including the Flp recombinase of *S. cerevisiae* (53), the R recombinase of *Zygosaccharomyces rouxi* (109), the prokaryotic β-recombinase (35), and the phage HK022 Int recombinase (82). Combining two of these systems, for example Cre and Flp, allows for strategies that would be difficult to achieve by either alone. For example, *Drosophila* with a single genomic insertion of a construct carrying two genes, one flanked by *loxP* sites, the other by FRT (Flp recognition targets) sites, can be mated with either *cre* flies or *FLP* flies to delete the undesired gene or allele and thus generate progeny that are identical except for the gene inserted at that locus (144). In mice a similar combination Cre-Flp strategy has been used to generate an allelic series at the FGF8 locus (104).

Recombination-Activated Gene Expression

It was not obvious initially that a prokaryotic DNA recombinase like Cre would be capable of recognizing and recombining *loxP* sites placed onto a chromosome in a mammalian cell or other higher eukaryote. Genes placed into mammalian cells are often subject to transcriptional silencing and other position effects on gene expression, perhaps imposed by chromatin (a distinctly unprokaryotic environment) or some other mechanism, and these same factors could well have posed a significant obstacle to Cre finding and recombining chromosomally located *loxP* sites. To detect even low-efficiency recombination, a selectable reporter gene cassette was designed that could be activated only if Cre-mediated DNA recombination were to remove a *loxP*-flanked inhibitory DNA sequence (the *LEU2* gene of yeast) interposed between a promoter and the *neo* structural gene (134). In cultured murine cells this recombination-activated gene expression assay showed that Cre-mediated recombination at a chromosomal target was efficient; *neo* activation occurred in nearly all cells expressing Cre recombinase. The efficiency of Cre-mediated chromosomal recombination was confirmed in transgenic tobacco plants (108) with a similar gene activation strategy using the polyadenylation site of the ribulose bisphosphate decarboxylase/oxygenase gene to conditionally block *neo* activation.

Molecular switches based on Cre-mediated gene activation provide a powerful method for genetically manipulating gene expression in gene-modified mice. Because certain transgenes may reduce fertility, be lethal to embryos, or otherwise decrease the ability to thrive, mice carrying such deleterious transgenes may not be able to propagate or even to survive to adulthood. Cre-mediated gene activation provides both a strategy to propagate such mice and the ability to target transgene activation to the tissue and developmental time desired (Fig. 3). Insertion of a *loxP*-flanked block to gene expression called STOP (consisting of the SV40 polyadenylation signal, a spurious RNA splice donor site, and an efficient but inauthentic AUG translational start that acts as a ribosome trap) between the αA-crystallin promoter and the SV40 T-Ag gene rendered tumor formation in the lens in these transgenics completely dependent on mating with a second *cre* transgenic mouse (86). Either widespread *cre* expression from the CMV promoter or lens-specific expression from the αA-crystallin promoter activated lens-targeted tumor formation in double transgenics. In a similar strategy, D1 receptor-positive neurons in knockin transgenic mice have been

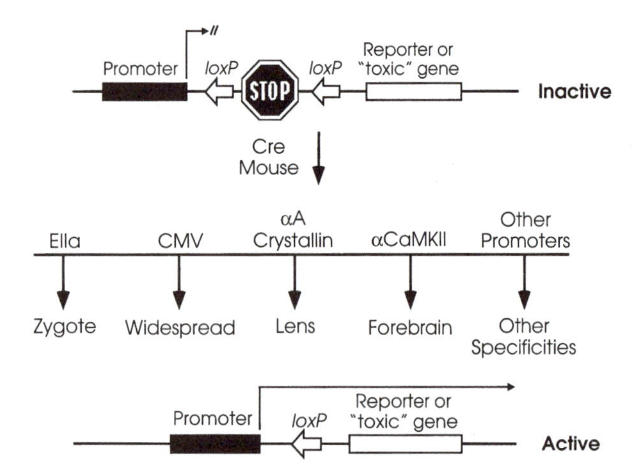

Figure 3. Conditional Cre-mediated gene activation. Mice carrying a *loxP*[2] STOP cassette (GenBank accession no. U51223) inserted between a promoter and a reporter or potentially lethal gene are mated with *cre* transgenic mice having expression under the control of promoters with the desired spatial-temporal pattern. Double transgenic progeny are thereby produced in which the STOP cassette has been removed in the desired spatial-temporal manner. Example *cre* transgenics use the adenovirus EIIa (85), CMV or cytomegalovirus major immediate early (86), and α-CaMKII or α-calcium-calmodulin-dependent kinase II (158) promoters.

made quiescent for expression of the diphtheria toxin A-chain gene (37). When mated with the zygote-specific EIIa-*cre* transgenic mouse, however, recombination activated toxin expression, precisely ablating D1 receptor-positive neurons and generating mice with pronounced movement disorders related to those seen in idiopathic Parkinson's disease.

Use of Cre-mediated recombination activation of gene expression for conditional gene expression is predicated on the notion that the *loxP*-flanked STOP sequence can be placed in an innocuous location within a gene, for example, in the 5′-untranslated region or in an intron. In particular, the intention of the strategy is both to allow full restoration of wild-type gene activity by Cre-mediated recombination, that is, after STOP excision the remaining *loxP* site does not interfere with gene expression, and to convert the target gene to a complete null allele when STOP is present. This latter intent may not always be fulfilled, however, depending on where the *loxP*-flanked STOP sequence is located. Placement of the STOP cassette within intron 22 of a type I collagen gene in mice resulted in an unexpected dominant mutation reminiscent of lethal type II osteogenesis imperfecta (45). This occurred from aberrant splicing events involving STOP that resulted in an in-frame insertion of noncollagen sequences into the mature mRNA. Although in this case the unexpected and aberrant splicing at STOP fortuitously gave rise to a useful murine model of osteogenesis imperfecta, caution should be exercised when placing the STOP cassette to ensure that it produces the desired null mutation.

The strategy shown in Fig. 3 also provides a design for recombination-reporter mice that allows the functional evaluation of *cre* transgenic mice for both spatial and temporal specificity. A transgenic mouse carrying a chicken β-actin promoter-*loxP*-chloramphenicol acetyltransferase (CAT) gene-*loxP*-*lacZ* gene construct activated β-galactosidase expression after Cre-mediated DNA recombination (9). A similar reporter mouse, having a chicken β-actin promoter separated from a *lacZ* reporter gene by a *loxP*[2] STOP cassette, helped identify a line of αCaNKII-*cre* transgenic mice that targeted expression almost exclusively to CA1 neurons in the hippocampus (158), and thus showed the suitability of that *cre* transgenic for a subsequent gene ablation experiment. The utility of such reporter mice critically depends on the ability of the reporter gene to be expressed ubiquitously (4). Because standard transgenic mice are notorious for showing unreliable position-dependent gene expression, several mouse lines have been generated by homologous recombination in ES cells to place knockin reporter constructs at the ubiquitously expressed ROSA26 locus. In one instance the *lacZ* reporter is activated upon Cre-mediated deletion of the *loxP*[2] STOP cassette (98) and in another by deletion of a *loxP*-flanked cassette having four repeats of the SV40 polyadenylation signal (146). Because absence of reporter gene expression in a particular cell after mating of a reporter mouse with a *cre* mouse could mean either that no recombination had occurred in that cell or that the promoter driving the reporter gene is simply not active in that cell, double reporter mice have been developed (96). These mice express *lacZ* before recombination, but delete *lacZ* and express the gene for human placental alkaline phosphatase after Cre-mediated recombination and thus, in principle, provide an unambiguous assessment of recombination status for each individual cell. An alternative to the use of recombination-activated *lacZ* as a reporter for Cre activity in a cell is the recombination-activated expression of GFP of *A. victoria* (27). Transgenic mice carrying an enhanced GFP gene separated from the very strong CAG promoter (chicken β-actin promoter with the CMV-IE enhancer) by a *loxP*-flanked CAT gene show nearly ubiquitous expression of GFP after Cre-mediated DNA recombination and so should also become useful as recombination reporter mice (75).

Activation of a reporter gene in a cell by Cre-mediated DNA recombination indicates that Cre is expressed in that same cell or, alternatively, that Cre had been expressed previously, either in that cell or in its precursor. Because this allows a cell and all of its descendents to be marked by the recombination event, Cre-mediated tagging of cells has been proposed as a method for cell lineage analysis and fate mapping (134). With the use of recombination reporter mice, as described above, this has now become a powerful method for mammalian cell lineage analysis. Cre-based tagging has been used to fate-map cells originating in the dorsal midbrain-hindbrain constriction (174), to investigate the lineage relationships between glucagon gene-expressing and insulin gene-expressing cells in the islets of Langerhans (62, 63), to analyze neural crest cell differentiation (26, 39, 74, 170), and to identify distinct lineage boundaries in mouse ectoderm before limb bud outgrowth (78). These same considerations have also been embraced in a retroviral gene trap vector to identify transiently expressed genes in developing embryos (155). Insertion of a U3*cre* retroviral vector into even a transiently expressed gene leads to recombinational activation of *lacZ* expression at a reporter locus, thus marking these cells so that they can be traced during embryogenesis. Incorporating recombination-based lineage-tracing methodology into mice that are also mutant for defined lineage-determining factors will almost certainly help to provide additional insight into the specification of cell fate in the developing embryo.

Conditional Gene Ablation

The generation of simple null mutations in any gene in the mouse has become a relatively straightforward procedure. Standard molecular biological techniques are used to modify cloned mouse genomic DNA, both to make a null mutation in the gene, usually a deletion or an inactivating insertion, and to incorporate a selectable marker in the targeting construct. The modified region is flanked by long stretches of DNA homologous to the genomic locus to be targeted, and then homologous recombination in totipotent murine ES cells is used to replace one or both copies of the endogenous gene with the mutated one. To generate a mutant mouse the modified ES cells are injected into mouse blastocysts to produce a chimeric animal that is then bred to give germline transmission of the modified chromosome.

A major concern of genetic analysis in all metazoans is that the role of a particular gene may differ from one type of cell to another. The same gene may have no role in one cell type (or even be expressed), have a distinct role in another cell type, and have yet another function in a third cell type. Moreover, a gene may be essential for embryogenesis but have a separate and distinct role in a particular adult tissue. Characterization of that role would be difficult unless a conditional mutagenesis or other strategy to circumvent the embryonic lethality existed. Cre-mediated recombination provides a conceptually simple but powerful strategy that addresses these issues by allowing the conditional generation of either loss-of-function or gain-of-function mutations in any gene.

Conditional ablation of an endogenous gene in the mouse was used first to better understand the role of DNA β-polymerase in T cells (57). The promoter and first exon of the endogenous mouse DNA β-polymerase gene (*pol*β) were flanked by directly repeated *loxP* sites, colloquially referred to as a flox-ed (*fl*anked by *lox*) construct, and these mice were then crossed with a second genetically manipulated mouse, a *cre* transgenic, that expressed the *cre* gene specifically in T cells (110). This led to the generation of "doubly modified" mice in which the *pol*β gene was specifically inactivated in 40% of the T-cell population but was completely untouched in all other tissues. Incomplete DNA excision in the T-cell population in these mice may have arisen from low-level *cre* expression, an occurrence which most likely could be remedied by using a "Kozak"-modified *cre* gene (58, 83, 137) that promotes efficient translation initiation in eukaryotes. This tissue-specific knockout strategy is one that can be generalized to any gene that can be modified by homologous targeting in ES cells. It is a strategy of considerable power, because the same *loxP*-modified mouse can be crossed to each of several different *cre* transgenic mice, each expressing recombinase under the control of a particular desired promoter and each with its own distinct pattern of tissue and developmental specificity. Such a binary strategy allows the targeted ablation of any *loxP*-modified gene in any cell type, with both temporal and spatial specification dependent only on the precision with which Cre activity can be regulated.

Numerous conditional knockout mutations have been generated in mice with the use of Cre-mediated excision, so the utility of this approach has been confirmed many times over. In one dramatic example conditional mutagenesis provided confirmatory evidence for the intimate involvement of an essential subunit of the *N*-methyl-D-aspartate receptor (NMDAR1) in the development of spatial memory in the mouse (159). Neonatal lethality from a complete *NMDAR1* knockout was avoided by using mice having a *loxP*-flanked *NMDAR1* gene; these mice were phenotypically normal. To obtain selective ablation of the *NMDAR1* gene in CA1 hippocampal neurons, the *loxP*-tagged mice were mated with an αCaMKII-*cre* transgenic mouse in which *cre* expression is directed postnatally and exclusively to CA1 neurons in the hippocampus by the α-calcium-calmodulin-dependent kinase II promoter. Gene ablation was confined to CA1 neurons and resulted in adult mice with no obvious abnormalities, but with a pronounced deficit in spatial memory. An important factor in ensuring that *NMDAR1* was ablated specifically in CA1 neurons, and not elsewhere, was the careful prior characterization of the Cre expression profile of the *cre* transgenic mouse using both a specific antibody (140) and a recombination reporter mouse (158), as described above.

Partial loss-of-function mutations can also be generated by Cre-mediated excision. For example, both mild and severe models of Pompe syndrome in mice were developed by a single gene-targeting strategy (117). Insertion of *neo* into exon 6 of the gene for acid α-glucosidase disrupts the gene and thus mimics similar chain termination mutants in patients with Pompe syndrome. By also flanking the mutagenized exon 6 with *loxP* sites, these same gene-targeted mice were mated with a *cre* transgenic mouse to delete exon 6 and thus mimic a milder form of Pompe syndrome arising from a splicing defect that results in exon 6 skipping in patients. In an alternative strategy, because the *neo* gene can be used to dampen target gene expression (101), although possibly also adjacent genes, a *loxP*-flanked *neo* was used to generate a hypomorphic mutation in the N-*myc* proto-oncogene (106). Because Cre-mediated excision can then be used to regain N-*myc* activity, this strategy allows tissue-specific restoration of gene activity in animals hypomorphic for a desired gene.

Because Cre-mediated deletion is a binary strategy, the mouse carrying the *lox*-tagged locus can easily be used not only for Cre-mediated gene ablation in each of several distinct tissues (or developmental times), but also for generation of animals or plants (143) mosaic for the null mutation. Mice having graded percentages of recombinant and nonrecombinant cells have been generated by mating *loxP*-modified mice with EIIa-*cre* mice (85). Because these *cre* mice provide Cre activity only in early embryos, they allow a burst of recombination before the establishment of distinct lineages and, importantly, no subsequent recombination, so the percentage of recombinant cells in the embryo is frozen. Generation of mice mosaic for knockout of the essential insulin-like growth factor 1 gene (*igf1*) showed that mice having a low percentage of IGF-1$^+$ cells were viable but were growth retarded, with the level of retardation corresponding inversely with the percentage of IGF-1$^+$ cells (94). In contrast, a liver-specific knockout of *igf1* reduced serum levels of IGF-1 dramatically, but did not result in growth retardation, thus providing direct evidence of a paracrine or autocrine role for IGF-1 (169).

A nestin-*cre* transgenic mouse is among other *cre* mice that have been suggested to be useful for generating mosaic knockout mice (18). Although nestin expression is usually restricted to the developing nervous system, nestin-*cre* transgenic mice were generated that express *cre* in a wide variety of tissues, which suggests that these *cre* transgenic mice may also be useful in producing mosaic knockout mice when mated to mice having a *loxP*-flanked gene of interest. In fact, any *cre* transgenic mouse having incomplete penetrance of expression would lead to a mosaic deletion phenotype. This approach with a tissue-specific promoter driving *cre* could also permit generation of tissue-specific mosaics. One caveat of such approaches, of course, is that any cell that activates expression of the *cre* gene as the animal ages would commit to deletion at the *loxP*-tagged target only after age-related activation. Such tissues would exhibit an age-dependent increase of cells having the Cre-mediated deletion.

Although recessive loss-of-function mutations are the most common type of conditional mutation that has been generated by Cre-mediated excision of endogenous DNA sequences, dominant mutations can also be generated. For example, in melanoma and colorectal cancer cell lines several stabilizing β-catenin mutations have been found at N-terminal threonine and serine residues in exon 3. These mutations block phosphorylation and lead to down-regulation of β-catenin. Cre-mediated recombination, targeted specifically to the intestine of mice, was used to delete a *loxP*-flanked exon 3 of the β-catenin gene. This resulted in a dominant mutation that produces intestinal polyposis (60), thus providing genetic evidence for Wnt signaling in intestinal and colonic tumors.

Inducible Recombination

The power of Cre-mediated conditional mutagenesis depends both on the penetrance of recombination in the target tissue and on the temporal and spatial precision with which that recombination can be directed in the organism. Although developmentally regulated and tissue-specific promoters can be used to direct *cre* expression to a particular stage of embryogenesis and/or a particular cell type, it would be extremely useful to be able to simply turn on recombination by an external stimulus, such as a pharmacological agent, over which the experimentalist has complete control. To this end, two types of approaches have been investigated: inducible promoters that regulate *cre* gene expression, and variant Cre proteins whose activity in eukaryotes is inducible by a small molecule.

Initial work showed that mammalian cells carrying the *cre* gene under the control of the metallothionein-I promoter could be rapidly induced with heavy metals to give Cre-mediated recombination (136). However, low-level expression from this promoter in the absence of inducers prevents it from being generally useful for inducible recombination in genetically engineered mice. The first successful approach to achieving the on/off expression control needed for inducible recombination-mediated gene ablation in mice was the use of the interferon-inducible Mx promoter to drive *cre* expression (84). Doubly modified mice, obtained by mating an Mx-*cre* transgenic mouse with mice having a *loxP*-flanked DNA β-polymerase gene, showed little or no recombination until induced. However, administration of either interferon or poly(I-C) induced Cre-mediated recombination in most cells of every organ examined. Use of the Mx-*cre* mouse thus provides a convenient strategy for inducing gene ablation (or activation) by allowing recombination by means of a simple dosing of a mouse at a desired time with a pharmacological agent.

Several other promoters have been enlisted to provide inducible Cre production in mice. In *Drosophila melanogaster* heat shock promoters have provided tight regulation of inducible expression of the Flp recombinase (53). However, when a similar strategy was used to regulate Cre recombinase in mice, *cre* expression occurred in early embryos without heat shock (36). Nevertheless, because this expression seems to be stochastic, relatively "unrecombined" mice can be identified by tail-snip analysis and then subjected to heat shock to induce DNA recombination. Moreover, even with low-level uninduced recombination, these mice may be useful for generating mosaics having different percentages of DNA recombination in their organs. A second strategy for regulating *cre* expression in mice is based on the successful adaptations of the prokaryotic tetracycline repressor-operator system to regulate gene expression in eukaryotes (46, 51, 54, 55, 142). In these systems the tetracycline repressor is used to bind to operator sequences placed into a eukaryotic promoter to give negative regulation, or fused to the VP16 transcriptional activator to promote expression at an operator-containing "minimal" promoter. In both cases the addition of tetracycline reverses gene repression or gene activation, respectively, by abolishing DNA binding. In the reverse tet transactivator (rtTA) variation of this strategy, tetracycline is required for DNA binding of the repressor-VP16 fusion protein and so gene expression can be activated by addition of tetracycline or doxycycline (55).

Of the tetracycline-regulated systems, the rtTA system is particularly appealing for controlling *cre* expression, not only because it exhibits tight regulation by addition rather than by removal of tetracycline, but also because the rtTA regulator can in principle be expressed in a tissue-specific manner so that induction of recombination will also be directed in that same tissue-specific fashion. That this can be achieved was demonstrated by constructing transgenic mice having both rtTA under the control of either the retinoblastoma promoter or the whey acidic acid protein promoter and also an rtTA promoter to regulate the *cre* gene (161). With both of these, two transgenic lines were crossed with a *lacZ* recombination reporter mouse and induced with doxycycline, and they each gave the spatial and temporal pattern of Cre-mediated recombination expected from the promoter-driving rtTA expression. Mice have also been generated with the rtTA controlled by the fatty acid-binding promoter, which is expressed predominantly in the gut (127). Tri-transgenic mice having this construct, the *cre* gene under the control of an rtTA-regulatable promoter and also a recombination reporter gene, received doxycycline orally for 4 days. Analysis showed that recombination was gut specific and that epithelial progenitor cells had undergone recombination, but that no recombination had occurred in the absence of induction.

As an alternative to transcriptional regulation of the *cre* gene, control of Cre activity itself in eukaryotes can be achieved by fusing Cre to a steroid receptor ligand binding domain (LBD) (114). In the absence of a steroid inducer, LBD fusion proteins are sequestered from the nucleus, rendering them inactive until a steroid inducer is provided and nuclear localization is restored. There are two perceived advantages to this approach: (i) such regulation of activity would be more rapid than regulation of transcription and (ii) the construction of engineered mice would be simplified because the need to generate tri-transgenic mice (regulator, *cre* gene, and *loxP*-modified locus) would be eliminated. Fusion of the LBD of the estrogen receptor (ER) to the C terminus of Cre results in a fusion protein that retains recombinase activity at *loxP* sites in cultured mammalian cells in the presence of the estrogen estradiol, but which is inactive for recombination in the absence of steroid (103).

To eliminate unwanted induction of Cre activity from endogenous estrogens in mice, Cre fusions have been made to mutated versions of both the mouse (172) and human (41) LBD that are insensitive to endogenous steroids but which are still sensitive to the synthetic steroid tamoxifen. No background recombination was detected at a *loxP*-flanked reporter locus in transgenic mice expressing this Cre-ER^T tamoxifen-sensitive chimeric protein under the control of the CMV promoter (21, 41), confirming that the fusion was not responsive to endogenous steroids. However, animals given 1 mg of tamoxifen daily for 5 days showed up to 50% recombination in a variety of organs. To avoid the possibility of undesired tamoxifen-induced toxicity or other side effects, Cre has also been fused to a multiply mutated estrogen receptor LBD (ER^{T2}) that shows markedly enhanced sensitivity to the synthetic steroid 4-hydroxytamoxifen (OHT) but that does not bind estradiol (42). With the use of the keratin K5 promoter to direct expression of the transgene to epidermal proliferating keratinocytes, mice with the Cre-ER^{T2} chimeric protein required ~10-fold less OHT than did mice expressing Cre-ER^T (72) for induction of DNA recombination after a 5-day intraperitoneal treatment regimen. Use of this multiply mutated LBD may be particularly useful in inducing conditional gene activation or inactivation in developing embryos. Although a single 1-mg dose of tamoxifen to a pregnant transgenic mouse carrying a tamoxifen-sensitive Cre-ER^{TM} fusion under the control of the *Wnt* promoter induced high levels of recombination within 24 h at a reporter locus in developing embryos, higher doses or multiple doses resulted in significant embryonic lethality at term (33), most likely because of the antiestrogen effects of tamoxifen.

Topical treatment with steroid provides a second strategy to reduce potential complicating toxicity from systemic steroid treatment. Transgenic mice expressing a $CreER^{tam}$ tamoxifen-sensitive fusion protein in basal keratinocytes using the keratin K14 promoter were induced to give high levels of DNA recombination after treatment either orally or topically with tamoxifen (163). An added experimental benefit of the topical treatment strategy is that mock application to a second patch of skin provides a same-animal control for the steroid-induced recombinant skin patch under investigation.

The LBDs of several other steroid receptors have also been used to regulate Cre activity in eukaryotic cells. Just as for the estrogen receptor (172), the LBD domain of the progesterone receptor can be fused to either the amino or carboxyl terminus of Cre to give a fusion protein that retains recombinase activity in vivo, although carboxyl-terminal fusions seem to retain greater activity (77). Fusion of Cre to a mutated progesterone receptor LBD that no longer binds progesterone, but which is still sensitive to the synthetic steroid RU486, places Cre activity under control of the RU486 ligand. Transgenic mice expressing such a chimeric Cre protein in the brain showed little or no DNA recombination at a reporter locus until induced with RU486, whereupon efficient DNA recombination occurred specifically in the target regions of the brain (76, 160). In a similar vein, Cre was also fused to the LBD of a mutant glucocorticoid receptor so that the resulting fusion protein activated DNA recombination in cultured cells in response to dexamethasone, RU486, and other synthetic ligands, but was unresponsive to physiological concentrations of naturally occurring ligands of the glucocorticoid receptor (20).

Targeted DNA Integration

A confounding problem in molecular manipulation of mammalian cells is that stable integration of exogenous DNA into the mammalian genome almost always occurs in an uncontrolled fashion and only seldom targets a specific locus by homologous recombination. The resulting variability in copy number and genome position between different transformants results in variable gene expression between different cell line isolates, and in transgenic mice this can result in unexpected and variable patterns of gene expression from an injected transgene. Use of targeted genomic integration of DNA by Cre DNA recombinase may provide a partial solution to this problem.

Initial work showed that once a *loxP* site has been placed into the genome of cultured mammalian cells, transient Cre expression (from a cotransformed *cre* expression plasmid) can direct the targeted integration of a *loxP*-containing plasmid to a genomic *loxP* site (137). Because this approach results in *loxP*-targeted integration at a frequency of only about 10 to 25% of random, illegitimate DNA integration, a promoter trap method was used in conjunction with a promoterless *loxP*-selectable marker on the targeting plasmid so that site-specific targeting to a chromosomally situated promoter *loxP* target could be selected directly. This resulted in more than 90% of the selected transformants being single-copy integrants of the targeting plasmid specifically integrated at the chromosomal *loxP* target, with only occasional recovery of "double" plasmid integrants (137). A variation of this strategy using an ATG-less *loxP-neo* fusion gene as the chromosomal target, where the *loxP*-targeting vector provides both the promoter and the ATG start, ensured that nearly 100% of the selected $G418^R$ colonies obtained were single-copy integrants of the targeting vector (48). Because the level of expression of a Cre-targeted *lacZ* reporter gene was nearly identical in independent integrants (48), the same strategy was used to construct isogenic cell lines expressing different alleles or isoforms of an angiotensin II AT_I receptor transgene at approximately the same level to facilitate pharmacological evaluation (133). Site-specific targeting of DNA to a chromosomal target does not itself abolish position effects on gene expression. Site-specific targeting only guarantees that transgenes targeted to a particular *loxP*-tagged chromosomal locus site will be single-copy and will experience the same chromosomal environment. Site-specific targeting thus eliminates the variability of expression that comes from transgenes being integrated at different chromosomal positions, but can not negate the influence that a particular chromosomal environment might have in influencing expression of the integrated transgene.

To better ensure that a Cre-targeted transgene experiences a "known" chromosomal environment, a single *loxP* site can be installed in a preselected locus by homologous targeting in murine ES cells. Installation of the *loxP* target at the mammary gland-specific whey acidic protein (WAP) locus allowed subsequent Cre-mediated targeting of a *loxP*-containing plasmid to that locus at a frequency of 23% compared with random DNA integration (125). Similarly, installation of a *loxP* site at the genomic locus for β-casein, another milk protein, in ES cells also allowed Cre-mediated integrative targeting, using a gene trap or split gene strategy to enrich for site-specific integrants (80). A lower frequency of site-specific to random integration was seen at the β-casein locus (~1%), possibly because a somewhat more sophisticated strategy was used to prevent chromosomal integration of accompanying plasmid backbone sequences. In this case the targeting plasmid was designed to have the selection cassette flanked by directly repeated *loxP* sites. In this strategy Cre first excises the cassette from the transfected plasmid by excisive recombination between the two directly repeated *loxP* sites and then integrates it into the genome in a second site-specific DNA recombination step (81, 137). An important feature of Cre-mediated integration is that very large DNA inserts can be efficiently integrated at a chromosomal *loxP* target. Efficient integration of both a 230-kb *loxP*-equipped BAC into a chromosomal *loxP* target in plants (29) and circularized YACs into various chromosomal *loxP* targets in ES cells (24) has been demonstrated recently. Because ES cells can be injected into murine blastocysts to generate genetically modified mice, Cre-mediated integrative targeting in ES cells may well provide a useful method for generating transgenic animals with more predictable patterns of expression.

Several strategies to increase the frequency of Cre-mediated targeting of DNA to a chromosomal target have been investigated. Because integration of a *loxP*-containing plasmid into a chromosomal *loxP* site necessarily results in the integrated plasmid being bounded by two directly repeated *loxP* sites, continued presence of Cre will quickly excise the plasmid. Transient expression from a cotransfected *cre* expression plasmid allows integrants to be trapped in the chromosome as the level of the recombinase in the cells wanes; that is, several cycles of integration and excision probably occur before the concentration of Cre declines to a level no longer sufficient for DNA recombination. Thus, an alternative strategy that eliminates continued production of Cre in the cell would be to cotransfect the targeting plasmid directly with purified Cre protein (15). This strategy did yield site-specific integrants, but not at an appreciably higher frequency. *loxP* sites at different chromosomal positions displayed Cre-mediated targeting frequencies that differed by as much as 10-fold, suggesting that improvements in targeting strategies should be evaluated at the same chromosomal position to avert position-imposed targeting variability.

Unlike many of the other members of the Int family of DNA recombinases, the recombination event catalyzed by Cre between two *loxP* sites is fully reversible by a second round of Cre-mediated recombination. Because multiple rounds of recombination are permitted in vivo, such reversibility favors intramolecular excisive recombination over intermolecular integrative recombination. Thus, one strategy to bias Cre-mediated recombination toward integration is to design the recombination reaction so that after integration one or both of the *lox* sites flanking the inserted DNA are no longer good substrates for Cre excision. Biochemical characterization of the *loxP* site had shown that mutation of the outermost 4 to 8 bp of either arm or inverted repeat element of the site (Fig. 1) does not block DNA recombination with an unmutated *loxP* site, but that such mutated sites do not recombine with each other (1, 150). Cooperative binding of Cre to each inverted repeat of a full *loxP* site presumably compensates for impaired binding at the mutated site during synapsis. Sites that are doubly mutated on both arms, however, would be expected to be incapable of recombination with an unmutated *loxP* site because they bind Cre poorly. Because such doubly mutated sites would be generated by recombination between a left-arm mutated site and a right-arm mutated site, integration might be favored by a strategy in which the chromosomal target *lox* site is mutated in one arm and the targeting vector is mutated in the other (Fig. 4A). Recombination would generate an insert flanked by one doubly mutated site and one wild-type site, thus locking the insert into the chromosome by generating a poor excision substrate.

A strategy along these lines met with mixed success using Flp recombinase in bacteria (70), but was successful with Cre, in both plants (6) and murine embryonic stem cells (10). With a mutant *lox* site with 4 bp removed from the right arm as a chromosomal target and a targeting vector with a similar left-arm mutant *lox* site, site-specific integration at several randomly chosen sites was 3- to 12-fold higher than targeting at several different *loxP* sites randomly placed into the genome. In ES cells the percentage of site-specific integrants among all drug-resistant selected colonies varied depending on the particular ES *lox* target line used, but was generally between 1 and 8%, with one line showing targeting as high as 16% of all selected colonies (10).

An alternative strategy to enhance Cre-mediated placement of DNA into a eukaryotic genome is shown in Fig. 4B, and is similar to a combinatorial infection and in vivo recombination strategy for generating large phage antibody display libraries in *E. coli* (167). In this strategy the chromosomal target consists of two heterospecific *lox* sites, *loxP* and *lox511* (Fig. 1), flanking either a reporter gene or a negative selection marker. Because these two *lox* have different spacer sequences, they recombine efficiently only with *lox* sites having the same spacer specificity (68). The targeting vector carries the same two heterospecific *lox* sites flanking the gene to be inserted and/or a positive selection marker gene. Cre-mediated recombination between plasmid-borne and chromosomal *lox* sites of similar specificities then results in what is effectively a double-crossover event, replacing the *lox*-flanked chromosomal interval precisely with the corresponding DNA segment on the plasmid. One appealing aspect of this double *lox* strategy is that no vector backbone sequences are integrated by site-specific targeting.

With use of a *lox-neo* split gene selection approach, direct comparison of single *lox* and double *lox* targeting strategies at exactly the same chromosomal site in murine 3T3 cells showed that double *lox* targeting was both 40-fold more efficient than single *lox* targeting and also that targeting required far less recombinase (17). In fact, site-specific double *lox* targeting was 3-fold more frequent than random DNA integration, with a frequency of ~1% of all surviving cells. Moreover, elimination of chromosomal plasmid sequences by replacement targeting allowed better inducibility of a p53 promoter-reporter gene. A similar

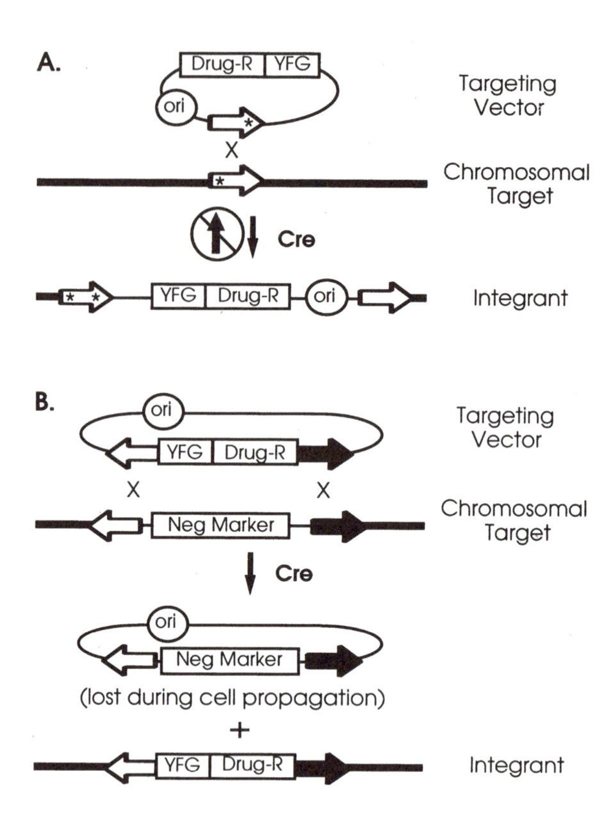

Figure 4. Cre-mediated chromosomal targeting in ES cells. The thin lines represent the transfected plasmid DNA; thick lines represent chromosomal DNA sequences. The desired transgene is designated YFG (*y*our *f*avorite *g*ene) and the plasmid replication origin (ori) is represented by the circle. (A) Mutated *lox* site strategy (10). Mutation (a 4-bp deletion) of *loxP* in one of the inverted repeat elements is represented by an asterisk. After intermolecular recombination, one of the product *lox* sites carries mutations in both arms or inverted repeat elements and is thus blocked from participating in Cre-mediated reversal of integration. (B) Double *lox* replacement recombination (147). Different heterospecific *lox* sites having non-identical spacer sequences are indicated by open (*loxP*) and filled (*lox511*) large arrows. Alternatively, two *loxP* sites in inverted configuration have been used (44). Double-crossover recombination between sites on the incoming plasmid and the resident chromosomal sites replaces the *lox*-bounded chromosomal interval with that on the plasmid. Integration can be selected by incorporation of a marker on the plasmid into the chromosome, or by negative selection for loss of a negative selectable marker from the chromosome.

double *lox* strategy was used to generate a series of allelic globin enhancer reporter gene constructs identically targeted to an arbitrarily chosen locus in mouse erythroleukemia MEL cells (19). Again, the incidence of site-specific targeting, between 0.2 and 0.5% of surviving cells, was high. To allow facile removal of the selectable marker, the selection cassette was flanked by directly repeated FRT sites so that it could be removed by Flp recombinase in a subsequent round of DNA transfection.

In ES cells, gene knockins by homologous recombination allow placement of a desired transgene or allelic variant at a desired locus, thus minimizing or eliminating position effects on gene expression by relying on endogenous surrounding regulatory elements, but the frequency of targeting by homologous recombination is often quite low. After installation of the double *lox* cassette at the asialoglycoprotein receptor gene locus in mouse ES cells by homologous recombination, subsequent knockin of any gene by double *lox* targeting occurred at a frequency 60 to 500 times greater than that usually achievable by homologous recombination and three times that of nontargeted random integration of DNA into the genome (147). Efficient recovery of knockins was achieved even in the absence of a positive selectable marker on the targeting construct by using ganciclovir to select for eviction of the herpes simplex virus TK gene and its replacement by the desired knockin (Fig. 4B). The frequency of site-specific targeting was sufficiently high that knockins could be identified by simply screening nonselected colonies obtained from a population of cells enriched by FACS to be fluorescent from having been cotransfected with a GFP-expressing construct. Moreover, the frequency of site-specific targeting was insensitive to the transcriptional status of the locus; that is, the frequency of targeting was the same whether or not an actively expressed gene was present at the chromosomal target. In similar experiments, a high frequency of Cre-mediated targeting at randomly placed targets in ES cells was observed with a closely related strategy that used two identical *loxP* sites but which were inverted (44). Because the *loxP* sites are inverted in this strategy there is no deletion of the intervening genomic interval, although inversion can occur. This inverted configuration also allows Cre-mediated double-crossover recombination to proceed with a transfected targeting plasmid having similarly configured *loxP* sites. In this case, inserts were obtained in both possible orientations.

In all of these integration strategies a *lox* target must first be placed into the genome either by homologous targeting or randomly, for example, by gene trap vectors (11, 47, 61). Might a *lox*-like sequence(s) already exist in the genome that can be targeted efficiently by Cre recombinase? Recently, so-called pseudo-*lox* sites have been identified from mammalian cells that are efficient substrates for Cre-mediated recombination in bacteria, and that support Cre-mediated integrative targeting when present on an extrachromosomal element in cultured mammalian cells (157). Regardless of whether these sites prove to be located at a useful (noncoding, nonsilenced) chromosomal position, their existence may be relevant to the development of double *lox*-targeting strategies. Because a variety of functional heterospecific *lox* sites with altered spacers exist (68, 89, 129, 131), it is likely that several of these will be useful for double *lox* targeting, as long as they do not also recombine with endogenous *lox*-like sequences. Curiously, at several spacer *lox* mutants, recombination in vitro with a modified Cre recombinase in a crude mammalian extract (89) occurred more readily than had been seen previously with purified wild-type Cre recombinase (68). Although this may simply be due to differences in the in vitro assay conditions, these results do raise the possibility, however unlikely, that Cre-mediated recombination in eukaryotic cells might be somewhat more tolerant of sequence differences in the *lox* site. If this were to be the case, such permissivity could restrict the choice of a useful heterospecific *lox* site for double *lox* targeting.

Several important experimental organisms, notably *Drosophila,* have remained refractory to targeted modification of the genome by homologous recombination with methods developed for yeast and mice. Here too, site-specific recombination has provided assistance in successfully developing a homologous targeting strategy (122). The I-SceI nuclease of yeast has an 18-bp specificity (30) that can be harnessed in eukaryotic cells to generate recombinogenic DNA ends. Such I-SceI-generated ends promote homologous DNA recombination and targeting when the I-SceI cut site is placed within a region of homology (124). To develop a homologous targeting system in *Drosophila,* the I-SceI cut site was placed within a cloned region of homology flanked by directly repeated FRT (Flp recognition target) sites, and the entire assemblage was then placed into the fly genome by P-element insertion (122). In flies carrying this construct, and also inducible genes for both the Flp site-specific DNA recombinase and the I-SceI nuclease, induction of both enzymes should lead to release of the FRT-flanked targeting construct from the genome and cutting of the excised circle to allow homologous targeting. Induction of such flies early in development did lead to easily detected homologous recombination at the endogenous target locus. In the female germ line homologous events occurred at a frequency of about 1 in 500 gametes, although homologous recombination was considerably lower in the male germ line. Strategies such as this, using one of the site-specific DNA recombinases such as Cre or Flp and a rare-cutting nuclease, allow extension of homologous targeting strategies to a greater number of eukaryotes and also thereby permit the precise implementation of the site-specific recombinase-mediated targeting strategies discussed above.

CHROMOSOME ENGINEERING

Chromosome Translocation

Both the alacrity with which site-specific recombinases catalyze interchromosomal DNA recombination in yeast (100, 129) and the demonstration that in higher eukaryotes Cre efficiently integrates transfected DNA into a chromosomal target by intermolecular site-specific DNA recombination argue strongly that Cre could be used as an engine to generate precisely designed chromosomal translocations in genetically engineered animals and plants. This has been true for Cre, and also for the related Flp recombinase (52). Cre-mediated translocation between nonhomologous chromosomes in tobacco was directly selected at a frequency of a few percent from among the progeny of a cross between two plants having randomly placed *loxP* sites, where recombination between a *lox*-containing drug-resistant gene of one parental plant and a promoter-*lox-cre* cassette of the second parent reconstructs a functional drug selection marker (115). The Cre-mediated translocations in these plants appeared to be balanced and reciprocal, and they not only occurred in somatic tissue but also were transmitted through seed formation to the next generation.

To generate precise chromosome translocations, *loxP* sites have been installed at defined sites on mouse chromosomes by homologous DNA recombination in ES cells to mimic human chromosome-translocation associated with leukemia and other neoplasms. After targeting of complementary halves of a *lox-Hprt* fusion gene to the *IgH* and c-*myc* loci in *Hprt*$^-$ ES cells, transfection with a *cre* expression plasmid generated Hprt$^+$ colonies carrying a precise t(12;15) translocation at an overall frequency of 5×10^{-8} (145). In a second report, *loxP* sites were placed within the introns of the *Dek* (chromosome 13) and *Can* (chromosome 2) genes by homologous targeting. Colonies obtained after transfection with a *cre* expression plasmid were then simply screened by PCR for the predicted recombination event and translocations were identified at an overall frequency of 2×10^{-5} to 5×10^{-5} (162). Because not all ES cells were transfected with the *cre* plasmid, the actual frequency of Cre catalyzing reciprocal chromosome translocation events is undoubtedly at least 10-fold higher. Still, the difference in frequency between the two different translocations could indicate differential chromosome accessibility to recombination, or may simply derive from the selection/detection method used, as described in the next section.

Deletion, Duplication, Inversion, and Chromosome Loss

Precise engineering of individual chromosomes themselves, with production of defined chromosomal inversions, deletions, and duplications, is now possible using Cre recombinase. Recombination between directly repeated *loxP* sites on the same chromosome will lead either to simple deletion of the intervening DNA after intramolecular recombination or, if recombination occurs in mitotic cells between centromere proximal and distal *loxP* sites on sister chromatids, to a balanced chromosome deletion and duplication. Similarly, recombination between *loxP* sites in an inverted configuration on the same chromosome will lead to a chromosomal inversion. In this case, recombination between oppositely oriented recombination sites on sister chromatids will generate dicentric and acentric products that are unstable to mitosis and can lead to chromosome loss (40). Indeed, the presence of inverted *loxP* sites on the same chromosome leads to loss of that chromosome in embryos expressing Cre recombinase (90). This may become a useful technique for producing partial monosomies in mice.

Both inversions and deletions were selected at the *HoxB-wnt3* interval on chromosome 11 with use of a *loxP-Hprt* gene fusion selection strategy (119). Relatively short inversions and deletions, between 90 kb and 1 Mb, were selected in ES cells at a frequency of either 1×10^{-7} or 1×10^{-5} to 5×10^{-5}, depending on the particular ES clone, after transfection with a *cre* expression plasmid. Clones showing the lower frequency of recombination were those in which the *loxP* sites were located in *trans* on chromosome homologs, whereas ES cells having the *loxP* sites situated in *cis* on the same chromosome exhibited the higher recombination frequency. At distances of greater than 3 to 4 cM, recombination was still lower, rivaling or surpassing those reported for chromosome translocations. Even with such a low-frequency recombination, Cre-mediated segmental aneuploidy on chromosome 11 allowed the generation of inversions, deletions, and duplications over a 22-cM interval (95), including several short, 3 to 4 cM, deficiencies that were lethal to the embryo. In contrast, Cre-mediated deletions of a 200-kb interval at the amyloid precursor protein (APP) locus occurred at a somewhat higher frequency, and were recovered by selecting for loss of a herpes simplex virus TK gene positioned between two *loxP* sites that had been targeted to flank the APP gene after two rounds of homologous recombination (93). Cre-mediated recombination was sufficiently high that the second round of homologous recombination for placement of the second *loxP* site could be combined with cotransfection of a *cre* expression vector, followed by simultaneous selection for both homologous targeting and the Cre-mediated deletion event. Because homologous recombination is itself a rare event, the ability to combine both recombination events in a single selection strategy argues that Cre-mediated recombination had occurred at an appreciable frequency.

These results suggested that certain Cre-mediated long-distance recombination events, perhaps depending on the locus or the selection scheme, occurred significantly less frequently than others, even though the physical distances were approximately the same. This seeming mystery was solved by the discovery that the split gene *lox-Hprt* cassettes used to select for Cre-mediated recombination were mutated (173). Thus, functional *Hprt* gene reconstruction using these older constructs required not only a site-specific DNA recombination event, but also a low-frequency gene conversion event. Reanalysis of the Cre-mediated deletion frequency on chromosome 11 in mouse ES cells using unmutated split gene *lox-Hprt* cassettes showed that the actual frequency of generating a 2-cM deletion when *loxP* sites were located on the same chromosome was quite high, about 11%. As before, interhomolog recombination in *trans* to generate a balanced deletion duplication was lower, about 1×10^{-5}. Even very large deletions of 60 cM occurred at an experimentally accessible frequency of about 1×10^{-6}. These results strongly support the use of Cre/*loxP* recombination strategies for making precisely engineered chromosomes for genetic analysis in higher eukaryotes.

Several approaches have been investigated to further expand the utility of Cre recombinase for engineering chromosomes. To make nested deletions or inversions having one predefined anchor point in plants, the Cre/*loxP* system was combined with the Ds-transposable element in tobacco (102), *Arabidopsis* (111), and tomato (152). In this strategy a Ds-*lox* element is first inserted into the genome using T-DNA or other standard gene transfer method, Ac transposase activity is then provided so that the Ds-*lox* element transposes to a new site on the same chromosome, and the resulting plant is mated with a Cre plant to catalyze DNA recombination between the Ds-*lox* at the original insertion site and the transposed Ds-*lox* element. Cre-mediated recombination was readily detected between sites up to 16.5 cM apart from each other (111). Moreover, the resulting *loxP*-flanked interval can be released from genomic DNA by performing Cre-mediated recombination with *loxP* oligonucleotides in vitro (116, 152). This latter strategy should help facilitate both the long-range mapping of chromosomes and aid in cloning specific chromosomal regions.

A somewhat similar transposon strategy allows the facile generation of nested deletions in ES cells (153). Homologous recombination in ES cells was used to place a *lox-Hprt* cassette at a desired locus and these cells were then infected with a recombinant retrovirus carrying the other half of the *lox-Hprt* split gene cassette. Unlike Ds transposition, the sites at which retrovirus insertion occurs are not necessarily on the same chromosome and *cis* to the desired *lox* anchor point. Still, the bias of Cre-mediated recombination to efficiently recombine *lox* sites on the same chromosome allowed the ready recovery of nested chromosomal deletions having one predetermined endpoint and one randomly chosen endpoint from retroviral insertion. The ease with which such deletions were recovered promises to aid greatly both in the functional mapping of important genes by complementation and in the generation of a physical map in mice.

CONCLUDING REMARKS

Advances in the genetic manipulation of model organisms have consistently led to deeper appreciation and understanding of their basic biology and to insights into the etiology of human disease. A key advance certainly in bacterial genetics was the ability to easily transfer specific genetic information from one bacterium to another. In *E. coli* both conjugation and transduction by phage P1 became, and still remain, important tools for gene transfer, gene mapping, and strain construction, and have been exploited to not only answer fundamental questions in virtually all aspects of bacterial biology, but also to engineer bacteria for a variety of industrial and commercial applications. During the past decade the astonishing facility with which bacterial genomes can be manipulated has been extended to the genomes of a variety of higher eukaryotes, including that of the common mouse, a model organism that has now become essential for understanding mammalian gene function. Bacteriophage P1 has again provided an important tool for genetic manipulation, that tool being the Int family site-specific DNA recombinase Cre.

Cre and other Int family members will undoubtedly be joined by other recombinases, for example, members of the resolvase family (56, 97, 156), in developing increasingly powerful strategies for genomic manipulation and chromosome engineering in a wide range of both prokaryotic and eukaryotic organisms. In conjunction with homologous recombination strategies and transposon methods these tools have already led to more precise design of model organisms, with built-in switches for turning genes on or off. Moreover, many of these strategies will be adapted not only to engineer industrially and commercially valuable plants and animals, but also to develop novel gene therapeutic agents for human disease.

REFERENCES

1. **Abremski, K., and R. Hoess.** 1985. Phage P1 Cre-*loxP* site-specific recombination: effects of DNA supercoiling on catenation and knotting of recombinant products. *J. Mol. Biol.* **184:**211–220.
2. **Abremski, K., R. Hoess, and N. Sternberg.** 1983. Studies on the properties of P1 site-specific recombination: evidence for topologically unlinked products following recombination. *Cell* **32:**1301–1311.
3. **Abuin, A., and A. Bradley.** 1996. Recycling selectable markers in mouse embryonic stem cells. *Mol. Cell. Biol.* **16:** 1851–1856.
4. **Akagi, K., V. Sandig, M. Vooijs, M. Van der Valk, M. Giovannini, M. Strauss, and A. Berns.** 1997. Cre-mediated somatic site-specific recombination in mice. *Nucleic Acids Res.* **25:**1766–1773.
5. **Alani, E., L. Cao, and N. Kleckner.** 1987. A method for gene disruption that allows repeated use of URA3 selection in the construction of multiply disrupted yeast strains. *Genetics* **116:**541–545.
6. **Albert, H., E. C. Dale, E. Lee, and D. W. Ow.** 1995. Site-specific integration of DNA into wild-type and mutant *lox* sites placed in the plant genome. *Plant J.* **7:**649–659.
7. **Anton, M., and F. L. Graham.** 1995. Site-specific recombination mediated by an adenovirus vector expressing the Cre recombinase protein: a molecular switch for control of gene expression. *J. Virol.* **69:**4600–4606.
8. **Aoki, K., C. Barker, X. Danthinne, M. J. Imperiale, and G. J. Nabel.** 1999. Efficient generation of recombinant adenoviral vectors by Cre-lox recombination *in vitro*. *Mol. Med.* **5:** 224–231.
9. **Araki, K., M. Araki, J.-I. Miyazaki, and P. Vassalli.** 1995. Site-specific recombination of a transgene in fertilized eggs by transient expression of Cre recombinase. *Proc. Natl. Acad. Sci. USA* **92:**160–164.
10. **Araki, K., M. Araki, and K. Yamamura.** 1997. Targeted integration of DNA using mutant lox site in embryonic stem cell. *Nucleic Acids Res.* **25:**868–872.
11. **Araki, K., T. Imaizumi, T. Sekimoto, K. Yoshinobu, J. Yoshimuta, M. Akizuki, K. Miura, M. Araki, and K. Yamamura.** 1999. Exchangeable gene trap using the Cre/mutated *lox* system. *Cell. Mol. Biol.* **45:**737–750.
12. **Artelt, P., R. Grannemann, C. Stocking, J. Friel, J. Bartsch, and H. Hauser.** 1991. The prokaryotic neomycin-resistance-encoding gene acts as a transcriptional silencer in eukaryotic cells. *Gene* **99:**249–254.
13. **Austin, S., M. Ziese, and N. Sternberg.** 1981. A novel role for site-specific recombination in maintenance of bacterial replicons. *Cell* **25:**729–736.
14. **Backman, K., M. J. O'Conner, A. Maruya, and M. Erfle.** 1984. Use of synchronous site specific recombination in vivo to regulate gene expression. *Biotechnology* **2:**1045–1049.
15. **Baubonis, W., and B. Sauer.** 1993. Genomic targeting with purified Cre recombinase. *Nucleic Acids Res.* **21:**2025–2029.
16. **Berghella, L., L. De Angelis, M. Coletta, B. Berarducci, C. Sonnino, G. Salvatori, C. Anthonissen, R. Cooper, G. S. Butler-Browne, V. Mouly, G. Ferrari, F. Mavilio, and G. Cossu.** 1999. Reversible immortalization of human myogenic cells by site-specific excision of a retrovirally transferred oncogene. *Hum. Gene Ther.* **10:**1607–1617.
17. **Bethke, B., and B. Sauer.** 1997. Segmental genomic replacement by Cre-mediated recombination: genotoxic stress activation of the p53 promoter in single-copy transformants. *Nucleic Acids Res.* **25:**2828–2834.
18. **Betz, U. A. K., C. A. J. Voßhenrich, K. Rajewsky, and W. Müller.** 1996. Bypass of lethality with mosaic mice generated by Cre-*loxP*-mediated recombination. *Curr. Biol.* **6:** 1307–1316.
19. **Bouhassira, E. E., K. Westerman, and P. Leboulch.** 1997. Transcriptional behavior of LCR enhancer elements integrated at the same chromosomal locus by recombinase-mediated cassette exchange. *Blood* **90:**3332–3344.
20. **Brocard, J., R. Feil, P. Chambon, and D. Metzger.** 1998. A chimeric Cre recombinase inducible by synthetic, but not by natural ligands of the glucocorticoid receptor. *Nucleic Acids Res.* **26:**4086–4090.
21. **Brocard, J., X. Warot, O. Wendling, N. Messaddeq, J. L. Vonesch, P. Chambon, and D. Metzger.** 1997. Spatio-temporally controlled site-specific somatic mutagenesis in the mouse. *Proc. Natl. Acad. Sci. USA* **94:**14559–14563.
22. **Brunelli, J. P., and M. L. Pall.** 1993. A series of yeast/*Escherichia coli* lambda expression vectors designed for directional cloning of cDNAs and cre/lox-mediated plasmid excision. *Yeast* **9:**1309–1318.
23. **Burgess, S. M., and N. Kleckner.** 1999. Collisions between yeast chromosomal loci *in vivo* are governed by three layers of organization. *Genes Dev.* **13:**1871–1883.
24. **Call, L. M., C. S. Moore, G. Stetten, and J. D. Gearhart.** 2000. A Cre-lox recombination system for the targeted integration of circular yeast artificial chromosomes into embryonic stem cells. *Hum. Mol. Genet.* **9:**1745–1751.
25. **Castilho, B. A., P. Olfson, and M. J. Casadaban.** 1984. Plasmid insertion mutagenesis and *lac* gene fusion with mini-Mu bacteriophage transposons. *J. Bacteriol.* **158:**488–495.
26. **Chai, Y., X. Jiang, Y. Ito, P. Bringas, Jr., J. Han, D. H. Rowitch, P. Soriano, A. P. McMahon, and H. M. Sucov.** 2000. Fate of the mammalian cranial neural crest during tooth and mandibular morphogenesis. *Development* **127:**1671–1679.
27. **Chalfrie, M., Y. Tu, G. Euskirchen, W. W. Ward, and D. C. Prasher.** 1994. Green fluorescent protein as a marker for gene expression. *Science* **263:**802–805.
28. **Chesney, R. H., J. R. Scott, and D. Vapnek.** 1979. Integration of the plasmid prophages P1 and P7 into the chromosome of *Escherichia coli*. *J. Mol. Biol.* **130:**161–173.
29. **Choi, S., D. Begum, H. Koshinsky, D. W. Ow, and R. A. Wing.** 2000. A new approach for the identification and cloning of genes: the pBACwich system using Cre/lox site-specific recombination. *Nucleic Acids Res.* **28:**e19.
30. **Colleaux, L., L. D'Auriol, F. Galibert, and B. Dujon.** 1988.

Recognition and cleavage site of the intron-encoded omega transposase. *Proc. Natl. Acad. Sci. USA* **85**:6022–6026.

31. **Dale, E. C., and D. W. Ow.** 1991. Gene transfer with subsequent removal of the selection gene from the host genome. *Proc. Natl. Acad. Sci. USA* **88**:10558–10562.
32. **Dale, E. C., and D. W. Ow.** 1990. Intra- and intermolecular site-specific recombination in plant cells mediated by bacteriophage P1 recombinase. *Gene* **91**:79–85.
33. **Danielian, P. S., D. Muccino, D. H. Rowitch, S. K. Michael, and A. P. McMahon.** 1998. Modification of gene activity in mouse embryos in utero by a tamoxifen-inducible form of Cre recombinase. *Curr. Biol.* **8**:1323–1326.
34. **de Wit, T., D. Drabek, and F. Grosveld.** 1998. Microinjection of cre recombinase RNA induces site-specific recombination of a transgene in mouse oocytes. *Nucleic Acids Res.* **26**: 676–678.
35. **Díaz, V., F. Rojo, C. Martínez-A., J. C. Alonso, and A. Bernad.** 1999. The prokaryotic β-recombinase catalyzes site-specific recombination in mammalian cells. *J. Biol. Chem.* **274**: 6634–6640.
36. **Dietrich, P., I. Dragatsis, S. Xuan, S. Zeitlin, and A. Efstratiadis.** 2000. Conditional mutagenesis in mice with heat shock promoter-driven cre transgenes. *Mamm. Genome* **11**: 196–205.
37. **Drago, J., P. Padungchaichot, J. Y.-F. Wong, A. J. Lawrence, J. F. McManus, S. H. Sumarsono, A. L. Natoli, M. Lakso, N. Wreford, H. Westphal, I. Kola, and D. I. Finkelstein.** 1998. Targeted expression of a toxin gene to D1 dopamine receptor neurons by Cre-mediated site-specific recombination. *J. Neurosci.* **18**:9845–9857.
38. **Elledge, S. J., J. T. Mulligan, S. W. Ramer, M. Spottswood, and R. W. Davis.** 1991. λYES: a multifunctional cDNA expression vector for the isolation of genes by complementation of yeast and *Escherichia coli* mutations. *Proc. Natl. Acad. Sci. USA* **88**:1731–1735.
39. **Epstein, J. A., J. Li, D. Lang, F. Chen, C. B. Brown, F. Jin, M. M. Lu, M. Thomas, E. Liu, A. Wessels, and C. W. Lo.** 2000. Migration of cardiac neural crest cells in Splotch embryos. *Development* **127**:1869–1878.
40. **Falco, S. C., Y. Li, J. R. Broach, and D. Botstein.** 1982. Genetic properties of chromosomally integrated 2 mu plasmid DNA in yeast. *Cell* **29**:573–584.
41. **Feil, R., J. Brocard, B. Mascrez, M. LeMeur, D. Metzger, and P. Chambon.** 1996. Ligand-activated site-specific recombination in mice. *Proc. Natl. Acad. Sci. USA* **93**: 10887–10890.
42. **Feil, R., J. Wagner, D. Metzger, and P. Chambon.** 1997. Regulation of Cre recombinase activity by mutated estrogen receptor ligand-binding domains. *Biochem. Biophys. Res. Commun.* **237**:752–757.
43. **Feiring, S., C. G. Kim, E. M. Epner, and M. Groudine.** 1993. An "in-out" strategy using gene targeting and FLP recombinase for the functional dissection of complex DNA regulatory elements: analysis of the β-globin locus control region. *Proc. Natl. Acad. Sci. USA* **90**:8469–8473.
44. **Feng, Y.-Q., J. Seibler, R. Alami, A. Eisen, K. Westerman, P. Leboulch, S. Feiring, and E. E. Bouhassira.** 1999. Site-specific chromosomal integration in mammalian cells: highly efficient CRE recombinase-mediated cassette exchange. *J. Mol. Biol.* **292**:779–785.
45. **Forlino, A., F. D. Porter, E. J. Lee, H. Westphal, and J. C. Marini.** 1999. Use of the Cre/lox recombination system to develop a non-lethal knock-in murine model for osteogenesis imperfecta with an alpha(I) G349C substitution. Variability in phenotype in BrtlIV mice. *J. Biol. Chem.* **274**: 37923–37931.
46. **Freundlieb, S., C. Schirra-Muller, and H. Bujard.** 1999. A tetracycline controlled activation/repression system with increased potential for gene transfer into mammalian cells. *J. Gene Med.* **1**:4–12.
47. **Fukushige, S., and J. E. Ikeda.** 1996. Trapping of mammalian promoters by Cre-lox site-specific recombination. *DNA Res.* **3**:73–80.
48. **Fukushige, S., and B. Sauer.** 1992. Targeted genomic integration with a positive selection lox recombination vector allows highly reproducible gene expression in mammalian cells. *Proc. Natl. Acad. Sci. USA* **89**:7905–7909.
49. **Gage, P. J., B. Sauer, M. Levine, and J. C. Glorioso.** 1992. A cell-free recombination system for site-specific integration of multigenic shuttle plasmids into the herpes simplex virus type 1 genome. *J. Virol.* **66**:5509–5515.
50. **Gagneten, S., Y. Le, J. Miller, and B. Sauer.** 1997. Brief expression of a GFP*cre* fusion gene in embryonic stem cells allows rapid retrieval of site-specific genomic deletions. *Nucleic Acids Res.* **25**:3326–3331.
51. **Gatz, C., and P. H. Quail.** 1988. Tn10-encoded tet repressor can regulate an operator-containing plant promoter. *Proc. Natl. Acad. Sci. USA* **85**:1394–1397.
52. **Golic, K.** 1991. Site-specific recombination between homologous chromosomes in *Drosophila*. *Science* **252**:958–961.
53. **Golic, K. G., and S. Lindquist.** 1989. The FLP recombinase of yeast catalyzes site-specific recombination in the Drosophila genome. *Cell* **59**:499–509.
54. **Gossen, M., and H. Bujard.** 1992. Tight control of gene expression in mammalian cells by tetracycline-responsive promoters. *Proc. Natl. Acad. Sci. USA* **89**:5547–5551.
55. **Gossen, M., S. Freundlieb, G. Bender, G. Muller, W. Hillen, and H. Bujard.** 1995. Transcriptional activation by tetracyclines in mammalian cells. *Science* **268**:1766–1769.
56. **Groth, A. C., E. C. Olivares, B. Thyagarajan, and M. P. Calos.** 2000. A phage integrase directs efficient site-specific integration in human cells. *Proc. Natl. Acad. Sci. USA* **97**: 5995–6000.
57. **Gu, H., J. D. Marth, P. C. Orban, H. Mossmann, and K. Rajewsky.** 1994. Deletion of a polymerase beta gene segment in T cells using cell type-specific gene targeting. *Science* **265**: 103–106.
58. **Gu, H., Y. R. Zou, and K. Rajewsky.** 1993. Independent control of immunoglobulin switch recombination at individual switch regions evidenced through Cre-*loxP*-mediated gene targeting. *Cell* **73**:1155–1164.
59. **Guldener, U., S. Heck, T. Fielder, J. Beinhauer, and J. H. Hegemann.** 1996. A new efficient gene disruption cassette for repeated use in budding yeast. *Nucleic Acids Res.* **24**: 2519–2524.
60. **Harada, N., Y. Tomai, T. Ishikawa, B. Sauer, K. Takaku, M. Oshima, and M. M. Taketo.** 1999. Intestinal polyposis in mice with a dominant stable mutation of the β-catenin gene. *EMBO J.* **18**:5931–5942.
61. **Hardouin, N., and A. Nagy.** 2000. Gene-trap-based target site for Cre-mediated transgenic insertion. *Genesis* **26**: 245–252.
62. **Herrera, P. L.** 2000. Adult insulin- and glucagon-producing cells differentiate from two independent cell lineages. *Development* **127**:2317–2322.
63. **Herrera, P. L., L. Orci, and J. D. Vassalli.** 1998. Two transgenic approaches to define the cell lineages in endocrine pancreas development. *Mol. Cell. Endocrinol.* **140**:45–50.
64. **Hoekstra, M. F., D. Burbee, J. Singer, E. Mull, E. Chiao, and F. Heffron.** 1991. A Tn3 derivative that can be used to make short in-frame insertions within genes. *Proc. Natl. Acad. Sci. USA* **88**:5457–5461.

65. **Hoekstra, M. F., H. S. Seifert, J. Nickoloff, and F. Heffron.** 1991. Shuttle mutagenesis: bacterial transposons for genetic manipulation in yeast. *Methods Enzymol.* **194:**329–342.
66. **Hoess, R. H., and K. Abremski.** 1990. The Cre-*lox* recombination system, p. 99–109. *In* F. Eckstein and D. M. J. Lilley (ed.), *Nucleic Acids and Molecular Biology*, vol. 4. Springer-Verlag KG, Berlin, Germany.
67. **Hoess, R. H., and K. Abremski.** 1984. Interaction of the bacteriophage P1 recombinase Cre with the recombining site *loxP*. *Proc. Natl. Acad. Sci. USA* **81:**1026–1029.
68. **Hoess, R. H., A. Wierzbicki, and K. Abremski.** 1986. The role of the spacer region in P1 site-specific recombination. *Nucleic Acids Res.* **14:**2287–2300.
69. **Hoess, R. H., M. Ziese, and N. Sternberg.** 1982. P1 site-specific recombination: nucleotide sequence of the recombining sites. *Proc. Natl. Acad. Sci. USA* **79:**3398–3402.
70. **Huang, L.-C., E. A. Wood, and M. M. Cox.** 1991. A bacterial model system for chromosomal targeting. *Nucleic Acids Res.* **19:**443–448.
71. **Iida, S.** 1984. Bacteriophage P1 carries two related sets of genes determining its host range in the invertible C segment of its genome. *Virology* **134:**421–434.
72. **Indra, A. K., X. Warot, J. Brocard, J. M. Bornert, J. H. Xiao, P. Chambon, and D. Metzger.** 1999. Temporally-controlled site-specific mutagenesis in the basal layer of the epidermis: comparison of the recombinase activity of the tamoxifen-inducible Cre-ER(T) and Cre-ER(T2) recombinases. *Nucleic Acids Res.* **27:**4324–4327.
73. **Ioannou, P. A., C. T. Amemiya, J. Garnes, P. M. Kroisel, H. Shizuya, C. Chen, M. A. Batzer, and P. J. de Jong.** 1994. A new bacteriophage P1-derived vector for the propagation of large human DNA fragments. *Nat. Genet.* **6:**84–89.
74. **Jiang, X., D. H. Rowitch, P. Soriano, A. P. McMahon, and H. M. Sucov.** 2000. Fate of the mammalian cardiac neural crest. *Development* **127:**1607–1614.
75. **Kawamoto, S., H. Niwa, F. Tashiro, S. Sano, G. Kondoh, J. Takeda, K. Tabayashi, and J.-I. Miyazaki.** 2000. A novel reporter mouse strain that expresses enhanced green fluorescent protein upon Cre-mediated recombination. *FEBS Lett.* **470:**263–268.
76. **Kellendonk, C., F. Tronche, E. Casanova, K. Anlag, C. Opherk, and G. Schutz.** 1999. Inducible site-specific recombination in the brain. *J. Mol. Biol.* **285:**175–182.
77. **Kellendonk, C., F. Tronche, A.-P. Monaghan, P.-O. Angrand, and F. Stewart.** 1996. Regulation of Cre recombinase activity by the synthetic steroid RU486. *Nucleic Acids Res.* **24:**1404–1411.
78. **Kimmel, R. A., D. H. Turnbull, V. Blanquet, W. Wurst, C. A. Loomis, and A. L. Joyner.** 2000. Two lineage boundaries coordinate vertebrate apical ectodermal ridge formation. *Genes Dev.* **14:**1377–1389.
79. **Kobayashi, N., T. Fujiwara, K. A. Westerman, Y. Inoue, M. Sakaguchi, H. Noguchi, M. Miyazaki, J. Cai, N. Tanaka, I. J. Fox, and P. Leboulch.** 2000. Prevention of acute liver failure in rats with reversibly immortalized human hepatocytes. *Science* **287:**1258–1262.
80. **Kolb, A. F., R. Ansell, J. McWhir, and S. G. Siddell.** 1999. Insertion of a foreign gene into the beta-casein locus by Cre-mediated site-specific recombination. *Gene* **227:**21–31.
81. **Kolb, A. F., and S. G. Siddell.** 1997. Genomic targeting of a bicistronic DNA fragment by Cre-mediated site-specific recombination. *Gene* **203:**209–216.
82. **Kolot, M., N. Silberstein, and E. Yagil.** 1999. Site-specific recombination in mammalian cells expressing the Int recombinase of bacteriophage HK022. *Mol. Biol. Rep.* **26:**207–213.
83. **Kozak, M.** 1986. Point mutations define a sequence flanking the AUG initiator codon that modulates translation by eukaryotic ribosomes. *Cell* **44:**283–292.
84. **Kühn, R., F. Schwenk, M. Aguet, and K. Rajewsky.** 1995. Inducible gene targeting in mice. *Science* **269:**1427–1429.
85. **Lakso, M., J. G. Pichel, J. R. Gorman, B. Sauer, Y. Okamoto, E. Lee, F. W. Alt, and H. Westphal.** 1996. Efficient in vivo manipulation of mouse genomic sequences at the zygote stage. *Proc. Natl Acad. Sci. USA* **93:**5860–5865.
86. **Lakso, M., B. Sauer, B. Mosinger, Jr., E. J. Lee, R. W. Manning, S.-H. Yu, K. L. Mulder, and H. Westphal.** 1992. Targeted oncogene activation by site-specific recombination in transgenic mice. *Proc. Natl. Acad. Sci. USA* **89:**6232–6236.
87. **Le, Y., S. Gagneten, D. Tombaccini, B. Bethke, and B. Sauer.** 1998. Identification of nuclear targeting determinants in the Cre recombinase of phage P1. *Nucleic Acids Res.* **24:**4703–4709.
88. **Le, Y., J. L. Miller, and B. Sauer.** 1999. GFP*cre* fusion vectors with enhanced expression. *Anal. Biochem.* **270:**334–336.
89. **Lee, G., and I. Saito.** 1998. Role of nucleotide sequences of loxP spacer region in Cre-mediated recombination. *Gene* **216:**55–65.
90. **Lewandoski, M., and G. R. Martin.** 1997. Cre-mediated chromosome loss in mice. *Nat. Genet.* **17:**223–225.
91. **Lewandoski, M., K. M. Wassarman, and G. R. Martin.** 1997. Zp3-cre, a transgenic mouse line for the activation or inactivation of loxP-flanked target genes specifically in the female germ line. *Curr. Biol.* **7:**148–151.
92. **Li, L., and S. N. Cohen.** 1996. Tsg101: a novel tumor susceptibility gene isolated by controlled homozygous functional knockout of allelic loci in mammalian cells. *Cell* **85:**319–329.
93. **Li, Z.-W., G. Stark, J. Götz, T. Rülicke, U. Müller, and C. Weissmann.** 1996. Generation of mice with a 200-kb amyloid precursor protein gene deletion by Cre recombinase-mediated site-specific recombination in embryonic stem cells. *Proc. Natl. Acad. Sci. USA* **93:**6158–6162.
94. **Liu, J.-L., A. Grinberg, H. Westphal, B. Sauer, D. Accili, M. Karas, and D. LeRoith.** 1998. Insulin-like growth factor-I affects perinatal lethality and postnatal development in a gene dosage-dependent manner: manipulation using the Cre/loxP system in transgenic mice. *Mol. Endocrinol.* **12:**1452–1462.
95. **Liu, P., H. Zhang, A. McLellan, H. Vogel, and A. Bradley.** 1998. Embryonic lethality and tumorigenesis caused by segmental aneuploidy on mouse chromosome 11. *Genetics* **150:**1155–1168.
96. **Lobe, C. G., K. E. Koop, W. Kreppner, H. Lomeli, M. Gertsenstein, and A. Nagy.** 1999. Z/AP, a double reporter for cre-mediated recombination. *Dev. Biol.* **208:**281–292.
97. **Maeser, S., and R. Kahmann.** 1991. The Gin recombinase of bacteriophage Mu can catalyse site-specific recombination in plant protoplasts. *gMol. Gen. Genet.* **230:**170–176.
98. **Mao, X., Y. Fujiwara, and S. H. Orkin.** 1999. Improved reporter strain for monitoring Cre recombinase-mediated DNA excisions in mice. *Proc. Natl. Acad. Sci. USA* **96:**5037–5042.
99. **Marth, J. D.** 1996. Recent advances in gene mutagenesis by site-directed recombination. *J. Clin. Invest.* **97:**1999–2002.
100. **Matsuzaki, H., R. Nakajima, J. Nishiyama, H. Araki, and Y. Oshima.** 1990. Chromosome engineering in *Saccharomyces cerevisiæ* by using a site-specific recombination system of a yeast plasmid. *J. Bacteriol.* **172:**610–618.
101. **McDevitt, M. A., R. A. Shivdasani, Y. Fujiwara, H. Yang, and S. H. Orkin.** 1997. A "knockdown" mutation created by cis-element gene targeting reveals the dependence of erythroid cell maturation on the level of transcription factor GATA-1. *Proc. Natl. Acad. Sci. USA* **94:**6781–6785.
102. **Medberry, S. L., E. Dale, M. Qin, and D. W. Ow.** 1995.

Intra-chromosomal rearrangements generated by Cre-*lox* site-specific recombination. *Nucleic Acids Res.* 23:485–490.

103. **Metzger, D., J. Clifford, H. Chiba, and P. Chambon.** 1995. Conditional site-specific recombination in mammalian cells using a ligand-dependent chimeric Cre recombinase. *Proc. Natl. Acad. Sci. USA* **92**:6991–6995.
104. **Meyers, E. N., M. Lewandoski, and G. R. Martin.** 1998. An Fgf8 mutant allelic series generated by Cre- and Flp-mediated recombination. *Nat. Genet.* **18**:136–141.
105. **Mozo, T., and P. J. J. Hooykaas.** 1992. Design of a novel system for the construction of vectors for Agrobacterium-mediated plant transformation. *Mol. Gen. Genet.* **236**:1–7.
106. **Nagy, A., C. Moens, E. Ivanyi, J. Pawling, M. Gertsenstein, A. K. Hadjantonakis, M. Pirity, and J. Rossant.** 1998. Dissecting the role of N-myc in development using a single targeting vector to generate a series of alleles. *Curr. Biol.* **8**: 661–664.
107. **Nunes-Düby, S. E., H. J. Kwon, R. S. Tirumalai, T. Ellenberger, and A. Landy.** 1998. Similarities and differences among 105 members of the Int family of site-specific recombinases. *Nucleic Acids Res.* **26**:391–406.
108. **Odell, J., P. Caimi, B. Sauer, and S. Russell.** 1990. Site-directed recombination in the genome of transgenic tobacco. *Mol. Gen. Genet.* **223**:369–378.
109. **Onouchi, H., K. Yokoi, C. Machida, H. Matsuzaki, Y. Oshima, K. Matsuoka, K. Nakamura, and Y. Machida.** 1991. Operation of an efficient site-specific recombination system of *Zygosaccharomyces rouxii* in tobacco cells. *Nucleic Acids Res.* **19**:6373–6378.
110. **Orban, P. C., D. Chui, and J. D. Marth.** 1992. Tissue- and site-specific DNA recombination in transgenic mice. *Proc. Natl. Acad. Sci. USA* **89**:6861–6865.
111. **Osborne, B. I., U. Wirtz, and B. Baker.** 1995. A system for insertional mutagenesis and chromosomal rearrangement using the Ds transposon and Cre-*lox*. *Plant J.* **7**:687–701.
112. **Palazzolo, M. J., B. A. Hamilton, D. Ding, C. H. Martin, D. A. Mead, R. C. Mierendorf, K. V. Raghavan, E. M. Meyerowitz, and H. D. Lipshitz.** 1990. Phage lambda cDNA cloning vectors for subtractive hybridization, fusion-protein synthesis and Cre-*loxP* automatic plasmid subcloning. *Gene* **88**:25–36.
113. **Peakman, T. C., R. A. Harris, and D. R. Gewert.** 1992. Highly efficient generation of recombinant baculoviruses by enzymatically mediated site-specific *in vitro* recombination. *Nucleic Acids Res.* **20**:495–500.
114. **Picard, D.** 1994. Regulation of protein function through expression of chimaeric proteins. *Curr. Opin. Biotechnol.* **5**: 511–515.
115. **Qin, M., C. Bayley, T. Stockton, and D. W. Ow.** 1994. Cre recombinase-mediated site-specific recombination between plant chromosomes. *Proc. Natl. Acad. Sci. USA* **91**: 1706–1710.
116. **Qin, M., E. Lee, T. Zankel, and D. W. Ow.** 1995. Site-specific cleavage of chromosomes *in vitro* through Cre-lox recombination. *Nucleic Acids Res.* **23**:1923–1927.
117. **Raben, N., K. Nagaraju, E. Lee, P. Kessler, B. Byrne, L. Lee, M. LaMarca, C. King, J. Ward, B. Sauer, and P. Plotz.** 1998. Targeted disruption of the acid α-glucosidase gene in mice causes an illness with the critical features of both infantile and adult human Glycogen Storage Disease Type II. *J. Biol. Chem.* **273**:19086–19092.
118. **Rajewsky, K., H. Gu, R. Kühn, U. A. Betz, W. Müller, J. Roes, and F. Schwenk.** 1996. Conditional gene targeting. *J. Clin. Invest.* **98**:600–603.
119. **Ramírez-Solis, R., P. Liu, and A. Bradley.** 1995. Chromosome engineering in mice. *Nature* (London) **378**:720–724.
120. **Rijli, F. M., P. Dollé, V. Fraulob, M. Lemeur, and P. Chambon.** 1994. Insertion of a targeting construct in a *Hoxd-10* allele can influence the control of *Hoxd-9* expression. *Dev. Dyn.* **201**:366–377.
121. **Ripoll, P. J., A. Cowper, S. Salmeron, P. Dickinson, D. Porteous, and B. Arveiler.** 1998. A new yeast artificial chromosome vector designed for gene transfer into mammalian cells. *Gene* **210**:163–172.
122. **Rong, Y. S., and K. G. Golic.** 2000. Gene targeting by homologous recombination in *Drosophila*. *Science* **288**: 1973–1975.
123. **Ross-Macdonald, P., A. Sheehan, G. S. Roeder, and M. Snyder.** 1997. A multipurpose transposon system for analyzing protein production, localization, and function in *Saccharomyces cerevisiae*. *Proc. Natl. Acad. Sci. USA* **94**:190–195.
124. **Rouet, P., F. Smith, and M. Jasin.** 1994. Expression of a site-specific endonuclease stimulates homologous recombination in mammalian cells. *Proc. Natl. Acad. Sci. USA* **91**: 6064–6068.
125. **Rucker, E. B., and J. A. Piedrahita.** 1997. Cre-mediated recombination at the murine whey acidic protein (mWAP) locus. *Mol. Reprod. Dev.* **48**:324–331.
126. **Russell, S. H., J. L. Hoopes, and J. T. Odell.** 1992. Directed excision of a transgene from the plant genome. *Mol. Gen. Genet.* **234**:49–59.
127. **Saam, J. R., and J. I. Gordon.** 1999. Inducible gene knockouts in the small intestinal and colonic epithelium. *J. Biol. Chem.* **274**:38071–38082.
128. **Sauer, B.** 1987. Functional expression of the *cre-lox* site-specific recombination system in the yeast *Saccharomyces cerevisiae*. *Mol. Cell. Biol.* **7**:2087–2096.
129. **Sauer, B.** 1992. Identification of cryptic *lox* sites in the yeast genome by selection for chromosome translocations that confer multiple drug resistance. *J. Mol. Biol.* **223**:911–928.
130. **Sauer, B.** 1998. Inducible gene targeting in mice using the Cre/*lox* system. *Methods* **14**:381–392.
131. **Sauer, B.** 1996. Multiplex Cre/*lox* recombination permits selective site-specific DNA targeting to both a natural and an engineered site in the yeast genome. *Nucleic Acids Res.* **24**: 4608–4613.
132. **Sauer, B.** 1994. Recycling selectable markers in yeast. *BioTechniques* **16**:1086–1088.
133. **Sauer, B., W. Baubonis, S. Fukushige, and L. Santomenna.** 1992. Construction of isogenic cell lines expressing human and rat angiotensin II AT_1 receptors by Cre-mediated site specific recombination. *Methods* **4**:143–149.
134. **Sauer, B., and N. Henderson.** 1989. Cre-stimulated recombination at *loxP* sites placed into the genome of mammalian cells. *Nucleic Acids Res.* **17**:147–161.
135. **Sauer, B., and N. Henderson.** 1988. The cyclization of linear DNA in *Escherichia coli* by site-specific recombination. *Gene* **70**:331–341.
136. **Sauer, B., and N. Henderson.** 1988. Site-specific DNA recombination in mammalian cells by the Cre recombinase of bacteriophage P1. *Proc. Natl. Acad. Sci. USA* **85**:5166–5170.
137. **Sauer, B., and N. Henderson.** 1990. Targeted insertion of exogenous DNA into the eukaryotic genome by the Cre recombinase. *New Biol.* **2**:441–449.
138. **Sauer, B., M. Whealy, A. Robbins, and L. Enquist.** 1987. Site-specific insertion of DNA into a pseudorabies virus vector. *Proc. Natl. Acad. Sci. USA* **84**:9108–9112.
139. **Schwenk, F., U. Baron, and K. Rajewsky.** 1995. A *cre*-transgenic mouse strain for the ubiquitous deletion of *loxP*-flanked gene segments including deletion in germ cells. *Nucleic Acids Res.* **23**:5080–5081.
140. **Schwenk, F., B. Sauer, N. Kukoc, R. Hoess, W. Müller, C. Kocks, R. Kühn, and K. Rajewsky.** 1997. Generation of Cre

recombinase-specific monoclonal antibodies to characterize the pattern of Cre expression in *cre*-transgenic mouse strains. *J. Immunol. Methods* **207**:203–212.
141. **Scott, J. R.** 1968. Genetic studies on bacteriophage P1. *Virology* **36**:564–574.
142. **Shockett, P., M. Difilippantonio, N. Hellman, and D. G. Schatz.** 1995. A modified tetracycline-regulated system provides autoregulatory, inducible gene expression in cultured cells and transgenic mice. *Proc. Natl. Acad. Sci. USA* **92**: 6522–6526.
143. **Sieburth, L. E., G. N. Drews, and E. M. Meyerowtz.** 1998. Non-autonomy of AGAMOUS function in flower development: use of Cre/*loxP* method for mosaic analysis in *Arabidopsis. Development* **125**:4303–4312.
144. **Siegal, M. L., and D. L. Hartl.** 1996. Transgene coplacement and high efficiency site-specific recombination with the Cre/loxP system in Drosophila. *Genetics* **144**:715–726.
145. **Smith, A. J. H., M. A. De Sousa, B. Kwabi-Addo, A. Heppell-Parton, H. Imprey, and P. Rabbitts.** 1995. A site-directed chromosomal translocation induced in embryonic stem cells by Cre-*loxP* recombination. *Nat. Genet.* **9**:376–385.
146. **Soriano, P.** 1999. Generalized *lacZ* expression with the ROSA26 Cre reporter strain. *Nat. Genet.* **21**:70–71.
147. **Soukharev, S., J. L. Miller, and B. Sauer.** 1999. Segmental genomic replacement in embryonic stem cells by double *lox* targeting. *Nucleic Acids Res.* **27**:e21.
148. **Sternberg, N.** 1990. Bacteriophage P1 cloning system for the isolation, amplification, and recovery of DNA fragments as large as 100 kilobase pairs. *Proc. Natl. Acad. Sci. USA* **87**: 103–107.
149. **Sternberg, N.** 1978. Demonstration and analysis of P1 site-specific recombination using λ-P1 hybrid phages constructed in vitro. *Cold Spring Harbor Symp. Quant. Biol.* **43**: 1143–1146.
150. **Sternberg, N., R. Hoess, and K. Abremski.** 1983. The P1 *lox*-Cre site specific recombination system: properties of *lox* sites and biochemistry of *lox*-Cre interactions, p. 671–684. *In* N. R. Cozzarelli (ed.), *UCLA Symposia on Molecular and Cellular Biology,* vol. 10. *Mechanisms of DNA Replication and Recombination.* Alan R. Liss, New York, N.Y.
151. **Sternberg, N., B. Sauer, R. Hoess, and K. Abremski.** 1986. Bacteriophage P1 *cre* gene and its regulatory region: evidence for multiple promoters and for regulation by DNA methylation. *J. Mol. Biol.* **187**:197–212.
152. **Stuurman, J., M. J. de Vroomen, H. J. Nijkamp, and M. J. van Haaren.** 1996. Single-site manipulation of tomato chromosomes *in vitro* and *in vivo* using Cre-*lox* site-specific recombination. *Plant Mol. Biol.* **32**:901–913.
153. **Su, H., X. Wang, and A. Bradley.** 2000. Nested chromosomal deletions induced with retroviral vectors in mice. *Nat. Genet.* **24**:92–95.
154. **Taniguchi, M., M. Sanbo, S. Watanabe, I. Naruse, M. Mishina, and T. Yagi.** 1998. Efficient production of Cre-mediated site-directed recombinants through the utilization of the puromycin resistance gene, pac: a transient gene-integration marker for ES cells. *Nucleic Acids Res.* **26**:679–680.
155. **Thorey, I. S., K. Muth, A. P. Russ, J. Otte, A. Reffelmann, and H. von Melchner.** 1998. Selective disruption of genes transiently induced in differentiating mouse embryonic stem cells by using gene trap mutagenesis and site-specific recombination. *Mol. Cell. Biol.* **18**:3081–3088.
156. **Thorpe, H. M., and M. C. Smith.** 1998. In vitro site-specific integration of bacteriophage DNA catalyzed by a recombinase of the resolvase/invertase family. *Proc. Natl. Acad. Sci. USA* **95**:5505–5510.
157. **Thyagarajan, B., M. J. Guimarães, A. C. Groth, and M. P. Calos.** 2000. Mammalian genomes contain active recombinase recognition sites. *Gene* **244**:47–54.
158. **Tsien, J. Z., D. F. Chen, D. Gerber, C. Tom, E. H. Mercer, D. J. Anderson, M. Mayford, E. R. Kandel, and S. Tonegawa.** 1996. Subregion- and cell type-restricted gene knockout in mouse brain. *Cell* **87**:1317–1326.
159. **Tsien, J. Z., P. T. Huerta, and S. Tonegawa.** 1996. The essential role of hippocampal CA1 NMDA receptor-dependent synaptic plasticity in spatial memory. *Cell* **87**:1327–1338.
160. **Tsujita, M., H. Mori, M. Watanabe, M. Suzuki, J. Miyazaki, and M. Mishina.** 1999. Cerebellar granule cell-specific and inducible expression of Cre recombinase in the mouse. *J. Neurosci.* **19**:10318–10323.
161. **Utomo, A. R., A. Y. Nikitin, and W. H. Lee.** 1999. Temporal, spatial, and cell type-specific control of Cre-mediated DNA recombination in transgenic mice. *Nat. Biotechnol.* **17**: 1091–1096.
162. **van Deursen, J., M. Fornerod, B. van Rees, and G. Grosveld.** 1995. Cre-mediated site-specific translocation between nonhomologous mouse chromosomes. *Proc. Natl. Acad. Sci. USA* **92**:7376–7380.
163. **Vasioukhin, V., L. Degenstein, B. Wise, and E. Fuchs.** 1999. The magical touch: genome targeting in epidermal stem cells induced by tamoxifen application to mouse skin. *Proc. Natl. Acad. Sci. USA* **96**:8551–8556.
164. **Vidal, F., J. Sage, F. Cuzin, and M. Rassoulzadegan.** 1998. Cre expression in primary spermatocytes: a tool for genetic engineering of the germ line. *Mol. Reprod. Dev.* **51**:274–280.
165. **Walker, D. H., Jr., and J. T. Walker.** 1975. Genetic studies of coliphage P1. I. Mapping by use of prophage deletions. *J. Virol.* **16**:525–534.
166. **Wang, P., M. Anton, F. L. Graham, and S. Bacchetti.** 1995. High frequency recombination between *loxP* sites in human chromosomes mediated by an adenovirus vector expressing Cre recombinase. *Somat. Cell Mol. Genet.* **21**:429–441.
167. **Waterhouse, P., A. D. Griffiths, K. S. Johnson, and G. Winter.** 1993. Combinatorial infection and in vivo recombination: a strategy for making large phage antibody repertoires. *Nucleic Acids Res.* **21**:2265–2266.
168. **Westerman, K. A., and P. Leboulch.** 1996. Reversible immortalization of mammalian cells mediated by retroviral transfer and site-specific recombination. *Proc. Natl. Acad. Sci. USA* **93**:8971–8976.
169. **Yakar, S., J.-L. Liu, B. Stannard, A. Butler, D. Accili, B. Sauer, and D. LeRoith.** 1999. Normal growth and development in the absence of hepatic insulin-like growth factor I. *Proc. Natl. Acad. Sci. USA* **96**:7324–7329.
170. **Yamauchi, Y., K. Abe, A. Mantani, Y. Hitoshi, M. Suzuki, F. Osuzu, S. Kuratani, and K. Yamamura.** 1999. A novel transgenic technique that allows specific marking of the neural crest cell lineage in mice. *Dev. Biol.* **212**:191–203.
171. **Yarmolinsky, M. B., and N. Sternberg.** 1988. Bacteriophage P1, p. 291–438. *In* R. Calendar (ed.), *The Bacteriophages,* vol. 1. Plenum Press, New York.
172. **Zhang, Y., C. Riesterer, A. M. Ayrall, F. Sablitzky, T. D. Littlewood, and M. Reth.** 1996. Inducible site-directed recombination in mouse embryonic stem cells. *Nucleic Acids Res.* **24**:543–548.
173. **Zheng, B., M. Sage, E. A. Sheppeard, V. Jurecic, and A. Bradley.** 2000. Engineering mouse chromosomes with Cre-*loxP:* range, efficiency, and somatic applications. *Mol. Cell. Biol.* **20**:648–655.
174. **Zinyk, D. L., E. H. Mercer, E. Harris, D. J. Anderson, and A. L. Joyner.** 1998. Fate mapping of the mouse midbrain-hindbrain constriction using a site-specific recombination system. *Curr. Biol.* **8**:665–668.

Mobile DNA II
Edited by N. L. Craig et al.

Chapter 5

Transposable Elements as Sources of Genomic Variation

MARGARET G. KIDWELL AND DAMON R. LISCH

INTRODUCTION

Transposable elements are potent mutagenic agents with the potential to cause a variety of changes in the genomes of their host organisms. Because an increasing number of new transposable element systems have been described during the past 30 years, one of the most intriguing observations has been the diversity, complexity, and ubiquity of the molecular and phenotypic variation produced by their activity. The effects of this variation can range from knockouts of host gene function to subtle changes in tissue specificity. Although often described as a consequence of unbridled selfishness, an opinion increasingly expressed in recent years is that the large fraction of transposable elements in human and other genomes may be a manifestation of the evolutionary benefits of genomic flexibility (e.g., reference 104).

Their widespread application and use in laboratory experimentation and their use as markers for genetic and phylogenetic analyses have demonstrated most dramatically the potency of transposable elements as mutagenic agents. However, transposable elements are also active as spontaneous mutator mechanisms, providing a broad range of genetic variation in natural populations of virtually all species that have been examined. Although the diversity of their effects is now well accepted, very few good estimates of the relative importance of transposable element-induced variation are available for natural populations.

This chapter aims to provide a general description of the types of genetic variation caused by transposable elements in animals and plants and to examine this variation within an evolutionary framework. Because the scope of the topic is so vast, no attempt

Margaret G. Kidwell • Department of Ecology and Evolutionary Biology, The University of Arizona, Tucson, AZ 85721. **Damon R. Lisch** • Department of Plant and Microbial Biology, University of California, Berkeley, Berkeley, CA 94720.

is made to be comprehensive, and examples are biased toward well-studied model organisms, such as maize and *Drosophila*. Because most of the other chapters in this book, in one way or another, address the variation in the size, frequency, copy number, and other properties of transposable elements, we only include a brief reference to these topics here. Rather, we focus on the variation induced by transposable elements in their host organisms. However, as seen, it is often not possible to draw a clear distinction between the two.

Extent, Frequency, and Significance of Transposable Element-Induced Variation

The host variation associated with transposable elements can result from several interconnected aspects of transposable element activity. The variation may result directly from transposable element-induced "mutations" of host genes. Alternatively, because transposable elements represent a rich source of enzymatic and regulatory diversity, it may be the result of the co-option, over evolutionary time, of the sequences and enzymatic machinery of the elements themselves. Finally, variation can result from the evolution, or elaboration, of mechanisms to control the activity of transposable elements.

Estimates of the frequencies of new transposable element-induced mutations have been made under laboratory conditions and varied over an enormous range. An active system of transposons can be highly mutagenic. In maize, for example, the *Mutator* system can by itself increase overall mutation frequencies to more than 50 times that of background (213), and the transposition frequency of *P* elements in the progeny of *Drosophila* hybrid dysgenic crosses has been estimated to be 0.25 per element per generation (74). These mutagenic lines are, however, the exception. Apparently, there is wide variation in *Drosophila* transposable element-induced mutation rates, depending on many factors (197, 198).

Similar to newly induced mutations produced by other mechanisms, those caused by transposable elements are subject to strong purifying selection in natural populations. For both these reasons, questions of special interest and relevance for evolution are, "What fraction of the spontaneous mutations observed in natural populations of organisms is attributable to transposable elements? And how many of these become fixed in populations due to positive selection?" Unfortunately, reliable answers are currently hard to find. The availability of sequence information from genome projects is already starting to improve this situation, however. Comparisons among well-characterized genes in several related species help to determine to what extent transposons have contributed to the evolution of new phenotypes. The changes in the regulatory regions of these genes are of particular interest.

In some model organisms, the fraction of transposable element-induced mutations has been estimated to be very high. In *Drosophila melanogaster*, for example, the total genomic rate of transposition, 0.37 per generation (198), is estimated to be the same order of magnitude as the spontaneous nucleotide mutation rate (86). Consistent with this estimate, approximately 50% to 80% of mutations in this species have been estimated to be transposable element-induced (6, 216). In humans, however, a recent estimate of the fraction of mutations that was transposable element-induced was surprisingly low, especially given the fraction of the genome that comprises transposable elements. Kazazian and Moran (128) reported that in humans, only 0.17% of all mutations was caused by long interspersed elements (LINEs), elements that make up a large proportion of the transposable element complement of the human genome. This contrasted with a much higher estimate of 10.64% in the mouse (B. A. Taylor, cited by Kazazian and Moran [128]). These highly variable percentages underscore the urgent need for better estimates of the frequency of transposable element-induced mutations relative to mutations induced by other mechanisms in natural populations from a variety of species, in a range of conditions. Not until we have such estimates will informed judgments about the relative importance of the broad array of transposable element-induced variation be possible.

Much of transposable element activity is likely to be episodic and conditional. Much of the mutagenic potential of transposable elements is usually held in check by various host and element-encoded proteins. However, under the appropriate circumstances, transposition frequencies can increase by orders of magnitude. Thus, current levels of transposition in any given species at any given time may not be representative of transposition frequencies that have occurred in the past, or may be likely to occur in the future.

Transposable Elements as Mutator Mechanisms

Properties that distinguish transposable element-induced variation from those produced by other mutational mechanisms are (i) their potency, (ii) their broad spectrum of induced mutations, and (iii) the degree to which new mutations are regulated by both the host and the transposable element itself. In addition, it is possible that during the course of coevolution of these elements with their host genomes, they may become chromatin-regulated "mutator genes" responsible for targeting insertion mutations to specific chromosomal regions (123).

Although the frequency of mutations induced by transposable elements varies widely depending on many factors, there is evidence that their potency can equal or exceed that of strong chemical mutagens and radiation. This has been documented for comparisons of P-M mutagenesis with mutagenesis induced by ethyl methane sulfonate (EMS) and X-irradiation on quantitative variation in *D. melanogaster*. With respect to bristle number and viability, the effects of *P*-element insertions were on average larger than the effects inferred from spontaneous mutation (167). Hybrid dysgenic crosses are capable of inducing polygenic mutational variance at levels equivalent to the heritability of these traits from natural populations (148). A single generation of exposure to *P*-element mutagenesis was estimated to be 100 times more powerful than 1000r X-rays in inducing variation for abdominal bristle counts (148, 172). This level of potency seems to be unprecedented among mutagenic agents. In a similar way, when the proportions of experimentally induced recessive lethal mutations were compared for *P* and EMS mutagenesis, the mutational effects of transposition were estimated to be equivalent to a dose of 1.0×10^{-3} to 2.5×10^{-3} M EMS (171).

Transposable elements like other mutators can produce major effects on phenotypic traits or small silent changes that are detectable only at the DNA sequence level. They produce their mutagenic effects not simply on initial insertion into host DNA, but also when they excise imprecisely, leaving either no identifying sequence or only small "footprints" of their previous presence. Because of their rapid divergence, after a while it is often difficult to identify sequences having a transposable element origin.

The ability of transposable elements to regulate themselves is a property not shared with physical and chemical mutagenic agents. Self-regulation is often a property that has presumably evolved to modulate the extent of damage to the host species. Another interesting aspect of the mutator role of transposable elements, particularly in plant evolution, is their relative quiescence during normal development and their activation by stresses, including wounding, pathogen attack, and cell culture (265). The low level of transposon activity under normal circumstances can be interpreted as a mechanism to ensure against massive, potentially catastrophic, expression that would lower the fitness of both the host and the element. The conditional activation of transposable elements can be seen in this context as a breakdown of the normal regulation of transposable elements due to stress, or an attempt by the transposable elements to ensure representation among the reduced number of surviving progeny of a stressed plant. In either event, the effect is the production in larger numbers of transposable element-induced mutations after stress, which could produce potentially more fit host variants. Transposable elements are, then, tightly regulated and conditionally expressed mutagenic agents.

VARIATION AT THE SUBGENOMIC LEVEL

Variation Mediated by Insertions

Transposable elements may affect host gene expression by insertion into, and disruption of, coding sequences, and insertion into nontranscribed or nontranslated regions of genes, potentially resulting in immediate or delayed changes of gene regulation. Insertions may also place a gene under the control of the regulatory sequences of the transposable element, as described in a later section. In plants and other organisms in which transposition is not restricted to the germ line, the opportunity for somatic activity of transposable elements provides the opportunity for transient phenotypic variability that can sometimes be spectacular, as well as germ line changes that are potentially of evolutionary importance.

Mutation by insertion into gene coding regions

Insertions by transposable elements into coding regions often disrupt gene function, a property that led to the discovery of transposable elements more than 50 years ago. A mutable allele of the *waxy* gene, *wx-m9*, first isolated by McClintock in 1951 (178), was caused by an insertion of a *Ds* element into the tenth exon of the gene (83). Similarly, a series of null mutations at the *white* locus in *Drosophila* led to the discovery of *P* elements (214). In humans, some diseases are produced by the insertion of transposable elements in exons, for instance, from the insertion of an *Alu* element into the coding region of the porphobilinogen deaminase (*PBGD*) gene resulting in acute intermittent porphyria (AIP) (190), and in another case, from the insertion of an *L1* element into the factor VIII gene resulting in hemophilia A (129). In addition to heritable diseases, transposon insertion has also been associated with carcinogenesis (183).

To the extent that transposable elements can produce null mutations, they differ little in kind (although not always in frequency) from other mutagens and would be expected to be selectively advantageous to the same extent as any null mutation. However, even in those cases in which a transposon inserts directly into a coding sequence, the knockout can be attenuated, sometimes in subtle ways. In the nematode

Caenorhabditis elegans, for instance, all eight independent insertions of *Tcl* into exons were spliced out of mature mRNAs with high efficiency, suggesting that many or even most insertions of this element are silent (215). However, that kind of silence can be conditional and variations in the efficiency of splicing can introduce new forms of tissue specificity. In maize, splicing of a retroelement element from the *wxG* allele is 30 times more efficient in the pollen than it is in the endosperm (173). Splicing of transposable elements has also been observed in *Drosophila* (92) and mammals (144). Because they are largely invisible to geneticists, spliced transposons may be more common than was once suspected. Even when spliced, however, introns are sometimes the site of regulatory sequences. In these cases, transposable element insertions into introns can affect gene regulation in surprising ways. For instance, the insertion of *Mu* elements into an intron of the *knotted* locus in maize induces ectopic expression of the gene, suggesting that the intron carries sequences usually required to repress expression of the gene in certain tissues (98) (Fig. 1). Similarly, in *Antirrhinum,* complementary floral homeotic phenotypes result from opposite orientations of a *Tam3* transposon in an intron of the *ple* gene (23) (Fig. 2).

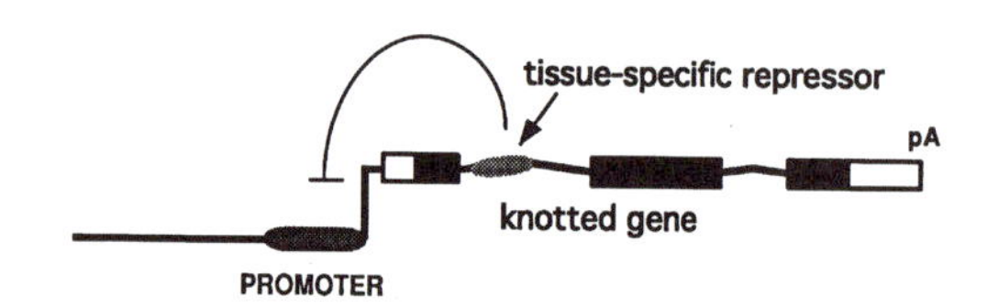

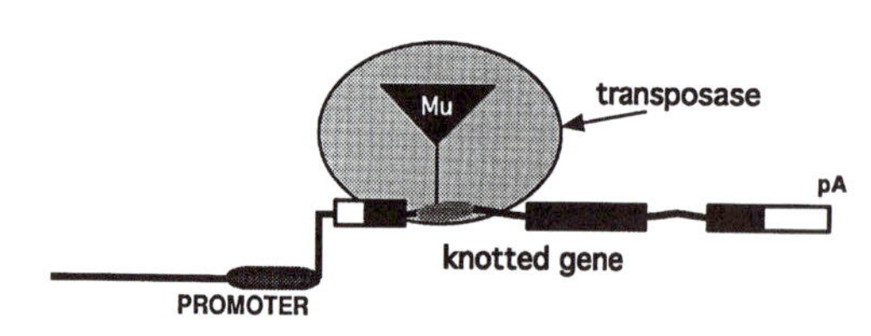

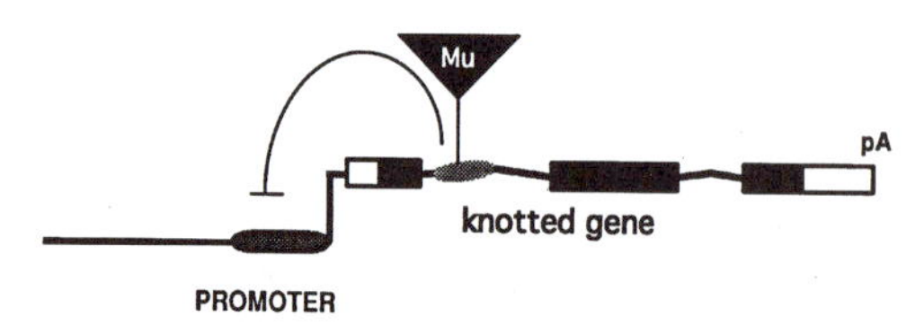

Figure 1. *Mu* transposon insertion into an intron causes ectopic expression of the *knotted1* gene in maize. In the presence of the transposase, function of a putative leaf-specific repressor in the intron is blocked. In the absence of transposase, the repressor functions normally and expression in the leaf is blocked. Thus, in this case, the transposase becomes part of a regulatory pathway in plant development. Adapted from *Genetics* (98) with permission from the publisher.

In some cases, partial splicing of transposable elements can lead to the formation of new genes, as observed in the *Drosophila montium* species group (196). The exon of a new gene was constructed from the genomic flanking sequence of a *P*-element sequence, and the first half of the intron is composed of a genomic flanking sequence and the second half is composed of a *P* sequence. From sequence data, it is speculated that cryptic acceptor and donor splicing sites present on the *P* element and its flanking sequences have been under selective constraints, which has led to the emergence of a new intron.

Insertions into noncoding regions of genes

In addition to affecting host variation by inserting into coding regions of genes, transposable elements are responsible for a variety of phenotypic effects in noncoding regions of the genome. Many examples of these types of insertions are described in later sections of this chapter. (See particularly the sections on targeting and host recruitment of transposable element functions.) Some of the phenotypic effects produced by insertions into noncoding regions are direct, but others are more subtle, nuanced, and sophisticated. Some insertions may even evolve over time in the host genome to become co-opted for use by the host organism. Estimates of the extent to which observed phenotypic variability is associated with transposable element insertions in noncoding regions of genes are very limited. However, two recent studies of plants have concluded that transposable elements make a surprisingly large contribution to variation in the phenotypic traits examined. Most of the phenotypic variation was caused by insertions into noncoding DNA.

A study of 18 alleles at the *b* locus in maize (223) presented evidence suggesting that major changes in gene expression can result from large-scale changes in DNA sequence and structure that are likely to have been mediated by transposable elements. The *b* locus encodes a transcription factor that regulates the expression of genes responsible for purple anthocyanin pigment production. Five of 18 alleles examined at the *b* locus appear to have arisen relatively recently and are associated with distinct tissue-specific patterns of expression. The presence of a transposable element insertion is responsible for the unique expression of one of these five alleles and is correlated with the phenotype of two others.

In another plant, morning glory, it is possible to associate discrete floral color phenotypes with individual genes (Color Plate 7 [see color insert]). In a

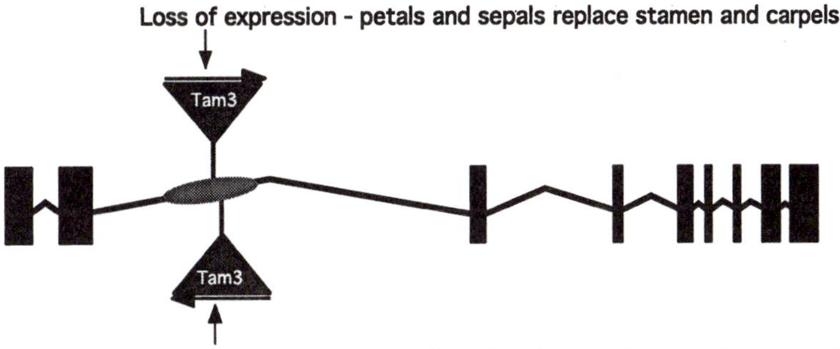

Figure 2. Insertions of a *Tam3* element into the first intron of the *plena* locus in *Antirrhinum majus* have opposite effects depending on the orientation of the transposon. In one orientation the result of an insertion is a recessive homeotic conversion of sex organs to sterile perianth organs because of loss of expression in the inner two whorls of the flower. Insertion in the opposite orientation leads to a semidominant conversion of sterile organs into sex organs caused by ectopic expression of plena in the outer two whorls of the flower and vegetative organs for promoting sex organ development within the context of the flower. Adapted from *Cell* (23) with permission from the publisher.

study of variation of flower color, all except one of the phenotypes analyzed in the Japanese morning glory, *Ipomoea nil,* and its close relative, *Ipomoea purpurea,* were the result of transposon insertions (45). A large proportion of these insertions was in introns or the 5′ flanking regions of genes in the flavonoid biosynthetic pathway. At present, it is not known whether the results of these studies are typical of other loci, characters, and species, or whether they are exceptional, and if exceptional, for what reason. One possibility is that there is a relationship between level of expression and liability to insertional events.

Variation Mediated by Excisions and Transposase-Mediated Deletions

In addition to their ability to disrupt or alter gene function in various ways upon insertion, transposable elements that move by a cut-and-paste mechanism often produce further variation when they excise. This is because after repair the original sequence is seldom restored precisely as it existed previously. There is evidence in many eukaryotes that transposon excision from a given site can generate a high degree of variation in DNA sequence and phenotype. The excision process may result in either the addition of new sequences or deletion of host sequences, flanking the insertion site (218). There is also evidence that new variation is not completely randomly determined. For example, examination of numerous excision products of the *Ds*-like transposon *Ascot-1* in a spore-color gene of the fungus *Ascobolus immersus* (47) revealed small palindromic nucleotide additions in many of them. Evidence was obtained for intermediate hairpin formation during the excision reaction and for strong biases in the subsequent processing steps at the empty site. Several parallels are suggested between the excision reaction of these and other *hAT* transposons, on the one hand, and V(D)J recombination, on the other. However, whereas the latter is a somatic process with little direct evolutionary consequences, small palindromic additions in the excision of *hAT* transposons can occur during meiosis with the potential to contribute significantly to germ line variability and the evolution of phenotypic traits.

Because of the clonal development of plants, sectors of revertant tissue can result from the somatic excision of transposable elements and restoration of gene activity (reviewed in reference 264). When transposable elements excise from protein coding regions, selection of phenotypic revertants favors those excisions that produce transposon footprints of 3 or 6 bp, because these restore the original reading frame. Thus, clonal selection during plant development can, in some circumstances, "test" excision events for their degree of viability (or even altered function) before they are transmitted to progeny.

An interesting contrast is observed between the rarity of precise excisions of the *Ac* element in maize (9) and the frequent occurrence of those of *D. melanogaster P* elements (74). Both *Ac* and *P* are DNA transposons, but they (or their hosts) use different mechanisms of repair, which seems to explain their differences in observed excision frequency. In maize, *Ac* and *Ds* elements are excised by staggered cuts in the 8-bp target-site duplication and repaired to variable extents by exonuclease activity, and in only a low number of cases, is the repair complete. However, in *Drosophila,* precise excision of *P* elements results from double-strand gap repair using the wild-type homologue to restore the excision site (76). Thus, in the *P* element, precise excision is a direct expression of the relatively frequent presence of wild-type homologues for use as repair templates, whereas the low frequency of precise *Ac* excision reflects the relatively low frequency of homologue-mediated repair.

Transposase-mediated deletions within an element are likely to inactivate the element. Such deletions can also produce new unstable phenotypes, or lead to new forms of element regulation. The *Drosophila P* element provides a good example of a deletion leading to an unstable phenotype (58). When a transformed strain that carried a *rosy* transposon inserted in or near the X chromosome *scalloped* locus was destabilized by P-M hybrid dysgenesis, mutations were recovered that exhibited alterations in the *rosy* and/or *scalloped* phenotypes. The most common events were DNA excisions wholly internal to the transposon and representing sections of *rosy* DNA. In addition to loss of *rosy* locus function, such excisions affected expression of the *scalloped* locus as well (58).

P elements also provide several examples of new forms of element regulation as the result of internal deletion (75, 187). The *KP* element, for example, a nonautonomous *P* element with a large internal deletion, is frequently present in one or more copies in natural populations of *D. melanogaster* and represses the activity of *P* elements (20). The element encodes a 207-amino-acid protein that can dimerize and interacts with multiple sites on *P*-element DNA (151). Similarly, a deletion derivative of the *Spm* element in maize can negatively regulate full-length *Spm* elements in *trans* (55).

Quantitative Trait Variation

Variation in quantitative genetic traits is of importance in providing the potential for artificial selection in animal and plant breeding and also in providing the variation in many fitness traits on which natural selection acts in both artificial and natural populations. There is now evidence that transposable elements are important sources of this variation. *P* elements in *D. melanogaster* provide some of the most detailed and comprehensive examples of the potency of transposable element insertions in generating new quantitative variation. In a series of large-scale experiments, Mackay and colleagues (5, 54, 167, 170) have shown that *P*-element insertions induced by hybrid dysgenesis can provide statistically significant changes in both the mean and variance of a wide range of quantitative genetic traits, including those affecting fitness, morphology, and behavior.

With respect to fitness traits, *P* elements reduced average homozygous and heterozygous viability by 12.2% and 5.5% per insert, respectively (170). Recessive lethal mutations were induced at an average rate of 3.8% per insert (170). Well-documented morphological traits affected by *P*-element mutagenesis include bristle numbers and body size (54, 167, 170). For example, 50 *P*-element inserts tagging quantitative trait loci (QTLs) with large effects on bristle number were mapped cytogenetically (167). Effects of *P*-element inserts on odor-guided behavior were documented by Anholt and colleagues (5). Different degrees of avoidance response to benzaldehyde were observed in 14 of 379 homozygous insert lines evaluated quantitatively. Smell impaired (*smi*) loci were identified in these lines which mapped to different cytological locations. It was estimated through enhancer trap analysis that expression of at least 10 *smi* genes is controlled by olfactory tissue-specific promoter/enhancer elements.

Transposition of Ty1 elements is responsible for much of the genetic variation observed in populations of the yeast, *Saccharomyces cerevisiae* (268). Wilke and Adams measured the distributions of stationary-phase densities and found that, although many Ty1 insertions have negative effects on fitness, some may also have positive effects (268). They concluded that Ty1 transposition events were directly responsible for the production of adaptive mutations in the clones that predominated in the populations they studied.

Sterility Associated with Hybrid Dysgenesis

The partial or complete sterility associated with several systems of hybrid dysgenesis in *Drosophila* provides an interesting aspect of variation associated with transposable element activity. Two broad types of sterility have been described (24). In *D. melanogaster,* the *I* element, a LINE-like retrotransposon, is responsible to a variable degree for a type of sterility, known as SF sterility, in the I-R system of hybrid dysgenesis. This is manifested as the death of embryos in eggs laid by F1 females produced by dysgenic crosses. The intensity of the syndrome in the dysgenic cross essentially depends on the reactivity level of the R females, which is ultimately controlled by still unresolved polygenic chromosomal determinants (117). The second type of sterility, GD sterility, is associated with the P-M (132) and H-E (21) systems of hybrid dysgenesis in *D. melanogaster* and *D. simulans* (57) and the multielement system of hybrid dysgenesis in *Drosophila virilis* (165). This sterility is manifested as unilateral or bilateral gonadal dysgenesis in the F1 female and male progeny of hybrid dysgenic crosses (74). Although male sterility can be caused by *P*-element insertion into specific Y chromosome loci encoding sterility factors (274), the underlying mechanism of GD sterility seems likely to be dominant lethal chromosome breakage, which kills the germ cells before adult gonadal development (74).

VARIATION AT THE GENOMIC LEVEL

The ability of transposable elements to induce relatively small-scale changes in both the size and arrangement of host genomes, including inversions, translocations, deletions, and additions, has been appreciated for some time. Indeed, Barbara McClintock saw genome restructuring mediated by transposable element activity as an essential component of the host's response to stress (179). However, only recently has evidence become available for a role of transposable elements in mediating large-scale changes in genome structure, size, and organization, in a variety of organisms. Transposable elements are viewed as one component of natural genetic engineering systems that are capable of acting not just on one site at a time, but on the genome as a whole (226). Several ways that transposable elements influence genome structure are now described briefly.

Transposable Elements and Genome Size

Genome size among prokaryotes varies relatively little, but that among eukaryotic species reveals a difference of greater than 80,000-fold between the smallest and largest genomes (157). This large variation in genome size is not closely correlated with genome complexity, and thus gives rise to the C-value paradox (the lack of correlation between genome size and organismal complexity). Among eukaryotes, there are interesting patterns of genome size variability. Among birds, mammals, and teleost fish, genome size variation is relatively small, suggesting that changes occur by gradual accretion of small blocks of DNA (99). However, the large variation in genome size observed among invertebrates and plants suggests that quantum jumps make an important contribution. Eukaryotic transposable elements represent a highly variable proportion of the nuclear genomes of their hosts ranging from a low percentage in some small genomes such as yeast, to a huge majority of the whole genome in some plants (Fig. 3). In fact, transposable elements make up the major type of identified nongenic DNA in every plant species (15). Although in grass genomes, differential amplification of retrotransposons largely accounts for the C-value paradox (263), other factors than transposable elements seem to have a greater influence on genome size differences in other organisms.

Cytological, physiological, and geographical correlates of genome size have been recorded (reviewed in reference 99). Apparently, the size of some transposable element families can wax and wane quite rapidly. Consequently, the proportion of the genome occupied by transposable elements is not constant in the same species, and sometimes not even in the same local population. This has recently been dramatically illustrated by the variability in the share of the genome of wild barley, *Hordeum spontaneum,* occupied by the *BARE-1* retrotransposon in different ecological regions of the Evolution Canyon in Israel (125). The copy number of full-length elements varied across an almost threefold range from 8.3×10^3 to 22.1×10^3 per haploid genome equivalent, among individuals at a single microsite. The data support an association between *BARE-1* copy number and the ecogeography of the canyon. Copy number is higher in the upper, drier sites of the canyon that present greater stress to the plants. The implication is that the proliferation of this element may contribute to genome size evolution within and among local populations. It is unknown at this time whether the associations described are correlative or causal, and whether *BARE-1* copy number variation is a direct or an indirect response to environmental variation.

Nonrandom Distribution and Genome Partitioning

Diverse nonrandom distribution patterns have been reported for many transposable elements. In addition to the general propensity of some elements to jump into closely linked sites, two broad types of underlying mechanisms may be responsible for the patterns observed: target-site preference during transposition and subsequent natural selection. In practice, it is likely to be difficult to disentangle these two types of mechanisms in any particular case. Indeed, it has frequently been argued that target-site selectivity at the molecular level has evolved because it both favors element propagation and works toward the optimization of the relationship between element and host (e.g., reference 52). Although those same selective forces are working on all elements in any given genome, it is becoming increasingly evident that different element classes can inhabit very different ecological niches within the host genome.

The role of natural selection

Nonrandom distributions of transposable elements may be explained, in part, by natural selection which acts both through the differential fitness of transposable element-induced insertions into gene coding regions, and other mutations, and through the consequences of transposable element-induced recombination. Ectopic recombination between copies of the same transposable element family, located in nonhomologous genomic locations, can lead to duplications, deficiencies, and new linkage relationships, often with consequent fitness reduction. This can potentially lead to selection against increased copy number and is considered by some (e.g., reference 38) to

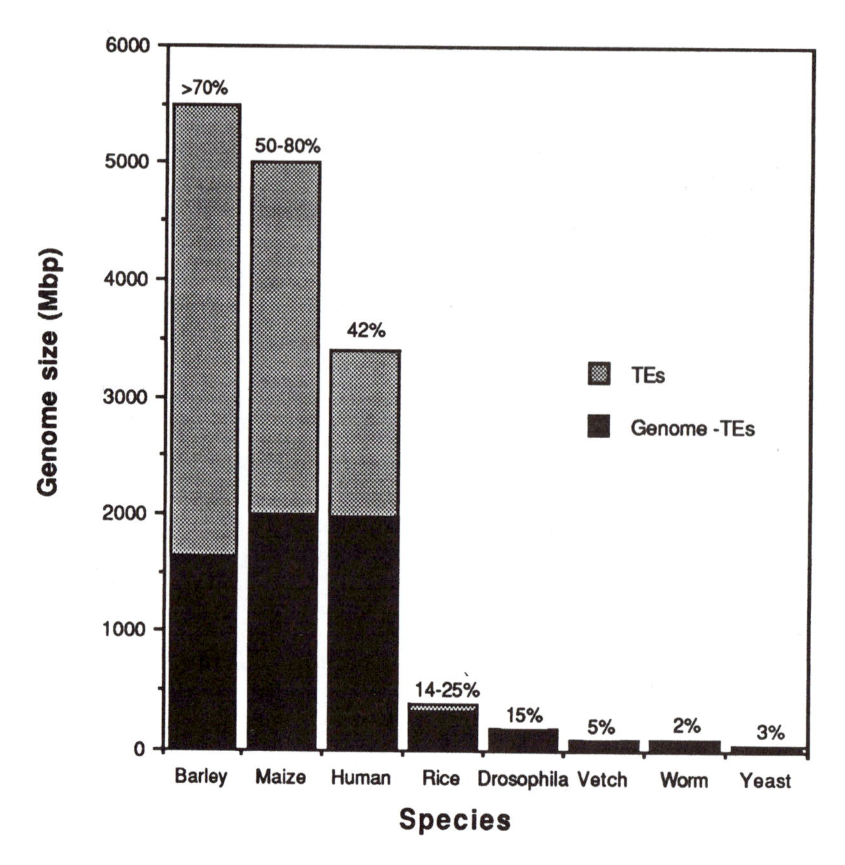

Figure 3. A histogram showing total genome sizes and the percentage of those sizes occupied by transposable elements (TEs) for eight species: barley (*Hordeum* species) (256), maize (*Zea mays*) (217), human (*Homo sapiens*) (233), rice (*Oryza sativa*) (244), fruit fly (*D. melanogaster*) (17), vetch (*A. thaliana*) (150), worm (*C. elegans*) (261), and yeast (*S. cerevisiae*) (135). Genome sizes for each of these species except barley and maize are taken from http://www.cbs.dtu.dk/databases/DOGS/.

be a more important factor in containing transposable element copy number than selection against insertion mutations (17). The theory predicts that because of their propensity for ectopic recombination, transposable elements will be found more frequently in chromosomal regions having lower recombination, such as in centromeric and telomeric heterochromatin, than in other chromosomal regions (149).

Negative, purifying selection is known to act against many insertions in gene coding regions, and against many of the products of transposable element-induced recombination (39). Selection may also be active in removing insertions that result in the deregulation of transcription because of the presence of promoter elements within many transposons. In mammalian genomes, selection may also have been prominent in weeding out premature transcription termination caused by the insertion of *L1* and long terminal repeat (LTR) elements into introns. Because of the huge area they offer for insertion, introns are particularly likely to be the recipients of such insertions (233). This may have been the reason for the evolution of a weak termination site in the 3′-flanking region of current *L1* transcripts (188), which would be expected to minimize the effects of insertions of *L1* elements into introns.

The distribution of transposable elements may also be influenced by positive selection. For example, although *Alu* elements in mammals prefer integrating into the same AT-rich (gene-poor) targets as *L1* elements, the observed distributions show a GC-rich (gene-rich) bias. A possible interpretation is that these elements originally insert randomly and later are preferentially fixed in GC-rich regions by positive selection, consistent with their postulated function as stress genes (219, 233).

Target-site preference during transposition

Target-site selection can occur through several molecular mechanisms, including (i) target selection by the interaction of transposase either with the target DNA or with accessory proteins and (ii) selective recognition of a target site by an element-encoded nuclease to produce a priming site for reverse transcription (52). Different families of elements appear to use different mechanisms for selecting target sites. For example, the *C. elegans* elements *Tc1* and *Tc3* display marked preferences for target sites that share a common consensus sequence. The key determinant in target-site selection, in this case, seems to be the interaction of transposase with these sites. In contrast, *nonLTR* or *PolyA*$^+$ retroposons transpose by cleavage of the DNA to produce a priming site for reverse transcriptase (RT). In some of these elements, the target site is highly specific, reflecting the specificity of the element-encoded nuclease (52). Examples of retroposons with high target-site specificity are the *R1* and *R2* families of retroposons in numerous insect species, including *D. melanogaster* and the silkworm, *Bombyx mori.* These elements integrate independently at specific sites in the *28S rRNA* genes. The frequency of insertions ranges from a few percent to greater than 50% of the *28S* genes (115). Approximately 25 copies of *R2* exist within the *B. mori* genome, of which at least 20 are located at a precise location within otherwise typical rDNA units. The precise location of the element within the genome implies that its transposition must occur with remarkable insertion sequence specificity (116, 271). The high level of sequence homogeneity between copies of each element within the same species is consistent with the high turnover rate expected to result from these processes (116).

Accumulation in gene-rich regions. Broad patterns of nonrandom distribution have been recorded in many organisms often related to the type of element involved. For example, in plants, all DNA-transposable elements exhibit preferential insertion and/or retention within euchromatic regions of the genome that are genetically active and unmethylated (15). An example of elements that specifically target gene sequences is the *Mu* elements in maize, which may be targeting the open chromatin configuration and local concentration of transcription factors around genes, but at the cost of increased selection against new insertions (53). Some LTR retrotransposons in plants also share this pattern (93). This type of element tends to have low copy numbers per genome, perhaps because of greater selection against unrestricted transposition in gene-rich regions.

Although, in plants, miniature inverted repeat transposons (MITEs) are present in high copy numbers (266), they are found almost exclusively in or near host genes, suggesting their exclusion from the specialized domain inhabited by retrotransposons (275). For example, in domesticated rice (*O. sativa*), 32 common sequences were shown by a computer-based search to belong to nine putative mobile-element families (33). Five of the nine families were first discovered through this computer search and four of these five had characteristics of MITEs. As in plants, MITEs present in the yellow fever mosquito, *Aedes aegypti* (252), are also closely associated with genes and are flanked by TAA target duplications. However, MITEs in *C. elegans,* and in the malaria mosquito *Anopheles gambiae,* have relatively low genomic copy numbers and tend to avoid gene-rich regions, thus showing marked differences in copy number and distribution patterns from those seen in plants and *A. aegypti* (see below).

In mammals, short interspersed elements (SINEs) are exceptional among transposable elements in being found preferentially in GC-rich, gene-rich, regions of the genome (233). In some SINE families, this seems more likely to be the result of positive selection than integration preference (233; also see above).

Targeting to chromatin domains. Different families of transposable elements in the yeast, *S. cerevisiae,* demonstrate interesting differences in insertional specificity. For example, the Ty1, Ty2, Ty3, and Ty4 retroelement families target the upstream region of the RNA polymerase III-transcribed genes (135). The yeast elements Ty1, Ty3, and Ty5 display distinct patterns of target-site selectivity, but all are likely to be mediated by collaborations between Ty-encoded and host-encoded proteins that are involved in chromatin restructuring (52). As discussed in a later section, exclusive targeting to the telomeres of chromosomes is seen by the *Het-A* and *TART* elements in *Drosophila* (18, 153, 202, 203), and by the *copia* retrotransposons found in *Allium cepea* (204). Two retroposons, *Zepp* from *Chlorella* (106) and *SART1* from the silkworm, *B. mori,* are inserted preferentially into telomeric repeats (241).

Accessibility and target-site preference. *P* elements in *Drosophila* insert into potential regulatory regions of genes, rather than exons, as shown by a large number of independent new insertions (235). In contrast to the strong bias in favor of insertion into the 5′ end of genes and especially the 5′-untranslated regions, only a small minority of *P* elements were observed to have inserted in coding sequences, and these were in the 5′ portion of them (Fig. 4). The true number of insertions into potential regulatory regions was actually underestimated in this study, because insertions that did not produce a mutant phenotype would be missed. This target preference of *P* elements is supported by the remarkable observation that in another experiment, more than 65% of the 500 independent *P*-element enhancer trap insertions were expressed in a spatially and temporally restricted fashion (14). A more recent study to determine the targeting mechanism involved (159) found that most *P*-element insertions were integrated within a few hundred bases of

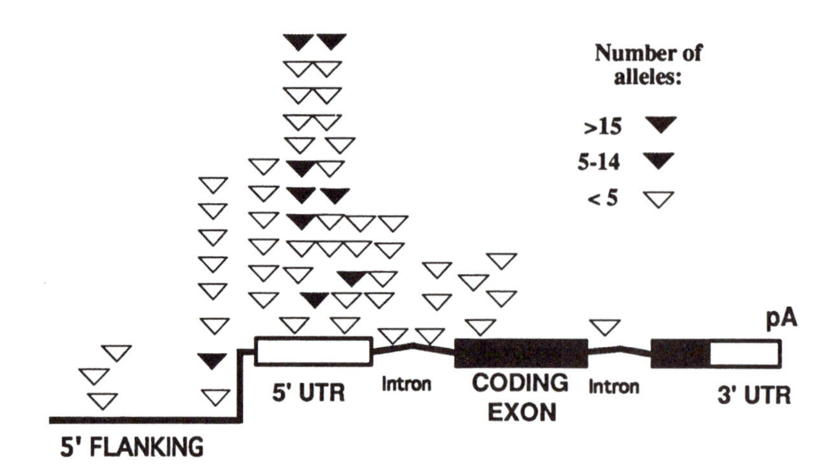

Figure 4. Preferential insertion of *P* elements near the 5′ end of transcription units. The insertion sites of 56 different mutagenic *P* elements are shown. Each insertion site is plotted with respect to a simplified standard gene containing one intron before, and one intron after, the AUG initiation site. Reprinted from the *Proceedings of the National Academy of Sciences USA* (235) with permission from the publisher.

the transcription start site of genes. Chromatin accessibility therefore seemed to be a major factor in site preference. However, within these open regions, it seemed that DNA structural features were favored at the site of insertion, rather than a specific sequence motif.

Accumulation in nongenic regions. In plants, the most abundant LTR retrotransposons, sometimes described as "intergenic LTR retrotransposons," have a very different lifestyle from DNA elements in the same species that tend to target genic regions. These LTR retrotransposons are found most frequently in methylated, presumably locally heterochromatic regions, often in nested clusters (217). They have a fivefold preference for insertion into the LTR, relative to any other part of the target retrotransposon (15). The region between maize genes is much less recombinogenic than are the genes themselves, suggesting that recombination between these elements (which is likely to be disastrous for the host) is suppressed (43). Another example of preferential integration into other transposable elements is provided by the *Zepp* retroelement in *Chlorella,* which appears to transpose preferentially into itself in telomeric heterochromatin (106). Transposons can also accumulate in other forms of repetitive DNA. *Grande* elements in *Zea* species, which are also often inserted into transposons, were preferentially integrated into certain sites of the 180-bp repeat in heterochromatic knobs (2). The absence of any phenotypic mutations caused by transposable elements in the slime mold, *Dictyostelium discoideum,* was ascribed to the preferential insertion of these elements into one another (88).

Transposable elements examined in mammalian genomes have also frequently been found to be nonrandomly distributed (267). For example, LINEs in human and mouse tend to accumulate preferentially in the G-banding regions of the chromosomes, which correspond to the AT-rich, gene-poor regions of the genome, and SINEs occur preferentially in R-banding regions, which correspond to the GC-rich, gene-rich regions of the genome. LINEs are also more abundant in both sex chromosomes that are rich in heterochromatin.

In *C. elegans,* the four MITE families (*Cele1, Cele2, Cele14,* and *Cele42*) displayed distinct chromosomal distribution profiles (239). One family clusters near the ends of autosomes, whereas another is more evenly distributed. The distribution patterns of each family are conserved on all five *C. elegans* autosomes, suggesting chromosomal polarity (239). In *A. gambiae,* the flanking sequences of most MITEs have, surprisingly, significantly higher AT content than either the genome average or the average of transcribed genes (252a). These MITEs tend to be associated with each other, but avoid gene-rich regions. It is not yet known whether the two distinct patterns of copy number and genomic distribution in MITEs are a reflection of different genomic architectures, or whether they are a result of different types of interactions between MITEs and their host genomes.

Genome partitioning

We previously postulated (133) that there are at least two types of elements that occupy two very different niches in the ecology of the genome. The first type preferentially inserts into regions distant from host gene sequences, such as heterochromatin, or the regions between genes. The second type inserts preferentially into, or near, single-copy sequences. The formation of these niches, and their exploitation by transposons, is likely to be a result of a long-term interaction between elements and their hosts, reflecting a combination of integration specificity and natural selection. Heterochromatin proteins can recognize and silence transposable elements (80), some of which target heterochromatin for insertion (247). Thus, the evolution of heterochromatin could have led to a self-perpetuating expansion of domains rich in transposable elements.

Sectorial mutagenesis of chromosomes

It has been proposed that during their coevolution with host genomes transposable elements may selectively target insertion mutations to particular chromosomal regions, or groups of genes (123). This targeting is envisaged to be achieved by "mutator genes" derived from wild-type transposable elements by natural selection. Such variants of autonomous elements would evolve to proliferate nonautonomous elements, rather than themselves, and target them to particular chromosomal niches. Such niches might be determined by accessibility of chromatin to insertion during the period of expression of the corresponding mutator gene (123).

Genome Structure and Reorganization

The ability of transposable elements to induce chromosomal rearrangements, such as deletions, duplications, inversions, and reciprocal translocations, has for some time been considered to provide the potential for both small- and large-scale genome reorganization. In fact, a study of mobilization rates of nine transposable elements in *D. melanogaster* indicated that most changes in restriction patterns were consistent with rearrangements, rather than with true transposition (62, 63). Further, clustering of movement across transposon families was observed, suggesting that transposition of different families may be nonindependent. Although several early studies of the breakpoints of chromosomal rearrangements in some *Drosophila* species failed to find evidence that transposable elements were implicated in their formation, transposable elements are undoubtedly responsible for a significant proportion of the observed karyotypic variation found among many groups. For example, it is claimed that retrotransposons are one of the major forces influencing the structure of plant genomes (145). However, reliable estimates of the fraction of all rearrangements accounted for by transposable elements are currently not available in any species.

Recombination

Two mechanisms are considered most likely to be responsible for transposable element-induced karyotypic changes. The best known mechanism is ectopic recombination, in which homologous recombination occurs between multiple copies of a transposable element present in a genome. It has been claimed that this mechanism is a significant source of the molecular rearrangements causing human disease (211). A second mechanism for inducing genomic rearrangements is alternative transposition of class II elements in bacteria, plants, and animals (97). In contrast to the traditional mode of "cut-and-paste" transposition that involves the synapsis of complementary ends of a single transposable element, alternative transposition involves the synapsis of complementary ends of different transposable elements on the same, or different, chromosomes. The hybrid element produced excises and reinserts in a new genomic location, as occurs in traditional transposition. This mode of transposition provides the potential for producing many different types of chromosomal rearrangements (97).

The numerous SINEs and LINEs present in primate genomes provide a good substrate and many opportunities for unequal homologous recombination. These events often occur intrachromosomally, resulting in deletion or duplication of exons in a gene, but they also can occur interchromosomally, causing more complex chromosomal abnormalities. Thirty-three cases of germ line genetic diseases and 16 cases of cancer were reported to have been caused by unequal homologous recombination between *Alu* repeats (59). An example of unequal recombination between LINE elements, in humans, is provided by a 7.5-kb deletion of genomic DNA that results from unequal exchange between two neighboring *LINE-1* sequences, giving rise to glycogen storage disease through phosphorylase kinase deficiency (35).

Some products of ectopic exchange survive natural selection and can result in genome reorganization. Exceptionally detailed examples of the role of transposon-mediated rearrangements and transpositions are seen in recent studies of the evolution of the human Y chromosome (146, 147, 220). Many blocks of DNA having sequence similarity are shared by the human X and Y chromosomes, and transposable elements seem to have played a role in mediating some of the major restructuring that has taken place during the last 300 million years (146). For example, an inversion mediated by recombination between *LINE-1*

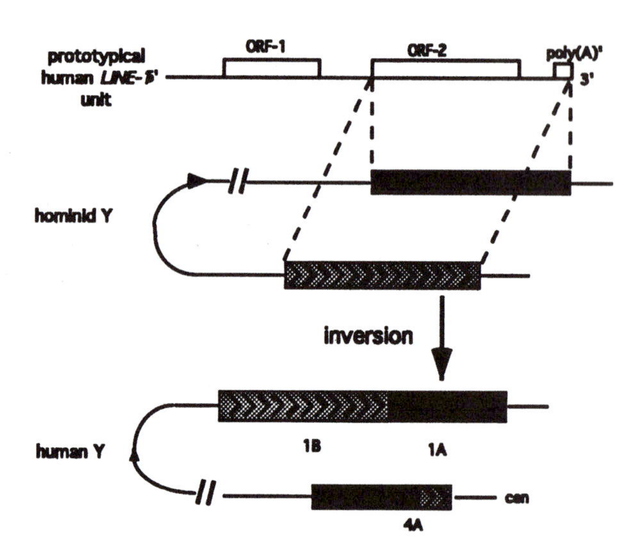

Figure 5. A Y chromosome inversion generated by recombination between LINE elements in hominids. At the top is depicted a schematic representation of a prototypical, full-length human LINE element showing the location of two long open reading frames (ORFs) and the polyadenylated tail. Below that is shown a model of an inversion between two LINE elements drawn out of register. Reprinted from *Human Molecular Genetics* (220) with permission from the publisher.

elements (Fig. 5) occurred in the Y chromosome after transposition of an approximately 4-Mb segment from the X to the Y chromosome. Both the X-Y transposition and the subsequent LINE-mediated Y inversion occurred after the divergence of hominid and chimpanzee lineages, but before the radiation of extant human populations (220). In addition, both retroposition and transposition of genomic DNA harboring intact transcription units (147) have mediated the gene content of the human Y chromosome.

Examples of transposable element-induced chromosomal rearrangements are found in many organisms. The five Ty retrotransposon families in *S. cerevisiae* have been important in the formation of chromosomal rearrangements (135). Evidence for recombination was present near many of the elements studied, and 5'- and 3'-flanking sequences were often duplicated and rearranged among multiple chromosomes. This suggests that recombination between retrotransposons has occurred relatively frequently and has had an important influence on both genome organization and the evolution of the Ty family of elements itself (122). Exceptional levels of karyotypic variation have also been recorded in several filamentous fungal pathogens, likely attributable to transposable element-induced ectopic recombination (56).

Translocations and inversions

There are several possible mechanisms by which a transposable element-mediated chromosomal inversion might be formed, including ectopic recombination between two inverted transposable element copies and breakage at or near the site of insertion of two elements in the same orientation, with subsequent inversion of the intervening sequences. Polymorphic inversions are very common in *Drosophila,* and evidence for the implication of transposable elements has been found for many of these. For example, many breakpoints in hybrid dysgenesis-induced chromosomal rearrangements occurred at, or very near to, the sites of *P* and *hobo* element insertions (77, 160). More than half of newly induced chromosomal rearrangements induced by hybrid dysgenesis in *D. virilis* were also found to contain *Penelope* and *Ulysses* transposable elements in the same chromosomal subsections as their breakpoints (79). Many of these breakpoints also coincide with chromosomal locations of *Penelope* and *Ulysses* insertions in the parental strains and with breakpoints of inversions previously established for other species of the *virilis* group. These observations suggest that transposable elements may have played an important role in the evolution of the *D. virilis* species group. In *Drosophila buzzatii,* the breakpoints of the cosmopolitan inversion 2j contain large insertions corresponding to a transposable element (36). The inversion arose by ectopic recombination between two copies of the transposon located in opposite orientations. Target-site duplications generated upon insertion apparently were exchanged during the inversion event.

In the mosquito *Anopheles arabiensis,* the *Odysseus* transposable element was recently discovered by detailed molecular analysis to lie at the distal breakpoint of a polymorphic inversion (175). Transposable elements are more likely to be associated with endemic inversions than cosmopolitan ones, because transposable elements tend to be lost over time from inversion breakpoints (77, 175). In hominids, there is evidence that an inversion on the Y chromosome was mediated by recombination between *LINE-1* elements in otherwise nonhomologous regions (220). This inversion appears to have occurred after the divergence of the hominid and chimpanzee lineages but before the radiation of extant human populations.

Gene duplication and pseudogene formation

Transposable elements play a significant role in genome reorganization by promoting the insertion of other gene sequences and the formation of pseudogenes (30). For example, Esnault et al. (78) recently demonstrated that LINEs can mobilize transcribed DNA not associated with their own sequences by a process involving the diversion of the LINE enzymatic machinery by the corresponding mRNA transcripts. This process results in the retroposition of the transcribed gene to form new genes that have the properties of processed pseudogenes, such as the loss of intron and promoter, the acquisition of a poly(A) tail, and the presence of target-site duplications of varying lengths. The formation of processed pseudogenes by LINE-mediated retroposition may thus be an important mechanism for genome duplication and genome remodeling. There is evidence that, among the retroelements, pseudogene formation may be specific to LINEs (78).

The presence of large amounts of RT has facilitated the production of numerous intronless pseudogenes in many organisms (32). In some cases, these retrotransposed genes have acquired new functions, either as a direct consequence of mutation during retrotransposition or because of the acquisition of new regulatory functions at their new position or both. A recent example is provided by *Cg*, the gene encoding the Cg catalytic subunit of cAMP-dependent protein kinase in humans (210).

Transduction and exon shuffling

Some transposable elements can provide a vehicle for the mobilization of flanking nucleotide sequences accompanying aberrant transposition events. In addition to nongenic sequences, gene sequences, such as exons or promoters, can sometimes be transduced and inserted into other existing genes (29). This may provide a general mechanism for the evolution of new genes. Examples of the transduction of downstream sequences by endogenous *L1* elements have been documented in humans (e.g., references 110 and 182). An example of so-called "exon retroshuffling," i.e., exon shuffling mediated by a retroelement, was recently reported to have occurred in human tissue culture cells (188). *L1*s inserted into transcribed genes retrotransposed sequences derived from their 3′ flanks to new genomic locations. The processing machinery at the 3′ end may occasionally bypass the *L1* polyadenylation signal and use a second downstream polyadenylation site instead, with the possibility of producing chimeric proteins.

The potential significance of transposable element-mediated transduction for host evolution was also recently highlighted with use of a bioinformatics approach (205). Computational analysis of human genome sequences indicated that at least 10% of 129 full-length active *L1* elements have an associated putative 3′-transduced segment. Extrapolating this result to the whole genome, at least 1% of the human genome may represent sequences transduced by *L1* elements. A similar conclusion was reached when 23% of the flanking DNA of 66 previously uncharacterized *L1* sequences, selected from the GenBank database, was found to contain transduced sequences (95). Two of 24 mice *L1*s were also found to have long, transduced sequences associated with them. Short pericentromeric repeats on both human and other primate chromosomes may be involved in the transposition of duplicated exons, or even entire genes, to new chromosomal locations (73).

Examples of transposable element-associated transduction have also been reported in plants. For example, among the *Mu* elements of maize, the nonautonomous element *Mu1* is homologous to a host gene, *MRS-A* (*Mu*-related sequence) (243). Recent database searches, with other nonautonomous *Mu* elements, reveal that all of them seem to contain transduced host sequences flanked by conserved terminal inverted repeats (D. R. Lisch, unpublished results). A specialized transducing transposon, *Tpn1*, has been recognized in the Japanese morning glory (242). *Tpn1*, a class II element, is found in the second intron of the *DFR-B* gene for flower pigmentation, in the mutable line "flecked" that shows variegation for flower color. In this location the element has apparently captured a genomic DNA segment containing at least four host exon sequences.

Retroelements in plants have also been observed to transduce. The *Bs1* retrotransposon family is represented by one to five copies in maize and is also found in its close relative, teosinte. About one-fifth of the *Bs1* element is made up of a sequence that appears to have been derived from the cDNA of part of a host gene (*pma*) encoding plasma membrane proton-translocating ATPase (34). These observations have possible relevance to theories about the origin of retroviruses. The ability of retrotransposons to transduce host sequences is a key requirement of the hypothesis that retroviruses evolved from ancient retrotransposons after the transduction of an *env*-like cellular gene (66).

Genome reorganization in ciliated protozoa

The programmed somatic excision of interstitial segments of DNA and the rejoining of flanking sequences occur in a variety of organisms ranging from bacteria to humans. It has been speculated that at least some excised DNA sequences are derived from viruses or transposable elements whose activities have coevolved to adapt to the normal differentiation pathways of their hosts (141). Ciliated protozoa provide a fascinating example of the co-option of transposable elements for a host function. These organisms contain two types of nuclei: a micronucleus and a macronucleus. The macronucleus provides the nuclear RNA for vegetative growth (207). The micronucleus consists of conventional eukaryotic chromosomes and can be viewed as the "germ line" nucleus (140). It is transcriptionally inactive during vegetative propagation, but is active during sexual reproduction. The genome of the macronucleus consists of a subset of the micronucleus DNA that is extensively rearranged by chromosome fragmentation, de novo telomere synthesis, and DNA amplification. During the reorganization process, large numbers of interstitial segments of DNA (internal eliminated sequences [IES]) are exactly excised from the somatic nucleus through DNA breaking, rejoining, and splicing. There is evidence that transposable elements are a major component of the highly variable structures of IESs (140, 221, 222). It has been proposed (141) that transposon invasions have generated the ciliate DNA excision phenomena. Further, it has been suggested that short IESs are ancient transposon IESs that have shrunk by loss of internal sequences unnecessary for host developmental excision (141).

TRANSPOSON-INDUCED VARIATION AND HOST EVOLUTION

Germ Line Restriction of Transposition

The restriction of transposition to the germ line appears to be a highly efficient strategy for transposable elements to ensure transmission to the next generation, yet minimize damage to the host. The adoption of such a strategy, however, does not distinguish between strictly parasitic and symbiotic elements and appears to have been adopted by a broad range of elements. Whether or not transposition occurs in somatic cells has clear implications for the patterns of transposable element-induced variability. In plants and other organisms that do not sequester germ line cells early in development, this option is more limited as a strategy than in those organisms that do so, but not impossible. For example, one of the two genes encoded by the *MuDR* transposon in maize is specifically expressed only in those cells that are potentially transmitted to the next generation. Expression of this gene is down-regulated in terminally differentiated tissue and markedly up-regulated in floral tissues (65). The *Mu* transposons of the *Robertson's Mutator* transposable element system in maize are unusual, when compared with the other known plant transposon systems, because the excision of these elements occurs late in somatic tissues and, in contrast to other elements such as *Ac* and *Spm,* while new mutations are common, heritable revertants are very rare (68).

Numerous examples of restriction of transposition in the germ line in animals exist. In the *P* element in *D. melanogaster,* the 2-3 intron is spliced only in germ cells. The absence of transposase in somatic cells results from the production of a truncated 66-kDa protein because of a premature stop codon after a frameshift. Two different proteins encoded by host genes have been implicated in this somatic inhibition (227, 228). Expression of active *LINE-1* sequences also appears to be favored in cells of germ line origin in human (230) and mouse (250) genomes. Also, the expression of intracisternal A-particles (IAPs) in the mouse genome was found to be essentially restricted to the male germ line (69). Transposition of SINEs in mammals, however, is not restricted to the germ line, but also occurs in somatic cells. This property may be essential for a possible function of SINEs in mediating responses to cellular stress (163, 219).

Host Recruitment of Transposable Element Functions

In contrast to the prevailing selfish DNA paradigm of the past 20 years, a minority opinion has maintained that at least some transposable element sequences have become "useful" for some functions of their hosts. This idea has been expressed in varying terminology. Several examples of transposable element-induced variation have been cited as exaptations (i.e., features that have previously evolved for functions other than they are now fulfilling, or for no function at all, but which have been co-opted for a new use [31]). The term "molecular domestication" has been applied more recently to various transposable element sequences in *Drosophila* and humans that have been exapted for host functions (184). Britten (25–27) has accumulated evidence for the exaptation of transposable elements for a role in host regulatory evolution. Transposable elements have also been viewed as drive mechanisms in the evolution of eukaryotic enhancers (181). We earlier argued (134) that from a long-term evolutionary perspective, transposable elements might best be considered within a parasite-mutualist continuum.

We currently have no good estimates of the relative frequency of exaptation of transposon functions and sequences. Therefore, conclusions about their relative significance are premature. Nevertheless, molecular domestication and exaptation provide examples of some of the more remarkable, and perhaps unexpected, ways that transposons can act as a source of variation. Transposable elements represent (among other things) a pool of rapidly diversifying enzymatic and regulatory functions. Because their proper function is not required for host viability, these mechanisms are free of many of the selective constraints that inhibit rapid variation of host genes. Thus, transposable elements are apt to evolve classes of enzymatic and regulatory machinery that would not necessarily have arisen in their absence. In addition, because they are mobile, transposable elements often move regulatory elements into proximity with host genes. Given the opportunistic nature of evolution, some of this machinery inevitably will be used by the host organism. Although only a limited number of documented instances of domestication exist, these cases suggest that transposons can take on important roles in such fundamental processes as chromosome replication, immunological activity, and centromere function. It may turn out that the recruitment of transposable elements, or their derivatives, is more than a sporadic episode in genome evolution (185).

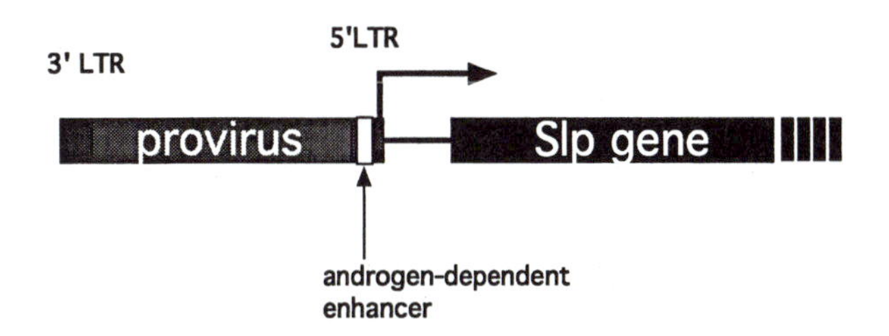

Figure 6. Association of a retrovirus-related element with androgen regulation of the sex-limited protein (*Slp*) gene in mouse. Adapted from *Cell* (236) with permission from the publisher.

Transposable elements in host gene regulation

There are several ways that transposable elements can affect gene regulation. A new insertion can either disrupt an existing host regulatory element, or it may contribute its own regulatory sequences to a host gene in which it has inserted in an appropriate location. Inserted sequences may also provide the potential for future evolution of regulatory sequences. The preference of several transposable elements for insertion in genic regions makes these elements good candidates for co-option as regulatory elements. Good examples are MITES in the yellow fever mosquito, *A. aegypti* (252), and several plants, including maize (275), *Arabidopsis* (84), and rice (33). Relatively small elements, such as SINEs and MITEs, seem to be particularly likely to assume such a role, but larger elements may sometimes act in a similar way. LTR retrotransposons carry a promoter within each LTR providing the potential for contributing to the potential for new patterns of host gene expression. Many transposable elements can act as movable carriers of regulatory elements, such as promoters or enhancers, and integrate into or near genes. They can thus contribute to the functional diversification of genes by supplying *cis*-regulatory domains and have the potential to alter spatial and tissue-specific expression patterns. A second possibility is that, after their initial insertion into or near genic regions, and after lying immobilized and dormant, perhaps for some time, transposable elements can mutate to be able to regulate nearby genes. The most common regulatory functions served by co-opted transposable elements include those of transcriptional enhancers, reducers, modulators, and polyadenylation signals (30).

Britten has used stringent criteria to identify strong cases in which transposable elements have been recruited for host gene regulation (25–27). The identification of such cases requires a long-term evolutionary perspective and the divergence of such sequences, over time, is likely to make many no longer identifiable as former transposable elements. The number of cases he has identified is small but growing. In addition to the plant MITE examples discussed above, he includes cases involving *Alu*-containing T cell-specific enhancers in the human *CD8a* gene (101), the association of a retrovirus-related element with androgen regulation of the sex-limited protein (*Slp*) gene in mouse (236) (Fig. 6), and inverted repeats in the *CyI-IIa* actin gene of sea urchin (4).

Brosius (30) provides an extensive list of possible examples of regulatory elements derived from transposable elements in vertebrates. Most of the examples concern SINEs in mouse, rat, rabbit, and human, but a few examples of *MER* elements are also included. Some examples are well documented, and others are variably problematical.

Insertion into regulatory regions of genes. The mouse provides an interesting example of a regulatory signal provided by a LINE inserted millions of years ago (102). The mouse thymidylate synthase (TS) mRNA is unusual in that the poly(A) tail is added at the translation stop codon. The second of at least two sequences that are required for efficient polyadenylation at the stop codon is a stretch of 14 consecutive uridylate residues, 32 nucleotides downstream of the stop codon. This U-rich region is absent from the human TS gene, which explains why the human TS mRNA is not polyadenylated at the stop codon even though the mouse and human genes are otherwise almost identical through this region. The U-rich region in mouse corresponds to the 3′ end of a 360-nucleotide *L1* element that was inserted in the opposite orientation to the gene more than 5 million years ago. Thus the polyadenylation signal of the present mouse TS gene was created by the transposition of a LINE downstream of a cryptic polyadenylation signal. The apolipoprotein (a) gene in humans (272) provides another example of a LINE that became an integral part of the regulation of a host gene. An enhancer of this gene, located 20 kb upstream of the start of transcription, is located entirely within a LINE. The enhancer element functions in either orientation to confer a greater than 10-fold increase in the activity of the apo(a) minimal promoter in cultured hepatocyte cells.

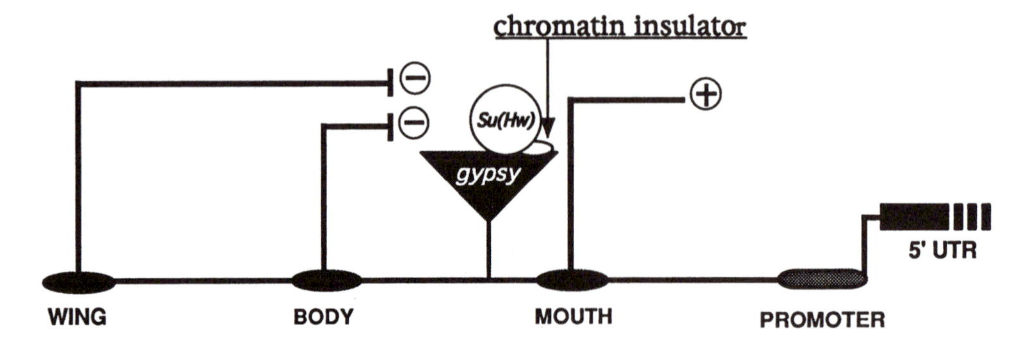

Figure 7. A gypsy insertion blocks a subset of upstream enhancers of the yellow gene in *D. melanogaster*. The gypsy transposon carries a chromatin insulator which binds to the protein product of *Su(Hw)* (circle). Because this insulator acts in a polar fashion, when the element is inserted between the promoter and the enhancers specific for wing and body expression, the yellow gene only expresses in the mouth, whose enhancer remains unaffected. Adapted from *Trends in Genetics* (50) with permission from the publisher.

Rewiring of regulatory networks. Another excellent example of this type is the insertion of gypsy into the 5′ upstream region of the yellow gene in *Drosophila,* which has caused a loss of expression of the yellow gene in specific tissues (50) (Fig. 7). The loss of expression in some tissues, and not others, in this case is the result of the interaction of the element, tissue-specific enhancers upstream of the element, and specific host factors. In *Antirrhinum,* a *Tam3* element was observed to insert into a region 5′ of the *niv* gene, which is involved in the synthesis of anthocyanin pigments. The initial insertion was observed to down-regulate expression of the gene. However, a series of rearrangements mediated by this element resulted in a change in the level and tissue specificity of expression of *niv*. The net effect is a new and novel distribution of anthocyanin pigment in the flower tube (162). This kind of mutation exemplifies the potential for transposable element-mediated "rewiring" of regulatory networks, in this case by bringing new regulatory sequences in proximity to exonic sequences via an inversion, followed by an imprecise excision event.

Transposable element activity can result in even more complex rearrangements that can have effects on gene regulation. In maize, the insertion of a *Mu* (Mutator) element into the TATA box of alcohol dehydrogenase (*Adh1*) changed the tissue specificity of RNA expression (143) (Fig. 8). Excision of this element caused a complex series of duplications and inversions whose net effect was to cause additional changes in tissue specificity (142) (Fig. 8). Kloeckener-Gruissem and Freeling suggest that this kind of "promoter scrambling" may represent a more general process by which transposons produce variants of a type not produced by other mechanisms. Further, in this case, like that of the *Tam* element in the *niv* gene of *Antirrhinum,* the transposable element footprint left behind after the element has generated the new mutation is small enough to be invisible to transposable element probes.

Another interesting example of insertional regulation of the *Adh* gene was reported in *D. melanogaster* (270). In this case, a *KP* element was inserted in the same orientation as *Adh* transcription, but five nucleotides upstream of the distal transcription start site. ADH activity was reduced to less than 12% of that of the controls in adult flies having either one or two copies of the allele. This reduction was apparently

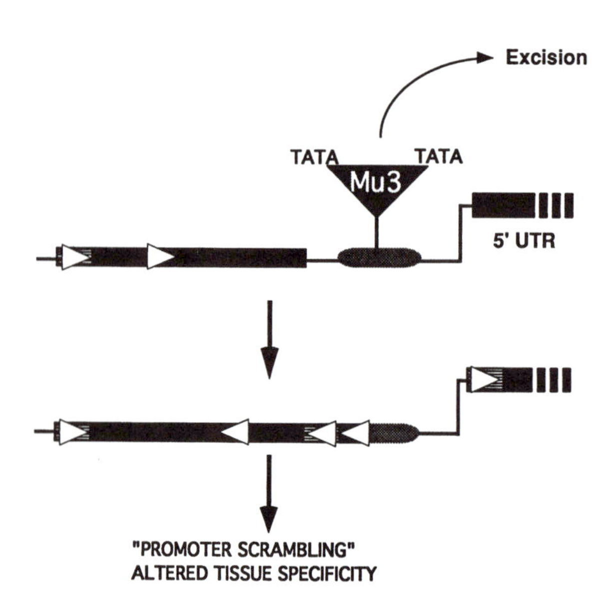

Figure 8. Promoter scrambling in plants. The original insertion of a *Mu3* transposon into the TATA box of the *Adh-1* gene caused a tissue-specific change in expression. Subsequent excision and promoter "scrambling" caused additional changes in tissue specificity. Reprinted from the *Proceedings of the National Academy of Sciences USA* (142) with permission from the publisher.

caused by reduced levels of transcription from the distal promoter. In third-instar larvae, the *KP*-element insertion affected the switch from the proximal to the distal promoter (270).

Changes in *cis*-regulatory regions of duplicated genes. It has been speculated for some time that changes in *cis*-regulatory regions of duplicated genes may be more important for the evolution and divergence of functional and morphological characters than mutations in coding sequences (e.g., references 28 and 137). Only recently has evidence started to accumulate to support this hypothesis. In *Drosophila,* for instance, the three homeotic genes, paired (*prd*), gooseberry (*gsb*), and gooseberry neuro (*gsbn*), have evolved from a single ancestral gene after gene duplication. They now have distinct developmental functions during embryogenesis. The three corresponding proteins, PRD, GSB, and GSBN, are transcription factors. Li and colleagues demonstrated that the three proteins are interchangeable with respect to their regulatory function (158) and that their distinct developmental functions are a consequence of changes in the regulatory sequences rather than in the proteins themselves. Because they lend themselves so well to changes in the architecture of promoter regions, it is likely that transposable element mutations have been involved in this kind of regulatory evolution (85).

Centromere structure and function

The centromere is a specialized domain of the chromosome that mediates the appropriate segregation of homologous and sister chromatids during cell division. Considering their ancient and conserved function, centromeres show a remarkable degree of sequence variation, both among different chromosomes within a species and between species. As an increasing number of centromeric sequences have become available, it has become apparent that transposons make up a significant fraction of many of them. In *Drosophila,* the centromere of minichromosome Dp1187, which is encompassed by 420 kb of centric heterochromatin, is made up primarily of satellites and complete transposable elements (238). A centromere of the yeast *Schizosaccharomyces pombe* is composed of a centromere-specific, nonrepetitive AT-rich central core, flanked by several classes of centromeric-specific repetitive elements that bind to a transposon homologue (44, 89). Also, a centromeric region of one *Neurospora crassa* chromosome contains several transposable elements surrounded by a sea of simple sequence repeats (37). Notably, one transposon in this species *Tcen* was found exclusively at the centromere. Mammalian centromeric regions also seem to be enriched in transposable element sequences as well (11, 138, 253). In plants, complete sequences of many *Arabidopsis* centromeric regions are available, although the minimal functional requirements for plant centromeres are still not known (49). As in mammals, these centromeric regions are composed primarily of satellite and transposable elements. In the cereals, several centromeric repetitive sequences have been identified (2, 64). One element, present in the centromeres of sorghum, rice, maize, and the Triticeae, is a retrotransposon of the *Ty3/Gypsy* class of retroelements (118).

What the presence of transposable elements within centromeres means with respect to function and structure remains to be seen. Although the presence of heterochromatin and various transposable elements is often associated with centromeric function, it is clear that perfectly functional centromeres can exist in the absence of either of these features. *S. cerevisiae,* for instance, has a very short centromere that lacks extensive satellite DNA or transposons. A region of only 125 bp has been determined to be sufficient for centromeric function (51). Nevertheless, there are indications that, at least in species having "regional" as opposed to "point" centromeres, transposon-related sequences may be playing important roles in centromere function. In mammals, centromeric repeat binding proteins bind repetitive motifs within centromeric regions. One of these proteins, CENP-B, the 80-kDa human centromere autoantigen recognized by patients with anticentromere antibodies, is clearly related to the *pogo* transposable element of *Drosophila* and the *Tigger* transposon in humans (70). Homologues of CENP-B are found in species as diverse as fission yeast (152) and *Arabidopsis thaliana* (161). Although sequences similar to those bound by CENP-B in humans (CENP-B boxes) are found in centromeric regions of several organisms, they are apparently much less widespread than CENP-B itself (reviewed in references 52 and 138). The *Tigger 2* terminal inverted repeat, which is presumably interacting with the transposase, contains a close match to the CENP-B box. Thus, it seems that at least one function of the *Tigger*-like transposon, i.e., the binding of a specific DNA sequence, has been retained by CENP-B (138).

In some genetic backgrounds, mice that are mutant for the *Cenp-b* locus show significant reduction in fertility, perhaps because of disruptions in the normal development of the uterine epithelium, a process that requires extensive mitotic division (91). The observed dependence on genetic background in mice may be related to redundancy of function. In support of this hypothesis, in fission yeast, when two redundant homologues of CENP-B were mutated, the resulting strains were severely compromised in growth rate and morphology, and they exhibited marked chromosome missegregation (12). These data suggest that CENP-B, which is clearly related to an ancient and widespread family of transposable elements, can play an important role in host chromosome segregation. Not all centromeres in humans require CENP-B function, however. The Y chromosome, which lacks CENP-B boxes, shows no evidence for binding to CENP-B, and a neocentromere (a noncentromeric region of a chromosome that can act as a centromere) lacks the binding motif for CENP-B (reviewed in reference 138). Thus, although CENP-B and its cognate binding site may play a role in the activity of some centromeres, it is not necessarily an essential component of all functional centromeres. The same can be said of nearly any specific sequence, which has led to the suggestion that general structural (i.e., secondary or tertiary) features are important, as well as epigenetically regulated chromatin level features (126). The extent to which transposable elements play a role in the formation of such features is as yet unknown. However, it has been suggested (138) that CENP-B nicks centromeric DNA, resulting in the expansion of centromeric repeats through recombination and subsequent formation of a secondary structure that is more competent to act as a centromere.

Telomeres

Although most organisms examined to date use telomerase to prevent the loss of sequences at the ends of chromosomes after DNA replication, Diptera appear to use at least two non-LTR class retroelements. *HetA* and *TART* elements transpose specifically to the ends of *D. melanogaster* chromosomes and are present in tandem arrays at the ends of these chromosomes in place of the more typical telomere repeat sequences (153). It is the serial addition of these elements to the ends of dipteran chromosomes that maintains their length after DNA replication. Unlike the Diptera, most close relatives have both telomerase activity and more typical telomeric repeats (18). Thus, it has been postulated that the use of these retroelements is a recently derived characteristic that has taken the place of the ancestral use of telomerase. Some other insects, such as *B. mori,* also have telomere-specific non-LTR retrotransposons. However, rather than targeting the poly(A) tail of a previously inserted retroelement, these elements target the canonical (TTAGG)n repeat at the ends of *Bombyx* telomeres (200, 241). One could imagine a scenario in which, in an ancestor of the drosophilids, elements that originally targeted a telomere repeat shifted targets to themselves. If the transposition frequency was high enough to replace the net loss of chromosomal material after DNA replication, telomerase could have become redundant and lost. Subsequent selection on the frequency and specificity of *Het-A* and *TART* activity would ensure that these elements would act as well-behaved cellular components, rather than as benign parasites.

Stress-induced responses

All SINEs examined in humans and mice unexpectedly accumulate in gene-rich regions (233). They also exhibit high transcription rates in somatic cells under stress. *Alu* RNA is increased by cellular insults, and dispersed *Alu*s exhibit remarkable tissue-specific differences in the level of their 5-methylcytosine content (219). Differences in *Alu* methylation in the male and female germ lines might suggest that *Alu* DNA may be involved in either the unique chromatin organization of sperm or signaling events in the early embryo (219). However, the possibility that GC-rich DNA may be linked to the striking hypomethylation of SINEs in the germ line, although true for *Alu*s, does not hold for other SINEs that have very few or no CpGs. Transcript levels in cells under stress, but not those of related RNA molecules, can antagonize the activation of double-stranded RNA-induced kinase PKR and increase translation of a reporter gene (41). Thus, it is possible that increased transcription of *Alu*s saturates and inhibits PKR and may induce protein translation. It has been proposed (233) that *Alu*s may represent a class of short RNAs of unspecific sequence that can quickly and transiently suppress PKR inhibition of translation after stress. Other requirements for such a molecule are a stable secondary structure and freedom to bind PKR in an inactive monomer. The open chromatin configuration of gene-rich regions is expected to facilitate this function which may be performed by mammalian-wide interspersed repeats (MIRs) or other, species-specific SINEs in various mammals. Similar observations of stress-induced high transcription rates of SINEs have been reported in insects. For example, the level of transcripts of *Bml,* a SINE found in the silk worm, *B. mori* (136), increases in response to either heat shock, inhibiting protein synthesis by cycloheximide or viral infection. Posttranscriptional events may partially account for stress-induced increases in *Bm1* RNA abundance. In light of these results, it has been proposed (136) that SINE RNAs may serve a role in the cell stress response that predates the divergence of insects and mammals implying that SINEs may have been exapted as a class of cell stress genes.

Molecular domestication in *Drosophila*

Although they have no known function, there is strong circumstantial evidence that some derivatives of *P* elements have become domesticated (184–186). Some of these elements, found in the stationary *P* element-related gene clusters of *Drosophila guanche, Drosophila madeirensis,* and *Drosophila subobscura* lack inverted terminal repeats, suggesting that they are no longer capable of transposition. However, they are well conserved within groups, particularly in their N-terminal domains. Their promoters are made up of additional insertions of small, MITE-like elements. These new promoters are much stronger than are those of *P* elements. Because they are so well conserved and are clearly no longer transposable elements in the strict sense of the term, it has been argued that they must be under selective constraints because of some function that they have acquired that provides some benefit to their host.

An internally deleted immobile *P* sequence cloned from *Drosophila tscasi,* a species of the *D. montium* subgroup (195), provides evidence that molecular domestication of a transposable element family may recur in a host lineage. This element produces a polyadenylated RNA that also has a coding capacity for a 66-kDa "repressor-like" protein. This domestication event has included the evolution of a new promoter and a new intron (196). Although the immobile *obscura* and *montium P* sequences were derived from the same ancestral mobile *P*-element family, the structures of the flanking regions of these sequences indicate that they were produced by two completely independent evolutionary events.

Miller and colleagues (185) maintain that these examples of domestication represent more than just isolated examples of an infrequent process. Rather, they argue, the adaptive integration of a short piece of autonomous DNA into a complex regulatory network that occurs in molecular domestication might be considered to represent an evolutionary process that has been common in the evolution of increasingly complex forms of life.

Molecular domestication in humans

Twenty human examples of transposable elements, which have been co-opted for host function, have been listed by Smit (233). At least seven of these are chimeras with C-terminal transposase-derived domains. These include interesting examples in which cellular genes, such as genes derived from human endogenous retroviruses (HERVs), encompass complete transposable element genes. There is also evidence for the recruitment of several SINEs in vertebrates (30, 233), including BC200, a brain-specific RNA that is part of a ribonucleoprotein complex that is preferentially located in primate dendrites. The gene encoding BC200 RNA arose from a monomeric *Alu* element at least 35 to 55 million years ago (231) and has subsequently been recruited for a function in the nervous system. The brain-specific expression pattern of BC200 RNA and its presence as a ribonucleoprotein particle (RNP) are conserved in Old World and New World monkeys and appear to be under positive selective pressure. The C gamma (*Cg*) gene appears to be a retroposon that is expressed specifically in the testis. Apparently it is a functional retrotransposed portion of the homologous *Ca* gene. The *Cg* gene lacks introns, contains a remnant of a poly(A) tail, and is flanked by direct repeats. Given that this gene can be translated in vitro (90), is expressed in a highly regulated manner (13), and is conserved in monkeys and humans, it is very likely to produce a functional protein. Because the start of transcription of this gene is likely to be located outside the retrotransposed portion of the gene (in an *Alu* element), it seems that its tissue specificity arose because of borrowed signals from its new position.

Other exaptation examples of particular interest include the role of *L1* elements in X-inactivation, the origin of antigen-specific immunity, and the evolution of centromeres and telomeres.

***LINE-1* elements in mammalian X inactivation.** It has been proposed (169) that *LINE-1* elements act as booster elements to promote the spread of Xist mRNA, the agent responsible for X inactivation. Nonrandom properties of *L1* distribution (7, 168) provide evidence that *L1* elements may serve as DNA signals to propagate X inactivation along the chromosome. These include the twofold enrichment for *L1* elements on the X chromosome, with the greatest enrichment for a restricted subset of *LINE-1* elements that were active 100 million years ago. Analysis of the distribution of *L1* elements along the X chromosome revealed a significant enrichment of these elements at the center of X inactivation, and significantly fewer *L1* elements were found in genic regions that escape X inactivation.

Immunoglobulin diversity. To respond to invasive foreign organisms, the vertebrate immune system must generate a massive degree of variation during the life of a single individual. In retrospect, it is perhaps not surprising that transposable elements, with their demonstrated capacity to generate variation, would have been co-opted for this purpose. In humans, immunoglobulin diversity is generated via sequence-specific recombination between three different domains: V, D, and J (reviewed in reference 92a; see also chapter 10). Because many different copies of each domain can be used, and because the joining reaction is somewhat imprecise, this reaction can produce a vast number of unique immunoglobulins. This recombination pathway requires the activity of two enzymes, RAG1 and RAG2, which are encoded by two genes that are located quite close to each other in reverse orientation (199). There is strong evidence that the genes that code for these enzymes are derived from transposons. The mechanism of V(D)J recombination bears striking similarities to transposable element transposition (reviewed in reference 154), which leads to speculation that the two mechanisms may be related in an evolutionary sense (234). More recent work has demonstrated that RAG1 and RAG2 can act as legitimate transposases in vitro, catalyzing the transposition of sequences flanked by the recombination signal sequences (1, 108). The origin of the putative RAG ancestral transposon is unknown, and there are no known transposons that can serve as a "missing link" in the evolution of RAG1/RAG2 recombination. However, it has been hypothesized that a transposon was transferred to the vertebrate lineage and co-opted for use in V(D)J recombination sometime before the evolution of bony fishes some 450 million years ago.

Resistance to viral infection. Viral resistance in mice is often genetically determined. Recent work suggests that at least two genes involved in viral resistance are themselves derived from viruses. The *Fv1* (Friend virus susceptibility 1) gene is responsible for resistance to Friend virus in mice. It acts after viral entry into the cell but before integration of the virus into the host genome (207a) and is clearly related to a retroviral gag gene (16). The apparent derivation of *Fv1* from a viral genome suggests that a viral gene has been co-opted for use in viral resistance, most likely by binding some component of the viral integration complex. Similarly, *Fv4,* a gene that confers resistance to murine leukemia virus (MuLV), is an endogenous defective version of the infective exogenous virus. Mice that carry *Fv4* secrete an *env*-derived surface peptide in their serum that can block the entry of that virus (193). A very different form of viral resistance has also been documented in mice (reviewed in reference 164). Mouse mammary tumor virus (MMTV) exists in both exogenous and endogenous forms. The integrated forms of both produce a super antigen (SAG) encoded by an ORF in their LTRs. Expression of SAG facilitates transient proliferation and subsequent infection of lymphocytes, which are transported to progeny via the mammary glands. High levels of expression of endogenous SAG before infection lead to the proliferation and subsequent loss of the population of T and B cells that are usually infected by exogenous MMTV, which makes mice carrying the endogenous virus resistant to infection from related exogenous viruses (reference 94 and references therein). It has been suggested that the presence of SAG-expressing endogenous elements in the absence of cognate exogenous viruses may represent cases in which the exogenous virus has been eliminated altogether because of the protective activity of the endogenous variant (164).

A functional role for endogenous retroviruses? Endogenous retroviruses are a ubiquitous component of many mammalian genomes. In humans, for instance, they have been estimated to make up 1% of the human genome (164). The vast majority of these elements do not contain intact reading frames and are thought to be extinct relics of previous expansions of active families of elements. However, a few elements retain intact reading frames and are expressed in specific tissues at specific times during development. Although no defined role has been assigned to these elements, the specificity of their expression and conservation of specific domains suggest that they may have one. *HERV-R* (*ERV-3*), for instance, is expressed at high levels in the placenta, a tissue with pronounced immunosuppressive properties (22). This single copy element expresses an Env protein that promotes immunosuppression (42). On the basis of these findings, it has been suggested that *HERV-R* constitutes a domesticated retrovirus that functions to permit, or at least facilitate, vivipary in mammals (255). Recent data on extant polymorphism within human *HERV-R* sequences cast some doubt on this hypothesis, however, because several individuals carried *HERV-R Env*-coding sequences that carried stop codon, suggesting that the protein is not necessary for normal reproduction in humans, or at least that this gene is redundant in function (208).

Host Responses: Epigenetic Regulation of Transposable Elements

Any nucleus can be viewed as a high concentration of valuable replication machinery. Therefore, it seems inevitable that many units of selection will have learned to parasitize that machinery. To the extent that this produces negative selective pressure, this parasitism would be expected to lead to the evolution of mechanisms to mitigate its effects. To the extent that these mechanisms have utility for the host, particularly with respect to subtle and reversible modulation of gene expression, they are then themselves subject to positive selection. Thus, one of the most important consequences of transposable element activity may be the response that it provokes in host organisms and the subsequent exaptation of those response pathways for other purposes (177).

A growing body of research suggests that the epigenetic regulation of transposable elements is intimately tied to that of host genes. These mechanisms include (but are not limited to) modulation of transcription through chromatin configuration, histone acetylation, and changes in methylation status, as well as posttranscriptional regulation, such as selective degradation of aberrant RNA, and various combinations thereof (reviewed in reference 269). Because the details of the specific modes of action of each of these mechanisms are covered extensively elsewhere, we will focus here on the evidence that host and parasite may be subject to a similar suite of epigenetic regulatory pathways. The origin of these pathways is not known. However, because many of them can be provoked to respond to the consequences of transposable element activity (such as changes in copy number and aberrant RNAs), it is quite possible that they arose originally as a means to repress transposable element activity (105).

Posttranscriptional gene silencing

Posttranscriptional gene silencing was first identified in plants (191, 254) in which it was discovered that introduced transgenes and homologous host genes would regularly become silenced. Because of the similarities between mechanisms for posttranscriptional gene silencing and repression of viral proliferation in plants (e.g., reference 260), including the presence of at least one mutant, *sgs2,* that affects both phenomena (189), it has been suggested that posttranscriptional gene silencing in plants evolved as a defense against viral infection (176).

Regulation of repetitive DNA via degradation of specific species of RNA has since been identified in both vertebrate (259a) and invertebrate (87a) animals as well. In animals, the most extensive work on posttranscriptional gene silencing has been done in *C. elegans,* where it is called double-stranded RNA-mediated interference (RNAi) (reviewed in reference 111). RNAi was first identified as a consequence of attempts to use antisense RNA to silence genes. It was found that constructs that encoded double-stranded RNA were as effective as antisense constructs at silencing native genes (100). Remarkably, RNAi can be induced via ingestion (248).

Mutagenesis experiments have identified several genes that are involved in regulating this process in *C. elegans* (131, 240). In addition to their effects on RNAi, several of these mutants were found to activate otherwise quiescent transposable elements, suggesting that one of the normal functions of RNAi is the regulation of transposable elements. One of the genes that inhibit RNAi, *rde-1,* has homologues in *Drosophila* and *Arabidopsis* that are involved in the maintenance of pluripotent stem cells (240). Another gene that affects both RNAi and transposon silencing, *mut-7,* encodes a protein that is similar to RNA helicases and RNase D. In addition to its effects on RNAi and transposon activity, *mut-7* mutant exhibits temperature-sensitive effects on fertility and sex ratio (131), suggesting that a mechanism that can regulate transgenes and transposons is also used by the host for other purposes. There is additional evidence that genes regulating RNAi may also be involved in host gene regulation. *ego-1,* a gene that is required for oogenesis in *C. elegans,* also exhibits deficiencies in RNAi. This gene is homologous to RNA-dependent RNA polymerase (232). So is *qde-1* in *N. crassa,* a gene that is involved in quelling, a mechanism for silencing repetitive sequences in this fungus (46), and *sgs2,* a gene required for posttranscriptional gene silencing of transgenes and viruses in *Arabidopsis* (189). The fact that mutations in the genes in *Neurospora* and *Arabidopsis* do not have any obvious phenotype, apart from a lack of silencing, suggests that silencing of repetitive elements may be the more ancient function of this class of genes.

Methylation

A distinct but related phenomenon involves methylation of cytosine residues. Methylation is often associated with transcriptional silencing (246), which can be reversed by the application of chemical agents that block methylation (121). Although a convenient marker for that inactivation, however, methylation often is caused by (225) or causes (72, 192) additional changes, such as in chromatin configuration or histone acetylation. Further, posttranscriptional silencing, discussed above, is often associated with DNA methylation, perhaps because of an interaction between aberrant RNA species and DNA (reviewed in reference 229). Thus, the direct causal relationship between inactivation and methylation is sometimes obscure. Methylation is clearly a component of gene-silencing mechanisms in many organisms, however.

Methylation is not universally required for eukaryotic development. Some eukaryotes, such as *Drosophila,* do without it altogether. However, among those that do methylate their DNA, this reversible modification apparently plays an important role in host gene regulation. In humans, genetic imprinting, which involves the specific inactivation of an allele from one parent, is often associated with methylation of the inactive allele (155), and failures in imprinting are associated with several diseases (209). Imprinting has also been observed in plants, where it is regulated by genes that affect global methylation (257).

Methylation is required for X chromosome inactivation in mammals (201), a process necessary to express X-linked genes at the appropriate levels in females. Mice that are homozygous for a mutation (*Dnmt*) in a mammalian methyltransferase die soon after gastrulation, implying that the methylation function is essential for normal mammalian development (156). Similarly, *Arabidopsis* plants homozygous for an antisense version of a plant methyltransferase cDNA (METI) exhibit several developmental abnormalities (87), as do those that are mutant for *ddm1,* a chromatin-remodeling gene involved in global methylation (124).

The data quite clearly suggest that methylation and associated changes in chromatin configuration play an important role in the normal development of many species. However, it is also clear that methylation plays an equally important role in the regulation of repetitive sequences. In both mammals and plants the vast majority of methylated DNA is within transposable elements (217, 273), and there is a long-established link between transposon activity and methylation. Maize elements, for instance, undergo reversible methylation associated with changes in their activity (8; see chapter 42). Similarly, in humans, cytosine methylation is involved in the repression of the IAP endogenous retrovirus (259). Indeed, methylation of transposons (in hosts that methylate their DNA) is the most reliable single predictor of that activity (174). In *Aspergillus,* a highly efficient system operates to methylate repetitive sequences. The process known as methylation induced premeiotically (MIP) induces heavy methylation of repetitive DNA (212) and is hypothesized to have evolved to silence transposable elements (224).

Experimental evidence also exists that methylation of transposable elements and changes in host gene regulation can be linked. Transposable elements in the promoters of maize genes, for instance, can be epigenetically modified, and those modifications can affect expression of the host gene (82). Further, mutations that affect paramutation in maize, which involves the heritable epigenetic alteration of host gene expression, also affect methylation of the *Mutator* system of transposons (D. R. Lisch et al., unpublished data). Similarly, methylation near a *doppia* transposon is also associated with changes in expression of the paramutable *r* locus in maize (258). Finally, mutations that affect levels of methylation in *Arabidopsis* can affect both host gene expression (114) and transposable element activation (109).

Given the consistent links between transposable element silencing and DNA methylation, it has been suggested that the regulation of repetitive DNA is the primary role of DNA methylation (10, 273). These authors argue that the vast majority of potentially methylated sequences are with transposons, not host genes. A similar argument has been made to explain imprinting in humans. The parent conflict model asserts that silencing of imprinted genes represents the outcome of a conflict between maternal and paternal genes (reviewed in reference 113). In addition, genes subject to imprinting have some interesting similarities to transposable elements, including a few or no introns and an enrichment of repetitive sequences (273). According to this view, methylation of these genes has more in common with the silencing of "selfish" transposons than it does with intrinsically important developmental regulation of host genes.

Needless to say, the idea that methylation is primarily a means to control selfish DNA has been a controversial one. It has been argued that, given the numerous transposable elements in hosts that use methylation, this mechanism has hardly been effective (19), and that many transposons are hypomethylated at exactly the time during mammalian development (such as the male germ cell) that one might expect them to be silenced if heritable insertions are to be prevented (119). However, in maize, those elements that are most heavily and consistently methylated are the least mutagenic (145), and very few mutations in humans are attributable to transposable element insertions (127). Given the vast number of elements present in the genomes of both of these species, the number of new mutations under normal circumstances is remarkably low, suggesting that mechanisms to repress them, such as methylation, are quite efficient.

Heterochromatin

Heterochromatin represents a significant portion of the eukaryotic genome. As much as 30% of the *D. melanogaster* genome, for instance, is made up of heterochromatin, as is 15% of the human genome (120). As we have suggested, heterochromatin is rich in satellite DNA and transposable elements and relatively poor in host gene sequences. The presence of transposable elements in heterochromatin has been documented in species as diverse as *D. melanogaster* (206), hamster (245), *Arabidopsis* (48), and maize (2, 3). Recombination within heterochromatin is markedly reduced (reviewed in reference 103) and, as discussed above, it has been suggested that transposons are often overrepresented in heterochromatin because selection has preferentially removed elements from euchromatic sites where ectopic recombination would select for their loss (39). Although this is a reasonable interpretation that has some experimental validation (e.g., reference 249), several experimental results suggest that other processes are involved as well.

First, it is clear that within heterochromatin there is a great deal of variation in distribution of elements. For instance, Pimpinelli and coworkers (206) found that each of several classes of elements inhabits a specific domain within *Drosophila* heterochromatin. Differential selection due to differences in recombination frequency would not be expected to produce this pattern, because recombination is suppressed throughout heterochromatin. Second, some elements, such as the *I* elements of *D. melanogaster* (60) and the Ty5 elements in yeast (135), as discussed above, clearly target heterochromatin. De novo insertion of *I* elements into a heterochromatic gene was an order of magnitude higher than into a euchromatic gene on the same chromosome (60) and Ty5 elements transpose exclusively to subtelomeric locations (276).

Finally, there is evidence that the presence of transposons themselves can contribute to the formation of heterochromatin. For example, it has been demonstrated that gene silencing and heterochromatin formation can be induced by transposable elements (67, 80). Further, Steinemann and coworkers (237) present compelling evidence that the first step in Y chromosome degeneration is driven by the accumulation of transposable elements, especially retrotransposons. They suggest that an enrichment of these elements along an evolving Y chromosome could account for the switch from a euchromatic to a heterochromatic chromatin structure. This kind of experiment has led to the suggestion that the evolution of heterochromatin may have been a response to invasive transposable elements. Further, the observed reduction in recombination within heterochromatin may be a strategy evolved by the host to reduce both ectopic recombination between transposable elements and expression and consequent new transpositions of those elements. Specific targeting of transposable elements into heterochromatin may be the result of subsequent selection against insertions that reduce host fitness.

Although the repression of transposable element transcription by heterochromatin can reasonably be seen as a means by which transposons can be held in check by the host, and targeting of heterochromatin by transposons can be seen as a mechanism by which transposable elements limit the damage they do to their host, targeting begs the question as to how these elements avoid extinction due to transcriptional repression. After all, a transposable element can only thrive because of replication advantage, an advantage it forgoes if it transposes into a permanently silenced state. Work with the yeast Ty5 element suggests a way that this paradox can be resolved. As discussed above, the Ty5 element targets inactive regions (silent chromatin) at the HM loci and is usually transcriptionally repressed. However, Ty5 transcription is induced 20-fold by pheromone treatment, leading to subsequent element transposition. This transcriptional activation is mediated by the presence of pheromone response elements in the LTRs of Ty5 (130). Thus, chromatin repression of this element can be reversed at precisely the time (during mating) that selection at the transposon level might be expected to favor element amplification.

Although one of its roles may be to regulate transposable element activity, there is strong evidence that heterochromatin can influence normal gene expression, and that the same proteins that are involved in the formation and maintenance of heterochromatin are also involved in the appropriate expression of several host genes. Position effect variegation (PEV), first discovered in *Drosophila,* occurs when a gene is put in proximity with a block of heterochromatin (reviewed in reference 262). This phenomenon has also been observed in species as diverse as mice (61) and yeast (96, 194), suggesting a general propensity of heterochromatin to silence nearby genes. Other epigenetically regulated phenomena, such as homeotic gene regulation in *Drosophila* and silencing in yeast, are regulated by some of the same or homologous proteins. Some genes that regulate PEV, such as *Su(var)3–9,* contain domains similar to those that regulate homeotic genes in *Drosophila* (81, 251), and at least one gene in yeast that is essential for silencing of centromeres and the mating-type loci (112). This same gene, *Clr4,* is also involved in the proper functioning of the centromere, supporting the hypothesis that centromere function involves chromatin configuration (reviewed in reference 40).

In *Drosophila,* some host genes are located within heterochromatin. When translocations move these loci into proximity to euchromatin, their expression becomes variegated (reviewed in references 71 and 107). Consistent with these results, these genes seem to require at least one suppressor of variegation to function, suggesting that they are specifically adapted to the heterochromatic niche (166). This suggests that heterochromatin can be maintained because of selection in favor of the expression of these heterochromatin host genes as well as against the proliferation of transposable elements located within the heterochromatin.

Recent work on both plants and animals suggests

that host gene regulation leading to the appropriate unfolding of developmental programs and transposable element regulation are inextricably linked. Similarly, there is little doubt that the influence of heterochromatin on transposable element activity and host gene regulation are functionally linked, and it seems likely that the behavior and distribution of transposable elements can influence the position, frequency, and regulation of heterochromatin, which, in turn, can influence host gene expression. In these and other examples, epigenetic regulation represents a suite of mechanisms that can modulate the expression of a variety of genes, including those of transposable elements. Whether or not transposable elements provided the impetus for the origin of such mechanisms, it seems likely that the competition between host and parasite has contributed to the continuing evolution of both, providing opportunities for the exaptation of capabilities that may have come about in response to the activities of invasive mobile DNA.

A Matter of Perspective

The relative role of transposable elements in host evolution is a matter of considerable controversy at present. Given the remarkable array of variation produced by transposable element activity, it is tempting to assume that transposable elements have systematically contributed to the evolution, or even the "evolvability" (139) of their hosts. This perspective has been expressed in several recent reviews (27, 30, 180). However, it should be recognized that there is a discrepancy between what can occur (as determined by geneticists working on a particular experimental system) and what has occurred (as determined by evolutionary biologists by comparing many such systems). Of necessity, genetic evidence is anecdotal, and it is tempting for molecular biologists and geneticists to emphasize the importance of mechanisms that can produce dramatic, rapid changes in phenotype over those that produce gradual changes in the frequency of multiple alleles. It is a reasonable criticism of this approach that it disregards what has actually happened in favor of what could have happened. In this sense, an historical perspective is lacking, and there is a strong tendency to overgeneralize from possibly rare events. It is hardly an exaggeration to say that, among biological systems, it is possible to find an example to fit nearly any perceived rule, and it certainly matters how typical any given example has actually been.

On the other hand, many classically trained evolutionary biologists and ecologists tend to overlook a revolution in our understanding of how molecular evolution can occur. Molecular and genetic analyses of numerous systems suggest that the process of mutagenesis can be more subtle, nuanced, and sophisticated than would have been believed when the neo-Darwinian synthesis was first formulated more than 50 years ago. Of course an understanding of the ways in which allele frequencies change in response to selective pressure has to form the core of our understanding of how evolutionary change occurs. However, the constraints and opportunities intrinsic to genomes determine what kinds of alleles are available, and hence, how genomes can respond to those selective pressures.

SUMMARY AND CONCLUSIONS

Transposable elements provide sources of variation in two main ways. First, as a result of their activity, they induce new variation directly through mutations and genomic rearrangements. However, because of natural and artificial selection, the standing variation in populations reflects only a fraction of the original variation induced by these elements. The co-option of transposable element sequences and enzymatic machinery by their host organisms provides a second way that transposable elements contribute to variation throughout evolutionary time. However, such co-opted sequences and mechanisms are likely to be increasingly difficult to identify with the passage of time. Therefore, the amount of variation of this type that can be observed is likely to be a gross underestimate of the true frequency, but improved identification techniques are already starting to improve this situation. Studies to estimate the frequency of both these broad types of variation are urgently needed to determine the importance of transposable elements as agents for the generation of phenotypic variation in natural populations.

At the subgenomic level, mutations can result from transposable element insertion in coding or noncoding regions, or when insertions produced by elements of the DNA type excise imprecisely. Quantitative variation determined by multiple loci can be molded by transposable elements, as well as traits determined by single genes. At the genomic level, transposable element activity can result in both small and large changes in both the size and arrangement of host genomes. These changes can occur through several mechanisms, including homologous recombination between elements located at different genomic locations, duplications, deletions, inversions, translocations, and transduction. Effects of transposable elements on genome restructuring can take place during the life span of a single individual, such as the programmed somatic excision of interstitial DNA segments in ciliated protozoa, or over millions of years, such as the doubling of the maize genome because of retroelement expansion. Some fraction of transposable element-mediated restructuring also seems to occur in response to environmental stress.

Nonrandom genomic distribution is a common attribute of transposable element insertions. Target-site preference on insertion and natural selections both play a role in determining the distribution patterns observed. Preference for DNA sites may vary widely in precision. Highly specific site preference is favored by a few transposable elements, whereas others target, or avoid, broad chromatin domains or genomic regions. Some transposable elements prefer to insert into one another, which results in clustering in broad regions such as heterochromatin. The functional and structural significance of certain regions of high transposable element density, such as centromeres, is not yet understood.

There are many ways that transposable elements can be domesticated or exapted for the functions of the host with the passage of evolutionary time. Although the number of possible examples of cases of this type has recently grown rapidly, there are no good estimates of their frequency or significance relative to other mutation mechanisms. A role in gene regulation, including the rewiring of regulatory networks, is a potentially significant way that transposable elements can positively contribute to host evolution. Other possible exaptations include, but are by no means limited to, telomeric structural features, stress-induced responses, large-scale DNA sequence inactivation, and immunological diversity. There is good evidence that transposable elements and at least some host genes may be subject to similar suites of regulatory pathways, including posttranscriptional silencing, methylation, and the formation of relatively inert heterochromatin.

The complexity of possible interactions between transposable elements and their hosts is becoming increasingly appreciated, but a full understanding of these interactions will take some time to be realized. Although evidence for the primary parasitic nature of many transposable elements is compelling, especially in the early stages of their invasion of host genomes, over time there seems to be a tendency for the relationship to become more benign, even mutualistic. Some features of both transposable elements and hosts suggest coadaptations to mitigate the reduction of fitness expected from unfettered transposition, and to provide a wide range of new variations on which natural selection can act. It is possible, then, that the dynamic instability conferred by the initially selfish behavior of transposable elements has paradoxically contributed in significant ways to the capacity of complex eukaryotes to evolve.

Acknowledgments. We thank Andrew Holyoake for valuable suggestions on the manuscript.

This work was supported by National Science Foundation grant DEB 9815754.

REFERENCES

1. **Agrawal, A., Q. M. Eastman, and D. G. Schatz.** 1998. Transposition mediated by *RAG1* and *RAG2* and its implications for the evolution of the immune system. *Nature* **394:** 744–751.
2. **Ananiev, E. V., R. L. Phillips, and H. W. Rines.** 1998. Chromosome-specific molecular organization of maize (*Zea mays* L.) centromeric regions. *Proc. Natl. Acad. Sci. USA* **95:** 13073–13078.
3. **Ananiev, E. V., R. L. Phillips, and H. W. Rines.** 1998. Complex structure of knob DNA on maize chromosome 9. Retrotransposon invasion into heterochromatin. *Genetics* **149:** 2025–2037.
4. **Anderson, R., R. J. Britten, and E. H. Davidson.** 1994. Repeated sequence target sites for maternal DNA-binding proteins in genes activated in early sea urchin development. *Dev. Biol.* **163:**11–18.
5. **Anholt, R. R., R. F. Lyman, and T. F. Mackay.** 1996. Effects of single *P*-element insertions on olfactory behavior in *Drosophila melanogaster. Genetics* **143:**293–301.
6. **Ashburner, M.** 1992. *Drosophila, a Laboratory Handbook.* Cold Spring Harbor Laboratory Press, Cold Spring Harbor, N.Y.
7. **Bailey, J. A., L. Carrel, A. Chakravarti, and E. E. Eichler.** 2000. Molecular evidence for a relationship between *LINE-1* elements and *X* chromosome inactivation: The Lyon repeat hypothesis. *Proc. Natl. Acad. Sci. USA* **97:**6634–6639.
8. **Banks, J. A., P. Masson, and N. Fedoroff.** 1988. Molecular mechanisms in the developmental regulation of the maize *Suppressor-mutator* transposable element. *Genes Dev.* **2:** 1364–1380.
9. **Baran, G., C. Echt, T. Bureau, and S. Wessler.** 1992. Molecular analysis of the maize *wx-B3* allele indicates that precise excision of the transposable *Ac* element is rare. *Genetics* **130:** 377–384.
10. **Barlow, D. P.** 1993. Methylation and imprinting: from host defense to gene regulation? *Science* **260:**309–310.
11. **Barry, A. E., E. V. Howman, M. R. Cancilla, R. Saffery, and K. H. Choo.** 1999. Sequence analysis of an 80 kb human neocentromere. *Hum. Mol. Genet.* **8:**217–227.
12. **Baum, M., and L. Clarke.** 2000. Fission yeast homologs of human CENP-B have redundant functions affecting cell growth and chromosome segregation. *Mol. Cell. Biol.* **20:** 2852–2864.
13. **Beebe, S. J., P. Salomonsky, T. Jahnsen, and Y. Li.** 1992. The C gamma subunit is a unique isozyme of the cAMP-dependent protein kinase. *J. Biol. Chem.* **267:**25505–25512.
14. **Bellen, H., C. J. O'Kane, C. Wilson, U. Grossniklaus, R. K. Pearson, and W. J. Gehring.** 1989. *P*-element-mediated enhancer detection: a versatile method to study development in *Drosophila. Genes Dev.* **3:**1288–1300.
15. **Bennetzen, J. L.** 2000. Transposable element contributions to plant gene and genome evolution. *Plant Mol. Biol.* **42:** 251–269.
16. **Best, S., P. Le Tissier, G. Towers, and J. P. Stoye.** 1996. Positional cloning of the mouse retrovirus restriction gene *Fv1. Nature* **382:**826–829.
17. **Biemont, C., A. Tsitrone, C. Vieira, and C. Hoogland.** 1997.

Transposable element distribution in *Drosophila. Genetics* **147**:1997–1999.

18. **Biessmann, H., and J. M. Mason.** 1997. Telomere maintenance without telomerase. *Chromosoma* **106**:63–69.
19. **Bird, A.** 1997. Does DNA methylation control transposition of selfish elements in the germline? *Trends Genet.* **13**: 469–472.
20. **Black, D. M., M. S. Jackson, M. G. Kidwell, and G. A. Dover.** 1987. *KP* elements repress *P*-induced hybrid dysgenesis in *Drosophila melanogaster. EMBO J.* **6**:4125–4135.
21. **Blackman, R., R. Grimaila, M. Koehler, and W. Gelbart.** 1987. Mobilization of *hobo* elements residing within the decapentaplegic gene complex: suggestion of a new hybrid dysgenesis system in *Drosophila melanogaster. Cell* **49**:497–505.
22. **Boyd, M. T., C. M. Bax, B. E. Bax, D. L. Bloxam, and R. A. Weiss.** 1993. The human endogenous retrovirus *ERV-3* is upregulated in differentiating placental trophoblast cells. *Virology* **196**:905–909.
23. **Bradley, D., R. Carpenter, H. Sommer, N. Hartley, and E. Coen.** 1993. Complementary floral homeotic phenotypes result from opposite orientations of a transposon at the plena locus of Antirrhinum. *Cell* **72**:85–95.
24. **Bregliano, J. C., and M. G. Kidwell.** 1983. Hybrid dysgenesis determinants, p. 363–410. *In* J. A. Shapiro (ed.), *Mobile Genetic Elements.* Academic Press, New York, N.Y.
25. **Britten, R. J.** 1996. Cases of ancient mobile element DNA insertions that now affect gene regulation. *Mol. Phylogenet. Evol.* **5**:13–17.
26. **Britten, R. J.** 1996. DNA sequence insertion and evolutionary variation in gene regulation. *Proc. Natl. Acad. Sci. USA* **93**: 9374–9377.
27. **Britten, R. J.** 1997. Mobile elements inserted in the distant past have taken on important functions. *Gene* **205**:177–182.
28. **Britten, R. J., and E. H. Davidson.** 1971. Repetitive and nonrepetitive DNA sequences and a speculation on the origin of evolutionary novelty. *Q. Rev. Biol.* **46**:111–138.
29. **Brosius, J.** 1999. Genomes were forged by massive bombardments with retroelements and retrosequences. *Genetica* **107**: 209–238.
30. **Brosius, J.** 1999. RNAs from all categories generate retrosequences that may be exapted as novel genes or regulatory elements. *Gene* **238**:115–134.
31. **Brosius, J., and S. J. Gould.** 1992. On "genomenclature": a comprehensive (and respectful) taxonomy for pseudogenes and other "junk DNA." *Proc. Natl. Acad. Sci. USA* **89**: 10706–10710.
32. **Brosius, J., and H. Tiedge.** 1995. Reverse transcriptase: mediator of genomic plasticity. *Virus Genes* **11**:163–179.
33. **Bureau, T. E., P. C. Ronald, and S. R. Wessler.** 1996. A computer-based systematic survey reveals the predominance of small inverted-repeat elements in wild-type rice genes. *Proc. Natl. Acad. Sci. USA* **93**:8524–8529.
34. **Bureau, T. E., S. E. White, and S. R. Wessler.** 1994. Transduction of a cellular gene by a plant retroelement. *Cell* **77**: 479–480.
35. **Burwinkel, B., and M. W. Kilimann.** 1998. Unequal homologous recombination between LINE-1 elements as a mutational mechanism in human genetic disease. *J. Mol. Biol.* **277**: 513–517.
36. **Caceres, M., J. M. Ranz, A. Barbadilla, M. Long, and A. Ruiz.** 1999. Generation of a widespread *Drosophila* inversion by a transposable element. *Science* **285**:415–418.
37. **Cambareri, E. B., R. Aisner, and J. Carbon.** 1998. Structure of the chromosome VII centromere region in *Neurospora crassa:* degenerate transposons and simple repeats. *Mol. Cell. Biol.* **18**:5465–5477.
38. **Charlesworth, B., C. H. Langley, and P. D. Sniegowski.** 1997. Transposable element distributions in Drosophila. *Genetics* **147**:1993–1995.
39. **Charlesworth, B., P. Sniegowski, and W. Stephan.** 1994. The evolutionary dynamics of repetitive DNA in eukaryotes. *Nature* **371**:215–220.
40. **Choo, K. H.** 2000. Centromerization. *Trends Cell Biol.* **10**: 182–188.
41. **Chu, W. M., R. Ballard, B. W. Carpick, B. R. Williams, and C. W. Schmid.** 1998. Potential Alu function: regulation of the activity of double-stranded RNA-activated kinase PKR. *Mol. Cell. Biol.* **18**:58–68.
42. **Cianciolo, G. J., T. D. Copeland, S. Oroszlan, and R. Snyderman.** 1985. Inhibition of lymphocyte proliferation by a synthetic peptide homologous to retroviral envelope proteins. *Science* **230**:453–455.
43. **Civardi, L., Y. Xia, K. J. Edwards, P. S. Schnable, and B. J. Nikolau.** 1994. The relationship between genetic and physical distances in the cloned a1-sh2 interval of the *Zea mays* L. genome. *Proc. Natl. Acad. Sci. USA* **91**:8268–8272.
44. **Clarke, L., H. Amstutz, B. Fishel, and J. Carbon.** 1986. Analysis of centromeric DNA in the fission yeast *Schizosaccharomyces pombe. Proc. Natl. Acad. Sci. USA* **83**:8253–8257.
45. **Clegg, M. T., and M. L. Durbin.** 2000. Flower color variation: a model for the experimental study of evolution. *Proc. Natl. Acad. Sci. USA* **97**:7016–7023.
46. **Cogoni, C., and G. Macino.** 1999. Gene silencing in *Neurospora crassa* requires a protein homologous to RNA-dependent RNA polymerase. *Nature* **399**:166–169.
47. **Colot, V., V. Haedens, and J. L. Rossignol.** 1998. Extensive, nonrandom diversity of excision footprints generated by *Ds*-like transposon *Ascot-1* suggests new parallels with V(D)J recombination. *Mol. Cell. Biol.* **18**:4337–4346.
48. **Consortium, A. S.** 2000. The complete sequence of a heterochromatic island from a higher eukaryote. The Cold Spring Harbor Laboratory: Washington University Genome Sequencing Center, and PE Biosystems Arabidopsis Sequencing Consortium. *Cell* **100**:377–386.
49. **Copenhaver, G. P., and D. Preuss.** 1999. Centromeres in the genomic era: unraveling paradoxes. *Curr. Opin. Plant Biol.* **2**:104–108.
50. **Corces, V. G., and P. K. Geyer.** 1991. Interactions of retrotransposons with the host genome. *Trends Genet.* **7**:86–90.
51. **Cottarel, G., J. H. Shero, P. Hieter, and J. H. Hegemann.** 1989. A 125-base-pair CEN6 DNA fragment is sufficient for complete meiotic and mitotic centromere functions in *Saccharomyces cerevisiae. Mol. Cell. Biol.* **9**:3342–3349.
52. **Craig, N. L.** 1997. Target site selection in transposition. *Annu. Rev. Biochem.* **66**:437–474.
53. **Cresse, A. D., S. H. Hulbert, W. E. Brown, J. R. Lucas, and J. L. Bennetzen.** 1995. Mu1-related transposable elements of maize preferentially insert into low copy number DNA. *Genetics* **140**:315–324.
54. **Currie, D. B., T. F. Mackay, and L. Partridge.** 1998. Pervasive effects of *P* element mutagenesis on body size in *Drosophila melanogaster. Genet. Res.* **72**:19–24.
55. **Cuypers, H., S. Dash, P. A. Peterson, H. Saedler, and A. Gierl.** 1988. The defective *En-I102* element encodes a product reducing the mutability of the *En-Spm* system of *Zea mays. EMBO J.* **7**:2953–2960.
56. **Daboussi, M. J.** 1997. Fungal transposable elements and genome evolution. *Genetica* **100**:253–260.
57. **Daniels, S. B., A. Chovnick, and M. G. Kidwell.** 1989. Hybrid dysgenesis in *Drosophila simulans* lines transformed with autonomous *P* elements. *Genetics* **121**:281–291.
58. **Daniels, S. B., M. McCarron, C. Love, and A. Chovnick.**

1985. Dysgenesis-induced instability of rosy locus transformation in *Drosophila melanogaster:* analysis of excision events and the selective recovery of control element deletions. *Genetics* **109:**95–117.

59. **Deininger, P. L., and M. A. Batzer.** 1999. Alu repeats and human disease. *Mol. Genet. Metab.* **67:**183–193.
60. **Dimitri, P., B. Arca, L. Berghella, and E. Mei.** 1997. High genetic instability of heterochromatin after transposition of the LINE-like I factor in *Drosophila melanogaster. Proc. Natl. Acad. Sci. USA* **94:**8052–8057.
61. **Dobie, K., M. Mehtali, M. McClenaghan, and R. Lathe.** 1997. Variegated gene expression in mice. *Trends Genet.* **13:** 127–130.
62. **Dominguez, A., and J. Albornoz.** 1996. Rates of movement of transposable elements in *Drosophila melanogaster. Mol. Gen. Genet.* **251:**130–138.
63. **Dominguez, A., and J. Albornoz.** 1999. Structural instability of 297 element in *Drosophila melanogaster. Genetica* **105:** 239–248.
64. **Dong, F., J. T. Miller, S. A. Jackson, G. L. Wang, P. C. Ronald, and J. Jiang.** 1998. Rice (*Oryza sativa*) centromeric regions consist of complex DNA. *Proc. Natl. Acad. Sci. USA* **95:**8135–8140.
65. **Donlin, M. J., D. Lisch, and M. Freeling.** 1995. Tissue-specific accumulation of MURB, a protein encoded by MuDR, the autonomous regulator of the Mutator transposable element family. *Plant Cell* **7:**1989–2000.
66. **Doolittle, R. F., D. F. Feng, M. S. Johnson, and M. A. McClure.** 1989. Origins and evolutionary relationships of retroviruses. *Q. Rev. Biol.* **64:**1–30.
67. **Dorer, D. R., and S. Henikoff.** 1994. Expansions of transgene repeats cause heterochromatin formation and gene silencing in *Drosophila. Cell* **77:**993–1002.
68. **Doseff, A., R. Martienssen, and V. Sundaresan.** 1991. Somatic excision of the *Mu1* transposable element of maize. *Nucleic Acids Res.* **19:**579–584.
69. **Dupressoir, A., and T. Heidmann.** 1996. Germ line-specific expression of intracisternal A-particle retrotransposons in transgenic mice. *Mol. Cell. Biol.* **16:**4495–4503.
70. **Earnshaw, W. C., K. F. Sullivan, P. S. Machlin, C. A. Cooke, D. A. Kaiser, T. D. Pollard, N. F. Rothfield, and D. W. Cleveland.** 1987. Molecular cloning of cDNA for CENP-B, the major human centromere autoantigen. *J. Cell Biol.* **104:** 817–829.
71. **Eberl, D. F., B. J. Duyf, and A. J. Hilliker.** 1993. The role of heterochromatin in the expression of a heterochromatic gene, the rolled locus of *Drosophila melanogaster. Genetics* **134:** 277–292.
72. **Eden, S., T. Hashimshony, I. Keshet, H. Cedar, and A. W. Thorne.** 1998. DNA methylation models histone acetylation. *Nature* **394:**842.
73. **Eichler, E. E., M. L. Budarf, M. Rocchi, L. L. Deaven, N. A. Doggett, A. Baldini, D. L. Nelson, and H. W. Mohrenweiser.** 1997. Interchromosomal duplications of the adrenoleukodystrophy locus: a phenomenon of pericentromeric plasticity. *Hum. Mol. Genet.* **6:**991–1002.
74. **Engels, W. R.** 1989. P elements in *Drosophila,* p. 437–484. *In* D. E. Berg and M. M. Howe (ed.), *Mobile DNA.* American Society for Microbiology, Washington, D.C.
75. **Engels, W. R.** 1996. *P* elements in *Drosophila,* p. 103–123. *In* H. Saedler and A. Gierl (ed.), *Transposable Elements.* Springer-Verlag, Berlin, Germany.
76. **Engels, W. R., D. M. Johnson-Schlitz, W. B. Eggleston, and J. Sved.** 1990. High-frequency *P* element loss in *Drosophila* is homolog dependent. *Cell* **62:**515–525.
77. **Engels, W. R., and C. R. Preston.** 1984. Formation of chromosome rearrangements by *P* factors in *Drosophila. Genetics* **107:**657–678.
78. **Esnault, C., J. Maestre, and T. Heidmann.** 2000. Human *LINE* retrotransposons generate processed pseudogenes. *Nat. Genet.* **24:**363–367.
79. **Evgen'ev, M. B., H. Zelentsova, H. Poluectova, G. T. Lyozin, V. Veleikodvorskaja, K. I. Pyatkov, L. Zhivotovski, and M. G. Kidwell.** 2000. Mobile elements and chromosomal evolution in the *virilis* group of *Drosophila. Proc. Natl. Acad. Sci. USA* **97:**11337–11342.
80. **Fanti, L., D. R. Dorer, M. Berloco, S. Henikoff, and S. Pimpinelli.** 1998. Heterochromatin protein 1 binds transgene arrays. *Chromosoma* **107:**286–292.
81. **Farkas, G., J. Gausz, M. Galloni, G. Reuter, H. Gyurkovics, and F. Karch.** 1994. The *Trithorax*-like gene encodes the *Drosophila* GAGA factor. *Nature* **371:**806–808.
82. **Fedoroff, N.** 1989. Maize transposable elements, p. 375–411. In D. E. Berg and M. M. Howe (ed.), *Mobile DNA.* American Society for Microbiology, Washington, D.C.
83. **Fedoroff, N., S. Wessler, and M. Shure.** 1983. Isolation of the transposable maize controlling elements *Ac* and *Ds. Cell* **35:**235–242.
84. **Feschotte, C., and C. Mouches.** 2000. Evidence that a family of miniature inverted-repeat transposable elements (MITEs) from the *Arabidopsis thaliana* genome has arisen from a pogo-like DNA transposon. *Mol. Biol. Evol.* **17:**730–737.
85. **Fincham, J. R. S., and G. R. K. Sastry.** 1974. Controlling elements in maize. *Annu. Rev. Genet.* **8:**15–50.
86. **Finnegan, D. J.** 1992. Transposable elements, p. 1096–1107. *In* D. L. Lindsley and G. Zimm (ed.), *The Genome of* Drosophila melanogaster. Academic Press, New York, N.Y.
87. **Finnegan, E. J., W. J. Peacock, and E. S. Dennis.** 1996. Reduced DNA methylation in *Arabidopsis thaliana* results in abnormal plant development. *Proc. Natl. Acad. Sci. USA* **93:** 8449–8454.

87a. **Fire, A., S. Xu, M. K. Montgomery, S. A. Kostas, S. E. Driver, and C. C. Mello.** 1998. Potent and specific genetic interference by double-stranded RNA in *Caenorhabditis elegans. Nature* **391:**806–811.

88. **Firtel, R. A.** 1989. Mobile elements in the cellular slime mold *Dictyostelium discoideum,* p. 557–566. *In* D. E. Berg and M. M. Howe (ed.), *Mobile DNA.* American Society for Microbiology, Washington, D.C.
89. **Fishel, B., H. Amstutz, M. Baum, J. Carbon, and L. Clarke.** 1988. Structural organization and functional analysis of centromeric DNA in the fission yeast *Schizosaccharomyces pombe. Mol. Cell. Biol.* **8:**754–763.
90. **Foss, K. B., B. Landmark, B. S. Skalhegg, K. Tasken, E. Jellum, V. Hansson, and T. Jahnsen.** 1994. Characterization of in-vitro-translated human regulatory and catalytic subunits of cAMP-dependent protein kinases. *Eur. J. Biochem.* **220:** 217–223.
91. **Fowler, K. J., D. F. Hudson, L. A. Salamonsen, S. R. Edmondson, E. Earle, M. C. Sibson, and K. H. Choo.** 2000. Uterine dysfunction and genetic modifiers in centromere protein B-deficient mice. *Genome Res.* **10:**30–41.
92. **Fridell, R. A., A. M. Pret, and L. L. Searles.** 1990. A retrotransposon *412* insertion within an exon of the *Drosophila melanogaster vermilion* gene is spliced from the precursor RNA. *Genes Dev.* **4:**559–566.

92a. **Fugmann, S. D., A. I. Lee, P. E. Shockett, I. J. Villey, and D. G. Schatz.** 2000. The RAG proteins and V(D)J recombination: complexes, ends, and transposition. *Annu. Rev. Immunol.* **18:**495–527.

93. **Garber, K., I. Bilic, O. Pusch, J. Tohme, A. Bachmair, D. Schweizer, and V. Jantsch.** 1999. The Tpv2 family of retro-

transposons of *Phaseolus vulgaris:* structure, integration characteristics, and use for genotype classification. *Plant Mol. Biol.* **39:**797–807.

94. **Golovkina, T. V., A. Chervonsky, J. P. Dudley, and S. R. Ross.** 1992. Transgenic mouse mammary tumor virus superantigen expression prevents viral infection. *Cell* **69:**637–645.
95. **Goodier, J. L., E. M. Ostertag, and H. H. Kazazian, Jr.** 2000. Transduction of 3′-flanking sequences is common in *L1* retrotransposition. *Hum. Mol. Genet.* **9:**653–657.
96. **Gottschling, D. E., O. M. Aparicio, B. L. Billington, and V. A. Zakian.** 1990. Position effect at *S. cerevisiae* telomeres: reversible repression of Pol II transcription. *Cell* **63:**751–762.
97. **Gray, Y. H.** 2000. It takes two transposons to tango: transposable-element-mediated chromosomal rearrangements. *Trends Genet.* **16:**461–468.
98. **Greene, B., R. Walko, and S. Hake.** 1994. Mutator insertions in an intron of the maize *knotted1* gene result in dominant suppressible mutations. *Genetics* **138:**1275–1285.
99. **Gregory, T. R., and P. D. Hebert.** 1999. The modulation of DNA content: proximate causes and ultimate consequences. *Genome Res.* **9:**317–324.
100. **Guo, S., and K. J. Kemphues.** 1995. par-1, a gene required for establishing polarity in *C. elegans* embryos, encodes a putative Ser/Thr kinase that is asymmetrically distributed. *Cell* **81:**611–620.
101. **Hambor, J. E., J. Mennone, M. E. Coon, J. H. Hanke, and P. Kavathas.** 1993. Identification and characterization of an *Alu*-containing, T-cell-specific enhancer located in the last intron of the human *CD8 alpha* gene. *Mol. Cell. Biol.* **13:** 7056–7070.
102. **Harendza, C. J., and L. F. Johnson.** 1990. Polyadenylylation signal of the mouse thymidylate synthase gene was created by insertion of an *L1* repetitive element downstream of the open reading frame. *Proc. Natl. Acad. Sci. USA* **87:** 2531–2535.
103. **Harper, L. C., and W. Z. Cande.** Mapping a new frontier; development of integrated cytogenetic maps in plants. *Func. Integr. Genomics,* in press.
104. **Henikoff, S., E. A. Greene, S. Pietrokovski, P. Bork, T. K. Attwood, and L. Hood.** 1997. Gene families: the taxonomy of protein paralogs and chimeras. *Science* **278:**609–614.
105. **Henikoff, S., and M. A. Matzke.** 1997. Exploring and explaining epigenetic effects. *Trends Genet.* **13:**293–295.
106. **Higashiyama, T., Y. Noutoshi, M. Fujie, and T. Yamada.** 1997. *Zepp,* a *LINE*-like retrotransposon accumulated in the *Chlorella* telomeric region. *EMBO J.* **16:**3715–3723.
107. **Hilliker, A. J., R. Appels, and A. Schalet.** 1980. The genetic analysis of *D. melanogaster* heterochromatin. *Cell* **21:** 607–619.
108. **Hiom, K., M. Melek, and M. Gellert.** 1998. DNA transposition by the RAG1 and RAG2 proteins: a possible source of oncogenic translocations. *Cell* **94:**463–470.
109. **Hirochika, H., H. Okamoto, and T. Kakutani.** 2000. Silencing of retrotransposons in *Arabidopsis* and reactivation by the *ddm1* mutation. *Plant Cell* **12:**357–369.
110. **Holmes, S. E., B. A. Dombroski, C. M. Krebs, C. D. Boehm, and H. H. Kazazian, Jr.** 1994. A new retrotransposable human L1 element from the *LRE2* locus on chromosome *1q* produces a chimaeric insertion. *Nat. Genet.* **7:**143–148.
111. **Hunter, C. P.** 2000. Gene silencing: shrinking the black box of RNAi. *Curr. Biol.* **10:**137–140.
112. **Ivanova, A. V., M. J. Bonaduce, S. V. Ivanov, and A. J. Klar.** 1998. The chromo and SET domains of the Clr4 protein are essential for silencing in fission yeast. *Nat. Genet.* **19:** 192–195.
113. **Iwasa, Y.** 1998. The conflict theory of genomic imprinting: how much can be explained? *Curr. Top. Dev. Biol.* **40:** 255–293.
114. **Jacobsen, S. E., H. Sakai, E. J. Finnegan, X. Cao, and E. M. Meyerowitz.** 2000. Ectopic hypermethylation of flower-specific genes in *Arabidopsis. Curr. Biol.* **10:**179–186.
115. **Jakubczak, J. L., W. D. Burke, and T. H. Eickbush.** 1991. Retrotransposable elements *R1* and *R2* interrupt the rRNA genes of most insects. *Proc. Natl. Acad. Sci. USA* **88:** 3295–3299.
116. **Jakubczak, J. L., Y. Xiong, and T. H. Eickbush.** 1990. Type I (R1) and type II (R2) ribosomal DNA insertions of *Drosophila melanogaster* are retrotransposable elements closely related to those of *Bombyx mori. J. Mol. Biol.* **212:**37–52.
117. **Jensen, S., L. Cavarec, M. P. Gassama, and T. Heidmann.** 1995. Defective *I* elements introduced into *Drosophila* as transgenes can regulate reactivity and prevent I-R hybrid dysgenesis. *Mol. Gen. Genet.* **248:**381–390.
118. **Jiang, J., S. Nasuda, F. Dong, C. W. Scherrer, S. S. Woo, R. A. Wing, B. S. Gill, and D. C. Ward.** 1996. A conserved repetitive DNA element located in the centromeres of cereal chromosomes. *Proc. Natl. Acad. Sci. USA* **93:**14210–14213.
119. **Joblonka, E., and M. J. Lamb.** 1995. *Epigenetic Inheritance and Evolution.* Oxford Univ. Press, Oxford, UK.
120. **John, B.** 1988. The biology of heterochromatin, p. 1–128. *In* R. S. Verma (ed.), *Heterochromatin, Molecular and Structural Aspects.* Cambridge University Press, Cambridge, UK.
121. **Jones, P. A.** 1985. Altering gene expression with 5-azacytidine. *Cell* **40:**485–486.
122. **Jordan, I. K., and J. F. McDonald.** 1998. Evidence for the role of recombination in the regulatory evolution of *Saccharomyces cerevisiae Ty* elements. *J. Mol. Evol.* **47:**14–20.
123. **Jurka, J., and V. V. Kapitonov.** 1999. Sectorial mutagenesis by transposable elements. *Genetica* **107:**239–248.
124. **Kakutani, T., J. A. Jeddeloh, S. K. Flowers, K. Munakata, and E. J. Richards.** 1996. Developmental abnormalities and epimutations associated with DNA hypomethylation mutations. *Proc. Natl. Acad. Sci. USA* **93:**12406–12411.
125. **Kalendar, R., J. Tanskanen, S. Immonen, E. Nevo, and A. H. Schulman.** 2000. Genome evolution of wild barley (*Hordeum spontaneum*) by *BARE-1* retrotransposon dynamics in response to sharp microclimatic divergence. *Proc. Natl. Acad. Sci. USA* **97:**6603–6607.
126. **Karpen, G. H., and R. C. Allshire.** 1997. The case for epigenetic effects on centromere identity and function. *Trends Genet.* **13:**489–496.
127. **Kazazian, H. H., Jr.** 1999. An estimated frequency of endogenous insertional mutations in humans. *Nat. Genet.* **22:**130.
128. **Kazazian, H. H., Jr., and J. V. Moran.** 1998. The impact of *L1* retrotransposons on the human genome. *Nat. Genet.* **19:** 19–24.
129. **Kazazian, H. H., Jr., C. Wong, H. Youssoufian, A. F. Scott, D. G. Phillips, and S. E. Antonarakis.** 1988. Haemophilia A resulting from de novo insertion of *L1* sequences represents a novel mechanism for mutation in man. *Nature* **332:**164–166.
130. **Ke, N., P. A. Irwin, and D. F. Voytas.** 1997. The pheromone response pathway activates transcription of *Ty5* retrotransposons located within silent chromatin of *Saccharomyces cerevisiae. EMBO J.* **16:**6272–6280.
131. **Ketting, R. F., T. H. Haverkamp, H. G. van Luenen, and R. H. Plasterk.** 1999. *Mut-7* of *C. elegans,* required for transposon silencing and RNA interference, is a homolog of Werner syndrome helicase and RNaseD. *Cell* **99:**133–141.
132. **Kidwell, M. G., J. F. Kidwell, and J. A. Sved.** 1977. Hybrid dysgenesis in *Drosophila melanogaster:* a syndrome of aberrant traits including mutation, sterility & male recombination. *Genetics* **36:**813–833.

133. **Kidwell, M. G., and D. Lisch.** 1997. Transposable elements as sources of variation in animals and plants. *Proc. Natl. Acad. Sci. USA* **94:**7704–7711.
134. **Kidwell, M. G., and D. R. Lisch.** 2000. Transposable elements and host genome evolution. *Trends Ecol. Evol.* **15:** 95–99.
135. **Kim, J. M., S. Vanguri, J. D. Boeke, A. Gabriel, and D. F. Voytas.** 1998. Transposable elements and genome organization: a comprehensive survey of retrotransposons revealed by the complete *Saccharomyces cerevisiae* genome sequence. *Genome Res.* **8:**464–478.
136. **Kimura, R. H., P. V. Choudary, and C. W. Schmid.** 1999. Silk worm *Bm1* SINE RNA increases following cellular insults. *Nucleic Acids Res.* **27:**3380–3387.
137. **King, M. C., and A. C. Wilson.** 1975. Evolution at two levels in humans and chimpanzees. *Science* **188:**107–116.
138. **Kipling, D., and P. E. Warburton.** 1997. Centromeres, CENP-B and *Tigger* too. *Trends Genet.* **13:**141–145.
139. **Kirschner, M., and J. Gerhart.** 1998. Evolvability. *Proc. Natl. Acad. Sci. USA* **95:**8420–8427.
140. **Klobutcher, L. A., and G. Herrick.** 1995. Consensus inverted terminal repeat sequence of Paramecium IESs: resemblance to termini of *Tc1*-related and *Euplotes Tec* transposons. *Nucleic Acids Res.* **23:**2006–2013.
141. **Klobutcher, L. A., and G. Herrick.** 1997. Developmental genome reorganization in ciliated protozoa: the transposon link. *Proc. Nucleic Acid Res. Mol. Biol.* **56:**1–62.
142. **Kloeckener-Gruissem, B., and M. Freeling.** 1995. Transposon-induced promoter scrambling: a mechanism for the evolution of new alleles. *Proc. Natl. Acad. Sci. USA* **92:** 1836–1840.
143. **Kloeckener-Gruissem, B., J. M. Vogel, and M. Freeling.** 1992. The TATA box promoter region of maize *Adh1* affects its organ-specific expression. *EMBO J.* **11:**157–166.
144. **Kobayashi, S., T. Hirano, M. Kakinuma, and T. Uede.** 1993. Transcriptional repression and differential splicing of FAS mRNA by early transposon (ETn) insertion in autoimmune LPR mice. *Biochem. Biophys. Res. Commun.* **191:**617–624.
145. **Kumar, A., and J. L. Bennetzen.** 1999. Plant retrotransposons. *Annu. Rev. Genet.* **33:**479–532.
146. **Lahn, B. T., and D. C. Page.** 1999. Four evolutionary strata on the human X chromosome. *Science* **286:**964–967.
147. **Lahn, B. T., and D. C. Page.** 1999. Retroposition of autosomal mRNA yielded testis-specific gene family on human *Y* chromosome. *Nat. Genet.* **21:**429–433.
148. **Lai, C., and T. F. Mackay.** 1993. Mapping and characterization of *P*-element-induced mutations at quantitative trait loci in *Drosophila melanogaster. Genet. Res.* **61:**177–193.
149. **Langley, C. H., E. Montgomery, R. Hudson, N. Kaplan, and B. Charlesworth.** 1988. On the role of unequal exchange on the containment of transposable element copy number. *Genet. Res.* **52:**223–235.
150. **Le, Q. H., S. Wright, Z. Yu, and T. Bureau.** 2000. Transposon diversity in *Arabidopsis thaliana. Proc. Natl. Acad. Sci. USA* **97:**7376–7381.
151. **Lee, C. C., Y. M. Mul, and D. C. Rio.** 1996. The *Drosophila P*-element *KP* repressor protein dimerizes and interacts with multiple sites on *P*-element DNA. *Mol. Cell. Biol.* **16:** 5616–5622.
152. **Lee, J. K., J. A. Huberman, and J. Hurwitz.** 1997. Purification and characterization of a CENP-B homologue protein that binds to the centromeric K-type repeat DNA of *Schizosaccharomyces pombe. Proc. Natl. Acad. Sci. USA* **94:** 8427–8432.
153. **Levis, R. W., R. Ganesan, K. Houtchens, L. A. Tolar, and F. M. Sheen.** 1993. Transposons in place of telomeric repeats at a *Drosophila* telomere. *Cell* **75:**1083–1093.
154. **Lewis, S. M., and G. E. Wu.** 1997. The origins of V(D)J recombination. *Cell* **88:**159–162.
155. **Li, E., C. Beard, and R. Jaenisch.** 1993. Role for DNA methylation in genomic imprinting. *Nature* **366:**362–365.
156. **Li, E., T. H. Bestor, and R. Jaenisch.** 1992. Targeted mutation of the DNA methyltransferase gene results in embryonic lethality. *Cell* **69:**915–926.
157. **Li, W.-H.** 1997. *Molecular Evolution.* Sinauer Associates Inc., Sunderland, Mass.
158. **Li, X., and M. Noll.** 1994. Evolution of distinct developmental functions of three *Drosophila* genes by acquisition of different cis-regulatory regions. *Nature* **367:**83–87.
159. **Liao, G. C., E. J. Rehm, and G. M. Rubin.** 2000. Insertion site preferences of the *P* transposable element in *Drosophila melanogaster. Proc. Natl. Acad. Sci. USA* **97:**3347–3351.
160. **Lim, J. K., and M. J. Simmons.** 1994. Gross chromosome rearrangements mediated by transposable elements in *Drosophila melanogaster. Bioessays* **16:**269–275.
161. **Lin, X., S. Kaul, S. Rounsley, T. P. Shea, M. I. Benito, C. D. Town, C. Y. Fujii, T. Mason, C. L. Bowman, M. Barnstead, T. V. Feldblyum, C. R. Buell, K. A. Ketchum, J. Lee, C. M. Ronning, H. L. Koo, K. S. Moffat, L. A. Cronin, M. Shen, G. Pai, S. Van Aken, L. Umayam, L. J. Tallon, J. E. Gill, J. C. Venter, et al.** 1999. Sequence and analysis of chromosome 2 of the plant *Arabidopsis thaliana. Nature* **402:**761–768.
162. **Lister, C., D. Jackson, and C. Martin.** 1993. Transposon-induced inversion in *Antirrhinum* modifies *nivea* gene expression to give a novel flower color pattern under the control of cycloidearadialis. *Plant Cell* **5:**1541–1553.
163. **Liu, W. M., W. M. Chu, P. V. Choudary, and C. W. Schmid.** 1995. Cell stress and translational inhibitors transiently increase the abundance of mammalian *SINE* transcripts. *Nucleic Acids Res.* **23:**1758–1765.
164. **Lower, R., J. Lower, and R. Kurth.** 1996. The viruses in all of us: characteristics and biological significance of human endogenous retrovirus sequences. *Proc. Natl. Acad. Sci. USA* **93:**5177–5184.
165. **Lozovskaya, E. R., V. S. Scheinker, and M. B. Evgen'ev.** 1990. A hybrid dysgenesis syndrome in *Drosophila virilis. Genetics* **126:**619–623.
166. **Lu, B. Y., P. C. R. Emtage, B. J. Duyf, A. J. Hilliker, and J. C. Eissenberg.** 2000. Heterochromatin protein 1 is required for the normal expression of two heterochromatin genes in *Drosophila. Genetics* **155:**699–708.
167. **Lyman, R. F., F. Lawrence, S. V. Nuzhdin, and T. F. Mackay.** 1996. Effects of single *P*-element insertions on bristle number and viability in *Drosophila melanogaster. Genetics* **143:** 277–292.
168. **Lyon, M. F.** 2000. *LINE-1* elements and *X* chromosome inactivation: a function for "junk" DNA? *Proc. Natl. Acad. Sci. USA* **97:**6248–6249.
169. **Lyon, M. F.** 1998. X-chromosome inactivation: a repeat hypothesis. *Cytogenet. Cell Genet.* **80:**133–137.
170. **Mackay, T. F., R. F. Lyman, and M. S. Jackson.** 1992. Effects of *P* element insertions on quantitative traits in *Drosophila melanogaster. Genetics* **130:**315–332.
171. **Mackay, T. F. C.** 1986. Transposable element-induced fitness mutations in *Drosophila melanogaster. Genet. Res.* **48:** 77–87.
172. **Mackay, T. F. C.** 1987. Transposable element-induced polygenic mutations in *Drosophila melanogaster. Genet. Res.* **49:** 225–233.
173. **Marillonnet, S., and S. R. Wessler.** 1997. Retrotransposon

insertion into the maize waxy gene results in tissue-specific RNA processing. *Plant Cell* 9:967–978.

174. **Martienssen, R. A.** 1996. Epigenetic silencing of *Mu* transposable elements in maize, p. 593–608. *In* V. E. A. Russo, R. A. Martienssen, and A. D. Riggs (ed.), *Epigenetic Mechanisms of Gene Regulation.* Cold Spring Harbor Laboratory Press, Cold Spring Harbor, N.Y.
175. **Mathiopoulos, K. D., A. della Torre, V. Predazzi, V. Petrarca, and M. Coluzzi.** 1998. Cloning of inversion breakpoints in the *Anopheles gambiae* complex traces a transposable element at the inversion junction. *Proc. Natl. Acad. Sci. USA* **95:**12444–12449.
176. **Matzke, M. A., and A. J. Matzke.** 1998. Epigenetic silencing of plant transgenes as a consequence of diverse cellular defense responses. *Cell. Mol. Life Sci.* **54:**94–103.
177. **Matzke, M. A., M. F. Mette, W. Aufsatz, J. Jakowitsch, and A. J. Matzke.** 1999. Host defenses to parasitic sequences and the evolution of epigenetic control mechanisms. *Genetica* **107:**271–287.
178. **McClintock, B.** 1951. Chromosome organization and genic expression. *Cold Spring Harbor Symp. Quant. Biol.* **16:** 13–47.
179. **McClintock, B.** 1978. Mechanisms that rapidly reorganize the genome. *Stadler Symp.* **10:**25–47.
180. **McDonald, J. F.** 1998. Transposable elements, gene silencing and macroevolution. *Trends Ecol. Evol.* **13:**94–95.
181. **McDonald, J. F., L. V. Matyunina, S. Wilson, I. K. Jordan, N. J. Bowen, and W. J. Miller.** 1997. LTR retrotransposons and the evolution of eukaryotic enhancers. *Genetica* **100:** 3–13.
182. **McNaughton, J. C., G. Hughes, W. A. Jones, P. A. Stockwell, H. J. Klamut, and G. B. Petersen.** 1997. The evolution of an intron: analysis of a long, deletion-prone intron in the human *dystrophin* gene. *Genomics* **40:**294–304.
183. **Miki, Y., I. Nishisho, A. Horii, Y. Miyoshi, J. Utsunomiya, K. W. Kinzler, B. Vogelstein, and Y. Nakamura.** 1992. Disruption of the *APC* gene by a retrotransposal insertion of *L1* sequence in a colon cancer. *Cancer Res.* **52:**643–645.
184. **Miller, W. J., S. Hagemann, E. Reiter, and W. Pinsker.** 1992. *P*-element homologous sequences are tandemly repeated in the genome of *Drosophila guanche. Proc. Natl. Acad. Sci. USA* **89:**4018–4022.
185. **Miller, W. J., J. F. McDonald, D. Nouaud, and D. Anxolabehere.** 1999. Molecular domestication: more than a sporadic episode in evolution. *Genetica* **107:**197–207.
186. **Miller, W. J., J. F. McDonald, and W. Pinsker.** 1997. Molecular domestication of mobile elements. *Genetica* **100:** 261–270.
187. **Misra, S., and D. C. Rio.** 1990. Cytotype control of *Drosophila P* element transposition: the 66 kd protein is a repressor of transposase activity. *Cell* **62:**269–284.
188. **Moran, J. V., R. J. DeBerardinis, and H. H. Kazazian, Jr.** 1999. Exon shuffling by *L1* retrotransposition. *Science* **283:** 1530–1534.
189. **Mourrain, P., C. Beclin, T. Elmayan, F. Feuerbach, C. Godon, J. B. Morel, D. Jouette, A. M. Lacombe, S. Nikic, N. Picault, K. Remoue, M. Sanial, T. A. Vo, and H. Vaucheret.** 2000. *Arabidopsis SGS2* and *SGS3* genes are required for posttranscriptional gene silencing and natural virus resistance. *Cell* **101:**533–542.
190. **Mustajoki, S., H. Ahola, P. Mustajoki, and R. Kauppinen.** 1999. Insertion of *Alu* element responsible for acute intermittent porphyria. *Hum. Mutat.* **13:**431–438.
191. **Napoli, C., C. Lemieux, and R. Jorgensen.** 1990. Introduction of a chimeric *chalcone synthase* gene into petunia results in reversible co-suppression of homologous genes in trans. *Plant Cell* **2:**279–289.
192. **Ng, H. H., P. Jeppesen, and A. Bird.** 2000. Active repression of methylated genes by the chromosomal protein MBD1. *Mol. Cell. Biol.* **20:**1394–1406.
193. **Nihrane, A., I. Lebedeva, M. S. Lyu, K. Fujita, and J. Silver.** 1997. Secretion of a murine retroviral *Env* associated with resistance to infection. *J. Gen. Virol.* **78:**785–793.
194. **Nimmo, E. R., G. Cranston, and R. C. Allshire.** 1994. Telomere-associated chromosome breakage in fission yeast results in variegated expression of adjacent genes. *EMBO J.* **13:**3801–3811.
195. **Nouaud, D., and D. Anxolabéhère.** 1997. P element domestication: a stationary truncated *P* element may encode a 66-kDa repressor-like protein in the *Drosophila montium* species subgroup. *Mol. Biol. Evol.* **14:**1132–1144.
196. **Nouaud, D., B. Boeda, L. Levy, and D. Anxolabéhère.** 1999. A *P* element has induced intron formation in *Drosophila. Mol. Biol. Evol.* **16:**1503–1510.
197. **Nuzhdin, S. V., and T. F. Mackay.** 1994. Direct determination of retrotransposon transposition rates in *Drosophila melanogaster. Genet. Res.* **63:**139–144.
198. **Nuzhdin, S. V., E. G. Pasyukova, and T. F. Mackay.** 1997. Accumulation of transposable elements in laboratory lines of *Drosophila melanogaster. Genetica* **100:**167–175.
199. **Oettinger, M. A., D. G. Schatz, C. Gorka, and D. Baltimore.** 1990. RAG-1 and RAG-2, adjacent genes that synergistically activate V(D)J recombination. *Science* **248:**1517–1523.
200. **Okazaki, S., H. Ishikawa, and H. Fujiwara.** 1995. Structural analysis of TRAS1, a novel family of telomeric repeat-associated retrotransposons in the silkworm, *Bombyx mori. Mol. Cell. Biol.* **15:**4545–4552.
201. **Panning, B., and R. Jaenisch.** 1998. RNA and the epigenetic regulation of X chromosome inactivation. *Cell* **93:**305–308.
202. **Pardue, M. L., O. N. Danilevskaya, K. Lowenhaupt, F. Slot, and K. L. Traverse.** 1996. *Drosophila* telomeres: new views on chromosome evolution. *Trends Genet.* **12:**48–52.
203. **Pardue, M. L., and P. G. DeBaryshe.** 1999. *Drosophila* telomeres: two transposable elements with important roles in chromosomes. *Genetics* **107:**189–196.
204. **Pearce, S. R., U. Pich, G. Harrison, A. J. Flavell, J. S. Heslop-Harrison, I. Schubert, and A. Kumar.** 1996. The *Ty1-copia* group retrotransposons of *Allium cepa* are distributed throughout the chromosomes but are enriched in the terminal heterochromatin. *Chromosome Res.* **4:**357–364.
205. **Pickeral, O. K., W. Makaowski, M. S. Boguski, and J. D. Boeke.** 2000. Frequent human genomic DNA transduction driven by *LINE-1* retrotransposition. *Genome Res.* **10:** 411–415.
206. **Pimpinelli, S., M. Berloco, L. Fanti, P. Dimitri, S. Bonaccorsi, E. Marchetti, R. Caizzi, C. Caggese, and M. Gatti.** 1995. Transposable elements are stable structural components of *Drosophila melanogaster* heterochromatin. *Proc. Natl. Acad. Sci. USA* **92:**3804–3808.
207. **Prescott, D. M.** 1994. The DNA of ciliated protozoa. *Microbiol. Rev.* **58:**233–267.

207a. **Pryciak, P. M., and H. E. Varmus.** 1992. Fv-1 restriction and its effects on murine leukemia virus integration in vivo and in vitro. *J. Virol.* **66:**5959–5966.

208. **Rasmussen, H. B., and J. Clausen.** 1998. Large number of polymorphic nucleotides and a termination codon in the *env* gene of the endogenous human retrovirus ERV3. *Dis. Markers* **14:**127–133.
209. **Reik, W., and E. R. Maher.** 1997. Imprinting in clusters: lessons from Beckwith-Wiedemann syndrome. *Trends Genet.* **13:**330–334.

210. **Reinton, N., T. B. Haugen, S. Orstavik, B. S. Skalhegg, V. Hansson, T. Jahnsen, and K. Tasken.** 1998. The gene encoding the C gamma catalytic subunit of cAMP-dependent protein kinase is a transcribed retroposon. *Genomics* **49:** 290–297.

211. **Reiter, L. T., T. Liehr, B. Rautenstrauss, H. M. Robertson, and J. R. Lupski.** 1999. Localization of *mariner* DNA transposons in the human genome by PRINS. *Genome Res.* **9:** 839–843.

212. **Rhounim, L., J. L. Rossignol, and G. Faugeron.** 1992. Epimutation of repeated genes in *Ascobolus immersus. EMBO J.* **11:**4451–4457.

213. **Robertson, D. S.** 1978. Characterization of a mutator system in maize. *Mutat. Res.* **51:**21–28.

214. **Rubin, G. M., M. G. Kidwell, and P. M. Bingham.** 1982. The molecular basis of P-M hybrid dysgenesis: the nature of induced mutations. *Cell* **29:**987–994.

215. **Rushforth, A. M., and P. Anderson.** 1996. Splicing removes the *Caenorhabditis elegans* transposon *Tc1* from most mutant pre-mRNAs. *Mol. Cell. Biol.* **16:**422–429.

216. **Sankaranarayanan, K.** 1988. Mobile genetic elements, spontaneous mutations, amd the assessment of genetic radiation hazards in man. *In* M. E. Lambert, J. F. McDonald, and I. B. Weinstein (ed.), *Eukaryotic Transposable Elements as Mutagenic Agents.* Cold Spring Harbor Laboratory, Cold Spring Harbor, N.Y.

217. **SanMiguel, P., A. Tikhonov, Y. K. Jin, N. Motchoulskaia, D. Zakharov, A. Melake-Berhan, P. S. Springer, K. J. Edwards, M. Lee, Z. Avramova, and J. L. Bennetzen.** 1996. Nested retrotransposons in the intergenic regions of the maize genome. *Science* **274:**765–768.

218. **Schiefelbein, J. W., D. B. Furtek, H. K. Dooner, and O. E. Nelson, Jr.** 1988. Two mutations in a maize bronze-1 allele caused by transposable elements of the *Ac-Ds* family alter the quantity and quality of the gene product. *Genetics* **120:** 767–777.

219. **Schmid, C. W.** 1998. Does *SINE* evolution preclude *Alu* function? *Nucleic Acids Res.* **26:**4541–4550.

220. **Schwartz, A., D. C. Chan, L. G. Brown, R. Alagappan, D. Pettay, C. Disteche, B. McGillivray, A. de la Chapelle, and D. C. Page.** 1998. Reconstructing hominid *Y* evolution: *X*-homologous block, created by *X-Y* transposition, was disrupted by *Yp* inversion through *LINE-LINE* recombination. *Hum. Mol. Genet.* **7:**1–11.

221. **Seegmiller, A., and G. Herrick.** 1998. A short internal eliminated sequence with central conserved sequences interrupting the *LA-MSC* gene of the 81 locus in the hypotrichous ciliates *Oxytricha fallax* and *O. trifallax. J. Eukaryot. Microbiol.* **45:** 55–58.

222. **Seegmiller, A., K. R. Williams, R. L. Hammersmith, T. G. Doak, D. Witherspoon, T. Messick, L. L. Storjohann, and G. Herrick.** 1996. Internal eliminated sequences interrupting the *Oxytricha* 81 locus: allelic divergence, conservation, conversions, and possible transposon origins. *Mol. Biol. Evol.* **13:**1351–1362.

223. **Selinger, D. A., and V. L. Chandler.** 1999. Major recent and independent changes in levels and patterns of expression have occurred at the *b* gene, a regulatory locus in maize. *Proc. Natl. Acad. Sci. USA* **96:**15007–15012.

224. **Selker, E. U.** 1997. Epigenetic phenomena in filamentous fungi: useful paradigms or repeat-induced confusion? *Trends Genet.* **13:**296–301.

225. **Selker, E. U.** 1998. Trichostatin A causes selective loss of DNA methylation in *Neurospora. Proc. Natl. Acad. Sci. USA* **95:**9430–9435.

226. **Shapiro, J. A.** 1999. Transposable elements as the key to a 21st century view of evolution. *Genetica* **107:**171–179.

227. **Siebel, C. W., A. Admon, and D. C. Rio.** 1995. Soma-specific expression and cloning of PSI, a negative regulator of *P* element pre-mRNA splicing. *Genes Dev.* **9:**269–283.

228. **Siebel, C. W., L. D. Fresco, and D. C. Rio.** 1992. The mechanism of somatic inhibition of *Drosophila P*-element pre-mRNA splicing: multiprotein complexes at an exon pseudo-5′ splice site control U1 snRNP binding. *Genes Dev.* **6:** 1386–1401.

229. **Sijen, T., and J. M. Kooter.** 2000. Post-transcriptional gene-silencing: RNAs on the attack or on the defense? *Bioessays* **22:**520–531.

230. **Singer, M. F., V. Krek, J. P. McMillan, G. D. Swergold, and R. E. Thayer.** 1993. *LINE-1:* a human transposable element. *Gene* **135:**183–188.

231. **Skryabin, B. V., J. Kremerskothen, D. Vassilacopoulou, T. R. Disotell, V. V. Kapitonov, J. Jurka, and J. Brosius.** 1998. The *BC200* RNA gene and its neural expression are conserved in *Anthropoidea* (primates). *J. Mol. Evol.* **47:** 677–685.

232. **Smardon, A., J. M. Spoerke, S. C. Stacey, M. E. Klein, N. Mackin, and E. M. Maine.** 2000. EGO-1 is related to RNA-directed RNA polymerase and functions in germ-line development and RNA interference in *C. elegans. Curr. Biol.* **10:** 169–178.

233. **Smit, A. F.** 1999. Interspersed repeats and other mementos of transposable elements in mammalian genomes. *Curr. Opin. Genet. Dev.* **9:**657–663.

234. **Spanopoulou, E., F. Zaitseva, F. H. Wang, S. Santagata, D. Baltimore, and G. Panayotou.** 1996. The homeodomain region of *Rag-1* reveals the parallel mechanisms of bacterial and V(D)J recombination. *Cell* **87:**263–276.

235. **Spradling, A. C., D. M. Stern, I. Kiss, J. Roote, T. Laverty, and G. M. Rubin.** 1995. Gene disruptions using P transposable elements: an integral component of the *Drosophila* genome project. *Proc. Natl. Acad. Sci. USA* **92:**10824–10830.

236. **Stavenhagen, J. B., and D. M. Robins.** 1988. An ancient provirus has imposed androgen regulation on the adjacent mouse sex-limited protein gene. *Cell* **55:**247–254.

237. **Steinemann, M., and S. Steinemann.** 1998. Enigma of Y chromosome degeneration: neo-*Y* and neo-*X* chromosomes of *Drosophila miranda* a model for sex chromosome evolution. *Genetica* **103:**409–420.

238. **Sun, X., J. Wahlstrom, and G. Karpen.** 1997. Molecular structure of a functional *Drosophila* centromere. *Cell* **91:** 1007–1019.

239. **Surzycki, S. A., and W. R. Belknap.** 2000. Repetitive-DNA elements are similarly distributed on *Caenorhabditis elegans* autosomes. *Proc. Natl. Acad. Sci. USA* **97:**245–949.

240. **Tabara, H., M. Sarkissian, W. G. Kelly, J. Fleenor, A. Grishok, L. Timmons, A. Fire, and C. C. Mello.** 1999. The rde-1 gene, RNA interference, and transposon silencing in *C. elegans. Cell* **99:**123–132.

241. **Takahashi, H., S. Okazaki, and H. Fujiwara.** 1997. A new family of site-specific retrotransposons, SART1, is inserted into telomeric repeats of the silkworm, *Bombyx mori. Nucleic Acids Res.* **25:**1578–1584.

242. **Takahashi, S., Y. Inagaki, H. Satoh, A. Hoshino, and S. Iida.** 1999. Capture of a genomic HMG domain sequence by the *En/Spm*-related transposable element *Tpn1* in the Japanese morning glory. *Mol. Gen. Genet.* **261:**447–451.

243. **Talbert, L. E., and V. L. Chandler.** 1988. Characterization of a highly conserved sequence related to *mutator* transposable elements in maize. *Mol. Biol. Evol.* **5:**519–529.

244. **Tarchini, R., P. Biddle, R. Wineland, S. Tingey, and A. Rafal-**

ski. 2000. The complete sequence of 340 kb of DNA around the rice Adh1-adh2 region reveals interrupted colinearity with maize chromosome 4. *Plant Cell* **12:**381–391.

245. **Taruscio, D., and L. Manuelidis.** 1991. Integration site preferences of endogenous retroviruses. *Chromosoma* **101:** 141–156.
246. **Tate, P. H., and A. P. Bird.** 1993. Effects of DNA methylation on DNA-binding proteins and gene expression. *Curr. Opin. Genet. Dev.* **3:**226–231.
247. **Terrinoni, A., C. D. Franco, P. Dimitri, and N. Junakovic.** 1997. Intragenomic distribution and stability of transposable elements in euchromatin and heterochromatin of *Drosophila melanogaster:* non-LTR retrotransposon. *J. Mol. Evol.* **45:** 145–153.
248. **Timmons, L., and A. Fire.** 1998. Specific interference by ingested dsRNA. *Nature* **395:**854.
249. **Torti, C., L. M. Gomulski, D. Moralli, E. Raimondi, H. M. Robertson, P. Capy, G. Gasperi, and A. R. Malacrida.** 2000. Evolution of different subfamilies of *mariner* elements within the medfly genome inferred from abundance and chromosomal distribution *Chromosoma* **108:**523–532.
250. **Trelogan, S. A., and S. L. Martin.** 1995. Tightly regulated, developmentally specific expression of the first open reading frame from *LINE-1* during mouse embryogenesis. *Proc. Natl. Acad. Sci. USA* **92:**1520–1524.
251. **Tschiersch, B., A. Hofmann, V. Krauss, R. Dorn, G. Korge, and G. Reuter.** 1994. The protein encoded by the *Drosophila* position-effect variegation suppressor gene *Su(var)3–9* combines domains of antagonistic regulators of homeotic gene complexes. *EMBO J.* **13:**3822–3831.
252. **Tu, Z.** 1997. Three novel families of miniature inverted-repeat transposable elements are associated with genes of the yellow fever mosquito, *Aedes aegypti. Proc. Natl. Acad. Sci. USA* **94:**7475–7480.

252a. **Tu, Z.** 2001. Eight novel families of miniature inverted repeat transposable elements in the African malaria mosquito, *Anopheles gambiae. Proc. Natl. Acad. Sci. USA* **98:** 1699–1704.

253. **Tyler-Smith, C., R. J. Oakey, Z. Larin, R. B. Fisher, M. Crocker, N. A. Affara, M. A. Ferguson-Smith, M. Muenke, O. Zuffardi, and M. A. Jobling.** 1993. Localization of DNA sequences required for human centromere function through an analysis of rearranged Y chromosomes. *Nat. Genet.* **5:** 368–375.
254. **van der Krol, A. R., L. A. Mur, M. Beld, J. N. Mol, and A. R. Stuitje.** 1990. Flavonoid genes in petunia: addition of a limited number of gene copies may lead to a suppression of gene expression. *Plant Cell* **2:**291–299.
255. **Venables, P. J., S. M. Brookes, D. Griffiths, R. A. Weiss, and M. T. Boyd.** 1995. Abundance of an endogenous retroviral envelope protein in placental trophoblasts suggests a biological function. *Virology* **211:**589–592.
256. **Vicient, C. M., A. Suoniemi, K. Anamthawat-Jonsson, J. Tanskanen, A. Beharav, E. Nevo, and A. H. Schulman.** 1999. Retrotransposon *BARE-1* and its role in genome evolution in the genus *Hordeum. Plant Cell* **11:**1769–1784.
257. **Vielle-Calzada, J. P., J. Thomas, C. Spillane, A. Coluccio, M. A. Hoeppner, and U. Grossniklaus.** 1999. Maintenance of genomic imprinting at the *Arabidopsis medea* locus requires zygotic DDM1 activity. *Genes Dev.* **13:**2971–2982.
258. **Walker, E. L.** 1998. Paramutation of the r1 locus of maize is associated with increased cytosine methylation. *Genetics* **148:**1973–1981.
259. **Walsh, C. P., J. R. Chaillet, and T. H. Bestor.** 1998. Transcription of *IAP* endogenous retroviruses is constrained by cytosine methylation. *Nat. Genet.* **20:**116–117.

259a. **Wargelius, A., S. Ellingsen, and A. Fjose.** 1999. Double-stranded RNA induces specific developmental defects in zebrafish embryos. *Biochem. Biophys. Res. Commun.* **263:** 156–161.

260. **Waterhouse, P. M., M. W. Graham, and M. B. Wang.** 1998. Virus resistance and gene silencing in plants can be induced by simultaneous expression of sense and antisense RNA. *Proc. Natl. Acad. Sci. USA* **95:**13959–13964.
261. **Waterston, R., and J. Sulston.** 1995. The genome of *Caenorhabditis elegans. Proc. Natl. Acad. Sci. USA* **92:** 10836–10840.
262. **Weiler, K. S., and B. T. Wakimoto.** 1995. Heterochromatin and gene expression in *Drosophila. Annu. Rev. Genet.* **29:** 577–605.
263. **Wendel, J. F., and S. R. Wessler.** 2000. Retrotransposon-mediated genome evolution on a local ecological scale. *Proc. Natl. Acad. Sci. USA* **97:**6250–6252.
264. **Wessler, S. R.** 1988. Phenotypic diversity mediated by the maize transposable elements *Ac* and *Spm. Science* **242:** 399–405.
265. **Wessler, S. R.** 1996. Turned on by stress. Plant retrotransposons. *Curr. Biol.* **6:**959–961.
266. **Wessler, S. R., T. E. Bureau, and S. E. White.** 1995. LTR-retrotransposons and *MITEs:* important players in the evolution of plant genomes. *Curr. Opin. Genet. Dev.* **5:**814–821.
267. **Wichman, H. A., R. A. Van den Bussche, M. J. Hamilton, and R. J. Baker.** 1992. Transposable elements and the evolution of genome organization in mammals. *Genetica* **86:**287–293.
268. **Wilke, C. M., and J. Adams.** 1992. Fitness effects of *Ty* transposition in *Saccharomyces cerevisiae. Genetics* **131:**31–42.
269. **Wolffe, A. P., and M. A. Matzke.** 1999. Epigenetics: regulation through repression. *Science* **286:**481–486.
270. **Wu, Y. H., A. V. Wilks, and J. B. Gibson.** 1998. A *KP* element inserted between the two promoters of the alcohol dehydrogenase gene of *Drosophila melanogaster* differentially affects expression in larvae and adults. *Biochem. Genet.* **36:** 363–379.
271. **Xiong, Y., and T. H. Eickbush.** 1988. The site-specific ribosomal DNA insertion element *R1Bm* belongs to a class of non-long-terminal-repeat retrotransposons. *Mol. Cell. Biol.* **8:**114–123.
272. **Yang, Z., D. Boffelli, N. Boonmark, K. Schwartz, and R. Lawn.** 1998. Apolipoprotein(a) gene enhancer resides within a LINE element. *J. Biol. Chem.* **273:**891–897.
273. **Yoder, J. A., C. P. Walsh, and T. H. Bestor.** 1997. Cytosine methylation and the ecology of intragenomic parasites. *Trends Genet.* **13:**335–340.
274. **Zhang, P., and R. L. Stankiewicz.** 1998. Y-Linked male sterile mutations induced by *P* element in *Drosophila melanogaster. Genetics* **150:**735–744.
275. **Zhang, Q., J. Arbuckle, and S. R. Wessler.** 2000. Recent, extensive, and preferential insertion of members of the miniature inverted-repeat transposable element family *Heartbreaker* into genic regions of maize. *Proc. Natl. Acad. Sci. USA* **97:**1160–1165.
276. **Zou, S., N. Ke, J. M. Kim, and D. F. Voytas.** 1996. The *Saccharomyces* retrotransposon *Ty5* integrates preferentially into regions of silent chromatin at the telomeres and mating loci. *Genes Dev.* **10:**634–645.

II. CONSERVATIVE SITE-SPECIFIC RECOMBINATION: REARRANGEMENTS THAT INVOLVE COVALENT PROTEIN-DNA INTERMEDIATES

Mobile DNA II
Edited by N. L. Craig et al.

Chapter 6

A Structural View of Tyrosine Recombinase Site-Specific Recombination

GREGORY D. VAN DUYNE

INTRODUCTION

DNA rearrangements catalyzed by site-specific recombinases from the tyrosine recombinase family are used to accomplish a variety of important biological functions (reviewed in references 41, 70, 90, and 94). The integration of the bacteriophage λ genome into the *Escherichia coli* chromosome has served as the paradigm for this process after being proposed by Campbell nearly 40 years ago (16). In addition to the integration and excision of viral genomes into and out of host chromosomes, site-specific recombination is used to resolve multimeric plasmids and chromosomes to monomers in order to ensure faithful partitioning upon cell division, to amplify the copy number of yeast 2μm circles by inverting DNA segments during replication, and to switch on and off the expression of genes by inversion of DNA segments. In all of these cases, recombinase enzymes recognize specific DNA sequences and catalyze DNA rearrangements only between those sequences.

The λ integrase family of site-specific recombinases, also referred to as the tyrosine recombinases (91), includes over 100 members identified based on sequence similarity (75). The most well-studied examples include, in addition to the integrase protein from bacteriophage λ (55), the bacterial XerC and XerD recombinases (90), Cre recombinase from bacteriophage P1 (1), and the Flp recombinase from the *Saccharomyces cerevisiae* 2μm circle (85). The tyrosine recombinases carry out site-specific recombination in a stepwise manner, exchanging one pair of DNA strands to form a Holliday junction intermediate and then resolving the Holliday junction to products by exchange of the second pair of strands. The emergence of structural data for the tyrosine recombinases has been a relatively recent event, beginning with the report of the λ integrase catalytic domain crystal structure early in 1997 (53). Since then, structures of the HP1 integrase catalytic domain (45), the XerD protein (96), and three Cre recombinase-DNA complexes

Gregory D. Van Duyne • Department of Biochemistry and Biophysics and Howard Hughes Medical Institute, University of Pennsylvania School of Medicine, Philadelphia, PA 19104.

Table 1. Structural models in the tyrosine recombinase and topoisomerase Ib families

System	Structure	PDB[a] code	Reference
λ Integrase	Int Y342F catalytic domain	1AE9	53
HP1 integrase	HP1 Int catalytic domain	1AIH	45
XerCD	XerD (full length)	1AOP	96
Cre	Cre-*loxA* covalent complex	1CRX	39
Cre	Cre R173K-HJ1 complex	3CRX	32
Cre	Cre-HJ2 complex	2CRX	32
Cre	Cre R173K-*loxS* synaptic complex	4CRX	38
Cre	Cre Y324F-*loxS* synaptic complex	5CRX	38
Flp	F1p-HJ complex	1FL0	21
Vaccinia virus topoisomerase Ib	Topoisomerase catalytic domain	1A41	22
Human topoisomerase Ib	Topoisomerase-DNA covalent complex	1A31	80
Human topoisomerase Ib	Topoisomerase Y723F-DNA complex	1A35	80
Human topoisomerase Ib	70-KDa topoisomerase Y723F-DNA complex	1A36	95

[a] PDB, Protein Data Bank.

(32, 38, 39) have been described. Most recently, the structure of an Flp recombinase-DNA complex has been determined (21). The resulting structural models for the tyrosine recombinase family (summarized in Table 1) are the subject of this review, with a focus on mechanistic aspects of the reaction, and in particular how the available protein and protein-DNA structural models influence current views of each step in the recombination pathway. Much of the discussion and figures will make use of structures in the Cre-*loxP* system, for which models of the recombinase-DNA complexes for three intermediates in the reaction pathway are available.

SIMPLE AND COMPLEX RECOMBINATION SYSTEMS

The DNA recombination sequences where recombinase binding and strand exchange take place are referred to as the core recombination sites, illustrated in Fig. 1a. These sites are composed of two recombinase binding elements (RBEs) arranged as inverted repeats surrounding a central strand exchange or crossover region. Two recombinase subunits bind to each core site (one to each RBE), often with a high level of cooperativity (11, 83). The phosphoryl transfer strand exchange chemistry between substrates occurs within the crossover region, a relatively short segment of DNA (6 to 8 bp) compared to the sequences required for a complete recombination event. Cleavage occurs at the border between the RBE and the 5′ end of the crossover region. The crossover sequence is asymmetric and therefore provides directionality to the site. For the simplest systems, exemplified by the Cre and Flp recombinases, the core recombination site is sufficient for the complete recombination reaction (2, 85).

The resolution of ColE1 dimers by XerC-XerD represents a more complex tyrosine recombinase system. The full recombination sites required in this case (called *cer;* Fig. 1b) are over 200 bp long and contain ~190 bp of accessory sequences adjacent to the core recombination site (97). The accessory sequences contain binding sites for two proteins, the arginine repressor (ArgR) and leucine aminopeptidase (PepA). Both ArgR and PepA are required for this recombination reaction, which occurs only on supercoiled substrates containing directly repeated *cer* sites and is exclusively intramolecular. These requirements together provide a topological filter that ensures that only the resolution reaction to produce plasmid monomers, and not the dimerization-integration reaction between plasmid monomers, occurs efficiently in vivo (4, 24).

An even more complex system is exemplified by the λ integrase integration reaction, in which the two DNA substrate recombination sites are quite different. The phage attachment site (*attP*) is 240 bp long and contains binding sites for the integration host factor (IHF) protein, as well as "arm-binding sites" for auxiliary DNA-binding domains present in the integrase protein (55). Thus, Int binds both to the core recombination site and to the arm sites, with IHF serving as an obligatory accessory protein that provides the sharp DNA bends necessary for intasome assembly (69). The bacterial attachment site (*attB*) consists of only the relatively short core sequence (Fig. 1c).

OLDER AND NEWER TYROSINE RECOMBINASE MECHANISMS

Tyrosine recombinases use topoisomerase I chemistry to cleave and religate DNA strands during recombination. A historical model for the recombina-

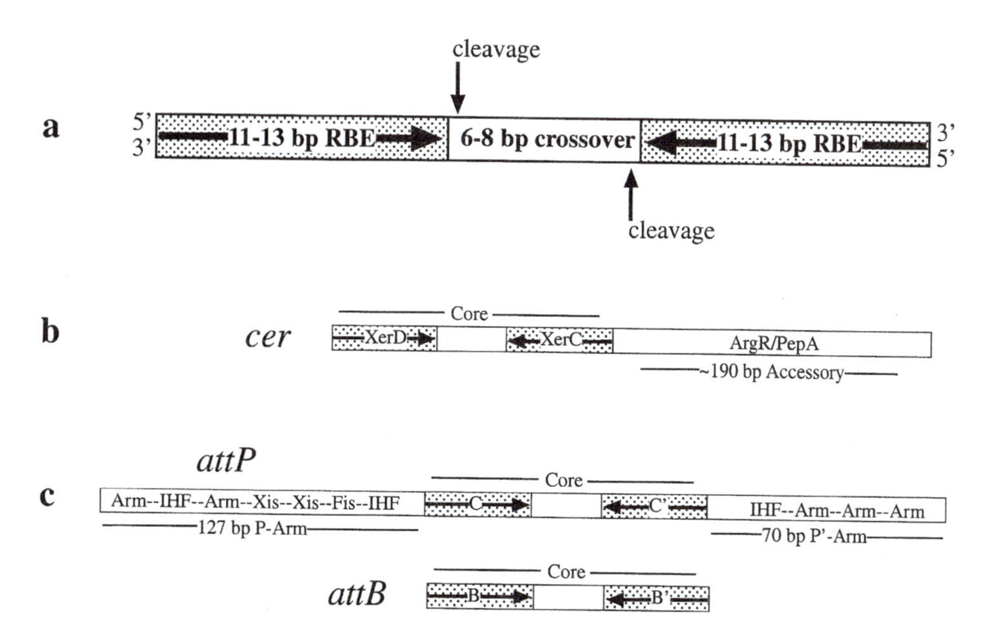

Figure 1. Organization of the recombination sites. (a) The simplest sites, such as those found in the Cre-*loxP* system (48), are composed of two 11- to 13-bp RBEs arranged as inverted repeats around a central crossover region. Cleavage of the sites occurs at the borders between the crossover region and the RBEs. (b) Resolution of ColE1 plasmid multimers by XerC and XerD recombinases occurs at *cer* sites. Nearly 200 bp of accessory sequences located adjacent to the core site are bound by the ArgR and PepA proteins when two *cer* sites are arranged in direct repeat on a supercoiled substrate (24). (c) The integration reaction of bacteriophage λ integrase requires a complex phage site (*attP*) composed of a core site and flanking accessory arms and a bacterial site (*attB*) composed of only the core site. The P and P′ arms of *attP* contain binding sites for the arm-binding domains of integrase protein, for IHF, and for two other accessory proteins (Xis and Fis) which stimulate the excision reaction (55).

tion pathway is shown in Fig. 2, which illustrates the stepwise nature of the reaction. This view of the reaction (and all views discussed in this chapter) involves only the core recombination sites and the core-binding components of the recombinase enzymes. For simple systems (such as Flp and Cre) recombining linear substrates, such a model represents all of the interacting components. For more complex systems, there are additional factors that play architectural and/or energetic roles that could influence the various steps in the recombination pathway. For reviews of the λ integrase family recombination mechanism as discussed prior to 1995, see references 26, 55, 56, 70, and 93.

In the model shown in Fig. 2, two recombinase-bound core sites associate to form a recombination synapse where the sites are aligned in a parallel orientation. Two of the four recombinase subunits in the synapse are activated to cleave the top strands of the DNA substrates using conserved tyrosine side chains as nucleophiles, forming covalent 3′-phosphotyrosine linkages to DNA and releasing free 5′-hydroxyl groups. The 5′-hydroxyl groups can either reseal the nicks to restore the original substrates and complete one round of topoisomerase I-like cleavage and religation, or the 5′-hydroxyl groups can attack the 3′-phosphotyrosine linkages of the partner substrates, resulting in exchange of the top strands and formation of a Holliday junction (HJ) intermediate. After branch migration through the crossover region, the bottom strands of the HJ intermediate are then cleaved and exchanged by the second pair of recombinase subunits to form recombinant products.

The tyrosine recombinases display a strong requirement for sequence identity in the crossover region between recombining sites. Early models have explained how each site senses homology with its partner by requiring the HJ intermediate to branch migrate through this region (102). Mismatches in sequence would provide energetic barriers to this process and ensure that only identical crossover regions could be efficiently recombined. In addition to providing a method for testing homology, the branch migration mechanism also provided an allosteric model for activation of the recombinase subunits. The initial cleavage events in this model occur at one end of the crossover region by the pair of subunits bound to the RBEs at the same end. Branch migration then places the second pair of recombinase subunits in a similar situation at the opposite end of the crossover regions, where they could be activated to continue the reaction and process the second pair of DNA substrate strands.

The branch migration model is satisfying in terms

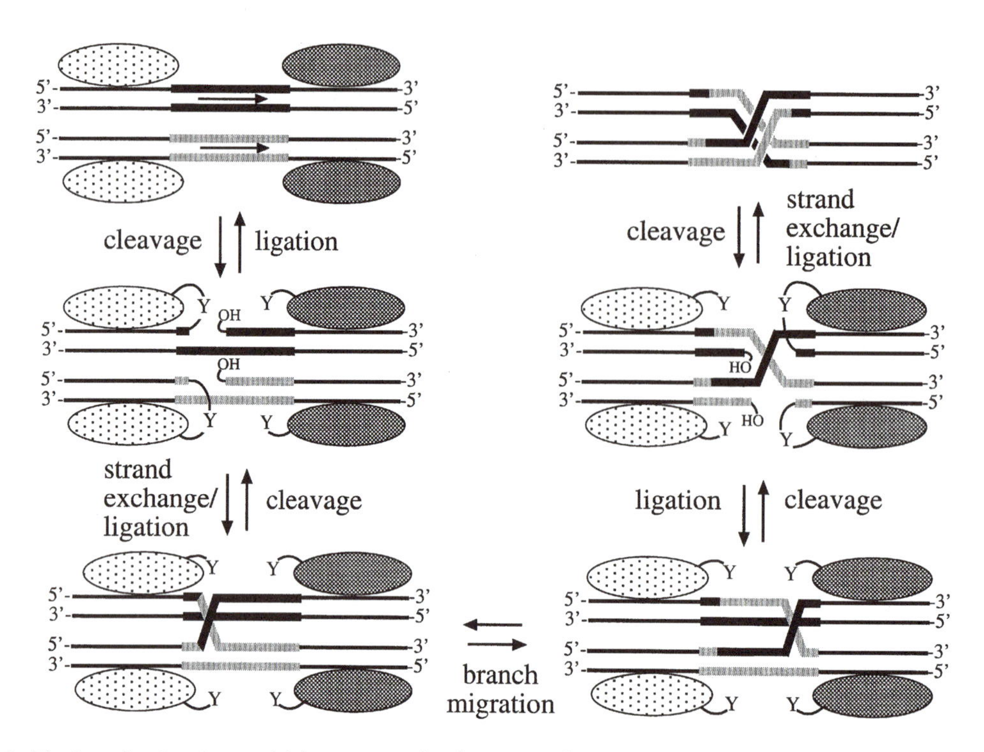

Figure 2. The branch migration model for integrase family site-specific recombination (26). Two recombinase-bound sites associate to form a recombination synapse (top left panel). Two subunits cleave the top strands of the substrates with conserved tyrosine side chains to form 3′-phosphotyrosine linkages and release free 5′-hydroxyl groups (middle left panel). The 5′-hydroxyl groups undergo intermolecular attack of the partner phosphotyrosine to complete the exchange of one pair of DNA strands between the two substrates and form an H-J intermediate (bottom left panel). The branch point of the junction starts at the site of initial strand exchange and then migrates through the crossover region to the second set of cleavage sites. The second pair of subunits are then activated and/or positioned for cleavage of the bottom substrate strands, which are exchanged to form recombinant products. Heterology between crossover sequences could block efficient branch migration and prevent the reaction from proceeding to the second strand exchange. For simplicity, the DNA sites are shown associating in a parallel orientation in this figure. For a schematic representation of the same reaction in an antiparallel alignment, see reference 5.

of explaining the requirements for homology and the observed reversibility of the recombination pathway but is difficult to reconcile with the motions required of the recombinase subunits and DNA arms of the HJ in three dimensions (95). In order to branch migrate 6 to 8 bp, the recombinase-bound arms of the HJ intermediate would need to undergo extensive rotations, requiring that any protein-protein interfaces formed in the synapse be disrupted. In the mid-1990s, experiments from several laboratories using synthetic HJ substrates supported the idea that the branch point of the HJ intermediate does not need to be located at the site of cleavage in order to resolve the junction to products (5, 27, 59, 71). In addition, evidence strongly indicated that efficient strand ligation depends on sequence homology adjacent to the point of ligation, a finding that was also at odds with the older branch migration model (15, 57). Nunes-Düby et al. therefore proposed an alternative "strand-swapping–isomerization" model for integrase family site-specific recombination (71). In this model, strand exchange occurs following cleavage of the site by melting two to three bases from their complementary strand and annealing the bases to the corresponding complementary strand in the recombining partner. Those bases are then effectively tested for homology with the partner by Watson-Crick base pairing and subsequent ligation.

In the strand-swapping–isomerization model, the HJ intermediate undergoes a change in quaternary structure which switches the roles of the protein subunits and the DNA strands, setting the stage for a second strand swap. A local branch migration of 1 to 2 bp tests homology between the central base pair(s) in the crossover region, with the 3-bp flanking sequences tested via strand swaps. A helical restacking event in which the duplex arms of the junction exchange stacking partners is coupled to limited branch migration and provides the allosteric signals required to commit the system to resolving only one pair of strands in a given isomeric configuration. Closely related versions of this mechanism have also been described for XerCD (7), and strand-swapping mechanisms with alternative HJ isomerization schemes have been discussed for XerCD (40), for Flp (62, 100), and for Cre (32, 33, 39).

STRUCTURES OF THE RECOMBINASES

Before three-dimensional structural data became available, most of what was known about protein structures in the tyrosine recombinase family came from sequence comparisons and limited proteolysis experiments. The tyrosine recombinases share limited sequence similarity overall, although it is now clear that family members have a common catalytic domain fold (30, 75). Early sequence analyses revealed four strictly conserved amino acids required for catalysis: the Arg-His-Arg catalytic triad and the tyrosine nucleophile (3, 9). All four residues were found to be in the C-terminal halves of the recombinase sequences, which showed more sequence similarity than the N-terminal halves (Fig. 3a).

Limited proteolysis experiments in the Cre and Flp systems identified probable domain boundaries, which divided the recombinases into an ~13-kDa amino-terminal domain and a larger carboxy-terminal catalytic domain of 25 kDa for Cre and 32 kDa for Flp (46, 78). The Cre C-terminal domain was shown to contact the outermost region of the RBE, with the remaining contacts proximal to the crossover region occurring only in the presence of the full-length protein. The region N terminal to the catalytic domain for λ-Int is further subdivided into a domain thought to serve the same functional role as the Cre, Flp, and XerCD N-terminal domains, plus an extra DNA-binding domain at the extreme N terminus that is re-

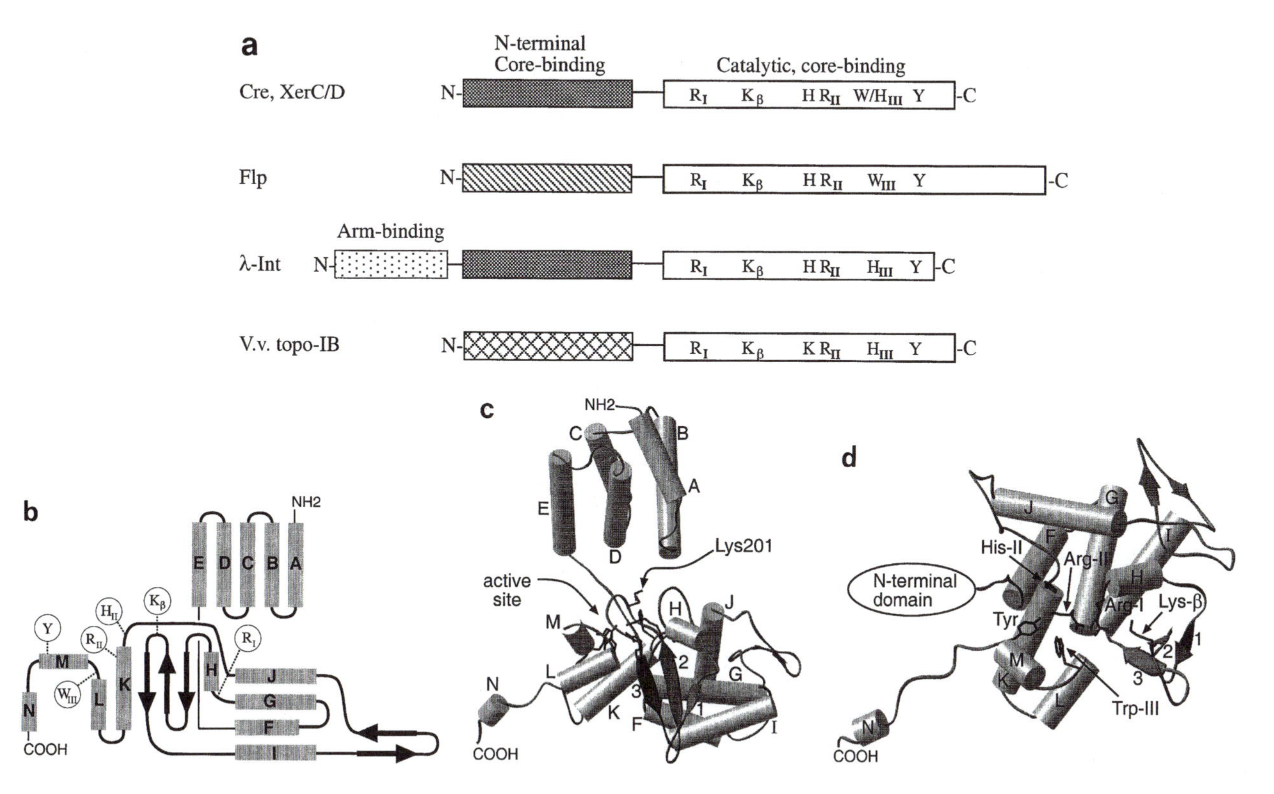

Figure 3. Domain structure of the tyrosine recombinases. (a) The tyrosine recombinases and the eukaryotic topoisomerase Ib family (represented here by the vaccinia virus enzyme) share conserved C-terminal catalytic domains but have more-divergent N-terminal domains. The arrangement of conserved catalytic residues in the C-terminal domains is indicated (see definitions in Table 2). V.v. topo-IB, vaccinia virus topoisomerase Ib. (b) Folding pattern of Cre recombinase. Helices A to E comprise the N-terminal domain and the remainder makes up the catalytic domain. A similar fold has been observed for the catalytic domains of λ integrase, HP1 integrase, and XerD recombinase, and this fold forms a subset of the eukaryotic topoisomerase Ib and the yeast (Flp) recombinase catalytic domains (see Table 1). The locations of conserved catalytic residues are indicated. (c) Ribbon-cylinder representation of a Cre recombinase subunit in the DNA-bound form, with the DNA removed from the drawing. Helices are labeled as in panel b, and active-site residues are drawn as black sticks. The location of Lys201 (Lys-β) is indicated. (d) Structure of the Cre recombinase catalytic domain, viewed from the amino-terminal domain (from the top as drawn in panel c). Active-site residues are indicated by arrows. Ribbon-cylinder illustrations in this chapter were produced with the program RIBBONS (18).

sponsible for contacting the arm sites in the accessory region of the λ recombination sites (98, 99).

So far, X-ray crystal structures for five λ integrase family site-specific recombinases have been reported (see Table 1 for a summary). In two cases (λ-Int and HP1-Int), the structures are of the isolated catalytic domains (45, 53). The structure of XerD is of the full-length protein (96). The structures of Cre and Flp recombinases are bound to DNA substrates (32, 38, 39, 81). In addition, the structures of two eukaryotic type IB topoisomerases have been determined, illustrating that the type IB topoisomerases and the λ integrase family recombinases share a common catalytic domain and a common enzymology of strand cleavage and ligation (22, 80, 95). A number of excellent minireviews and commentaries provide a mechanistic perspective on these new contributions to the field (36, 54, 63, 91, 105).

In all cases, the structures that have been determined indicate a common catalytic domain fold (Fig. 3). In the case of Flp, the C-terminal domain is somewhat larger, but the core of the domain is similar to the other known catalytic domain structures. Of the N-terminal domain structures determined thus far, only those of Cre and XerD closely resemble one another. Both contain a loosely arranged bundle of α-helices with small to moderate-size connecting loops. The N-terminal domains from Flp and from the vaccinia virus and human topoisomerases contain a substantial fraction of β-structure yet appear unrelated to one another and are different from the equivalent domains in Cre and XerD.

RECOMBINASE BINDING TO DNA

The complexes of Cre recombinase with DNA substrates revealed an extensive protein-DNA interface between each protein subunit and its contacted RBE (32, 38, 39). The two domains of Cre form a "C-shaped clamp" which grasps the DNA from opposite sides, where the N-terminal domain interacts primarily with the major groove proximal to the crossover region on one face and the C-terminal domain interacts with successive minor, major, and minor grooves on the opposite face (Color Plate 8 [see color insert]). A similar organization has been observed in the Flp-DNA complex (21) and in the human topoisomerase-DNA complexes (80), despite clear differences in the structures of the proteins N-terminal to the conserved core catalytic domains. Biochemical evidence in the Xer and λ Int systems, together with the available structural models, indicates that this recombinase-DNA organization is likely to be a common feature of the tyrosine recombinases (44, 99).

In addition to the extensive protein-DNA interface formed between Cre and its recombination site *loxP*, a substantial protein-protein interface is formed between the two subunits bound to RBEs on the same site (Color Plate 8). On one side of the DNA, the two N-terminal domains interact with one another primarily through helix-helix contacts, but they do not interact substantially with the catalytic domains. On the opposite side, the two catalytic domains interact via an exchange of helices located at the carboxyl termini. The C-terminal helix in one Cre subunit (helix N in Fig. 3 and Color Plate 8) buries its hydrophobic surface in an "acceptor pocket" on the adjacent subunit. This network of contacts between N- and C-terminal domains in DNA-bound Cre subunits contributes to a large cooperativity of binding to the *loxP* site (1, 83). Cooperative binding of the recombinases has also been noted in the XerCD system (11, 14).

The C-terminal helix exchange is not reciprocal between subunits bound to the same DNA site in the Cre duplex DNA crystal structures because a second Cre-*lox* complex interacts to form a synaptic complex and the helix exchange is cyclic among the four subunits. In the HP1-Int catalytic domain structure, however, a C-terminal helix swap occurs in a reciprocal fashion, leading to an HP1-Int dimer (45). An HP1-like reciprocal exchange of C-terminal helices between DNA-bound Cre subunits appears to be precluded for geometric reasons if the *loxP* site is bent even slightly. Modeling exercises indicate that even in the case of an unbent *loxP* site, mutual exchange of C-terminal helices would be difficult and may require partial unfolding or repositioning of the penultimate helix (helix M) in order to form the correct helix-acceptor interactions (37).

Although Cre binding to the *loxP* site in the context of a synaptic complex indicates a sharp bend of ~75° (38), a detailed study of how Cre bends the isolated *loxP* site has not yet been reported. In contrast, several studies have described bending in the Flp system (66, 67, 86, 87). Based on electrophoretic mobility assays, Flp bends its recombination site *frt* by 140°, a distortion in excess of that seen in the Cre-DNA crystal structures. It remains to be seen if the difference is due to limitations of the bending assays, differences between Flp and Cre, or differences between synapsed and unsynapsed recombinase-DNA complexes.

The helix-swapping arrangement observed in the Cre-DNA structures provides not only a model for understanding cooperative binding, bending, and synapsis of *loxP* sites but also a provocative mechanistic argument for control of the cleavage reaction. The conserved tyrosine nucleophile (Tyr324) is located close to the peptide linker (Arg326-Gly333) that connects helix N to the rest of the domain. Changes in quaternary structure of the recombination complex could therefore be sensed by the tyrosine nucleophile through this peptide linker, providing a stereochemical coupling between the overall structure of the recombination assembly and positioning of a catalytic amino acid. Although it appears that the XerC-XerD system may use a similar ball-and-socket connectivity between catalytic domains in the recombination complex (92), it is not clear that the same structural features exist in the λ-Int or Flp systems. For λ-Int, the C terminus is shorter than Cre and XerCD and there may be no corresponding C-terminal helix. For Flp, the catalytic domain is substantially larger and the intersubunit contacts are different, as described later in this chapter.

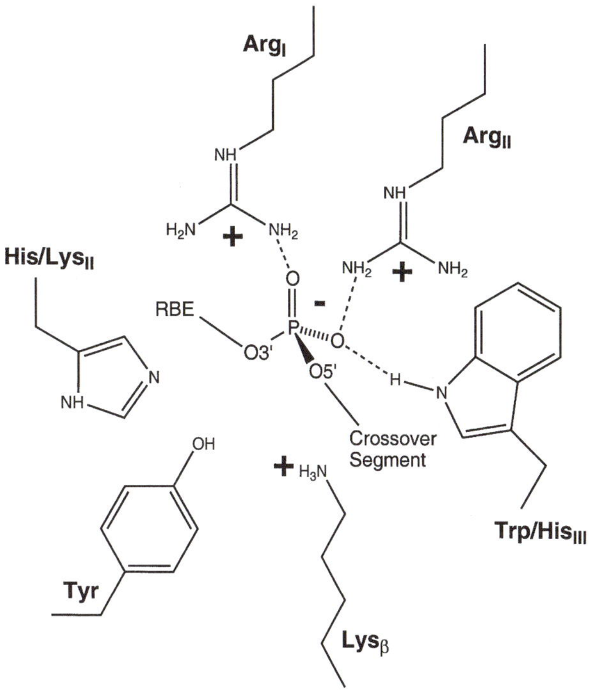

RECOMBINASE ACTIVE SITES

The active-site model that has emerged from the integrase family catalytic domain structures, from the eukaryotic topoisomerase Ib catalytic domain structures, and from an analysis of three intermediates in the Cre-DNA pathway is illustrated in Fig. 4, with a summary of the important residues in Table 2. The previously identified Arg-I, His-II, and Arg-II side chains coordinate the scissile phosphates during recombination, while the conserved tyrosine side chain is positioned near the scissile phosphate, poised for nucleophilic attack. In addition to these four residues, two new participants in the enzyme active site have been identified. A tryptophan (His or Trp-III) forms a hydrogen bond to the scissile phosphate via the indole nitrogen in Cre, whereas a histidine side chain occupies this position in the sequences of λ-Int, HP1-Int, XerD, and the vaccinia virus and human topoisomerases. In the HP1-Int catalytic domain structure, His-III (His306) is hydrogen bonded to a sulfate ion that appears to play the role of a surrogate phosphate (45), and the human topoisomerase-DNA complexes show that the corresponding histidine also coordinates the scissile phosphate (80). A comparison of tyrosine recombinase sequences indicates that this His/Trp hy-

Figure 4. Schematic representation of the tyrosine recombinase active site. See Table 2 for the residue identities in each of the well-studied systems. The RBE flanks the 3′ side of the scissile phosphate and the central crossover strand flanks the 5′ side.

Table 2. Catalytic residues in the tyrosine recombinases and topoisomerases Ib

Protein	Catalytic residue[a]					
	Arg-I	Lys-β	His/Lys-II	Arg-II	His/Trp-III	Tyr
Cre	Arg173	Lys201	His289	Arg292	Trp315	Tyr324
λ Integrase	Arg212	Lys235	His308	Arg311	His333	Tyr342
HP1 integrase	Arg207	Lys230	His280	Arg283	His306	Tyr315
XerC	Arg148	Lys172	His240	Arg243	His266	Tyr275
XerD	Arg148	Lys172	His244	Arg247	His270	Tyr279
Flp	Arg191	Lys223	His305	Arg308	Trp330	Tyr343
Vaccinia virus topoisomerase	Arg130	Lys167	Lys220	Arg223	His265	Tyr274
Human topoisomerase Ib	Arg488	Lys532	Lys587	Arg590	His632	Tyr723

[a]Arg-I, His-II, and Arg-II are the conserved RHR triad of residues, labeled based on their positions in conserved motifs I and II (3). Lys-β is the conserved lysine in the loop between β-strands 2 and 3. His/Trp-III are active site residues that coordinate the scissile phosphate during catalysis.

drogen bond is likely to be conserved throughout the protein family, although an Flp-DNA complex crystal structure does not show this active-site configuration (21). As a result of a base pairing rearrangement following cleavage of a nicked suicide substrate by Flp, an alternative 5′-hydroxyl group has shifted into the active site and assumed a position where the normal 5′-hydroxyl would be expected to attack the phosphotyrosine linkage in the ligation step of the reaction. In the active sites where hydrolysis of the 3′-phosphotyrosine bond has led to formation of a 3′-phosphate, Trp-III is observed hydrogen bonded instead to the free 5′-hydroxyl group.

The second previously unidentified active site component is a lysine residue located in the loop between β-strands 2 and 3 (referred to here as Lys-β). Structural and biochemical studies in the vaccinia virus topoisomerase system were key to realizing the importance of this residue in the site-specific recombinases (22, 104). In the vaccinia virus enzyme, mutation of the corresponding residue (Lys165) results in a loss of 10^4-fold in catalytic efficiency (104). The alanine mutants of this lysine in XerD (17) and Cre (F. Guo and G. Van Duyne, unpublished data) are each defective in recombination. In the covalent Cre-DNA and human topoisomerase Ib DNA structures, this lysine contacts a base adjacent to the cleaved phosphate in the minor groove (39, 80). In the synaptic Cre-DNA complexes and in the Cre-HJ complex structures, Lys-β is not well ordered, most likely indicating a high degree of mobility.

An important characteristic of the site-specific recombinase active sites based on structures with and without bound substrate is the mobility of the tyrosine nucleophile. In Cre, Y324 is located on the penultimate C-terminal helix (helix M) (Fig. 3), which is positioned to move in and out of the active site pocket in a direction parallel to its helical axis. In the characterized Cre active sites, this tyrosine either is covalently linked to the scissile phosphate (in the active subunits of the covalent complex) or is hydrogen bonded to the 5′-adjacent phosphate, where the tyrosine hydroxyl lies roughly midway between the two phosphate groups. In the HP1-Int, λ-Int, and XerD recombinase structures (which are not bound to substrate), the tyrosine nucleophile adopts a much broader ensemble of positions that are quite distinct from one another. Part of the reason for these conformational differences may simply reflect the need to assemble the active site on the DNA substrate. Upon substrate binding, these structural motifs are likely to adopt conformations similar to those that have been observed in Cre or Flp. The distribution of tyrosine positions in the structural models reported in the absence of substrate, along with the distribution of tyrosine positions observed in the three reaction intermediates in the Cre-*loxP* pathway, demonstrates an inherent flexibility in the region of the recombinases that contain this important residue.

THE SYNAPTIC RECOMBINATION COMPLEX

In order for recombination to occur, two DNA sites must associate to form a synaptic complex, within which the cleavage and strand exchange reactions take place. Figure 2 implies that in early models of the integrase family recombination mechanism the sites come together in a parallel alignment. Here, I define a parallel alignment of sites as one in which the DNA helical axes of the crossover regions form an angle less than 90° with respect to the crossover sequence directions. In the more recent strand-swapping–isomerization model, the issue of site alignment was avoided altogether and the schematic representation of the proposed mechanism had sites aligned in an orthogonal orientation (74). Neither the branch migration nor the strand-swapping mechanism in principle requires a strict parallel or antiparallel alignment of sites during synapsis, however, since they both allow extensive motions of the DNA arms and bound recombinases during the recombination process.

The only crystallographic data describing synapsed DNA sites in the tyrosine recombinase family come from the Cre-*loxP* system. In the structures of two Cre mutants bound to symmetrized *loxP* sites (38), the Cre-bound DNA duplexes are bent sharply within the crossover regions and are brought together with all four arms (defined as the DNA segments on either sides of the bends) lying nearly in the same plane (Color Plate 9 [see color insert]). The four recombinase monomers create a pseudo-fourfold-symmetric network of protein-protein interactions responsible for holding the synapse together, with a set of interactions between the N-terminal domains on one side of the DNA plane and an independent set of interactions between C-terminal domains on the opposite side of the DNA substrate plane. The complex is strictly twofold symmetric overall, with the dyad of symmetry passing through the center of the complex in a direction perpendicular to the plane of the DNA substrates. A similar twofold-symmetric arrangement of recombinase and substrate is present in the structure of the cleaved covalent Cre-DNA intermediate (39) and in the structures of the Cre- and Flp-bound Holliday junction intermediates (32, 81). Although the DNA sites in these cases are symmetric and therefore have no sequence-derived directionality that would specify parallel versus antiparallel alignment, the twofold-symmetric structures strongly imply an antiparallel alignment of wild-type recombination sites. The three-dimensional mechanism proposed for the Cre-*loxP* pathway (discussed later in this chapter) in fact requires that the sites be aligned in an antiparallel fashion in order for productive strand exchange to occur in the absence of large structural rearrangements.

In addition to the Cre- and Flp-DNA crystal structures determined thus far, two other structural techniques have been used to probe tyrosine recombinase synaptic complexes. Huffman and Levene have used electron microscopy (EM) to image Flp-DNA synaptic complexes and concluded that a majority of the synaptic assemblies had *frt* sites aligned in parallel (49). This bias in alignment was dependent on Flp cleavage activity, and the source of the bias was suggested to be related to the structure of the Holliday intermediates formed. Lee et al. have attempted to explain these EM results by analyzing the resolution properties of tethered parallel HJs by Flp recombinase (61). Their conclusion was that parallel Flp-DNA complexes undergoing recombination might contain only three recombinase subunits and do not represent the normal Flp pathway. In separate experiments, the same laboratory showed that site alignment in the Flp system is antiparallel, based on a topological analysis of the recombination reaction (34).

Cassell et al. have used both atomic force microscopy and solution ligation assays to demonstrate that *attL* sites form synaptic complexes that have either tetrahedral or square-planar geometry and are skewed towards a parallel orientation (19). It is not yet clear how to reconcile the EM and atomic force microscopy data in the Flp and Int systems with crystallographic and biochemical data that indicate an antiparallel alignment of core recombination sites. In the Cre-DNA synaptic structures, we found no reason to conclude that parallel synapses should be significantly less stable on structural grounds alone, but rather that recombination in a parallel synapse should be blocked by sequence heterology.

The nature of the protein-protein contacts in the Cre-DNA synaptic complexes, and more recently in the Flp-HJ complex, has provided a great deal of insight into many aspects of the recombination process. A striking feature of these structures is the types and the extents of the protein-protein interfaces. In the Cre tetramer, the self-association of N-terminal domains alone buries ~4,000 Å^2 of solvent-accessible surface area, with the interactions between C-terminal domains burying a further ~7,000 Å^2. The N-terminal domains interact primarily with one another (and with the DNA) to form a nearly fourfold-symmetric tetramer via helix-helix and loop-helix interactions. As a consequence of this symmetry, the interactions between two N-terminal domains bound to the same site are nearly identical to the synaptic interactions between N-terminal domains on different sites. This arrangement of N-terminal domains appears to be capable of accommodating the synapsed substrates, the cleaved substrates, and the HJ intermediate in both halves of the pathway with virtually no changes in quaternary structure. The N-terminal domains therefore form a relatively rigid structure that persists throughout the recombination process.

As discussed earlier, the C-terminal domains of Cre interact with one another through a cyclic exchange of C-terminal helices, where a short helix at the C terminus interacts extensively with the catalytic domain of the neighboring subunit. Spiers and Sherratt have analyzed these regions in *E. coli* XerC and XerD by mutagenesis and found that mutations in either the C-terminal helix or in the acceptor pocket result in dimished recombination efficiency, indicating that these interactions are probably not unique to the Cre-*loxP* system (93). Only a few residues preceding this helix is the catalytic tyrosine, prompting the suggestions that changes in this region of the recombinases could be part of the allosteric regulation mechanism for control of the recombinase subunits (39).

In the Flp-HJ complex (21), the nature of the domain-domain contacts in the recombinase tetramer is quite different. The N-terminal domains do not interact with one another at all, but instead interact with a helix that connects the N- and C-terminal domains of an adjacent subunit. This helix is not well packed against its catalytic domain but instead is extended away from the domain to provide a docking surface for the neighboring N-terminal domain. As in Cre, the intersubunit contacts are cyclic, forming a pseudo-fourfold-symmetric network of interactions within the recombinase tetramer. Flp recombinase has a larger C-terminal domain than what is found in Cre, XerCD, and λ-Int and does not form a cyclic exchange of C-terminal helices. Instead, Flp creates a cyclic exchange of the helices corresponding to helix M in Cre. These helices, which carry the tyrosine nucleophiles, are provided in *trans* to the neighboring subunits, where they participate in the formation of shared active sites between pairs of adjacent subunits. The mechanistic implications of this alternative arrangement are discussed below.

Given the similar folds of the N-terminal domains and the results of mutagenesis experiments on the C-terminal domains, it seems probable that the protein-protein contacts in a synaptic XerCD complex with *dif* sites (6-bp crossover) are similar to those seen in the Cre-*lox* structures. In both the XerC-

XerD-*dif* and Cre-*loxP* cases, there are no further accessory factors required to synapse the recombination sites and to begin recombination. The situation with XerCD acting on *cer* sites (8-bp crossover), however, is likely to be quite different. The additional two base pairs in the crossover region require both a change in distance (~7 Å) and a relative rotation (~68°) of the domains bound to their recognition elements, requiring that a somewhat different protein-protein interface be formed (12). Recombination at *cer* requires both supercoiled substrates and the ArgR and PepA proteins, which are perhaps able to serve their regulatory roles by promoting synapsis between sites that can associate only weakly on their own (12, 24).

CLEAVAGE AND STRAND EXCHANGE

Some of the most informative insights from structural models in the tyrosine recombinase system concern the geometry of the cleavage and strand exchange steps in the reaction. Our first snapshot of this process in action came from the crystal structure of Cre recombinase covalently bound to a suicide DNA substrate (39). This intermediate was formed using symmetrized *loxP* sites that contained nicks adjacent to the scissile phosphates. Cleavage by the recombinase proteins resulted in formation of 3′-phosphotyrosine linkages to the DNA and release of free cytidine molecules, which diffuse away from the active sites (72). The resulting complex was trapped at the strand exchange step, since the departing cytidine takes with it the 5′-hydroxyl required to either continue or reverse the strand exchange reaction and the remaining strand is too short on the 5′ end to reach either of the activated phosphotyrosines.

The covalent Cre-DNA complex revealed a synaptic assembly with an architecture nearly identical to that observed later in the precleavage synaptic complex (Color Plate 9). The recombinase proteins form a pseudo-fourfold-symmetric tetramer with the DNA duplex substrates in a distorted square planar arrangement. Two of the four protein subunits diagonally related in the complex have cleaved the DNA to form covalent bonds to the scissile phosphate (the green subunits in Color Plate 9), but the other two subunits have not cleaved the DNA. Remarkably, the 5′ ends of the DNA strands that would be transferred to the partner substrates in the normal reaction (if the 5′-cytidines were present) have partially melted away from their complementary strands, with their 5′-thymidine ends converging in the middle of the synaptic complex (Fig. 5a). The 5′-thymine bases of these strands actually make van der Waals contact with one another, perhaps explaining why this unusual DNA conformation could be visualized in the crystal structure.

It is tempting to think of this cleaved intermediate structure as a snapshot of the strand exchange process in action. In this context, the complex structure could equally well represent a trapped-strand transfer reaction in the HJ-forming step or in the HJ-resolving step. In other words, the model can equally well represent the strands going (from the duplex substrates) or coming (from the HJ intermediate to form products). One of the interesting questions concerning this view of the structural model is whether the observed protein-DNA architecture needs to undergo a significant change in quaternary structure in order to accommodate the transfer and ligation of DNA strands. Modeling exercises suggest that the intact strands (with the 5′-cytidine restored) would be able to reach the complementary strands of the partner substrate and form at least two Watson-Crick base pairs immediately adjacent to the scissile phosphate. This annealing process could occur by moving only the 5′ ends of the DNA strands and leaving the remainder of the protein-DNA assembly unchanged. With more-extensive movement of the DNA strands, a third base pair (and thus a fully base-paired HJ intermediate) can be formed (Fig. 5b).

A series of experiments in the λ-Int system have indicated that formation of base pairs, and therefore homology sensing, occurs prior to chemical ligation (15, 76). In support of these observations, access to the phosphotyrosine linkage is limited stereochemically in the structure of the Cre-DNA cleaved intermediate. In order for the 5′-hydroxyl of the transferring strand to be positioned properly for attack at the activated phosphate, the 5′ base needs to occupy a base pairing position. The covalent Cre-DNA structure is therefore consistent with the idea that a mismatched base adjacent to the scissile phosphate presents a stereochemical barrier to efficient ligation. Indeed, evidence for the importance of heterology adjacent to the scissile phosphate had also been demonstrated for the Flp system prior to the appearance of any structural models (57, 106).

The structure of the Cre-HJ intermediate (discussed below) indicates that three bases are exchanged in each strand of the duplex substrates, placing the branch point of the HJ formed at the center of the *loxP* site. This implies that exactly half of the 6-bp crossover site between scissile phosphates are exchanged in forming the HJ (shown in Fig. 5) and the remaining half are exchanged in the resolution of the HJ intermediate to products. Although this recom-

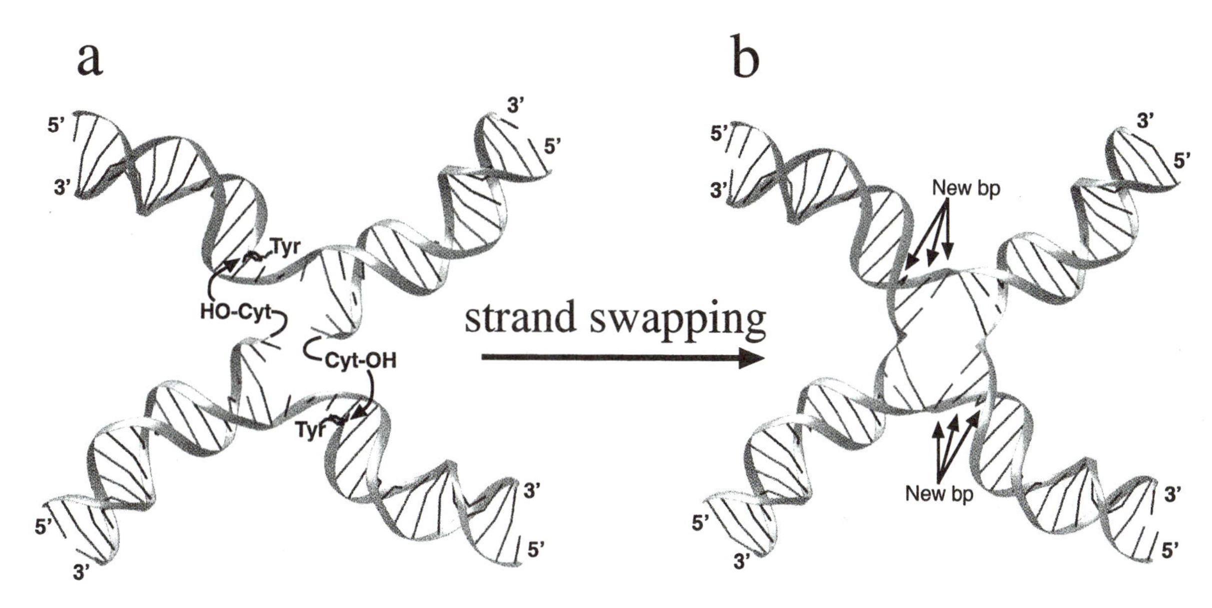

Figure 5. Exchange of DNA strands by strand swapping. (a) DNA structure in the covalent Cre-DNA intermediate (39). 5′-Cytidine residues were cleaved and allowed to diffuse away in order to trap this intermediate (72). These residues are shown modeled into the structure, with the direction of nucleophilic attack indicated by arrows. (b) DNA structure in the Cre-HJ intermediate (32). The formation of 3 bp with the partner substrates as a result of strand exchange is indicated by arrows. Only the phosphodiester backbone of the exchanged bases needs to move in order to form the HJ intermediate.

binase-DNA model may apply quite well to the XerC-XerD-*dif* recombination system, where there is also a 6-bp crossover (8), an interesting question arises with the λ-Int system, where the crossover site is 7 bp (55), or in the Flp-*frt* (86) and XerC-XerD-*cer* (91) systems, which have 8-bp crossovers. Two simple strand-swapping scenarios can be imagined that would accommodate the difference in bases to be exchanged: (i) 3 bp could be swapped at each step, with the remainder tested by branch migration of the HJ intermediate (74), or (ii) all bases could be exchanged by swapping. If all bases are homology tested by swapping without use of branch migration, then λ-Int would require either unequal swap sizes or else equal swaps of 3.5 bp with a symmetrical arrangement of unpaired bases at the center of the junction. Flp and XerCD-*cer* would require 4-bp swaps in the no-branch-migration model.

cis VERSUS *trans* CLEAVAGE

In 1992, Chen et al. reported that the Flp recombinase subunit does not cleave the scissile phosphate adjacent to the site to which it binds, but rather it cleaves the phosphate on the opposite end of the crossover region (20). This *trans* cleavage mechanism requires that the Flp active sites be formed from at least two subunits and that no less than two Flp monomers are required to form an active site that is competent for cleavage. A series of further experiments in the Flp system throughout the 1990s have corroborated this finding (28, 58, 103), but the same cannot be said of other tyrosine recombinase systems, where conflicting results were published. Han et al. reported in 1993 that λ-Int also cleaves in *trans* (42), but this result was soon challenged with the demonstration of *cis* cleavage by λ-Int of bispecific DNA substrates designed from the closely related λ and HK022 bacteriophage integrase systems (73). In the Xer system, where the distinct DNA-binding specificities of the two recombinases provide a convenient experimental framework for performing complementation experiments, *cis* cleavage was clearly demonstrated (8). Finally, Shaikh and Sadowski reported in 1997 that Cre recombinase uses a *trans* cleavage mechanism (90). A number of explanations have been offered for the different observations among the well-studied tyrosine recombinases (13, 50, 73).

The Cre-DNA and Flp-DNA crystal structures clearly demonstrate that there are two cleavage lifestyles in the tyrosine recombinase family. The Cre-DNA crystal structures indicate *cis* cleavage, with all amino acids that form the active site around a given scissile phosphate coming from the protein subunit that is bound adjacent to that phosphate (39). The Cre-DNA structures also provide a rationale for how *trans* cleavage might occur in a suitably perturbed Cre system or in the Flp system (33). In contrast, the Flp-HJ complex clearly illustrates *trans* cleavage and formation of a shared active site between subunits. The two active site architectures are illustrated schematically in Color Plate 10 (see color insert). In Cre, the active-site residues Arg-I, Lys-β, His-II, Arg-II, and Trp-III (RKHRW) come from the catalytic domain bound adjacent to the scissile phosphate, the tyrosine nucleophile comes from helix M of the same domain, and the C-terminal helix N is donated in *trans* to the adjacent subunit. In Flp, the active-site RKHRW residues also come from the *cis*-bound catalytic domain, but the tyrosine nucleophile is donated from the helix M equivalent in *trans* from the adjacent subunit. It has been noted that the linker between helix L and helix M is longer in Flp than in Cre, which presumably allows helix M to reach the neighboring active sites in the yeast recombinases but not in the bacterial systems (13, 33).

It is difficult to interpret the crystal structures of the λ- and HP1-Int catalytic domains and the XerD protein in terms of *cis* and *trans* cleavage modes in the absence of DNA substrates; however, there are some important indications from two of the structures. The HP1-Int structure shows the tyrosine nucleophile bound to a sulfate ion in the active site, strongly suggestive of *cis* cleavage geometry (45). The XerD structure shows the catalytic tyrosine hydrogen-bonded to the protein backbone adjacent to the *cis* active site, also suggestive of *cis* cleavage geometry (96). The biochemical evidence in the λ-Int and Xer systems also seems to indicate that, like Cre, they cleave their substrates in *cis* (8, 73).

WHICH STRAND IS CLEAVED FIRST DURING SITE-SPECIFIC RECOMBINATION?

In principle, either of the DNA strands in the two duplex substrates could be cleaved and exchanged first in the reaction pathway. From a structural point of view, it is useful to consider whether there are differences between the DNA strands and/or the active sites in the precleavage synaptic complex that would favor the initial cleavage and exchange of one strand over the other. In the Cre-DNA synapse formed with inactive Cre mutants and symmetrized *loxP* sites (38), there is a clear distinction between the two strands (Color Plate 11 [see color insert]). The strand that corresponds to the one caught in the act of swapping in Fig. 5 has a helicoidal trajectory that takes it through the middle of the open strand exchange region between duplex substrates. This strand is also relatively free from recombinase-backbone contacts that might limit its mobility. In contrast, the complementary strand is embedded in a more extensive protein interface, which is primarily composed of sugar and phosphate contacts located on the side of the DNA duplex pointing away from the strand exchange region. If this strand were to be exchanged first, a more complicated unraveling from the duplex and dissociation from recombinase interactions would be required.

In addition to differences in trajectories and protein contacts of the two DNA strands, there is a difference in the relative active site positioning in the precleavage complex. The two active sites that are "activated" for cleavage and strand exchange are 34 Å apart (measured as the distance between scissile phosphates). The inactive active sites where cleavage and strand exchange do not occur at this step of the reaction are 43 Å apart. This difference, combined with the nonideal positioning of the complementary DNA strands, may make the strand exchange event geometrically inaccessible for those strands even if cleavage were to occur at the alternative scissile phosphates.

The above analysis of the precleavage Cre-DNA structure suggests that the order of strand exchanges in the recombination pathway is determined once the synaptic complex is formed. However, two distinct antiparallel (and two distinct parallel) synapses are possible, which differ by the direction of bending of the sites (Color Plate 11). Recombination would therefore be expected to proceed with an opposite order of strand exchanges in these two cases (38). For Cre, a preference for cleaving at the G/C end of the *loxP* crossover has been reported (47), which suggests that one of the two synapses shown in Color Plate 11 is slightly favored over the other. In principle, a bending preference arising from the crossover sequence itself could be responsible for the observed bias, although there are no published experimental data to support this idea for the Cre-*loxP* system.

The Flp recombinase, which has a larger crossover compared to Cre, shows little or no preference for the order of strand exchange during recombination (60, 61). Although bending of the *frt* site by Flp has been well studied by the Sadowski laboratory (66, 67, 86, 87), there is no strong evidence that the bend occurs in a preferred direction, which is consistent with a lack of initial cleavage preference. The λ-Int and Xer systems, on the other hand, show a strong preference for cleaving one strand first during recombination (8, 51, 72). Since formation of the λ intasome requires a precise higher-order assembly of proteins and supercoiled *attP*-containing substrate (55), it is reasonable to assume that there are not two alternative intasomes that could proceed through the recombination pathway with an opposite order of strand exchange. Indeed, the Landy and Nash laboratories provided evidence over 10 years ago that it is the accessory region of *attP* which dictates the order of strand exchanges during λ recombination (51, 72). In the context of the Cre-DNA synapse models in Color Plate 11, this could be interpreted as a unique bend direction, and therefore a unique synaptic complex, with only one pair of strands poised for initial exchange.

Strong biochemical evidence supporting the relationship between bending of the core recombination site and strand cleavage preference in the synaptic complex comes from studies of bulged DNA substrates in the Cre and Flp systems. Lee et al. have shown that when recombination sites containing one wild-type strand and one strand with a bulge (extra nucleotides) located in the crossover region are processed by the appropriate recombinase, the strand harboring the bulge becomes preferentially selected for cleavage (60). The results of these experiments can be readily explained with respect to the Cre-DNA synaptic structures. Only one of the two crossover strands (the one destined for initial cleavage and exchange) could sterically accommodate a polynucleotide bulge, and the site is bent in the same direction as that expected for a bulge-induced bend.

The XerCD system also shows a strong preference for the order of strand exchanges during recombination, with XerC catalyzing strand exchange to form the HJ intermediate and XerD (or other factors) being responsible for HJ resolution (8, 90). In addition to the structural constraints imposed by accessory proteins and intrinsic preferences in the crossover DNA sequence, the fact that two distinct recombinases are used in this system provides a third source of bias in synapse formation. The interface formed between XerC and XerD bound to a recombination site is likely to differ from that between XerD and XerC bound to the same site, but bent in the opposite direction. This is most easily visualized in terms of the ball-and-socket cyclic helix swap between subunits in the Cre-DNA structures where the XerC ball fitting into the XerD socket is not identical to the XerD ball fitting into the XerC socket (see also Fig. 6 in reference 40). One of these types of interaction will provide intersubunit contacts on a single site and the other type of interaction will exist between synapsed duplexes at the start of the recombination pathway. An energetic preference for which interaction occurs in which context could therefore also impose a strong bias in the assembly of the XerCD-DNA synapse.

THE HOLLIDAY INTERMEDIATE

Much of the focus in tyrosine recombinase research in recent years has been on the Holliday intermediate of the reaction pathway. Many of the questions raised have dealt with the initial location of the branch point when the HJ is first formed, whether that branch point migrates during the recombination process, and if so how far it migrates. The branch migration issue is related to the requirement for homology between crossover sites and the need for any mechanism to account for this observation, but branch migration also could play a central role in the isomerization of the junction, allowing the reaction to switch from top-strand cleavage mode to bottom-strand cleavage mode and proceed to form recombinant products. Here, I summarize what we have learned from the crystal structures of recombinase-HJ intermediates and then relate the structural models to recent experimental data in the tyrosine recombinase family.

There are at present two crystal structures available representing the Cre-HJ intermediate (32) and one crystal structure of Flp recombinase bound to an HJ substrate with a 7-bp crossover site (21). In one of the Cre-bound junctions (HJ2 in reference 32), eight overlapping DNA strands are assembled to form an HJ with four nicks (missing phosphates) that is fourfold symmetric with respect to sequence symmetry and in principle is free to branch migrate within the crossover region. The branch point is located at the center of the crossover region in this structure. The other Cre-bound HJ (HJ1 in reference 32) is an immobile junction with the branch point fixed by design at the center of the crossover region. The structures of the two junctions are nearly superimposable, indicating that the presence of the nicks and the asymmetries imposed to form an immobile junction most likely do not distort the observed structures significantly from that of the true HJ intermediate in Cre-*loxP* site-specific recombination. The DNA structure in the Cre-bound HJ complex is shown in Fig. 6.

The Cre-bound HJ arms, defined here as the duplex DNA segments extending away from the branch point, lie nearly in the same plane, with a small amount of curvature within each arm that creates a slightly concave surface on the side of the junction plane where the recombinase catalytic domains bind. The arms adopt a distorted square planar arrangement that is strictly only twofold symmetric, with interarm angles of ~75° and ~105°. At the center of the junction, all bases are Watson-Crick paired, and all eight bases are completely unstacked and exposed to solvent. The nearly square planar arms and the complete unstacking at the branch point in the Cre-HJ complexes most closely resemble the square planar form of the HJ characterized in solution by Duckett et al. (29) and observed in the crystal structures with bound RuvA protein (43, 84). For the free HJ in solution, this junction conformer exists in the absence of divalent ions and low concentrations of monovalent ions, where phosphate-phosphate repulsions presumably lead to a fully extended fourfold-symmetric structure (65).

The alternative form of the HJ in solution is referred to as the stacked-X conformation, where the junction arms pair up to form fully stacked, continuous duplexes that cross over between stacks at the branch point. This conformer forms interstack angles of 60°, with twofold symmetry overall. The more stable form of the stacked-X junction is one in which the pair of strands that form a continous helical path through the stacked duplex arms are antiparallel with respect to one another. A given antiparallel stacked-X junction can still exist in one of two conformers, however, depending on which pair of strands form a "continuous" path and which pair crosses over between duplex stacks. It is this feature of the junction, where there is a clear stereochemical distinction between the DNA strands, that is shared by the Cre-HJ structures. For reviews of HJ structures and properties, see references 64, 65, and 88.

In addition to the unequal interarm angles, the geometry of the deoxyribose-phosphate backbone is quite different for the two pairs of DNA strands in the Cre-bound HJ. The strands that span the obtuse interarm angle form a more or less uninterrupted helicoidal trajectory through the branch point of the junction. The DNA strands that span the acute interarm angles have a rather sharp discontinuity at the branch point, where the branch point phosphate rotates towards the center of the junction, away from the path traced by the flanking phosphates. Thus, in the Cre-bound HJ, like in the stacked-X junction, there are a pair of continuous DNA strands and a pair of crossing DNA strands which are stereochemically distinct. It follows, therefore, that the recombinase active sites that surround the scissile phosphates on the continuous versus crossing strands in the junction are not stereochemically equivalent. As discussed below, the crossing strand active sites are activated for cleavage and the continuous strand active sites are prevented from cleaving the DNA. A similar HJ geometry has been observed in the Flp-DNA complex (see Color Plate 12) (21).

The properties of tyrosine recombinase HJ intermediates in solution have been studied for the λ-Int, Xer, and Flp systems. At the same time as structural models in the tyrosine recombinase family began to emerge, the Landy and Sherratt laboratories described two different, but complementary, experiments that addressed the relationship between junction conformation and strand cleavage specificity. In the λ-Int study, Azaro et al. manipulated the DNA sequences at the branch point of the junction and found that purines strongly favor adopting the crossing-strand configuration in the unbound stacked-X junction (10). This allowed the construction of immobile HJs that were substrates for resolution by λ-Int which adopt only a single-crossover isomeric form prior to λ-Int binding. In the presence of the integrase protein, the junction was resolved to form only one of two possible pairs of products, indicating that a conformational bias existed in the integrase-bound complex and that this bias is somehow related to that observed in the free junction. The strands cleaved by λ-Int in this case were the ones that adopt the crossed configuration in the free junction.

Arciszewska et al. used a different approach and arrived at a similar result (7). Using synthetic HJs that are substrates for Xer resolution, the junction arms were physically connected by a polythymidine tether, forcing the connected arms to make a more acute angle and thus biasing the conformation of the free junction towards one of the stacked-X crossover isomers. As in the λ-Int case, resolution of the tethered junctions by XerC-XerD produced primarily one set of products, corresponding to cleavage of the strands that are crossed in the free junction. Importantly, the same study also demonstrated using permanganate oxidation sensitivity of thymine bases that the recombinase-bound form of the junction is clearly distinct from the stacked-X conformation. Thymine bases at the branch point of the HJ complexes become sensitive to oxidation when XerC-XerD are bound, indicating that unstacking and exposure of bases to solvent occurs in the junction.

Only in the Flp system have the conformational properties of the recombinase-bound junctions been probed in detail (62). The conformational symmetry of the Flp-HJ complex has been analyzed using the electrophoretic methods introduced by Cooper and Hagerman (25). Rather than displaying a pattern of twofold symmetry as might be expected from the Cre-HJ structures and the λ-Int and Xer HJ resolution experiments discussed above, a square-planar, fourfold-symmetric conformation was most consistent with the Flp-DNA electrophoretic mobility data. One possible explanation for the high symmetry observed is that the Flp-HJ complex can interconvert readily between two dyad symmetric forms, resulting in a time-averaged mobility indicative of a fourfold-symmetric structure. In a more recent report, a series of resolution experiments using Flp recombinase and tethered junctions in both an antiparallel and a parallel orientation was described (61). For the antiparallel junctions, only modest resolution bias was observed as a result of tethering the junction arms and little or no bias was observed as a result of modifying the crossover region DNA sequence. These results appear to contrast with those observed in the λ-Int and Xer system and may represent differences between the yeast and bacterial recombinase subfamilies. Similar experiments have not yet been described for Cre recombinase.

ISOMERIZATION OF THE HJ INTERMEDIATE

It has been understood for many years that a change in the structure of the HJ intermediate in tyrosine recombinase recombination must serve the pivotal role of deciding whether the reaction should proceed forward to recombinant products or backwards to restore the original substrates (see discussion in references 26, 32, and 74). Early models of this isomerization step involved a migration of the junction branch point between the sites of cleavage within the crossover region (Fig. 2). The strands cleaved in resolution of the HJ would be those whose scissile phosphates were located at the junction branch point. In addition to providing a mechanism for reading homology between sites, this model provided a rationale for how pairs of active sites could be turned on and off in response to the quaternary structure of the recombination complex. The Cre-DNA complex structures, and in particular the structures of the Cre-HJ intermediates, have provided a different view of this central isomerization step in the recombination pathway, which along with biochemical studies from several laboratories has led to an alternative model (32, 33).

In the specific case of the Cre-bound HJ complex, there are two different ways one could draw the structure shown on the left of Fig. 6. The conformers differ by which arms form acute and which arms form obtuse angles and by which DNA strands adopt the crossing configuration with inverted phosphates at the branch point and which strands adopt the more continuous configuration. The Cre-*loxP* model for HJ isomerization shown in Fig. 6 involves a planar scissoring motion of the HJ arms which swap the pairs of interarm angles along with adjustments to the deoxyribose-phosphate backbone torsion angles that leads to a swap of branch point phosphate configurations. The analogy with crossover isomerization of the Holliday intermediate in homologous recombination is remarkable. In both cases, the generation of strand equivalence by quaternary changes in the junction structure is accomplished. It is not difficult to imagine that tethering the arms of the junction (7) or imposing torsional or bending constraints with accessory factors (31) could influence which isomer is energetically preferred or which isomer is more likely to be initially formed upon recombinase binding.

The effect of HJ isomerization on the catalytic domains of the recombinase subunits mirrors that of the DNA strands. Those pairs of adjacent subunits that are separated by a larger distance in one conformer are closer together in the alternative conformer. Accordingly, the nature of the interactions between the catalytic domains in these two distinct interfaces is exchanged between recombinase pairs. As discussed later in this chapter, these differences in domain-domain interfaces may relate directly to the positioning of active site residues and therefore in determining which active sites are activated for cleavage. The situation with the N-terminal domains of the Cre subunits is somewhat different. These domains main-

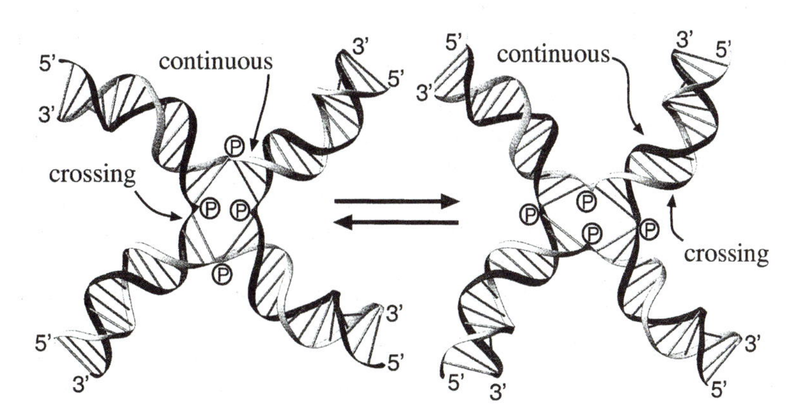

Figure 6. Isomerization model for the HJ intermediate in Cre-*loxP* site-specific recombination. The conformer on the left differs from that on the right by an exchange in the interarm angles and an exchange in the positions of the branch point phosphates. The stereochemical identities of the dark (crossing) strands on the left are identical to those of the light (crossing) strands on the right. Likewise, the light strands (continuous) on the left are equivalent to the dark (continuous) strands on the right. The strands labeled "crossing" are activated for cleavage and exchange in the two conformers.

tain a nearly fourfold-symmetric arrangement in both isomers, although the interactions formed with the crossing and continuous strands of the junction are not the same.

In terms of active versus inactive subunits in the Cre-HJ model, the subunits whose active sites are positioned on the crossing strands are predicted to be active for strand exchange and the pair of subunits on the continuous strands are predicted to be rendered inactive. These predictions follow from a comparison with the cleaved Cre-DNA intermediate (39), where the active site that has catalyzed formation of a covalent Cre-DNA linkage corresponds to the one containing the crossing-strand phosphates in the HJ complex. A similarity also exists at the quaternary structural level between the synapsed substrates and the HJ intermediate. Aside from differences in DNA structure at the center of the synaptic complex due to the strand exchange event, the structures of the synapsed sites, the covalent Cre-DNA intermediate, and the Cre-HJ intermediate are nearly superimposable, allowing for easy prediction of the active subunits in the two complexes where direct evidence of cleavage is not available.

It is tempting to conclude that the crossed strands cleaved in the λ-Int experiments with sequence-biased junctions and in the XerCD experiments with tethered junctions correspond to the strands we have labeled crossing in the Cre-HJ structures. In particular, the resolution bias observed with tethered junction arms in the Xer system (7) could be readily explained with respect to Fig. 6 if the tethered duplex arms are simply forced to adopt acute interarm angles. The resolution bias observed by λ-Int in response to the choice of branch point sequence (10) is not so easily reconciled with the Cre-HJ models. There is no obvious reason why purines should be preferred in the crossed configuration in Fig. 6; the bases on the crossing and continuous strands are almost entirely solvent exposed at the branch point, and in neither case do the bases interact with other macromolecular components in the recombination complex. One possibility is that the branch point sequence in the λ-Int-bound junction does not in fact significantly bias the Int-bound HJ conformer, but the strongly biased conformer of the stacked-X junction that exists prior to Int binding is most readily opened to form only one of the two possible conformers, and resolution of the junction occurs before significant isomerization can occur. This type of kinetic effect would assume that cleavage and strand exchange are fast compared to the isomerization process.

It is clear from Fig. 6 that the proposed isomerization of the Cre-HJ intermediate does not involve migration of the branch point from the center of the crossover region. In fact, given the extent of the protein-protein interfaces (average surface size of ~3,400 Å^2 buried in each pair of adjacent subunits) in the Cre tetramer, it is difficult to imagine how this HJ intermediate could undergo any branch migration. A coordinated rotation of the protein-bound DNA arms to move the branch point would require that the observed protein-protein interface be disrupted and a new interface be formed. Even if the old and new interfaces were isoenergetic, considerable activation energy would no doubt be required to accomplish this type of rearrangement.

Despite the difficulties in accepting a branch migration model for Cre-*loxP* site-specific recombination based on the crystal structures of the reaction intermediates, the possibility is much more compelling for the λ-Int, Flp, and Xer systems. As discussed previously, the crossover regions in these systems lead to an intriguing set of questions concerning the number of nucleotides swapped at each strand exchange step of the reaction and the role, if any, of branch migration in further testing homology and activating or deactivating the recombinase active sites. In the λ-Int system, Nunes-Düby et al. have suggested that three bases could be swapped at the initial strand exchange step, placing the branch point one-half base pair from the center of the crossover region (71). Successful branch migration through this short distance would test the central base pair between substrate sites and activate the second pair of active sites to swap the second pair of DNA strands. The isomerization in this case would most likely involve a coordinated 17° rotation of the protein subunits and a minor remodeling of the intersubunit contacts in the complex.

An interesting alternative to the limited branch migration model for λ-Int is suggested by the Flp-HJ complex. Chen et al. created a Flp-HJ complex using nicked suicide substrates that had unexpectedly rearranged to yield an HJ with a 7-bp spacer rather than the 8-bp spacer found in the normal Flp reaction (Color Plate 12 [see color insert]) (21). In this complex, the branch point of the junction is exactly at the center of the crossover site, and the four central odd bases in the junction (all of which are adenine) stack against the junction arms but are unpaired. If the λ-Int HJ intermediate were to adopt such a configuration following an initial strand exchange of 3.5 bp, the resulting complex could in principle share properties of the Cre-HJ model, including a more subtle rearrangement of subunits during HJ isomerization.

The Flp recombinase system presents a related, but more symmetric version of the branch migration question. The initial strand exchange in this case could involve a 3-bp swap, with formation of an HJ intermediate with its branch point located 1 bp from the center of the crossover. Isomerization of the junction intermediate could then incorporate a 2-bp movement of the branch point, thereby activating the second strand exchange. An alternative to this mechanism is the simpler isomerization described for Cre, requiring a symmetric 4-bp strand swap to form and resolve the Flp-HJ intermediate. A generalized model for the tyrosine recombinase reaction has been described by Voziyanov et al. which incorporates the rotations and displacements required by 0-, 1-, or 2-bp branch migration models, with particular attention to the relative geometries of the bound recombinase subunits and the scissile phosphates as the reaction proceeds (101).

The XerCD system presents a number of mechanistic questions for the HJ intermediate, stemming from the different crossover sequences found in plasmid and chromosome substrates, from the various requirements for auxiliary proteins and auxiliary DNA sequences, and from the fact that two distinct recombinases are involved. Recombination at *dif* (6-bp crossover), for example, does not require accessory proteins and could in principle proceed in a three-dimensional framework similar to that observed in the Cre-DNA crystal structures (40). Recombination at *cer* (8-bp crossover), in contrast, requires two accessory proteins for recombination of supercoiled substrates containing sites arranged in direct repeat (24). The fact that initiation of recombination at *dif* requires only XerC and XerD suggests that the protein-protein interactions required for association of sites are relatively strong in this case, as in Cre and Flp. The weaker synaptic interactions that presumably make *cer* recombination PepA and ArgR-dependent are consistent with the idea that the larger crossover disrupts optimal intersubunit interactions and is also reflected in a lower cooperativity of recombinase binding to the core sites in these cases (discussed in reference 12).

The most simplistic view of XerCD recombination, given the available structural models and results of biochemical and mutagenesis studies, is that recombination at *dif* most closely resembles Cre-*loxP* recombination, with 3-bp swaps during strand exchange and a subtle isomerization of the HJ intermediate. This model must in some ways be an oversimplification, since resolution of the HJ intermediate at both *cer* and *dif* sites in vivo requires additional cellular factors to resolve the HJ intermediate produced by XerC-catalyzed strand exchange (12, 41). Recombination within the core site of *cer* presents the same questions as in the Flp system regarding the strand swap sizes and isomerization of the HJ intermediate. It remains to be seen whether these systems all use similar mechanisms or whether evolution has refined unique solutions to apparently related problems.

MECHANISM OF PHOSPHORYL TRANSFER

A distinguishing feature of the tyrosine recombinase mechanism of site-specific recombination is the stepwise exchange of DNA strands between recombining sites. One pair of strands is exchanged to form an HJ intermediate, which is then a substrate for exchange of the second pair of strands. This aspect of the reaction is quite different than found in the invertase/resolvase family of enzymes, where both pairs of strands are cleaved and exchanged in a more concerted overall reaction without formation of a branched intermediate (93). In order to understand the molecular basis of this stepwise process, it is important to understand the enzymology of the cleavage and ligation reactions that the recombinase subunits catalyze. Unfortunately, a detailed kinetic model for the wild-type and mutant recombinase enzymes is not yet available, although the rates for some specific steps in the Cre and Flp reactions have been investigated (68, 83, 101). More extensive data from the closely related vaccinia virus topoisomerase system have, however, shed considerable light on active site catalysis in the site-specific recombinases (see references in references 22 and 52).

A general model for acid-base catalysis of the phosphoryl transfer steps in the tyrosine recombinase cleavage reaction is shown in Fig. 7 (see also references 35 and 91). The model is an associative mechanism, with in-line attack of the scissile phosphate by tyrosine to form a trigonal bipyramidal transition state, followed by expulsion of the 5′-hydroxyl leaving group in the cleavage reaction. The ligation reaction is the reverse of the cleavage reaction. The Arg-I, His-II, Arg-II triad of conserved side chains, along with the Trp-III side chain observed in Cre (His in λ-Int, HP1-Int, and XerD), provides an array of hydrogen bond donors surrounding the scissile phosphate that nicely fulfills the requirement of stabilizing a pentacoordinate transition state of the phosphate and probably contributes an electrophilic catalysis component to the reaction.

The identities of the catalytic base and acid in Fig. 7 have not yet reached a consensus opinion in the field. Stewart et al., for example, have suggested that His632 (His/Trp-III equivalent [see Table 2]) could be the general acid responsible for protonating the 5′-OH leaving group in human topoisomerase I (95). Grainge and Jayaram have suggested that the corresponding Trp in the Flp and Cre enzymes could

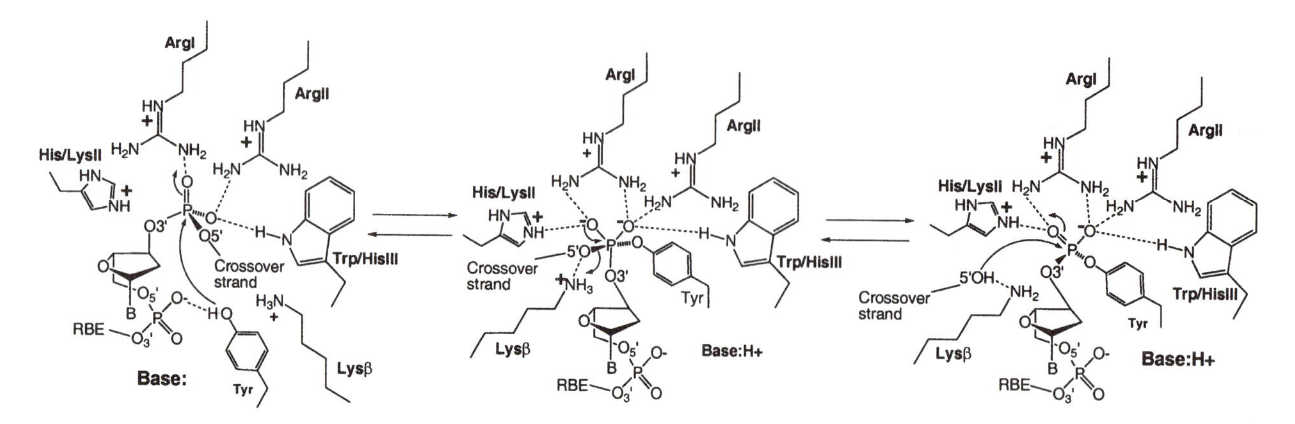

Figure 7. A cleavage mechanism for tyrosine recombinases. The general base responsible for accepting a proton from the tyrosine nucleophile during cleavage (indicated by "Base:") has not been clearly identified biochemically, but could be His/Lys-II. The general acid responsible for protonating the 5′-OH leaving group has been identified as Lys-β in the eukaryotic topoisomerases and is therefore likely to play the same role in the tyrosine recombinases as shown here (52). The ligation reaction is simply the reverse of that shown above, beginning with nucleophilic attack of the 3′-phosphotyrosine intermediate by a crossing strand 5′-hydroxyl group.

also perform this function (35). In contrast with these proposals, Shuman and colleagues have shown that while mutation of the equivalent histidine in vaccinia virus topoisomerase I (His265) to alanine results in a 100-fold decrease in catalytic efficiency, mutations to Asn or Gln (which could also hydrogen-bond to the scissile phosphate) restore near wild-type activity (79). Since Asn, Gln, and Trp are poor acids, the most likely role of His265 therefore appears to be coordination and stabilization of negative charge on the nonbridging phosphate oxygen during catalysis.

The structure of the vaccinia virus topoisomerase enzyme and the subsequent comparison to the tyrosine recombinases revealed that topoisomerase Lys167 (Lys-β) is invariant in both the topoisomerase Ib family and in the integrase family recombinases (22). Although it was known that Lys167 is essential for topoisomerase catalysis (104), little attention had been paid to the corresponding lysine in the recombinases. In fulfillment of the prediction by Cheng et al. that this lysine is also likely to be essential for the recombinases enzymes (22), the XerD Lys172 to Ala mutant was shown to be inactive (17) and we found that the Lys201-to-Ala Cre mutant was also inactive (F. Guo and G. Van Duyne, unpublished data). Wittschieben and Shuman originally proposed in a biochemical characterization of the topoisomerase Lys-β mutant that this residue may serve as the catalytic acid responsible for protonating the 5′-OH leaving group during the cleavage reaction (104). In the Cre-DNA covalent intermediate structure, Lys201 hydrogen bonds to the O-2 atom of Thy adjacent to the scissile phosphate on the noncleaved strand (39), prompting the suggestion that this conserved lysine may play a structural role in organizing the active site (35). Most recently, Krogh and Shuman have reported experimental evidence using 5′-bridging phosphorothiolate-containing substrates that Lys-β is in fact the general acid responsible for activating the 5′-hydroxyl leaving group during topoisomerase catalysis, as indicated in Fig. 7 (52).

On the other side of the scissile phosphate, there have also been multiple proposals for the identity of a catalytic base that accepts a proton from tyrosine during cleavage and returns the proton during ligation. Champoux and coworkers have suggested that a water molecule might serve as proton acceptor during cleavage by human topoisomerase I (95), whereas conserved His-II has been proposed to serve this function in the tyrosine recombinases (39, 91). The alanine mutant at this position in Flp recombinase is defective in the ligation step of the reaction but is competent in the cleavage reaction (77). In the eukaryotic topoisomerase I enzymes, the corresponding residue is a lysine (Table 2). Mutation of this residue in the vaccinia virus topoisomerase to alanine results in a modest 10-fold loss of activity (23).

The Cre-DNA complex structures show His289 to be ideally positioned to serve as both a proton acceptor during cleavage and a proton donor during ligation (32, 38, 39). When protonated, His289 could swing into the hydrogen bonding position, where it is observed coordinating the cleaved phosphate in the covalent Cre-DNA complex. In the precleavage Cre-DNA complexes, His289 is not hydrogen bonded to the tyrosine nucleophile hydroxyl group but could readily be so with a small rotation of side chain torsion angles as the tyrosine moves into cleavage position (38). In the human topoisomerase I-DNA structures, the corresponding lysine is appropriately positioned to play a similar role (80), but a larger movement of the side chain would be required in order to reach the tyrosine hydroxyl group. While it seems reasonable to assume that the conserved His/Lys-II residues perform the role of general base and acid in the recombinases and topoisomerases cleavage and ligation steps, convincing biochemical evidence is still lacking.

REGULATION OF ACTIVE SITES

The relatively simple question of what turns on one pair of recombinase subunits and what turns off the other pair during tyrosine recombinase site-specific recombination has been a difficult one to answer with the available structural data. The models that have been most recently discussed (33, 40, 100) involve changes in intersubunit interactions as the recombination reaction proceeds (Fig. 8). At the start of the reaction, synapsed sites form a pair of interfaces between recombinases bound to the same DNA substrate (referred to here as type I) and a pair of synaptic interfaces between subunits bound to different recombination sites (referred to here as type II). Following cleavage and strand exchange to form the Holliday intermediate, the HJ inherits the same arrangement of interfaces and the reaction can be reversed by cleavage of the same pair of phosphates to produce starting substrates. Upon isomerization of the Holliday intermediate, the type I interfaces become type II and the type II interfaces become type I. If we assign directionality to these interface labels (Fig. 8), then we can further specify that the subunit at the receiving end of the type II interface is active and the subunits at the receiving ends of the type I interfaces are inactive in the case of the Cre-*loxP* system. This view of the reaction emphasizes that it is the directional protein-protein interactions between adjacent subunits in the recombination complex that dictate the state of each subunit. Such a model would equally well describe isomerizations that may involve limited branch migration (i.e., for Flp and Int). For Cre recombinase, the type I and type II interfaces are very similar because the junction isomerization is subtle. For those systems that branch migrate, the interfaces would be expected to be more distinct, given the greater extent of confor-

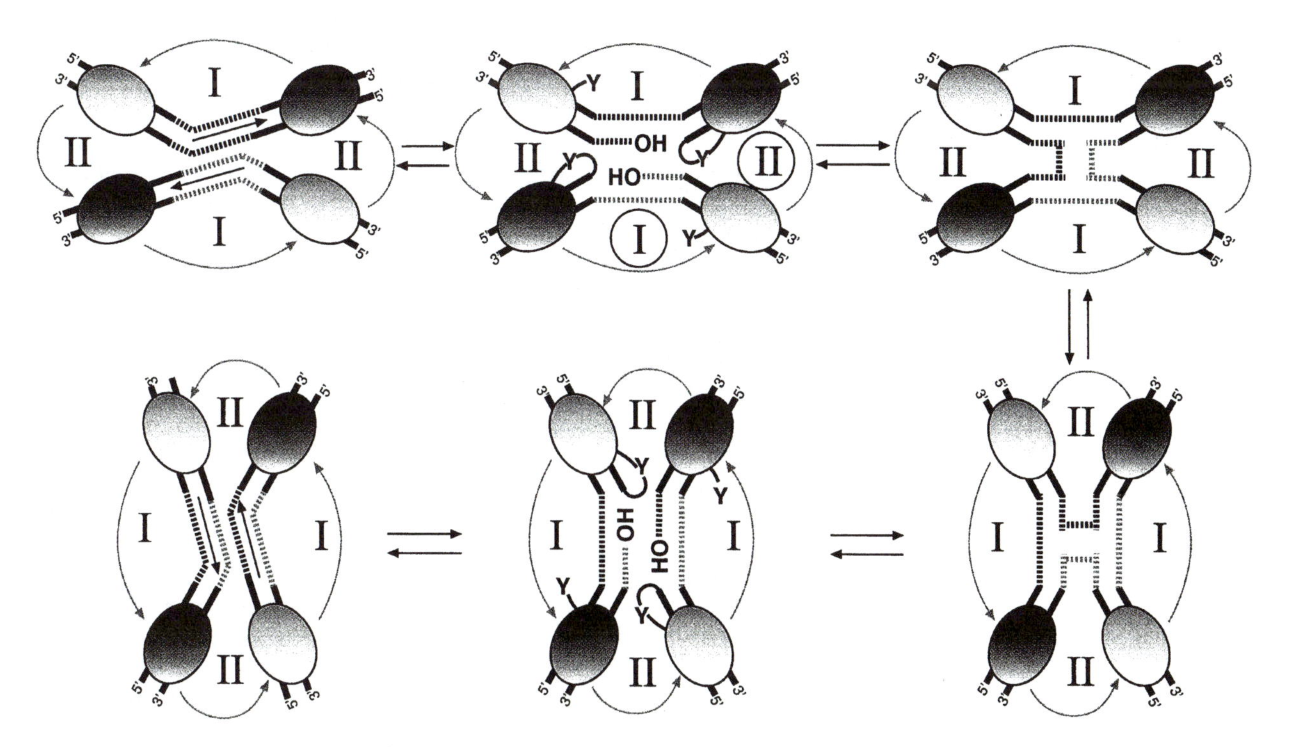

Figure 8. Strand-swapping model of Cre-*loxP* site-specific recombination. Dark gray subunits are active for cleavage in the top half of the pathway, and light gray subunits are active for cleavage in the bottom half of the pathway. The DNA substrates lie nearly in the same plane and undergo only a subtle scissoring motion at the HJ isomerization step of the reaction, which serves to switch the roles of the protein subunits and switch which strands are activated for exchange. The mechanism does not involve branch migration of the HJ intermediate. I and II refer to the type I and type II interfaces formed between protein subunits in the tetrameric recombination synapse, and curved gray arrows indicate the directionality of the interface (see text). The two circled interfaces are the ones illustrated in Color Plate 13.

mational changes required to move the junction branch point. Voziyanov et al. have presented a thorough geometric treatment of the tyrosine recombinase reaction pathway model that focuses on the relative rotation and positioning of the reaction components as the reaction proceeds (100).

The XerCD system represents a special case in thinking about intersubunit interactions. Since the two recombinases bound to each site are different, the type I and type II interfaces are by definition distinct, although the high degree of sequence similarity between the proteins implies that the interfaces will probably not differ greatly based on amino acid sequences alone. The power of the two-recombinase XerCD system has led to the most extensive biochemical investigation of intersubunit interactions among the tyrosine recombinases (6, 40, 92). Mutations in XerC not only are able to affect the catalytic state of the XerC subunit but can also dramatically influence that of the XerD subunits. Similarly, XerD mutants affect the activities of both XerD and XerC. These types of experiments strongly reinforce the general model in Fig. 8 and, most importantly, establish a specific framework for understanding these interactions on an amino acid level.

In attempting to dissect the differences between the type I and type II interfaces in the Cre-DNA structural models, we have employed the difference-distance matrix technique to compare active and inactive subunits (82). The results of this comparison for the precleavage Cre R173K-*loxS* complex indicate two regions of the structure that differ significantly between the two subunits. These regions correspond to the loop between β-strands 2 and 3, which contains Lys-β, and the C-terminal segment of the recombinases, which includes helix M, the M-N linker, and helix N. Other than in these two regions, the recombinase subunits in structures of the precleavage complex, the covalent intermediate, and the HJ intermediate superimpose quite well.

As indicated from the difference-distance matrix analysis, the helix M-helix N segments adopt very different conformations in the two types of interfaces. In the type I interface, helix N is donated to the catalytic domain of the inactive subunit and the linker between helix M and N is in an extended conformation (Color Plate 13a [see color insert]). In the type II interface, helix N is donated from the inactive to the active subunit and the helix M-helix N linker adopts a more compact conformation (Color Plate 13b). One likely source of the different peptide linker conformations in the two interfaces is the difference in intersubunit distances in the Cre-DNA complexes (25 Å between scissile phosphates in the type I interface versus 30 Å in the type II interface). Although the differences in linker conformations are dramatic, the linker amino acids themselves are not components of the active site, and any effects on catalysis are therefore expected to be indirect. A related observation from comparison of the C-terminal segments in the active versus inactive subunits in the three Cre-DNA reaction intermediates involves the positioning of the Tyr324 nucleophile. In each of the three reaction intermediates, the inactive Tyr324 does not contact the scissile phosphate but instead is hydrogen bonded to the adjacent phosphate on the 5′ side. This inactive tyrosine is linked via helix M to the compact linker peptide in the type II interface. Tyr324, helix M, the linker, and helix N have well-defined electron density for the type II interface in the Cre-DNA crystal structures, indicative of a single conformation with relatively low mobility of the peptide backbone. The active Tyr324 adopts a slightly different conformation in the structures of the three Cre-DNA intermediates. In the precleavage synaptic complex and in the HJ complex, the Tyr324 hydroxyl groups are intermediate between the scissile phosphate and the 5′-adjacent phosphate. These tyrosines are connected via helix M to the extended linker peptides in the type I interfaces. The electron densities of Tyr324, helix M, and the linker segment are much less well-defined than those in the type II interfaces, with higher thermal parameters indicative of main chain mobility in this region of the structure (37).

In terms of active-site regulation, these observations could be interpreted as follows. The Tyr324 nucleophile in the subunit rendered inactive in the recombination complex may have a restricted mobility because the compact and well-defined conformation of the helix M-N linker in the type II interface acts as a tether for helix M. On the other hand, helix M and Tyr324 may have much greater mobility in the active subunits because they are connected to an extended, flexible linker peptide in the type I interface. It is clear from comparison of the inactive and active sites in the Cre-DNA complexes that the Tyr324 hydroxyl group must move by ~6 Å in order to cleave the scissile phosphate. It should not be surprising, therefore, that helix M in the active subunit has electron density and thermal parameters characteristic of a mobile peptide segment. It seems likely that similar factors could exist in the XerCD system, where mutagenesis and biochemical studies indicate a similar ball-and-socket set of interactions between catalytic domains (92). In the λ-Int and Flp systems, the C-terminal regions are quite different (75). λ-Int may not have a helix N equivalent, and Flp has a more extensive C-terminal region in its catalytic domain in addition to using an alternative cleavage mode. The generality of helix M tethering as an active site regulation mechanism among the tyrosine recombinases therefore requires further structural and biochemical investigation.

The second difference between active and inactive subunits indicated from difference-distance matrix analysis is the conformation of the β2-β3 loop. In the Cre-DNA covalent complex structure, it was noted that this loop approaches the active site in the active subunit, where Lys201 contacts O-2 of Thy adjacent to the scissile phosphate on the noncleaved strand, but this loop is projected away from the active site in the inactive subunit (Color Plate 13) (39). Given the experimental evidence that Lys-β is the catalytic acid responsible for protonating the 5′-hydroxyl leaving group in the vaccinia virus topoisomerase (52), the conservation of structure and mechanism between the topoisomerase Ib and integrase family recombinase enzymes (22), and the fact that the corresponding residue is known to be essential in Cre, XerD (17), and presumably in all tyrosine recombinases, it seems plausible that the allosteric control of positioning this important residue could be used in the regulation of the recombinase active sites. Unfortunately, Lys201 is poorly ordered in the active sites of the precleavage Cre-DNA complex structures, and it has not been possible to visualize a direct lysine contact to the 5′-hydroxyl group in any structure thus far. In addition, the β2-β3 loop shows a wide range of conformations in the unbound tyrosine recombinase (45, 53, 96) and vaccinia virus topoisomerase structures (22). We can, however, infer from the position of the (weakly defined) β2-β3 loop in the precleavage Cre-DNA complexes and the Cre-HJ complexes that the observations made in the Cre-DNA covalent intermediate are likely to be true of all three reaction intermediate structures. In the active subunits, the β2-β3 loop is positioned such that Lys201 can access the active site and reach both the 5′-hydroxyl leaving group as well as the Thy–O-2 base in the minor groove. In the inactive subunits, the β2-β3 loop is positioned so as to preclude Lys201 from entering the active site region to as great an extent.

The relationship between the conformation of the β2-β3 loops and the type I and type II interfaces between recombinase subunits in the Cre-DNA structures is intriguing (Color Plate 13). The β2-β3 loop is located near the interface between adjacent subunits in the tetrameric Cre-DNA complex, but the active conformation of the loop is in the type II interface, not the type I interface that contains the active Tyr324 nucleophile. Likewise, the inactive conformation of the β2-β3 loop is located in the type I interface. The β3-strand in the type I interface in fact directly contacts the extended helix M-N linker, forming a pseudo-fourth β-strand in the inactive catalytic domain (Color Plate 13a). In the type II interface, contacts between β3 and the linker peptide are quite different (Color Plate 13b). While the differences between protein-protein contacts in the type I and type II interfaces could directly affect the positioning of the β2-β3 loop, a second possibility is that the DNA substrate influences the loop conformation.

Of the two differences between active and inactive subunits in the structures of Cre-DNA intermediates that give rise to the type I and type II interfaces, positioning of the β2-β3 loop appears to be the most plausible means of active-site regulation that could be generalized to other integrase family members. This structural feature is highly conserved in the tyrosine recombinases (75) and in the eukaryotic topoisomerase Ib enzymes (22). The homologue of Lys-β is essential for Xer and Cre recombinase catalysis and has been identified as the general acid responsible for promoting the 5′-hydroxyl leaving group during vaccinia virus topoisomerase catalysis (52). Given the extremely close similarity between the members of this superfamily of enzymes, it seems quite likely that this conserved lysine fulfills the same role in the integrase family of enzymes. By preventing Lys-β from accessing the active site in the type I interface or by promoting the entry of the β2-β3 loop into the minor groove of the DNA bound by the active subunit in the type II interface (or both), the recombination complex could ensure that only two subunits are active at any step in the recombination pathway. An interesting implication of the promotion of a Lys-β interaction via the synaptic type II interface would be that synapsis may be required in order to stimulate cleavage of the sites. This prerequisite could provide a safety measure to exclude unproductive and possibly dangerous cleavage reactions in the absence of an appropriate recombining partner.

SUMMARY

The convergence of biochemical, genetic, and structural studies in an area of biological research generally results in an enlightened view of the underlying molecular processes. This is clearly the case for the tyrosine recombinase family of site-specific recombination enzymes. Structural data for five of the integrase family recombinase systems and two eukaryotic topoisomerase Ib systems have provided a three-dimensional framework for understanding years of results obtained by a number of different laboratories. Biochemical studies in the vaccinia virus topoisomerase Ib enzyme have also contributed greatly to our understanding of catalysis in this superfamily. Much of the results from the topoisomerase studies seem likely to be directly applicable to the integrase family enzymes.

Our current understanding of the tyrosine recombinase site-specific recombination pathway in the three-dimensional sense has been guided primarily by structures of reaction intermediates in the Cre-*loxP* system. The most surprising feature of what we have learned from this system is that the reaction proceeds with a single protein-DNA architecture that requires no large changes in quaternary structure in converting substrates to products. A subtle isomerization of the nearly coplanar HJ between dyad-symmetric forms appears sufficient to act as the active site switch that triggers exchange of only one of the two pairs of DNA strands to produce either substrates or products. While this might appear to contrast sharply with the ideas of helical restacking discussed for λ-Int (10, 71) and for XerCD (7), it is worth noting that the Cre-HJ isomerization model could also be thought of as a kind of helical restacking process. In this case, however, both stacks are always discontinuous and the restacking event simply switches the angular relationships between the duplex arms. Our current model for the Cre-*loxP* recombination pathway is outlined in Fig. 8. This mechanism requires antiparallel synapsis of the sites in order to lead to productive strand exchange. The strand exchange process occurs by a strand-swapping mechanism similar to that proposed in 1995 by Nunes-Düby et al. (74). Three bases are tested for homology with the recombining site and exchanged to form the Holliday intermediate, and the remaining three bases in the crossover region are tested during the second strand exchange step.

Perhaps the most compelling questions remaining at the structural and mechanistic level for the tyrosine recombinases involve the detailed stereochemical interfaces between subunits that regulate the recombination process. While Cre and XerCD have similar N-terminal domain structures and might be expected to form similar protein-protein contacts in the recombination assembly, it is quite clear that these contacts have not been conserved in Flp recombinase (21). In the case of XerCD acting on plasmid sites, the nature of protein-protein contacts is expected to be quite different from that found on *dif* sites because of the increase in crossover length. The "loss" of intimate contacts capable of mediating the recombination reaction in the absence of accessory factors may have been replaced to some extent in these systems by a topological regulatory mechanism that is assembled only in an intramolecular or direct repeat context (12, 24). If the scheme outlined here for regulating stepwise cleavage and exchange of strands is general for the tyrosine recombinases, then we should expect to see similar types of protein-protein contacts in the crucial regulatory interfaces for other family members.

A second question regards isomerization of the HJ intermediate in tyrosine recombinase recombination. While the isomerization appears to be quite subtle in the Cre system, requiring only small changes in quaternary structure and no branch migration of the junction, the same need not be true of Flp, Int, and XerCD-*cer*. In the Flp system, both a limited branch migration model and the simpler 4-bp swap model would seem to be possible (100). In the XerCD–8bp crossover and λ-Int systems, the formation of specific nucleoprotein architectures adjacent to the recombination sites leads to interesting possibilities. The extent to which these assemblies might inhibit, promote, or simply tolerate a limited branch migration process during recombination (31) will be an important issue in understanding the more complex tyrosine recombination systems in the future. In the realm of structural biology, a three-dimensional experimental model of the λ intasome or the XerCD-PepA-ArgR-*cer* assembly will be tantalizing goals for the years to come.

Acknowledgments. I gratefully acknowledge the valuable comments, discussions, and support received from members of the site-specific recombination community during my laboratory's work in the Cre-*loxP* system and Phoebe Rice for supplying Color Plate 12 and results prior to publication. Feng Guo and Michael Gopaul carried out the structural studies in the Cre-*loxP* system described in this chapter.

The National Institutes of Health provided financial support for this work.

REFERENCES

1. **Abremski, K., and R. Hoess.** 1984. Bacteriophage P1 site-specific recombination. Purification and properties of the Cre recombinase protein. *J. Biol. Chem.* **259:**1509–1514.
2. **Abremski, K., R. Hoess, and N. Sternberg.** 1983. Studies on the properties of P1 site-specific recombination: evidence for topologically unlinked products following recombination. *Cell* **32:**1301–1311.
3. **Abremski, K. E., and R. H. Hoess.** 1992. Evidence for a second conserved arginine residue in the integrase family of recombination proteins. *Protein Eng.* **5:**87–91.
4. **Alén, C., D. J. Sherratt, and S. D. Colloms.** 1997. Direct interaction of aminopeptidase A with recombination site DNA in Xer site-specific recombination. *EMBO J.* **16:** 5188–5197.
5. **Arciszewska, L., I. Grainge, and D. Sherratt.** 1995. Effects of Holliday junction position on Xer-mediated recombination in vitro. *EMBO J.* **14:**2651–2660.
6. **Arciszewska, L. K., R. A. Baker, B. Hallet, and D. J. Sherratt.** 2000. Coordinated control of XerC and XerD catalytic activities during Holliday junction resolution. *J. Mol. Biol.* **299:** 391–403.
7. **Arciszewska, L. K., I. Grainge, and D. J. Sherratt.** 1997. Action of site-specific recombinases XerC and XerD on tethered Holliday junctions. *EMBO J.* **16:**3731–3743.
8. **Arciszewska, L. K., and D. J. Sherratt.** 1995. Xer site-specific recombination in vitro. *EMBO J.* **14:**2112–2120.

9. **Argos, P., A. Landy, K. Abremski, J. B. Egan, L. E. Haggard, R. H. Hoess, M. L. Kahn, B. Kalionis, S. V. Narayana, L. D. Pierson, et al.** 1986. The integrase family of site-specific recombinases: regional similarities and global diversity. *EMBO J.* **5**:433–440.
10. **Azaro, M. A., and A. Landy.** 1997. The isomeric preference of Holliday junctions influences resolution bias by lambda integrase. *EMBO J.* **16**:3744–3755.
11. **Blakely, G., G. May, R. McCulloch, L. K. Arciszewska, M. Burke, S. T. Lovett, and D. J. Sherratt.** 1993. Two related recombinases are required for site-specific recombination at dif and cer in E. coli K12. *Cell* **75**:351–361.
12. **Blakely, G., and D. Sherratt.** 1996. Determinants of selectivity in Xer site-specific recombination. *Genes Dev.* **10**: 762–773.
13. **Blakely, G. W., and D. J. Sherratt.** 1996. *cis* and *trans* in site-specific recombination. *Mol. Microbiol.* **20**:233–238.
14. **Blakely, G. W., and D. J. Sherratt.** 1994. Interactions of the site-specific recombinases XerC and XerD with the recombination site dif. *Nucleic Acids Res.* **22**:5613–5620.
15. **Burgin, A. B., and H. A. Nash.** 1995. Suicide substrates reveal properties of the homology-dependent steps during integrative recombination of bacteriophage lambda. *Curr. Biol.* **5**: 1312–1321.
16. **Campbell, A.** 1962. Episomes. *Adv. Genet.* **11**:101–145.
17. **Cao, Y., and F. Hayes.** 1999. A newly identified, essential catalytic residue in a critical secondary structure element in the integrase family of site-specific recombinases is conserved in a similar element in eucaryotic type IB topoisomerases. *J. Mol. Biol.* **289**:517–527.
18. **Carson, M.** 1991. Ribbons 2.0. *J. Appl. Crystallogr.* **24**: 958–961.
19. **Cassell, G., R. Moision, E. Rabani, and A. Segall.** 1999. The geometry of a synaptic intermediate in a pathway of bacteriophage lambda site-specific recombination. *Nucleic Acids Res.* **27**:1145–1151.
20. **Chen, J. W., J. Lee, and M. Jayaram.** 1992. DNA cleavage in trans by the active site tyrosine during Flp recombination: switching protein partners before exchanging strands. *Cell* **69**:647–658.
21. **Chen, Y., U. Narendra, E. L. Iype, M. M. Cox, and A. P. Rice.** 2000. Crystal structure of a Flp recombinase-Holliday junction complex: assembly of an active oligomer by helix swapping. *Mol. Cell* **6**:885–897.
22. **Cheng, C., P. Kussie, N. Pavletich, and S. Shuman.** 1998. Conservation of structure and mechanism between eukaryotic topoisomerase I and site-specific recombinases. *Cell* **92**: 841–850.
23. **Cheng, C., L. K. Wang, J. Sekiguchi, and S. Shuman.** 1997. Mutational analysis of 39 residues of vaccinia DNA topoisomerase identifies Lys-220, Arg-223, and Asn-228 as important for covalent catalysis. *J. Biol. Chem.* **272**: 8263–8269.
24. **Colloms, S. D., J. Bath, and D. J. Sherratt.** 1997. Topological selectivity in Xer site-specific recombination. *Cell* **88**: 855–864.
25. **Cooper, J. P., and P. J. Hagerman.** 1987. Gel electrophoretic analysis of the geometry of a DNA four-way junction. *J. Mol. Biol.* **198**:711–719.
26. **Craig, N. L.** 1988. The mechanism of conservative site-specific recombination. *Annu. Rev. Genet.* **22**:77–105.
27. **Dixon, J. E., and P. D. Sadowski.** 1994. Resolution of immobile chi structures by the FLP recombinase of 2 microns plasmid. *J. Mol. Biol.* **243**:199–207.
28. **Dixon, J. E., A. C. Shaikh, and P. D. Sadowski.** 1995. The Flp recombinase cleaves Holliday junctions in trans. *Mol. Microbiol.* **18**:449–458.
29. **Duckett, D. R., A. I. Murchie, S. Diekmann, E. von Kitzing, B. Kemper, and D. M. Lilley.** 1988. The structure of the Holliday junction, and its resolution. *Cell* **55**:79–89.
30. **Esposito, D., and J. J. Scocca.** 1997. The integrase family of tyrosine recombinases: evolution of a conserved active site domain. *Nucleic Acids Res.* **25**:3605–3614.
31. **Franz, B., and A. Landy.** 1995. The Holliday junction intermediates of lambda integrative and excisive recombination respond differently to the bending proteins integration host factor and excisionase. *EMBO J.* **14**:397–406.
32. **Gopaul, D. N., F. Guo, and G. D. Van Duyne.** 1998. Structure of the Holliday junction intermediate in Cre-loxP site-specific recombination. *EMBO J.* **17**:4175–4187.
33. **Gopaul, D. N., and G. D. Van Duyne.** 1999. Structure and mechanism in site-specific recombination. *Curr. Opin. Struct. Biol.* **9**:14–20.
34. **Grainge, I., D. Buck, and M. Jayaram.** 2000. Geometry of site alignment during int family recombination: antiparallel synapsis by the Flp recombinase. *J. Mol. Biol.* **298**:749–764.
35. **Grainge, I., and M. Jayaram.** 1999. The integrase family of recombinase: organization and function of the active site. *Mol. Microbiol.* **33**:449–456.
36. **Grindley, N. D. F.** 1997. Site-specific recombination: synapsis and strand exchange revealed. *Curr. Biol.* **7**:R608–R612.
37. **Guo, F.** 2000. Ph.D. thesis. University of Pennsylvania, Philadelphia.
38. **Guo, F., D. N. Gopaul, and G. D. Van Duyne.** 1999. Asymmetric DNA bending in the Cre-loxP site-specific recombination synapse. *Proc. Natl. Acad. Sci. USA* **96**:7143–7148.
39. **Guo, F., D. N. Gopaul, and G. D. Van Duyne.** 1997. Structure of Cre recombinase complexed with DNA in a site-specific recombination synapse. *Nature* **389**:40–46.
40. **Hallet, B., L. K. Arciszewska, and D. J. Sherratt.** 1999. Reciprocal control of catalysis by the tyrosine recombinases XerC and XerD: an enzymatic switch in site-specific recombination. *Mol. Cell* **4**:949–959.
41. **Hallet, B., and D. J. Sherratt.** 1997. Transposition and site-specific recombination: adapting DNA cut-and-paste mechanisms to a variety of genetic rearrangements. *FEMS Microbiol. Rev.* **21**:157–178.
42. **Han, Y. W., R. I. Gumport, and J. F. Gardner.** 1993. Complementation of bacteriophage lambda integrase mutants: evidence for an intersubunit active site. *EMBO J.* **12**: 4577–4584.
43. **Hargreaves, D., D. W. Rice, S. E. Sedelnikova, P. J. Artymiuk, R. G. Lloyd, and J. B. Rafferty.** 1998. Crystal structure of *E. coli* RuvA with bound DNA Holliday junction at 6 A resolution. *Nat. Struct. Biol.* **5**:441–446.
44. **Hayes, F., and D. J. Sherratt.** 1997. Recombinase binding specificity at the chromosome dimer resolution site dif of Escherichia coli. *J. Mol. Biol.* **266**:525–537.
45. **Hickman, A. B., S. Waninger, J. J. Scocca, and F. Dyda.** 1997. Molecular organization in site-specific recombination: the catalytic domain of bacteriophage HP1 integrase at 2.7 A resolution. *Cell* **89**:227–237.
46. **Hoess, R., K. Abremski, S. Irwin, M. Kendall, and A. Mack.** 1990. DNA specificity of the Cre recombinase resides in the 25 kDa carboxyl domain of the protein. *J. Mol. Biol.* **216**: 873–882.
47. **Hoess, R., A. Wierzbicki, and K. Abremski.** 1990. Synapsis in the Cre-*lox* site-specific recombination system, p. 203–213. *In* R. H. Sarma and M. H. Sarma (ed.), *Structure and Methods*, vol. 1. Adenine Press, New York, N.Y.
48. **Hoess, R. H., and K. Abremski.** 1985. Mechanism of strand

cleavage and exchange in the Cre-lox site-specific recombination system. *J. Mol. Biol.* **181**:351–362.
49. **Huffman, K. E., and S. D. Levene.** 1999. DNA-sequence asymmetry directs the alignment of recombination sites in the FLP synaptic complex. *J. Mol. Biol.* **286**:1–13.
50. **Jayaram, M.** 1997. The cis-trans paradox of integrase. *Science* **276**:49–51.
51. **Kitts, P. A., and H. A. Nash.** 1988. Bacteriophage lambda site-specific recombination proceeds with a defined order of strand exchanges. *J. Mol. Biol.* **204**:95–107.
52. **Krogh, B. O., and S. Shuman.** 2000. Catalytic mechanism of DNA topoisomerase IB. *Mol. Cell* **5**:1035–1041.
53. **Kwon, H. J., R. Tirumalai, A. Landy, and T. Ellenberger.** 1997. Flexibility in DNA recombination: structure of the lambda integrase catalytic core. *Science* **276**:126–131.
54. **Landy, A.** 1999. Coming or going it's another pretty picture for the lambda-Int family album. *Proc. Natl. Acad. Sci. USA* **96**:7122–7124.
55. **Landy, A.** 1989. Dynamic, structural, and regulatory aspects of lambda site-specific recombination. *Annu. Rev. Biochem.* **58**:913–949.
56. **Landy, A.** 1993. Mechanistic and structural complexity in the site-specific recombination pathways of Int and FLP. *Curr. Opin. Genet. Dev.* **3**:699–707.
57. **Lee, J., and M. Jayaram.** 1995. Role of partner homology in DNA recombination. *J. Biol. Chem.* **270**:4042–4052.
58. **Lee, J., M. Jayaram, and I. Grainge.** 1999. Wild-type Flp recombinase cleaves DNA in trans. *EMBO J.* **18**:784–791.
59. **Lee, J., J. Lee, and M. Jayaram.** 1995. Junction mobility and resolution of Holliday structures by Flp site-specific recombinase. Testing partner compatibility during recombination. *J. Biol. Chem.* **270**:19086–19092.
60. **Lee, J., T. Tonozuka, and M. Jayaram.** 1997. Mechanism of active site exclusion in a site-specific recombinase: role of the DNA substrate in conferring half-of-the-sites activity. *Genes Dev.* **11**:3061–3071.
61. **Lee, J., G. Tribble, and M. Jayaram.** 2000. Resolution of tethered antiparallel and parallel holliday junctions by the Flp site-specific recombinase. *J. Mol. Biol.* **296**:403–419.
62. **Lee, J., Y. Voziyanov, S. Pathania, and M. Jayaram.** 1998. Structural alterations and conformational dynamics in Holliday junctions induced by binding of a site-specific recombinase. *Mol. Cell* **1**:483–494.
63. **Lilley, D. M.** 1997. Site-specific recombination caught in the act. *Chem. Biol.* **4**:717–720.
64. **Lilley, D. M. J.** 2000. Structures of helical junctions in nucleic acids. *Q. Rev. Biophys.* **33**:109–159.
65. **Lilley, D. M. J., and R. M. Clegg.** 1993. The structure of the four-way junction in DNA. *Annu. Rev. Biophys. Biomol. Struct.* **22**:299–328.
66. **Luetke, K. H., and P. D. Sadowski.** 1998. Determinants of the position of a Flp-induced DNA bend. *Nucleic Acids Res.* **26**:1401–1407.
67. **Luetke, K. H., and P. D. Sadowski.** 1995. The role of DNA bending in Flp-mediated site-specific recombination. *J. Mol. Biol.* **251**:493–506.
68. **Meyer-Leon, L., R. B. Inman, and M. M. Cox.** 1990. Characterization of Holliday structures in FLP protein-promoted site-specific recombination. *Mol. Cell. Biol.* **10**:235–242.
69. **Nash, H. A.** 1990. Bending and supercoiling of DNA at the attachment site of bacteriophage lambda. *Trends Biochem. Sci.* **15**:222–227.
70. **Nash, H. A.** 1996. Site-specific recombination: integration, excision, resolution, and inversion of defined DNA segments, p. 2363–2376. *In* F. C. Neidhardt et al. (ed.), Escherichia coli *and* Salmonella: *Cellular and Molecular Biology*. ASM Press, Washington, D.C.
71. **Nunes-Düby, S., M. A. Azaro, and A. Landy.** 1995. Swapping DNA strands and sensing homology without branch migration in lambda site-specific recombination. *Curr. Biol.* **5**:139–148.
72. **Nunes-Düby, S., L. Matsumoto, and A. Landy.** 1987. Site-specific recombination intermediates trapped with suicide substrates. *Cell* **50**:779–788.
73. **Nunes-Düby, S., R. S. Tirumalai, L. Dorgai, E. Yagil, R. A. Weisberg, and A. Landy.** 1994. Lambda integrase cleaves DNA in cis. *EMBO J.* **13**:4421–4430.
74. **Nunes-Duby, S. E., M. A. Azaro, and A. Landy.** 1995. Swapping DNA strands and sensing homology without branch migration in lambda site-specific recombination. *Curr. Biol.* **5**:139–148.
75. **Nunes-Duby, S. E., H. J. Kwon, R. S. Tirumalai, T. Ellenberger, and A. Landy.** 1998. Similarities and differences among 105 members of the Int family of site-specific recombinases. *Nucleic Acids Res.* **26**:391–406.
76. **Nunes-Duby, S. E., D. Yu, and A. Landy.** 1997. Sensing homology at the strand-swapping step in lambda excisive recombination. *J. Mol. Biol.* **272**:493–508.
77. **Pan, G., K. Luetke, and P. D. Sadowski.** 1993. Mechanism of cleavage and ligation by FLP recombinase: classification of mutations in FLP protein by in vitro complementation analysis. *Mol. Cell. Biol.* **13**:3167–3175.
78. **Pan, H., D. Clary, and P. D. Sadowski.** 1991. Identification of the DNA-binding domain of the FLP recombinase. *J. Biol. Chem.* **266**:11347–11354.
79. **Petersen, B. O., and S. Shuman.** 1997. Histidine 265 is important for covalent catalysis by vaccinia topoisomerase and is conserved in all eukaryotic type I enzymes. *J. Biol. Chem.* **272**:3891–3896.
80. **Redinbo, M. R., L. Stewart, P. Kuhn, J. J. Champoux, and W. G. Hol.** 1998. Crystal structures of human topoisomerase I in covalent and noncovalent complexes with DNA. *Science* **279**(5356):1504–1513.
81. **Rice, P.** 2000. Structure of the Flp-DNA complex. *Mol. Cell* **6**:885–897.
82. **Richards, F. M., and C. E. Kundrot.** 1988. Identification of structural motifs from protein coordinate data: secondary structure and first-level supersecondary structure. *Proteins* **3**:71–84.
83. **Ringrose, L., V. Lounnas, L. Ehrlich, F. Buchholz, R. Wade, and A. F. Stewart.** 1998. Comparative kinetic analysis of FLP and Cre recombinases: mathematical models for DNA binding and recombination. *J. Mol. Biol.* **284**:363–384.
84. **Roe, S. M., T. Barlow, T. Brown, M. Oram, A. Keeley, I. R. Tsaneva, and L. H. Pearl.** 1998. Crystal structure of an octameric RuvA-Holliday junction complex. *Mol. Cell* **2**:361–372.
85. **Sadowski, P. D.** 1995. The Flp recombinase of the 2-microns plasmid of Saccharomyces cerevisiae. *Prog. Nucleic Acids Res. Mol. Biol.* **51**:53–91.
86. **Schwartz, C. J., and P. D. Sadowski.** 1990. FLP protein of 2 mu circle plasmid of yeast induces multiple bends in the FLP recognition target site. *J. Mol. Biol.* **216**:289–298.
87. **Schwartz, C. J., and P. D. Sadowski.** 1989. FLP recombinase of the 2 microns circle plasmid of Saccharomyces cerevisiae bends its DNA target. Isolation of FLP mutants defective in DNA bending. *J. Mol. Biol.* **205**:647–658.
88. **Seeman, N. C., and N. R. Kallenbach.** 1994. DNA branched junctions. *Annu. Rev. Biophys. Biomol. Struct.* **23**:53–86.
89. **Shaikh, A. C., and P. D. Sadowski.** 1997. The Cre recombi-

nase cleaves the lox site in trans. *J. Biol. Chem.* **272:** 5695–5702.

90. **Sherratt, D. J., L. K. Arciszewska, G. Blakely, S. Colloms, K. Grant, N. Leslie, and R. McCulloch.** 1995. Site-specific recombination and circular chromosome segregation. *Philos. Trans. R. Soc. Lond. Ser. B* **347**(1319):37–42.
91. **Sherratt, D. J., and D. B. Wigley.** 1998. Conserved themes but novel activities in recombinases and topoisomerases. *Cell* **93:**149–152.
92. **Spiers, A. J., and D. J. Sherratt.** 1999. C-terminal interactions between the XerC and XerD site-specific recombinases. *Mol. Microbiol.* **32:**1031–1042.
93. **Stark, W. M., M. R. Boocock, and D. J. Sherratt.** 1992. Catalysis by site-specific recombinases. *Trends Genet.* **8:**432–439.
94. **Stark, W. M., D. J. Sherratt, and M. R. Boocock.** 1989. Site-specific recombination by Tn3 resolvase: topological changes in the forward and reverse reactions. *Cell* **58:**779–790.
95. **Stewart, L., M. R. Redinbo, X. Qiu, W. G. J. Hol, and J. J. Champoux.** 1998. A model for the mechanism of human topoisomerase I. *Science* **279:**1534–1540.
96. **Subramanya, H. S., L. K. Arciszewska, R. A. Baker, L. E. Bird, D. J. Sherratt, and D. B. Wigley.** 1997. Crystal structure of the site-specific recombinase, XerD. *EMBO J.* **16:** 5178–5187.
97. **Summers, D. K., and D. J. Sherratt.** 1988. Resolution of ColE1 dimers requires a DNA sequence implicated in the three-dimensional organization of the cer site. *EMBO J.* **7:** 851–858.
98. **Tirumalai, R. S., E. Healey, and A. Landy.** 1997. The catalytic domain of lambda site-specific recombinase. *Proc. Natl. Acad. Sci. USA* **94:**6104–6109.
99. **Tirumalai, R. S., H. J. Kwon, E. H. Cardente, T. Ellenberger, and A. Landy.** 1998. Recognition of core-type DNA sites by lambda integrase. *J. Mol. Biol.* **279:**513–527.
100. **Voziyanov, Y., S. Pathania, and M. Jayaram.** 1999. A general model for site-specific recombination by the integrase family recombinases. *Nucleic Acids Res.* **27:**930–941.
101. **Waite, L. L., and M. M. Cox.** 1995. A protein dissociation step limits turnover in FLP recombinase-mediated site-specific recombination. *J. Biol. Chem.* **270:**23409–29414.
102. **Weisberg, R. A., L. W. Enquist, C. Foeller, and A. Landy.** 1983. A role for DNA homology in site-specific recombination: the isolation and characterization of a site affinity mutant of coliphage lambda. *J. Mol. Biol.* **170:**319–342.
103. **Whang, I., J. Lee, and M. Jayaram.** 1994. Active-site assembly and mode of DNA cleavage by Flp recombinase during full-site recombination. *Mol. Cell. Biol.* **14:**7492–7498.
104. **Wittschieben, J., and S. Shuman.** 1997. Mechanism of DNA transesterification by vaccinia topoisomerase: catalytic contributions of essential residues Arg-130, Gly-132, Tyr-136 and Lys-167. *Nucleic Acids Res.* **25:**3001–3008.
105. **Yang, W., and K. Mizuuchi.** 1997. Site-specific recombination in plane view. *Structure* **5:**1401–1406.
106. **Zhu, X. D., G. Pan, K. Luetke, and P. D. Sadowski.** 1995. Homology requirements for ligation and strand exchange by the FLP recombinase. *J. Biol. Chem.* **270:**11646–11653.

Mobile DNA II
Edited by N. L. Craig et al.

Chapter 7

λ Integrase and the λ Int Family

Marco A. Azaro and Arthur Landy

Marco A. Azaro and Arthur Landy • Department of Molecular Biology, Cell Biology, and Biochemistry, Brown University, Providence, RI 02912.

INTRODUCTION

This chapter starts with Allan Campbell's insightful proposal for the pathway by which the chromosome of bacteriophage λ is integrated into and excised from the chromosome of its *Escherichia coli* host (28). A combination of sophisticated genetics and clever geneticists established the basic elements of the, then novel, site-specific recombination pathway even before its broader significance was fully appreciated (89, 91). However, the route to a detailed biochemical understanding of this reaction was not opened until Howard Nash and his colleagues purified the λ-encoded integrase (Int) and its *E. coli*-encoded accessory protein, integration host factor (IHF) (101, 149). Delineation of the DNA substrates, called *att* sites, provided unexpected insights and provoked the exploration of additional levels of complexity (115). While this review is primarily about the integration-excision reaction of λ Int, outlined in Fig. 1, it will necessarily draw on results and insights from other well-characterized λ Int family members since many important ideas about the λ Int reaction come from them.

Concomitant with the early studies on λ site-specific recombination, work in other systems began to reveal a variety of pathways, and several different mechanisms, devoted to the rearrangement of DNA. Somewhat later, the outlines of a rapidly growing λ Int family became apparent (7). Now that a great deal is known about the relevant molecular mechanisms, membership in the family has taken on a mechanistic, rather than a biological, definition. As we shall discuss below, the biology of the family ranges far and wide, but its chemistry is tightly defined by two features: the first is the ability to cleave and reseal DNA in the absence of high-energy cofactors via a 3′-phosphotyrosine intermediate, and the second is the presence of a small number of highly conserved residues in the catalytic domain. Besides these two features, the λ Int family reactions encompass a wide range of biological functions, global protein structures, reaction products, and dependencies on accessory proteins. Many of the earlier reviews on the λ Int family include information and perspectives that are still relevant but might not, or could not, be covered here (44, 84, 113, 114, 136, 146, 184, 205, 224). Although the λ Int family is sometimes referred to as the tyrosine recombinase family, we shall use the former term for historical continuity and because it avoids minor confusions. For example, some λ Int family members are not recombinases, and some virally encoded site-specific re-

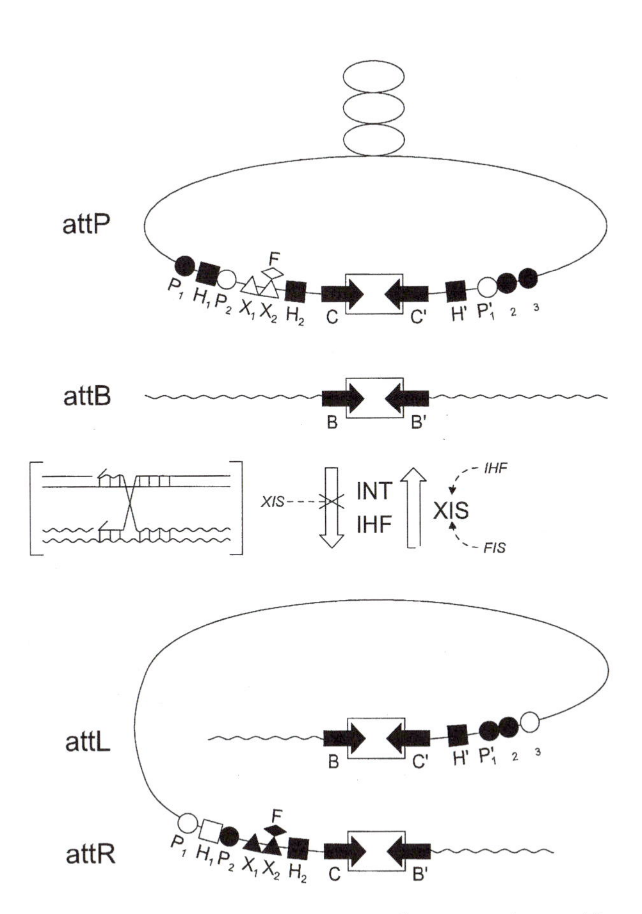

Figure 1. Integrative and excisive recombination pathways. The protein binding sites for arm-type Int (○), core-type Int (⇒), IHF (□), Xis (△), and Fis (◇) are indicated by filled symbols when that site is occupied by its cognate protein to make a competent recombination partner for integrative (⇓) or excisive (⇑) recombination. Required proteins (Int, IHF, and Xis) are in boldface type, and proteins that inhibit (Xis and IHF) or enhance (Fis) the indicated reactions are in italics. The Holliday junction intermediate (square brackets) results from the reciprocal single-strand swap of the 5′ ends liberated by Int cleavage. The three bases of swapped top strands are shown just prior to ligation and formation of the four-way junction. This is followed by conformational changes that allow cleavage, swapping, and ligation of the bottom strands, i.e., resolution of the Holliday junction to yield recombinant products.

combinases are not λ Int family members. The other major family of energy-independent site-specific recombinases is the resolvase/invertase family, which uses a serine nucleophile and a different collection of active-site residues to cleave and religate DNA (see chapters 13 and 14).

DIFFERENT LEVELS OF λ Int FAMILY COMPLEXITY

The λ Int family can be divided into four groups, or branches, on the basis of reaction complexity, which is also manifested by their respective DNA target sites (Fig. 2). The best-studied exemplars of the first family branch are the vaccinia virus and human type IB topoisomerases, which function as monomers because their primary role requires the cleavage and religation of a single DNA strand. This contrasts with the other three branches of the family, which are all populated by recombinases that operate in groups of four, one for each of the DNA strands that must be cleaved and religated. We include the type IB eukaryotic DNA topoisomerases in the λ Int family because they clearly meet the family definition articulated above and there is nothing in their enzymology that suggests they do not belong. Studies on these topoisomerases have been extremely valuable in understanding the chemistry and enzymology of the DNA cleavage-ligation reaction that they share with other λ Int family members (see "Catalysis," below).

The second family branch consists of the recombinases that function without accessory proteins. The best studied archetypes of this group are the Cre recombinase of *E. coli* bacteriophage P1 and the Flp recombinase encoded by the 2μm plasmid of *Saccharomyces cerivisiae.* Proteins in these two branches of the λ Int family that function without the need for accessory proteins, not surprisingly, have been the first to yield cocrystals with their cognate DNA targets. These structures have had a major impact on our views and understanding of the entire family.

The third family group comprises those recombinases which require or exploit accessory proteins to impose topological specificity, and possibly also regulation. The best-studied member of this group is the XerCD system of *E. coli,* which on some substrates requires the proteins PepA and ArgR. In addition to the great biological importance of this system and its many relatives in bacterial chromosome segregation, the XerCD pathway has the experimental advantage of using two slightly different recombinases for the front and back halves of the recombination reaction.

The fourth level of familial complexity consists of those recombinases that are heterobivalent, in addition to requiring accessory proteins. Many of these recombinases catalyze the integration and excision of viral chromosomes into and out of their host's chromosomes, such as the well-studied integrases of *E. coli* phage λ, *Haemophilus* phage HP1, and *Mycobacterium* phage L5. However, other members of this group are found in conjugative plasmids such as Tn*916* and Tn*544*.

FAMILY BIOLOGY

Linking Number Control

In terms of the DNA transactions they mediate, the eukaryotic type I topoisomerases (type IB topoisomerases) are the least complex members of the λ Int family. This branch of the family includes the nuclear topoisomerase I found in all eukaryotic cells and the topoisomerases encoded by vaccinia virus and other cytoplasmic poxviruses. These enzymes transiently break one strand of duplex DNA and thereby allow for single-step changes in the linking number. They relax both negative and positive supercoils and thereby play critical roles in DNA replication, transcription, and recombination (for reviews see references 31, 198, and 222). The prokaryotic type I topoisomerases (type IA) do not belong to the λ Int family: they require magnesium, they form a covalent intermediate with the 5′ end of the cleaved DNA strand, and they seem to share no sequence or structural similarity with the type IB enzymes. Even further removed are the type II topoisomerases, which are dimeric and effect linking number changes in steps of two by passing a region of duplex DNA through a double-strand break. Although the type IB topoisomerases can transfer the covalent DNA cleavage intermediate to a new DNA strand to generate recombinant molecules and do catalyze illegitimate recombination in vivo, this is not regarded as their primary function (23, 196).

Whereas cellular type IB topoisomerases have a loose preference for cleaving just after a four-base motif, 5′(A/T)(G/C)(A/T)T, the enzyme encoded by vaccinia virus and other poxviruses cleaves specifically after the sequence 5′(T/C)CCTT, with the 3′ T forming the target for the catalytic Tyr (95, 197, 199). The viral enzymes, at approximately 325 amino acids, are not much larger than the smallest family members (FimB and FimE weigh in at approximately 200 amino acids), whereas the cellular topoisomerases are larger (approximately 760 to 1,020 amino acids) and have a more complex domain structure (174). Crystal structures have been solved for both the human and vaccinia virus topoisomerases and also a cocrystal structure of the human enzyme complexed with a

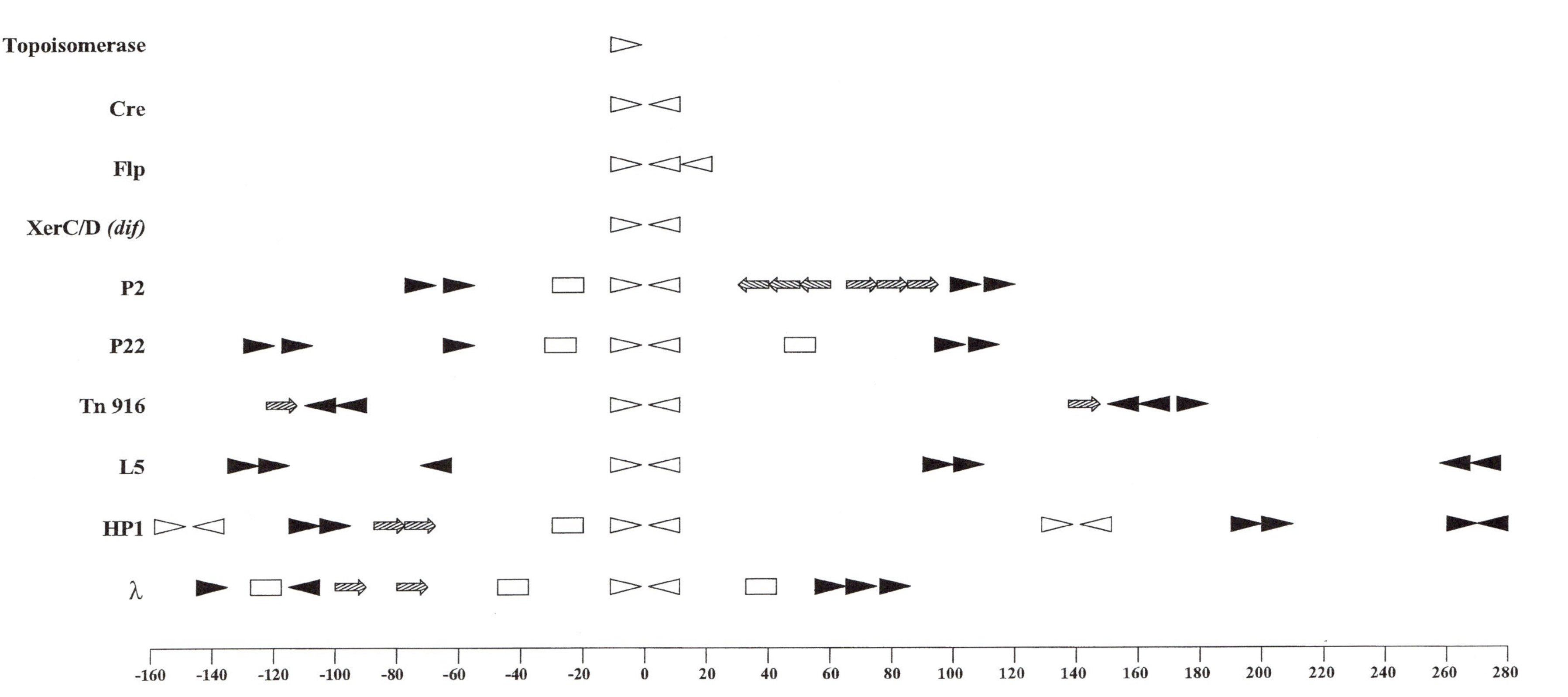

Figure 2. Hierarchy of λ Int family attachment sites increasing in complexity from top to bottom. The *attP* sites of the heterobivalent recombinases (P2, P22, Tn*916*, L5, HP1, and λ) are not ordered on the basis of organizational complexity. Nevertheless, their arrangement captures the variability in number, spacing, orientation, and nature of different DNA binding sites within each site. Each *att* site map (except that for topoisomerase) is centered about its respective core region, which consists of the 6- to 8-bp overlap region (i.e., the locus of catalysis and strand exchange) and flanking inverted-repeat core-type protein binding sites (open arrowheads). Topoisomerase functions as a monomer, and therefore just one core site is indicated. Cre exemplifies the simplest recombinases in that it requires only a pair of inverted core sites (per partner). Although Flp, like Cre, needs only a single pair of inverted core sites (per partner), its biological *att* sites contain a third core site adjacent to the functional pair. An inverted pair of core sites is sometimes sufficient for XerCD-mediated recombination (e.g., at *dif* sites); however, for other substrates, such as *cer,* XerCD requires participation of the accessory proteins ArgR and PepA (sites not shown). The remaining maps represent the *attP* sites of recombinases that are heterobivalent and also require accessory proteins. The intrinsically asymmetric integrase arm-type sites and excisive factor binding sites (Xis or cox sites) are indicated with black arrowheads and hatched arrows, respectively, while IHF binding sites are denoted with open boxes. The Xis site of L5 has not been mapped and is not indicated. The heterobivalent integrase *att* site maps are ordered with the P arms to the left and the P′ arms to the right of the origin, except HP1, which has been reversed. The Tn*916* map depicts the structure of the circular transposon intermediate postexcision and preintegration. Note that the L5 *attP* is interchangeable with the D29 *attP* and likewise the HK022 *attP* is interchangeable with the λ *attP*.

DNA substrate (36, 175, 207). These structures and especially the thorough mutational and biochemical analyses of the vaccinia virus have been extremely important in working out the λ Int family mechanisms of DNA cleavage and ligation (see below).

Gene Expression

The best-studied example of regulation of gene expression by a λ Int family member is the FimB- and FimE-dependent site-specific DNA inversion system that controls phase variation of type 1 fimbriation in *E. coli* (1, 25). Type 1 fimbriae are thin protein surface polymers that mediate adhesion to mannosides and enable *E. coli* to colonize epithelial surfaces. The ability to shift between a fimbriate and nonfimbriate state (phase variation) is thought to help cells avoid the host immune system. The frequency of phase variation is on the order of 1 per 1,000 cells per generation (58) and results from the inversion of a 314-bp DNA element known as the *fim* switch (*fimS*) (1). The orientation of the *fim* switch determines whether or not an encoded promoter is able to drive expression of *fimA*, which encodes the major fimbrial subunit (for a review, see reference 88). The FimB and FimE recombinases are extremely interesting from several points of view (in addition to any pathogenicity-related perspectives). At approximately 200 amino acids they are the smallest members of the λ Int family. The two recombinases, which are approximately 50% identical, are encoded by tandemly arranged genes adjacent to the *fimB* element (105). FimB catalyzes recombination of either orientation of *fimS*, while FimE can only catalyze recombination from the on to off orientation (70, 130). The latter limitation has been explained, in part, by a reduced binding affinity of FimE for one of the four possible recombination targets (inversion generates slightly different junction sequences in each of the four recombination targets) (111). Both FimB and FimE require the accessory proteins IHF and Lrp (leucine-responsive regulatory protein) for efficient inversion of *fimS* (20, 53, 59). While IHF has turned up as a frequently used accessory protein, especially by the virally encoded heterobivalent recombinases, it is evident that a number of different DNA bending proteins have also been exploited by various λ Int family members to facilitate, direct, or regulate their respective reactions. The location of the IHF and Lrp binding sites within the 300-bp invertible segment suggests that active DNA bending may be required for synapsis across such a short stretch of DNA. Alternatively, or additionally, the bending proteins may prevent intermolecular recombination through a topological filter, similar to the use of ArgR and PepA in the XerCD pathway.

Chromosome Segregation

Together with the biological function of chromosome segregation we include plasmid copy number control, since it works toward the same goal. This is a rich area for the λ Int family on several different levels. Three of the most extensively characterized family members are devoted to this purpose: Cre (see chapter 6), Flp (see chapter 11), and XerCD (see chapter 8). Two of these recombinases (Cre and XerCD) reduce dimeric genomes to monomers and therefore are functionally resolvases. By the same criteria Flp is functionally an invertase that works during DNA replication of the 2μm plasmid in *S. cerevisiae* to reverse one of the two oppositely oriented replication forks so they move in the same direction (69, 221). The resulting double rolling circle intermediate generates multimeric-size plasmids that are subsequently resolved to monomers. The consequent high copy numbers of 2μm plasmid favor their transmission to both daughter cells. For the full story on each of these three family members the reader is referred to their respective chapters in this volume.

A less well characterized but particularly interesting family member with resolvase function is the TnpI recombinase encoded by the *Bacillus* transposons Tn*5401* and Tn*4430* (15, 127). Recombinases with related sequences and possibly related function are also encoded by transposons Tn*5041*, Tn*4652*, and Tn*4556* (N. Grindley, personal communication). A particularly interesting feature described for the TnpI recombinase of Tn*5401* is its ability to act as a negative transcription regulator of another gene (the *tnpA* transposase) and also as a modulator of TnpA binding to its DNA target sites (15).

Chromosome Evolution

Integron organization

Integrons are the medically most important members of the λ Int family. It is therefore especially unfortunate that they are also the family members about which we know the least, a deficiency attributable to the evolutionary time scale on which they operate. Integrons are DNA elements that accumulate and disseminate bacterial genes that are often (but not only) related to antibiotic resistance, pathogenesis, and the ability to survive noxious environmental agents. More precisely, they can be considered gene expression elements that accumulate open reading frames (gene cassettes) and convert them to functional genes. Their essential components are a site-specific recombinase (*intI*) and an associated primary recombination target, or *att* site (*attI*). The other integron feature is a strong promoter (P_{ANT}) oriented and positioned with respect to *attI* so as to transcribe any gene cassettes that have recombined into it. The reason gene cassettes are able to recombine into *attI* is that they carry their own *att* site, which is called either *attC* or 59-base element, whose origin will be discussed below. We shall use the former term because it emphasizes function rather than structure. Some of the disagreement about nomenclature is historical (R. Hall, personal communication), and some of it reflects the fact that there is great variability in architecture and length (from approximately 60 to 140 bp) among the *attC* sequences associated with different gene cassettes. (For reviews, see references 64, 173, 182, and 208).

There are presently four classes of integron integrases, IntI1 to -4, sharing approximately 50% identity. Almost all characterization to date has been of the IntI1 integron, using either an in vivo mating-out assay (160) or purified IntI1 integrase and fusion proteins (41, 65, 76, 77). Membership of the 337-residue IntI1 protein in the λ Int family (160) has been legitimized by the phenotypes of site-directed mutants at key family residues (41, 77). One of the factors complicating the analysis of the integrons is that in vivo IntI1 can catalyze recombination between any pair of *att* sites as well as between an *att* site and a secondary, or degenerate, *att* site. While this kind of reaction flexibility is characteristic of almost all λ Int family members, it is more confusing in the case of the integrons because even the "primary" reaction pathways have a low efficiency. A further impediment to understanding these reactions is the inability to demonstrate any in vitro activities for IntI1 to date.

Integron *att* sites

In contrast to the IntI1 integrase, which has not suggested any striking deviations from its cousins, the integron recombination targets, *attI1* and *attC*, present several new variations on λ Int family themes (41, 65, 76, 166). The first unusual feature is that *attI1* and *attC* look very different from each other. While this asymmetry in recombination targets is a common feature of the heterobivalent recombinases, there is no suggestion of a second DNA binding domain or a second consensus recognition sequence for IntI1. The *attI1* site contains four binding sites for IntI1. Two of these are arranged as inverted repeats separated by a 7-bp spacer, i.e., like a typical core region. If *attI1* behaved like other λ Int family recombinases, cleavage by IntI1 would occur at the two boundaries of the 7-bp spacer region. However, in vivo evidence thus far indicates that the crossover occurs at a unique position in the core region, at one boundary of the 7-bp spacer (87, 208). Indeed, recombination between *att* sites with identical 7-bp spacer regions is not any more efficient than that for two sites differing at five positions (208). This is in contrast to most other λ Int family recombinases but similar to one subgroup, exemplified by the Tn*916*- and Tn*1545*-like transposases (190).

Among the various interpretations of this result are suggestions that (i) strand exchange occurs at the same site on both strands, (ii) the Holliday junction generated by a single strand exchange is resolved by a generic Holliday junction-resolving enzyme (as is the case for XerC- and XerD-catalyzed recombination between *cer* sites under certain conditions [43, 131]), and (iii) the Holliday junction serves as a template for DNA synthesis. Deciding among these and other possible explanations will probably require the development of an in vitro reaction.

The second unusual feature of *attI1* is the presence of two additional IntI1 binding sites arranged as direct repeats and separated by approximately 20 bp, that are located 10 bp from the core region (76, 87, 166). It remains to be determined how the two directly repeated "non-core" sites, distant from the site of strand exchange, actually function to enhance recombination. Similar arrangements of additional recombinase binding sites are found in other λ Int family reactions, such as the FRT sites recognized by Flp recombinase and the resolution site recognized by TnpI recombinase (14).

The structure of *attC* is even more unusual than that of its *attI1* partner and even less well understood. Since binding of IntI1 to *attC* in vitro is very weak (65, 76) a major source of interaction data comes from in vivo mutational studies (208). The highly variable *attC* genes (length ranges from 60 to 140 bp) combine with the 3′ end of an adjacent coding segment to create a gene cassette. Of the more than 100 gene cassettes identified, each one carries a different version of *attC* that is defined primarily by a pair of core regions at its outer ends (this distance may range from 20 to 100 bp). The core region distal to the coding sequence binds IntI1 with higher affinity (binding to the proximal *att* site is not detectable) and is the site of strand exchange (65, 76, 208). More precisely, exchange takes place on the right side of the 6- to 7-bp spacer region, in the same relative location as described for *attI1*. The difference in architecture between *attI1* and *attC* is very likely to be a significant factor in favoring *attI* × *attC* recombinations over *attC* × *attC*. This has the effect of directing newly integrating cassettes to the *attI1* site (adjacent to P_{ANT}), as opposed to the many *attC* sites that are already present in the integron.

We do not yet know anything about the initial

formation of gene cassettes or the initial union of *attC* genes with the 3′ termini of coding regions. It has been suggested that, since the majority of gene cassettes contain promoterless open reading frames with an *attC* at the 3′ end, they may have arisen via reverse transcription from mRNA (173). This is an appealing proposal that nevertheless would require some additional explanation for the small number of gene cassettes that do have their own promoter or have their *attC* at the 5′ end of the open reading frame (64, 182). A number of important insights have emerged from the finding that *Vibrio cholerae,* a significant pathogen for humans and animals, has very large clusters of genes in integron-like elements (129, 173). Indeed, the integrase (IntI4) associated with these islands of genes dedicated to pathogenicity can mediate recombination into a class 1 integron (129). The *Vibrio*-associated *attC* genes, referred to as *Vibrio cholerae* repeated sequences, are extremely similar (some are virtually identical) to those of class 1 *attC* genes. It is estimated that there are 60 to 100 different genes accumulated in one *Vibrio* "super integron," and it appears from old *Vibrio* isolates that some *V. cholerae* repeated-sequence cassettes predate the introduction of antibiotics and were probably acquired from other (non-*Vibrio*) microbes (129). Understanding the origin and formation of integrons and gene cassettes is an important and interesting challenge.

Conjugative Transposition

Another λ Int family branch of great medical importance contains those recombinases encoded by conjugative transposons. These are DNA elements that are excised from a chromosome to form circular intermediates which, in turn, can be transferred to another cell, where they may integrate into a new chromosome. Some of these circular intermediates have been shown to be capable of replication in some strains, but transmissibility is their essential feature. Indeed, conjugative transposons are much like lysogenic phage with the obvious exception that they are transferred by conjugation rather than by viral infection. In fact, two very well studied integrases, encoded by the Tn*916* and Tn*1545* transposons, are heterobivalent DNA binding proteins just like the viral integrases. Furthermore, like the viral integrases they use a transposon-encoded DNA bending accessory protein (Xis) for excision (172, 190, 209; also see chapter 10). Whereas some of the transposons integrate at relatively specific sites in the chromosome, others, like Tn*916* and Tn*1545,* are quite promiscuous in their site selection (187, 190). The molecular basis of DNA target site promiscuity is one of the extremely interesting features of the conjugative transposons that distinguishes them (along with the integrons) from many other λ Int family members. Unlike most family members, these recombinases do not require DNA homology in the short (6- to 8-bp) "overlap" region that becomes heteroduplex after recombination (see below; also see chapter 10).

The conjugative transposons are undoubtedly a significant vector in the horizontal and interspecies spread of antibiotic resistance genes. It is therefore not unreasonable to speculate that they may also be involved in the spread of other genes, including those conferring pathogenicity (186).

Viral Integration and Excision

In contrast to virulent viruses which must destroy their host cell in order to propagate, temperate viruses have the added capacity to propagate in a form that allows concurrent multiplication of virus and host cell. One of the popular components among the strategies for achieving this benign state is the integration (and appropriately timed excision) of the viral chromosome at a specific locus on the chromosome of the infected cell. Execution of this site-specific recombination of temperate viruses calls upon one of the largest branches of the λ Int family. It is fitting that this branch of the family is numerically so well represented, since the eponymous bacteriophage λ integrase was the first family member to be characterized genetically and biochemically.

The two features most clearly differentiating this branch from other family members are a recombinase that is a heterobivalent DNA binding protein and a highly evolved mechanism for controlling directionality via a virally encoded accessory excision protein (often, but not always, called Xis). Obviously, the excision protein greatly stimulates recombination in the excisive direction and, where studied, also inhibits recombination in the integrative direction. Several other features generally characteristic of the viral recombinase pathways are only briefly mentioned here, as they will be elaborated further below. During integrative recombination the virally encoded recombinases direct recombination to a specific target site on the host chromosome (called *attB*). This contrasts with the functionally very similar recombinases of conjugative transposons, which are quite promiscuous in their choice of chromosomal integration sites. The viral integrative recombination pathway is notably asymmetric in the complexity of the two recombination targets, with the site on the viral chromosome (called *attP*) being much more complex than the simple *attB* site on the host chromosome. This asymmetry is much reduced in the excisive recombination reaction that involves two hybrid sites, *attL* and *attR.* The viral recombination reactions all require (or utilize) accessory DNA bending proteins to varying extents and with varying degrees of specificity, and in all cases these proteins bind to the viral DNA sequences (arms) of *attP*: *attL* and *attR*. Target sequences contributed by the host chromosome do not contain sites for accessory proteins.

It is worth noting that of the virally encoded recombinases that execute site-specific integration and excision with a directionality imposed by a small accessory protein, not all are members of the λ Int family. For example, integrases encoded by the lactococcal phage TP901-1 and the actinophages R4 and ϕC31 comprise a subgroup of the resolvase/invertase family based on their primary amino acid sequences and the mechanisms of DNA cleavage (22, 110, 128, 216).

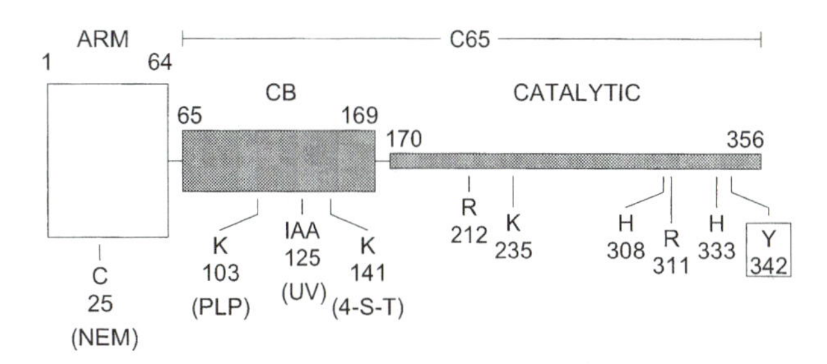

Figure 3. Domain structure and selected residues of λ Int. The heights of the boxes approximate the different affinities of the three domains for their respective DNA targets. The amino-terminal domain (residues 1 to 64) binds to arm-type sites and includes a single NEM-reactive cysteine (C25). The central CB domain (residues 65 to 169) binds to core-type sites and contains residues in close proximity to DNA, as determined by DNA-sensitive pyridoxal 5′-phosphate (PLP) reactivity with Lys 103, UV zero-length cross-linking to Ala 125-Ala 126, and photoactivated cross-linking of 4-thio-T to Lys 141. The catalytic domain (residues 170 to 356) contains the most highly conserved λ Int family residues, including the catalytic pentad Arg 212-Lys 235-His 308-Arg 311-His 333 and the active-site nucleophile Tyr342.

λ INTEGRASE

Domains and Structure

λ Int is a basic protein with an M_r of 40,330 that has topoisomerase activity and that carries out the cleavage and religation of *att* site DNA via its absolutely conserved Tyr 342 residue (for reviews, see references 113, 144, and 146). It is a heterobivalent protein that can be subdivided into three functional and structural domains based on collected evidence from recent experiments: an amino-terminal region (residues 1 to 64) that recognizes arm-type sequences with high affinity, a central core-binding (CB) domain (residues 65 to 169) that contributes to specific recognition of core-type sequences at sites of strand exchange and possibly participates in protein-protein interactions, and a carboxy-terminal minimal catalytic domain (residues 170 to 356) that contains the requisite machinery for executing DNA strand cleavage and ligation (Fig. 3) (217).

Using C65 (residues 65 to 356) as a monovalent analog to probe Int-DNA interactions, several modification strategies were employed to show that the CB domain contains critical determinants of core-type DNA binding: zero-length cross-linking with UV light identified Ala 125 and Ala 126 as contacting residues, substrate-sensitive modification with pyridoxal 5′-phosphate identified Lys 103 (218), and photoactivated cross-linking to 4-thio-T identified Lys 141 (M. Kovach and A. Landy, unpublished data). Furthermore, a cloned polypeptide encompassing residues 65 to 169 was shown to bind autonomously and specifically to core-type DNA.

Notwithstanding the unambiguous and direct participation of the CB domain in binding and the relative inability of the C170 catalytic domain to form electrophoretically stable complexes, the CB domain cannot be the sole determinant in core-type binding. This is based on several lines of evidence: the catalytic domain encompasses the Int family's conserved active site residues, which, when mutated, cause significant defects in DNA cleavage and/or ligation, and in λ, such changes are shown to greatly reduce affinity for core-type interactions (68, 86, 126, 165, 228). Work by Dorgai et al. (52) has shown that substitution of just five residues in λ Int to the corresponding amino acids from the closely related HK022 integrase is sufficient to transform core-binding specificity from λ- to HK022-type. Only one of these, Asn 99, lies in the CB domain, while the remainder reside in the catalytic domain (Ser 282, Gly 283, Arg 287, and Glu 319). But, most persuasively, the Cre cocrystal structures show that each of the active site residues corresponding to R212, H308, R311, and H333 in λ makes at least one hydrogen bond to the scissile phosphate. We believe, however, that there is no real dichotomy between the localization of several core specificity mutations to the catalytic domain and, separately, the localization of substantive core-binding interactions to the CB domain. It should be noted that the former were identified on the basis of recombination defects and the latter were identified on the less stringent criterion of forming electrophoretically stable complexes. Perhaps the protein-DNA "fit" required for efficient catalysis is more restrictive than that required for stable binding.

The C65 domain of λ Int is analogous to the

monovalent Cre and Xer recombinases that naturally lack an auxiliary DNA-binding module. In fact the CB and catalytic domains of λ Int appear to reflect the N- and C-terminal domain organization of XerD and Cre. The observation that the catalytic domain of Cre binds autonomously to core-type sites but the same domain of λ Int does not may be a consequence of differential distribution of binding energy between the CB and catalytic domains of each protein. In this regard, it is worth noting that λ Int and Cre represent two extremes in the Int family of recombinases; Cre is a monovalent protein that has apparently evolved strong core interactions, enabling it to recombine simple *att* sites composed only of core-type sites, whereas λ Int is a bivalent protein that has evolved weak core interactions, presumably to render it dependent on arm binding. Recent evidence suggests that λ Int may have gone even further in evolving a dependence on arm binding. Whereas full Int binds very poorly to core-type DNA sites, C65 (lacking the N-terminal domain) binds very well (189). On the other hand, adding back the N-terminal domain in solution stimulates C65 binding. Thus, the N-terminal domain can either be an inhibitor (in *cis* with respect to C65) or a stimulator (mimicked by being in *trans* with respect to C65) of C65 binding to core-type DNA sites. It has been suggested that arm binding to the N-terminal domain may help trigger the switch between these two modes.

The structure of the C170 catalytic domain of λ Int was solved by X-ray crystallography at 1.9-Å resolution to reveal a globular protein with a mixed α-β structure consisting of seven α helices and seven β-strands (Color Plate 14 [see color insert]). An unusual α-helical bundle that consisted of a core of three helices (αA, αB, and αC) circumscribed by a ring formed by αD, αE, and αF identified a new protein fold (112).

Mutants

Han and colleagues employed powerful in vivo and in vitro approaches to identify four classes of Int mutants (86). Class I mutants, which populate the region spanning residues 80 to 168, have the property of binding more efficiently to P′1 substrates than P′123 substrates (Fig. 1). Han et al. speculate that these mutations may have deleterious consequences on protein-protein interactions and thereby reduce the cooperative binding typically seen by λ Int with P′123. Combined with the observation that an overlapping set of mutants, characterized by their inability to diminish in vitro topoisomerase function, also maps to this region, they suggest that this domain seems not to be critical for catalysis. Notice that this region is contained entirely within the CB domain (see above), which has been separately implicated in core binding (217, 218).

Class II and III mutants, which are defined by a reduced ability to bind to either P′123 and P′1 or P′123, respectively, map to the region encompassed by residues 217 to 332, deep within the catalytic domain. Although far removed from the arm-binding domain (residues 1 to 64) with respect to primary sequence, this region may, the authors reason, contribute in a subtle way to arm binding, possibly by some physical interaction. They also provide evidence that R212 and neighbors (G214 and C217) and R311 are important catalytic residues, consistent with previous studies that showed that the corresponding residues in Flp are directly implicated in catalysis.

Interestingly, they reisolated a nonsense mutant, W350 (16), that is missing the carboxy-terminal seven amino acids and that behaves like Int in most assays but has enhanced topoisomerase activity. The authors surmise that this short region may be important for modulating the cleavage-religation activity of λ Int and that its absence may result in uncoupled nicking and religation that contrasts with the controlled relaxation of supercoils (one supercoil relaxed per cycle) previously demonstrated for wild-type Int (101).

Specificity Determinants

λ and HK022 are closely related integrases that have identical amino acid sequences over the N-terminal 55 residues and share approximately 70% homology over the remainder of their lengths. Interestingly, although they recognize identical arm-type sequences and have similarly arranged *att* sites, neither is capable of recombining the other's substrates. It appears that the critical DNA-based specificity differences reside in three of the four core sequences that surround the region of strand exchange (see also "DNA Specificity Determinants," below). In an attempt to identify the protein-based determinants of specificity Weisberg and colleagues undertook a two-pronged approach: they isolated and characterized chimeric recombinases with altered specificities (232) and separately mutagenized λ Int to identify substitutions that endow it with HK022 specificity (52). They identified a minimum of five λ-to-HK substitutions that are sufficient to obliterate λ specificity and confer virtually wild-type HK022 specificity. Two of these mutations (N99D and E319R) seem to act primarily to broaden specificity; that is, they allow λ Int to efficiently recombine HK022 sites while retaining the ability to recombine its own sites. The remaining mutations (S282P, G283K, and R287K) seem to function by preventing λ-specific recombination. Based on their individual phenotypes the authors inferred that the N99D, S282P, G283K, and R287K mutations have a direct influence on protein-DNA contacts, whereas the E319R conversion alters a catalytic step. Notice that only one of the proposed direct contacts resides in the CB domain, while the other three lie in a cluster in the catalytic domain. As we suggested earlier, this bias may be due to the fact that recombination was used as a probe for specificity changes versus core site affinity.

Based on the above data Weisberg and colleagues suggest a plausible model for the evolution of altered specificity. Whereas a mutation that simultaneously increases recognition of a new site and decreases recognition of an old site may be disadvantageous, a multistep change whereby specificity is first broadened and then narrowed may allow a protein to distinguish a new target without making itself obsolete in the process.

attB

Overlap Region and DNA-DNA Homology

Integration of bacteriophage λ into its *E. coli* host chromosome occurs primarily at a single preferred site, an ~21-bp element called *attB* which consists of three distinct elements: two core-binding sites, B and B′ (each of which is a locus of interaction with the weaker of the two heterobivalent binding activities of λ Int [see "Domains and Structure," above]), which lie in near perfect dyad symmetry surrounding and partially overlapping a central 7-bp overlap region. These three elements comprise what is called the core region. During recombination this *attB* is captured by a preorganized nucleoprotein complex (180) consisting of the ~234-bp phage attachment site (*attP*), IHF, and λ Int and is believed to be aligned in antiparallel with its analogous DNA element (204, 206), which also contains a core region (two inverted core-binding sites [C and C′] and a central overlap region). The overlap regions of these two *att* sites must be perfectly homologous for efficient recombination to occur.

This stringent requirement for homology is the most striking difference between the reaction catalyzed by λ Int and that catalyzed by the closely related integrases of conjugative transposons such as Tn*916* and Tn*1545* (190, 220), where the degree of homology between two recombining overlap regions has little effect on target site selection or orientation of insertion. This has been perceived as attesting to an essential mechanistic difference between the two reactions. However, it should be pointed out that strand exchange mediated by λ Int can, albeit infrequently, tolerate mismatches in the reaction products at secondary *att* sites (see below). One might speculate that the reactions catalyzed by Tn*916* and Tn*1545* utilize a recalibrated version of the strand swapping model, where the homology-sensing apparatus has been confounded and mismatches can be readily accommodated.

Core-Type Int Binding Sites

The four core sites of λ (B, B′, C, and C′) were originally identified by a combination of sequencing (115) and methylation protection experiments (181), yielding a consensus core-binding sequence of 5′CAACTTNNT, which corresponds precisely to C′. Although these four sites are similar at the sequence level, there is a clear hierarchy of core-binding affinity, with C′ exhibiting the strongest affinity, C and B exhibiting equal intermediate affinities, and B′ exhibiting the weakest affinity. Interestingly, the probing experiments reveal that all but one of the apparent protein-DNA contacts lie along a single face of the DNA double helix. The staggered cleavage sites also lie on this face and are separated by the width of the minor groove. The dyad symmetrical arrangement of binding sites, which is believed to allow each bound Int to relate to its respective cleavage site in a similar manner, turns out to be a general feature of lambdoid phage recombination systems (29).

Chromosomal Targets

The above discussion cannot easily be extended to the *attB* sites of other temperate viruses without some brief reference to secondary *att* sites. These are sites utilized for λ integration (at greatly reduced efficiencies) when the canonical *attB* is deleted. They are degenerate facsimiles whose relative recombination efficiencies are determined by the extent to which they emulate the spacing and sequence of the overlap region and core-type Int binding sites of the canonical *attB* (181, 224). Such sites are presumably fodder for the evolution of new high-efficiency *att* sites with different sequences and correspondingly evolved integrases.

Whereas λ *attB* is located in an intergenic region between the *gal* and *bio* genes, such an unobtrusive location is not the general rule for the *attB* genes of other phage. Indeed, it appears that the majority, or at least a large fraction, of the *attB* genes in other recombination pathways are located within tRNA genes or coding regions (29, 176). The large number of *attB* sites mapped to tRNA genes is spread out over many species in both gram-positive and gram-negative bacteria. It has been suggested that the popularity of this destination reflects (i) the location of some evolutionarily successful ancestral phage, (ii) the felicitous structural features of the tRNA gene with inverted repeats separated by a 7-bp spacer (e.g., anticodon loop), or (iii) the conservation of tRNA sequences (and therefore potential target sites) in different bacterial hosts. It should be noted that for the majority of these *attB* genes, phage integration does not inactivate the target gene. As a result of the obvious selective pressure against such a calamity, each phage carries a duplication of a portion of the interrupted gene so that integration restores its integrity.

DNA Specificity Determinants

Despite their thematic similarities, the nucleotide diversity of core sequences is exceptional. In fact it is only with the λ and HK022 site-specific recombination systems that we see specificities that, while distinct, are closely related. This close kinship and minimal cross reactivity have permitted enlightening studies on the mechanism and nature of evolutionary changes in specificity of Int-DNA interactions—primarily with regard to core site interactions, since it is here that the major determinants appear to reside (reviewed in reference 226). The amino acid and nucleotide residues responsible for distinction have been variously identified by genetic selections, construction of chimeric integrases (via recombination or site-directed mutagenesis), and alteration of core sites (37, 52, 142, 232).

Particularly dramatic was the demonstration that just five HK022-type amino acid substitutions into λ Int are sufficient to convert recombination specificity from exclusively λ type to exclusively HK022 type (52) (see also "λ Integrase," above). The diminished performance of class II mutants (grouped on the basis of narrowed specificity) was reasoned to be due to the creation of negative interactions between the substituted residues and λ core sites only. Support for this conjecture came from complementary experiments with altered core sites: base substitutions at three positions in λ *attB* remove the barrier to recombination by HK and the class II mutants of λ, without preventing recombination by wild-type λ Int, and conversely, base substitutions at two positions in the HK022 core removed or reduced the barrier to recombination by λ Int without preventing recombination by HK022 (51). Correlating these substitution positions with the analogous residues in the crystal structures of λ Int and Cre-*lox*, Weisberg and colleagues suggest that four of the five are poised to make direct DNA contacts (226). The core site specificity mutants reside almost exclusively in the C and B′ sites and fall into two categories that reflect the organization of the Int mutants: one type allows recombination by a noncognate Int (i.e., relaxes specificity), while the second type prevents recombination by a particular Int (i.e., narrows specificity). Mutants of HK022 that were isolated on the basis of enhanced affinity for λ B′ sites, surprisingly, did not acquire the ability to recombine λ sites (37).

STRAND SWAPPING

DNA Homology

λ and most of its siblings require perfect sequence homology between partner overlaps for recombination to occur. Many sequence modifications can be accommodated into an overlap as long as identical modifications are introduced into the partner (224). It was clear even in the nascent stages of λ research that implicit in this stringent homology requirement lay a mechanistic scheme of site-specific recombination and that a thorough understanding of this prerequisite was mandatory.

Early models that invoked a four-stranded intermediate or a concerted cleavage-exchange mechanism were dismissed in favor of an alternative scheme that invoked stepwise pairs of DNA cleavages and exchanges. These events respectively create and then resolve a Holliday junction intermediate, which itself must branch migrate across the entire overlap region (113). Indeed a host of studies were all consistent with a model in which homology-dependent branch migration served to translate the branch point from its locus of synthesis to its locus of resolution (reviewed in references 114 and 146). Since these experiments did not eliminate other interpretations, finer dissection of the reaction was required.

Resolution assays performed with immobile Holliday junctions suggested that the optimal position of formation and resolution of the branch point is not adjacent to the cleavage sites, but instead is close to the center of the 7-bp overlap (Fig. 4) (154). Particularly striking was the demonstration that only a minor shift about the central base pair of the overlap was sufficient to cause a dramatic reversal in bias. Thus, the homology requirement originally perceived to potentiate branch migration is now viewed as permitting two pairs of reversible swaps of 2 to 3 nt that assay the relative match between a potential partner and an existing partner. These swapping events were suggested to be linked by a central isomerization step which might involve branch migration of 1 to 3 bp and other structural rearrangements. This model is supported by experiments in the related systems of XerCD and FLP (4, 118, 119). Burgin and Nash (24) employed an ingenious cleavage site modification (5′-bridging phosphorothiolate) to monitor the relation-

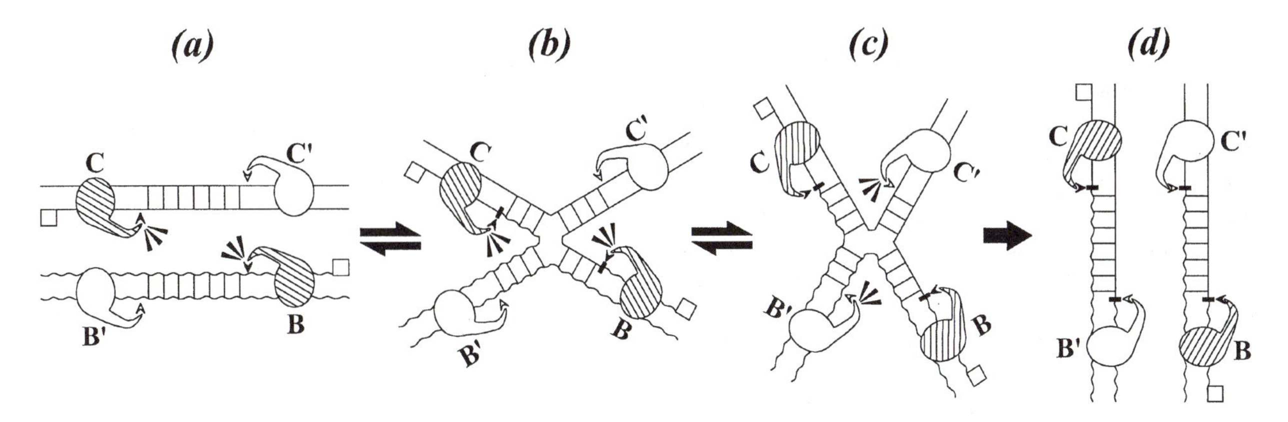

Figure 4. Scheme of Holliday junction formation and resolution at the core sites of *attP* (COC′) and *attB* (BOB′) during integrative recombination. (a) The core sites are shown synapsed in antiparallel within a tetrameric arrangement of λ integrases. The 5′ terminus of the top strand from each partner substrate is indicated with an open square, and the top-strand and bottom-strand cleavage sites are represented with filled and open arrowheads, respectively. The integrases poised to cleave the top strands in *cis* (i.e., bound at C and B) are displayed with hatched ovals, while the integrases poised to cleave bottom strands in *cis* (i.e., bound at C′ and B′) are displayed with open ovals. The catalytically active Ints (i.e., at C and B) (accented with triple arrowheads) cleave the top strand of each partner (a), generating a transient enzyme-Tyr-3′-phosphodiester intermediate (not shown), and permit a reciprocal swap of 5′OH-terminated single-strand segments (each approximately 3 nucleotides long that, in turn, displace the covalently bound integrases and create a recombinant joint (black bar). The resulting Holliday junction (b) most closely resembles a top-strand crossed isomer. In this structure the integrases bound to C and B are still primed for cleavage and may reverse the first strand exchange event. The transition from panel b to panel c represents the isomerization of the tetramer-Holliday junction complex required to prime the integrases bound at C′ and B′ to cleave the bottom strands: the arms of the Holliday junction shift in a scissor-like fashion so that the top strands now subtend an obtuse angle while the bottom strands subtend an acute angle, and the branch point shifts approximately one position closer to the bottom strand cleavage sites. (d) The primed integrases execute the second pair of reciprocal swaps to generate the recombinant products *attR* (COB′) and *attL* (BOC′) by a step drawn here as primarily unidirectional.

ship between homology-sensing and strand exchange events; they found that bottom-strand cleavage is strongly dependent on the presence of homology between the first 3 bp of partner overlaps (24).

Holliday Junction Isomerization

The dramatic shift in resolution bias caused by just a single base pair relocation of the branch point (see above) strongly implied that resolution of the top and bottom strands might be symmetric. It was subsequently shown that besides this positional symmetry there appears to be structural symmetry related to each cleavage and exchange event (5, 10).

The structure of immobile Holliday junctions in solution has been extensively characterized (reviewed in references 124 and 191) and has been shown to depend both on the ionic conditions and on the sequence at the branch point (55, 56). Cations (such as Mg^{2+} or Na^{+}) promote a pairwise stacking of the helical arms and rotation into an antiparallel X structure, where two strands become continuous with each helical axis and two strands bend sharply and cross between the two stacked helices. Immobile four-way junctions choose one of two possible isomers of the stacked structure, based on their relative stability (55). It was demonstrated that there is a strong correlation between the preferred isomeric structure of a naked Holliday junction substrate and the ability of λ Int to cleave, exchange, and ligate the crossed pair of strands (10). A more direct approach involved fixing artificial Holliday junction substrates into a particular isomeric form with a 9-nucleotide tether. With these physically constrained substrates it was shown that XerD shows a strong proclivity to cleave strands that are crossed, while XerC is more ambivalent (5).

A general prediction of both of these experiments is that there exist stereochemical characteristics that distinguish a λ Int-Holliday junction complex that is primed to cleave its top strands from one that is primed to cleave its bottom strands. More precisely, it was suggested that the former would possess features that more closely approximate a Holliday junction isomer with its top strands crossed than one with its bottom strands crossed. What was unknown at the time was to what extent the DNA framework of the recombination complex would resemble the extreme twofold symmetry of a fully stacked Holliday junction or the ambivalent fourfold symmetry of a square planar Holliday junction. Enter the Cre-*lox* cocrystal structures (74, 79, 80).

The crystal structure of the Cre-*loxP* site-specific recombination complex reveals a twofold-symmetric Holliday junction substructure that possesses features of both the stacked and square planar Holliday junction models while remaining distinct in detail. Surprisingly, the twofold symmetry derives from a small but significant deviation from an open square planar conformation. The strands that are primed for cleavage are indeed the ones that are crossing (i.e., they subtend an acute angle of ~76°), while the strands that are immune from cleavage comprise the continuous strands (i.e., they subtend an obtuse angle of ~101°).

All of the Cre cocrystals suggest a much more subtle protein-mediated isomerization than previously imagined; indeed, it may entail little or no branch migration at all. Instead, isomerization would convert crossed strands to continuous strands, and vice versa, rendering opposite pairs of strands susceptible to cleavage with minimal changes in protein-protein and protein-DNA contacts. Another striking feature of these structures is the gaping "strand exchange cavity"; given the requirement for large movements of DNA demanded by a strand-swapping model it is notable that very few protein-DNA contacts are made in this region. In fact, within the four base pairs adjacent to the branch point there are no direct contacts made by the proteins to bases in either the major or minor groove. This is consistent with the observation that no specific overlap sequence is dictated.

CATALYSIS

Chemistry

It is widely accepted that the integrase family of enzymes catalyzes recombination by carefully coordinated transesterifications of the target DNA backbone. The chemistry of the reaction is initiated by the hydroxyl moiety of an absolutely conserved Tyr (residue 342 in λ Int) which executes a nucleophilic attack on a target phosphate that has been rendered scissile by a constellation of conserved basic residues; all of this is done in the absence of exogenous energy cofactors or metal cations. A single cleavage event generates a covalent DNA–3′-phosphotyrosyl-enzyme intermediate and a freed 5′OH-terminated DNA strand. This reaction may be reversed to regenerate the free Tyr-OH and religate the DNA backbone, or alternatively, the phosphotyrosyl bond may be attacked by an extrinsic nucleophile. During recombination and strand exchange this external component comes from a 5′OH-terminated DNA strand that has been liberated in an identical manner from a partner substrate. Several of the conserved integrase residues are postulated to assist catalysis by stabilizing the negative charges that develop on the equatorial oxygens of the pentacoordinate phosphorane intermediate that arises during nucleophilic attack by Tyr, while other conserved residues are suggested to promote catalysis via a general acid-base mechanism.

Although the members of the λ Int family of recombinases are highly diverged in their primary amino acid sequence, it has been shown that a handful of basic residues, collectively referred to as the RHR triad (R212, H308, and R311 in λ Int), are highly conserved (3, 19, 62, 155) and that each of these residues is important for catalysis (32, 63, 68, 120, 163). Mutational analysis and phosphate interference footprinting of various recombinases strongly suggest that the RHR triad activates the scissile phosphate for cleavage (32, 33).

Interestingly, it seems that in some recombinases the His of the RHR triad can be dispensed with at certain catalytic steps. For example in the XerCD system the two Arg residues are required for strand cleavage while the His is required for strand joining (6, 18). Likewise, both Arg residues are required for transesterification by Flp recombinase but the His of the conserved triad is not required for strand cleavage (164, 165). This hints at a more modest contribution of the His on catalysis (see below). Analysis of these issues, and others pertaining to strand transfer, should be greatly aided by the synthesis and exploitation of a chromogenic DNA substrate that allows ligation to be studied in real time and in the absence of the competing cleavage reaction (230).

Crystal Structures

These last few years the site-specific recombination field has witnessed a remarkable proliferation of informative crystal structures: the isolated catalytic domain of λ Int (112), the dimerized catalytic domain HP1 integrase (92), the fully intact XerD protomer (210), several striking cocrystals of Cre recombinase complexed with substrate at different stages of catalysis (74, 79, 80), and most recently Flp engaged with a Holliday junction substrate (35; see also chapters 6 and 12). Catalytic residues previously identified by mutation-function analysis are found in close collusion: for example, members of the RHR triad are all located within 7 Å of each other in a shallow basic groove on the surface of λ Int, which also encompasses other invariant residues including His333 (Color Plate 14). The unliganded structures of HP1 and XerD assume very similar secondary and tertiary structures to λ Int and place their conserved RHR triads in corresponding positions. An alignment of 66 members of the integrase family shows that the invariant residues are all located on the proposed DNA interaction surface revealed by the structure of λ Int (112). This suggests that the tertiary structure of the Int active site may be recapitulated in the family at large.

One surprising feature of the unliganded Int structures (λ Int, HP1, and XerD) is the variability of their extreme C termini which contain the catalytic Tyr. In λ Int, residues 334 to 356 form a flexible loop that is completely disordered in one of the crystalline asymmetric units and is only partially disordered in the other, where the final 15 residues form two extra β sheets. In either case the Tyr is physically removed from the other catalytic residues. Amino acids 307 to 337 of HP1 form an extended structure that protrudes from the surface and which contains two short helices. This region seems to have been important for crystal contacts, and Hickman et al. (92) suggest that it may represent a protein dimer interface. The C terminus of XerD (residues 271 to 298), on the other hand, forms a turn followed by a long α helix that contains the Tyr, and the last six residues are disordered. In fact, even though the catalytic Tyr of XerD is relatively close to the active site, it is actually buried in the protein in such a manner that would not permit direct attack of the scissile phosphate. However, the authors point out that a small alteration in the position of the terminal helix would be sufficient to reorient it. It has been suggested that this variability in C-terminal configurations reflects different modes of active site assembly proposed for some λ Int family members (see "*cis* versus *trans* DNA Cleavage," below).

Perhaps most exciting of all are the striking Cre-*lox* and Flp-FRT cocrystal structures; collectively they manage to answer questions that have occupied researchers for years. They provide structural models for synapsis in site-specific recombination and provide gross and fine mechanistic insights that the unliganded proteins cannot. We shall focus on the Cre-*lox* structures since they provide the closest analog for λ Int. Common to all of these structures is a pseudofourfold tetramer of Cre monomers intimately networked in a cyclic fashion. The bound DNA substrate is disposed to resemble a square planar Holliday junction, regardless of its original form. The protein-DNA complex surrounds a strand exchange cavity which accommodates the central 6-bp overlap. Released DNA strands can apparently exchange with ease in this architecture without gross alterations in quaternary structure or steric interference.

The first reported structure actually shows Cre covalently linked to the substrate by a phosphotyrosine linkage. The tetrameric complex contains two distinct active sites: one diagonally opposed pair is covalently linked and primed for cleavage, whereas the second pair is disabled by a subtle rearrangement of critical residues. The catalytic sites contain essential residues RHRY and W (the corresponding residue in λ is a His 333), which are all provided by a single Cre monomer, consistent with a *cis* mechanism. Both Arg residues and Trp are engaged in H-bonding interactions with the nonbridging oxygens of the scissile phosphate in both active sites, whereas the His is H-bonded to the S′p oxygen in the covalent active site but is shifted away in the noncovalent site. Guo et al. (79) observe that the active site His, Arg, and Trp side chains form a positive "proton cradle" to accommodate the pentacoordinate phosphorane intermediate (Fig. 5).

Enzymology

Another exciting development was the confirmation of membership of the eukaryotic type IB topoisomerases in the λ Int family. Not only do these enzymes carry out a similar repertoire of activities and share conserved catalytic residues, but they also possess extensive structural similarity. Superimposing the crystal structure of the catalytic domain of vaccinia virus topoisomerase on the coordinates of the previously reported HP1 and Cre recombinase structures, it is apparent that the order and topology of the secondary and tertiary structural elements are strikingly similar (36). The conserved structural domain spans approximately 7 to 10 α-helices and a three-stranded β-sheet, and all critical catalytic residues occupy homologous positions.

From a practical point of view, it is further justification that the λ field should avail itself of the vast corpus of elegant work performed by Stuart Shuman and colleagues (for a review, see reference 198). They have undertaken an extensive mutation-function analysis of vaccinia virus topoisomerase (142 out of 314 residues were altered) and have identified 7 essential and 3 important residues, most of which correspond to the catalytic residues in λ Int, etc. The 7 essential positions were subjected to Ala scanning mutagenesis and careful kinetic analyses that provided valuable information regarding their individual enzymic rate enhancements and their possible mode of action.

Of the $\sim10^9$-fold rate enhancement provided by the topoisomerase, Arg 130 and Arg 223 each contribute $\sim10^4$- to 10^5-fold, the largest individual contributions witnessed. They propose that Arg 130 makes bidentate contacts to scissile phosphate nonbridging oxygens while Arg 223 makes only monodentate contacts. This is based on the observation that only the latter residue can be complemented by replacement with Lys. Arg 173 of Cre (analogous to Arg 130) does indeed simultaneously contact two ox-

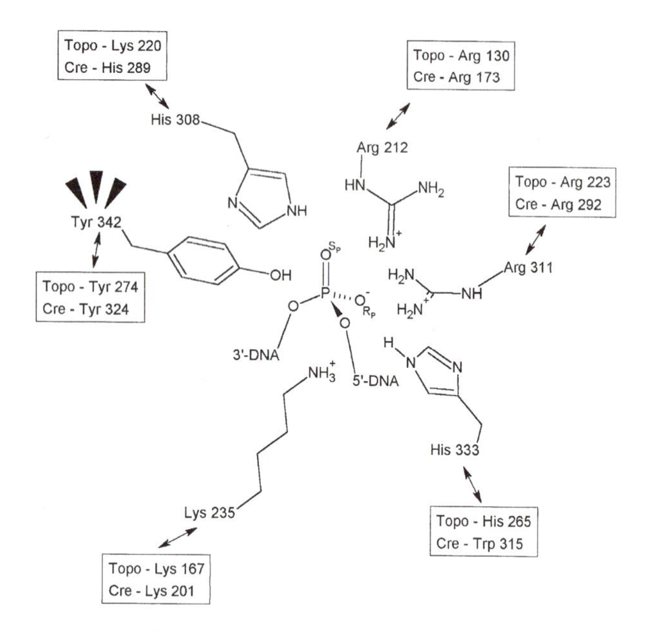

Lambda	Arg 212	Lys 235	His 308	Arg 311	His 333	Tyr 342
HP1	Arg 207	Lys 230	His 280	Arg 283	His 306	Tyr 315
XerD	Arg148	Lys 172	His 244	Arg 247	His 270	Tyr 279
Flp	Arg 191	Lys 223	His 305	Arg 308	Trp 330	Tyr 343

Figure 5. Schematic representation of active-site catalytic residues from λ Int surrounding the scissile phosphate of a substrate DNA. This model is based on, and representative of, a synthesis of structural and functional data derived from many systems (see text for details and references). For each λ Int residue there is also indicated the corresponding residue from vaccinia virus topoisomerase (Topo), Cre, HP1, XerD, and Flp. The homologous residues of Topo and Cre are boxed and juxtaposed to their λ counterparts since they are referred to extensively in the text. The diagrammed positions approximate the relative disposition of each amino acid to the element with which it is expected to interact. Members of the canonical Arg-His-Arg triad are Arg 212, His 308, and Arg 311. Arg 212 is shown poised to make bidentate H bonds with the nonbridging oxygens of the scissile phosphate, while His 308 and Arg 311 are both oriented to make only monovalent contacts. His 333 most likely interacts with a nonbridging oxygen and probably does not act as a general acid catalyst to assist in 5′OH release. The nucleophilic Tyr 342 is accented with triple arrowheads and is shown poised to execute an inline attack of the scissile phosphate to generate a covalent Tyr-3′-phosphate linkage and a freed 5′OH. Note that the Tyr may be shifted away from its target when the catalytic site is inactive. Lys 235 is a putative general acid catalyst and is shown prepared to donate a proton to the 5′ oxygen. The chirality of the nonbridging oxygens is indicated; note that R_p and S_p become R'_p and S'_p in the covalent complex.

ygens in the covalently complexed active site. H265 of vaccinia virus topoisomerase is profferred to also contact a nonbridging oxygen based on its ability to be complemented with Gln or Asn. Consistent with this suggestion is the observation that the corresponding residues of human topoisomerase I and Cre do precisely this.

Identification of residues that could act as general acids or bases in catalysis has been less fruitful. However, in a recent study, Lys 167 of vaccinia virus topoisomerase is strongly implicated as being a general acid catalyst (109). Conversion of this residue to Arg restores partial function commensurate with a predicted 2-log difference in pK_a between Lys and Arg (229). K167 of vaccinia virus topoisomerase and K235 of λ Int are both located atop flexible loops that may undergo conformational changes upon substrate binding, bringing them into proximity with other catalytic residues. Unpublished data from our laboratory indicate that this residue is indeed essential for catalysis by λ Int (J. Liu and S. E. Nunes-Düby, unpublished results), while Van Duyne and colleagues cite unpublished evidence that the equivalent residue of Cre (K201) is also required for catalysis (80). It is interesting that in the covalent Cre-*lox* cocrystal, this residue is seen interacting in the minor groove with the +1 base on the scissile strand and the +2 base of the nonscissile strand. Consequently, it is close enough to the scissile group to easily imagine that it could interact transiently with the catalytic apparatus.

In light of these observations it is interesting to reflect on the relative conservation of this previously underappreciated Lys residue (K235) to that of the central His of the well-known RHR triad (H311). In λ Int family alignments it can be seen by careful inspection that the Lys is even more conserved than the His; this implies that the former residue may play a more critical role in catalysis (19, 62, 155). Indeed, Shuman and colleagues find that the corresponding His of vaccinia virus topoisomerase contributes only modestly to the total catalytic rate enhancement (approximately 10^1) whereas the corresponding Lys provides a 10^4-fold enhancement.

cis VERSUS *trans* DNA CLEAVAGE

While the λ Int family recombinases appear to utilize a common chemical mechanism to execute DNA cleavage and religation (see "Catalysis," above), there is compelling evidence that they differ in the combinatorial organization of active sites: in some cases the same monomer that activates a scissile phosphate for attack also provides the nucleophilic Tyr for cleavage in *cis,* whereas in other cases this Tyr is provided in *trans* by a separate protomer. It was not clear until recently whether this disparity, referred to as the "*cis/trans* paradox" (96, 202), represented an essential difference between integrase family members or a misinterpretation of one or more sets of data. Indeed, this topic has been further complicated not only by the apparent differences between certain integrases but, more perplexingly, by conflicting results obtained within some systems. The recent crystallographic structures (particularly those which show recombinases interacting with substrate) have provided a new and revealing perspective on this topic and indicate where essential or perceived differences may lie.

Biochemical experiments with Flp recombinase on various substrates (half sites, full sites, Y junctions, or Holliday junctions) consistently and convincingly reveal *trans* cleavage only (33, 34, 116, 117). These solution studies are beautifully supported by the recently solved structure of a Flp recombinase tetramer bound to a Holliday junction—indeed the helix harboring the catalytic Tyr in each monomer is domain swapped with the neighboring monomer in a cyclic exchange previously suggested by Guo et al. (79). On the other hand, experiments conducted with the XerC and XerD recombinases of *E. coli* demonstrate that the DNA substrate is cleaved exclusively in *cis* (6).

Matters become less clear-cut when one considers the biochemical and structural data that exist for λ and Cre. Catalytic complementation assays (similar to those employed for Flp) between two different active-site mutants of λ Int have suggested that cleavage occurs in *trans* (85), whereas experiments that exploited the unique core-binding specificities of the closely related HK022 Int and λ Int indicate a *cis* cleavage mechanism (157). The unliganded monomeric structure of λ is, unfortunately, equally ambivalent regarding this issue; the nucleophilic Tyr (Phe in the crystal structure) is located on a flexible loop approximately 17 residues long which, when docked with an idealized *att* site substrate, can readily be modeled in a *cis* or *trans* attacking mode. One impatiently waits to see which mechanism prevails in a recombinogenic protein-DNA crystal structure.

Initial biochemical experiments addressing the active-site assembly of Cre recombinase suggested a *trans* interaction (194), but the appearance of the Cre-*lox* cocrystal structure shortly thereafter unequivocally displayed a *cis* mechanism. Revisiting the *cis-trans* issue with a modified assay and noting that their earlier study, which employed an active-site complementation type assay, did not actually rule out the possibility of *cis* cleavage, they find that Cre does indeed cleave in *cis,* and that the *cis* versus *trans* determinant resides in the C terminus (consistent with the Cre and Flp cocrystal structures) based on construction and analysis of Cre-Flp hybrids (195).

TOPOLOGY

A unified topological mechanism of site-specific recombination by λ integrase family members has remained elusive until fairly recently, in contrast to the well-characterized mechanisms for several enzymes of the resolvase/invertase family (203). Many of the early studies in λ site-specific recombination research addressed the gross topological features of the recombination reaction, but over the years inquiry gradually shifted towards finer mechanistic points, leaving a wealth of topological information that lacked a consistent theme. Understanding the nature of productive synapse assembly and product generation in the resolvase/invertase class of proteins has been facilitated by a number of advantageous proprietary features. These enzymes tend to specialize in one particular mode of recombination (i.e., inversion, deletion, or fusion) that typically takes place in a highly defined protein-DNA structure in which a fixed number of nodes or supercoils have been trapped. Strand exchange in such a bounded architecture yields products with different but also well-defined topologies (203). Additionally, elucidation of the topological mechanism has been greatly facilitated by the phenomenon of processive recombination, whereby an initial synaptic intermediate is converted into a unique ensemble of topological products by multiple rounds of exchange (99, 223). Int family members, however, do not mediate this process. A topological scheme for recombination by the λ family of enzymes must take into account the much more diverse range of activities that its members are known to mediate (integrases can recombine directly or inversely repeated sites in supercoiled, nicked, or linear substrates), as well as the observation that synapsis appears to proceed via random site collision (171, 201).

The first significant contribution to understanding λ recombination was a prescient topological prediction made by Allan Campbell (28). He proposed that the λ phage DNA integrates into the *E. coli* genome as a covalently closed circular molecule, creating a circular cointegrate. It was later shown experimentally that the *attP* locus must be situated on a circular molecule that is negatively supercoiled to permit efficient recombination in vivo or under high ionic conditions in vitro (for a review, see reference 145). In contrast to integration, supercoiling is not required for *attL* by *attR* excisive recombination in vitro (170) even though under normal in vivo conditions excision is an intramolecular reaction.

The next set of significant observations and interpretations concerned the surprising topological complexity of in vitro recombination products. Integrative recombination between intramolecular direct repeats of *attP* and *attB* on a supercoiled substrate resulted in catenated circles, the combined superhelicity of which totaled that of the parent molecule (137). This containment of supercoiling through recombination suggested that the reaction must proceed as a highly coupled process which does not permit free rotation around single-strand breaks. It also suggested that synapsis probably takes place by three-dimensional diffusion and random collision versus one-dimensional tracking.

Further analysis of knotted or catenated product formation by intramolecular recombination between inverted or direct repeats on supercoiled or relaxed substrates offered further mechanistic insights. Nash and colleagues noted that integrative recombination is characterized by a unique change in linking number (i.e., relaxation of two supercoiled nodes) (148) and the excess knots it produces on relaxed substrates are trefoils of unique topology (78, 171). They interpreted this to reflect a consistent mechanism of synapsis and suggested that it could be explained by a nucleosome-type wrapping of *attP* around the recombinase proteins at synapsis. They also found that although excision consistently generated topologically simpler products than did integration, recombination between intramolecular inverted repeats of *attL* and *attR* gives a change in linking number of +2 or −2 (148); this disparity was not explained until much later (46).

The first complete topological model recombination described for a λ Int family member was based on an analysis of XerCD-mediated monomerization of plasmid substrates (42). This system is unusually stringent in that recombination only takes place between core sites that are direct repeats on the same molecule and that all reaction products display a specific topology; in this sense it resembles a "topological filter" mechanism, initially described for the resolvase/invertase family (204, 206).

Deciphering a complete model for λ Int recombination, however, would require one to address the additional complexity of variable synapse topology. This problem was elegantly resolved by Crisona et al. (46) who employed a mathematical method called "tangles"; this technique permits one not only to describe the path of any DNA molecule in a two-dimensional projection but also to abstract any essential topological features from those that are randomly generated. The authors note that both types of nodes contribute to the complexity of the products but that only the former are illustrative of the mechanism and represent the invariant features of synaptic complex formation.

By separately considering Int-mediated excision, integration, and Cre and Flp recombination they succeeded in deducing a unified model for Int family recombination. They find that all Int reactions are chiral; that is, the product generated by strand exchange is one of two possible enantiomers. They interpret that this reflects a right-handed crossing within or between recombining sites that actually favors formation of a right-handed Holliday junction intermediate; this is presumed to be enacted by an asymmetric protein-mediated organization of recombinase sites that is maintained through the formation and resolution of a Holliday junction. They also find that integrative recombination does indeed differ from excisive recombination (171; see above) in that the synaptic complex for the latter reaction traps an additional (−) plectonemic supercoil with a relaxed substrate or two additional (−) plectonemic supercoils with a supercoiled substrate. They explain that the +2 *or* −2 changes in linking number observed for intramolecular excision on relaxed circular substrates observed by Pollock and Nash (171) is not due to formation of alternative synaptic complexes but is simply due to different distal wrapping of the plasmid. As one might expect, all of these assemblies appear to differ from the topological filter-type structures favored by the resolvase/invertase family of enzymes and Xer-mediated recombination at directly repeated sites on plasmid substrates.

IHF

Biology

IHF was originally identified in *E. coli* extracts as a cellular function essential for both in vivo and in vitro λ site-specific integrative and excisive recombination (133, 135), but it is now known to participate in several other cellular processes, such as gene expression, transposition, phage packaging, and plasmid replication (for reviews, see references 67, 73, and 147). It is a heterodimeric protein consisting of two highly basic polypeptides, α and β, with predicted molecular masses of 11,200 and 10,580 Da, respectively. These two subunits share significant homology to each other (~30%) and also to the family of type II DNA-binding proteins that includes major histone-like proteins of *E. coli* such as HU (reviewed in reference 54). The α subunit is encoded by the *himA* gene, which maps to min 37 of the *E. coli* chromosome, while the β subunit is encoded by the *himD* (or *hip*) gene, which maps to min 20.

Although the absolute amount of intracellular IHF is quite high (~7 to 14 μM in exponential-phase cells [49, 149]), in vivo footprinting of cognate IHF binding sites demonstrates that the effective cellular concentration in exponential-phase cells is just 0.7 nM (141, 234). The general consequence is that low-affinity sites are partially occupied and high-affinity sites are semisaturated. Evidently the vast majority of

IHF dimers are sequestered via nonspecific interactions with chromosomal DNA, possibly suggesting a role in global DNA compaction and condensation. Equally important, the relative abundance of IHF changes with the physiological state of the cell, increasing approximately five- to sevenfold over a 6-h span after entry into stationary phase (49, 141) and decreasing when stationary-phase cells are diluted into fresh medium and cell mass begins to double (26). This downshift in concentration is probably not due to increased protein degradation, since IHF is generally observed to be metabolically stable in its dimeric form (75, 150), but instead appears to be a consequence of increased transcription of the individual subunits upon entry into stationary phase (9) and arrested transcription upon entry into exponential phase.

There is good indication that IHF participates in its own regulation. First, *himA* transcription appears to be repressed by IHF (132, 134), and second, strains that produce inactive heterodimers show an increase in abundance of α and β subunits (75). Despite the numerous physiological alterations effected by changes in its concentration, strains lacking IHF are viable. The lack of severe growth defects or stationary-phase recovery problems in such IHF-deficient strains leads to the general impression that IHF's role in growth phase regulation is either redundant, subtle, or both (49). However, there is evidence that IHF may play an essential role in cell starvation survival: not only is IHF critical for induction of 14 proteins from the glucose starvation stimulon but *himA* mutants appear to be severely compromised in their ability to survive glucose starvation (158).

Structure

Before the publication of the IHF-DNA cocrystal (178), our understanding of the structure and mode of interaction of IHF with DNA was based on a model proposed by Nash and colleagues. It embodied a creative combination of crystal structure data from the unliganded homodimeric HU protein of *Bacillus stearothermophilus* (211) and inferences drawn from a variety of biochemical experiments (147). A logical and satisfying culmination of these efforts was the appearance of the revealing IHF-DNA cocrystal structure (178). The structure is strikingly similar to that of HU and reveals a compact globular domain, primarily α-helical, consisting of symmetrically intertwined α and β subunits, nestled into the concave surface of a severely bent (>160°) H′-containing duplex (one strand of this duplex is nicked). The two arms that were unresolved in the HU structure track along the minor groove around to the convex surface, as predicted by the model of Nash and colleagues (227, 233); these clasping arms bury into the minor groove, widening it and bending the DNA. The protein-DNA interface is extensive (>4,600 Å^2 buried) and results primarily from minor groove contacts, consistent with previous footprinting data (233). Concern that the presence of a nick in one strand of the crystallization substrate might alter the conformation of the complex was allayed by a recent FRET analysis (125).

Particularly outstanding are the presence of two sharp kinks separated by 9 bp and symmetrically disposed around the functional bend center that result from the intercalation of a conserved proline residue located at the tip of each arm of the dimer. These hydrophobic residues interrupt the base stacking and wedge open the minor groove, apparently to provide the extra motivation needed to wrap the DNA around the globular core. Other "big benders" (reviewed in reference 219) use a similar mechanism of concentrating a large portion of their net induced bend at a such a kink. The second major mechanism of bend stabilization involves an extensive network of electrostatic interactions between the core of the heterodimeric IHF and the phosphate backbone on the concave surface of the DNA hairpin. This has the general effect of neutralizing the repulsive forces between the phosphates on the narrowed major groove.

An emphatic message from the IHF-DNA cocrystal structure is that the sequence preference displayed by IHF cannot be explained by specific side chain-to-DNA base contacts. In fact only three residues make direct contacts to the edges of DNA bases, and of these, only one (Arg 46 of the β subunit) is conserved. Instead, IHF seems to exert its preference by "indirect readout"; that is, its selection appears to be based on the sequence dependent distortability of substrate DNA. For example, IHF displays a strong preference for stretches of A-T base pairs to the "left" of the consensus sequence (72); introduction of an A-tract into a recognition site that does not normally contain one has been shown to increase IHF's affinity for it (83). The A-tract segment of the IHF-DNA complex manifests all of the same exceptional features that were previously characterized in free DNA (47, 48, 151): it is remarkably straight, it has a narrowed minor groove, and it displays high propeller twist between the bases. It has been suggested (177) that these idiosyncratic features are important for making critical contacts, such as the "tripartite clamp" (178), which comprises a network of interactions that span the narrowed minor groove.

The average twist angle in the bound DNA (~33.3°/bp) is very similar to that of average B-DNA with the topological consequence that it is not significantly underwound by IHF. The ends of the bent

DNA hairpin lie approximately in the same plane, subtending a dihedral angle of only 10 to 15° whose handedness is consistent with the wrap of negatively supercoiled DNA. This minor out-of-plane deviation means that IHF's overall effect on writhe, or the formation of a three-dimensional framework, will depend substantially on its phasing with any other bending modules.

Architectural Elements

The three IHF sites of *attP* (H1 and H2 of the P arm and H′ of the P′ arm) are found interspersed between the various core-binding and arm-binding sites of λ Int (45; see "*attP*," below). This arrangement, plus the observation that the λ *att* site forms a higher-order nucleoprotein complex (termed the intasome), led to speculations that IHF might assist in the folding of this structure (179). In support of this thinking, the only well-established function of IHF known at the time was its ability to induce extreme DNA bending (for reviews, see references 113 and 147).

Persuasive evidence of a primarily architectural role in the assembly of recombinogenic nucleoprotein complexes came from "bend swap" experiments (71, 147, 192) where one or more IHF sites were replaced by unrelated DNA-bending modules (either intrinsically bent DNA or different DNA bending proteins [e.g., HU, HMG1/2, etc.]). Although able to complement the lack of IHF-induced bending, none of the chimeric constructions performed as efficiently as the wild-type arrangement. A specific architectural role was demonstrated by the observation that IHF bending could stimulate Int binding and cleavage at low-affinity core sites (138). This activity depended on the presence of a high-affinity arm-binding site that could effectively recruit an Int protomer from solution and direct the protein's core-binding domain to its cognate site by appropriate bending.

The inability of the bend swap chimeras to achieve wild-type efficiency implied that either IHF performed some secondary function besides bending or that only IHF alone could provide the requisite precision. Evidence for the latter case derived from the observation that a wrongly positioned IHF bend (of just 1 bp), in a loop of constant length between C′ and P′1 of *attL*, could severely reduce excisive recombination efficiency (156).

Xis

Xis is a small basic protein (72 amino acids) encoded by the phage genome that is essential for excisive recombination but inhibitory towards integrative recombination (2, 143). It does not manifest any detectable catalytic activity; instead it binds and bends DNA in a site-specific fashion (two cooperatively bound Xis protomers induce a bend of ≥140° [212]). Thus, like IHF, it appears to act as an essential architectural element in the assembly of competent or incompetent nucleoprotein complexes (depending on the circumstances); however, it also engages in several cooperative protein-protein interactions (so far not seen for IHF) that are important for its function.

Xis recognizes two 13-bp sequences, designated X1 and X2, that are located in the P arm of *attP* (or *attR*) and are, curiously, arranged as tandem direct repeats (235). Consequently, two protomers will bind cooperatively in a head-to-tail orientation, distinguishing Xis from the vast majority of homodimeric DNA-binding proteins that interact in a dyad symmetrical head-to-head mode (27). Partially overlapping X2 is a recognition element (F) for the unrelated host protein Fis (factor for inversion stimulation) that can efficiently substitute for an Xis at X2 (binding to F or X2 is mutually exclusive) and cooperate with an Xis bound to X1 (214). Together, Xis, Int, and IHF stimulate excision by assembling a recombinogenic intasome on *attR* (17) that can interact productively with an *attL* intasome (Fig. 1). Of particular interest are the cooperative interactions between a pair of Xis molecules bound at X1 or X2 and an Int protomer bound via its arm domain to the adjacent P2 sequence. Although an Int bound to this site is not required for excision or integration, it can enhance Xis binding by approximately 16-fold, thus decreasing the concentration required to stimulate the former and inhibit the latter process (26, 214).

It is proposed that Xis binding inhibits integration by effectively converting a recombinogenic *attP* intasome into a specific looped structure that precludes the normal productive interactions that would lead to the first strand exchange event (139). More precisely, it appears that Xis prevents Int and IHF from occupying their requisite sites via competitive interactions that influence sites far removed from X1 and X2 on the primary DNA sequence (e.g., P1 and H1 of the P arm and P′3 of the P′ arm [which is ~140 bp away]). Thus, Xis binding can be viewed as a potential regulatory switch that ensures the appropriate equilibrium between lysogeny and lytic growth (see "*attP*" and "Directionality and Regulation," below). Further support for this notion comes from the observation that although Xis is relatively thermostable in vitro, it is rapidly degraded in vivo (225). This contrasts with the comparatively higher stability of Int ($t_{1/2} \cong 60$ min) and would be expected of a potential regulatory factor. A recent study suggests that Xis is degraded in vivo by two ATP-dependent proteases

(Lon and FtsH [HflB]) (121). Interestingly, the same study finds that increasing the intracellular concentration of Xis in wild-type cells, or in cells harboring disabled proteases, can inhibit integrative recombination and promote the curing of established lysogens. The latter observation demonstrates that the residual level of Int is sufficient to mediate excision and probably is not a source of regulation.

At present, there exists no structural data for Xis (i.e., in the form of a crystal or nuclear magnetic resonance structure) although a number of recent studies have combined structural inferences derived from computer-based secondary-structure predictions with functional assays to dissect and localize the various activities of Xis. A truncated Xis (residues 1 to 53) retains DNA binding and cooperative interactions with Fis (making it sufficient to stimulate excision) but lacks cooperativity with Int, thus implicating the missing C-terminal portion in this activity (153). An amphipathic helix of Xis (residues 57 to 65) appears to be the primary target of interaction with Int (231), while a second amphipathic helix (residues 18 to 28) plus the region around residue 40 appear to contain the DNA binding activity (39). It remains to be seen precisely where the Fis interaction domain dwells.

Fis

Structure

Fis is a small (98-amino-acid), basic, heat-stable protein that binds and bends DNA as a homodimer and is encoded by a gene mapping to 72.5 min on the *E. coli* chromosome (see chapter 13). Crystal structures of dimeric Fis reveal a compact ellipsoid core, consisting of four α-helices that terminate in a canonical helix-turn-helix motif in each subunit (108, 238). The N-terminal 25 amino acids of each subunit form a highly flexible β-hairpin arm that protrudes >20 Å from the protein core (185) that, interestingly, can be converted to an α-helical structure by mutating Pro 26 (which forms a hinge at the base of this structure) to Ala. As an alternative to utilizing the highly degenerate 15-bp consensus sequence (93), which discards important information about the observed frequency of bases, a compelling case has been made for using information analysis (90) (which employs a graphical system that retains the subtleties of sequence data) to identify potential Fis sites.

For Fis to dock with its cognate DNA substrate, its two helix-turn-helix moieties (separated by just 25 Å) must contact two adjacent major grooves separated by 34 Å (according to methylation interference footprinting). Clearly, the DNA and/or protein must undergo substantial distortion to achieve union. A computer modeling exercise that docked the unliganded structure of Fis (which was kept relatively rigid) to a yielding DNA substrate resulted in a bend of ~60° (188), which is in general accord with the biochemical data (161, 162, 169). It is interesting that Fis bending can be quite variable—ranging from ~50° to ~90°—due apparently to differential interactions with bases flanking the recognition sequence (161, 162).

Biology

Fis is implicated in an impressive array of functions that relate not only to its capacity to site-specifically bind and bend DNA but also to its ability to participate in a variety of protein-protein interactions. Nevertheless, it was first identified as a factor required to stimulate site-specific DNA inversion in the Hin system of *Salmonella enterica* serovar Typhimurium (98), the Gin system of bacteriophage Mu (106), and the Cin system of P1 bacteriophage (81).

Fis's role in λ site-specific recombination was first identified from a screen of *E. coli* proteins by its ability to stimulate excision ~20-fold at limiting Xis concentrations (213). It accomplishes this by substituting for an Xis protomer at X2 in *attR* and cooperatively assisting in the binding of an Xis to X1 (approximately fourfold) (214), presumably via a different contact surface than the one that activates Hin cleavage, since Fis mutants lacking the N-terminal β-hairpin still stimulate excision (107, 159). Not only is the net bend induced by the Xis-Fis pair equivalent to that of the Xis-Xis pair (212), but so is its influence on the efficiency and directionality of the recombination reaction. Fis has no effect at saturating Xis in vitro, although evidently, Xis is not saturating in vivo since Fis manifests a significant stimulation of excision. Additionally, Fis has been shown to stimulate integration approximately twofold and to enhance the maintenance of lysogeny independent of its interaction with the F site in *attP* and *attR* (11, 12).

It is tempting to speculate that Fis sensitivity was acquired by the λ system to provide a sensor of host physiology and thus a competitive advantage to the phage's survival (see "Directionality and Regulation," below). Fis concentration varies dramatically with the growth phase of the cell and changes with nutrient conditions; its concentration is very low to undetectable during stationary phase or slow growth but increases to approximately 25,000 to 50,000 dimers per cell when shifted to rich medium (i.e., during early exponential phase). Levels then decrease to ~1% of peak levels as cells enter stationary phase again (13, 152). Consistent with this variation, the F site of *attP* and *attR* is occupied during exponential phase (i.e., when Fis typically manifests its stimulatory effect in vivo) and is vacant during stationary phase (213).

attP

Int Binding Sites

The 240-bp λ phage *att* site consists of an *attB*-like core region with flanking arms, called P and P′ (Fig. 1). The core region is where all of the chemistry and strand exchange of recombination takes place, as described above for *attB*, while the P and P′ arms contain a variety of protein binding sites that govern the directionality and efficiency of recombination. As a result of integrative recombination between *attP* (POP′) and *attB* (BOB′), the complexity of *attP* is distributed to the left and right prophage *att* sites, *attL* (BOP′) and *attR* (POB′), which are themselves potential substrates for excisive recombination.

One of the striking features of the P and P′ arms is that they contain a class of arm-type Int binding sites that is distinct from the core-type Int binding sites where DNA cleavage and ligation are executed. Whereas the latter are recognized by the large carboxy-terminal domain of Int, the former are recognized by the small amino-terminal domain (113). The five arm-type Int binding sites (P1 and P2 in the P arm and P′1, P′2, and P′3 in the P′ arm) have a consensus recognition sequence of seven contiguous base pairs. Their approximate binding affinities for Int suggest that P′1 and P1 have equal affinities, greater than that of P2 (the affinities for P′2 and P′3 not having been examined in isolation). From experiments with individually mutated arm-type sites and *att* sites with resected arms it has been shown that the P1, P′2, and P′3 sites are all required for integrative recombination, while occupancy of the P2 site is actually detrimental. The apparent dispensability of P′1 for integrative recombination is subject to the reservation that applies to all point mutants without a phenotype in highly cooperative systems. In excisive recombination the two most convincing observations are the strong requirement for P′1 and the dispensability of P1.

IHF Binding Sites

Interspersed between the arm-type and core-type Int binding sites are three binding sites for the accessory DNA bending protein IHF: H1 and H2 in the P arm and H′ in the P′ arm (45). As discussed above, the cocrystal structure of IHF bound to one of its cognate sites has dramatically confirmed a great deal of biochemistry (178) and has illuminated the structural basis for how this protein introduces such sharp and precise bends in DNA (see "IHF," above). Indeed, precision at the level of 1 bp is required for the proper relative positioning of arm- and core-type Int sites in order to assemble a functional recombinogenic complex (156).

Differential Use of Binding Sites

Despite the gross similarity of the three IHF sites in the extent of induced DNA bending, the three sites are not functionally equivalent in recombination. Whereas H1 is required for integrative but not excisive recombination, H2 and H′ are required for both reactions (215). Between the P2 Int site and the H2 IHF sites are the two binding sites for Xis, X1 and X2, and overlapping the latter is the binding site for Fis, F. As discussed above, each of the accessory proteins is a DNA bending protein that plays a role in the formation of the distinctly different sets of higher-order complexes required for integrative and excisive recombination, respectively (see "Xis" and "Fis," above). Analysis has been easier for the complexes involved in excisive as opposed to integrative recombination for two technical reasons: the requirement for supercoiling of the *attP* partner in integrative recombination and the dispersion of the higher-order complexity between the two partners of excisive recombination.

Higher-Order Complexes

Specific interactions supporting the recombinogenic complexes have been deduced by correlation of recombination with multiprotein nuclease protection assays, competition binding studies, site-directed mutagenesis of protein binding sites, studies with suicide recombination substrates, gel shift assays, and an in vitro genetic test for synergism (103, 113). One picture that has emerged for the *attL-attR* synaptic complex draws on a basic motif of protein-induced bends that enable bivalent Int molecules to form intra- and intermolecular bridges between appropriate pairs of arm- and core-type sites. In this model one Int forms an intramolecular bridge between P′1 and C′ of *attL* and a second Int forms an intermolecular bridge between P2 of *attR* and B of *attL*. The Int bound at P′2 of *attL* could form an intermolecular bridge with either C or B′ of *attR*. The authors suggest that the former might be more likely since the B′ site is not required for synapsis or top-strand exchange and the P′2 and C sites are both required for these early steps in recombination. If this were the case, the B′ site could recruit an Int from solution, possibly even at a later stage of the reaction (103).

The results of these experiments point to the importance of the P′1-C′ bridge on *attL*. The data show this to be the strongest of the Int bridges, and it is also the only bridge that is intramolecular. Thus, as in integrative recombination, which involves a complex *attP* and a simple *attB*, there is also a functional disparity between the recombination partners in excision. It seems that the primary structures of recombination are located on the *attL* arm and the modulating structures, such as the F and H1 sites (see below), are located on the *attR* arm. A similar distribution of functions was also indicated by studying the effect of Xis and IHF on the direction of resolution of preformed synthetic *att* site Holliday junctions containing either the P or P′ arms, or both (66). The patterns of resolution showed the P′ arm largely governed the efficiency of the reaction while Xis and the P arm primarily influenced the direction of resolution, i.e., which pair of strands was cleaved. The picture of the excisive recombinogenic complexes is far from complete, and that of the integrative complexes has not even been started. The same general design principles elucidated for the λ pathway also underlie two different patterns of Int-arm-core bridges proposed for the Tn*916* and L5 recombination pathways (97, 168).

Family Considerations

A particularly interesting and puzzling aspect of the λ P and P′ arm structures is their relationship to the analogous DNA sequences in the *att* sites of other heterobivalent λ Int family members. Each of these other recombination pathways also makes use of accessory DNA bending proteins interspersed between arm and core-type Int binding sites. However, the nature, number, spacing, and orientation of these sites can be quite different from system to system (Fig. 2). Whereas the basic theme of the core region and its core-type Int sites is conserved, the disposition and orientation of the arm-type Int sites vary considerably. One of the most dramatic differences in Int binding patterns is found in the HP1 *attP* site, where the arm-type Int sites are joined by additional core-type Int binding sites in the P and P′ arms (82).

In comparison to the three IHF sites of λ *attP*, HP1 and P22 have only two and φ80 and P2 appear to have only one (45, 94, 122, 200, 236). L5 has a requirement for a mycobacterial IHF but it does not bind specifically to the L5 *attP* DNA in the absence of Int (167). The *fimB-fimE* system utilizes IHF, but these recombinases are not heterobivalent. Lrp is thought to introduce DNA bends that promote synapsis between the target sites flanking the invertible DNA segment (21). At the opposite end of the spectrum from λ Int is the Int from *Lactobacillus* phage mv4, which does not seem to have any requirement for IHF, despite the fact that it has a large *attP* site with dispersed arm-type Int binding sites (8). This observation must be considered preliminary because the mv4 Int has not been purified. Furthermore, there is already the example of the P22 *attP* which has well-defined IHF sites but seems to function in vivo without IHF, possibly using some other protein(s) in its place (38).

A similar range of differences is observed for the accessory proteins that are Xis-like in function. In the case of Tn*916*, the Xis protein, which is very similar in size to λ Xis, has binding sites on both sides of the core region. Only one of these sites is between arm- and core-type Int binding sites, and the other site is postulated to function primarily via a stimulation of Int binding (183). The three related recombination systems of HP1, P2, and 186 all stimulate excision via a "cox" protein, which is quite different from λ Xis in its structure and mode of DNA binding (50, 61, 237). In all three phages the cox protein also functions as a regulator of phage gene transcription. Despite their differences, all of these proteins, including the Xis of phage L5 (123), are functionally similar to λ Xis in their ability to inhibit integrative as well as profoundly stimulate excisive recombination.

We consider two extreme views of the extent of diversity among the *attP* sites of different heterobivalent λ Int family members. One view minimizes the differences and considers them to be more apparent than real. According to this view, the different linear arrays of protein binding sites are all compatible with basically similar higher-order structures. According to the second view, each (or many) of the different linear arrays is associated with a different higher-order structure. Even between these two simplified opposing views there remains the question, "What structural features are regarded as being substantially, versus superficially, different?"

PATHWAY FLEXIBILITY

It has been known for quite a while that λ Int mediates efficient recombination between two *attL* sites in vivo (57, 100). Motivated by a desire to understand the nature of this phenomenon, Howard Nash and colleagues reconstituted *attL* × *attL* recombination in vitro to find that the reaction falls into two distinct categories that differ in protein factor requirement, binding site occupancy, rate and extent of reaction, and most likely complex geometry too—hence their graphic sobriquets: "straight-L" and "bent-L" recombination.

The straight-L reaction is distinguished by the

ability of λ Int to align and hold together two DNA molecules, forming stable non-covalent complexes (BMC) that support a low level (~5%) of recombination (193). Whereas BMC formation clearly requires both core sites (B and C′ in this instance), the data for the arm sites are ambiguous. Methylation interference footprinting suggests that no single arm site (out of two pairs of P′123) is absolutely required, while deletion analysis indicates that several sites need to be occupied during the reaction and that P′1 is perhaps the most important arm site. The observation that IHF inhibits formation of the straight-L complex was a sign that this reaction was not commensurate with previous in vivo data; the *attL* × *attL* reaction is stimulated approximately 5- to 10-fold in an IHF-competent host versus one that is lacking IHF and can reach levels of recombination comparable to integration and excision (50 to 70%). This cannot be accounted for by the sluggish, low level of recombination manifested by the straight-L reaction.

The pivotal finding was that IHF can stimulate homology-dependent recombination between *attL* partners that are modified to be defective in P′1 binding. This reaction is referred to as bent-L recombination to accentuate the critical role of IHF participation. The most interesting aspects of bent-L recombination reaction are best viewed in contrast to the complementary features of the excisive *attL* complex (Fig. 6) (102). First of all, while a P′1 binding site is deleterious for bent-L recombination in vitro, both of the remaining sites (P′2 and P′3) are required. This differs from the excisive reaction, which requires P′1 but can dispense with P′3. Secondly, although both reactions require IHF, the nature of its interaction with each complex seems to be different. This is based on the observation that certain nonspecific DNA bending proteins (HU and HMG1/2) can stimulate *attL* intasome formation and excisive recombination in lieu of IHF. The bent-L reaction, on the other

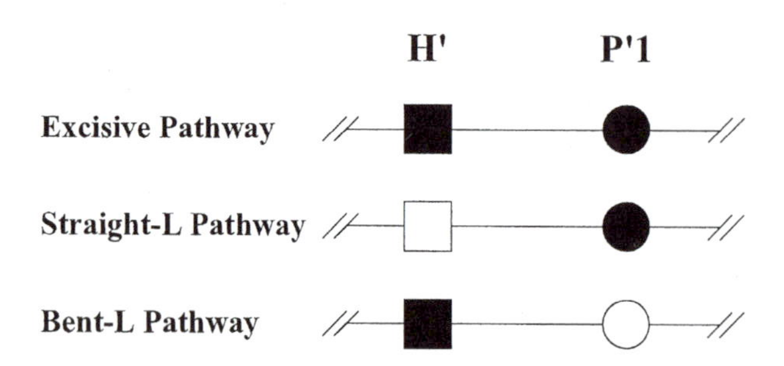

Figure 6. Differential occupancy of H′ and P′1 sites of *attL* that characterize various recombination pathways. Occupied sites are indicated by filled symbols, while vacant sites are indicated by open symbols.

hand, cannot utilize these proteins to form complexes or to sponsor recombination.

Although this model reaction seems to account for previously documented *attL* × *attL* recombination in vivo, one discrepancy remains unresolved: bent-L recombination cannot take place between wild-type *attL* substrates in vitro, while it appears to take place between wild-type or P′1 mutated sites in vivo. Nash and colleagues speculate that there is an as-yet-unidentified factor that permits bent-L recombination in vivo by, for example, mitigating the deleterious effect of P′1 occupancy on the reaction seen in vitro.

One promising approach to understanding the basis of this flexibility in recombination has recently been reported by Anca Segall and her colleagues (30, 104). They have shown that the generically useful combinatorial peptide libraries can be used to identify specific hexapeptide sequences that differentially inhibit the various λ Int pathways. It is reasonable to expect that this approach will also have great potential in dissecting out other aspects of the recombination pathway(s).

It is interesting that the flexibility in the way different pairs of *att* sites recombine is influenced not only by mutations in the *att* sites, as described above, but also by mutations in the integrase protein. This was first shown for an Int point mutant (E174K) that could execute excisive recombination in the absence of Xis (60). Most recently the same mutation, alone or in combination with E218K, has been shown to lead to deletions between substrates that would normally lead to inversion (40).

DIRECTIONALITY AND REGULATION

The critical importance for temperate phages to regulate their integration and excision reactions is evidenced by the many different levels at which regulation is exerted and the number of inputs to which it responds. A common mode of regulation is of course gene expression (in this case) of the *int* and *xis* genes. Although it is beyond the scope of this chapter to explore the elegant and well-studied control circuits of bacteriophage λ, we note that the expression of *int* and *xis* is influenced by several different types of regulation (89, 91). Their expression is under the negative control of the cI repressor acting at O_L, positive control of cII acting on pI, differential control due to the fact that *xis* and *int* overlap out of frame, and "retroregulation" by the *cis*-acting downstream *sib* sequences. The last feature is particularly interesting and relevant to recombination genes because it is position sensitive. The regulatory *sib* sequences are distal to the *attP* site in the viral chromosome so they are separated from the *int* gene, and rendered nonfunctional, in the lysogen (140).

In addition to the modulation of intracellular Int and Xis concentrations, there are also complex interactions between these factors and their substrates that determine their relative ability to execute integration or excision (113). The sharpest insights about regulation at this level have come from in vitro experiments, but extrapolation to the in vivo situation has not always been straightforward (see below). While Xis protein is required for excisive recombination under physiological conditions, it is inhibitory for integrative recombination. Indeed, it is the main switch driving excisive versus integrative recombination. However, there are additional switches influencing the likelihood that excision will occur. One of these is the H1 site, which must be occupied by IHF for integrative recombination and must be vacant for excisive recombination. It has been shown in vitro that the relative levels of IHF, Int, Xis, and Fis influence differential occupancy of this site. The significance of the H1 switch is that high levels of IHF have the potential to inhibit excisive recombination. In contrast, Fis, the other host-encoded accessory protein, has the potential to stimulate excision when it is present at high levels (and Xis is limiting). Thus, the modulating effects of both host proteins are themselves modulated by the levels of Int and Xis: high Int or high Xis overcomes the IHF inhibition, and high levels of Xis render excision insensitive to Fis.

How do these modulating effects by two host-encoded proteins (which were determined from in vitro experiments) extrapolate to the in vivo situation? There are two different answers to this question. The first is that they make a great deal of biological sense. IHF levels rise as cells enter stationary phase (26, 49), and it makes sense to reduce the likelihood of excision at a time when intracellular resources may limit a phage burst and the extracellular environment may not afford many growing cells to infect. Fis on the other hand undergoes a rapid burst of synthesis just as cells are commencing exponential growth (13), a time when it makes great sense to favor excision and the transition to lytic growth. The second answer to the question is that in vivo experiments do not support the scenarios of modulation extrapolated from in vitro experiments, and it has been suggested that Fis and IHF are used by λ because of their DNA-bending properties, rather than their expression patterns (12, 146). We confess to a bias that gives more weight to the in vitro experiments and the likelihood that the expression patterns of IHF and Fis are not coincidence or incidental to their role in recombination. Both of these proteins influence many pathways, including additional λ pathways. It may therefore be difficult to deconvolute these networks enough to uncover low-frequency events that are nevertheless profoundly important to a phage population (such as in spontaneous phage production that occurs at 1 in 10^3 to 10^5 cell generations).

CONCLUDING REMARKS

The range and variety of biological functions and reaction mechanisms encompassed by the λ Int family are exemplified by a large number of systems at overlapping stages of development. Indeed, studies throughout the family have complemented each other so well precisely because the high degree of relatedness among family members is counterbalanced by a wide spectrum of properties and activities. Unfortunately, limitations of space in this article have forced us to leave out many interesting segments of this rich diversity.

This intellectual reciprocity is easily seen by a quick look at some of the most "mature" λ Int family members. The unadorned minimal Cre and type I topoisomerase reactions have greatly facilitated high-resolution studies of catalysis and strand exchange. The dual-protein XerC and XerD system has helped to differentiate the front and back halves of the central reaction and the action of topological filters. The *trans*-cleaving Flp protein has highlighted the concept of composite and interacting active sites. The promiscuous Tn*916*-Tn*1545* recombinases have challenged the role of overlap region homology. And the genetically gifted (well-endowed) λ pathway has helped guide the way to understanding the directionality, regulation, and higher-order structures of the viral and heterobivalent recombinases.

Even among the heterobivalent recombinases there is a level of diversity that we think constitutes one of the most interesting questions for this branch of the family. Does the variety of arrangements of protein binding sites in the P and P′ arms of different *att* sites (as discussed above) reflect a similar variety of higher-order structures? Or, does it reflect a degree of "sloppiness" that can be accommodated in the assembly of a small number of basically similar higher-order structures? If the latter is true, does the concept of "approximate structures" only apply to a collection of pathways, each of which has its own precisely ordered structures, or does it also apply to individual pathways? The present data relevant to these questions do not fit a simple or single answer.

The last several years have witnessed great strides in our understanding of both the λ integration-excision pathway and the entire λ Int family. These advances include a sharpened view of strand exchange and how homology is sensed, high-resolution insights regarding protein active sites and the chemistry of DNA cleavage and ligation, models for the higher-order structures in several systems, a developing functional anatomy of some of the proteins, increases in the diversity and breadth of the family, an improved understanding of how different family members function in their respective biological contexts, and the development of several family members as investigative and manipulative tools in other areas of biology and molecular technology. Each of these recent strides has served to increase our expectations for the next step toward elucidating the details of family principles and uncovering the diversity among individual family members.

Acknowledgments. We thank our many colleagues who provided us with preprints and reprints. We are grateful to our coworkers at Brown University and in the Ellenberger laboratory at Harvard Medical School for intellectual stimulation and critical feedback. We thank Joan Boyles for assistance in preparing the manuscript.

We thank the National Institutes of Health (grants GM62723 and GM33928) for support of the work carried out here.

REFERENCES

1. **Abraham, J. M., C. S. Freitag, J. R. Clements, and B. I. Eisenstein.** 1985. An invertible element of DNA controls phase variation of type 1 fimbriae of *Escherichia coli. Proc. Natl. Acad. Sci. USA* **82:**5724–5727.
2. **Abremski, K. E., and S. Gottesman.** 1982. Purification of the bacteriophage λ *xis* gene product required for λ excisive recombination. *J. Biol. Chem.* **257:**9658–9662.
3. **Abremski, K. E., and R. H. Hoess.** 1992. Evidence for a second conserved arginine residue in the integrase family of recombination proteins. *Protein Eng.* **5:**87–91.
4. **Arciszewska, L., I. Grainge, and D. Sherratt.** 1995. Effects of Holliday junction position on Xer-mediated recombination in vitro. *EMBO J.* **14:**2651–2660.
5. **Arciszewska, L. K., I. Grainge, and D. J. Sherratt.** 1997. Action of site-specific recombinases XerC and XerD on tethered Holliday junctions. *EMBO J.* **16:**3731–3743.
6. **Arciszewska, L. K., and D. J. Sherratt.** 1995. Xer site-specific recombination in vitro. *EMBO J.* **14:**2112–2120.
7. **Argos, W., A. Landy, K. Abremski, J. B. Egan, E. Haggård-Ljungquist, R. H. Hoess, M. L. Kahn, W. Kalionis, S. V. L. Narayana, L. S. I. Pierson, N. Sternberg, and J. M. Leong.** 1986. The integrase family of site-specific recombinases: regional similarities and global diversity. *EMBO J.* **5:**433–440.
8. **Auvray, F., M. Coddeville, G. Espagno, and P. Ritzenthaler.** 1999. Integrative recombination of *Lactobacillus delbrueckii* bacteriophage mv4: functional analysis of the reaction and structure of the *attP* site. *Mol. Gen. Genet.* **262:**355–366.
9. **Aviv, M.** 1994. Expression of the genes coding for the Escherichia coli integration host factor are controlled by growth phase, rpoS, ppGpp and by autoregulation. *Mol. Microbiol.* **14:**1021–1031.
10. **Azaro, M. A., and A. Landy.** 1997. The isomeric preference of Holliday junctions influences resolution bias by λ integrase. *EMBO J.* **16:**3744–3755.
11. **Ball, C. A., and R. C. Johnson.** 1991. Efficient excision of phage λ from the *Escherichia coli* chromosome requires the Fis protein. *J. Bacteriol.* **173:**4027–4031.
12. **Ball, C. A., and R. C. Johnson.** 1991. Multiple effects of Fis on integration and the control of lysogeny in phage λ. *J. Bacteriol.* **173:**4032–4038.
13. **Ball, C. A., R. Osuna, K. C. Ferguson, and R. C. Johnson.** 1992. Dramatic changes in Fis levels upon nutrient upshift in *Escherichia coli. J. Bacteriol.* **174:**8043–8056.
14. **Baum, J. A.** 1995. TnpI recombinase: identification of sites within Tn*5401* required for TnpI binding and site-specific recombination. *J. Bacteriol.* **177:**4036–4042.
15. **Baum, J. A., A. J. Gilmer, and A.-M. L. Mettus.** 1999. Multiple roles for TnpI recombinase in regulation of Tn*401* transposition in *Bacillus thuringiensis. J. Bacteriol.* **181:** 6271–6277.
16. **Bear, S. E., J. B. Clemens, L. W. Enquist, and R. J. Zagursky.** 1987. Mutational analysis of the lambda *int* gene: DNA sequence of dominant mutations. *J. Bacteriol.* **169:**5880–5883.
17. **Better, M., S. Wickner, J. Auerbach, and H. Echols.** 1983. Role of the Xis protein of bacteriophage λ in a specific reactive complex at the *attR* prophage attachment site. *Cell* **32:** 161–168.
18. **Blakely, G., G. May, R. McCulloch, L. K. Arciszewska, M. Burke, S. T. Lovett, and D. J. Sherratt.** 1993. Two related recombinases are required for site-specific recombination at dif and cer in E. coli K12. *Cell* **75:**351–361.
19. **Blakely, G. W., and D. J. Sherratt.** 1996. *cis* and *trans* in site-specific recombination. *Mol. Microbiol.* **20:**234–237.
20. **Blomfield, I. C., P. J. Calie, K. J. Eberhardt, M. S. McClain, and B. I. Eisenstein.** 1993. Lrp stimulates phase variation of type 1 fimbriation in *Escherichia coli* K-12. *J. Bacteriol.* **175:** 27–36.
21. **Blomfield, I. C., D. H. Kulasekara, and B. I. Eisenstein.** 1997. Integration host factor stimulates both FimB- and FimE-mediated site-specific DNA inversion that controls phase variation of type 1 fimbriae expression in *Escherichia coli. Mol. Microbiol.* **23:**705–717.
22. **Breuner, A., L. Brondsted, and K. Hammer.** 1999. Novel organization of genes involved in prophage excision identified in the temperate lactococcal bacteriophage TP901-1. *J. Bacteriol.* **181:**7291–7297.
23. **Bullock, P., J. J. Champoux, and M. Botchan.** 1985. Association of crossover points with topoisomerase I cleavage sites: a model for nonhomologous recombination. *Science* **230:** 954–958.
24. **Burgin, A. B., and H. A. Nash.** 1995. Suicide substrates reveal properties of the homology-dependent steps during integrative recombination of bacteriophage λ. *Curr. Biol.* **5:** 1312–1321.
25. **Burns, L. S., S. G. J. Smith, and C. J. Dorman.** 2000. Interaction of the FimB integrase with the *fimS* invertible DNA element in *Escherichia coli* in vivo and in vitro. *J. Bacteriol.* **182:** 2953–2959.
26. **Bushman, W., J. F. Thompson, L. Vargas, and A. Landy.** 1985. Control of directionality in lambda site-specific recombination. *Science* **230:**906–911.
27. **Bushman, W., S. Yin, L. L. Thio, and A. Landy.** 1984. Determinants of directionality in lambda site-specific recombination. *Cell* **39:**699–706.
28. **Campbell, A. M.** 1962. Episomes, p. 101–145. *In* E. W. Caspari (ed.), *Advances in Genetics,* 1st ed. Academic Press, New York, N.Y.
29. **Campbell, A. M.** 1992. Chromosomal insertion sites for phages and plasmids. *J. Bacteriol.* **174:**7495–7499.
30. **Cassell, G., M. Klemm, C. Pinilla, and A. Segall.** 2000. Dissection of bacteriophage λ site-specific recombination using synthetic peptide combinatorial libraries. *J. Mol. Biol.* **299:** 1193–1202.

31. **Champoux, J. J.** 1998. Domains of human topoisomerase I and associated functions. *Prog. Nucleic Acid Res.* **60:** 111–132.
32. **Chen, J.-W., B. R. Evans, S.-H. Yang, H. Araki, Y. Oshima, and M. Jayaram.** 1992. Functional analysis of box I mutations in yeast site-specific recombinases F1p and R: pairwise complementation with recombinase variants lacking the active-site tyrosine. *Mol. Cell. Biol.* **12:**3757–3765.
33. **Chen, J.-W., J. Lee, and M. Jayaram.** 1992. DNA cleavage in trans by the active site tyrosine during F1p recombination: switching protein partners before exchanging strands. *Cell* **69:**647–658.
34. **Chen, J.-W., S.-H. Yang, and M. Jayaram.** 1993. Tests for the fractional active site model in F1p site-specific recombination: assembly of a functional recombination complex in half-site and full-site strand transfer. *J. Biol. Chem.* **268:** 14417–14425.
35. **Chen, Y., U. Narendra, L. E. Iype, M. M. Cox, and P. A. Rice.** 2000. Crystal structure of a F1p recombinase-Holliday junction complex: assembly of an active oligomer by helix swapping. *Mol. Cell* **6:**885–897.
36. **Cheng, C., P. Kussie, N. Pavletich, and S. Shuman.** 1998. Conservation of structure and mechanism between eukaryotic topoisomerase I and site-specific recombinases. *Cell* **92:** 841–850.
37. **Cheng, Q., B. M. Swalla, M. Beck, R. Alcaraz, Jr., R. I. Gumport, and J. F. Gardner.** 2000. Specificity determinants for bacteriophage Hong Kong 022 integrase: analysis of mutants with relaxed core-binding specificities. *Mol. Microbiol.* **36:** 424–436.
38. **Cho, E. H., C.-E. Nam, R. Alcaraz, Jr., and J. F. Gardner.** 1999. Site-specific recombination of bacteriophage P22 does not require integration host factor. *J. Bacteriol.* **181:** 4245–4249.
39. **Cho, E. H., R. Alcaraz, Jr., R. I. Gumport, and J. F. Gardner.** 2000. Characterization of bacteriophage lambda excisionase mutants defective in DNA binding. *J. Bacteriol.* **182:** 5807–5812.
40. **Christ, N., and P. Dröge.** 1999. Alterations in the directionality of λ site-specific recombination catalyzed by mutant integrases in vivo. *J. Mol. Biol.* **288:**825–836.
41. **Collis, C. M., M.-J. Kim, H. W. Stokes, and R. M. Hall.** 1998. Binding of the purified integron DNA integrase IntI1, to integron- and cassette-associated recombination sites. *Mol. Microbiol.* **29:**477–490.
42. **Colloms, S. D., J. Bath, and D. J. Sherratt.** 1997. Topological selectivity in xer site-specific recombination. *Cell* **88:** 855–864.
43. **Colloms, S. D., R. McCulloch, K. Grant, L. Neilson, and D. J. Sherratt.** 1996. Xer-mediated site-specific recombination in vitro. *EMBO J.* **15:**1172–1181.
44. **Craig, N. L.** 1988. The mechanism of conservative site-specific recombination. *Annu. Rev. Genet.* **22:**77–105.
45. **Craig, N. L., and H. A. Nash.** 1984. *E. coli* integration host factor binds to specific sites in DNA. *Cell* **39:**707–716.
46. **Crisona, N. J., R. L. Weinberg, B. J. Peter, D. W. Sumners, and N. R. Cozzarelli.** 1999. The topological mechanism of phage λ integrase. *J. Mol. Biol.* **289:**747–775.
47. **DiGabriele, A. D., and T. A. Steitz.** 1993. A DNA dodecamer containing an adenine tract crystallizes in a unique lattice and exhibits a new bend. *J. Mol. Biol.* **231:**1024–1039.
48. **DiGabriele, A. D., M. R. Sanderson, and T. A. Steitz.** 1989. Crystal lattice packing is important in determining the bend of a DNA dodecamer containing an adenine tract. *Proc. Natl. Acad. Sci. USA* **86:**1816–1820.
49. **Ditto, M. D., D. Roberts, and R. A. Weisberg.** 1994. Growth phase variation of integration host factor level in *Escherichia coli. J. Bacteriol.* **176:**3738–3748.
50. **Dodd, I. B., M. R. Reed, and J. B. Egan.** 1994. The cro-like Ap1 repressor of coliphage 186 is required for prophage excision and binds near the phage attachment site. *Mol. Microbiol.* **10:**1139–1150.
51. **Dorgai, L., S. Sloan, and R. A. Weisberg.** 1998. Recognition of core binding sites by bacteriophage integrases. *J. Mol. Biol.* **277:**1059–1070.
52. **Dorgai, L., E. Yagil, and R. A. Weisberg.** 1995. Identifying determinants of recombination specificity: construction and characterization of mutant bacteriophage integrases. *J. Mol. Biol.* **252:**178–188.
53. **Dorman, C. J., and C. F. Higgins.** 1987. Fimbrial phase variation in *Escherichia coli:* dependence on integration host factor and homologies with other site-specific recombinases. *J. Bacteriol.* **169:**3840–3843.
54. **Drlica, K., and J. Rouviere-Yaniv.** 1987. Histone-like proteins of bacteria. *Microbiol. Rev.* **51:**301–319.
55. **Duckett, D. R., A. I. H. Murchie, S. Diekmann, E. von Kitzing, B. Kemper, and D. M. J. Lilley.** 1988. The structure of the Holliday junction, and its resolution. *Cell* **55:**79–89.
56. **Duckett, D. R., A. I. H. Murchie, and D. M. J. Lilley.** 1990. The role of metal ions in the conformation of the four-way DNA junction. *EMBO J.* **9:**583–590.
57. **Echols, H.** 1970. Integrative and excisive recombination by bacteriophage λ: evidence for an excision-specific recombination protein. *J. Mol. Biol.* **47:**575–583.
58. **Eisenstein, B.** 1981. Phase variation of type 1 fimbriae in *Escherichia coli* is under transcriptional control. *Science* **214:** 337–339.
59. **Eisenstein, B. I., D. S. Sweet, V. Vaughn, and D. I. Friedman.** 1987. Integration host factor is required for the DNA inversion that controls phase variation in Escherichia coli. *Proc. Natl. Acad. Sci. USA* **84:**6506–6510.
60. **Enquist, L. W., and R. A. Weisberg.** 1984. An integration-proficient *int* mutant of bacteriophage λ. *Mol. Gen. Genet.* **195:**62–69.
61. **Esposito, D., and J. J. Scocca.** 1997. Purification and characterization of HP1 cox and definition of its role in controlling the direction of site-specific recombination. *J. Biol. Chem.* **272:**8660–8670.
62. **Esposito, D., and J. J. Scocca.** 1997. The integrase family of tyrosine recombinases: evolution of a conserved active site domain. *Nucleic Acids Res.* **25:**3605–3614.
63. **Evans, B. R., J. W. Chen, R. L. Parsons, T. K. Bauer, D. B. Teplow, and M. Jayaram.** 1990. Identification of the active site tyrosine of FLP recombinase. *J. Biol. Chem.* **265:** 18504–18510.
64. **Fluit, A. C., and F. J. Schmitz.** 1999. Class 1 integrons, gene cassettes, mobility, and epidemiology. *Eur. J. Clin. Microbiol. Infect. Dis.* **18:**761–770.
65. **Francia, M. V., J. C. Zabala, F. de la Cruz, and J. M. G. Lobo.** 1999. The IntI1 integron integrase preferentially binds single-stranded DNA of the *attC* site. *J. Bacteriol.* **181:** 6844–6849.
66. **Franz, B., and A. Landy.** 1995. The Holliday junction intermediates of λ integrative and excisive recombination respond differently to the bending proteins, IHF and Xis. *EMBO J.* **14:**397–406.
67. **Friedman, D. I.** 1988. Integration host factor: a protein for all reasons. *Cell* **55:**545–554.
68. **Friesen, H., and P. D. Sadowski.** 1992. Mutagenesis of a conserved region of the gene encoding the FLP recombinase of *Saccharomyces cerevisiae. J. Mol. Biol.* **225:**313–326.
69. **Futcher, A. B.** 1986. Copy number amplification of the 2

micron M circle plasmid of *Saccharomyces cerevisiae*. *J. Theor. Biol.* **119:**197–204.
70. **Gally, D. L., J. Leathart, and I. C. Blomfield.** 1996. Interaction of FimB and FimE with the *fim* switch that controls the phase variation of type 1 fimbriae in *Escherichia coli* K-12. *Mol. Microbiol.* **21:**725–738.
71. **Goodman, S. D., S. C. Nicholson, and H. A. Nash.** 1992. Deformation of DNA during site-specific recombination of bacteriophage λ: replacement of IHF protein by HU protein or sequence-directed bends. *Proc. Natl. Acad. Sci. USA* **89:** 11910–11914.
72. **Goodrich, J. A., M. L. Schwartz, and W. R. McClure.** 1990. Searching for and predicting the activity of sites for DNA binding proteins: compilation and analysis of the binding sites for *Escherichia coli* integration host factor (IHF). *Nucleic Acids Res.* **18:**4993–5000.
73. **Goosen, N., and P. van de Putte.** 1995. The regulation of transcription initiation by integration host factor. *Mol. Microbiol.* **16:**1–7.
74. **Gopaul, D. N., F. Guo, and G. D. Van Duyne.** 1998. Structure of the Holliday junction intermediate in Cre-*loxP* site-specific recombination. *EMBO J.* **17:**4175–4187.
75. **Granston, A. E., and H. A. Nash.** 1993. Characterization of a set of integration host factor mutants deficient for DNA binding. *J. Mol. Biol.* **234:**45–59.
76. **Gravel, A., B. Fournier, and P. H. Roy.** 1998. DNA complexes obtained with the integron integrase Intl at the *att*1 site. *Nucleic Acids Res.* **26:**4347–4355.
77. **Gravel, A., N. Messier, and P. H. Roy.** 1998. Point mutations in the integron integrase IntI1 that affect recombination and/ or substrate recognition. *J. Bacteriol.* **180:**5437–5442.
78. **Griffith, J. D., and H. A. Nash.** 1985. Genetic rearrangement of DNA induces knots with a unique topology: implication for the mechanism of synapsis and crossing-over. *Proc. Natl. Acad. Sci. USA* **82:**3124–3128.
79. **Guo, F., D. N. Gopaul, and G. D. Van Duyne.** 1997. Structure of Cre recombinase complexed with DNA in a site-specific recombination synapse. *Nature* **389:**40–46.
80. **Guo, F., D. N. Gopaul, and G. D. Van Duyne.** 1999. Asymmetric DNA-bending in the Cre-loxP site-specific recombination synapse. *Proc. Natl. Acad. Sci. USA* **96:**7143–7148.
81. **Haffter, P., and T. A. Bickle.** 1987. Purification and DNA-binding properties of FIS and CIN, two proteins required for the bacteriophage P1 site-specific recombination system, *cin*. *J. Mol. Biol.* **198:**579–587.
82. **Hakimi, J. M., and J. J. Scocca.** 1994. Binding sites for bacteriophage HP1 integrase on its DNA substrates. *J. Biol. Chem.* **269:**21340–21345.
83. **Hales, L. M., R. I. Gumport, and J. F. Gardner.** 1996. Examining the contribution of a dA + dT element to the conformation of *Escherichia coli* integration host factor-DNA complexes. *Nucleic Acids Res.* **24:**1780–1786.
84. **Hallet, B., and D. J. Sherratt.** 1997. Transposition and site-specific recombination: adapting DNA cut-and-paste mechanisms to a variety of genetic rearrangements. *FEMS Microbiol. Rev.* **21:**157–178.
85. **Han, Y. W., R. I. Gumport, and J. F. Gardner.** 1993. Complementation of bacteriophage lambda integrase mutants: evidence for an intersubunit active site. *EMBO J.* **12:** 4577–4584.
86. **Han, Y. W., R. I. Gumport, and J. F. Gardner.** 1994. Mapping the functional domains of bacteriophage lambda integrase protein. *J. Mol. Biol.* **235:**908–925.
87. **Hansson, K., O. Skold, and L. Sundstrom.** 1997. Non-palindromic *att*1 sites of integrons are capable of site-specific recombination with one another and with secondary targets. *Mol. Microbiol.* **26:**441–453.
88. **Henderson, I. R., P. Owen, and J. P. Nataro.** 1999. Molecular switches—the ON and OFF bacterial phase variation. *Mol. Microbiol.* **33:**919–932.
89. **Hendrix, R. W., J. W. Roberts, F. W. Stahl, and R. A. Weisberg (ed.).** 1983. *Lambda II*. Cold Spring Harbor Laboratory, Cold Spring Harbor, N.Y.
90. **Hengen, P. N., S. L. Bartram, L. E. Stewart, and T. D. Schneider.** 1997. Information analysis of Fis binding sites. *Nucleic Acids Res.* **25:**4994–5002.
91. **Hershey, A. D. (ed.).** 1971. *The Bacteriophage Lambda*. Cold Spring Harbor Laboratory, Cold Spring Harbor, N.Y.
92. **Hickman, A. B., S. Waninger, J. J. Scocca, and F. Dyda.** 1997. Molecular organization in site-specific recombination: the catalytic domain of bacteriophage HP1 integrase at 2.7 Å resolution. *Cell* **89:**227–237.
93. **Hubner, P., and W. Arber.** 1989. Mutational analysis of a prokaryotic recombinational enhancer element with two functions. *EMBO J.* **8:**577–585.
94. **Hwang, E. S., and J. J. Scocca.** 1990. Interaction of integration host factor from *Escherichia coli* with the integration region of the *Haemophilus influenzae* bacteriophage HP1. *J. Bacteriol.* **172:**4852–4860.
95. **Jaxel, C., G. Capranico, D. Kerrigan, K. W. Kohn, and Y. Pommier.** 1991. Effect of local DNA sequence on topoisomerase I cleavage in the presence or absence of camptothecin. *J. Biol. Chem.* **266:**20418–20423.
96. **Jayaram, M.** 1997. The cis-trans paradox of integrase. *Science* **276:**49–51.
97. **Jia, Y., and G. Churchward.** 1999. Interactions of the integrase protein of the conjugative transposon Tn*916* with its specific DNA binding sites. *J. Bacteriol.* **181:**6114–6123.
98. **Johnson, R. C., M. F. Bruist, and M. I. Simon.** 1986. Host protein requirements for in vitro site-specific DNA inversion. *Cell* **46:**531–539.
99. **Kanaar, R., A. Klippel, E. Shekhtman, J. M. Dungan, R. Kahmann, and N. R. Cozzarelli.** 1990. Processive recombination by the phage Mu Gin system: implications for the mechanisms of DNA strand exchange, DNA site alignment, and enhancer action. *Cell* **62:**353–366.
100. **Kikuchi, A., E. Flamm, and R. A. Weisberg.** 1985. An *Escherichia coli* mutant unable to support site-specific recombination of bacteriophage λ. *J. Mol. Biol.* **183:**129–140.
101. **Kikuchi, Y., and H. A. Nash.** 1978. The bacteriophage λ *int* gene product. *J. Biol. Chem.* **253:**7149–7157.
102. **Kim, S., L. Moitoso de Vargas, S. E. Nunes-Düby, and A. Landy.** 1990. Mapping of a higher order protein-DNA complex: two kinds of long-range interactions in λ attL. *Cell* **63:** 773–781.
103. **Kim, S.-H., and A. Landy.** 1992. Lambda Int protein bridges between higher order complexes at two distant chromosomal loci attL and attR. *Science* **256:**198–203.
104. **Klemm, M., C. Cheng, G. Cassell, S. Shuman, and A. M. Segall.** 2000. Peptide inhibitors of DNA cleavage by tyrosine recombinases and topoisomerases. *J. Mol. Biol.* **299:** 1203–1216.
105. **Klemm, P.** 1986. Two regulatory *fim* genes, *fim*B and *fim*E, control the phase variation of type 1 fimbriae in *Escherichia coli*. *EMBO J.* **5:**1389–1393.
106. **Koch, C., and R. Kahmann.** 1986. Purification and properties of the *Escherichia coli* host factor required for inversion of the G segment in bacteriophage Mu. *J. Biol. Chem.* **261:** 15673–15678.
107. **Koch, C., O. Ninnemann, H. Fuss, and R. Kahmann.** 1991. The N-terminal part of the E. coli DNA binding protein FIS

is essential for stimulating site-specific DNA inversion but is not required for specific DNA binding. *Nucleic Acids Res.* **19:**5915–5922.

108. **Kostrewa, D., J. Granzin, C. Koch, H.-W. Choe, J. Labahn, R. Kahmann, and W. Saenger.** 1991. Three-dimensional structure of the *E. coli* DNA-binding protein FIS. *Nature* **349:** 178–180.
109. **Krogh, B. O., and S. Shuman.** 2000. Catalytic mechanism of DNA topoisomerase IB. *Mol. Cell* **5:**1035–1041.
110. **Kuhstoss, S., and R. N. Rao.** 1991. Analysis of the integration function of the streptomycete bacteriophage phi C31. *J. Mol. Biol.* **222:**897–908.
111. **Kulasekara, H. D., and I. C. Blomfield.** 1999. The molecular basis for the specificity of *fimE* in the phase variation of type 1 fimbriae of *Escherichia coli* K-12. *Mol. Microbiol.* **31:** 1171–1181.
112. **Kwon, H. J., R. S. Tirumalai, A. Landy, and T. Ellenberger.** 1997. Flexibility in DNA recombination: structure of the λ integrase catalytic core. *Science* **276:**126–131.
113. **Landy, A.** 1989. Dynamic, structural, and regulatory aspects of lambda site-specific recombination. *Annu. Rev. Biochem.* **58:**913–949.
114. **Landy, A.** 1993. Mechanistic and structural complexity in the site-specific recombination pathways of Int and FLP. *Curr. Biol.* **3:**699–707.
115. **Landy, A., and W. Ross.** 1977. Viral integration and excision: structure of the lambda *att* sites. *Science* **197:**1147–1160.
116. **Lee, J., and M. Jayaram.** 1993. Mechanism of site-specific recombination: logic of assembling recombinase catalytic site from fractional active sites. *J. Biol. Chem.* **268:** 17564–17570.
117. **Lee, J., and M. Jayaram.** 1995. Functional roles of individual recombinase monomers in strand breakage and strand union during site-specific DNA recombination. *J. Biol. Chem.* **270:** 23203–23211.
118. **Lee, J., and M. Jayaram.** 1995. Junction mobility and resolution of Holliday structures by Flp site-specific recombinase. *J. Biol. Chem.* **270:**19086–19092.
119. **Lee, J., and M. Jayaram.** 1995. Role of partner homology in DNA recombination. *J. Biol. Chem.* **270:**4042–4052.
120. **Lee, J., M. C. Serre, S.-H. Yang, I. Whang, H. Araki, Y. Oshima, and M. Jayaram.** 1992. Functional analysis of box II mutations in yeast site-specific recombinases FLP and R. *J. Mol. Biol.* **228:**1091–1103.
121. **Leffers, G. G., Jr., and S. Gottesman.** 1998. Lambda Xis degradation in vivo by Lon and FtsH. *J. Bacteriol.* **180:** 1573–1577.
122. **Leong, J., S. Nunes-Düby, C. Lesser, P. Youderian, M. M. Susskind, and A. Landy.** 1985. Primary structure of the ϕ and P22 attachment sites and their interactions with E. coli integration host factor. *J. Biol. Chem.* **260:**4468–4477.
123. **Lewis, J. A., and G. F. Hatfull.** 2000. Identification and characterization of mycobacteriophage L5 excisionase. *Mol. Microbiol.* **35:**350–360.
124. **Lilley, D. M. J., and R. M. Clegg.** 1994. The structure of the four-way junction in DNA. *Annu. Rev. Biophys. Biomol. Struct.* **22:**299–328.
125. **Lorenz, M., A. Hillisch, S. D. Goodman, and S. Diekmann.** 1999. Global structure similarities of intact and nicked DNA complexed with IHF measured in solution by fluorescence resonance energy transfer. *Nucleic Acids Res.* **27:**4619–4625.
126. **MacWilliams, M. P., R. I. Gumport, and J. F. Gardner.** 1996. Genetic analysis of the bacteriophage λ *att*L nucleoprotein complex. *Genetics* **143:**1069–1079.
127. **Mahillon, J., and D. Lereclus.** 1988. Structural and functional analysis of Tn*4430:* identification of an integrase-like protein involved in the co-integrate-resolution process. *EMBO J.* **7:** 1515–1526.
128. **Matsuura, M., T. Noguchi, D. Yamaguchi, T. Aida, M. Asayama, H. Takahashi, and M. Shirai.** 1996. The *sre* gene (ORF469) encodes a site-specific recombinase responsible for integration of the R4 phage genome. *J. Bacteriol.* **178:** 3374–3376.
129. **Mazel, D., B. Dychinco, V. A. Webb, and J. Davies.** 1998. A distinctive class of integron in the *Vibrio cholerae* genome. *Science* **280:**605–608.
130. **McClain, M. S., I. C. Blomfield, and B. I. Eisenstein.** 1991. Roles of *fimB* and *fimE* in site-specific DNA inversion associated with phase variation of type 1 fimbriae in *Escherichia coli. J. Bacteriol.* **173:**5308–5314.
131. **McCulloch, R., L. W. Coggins, S. D. Colloms, and D. J. Sherratt.** 1994. Xer-mediated site-specific recombination at *cer* generates Holliday junctions in vitro. *EMBO J.* **13:** 1844–1855.
132. **Mechulam, Y., S. Blanquet, and G. Fayat.** 1987. Dual level control of the *Escherichia coli pheST-himA* operon expression. *J. Mol. Biol.* **197:**453–470.
133. **Miller, H. I., A. Kikuchi, H. A. Nash, R. A. Weisberg, and D. I. Friedman.** 1979. Site-specific recombination of bacteriophage λ: the role of host gene products. *Cold Spring Harbor Symp. Quant. Biol.* **43:**1121–1126.
134. **Miller, H. I., M. Kirk, and H. Echols.** 1981. SOS induction and autoregulation of the *him*A gene for site-specific recombination in *Escherichia coli. Proc. Natl. Acad. Sci. USA* **78:** 6754–6758.
135. **Miller, H. I., and H. A. Nash.** 1981. Direct role of the *him*A gene product in phage λ integration. *Nature* **290:**523–526.
136. **Mizuuchi, K.** 1992. Polynucleotidyl transfer reactions in transpositional DNA recombination. *J. Biol. Chem.* **267:** 21273–21276.
137. **Mizuuchi, K., M. Gellert, R. A. Weisberg, and H. A. Nash.** 1980. Catenation and supercoiling in the products of bacteriophage λ integrative recombination in vitro. *J. Mol. Biol.* **141:**485–495.
138. **Moitoso de Vargas, L., S. Kim, and A. Landy.** 1989. DNA looping generated by the DNA-bending protein IHF and the two domains of lambda integrase. *Science* **244:**1457–1461.
139. **Moitoso de Vargas, L., and A. Landy.** 1991. A switch in the formation of alternative DNA loops modulates λ site-specific recombination. *Proc. Natl. Acad. Sci. USA* **88:**588–592.
140. **Montanez, C., J. Bueno, U. Schmeissner, D. L. Court, and G. Guarneros.** 1986. Mutations of bacteriophage lambda that define independent but overlapping RNA processing and transcription termination sites. *J. Mol. Biol.* **191:**29–37.
141. **Murtin, C., M. Engelhorn, J. Geiselmann, and F. Boccard.** 1998. A quantitative UV laser footprinting analysis of the interaction of IHF with specific binding sites: re-evaluation of the effective concentration of IHF in the cell. *J. Mol. Biol.* **284:**949–961.
142. **Nagaraja, R., and R. A. Weisberg.** 1990. Specificity determinants in the attachment sites of bacteriophage HK022 and λ. *J. Bacteriol.* **172:**6540–6550.
143. **Nash, H. A.** 1975. Integrative recombination of bacteriophage lambda DNA in vitro. *Proc. Natl. Acad. Sci. USA* **72:** 1072–1076.
144. **Nash, H. A.** 1981. Integration and excision of bacteriophage lambda: the mechanism of conservative site-specific recombination. *Annu. Rev. Genet.* **15:**143–167.
145. **Nash, H. A.** 1990. Bending and supercoiling of DNA at the attachment site of bacteriophage lambda. *Trends Biochem. Sci.* **15:**222–227.
146. **Nash, H. A.** 1996. Site-specific recombination: integration,

excision, resolution, and inversion of defined DNA segments, p. 2363–2376. *In* F. C. Neidhardt, R. Curtiss III, J. L. Ingraham, E. C. C. Lin, K. B. Low, B. Magasanik, W. S. Reznikoff, M. Riley, M. Schaechter, and H. E. Umbarger (ed.), Escherichia coli *and* Salmonella typhimurium, 2nd ed. ASM Press, Washington, D.C.

147. **Nash, H. A.** 1996. The HU and IHF proteins: accessory factors for complex protein-DNA assemblies, p. 149–179. *In* E. C. C. Lin and A. S. Lynch (ed.), *Regulation of Gene Expression in* Escherichia coli. R. G. Landes Company, Austin, Tex.
148. **Nash, H. A., and T. J. Pollock.** 1983. Site-specific recombination of bacteriophage lambda: the change in topological linking number associated with exchange of DNA strands. *J. Mol. Biol.* **170:**19–38.
149. **Nash, H. A., and C. A. Robertson.** 1981. Purification and properties of the *Escherichia coli* protein factor required for λ integrative recombination. *J. Biol. Chem.* **256:**9246–9253.
150. **Nash, H. A., C. A. Robertson, E. Flamm, R. A. Weisberg, and H. I. Miller.** 1987. Overproduction of *Escherichia coli* integration host factor, a protein with nonidentical subunits. *J. Bacteriol.* **169:**4124–4127.
151. **Nelson, H. C. M., J. T. Finch, F. L. Bonaventura, and A. Klug.** 1987. The structure of an oligo(dA) oligo(dT) tract and its biological implications. *Nature* **330:**221–226.
152. **Nilsson, L., H. Verbeek, E. Vijgenboom, C. van Drunen, A. Vanet, and L. Bosch.** 1992. FIS-dependent *trans* activation of stable RNA operons of *Escherichia coli* under various growth conditions. *J. Bacteriol.* **174:**921–929.
153. **Numrych, T. E., R. I. Gumport, and J. F. Gardner.** 1992. Characterization of the bacteriophage lambda excisionase (Xis) protein: the C-terminus is required for Xis-integrase cooperativity but not for DNA binding. *EMBO J.* **11:**3797–3806.
154. **Nunes-Düby, S., M. A. Azaro, and A. Landy.** 1995. Swapping DNA strands and sensing homology without branch migration in λ site-specific recombination. *Curr. Biol.* **5:**139–148.
155. **Nunes-Düby, S., R. S. Tirumalai, H. J. Kwon, T. Ellenberger, and A. Landy.** 1998. Similarities and differences among 105 members of the Int family of site-specific recombinases. *Nucleic Acids Res.* **26:**391–406.
156. **Nunes-Düby, S. E., L. I. Smith-Mungo, and A. Landy.** 1995. Single base-pair precision and structural rigidity in a small IHF-induced DNA loop. *J. Mol. Biol.* **253:**228–242.
157. **Nunes-Düby, S. E., R. S. Tirumalai, L. Dorgai, R. Yagil, R. Weisberg, and A. Landy.** 1994. λ Integrase cleaves DNA in *cis*. *EMBO J.* **13:**4421–4430.
158. **Nystrom, T.** 1995. Glucose starvation stimulon of *Escherichia coli:* role of integration host factor in starvation survival and growth phase-dependent protein synthesis. *J. Bacteriol.* **177:**5707–5710.
159. **Osuna, R., S. E. Finkel, and R. C. Johnson.** 1991. Identification of two functional regions in Fis: the N-terminus is required to promote Hin-mediated DNA inversion but not λ excision. *EMBO J.* **10:**1593–1603.
160. **Ouellette, M., and P. H. Roy.** 1987. Homology of ORFs from Tn*2603* and from R46 to site-specific recombinases. *Nucleic Acids Res.* **15:**10055.
161. **Pan, C. Q., J. A. Feng, S. E. Finkel, R. Landgraf, D. Sigman, and R. C. Johnson.** 1994. Structure of the *Escherichia coli* Fis-DNA complex probed by protein conjugated with 1,10-phenanthroline copper(I) complex. *Proc. Natl. Acad. Sci. USA* **91:**1721–1725.
162. **Pan, C. Q., S. E. Finkel, S. E. Cramton, J. -A. Feng, D. S. Sigman, and R. C. Johnson.** 1996. Variable structures of Fis-DNA complexes determined by flanking DNA-protein contacts. *J. Mol. Biol.* **264:**675–695.
163. **Pargellis, C. A., S. E. Nunes-Düby, L. Moitoso de Vargas, and A. Landy.** 1988. Suicide recombination substrates yield covalent λ integrase-DNA complexes and lead to identification of the active site tyrosine. *J. Biol. Chem.* **263:**7678–7685.
164. **Parsons, R. L., B. R. Evans, L. Zheng, and M. Jayaram.** 1990. Functional analysis of Arg-308 mutants of Flp recombinase. Possible role of Arg-308 in coupling substrate binding to catalysis. *J. Biol. Chem.* **265:**4527–4533.
165. **Parsons, R. L., P. V. Prasad, R. M. Harshey, and M. Jayaram.** 1988. Step-arrest mutants of FLP recombinase: implications for the catalytic mechanism of DNA recombination. *Mol. Cell. Biol.* **8:**3303–3310.
166. **Partridge, S. R., G. D. Recchia, C. Scaramuzzi, C. M. Collis, H. W. Stokes, and R. M. Hall.** 2000. Definition of the *attI1* site of class 1 integrons. *Microbiology* **146:**2855–2864.
167. **Pedulla, M. L., M. H. Lee, D. C. Lever, and G. F. Hatfull.** 1996. A novel host factor for integration of mycobacteriophage L5. *Proc. Natl. Acad. Sci. USA* **93:**15411–15416.
168. **Pena, C. E. A., M. Kahlenberg, and G. F. Hatfull.** 2000. Assembly and activation of site-specific recombination complexes. *Proc. Natl. Acad. Sci. USA* **97:**7760–7765.
169. **Perkins-Balding, D., D. P. Dias, and A. C. Glasgow.** 1997. Location, degree, and direction of DNA bending associated with the Him recombinational enhancer sequence and Fis-enhancer complex. *J. Bacteriol.* **179:**4747–4753.
170. **Pollock, T. J., and K. Abremski.** 1979. DNA without supertwists can be an in vitro substrate for site-specific recombination of bacteriophage λ. *J. Mol. Biol.* **131:**651–654.
171. **Pollock, T. J., and H. A. Nash.** 1983. Knotting of DNA caused by a genetic rearrangement: evidence for a nucleosome-like structure in site-specific recombination for bacteriophage lambda. *J. Mol. Biol.* **170:**1–18.
172. **Poyart-Salmeron, C., P. Trieu-Cuot, C. Carlier, and P. Courvalin.** 1989. Molecular characterization of two proteins involved in the excision of the conjugative transposon Tn1545: homologies with other site-specific recombinases. *EMBO J.* **8:**2425–2433.
173. **Recchia, G. D., and R. M. Hall.** 1997. Origins of the mobile gene cassettes found in integrons. *Trends Genet.* **5:**389–394.
174. **Redinbo, M. R., J. J. Champoux, and W. G. J. Hol.** 2000. Novel insights into catalytic mechanism from a crystal structure of human topoisomerase I in complex with DNA. *Biochemistry* **39:**6832–6840.
175. **Redinbo, M. R., L. Stewart, P. Kuhn, J. J. Champoux, and W. G. J. Hol.** 1998. Crystal structures of human topoisomerase I in covalent and noncovalent complexes with DNA. *Science* **279:**1504–1513.
176. **Reiter, W. D., P. Palm, and S. Yeats.** 1989. Transfer RNA genes frequently serve as integration sites for prokaryotic genetic elements. *Nucleic Acids Res.* **17:**1907–1914.
177. **Rice, P. A.** 1997. Making DNA do a U-turn: IHF and related proteins. *Curr. Opin. Struct. Biol.* **7:**86–93.
178. **Rice, P. A., S.-W. Yang, K. Mizuuchi, and H. A. Nash.** 1996. Crystal structure of an IHF-DNA complex: a protein-induced DNA U-turn. *Cell* **87:**1295–1306.
179. **Richet, E., P. Abcarian, and H. A. Nash.** 1986. The interaction of recombination proteins with supercoiled DNA: defining the role of supercoiling in lambda integrative recombination. *Cell* **46:**1011–1021.
180. **Richet, E., P. Abcarian, and H. A. Nash.** 1988. Synapsis of attachment sites during lambda integrative recombination involves capture of a naked DNA by a protein-DNA complex. *Cell* **52:**9–17.
181. **Ross, W., and A. Landy.** 1983. Patterns of λ Int recognition in the regions of strand exchange. *Cell* **33:**261–272.

182. **Rowe-Magnus, D. A., and D. Mazel.** 1999. Resistance gene capture. *Curr. Opin. Microbiol.* **2:**483–488.
183. **Rudy, C. K., J. R. Scott, and G. Churchward.** 1997. DNA binding by the Xis protein of the conjugative transposon Tn*916*. *J. Bacteriol.* **179:**2567–2572.
184. **Sadowski, P. D.** 1993. Site-specific genetic recombination: hops, flips and flops. *FASEB J.* **7:**760–767.
185. **Safo, M. K., W.-Z. Yang, L. Corselli, S. E. Cramton, H. S. Yuan, and R. C. Johnson.** 1997. The transactivation region of the Fis protein that controls site-specific DNA inversion contains extended mobile β-hairpin arms. *EMBO J.* **16:** 6860–6873.
186. **Salyers, A., N. Shoemaker, G. Bonheyo, and J. Frias.** 1999. Conjugative transposons: transmissible resistance islands, p. 331–345. *In* J. B. Kaper and J. Hacker (ed.), *Pathogenicity Islands and Other Mobile Virulence Elements.* American Society for Microbiology, Washington, D.C.
187. **Salyers, A. A., N. B. Shoemaker, A. M. Stevens, and L.-Y. Li.** 1995. Conjugative transposons: an unusual and diverse set of integrated gene transfer elements. *Microbiol. Rev.* **59:** 579–590.
188. **Sandmann, C.** 1996. Structure model of a complex between the factor for inversion stimulation (FIS) and DNA: modeling protein-DNA complexes with dyad symmetry and known protein structures. *Proteins* **25:**486–500.
189. **Sarkar, D., M. Radman-Livaja, and A. Landy.** 2001. The small DNA binding domain of λ integrase is a context-sensitive modulator of recombinase functions. *EMBO J.* **20:** 1203–1212.
190. **Scott, J. R., and G. G. Churchward.** 1995. Conjugative transposition. *Annu. Rev. Microbiol.* **49:**367–397.
191. **Seeman, N. C., and N. R. Kallenbach.** 1994. DNA branched junctions. *Annu. Rev. Biophys. Biomol. Struct.* **23:**53–86.
192. **Segall, A. M., S. D. Goodman, and H. A. Nash.** 1994. Architectural elements in nucleoprotein complexes: interchangeability of specific and non-specific DNA binding proteins. *EMBO J.* **13:**4536–4548.
193. **Segall, A. M., and H. A. Nash.** 1993. Synaptic intermediates in bacteriophage lambda site-specific recombination: integrase can align pairs of attachment sites. *EMBO J.* **12:** 4567–4576.
194. **Shaikh, A. C., and P. D. Sadowski.** 1997. The Cre recombinase cleaves the *lox* site in *trans*. *J. Biol. Chem.* **272:** 5695–5702.
195. **Shaikh, A. C., and P. D. Sadowski.** 2000. Chimeras of the Flp and Cre recombinases: tests of the mode of cleavage by Flp and Cre. *J. Mol. Biol.* **302:**27–48.
196. **Shuman, S.** 1989. Vaccinia DNA topoisomerase I promotes illegitimate recombination in *Escherichia coli*. *Proc. Natl. Acad. Sci. USA* **86:**3489–3493.
197. **Shuman, S.** 1991. Site-specific interaction of vaccinia topoisomerase I with duplex DNA. Minimal DNA substrate for strand cleavage in vitro. *J. Biol. Chem.* **266:**11372–11379.
198. **Shuman, S.** 1998. Vaccinia virus DNA topoisomerase: a model eukaryotic type IB enzyme. *Biochim. Biophys. Acta* **1400:**321–337.
199. **Shuman, S., E. M. Kane, and S. G. Morham.** 1989. Mapping the active-site tyrosine of vaccinia virus DNA topoisomerase I. *Proc. Natl. Acad. Sci. USA* **86:**9793–9797.
200. **Smith-Mungo, L., I. T. Chan, and A. Landy.** 1994. Structure of the P22 att site: conservation and divergence in the lambda motif of recombinogenic complexes. *J. Biol. Chem.* **269:** 20798–20805.
201. **Spengler, S. J., A. Stasiak, and N. R. Cozzarelli.** 1985. The stereostructure of knots and catenanes produced by phage λ integrative recombination: implications for mechanism and DNA structure. *Cell* **42:**325–334.
202. **Stark, W. M., and M. R. Boocock.** 1995. Gatecrashers at the catalytic party. *Trends Genet.* **11:**121–123.
203. **Stark, W. M., and M. R. Boocock.** 1995. Topological selectivity in site-specific recombination, p.101–129. *In* D. Sherratt (ed.), *Mobile Genetic Elements.* Oxford University Press, New York, N.Y.
204. **Stark, W. M., M. R. Boocock, and D. J. Sherratt.** 1989. Site-specific recombination by Tn3 resolvase. *Trends Genet.* **5:** 304–309.
205. **Stark, W. M., M. R. Boocock, and D. J. Sherratt.** 1992. Catalysis by site-specific recombinases. *Trends Genet.* **8:**432–439.
206. **Stark, W. M., D. J. Sherratt, and M. R. Boocock.** 1989. Site-specific recombination by Tn3 resolvase: topological changes in the forward and reverse reactions. *Cell* **58:**779–790.
207. **Stewart, L., M. R. Redinbo, X. Qiu, W. G. J. Hol, and J. J. Champoux.** 1998. A model for the mechanism of human topoisomerase I. *Science* **279:**1534–1541.
208. **Stokes, H. W., D. B. O'Gorman, G. D. Recchia, M. Parsekhian, and R. M. Hall.** 1997. Structure and function of 59-base element recombination sites associated with mobile gene cassettes. *Mol. Microbiol.* **26:**731–745.
209. **Su, Y. A., and D. B. Clewell.** 1993. Characterization of the left 4kb of conjugative transposon Tn916: determinants involved in excision. *Plasmid* **30:**234–250.
210. **Subramanya, H. S., L. K. Arciszewska, R. A. Baker, L. E. Bird, D. J. Sherratt, and D. B. Wigley.** 1997. Crystal structure of the site-specific recombinase, XerD. *EMBO J.* **16:** 5178–5187.
211. **Tanaka, I., K. Appelt, J. Dijk, S. W. White, and K. S. Wilson.** 1984. 3-Å resolution structure of a protein with histone-like properties in prokaryotes. *Nature* **310:**376–381.
212. **Thompson, J. F., and A. Landy.** 1988. Empirical estimation of protein-induced DNA bending angles: applications to λ site-specific recombination complexes. *Nucleic Acids Res.* **16:** 9687–9705.
213. **Thompson, J. F., L. Moitoso de Vargas, C. Koch, R. Kahmann, and A. Landy.** 1987. Cellular factors couple recombination with growth phase: characterization of a new component in the λ site-specific recombination pathway. *Cell* **50:** 901–908.
214. **Thompson, J. F., L. Moitoso de Vargas, S. E. Skinner, and A. Landy.** 1987. Protein-protein interactions in a higher-order structure direct lambda site-specific recombination. *J. Mol. Biol.* **195:**481–493.
215. **Thompson, J. F., D. Waechter-Brulla, R. I. Gumport, J. F. Gardner, L. Moitoso de Vargas, and A. Landy.** 1986. Mutations in an integration host factor-binding site: effect on lambda site-specific recombination and regulatory implications. *J. Bacteriol.* **168:**1343–1351.
216. **Thorpe, H. M., and M. C. M. Smith.** 1998. In vitro site-specific integration of bacteriophage DNA catalyzed by a recombinase of the resolvase/invertase family. *Proc. Natl. Acad. Sci. USA* **95:**5505–5510.
217. **Tirumalai, R. S., E. Healey, and A. Landy.** 1997. The catalytic domain of λ site-specific recombinase. *Proc. Natl. Acad. Sci. USA* **94:**6104–6109.
218. **Tirumalai, R. S., H. Kwon, E. Cardente, T. Ellenberger, and A. Landy.** 1998. The recognition of core-type DNA sites by λ integrase. *J. Mol. Biol.* **279:**513–527.
219. **Travers, A.** 1997. DNA-protein interactions: IHF—the master bender. *Curr. Biol.* **7:**R252–R254.
220. **Trieu-Cuot, P., C. Poyart-Salmeron, C. Carlier, and P. Courvalin.** 1994. Sequence requirements for target activity in site-

specific recombination mediated by the Int protein of transposon Tn*1545*. *Mol. Microbiol.* **8:**179–185.
221. **Volkert, F. C., and J. R. Broach.** 1986. Site-specific recombination promotes plasmid amplification in yeast. *Cell* **46:** 541–550.
222. **Wang, J. C.** 1996. DNA topoisomerases. *Annu. Rev. Biochem.* **65:**635–692.
223. **Wasserman, S. A., J. M. Dungan, and N. R. Cozzarelli.** 1985. Discovery of a predicted DNA knot substantiates a model for site-specific recombination. *Science* **229:**171–174.
224. **Weisberg, R., and A. Landy.** 1983. Site-specific recombination in phage lambda, p. 211–250. *In* R. W. Hendrix, J. W. Roberts, F. W. Stahl, and R. A. Weisberg (ed.), *Lambda II.* Cold Spring Harbor Laboratory, Cold Spring Harbor, N.Y.
225. **Weisberg, R. A., and M. E. Gottesman.** 1971. The stability of Int and Xis functions, p. 489–500. *In* A. D. Hershey (ed.), *The Bacteriophage Lambda.* Cold Spring Harbor Laboratory, Cold Spring Harbor, N.Y.
226. **Weisberg, R. A., M. E. Gottesman, R. W. Hendrix, and J. W. Little.** 1999. Family values in the age of genomics: comparative analyses of temperate bacteriophage HK022. *Annu. Rev. Genet.* **33:**565–602.
227. **White, S. W., K. Appelt, K. S. Wilson, and I. Tanaka.** 1989. A protein structural motif that bends DNA. *Proteins* **5:** 281–288.
228. **Wierzbicki, A., M. Kendall, K. Abremski, and R. Hoess.** 1987. A mutational analysis of the bacteriophage P1 recombinase Cre. *J. Mol. Biol.* **195:**785–794.
229. **Wittschieben, J., and S. Shuman.** 1997. Mechanism of DNA transesterification by vaccinia topoisomerase: catalytic contributions of essential residues Arg-130, Gly-132, Tyr-136 and Lys-167. *Nucleic Acids Res.* **25:**3001–3008.
230. **Woodfield, G., C. Cheng, S. Shuman, and A. B. Burgin.** 2000. Vaccinia topoisomerase and Cre recombinase catalyze direct ligation of activated DNA substrates containing a 3′-*para*-nitrophenyl phosphate ester. *Nucleic Acids Res.* **28:** 3323–3331.
231. **Wu, Z., R. I. Gumport, and J. F. Gardner.** 1998. Defining the structural and functional roles of the carboxyl region of the bacteriophage lambda excisionase (Xis) protein. *J. Mol. Biol.* **281:**651–661.
232. **Yagil, E., L. Dorgai, and R. Weisberg.** 1995. Identifying determinants of recombination specificity: construction and characterization of chimeric bacteriophage integrases. *J. Mol. Biol.* **252:**163–177.
233. **Yang, C.-C., and H. A. Nash.** 1989. The interaction of E. coli IHF protein with its specific binding sites. *Cell* **57:**869–880.
234. **Yang, S., and H. A. Nash.** 1995. Comparison of protein binding to DNA in vivo and in vitro: defining an effective intracellular target. *EMBO J.* **14:**6292–6300.
235. **Yin, S., W. Bushman, and A. Landy.** 1985. Interaction of λ site-specific recombination protein Xis with attachment site DNA. *Proc. Natl. Acad. Sci. USA* **82:**1040–1044.
236. **Yu, A., and E. Haggård-Ljungquist.** 1993. Characterization of the binding sites of two proteins involved in the bacteriophage P2 site-specific recombination system. *J. Bacteriol.* **175:**1239–1249.
237. **Yu, A., and E. Haggård-Ljungquist.** 1993. The Cox protein is a modulator of directionality in bacteriophage P2 site-specific recombination. *J. Bacteriol.* **175:**7848–7855.
238. **Yuan, H. S., S. E. Finkel, J.-A. Feng, M. Kaczor-Grzeskowiak, R. C. Johnson, and R. E. Dickerson.** 1991. The molecular structure of wild-type and a mutant Fis protein: relationship between mutational changes and recombinational enhancer function or DNA binding. *Proc. Natl. Acad. Sci. USA* **88:** 9558–9562.

Mobile DNA II
Edited by N. L. Craig et al.

Chapter 8

Xer Site-Specific Recombination: Promoting Chromosome Segregation

FRANÇOIS-XAVIER BARRE AND DAVID J. SHERRATT

INTRODUCTION

Xer site-specific recombination was discovered in 1984 through its role in converting multimers of ColE1-related multicopy plasmids to monomers and hence ensuring their stable inheritance within *Escherichia coli* (77). Multimers can arise through homologous recombination and during rolling circle replication, which can occur during conjugal transfer and sometimes during vegetative replication. The ColE1 Xer recombination site, *cer,* and the chromosomal genes whose functions are required for Xer recombination on ColE1 plasmids were subsequently identified by genetic means (11, 20, 72–74, 78). The first two genes to be identified, *argR* and *pepA,* encoded proteins that were not recombinases, but a previously identified transcriptional repressor and an aminopeptidase, respectively (72, 74). These are now known to be essential accessory proteins, which interact with accessory DNA sequences located within *cer* and are necessary for synapse formation and the direction of recombination outcome. The third gene to be identified, *xerC* (20), was immediately assigned a direct role in recombination by its homology to the bacteriophage λ integrase gene, a member of the tyrosine recombinase family of site-specific recombinases (28). Subsequently, it was shown that the XerC recombinase is required for normal chromosome segregation at cell division, apparently by converting chromosome dimers to monomers through its action at a chromosomal recombination site, *dif,* located in the replication terminus region of the *E. coli* chromosome (10, 16, 43). Indeed, Xer recombination seems to be a conserved feature of most organisms that contain circular chromosomes, and presumably functions ubiquitously in converting chromosomal dimers to monomers (58). Subsequently, a second recombinase related to XerC, XerD, was found to be necessary for Xer recombination at sites present in plasmids or on the chromosome (11). This feature differentiates the Xer recombination from most other site-specific recombination reactions that use a single recombinase, and makes it an excellent model for studying the molecular interplay among the proteins during site-specific recombination. Thus, research on Xer recombination has been devoted to: (i) characterizing the recombination reaction at the molecular level; (ii) understanding how the system converts plasmid and chromosomal dimers to monomers (and not monomers to dimers); and (iii) understanding how Xer re-

François-Xavier Barre and David J. Sherratt • Division of Molecular Genetics, Department of Biochemistry, University of Oxford, South Parks Rd., Oxford OX1 3QU, United Kingdom.

combination at chromosomal *dif* is integrated into other aspects of DNA metabolism and processing during the bacterial cell cycle. Enough progress has been made in each of these areas of research, so that it is now possible to integrate the particular mechanistic features of Xer recombination into our understanding of its cellular roles.

THE XER SITE-SPECIFIC RECOMBINATION MACHINE

XerC and XerD Are Members of the Tyrosine Recombinase Family

Both XerC and XerD are identified as members of the tyrosine recombinase family because of the characteristic RKHRHY signature of active site residues (Color Plate 15A [see color insert]) (28, 65). The Xer recombinases work on several different recombination sites, the best studied of which are the *cer* (78) and *psi* (23) sites of plasmid ColE1 and pSC101, respectively, and the *dif* (10, 43) chromosomal site (Color Plate 15B). These sites share a highly homologous region, the core recombination site, which is organized identically to the core recombination sites of other tyrosine recombinases. The core site contains two 11-bp inverted recombinase binding sites separated by a central region at the outer boundaries of which recombination occurs. *dif* and *psi* have 6-bp central regions, whereas *cer* has an 8-bp central region. Within the recombinase binding sites, the 5 bp closest to the central region are palindromic in *dif*, whereas 5 of the 6 outer base pairs are not, and ensure the specificity of binding of XerD to the right side of *dif* and XerC to the left side (11, 35). Although a *dif* core site is fully functional, functional *cer* and *psi* sites consist of a core site plus adjacent accessory sequences (see below).

The use of two recombinases, XerC and XerD, by the Xer recombination machine is unusual but not unique. For example, FimB and FimE of *E. coli* mediate an inversion gene switch that regulates expression of type I fimbriae (39), whereas in *Staphylococcus aureus*, two related tyrosine recombinases of Tn*554* mediate promiscuous site-specific recombination (50). Although *E. coli* XerC and XerD share only 37% identity, they are the closest relatives to each other among the highly diverged tyrosine recombinase family. Similarly, in other eubacteria, XerC-XerD homologues are readily identified by their homology to each other and to their *E. coli* recombinases (34, 51, 58, 63). The genes encoding XerC and XerD are invariably located in different regions of the chromosome, where they often lie adjacent to and may be coexpressed with genes involved in recombination, repair, and cellular response to stress (58).

The crystal structure of XerD (Color Plate 15C) (76) is strikingly similar to that of Cre and the other tyrosine recombinases that have had structures determined at atomic resolution, despite high sequence divergence (28, 65). Indeed, tyrosine recombinases and type IB topoisomerases of eukaryotes, which share no detectable sequence similarity, share the same overall structure and catalytic mechanism (41, 59, 65). Correspondingly, both XerC and XerD can act as site-specific type I topoisomerases by relaxing plasmid DNA containing their binding site (21). It therefore seems likely that the XerC structure will be similar to that of XerD. Indeed, the predicted secondary structure of XerC fits well with the known structure of XerD (Color Plate 15A).

Like Cre, the XerC and XerD proteins can be divided into two domains that make a C-shaped clamp into which the DNA is bound (shown schematically in Fig. 1) (29). It is intriguing that the FimBE recombinases lack an N-terminal domain. Based on analogy with Cre, α-helices B and D of the N-terminal domain are expected to interact with the major groove formed by the inner palindromic nucleotides of the recombinase binding sites. The N-terminal domain is also expected to contain determinants for protein-protein interactions, whereas the larger, more conserved C-terminal domain contains the catalytic residues and determinants for specific DNA binding and protein-protein interactions. The residues proposed to be involved in specific major groove DNA binding are located within α-helices G and J, with R221 and Q222 being implicated in providing specificity (Color Plate 15A) (76). XerD shows a stronger affinity for its binding site than XerC, each of them binding more tightly to its site in the presence of the partner (11, 13, 66).

The Recombination Mechanism

The recombination mechanism of tyrosine recombinases has been well documented (29, 30, 32) and can be divided into the five steps shown schematically in Fig. 1. A key feature of this mechanism is that DNA strands are exchanged in two pairs with a Holliday junction (HJ)-containing molecule as an essential reaction intermediate. The two pairs of strand exchanges are separated in time and space. Consequently, in the recombining nucleoprotein complex only two of the four recombinase molecules are catalytically active at any one time, a reciprocal switch between activity and inactivity occurring for each pair of recombinases during a change in conformation of the 2-fold symmetrical HJ intermediate. This switch is facilitated by the cyclic interactions that link the

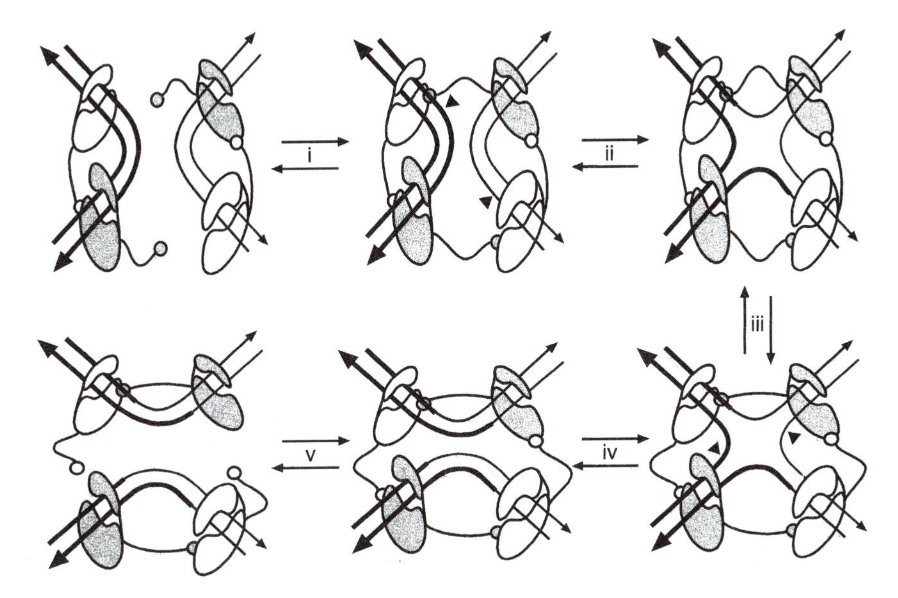

Figure 1. The site-specific recombination reaction. The XerC and XerD recombinases are represented by the white and grey C-shaped molecules, respectively, the N terminus lying on top of the DNA and the C terminus lying below. The DNA strands of each of the two recombination sites are represented by arrows pointing in the 5′ to 3′ direction. The strands of one of the sites are represented by thick lines, and the strands of the other site are represented by thin lines. Each site contains a black strand and a grey strand. Black strands are processed by XerC, and grey strands are processed by XerD. Note that the two synapsed sites are represented as being antiparallel because the black strands (and the grey strands) from the two sites run in opposite directions. Black and grey triangles indicate the positions of cutting and rejoining by the XerC and XerD recombinases, respectively. Active recombinases have an extended C-terminal tail. (i) synapsis; (ii) XerC strand exchange; (iii) HJ conformation change; (iv) XerD strand exchange; (v) synapsis breakage.

four recombinase molecules in a recombining tetramer (29, 30, 32). In this chapter, we focus on those features of the Xer system that differentiate it from the related site-specific recombination systems, Cre-*loxP*, λ Int-*attP*/*attB*, and Flp-*frt*, which are discussed in chapters 4, 7, and 12. Chapters 9 and 10 deal with promiscuous site-specific recombination mediated by integrons and conjugative transposons, respectively.

Synapsis

In Cre-*loxP* recombination, interactions between the recombinases and between the recombinases and DNA are sufficient to form a productive synaptic complex. This is likely to be the case for Xer recombination at *dif*, because a *dif* core site is fully functional in vivo as assayed by normal chromosome segregation and its ability to be a substrate for recombination when present in a plasmid (11). In contrast to *dif*, productive synapsis of *psi* and *cer* sites seems to require the interactions of specific accessory proteins with accessory DNA, located adjacent to the XerC binding site of the recombination core site (see later; 17, 23, 72, 73, 78). Apparently, accessory components are required because XerCD binds less tightly to duplex *cer* and *psi* than to *dif* (9). A consequence of this requirement for accessory factors is that it provides a mechanism for resolution selectivity, the process that limits recombination to intramolecular events between directly repeated sites.

Strand exchange

Tyrosine recombinases cleave and rejoin DNA strands through the formation of a transient 3′-phosphotyrosyl DNA-protein covalent intermediate involving the conserved tyrosine as the catalytic nucleophile. The site of recombinase-mediated cleavage is at the boundary of the central region and the binding site of the cleaving recombinase (Fig. 1). The five other conserved residues, RKHRH, are thought to be involved in stabilizing the transition state and/or in general acid-base catalysis (65). In the second catalytic step, the 5′-OH of the partner strand, released during the initial cleavage, acts as a nucleophile to attack the 3′-phosphotyrosyl bond of the HJ intermediate, thereby completing strand exchange. Current evidence indicates that the conserved K contained within the β-hairpin loop (highlighted in Color Plate 15A) is the acid that protonates the 5′-OH leaving group during the initial nucleophilic attack, whereas a water molecule may be the initial base that facilitates nucleophilic attack by the catalytic tyrosine (40, 41, 59). The XerCD recombinases, like Cre, contain a functional catalytic unit within a single recombinase molecule. In contrast, Flp assembles a single active site from two molecules because the cyclic interactions that join the four molecules in a recombining tetramer involve not the C-terminal tail of the recombinase as in Cre, but a loop that inserts the tyrosine of one molecule into the active site region of its partner (Fig. 1) (15, 29).

Experimental data from in vivo and in vitro experiments show that XerC usually initiates catalysis on synapsed *psi, cer,* or *dif* sites to form a HJ (12, 19, 51). Nevertheless, a low level of recombination in vivo has been observed on supercoiled plasmids carrying two *dif* sites in strains expressing a catalytically inactive form of XerC, suggesting that XerD may be able to initiate catalysis on *dif* under some conditions (33). It seems likely that XerC (Fig. 1) is initially active because the DNA adopts an asymmetric conformation in the synapsed nucleoprotein complex, distortion occurring at the boundary of the XerD binding site and the central region on XerCD binding. This view is based on the structure of two synapsed *loxP* sites bound by Cre. Substantial asymmetric distortion of the DNA duplexes at the boundary of the central region and of one of the recombinase binding sites seems to direct the initial strand exchange to the boundary of the central region distal to the kink, and therefore it directs which pair of recombinase molecules will initiate strand exchange (29).

Once the initial HJ intermediate has formed, three outcomes of the reaction are possible. The initial HJ intermediate can be resolved back to substrate by a second pair of XerC-mediated strand exchanges (Fig. 1, step ii). This reaction occurs efficiently on synthetic HJs in vitro and may be a frequent event during in vivo Xer recombination at *dif* (see below). Alternatively, the second pair of recombinases may become activated, as a consequence of a conformational change from one 2-fold symmetrical HJ intermediate to the other (Fig. 1, step iii). This conformational change of the HJ intermediate was initially proposed to be central to the reaction mechanism because of symmetry considerations (68), a view supported by biochemical and structural data that show that tyrosine recombinases act on the "crossing strands" of the HJ, i.e., the pair of strands that subtend a more acute angle at the HJ branchpoint (3, 5). The exchange of the second pair of strands by XerD resolves the HJ into complete recombinant products (step iv) that can then dissociate (step v) during recombination at *psi* and *dif*. A third possible reaction outcome is observed at *cer* in vivo and in vitro. The HJ intermediates formed by XerC accumulate, being neither resolved by XerD nor efficiently converted back to substrate (12, 19, 47). This is almost certainly a consequence of the particular sequence of the *cer* core site, and its 8-bp central region (Color Plate 15B). The in vivo resolution process, which seems to be RuvC independent, is still unknown (47), although replication of the HJ-containing intermediate seems a likely candidate. Synthetic HJs containing *dif* or *psi* core sites can be recombined (albeit inefficiently) in vitro by XerD, unlike *cer* HJs, although all three types of HJ are recombined efficiently by XerC (4, 33).

Coordination of the Recombination Reaction

Structural, biochemical, and genetic experiments have contributed to our understanding of the ways in which the activities of the four recombinase molecules of a recombining complex are coordinated so that the two pairs of strand exchanges occur at the correct time and in the correct place. Structures of Cre bound to DNA show four recombinase molecules interacting in a cyclic manner on synapsed duplex DNA or a HJ intermediate (Fig. 2) (29, 30, 32). In these interactions, each monomer has its extreme C terminus (the donor region) contacting the neighboring partner recombinase close to its active site (the acceptor region). Synapsis yields an asymmetric complex in which the two members of each active pair of recombinases have their C-terminal tails in a more extended conformation, thereby helping to position their tyrosine nucleophiles next to the scissile phosphate of the crossing strands. Small movements in DNA and protein allow the change in conformation that reciprocally activates the second pair of recombinases, while inactivating the other recombinase pair.

Genetic and biochemical experiments with XerCD support this general model. Specific mutations in either the presumptive donor or acceptor regions of XerCD, and indeed at other specific positions, lead to mutant phenotypes that exhibit either reciprocal stimulation of catalysis by partner and impairment of catalysis by self (the SPIS phenotype), or reciprocal impairment of partner catalysis and stimulation of catalysis by self (the IPSS phenotype) on synthetic HJ substrates (2, 33, 66). Furthermore, changing the purine content of a strand in the region of the HJ branchpoint can modulate recombination outcome. Purine richness of a strand predisposes it to adopt the "crossing" configuration, and therefore to be a substrate for XerCD-mediated strand exchange (2, 5, 27). For example, a *cer*-containing HJ, which contains a purine-rich XerC substrate strand, undergoes only XerC-mediated strand exchange, whereas a *dif* HJ, with alternating purines and pyrimidines in its central region, exhibits a little XerD-mediated strand exchange in addition to the efficient XerC-mediated strand exchange. A *psi* HJ, which is purine-rich in its XerD substrate strand, exhibits a higher level of catalysis by XerD, although XerC-mediated catalysis still predominates. Therefore, the Xer recombination system is "set" to undergo XerC-mediated strand exchange predominantly, presumably because the nature of the recombinases and their macromolecular interactions directs the favored HJ conformation. By using molecular tethers or bulges it is possible to overcome this natural tendency and produce substrates that undergo XerD strand exchange preferentially. The SPIS/IPSS mutant phenotypes are modulated by the nature of the HJ substrate, the switch phenotypes being strongest on HJ substrates that are not too strongly constrained to adopt one particular function (2, 4, 31, 33).

A major conclusion from these experiments is that the relative concentrations of the two HJ intermediate forms determine the relative levels of catalysis mediated by XerC and XerD. The use of two recombinases in the Xer system not only ensures that XerC will initiate catalysis, but provides an editing mechanism; catalysis by XerD is only favored on substrates that can adopt a conformation in which the XerD substrate strands are crossed. Even then, additional factors are required to overcome the thermodynamic and/or kinetic barriers required to form the appropriate substrate for catalysis by XerD (see below).

STABLE INHERITANCE OF MULTICOPY PLASMIDS

Mechanistic studies of recombination at plasmid *cer* and *psi* have played a central role in the molecular characterization of the Xer recombination machine. The *cer* and *psi* sites consist of a *dif*-like core site and approximately 178 bp (*cer*) or 158 bp (*psi*) of accessory sequence adjacent to the XerC binding site. Reduction of the *cer* accessory sequences to 174 bp leads to a reduction in function (78), whereas reduction of the *psi* accessory sequences to 127 bp leads to a complete loss of activity (S. D. Colloms, unpublished data).

In addition to XerC and XerD, recombination at *psi* or *cer* depends on two other host-encoded proteins: PepA and ArgR at *cer* (72, 74) and PepA and phosphorylated ArcA at *psi* (17). Those proteins have an architectural role and have no direct catalytic role in the recombination reaction (14, 46). All three proteins are known transcription factors with specific DNA-binding properties. ArgR binds as a hexamer to a single ArgR box within *cer* (Color Plate 15C) (14), whereas PepA binds as a hexamer to synapsed *cer* accessory sequences on both sides of the ArgR binding site (Color Plate 15B) (1). ArcA-P binds to the *psi* accessory sequences at a place similar to that bound by ArgR in *cer* (Color Plate 15B) (17). The accessory sequences and proteins are required for synapse formation and stabilization, and at least in part compensate for the weak interactions of XerCD with their *cer* and *psi* core recombination sites.

The accessory sequences and proteins direct the formation of a productive synapse that has a precise geometry, which is reflected in the precise topology of the products of recombination at *cer* and *psi* (Fig. 2A). The *psi* recombination product consists of two right-handed catenated circles containing four interlinks, and with the *psi* sites oriented antiparallel with respect to each other (−4A catenanes; Fig. 2A) (18). The *cer* recombination product is a HJ in which the recombined strands form catenated rings with a topology identical with that of the *psi* product, whereas the unrecombined strand remains as an unknotted circle. On the basis of this topological information and on the known binding of the accessory proteins to the accessory sequences, a model for the synaptic structure was proposed (Fig. 2B) (1, 7, 18). The crystal structure of PepA (75) helped refine the model for the synapse at *cer;* two PepA hexamers are sandwiched around a hexamer of ArgR and the recombination sites interwrap around accessory proteins as a right-handed superhelix that contains three negative superhelical turns for recombination sites that are viewed

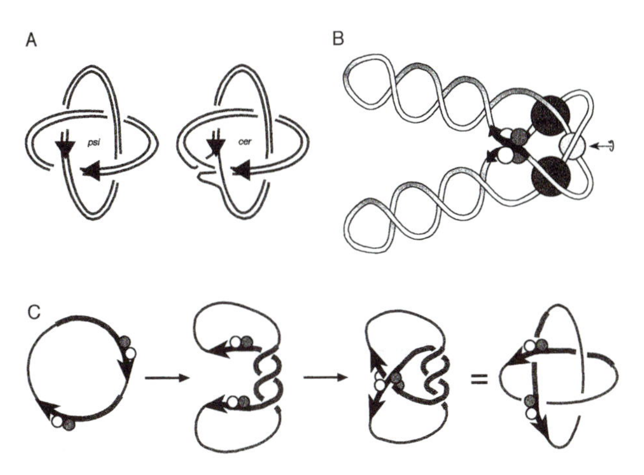

Figure 2. Topology of recombination on plasmid substrates. (A) The recombination products of *psi* and *cer* substrates have a defined topology. Plasmid DNA is represented by its two strands. The oriented core recombination sites are indicated by black triangles. (B) A model for synapsis at *cer* or *psi*. XerC and XerD are represented as small grey and white circles, PepA molecules are shown as large black circles, and ArgR or P-ArcA (for *cer* and *psi,* respectively) is shown as a medium-sized grey circle. The core recombination sites are shown in parallel, as indicated by the black arrows. (C) The *psi* reaction, from substrate (left) to product (right).

as being aligned in parallel. Given our knowledge of the structure of synapsed recombination sites, random projections onto a plane of aligned recombination sites will show them most often as antiparallel to each other (see Fig. 1). This translates into a synaptic structure in which four negative supercoils are entrapped (three being bound by accessory proteins and the fourth being "free"). (As the view of a pair of aligned core sites changes from antiparallel to parallel, the fourth free node disappears.) A similar complex is thought to be formed between the *psi* accessory sequence, PepA and ArcA-P (17). The absolute requirement for proteins and sequences that impose a synapse with a defined topology provides a topological filter mechanism for resolution selectivity, the process that limits Xer recombination on *cer* and *psi* to intramolecular events between directly repeated sites. Synapses of the observed defined topology can only readily form on such molecules (Fig. 2C) (67). The precise synaptic structure can also influence the order of strand exchange (M. Bregu, S. D. Colloms, and D. J. Sherratt, unpublished data).

The accessory factors not only play a role in the assembly of a synapse with precise geometry, but also in activating the second pair of strand exchange by XerD. Indeed, in the absence of PepA, supercoiling or catenation, XerCD convert this HJ intermediate back to substrate (M. Robertson and D. J. Sherratt, unpublished data). Thus, the catenated HJ intermediate formed by XerC on supercoiled *psi*-containing substrates in vitro requires PepA, and presumably the associated synapse geometry to promote the conformational change required for catalysis by XerD. These observations are in agreement with those that show that synthetic *psi*-containing HJ substrates are recombined much more efficiently by XerC than by XerD.

CHROMOSOME SEGREGATION

The discovery and initial characterization of Xer recombination through its role dimer resolution and consequent stable plasmid inheritance immediately raised the possibility that the primary role of this recombination system might be to resolve chromosome dimers introduced by homologous recombination. The identification of a functional Xer recombination site, *dif,* located in the replication terminus region of the chromosome, and the demonstration that Xer-defective cells form filaments with aberrant nucleoids, which are unable to partition correctly (10, 16, 43), provided experimental support for this view. Furthermore, the phenotype of Xer mutants is suppressed in strains deficient in homologous recombination (10).

In addition to *E. coli,* the phenotype of Xer$^-$ mutants has been reported for *Pseudomonas aeruginosa* (38), *Haemophilus influenzae* (51), and *Bacillus subtilis* (63). The precise Xer mutant phenotype in these bacteria, although consistent with a role for Xer recombination in the resolution of chromosome dimer, differs somewhat between species. For example, *H. influenzae* cells lacking the Xer recombinases form long chains (52), whereas the phenotype is similar but milder for mutation in *ripX* (the *xerD* homologue) of *B. subtilis* (63). Surprisingly, in *B. subtilis* the mutant phenotype is not suppressed by a *recA* mutation. With use of time-lapsed microscopy, Hendricks et al. (37) have helped refine our understanding of the Xer$^-$ phenotype in *E. coli;* the filaments present in Xer$^-$ cultures derive in a clonal manner from dividing cells in which the chromosome has been bisected by the septum. The SOS system, probably induced by the chromosome breakage, is responsible for the elongation of these cells.

In cells containing circular chromosomes, chromosome dimers can arise by RecA-dependent homologous recombination between sister chromatids (Fig. 3). Indeed, the Xer recombination machinery appar-

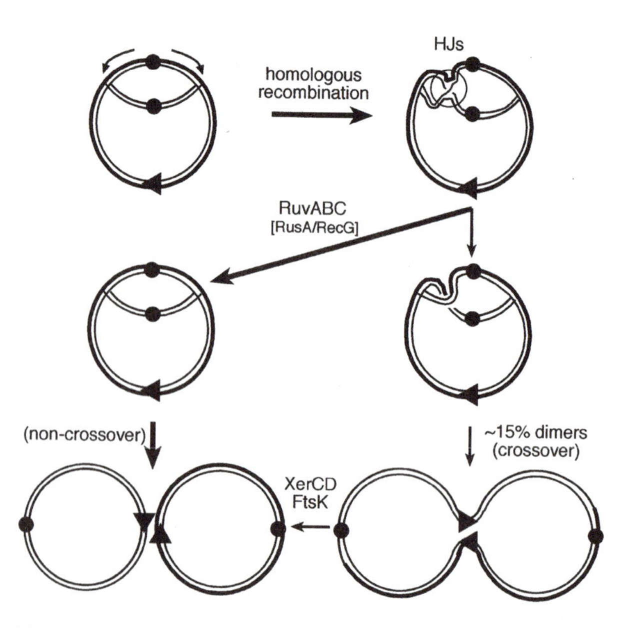

Figure 3. Dimer formation by homologous recombination. Chromosomal DNA is depicted by black and grey double strands. To differentiate the two sister chromatids arising by replication, one of the parental black strands is depicted by a thicker line. The origin of replication is shown as a black circle, and *dif* is represented by a black triangle. The thickness of the arrows follows the relative frequency of the events shown. HJs made by homologous recombination can be resolved to crossover or noncrossover events. RuvABC (and RusA/RecG) preferentially resolves them to noncrossover events. However, crossover events still occur. Consequently, dimers are formed. The Xer recombination system ensures their conversion to monomers.

ently is present in most eubacteria that have a circular chromosome and a full complement of homologous recombination (34, 58). Xer homologues have not been reported for bacteria with linear chromosomes. The current view is that a major role of homologous recombination is to allow the reassembly of functional replication forks that have broken or stalled, either as a consequence of DNA breaks or lesions (24) or because of stalled transcription machinery (48). Resolution of the HJ recombination intermediates can lead to either noncrossover or crossover products, the latter leading to sister chromatid exchange (SCE). In cells with circular chromosomes, odd numbers of SCEs lead to the formation of chromosome dimers.

On the basis of the assumption that all chromosome dimers are resolved at *dif*, chromosome dimer formation has been indirectly quantified by monitoring Xer recombinational exchanges at *dif* with a density label assay (70). Dimer formation reaches 15% in wild-type cells and depends on homologous recombination mediated by RecA using both the RecFOR and the RecBCD pathways (70, 71). Mutation in either of these pathways leads to about a 50% decrease in the number of Xer recombinational exchanges at *dif*, whereas mutational ablation of both pathways almost abolishes Xer recombinational exchanges at *dif*. This estimate of the frequency of SCEs that lead to dimers fits well with the general phenotype of Xer-defective cells. Indeed, the slow cell growth of Xer mutants is consistent with 15% of divisions giving no viable progeny (56).

If homologous recombination intermediates are resolved to crossover and noncrossover products with equal frequency, then the values above indicate that homologous recombination events occur in about 30% of chromosome replications. However, recent data indicate that there may be a strong bias toward noncrossover events, thereby minimizing chromosomal dimer formation. This bias seems to occur at the level of HJ resolution and requires RuvABC function (Fig. 3) (49, 80). In the absence of Ruv, this bias is lost and the action of Xer recombination becomes important for survival, particularly under conditions in which recombination events at the replication fork are increased. For example, although Xer$^-$ cells are not significantly UV^S, Ruv$^-$ Xer$^-$ cells are much more UV^S than Ruv$^-$ cells alone (49). The activation of expression of the cryptic Rus resolvase can re-exert a RecG-dependent bias in the formation of noncrossover products (49). These results suggest that homologous recombination events usually occur in greater than 30% of replication events and may occur in most wild-type cell generations. A major function of HJ-resolving enzymes like bacterial RuvABC and yeast mitochondrial CCE1 and its eukaryote mitochondrial homologues may be to bias HJ resolution and thereby minimize circular dimer formation (49). *B. subtilis* cells, which exhibit a particularly mild phenotype when Xer is inactivated, do not obviously possess a functional HJ resolvase such as RuvC (64). This is a puzzling feature of a species that possesses a very efficient transformation system, and indicates that there may be major differences between the replication and homologous recombination processes of *E. coli* and *B. subtilis*.

Converting Chromosome Dimers to Monomers

The precise mechanism by which the Xer site-specific recombination system differentiates sister *dif* sites carried by a chromosome dimer from sites carried by monomers, and thus converts dimers to monomers rather than creating dimers from monomers, remains to be determined.

It has been argued that there may be no directional control of Xer recombination at *dif* for its normal cellular role to be mediated (10, 43). For example, the Xer recombinases could catalyze a complete recombination reaction between *dif* sites, irrespective of their location, and so could create and resolve chromosome dimers. In one version of this model, when the two *dif* sites are present on separate monomers, the condensation and segregation of the two sister chromatids in the daughter cells lead to their separation before an opportunity for Xer recombination at *dif* arises. When the two *dif* sites are carried by a dimer, the Xer system can resolve the dimer into monomers before segregation can occur. If such a hypothesis were correct, replacing the Xer system by other site-specific recombination systems would correct the phenotype of Xer-defective cells. In favor of this hypothesis, it was found that adding back a *psi* core site to a *dif* deleted strain (22), or even a *loxP* site along with the Cre recombinase (44), alleviates the Xer mutant phenotype. However, the complementation is only partial (F.-X. Barre and D. J. S. Sherratt, unpublished data), raising the likelihood that a more sophisticated mechanism is operating. Consistent with this, the inability of *E. coli* XerCD to mediate a complete reaction at *dif* in vitro on supercoiled plasmids could be attributed to the requirement for some cellular factors involved in the control of Xer recombination in vivo. Furthermore, this control must be different from the accessory factor-dependent topological filter that restricts recombination to intramolecular events at *cer* and *psi*. A minimal chromosomal *dif* core site is functional for in vivo recombination, even when it is bounded by a 173-kb deletion of the normal flanking chromosomal DNA (Fig. 4) (79). Finally, replacing *E. coli* XerCD by *H. influenzae*

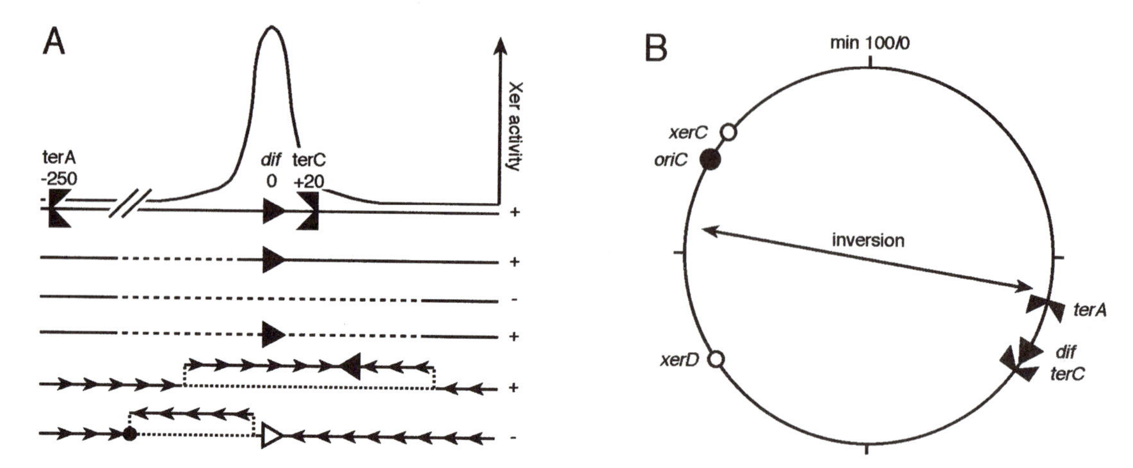

Figure 4. The *dif* activity zone. (A) The DAZ corresponds to ~30 kb around the natural position of *dif* (22, 42, 79). The Xer activity of *dif* sites away from their original position is represented on top of the drawing (56). Various deletions (dashed lines) and inversions around *dif* are also shown. Their effect on *dif* activity is indicated on the right by a plus or a minus sign. The simplest explanation for these complicated data is that there are sequence elements all over the chromosome that point toward *dif*, and that to be active, a *dif* site must be placed at the junction between those elements (22, 42, 56). It is believed that those sequence elements are smaller than 8 bp (F. Cornet, unpublished data). (B) Map of the *E. coli* chromosome, showing the position of the *xerC* and *xerD* genes, the position of *oriC* and *dif*, and the position of two replication terminator sites. There is absolutely no link between timing and/or termination of replication and *dif* activity. Indeed, a strain carrying a large inversion of the chromosome (the inverted region is indicated with an arrow), displacing *dif* close to oriC, does not present with the classical filamentous phenotype of *dif* strains (22). In this strain, replication of *dif* occurs rapidly after initiation, and termination of replication is outside the zone for competency of *dif* (22).

XerC and *E. coli* XerD leads to recombination between *dif* sites present in a supercoiled plasmid, producing products of mixed topology. This observation again argues that there is a normal cellular control of Xer recombination at *dif*, and that this can be overcome by using a heterologous mixture of recombinases (51).

Temporal and Spatial Regulation of Xer Recombination at *dif*

Hints on what might regulate the Xer recombination machine at the chromosomal *dif* sites have been given by its dependence on four major cellular factors: the position of the *dif* site in the chromosome, homologous recombination, cell division, and the septum-located protein, FtsK.

Combining all the available evidence, we propose the following model (Fig. 5). XerCD are sufficient to synapse two *dif* sites irrespective of their position and relative orientation. In general, the opportunities for synapsis of two chromosomal *dif* sites occur most readily immediately after replication, before the two daughter chromatids begin to be segregated. In most positions of the chromosome, a very small window of opportunity for such synapsis exists because of rapid separation of the two sister *dif* sites after replication. In the normal replication terminus position, the replicated *dif* sites remain in proximity for a longer period, thereby increasing the opportunity for synapsis. Supporting this view, we have shown that the level of HJs at *dif* in vivo in the chromosome is higher when *dif* is located in the replication terminus region rather than near the origin (6).

Xer recombination at *dif* requires cell division and the C-terminal domain of the septum-located protein FtsK ($FtsK_C$) (57, 69). FtsK is a 1329 amino acid protein divided into three separate domains. The 200-aa N-terminal domain is necessary for cell division (8, 25, 26) and localizes to the septum in an FtsZ-dependent manner (81, 83). The 500-aa $FtsK_C$ domain is highly homologous to the C-terminal domain of the SpoIIIE protein from *B. subtilis* (82) and is implicated in chromosome segregation (45, 84). Therefore, FtsK seems well suited to coordinate cell division with chromosome segregation. The central region is characterized by its high content in proline and glutamine, and by the presence of seven repeats of a 10-aa motif (8). Cells lacking $FtsK_C$ form filaments and chains with mispositioned nucleoids (45, 57, 84) and are defective in Xer recombination at chromosomal and plasmid *dif* (57, 69), but not at *cer* and *psi* sites (57). Most, if not all, of the chromosome segregation defect is suppressed in a *recA* background (57), consistent with a major part of the chromosome segregation function of $FtsK_C$ being in its role in promoting Xer recombination at *dif*.

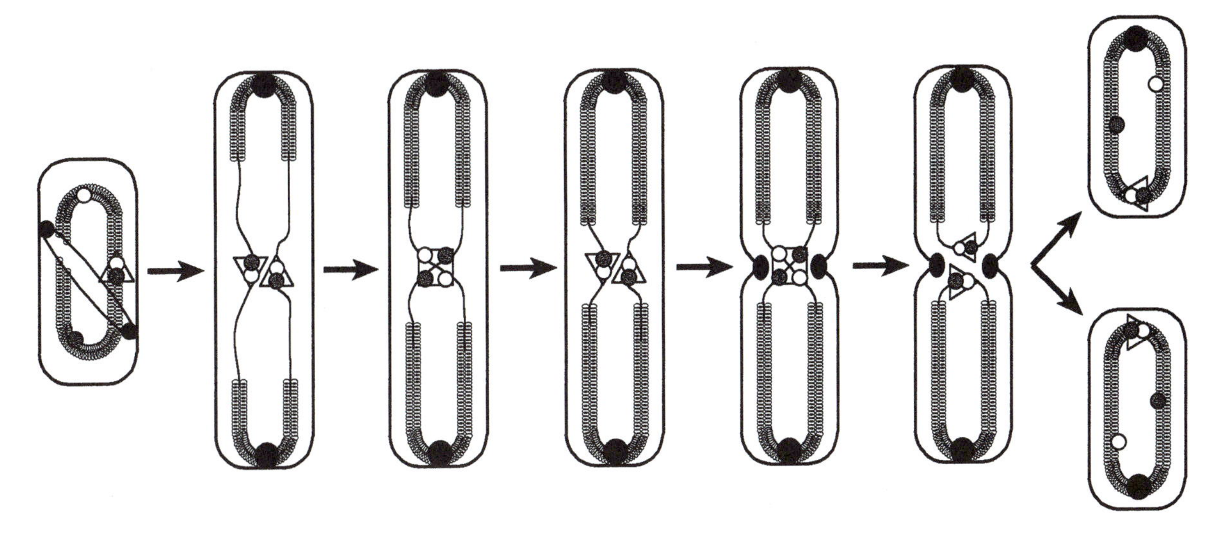

Figure 5. A model for the regulation of Xer recombination at *dif*. Just after initiation of replication, the two newly replicated origins (black circles) are segregated away from midcell. In contrast, the *dif* region (white triangles) stays close to midcell after replication. Just after being replicated, the sister *dif* sites can be synapsed by the Xer recombinases (grey and white circle). XerC catalyzes HJ formation, and in the absence of $FtsK_C$, resolves them back to substrate. Chromosome dimers block segregation, forcing septum closure around the synapsed *dif* sites. $FtsK_C$ (black ellipses at midcell) can then activate the complete recombination reaction.

To determine in which step of the Xer site-specific recombination reaction $FtsK_C$ is involved, we have monitored the level of HJs formed by Xer at *dif* in an $FtsK_C^-$ strain. We found that $FtsK_C$ is not required for HJ formation (6). We therefore propose that Xer recombination at *dif* is regulated at the stage of the HJ conformation change (Fig. 1, step iii). The chromosomal *dif* sites could be acted on by the Xer recombinases just after replication to form an XerC-catalyzed HJ intermediate. In the absence of the factors required to promote the conformational change to a substrate for XerD-mediated catalysis, XerC would be able to catalyze cycles of HJ formation and resolution that would lead to no complete recombinant products. Such XerC-mediated cycling could continue until segregation of the nucleoids breaks the synaptic complex. When a chromosome dimer with correctly positioned *dif* sites is present, the septum would close on the Xer synaptic complex, bringing in FtsK and any other accessory factors needed to allow the reaction to proceed to completion (Fig. 5). This model can be easily fitted to take into account the particular characteristics of the Xer recombinases, XerC being able to perform strand exchange between *dif* sites, but XerD needing a conformational change in the structure of the HJ to proceed. We propose that FtsK facilitates this conformational change in the HJ intermediate directly and therefore it performs a function equivalent to that of PepA in recombination at *psi*. However, although PepA functions through its interactions with specific accessory sequences, FtsK must mediate its effect directly or indirectly by either interacting with nonspecific DNA sequences or the synaptic complex itself. Steiner and Kuempel have argued against this model because they did not detect strand exchanges at *dif* in the absence of cell division (70). However, the level of HJs at *dif* is too low to be detected by the density label assay they used. We could detect and quantitate the level of HJs at *dif* by using a more sensitive technique of hybridization after two-dimensional gel electrophoresis (6, 61, 62). Consistent with a model in which XerC can continuously form and resolve HJ intermediates in the absence of the conditions needed for complete dimer resolution, we find that although XerCD are unable to mediate a complete Xer recombination reaction in vitro between *dif* sites on a supercoiled plasmid, substantial levels of HJ intermediates can be observed under appropriate conditions (6).

The model presented in Fig. 5 also provides explanations for the dependence of Xer recombination at *dif* on the position of *dif* on the chromosome. Indeed, for full activity, the *dif* site must be located in a ~30-kb zone, the *dif* activity zone (DAZ) (Fig. 4A, B) (22, 42, 56, 79). Whereas displacing *dif* 150 kb away from its natural position drastically diminishes Xer recombination at the site, we found that it had very little or no effect on HJ formation (6). We propose that the DAZ is involved in proper positioning of *dif* close to $FtsK_C$ at cell division. Indeed, the DAZ has been located close to midcell before cell division, in contrast to regions close to the origin (53–55).

Moreover, this characteristic seems conserved in cells that carry a large chromosomal inversion that moves DAZ close to *oriC* (55), a situation that can maintain functional *dif* activity (22). A septal location of DAZ at cell division would seem to be optimal, because it allows Xer recombination to act only when the DNA has been almost completely segregated into the two daughter cells. If Xer recombination acted earlier, late homologous recombination events could still occur and create chromosome dimers that would not be resolved by Xer.

Complete Xer recombinant product at chromosomal or plasmid *dif* also depends on a functional homologous recombination system, which is necessary to generate dimeric chromosomes. Furthermore, the levels of Xer recombinational exchanges at *dif* seem to be proportional to the amounts of ongoing homologous recombination (57, 70, 71). In contrast, homologous recombination is not needed for HJ formation and the level of HJ at *dif* is not modulated by conditions that change homologous recombination (6). As for the DAZ, we propose that homologous recombination helps position the chromosomal *dif* site close to the C terminus of FtsK. Indeed, the time between the termination of replication and the completion of cell division, the D period, has been estimated to be 20 min in standard *E. coli* cultures (36). During this period, the Xer recombinases can form HJs at *dif,* but the process cannot go to completion because $FstK_C$ is missing. Thus, there is ample time to segregate monomeric chromosomes away from midcell before septum closure. In contrast, a chromosome dimer cannot be segregated, forcing septum closure around the DAZ region and allowing FtsK-dependent Xer recombination.

With use of the same kind of argument, the low efficiency of Xer recombination between directly repeated plasmid *dif* sites, in the order of 30% resolution in an overnight culture in wild-type cells (57), can be attributed to the small fraction of plasmids that are close to the septum, and therefore accessible to $FtsK_C$, at each cell division. However, it remains unclear why the Xer recombination between plasmid *dif* sites should depend on RecA. Note that the dependence on homologous recombination functions for Xer recombination at plasmid *dif* is a requirement for conditions that generate chromosomal dimers rather than for plasmid dimers per se, because *recF* mutations, which almost completely abolish plasmid homologous recombination, have no effect on Xer-dependent recombination at plasmid *dif* (57). It is possible that the presence of a chromosome dimer in the cell increases the time during which $FtsK_C$, and/or a $FtsK_C$-dependent factor necessary for Xer recombination, is accessible to the plasmid, but we think that the presence of a chromosome dimer might also provide a signal to activate the FtsK protein.

CONCLUDING REMARKS

Studies of Xer site-specific recombination have revealed new mechanistic details of the tyrosine recombinase mechanism. In particular, the use of two recombinases that bind to related yet different half sites has provided a powerful tool for determining the precise role of different molecules as the recombination reaction proceeds. Genetic and biochemical studies have also revealed new insight into how XerC-XerD interactions and recombination site core sequence coordinate and control recombination outcome. It seems that the use of two recombinases by the Xer system has evolved to ensure that a complete recombination reaction is completed only when very special conditions are met. Asymmetry imposed by having the two recombinases provides an opportunity to form HJ intermediates that are unable to undergo the conformational change needed for the second pair of strand exchanges. This checkpoint then allows other conditions and factors to facilitate this change and thereby allows Xer recombination to mediate its specific biological roles.

The ability of Xer recombination to exhibit resolution selectivity, the process that limits recombination to intramolecular events between two directly repeated identical *cer* or *psi* recombination sites and thereby converts dimers (and other multimers) exclusively to monomers, demonstrates how precise synapse topology and the topological filter mechanism has evolved to act in unrelated specialized recombination systems. Furthermore, these studies have revealed novel architectural roles for proteins, like PepA, that have other quite different biochemical functions.

Uncovering the relationships between Xer recombination and chromosome segregation and how these relate to cell division, homologous recombination, and DNA replication is revealing the sophisticated methods that have evolved in bacteria to ensure the stable duplication and inheritance of the genetic material. Finally, the characterization of a strategy to ensure that Xer recombination at chromosomal *dif* is restricted to converting chromosome dimers present at cell division to monomers reveals one new way in which bacteria can process temporal and spatial information.

Acknowledgments. F.-X.B. was supported by ARC and EMBO fellowships. D.J.S. is supported by the Wellcome Trust.

We thank M. Chandler, F. Cornet, and H. Ferreira for helpful discussions.

REFERENCES

1. **Alen, C., D. J. Sherratt, and S. D. Colloms.** 1997. Direct interaction of aminopeptidase A with recombination site DNA in Xer site-specific recombination. *EMBO J.* **16:**5188–5197.
2. **Arciszewska, L. K., R. A. Baker, B. Hallet, and D. J. Sherratt.** 2000. Coordinated control of XerC and XerD catalytic activities during Holliday junction resolution. *J. Mol. Biol.* **299:** 391–403.
3. **Arciszewska, L. K., I. Grainge, and D. J. Sherratt.** 1997. Action of site-specific recombinases XerC and XerD on tethered Holliday junctions. *EMBO J.* **16:**3731–3743.
4. **Arciszewska, L. K., and D. J. Sherratt.** 1995. Xer site-specific recombination in vitro. *EMBO J.* **14:**2112–2120.
5. **Azaro, M. A., and A. Landy.** 1997. The isomeric preference of Holliday junctions influences resolution bias by lambda integrase. *EMBO J.* **16:**3744–3755.
6. **Barre, F.-X., M. Aroyo, S. D. Colloms, A. Helfrich, F. Cornet, and D. J. Sherratt.** 2000. FtsK functions in the processing of a Holliday junction intermediate during bacterial chromosome segregation. *Genes Dev.* **14:**2976–2988.
7. **Bath, J., D. J. Sherratt, and S. D. Colloms.** 1999. Topology of Xer recombination on catenanes produced by lambda integrase. *J. Mol. Biol.* **289:**873–883.
8. **Begg, K. J., S. J. Dewar, and W. D. Donachie.** 1995. A new *Escherichia coli* cell division gene, ftsK. *J. Bacteriol.* **177:** 6211–6222.
9. **Blake, J. A., N. Ganguly, and D. J. Sherratt.** 1997. DNA sequence of recombinase-binding sites can determine Xer site-specific recombination outcome. *Mol. Microbiol.* **23:**387–398.
10. **Blakely, G., S. Colloms, G. May, M. Burke, and D. Sherratt.** 1991. *Escherichia coli* XerC recombinase is required for chromosomal segregation at cell division. *New Biol.* **3:**789–798.
11. **Blakely, G., G. May, R. McCulloch, L. K. Arciszewska, M. Burke, S. T. Lovett, and D. J. Sherratt.** 1993. Two related recombinases are required for site-specific recombination at *dif* and *cer* in *E. coli* K12. *Cell* **75:**351–361.
12. **Blakely, G. W., A. O. Davidson, and D. J. Sherratt.** 2000. Sequential strand exchange by XerC and XerD during site-specific recombination at *dif*. *J. Biol. Chem.* **275:**9930–9936.
13. **Blakely, G. W., and D. J. Sherratt.** 1996. Cis and trans in site-specific recombination. *Mol. Microbiol.* **20:**234–237.
14. **Burke, M., A. F. Merican, and D. J. Sherratt.** 1994. Mutant *Escherichia coli* arginine repressor proteins that fail to bind L-arginine, yet retain the ability to bind their normal DNA-binding sites. *Mol. Microbiol.* **13:**609–618.
15. **Chen, Y., U. Narendra, L. E. Iype, M. M. Cox, and P. A. Rice.** 2000. Crystal structure of a Flp recombinase-holliday junction complex: assembly of an active oligomer by helix swapping. *Mol. Cell* **6:**885–897.
16. **Clerget, M.** 1991. Site-specific recombination promoted by a short DNA segment of plasmid R1 and by a homologous segment in the terminus region of the *Escherichia coli* chromosome. *New Biol.* **3:**780–788.
17. **Colloms, S. D., C. Alen, and D. J. Sherratt.** 1998. The ArcA/ArcB two-component regulatory system of *Escherichia coli* is essential for Xer site-specific recombination at psi. *Mol. Microbiol.* **28:**521–530.
18. **Colloms, S. D., J. Bath, and D. J. Sherratt.** 1997. Topological selectivity in Xer site-specific recombination. *Cell* **88:**855–864.
19. **Colloms, S. D., R. McCulloch, K. Grant, L. Neilson, and D. J. Sherratt.** 1996. Xer-mediated site-specific recombination in vitro. *EMBO J.* **15:**1172–1181.
20. **Colloms, S. D., P. Sykora, G. Szatmari, and D. J. Sherratt.** 1990. Recombination at ColE1 *cer* requires the *Escherichia coli xerC* gene product, a member of the lambda integrase family of site-specific recombinases. *J. Bacteriol.* **172:** 6973–6980.
21. **Cornet, F., B. Hallet, and D. J. Sherratt.** 1997. Xer recombination in *Escherichia coli*. Site-specific DNA topoisomerase activity of the XerC and XerD recombinases. *J. Biol. Chem.* **272:** 21927–21931.
22. **Cornet, F., J. Louarn, J. Patte, and J. M. Louarn.** 1996. Restriction of the activity of the recombination site *dif* to a small zone of the *Escherichia coli* chromosome. *Genes Dev.* **10:** 1152–1161.
23. **Cornet, F., I. Mortier, J. Patte, and J. M. Louarn.** 1994. Plasmid pSC101 harbors a recombination site, *psi*, which is able to resolve plasmid multimers and to substitute for the analogous chromosomal *Escherichia coli* site *dif*. *J. Bacteriol.* **176:**3188–3195.
24. **Cox, M. M., M. F. Goodman, K. N. Kreuzer, D. J. Sherratt, S. J. Sandler, and K. J. Marians.** 2000. The importance of repairing stalled replication forks. *Nature* **404:**37–41.
25. **Diez, A. A., A. Farewell, U. Nannmark, and T. Nystrom.** 1997. A mutation in the *ftsK* gene of *Escherichia coli* affects cell-cell separation, stationary-phase survival, stress adaptation, and expression of the gene encoding the stress protein UspA. *J. Bacteriol.* **179:**5878–5883.
26. **Draper, G. C., N. McLennan, K. Begg, M. Masters, and W. D. Donachie.** 1998. Only the N-terminal domain of FtsK functions in cell division. *J. Bacteriol.* **180:**4621–4627.
27. **Duckett, D. R., A. I. Murchie, and D. M. Lilley.** 1995. The global folding of four-way helical junctions in RNA, including that in U1 snRNA. *Cell* **83:**1027–1036.
28. **Esposito, D., and J. J. Scocca.** 1997. The integrase family of tyrosine recombinases: evolution of a conserved active site domain. *Nucleic Acids Res.* **25:**3605–3614.
29. **Gopaul, D. N., and G. D. Duyne.** 1999. Structure and mechanism in site-specific recombination. *Curr. Opin. Struct. Biol.* **9:**14–20.
30. **Gopaul, D. N., F. Guo, and G. D. Van Duyne.** 1998. Structure of the Holliday junction intermediate in Cre-loxP site-specific recombination. *EMBO J.* **17:**4175–4187.
31. **Grainge, I., and D. J. Sherratt.** 1999. Xer site-specific recombination. DNA strand rejoining by recombinase XerC. *J. Biol. Chem.* **274:**6763–6769.
32. **Guo, F., D. N. Gopaul, and G. D. Van Duyne.** 1999. Asymmetric DNA bending in the Cre-loxP site-specific recombination synapse. *Proc. Natl. Acad. Sci. USA* **96:**7143–7148.
33. **Hallet, B., L. K. Arciszewska, and D. J. Sherratt.** 1999. Reciprocal control of catalysis by the tyrosine recombinases XerC and XerD: an enzymatic switch in site-specific recombination. *Mol. Cell* **4:**949–959.
34. **Hayes, F., S. A. Lubetzki, and D. J. Sherratt.** 1997. *Salmonella typhimurium* specifies a circular chromosome dimer resolution system which is homologous to the Xer site-specific recombination system of *Escherichia coli*. *Gene* **198:**105–110.
35. **Hayes, F., and D. J. Sherratt.** 1997. Recombinase binding specificity at the chromosome dimer resolution site *dif* of *Escherichia coli*. *J. Mol. Biol.* **266:**525–537.
36. **Helmstetter, C. E.** 1996. Timing of synthetic activities in the cell cycle, p. 1627–1639. *In* F. C. Neidhardt et al. (ed.), Escherichia coli *and* Salmonella: *Cellular and Molecular Biology,* 2nd ed., vol. 2. American Society for Microbiology, Washington, D.C.
37. **Hendricks, E. C., H. Szerlong, T. Hill, and P. Kuempel.** 2000. Cell division, guillotining of dimer chromosomes and SOS induction in resolution mutants (*dif, xerC and xerD*) of *Escherichia coli*. *Mol. Microbiol.* **36:**973–981.
38. **Hofte, M., Q. Dong, S. Kourambas, V. Krishnapillai, D. Sher-**

ratt, and M. Mergeay. 1994. The sss gene product, which affects pyoverdin production in Pseudomonas aeruginosa 7NSK2, is a site-specific recombinase. *Mol. Microbiol.* 14: 1011–1020.
39. Klemm, P. 1986. Two regulatory *fim* genes, *fimB* and *fimE*, control the phase variation of type 1 fimbriae in *Escherichia coli*. *EMBO J.* 5:1389–1393.
40. Krogh, B. O., and S. Shuman. 2000. Catalytic mechanism of DNA topoisomerase IB. *Mol. Cell* 5:1035–1041.
41. Krogh, B. O., and S. Shuman. 2000. DNA strand transfer catalyzed by vaccinia topoisomerase: peroxidolysis and hydroxylaminolysis of the covalent protein-DNA intermediate. *Biochemistry* 39:6422–6432.
42. Kuempel, P., A. Hogaard, M. Nielsen, O. Nagappan, and M. Tecklenburg. 1996. Use of a transposon (Tn*dif*) to obtain suppressing and nonsuppressing insertions of the *dif* resolvase site of *Escherichia coli*. *Genes Dev.* 10:1162–1171.
43. Kuempel, P. L., J. M. Henson, L. Dircks, M. Tecklenburg, and D. F. Lim. 1991. *dif*, a recA-independent recombination site in the terminus region of the chromosome of *Escherichia coli*. *New Biol.* 3:799–811.
44. Leslie, N. R., and D. J. Sherratt. 1995. Site-specific recombination in the replication terminus region of *Escherichia coli*: functional replacement of *dif*. *EMBO J.* 14:1561–1570.
45. Liu, G., G. C. Draper, and W. D. Donachie. 1998. FtsK is a bifunctional protein involved in cell division and chromosome localization in *Escherichia coli*. *Mol. Microbiol.* 29:893–903.
46. McCulloch, R., M. E. Burke, and D. J. Sherratt. 1994. Peptidase activity of *Escherichia coli* aminopeptidase A is not required for its role in Xer site-specific recombination. *Mol. Microbiol.* 12:241–251.
47. McCulloch, R., L. W. Coggins, S. D. Colloms, and D. J. Sherratt. 1994. Xer-mediated site-specific recombination at *cer* generates Holliday junctions in vivo. *EMBO J.* 13:1844–1855.
48. McGlynn, P., and R. G. Lloyd. 2000. Modulation of RNA polymerase by (p)ppGpp reveals a RecG-dependent mechanism for replication fork progression. *Cell* 101:35–45.
49. Michel, B., G. D. Recchia, M. Penel-Colin, S. D. Ehrlich, and D. J. Sherratt. 2000. Resolution of Holliday junctions by RuvABC prevents dimer formation in *rep* mutants and UV irradiated cells. *Mol. Microbiol.* 37:181–191.
50. Murphy, E. 1989. Transposable elements in gram-positive bacteria, p. 269–288. *In* D. E. Berg and M. M. Howe (ed.), *Mobile DNA*. American Society for Microbiology, Washington, D.C.
51. Neilson, L., G. Blakely, and D. J. Sherratt. 1999. Site-specific recombination at *dif* by *Haemophilus influenzae* XerC. *Mol. Microbiol.* 31:915–926.
52. Neilson, L. M. 1999. *Xer Site-Specific Recombination at* dif *by* Haemophilus influenzae *XerC*. D.Phil. dissertation. Oxford University, Oxford, United Kingdom.
53. Niki, H., and S. Hiraga. 1997. Subcellular distribution of actively partitioning F plasmid during the cell division cycle in *E. coli*. *Cell* 90:951–957.
54. Niki, H., and S. Hiraga. 1998. Polar localization of the replication origin and terminus in *Escherichia coli* nucleoids during chromosome partitioning. *Genes Dev.* 12:1036–1045.
55. Niki, H., Y. Yamaichi, and S. Hiraga. 2000. Dynamic organization of chromosomal DNA in *Escherichia coli*. *Genes Dev.* 14:212–223.
56. Perals, K., F. Cornet, Y. Merlet, I. Delon, and J. M. Louarn. 2000. Functional polarization of the *Escherichia coli* chromosome terminus: the *dif* site acts in chromosome dimer resolution only when located between long stretches of opposite polarity. *Mol. Microbiol.* 36:33–43.
57. Recchia, G. D., M. Aroyo, D. Wolf, G. Blakely, and D. J. Sherratt. 1999. FtsK-dependent and -independent pathways of Xer site-specific recombination. *EMBO J.* 18:5724–5734.
58. Recchia, G. D., and D. J. Sherratt. 1999. Conservation of Xer site-specific recombination genes in bacteria. *Mol. Microbiol.* 34:1146–1148.
59. Redinbo, M. R., J. J. Champoux, and W. G. Hol. 2000. Novel insights into catalytic mechanism from a crystal structure of human topoisomerase I in complex with DNA. *Biochemistry* 39:6832–6840.
60. Rost, B. 1996. PHD: predicting one-dimensional protein structure by profile based neural networks. *Methods Enzymol.* 266: 525–539.
61. Schwacha, A., and N. Kleckner. 1994. Identification of joint molecules that form frequently between homologs but rarely between sister chromatids during yeast meiosis. *Cell* 76:51–63.
62. Schwacha, A., and N. Kleckner. 1995. Identification of double Holliday junctions as intermediates in meiotic recombination. *Cell* 83:783–791.
63. Sciochetti, S. A., P. J. Piggot, D. J. Sherratt, and G. Blakely. 1999. The *ripX* locus of *Bacillus subtilis* encodes a site-specific recombinase involved in proper chromosome partitioning. *J. Bacteriol.* 181:6053–6062.
64. Sharples, G. J., S. M. Ingleston, and R. G. Lloyd. 1999. Holliday junction processing in bacteria: insights from the evolutionary conservation of RuvABC, RecG, and RusA. *J. Bacteriol.* 181:5543–5550.
65. Sherratt, D. J., and D. B. Wigley. 1998. Conserved themes but novel activities in recombinases and topoisomerases. *Cell* 93: 149–152.
66. Spiers, A. J., and D. J. Sherratt. 1999. C-terminal interactions between the XerC and XerD site-specific recombinases. *Mol. Microbiol.* 32:1031–1042.
67. Stark, W. M., and M. R. Boocock. 1995. Topological selectivity in site-specific recombination, p. 101–129. *In* D. J. Sherratt (ed.), *Mobile Genetic Elements*. IRL Press, Oxford, United Kingdom.
68. Stark, W. M., D. J. Sherratt, and M. R. Boocock. 1989. Site-specific recombination by Tn3 resolvase: topological changes in the forward and reverse reactions. *Cell* 58:779–790.
69. Steiner, W., G. Liu, W. D. Donachie, and P. Kuempel. 1999. The cytoplasmic domain of FtsK protein is required for resolution of chromosome dimers. *Mol. Microbiol.* 31:579–583.
70. Steiner, W. W., and P. L. Kuempel. 1998. Cell division is required for resolution of dimer chromosomes at the *dif* locus of *Escherichia coli*. *Mol. Microbiol.* 27:257–268.
71. Steiner, W. W., and P. L. Kuempel. 1998. Sister chromatid exchange frequencies in *Escherichia coli* analyzed by recombination at the *dif* resolvase site. *J. Bacteriol.* 180:6269–6275.
72. Stirling, C. J., S. D. Colloms, J. F. Collins, G. Szatmari, and D. J. Sherratt. 1989. *xerB*, an *Escherichia coli* gene required for plasmid ColE1 site-specific recombination, is identical to *pepA*, encoding aminopeptidase A, a protein with substantial similarity to bovine lens leucine aminopeptidase. *EMBO J.* 8: 1623–1627.
73. Stirling, C. J., G. Stewart, and D. J. Sherratt. 1988. Multicopy plasmid stability in *Escherichia coli* requires host-encoded functions that lead to plasmid site-specific recombination. *Mol. Gen. Genet.* 214:80–84.
74. Stirling, C. J., G. Szatmari, G. Stewart, M. C. Smith, and D. J. Sherratt. 1988. The arginine repressor is essential for plasmid-stabilizing site-specific recombination at the ColE1 *cer* locus. *EMBO J.* 7:4389–4395.
75. Strater, N., D. J. Sherratt, and S. D. Colloms. 1999. X-ray structure of aminopeptidase A from *Escherichia coli* and a model for the nucleoprotein complex in Xer site-specific recombination. *EMBO J.* 18:4513–4522.

76. **Subramanya, H. S., L. K. Arciszewska, R. A. Baker, L. E. Bird, D. J. Sherratt, and D. B. Wigley.** 1997. Crystal structure of the site-specific recombinase, XerD. *EMBO J.* **16:**5178–5187.
77. **Summers, D. K., and D. J. Sherratt.** 1984. Multimerization of high copy number plasmids causes instability: ColE1 encodes a determinant essential for plasmid monomerization and stability. *Cell* **36:**1097–1103.
78. **Summers, D. K., and D. J. Sherratt.** 1988. Resolution of ColE1 dimers requires a DNA sequence implicated in the three-dimensional organization of the *cer* site. *EMBO J.* **7:**851–858.
79. **Tecklenburg, M., A. Naumer, O. Nagappan, and P. Kuempel.** 1995. The *dif* resolvase locus of the *Escherichia coli* chromosome can be replaced by a 33-bp sequence, but function depends on location. *Proc. Natl. Acad. Sci. USA* **92:**1352–1356.
80. **van Gool, A. J., N. M. Hajibagheri, A. Stasiak, and S. C. West.** 1999. Assembly of the *Escherichia coli* RuvABC resolvasome directs the orientation of holliday junction resolution. *Genes Dev.* **13:**1861–1870.
81. **Wang, L., and J. Lutkenhaus.** 1998. FtsK is an essential cell division protein that is localized to the septum and induced as part of the SOS response. *Mol. Microbiol.* **29:**731–740.
82. **Wu, L. J., P. J. Lewis, R. Allmansberger, P. M. Hauser, and J. Errington.** 1995. A conjugation-like mechanism for prespore chromosome partitioning during sporulation in *Bacillus subtilis. Genes Dev.* **9:**1316–1326.
83. **Yu, X. C., A. H. Tran, Q. Sun, and W. Margolin.** 1998. Localization of cell division protein FtsK to the *Escherichia coli* septum and identification of a potential N-terminal targeting domain. *J. Bacteriol.* **180:**1296–1304.
84. **Yu, X. C., E. K. Weihe, and W. Margolin.** 1998. Role of the C terminus of FtsK in *Escherichia coli* chromosome segregation. *J. Bacteriol.* **180:**6424–6428.

Mobile DNA II
Edited by N. L. Craig et al.

Chapter 9

Gene Acquisition in Bacteria by Integron-Mediated Site-Specific Recombination

Gavin D. Recchia and David J. Sherratt

INTRODUCTION

Microbial genomes are dynamic mosaic structures. The fluidity of genome components and their relative arrangements allows bacteria to respond rapidly to changing environmental conditions and to new environmental challenges. The dynamic nature of microbial genomes can be attributed to the ability of bacteria to share genetic information by horizontal gene transfer systems and to the variety of homologous and nonhomologous recombinational processes that integrate, excise, and translocate genes into and from homologous, homeologous, and nonhomologous loci (62). One of the most remarkable examples of the impact of these processes has been the bacterial response to use of antibiotics.

Historical Perspective: the Antibiotic Resistance Phenomenon

In the 1970s transposable elements were identified as major contributors to bacterial genomic fluidity by virtue of their possession of antibiotic resistance genes (9, 43). Transposable elements can be disseminated within and between bacterial species by horizontal gene transfer systems such as conjugation, bacteriophage-mediated transduction, and genetic transformation. Subsequently, other homologous and site-specific recombinational processes have also generated novel gene combinations at new genetic loci, and again these novel genetic combinations have often been subject to widespread horizontal transfer.

In the relatively short time since the introduction of antibiotic therapy, resistance to virtually all antibiotics has appeared in pathogenic organisms (22). Most resistance genes clearly have not evolved since the advent of antibiotic therapy, nor have they evolved de novo in pathogenic bacteria (23), but rather they have been acquired by horizontal transfer. The development of DNA sequencing revealed that many of the resistance genes in gram-negative pathogens reside in a similar genetic context. In particular, their flanking sequences were similar or identical, which suggested the existence of a specific mechanism for the dissemination of these genes (13, 39, 55, 63, 74, 82, 88). This led to the discovery of a RecA-independent site-specific recombination system (55), encoded by genetic elements termed integrons (82). A minimal integron consists of a gene, *intI*, encoding a site-specific recombinase of the tyrosine recombinase family and an adjacent recombination site, *attI*. Integrons may also contain one or more gene cassettes

Gavin D. Recchia • Freehills Carter Smith Beadle, Sydney, New South Wales 2000, Australia. **David J. Sherratt** • Division of Molecular Genetics, Department of Biochemistry, University of Oxford, Oxford OX1 3QU, United Kingdom.

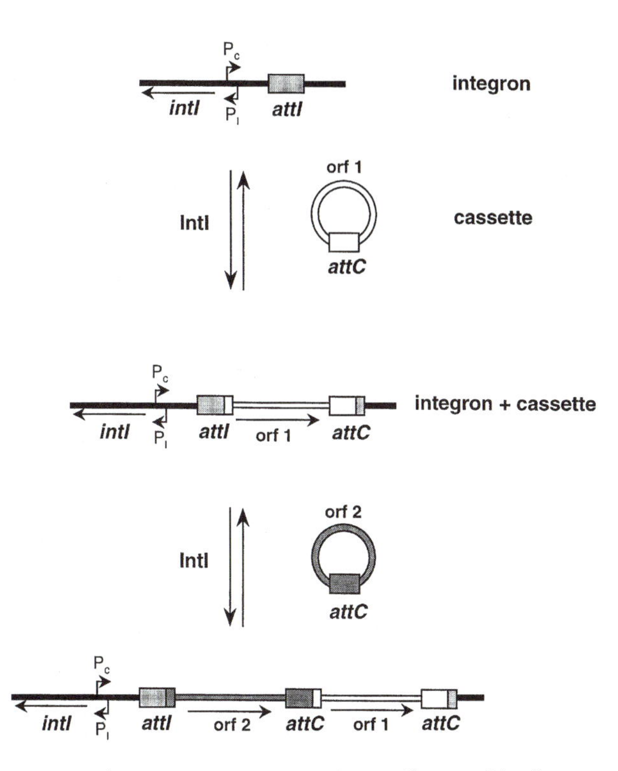

Figure 1. The integron-gene cassette site-specific recombination system. The *attI* recombination site into which cassettes are integrated is represented by a filled box, whereas the *attC* site of the gene cassette is shown as an open box. The *intI* gene encoding a tyrosine recombinase is transcribed by the promoter P_I, while promoter P_c directs transcription of integrated gene cassettes. Phenotypic markers are designated orf1 and orf2. A schematic model for the successive integration of two circular cassettes into the *attI* site of an integron is shown. Incoming cassettes are preferentially integrated at the *attI* site (17, 66).

integrated at *attI* (Fig. 1) (35, 68). Gene cassettes contain a recombination site that we designate *attC* and antibiotic resistance or other phenotypically detectable genes. They represent a vast mobile gene pool that can be mobilized by integron-mediated recombination. Gene cassettes can be assembled into tandem arrays and rearranged within integrons, and subsequently disseminated throughout a bacterial population by horizontal transfer mechanisms. Together integrons and gene cassettes constitute a two-component site-specific recombination system for gene acquisition and dissemination. Gene cassettes are the mobile elements and integrons the recombinase-encoding master elements into which cassettes are integrated and excised. Gene cassettes are not the only antibiotic resistance mobile elements that move by site-specific recombination. Conjugative transposons, which are important determinants of antibiotic resistance, particularly in gram-positive bacteria (73, 75), are also mobilized by the action of tyrosine recombinases, presumably by a site-specific recombination mechanism similar to that which mobilizes gene cassettes (see chapter 10). Conjugative transposons are autonomous genetic elements that not only encode the recombinase responsible for their mobility, but also mobilize the recombinase gene along with the phenotypic marker that they carry. In contrast, gene cassettes are nonautonomous genetic elements because they rely on the nonmobilizable recombinase gene of the integron for their mobility.

Functions Encoded by Gene Cassettes

The number of resistance genes identified as being cassette-associated has continued to grow. Outbreaks of nosocomial infections caused by gram-negative pathogens have increasingly been attributed to the presence of integrons (5, 25, 44, 72), and significant reservoirs of multiresistant integron-containing strains have been found in animal populations such as poultry (7) and swine (86). More than 60 antibiotic resistance gene cassettes have been identified to date, encoding resistance to aminoglycosides, β-lactams, chloramphenicol, trimethoprim, erythromycin, streptothricin, rifampin, and a variety of antiseptics and disinfectants (58, 71). For some antibiotic resistance protein families, for example, class B chloramphenicol acetyltransferases, class D β-lactamases, and class B dihydrofolate reductases, most known genes are contained within mobile gene cassettes.

The bias toward identifying antibiotic resistance determinants in gene cassettes is almost certainly the consequence of nonrandom sampling, because most studies have focused on clinical strains. We now know that cassette-encoded functions are not limited solely to antibiotic resistance, and the role of integrons is not confined to the acquisition and dissemination of resistance genes. Other cassettes encode a diverse range of functions including toxins, lipoproteins, restriction-modification systems, and pathogenicity determinants (71). In principle, any gene can be packaged and mobilized as a cassette.

Distribution and Mobility

Integrons and gene cassettes are widely distributed in gram-negative bacteria. They have been found in more than a dozen genera of the *Enterobacteriaceae* (66) and in several members of the *Vibrionaceae* (16, 53, 58), various pseudomonads (58, 90), *Acinetobacter* (76), and xanthomonads (71). Many of these organisms are opportunistic pathogens of humans, reflecting a sampling bias because most integrons studied have been obtained from the clinical environment, whereas others are soil and marine organisms. However, integrons are not restricted to the gram-negative γ-proteobacteria, and are likely to be widespread in the natural environment. An integron has recently been isolated from the gram-positive organism *Corynebacterium glutamicum* (60) and an integron remnant is present in Tn*610* from *Mycobacterium fortuitum* (54).

The success of integrons and gene cassettes in penetrating diverse bacterial populations is largely attributable to the variety of mechanisms available for their mobility and transfer. Gene cassettes can be recruited individually and shuffled both between and within integrons by site-specific recombination reactions that exhibit some promiscuity. In turn, many integrons are located within transposons that are typically found on broad host range conjugative plasmids. An example of the hierarchical nature of these mobility mechanisms is provided by the composite transposon Tn*21* (51). Tn*21* and its close relatives are widespread in clinical environments and carry multiple antibiotic resistance gene cassettes contained within an integron. Tn*21* is located within another Tn*9*-like composite transposon, which in turn is found on the conjugative plasmid NR1. Indeed, all known type 1 integrons reside in Tn*5090*/Tn*402* or truncates thereof; Tn*21* contains one of these truncated forms (87). Thus site-specific recombination assembles arrays of cassettes within integrons, whereas classical transposition and conjugation act as powerful tools for the horizontal transfer of these arrays (87).

The purpose of this chapter is to review briefly the properties of integrons and gene cassettes, and then attempt to provide a mechanistic framework for integron-mediated gene cassette mobility using the typical tyrosine recombinase mechanism as a perspective.

THE RECOMBINATION REACTION

The Tyrosine Recombinase Paradigm

The tyrosine recombinase family is characterized both by the conservation of key catalytic residues and a shared mechanism of recombination. All well-studied members of the family perform recombination by using the same series of reactions (31, 32, 41; see also chapters 5–8). The recombination reaction involves two sequential strand cleavage and rejoining reactions that are separated in time and space (Fig. 2A). The first pair of strand exchanges generates a Holliday junction (HJ) intermediate, which then undergoes a conformational change that allows it to be resolved by the second pair of strand exchanges to produce recombinant product. Tyrosine recombinases typically recognize a ~30-bp core recombination site that binds two molecules of recombinase at two inversely oriented binding sites separated by a 6- to 8-bp central region (Fig. 2B). (Note that central regions have also been called "overlap" regions and "coupling" sequences.) Recombination occurs at the boundaries of this central region on each strand (32, 41). A recombining nucleoprotein complex contains four recombinase molecules, two of which are catalytically active and two of which are catalytically inactive at any given time. During a complete recombination reaction, a reciprocal switch in recombinase activity occurs during the change in conformation of the HJ intermediate (2, 31, 40).

Site-specific recombination reactions mediated by the highly characterized λ Int-*attP*/*attB*, Cre-*loxP*, Flp-*frt*, and XerCD-*dif*/*cer*/*psi* systems usually occur between core sites that are identical or almost identical. In particular, the participating central region sequences are always identical and mutations that create heterology between the participating duplexes can drastically reduce recombination in vivo and in vitro. Nevertheless, low levels of recombination have also been observed between a normal core site and related, yet diverged secondary sites. For example, bacteriophage λ can use its *attP* site not only to integrate into its normal chromosomal locus, *attB*, but also to integrate at low frequency into other secondary attachment sites (91). In in vitro recombination reactions, introduction of heterology can impair the initial recombinase-mediated cleavage or cause the reaction to stall at the HJ intermediate stage, rejoining being inhibited. Nevertheless, some level of rejoining can occur between central regions that differ in sequence or length to generate rejoined mismatch-containing products (2–4, 49, 92). These observations are relevant to the process of integron-mediated recombination because it involves interactions between recombination sites that contain heterologous central regions (see below).

Regulation of recombination outcome can also occur at the HJ intermediate during recombination between identical core sites. For example, in Xer site-specific recombination, the HJ intermediate can resolve back to substrate, be converted to complete recombinant product, or be the final recombinase-mediated reaction product, depending on the particular sites undergoing recombination and on the reaction conditions (see chapter 8).

Many site-specific recombination systems use additional accessory proteins to regulate the recombination reaction (Fig. 2B). These proteins interact with specific accessory sequences, usually adjacent to the

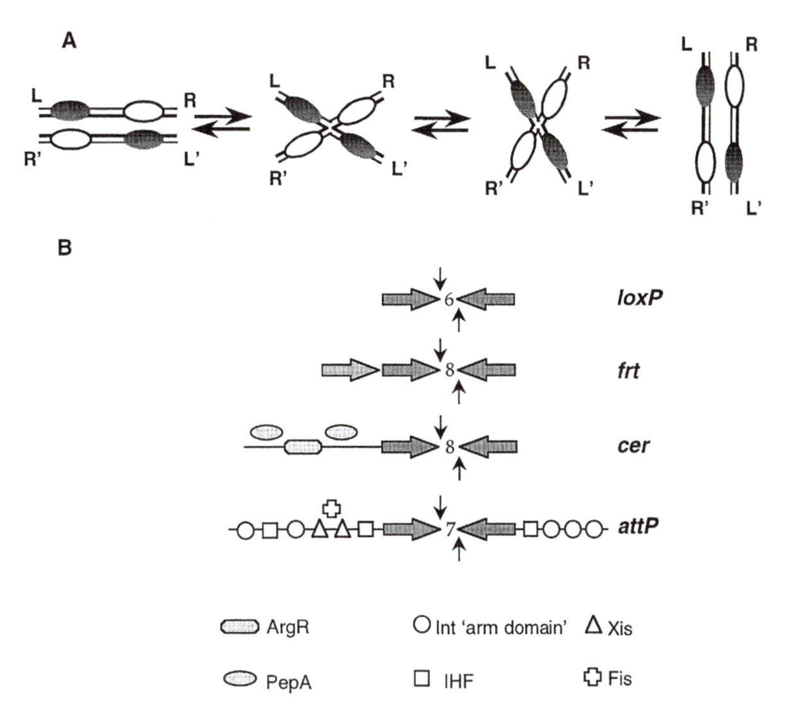

Figure 2. (A) Schematic representation of reactions catalyzed by tyrosine recombinases. Two recombinase molecules (ovals) are bound to each core recombination site. Recombination core sites synapse in an antiparallel orientation. In the first pair of strand exchanges two active recombinase monomers (unshaded) catalyze strand breakage and rejoining reactions of the thicker strands to generate a 2-fold symmetrical HJ intermediate, which then undergoes a conformational change to form a substrate for the second pair of strand exchanges. Tyrosine recombinases act on the "crossing" strands, the more acute pair of strands within the HJ. During the second pair of strand exchanges, the shaded recombinase monomers are active and catalyze breakage-rejoining of the thinner strands to generate recombinant products. (B) Structure of representative core recombination sites (not to scale). The *loxP* site is recognized by the phage P1 encoded Cre protein, the *frt* site by Flp encoded by the yeast 2μm plasmid, the *cer* site by XerCD, and the *attP* site by λ Int. Recombinase binding sites are represented by shaded horizontal arrows separated by a central region of 6 to 8 bp. The positions of strand exchange at each end of the central region are indicated by vertical arrows. Additional recombinase binding sites and binding sites for accessory proteins are also indicated.

core recombination site, and have an architectural role in the formation of the nucleoprotein-recombining complex (41, 48). Such complexes can provide selectivity to the recombination reaction, in particular restricting it to particular configurations of recombination sites (67, 80, 81). Accessory proteins can be encoded by the host, specialized proteins encoded by the element, or additional molecules of the recombinase itself. For example, site-specific recombinases of both the serine recombinase and tyrosine recombinase family can act as accessory proteins by binding to accessory sequences and to the core recombination sites, thereby facilitating synapsis and directing recombination outcome.

Because the integron-encoded IntI recombinase proteins possess all of the conserved sequence motifs typical of the tyrosine recombinase family, it seems inevitable that IntI recombinases will share the same catalytic mechanism as that mediated by Cre and the other well-characterized members of the family. Eukaryote type IB topoisomerases, which are far more diverged in primary sequence from Cre than the IntI proteins, share almost identical tertiary structures with Cre, as well as the same catalytic mechanism (24, 47, 79).

Biochemical studies of IntI are in their infancy, and IntI-mediated recombination has not been successfully reconstituted in vitro. Nevertheless, the available evidence suggests that integron-mediated recombination will conform to the tyrosine recombinase paradigm, although it may use special features to allow it to recombine pairs of sequences that are more diverged in primary sequence than the typical tyrosine recombinase and, in particular, contain heterologous central regions. In the subsequent presentation we attempt to fit the experimental data on integrons and their cassettes into the tyrosine recombinase mechanism paradigm.

Components of the Integron-Gene Cassette Recombination System

Integrons and their recombinases

Integrons are found in a variety of genetic contexts. Some are chromosomal, whereas others are located within transposable elements and/or conjugative plasmids. However, the precise genetic surroundings of most integrons have not been determined; consequently, unlike transposable elements, it is not possible to define integrons by virtue of specific physical boundaries or a precise sequence composition. Therefore, integrons have been defined in functional terms, as gene capture elements that encode the determinants for mobilization and expression of gene cassettes (36, 71, 82). These determinants are a recombinase gene, *intI*, an adjacent recombination site, *attI*, into which gene cassettes are integrated, and divergent promoters, located between *intI* and *attI*, one, P_I, for the expression of IntI and the other, P_c, for cassette gene expression, because most gene cassettes do not contain their own promoter (Fig. 1) (12, 19, 50). Each *intI* gene is associated with a unique *attI* site. Four IntI recombinases, sharing 42 to 58% amino acid sequence identity and similar in size (317 to 340 amino acid residues) to most other tyrosine recombinases, have been identified (Fig. 3A) (1, 16, 24, 58, 61, 64, 82; A. Pelletier and P. H. Roy, personal communication). They, like the best-characterized recombinase, Cre (see chapters 5, 30, and 31), are therefore expected to contain the two domains that form a C-shaped clamp around recombination site DNA. IntI1, IntI3, and *Vch* IntI have each been demonstrated to be functional recombinases, whereas IntI2 may be defective because the gene appears to be interrupted by a premature termination codon (Pelletier and Roy, personal communication). All IntI recombinases possess the features characteristic of the tyrosine recombinase family including the conserved catalytic residues, RKHRHY (24, 61). Consistent with findings for other family members, mutations in at least some of these residues in IntI1 reduce recombination in vivo to negligible levels (34, 45). These catalytic mutations give proteins that retain in vitro DNA-binding activity.

IntI recombinases contain a novel stretch of 20 to 22 amino acids, with a conserved motif of 15 predominantly nonpolar amino acids (Fig. 3B) (A. J. Holmes, personal communication). The function of this motif, present in the C-terminal catalytic domain, remains unclear. Phylogenetic analysis shows that the IntI recombinases are most similar to XerCD recombinases, Tn*4430* and Tn*5401* resolvases, and the resolvase of an uncharacterized transposon of *Weeksella zoohelcum* (G. D. Recchia, unpublished observations). Furthermore, whereas all tyrosine recombinases contain a catalytic K residue in a conserved β-hairpin turn (10, 14, 15, 46, 79, 84), the precise peptide sequence within which this is contained, although highly variable in the family as a whole, is highly conserved between the four IntIs, XerCD, and the Tn*5401*, Tn*4430*, and *W. zoohelcum* recombinases (GKGXKXR) (Fig. 3C). The loop containing the catalytic K is highly mobile and can be positioned in the minor groove close to the scissile phosphate (10, 84). The K is implicated as the acid in the first catalytic step, helping to protonate the 5′-OH leaving group (46). Because of these similarities, we view these recombinases as a functionally related subfamily of tyrosine recombinases that recognize and act on a recombination core site of related sequence (for example, see Fig. 5B).

Integrons are divided into types on the basis of the sequence of the recombinase, although all are likely to function in a similar way. All members of a given integron type encode the same IntI protein and include the same *attI* site sequence. They differ only in the genetic context in which they are located and in the cassettes they encode. Members of a single type may have slight variations in the sequence surrounding the P_c promoter. The most common and widely studied type of integron, the type 1 integron (sometimes referred to as the class 1 integron), encodes the IntI1 recombinase and contains the *attI1* recombination site. Integrons of this type are widespread in clinical isolates, are typically located within transposons or on conjugative plasmids, and contain one or more antibiotic resistance gene cassettes (37). Most experimental evidence, outlined below, is derived from studies of type 1 integrons. Type 2 integrons (encoding IntI2 and *attI2*) are only found in Tn7 and its close relatives (89; Pelletier and Roy, personal communication). Type 3 integrons have to date only been detected in clinical isolates from Japan (1, 77). The fourth characterized integron type is a chromosomal "superintegron" containing more than 179 cassettes found in the genome of *Vibrio cholerae* (*Vch* integron, previously designated class 4) (16, 58). Recently, more than a dozen more superintegrons in the genomes of a variety of species have been detected (53, 58, 71, 90; D. A. Rowe-Magnus and D. Mazel, personal communication).

Gene cassettes and their mobility

Gene cassettes usually include a single gene with very little extraneous noncoding DNA. Most include a suitable ribosome binding site, but no functional promoter. The specific recombination site, *attC*, associated with a particular gene cassette is unique to that

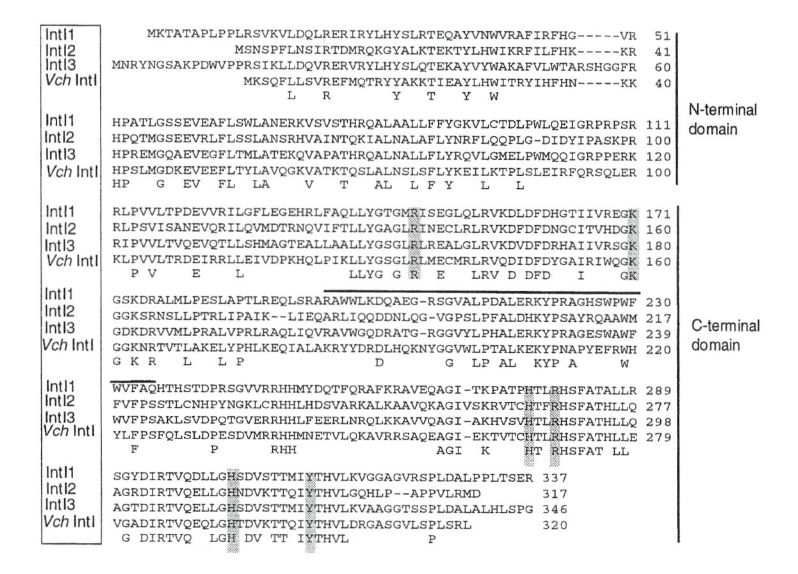

B

IntI1	RAWWLKDQAEGRSGVALPDALERKYPRAGHSWPWFWVFAQ
IntI2	RLIQQDDNLQG.VGPSLPFALDHKYPSAYRQAAWMFVFPS
IntI3	RAVWGQDRATGRGGVYLPHALERKYPRAGESWAWFWVFPS
Vch IntI	RYYDRDLHQKNYGGVWLPTALKEKYPNAPYEFRWHYLFPS
Eco XerD	RPWLLNGVSI DVLFPS
Tn*4430* TnpI	KTYSTAHES PYLFIS
Tn*5401* TnpI	EISQ SYLFPS
λ Int	KEILGG ETIIAS
Cre	GVADD PNNYLFC

C

Eco XerC	G	**K**	G	S	K	E	R
Eco XerD	G	**K**	G	N	K	E	R
IntI1	G	**K**	G	S	K	D	R
IntI2	G	**K**	G	G	K	S	R
IntI3	G	**K**	G	D	K	D	R
Vch IntI	G	**K**	G	G	K	N	R
Tn*5401* TnpI	G	**K**	R	N	K	Q	R
Tn*4430* TnpI	G	**K**	G	G	K	Q	R
Wzo TnpI	G	**K**	G	K	K	D	R

Figure 3. Sequences of integron-encoded recombinases. (A) Alignment of four characterized IntI proteins. Conserved amino acid residues are shown below the alignment. The conserved catalytic RKHRHY residues of tyrosine recombinases and type IB topoisomerases are shaded. The horizontal line above the IntI1 sequence indicates that region shown in panel B. Sequences are derived from NCBI entries AAB59081 (IntI1), L10818 (IntI2), BAA08929 (IntI3), and AAD53319 (*Vch* IntI, previously IntI4). The complete IntI2 sequence is obtained by removing an internal premature termination codon at position 179. (B) A conserved IntI-specific motif. Alignment of sequences from integron recombinases (highlighted in panel A) with corresponding regions of five other representative tyrosine recombinases: XerD from *E. coli* (P21891), TnpI from Tn*4430* (P10020), TnpI from Tn*5401* (AAA64588), phage λ Int (P03700), and the phage P1 encoded Cre (CAA27178). Amino acids completely conserved in IntI proteins but absent from other tyrosine recombinases (Holmes, personal communication) are shaded. This integron recombinase-specific motif lies at a position equivalent to that between α-helices I and J of *Eco* XerD (84). (C) Comparisons of recombinase sequences derived from sequence alignment in reference 61. Conserved residues are shaded. The sequences shown correspond to part of the conserved motif III of Xer recombinases which contains the catalytic lysine that is present in all tyrosine recombinases and type IB topoisomerases (15, 47).

cassette. Nevertheless, the conserved consensus sequences and overall organization are retained. Cassettes are usually integrated in only one of the two possible orientations within an integron, that orientation allowing for the expression of genes from P_c. This orientation specificity appears to be determined by polarity within the recombination sites, which determines how they align and recombine. Cassettes that rely on the integron promoter, P_c, for their phenotypic gene expression must integrate in an orientation compatible with expression if their mobilization is to be detected. Cassettes can also exist as free nonreplicative and transcriptionally silent, covalently closed circular molecules (17, 18).

Integration of a gene cassette into an integron frequently involves site-specific recombination between the *attI* site of an integron and the *attC* site of a free, circular cassette (see Fig. 1) (17). Cassette excision is the reverse of integration and can involve recombination between *attI* and *attC* or between two *attC* sites. Free simple or compound circular cassettes are regenerated by the excision reaction (18), and these are likely to be important intermediates in the mobility process. Cassettes may also be exchanged between integrons by a cointegration followed by resolution at a different recombination site (36). Table 1 lists the recombination events catalyzed by IntI1. Both inter- and intramolecular recombination events can occur without apparent preference for either resolution or integration. For cassette integration and excision, recombination between one *attI* site and one *attC* site or between two *attC* sites is the most biologically important reaction. In in vivo integration experiments, *attI* × *attC* recombination occurs at a higher frequency than *attC* × *attC* recombination (17, 66). Thus, in an integron with existing cassettes, an incoming cassette is most likely to be integrated immediately adjacent to the *attI* site, thereby ensuring maximal expression from the P_c promoter (Fig. 1). Recombination between two *attI* sites also occurs (42, 65, 66), although at a substantially lower frequency.

In addition to recombining *attI* and *attC* sites, IntI1 can also recombine, at low efficiency, these sites with secondary sites, which share very little sequence identity with either *attI* or *attC*. More than 50 secondary site sequences have been sequenced (26–28, 66, 70). Recombination between one specific site and one secondary site can give rise to novel biologically significant acquisition of gene cassettes by genomes that do not possess an integron. For example, an integron in one plasmid can supply the recombinase required for the excision of a cassette from that integron and its subsequent insertion into either the host chromosome or another resident plasmid that does not contain an integron. Most cassette insertions outside the integron context are likely to be stable because the hybrid sites that flank the cassette after the insertion event would be inactive (69). Several antibiotic resistance gene cassettes inserted at nonspecific locations have been discovered (69, 76; H.-K. Young and A. Mathers, GenBank accession no. Z50805). In each case the cassette has inserted downstream of an existing promoter that drives expression of the cassette-associated genes.

Integron-mediated gene cassette mobilization is the most promiscuous site-specific recombination system known. An enormous range of specific recombination sites of different sequences, including hundreds of cassette-associated *attC* sites, at least four *attI* sites, and potentially hundreds of secondary sites, are recognized by at least four recombinase proteins.

Recombination site organization

The *attI* site. Although the four known *attI* sites share only limited sequence identity with each other, they appear to have the same general organization (Fig. 4) (20, 38). Each IntI recombinase appears to preferentially recognize its own cognate *attI* site (38). A region of 65 to 70 bp is required for maximal *attI1* site activity in reactions with *attC* in vivo (42, 65, 70). However, in *attI1* × *attI1* recombination one full site

Table 1. IntI1-catalyzed recombination events

Participating sites	Recombination events		References
	Integrative	Excisive	
attI × *attC*	Yes	Yes	17, 18, 38, 42, 56, 66, 70
attC × *attC*	Yes	Yes	18, 35, 56, 83
attI × *attI*	Yes	Unknown	42, 65, 66
attC × 2°	Yes[a]	Unknown	26–28, 66, 70
attI × 2°	Yes[a]	Unknown	42, 66

[a]In most cases it is unlikely that integrative recombination between *attI* or *attC* and a secondary (2°) site would regenerate a functional hybrid site capable of undergoing further rounds of IntI-mediated recombination (69).

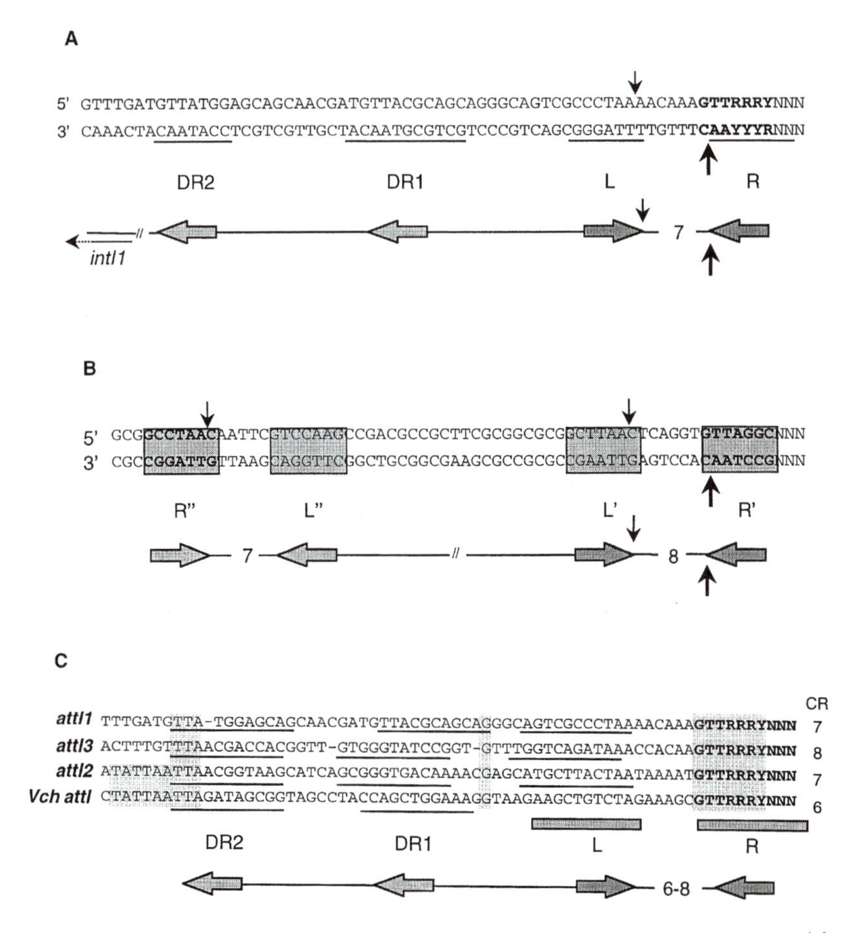

Figure 4. Recombination site organization. (A) Sequence of the *attI1* site (from type 1 integrons) required for efficient in vivo recombination. The 7-bp conserved sequence at the right end is shown in bold. The location of the normal recombination crossover position within this conserved sequence is marked by a thick vertical arrow. The thin vertical arrow indicates the location of recombination crossover events identified in rare recombinants and defines the second recombinase-mediated cleavage site. The extent of protected regions in methylation interference assays (33) is indicated by horizontal lines, corresponding to regions DR1 and DR2 and containing directly repeated accessory recombinase binding sites that contain the consensus DNA binding sequence (Fig. 5). The regions marked L and R represent recombinase binding sites within the core recombination site. Below the sequence is a schematic representation of *attI1* site organization. Recombinase binding sites are shown as shaded arrows. The number indicates the length of the predicted central region between inversely oriented IntI binding sites. The orientation of the *intI1* gene is also shown. Note that recombinase-mediated cleavage is presumed to occur on the strands and positions indicated by vertical arrows by analogy to the characterized tyrosine recombination systems and the known strand cleavage positions, 3′ of an A residue, in Xer recombination (78). (B) *attC* site organization. Sequence of the *aadB attC* sequence. The 7-bp inverted repeat conserved sequences found at each end are shown in bold. A major recombination crossover position is indicated by a thick vertical arrow. Alternative crossover positions in rare recombinants are indicated by thin vertical arrows. Sequences conserved between more than 60 *attC* sequences and representing the inner portion of putative IntI binding sites (83) are shaded. Invariably, the DNA sequence between L′ and L″ shows inverted repeat character. A schematic representation of *attC* organization is shown below. The break between L′ and L″ indicates that the length of this sequence can be longer than that of the *aadB attC* sequence shown here. Features are as in panel A. (C) Comparison of four known *attI* sequences. Regions of extensive conservation are highlighted, with recombinase binding sites indicated by underlining. A schematic of the proposed recombinase binding sites is shown below. CR indicates the central region. Sequences are derived from database entries referred to in Fig. 3A. Features are as in panel A.

can recombine with a minimal *attI1* partner consisting of only a putative core site (42). A complete site seems to consist of a recombination core site, containing two inverted recombinase binding sites (Fig. 4, L and R), separated by a 7-bp central region in *attI1* (*attI2, attI3,* and *Vch attI* appear to have, 7-, 8-, and 6-bp central regions, respectively [Fig. 4C]). Adjacent to the recombination core site are two more directly repeated recombinase binding sites, DR1 and DR2. We believe that recombinase binding to these provides an accessory function role. This organization is remarkably similar to that of the tyrosine recombination sites of Tn*4430,* Tn*5401,* and a *W. zoohelcum* transposon (Fig. 5) (8, 11, 52). In an IntI-mediated recombination reaction, recombinase binding site R is swapped with the equivalent site (R′) of *attC* as expected for a tyrosine recombinase reaction (35, 42, 70, 83).

In vitro gel shift assays using purified IntI1 and fragments containing the entire *attI1* region generate four protein-DNA complexes (20, 33). Analysis of these complexes by DNase I footprinting, methylation interference, and potassium permanganate interference reveals the presence of the four IntI binding regions (Fig. 4A). Site-directed mutagenesis of nucleotides within DR1 and DR2 reduces the levels of in vivo recombination with *attC*, demonstrating their importance in IntI-mediated recombination (38). Whether binding of four molecules of IntI1 is required for recombination in vitro remains to be determined.

The presumptive core site was identified by conserved sequence motifs and by identification of recombinant junctions from DNA sequencing studies (35, 70). A conserved 7-bp sequence of consensus GTTRRRY is found in all *attI* and *attC* sites. Point mutations within this sequence result in drastically reduced levels of recombination in vivo (42, 83). We believe that the G is part of the central region, whereas the other conserved residues appear to form part of the right-hand recombinase binding site, R (or R′ in *attC*). By analogy with the other tyrosine recombinases, we expect that recombinase-mediated cleavage occurs between the A and C on the opposite strand, the nucleotides complementary to the initial GT of the conserved sequence (Fig. 4; see below). The left-hand recombinase binding site (L) of the *attI* recombination core is less highly conserved than R; nevertheless, it retains the overall consensus in its inner portion. The inner nucleotides of the recombinase binding sites would be expected to be contacted by the recombinase N-terminal domain (30). Typical tyrosine recombinase binding sites consist of 11 to 13

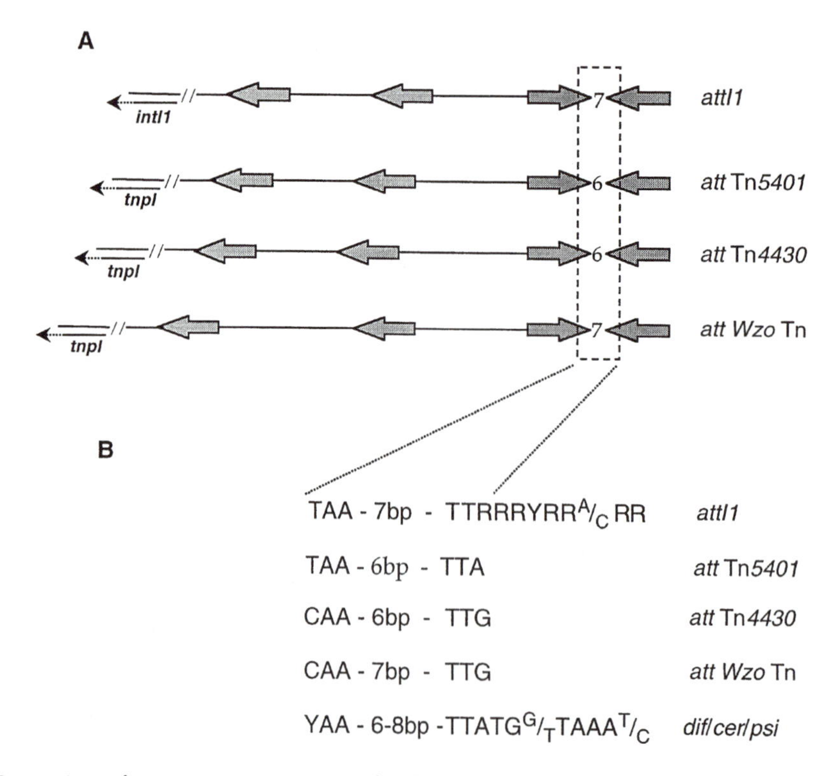

Figure 5. (A) Comparison of *attI1* site organization in related systems. Recombination site features are depicted as in Fig. 4. The relative orientations of the respective recombinase genes are shown. The Tn*5401* recombination site organization is derived from information in reference 8. Note that, based on analogy to characterized systems, the central region in this site is assumed to be 6 bp. The Tn*4430* recombination site organization is derived from reference 52 and V. Vanhooff and B. Hallet (personal communication). The length of central region in this site has been determined experimentally (Vanhooff and Hallet, personal communication). The uncharacterized recombination site found in a putative transposon from *W. zoohelcum* (11) was determined from analysis of the sequence in GenBank database entry U14952. (B) Comparison of the sequences immediately adjacent to the central regions (and hence strand cleavage positions) in the recombination sites shown in panel A with those of sites recognized by the XerC and XerD recombinases (*dif/cer/psi*) (78). The central regions of each site are AT rich. A conserved consensus for the presumptive complete 11-bp IntI recombinase binding sites in *attI* and *attC* is presented and compared with the remarkably similar sequence bound by XerD. The consensus is derived from the information in Fig. 4.

bp, the inner nucleotides of which at least show dyad symmetry, as here. The outer nucleotides appear to contact a DNA binding domain contained within the C-terminal domain of the protein, and at least in XerCD provide sequence recognition specificity (84). Examination of the available *attI* sequences (Fig. 4C) also reveals a well-conserved consensus for the outer parts of the IntI binding sites; this consensus is also present in *attC* sites (below) and is remarkably similar to the highly conserved XerD binding site (Fig. 5B). We therefore conclude that the complete *attI* binding site is a conserved sequence of 11 bp, an organization similar to that of other tyrosine recombinases.

The *attC* site. Recombination sites associated with cassettes found in multiresistant integrons (types 1, 2, and 3) have historically been referred to as "59-base elements" despite a substantial variation in length (35, 82). Those associated with the *Vch* super-integron have been designated VCRs (6, 16, 58). VCRs display a higher degree of conservation, in sequence and length, than 59-base elements, but despite differences, all of these sites share common features and all appear to be recognized by the same recombinases. For example, both IntI1 and IntI4 are able to catalyze recombination with 59-base and VCR elements (58). Therefore, we use the term *attC* to describe all gene cassette-associated recombination sites. The best-studied *attC* sites are those associated with cassettes encoding antibiotic resistance determinants. One such representative element, the *aadB attC* site, which has been the most extensively characterized and conforms closely to the *attC* consensus (83), is shown in Fig. 4B.

Characterized *attC* sites range from 57 to 141 bp in length and appear to contain a typical core recombination site plus two further inverted recombinase binding sites (Fig. 4B). A priori, one might consider this a second alternative core site or as two inverted accessory recombinase binding sites. The conserved left- and right-hand putative recombinase binding sites (L′ and R′, respectively) are separated by an 8-bp central region. The sequence of recombinant junctions supports the view that this corresponds to the central region (35, 66, 83). Mutations within the conserved GTT sequence (the G belonging to the central region and TT being part of R′) drastically reduce recombination activity in vivo, whereas changes to the AAC (AA being part of L′ and C part of the central region) reduce activity to a lesser extent (83). This core site appears to be able to recombine with *attI* sites having central regions of 6, 7, or 8 bp. The other pair of recombinase binding sites in *attC* may also very rarely be active as a core recombination site containing a 7-bp central region, because recombinant junctions between the putative recombinase binding site and central region have not often been found (Fig. 4) (83). We therefore tentatively label the binding sites as R″ for that which has the perfect consensus and L″. Use of this presumptive core would integrate the cassette in the opposite direction as compared with an integration event using the L′-R′ core; therefore, a measurement of the true frequency of recombination using this site would demand that the selective cassette marker be expressed independently of the normal integron promoter.

Preliminary in vivo studies suggest that both halves of the *attC* site, and thus all four putative IntI binding sites, are required for efficient *attC* activity (56), suggesting that even if both pairs of recombinase binding sites can act as core sites, the nonutilized core is providing accessory factor function in any given reaction. In vitro gel retardation experiments have shown only very weak binding of IntI1 to both *aadB attC* (20) and *aadA2 attC* (33). Two complexes were identified using the entire *aadB attC;* however, protein-DNA interactions appeared to be too weak to allow their detailed analysis (20). In *aadA2 attC,* a fragment containing only the left half of the site (containing R″ and L″) generated two weak complexes with IntI1 (33). The right half of an *attC* site presumably must also bind two molecules of IntI1 given our knowledge of the recombination reaction, but this still must be demonstrated. Putative recombinase binding sites within *attC* exhibit the same conserved consensus sequence for a complete 11-bp IntI binding site as those present in *attI* (Fig. 5B).

Despite considerable sequence divergence between sites, each *attC* sequence retains the ability to internally base pair and potentially form cruciform structures (35, 83). There has been much conjecture about whether this feature is important in the recombination reaction. Recently, Francia et al. (29) demonstrated binding of purified IntI1 to single-stranded *attC* DNA, but not to a double-stranded *attC* site. This was interpreted as suggesting a role for extruded *attC* cruciforms in the recombination mechanism (29). Extrusion of a cruciform to form a Holliday junction centered between L′ and L″ (Fig. 4B) would bring distant putative recombination half-sites into the branch-point position where they could be a substrate for IntI-mediated resolution, thereby generating two hairpin-containing molecules. It is unclear how these could be processed productively and what the role of these unprocessed or processed cruciforms would be. Therefore, for the moment, we confine our mechanistic models to those exhibited by the more characterized tyrosine recombinases.

The Recombination Reaction

IntI is the only protein that has been demonstrated to be required for integron-mediated, site-specific recombination (55). No recognition sites for DNA-binding proteins required in other systems such as integration host factor or factor for inversion stimulation (41, 48) have been identified adjacent to any IntI recombination site. Furthermore, there is no evidence for an integron-encoded excisionase, a contrast with some conjugative transposons (75). The fact that integrons are found in a diverse range of both gram-negative and gram-positive bacterial species is consistent with this system requiring no other cellular protein requirements for synapse assembly and strand exchange. If host accessory factors are required, then it is clear that they must be highly conserved through bacterial evolution.

Consistent with the lack of host accessory proteins is the apparent requirement of recombinase molecules as accessory proteins, in addition to those that execute breakage and rejoining at the recombination core site. The organization of the recombination sites and the observation that apparent accessory recombinase binding sites are not only present, but required for maximal recombination, indicates that they are a usual and necessary part of the recombination machinery. As noted earlier, strong similarities exist between the arrangement of recombinase binding sites in *attI1* and those of sites in several class II transposons involved in cointegrate resolution (Fig. 5A). Like *attI1*, these recombination sites in Tn*5401* (8), Tn*4430* (52), and an uncharacterized transposon of *W. zoohelcum* (11) are located adjacent to the tyrosine recombinase gene and in the same relative orientation. Furthermore, they each include two directly oriented recombinase binding sites (equivalent to DR1 and DR2 in *attI1*) upstream of the recombination core site (Fig. 5A). Closer examination of the core recombination site reveals further similarities between these sites and *attI1*. In each case the inner nucleotides of the recombinase binding site and the central region have the same consensus sequence (YAA–6- to 8-bp central region–TTR [Fig. 5B]). The central regions of each site are AT rich (not shown). The same consensus sequence is present in the Xer-specific recombination sites *dif, cer, psi,* and their relatives (78), and the full 11-bp XerD binding site has a consensus that is remarkably similar to that of *attI*.

In contrast to the strategy that appears to be used by the IntI recombinases in which the same DNA recognition sequences are used in both accessory sequences and the recombinase core site, some other tyrosine recombinase systems use accessory sites that bind additional recombinase molecules by using distinct DNA binding domains, for example, λ Int-*attP* (and its relatives), and the recombination sites used by the tyrosine recombinases of the conjugative transposons Tn*916* and Tn*1545* (48, 75). It remains to be seen whether the accessory sequences and proteins used in integron cassette movement and conjugative transposition function only to facilitate synapsis, or whether they also direct selectivity in recombination outcome, for example, by minimizing *attI* × *attI* recombination, or by promoting intermolecular recombination events preferentially.

An analysis of the available data makes us confident that IntI recombinase will initiate recombination and form a HJ intermediate, like the typical tyrosine recombinases. The likely core recombination sites and strand exchange positions are indicated in Fig. 4. The question then is whether the reaction halts at the HJ intermediate stage, as apparently for XerCD-mediated recombination at *cer* in vivo and in vitro (21, 59), leaving cellular proteins to produce recombinant product, or whether a second pair of strand exchanges occurs to form complete recombinant product. The Xer recombination system can stop at the HJ intermediate stage or progress to complete recombinant product depending on the nature of the recombination site (see chapter 8). Perhaps IntI-mediated recombination can act similarly.

As indicated earlier, during recombination catalyzed by most tyrosine recombinases, the central region sequences undergoing recombination are identical and introduction of heterology between the sequences can result in a dramatically reduced recombination frequency. Xer recombination is tolerant of mismatches in the recombinase binding sites, but introduction of heterology into the central region reduces the level of recombination (2, 3, 59; M. R. Robertson and D. J. Sherratt, unpublished observations). In a similar manner, in recombination mediated by Flp, central region heterology can impair strand cleavage or strand rejoining (4, 49, 92). In contrast, IntI-mediated recombination yields products at the same frequency irrespective of whether the presumptive central regions are homologous or heterologous (83). Nearly all *attC* sites differ in the sequence of their proposed central regions (83), and thus, in normal circumstances, IntI recombinases would seldom perform recombination between identical sites. The only other site-specific recombination system that appears to accommodate a similar level of nonhomology in the central region is that encoded by the conjugative transposons Tn*916* and Tn*1545* (75). Here, as with IntI recombination, the central regions (coupling sequences) of recombining sites are rarely identical. It has been proposed that a lack of requirement for partner site homology in these systems may make the HJ intermediate generated by a single-strand exchange a stable product, preventing its resolution back to substrate by the recombinase and instead allowing its resolution by some other mechanism (41, 59).

If central region heterology or some other feature of the IntI recombination system causes the recombination reaction to stop after HJ formation, the HJ intermediate could be replicated to form both substrate and recombinant product, an obvious advantage for a biological system whose function is to disseminate genes (83). Alternatively, the HJ intermediate could be resolved by a cellular protein, such as the HJ resolvase RuvC, after which the mismatch could be replicated or repaired. Note that RuvC-mediated cleavage of HJ intermediates formed by XerCD at *cer* does not appear to be a major pathway of resolution (59), and resolution through replication is currently favored (see chapter 8). A replication pathway can, in principle, readily process heterology in the central region of the HJ intermediate, with novel recombinant joints corresponding to just one of the central region boundaries being preferentially recovered, if reaction forms a HJ intermediate by the exchange of a specific pair of strands. Preferential recovery of a single recombinant joint is observed after IntI-mediated recombination (35, 66, 83), just as observed in the pioneering studies of Weisberg et al. (91) that analyzed the products of λ Int-mediated recombination in vivo between recombination sites containing heterologous central regions.

A mismatched recombinant can be generated if IntI catalyzes both pairs of strand exchanges at the predicted cleavage sites. Such mismatches could be replicated or repaired. There is no intrinsic barrier to generating mismatches during strand exchange mediated by tyrosine recombinase, even when the central regions are of different sizes, although the final rejoining step can be slow (2, 3, 85). Bias in the direction of any mismatch repair could also lead to apparent recovery of recombinant products having a single novel recombinant joint, despite recombinase-mediated cleavage at both ends of the central region. For example, if a second strand exchange step is initiated on the HJ intermediate, but rejoining is unfavorable, mismatch repair directed by limited 5′-OH exonucleolytic degradation of each of the mismatched central regions, followed by resynthesis, could occur to generate a product with a single novel joint corresponding to the position of the first pair of strand exchanges. Whether removal of the covalently bound protein by hydrolysis or esterification would be required to complete repair, or whether the repaired homologous 5′ terminus could attack the 3′-phosphoprotein to rejoin the DNA is not known.

Because the nucleotides immediately on either side of the presumed cleavage sites of *attI* and *attC* are conserved, we see no a priori reason why strand exchanges at either junction should not occur, raising the possibility that two pairs of strand exchanges can occur in a normal reaction between *attI* and *attC*. Nevertheless, heterology in the central region could limit the reaction to one pair of strand exchanges, and the strong purine bias in the strands of many of the presumptive central regions suggests that one pair of strand exchanges may be favored over the other (2). In conclusion, established tyrosine recombinase biology can be used to explain most if not all of the observations with integron-mediated site-specific recombination. Several pathways for which there are known precedents, particularly in the closely related Xer site-specific recombination system, can generate the observed products. It remains to be determined whether some or all of these pathways usually operate. It seems likely that the site-specific recombination systems used during conjugative transposition (see chapter 10) will share the same general features.

CONCLUDING REMARKS

The ideas discussed here provide a reasonable framework for understanding the IntI-mediated, site-specific recombination mechanism. Only further experimental data will confirm or refute these ideas. An understanding of the mechanistic details of IntI recombination has been hampered by the low occurrence of recombination and the lack of an in vitro system for studying the reaction. Nevertheless, experimental data from a range of tyrosine recombinase systems, and the data discussed here for integron-mediated recombination, reinforce our belief that the mechanism of mobilization of gene cassettes by integrons is fundamentally the same as that mediated by the more typical tyrosine recombinases. For example, tyrosine recombinases sometimes stall at the HJ intermediate stage, irrespective of the presence or absence of central region homology, providing an opportunity for resolution by replication. Even when central region heterology is present, they can sometimes complete both pairs of strand exchanges, although rejoining tends to be slow, which thereby provides an opportunity for directed mismatch repair. The similarity of integron-encoded recombinases with XerCD is interesting because XerCD seems to be good at stalling reaction at the HJ intermediate stage and at dealing with central region heterology. Because almost all bacteria with circular chromosomes possess XerCD homologs, we speculate that IntI genes of integrons may have evolved from XerCD-encoding genes. Indeed the ancestors of XerCD may have been the original tyrosine recombinases, functioning in ensuring effective duplication and segregation of chromosomes.

Acknowledgments. We thank Didier Mazel, Bernard Hallet, Paul Roy, and Andrew Holmes for sharing unpublished data and for helpful discussions. We also thank Gordon Churchward and Lars Sundström for helpful discussions.

G.D.R. has been supported by a CJ Martin Postdoctoral Fellowship from the National Health and Medical Research Council of Australia and by a research grant from the Australian Research Council. D.J.S. is supported by the Wellcome Trust.

REFERENCES

1. **Arakawa, Y., M. Murakami, K. Suzuki, H. Ito, R. Wacharotayankun, S. Ohsuka, N. Kato, and M. Ohta.** 1995. A novel integron-like element carrying the metallo-beta-lactamase gene *blaIMP*. *Antimicrob. Agents Chemother.* **39:**1612–1615.
2. **Arciszewska, L. K., R. A. Baker, B. Hallet, and D. J. Sherratt.** 2000. Coordinated control of XerC and XerD catalytic activities during Holliday junction resolution. *J. Mol. Biol.* **299:** 391–403.
3. **Arciszewska, L. K., and D. J. Sherratt.** 1995. Xer site-specific recombination in vitro. *EMBO J.* **14:**2112–2120.
4. **Azam, N., J. E. Dixon, and P. D. Sadowski.** 1997. Topological analysis of the role of homology in Flp-mediated recombination. *J. Biol. Chem.* **272:**8731–8738.
5. **Baggesen, D. L., D. Sandvang, and F. M. Aarestrup.** 2000. Characterization of *Salmonella enterica* serovar *typhimurium* DT104 isolated from Denmark and comparison with isolates from Europe and the United States. *J. Clin. Microbiol.* **38:** 1581–1586.
6. **Barker, A., C. A. Clark, and P. A. Manning.** 1994. Identification of VCR, a repeated sequence associated with a locus encoding a hemagglutinin in *Vibrio cholerae* O1. *J. Bacteriol.* **176:**5450–5458.
7. **Bass, L., C. A. Liebert, M. D. Lee, A. O. Summers, D. G. White, S. G. Thayer, and J. J. Maurer.** 1999. Incidence and characterization of integrons, genetic elements mediating multiple-drug resistance, in avian *Escherichia coli*. *Antimicrob. Agents Chemother.* **43:**2925–2929.
8. **Baum, J. A.** 1995. TnpI recombinase: identification of sites within Tn*5401* required for TnpI binding and site-specific recombination. *J. Bacteriol.* **177:**4036–4042.
9. **Berg, D. E.** 1989. Transposable elements in prokaryotes, p. 99–137. *In* S. B. Levy and R. V. Miller (ed.), *Gene Transfer in the Environment*. McGraw-Hill, New York, N.Y.
10. **Blakely, G. W., A. O. Davidson, and D. J. Sherratt.** 1997. Binding and cleavage of nicked substrates by site-specific recombinases XerC and XerD. *J. Mol. Biol.* **265:**30–39.
11. **Brassard, S., H. Paquet, and P. H. Roy.** 1995. A transposon-like sequence adjacent to the *Acc*I restriction-modification operon. *Gene* **157:**69–72.
12. **Bunny, K. L.** 1997. Expression of cassette-associated antibiotic resistance genes in class 1 integrons. Ph.D. dissertation. Macquarie University, Sydney, Australia.
13. **Cameron, F. H., D. J. Groot Obbink, V. P. Ackerman, and R. M. Hall.** 1986. Nucleotide sequence of the AAD(2″) aminoglycoside adenylyltransferase determinant *aadB*. Evolutionary relationship of this region with those surrounding *aadA* in R538-1 and *dhfrII* in R388. *Nucleic Acids Res.* **14:**8625–8635.
14. **Cao, Y., B. Hallet, D. J. Sherratt, and F. Hayes.** 1997. Structure-function correlations in the XerD site-specific recombinase revealed by pentapeptide scanning mutagenesis. *J. Mol. Biol.* **274:**39–53.
15. **Cao, Y., and Hayes, F.** 1999. A newly identified, essential catalytic residue in a critical secondary structure element in the integrase family of site-specific recombinases is conserved in a similar element in eucaryotic type IB topoisomerases. *J. Mol. Biol.* **289:**517–527.
16. **Clark, C. A., L. Purins, P. Kaewrakon, and P. A. Manning.** 1997. VCR repetitive sequence elements in the *Vibrio cholerae* chromosome constitute a megaintegron. *Mol. Microbiol.* **26:** 1137–1138.
17. **Collis, C. M., G. Grammaticopoulos, J. Briton, H. W. Stokes, and R. M. Hall.** 1993. Site-specific insertion of gene cassettes into integrons. *Mol. Microbiol.* **9:**41–52.
18. **Collis, C. M., and R. M. Hall.** 1992. Gene cassettes from the insert region of integrons are excised as covalently closed circles. *Mol. Microbiol.* **6:**2875–2885.
19. **Collis, C. M., and R. M. Hall.** 1995. Expression of antibiotic resistance genes in the integrated cassettes of integrons. *Antimicrob. Agents Chemother.* **39:**155–162.
20. **Collis, C. M., M.-J. Kim, H. W. Stokes, and R. M. Hall.** 1998. Binding of the purified integron DNA integrase IntI1 to integron- and cassette-associated recombination sites. *Mol. Microbiol.* **29:**477–490.
21. **Colloms, S. D., R. McCulloch, K. Grant, L. Nielson, and D. J. Sherratt.** 1996. Xer-mediated site-specific recombination in vitro. *EMBO J.* **15:**1172–1181.
22. **Davies, J.** 1994. Inactivation of antibiotics and the dissemination of resistance genes. *Science* **264:**375–382.
23. **Davies, J.** 1997. Origins, acquisition and dissemination of antibiotic resistance determinants. *Ciba Found. Symp.* **207:**15–27.
24. **Esposito, D., and J. J. Scocca.** 1997. The integrase family of tyrosine recombinases: evolution of a conserved active site domain. *Nucleic Acids Res.* **25:**3605–3614.
25. **Fluit, A. C., and F. J. Schmitz.** 1999. Class 1 integrons, gene cassettes, mobility, and epidemiology. *Eur. J. Clin. Microbiol. Infect. Dis.* **18:**761–770.
26. **Francia, M. V., P. Avila, F. de la Cruz, and J. M. García Lobo.** 1997. A hot spot in plasmid F for site-specific recombination mediated by Tn*21* integron integrase. *J. Bacteriol.* **179:** 4419–4425.
27. **Francia, M. V., F. de la Cruz, and J. M. García Lobo.** 1993. Secondary-sites for integration mediated by the Tn*21* integrase. *Mol. Microbiol.* **10:**823–828.
28. **Francia, M. V., and J. M. García Lobo.** 1996. Gene integration in the *Escherichia coli* chromosome mediated by Tn*21* integrase (Int21). *J. Bacteriol.* **178:**894–898.
29. **Francia, M. V., J. C. Zabala, F. de la Cruz, and J. M. García Lobo.** 1999. The IntI1 integron integrase preferentially binds single-stranded DNA of the *attC* site. *J. Bacteriol.* **181:** 6844–6849.
30. **Gopaul, D. N., F. Guo, and G. D. Van Duyne.** 1998. Structure of the Holliday junction intermediate in Cre-*loxP* site-specific recombination. *EMBO J.* **17:**4175–4187.
31. **Gopaul, D. N., and G. D. Van Duyne.** 1999. Structure and mechanism in site-specific recombination. *Curr. Opin. Struct. Biol.* **9:**14–20.
32. **Grainge, I., and Jayaram, M.** 1999. The integrase family of recombinases: organization and function of the active site. *Mol. Microbiol.* **33:**449–456.
33. **Gravel, A., B. Fournier, and P. H. Roy.** 1998. DNA complexes obtained with the integron integrase IntI1 at the *attI1* site. *Nucleic Acids Res.* **26:**4347–4355.
34. **Gravel, A., N. Messier, and P. H. Roy.** 1998. Point mutations in the integron integrase IntI1 that affect recombination and/or substrate recognition. *J. Bacteriol.* **180:**5437–5442.
35. **Hall, R. M., D. E. Brookes, and H. W. Stokes.** 1991. Site-specific insertion of genes into integrons: role of the 59-base

element and determination of the recombination cross-over point. *Mol. Microbiol.* **5**:1941–1959.

36. **Hall, R. M., and C. M. Collis.** 1995. Mobile gene cassettes and integrons: capture and spread of genes by site-specific recombination. *Mol. Microbiol.* **15**:593–600.
37. **Hall, R. M., and C. M. Collis.** 1998. Antibiotic resistance in gram-negative bacteria: the role of gene cassettes and integrons. *Drug Resist. Updates* **1**:109–119.
38. **Hall, R. M., C. M. Collis, M.-J. Kim, S. R. Partridge, G. D. Recchia, and H. W. Stokes.** 1999. Mobile gene cassettes and integrons in evolution. *Ann. NY Acad. Sci.* **870**:68–80.
39. **Hall, R. M., and C. Vockler.** 1987. The region of the IncN plasmid R46 coding for resistance to beta-lactam antibiotics, streptomycin/spectinomycin and sulphonamides is closely related to antibiotic resistance segments found in IncW plasmids and in Tn*21*-like transposons. *Nucleic Acids Res.* **15**: 7491–7501.
40. **Hallet, B., L. K. Arciszewska, and D. J. Sherratt.** 1999. Reciprocal control of catalysis by the tyrosine recombinases XerC and XerD: an enzymatic switch in site-specific recombination. *Mol. Cell* **4**:949–959.
41. **Hallet, B., and D. J. Sherratt.** 1997. Transposition and site-specific recombination: adapting DNA cut-and-paste mechanisms to a variety of genetic rearrangements. *FEMS Microbiol. Rev.* **21**:157–178.
42. **Hansson, K., O. Sköld, and L. Sundström.** 1997. Non-palindromic *attI* sites of integrons are capable of site-specific recombination with one another and with secondary targets. *Mol. Microbiol.* **26**:441–453.
43. **Hedges, R. W., and A. Jacob.** 1974. Transposition of ampicillin resistance from RP4 to other replicons. *Mol. Gen. Genet.* **132**:31–40.
44. **Jones, M. E., E. Peters, A.-M. Weersink, A. Fluit, and J. Verhoef.** 1997. Widespread occurrence of integrons causing multiple antibiotic resistance in bacteria. *Lancet* **349**:1742–1743.
45. **Kim, M.-J.** 1999. Class 1 and class 3 integron-associated site-specific recombination systems. Ph.D. dissertation. Macquarie University, Sydney, Australia.
46. **Krogh, B. O., and S. Shuman.** 2000. Catalytic mechanism of DNA topoisomerase IB. *Mol. Cell* **5**:1035–1041.
47. **Krogh, B. O., and S. Shuman.** 2000. DNA strand transfer catalyzed by vaccinia topoisomerase: peroxidolysis and hydroxylaminolysis of the covalent protein-DNA intermediate. *Biochemistry* **39**:6422–6432.
48. **Landy, A.** 1989. Dynamic, structural, and regulatory aspects of lambda site-specific recombination. *Annu. Rev. Biochem.* **58**:913–949.
49. **Lee, J., and M. Jayaram.** 1995. Role of partner homology in DNA recombination. *J. Biol. Chem.* **270**:4042–4052.
50. **Levesque, C., S. Brassard, J. Lapointe, and P. H. Roy.** 1994. Diversity and relative strength of tandem promoters for the antibiotic-resistance genes of several integrons. *Gene* **142**: 49–54.
51. **Liebert, C. A., R. M. Hall, and A. O. Summers.** 1999. Transposon Tn*21*, flagship of the floating genome. *Microbiol. Mol. Biol. Rev.* **63**:507–522.
52. **Mahillon, J., and D. Lereclus.** 1988. Structural and functional analysis of Tn*4430*: identification of an integrase-like protein involved in the co-integrate resolution process. *EMBO J.* **7**: 1515–1526.
53. **Manning, P. A., C. A. Clark, and T. Focareta.** 1999. Gene capture in *Vibrio cholerae*. *Trends Microbiol.* **7**:93–95.
54. **Martin, C., J. Timm, J. Rauzier, R. Gomez-Lus, J. Davies, and B. Gicquel.** 1990. Transposition of an antibiotic resistance element in mycobacteria. *Nature* **345**:739–743.
55. **Martinez, E., and F. de la Cruz.** 1988. Transposon Tn*21* encodes a RecA-independent site-specific integration system. *Mol. Gen. Genet.* **211**:320–325.
56. **Martinez, E., and F. de la Cruz.** 1990. Genetic elements involved in Tn*21* site-specific integration, a novel mechanism for the dissemination of antibiotic resistance genes. *EMBO J.* **9**:1275–1281.
57. **Mazel, D., and J. Davies.** 1999. Antibiotic resistance in microbes. *Cell. Mol. Life Sci.* **56**:742–754.
58. **Mazel, D., B. Dychinco, V. A. Webb, and J. Davies.** 1998. A distinctive class of integron in the *Vibrio cholerae* genome. *Science* **280**:605–608.
59. **McCulloch, R., L. W. Coggins, S. D. Colloms, and D. J. Sherratt.** 1994. Xer-mediated site-specific recombination at *cer* generates Holliday junctions in vivo. *EMBO J.* **13**:1844–1855.
60. **Nesvera, J., J. Hochmannova, and M. Patek.** 1998. An integron of class 1 is present on the plasmid pCG4 from gram-positive bacterium *Corynebacterium glutamicum*. *FEMS Microbiol. Lett.* **169**:391–395.
61. **Nunes-Düby, S. E., H. J. Kwon, R. S. Tirumalai, T. Ellenberger, and A. Landy.** 1998. Similarities and differences among 105 members of the Int family of site-specific recombinases. *Nucleic Acids Res.* **26**:391–406.
62. **Ochman, H., J. G. Lawrence, and E. A. Groisman.** 2000. Lateral gene transfer and the nature of bacterial innovation. *Nature* **405**:299–304.
63. **Ouellette, M., L. Bissonnette, and P. H. Roy.** 1987. Precise insertion of antibiotic resistance determinants into Tn*21*-like transposons: nucleotide sequence of the OXA-1 beta-lactamase gene. *Proc. Natl. Acad. Sci. USA* **84**:7378–7382.
64. **Ouellette, M., and P. H. Roy.** 1987. Homology of ORFs from Tn*2603* and from R46 to site-specific recombinases. *Nucleic Acids Res.* **15**:10055.
65. **Partridge, S. R., G. D. Recchia, C. Scaramuzzi, C. M. Collis, H. W. Stokes, and R. M. Hall.** 2000. Definition of the *attI1* site of class 1 integrons. *Microbiology* **146**:2855–2864.
66. **Recchia, G. D.** 1996. Mobile gene cassettes and integrons: evolutionary and recombinational studies. Ph.D. dissertation. Macquarie University, Sydney, Australia.
67. **Recchia, G. D., M. Aroyo, D. Wolf, G. Blakely, and D. J. Sherratt.** 1999. FtsK-dependent and -independent pathways of Xer site-specific recombination. *EMBO J.* **18**:5724–5734.
68. **Recchia, G. D., and R. M. Hall.** 1995. Gene cassettes: a new class of mobile element. *Microbiology* **141**:3015–3027.
69. **Recchia, G. D., and R. M. Hall.** 1995. Plasmid evolution by acquisition of mobile gene cassettes: plasmid pIE723 contains the *aadB* gene cassette precisely inserted at a secondary site in the incQ plasmid RSF1010. *Mol. Microbiol.* **15**:179–187.
70. **Recchia, G. D., H. W. Stokes, and R. M. Hall.** 1994. Characterisation of specific and secondary recombination sites recognised by the integron DNA integrase. *Nucleic Acids Res.* **22**: 2071–2078.
71. **Rowe-Magnus, D. A., A. M. Guerout, and D. Mazel.** 1999. Super-integrons. *Res. Microbiol.* **150**:641–651.
72. **Sallen, B., A. Rajoharison, S. Desvarenne, and C. Mabilat.** 1995. Molecular epidemiology of integron-associated antibiotic resistance genes in clinical isolates of enterobacteriaceae. *Microb. Drug Resist.* **1**:195–202.
73. **Salyers, A. A., N. B. Shoemaker, A. M. Stevens, and L. Y. Li.** 1995. Conjugative transposons: an unusual and diverse set of integrated gene transfer elements. *Microbiol. Rev.* **59**: 579–590.
74. **Schmidt, F. R., E. J. Nücken, and R. B. Henschke.** 1989. Structure and function of hot spots providing signals for site-directed specific recombination and gene expression in Tn*21* transposons. *Mol. Microbiol.* **3**:1545–1555.

75. **Scott, J. R., and G. G. Churchward.** 1995. Conjugative transposition. *Annu. Rev. Microbiol.* **49:**367–397.
76. **Segal, H., and B. G. Elisha.** 1997. Identification and characterization of an *aadB* gene cassette at a secondary site in a plasmid from *Acinetobacter. FEMS Microbiol. Lett.* **153:**321–326.
77. **Senda, K., Y. Arakawa, S. Ichiyama, K. Nakashima, H. Ito, S. Ohsuka, K. Shimokata, N. Kato, and M. Ohta.** 1996. PCR detection of metallo-beta-lactamase gene (*blaIMP*) in gram-negative rods resistant to broad-spectrum beta-lactams. *J. Clin. Microbiol.* **34:**2909–2913.
78. **Sherratt, D. J., L. Arciszewska, G. Blakely, S. Colloms, K. Grant, N. Leslie, and R. McCulloch.** 1995. Site-specific recombination and circular chromosome segregation. *Philos. Trans. R. Soc. Lond. B. Biol. Sci.* **347:**37–42.
79. **Sherratt, D. J., and D. B. Wigley.** 1998. Conserved themes but novel activities in recombinases and topoisomerases. *Cell* **93:** 149–152.
80. **Stark, W. M., and M. R. Boocock.** 1995. Topological selectivity in site-specific recombination, p. 101–129. *In* D. J. Sherratt (ed.), *Mobile Genetic Elements.* IRL Press, Oxford, United Kingdom.
81. **Steiner, W., G. Liu, W. D. Donachie, and P. Kuempel.** 1999. The cytoplasmic domain of FtsK protein is required for resolution of chromosome dimers. *Mol. Microbiol.* **31:**579–583.
82. **Stokes, H. W., and R. M. Hall.** 1989. A novel family of potentially mobile DNA elements encoding site-specific gene-integration functions: integrons. *Mol. Microbiol.* **3:**1669–1683.
83. **Stokes, H. W., D. B. O'Gorman, G. D. Recchia, M. Parsekhian, and R. M. Hall.** 1997. Structure and function of 59-base element recombination sites associated with mobile gene cassettes. *Mol. Microbiol.* **26:**731–745.
84. **Subramanya, H. S., L. K. Arciszewska, R. A. Baker, L. E. Bird, D. J. Sherratt, and D. B. Wigley.** 1997. Crystal structure of the site-specific recombinase, XerD. *EMBO J.* **16:**5178–5187.
85. **Summers, D. K.** 1989. Derivatives of ColE1 *cer* show altered topological specificity in site-specific recombination. *EMBO J.* **8:**309–315.
86. **Sunde, M., and H. Sørum.** 1999. Characterization of integrons in *Escherichia coli* of the normal intestinal flora of swine. *Microb. Drug Resist.* **5:**279–287.
87. **Sundström, L.** 1998. The potential of integrons and connected programmed rearrangements for mediating horizontal gene transfer. *APMIS Suppl.* **84:**37–42.
88. **Sundström, L., P. Rådström, G. Swedberg, and O. Sköld.** 1988. Site-specific recombination promotes linkage between trimethoprim- and sulfonamide resistance genes. Sequence characterization of *dhfrV* and *sulI* and a recombination active locus of Tn*21*. *Mol. Gen. Genet.* **213:**191–201.
89. **Sundström, L., P. H. Roy, and O. Sköld.** 1991. Site-specific insertion of three structural gene cassettes in transposon Tn*7*. *J. Bacteriol.* **173:**3025–3028.
90. **Vaisvila, R., R. D. Morgan, and E. A. Raleigh.** 1999. The *pacIR* gene resides within a potential superintegron in the *Pseudomonas alcaligenes* genome. IXth International Congress of Bacteriology and Applied Microbiology, Sydney, Australia.
91. **Weisberg, R. A., L. W. Enquist, C. Foeller, and A. Landy.** 1983. Role for DNA homology in site-specific recombination. The isolation and characterization of a site affinity mutant of coliphage lambda. *J. Mol. Biol.* **170:**319–342.
92. **Zhu, X.-D., G. Pan, K. Luetke, and P. D. Sadowski.** 1995. Homology requirements for ligation and strand exchange by the Flp recombinase. *J. Biol. Chem.* **270:**11646–11653.

Mobile DNA II
Edited by N. L. Craig et al.

Chapter 10

Conjugative Transposons and Related Mobile Elements

Gordon Churchward

Gordon Churchward • Department of Microbiology and Immunology, Emory University School of Medicine, Atlanta, GA 30322.

INTRODUCTION

Nomenclature

Conjugative transposons are genetic elements that move from the genome of a donor bacterium to the genome of a recipient bacterium by a process that requires intercellular contact. The first such elements were identified in the early 1980s (5, 22, 29, 84). The best-studied elements, the very closely related Tn*916* and Tn*1545* (16, 23), correspond to the accepted definition of a transposable element, because they move from one site in the genome of the donor to a different site in the genome of the recipient. However, as more related conjugal elements have been discovered and analyzed, it has become clear that many elements described as conjugative transposons integrate into unique target sites in the recipient that are identical with the site in which they reside in the donor, and thus do not necessarily correspond to the definition of a transposable element. Because of this phenomenon of site-specific integration, it has been proposed that such elements be called constins (an acronym for conjugative, self-transmissible, integrating element) (28). This would be a useful way of distinguishing between elements that integrate predominantly into single sites in their host's chromosome and elements such as Tn*916* that integrate into multiple sites and act as transposons.

To compound further the semantic confusion surrounding these elements, nearly all transposable elements use a transposase enzyme containing a DDE motif (19) that catalyzes cleavage and DNA strand transfer without the formation of covalent DNA-protein linkages. However, the movement of conjugative transposons and related elements is much more similar to the reactions catalyzed by site-specific recombinases. These enzymes form such linkages as intermediates in their cleavage and strand transfer reactions (see chapters 6 to 9, 11, and 12).

Scope

This chapter covers the conjugative transposons from gram-positive and gram-negative bacteria, beginning with a detailed description of the enterococcal and streptococcal transposons Tn*916* and Tn*1545* and the larger streptococcal element Tn*5253* and its components, Tn*5251* and Tn*5252*. This chapter also describes the conjugal transposons from *Lactococcus lactis* and *Bacteroides* and *Clostridium* spp. and mobilizable elements from *Bacteroides* and *Clostridium* spp. I then describe the element of *Vibrio cholerae* that encodes resistance to sulfamethoxazole and trimethoprim (SXT) that integrates into a unique target site in the genome of its host bacterium. I also detour into the world of integrating plasmids and the peculiar behavior of a class of elements originally described as conjugal plasmids assigned to the IncJ incompatibility group. Finally, I return to the gram-positive world and describe Tn*554* from *Staphylococcus aureus*.

Previous Reviews

The behavior of the conjugative transposons has been reviewed since the early 1980s. See those reviews for a more extensive discussion of the evidence relating to the behavior of these elements (10, 11, 13, 14, 40, 41, 56, 74, 76, 78, 79, 81, 90).

Structure

The general structure of these elements corresponds to that of other transposable elements, in the sense that the elements are either simple (composed of a single structural unit, e.g., Tn*916*) or complex (composed of units that are capable of independent movement, e.g., Tn*5253*). The smallest of the conjugative transposons is Tn*916*, 18.4 kb (21, 23). Thus, the conjugative transposons tend to be larger than many other transposable elements. They have a modular structure, containing gene segments required for DNA cleavage and strand transfer, conjugation, and containing antibiotic resistance or other accessory genes. They extend in size to upwards of 100 kb. The complete nucleotide sequence has not been reported for any of the larger elements, and so it is not clear what other functions they might encode.

Most conjugative transposons encode recombinases that belong to the lambda integrase family, but some elements found in species of clostridia encode a resolvase family recombinase (2, 102). Some elements also carry an adjacent gene, *xis,* which encodes a small basic protein that is analogous to lambda Xis protein.

The conjugation genes occupy a discrete segment of the transposons. Conjugation seems to occur by a mechanism similar to that of the well-known conjugative plasmids of *Escherichia coli.* For many transposons, an *oriT* sequence that is distinct from the ends of the integrated transposon has been identified, as have genes related to the *mob* genes encoding endonuclease relaxases that nick the DNA at the origin of DNA transfer (35). Other genes with similar functions, including those involved in partitioning DNA into bacterial cells or endospores and those conferring protection from restriction endonuclease cleavage, are also present (21).

General Mechanism of Movement

As described in more detail below, conjugative transposons and related elements move via a circular intermediate that is produced by excision of the integrated element from the donor chromosome (82). This circular intermediate then serves as a substrate for conjugation, and most likely a single DNA strand of the intermediate is transferred from donor to recipient with transfer initiating at *oriT* (80). Once in the recipient, the circular intermediate re-forms, and then integrates into the recipient genome.

Tn*916* AND Tn*1545*

Tn*916* and Tn*1545* were discovered in *Enterococcus faecalis* and *Streptococcus pneumoniae,* respectively (16, 23). Like many other conjugative transposons, they confer resistance to tetracycline upon their hosts through a *tet*M determinant that encodes a ribosome-binding protein. Tn*1545* also confers kanamycin and erythromycin resistance. These transposons are remarkably promiscuous. They have an exceptionally broad host range; they are able to transfer between many different gram-positive bacterial species and between gram-positive and gram-negative bacteria (3, 66). They encode putative antirestriction functions and possess a paucity of restriction enzyme cleavage sites (21). The *tet*M determinant is expressed in both gram-positive and gram-negative hosts. In addition, these elements do not confer immunity (in the sense of prophage immunity) upon their hosts. The presence of a resident element does not impede the entry and establishment of a second related element (61).

Structure

The genetic structure of Tn*916* is shown in Fig. 1 (21). The *int* and *xis* genes are located at the left end of the transposon, whereas the genes thought to be involved in conjugal transfer, *orf23* through *orf13,* are located at the right end (83). *orf23* encodes the putative Mob protein. With one exception (*orf9*) all genes are oriented from right to left. The origin of transfer, *oriT,* defined as a DNA sequence that, when cloned onto a plasmid, allows the plasmid to be mobilized in *trans* by Tn*916*, is located between *orf20* and *orf21* (32). The *tetM* gene is transcribed from P*tet* from right to left as shown (8, 95). The transcription of *tet*M is induced by tetracycline through an antitermination mechanism. Transcription from the *tet* promoter continues through *orf7* and *orf8,* which express two gene products that activate the downstream *orf7* promoter. Transcription from P*orf7* continues through the *xis* and *int* genes and, when the transposon is excised and is in its circular form, through the *tra* genes (8). There are also two other transcriptional promoters, P*xis* (8, 47) and P*int* (47), upstream of the *int* and *xis* genes, respectively.

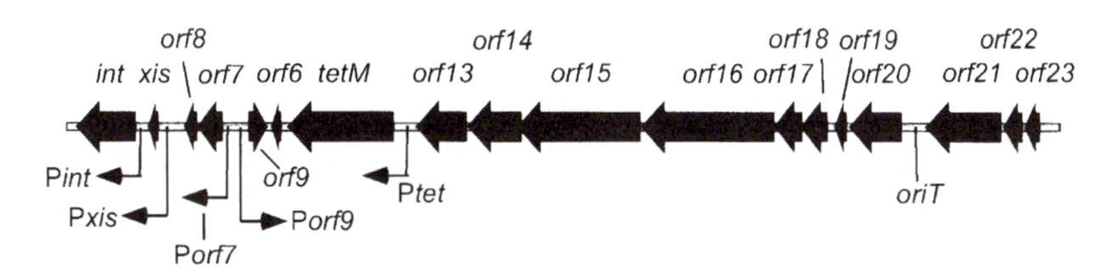

Figure 1. Genetic map of Tn*916*. The thin gray line represents transposon DNA and the thick black arrows represent open reading frames. Four short reading frames are not shown: *orf24,* to the right of *orf23; orf12,* to the right of *tetM; orf10,* to the right of *orf7;* and *orf5,* to the right of *xis*. The position of the origin of conjugal DNA transfer, *oriT,* is indicated between *orf21* and *orf20*. The thin arrows indicate five promoters and their direction of transcription.

Excision

Excision of Tn*916* requires Int and Xis (47, 65, 71, 94), although excision in the absence of Xis has been reported in *E. coli* (65). No host factors have been required for excision, and an excised transposon circle can be identified in vitro in reactions using a plasmid DNA substrate carrying the ends of Tn*916* flanking an antibiotic resistance marker and purified Int and Xis proteins (71). Int, like other similar recombinases, makes staggered cleavages at the ends of the element, which results in the formation of free 5′-OH groups and 3′-phosphotyrosine DNA-protein intermediates (43, 96). These cleavages are usually 6 bp apart (72). There is some variability in the site of cleavage at the right end of the element, with cleavage leaving either four or five T's (6). The resulting 5′-OH groups can then attack phosphotyrosine linkages at the other end of the element to promote DNA strand exchange.

Integrase Binding

Tn*916* Int is a bivalent DNA-binding protein that recognizes two distinct DNA sequences (38). The position of the Int binding sites is shown in Fig. 2. The C-terminal domain of Int, which binds to the ends of the transposon and flanking host DNA, contains the catalytic residues responsible for DNA cleavage. Two C-terminal domains bind each end of Tn*916* (33, 38). The N-terminal domain binds to short repeated sequences (AAGTGGTAAT), which are named DR-2 repeats (12) and located 150 bases from the left end and 90 bases from the right end of the transposon (39, 105). With the N-terminal domain fused to maltose-binding protein, two types of complex have been identified, which contain either two or four domains bound to each set of repeats (33). Presumably, in vivo four molecules of Int bind the transposon ends and internal repeats in a manner to bring together the ends of the transposon, in a way similar to that of other integrase enzymes. The exact structure of this presumed complex, especially which domain of Int binds to which segment of DNA, is not yet known. The significance of the Int N-terminal domain complexes containing four N-terminal domains is not clear. One difference between Tn*916* integrase and lambda integrase that may be significant for the formation of a recombination complex is that the apparent affinity of the C- and N-terminal domains is similar for Tn*916* integrase (38), whereas the apparent affinity of the C-terminal domain is much lower than that of the N-terminal domain for lambda integrase (48).

Xis Binding

Xis binds to sequences adjacent to the Int N-terminal domain repeats at both ends of the transposon (73) as shown in Fig. 2. The segment of DNA containing the Xis and Int binding sites from the right end of the transposon is bent, and Xis binding induces further bending of the DNA (D. Hinerfeld and G. Churchward, unpublished results). At the left end, where the Xis binding site is between the C- and N-terminal Int binding sites, such bending, if it occurs, could facilitate the binding of a single Int molecule forming a short loop. Mutations in the Xis binding site at the left end result in reduced levels of excision in vivo (25b). The situation at the right end of the transposon is different because the Xis binding site lies further to the interior of the transposon than the Int binding sites. The role of Xis binding at the right end of Tn*916* is discussed further below.

Mechanism of Excision

Our present model for Tn*916* excision is shown in Fig. 3. Excision of Tn*916* results in a circular intermediate form of the transposon where the transposon ends are separated by six nucleotides (6), reflecting the staggered cleavages made by Int. What sets Tn*916* excision apart from other integrase-catalyzed reactions is that the 6-bp regions between the transposon

Figure 2. Binding sites for Int and Xis at the ends of Tn*916*. The thick black line represents DNA from each end of the transposon. The diamonds labeled Int-C show where the C-terminal domains of Int bind to the transposon ends and flanking bacterial DNA. The black triangles labeled DR-2 and Int-N show the positions and relative orientation of binding sites for the N-terminal domain of integrase. The open triangles labeled Xis show the positions and relative orientation of binding sites for Xis.

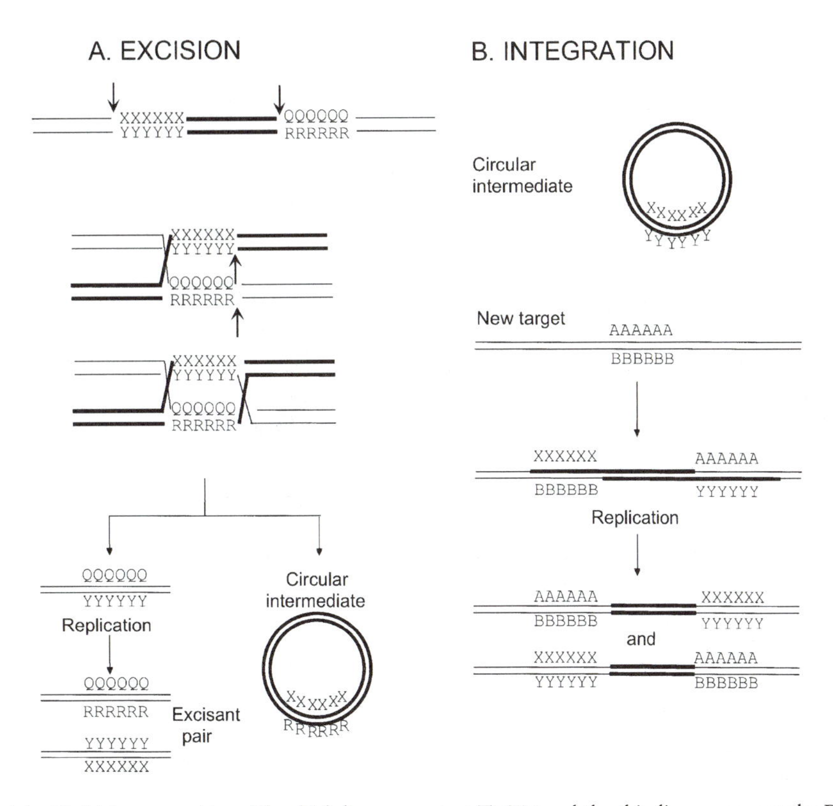

Figure 3. Model for Tn*916* transposition. The thick lines represent Tn*916* and the thin lines represent the DNA adjacent to the transposon. Coupling sequences are indicated by the hypothetical nucleotide pairs X-Y, Q-R, and A-B. (A) Cleavage of one DNA strand at each end of the transposon on the 5′ side of the coupling sequence, indicated by the vertical arrows, followed by DNA strand exchange leads to the formation of a Holliday junction intermediate. A second round of cleavage and DNA strand exchange results in the formation of an excised circular intermediate form of the transposon that contains a heteroduplex region formed from the base pairs originally present in the coupling sequences flanking the transposon in the donor. The reciprocal product can be processed by DNA replication to yield a pair of excisant molecules, each carrying one of the coupling sequences originally flanking the transposon. (B) After introduction of a single strand of the circular intermediate form of the transposon into the recipient, the complementary strand is synthesized to form a new intermediate with only one of the coupling sequences originally flanking the transposon in the donor. A recombination event similar to that shown in panel A results in the transposon being integrated into the recipient DNA where it is flanked on each side by a heteroduplex region composed of coupling sequence and target DNA. After replication, two DNA molecules are produced with sequences from the circular form of the transposon at either the left or the right of the integrated transposon.

ends, termed coupling sequences (equivalent to core or overlap sequences in other systems), can be heteroduplex in the excised transposon (6). These coupling sequences are nearly always different in the recombining transposon ends during excision or between the integrating transposon and its target. This is because the 6-bp coupling sequences flanking the ends of the transposon need not be homologous for recombination to occur, and in fact rarely are, because homology is not required when the transposon integrates (see below). We assume, though this has not been demonstrated experimentally, that excision involves two ordered rounds of cleavage and DNA strand exchange with the formation of a Holliday junction intermediate, as has been demonstrated for other similar recombinases (see chapters 6 to 9, 11, and 12).

Integration

Integration of Tn*916* requires Int but does not require Xis (47, 94), and, unlike lambda Xis, Tn*916* Xis does not inhibit integration in the recipient (36, 47). Integration is assumed to proceed in a manner similar to excision. As shown in Fig. 3, when Tn*916* enters a recipient by conjugation, the coupling sequence that is present at the left end of the transposon in the donor appears at either end of the integrated transposon in the recipients. This is evidence that only a single strand of the excised transposon circle carrying heteroduplex coupling sequences enters the recipient (80), where the complementary transposon strand is synthesized. A similar result would be obtained if directed mismatch repair of the heteroduplex occurred (44). In experiments in which a plasmid-expressing Int is used to complement an int^- transposon, it appears that a functional *int* gene is required only in the donor bacteria (4). This suggests that Int might be transferred from donor to recipient during conjugative transposition. There is no evidence for any process similar to transposition immunity, and a single recipient can harbor multiple transposon insertions (60). Whether these arise from a single recipient mating with several donors or are due to the introduction of several copies of the transposon from a single donor (see below) is not known.

Target Sites

Clear preferences for particular target sites exist, although the factors that make a particular site a preferred target remain unclear. In general, target sites are similar to the joined transposon ends of the circular intermediate, consisting of tracts of A's and tracts of T's separated by a variable sequence of six nucleotides, corresponding to the coupling sequences (30, 80). In one case, a sequence that in one orientation was identical with the joined transposon ends has acted as a preferred target when cloned into a plasmid, but insertions occur in both orientations and the sequence is not a preferred target when introduced into the bacterial chromosome (99). Int binds to preferred target sites with a higher affinity, but the differences in binding affinity are much smaller than the differences in frequency of target site utilization (39). There is no evidence that the transposon prefers to integrate into a primary target much more frequently than into other targets. No behavior analogous to that of the site affinity mutants of phage lambda (104) has been observed for Tn*916*. There is also no evidence that alterations in the coupling sequences lead to increases in preference for a particular target site. This is in marked contrast to the situation with phage λ, where a mutation in the core (coupling sequence) region can result in a switch in preference for integration (104).

Conjugal Transfer

As stated above, conjugal transfer seems to occur by a mechanism similar to that of the conjugal plasmids of enteric bacteria (32, 80). The *oriT* sequence shows similarities to the sites at which the Mob proteins of the F and R6K plasmids nick their conjugate plasmid *oriT* sequences. As described above, analysis of the inheritance of coupling sequences flanking the transposon in the donor bacterium suggests that only a single strand of the excised circular transposon is transferred into the recipient (80). If DNA transfer does occur by a mechanism similar to those described for the conjugal plasmids, then the introduction of multiple copies of the transposon into recipient bacteria could occur by rolling circle replication of the circular transposon to produce multimers. Two of the *orf* genes in the region of the transposon that are required for conjugative transposition, *orf13* and *orf15*, have been shown, when analyzed in *E. coli*, to contain membrane-spanning domains, and so are candidates for components of a hypothetical mating pore (J. B. Jewell, S. Loe, and R. E. Andrews, presentation, 1st International ASM Conference on Enterococci: Pathogenesis, Biology and Antibiotic Resistance, Banff, Alberta, Canada, 2000). There is also an *orf* gene, *orf21*, that is similar to the genes encoding the SpoIIIE and FtsK proteins, and the product of this gene might be involved in translocation of DNA from donor to recipient.

Gene Expression

Northern blot analysis shows that a transcript that begins at the *orf7* promoter can extend through *xis* and *int* and through all the transfer genes (8). Extension through the transfer genes depends on the presence of a functional *int* gene, implying that the transfer genes are only expressed when the transposon excises. The mechanism of transcriptional activation of this promoter is not known at present, but requires the products of *orf7* and *orf8*. These can be expressed from a tetracycline-inducible promoter upstream of the *tetM* gene, and tetracycline has been reported to cause modest increases in the frequency of conjugative transposition of Tn*916* (8, 18, 88). The *orf7* promoter is repressed by the product of *orf9*. Although Int and Xis can presumably be expressed from the *orf7* promoter, there is a functional promoter upstream of both *xis* and *int* (8, 47). The *xis* promoter does not appear to be regulated by any Tn*916* function, but recent evidence suggests that the *int* promoter may be up-regulated by a copy of Tn*916* in *trans* (D. Muller and G. Churchward, unpublished results).

Regulation of Conjugative Transposition

Conjugative transposition of Tn*916* is a relatively rare event, typically occurring at frequencies below 10^{-4} per donor bacterium. So far, only two factors have been reported to increase this frequency. One is addition of tetracycline, but increases in conjugative transposition frequency attributable to tetracycline are modest, and do not compare with the levels of induction of conjugative transposition of the *Bacteroides* conjugative transposons and related elements by tetracycline (8, 88, 93). The second is the nature of the coupling sequences flanking Tn*916* (31). The behavior of several insertions of Tn*916* from different donors into the same target site was analyzed, and the conjugative transposition frequencies of these inserted transposons varied over several orders of magnitude. The only difference between the donor strains should have been the sequence of the coupling sequences. This implies that the coupling sequences can have a profound effect on the frequency of conjugative transposition. How the coupling sequences might exert this effect is not known, but it is not due to an effect on the affinity of transposon ends with different coupling sequences for Int binding (63). In addition to these reported factors that affect transposon excision, we have found that mutations in the Xis binding site in the right end of the transposon that reduce binding of Xis cause elevated levels of excision in vivo (25b). DNase I protection experiments show that Xis competes with the binding of the N-terminal domain of Int to its adjacent binding site (C. Rudy, J. R. Scott, and G. Churchward, unpublished results). Thus, Xis not only is required for transposon excision but also serves to regulate the process.

Possible Regulatory Mechanism

Conjugative transposition consists of a minimum of three steps: excision, conjugal DNA transfer, and integration. One might suppose that all regulation occurs at the stage of excision (45) and is transcriptional, so that expression of Int and Xis leads to excision, which, in turn, results in expression of transfer functions leading to conjugation. If this were true, then one might expect that expression of Int and Xis would only occur in that small minority of cells that act as conjugal donors of Tn*916*. However, the *int, xis,* and *orf7* promoters are active in all cells (D. Muller and G. Churchward, unpublished results), yet excision is a rare event, making it unlikely that this simple model can account for regulation of Tn*916* transposition. Regulation of conjugal transposition may also occur at the level of conjugation. When Int and Xis are expressed from an efficient, exogenous promoter, excision occurs at a high level without a concomitant increase in conjugal transfer of Tn*916*, indicating that excision is necessary but not sufficient for conjugal transfer to occur (46). Markers in *cis* to the transposon are not mobilized (20), in agreement with the model that excision must occur before transfer can occur. However, it seems unlikely that the only mechanism that prevents the occurrence of Hfr-type transfer or mobilization of the plasmid on which Tn*916* resides before transposon excision is a requirement that the transposon ends be joined so that transcription of the transfer genes can occur. The transposon can insert into different regions, and if it transposes into a gene, then the transfer genes could be transcribed from an upstream, external promoter. However, we have recently observed that Int binds specifically to *oriT* (25a). If Int is present in all cells carrying the transposon, this binding possibly prevents transfer until the repression is relieved by formation of a recombination complex competent for excision. Such a mechanism would explain the coordinate activation that has been observed of multiple copies of Tn*916* present in a single donor cell (20).

Tn*5253*: A COMPOSITE OF Tn*5251* AND Tn*5252*

Tn*5253* was identified as an element conferring resistance to chloramphenicol and tetracycline in *S. pneumoniae* that was capable of conjugal transfer to a number of streptococci (5). Tn*5253* was subsequently found to be a composite of two elements, Tn*5251* and Tn*5252* (1, 67). Tn*5251* carries a *tet*M determinant, has a size similar to Tn*916*, and is almost identical in sequence with Tn*916* over a 4-kb region. Tn*5252* is a larger element (45 kb) and exemplifies the larger conjugative transposons found in gram-positive bacteria. Part of Tn*5252* has been sequenced, and it carries *int* and *xis* genes close to one end, a transfer region containing an *oriT* sequence and a relaxase protein that nicks *oriT* in vitro, a DNA cytosine methylase, and an operon that confers UV resistance on the host bacteria (34, 52, 77, 92, 100).

Excision and Integration

Two genes termed *orf1* and *orf2* are found close to one end of Tn*5252* (100). *orf2* encodes a small basic protein and *orf1* encodes an integrase family recombinase. The two *orf* genes are in an orientation (with respect to the ends of the transposon) similar to that of the *int* and *xis* genes of Tn*916*. Tn*5252* has a preferred target site in the pneumococcal genome. Integration presumably follows the phage lambda paradigm, although too few integration events have been analyzed to assess the degree of similarity between Tn*5252* and Tn*916*.

Tn*5276*: A CONJUGAL TRANSPOSON FROM *L. LACTIS*

Tn*5276* is a 70-kb conjugative transposon identified in *L. lactis* that encodes the lantibiotic nisin, as well as genes conferring the ability to ferment sucrose (68). Mechanistically it appears to be similar to Tn*916* (70).

Excision and Integration

Tn*5276* encodes a lambda integrase-type recombinase and a Xis-like protein (69). Expression of both proteins is required for efficient excision to occur. The junctions of insertions of Tn*5276* into two sites in the lactococcal chromosome have been analyzed, and the coupling sequences contain the hexanucleotide TTTTTG (70). However, the length of the coupling sequences cannot be determined exactly from the DNA sequence information that is available, and there is evidence for heterogeneity in the coupling sequences. The inheritance of the different sequences is consistent with the formation of heteroduplexes during excision and integration.

CTnDOT AND THE LARGE *BACTEROIDES* CONJUGATIVE TRANSPOSONS

Strains of *Bacteroides* species carry several large (<50 kb) conjugative transposons that belong to at least two distinct families, the best example being CTnDOT and its relatives (76, 90). As with the Tn*916* family, many of these transposons confer resistance to tetracycline by expression of a ribosome-binding protein, although the determinant is *tet*Q rather than *tet*M in this case (76, 91). None of these large transposons has had its DNA sequence reported in its entirety, so there is no information about the nature of the functions that are encoded in these transposons. Three features set this family of transposons apart from the Tn*916* family: conjugative transposition is induced by a factor of 10^3 to 10^4 by tetracycline (93), they can mobilize many smaller elements referred to as mobilizable transposons and nonreplicating *Bacteroides* units (NBUs), and these mobilizable transposons can mobilize genetic markers in *cis*.

Genetic Organization

As with the Tn*916* family, the structure of these transposons appears to be modular. A gene close to one end of CTnDOT encodes a lambda family integrase flanked by an open reading frame (*orf2*) encoding a small basic protein (9). However, *orf2* is not required for excision. The transfer genes, including regulatory genes and *oriT,* occupy about 16 kb in the middle of the transposon (37).

Excision

Excision of CTnDOT requires the activity of the transposon-encoded integrase and seems to occur by a mechanism similar to that of Tn*916* (9). PCR amplification products from reactions using primers that flank the joined transposon ends show that there is a circular intermediate, and that these intermediates contain both coupling sequences flanking the integrated transposon. It is not known if the circular forms of CTnDOT contain heteroduplex coupling sequences. Cleavage occurs 5 bp apart for CTnDOT and 4 or 5 bp apart for the related element CTnERL, depending on the site of integration of the transposon. Another relative, CTnXBU4422, can apparently produce excision products from the same insertion site with either 4- or 5-bp coupling sequences. Unlike Tn*916, int* and the adjacent *orf2* are not sufficient for excision to occur. It is not known what other genes might be required for excision of CTnDOT.

Integration

Integration of CTnDOT elements requires the joined transposon ends and *int* alone (9). As for Tn*916*, after integration events following conjugal transfer of the transposon, a single coupling sequence from the donor appears in the recipient at either end of the integrated element, indicating that a single strand of the circular excised transposon probably is transferred from donor to recipient. In one case, involving intracellular transposition of CTnXBU4422, the same donor coupling sequence was found in the integrated element nine of ten times, and this was interpreted as meaning that excision and integration involve blunt-end cleavage of donor and target sequences (76). In fact, similar biases have been observed in recombination of lambda phages carrying different core (coupling) sequences (104), and the inheritance of the coupling sequences observed with intracellular transposition of CTnXBU4422 seems very similar to integron recombination (see chapter 9). These results can be reconciled in several ways, but one attractive hypothesis is that sometimes CTnXBU4422 Int initiates recombination but cannot process a Holliday junction intermediate because of sequence heterology in the coupling sequences. Resolution of the intermediate is performed by other means, for example, DNA replication. Sherratt and Recchia discuss the effects of DNA heterology on integrase-mediated recombination reactions in more detail in chapter 9.

Target Sites

The target sites for integration of CTnDOT are related, and all contain a 10-bp sequence, GTTNNTTTGC, that is also present at the right end of the transposon. Apparently this sequence is responsible for the preferential integration of CTnDOT into about 7 sites in the *Bacteroides* chromosome (9).

Regulation of Conjugative Transposition

As stated above, tetracycline can stimulate conjugative transposition of CTnDOT by several orders of magnitude. This stimulation is the result of the action of three gene products encoded by the *rteA, rteB,* and *rteC* genes (75, 93). RteA and RteB have sequences similar to the proteins of two component regulatory signals, but rather than sensing tetracycline directly, their expression is increased by tetracycline induction of transcription from the upstream *tet*Q promoter (75). RteB is thought to activate expression of RteC by counteracting the activity of a transposon-encoded repressor, because transfer of a plasmid carrying the *tra* genes of CTnDOT is constitutive in the absence of both the *rte* genes and tetracycline (37). How RteC activates conjugative transposition of CTnDOT is unclear, although the regulated step appears to be excision of the transposon.

MOBILIZABLE TRANSPOSONS AND NBUs OF *BACTEROIDES*

Mobilizable transposons and NBUs are found in *Bacteroides* species, and are mobile genetic elements that are activated for excision and conjugally mobilized in the classic sense by the large conjugative Tc^R transposons such as CTnDOT (74, 76, 90). They can also be mobilized conjugally by incQ plasmids.

NBUs

There are three NBUs, NBU1, 2, and 3; all are approximately 10 to 12 kb in length. Only NBU1 has been studied in detail.

Excision and integration

Excision of NBU1 requires *intN1,* a gene encoding a lambda integrase family recombinase, and also a large stretch of contiguous DNA containing four open reading frames and *oriT* (86). Gene disruptions show that at least three of the four open reading frames are required for excision. A model has been proposed whereby the *oriT* region and associated proteins interact with the transposon ends and IntN1 to form an active recombination complex. Evidence for such interactions comes from experiments showing that, although the *oriT* region is required in *cis* for excision of NBU1, copies of the *oriT* region cloned in *trans* on a plasmid inhibit excision.

Integration of NBU1 requires *intN1* alone. Integration occurs preferentially into a single target site in the *Bacteroides* chromosome at the 3′ end of a leucine tRNA gene containing the 14-bp sequence CCGGGTCTGGGTAC that is present at one end of the element (87). Integration preserves the sequence of the tRNA gene. When mobilized into *E. coli,* NBU1 can integrate at low frequencies adjacent to several sites that show similarity to the 14-bp sequence (85). However, integration occurs 10- to 70-fold more frequently in *E. coli* when the primary *Bacteroides* integration sequence is present on a plasmid.

Regulation of transposition

Excision and transfer of NBU1 are induced by tetracycline in strains containing a large CTnDOT-type transposon (75). This induction requires *rteA* and *rteB* but, unlike the conjugative transposons, does not require *rteC.* Activation of excision is not caused by activation of *intN1* expression, because *intN1* is expressed constitutively in the absence of both tetracycline and *rteB.*

Tn*4555*

Tn*4555* is a 12.1-kb element encoding cefoxitin resistance whose movement can be induced by tetracycline in the presence of a large conjugative transposon (89).

Excision and integration

Excision and integration of Tn*4555* require a lambda family recombinase and occur by a mechanism similar to that of Tn*916* (97). The major difference from Tn*916* is that two proteins in addition to Int and Xis, TnpA and TnpC, are involved in the process (98). A mutation in *tnpC* results in reduced levels of integration, and *tnpC* is thought to play a role in the assembly of the integration complex. TnpA acts as a targeting protein. Tn*4555* usually integrates into a specific site in the *Bacteroides* chromosome, but, in the absence of TnpA, integration specificity is lost, and integration occurs into random sites. The mechanism by which TnpA directs integration into a specific site and whether there is any similarity to the targeting mechanism of Tn7 (see chapter 19) are not known.

Tn*4399*

Tn*4399* is a 9.6-kb conjugal mobilizing transposon isolated from *Bacteroides fragilis* (24). It contains a mobilization cassette that includes an *oriT* and two genes essential for transfer, *mocA* and *mocB* (53). The *mocA* gene is similar to other DNA relaxases encoded by conjugal elements. Both genes are required for nicking at *oriT*. The element can promote the mobilization of nonconjugal plasmids in *cis*, which indicates that excision is not required for transfer.

Excision and integration

The most remarkable feature of Tn*4399* transposition is the DNA sequence changes that accompany integration of the element (25). Three base pairs of the target are duplicated at each end of the integrated transposon, as would be expected if a transposase made staggered cleavages in the target DNA. However, an additional five nucleotides are inserted into the target between the end of the transposon and the repeated target sequence. This appears to be an unusual hybrid system having features found in classical transposition (target duplication) and conjugative transposition (extra bases brought in by the integrating transposon). However, the general features of this system can be explained by assuming that a transposase cleaves one end of Tn*4399* at the end of the transposon and five nucleotides from the transposon end at the other end, and strand exchange reactions occur to generate a circular transposon whose ends are separated by five nucleotides. If this circular form then is cleaved in a similar manner as during excision, on the same strand at the transposon end and five nucleotides from the transposon end on the other end (equivalent to a blunt-end cleavage leaving the five-nucleotide sequence attached to one end of the transposon), and the 3′-OH groups attack the target DNA making staggered cleavages 3 bp apart, the transposon will integrate into the target with a 3-bp duplication and with five nucleotides at one end of the transposon. Although such a scheme can explain the general features of Tn*4399* integration, it does not explain the specific sequence conservation observed in the 5-bp segments, or why four different 5-bp sequences were observed in transposition events occurring in a donor strain carrying three different insertions of Tn*4399* (25).

Tn*5397* AND OTHER CONJUGATIVE TRANSPOSONS OF CLOSTRIDIA

Two large conjugative transposons, Tn*5397* and Tn*5398*, have been identified in *Clostridium difficile* (49, 51). Tn*5397* encodes a *tetM* determinant, whereas Tn*5398* encodes erythromycin (MLS) resistance. Tn*5397* can transfer from *C. difficile* to *Bacillus subtilis* and then back to *C. difficile*. This transposon is very similar to Tn*916*, differing only in the nature of its recombinase (103) and in the presence of a group II intron in the reading frame analogous to *orf14* of Tn*916* (50).

Excision and Integration

In place of the *int* and *xis* genes present at one end of the Tn*916*, Tn*5397* encodes a resolvase protein, TndX (103). This protein is necessary and sufficient for excision and integration of Tn*5397*. In contrast to Tn*916* Int, the integrated transposon is flanked by a GA dinucleotide that is present in the target sequence. This is indicative of a resolvase-type recombination reaction with staggered cleavages 2 bp apart. Excision of Tn*5397* results in the formation of a circular excised form of the transposon, which in *C. difficile* integrates into a single site in the chromosome, and into multiple sites in the *B. subtilis* chromosome. It does not integrate into the *E. coli* chromosome, but will integrate into target sequences cloned from *C. difficile* or *B. subtilis*. These latter experiments indicate that the sole gram-positive determinant of target specificity is TndX.

Regulation of Conjugative Transposition

Little is known concerning the regulation of transposition of Tn*5397*, except for an interesting interaction between Tn*5397* and Tn*916* (103). When Tn*916* is introduced into *C. difficile*, it integrates into a single site in the chromosome. When Tn*916* enters a cell already containing a Tn*916* transposon, there is no inhibition of conjugation or integration, or loss of the resident transposon. When Tn*5397* is introduced into a strain carrying Tn*916*, 35% of the exconjugants carrying Tn*5397* lose the marker associated with Tn*916*. When the converse experiment is performed, with Tn*916* entering a cell containing Tn*5397*, 96% of the exconjugants lose the marker associated with Tn*5397*. Exconjugant strains carrying both transposons are stable through many subcultures. It therefore appears that, although the recombinase proteins of the two transposons are different, a transposon function is capable of activating excision of the resident transposon when Tn*916* or Tn*5397* enters the recipient. This "incompatibility" does not occur between Tn*916* transposons, but analysis of its molecular mechanism may yield important information concerning the regulation of conjugative transposition of Tn*916* and its relatives.

Tn*4451* AND Tn*4453*a AND Tn*4453*b: MOBILIZABLE TRANSPOSONS OF CLOSTRIDIA

The best-studied clostridial transposon is the 6-kb mobilizable transposon Tn*4551*, which, like the Tn*4453*a and Tn*4453*b elements, confers chloramphenicol resistance on its host (2, 42). These elements can be mobilized by broad host range plasmids such as RP4, and because they encode their own *oriT* and Mob proteins, can contribute to the conjugal mobilization of genetic elements into which they insert.

Excision and Integration

Tn*4451* was the first conjugative transposon or related mobilizable element shown to encode a resolvase-type recombinase (17). This recombinase also contains a domain with the conserved amino acid residues responsible for integrase-type catalysis of cleavage and strand exchange. A second gene, *tnpV*, encodes an Xis-like protein (17). Excision of this element requires *tnpX* but not *tnpV* and produces a circular form of the transposon. For Tn*4453*a, this has been shown to be a transposition intermediate (D. Lyras and J. I. Rood, personal communication). Analysis of the targets and integration products of Tn*4451* showed that the target contains a GA dinucleotide, and upon integration the element is flanked by GA dinucleotides. Consistent with this result, site-directed mutagenesis shows that the resolvase domain, rather than the integrase domain, is essential for excision and integration, and that modifications to the GA dinucleotides flanking the transposon can result in reduced levels of transposon excision.

Regulation of Transposition

The *tnpX* gene is transcribed from the end of the transposon toward the interior, and, as shown for many other transposons that form circular intermediates, the fusion of the ends of the element produces an active promoter that drives the expression of *tnpX* (D. Lyras and J. I. Rood, personal communication). Integration of the element separates the −10 and −35 elements of the promoter; so presumably, for excision to occur, transcription of *tnpX* must occur from a promoter resulting from the fusion of target and transposon sequences or from an upstream promoter.

THE SXT ELEMENT OF *V. CHOLERAE*

The SXT element is a 62-kb genetic element that encodes resistance to sulfamethoxazole and trimethoprim, chloramphenicol, and streptomycin that can conjugally transfer between different *V. cholerae* and *E. coli* strains (101).

Excision and Integration

The SXT element encodes an integrase family recombinase and a small *xis*-like open reading frame close to one end of the element (28). The integrase is essential for excision of the element, whereas the *xis*-like gene is not required. Excision results in the formation of a circular form of the element. Although the SXT element shares many features in common with Tn*916*, it integrates into a single site in the recipient chromosome in both *V. cholerae* and *E. coli* at the 5′ end of the *prfC* gene (encoding peptide chain release factor 3) (28). The integrated element restores the *prfC* open reading frame with alterations at the amino terminus, and the resulting allele is functional.

Transfer

In addition to its site-specific integration activity, the SXT element also differs from Tn*916* because it can mobilize adjacent genetic markers in *cis* by Hfr-type transfer (27). The conjugation functions of this element can be expressed in the integrated state, can mobilize RSF1010 and CloDF13 in *trans*, and can mobilize flanking chromosomal markers in *cis* in an Hfr-like manner.

THE IncJ "PLASMIDS" AND CTn*scr94*

There are seven related conjugal elements including R391 and R997, isolated from *Proteus*, *Pseudomonas*, and *Vibrio* species, that confer antibiotic and mercury resistance on their bacterial hosts (15). They were originally classified as plasmids belonging to the incompatibility group IncJ because of interactions that led to the loss of one element when a second was introduced into the host bacteria. Subsequently, R391 was found to integrate into the *E. coli* chromosome at a specific site between the *uxuA* and *serB* loci (98.0 to 99.5 min) (54, 55). Integration occurred in a *recA*-independent manner. One recent report shows that when one of these elements is introduced into *recA* bacteria carrying a resident element, the incoming element is maintained in plasmid form (62). This suggests that these elements can lead a bifunctional life either integrated into the chromosome or, under the influence of a resident element, as an autonomously replicating plasmid. Alternatively, one of the two elements might reversibly integrate and excise from the bacterial chromosome and be maintained by chromosomal replication during the integrated phase. More recently, it has been shown that the SXT element and R391 share an attachment site in the *prfC* gene and can form tandem arrays (25c).

CTn*scr94* is approximately 100 kb and encodes genes for a sucrose fermentation pathway (26). It integrates into two specific sites in *E. coli,* located in the 3′ ends of the *pheU* and *pheV* genes that encode a $tRNA^{Phe}$. Integration maintains a functional tRNA gene. The type of recombinase responsible for integration of CTn*scr94* is not known at present.

Tn*554:* A SITE-SPECIFIC TRANSPOSABLE ELEMENT

Tn*544* is a transposon from *S. aureus* that confers resistance to the macrolide-lincosamide-streotogramin B (MLS) antibiotics (56, 64). Its genetic organization is unusual because it carries two genes encoding integrase-family recombinases, *tnpA* and *tnpB* (7, 57). Both genes are required for transposition. A third gene, *tnpC,* is also required for transposition. Mutants defective in *tnpC* show reduced levels of transposition and lose the normal orientation specificity when inserting into the preferred chromosomal target site.

Excision and Integration

One unusual property of Tn*554,* compared with the other elements described in this chapter, is that it usually appears to transpose by a duplicative process rather than by excision from the donor molecule (59). Rather than causing target-site duplication, like most transposable elements, sequences from the donor accompany the transposon upon integration into the target (58). However, in contrast to Tn*916* and other elements discussed in this chapter, these sequences do not appear on either end of the integrated element. Rather, in the cases where donor and target sequences can be distinguished, the sequence at the left end of the integrated transposon always belongs to the target, whereas the sequence brought in with the transposon is always at the right end (58). This pattern of inheritance could be explained by directed correction of heteroduplexes formed during integration, assuming a mechanism similar to Tn*916*. However, it is also similar to the pattern of inheritance of sequences observed during integron recombination, suggesting that there may be similarities between the assembly of integrons and transposition of Tn*554*.

CONCLUDING REMARKS

This review has covered, at times superficially, the behavior of a bewildering variety of sometimes bewildering mobile genetic elements. However, it is possible to draw some general conclusions and to highlight some possible future directions for the study of these elements.

First, it appears that the mechanism of DNA cleavage and strand exchange for many of these elements occurs in a similar fashion. For most of the elements for which information is available, recombination is catalyzed by an integrase family recombinase, with varying degrees of insertion site specificity. For elements such as Tn*916* and CTnDOT, sequence heterology between donor and target sites can be accommodated by the recombinase complex and results in the formation of heteroduplex joints, which are repaired either by mismatch repair systems or by DNA replication. One can assume that recombination proceeds by an ordered pair of DNA strand exchanges via a Holliday junction intermediate. However, this has not been demonstrated for any of these systems, and such a demonstration remains a priority.

Second, an exciting future direction, which can just be glimpsed at the moment, concerns possible interactions between the transfer and recombination machineries of these elements. In both Tn*916* and NBU1, there are suggestions of direct interactions that may play important roles in regulating and coordinating excision and transfer of the elements. Hopefully, in a few years, this area will be a major part of the story of these elements.

Third, these elements have an extraordinary potential to promote genetic exchange between bacteria. Many of the conjugation systems of these elements are wildly promiscuous, capable of promoting conjugal transfer between bacteria belonging to different species and genera. The fact that many of the elements encode transfer systems that are expressed independently of excision of the element means that they can readily mobilize nonconjugal replicons into which they insert.

Finally, it should be obvious how little we know about these elements. The DNA sequences of the larger elements remain uncharted territory. Whereas some functions apart from antibiotic resistance that are encoded by the larger elements have been identified, the role these elements play in the biology of their bacterial hosts remains largely a mystery.

Acknowledgments. I thank all my colleagues who sent me information during the preparation of this article. I extend special thanks to Ruth Hall and Dave Sherratt for discussions of integron recombination and to June Scott for a long and fruitful collaboration in studying Tn*916*.

Work in my laboratory is currently supported by grant MCB-9876427 from the National Science Foundation.

REFERENCES

1. **Ayoubi, P., A. O. Kilic, and M. N. Vijayakumar.** 1991. Tn*5253*, the pneumococcal Ω (*cat tet*) BM6001 element, is a composite structure of two conjugative transposons, Tn*5251* and Tn*5252*. *J. Bacteriol.* **173:**1617–1622.
2. **Bannam, T. L., P. K. Crellin, and J. I. Rood.** 1995. Molecular genetics of the chloramphenicol-resistance transposon Tn*4551* from *Clostridium perfringens:* the TnpX site-specific recombinase excises a circular transposon molecule. *Mol. Microbiol.* **16:**535–551.
3. **Bertram, J., M. Stratz, and P. Durre.** 1991. Natural transfer of conjugative transposon Tn*916* between gram-positive and gram-negative bacteria. *J. Bacteriol.* **173:**443–448.
4. **Bringel, F. G., G. L. Van Alstine, and J. R. Scott.** 1992. Conjugative transposition of Tn*916:* the transposon *int* gene is required only in the donor. *J. Bacteriol.* **174:**4036–4041.
5. **Buu-Hoi, A., and T. Horodniceanu.** 1980. Conjugative transfer of multiple antibiotic resistance markers of *Streptococcus pneumoniae*. *J. Bacteriol.* **143:**313–320.
6. **Caparon, M. G., and J. R. Scott.** 1989. Excision and insertion of the conjugative transposon Tn*916* involves a novel recombination mechanism. *Cell* **59:**1027–1034.
7. **Carmo de Freire Bastos, M. D., and E. Murphy.** 1988. Transposon Tn*544* encodes three products required for transposition. *EMBO J.* **7:**2935–2941.
8. **Celli, J., and P. Trieu-Cuot.** 1998. Circularization of Tn*916* is required for expression of the transposon-encoded transfer functions: characterization of long tetracycline-inducible transcripts reading through the attachment site. *Mol. Microbiol.* **28:**103–117.
9. **Cheng, Q., B. J. Paszkeit, N. B. Shoemaker, J. Gardner, and A. A. Salyers.** 2000. Integration and excision of a *Bacteroides* conjugative transposon, CTnDOT. *J. Bacteriol.* **182:** 4035–4043.
10. **Clewell, D. B.** 1990. Movable genetic elements and antibiotic resistance in *Enterococci*. *Eur. J. Clin. Microbiol. Infect. Dis.* **9:**90–102.
11. **Clewell, D. B., and S. E. Flannagan.** 1993. The conjugative transposons of gram-positive bacteria, p. 369–393. *In* D. B. Clewell (ed.), *Bacterial Conjugation.* Plenum Publishing Corp., New York, N.Y.
12. **Clewell, D. B., S. E. Flannagan, Y. Ike, J. M. Jones, and C. Gawron-Burke.** 1988. Sequence analysis of termini of conjugative transposon Tn*916*. *J. Bacteriol.* **170:**3046–3052.
13. **Clewell, D. B., S. E. Flannagan, and D. D. Jaworski.** 1995. Unconstrained bacterial promiscuity: the Tn*916*-Tn*1545* family of conjugative transposons. *Trends Microbiol.* **3:** 229–236.
14. **Clewell, D. B., and C. Gawron-Burke.** 1986. Conjugative transposons and the dissemination of antibiotic resistance. *Annu. Rev. Microbiol.* **40:**635–659.
15. **Coetzee, N. A., N. Datta, and R. W. Hedges.** 1972. R-factors from *Proteus rettgeri*. *J. Gen. Microbiol.* **72:**543–555.
16. **Courvalin, P., and C. Carlier.** 1986. Transposable multiple antibiotic resistance in *Streptococcus pneumoniae*. *Mol. Gen. Genet.* **205:**291–297.
17. **Crellin, P. K., and J. I. Rood.** 1997. The resolvase/invertase domain of the site-specific recombinase TnpX is functional and recognizes a target sequence that resembles the junction of the circular form of the *Clostridium perfringens* transposon Tn*4451*. *J. Bacteriol.* **179:**5148–5156.
18. **Doucet-Populaire, F., P. Trieu-Cuot, I. Dosbaa, A. Andremont, and P. Courvalin.** 1991. Inducible transfer of conjugative transposon Tn*1545* from *Enterococcus faecalis* to *Listeria monocytogenes* in the digestive tracts of gnotobiotic mice. *Antimicrob. Agents Chemother.* **35:**185–187.
19. **Fayet, O., P. Ramond, P. Polard, M. F. Prere, and M. Chandler.** 1990. Functional similarities between the IS3 family of bacterial insertion elements. *Mol. Microbiol.* **4:**1771–1777.
20. **Flannagan, S. E., and D. B. Clewell.** 1991. Conjugative transfer of Tn*916* in *Enterococcus faecalis:* trans activation of homologous transposons. *J. Bacteriol.* **173:**7136–7141.
21. **Flannagan, S. E., L. A. Zitzow, Y. A. Su, and D. B. Clewell.** 1994. Nucleotide sequence of the 18-kb conjugative transposon Tn*916* from *Enterococcus faecalis*. *Plasmid* **32:**350–354.
22. **Franke, A. E., and D. B. Clewell.** 1981. Evidence for a chromosome-borne resistance transposon (Tn*916*) in *Streptococcus faecalis* that is capable of conjugative transfer in the absence of a conjugative plasmid. *J. Bacteriol.* **145:**494–502.
23. **Gawron-Burke, C., and D. B. Clewell.** 1982. A transposon in *Streptococcus faecalis* with fertility properties. *Nature* **300:** 281–284.
24. **Hecht, D. W., and M. H. Malamy.** 1989. Tn*4399*, a conjugal mobilizing transposon of *Bacteroides fragilis*. *J. Bacteriol.* **171:**3603–3608.
25. **Hecht, D. W., J. S. Thompson, and M. H. Malamy.** 1989. Characterization of the termini and transposition products of Tn*4399*, a conjugal mobilizing transposon of *Bacteroides fragilis*. *Proc. Natl. Acad. Sci. USA* **86:**5340–5344.

25a. **Hinerfeld, D., and G. Churchward.** 2001. Specific binding of integrase to the origin of transfer (*oriT*) of the conjugative transposon Tn*916*. *J. Bacteriol.* **183:**2947–2951.

25b. **Hinerfeld, D., and G. Churchward.** Xis protein of the conjugative transposon Tn*916* plays dual opposing roles in transposon excision. *Mol. Microbiol.*, in press.

25c. **Hochhut, B., J. W. Beaber, R. Woodgate, and M. K. Waldor.** 2001. Formation of chromosomal tandem arrays of the SXT element and R391, two conjugative chromosomally integrating elements that share an attachment site. *J. Bacteriol.* **183:** 1124–1132.

26. **Hochhut, B., K. Jahreis, J. W. Lengeler, and K. Schmid.** 1997. CTn*scr94*, a conjugative transposon found in enterobacteria. *J. Bacteriol.* **179:**2097–2102.
27. **Hochhut, B., J. Marrero, and M. K. Waldor.** 2000. Mobilization of plasmids and chromosomal DNA mediated by the SXT element, a constin found in *Vibrio cholerae* O139. *J. Bacteriol.* **182:**2043–2047.
28. **Hochhut, B., and M. K. Waldor.** 1999. Site-specific integration of the conjugal *Vibrio cholerae* SXT element into *prfC*. *Mol. Microbiol.* **32:**99–110.
29. **Horodniceanu, T., L. Bougueleret, and G. Bieth.** 1981. Conjugative transfer of multiple-antibiotic resistance markers in beta hemolytic group A, B, F and G streptococci in the absence of extrachromosomal deoxyribonucleic acid. *Plasmid* **5:**127–137.
30. **Hosking, S. L., M. E. Deadman, E. R. Moxon, J. F. Peden, and N. J. Saunders.** 1998. An in silico evaluation of Tn*916* as a tool for generalized mutagenesis in *Haemophilus influenzae* Rd. *Microbiology* **144:**2525–2530.
31. **Jaworski, D. D., and D. B. Clewell.** 1994. Evidence that coupling sequences play a frequency-determining role in conjugative transposition of Tn*916* in *Enterococcus faecalis*. *J. Bacteriol.* **176:**3328–3335.
32. **Jaworski, D. D., and D. B. Clewell.** 1995. A functional origin of transfer (*oriT*) on the conjugative transposon Tn*916*. *J. Bacteriol.* **177:**6644–6651.
33. **Jia, Y., and G. Churchward.** 1999. Interactions of the integrase protein of the conjugative transposon Tn*916* with its specific DNA binding sites. *J. Bacteriol.* **181:**6114–6123.
34. **Kilic, A. O., M. N. Vijayakumar, and S. F. al-Khaldi.** 1994. Identification and nucleotide sequence analysis of a transfer-related region in the streptococcal conjugative transposon Tn*5252*. *J. Bacteriol* **176:**5145–5150.

35. **Lanka, E., and B. M. Wilkins.** 1995. DNA processing reactions in bacterial conjugation. *Annu. Rev. Biochem.* **64:** 141–169.
36. **Leffers, G. G., and S. Gottesman.** 1998. Lambda Xis degradation in vivo by Lon and FtsH. *J. Bacteriol.* **180:**1573–1577.
37. **Li, L.-Y., N. B. Shoemaker, and A. A. Salyers.** 1995. Location and characteristics of the transfer region of a *Bacteroides* conjugative transposon and regulation of transfer genes. *J. Bacteriol.* **177:**4992–4999.
38. **Lu, F., and G. Churchward.** 1994. Conjugative transposition: Tn*916* integrase contains two independent DNA binding domains that recognize different DNA sequences. *EMBO J.* **13:** 1541–1548.
39. **Lu, F., and G. Churchward.** 1995. Tn*916* target DNA sequences bind the C-terminal domain of integrase protein with different affinities that correlate with transposon insertion frequency. *J. Bacteriol.* **177:**1938–1946.
40. **Lyras, D., and J. I. Rood.** 1997. Transposable genetic elements and antibiotic resistance determinants from *Clostridium perfringens* and *Clostridium difficile*, p. 73–92. *In* J. I. Rood, B. A. McClane, J. G. Songer, and R. W. Titball (ed.), *The* Clostridia: *Molecular Biology and Pathogenesis.* Academic Press, Inc., London, United Kingdom.
41. **Lyras, D., and J. I. Rood.** 2000. Clostridial genetics, p. 529–539. *In* V. A. Fischetti (ed.), *Gram-Positive Pathogens.* American Society for Microbiology, Washington, D.C.
42. **Lyras, D., C. Storie, A. S. Huggins, T. L. Bannam, and J. I. Rood.** 1998. Chloramphenicol resistance in *Clostridium difficile* is encoded on Tn*4453* transposons that are closely related to Tn*4451* from *Clostridium perfringens. Antimicrob. Agents Chemother.* **42:**1563–1567.
43. **Manganelli, R., S. Ricci, and G. Pozzi.** 1996. Conjugative transposon Tn*916:* evidence for excision with formation of 5′-protruding termini. *J. Bacteriol.* **178:**5813–5816.
44. **Manganelli, R., S. Ricci, and G. Pozzi.** 1997. The joint of Tn*916* circular intermediates is a homoduplex in *Enterococcus faecalis. Plasmid* **38:**71–78.
45. **Manganelli, R., L. Romano, S. Ricci, M. Zazzi, and G. Pozzi.** 1995. Dosage of Tn*916* circular intermediates in *Enterococcus faecalis. Plasmid* **34:**48–57.
46. **Marra, D., B. Pethel, G. Churchward, and J. R. Scott.** 1999. The frequency of conjugative transposition of Tn*916* is not determined by the frequency of excision. *J. Bacteriol.* **181:** 5414–5418.
47. **Marra, D., and J. R. Scott.** 1999. Regulation of excision of the conjugative transposon Tn*916*. *Mol. Microbiol.* **31:**609–621.
48. **Moitoso de Vargas, L., C. A. Pargellis, N. M. Hasan, E. W. Bushman, and A. Landy.** 1988. Autonomous DNA binding domains of lambda integrase recognize two different sequence families. *Cell* **54:**923–929.
49. **Mullaney, P., M. Wilks, and S. Tabaqchali.** 1995. Transfer of macrolide-lincosamide-streptogramin B (MLS) resistance in *Clostridium difficile* is linked to a gene homologous with toxin A and is mediated by a conjugative transposon Tn5398. *J. Antimicrob. Chemother.* **35:**305–315.
50. **Mullany, P., M. Pallen, M. Wilks, J. R. Stephen, and S. Tabaqchali.** 1996. A group II intron in a conjugative transposon from the gram-positive bacterium, *Clostridium difficile. Gene* **174:**145–150.
51. **Mullany, P., M. Wilks, I. Lamb, C. Clayton, B. Wren, and S. Tabaqchali.** 1990. Genetic analysis of a tetracycline resistance element from *Clostridium difficile* and its conjugal transfer to and from *Bacillus subtilis. J. Gen Microbiol.* **136:** 1343–1349.
52. **Munoz-Najar, U., and M. N. Vijayakumar.** 1999. An operon that confers UV resistance by evoking the SOS mutagenic response in streptococcal transposon Tn*5252*. *J. Bacteriol.* **181:**2782–2788.
53. **Murphy, C. G., and M. H. Malamy.** 1995. Requirements for strand- and site-specific cleavage within the *oriT* region of Tn*4399*, a mobilizing transposon from *Bacteroides fragilis. J. Bacteriol.* **177:**3158–3165.
54. **Murphy, D. B., and J. T. Pembroke.** 1995. Transfer of the IncJ plasmid R391 to recombination *E. coli. FEMS Microbiol. Lett.* **134:**153S–158S.
55. **Murphy, D. B., and J. T. Pembroke.** 1999. Monitoring of chromosomal insertions of the IncJ elements R391 and R997 in *Escherichia coli* K-12. *FEMS Microbiol. Lett.* **174:** 355–361.
56. **Murphy, E.** 1989. Transposable elements in gram-positive bacteria, p. 269–288. *In* D. E. Berg and M. M. Howe (ed.), *Mobile DNA.* American Society for Microbiology, Washington, D.C.
57. **Murphy, E., L. Huwyler, and M. de Carmo de Freire Bastos.** 1985. Transposon Tn*554:* complete nucleotide sequence and isolation of transposition-defective and antibiotic-sensitive mutants. *EMBO J.* **4:**3357–3365.
58. **Murphy, E., and S. Lofdahl.** 1984. Transposition of Tn*544* does not generate a target duplication. *Nature* **307:**292–295.
59. **Murphy, E., S. Phillips, I. Edelman, and R. P. Novick.** 1981. Tn*544:* isolation and characterization of plasmid insertions. *Plasmid* **5:**292–305.
60. **Norgren, M., and J. R. Scott.** 1991. The presence of conjugative transposon Tn916 in the recipient strain does not impede transfer of a second copy of the element. *J. Bacteriol.* **173:** 319–324.
61. **Norgren, M. G., and J. R. Scott.** 1991. Presence of the conjugative transposon Tn*916* in the recipient strain does not impede transfer of a second copy of the element. *J. Bacteriol.* **173:**319–324.
62. **Pembroke, J. T., and E. Murphy.** 2000. Isolation and analysis of a circular form of the IncJ conjugative transposon-like elements, R391 and R997: implications for IncJ incompatibility. *FEMS Microbiol. Lett.* **187:**133–138.
63. **Pethel, B., and G. Churchward.** 2000. Coupling sequences flanking Tn*916* do not determine the affinity of binding of integrase to the transposon ends and adjacent bacterial DNA. *Plasmid* **43:**123–129.
64. **Phillips, S., and R. P. Novick.** 1979. Tn*544*—a site-specific repressor-controlled transposon in *Staphylococcus aureus. Nature* **278:**476–478.
65. **Poyart-Salmeron, C., P. Trieu-Cuot, C. Carlier, and P. Courvalin.** 1989. Molecular characterization of two proteins involved in the excision of the conjugative transposon Tn*1545:* homologies with other site-specific recombinases. *EMBO J.* **8:**2425–2433.
66. **Poyart, C., J. Celli, and P. Trieu-Cuot.** 1995. Conjugative transposition of Tn*916*-related elements from *Enterococcus faecalis* to *Escherichia coli* and *Pseudomonas fluorescens. Antimicrob. Agents Chemother.* **39:**500–506.
67. **Proveddi, R., R. Manganelli, and G. Pozzi.** 1996. Characterization of conjugative transposon Tn*5251* of Streptococcus pneumoniae. *FEMS Microbiol. Lett.* **135:**231–236.
68. **Rauch, P. J., and W. M. de Vos.** 1992. Characterization of the novel nisin-sucrose conjugative transposon Tn*5276* and its insertion in *Lactococcus lactis. J. Bacteriol.* **174:** 1280–1287.
69. **Rauch, P. J., and W. M. de Vos.** 1994. Identification and characterization of genes involved in excision of the *Lactococcus lactis* conjugative transposon Tn*5276*. *J. Bacteriol.* **176:**2165–2171.
70. **Rauch, P. J. G.** 1996. *Transposition of the* Lactococcus lactis

Conjugative Transposon Tn5276: Identification of a Circular Intermediate and Analysis of Its Excision and Insertion Sites. Ph.D. dissertation. Netherlands Institute for Dairy Research, Ede, The Netherlands.

71. **Rudy, C., K. L. Taylor, D. Hinerfeld, J. R. Scott, and G. Churchward.** 1997. Excision of a conjugative transposon in vitro by the Int and Xis proteins of Tn*916*. *Nucleic Acids Res.* **25:**4061–4066.
72. **Rudy, C. K., and J. R. Scott.** 1996. Length of the coupling sequence of Tn*916*. *J. Bacteriol.* **176:**3386–3388.
73. **Rudy, C. K., J. R. Scott, and G. Churchward.** 1997. DNA binding by the Xis protein of the conjugative transposon Tn*916*. *J. Bacteriol.* **179:**2567–2572.
74. **Salyers, A. A., N. B. Shoemaker, and L.-Y. Li.** 1995. In the drivers seat: the *Bacteroides* conjugative transposons and the elements the mobilize. *J. Bacteriol.* **177:**5727–5731.
75. **Salyers, A. A., N. B. Shoemaker, and A. M. Stevens.** 1995. Tetracycline regulation of conjugal transfer genes, p. 393–400. *In* J. A. Hoch and T. J. Silhavy (ed.), *Two-Component Signal Transduction.* American Society for Microbiology, Washington, D.C.
76. **Salyers, A. A., N. B. Shoemaker, A. M. Stevens, and L.-Y. Li.** 1995. Conjugative transposons: an unusual and diverse set of integrated gene transfer elements. *Microbiol. Rev.* **59:** 579–590.
77. **Sampath, J., and M. N. Vijayakumar.** 1998. Identification of a DNA cytosine methyltransferase gene in conjugative transposon Tn*5252*. *Plasmid* **39:**63–76.
78. **Scott, J. R.** 1992. Sex and the single circle: conjugative transposition. *J. Bacteriol.* **174:**6005–6010.
79. **Scott, J. R.** 1993. Conjugative transposons, p. 597–614. *In* A. L. Sonenshein, J. A. Hoch, and R. Losick (ed.), Bacillus subtilis *and Other Gram-Positive Bacteria.* American Society for Microbiology, Washington, D.C.
80. **Scott, J. R., F. Bringel, D. Marra, G. Van Alstine, and C. K. Rudy.** 1994. Conjugative transposition of Tn*916:* preferred targets and evidence for conjugative transfer of a single strand and for a double-stranded circular intermediate. *Mol. Microbiol.* **11:**1099–1108.
81. **Scott, J. R., and G. G. Churchward.** 1995. Conjugative transposition. *Annu. Rev. Microbiol.* **49:**367–397.
82. **Scott, J. R., P. A. Kirchman, and M. G. Caparon.** 1988. An intermediate in the transposition of the conjugative transposon Tn*916*. *Proc. Natl. Acad. Sci. USA* **85:**4809–4813.
83. **Senghas, E., J. M. Jones, M. Yamamoto, C. Gawron-Burke, and D. B. Clewell.** 1988. Genetic organization of the bacterial conjugative transposon Tn*916*. *J. Bacteriol.* **170:**245–249.
84. **Shoemaker, N. B., M. D. Smith, and W. R. Guild.** 1980. DNase-resistant transfer of chromosomal *cat* and *tet* insertions by filter mating in pneumococcus. *Plasmid* **3:**80–87.
85. **Shoemaker, N. B., G.-R. Wang, and A. A. Salyers.** 1996. NBU1, a mobilizable site-specific integrated element from *Bacteroides* spp., can integrate nonspecifically in *Escherichia coli*. *J. Bacteriol.* **178:**3601–3607.
86. **Shoemaker, N. B., G.-R. Wang, and A. A. Salyers.** 2000. Multiple gene products and sequences required for excision of the mobilizable integrated *Bacteroides* element NBU1. *J. Bacteriol.* **182:**928–936.
87. **Shoemaker, N. B., G. R. Wang, and A. A. Salyers.** 1996. The *Bacteroides* mobilizable insertion element, NBU1, integrates into the 3′ end of a Leu-tRNA gene and has an integrase that is a member of the lambda integrase family. *J. Bacteriol.* **178:** 3594–3600.
88. **Showsh, S. A., and R. E. Andrews.** 1992. Tetracycline enhances Tn*916*-mediated conjugal transfer. *Plasmid* **28:** 213–224.
89. **Smith, C. J., and A. C. Parker.** 1993. Identification of a circular intermediate in the transfer and transposition of Tn*4555*, a mobilizable transposon from *Bacteroides* spp. *J. Bacteriol.* **175:**2682–2691.
90. **Smith, C. J., G. D. Tribble, and D. P. Bayley.** 1998. Genetic elements of *Bacteroides* species: a moving story. *Plasmid* **40:** 12–29.
91. **Speer, B. S., N. B. Shoemaker, and A. A. Salyers.** 1992. Bacterial resistance to tetracycline: mechanisms, transfer, and clinical significance. *Clin. Microbiol. Rev.* **5:**387–399.
92. **Srinivas, P., A. O. Kilic, and M. N. Vijayakumar.** 1997. Site-specific nicking *in vitro* at *ori*T by the DNA relaxase of Tn*5252*. *Plasmid* **37:**42–50.
93. **Stevens, A. M., N. B. Shoemaker, L.-Y. Li, and A. A. Salyers.** 1993. Tetracycline regulation of genes on *Bacteroides* conjugative transposons. *J. Bacteriol.* **175:**6134–6141.
94. **Storrs, M. J., C. Carlier, C. Poyart-Salmeron, P. Trieu-Cuot, and P. Courvalin.** 1991. Conjugative transposition of Tn*916* requires the excisive and integrative activities of the transposon-encoded integrase. *J. Bacteriol.* **173:**4347–4352.
95. **Su, Y. A., P. He, and D. B. Clewell.** 1992. Characterization of the tet(M) determinant of Tn*916:* evidence for regulation by transcription attenuation. *Antimicrob. Agents Chemother.* **36:**769–778.
96. **Taylor, K., and G. Churchward.** 1997. Specific DNA cleavage mediated by the integrase of conjugative transposon Tn*916*. *J. Bacteriol.* **179:**1117–1125.
97. **Tribble, G. D., A. C. Parker, and C. J. Smith.** 1997. The *Bacteroides* mobilizable transposon Tn*4555* integrates by a site-specific recombination mechanism similar to that of the gram-positive bacterial element Tn*916*. *J. Bacteriol.* **179:** 2731–2739.
98. **Tribble, G. D., A. C. Parker, and C. J. Smith.** 1999. Transposition genes of the bacteroides mobilizable transposon Tn*4555:* role of a novel targeting gene. *Mol. Microbiol.* **34:** 385–394.
99. **Trieu-Cuot, P., C. Poyart-Salmeron, C. Carlier, and P. Courvalin.** 1993. Sequence requirements for target activity in site-specific recombination mediated by the Int protein of transposon Tn*1545*. *Mol. Microbiol.* **8:**179–185.
100. **Vijayakumar, M. N., and S. Ayalew.** 1993. Nucleotide sequence analysis of the termini and chromosomal locus involved in site-specific integration of the streptococcal conjugative transposon Tn*5252*. *J. Bacteriol.* **175:**2713–2719.
101. **Waldor, M. K., H. Tschape, and J. J. Mekalanos.** 1996. A new type of conjugative transposon encodes resistance to sulfamthoxazole, trimethoprim and streptomycin in *Vibrio cholerae* O139. *J. Bacteriol.* **178:**4157–4165.
102. **Wang, H., A. P. Roberts, D. Lyras, J. I. Rood, M. Wilks, and P. Mullaney.** 2000. Characterization of the ends and target sites of the novel conjugative transposon Tn*5397* from *Clostridium difficile:* excision and circularization is mediated by the large resolvase, TndX. *J. Bacteriol.* **182:**3775–3783.
103. **Wang, H., A. P. Roberts, D. Lyras, J. I. Rood, M. Wilks, and P. Mullany.** 2000. Characterization of the ends and target sites of the novel conjugative transposon Tn*5397* from *Clostridium difficile:* excision and circularization is mediated by the large resolvase, TndX. *J. Bacteriol.* **182:**3775–3783.
104. **Weisberg, R. A., L. W. Enquist, C. Foeller, and A. Landy.** 1983. A role for DNA homology in site-specific recombination: the isolation and characterization of a site-affinity mutant of coliphage lambda. *J. Mol. Biol.* **170:**319–342.
105. **Wojciak, J. M., K. M. Connolly, and R. T. Clubb.** 1999. Solution structure of the Tn*916* integrase-DNA complex: specific binding using a three-strand beta-sheet. *Nat. Struct. Biol.* **6:**366–373.

Mobile DNA II
Edited by N. L. Craig et al.

Chapter 11

Site-Specific Recombination by the Flp Protein of *Saccharomyces cerevisiae*

Makkuni Jayaram, Ian Grainge, and Gena Tribble

INTRODUCTION

The Flp protein is a conservative site-specific recombinase. It mediates DNA rearrangements by catalyzing strand cleavage and rejoining between a pair of identical, relatively short (34 bp) target DNA segments. The reaction requires no exogenous high-energy source, and is performed without degradation or synthesis of DNA.

The natural source of Flp is the 2μm plasmid, a multicopy extrachromosomal element found in most common strains of *Saccharomyces* yeast (reviewed in reference 8) (see Color Plate 16A [see color insert]). The plasmid is a 6318-bp double-stranded circular DNA molecule containing two copies of a 599-bp sequence arranged in head-to-head (inverted) orientation (Color Plate 16B). The inverted repeat, which harbors the Flp recombination target (*FRT*) sites (Color Plate 16C), divides the plasmid genome into two unique domains. The Flp recombination system constitutes one part of a dual strategy by which stable high-copy number maintenance of the plasmid is achieved. A partitioning system consisting of two plasmid-encoded proteins, Rep1p and Rep2p, in conjunction with a *cis*-acting locus *STB* (also called *REP3*), ensures fairly even segregation of replicated plasmid molecules between mother and daughter at cell division. Plasmid duplication is performed by the cellular DNA replication apparatus, and is subject to the cell cycle restriction of one round of replication

Makkuni Jayaram, Ian Grainge, and Gena Tribble • Section of Molecular Genetics and Microbiology and Institute of Cellular and Molecular Biology, University of Texas at Austin, Austin, TX 78712.

initiation per S phase. During each cell cycle, the plasmid copy number is doubled with each molecule contributing equally to the doubling (149). If a random missegregation event causes the plasmid copy number to drop in a cell, it is quickly restored to normal by an amplification mechanism. Amplification is mediated by Flp recombination, and bypasses the normal regulation that proscribes more than one round of plasmid replication.

According to a widely accepted model proposed by Futcher (Color Plate 16D) (33), amplification depends on the asymmetric location of the replication origin with respect to the Flp recombination target (*FRT*) sites within the 2μm circle genome. During normal bidirectional replication, assuming equal rates of fork movement, the *FRT* site proximal to the origin would be duplicated before the distal site. Recombination between one of the two proximal sites with the distal one would invert the relative direction of fork movement. As a result, each of the two forks would traverse the circular template repeatedly, yielding multiple plasmid copies from a single initiation step at the origin. Amplification will be terminated when a second recombination event reinverts the forks and restores their bidirectional orientation. The tandemly repeated array of 2μm circles can be resolved into unit-length circles by Flp-mediated resolution between alternate copies of the *FRT* sites (which are in head-to-tail orientation). One key prediction of the model, that the absence of a functional recombination system would abolish amplification, has been verified (112, 138). However, the mechanics of amplification and the consequent "rolling circle intermediate" predicted by the model remain to be rigorously tested.

Plasmids similar to the 2μm circle in structural and functional organization have been identified in *Zygosaccharomyces* and *Kluyveromyces* yeasts (8). They harbor site-specific recombination systems analogous to Flp but with target specificities distinct from that of Flp. These plasmids have also maintained the asymmetry in the spacing between their replication origins and the target sites for recombination. This evolutionary conservation suggests that coupling of recombination to replication likely provides an effective common strategy for copy-number maintenance among the yeast plasmids. Although amplified DNA sequences found in eukaryotic chromosomes can be accommodated by mechanisms that use features of the Futcher model (23), there is so far no evidence for the involvement of site-specific recombination during their formation. The basic chemical mechanism of the Flp reaction is also used by recombinases of the integrase/tyrosine family (to which Flp belongs) for many other biological purposes: to mediate integration and excision of phage genomes in bacterial hosts, to select or regulate patterns of gene expression, and to ensure stable partitioning of circular genomes (44, 91).

In this chapter, we review the biochemical, mechanistic, and topological aspects of the Flp system with special emphasis on the contributions made by the Flp protein subunits and the target DNA to the individual steps of the recombination pathway. We illustrate how Flp recombination fits into the global physicochemical paradigm followed by the integrase/tyrosine family recombinases, while still retaining features that are strikingly different from other members of the family. We then discuss how the Flp active site provides a model for the emergence of complex active sites from elementary active sites under the functional constraints faced by biological catalysts during their evolution. Next, we allude to some of the applications of Flp in yeast and in extraneous host systems. We conclude by considering potential mechanisms for Flp regulation in vivo to rapidly commission or decommission the amplification machine as demanded by the copy-number status of the 2μm plasmid.

For aspects of the Flp reaction that are not covered in detail here, the reader is referred to earlier reviews (23, 58, 59, 114, 115). We acknowledge the early genetic contributions made by the laboratory of James Broach and subsequent biochemical advancements provided by the laboratories of Paul Sadowski and Michael Cox toward our current understanding of how Flp works.

THE Flp RECOMBINATION PATHWAY: CONTRIBUTIONS FROM THE PROTEIN AND DNA COMPONENTS OF THE REACTION

Like other members of the integrase/tyrosine family, Flp performs recombination in two steps: first, exchanging one pair of strands to form a Holliday junction intermediate; second, exchanging the second pair of strands to resolve the junction into reciprocal recombinants (Fig. 1). Each step of the reaction requires the cooperative action of four Flp monomers, with two bound on each of the two DNA substrates. In this section, we consider how the molecular recognition between Flp and DNA, together with the molecular interactions between Flp monomers bound on a single substrate or on partner substrates within the synaptic complex, gives rise to the sequential strand-exchange pathway of recombination.

Our discussion of the mechanism and dynamics of recombination is confined to the minimal *FRT* site (56, 108, 121) that is missing the binding element 1′b (see Color Plate 16B). This element does not directly

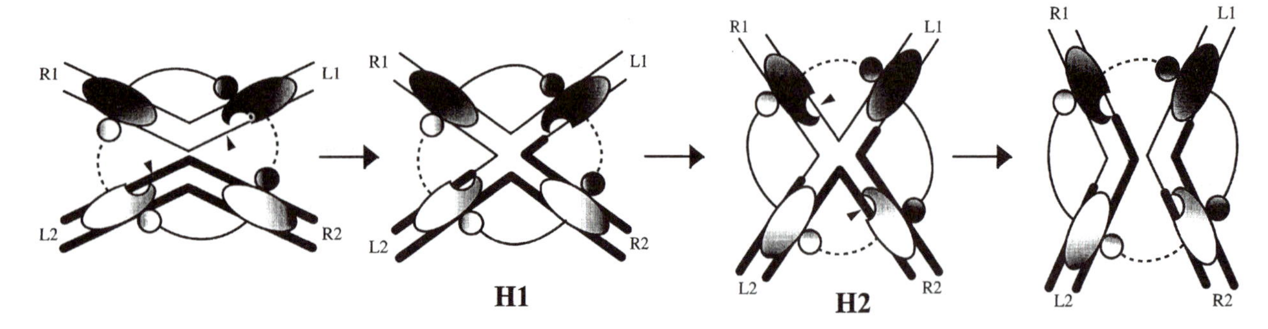

Figure 1. The Flp reaction is initiated within a synaptic structure containing two DNA substrates and the four Flp monomers bound to them. Exchange of the first pair of strands results in a Holliday intermediate (H1). Isomerization of the complex (H2) promotes the exchange of the second pair of strands. This diagram anticipates several aspects of the Flp reaction that will be clarified in the text. For example, the two substrates are bent identically and arranged in an antiparallel orientation (L, left; R, right). The interactions between recombinase monomers bound to each substrate (the two darkly shaded monomers in substrate 1, and the two lightly shaded monomers in substrate 2) are responsible for the first strand cleavage and exchange reaction. The Holliday intermediate is roughly square planar, but strictly only 2-fold symmetric. During the resolution step, the catalytic dimers are formed between Flp monomers bound on the left and right arms of partner substrates. Each active dimer is constituted by a darkly shaded and a lightly shaded monomer. Note that, throughout the reaction pathway, a cyclic peptide connectivity between the four Flp monomers is maintained. The catalytically active and inactive associations between pairs of Flp monomers are indicated by the solid and dashed arcs, respectively.

contribute to the recombination reaction in vivo or in vitro, but it could play a regulatory function in vivo (56), perhaps by acting as a loading site for Flp or, alternatively, by sequestering Flp from the recombination site. The minimal *FRT* represents the general organization of the core recombination sequences required by all integrase/tyrosine family recombinases (see chapters 6 to 10 and 12 to 14). The more complex members of the family may use two nonequivalent DNA elements (corresponding to 1a and 1′a in Color Plate 16B) to bind a recombinase dimer constituted by two distinct subunits (XerC/XerD [see chapter 8]) or harbor separate binding sites for accessory protein factors within their target sites (lambda Int and XerC/XerD; see chapters 7 and 8). The Cre recombinase of phage P1 and Flp are the two best understood simple members of the family, and comparisons between the two are frequently made to highlight mechanistic similarities and differences that exist within the family.

Functional Architecture of the Flp Recombination Complex: DNA-Protein and Protein-Protein Interactions

Interactions between Flp and the *FRT* site

The Flp protein is a monomer in solution, and associates with each 13-bp binding element of the *FRT* site as a monomer. The DNA-protein interactions are mediated through both major and minor grooves (9, 62, 100). The amino-terminal domain of Flp (approximately 13 kDa) is responsible for contacting the DNA segment immediately proximal to the scissile phosphodiester bond (98, 99, 101). The larger carboxyl-terminal domain (approximately 32 kDa), which harbors the essential catalytic residues, makes DNA contacts that are spread out over a larger portion of the binding element (99, 101, 134). The features of Flp-DNA interactions inferred from partial proteolysis experiments, DNA and protein footprinting analyses, and UV cross-linking assays (14, 99, 101, 134) are in good agreement with those revealed by the crystal structures of Cre and Flp in their DNA-bound forms (18, 47). A crescent-shaped protein scaffold formed by the amino-terminal and carboxyl-terminal domains of Flp with the DNA being cradled between the two seems to be a common DNA recognition motif within the integrase/tyrosine family. Dimerization of Flp occurs on the DNA by interactions between monomers bound on either side of the spacer. To accommodate the Flp-Flp interactions, the DNA deviates from linearity but remains essentially planar (see DNA bending, described below). Synapsis of two Flp-bound *FRT* sites is established by interactions between monomers across the binding arms of the DNA partners, resulting in a cyclic network of peptide interactions (Fig. 1; also see chapters 6 and 12). The overall geometry of this DNA-protein complex, which seems to be conserved within the integrase/tyrosine family, is well suited for conducting the chemical steps of recombination without large-scale distortions of the structure or disruptions of protein subunit contacts.

Bending of DNA by Flp

Binding of two Flp monomers to an *FRT* site results in a significant DNA bend, centered within the spacer region. The bending is required for establishing a functional interface between the monomers. Point mutants of Flp that bind DNA normally but cannot induce the bend (likely reflecting a weakening of the monomer-monomer interactions) have been isolated. Such mutants are defective in the chemical steps of recombination (12, 15, 66, 116, 117). By comparing the electrophoretic mobilities of Flp-bound DNA to bent DNA standards, the magnitude of the bend (referred to as the type II bend in the literature) was estimated to be approximately 140° (12, 117). The recent crystal structure of a Flp-DNA complex suggests that this value is likely an overestimate (18). The DNA bend in the structure corresponding to a cleavage-competent pair of Flp monomers is comparable with that seen in a similar Cre-DNA structure, approximately 80° (47). It is possible that the angle between adjacent DNA arms is altered when a Flp-bound *FRT* site associates with a second Flp-bound site to form the synaptic structure. This change in geometry may account for the discrepancy between the DNA bends seen in the crystal structure and estimated by gel retardation.

The bending of DNA is consistent with the assembly of the Flp-active site at the interface of the two monomers bound on either side of the spacer. The contributions made by the individual monomers toward the organization of a "shared active site" is discussed in detail later. When the geometric constraints normally posed by the spacer DNA are mitigated, either by splitting a full site into half sites (see Fig. 2 for the relevant features of a half site) or by providing a prebent full site, a bending-incompetent mutant becomes cleavage-proficient (12, 15, 75). Although the bend per se seems unlikely to impose direct strain on the labile phosphate as a requirement for strand cleavage, the bend directionality restricts cleavage to only one end of the spacer at a time (see below for details). Thus, the configuration of the DNA substrate is a principal determinant in charting the mechanistic course of the reaction.

Geometry of the Flp-DNA complex during strand cleavage: the principle of active-site exclusion

Solution studies have indicated that a Flp dimer bound to a minimal *FRT* site is chemically competent for DNA cleavage, even in the absence of synapsis between recombination partners (139). The dimer is functionally asymmetric as indicated by the lack of double cleavages within an *FRT* site and by the occurrence of only 50% of the possible cleavages in dimers of Flp-half *FRT* site complexes (Fig. 2A) (75, 85, 110). In principle, the Flp dimer has the potential to assemble two active sites simultaneously at either end of the spacer, because it has all of the catalytic functional groups to do so. However, the need to establish specific protein-protein contacts in the face of the geometry and stiffness of the spacer DNA makes this sterically impossible. Thus, only one of the two functional active sites can be arranged at a time, as would be consistent with the two-step, single strand-exchange recombination mechanism shown in Fig. 1. Two equivalent but asymmetric DNA bends directed opposite to each other, accommodating two distinct dimeric configurations of the recombinase, can provide the geometric basis for this active-site exclusion at a given spacer end (75, 83, 84). When DNA substrates are forced into assuming a single bend orientation using strategically positioned nucleotide bulges (36, 113) within the spacer, nearly all of the cleavages are confined to the strand containing the bulge (Fig. 2B). In the excluded active site, it is the cleavage nucleophile (Tyr343) of Flp that is misoriented, whereas the target phosphodiester is activated for cleavage. Thus, in a bulge-containing substrate occupied by Flp, the phosphodiester that is usually refractory to cleavage by Tyr343 can be cleaved by an exogenously added small nucleophile such as hydrogen peroxide (75). On the other hand, when the structural constraints of the spacer are minimized by eliminating normal base pairing, active-site exclusion can be overcome. In such a substrate, both ends of the spacer can be cleaved in a single molecule. The result of subsequent strand joining is a "reductional" recombination event that yields two hairpin products (Fig. 2C) (75).

Is the DNA bend used to promote strand exchange? The release of tension harnessed in the bend upon strand cleavage can, in principle, propel a segment of the spacer away from its original partner strand and toward the new partner (83, 140). A DNA "spring" may be established at one end of the spacer, if an intrinsic sequence-dependent bend at that position is opposed by the bend induced upon recombinase binding. The first pair of strand cleavages, followed by Holliday formation, would dissipate the accumulated torsional stress. After the second pair of cleavages, the broken strands would reseal in a "latch-like" manner to produce the recombinant products. Differences in the bend-induced tension at the two ends of a spacer (spring-like or latch-like) could impart a prescribed order of strand exchange to a recombination system. The order may be dictated by the sequence of the spacer favoring one type of DNA bend over the other. Alternatively, this asymmetry may be imparted by the DNA-protein and protein-protein in-

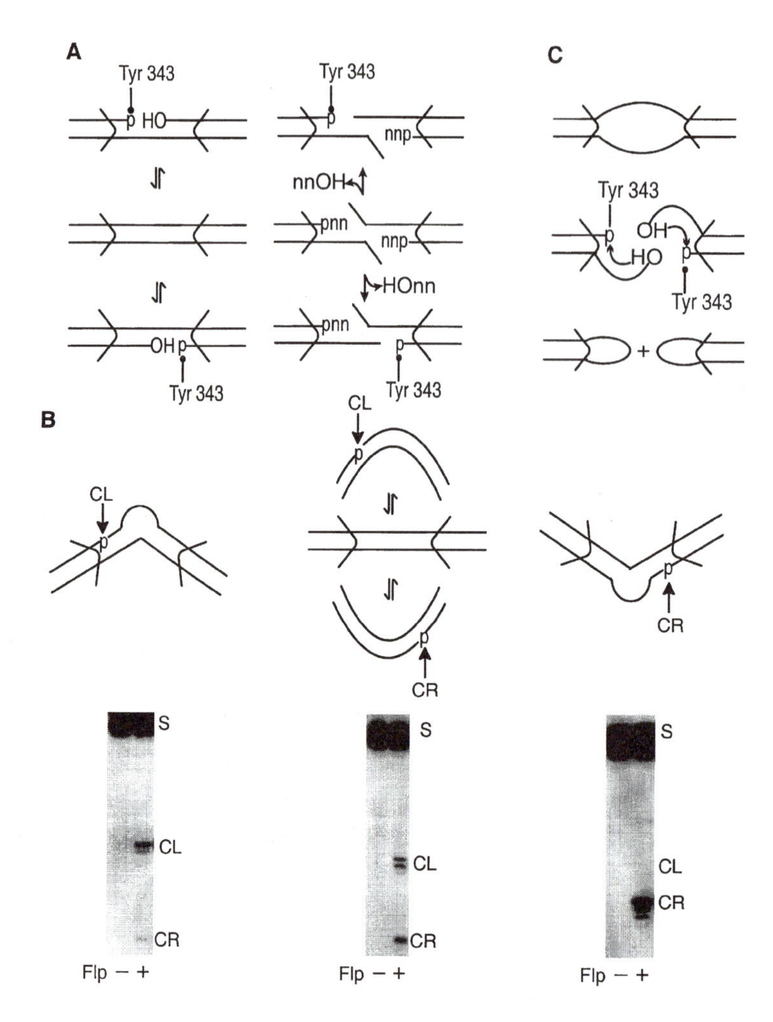

Figure 2. (A, left) A dimer of Flp bound to a full-site can assume one of two catalytic configurations. As a result, DNA cleavage can occur either at the left end of the spacer (top) or at the right end (bottom). In the cleaved complex, the Flp protein is linked to the 3′-phosphate end (p) by Tyr343, the cleavage nucleophile. (A, right) Analogous to the situation in the full-site, only half of the possible cleavages are observed within a dimer of a Flp-half-site complex (110). A typical half-site contains one Flp-binding element, only two or three spacer nucleotides (n) on the strand containing the labile phosphodiester, and a full complement of the spacer nucleotides on the other strand ending in a 5′-hydroxyl group. This hydroxyl may be blocked by a phosphate group to prevent a cleavage reaction from continuing through the strand-joining step. During cleavage of the half-site, the short di- or trinucleotide product (5′OHnn3′) diffuses away and cannot take part in strand joining. For simplicity, the occupancy of the binding elements in the full-site and half-site by Flp is not shown. (B) In full-sites containing nucleotide bulges within the spacer on the top strand (left) or the bottom strand (right), strand cleavage is biased nearly exclusively to one end of the spacer (indicated by the short vertical arrows). The bulge-containing strand is the cleavage-susceptible one. Nucleotide bulges were created by the inclusion of three consecutive A's on one strand with no complementary bases on the other. In the normal full-site (middle), cleavage can occur at either end of the spacer with roughly equal probability. The gel profile showing the Flp cleavage product(s) yielded by a substrate is shown below it. CL and CR stand for left and right cleavages, respectively. In the bulge-containing substrates, the cleavage bands are longer by 3 nt (because of the A3 bulge). (C) When the substrate has a fully mismatched spacer, both ends of the spacer may be cleaved. The 5′-hydroxyl groups at the cleaved spacer ends can take part in the indicated joining reactions to form hairpin products.

teractions contributed by accessory factors (in the case of the more complex members of the family). So far, there is no strong evidence indicating a defined order of strand exchange during the Flp reaction in vitro.

Chemistry and Stereochemistry of the Flp Reaction

Each strand exchange performed by Flp consists of two transesterification reactions, cleavage and rejoining. The reaction is chemically identical with that used by type IB topoisomerases. Cleavage occurs via nucleophilic attack by Tyr343 of Flp on the phosphodiester bond at one end of the spacer. The incoming nucleophile displaces one of the bridging phosphate oxygens to yield a 5′-hydroxyl group and a 3′-phosphotyrosyl bond. The strand-exchange reaction is chemically a reversal of the cleavage reaction, except that the 5′-hydroxyl group attacking the phosphotyrosyl bond is derived from the partner duplex that has also been cleaved at the same spacer end. As a result, a recombinant DNA strand is formed. Because a new phosphodiester bond is gained for one that is lost at both the cleavage and ligation steps, the overall bond energies are conserved during the reaction, so no high-energy cofactor such as ATP is required.

Active-site amino acids

Elucidation of the crystal structures of the integrase/tyrosine family recombinases (HP1 Int, lambda Int, and XerD), and in particular the cocrystals of Cre and Flp with DNA (18, 39, 47–49, 67, 131), has confirmed the identity of key catalytic residues in Flp inferred from mutational and biochemical analyses (13, 27, 32, 46, 74, 102, 103, 107). Despite the great diversity in their primary amino acid sequences, the solved structures reveal that the integrase/tyrosine family recombinases have nearly identical three-dimensional peptide folds, especially in their carboxyl-terminal domains. Comparison between Flp-DNA and Cre-DNA structures indicates that the peptide architecture of the Flp amino-terminal domain may differ significantly from other members of the family.

The Flp active site contains four residues, Arg191, His305, Arg308, and Tyr343, that are almost invariant throughout the family (26, 94). The Cre cocrystals also reveal a tryptophan (usually a histidine in other integrases but also tryptophan in Flp) which is hydrogen bonded to the scissile phosphate. Therefore the active-site motif can be generalized as an RHRH/W tetrad and a tyrosine nucleophile (Fig. 3). Not only are the reaction mechanisms similar for the integrase/tyrosine recombinases and type IB topoisomerases but their active sites, as seen by X-ray crystallography, are very similar too (19, 42, 92, 111, 126, 128). In the topoisomerases, the active site contains an R-K-R-H motif, which surrounds and activates the phosphodiester for cleavage compared with the R-H-R-H/W motif of tyrosine recombinases (Fig. 3).

The role of the arginines seems to be the same in topoisomerases and recombinases as has been seen in staphylococcal nuclease, namely to coordinate the nonbridging oxygens of the scissile phosphodiester and stabilize the pentavalent intermediate in the cleavage/ligation (Fig. 4) (31, 47, 111, 128). The role of the H, H/W residues (His305 and Trp330 in Flp) is less clear. The presence of two histidines (or a histidine and a tryptophan) around the active-site residues is reminiscent of the bovine pancreatic RNase catalytic site. During the RNase reaction, two histidines (His12 and His119) act reversibly as general acid and general base at the phosphate cyclization and hydrolysis steps (see reference 31). It is attractive to consider that the histidine or histidine/tryptophan pair may play a similar role in the recombinase active site, especially because the Flp active site harbors a cryptic RNA cleavage activity that follows the pancreatic RNase mechanism (145, 146). Recent evidence suggests, however, that an active-site histidine residue of vaccinia virus topoisomerase (corresponding to Trp330 of Flp) plays a role in stabilizing the negative charge on the transient pentavalent phosphate intermediate rather than acting as a general acid/base (129). This inference is based on the finding that the wild-type topoisomerase and a weakly active histidine mutant (H265A) display opposite stereochemical selectivity toward substrates containing a phosphorothioate substitution at the cleavage position. Whereas the wild-type enzyme prefers the *S*p form, the mutant prefers the *R*p form. The results would be consistent with the crystal structure of the homologous human type IB topoisomerase in which the H^{ϵ} of the corresponding histidine is in hydrogen bonding distance with the pro-*R*p oxygen (128). The stereochemistry of the Flp reaction is discussed below. Biochemical assays for cleavage alone (in a suicide substrate) or strand joining alone (in a precleaved substrate) have demonstrated that Arg191 and Arg308 of Flp are required at both steps (13, 32, 68, 97). His305 is essential for strand joining, whereas the cleavage nucleophile, Tyr343, is not required at this step. The potential role of His305 during strand cleavage is rather ambiguous. Full sites readily undergo cleavage in the absence of His305, yet half sites are not cleaved by Flp mutants lacking this residue. It is possible that, in addition to its direct catalytic role, His305 may indirectly contribute toward stabilizing the functional interaction between Flp monomers bound to two separate half sites. In one of the Cre structures (48), a well ordered water molecule is seen hydrogen bonded to several of the active-site residues, raising the possibility that acid/base catalysis may be mediated through water.

Are any further residues involved in active-site catalysis? A separate lysine, conserved in both the topoisomerase and integrase families, is located within a three β-sheet region in the solved crystal structures. In the human topoisomerase I and Cre cocrystals, this lysine interacts with the base immediately 5′ or 3′ of the cleavage site, respectively, through the minor groove. Mutations in this residue greatly reduce topoisomerase activity (10^3 to 10^4 times reduction in vaccinia virus topoisomerase I) (144), and this contact may be necessary for positioning of the active-site residues or could be directly involved in catalysis.

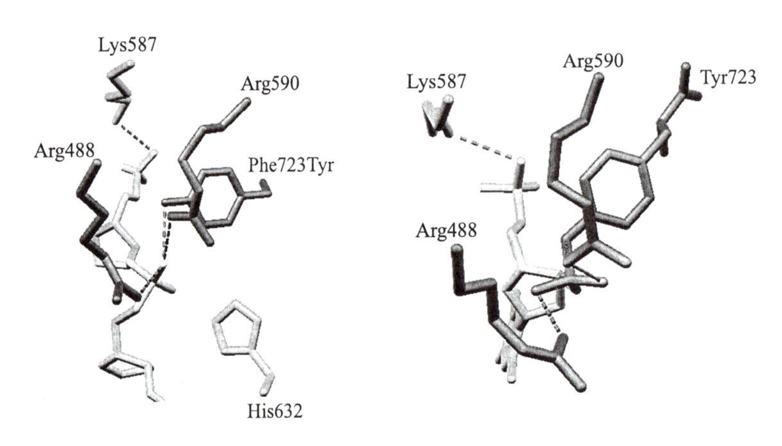

Figure 3. The disposition of the active-site residues in Cre and human topoisomerase I (topo I) cocrystals with DNA in the cleaved and uncleaved states is shown. Hydrogen bonds are represented by dashed lines. In the Cre structures the RHRW tetrad and the tyrosine nucleophile are indicated (47). For the topoisomerase I structure in the uncleaved state, the RKRH tetrad and the tyrosine nucleophile are depicted; however, in the cleaved form the histidine of the tetrad (His632) is disordered and is therefore not shown (111). The uncleaved complex was obtained with topoI mutated at the active-site tyrosine (Y723F). In the structure shown, a tyrosine is represented occupying exactly the same position as the phenylalanine and is denoted as Phe723Tyr.

Stereochemical course of recombination

Each of the two steps involved in the formation of a recombinant strand is thought to follow an in-line S_N2 type phosphoryl transfer reaction. The suspected mechanism has been tested by substituting either of the nonbridging oxygen atoms of the scissile phosphate by sulfur and examining the stereochemical consequence of the reaction. It is a well accepted tenet in enzymology that a single-step phosphoryl transfer (or an odd number of steps) causes inversion of phosphate configuration, whereas a two-step phosphoryl transfer (or an even number of steps) leads to retention of phosphate configuration. Phosphorothioate-substituted substrates have been used to establish direct attack by the 3′-hydroxyl of guanosine during the self-cleavage of the *Tetrahymena* rRNA intron (88). The method has been adapted and popularized in DNA phosphoryl transfer reactions by Mizuuchi and Adzuma (90) who demonstrated the inversion of phosphate configuration during the transfer of a "pre-nicked" Mu DNA end to a target phosphodiester by the phage Mu transposase. The Flp reaction was performed initially in a substrate containing the *R*p form of the phosphorothioate incorporated at the cleavage position using Klenow polymerase (123). Consistent with the proposed mechanism, the recombinant product shows retention of configuration as indicated by its cleavage susceptibility to snake venom and resistance to P1 nuclease, two stereo-specific nucleases (Fig. 5A). Subsequently, this result was confirmed by using both *R*p and *S*p phosphorothioate substrates

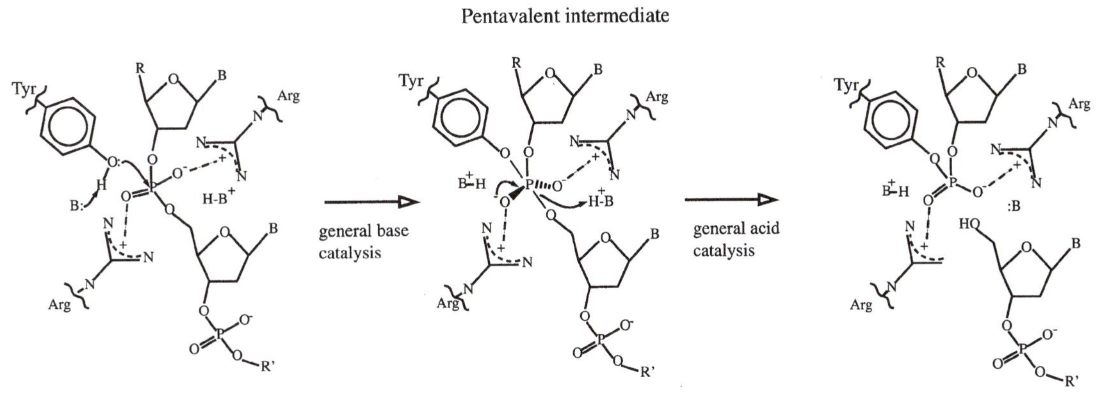

Pentavalent intermediate

general base catalysis

general acid catalysis

Figure 4. Plausible roles of the active-site residues during the cleavage reaction by an integrase/tyrosine recombinase. The two arginines and the tyrosine nucleophile are represented around the DNA. The histidine, or histidine-tryptophan, pair of the RHRH/W tetrad could act as general base (B:) and general acid ($B^{+}H$), respectively, for deprotonating the tyrosine nucleophile and protonating the leaving group at its 5′ end. If the recombination mechanism is an adaptation of the classical pancreatic RNase mechanism, the roles of the two histidines or the histidine-tryptophan pair would be reversed during strand joining. The general base would now abstract the proton from the 5′-hydroxyl group, and the general acid would protonate the tyrosine-leaving group. Alternatively, if the role of the histidines is only to help stabilize the pentavalent intermediate, then the general acid and general base may simply be water.

constructed with synthetic oligonucleotides (synthesized and purified by W. J. Stec and colleagues at the Polish Academy of Sciences, Lodz, Poland). The data for the *S*p form, for which Flp shows a large preference over the *R*p form (J. Lee and M. Jayaram, unpublished data), are shown in Fig. 5B. This stereochemical specificity is the same as that observed for the lambda Int protein (64, 90). A similar preference for the *S*p phosphorothioate over the *R*p form has also been observed for topoisomerase (129), lending further support to the functional and possibly evolutionary relatedness of these proteins.

The shared active site of Flp: the basis for DNA cleavage in *trans*

The first strong indication that Flp assembles a shared active site at the interface of two monomers was provided by the observation that a Flp variant containing a point mutation in the RHR triad (Arg 191, His305, and Arg308) can be catalytically complemented by the Tyr343 mutant (FlpY343F) in a half-site recombination reaction (16). No complementation was observed between two point mutants within the RHR triad. Because the formation of the hairpin product from a half site consists of one phosphodiester breakage followed by strand joining, it was concluded that the single active site required to perform this reaction is assembled by using the RHR triad of one Flp monomer and the catalytic tyrosine of the second monomer. Because the lifetime of the complex between a Flp monomer and a binding element is much longer than the time required for the chemical steps of recombination, it became possible to perform mixing reactions with substrates bound by specific Flp variants. By prebinding a triad mutant to one half site and the tyrosine mutant to a second half site, it could be shown that the cleavage reaction occurs only at the half site bound by the tyrosine mutant (Fig. 6A). Because Tyr343 is not essential for strand joining, the cleaved half site can go on to yield the hairpin recombinant. Two key predictions of the shared active-site model have been verified (17), and are shown diagrammatically in Fig. 6B. First, single, double, and triple mutants of the triad yield equivalent levels of complementation when paired with FlpY343F. Second, wild-type Flp, upon pairing with a triad, Tyr343 double mutant, is inactivated in the half-site reaction.

The half-site complementation assay has been used extensively to delineate amino acid residues that are required in *cis* or act in *trans* during strand cleavage by a Flp dimer (13, 68, 74, 97). The Flp related R recombinase encoded by a 2μm circle-like plasmid of *Zygosaccharomyces rouxii* also shares the RHR triad and the catalytic tyrosine between two monomers at the cleavage step (148). The shared active site seems to be a distinguishing feature of the yeast subfamily within the larger integrase/tyrosine family recombinases. Other well-studied non-yeast recombinases (lambda Int, Cre, XerC/XerD) derive all the active-site residues required for cleavage from a single monomer. However, as indicated by the crystal structures of Cre-DNA complexes (39, 47, 48), the alloste-

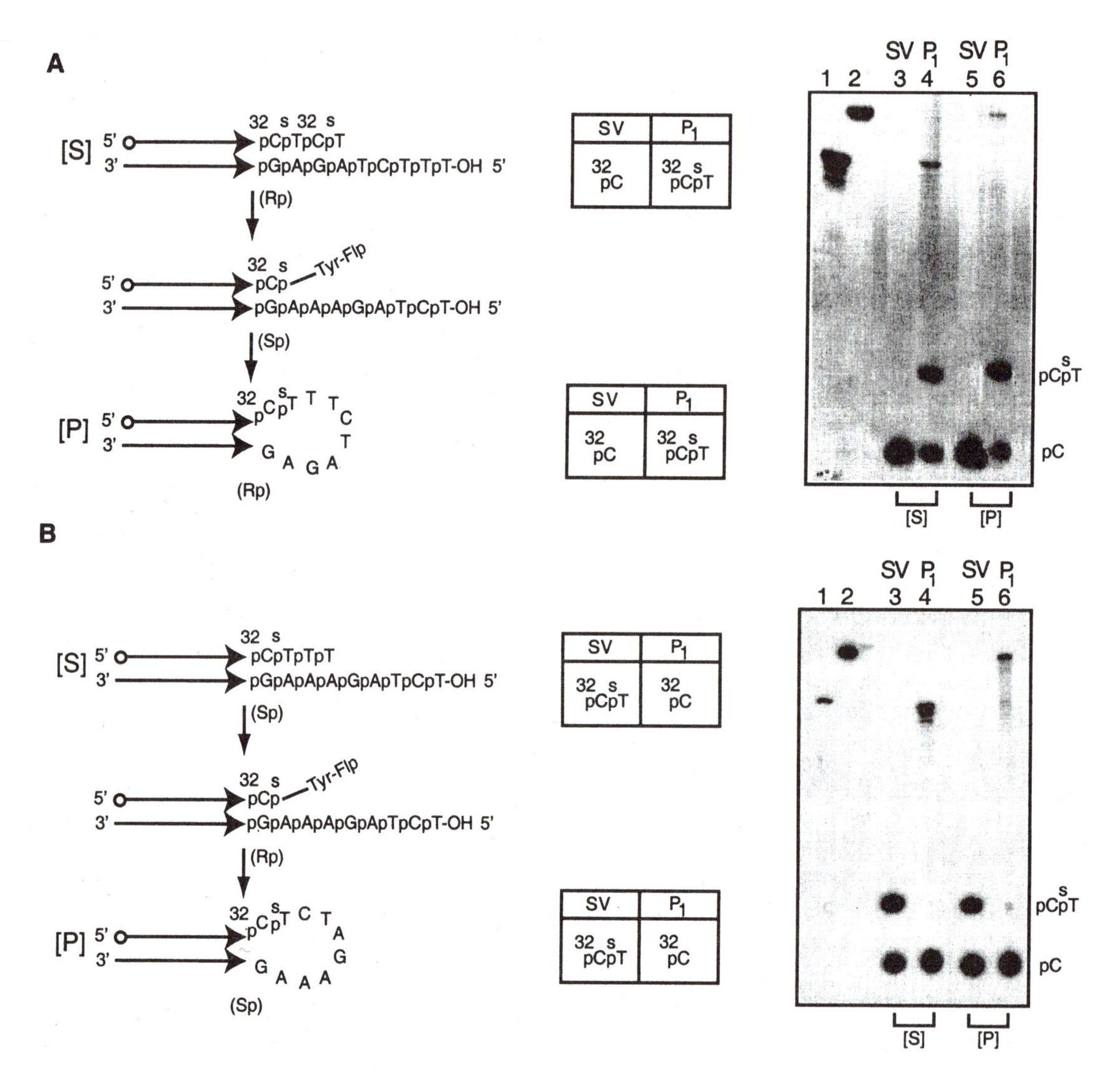

Figure 5. (A) The half-site substrate (S) containing the *Rp* form of the phosphorothioate at the scissile phosphodiester position (obtained by Klenow polymerase incorporation) should, in a two-step reaction, yield a hairpin recombinant (P) that retains the *Rp* configuration. The predicted diagnostic radioactive products of snake venom (SV) and P1 digestion are a dinucleotide and a mononucleotide, respectively, for both substrate and product. The experimental results shown at the right support this prediction. The mononucleotide band in the P1 digestion of the product indicates the level of phosphate contamination at the phosphorothioate position. The relative increase in this band in the P1 digestion of the substrate is caused by the second P^{32} position of the half-site. This labeled position is absent in the hairpin product. (B) The analysis is similar to that in A. The substrate containing the *Sp* form of the phosphorothioate at the cleavage position (preferentially cleaved by Flp over the *Rp* form) was prepared by ligating the tetranucleotide, 5′Cp^{s}TTT3′, phosphorylated at the 5′ end by P^{32} to the appropriate half-site oligonucleotide. The mononucleotide band in the snake venom digestion is caused by phosphate contamination at the phosphorothioate position.

ric activation of the "active monomer" is critically dependent on the interaction of its carboxyl-terminal region with the "inactive monomer." Thus, it is a fair generalization that the cleavage-competent entity is a recombinase dimer throughout the family.

The half-site recombination catalyzed by a pair of mutant Flp variants leads to the notion of *trans* DNA cleavage by Flp. That is, the tyrosine of a Flp monomer bound to a half site does not cleave the scissile phosphodiester adjacent to it (cleavage in *cis*); rather, it reaches across the spacer and cleaves the phosphodiester of the half site bound by a second Flp monomer (see Fig. 6A). Because the *trans* cleavage model is derived by the use of mutant Flp partners (and is at odds with the *cis* cleavage mode of Int, XerC/XerD, and Cre), it has been viewed with some skepticism. By appropriate experimental strategies, the occurrence of *trans* cleavage and the virtual ab-

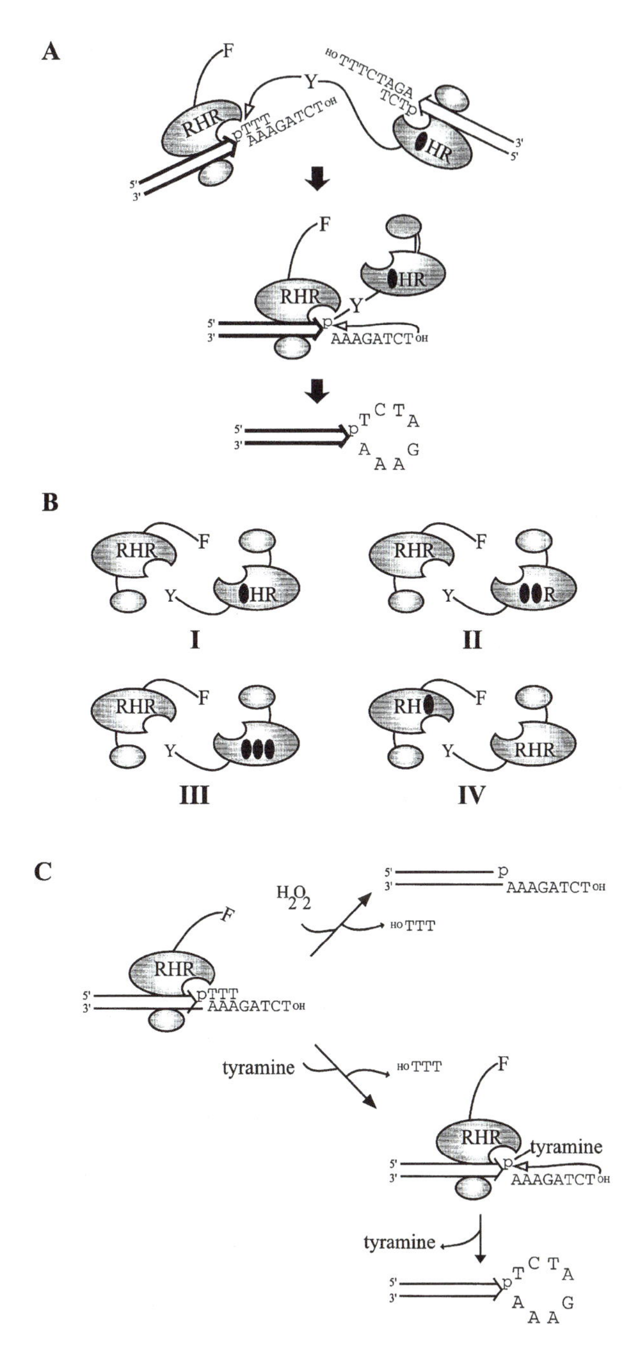

Figure 6. (A) In the complementation reaction shown, the left half site (bold lines) is bound by FlpY343F and the right half site (thin lines) by a triad mutant of Flp. A functional active site (RHR, Y) assembled at the left half site cleaves the scissile phosphodiester. FlpY343F mediates strand joining within the cleaved intermediate to form the hairpin. The right half site is inactive in this reaction. (B) One good active site (RHR, Y) can be derived from FlpY343F and a single (I), double (II), or triple (III) triad mutant of Flp. The catalytically inactive states in I, II, and III are (F, ●HR), (F, ●●R), and (F, ●●●), respectively. The dimer formed by wild-type Flp and a triad, Tyr343 double mutant shown in IV, cannot yield a good active site (RH●, Y in one case, and RHR, F in the other). (C) A half-site bound by FlpY343F can be cleaved by hydrogen peroxide or tyramine. The cleavage product from the tyramine reaction is active in the strand-joining step and yields the hairpin.

sence of *cis* cleavage have been demonstrated for a wild-type Flp-Flp (Y343F) dimer as well as a Flp dimer with all catalytic residues intact (73, 79). The crystal structure of the Flp-DNA complex also supports *trans* DNA cleavage (18).

Based on the active-site architecture of Flp, the strand cleavage and strand joining steps of recombination can be accommodated by a *cis* activation of phosphodiester *trans* nucleophilic attack mechanism (68, 69). The target phosphodiesters (the DNA phosphodiester or the phosphotyrosyl phosphodiester) are oriented in *cis* by the Flp monomer bound adjacent to it. The active nucleophiles (Tyr343 or the 5′-hydroxyl group) are then delivered in *trans,* either by a partner Flp monomer or by the cleaved DNA strand. Consistent with this scheme, Flp (Y343F) can perform a half-site recombination event in the presence of exogenously provided tyramine as a mimic of Tyr343 (68, 69). Similarly, Flp (Y343F) can perform strand cleavage at the scissile phosphodiester bond with hydrogen peroxide as the nucleophile to yield a product with a free 3′-phosphate end (63). Because the hydrogen peroxide reaction does not conserve the energy of the phosphodiester bond, the reaction cannot proceed to the hairpin product (Fig. 6C).

In summary, nearly all the contributions for transition-state stabilization at each chemical step of recombination are made by one Flp monomer. The function of the second monomer bound across the spacer is to deliver the tyrosine nucleophile in a precisely oriented manner. The spacer length of 8 bp ensures the correct spatial disposition of two Flp monomers to accomplish this active interaction between them. Changing the spacer length by the addition or deletion of 1 bp results in a large reduction in the efficiency of recombination between the mutant *FRT* and a wild-type partner (122). The defined spacer lengths in other recombination systems related to Flp (7 bp for lambda Int or 6 bp for Cre) also reflect the optimization of monomer-monomer interactions, even though they differ from Flp in the delivery of the cleavage nucleophile.

The types of *trans* cleavage during one round of Flp recombination

Having ruled out *cis* cleavage by Flp, the question arises: Which type(s) of *trans* cleavage does Flp use during recombination? In the context of a synapsed pair of DNA substrates (or a Holliday junction intermediate), a Flp monomer bound to a given DNA arm could, in principle, cleave in *trans* at three possible

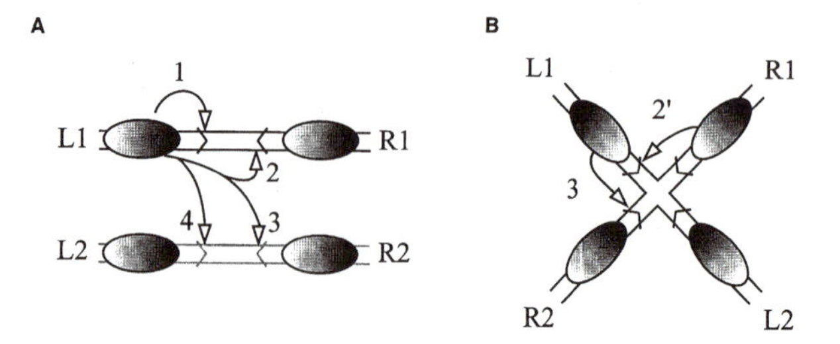

Figure 7. (A) The potential modes of cleavage by a given Flp monomer were originally defined with respect to a parallel arrangement of the DNA substrates. The left and right DNA arms are indicated by L and R, respectively, with the subscripts referring to substrate 1 and 2. The cleavage modes numbered 1 through 4 represent *cis*, *trans* horizontal, *trans* diagonal, and *trans* vertical, respectively. The curved arrows originating from the Flp monomer denote the delivery of Tyr343 nucleophile. (B) For an open-square arrangement of the DNA arms, which is close to what has been observed in Flp-bound Holliday junctions (77), the cleavage modes numbered 2′ and 3 would be *trans* horizontal and *trans* diagonal, respectively.

phosphodiester positions (Fig. 7). The corresponding three types of cleavage have been termed as *trans* horizontal, *trans* diagonal, and *trans* vertical. The nomenclature is based on the two DNA substrates being arranged in a parallel fashion (in the same left-to-right orientation). *trans* horizontal cleavage refers to cleavage within a single substrate, Tyr343 being donated across the spacer in a left-to-right (or right-to-left) orientation. *trans* diagonal and *trans* vertical cleavages refer to Tyr343 donation across partner substrates. *trans* diagonal (like *trans* horizontal) cleavage requires the catalytic cooperativity of Flp monomers bound on opposite sides of the spacer. Because the cleavage reaction requires only a dimer of Flp (139), *trans* horizontal and *trans* diagonal cleavages are functionally indistinguishable. Analyses of cleavage directionalities in a variety of substrate configurations have demonstrated that the cleavage mode in linear full-site substrates is *trans* horizontal (79). In half-site substrates or in synthetic Holliday junction substrates, the cleavage modes correspond to *trans* horizontal and *trans* diagonal (16, 25, 78). These results are consistent with a model in which Flp recombination is initiated by *trans* horizontal cleavage in the two DNA substrates to form the Holliday intermediate and terminated by *trans* diagonal cleavage to form the recombinant products (see below). *trans* vertical cleavage, in which Tyr343 from a Flp monomer bound to one DNA substrate cleaves the scissile phosphodiester on the second DNA substrate but on the same side of the spacer, has never been observed. Thus, catalytically productive interactions between Flp monomers bound to two left arms or two right arms of partner substrates can be ruled out.

Assembly of the shared active site in nonstandard substrates

Each of the two steps of strand exchange during Flp recombination requires the concerted action of two active sites. Because each active site is contributed by a dimer, one could theoretically derive a pair of functional active sites from a Flp dimer, trimer, or tetramer. As discussed earlier, because of the structural constraints imposed by the spacer DNA, a dimer is unable to assemble two active sites simultaneously. Two sets of experiments have demonstrated that a Flp trimer can provide two functional active sites. A synthetic Holliday junction containing only three Flp binding arms (the fourth being mutated drastically to eliminate binding) can be resolved into linear products by wild-type Flp (109). Similarly, an artificial three-armed Y substrate can be resolved into a linear and a hairpin product by Flp (78) (Fig. 8). The Y reaction is performed strictly by a Flp trimer with no contribution from a fourth Flp monomer in solution. A triad mutant of Flp, when paired with Flp(Y343F), is unable to execute this reaction. The two types of heterotrimers formed by the mutants will contain either two functional triads and one Tyr343 residue or one functional triad and two Tyr343 residues. Hence

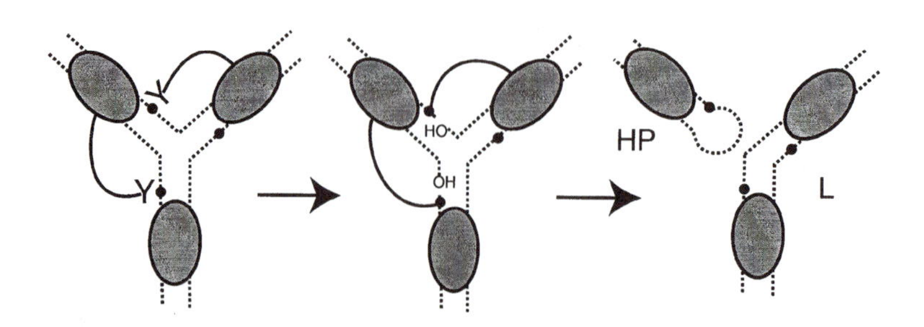

Figure 8. Two simultaneous strand cleavages can be mediated in a Y substrate by a Flp trimer. Subsequent strand joining yields the linear and hairpin recombinants (L and HP).

only one, but not two active sites, can be derived. Consistent with this logic, the mutant pair yields strand cleavage (from the single active site) in the Y substrate but no recombination. If a fourth Flp monomer from solution were a participant, assembly of two active sites and hence resolution of the Y substrate should have been possible. There is direct experimental evidence that two active sites can also be derived from a Flp tetramer. A triad mutant of Flp, in combination with Flp(Y343F), is able to resolve a synthetic Holliday junction into linear products (72). This is possible only by deriving the requisite triad pair from two monomers of Flp(Y343F) and the Tyr343 pair from two monomers of the triad mutant.

What then is the active unit of Flp during a normal recombination event? Is it a trimer or a tetramer? Qian and Cox (109) have proposed a model in which two sets of Flp trimers derived from the DNA-bound tetramer execute the Holliday formation and resolution steps. However, Lee and Jayaram (72) have demonstrated that, within an assembled Flp tetramer, the trimeric catalytic configuration is proscribed. Thus, all four Flp monomers participate at each step of the recombination reaction. However, as indicated in Fig. 1, each Flp monomer switches its catalytic association between two partner monomers during one round of recombination. The protein-DNA arrangements seen in the crystal structures of Cre and Flp support the tetramer (or dimer of dimers) model.

The arrangement of the Flp tetramer as two sets of active trimers in the Qian-Cox model (109) has a cyclic connectivity among the protein subunits that superficially resembles the cyclic protein interactions seen in the crystal structures of Cre and Flp (18, 47). However, from a functional perspective, the two arrangements are dissimilar. In the trimer model, the DNA partners are arranged in parallel, whereas their alignment in the structures is antiparallel.

THE GEOMETRY AND TOPOLOGY OF THE Flp RECOMBINATION COMPLEX

How do the two *FRT* sites align themselves within the synaptic structure: in parallel or antiparallel configuration? Given the overall near planarity of the Cre-DNA and Flp-DNA complexes, the distinction between the two possible arrangements is a valid one. The earlier results (75, 77) with Flp are most readily accommodated by an antiparallel geometry of the sites. For example, synthetic Holliday junctions that adopt a stacked X form in the presence of Mg^{2+} ions (80) are opened up into a nearly square planar configuration when bound by Flp (Fig. 9) (77). The two right arms (and the two left arms) flanking the spacer are coaxially directed to the diagonally opposite corners of the square. Based on the directed DNA bend for strand cleavage inferred from constrained

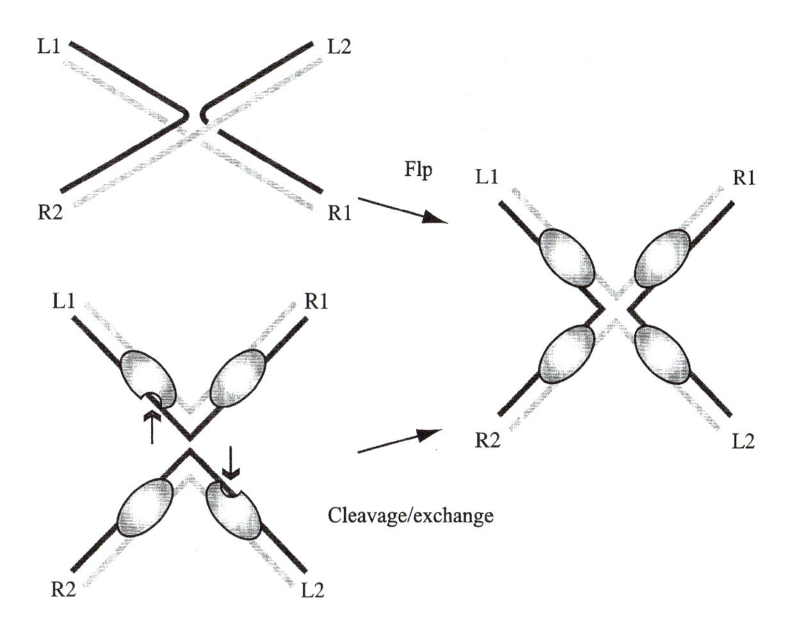

Figure 9. A free Holliday junction in the presence of divalent cations assumes the right-handed antiparallel stacked configuration (top left). When bound by Flp, the configuration switches to a nearly square planar form (right). In the context of recombination, the simplest scheme to produce this Holliday intermediate is to start with an antiparallel arrangement of the substrates (bottom left). The phosphodiester bonds involved in cleavage and exchange are indicated by the short vertical arrows.

DNA substrates (75), the square form of the Holliday is most easily arrived at from the antiparallel orientation of the *FRT* sites (Fig. 9).

Recombination initiated from antiparallel synapsis (as shown in Fig. 1) has been contradicted by results from electron microscopy of synaptic complexes formed by Flp. They seem to indicate that the productive form of the synapse has a parallel geometry. According to the results of Huffman and Levene (55), sites may come together in either orientation; yet, only the parallel synapse gives rise to Holliday junctions. Grainge et al. (41) have used a topological analysis to arrive at the geometry of the Flp synapse. In the most critical of their tests, they performed Flp recombination from a synaptic structure formed by the Tn*3* resolvase enzyme. These experiments were modeled after those performed by Kilbride et al. (61) demonstrating the ability of Cre to act from a preestablished resolvase synapse. Because three plectonemic negative supercoils are precisely trapped by resolvase, any additional nodes introduced by Flp for synapsis, or during strand exchange, can be indirectly arrived at by counting the knot nodes or catenane nodes in the products of recombination. For an inversion substrate (*FRT* sites in head-to-head orientation), the three-noded knot is uniquely enriched among the Flp recombination products when resolvase is present in the reaction (Fig. 10A). In the absence of resolvase, the product is largely the unknotted circle (which would migrate with the substrate band during electrophoresis), with some knotted products ranging from the simple trefoil knot to the more complex ones, but with odd numbers of nodes. For a deletion substrate (*FRT* sites in head-to-tail orientation), the "Flp alone" reaction produces mainly the unlinked circular products with a smattering of singly or multiply linked catenanes. In the presence of resolvase, however, there is a specific increase in the four-noded catenane product (Fig. 10B).

The formation of the trefoil and the 4-catenane during resolvase-assisted inversion and deletion reactions, respectively, mediated by Flp is most easily explained by antiparallel synapse of the *FRT* sites (Fig. 11). For the sites in head-to-head orientation, the three resolvase-trapped nodes would suffice to arrange them in the antiparallel mode. Because recombination by Flp from this synapse does not add or remove DNA crossings, the three synaptic nodes would be retained in the product as three-knot nodes. Because recombination results in relative inversion of DNA within the substrate, the sign of nodes would change from minus (in the synapse) to plus (in the product). For *FRT* sites in head-to-tail orientation, antiparallel arrangement would require an additional negative (because the substrate is negatively supercoiled) node, thus establishing a minus 4 synapse. The resultant product, with the same reaction mechanism which per se does not introduce any DNA crossings, would be a four-noded catenane.

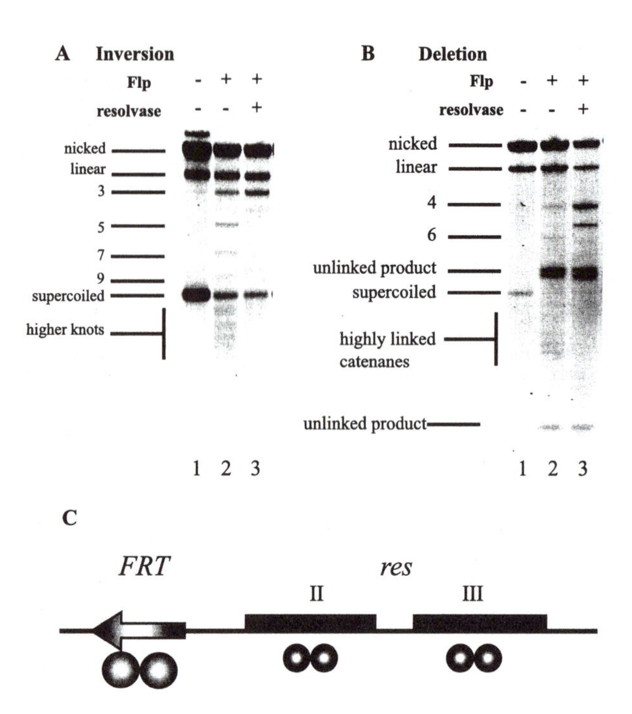

Figure 10. In the hybrid *FRT/res* plasmids, the accessory sites (*res* II/III) were used to build a topologically defined synapse, and recombination by Flp was conducted at the *FRT* sites. In both the inversion and deletion substrates, the *resII/III* sites were in direct repeat. The adjacent *FRT* sites were in either inverted (A) or direct (B) orientation. The substrate plasmid was reacted with Flp either with (lane 3) or without (lane 2) resolvase present, then nicked with DNase I and electrophoresed. An unknotted product resulting from Flp inversion (A) would comigrate with the substrate. The number of knot nodes (3, 5, etc.) and catenane nodes (4, 6, etc.) in the recombination products are indicated. (C) A representation of a single hybrid *FRT/res* site. The protein monomers that associate with the binding sites are represented as circles: light grey for Flp, dark grey for Tn*3* resolvase.

In principle, formation of the minus 4 catenane product is also satisfied by a minus 3 resolvase synapse, parallel arrangement of *FRT* sites, and a minus 1 crossing introduced by Flp during recombination (see the table in Fig. 11). However, if we apply this mechanism to the inversion reaction, the product should have been a 5-noded knot (minus 3 resolvase synapse, minus 1 additional node for parallel alignment, and minus 1 recombination crossing node). The trefoil/4-catenane combination of products, rather than the pentafoil/4-catenane combination or the trefoil/2-catenane combination (if the recombination crossing were plus 1 and not minus 1), strongly favors antiparallel site alignment within the synaptic complex.

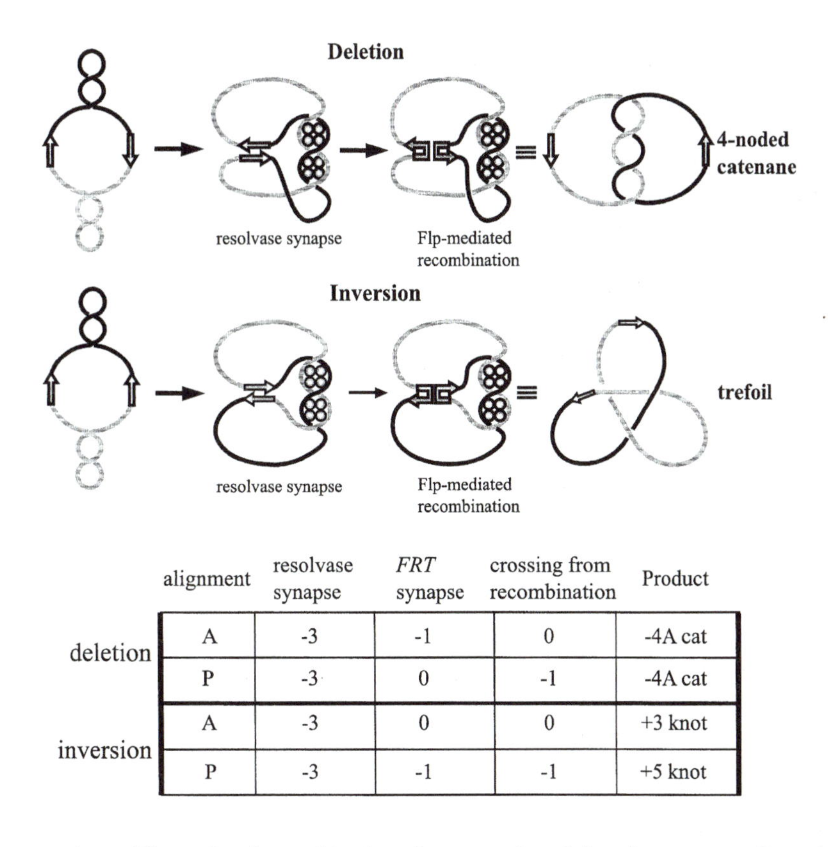

	alignment	resolvase synapse	*FRT* synapse	crossing from recombination	Product
deletion	A	-3	-1	0	-4A cat
	P	-3	0	-1	-4A cat
inversion	A	-3	0	0	+3 knot
	P	-3	-1	-1	+5 knot

Figure 11. Representations of Flp-mediated recombination adjacent to a 3-noded resolvase synapse. Recombination of direct-repeat *FRT* sites proceeding through antiparallel site alignment requires an extra DNA crossing, and the product is a 4-noded catenane. Recombination of the inverted *FRT* sites requires no additional crossing, and the product is a 3-noded knot. A possible pathway to the 4-noded catenane from a parallel synapsis of the direct *FRT* sites is outlined in the table. However, this pathway would predict a 5-noded (rather than 3-noded) knot for the inverted *FRT* sites. A and P indicate antiparallel and parallel alignment of sites.

The topological assays are supported by topoisomerase I relaxation assays or nick-ligation assays performed on inversion or deletion plasmids in the presence and absence of Flp(Y343F), the mutant Flp that binds the *FRT* site without cleaving it (41). After relaxation or ligation of the deletion substrate (*FRT* sites in head-to-tail orientation), the topoisomer distributions from the reactions with or without Flp(Y343F) are centered identically. For the inversion substrate (*FRT* sites in head-to-head orientation), the center of the topoisomer distribution is shifted by approximately minus 1 supercoil in the presence of Flp(Y343F). The results are interpreted to mean that Flp(Y343F) establishes a functional synapse by trapping zero DNA crossings between direct *FRT* sites and one negative crossing between inverted sites. The synapse geometry would therefore be antiparallel. The findings are surprising because they indicate that Flp senses site orientation before the chemical steps of recombination. Whereas protein-bound sites may associate in either orientation, the incorrect synapse is disfavored and likely dissociates rapidly. This selection may be imparted by the asymmetry in the sequence of the strand-exchange region (spacer) and by the directed bending of the *FRT* site bound by a Flp-dimer (75, 83).

The conclusions from the Grainge et al. (41) analyses agree with those arrived at by Crisona et al. (24) regarding the mode of synapsis during the lambda Int reaction. Crisona et al. (24) argue that a topological chirality is associated with the Int reaction (and perhaps the integrase/tyrosine family reactions in general). The DNA-crossing node introduced by Int to establish antiparallel synapse has a fixed sign: it is always a negative supercoil node. The finding by Grainge et al. (41) that nick ligation of a Flp inversion substrate in presence of Flp(Y343F) leads to a minus 1 (and not plus or minus 1) shift in topoisomer distribution would be consistent with the notion of topological chirality. However, the trivial possibility that the conditions used in the assay might have favored some underwinding of DNA cannot be ruled out at this time. Further experiments are needed to prove or disprove topological chirality during Flp recombination.

A MODEL FOR Flp RECOMBINATION: INTEGRATING ACTIVE-SITE ASSEMBLY, REACTION MECHANISM, AND THE GEOMETRY OF THE PROTEIN-DNA COMPLEX

All the physicochemical and geometric features of Flp recombination described so far fit nicely into the reaction scheme outlined in Fig. 1 (140). The synaptic structure is assembled by two Flp-bound *FRT* sites, bent identically and arranged in an antiparallel orientation within a nearly planar protein-DNA complex. Two active sites are assembled at the same ends of the spacer DNA, each one being derived from the two Flp monomers bound to the same DNA substrate. Strand cleavage and Holliday intermediate formation ensue. After this strand-exchange step, the reaction complex isomerizes to disassemble the first pair of active sites and assemble a new pair of active sites at the opposite spacer end. The catalytic contributions for the latter are made by Flp monomers bound on the DNA partners. The second cleavage and exchange step follows to resolve the Holliday intermediate.

Distinct DNA Cleavage Modes within a Common Architecture of the Integrase/Tyrosine Family Recombination Structures

The structures of the Cre recombination intermediates (39, 47, 48), together with that of the Flp-DNA complex (Color Plate 17 [see color insert]) (18), suggest that the pathway in Fig. 1 represents a general model for integrase/tyrosine family recombination. Because Cre and Flp cleave DNA in *cis* and *trans*, respectively, how does the same recombination model accommodate two distinct modes of DNA cleavage? Or, within a *cis* cleaving versus a *trans* cleaving recombinase dimer, what structural rules specify the monomer that provides the tyrosine nucleophile? The two possible alternatives, I and II or I and III, diagrammed in Fig. 12, schematically represent the *cis-trans* relationship. In the former case, equivalent recombinase monomers (shown in gray) provide the tyrosine, yielding cleavage at opposite ends of the spacer. In the latter case, nonequivalent monomers (gray during *cis* cleavage and white during *trans* cleavage) provide the tyrosine, causing cleavage to occur at the same end of the spacer. Tribble et al. (134) have addressed this issue by assaying DNA cleavage by Flp and Cre in synthetic substrates constrained in the

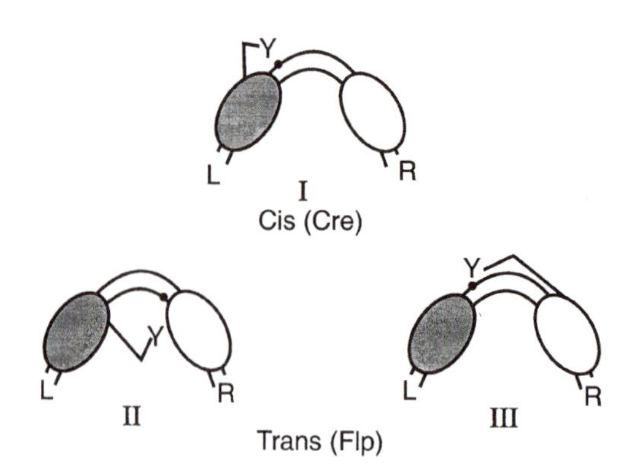

Figure 12. A recombinase dimer that cleaves in *cis* (I; Cre, for example) can, in principle, be related to one that cleaves in *trans* (Flp, for example) in one of two ways (II or III). The tyrosine (from the gray monomer) that cleaves at the left end of the spacer can be pushed to the right end in its functional orientation as shown in II. Alternatively, the tyrosine from the monomer at the right (white) can be moved to the left end as shown in III. The DNA configuration, indicated by the directed bend, is maintained the same in I, II, and III. The cleavage-susceptible phosphodiester positions are indicated by the black dots. Experimental evidence (134) relates Cre and Flp by I and III, respectively.

same bent orientation by strand-specific nucleotide bulges placed within the spacer (36, 75). They found that, for the same substrate geometry, the *cis* cleaving Cre and the *trans* cleaving Flp attack the scissile phosphodiester bond at the same spacer end. Thus, *cis* and *trans* cleavages by these two recombinases are related by the I/III states in Fig. 12 (and not the I/II states).

The Cre crystal structure reveals that it is the extreme carboxyl-terminal region of Cre consisting of the K, L, M, and N helices and the tail end of the protein that take part in the interactions responsible for the allosteric activation of one monomer within a dimer to cleavage competence (47). The perspective in Fig. 13A reveals the relevant portions of the carboxyl-terminal domain of Cre. The outline of the DNA backbone is included to indicate the direction of the DNA bend within the cleaved complex. A two-dimensional projection of the K-N helices and their connectivities is shown in the left panel of Fig. 13B. Gopaul and Van Duyne (40) have pointed out that, for a given recombinase subunit, an extensible tether between the L and M helices could, in principle, confer the *trans* configuration on M (which harbors the catalytic tyrosine), whereas the rest of the carboxyl-terminal peptide now folds back to produce the *cis* configuration of N. The strand cleavage patterns imposed by the spacer bulges in Cre and Flp reactions (134) are consistent with this model. In the representation of the Flp dimer in Fig. 13B (right), the gray L helix is connected to white M′-N′ helices, defining the peptide

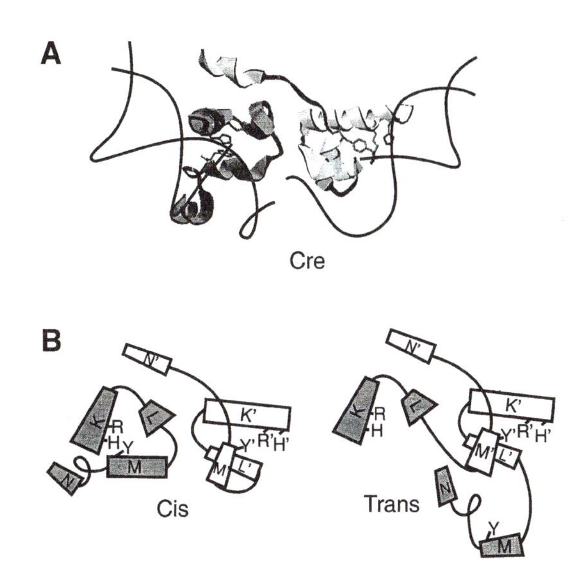

Figure 13. (A) Perspective of the cleaved Cre-DNA complex (47) as viewed from a vantage point on the carboxyl-terminal side of the protein. Cleavage has occurred at the right-hand side, executed in the *cis* mode by the Cre monomer in white. (B) The same perspective as in the top panel is redrawn at the left to represent the helices K-N and their linkages. At the right, the change in connectivity between the gray and white monomers to mediate *trans* cleavage is shown. Effectively, the gray L helix is linked to the white M′-N′ helices. Conversely, the white L′ helix is linked to the gray M-N helices. The new location of the M-N helices approximates the situation for the recombinase tetramer seen in DNA-protein cocrystal structures.

connectivity within a monomer of the *trans* cleaving recombinase. The M-N helices (now linked to the white L′ helix) are swung away toward the second substrate to designate their plausible location within a recombination complex containing a recombinase tetramer. Although the cyclic contact between adjacent recombinase monomers is counterclockwise for the Cre tetramer (Fig. 13B, left), it becomes clockwise for Flp (Fig. 13B, bottom right) to retain identity in strand cleavage. It seems plausible that a *trans* cleaving recombinase could have evolved differently, by maintaining the counterclockwise interaction seen in Cre, but adopting strand selectivity in cleavage opposite to that of Cre.

The evolution of two distinct DNA cleavage modes within the Int family recombinases by localized amino acid changes among its members does not contradict the rules of protein structure and folding. A relevant example is the conversion of staphylococcal nuclease from the native monomer form into a stable dimer by a short peptide deletion in a surface loop (43). The resultant stripping of the carboxyl-terminal α-helix from its normal position (*cis*) and its incorporation into an equivalent position on the adjoining monomer (*trans*) is analogous to the switching of peptide connectivities diagrammed in Fig. 13. More recently, it has been shown that interchanging two amino acid neighbors in the Arc repressor of phage P22 transforms the local peptide fold from a β strand into a right-handed helix (21). This structural alteration has little effect on the global protein stability or folding cooperativity.

The Significance of the Spacer Length and Sequence in Recombination

The length of the spacer sequence varies over a limited range within the integrase/tyrosine family. It is 8 bp for Flp, and 6 and 7 bp, respectively, for Cre and lambda Int. Several Flp spacers altered in sequence function efficiently in recombination as long as perfect homology with the partner spacer is maintained (122). Relatively weak reactivity of a subset of the variant spacers may be attributed to sequence-dependent alterations in DNA bendability or strand melting during recombination. The spacer length, unlike the spacer sequence, is almost inflexible. Recombination by Flp is reduced in 7- or 9-bp spacers, and is undetectable in <7- or >9-bp spacers.

As already indicated, the spacer DNA directly influences the course of the reaction by disallowing the simultaneous assembly of two active sites by a recombinase dimer. Furthermore, preferred bending of the spacer may impose a preferred sequence of strand exchange. An unsolved question concerns the relationship between the length of the spacer and its dynamics during the isomerization of the recombination complex to the Holliday resolving mode. When a cylindrical projection of the target site DNA is viewed down the helical axis, the scissile phosphates are nearly superimposed for the 6-bp spacer (Cre) and are offset by approximately 36° and 72°, respectively, for the 7-bp (Int) and 8-bp (Flp) spacers (Fig. 14A). The 6-bp spacer systems present the simplest case and suggest an elegant model of recombination supported by the Cre-DNA crystal structures. The first and second strand-exchange steps would swap 3 nucleotides (nt) each, with the branch point of the Holliday intermediate being placed at the center of the spacer. The required switch in the catalytic configurations for the two exchanges can be accomplished by scissors-like movement of the DNA arms, altering the included angle between them (Fig. 14B). What is the mechanism when the spacer length is increased to 7 or 8 bp? If the elegant symmetry and minimal movement of DNA or protein are to be retained for all Int-type recombinases, a mechanism that swaps exactly half the spacer segment at each cleavage/exchange step must be invoked. For a 7-bp spacer, this would mean

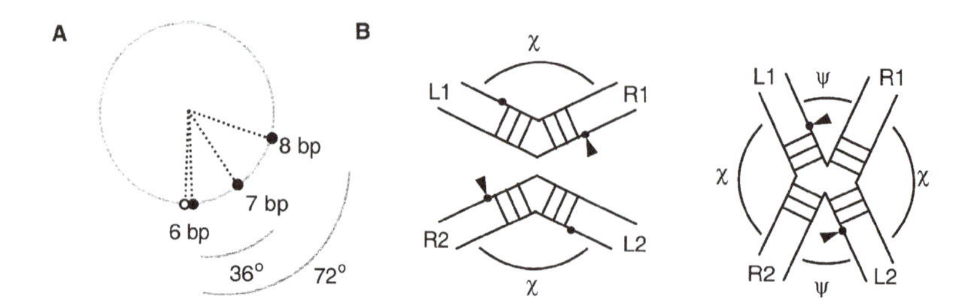

Figure 14. (A) In a cylindrical projection of the recombinase substrates with 6-, 7-, and 8-bp spacers as viewed along the helix axis, the scissile phosphodiesters at the near end are superimposed on each other (open circle). The relative angular displacements of those at the far end (filled circles) are indicated. (B) Recombinase monomers bound on L1/R1 and L2/R2 establish the included angle χ required for the first set of cleavages. The DNA dynamics associated with strand exchange (isomerization) establish χ between L1/R2 and L2/R1, promoting the second set of cleavages. The included angle gc; between two DNA arms represents the nonreactive configuration of the corresponding recombinase dimer.

that the central bases remain unpaired in the Holliday junction and become fully paired only after the resolution step. Perhaps the interactions between each recombinase and its target site have been evolutionarily optimized to accommodate this strand-swapping rule. A plausible alternative is that, in these systems, isomerization involves a higher degree of remodeling of the protein-DNA complex. In principle, branch migration of 1 or 2 bp and the concomitant DNA rotation of 36° or 72°, respectively, can accomplish this end for systems that exchange 7- and 8-bp spacers (25, 70, 71, 93). This would seem to require the disruption of some protein-protein interfaces and the subsequent formation of new but equivalent interfaces. Overall the process would be isoenergetic, and the energy barrier to the transition may not be high enough to slow down the reaction rate significantly. The disruption of contacts may be minimized by having the protein-protein interaction domains located on flexible linker regions. The finding that Flp, lambda Int, and XerC/XerD systems all appear to swap 3 nt of the spacer during strand exchange (3, 71, 93) would be consistent with the limited branch migration model.

Switching of Catalytic Association among Protein Subunits during Isomerization of the Recombination Complex

The isomerization step of recombination refers to the switching of the functional active sites from one end of the spacer to the other. During this step, catalytic interactions are broken between Flp monomers bound to the same substrate and established between those bound to partner substrates (which are now coalesced into the Holliday intermediate). The presence of two distinct forms of the Holliday intermediate along the reaction pathway was first suspected from kinetic evidence suggesting that the isomerization process is rate limiting at 0°C (89). What structural features of the Holliday junction are important in determining which two of the four strands will be cleaved? Note that in the conventional X form Holliday junction, one pair of strands is continuous (the helix axes being colinear), whereas the second pair is discontinuous or crossed (the helix axes being bent). The question of cleavage selectivity was addressed in the lambda Int and XerC/XerD systems with synthetic Holliday junctions in which particular strands were forced to be in the crossed configuration by sequence design or by the use of DNA tethers (4, 5). In these substrates, cleavage susceptibility seems to be dictated by the strand configuration at the branch point of the protein-free junction. It is the scissile phosphodiesters on the crossed strands that are targeted preferentially for cleavage. The isomerization step may then be described as a switch between two X forms of the Holliday junction in which the strand-crossing pattern is reversed with concomitant restacking of the helical arms (Fig. 15).

The helix-restacking model suggests a simple and elegant means for ensuring the forward progression of recombination, once the initial strand exchange has been completed. The strand-swapping mechanism, combined perhaps with the freedom for limited branch migration, could place the Holliday junction branch point within a sequence context that promotes the crossed configuration of the parental (unexchanged) strands. Their preferential cleavage susceptibility would direct the resolution reaction toward the recombinant products. Results with a series of synthetic Flp Holliday junctions, conformationally constrained by local sequences or by strand tethering, do not conform to this model (76). In these substrates, no strong resolution bias is observed in favor of the strands designed to assume the "crossed" configura-

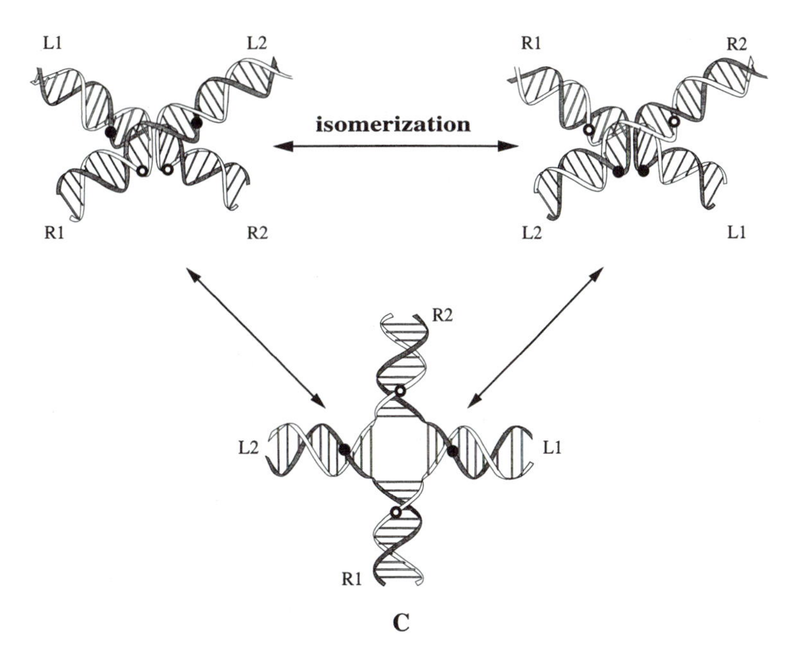

Figure 15. Isomerization between two alternative stacking patterns of the DNA arms within the Holliday junction can bring two scissile phosphodiesters into the proximal configuration, while simultaneously placing the other two in the distal configuration. The former pair is poised for cleavage and exchange, and the latter pair is in the nonreactive state. The functional outcome of this mode of isomerization is the same as that drawn in Fig. 1. In both cases, isomerization switches the active recombinase pairs between the L1/R1 plus L2/R2 set on the one hand and the R2/L1 and R1/L2 set on the other.

tion within the unbound junction. The Flp results can be reconciled with the lambda Int and XerC/XerD results in a less extreme form of the helix-restacking model, exemplified by the Cre crystal structures. In the nearly 4-fold symmetric, planar Cre-Holliday junction complex (39), the structural distinction between the scissile and nonscissile strands may be equated with the crossed and noncrossed (or continuous) strand arrangements, respectively. This distinction is also conspicuous in the uncleaved synaptic complex formed by a mutant Cre protein (48). In the "crossed" strands, the phosphate backbones at the center of the spacer region are directed away from the recombinase N-terminal binding surfaces toward the interior of the synapse and are relatively free from direct protein contacts. In the "continuous" strands, the central spacer backbones face the N-terminal recombinase binding surface and there are considerably more contacts, both direct and indirect, over the entire spacer length on this strand. The dynamics of isomerization in this revised model would be consistent with a reaction scheme (see Fig. 1) in which the prechemical and chemical steps of recombination are performed in a set of geometrically related and close to planar DNA-protein assemblies. Isomerization would be restricted to a relatively minor motion of the DNA arms in the unstacked Holliday junction. It should be noted from Fig. 1 and 15 that the outcome of either mode of isomerization is the same functionally, namely, a reconfiguration of the catalytic associations between a given recombinase monomer and two other monomers (one bound to the same substrate and the other bound to the second substrate).

The absence of strong resolution bias by Flp in tethered junctions or junctions with sequence-altered spacers (76) suggests that the binding energy derived from Flp-Holliday junction interactions is sufficient to overcome any bias in the direction of the isomerization step imposed by strand-crossing preferences.

THE EVOLUTION OF THE Flp ACTIVE SITE: CATALYTIC FLEXIBILITY AND FUNCTIONAL COMPLEXITY

The active site of Flp and related proteins designed to perform the complex operation of DNA recombination must couple elementary chemical mechanisms for strand cutting and joining to the conformational transitions required for exchanging strands between two double-helical partners. Thus, the retention of key catalytic motifs among enzymes that catalyze a variety of phosphoryl transfer reactions in nucleic acids (nucleases, topoisomerases, recombinases, and others) would not be surprising. The identification of Arg191-His305-Arg308 as a key catalytic triad in Flp (13, 32, 68, 74, 97, 102, 103) suggests plausible mechanistic similarities to pancreatic RNase and staphylococcal nuclease reactions (reviewed in reference 31). The step-arrest phenotypes, blockage of DNA cleavage or strand transfer or both, associated with point mutations in the triad cluster would be consistent with their role in stabilizing the pentaphosphate intermediate (or transition state), activating the nucleophile, and/or stabilizing the leaving group (74, 103). In accordance with this idea, Flp can be coaxed to perform reactions that are chemically related but not directly relevant to recombination. For example, a mutant of Flp lacking Tyr343 can mediate strand cleavage precisely at the normal scissile phosphodiester bond when provided with alternative nucleophiles such as the peroxide anion or a phenolate moiety (63, 68). Furthermore, in reactions that mimic the strand-joining step, glycerol and other polyhydric alcohols can take the place of the 5′-hydroxyl group of DNA (65).

Flp Harbors RNase and Topoisomerase Activities

With the help of mixed DNA-RNA substrates, Xu et al. (145, 146) have revealed two site-specific RNA cleavage activities of the Flp protein (Fig. 16). Flp-RNase I closely follows the normal recombination mechanism. It requires the active-site tyrosine, targets the same phosphodiester that takes part in DNA recombination, and seems to proceed by a phosphotyrosyl intermediate. By contrast, Flp-RNase II mimics the pancreatic RNase mechanism. It is independent of the active-site tyrosine and targets the phosphodiester immediately 3′ to the recombination phosphate. Sekiguchi and Shuman (120) have demonstrated an RNase activity in vaccinia virus topoisomerase I that is strikingly similar to Flp-RNase I. The only difference between the two is that the final product is a 3′-phosphate in Flp, whereas it is a 2′, 3′-cyclic phosphate in the topoisomerase. The activity equivalent to Flp-RNase II has not been reported in the vaccinia enzyme. Xu et al. (146) have also demonstrated an intrinsic topoisomerase activity associated with Flp. In appropriately designed DNA substrates, and under special reaction conditions, Flp can perform relaxation of negative supercoils. This cryptic topoisomerase activity must not be confused with the more obvious knotting reaction observed by Beatty et al. (6) during Flp recombination in negatively supercoiled substrates. Knotting is likely the result of random trapping of supercoils during *FRT* synapsis, which become fixed in the product after strand exchange.

That Flp can be induced to function as a topoisomerase and that both Flp and an authentic topoisomerase I contain intrinsic RNA cleavage functions suggest possible routes by which common catalytic motifs might be incorporated into functionally related active sites (11, 42, 120, 126, 146). Because enzymes have to perform numerous biological reactions with a limited set of useful functional groups, the evolutionary optimization of an active site for a certain chemical task may inadvertently include the catalytic design, albeit an inefficient one, for a related task as well. Because DNA breakage is inherent in the problem of recombination, the cryptic nuclease activity housed within the Flp active site is not surprising. Alternatively, evolution affords the opportunity to start with a preexisting catalytic strategy (even a clumsy

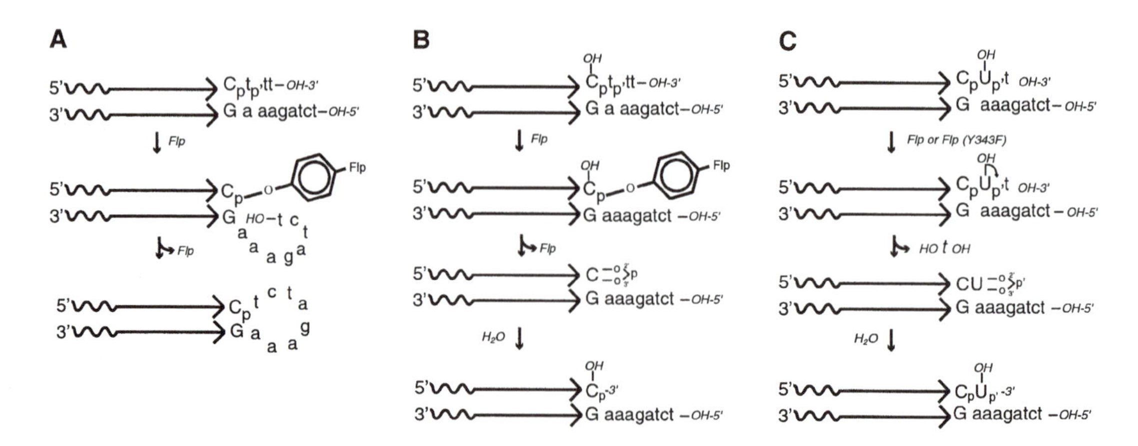

Figure 16. A half-site recombination mediated by Flp (A) is compared with Flp-RNase I (B) and Flp-RNase II (C). Flp RNase I, like the half-site reaction, is mediated by a "shared" active site and proceeds by Tyr343-mediated cleavage at the equivalent phosphodiester position. The only difference between the two is in the nucleophile used for the subsequent step: a 5′-hydroxyl group in panel A and a 2′-hydroxyl group in panel B. Flp-RNase II has a different target specificity and proceeds by a distinct mechanism that bypasses the tyrosine-mediated cleavage step.

one), and modify, refine, and expand it to meet the demands of novel, more complex biological reactions. In this view, an early step in the evolution of the recombinase must have been the formation of an elementary nuclease active site containing the RHR triad. The emergence of a precisely positioned protein nucleophile could then set the stage for the maturation of this proactive site into a "topoisomerase" and further into a "recombinase" active site. One obvious consequence of this catalytic design is that, within the "activated" active site, breakage at a scissile phosphodiester may be mediated by nonprotein nucleophiles as well: by an oriented hydroxyl group or a hydroxide ion derived from water (the nuclease active site), by a nucleophile derived from the nucleic acid substrate (the RNase active site), or even by an exogenously added nucleophile (hydrogen peroxide, tyramine, or phenol) that can gain access to the reaction center (63, 68).

The RNA cleavage reactions of recombinases and topoisomerases have also fueled speculations on the possible relevance of these activities in vivo (11, 120). The suggested roles include the triggering of repair pathways for the removal of ribonucleotides from DNA or involvement in RNA-splicing reactions. Whether these enzymes once performed reactions with which they are not presently identified, their active sites exemplify how novel and diverse activities can be derived from conserved chemical themes (126).

A previously unnoticed structural similarity between the integrase/tyrosine recombinases and the AraC family of transcriptional activators has come to light recently (35, 45). Several of the catalytic residues of the recombinases are located within two helix-turn-helix (HTH) motifs, suggesting that an ancient DNA-binding module might have duplicated and evolved to acquire enzymatic function. The tyrosine nucleophile for DNA breakage is harbored by the second helix of the carboxyl-terminal HTH in Cre. In the structures of other strand-breaking and -resealing enzymes (topoisomerase I, topoisomerase II, and archaeal topoisomerase VI), the catalytic tyrosine is present in an HTH region, although its position is not strictly conserved. These examples suggest parallel, independent evolution of homologous enzymatic domains with alternative placements of the catalytic residues.

The recently solved crystal structure of the *Nae*I restriction enzyme, an endonuclease that can be converted to a topoisomerase/recombinase by a leucine-to-lysine substitution, reveals a dimer with a bidomainal organization of each monomer subunit. The amino-terminal domain and the carboxyl-terminal domain contain potential DNA-binding motifs corresponding to the endonuclease and topoisomerase activities, respectively (52). The *Nae*I structure provides support for the evolutionary divergence of DNA-processing enzymes from a limited number of ancestral proteins or for the convergence of a finite set of catalytic motifs in them.

THE Flp ENZYME: A POOR CATALYST BUT AN OPTIMIZED FACILITATOR PROTEIN

The catalytic power of the Flp protein is quite weak compared with most other enzymes. For a long time, it was debated whether Flp and related recombinases were enzymes in the true sense, or rather acted stoichiometrically. Gates and Cox (34) have shown that Flp does turn over, although the turnover number is a sluggish 0.12 min^{-1} per Flp monomer. Furthermore, the dissociation of Flp from product is a slow process, occurring in a minimum of two steps (141). One or both of these steps seem to be rate limiting for Flp turnover under standard in vitro reaction conditions. Assuming that the nuclease active site was the progenitor of the recombinase active site, the price for the evolution of subunit cooperativity and novel chemical attributes required for coordinated cleavage and directed exchange of strands must have been paid for by a reduction in overall catalytic efficiency. A high turnover number would not be advantageous for Flp, whose proposed function is to trigger plasmid amplification by a recombination reaction that converts a bidirectional replication fork into two unidirectional forks on a circular template. Because a second recombination event would restore bidirectionality of the forks, the longer the interval between the two recombination events, the higher the degree of amplification. Flp is only one of many examples in biological catalysis in which catalytic prowess is sacrificed for the sake of a directed physiological outcome or purposes of fine-tuned regulation.

Our current model for Flp recombination (as drawn in Fig. 1) has one drawback. It does not explain how, after a recombination event, an immediate reversal of the reaction is prevented. Given that the dissociation of the protein from the products is slow, the recombination complex at the end of a reaction has the precise geometry to run the reaction backward. A defined order of strand exchange would provide a plausible solution to this problem, in which case the complex would have to be reconfigured to initiate the second round of reaction. At least with the minimal *FRT* sites used in standard in vitro reactions, no strong evidence for ordered strand exchange exists. However, it is possible that the situation may be different in the native 2μm circle that is packaged into minichromosomes and whose *FRT* sites include a third Flp-binding element (see Color Plate 16). Or, alternative mechanisms must exist to ensure that iteration of recombination does not negate amplification.

Flp APPLICATIONS

The Flp/*FRT* system has been used as a molecular tool to mediate directed genetic rearrangements in bacteria, fungi, plants, flies, and animals. Among the integrase/tyrosine family recombinases, Flp and Cre are the enzymes of choice for this purpose. They act on simple target sites, they mediate intramolecular inversions, deletions, or intermolecular exchanges, and the reactions require no cofactors and are indifferent to substrate topology. With higher cell systems, Cre has had greater popularity than Flp (7), probably because of occasional difficulties in Flp expression caused by cryptic poly(A) addition sites or splice sites within the Flp coding sequence (95).

Although the expression of active Flp in prokaryotes has been established for years (22), the use of Flp for routine genomic manipulations in bacteria is a more recent event. In *Escherichia coli* and in *Pseudomonas,* Flp/*FRT* constructs have been used for site-specific chromosomal insertion of DNA (53, 54), for targeted gene deletions in the chromosome (50, 118), and for expression of specific proteins from selected chromosomal locales (51). Other applications include excision of large chromosomal DNA circles for sequencing projects (104, 142, 143), rescue of pathogenicity islands (105), and the removal of exogenous sequences integrated into the chromosome (20, 87). Integration or excision events can be made virtually irreversible by suitably manipulating the two *FRT* sites taking part in the recombination (53, 106). The left Flp-binding element of one site and the right Flp-binding element of the other are mutated to significantly lower protein-DNA binding affinity. In the forward reaction, each *FRT* site is rescued by cooperativity from its normal binding element. However, the products of recombination are a wild-type *FRT* site and a doubly mutant *FRT* site. The doubly mutant *FRT* site is a very poor substrate for recombination.

An impressive application of Flp in eukaryotes has been the creation of mosaic flies in *Drosophila* by site-specific recombination between homologous chromosomes (37, 38, 133, 147). Analysis of these mosaics allows the tracking of cell lineages during development. The "Flp-out" technique in which recombination is used to abolish or activate gene expression, by excising the gene of interest or an intervening sequence within that gene, has also been used effectively in *Drosophila* developmental biology (37, 38, 130, 132). Similar experimental designs have been extended to plants and animals for the molecular analysis of development and for the creation of transgenics (2, 81, 86, 95, 96, 127). The isolation of a thermostable variant of Flp (10) and the design of ligand-dependent recombination (by using a hybrid Flp fused to the steroid hormone ligand binding domain, for example) (82) will further enhance the utility of Flp in genetic manipulations of higher systems.

Aside from its natural role in the physiology of the 2μm plasmid, Flp has been used to perform many artificial functions in its native host *Saccharomyces cerevisiae.* Flp-mediated unequal sister chromatid exchange via integrated copies of the 2μm circle inverted repeat results in dicentric chromosomes, whose breakage is repaired by copying the allelic information from the homologues (30). This Flp-induced instability forms the basis of a rapid method for mapping cloned

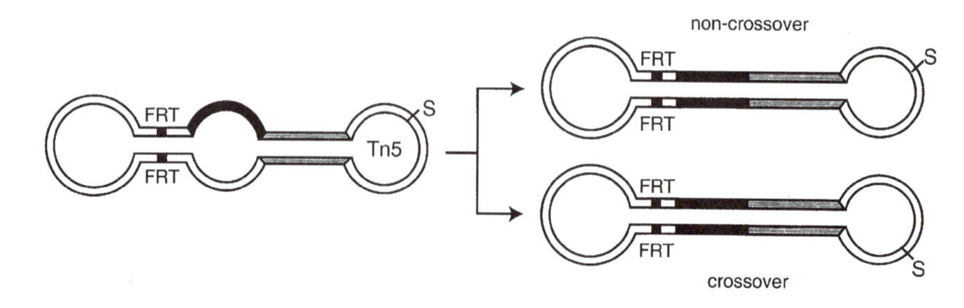

Figure 17. In a 2μm circle-derived hybrid plasmid, the Tn*5* insertion is shown as a lollipop at the right. DNA segments drawn as parallel lines represent the 2μm circle and Tn*5* repeats. A gene conversion event initiated by a Flp-induced break in one of the *FRT* sites (the example shown here corresponds to cutting off the bottom *FRT* site) can be terminated and resolved within the Tn*5* repeats. The resulting plasmids contain a large inverted repeat, made up of the 2μm circle repeat, the Tn*5* repeat, and the intervening gene-converted DNA. The plasmid products at the top and bottom are formed by the noncrossover and crossover modes of resolution, respectively. The parental and inverted configurations of Tn*5* are indicated by the relative locations of the *Sal*I restriction enzyme site (S).

yeast genes to the chromosomes from which they are derived (29). DNA breaks at the *FRT* sites caused by Flp are targets for host repair enzymes, leading to gene conversion events that are resolved with or without crossover of the flanking markers (57). Flp-mediated inversion in yeast of the Tn*5* transposon present on hybrid plasmids derived from the 2μm circle (60) (Fig. 17) is readily accommodated by a conversion event that is resolved within the Tn*5* repeats in the crossover mode (57). The original suspicion, based on local sequence similarities between the 2μm circle and Tn*5* repeats, that Flp might act directly on the Tn*5* repeats (60) was invalidated by the subsequent precise mapping of the *FRT* site. When cells are overdosed with *FRT* breaks by expressing a Flp variant that can cut but not rejoin DNA, Flp(H305L), from a strong promoter, the unrepaired DNA lesions cause the nicked plasmids to be lost at an easily detectable frequency (135). The [cir^0] strains thus produced have proven valuable in examining the role of 2μm circle-encoded components in plasmid-partitioning and copy-number control.

REGULATION OF Flp IN VIVO

Under conditions of steady-state copy number, the 2μm plasmid depends only on the partitioning system for its stable propagation. It is only when a missegregation event occurs that the Flp system is triggered into action to amplify the plasmid and reestablish normal copy number. Plasmid missegregation is quite infrequent. Density-shift experiments have shown that almost all plasmid molecules go through one and only one round of replication during one cell division cycle (149). Thus, the Flp-mediated amplification reaction is a safety device that is kept silent during normal growth. At least one mode of Flp regulation is mediated at the level of the *FLP* gene transcription. The negative regulator is believed to be a bipartite complex between the plasmid-coded Rep1 and Rep2 proteins (reviewed in reference 8). These two proteins show both self- and cross-associations in vivo and in vitro (1, 119, 136). The intracellular concentration of the Rep1-Rep2 complex would be determined by the balance among these interactions, and may provide an indirect readout of the plasmid copy number. In addition to the negative control of Flp gene expression, there seems to be a positive control as well, mediated by the plasmid-encoded Raf1 protein. The two alternative modes of regulation would permit the rapid commissioning and decommissioning of the amplification system as a function of the plasmid copy number (reviewed in reference 8).

In addition to the regulation of Flp expression, is there also regulation of Flp activity? The latter mechanism would be attractive for a rapid-response system, yet this question has not been addressed seriously. In principle, Flp can be kept quiescent by a reversible chemical modification or by a reversible association with a protein partner or an effector molecule. The active-site nucleophile for phosphodiester cleavage being a tyrosine, a combination of a tyrosine kinase and a tyrosine phosphotase could rapidly convert this residue from an inactive to active state or vice versa. Because the tyrosine takes part in a polynucleotidyl transfer reaction during recombination, it is not unreasonable to expect that it might serve as the target for phosphoryl transfer from a nucleoside triphosphate as well.

So far all the in vitro studies with Flp have been conducted with synthetic DNA substrates, plasmid molecules isolated from *E. coli* or linear DNA fragments derived from them. By contrast, the native substrate of Flp in yeast is a nuclear resident circular minichromatin packaged into a nucleosomal array. How does the protein interact with its target site in this nucleoprotein complex? And how does the chromatin structure affect the mechanics of the two-step strand-exchange reaction? The DNA region bracketing the *FRT* site is DNase I sensitive when chimeric plasmids containing a cloned copy of this region are isolated from yeast as nucleoprotein complexes (28). This observation suggests an altered pattern of chromatin organization in the vicinity of the *FRT* sites. If (as demanded by the Futcher model) plasmid amplification requires the coupling of DNA replication to recombination, the dynamics of chromatin reorganization or remodeling associated with the replication process may directly influence the recombination event.

Now that most interesting mechanistic questions of the Flp reaction have been solved satisfactorily, the next challenge is to understand how the reaction is regulated in the native physiological environment such that the copy number of the 2μm plasmid is maintained at its optimum value.

Acknowledgments. We acknowledge the contributions of several colleagues past and present to our efforts in exploring the mechanism of action of the Flp recombinase. We thank Michael Cox and Paul Sadowski for critically reading a nearly final version of this chapter. We are grateful to Phoebe Rice for providing the crystal structure of the Flp-DNA complex.

The work in the Jayaram laboratory on recombination has been supported over the years by awards from the National Institutes of Health, the National Science Foundation, the Robert F. Welch Foundation, the Texas Board for Coordinating Higher Education, and the Human Frontiers in Science Program.

ADDENDUM

While this chapter was being written, the crystal structure of the Flp-Holliday junction complex (18) was still unpublished and information was only available as a personal communication from the authors. The details revealed by the published structure, as they pertain to earlier solution studies, call for a few additional comments. The structure agrees with experiments with bulged DNA substrates (75) that revealed oriented DNA bend as the agent of half-of-the-sites activity in Flp (assembly of only one active site by a substrate-bound Flp dimer). Exclusion of one of the two possible active sites is caused by the misorientation of the catalytic tyrosine. The DNA geometry and cleavage positions within the crystal structure are in conformity with the finding that for a fixed substrate orientation, the *cis*-cleaving Cre and the *trans*-cleaving Flp would target the phosphodiester at the same side of the spacer (134). It is discernible from the structure how three monomers of Flp bound to an artificial three-armed DNA substrate (Y junction) could mediate its resolution into a linear product and a hairpin product, as had been demonstrated earlier (78). The structure also makes it clear that, in the context of a Flp tetramer bound to a Holliday junction, two active sites required for its resolution cannot be derived from a trimer but only from the whole tetramer acting as a dimer of dimers. The ability of the fourth monomer to preclude the trimer from assembling two active sites had also been surmised correctly from solution experiments (72).

A recent report by Shaikh and Sadowski (125) has used chimeric proteins containing amino- and carboxyl-terminal swaps between Flp and Cre to demonstrate that Cre donates its catalytic tyrosine in *cis* during strand cleavage, whereas Flp does so in *trans*. This finding now settles the uncertainty arising from a previous claim that Cre cleaves in *trans* (124), which is at odds with the crystal structure of the cleaved Cre-DNA complex (47). It is useful to point out that potential artifacts in the determination of *cis* or *trans* cleavage can be avoided by using a half-site complementation assay that scores not merely strand cleavage but also the subsequent strand-joining event (as done for Flp) (16). The rationale is that an RHR triad mutant of the recombinase is affected in both strand cleavage and joining steps in half-site reactions, whereas a tyrosine mutant is normal in the joining reaction. Thus, cleavage events in a half-site bound by a tyrosine mutant as a result of *trans* donation of tyrosine by a triad mutant will be processed efficiently through the joining step to yield hairpin recombinants. On the other hand, weak cleavage reactions mediated by a triad mutant of a *cis*-cleaving recombinase will not yield the corresponding hairpin product. By setting hairpin formation as the criterion for *trans* cleavage during complementation assays with half sites, we have not observed Cre to cleave in *trans*.

REFERENCES

1. **Ahn, Y. T., X. L. Wu, S. Biswal, S. Velmurugan, F. C. Volkert, and M. Jayaram.** 1997. The 2 micron-plasmid-encoded Rep1 and Rep2 proteins interact with each other and colocalize to the *Saccharomyces cerevisiae* nucleus. *J. Bacteriol.* **179:**7497–7506.
2. **Aladjem, M. I., L. L. Brody, S. O'Gorman, and G. M. Wahl.** 1997. Positive selection of FLP-mediated unequal sister chromatid exchange products in mammalian cells. *Mol. Cell. Biol.* **17:**857–861.
3. **Arciszewska, L., I. Grainge, and D. Sherratt.** 1995. Effects of Holliday junction position on Xer-mediated recombination *in vitro*. *EMBO J.* **14:**2651–2660.
4. **Arciszewska, L., I. Grainge, and D. Sherratt.** 1997. Action of site-specific recombinases XerC and XerD on tethered Holliday junctions. *EMBO J.* **16:**3731–3743.
5. **Azaro, M. A., and A. Landy.** 1997. The isomeric preference of Holliday junctions influences resolution bias by lambda integrase. *EMBO J.* **16:**3744–3755.
6. **Beatty, L. G., D. Babineau-Clary, C. Hogrefe, and P. D. Sadowski.** 1986. FLP site-specific recombinase of yeast 2-micron plasmid. Topological features of the reaction. *J. Mol. Biol.* **188:**529–544.
7. **Bethke, B. D., and B. Sauer.** 2000. Rapid generation of isogenic mammalian cell lines expressing recombinant transgenes by use of Cre recombinase. *Methods Mol. Biol.* **133:**75–84.
8. **Broach, J. R., and F. C. Volkert.** 1991. Circular DNA plasmids of yeast, p. 297–332. *In* J. R. Broach, J. R. Pringle, and E. W. Jones (ed.), *The Molecular and Cellular Biology of the Yeast* Saccharomyces, vol. 1. Cold Spring Harbor Press, New York.
9. **Bruckner, R. C., and M. M. Cox.** 1986. Specific contacts between the FLP protein of the yeast 2-micron plasmid and its recombination site. *J. Biol. Chem.* **261:**11798–11807.
10. **Buchholz, F., P. O. Angrand, and A. F. Stewart.** 1998. Improved properties of FLP recombinase evolved by cycling mutagenesis. *Nat. Biotechnol.* **16:**657–662.
11. **Burgin, A. B., Jr.** 1997. Can DNA topoisomerases be ribonucleases? *Cell* **91:**873–874.
12. **Chen, J. W., B. Evans, H. Rosenfeldt, and M. Jayaram.** 1992. Bending-incompetent variants of Flp recombinase mediate strand transfer in half-site recombinations: role of DNA bending in recombination. *Gene* **119:**37–48.
13. **Chen, J. W., B. R. Evans, S. H. Yang, H. Araki, Y. Oshima, and M. Jayaram.** 1992. Functional analysis of box I mutations in yeast site-specific recombinases Flp and R: pairwise complementation with recombinase variants lacking the active-site tyrosine. *Mol. Cell. Biol.* **12:**3757–3765.
14. **Chen, J. W., B. R. Evans, S. H. Yang, D. B. Teplow, and M. Jayaram.** 1991. Domain of a yeast site-specific recombinase (Flp) that recognizes its target site. *Proc. Natl. Acad. Sci. USA* **88:**5944–5948.
15. **Chen, J. W., B. R. Evans, L. Zheng, and M. Jayaram.** 1991. Tyr60 variants of Flp recombinase generate conformationally altered protein-DNA complexes. Differential activity in full-site and half-site recombinations. *J. Mol. Biol.* **218:**107–118.
16. **Chen, J.-W., J. Lee, and M. Jayaram.** 1992. DNA cleavage in *trans* by the active site tyrosine during Flp recombination: switching protein partners before exchanging strands. *Cell* **69:**647–658.
17. **Chen, J. W., S. H. Yang, and M. Jayaram.** 1993. Tests for the fractional active-site model in Flp site-specific recombination. Assembly of a functional recombination complex in half-site

and full-site strand transfer. *J. Biol. Chem.* **268:** 11417–14425.

18. **Chen, Y., U. Narendra, L. E. Iype, M. M. Cox, and P. A. Rice.** 2000. Crystal structure of a Flp recombinase-Holliday junction complex. Assembly of an active oligomer by helix swapping. *Mol. Cell* **6:**885–897.
19. **Cheng, C., P. Kussie, N. Pavletich, and S. Shuman.** 1998. Conservation of structure and mechanism between eukaryotic topoisomerase I and site-specific recombinases. *Cell* **92:** 841–850.
20. **Cherepanov, P. P., and W. Wackernagel.** 1995. Gene disruption in *Escherichia coli:* TcR and KmR cassettes with the option of Flp-catalyzed excision of the antibiotic-resistance determinant. *Gene* **158:**9–14.
21. **Cordes, M. H. J., N. P. Walh, C. J. McKnight, and R. T. Sauer.** 1999. Evolution of a protein fold *in vitro. Science* **284:** 325–327.
22. **Cox, M. M.** 1983. The FLP protein of the yeast 2-micron plasmid: expression of a eukaryotic genetic recombination system in *Escherichia coli. Proc. Natl. Acad. Sci. USA* **80:** 4223–4227.
23. **Cox, M. M.** 1989. DNA inversion in the 2μm plasmid of *Saccharomyces cerevisiae,* p. 661–670. *In* D. E. Berg and M. M. Howe (ed.), *Mobile DNA.* American Society for Microbiology, Washington, D.C.
24. **Crisona, N. J., R. L. Weinberg, B. J. Peter, D. W. Sumners, and N. R. Cozzarelli.** 1999. The topological mechanism of phage lambda integrase. *J. Mol. Biol.* **289:**747–775.
25. **Dixon, J. E., A. C. Shaikh, and P. D. Sadowski.** 1995. The Flp recombinase cleaves Holliday junctions in *trans. Mol. Microbiol.* **18:**449–458.
26. **Esposito, D., and J. J. Scocca.** 1997. The integrase family of tyrosine recombinases: evolution of a conserved active site domain. *Nucleic Acids Res.* **25:**3605–3614.
27. **Evans, B. R., J. W. Chen, R. L. Parsons, T. K. Bauer, D. B. Teplow, and M. Jayaram.** 1990. Identification of the active site tyrosine of Flp recombinase. Possible relevance of its location to the mechanism of recombination. *J. Biol. Chem.* **265:** 18504–18510.
28. **Fagrelius, T. J., A. D. Strand, and D. M. Livingston.** 1987. Changes in the DNase I sensitivity of DNA sequences within the yeast 2 micron plasmid nucleoprotein complex effected by plasmid-encoded products. *J. Mol. Biol.* **197:**415–423.
29. **Falco, S. C., and D. Botstein.** 1983. A rapid chromosome-mapping method for cloned fragments of yeast DNA. *Genetics* **105:**857–872.
30. **Falco, S. C., Y. Li, J. R. Broach, and D. Botstein.** 1982. Genetic properties of chromosomally integrated 2 mu plasmid DNA in yeast. *Cell* **29:**573–584.
31. **Fersht, A.** 1984. *Enzyme Structure and Function.* W. H. Freeman Company, New York.
32. **Friesen, H., and P. D. Sadowski.** 1992. Mutagenesis of a conserved region of the gene encoding the FLP recombinase of *Saccharomyces cerevisiae.* A role for arginine 191 in binding and ligation. *J. Mol. Biol.* **225:**313–326.
33. **Futcher, A. B.** 1986. Copy number amplification of the 2 micron circle plasmid of *Saccharomyces cerevisiae. J. Theor. Biol.* **119:**197–204.
34. **Gates, C. A., and M. M. Cox.** 1988. FLP recombinase is an enzyme. *Proc. Natl. Acad. Sci. USA* **85:**4628–4632.
35. **Gillette, W. K., S. Rhee, J. L. Rosner, and R. G. Martin.** 2000. Structural homology between MarA of the AraC family of transcriptional activators and the integrase family of site-specific recombinases. *Mol. Microbiol.* **35:**1582–1583.
36. **Gohlke, C., A. I. Murchie, D. M. Lilley, and R. M. Clegg.** 1994. Kinking of DNA and RNA helices by bulged nucleotides observed by fluorescence resonance energy transfer. *Proc. Natl. Acad. Sci. USA* **91:**11660–11664.
37. **Golic, K. G., and M. M. Golic.** 1996. Engineering the Drosophila genome: chromosome rearrangements by design. *Genetics* **144:**1693–1711.
38. **Golic, K. G., and S. Lindquist.** 1989. The FLP recombinase of yeast catalyzes site-specific recombination in the Drosophila genome. *Cell* **59:**499–509.
39. **Gopaul, D. N., F. Guo, and G. D. Van Duyne.** 1998. Structure of the Holliday junction intermediate in Cre-loxP site-specific recombination. *EMBO J.* **17:**4175–4187.
40. **Gopaul, D. N., and G. D. Van Duyne.** 1999. Structure and mechanism in site-specific recombination. *Curr. Opin. Struct. Biol.* **9:**14–20.
41. **Grainge, I., D. Buck, and M. Jayaram.** 2000. Geometry of site alignment during Int family recombination: antiparallel synapsis by the Flp recombinase. *J. Mol. Biol.* **298:**749–764.
42. **Grainge, I., and M. Jayaram.** 1999. The integrase family of recombinases: organization and function of the active site. *Mol. Microbiol.* **33:**449–456.
43. **Green, S. M., A. G. Gittis, A. K. Meeker, and E. E. Lattman.** 1995. One-step evolution of a dimer from a monomeric protein. *Nat. Struct. Biol.* **2:**746–751.
44. **Grindley, N. D.** 1997. Site-specific recombination: synapsis and strand exchange revealed. *Curr. Biol.* **7:**608–612.
45. **Grishin, N. V.** 2000. Two tricks in one bundle: helix-turn-helix gains enzymatic activity. *Nucleic Acids Res.* **28:** 2229–2233.
46. **Gronostajski, R. M., and P. D. Sadowski.** 1985. The FLP recombinase of the *Saccharomyces cerevisiae* 2 micron plasmid attaches covalently to DNA via a phosphotyrosyl linkage. *Mol. Cell. Biol.* **5:**3274–3279.
47. **Guo, F., D. N. Gopaul, and G. D. Van Duyne.** 1997. Structure of Cre recombinase complexed with DNA in a site-specific recombinase synapse. *Nature* **389:**40–46.
48. **Guo, F., D. N. Gopaul, and G. D. Van Duyne.** 1999. Assymetric DNA-bending in the Cre-loxP site-specific recombination synapse. *Proc. Natl. Acad. Sci. USA* **96:**7143–7148.
49. **Hickman, A. B., S. Waninger, J. J. Scocca, and F. Dyda.** 1997. Molecular organization in site-specific recombination: the catalytic domain of bacteriophage HP1 integrase at 2.7 A resolution. *Cell* **89:**227–237.
50. **Hoang, T. T., R. R. Karkhoff-Schweizer, A. J. Kutchma, and H. P. Schweizer.** 1998. A broad-host-range Flp-FRT recombination system for site-specific excision of chromosomally-located DNA sequences: application for isolation of unmarked *Pseudomonas aeruginosa* mutants. *Gene* **212:**77–86.
51. **Hoang, T. T., A. J. Kutchma, A. Becher, and H. P. Schweizer.** 2000. Integration-proficient plasmids for *Pseudomonas aeruginosa:* site-specific integration and use for engineering of reporter and expression strains. *Plasmid* **43:**59–72.
52. **Huai, Q., J. D. Colandene, Y. Chen, F. Luo, Y. Zhao, M. D. Topal, and H. Ke.** 2000. Crystal structure of NaeI—an evolutionary bridge between DNA endonuclease and topoisomerase. *EMBO J.* **19:**3110–3118.
53. **Huang, L. C., E. A. Wood, and M. M. Cox.** 1991. A bacterial model system for chromosomal targeting. *Nucleic Acids Res.* **19:**443–448.
54. **Huang, L. C., E. A. Wood, and M. M. Cox.** 1997. Convenient and reversible site-specific targeting of exogenous DNA into a bacterial chromosome by use of the FLP recombinase: the FLIRT system. *J. Bacteriol.* **179:**6076–6083.
55. **Huffman, K. E., and S. D. Levene.** 1999. DNA-sequence asymmetry directs the alignment of recombination sites in the FLP synaptic complex. *J. Mol. Biol.* **286:**1–13.
56. **Jayaram, M.** 1985. Two-micrometer circle site-specific re-

combination: the minimal substrate and the possible role of flanking sequences. *Proc. Natl. Acad. Sci. USA* **82:** 5875–5879.
57. **Jayaram, M.** 1986. Mating type-like conversion promoted by the 2 micron circle site-specific recombinase: implications for the double-strand-gap repair model. *Mol. Cell. Biol.* **6:** 3831–3837.
58. **Jayaram, M.** 1994. Mechanism of site-specific recombination: the Flp paradigm, p. 268–286. *In* F. Eckstein and D. M. J. Lilley (ed.), *Nucleic Acids and Molecular Biology*. Springer-Verlag, New York.
59. **Jayaram, M.** 1997. The cis-trans paradox of integrase. *Science* **276:**49–51.
60. **Jayaram, M., and J. R. Broach.** 1983. Yeast plasmid 2-micron circle promotes recombination within bacterial transposon Tn5. *Proc. Natl. Acad. Sci. USA* **80:**7264–7268.
61. **Kilbride, E., M. R. Boocock, and W. M. Stark.** 1999. Topological selectivity of a hybrid site-specific recombination system with elements from Tn3 res/resolvase and bacteriophage P1 loxP/Cre. *J. Mol. Biol.* **289:**1219–1230.
62. **Kimball, A. S., M. L. Kimball, M. Jayaram, and T. D. Tullius.** 1995. Chemical probe and missing nucleoside analysis of Flp recombinase bound to the recombination target sequence. *Nucleic Acids Res.* **23:**3009–3017.
63. **Kimball, A. S., J. Lee, M. Jayaram, and T. D. Tullius.** 1993. Sequence-specific cleavage of DNA via nucleophilic attack of hydrogen peroxide, assisted by Flp recombinase. *Biochemistry* **32:**4698–4701.
64. **Kitts, P. A., and H. A. Nash.** 1988. An intermediate in the phage lambda site-specific recombination reaction is revealed by phosphorothioate substitution in DNA. *Nucleic Acids Res.* **16:**6839–6856.
65. **Knudsen, B. R., K. Dahlstrom, O. Westergaard, and M. Jayaram.** 1997. The yeast site-specific recombinase Flp mediates alcoholysis and hydrolysis of the strand cleavage product: mimicking the strand-joining reaction with non-DNA nucleophiles. *J. Mol. Biol.* **266:**93–107.
66. **Kulpa, J., J. E. Dixon, G. Pan, and P. D. Sadowski.** 1993. Mutations of the FLP recombinase gene that cause a deficiency in DNA bending and strand cleavage. *J. Biol. Chem.* **268:**1101–1108.
67. **Kwon, H. J., R. Tirumalai, A. Landy, and T. Ellenberger.** 1997. Flexibility in DNA recombination: structure of the lambda integrase catalytic core. *Science* **276:**126–131.
68. **Lee, J., and M. Jayaram.** 1993. Mechanism of site-specific recombination. Logic of assembling recombinase catalytic site from fractional active sites. *J. Biol. Chem.* **268:** 17564–17570.
69. **Lee, J., and M. Jayaram.** 1995. Functional roles of individual recombinase monomers in strand breakage and strand union during site-specific DNA recombination. *J. Biol. Chem.* **270:** 23203–23211.
70. **Lee, J., and M. Jayaram.** 1995. Junction mobility and resolution of Holliday structures by Flp site-specific recombinase. Testing partner compatibility during recombination. *J. Biol. Chem.* **270:**19086–19092.
71. **Lee, J., and M. Jayaram.** 1995. Role of partner homology in DNA recombination. Complementary base pairing orients the 5′-hydroxyl for strand joining during Flp site-specific recombination. *J. Biol. Chem.* **270:**4042–4052.
72. **Lee, J., and M. Jayaram.** 1997. A tetramer of the Flp recombinase silences the trimers within it during resolution of a Holliday junction substrate. *Genes Dev.* **11:**2438–2447.
73. **Lee, J., M. Jayaram, and I. Grainge.** 1999. Wild-type Flp recombinase cleaves DNA in trans. *EMBO J.* **18:**784–791.
74. **Lee, J., M. C. Serre, S. H. Yang, I. Whang, H. Araki, Y. Oshima, and M. Jayaram.** 1992. Functional analysis of Box II mutations in yeast site-specific recombinases Flp and R. Significance of amino acid conservation within the Int family and the yeast sub-family. *J. Mol. Biol.* **228:**1091–1103.
75. **Lee, J., T. Tonozuka, and M. Jayaram.** 1997. Mechanism of active site exclusion in a site-specific recombinase: role of the DNA substrate in conferring half-of-the-sites activity. *Genes Dev.* **11:**3061–3071.
76. **Lee, J., G. Tribble, and M. Jayaram.** 2000. Resolution of tethered antiparallel and parallel holliday junctions by the Flp site-specific recombinase. *J. Mol. Biol.* **296:**403–419.
77. **Lee, J., Y. Voziyanov, S. Pathania, and M. Jayaram.** 1998. Structural alterations and conformational dynamics in Holliday junctions induced by binding of a site-specific recombinase. *Mol. Cell* **1:**483–493.
78. **Lee, J., I. Whang, and M. Jayaram.** 1996. Assembly and orientation of Flp recombinase active sites on two-, three-, and four-armed DNA substrates: implications for a recombination mechanism. *J. Mol. Biol.* **257:**532–549.
79. **Lee, J., I. Whang, J. Lee, and M. Jayaram.** 1994. Directed protein replacement in recombination full sites reveals *trans*-horizontal DNA cleavage by Flp recombinase. *EMBO J.* **13:** 5346–5354.
80. **Lilley, D. M., and R. M. Clegg.** 1993. The structure of the four-way junction in DNA. *Annu. Rev. Biophys. Biomol. Struct.* **22:**299–328.
81. **Lloyd, A. M., and R. W. Davis.** 1994. Functional expression of the yeast FLP/FRT site-specific recombination system in *Nicotiana tabacum*. *Mol. Gen. Genet.* **242:**653–657.
82. **Logie, C., and A. F. Stewart.** 1995. Ligand-regulated site-specific recombination. *Proc. Natl. Acad. Sci. USA* **92:** 5940–5944.
83. **Luetke, K. H., and P. D. Sadowski.** 1995. The role of DNA bending in Flp-mediated site-specific recombination. *J. Mol. Biol.* **251:**493–506.
84. **Luetke, K. H., and P. D. Sadowski.** 1998. DNA sequence determinant for Flp-induced DNA bending. *Mol. Microbiol.* **29:**199–208.
85. **Luetke, K. H., B. P. Zhao, and P. D. Sadowski.** 1997. Asymmetry in Flp-mediated cleavage. *Nucleic Acids Res.* **25:** 4240–4249.
86. **Lyznik, L. A., K. V. Rao, and T. K. Hodges.** 1996. FLP-mediated recombination of FRT sites in the maize genome. *Nucleic Acids Res.* **24:**3784–3789.
87. **Martinez-Morales, F., A. C. Borges, A. Martinez, K. T. Shanmugam, and L. O. Ingram.** 1999. Chromosomal integration of heterologous DNA in *Escherichia coli* with precise removal of markers and replicons used during construction. *J. Bacteriol.* **181:**7143–7148.
88. **McSwiggen, J. A., and T. R. Cech.** 1989. Stereochemistry of RNA cleavage by the Tetrahymena ribozyme and evidence that the chemical step is not rate-limiting. *Science* **244:** 679–683.
89. **Meyer-Leon, L., R. B. Inman, and M. M. Cox.** 1990. Characterization of Holliday structures in FLP protein-promoted site-specific recombination. *Mol. Cell. Biol.* **10:**235–242.
90. **Mizuuchi, K., and K. Adzuma.** 1991. Inversion of the phosphate chirality at the target site of Mu DNA strand transfer: evidence for a one-step transesterification mechanism. *Cell* **66:**129–140.
91. **Nash, H. A.** 1996. Site-specific recombination: integration, excision, resolution, and inversion of defined DNA segments, p. 2363–2376. *In* F. C. Neidhardt et al. (ed.), Escherichia coli *and* Salmonella: *Cellular and Molecular Biology*, vol. 2. American Society for Microbiology, Washington, D.C.

92. **Nash, H. A.** 1998. Topological nuts and bolts. *Science* **279:** 1490–1491.
93. **Nunes-Duby, S. E., M. A. Azaro, and A. Landy.** 1995. Swapping DNA strands and sensing homology without branch migration in lambda site-specific recombination. *Curr. Biol.* **5:** 139–148.
94. **Nunes-Duby, S. E., H. J. Kwon, R. S. Tirumalai, T. Ellenberger, and A. Landy.** 1998. Similarities and differences among 105 members of the Int family of site-specific recombinases. *Nucleic Acids Res.* **26:**391–406.
95. **O'Gorman, S., D. T. Fox, and G. M. Wahl.** 1991. Recombinase-mediated gene activation and site-specific integration in mammalian cells. *Science* **251:**1351–1355.
96. **O'Gorman, S., and G. M. Wahl.** 1997. Mouse engineering. *Science* **277:**1025.
97. **Pan, G., K. Luetke, and P. D. Sadowski.** 1993. Mechanism of cleavage and ligation by FLP recombinase: classification of mutations in FLP protein by in vitro complementation analysis. *Mol. Cell. Biol.* **13:**3167–3175.
98. **Pan, G., and P. D. Sadowski.** 1993. Identification of the functional domains of the FLP recombinase. Separation of the nonspecific and specific DNA-binding, cleavage, and ligation domains. *J. Biol. Chem.* **268:**22546–22551.
99. **Pan, H., D. Clary, and P. D. Sadowski.** 1991. Identification of the DNA-binding domain of the FLP recombinase. *J. Biol. Chem.* **266:**11347–11354.
100. **Panigrahi, G. B., L. G. Beatty, and P. D. Sadowski.** 1992. The FLP protein contacts both major and minor grooves of its recognition target sequence. *Nucleic Acids Res.* **20:** 5927–5935.
101. **Panigrahi, G. B., and P. D. Sadowski.** 1994. Interaction of the NH_2- and COOH-terminal domains of the FLP recombinase with the FLP recognition target sequence. *J. Biol. Chem.* **269:** 10940–10945.
102. **Parsons, R. L., B. R. Evans, L. Zheng, and M. Jayaram.** 1990. Functional analysis of Arg-308 mutants of Flp recombinase. Possible role of Arg-308 in coupling substrate binding to catalysis. *J. Biol. Chem.* **265:**4527–4533.
103. **Parsons, R. L., P. V. Prasad, R. M. Harshey, and M. Jayaram.** 1988. Step-arrest mutants of FLP recombinase: implications for the catalytic mechanism of DNA recombination. *Mol. Cell. Biol.* **8:**3303–3310.
104. **Posfai, G., M. Koob, Z. Hradecna, N. Hasan, M. Filutowicz, and W. Szybalski.** 1994. *In vivo* excision and amplification of large segments of the *Escherichia coli* genome. *Nucleic Acids Res.* **22:**2392–2398.
105. **Pósfai, G., M. D. Koob, H. A. Kirkpatrick, and F. R. Blattner.** 1997. Versatile insertion plasmids for targeted genome manipulations in bacteria: isolation, deletion, and rescue of the pathogenicity island LEE of the *Escherichia coli* O157:H7 genome. *J. Bacteriol.* **179:**4426–4428.
106. **Prasad, P. V., D. Horensky, L. J. Young, and M. Jayaram.** 1986. Substrate recognition by the 2 micron circle site-specific recombinase: effect of mutations within the symmetry elements of the minimal substrate. *Mol. Cell. Biol.* **6:** 4329–4334.
107. **Prasad, P. V., L. J. Young, and M. Jayaram.** 1987. Mutations in the 2-micron circle site-specific recombinase that abolish recombination without affecting substrate recognition. *Proc. Natl. Acad. Sci. USA* **84:**2189–2193.
108. **Proteau, G., D. Sidenberg, and P. Sadowski.** 1986. The minimal duplex DNA sequence required for site-specific recombination promoted by the FLP protein of yeast *in vitro*. *Nucleic Acids Res.* **14:**4787–4802.
109. **Qian, X. H., and M. M. Cox.** 1995. Asymmetry in active complexes of FLP recombinase. *Genes Dev.* **9:**2053–2064.
110. **Qian, X. H., R. B. Inman, and M. M. Cox.** 1990. Protein-based asymmetry and protein-protein interactions in FLP recombinase-mediated site-specific recombination. *J. Biol. Chem.* **265:**21779–21788.
111. **Redinbo, M. R., L. Stewart, P. Kuhn, J. J. Champoux, and W. G. Hol.** 1998. Crystal structures of human topoisomerase I in covalent and noncovalent complexes with DNA. *Science* **279:**1504–1513.
112. **Reynolds, A. E., A. W. Murray, and J. W. Szostak.** 1987. Roles of the 2 micron gene products in stable maintenance of the 2 micron plasmid of *Saccharomyces cerevisiae*. *Mol. Cell. Biol.* **7:**3566–3573.
113. **Rosen, M. A., L. Shapiro, and D. J. Patel.** 1992. Solution structure of a trinucleotide A-T-A bulge loop within a DNA duplex. *Biochemistry* **31:**4015–4026.
114. **Sadowski, P. D.** 1995. The Flp recombinase of the 2-micron plasmid of *Saccharomyces cerevisiae*. *Prog. Nucleic Acid Res. Mol. Biol.* **51:**53–91.
115. **Sadowski, P. D.** 1993. Site-specific genetic recombination: hops, flips, and flops. *FASEB J.* **7:**760–767.
116. **Schwartz, C. J., and P. D. Sadowski.** 1989. FLP recombinase of the 2 micron circle plasmid of *Saccharomyces cerevisiae* bends its DNA target. Isolation of FLP mutants defective in DNA bending. *J. Mol. Biol.* **205:**647–658.
117. **Schwartz, C. J., and P. D. Sadowski.** 1990. FLP protein of 2 micron circle plasmid of yeast induces multiple bends in the FLP recognition target site. *J. Mol. Biol.* **216:**289–298.
118. **Schweizer, H. P.** 1998. Intrinsic resistance to inhibitors of fatty acid biosynthesis in *Pseudomonas aeruginosa* is due to efflux: application of a novel technique for generation of unmarked chromosomal mutations for the study of efflux systems. *Antimicrob. Agents Chemother.* **42:**394–398.
119. **Scott-Drew, S., and J. A. Murray.** 1998. Localisation and interaction of the protein components of the yeast 2 micron circle plasmid partitioning system suggest a mechanism for plasmid inheritance. *J. Cell Sci.* **111:**1779–1789.
120. **Sekiguchi, J., and S. Shuman.** 1997. Site-specific ribonuclease activity of eukaryotic DNA topoisomerase I. *Mol. Cell* **1:** 89–97.
121. **Senecoff, J. F., R. C. Bruckner, and M. M. Cox.** 1985. The FLP recombinase of the yeast 2-micron plasmid: characterization of its recombination site. *Proc. Natl. Acad. Sci. USA* **82:** 7270–7274.
122. **Senecoff, J. F., and M. M. Cox.** 1986. Directionality in FLP protein-promoted site-specific recombination is mediated by DNA-DNA pairing. *J. Biol. Chem.* **261:**7380–7386.
123. **Serre, M. C., L. Zheng, and M. Jayaram.** 1993. DNA splicing by an active site mutant of Flp recombinase. Possible catalytic cooperativity between the inactive protein and its DNA substrate. *J. Biol. Chem.* **268:**455–463.
124. **Shaikh, A. C., and P. D. Sadowski.** 1997. The Cre recombinase cleaves the lox site in trans. *J. Biol. Chem.* **272:** 5695–5702.
125. **Shaikh, A. C., and P. D. Sadowski.** 2000. Chimeras of the Flp and Cre recombinases: tests of the mode of cleavage by Flp and Cre. *J. Mol. Biol.* **302:**27–48.
126. **Sherratt, D. J., and D. B. Wigley.** 1998. Conserved themes but novel activities in recombinases and topoisomerases. *Cell* **93:**149–152.
127. **Sonti, R. V., A. F. Tissier, D. Wong, J. F. Viret, and E. R. Signer.** 1995. Activity of the yeast FLP recombinase in Arabidopsis. *Plant Mol. Biol.* **28:**1127–1132.
128. **Stewart, L., M. R. Redinbo, X. Qiu, W. G. Hol, and J. J. Champoux.** 1998. A model for the mechanism of human topoisomerase I. *Science* **279:**1534–1541.
129. **Stivers, J. T., G. J. Jagadeesh, B. Nawrot, W. J. Stec, and S.**

Shuman. 2000. Stereochemical outcome and kinetic effects of Rp- and Sp-phosphorothioate substitutions at the cleavage site of vaccinia type I DNA topoisomerase. *Biochemistry* **39:** 5561–5572.

130. **Struhl, G., and K. Basler.** 1993. Organizing activity of wingless protein in Drosophila. *Cell* **72:**527–540.
131. **Subramanya, H. S., L. K. Arciszewska, R. A. Baker, L. E. Bird, D. J. Sherratt, and D. B. Wigley.** 1997. Crystal structure of the site-specific recombinase, XerD. *EMBO J.* **16:** 5178–5187.
132. **Sun, J., and J. Tower.** 1999. FLP recombinase-mediated induction of Cu/Zn-superoxide dismutase transgene expression can extend the life span of adult *Drosophila melanogaster* flies. *Mol. Cell. Biol.* **19:**216–228.
133. **Theodosiou, N. A., and T. Xu.** 1998. Use of FLP/FRT system to study Drosophila development. *Methods* **14:**355–365.
134. **Tribble, G., Y. T. Ahn, J. Lee, T. Dandekar, and M. Jayaram.** 2000. DNA recognition, strand selectivity, and cleavage mode during integrase family site-specific recombination. *J. Biol. Chem.* **275:**22255–22267.
135. **Tsalik, E. L., and M. R. Gartenberg.** 1998. Curing *Saccharomyces cerevisiae* of the 2 micron plasmid by targeted DNA damage. *Yeast* **14:**847–852.
136. **Velmurugan, S., Y. T. Ahn, X. M. Yang, X. L. Wu, and M. Jayaram.** 1998. The 2 micrometer plasmid stability system: analyses of the interactions among plasmid- and host-encoded components. *Mol. Cell. Biol.* **18:**7466–7477.
137. **Velmurugan, S., X. M. Yang, C. S. Chan, M. Dobson, and M. Jayaram.** 2000. Partitioning of the 2-micron circle plasmid of *Saccharomyces cerevisiae*. Functional coordination with chromosome segregation and plasmid-encoded rep protein distribution. *J. Cell Biol.* **149:**553–566.
138. **Volkert, F. C., and J. R. Broach.** 1986. Site-specific recombination promotes plasmid amplification in yeast. *Cell* **46:** 541–550.
139. **Voziyanov, Y., J. Lee, I. Whang, J. Lee, and M. Jayaram.** 1996. Analyses of the first chemical step in Flp site-specific recombination: synapsis may not be a pre-requisite for strand cleavage. *J. Mol. Biol.* **256:**720–735.
140. **Voziyanov, Y., S. Pathania, and M. Jayaram.** 1999. A general model for site-specific recombination by the integrase family recombinases. *Nucleic Acids Res.* **27:**930–941.
141. **Waite, L. L., and M. M. Cox.** 1995. A protein dissociation step limits turnover in FLP recombinase-mediated site-specific recombination. *J. Biol. Chem.* **270:**23409–23414.
142. **Wild, J., Z. Hradecna, G. Posfai, and W. Szybalski.** 1996. A broad-host-range *in vivo* pop-out and amplification system for generating large quantities of 50- to 100-kb genomic fragments for direct DNA sequencing. *Gene* **179:**181–188.
143. **Wild, J., M. Sektas, Z. Hradecna, and W. Szybalski.** 1998. Targeting and retrofitting pre-existing libraries of transposon insertions with FRT and oriV elements for *in vivo* generation of large quantities of any genomic fragment. *Gene* **223:** 55–66.
144. **Wittschieben, J., and S. Shuman.** 1997. Mechanism of DNA transesterification by vaccinia topoisomerase: catalytic contributions of essential residues Arg-130, Gly-132, Tyr-136 and Lys-167. *Nucleic Acids Res.* **25:**3001–3008.
145. **Xu, C. J., Y. T. Ahn, S. Pathania, and M. Jayaram.** 1998. Flp ribonuclease activities. Mechanistic similarities and contrasts to site-specific DNA recombination. *J. Biol. Chem.* **273:** 30591–30598.
146. **Xu, C. J., I. Grainge, J. Lee, R. M. Harshey, and M. Jayaram.** 1998. Unveiling two distinct ribonuclease activities and a topoisomerase activity in a site-specific DNA recombinase. *Mol. Cell* **1:**729–739.
147. **Xu, T., and S. D. Harrison.** 1994. Mosaic analysis using FLP recombinase. *Methods Cell Biol.* **44:**655–681.
148. **Yang, S. H., and M. Jayaram.** 1994. Generality of the shared active site among yeast family site-specific recombinases. The R site-specific recombinase follows the Flp paradigm. *J. Biol. Chem.* **269:**12789–12796.
149. **Zakian, V. A., B. J. Brewer, and W. L. Fangman.** 1979. Replication of each copy of the yeast 2 micron DNA plasmid occurs during the S phase. *Cell* **17:**923–934.

Mobile DNA II
Edited by N. L. Craig et al.

Chapter 12

Theme and Variation in Tyrosine Recombinases: Structure of a Flp-DNA Complex

Phoebe A. Rice

INTRODUCTION

Flp, encoded by the 2μm plasmid of the yeast *Saccharomyces cerevisiae,* is one of only a few known eukaryotic members of the λ integrase or tyrosine-based family of site-specific DNA recombinases (reviewed in references 14 and 37 and chapter 6). These enzymes all act at specific DNA sequences to catalyze inversions, deletions, or insertions of DNA segments. Flp-mediated DNA inversion allows amplification of the 2μm plasmid copy number during replication (reviewed in reference 47 and chapter 11). Flp is an outlier within this large and divergent family of enzymes: a PSI-BLAST data bank search for sequences related to one of the prokaryotic integrases retrieves hundreds of sequences, but Flp is not among them (2). Likewise, a search using Flp as the query retrieves only a few closely related recombinases from other fungal plasmids. Nevertheless, the recently determined structure of a Flp-DNA complex confirms that its catalytic domain closely resembles that of other family members, although the active site itself is assembled differently, as discussed below (11). The overall geometry of the synaptic complex formed by Flp with its DNA substrates is also similar to that of the prokaryotic recombinase Cre (see chapter 6), but the protein-protein contacts among monomers within the complex differ greatly. Flp is thus an interesting example of how general mechanistic unity can be conserved within a family of enzymes while allowing for substantial structural variability.

The reaction pathway proposed for this family of recombinases is shown in Fig. 1. The DNA recombination sites vary in complexity, but always include two binding sites for the recombinase that are inverted about a 6- to 8-bp spacer. Two such sites are synapsed by a tetramer of the recombinase. The catalytic activity of the recombinase monomers is carefully coordinated such that only two of four monomers are active at any given time, and double-strand breaks are never introduced into the DNA. One strand from each duplex is thus cleaved, by direct attack of an active site tyrosine, creating a transient covalent 3′-phosphotyrosyl linkage. The free 5′ ends are exchanged, and the 5′ hydroxyls attack the opposite phosphotyrosyl linkage, forming a Holliday-junction intermediate. After an isomerization step that activates the alternate pair of monomers, a second and similar set of cleavage and strand-exchange steps completes the recombination reaction.

Recombination is efficient only if the spacers of the two substrate DNAs are identical in sequence. Also, when the spacer sequence is asymmetric, only one of two possible synaptic orientations (antiparallel rather than parallel) can produce Watson-Crick base-paired products. Because the energy of the initial

Phoebe A. Rice • Department of Biochemistry and Molecular Biology, The University of Chicago, 920 E. 58th St., Chicago, IL 60637.

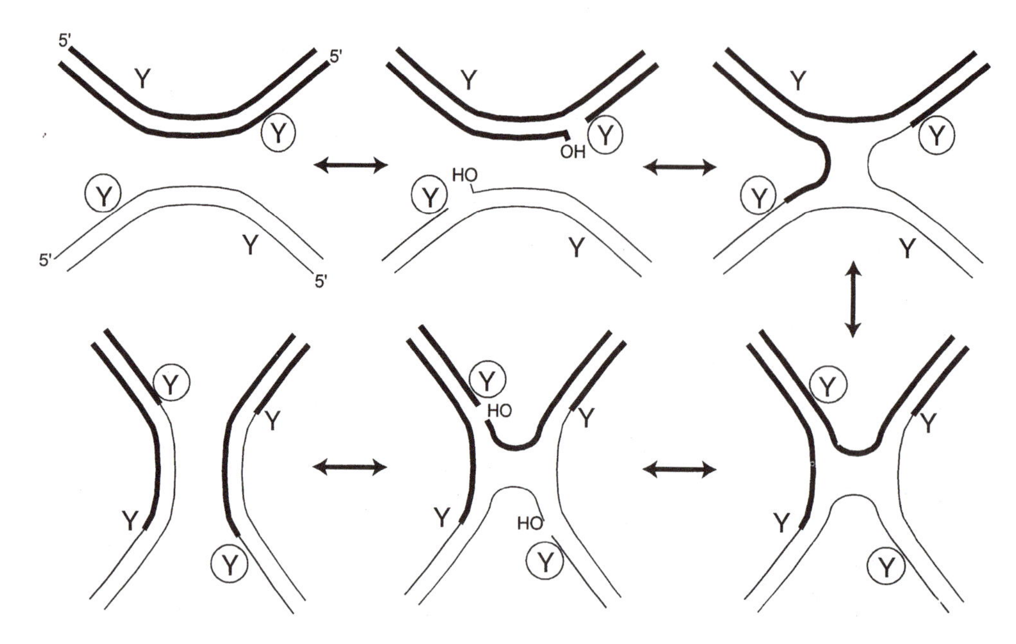

Figure 1. Proposed mechanism of site-specific recombination catalyzed by the tyrosine recombinases. Each recombinase monomer is represented by a single Y. Circles indicate the two active tyrosines, which form transient 3′ phosphotyrosine linkages with the DNA. The central isomerization step (right) activates the alternate pair of active sites, allowing for exchange of the second pair of DNA strands.

phosphodiester bond is stored in the phosphotyrosine intermediate, reactions that would result in mismatches can simply proceed backward to reseal the substrate duplexes.

The reaction mechanism as drawn is the distillation of years of work by many laboratories, which is beyond the scope of this chapter to review comprehensively. For a historical discussion of mechanistic models and the data supporting them, the reader is referred to other chapters in this book and references therein. The model in Fig. 1 is based in part on (i) observations that the ligation and/or subsequent isomerization steps of these reactions are particularly sensitive to the ability of the first two to three bases of the incoming strand to form Watson-Crick base pairs (3, 9, 32, 59), (ii) the sequential strand-swapping model of Nunes-Duby et al. (36), (iii) the suggestion of Stark et al. (49) that the square planar conformation of a Holliday junction is particularly convenient for strand exchange, and (iv) the crystal structure of Cre synaptic complex (20).

Much effort has been devoted to structural studies of this family of recombinases. In addition to the structure of the Flp-DNA complex that will be the focus of this chapter, crystal structures of four members of this family have been determined: the core domains of the λ (30) and HP1 (23) phage integrases, intact *Escherichia coli* XerD (52), and intact bacteriophage P1 Cre. Cre has been crystallized with a variety of DNA substrates, representing different steps along the reaction pathway shown in Fig. 1 (18, 20, 21).

The structurally conserved active-site-containing domain of these enzymes is also found in the eukaryotic type IB topoisomerases (12, 46). This reflects the fact that these topoisomerases also form a 3′-phosphotyrosyl intermediate, although they function as monomers and religate the original DNA strand after allowing torsional relaxation. The helical portion of this fold resembles two helix-turn-helix motifs and is also found in the AraC family of transcriptional regulators (16).

In addition to the nucleophilic tyrosine, the active sites of these recombinases and the type IB topoisomerases contain several other highly conserved residues (1, 6; reviewed in reference 19). Three of these, R191, H305, and R308 in Flp, have been demonstrated to be catalytically important in Flp and several other systems, although H305 is more important for religation than cleavage, and the equivalent lysine is not essential in the topoisomerases. These residues presumably localize the scissile phosphate and stabilize the transition state. The second arginine also binds a water molecule in human topoisomerase I that was suggested to act as the general base that accepts a proton from the attacking tyrosine (45). Structural studies have highlighted two additional conserved residues: a lysine and a tryptophan that is often replaced by histidine. This lysine in vaccinia virus topoisomerase has been shown to be the general acid that protonates the leaving 5′ hydroxyl (27). Sequence comparisons had tentatively identified these residues as K223 and W330 in Flp (37), which the Flp structure now confirms (11).

The active site of Flp has been known for some time to differ from that of its relatives in one very important aspect: in contrast to other tyrosine recombinases, including the λ and HP1 phage integrases (23, 39), Cre (20), and XerC and D (5, 7), the active sites within Flp complexes are composed of amino acid residues from two different monomers (references 10 and 33 and references therein). Jayaram and colleagues showed that the active site tyrosine (Y343) is donated not by the monomer bound adjacent to the cleaved phosphodiester (called cleavage in *cis*), but by an adjacent monomer within the complex (cleavage in *trans*). This arrangement appears to be unique to Flp and closely related recombinases from other yeast plasmids (57).

Another feature that varies among members of the tyrosine recombinase family is the length of the DNA spacer between cleavage sites. The Cre recombination site, loxP, has a 6-bp spacer, and recombination is very inefficient at sites with a 7-bp spacer (24, 50). The natural recombination sites for phage λ integrase have a 7-bp spacer, while *E. coli* XerC and XerD sites are found with both 6- and 8-bp spacers. Flp's natural recombination target (FRT) has an 8-bp spacer, but it readily recombines artificial sites with 7- or 9-bp spacers (48). Changing the length of the DNA in the center of the complex would presumably change the relative orientations of the four monomers, raising the question of how the half-of-the-sites activity is maintained in such a flexible system.

This chapter is intended to briefly describe the structure of a Flp-Holliday junction complex and its biochemical implications, and to explore its similarities to and differences from other tyrosine recombinases and the type IB topoisomerases.

OVERALL STRUCTURE OF THE Flp-DNA COMPLEX

Flp was co-crystallized with a binding site that was designed as a suicide inhibitor (38), but the DNA rearranged to form a Holliday junction with a 7-base spacer and nicks at the 3′ side of each scissile phosphate, as shown in Fig. 2 (11). The first 3 bases of

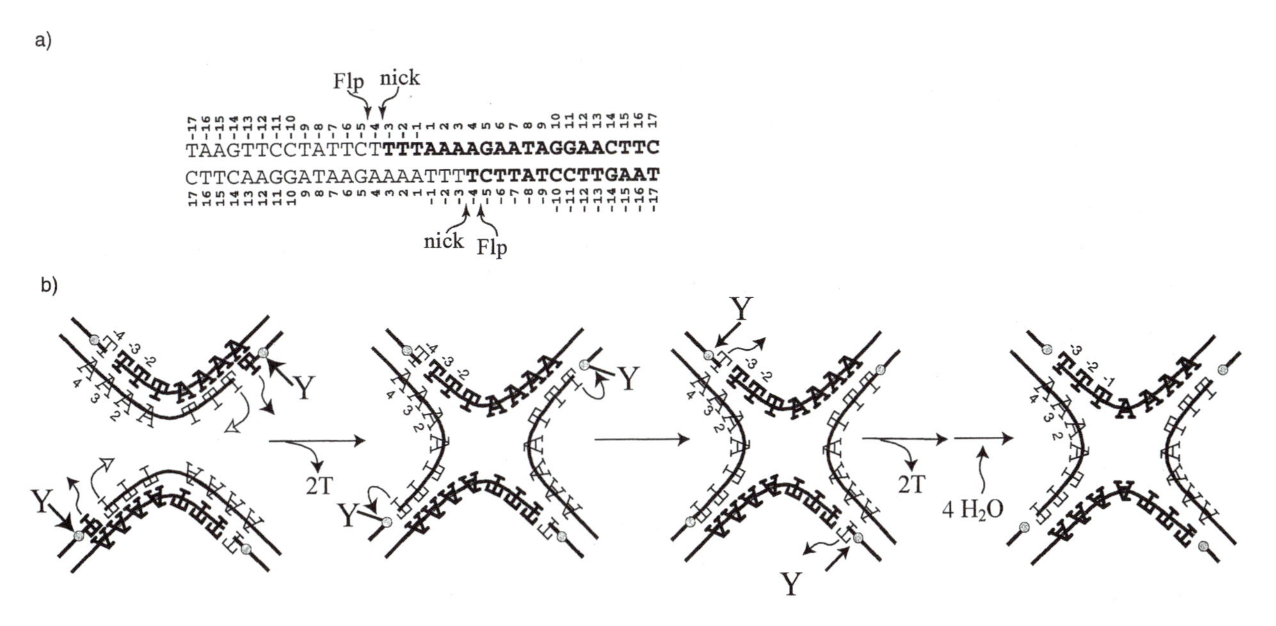

Figure 2. DNA substrate in the Flp crystal structure. (a) The symmetrized Flp binding site used in crystallization, numbered from center. Two identical half-sites (plain and bold text) anneal to form the duplex shown, with nicks at the junction. Flp cleaves DNA between nucleotides −4 and −5, one base away from the nick. (b) Pathway for formation of the Holliday junction present in the crystals. Two duplexes are synapsed by a tetramer of Flp protein. Only the sequence at the center is shown, and small circles represent the scissile phosphates. Due to the nicks, T-4 is lost into solution upon attack of the tyrosine. Since the spacer sequence used allows slippage of the base pairing, T-3 of the opposite duplex can fit into the pocket vacated by T-4, and its 5′ hydroxyl can attack the phosphotyrosine intermediate. After isomerization, the same sequence of events can occur at the other pair of active sites, resulting in a Holliday junction with a central unpaired A. All four scissile bonds were hydrolyzed at some point before data collection. This probably occurred because the Holliday junction was in equilibrium with the phosphotyrosine intermediate, which is subject to slow hydrolysis (26).

the overhanging bottom strand are base paired with their recombinant partners, although the geometry of the centermost base pairs is strained. The central A (no. 1) is not paired and would form an A:A mismatch if it were. Instead it stacks on the run of A's to its 3′ side.

Flp binds the DNA Holliday junction in a roughly square planar conformation, similar to that seen in the Cre complexes (see chapter 6), and as expected from solution studies of Flp (34). The complex has nearly perfect twofold symmetry, but only approximate fourfold symmetry. This corresponds to the stepwise nature of the reaction pathway, which implies that only two of four monomers should be in an active conformation at any given time. Since each of the four Flp monomers is in a different crystal-packing environment, the observed symmetry must be intrinsic to the complex rather than the result of crystal-packing forces. In contrast to Cre, and as discussed in more detail below, the Flp-DNA complex as a whole is a rather loosely packed, flexible entity. This reflects flexibility within individual monomers rather than a lack of contacts between monomers.

STRUCTURE OF A SINGLE PROTOMER

The Flp monomer is composed of two compact domains and two protruding segments that pack in *trans* onto neighboring protomers within the tetramer (Color Plate 18 [see color insert]). The first such segment connects the two domains and contains a single amphipathic helix, and the second contains helix M and the active site Y343. Within the complex, the monomer forms a ring that fully encircles the bound DNA.

The active site lies in the C-terminal domain. Much of this domain is structurally homologous to the catalytic domains of the other tyrosine recombinases as well as the type IB topoisomerases (55). However, while helix N is domain-swapped in the Cre and HP1 integrase structures, in Flp it is helix M that is swapped (the Flp secondary-structure elements have been named to correspond to those of Cre in this region). Since helix M contains the nucleophilic tyrosine, this leads to a major mechanistic difference between these enzymes: a single active site in a Flp complex is assembled in *trans* with catalytic residues donated by two different monomers, as predicted by Jayaram and colleagues (33 and references therein). Except for this difference in topology, the active site closely resembles that of Cre and other related enzymes.

The C-terminal domain of Flp contains additional secondary-structure elements not found in its prokaryotic relatives. Helix N lies at the C terminus of Cre and HP1 integrase, and by sequence comparison, probably many other tyrosine recombinases (37). However, Flp continues for another 60 residues after helix N. This extension forms a beta hairpin and three short helices that pack onto the opposite face of the C-terminal domain from the DNA. The hairpin makes some protein-protein contacts within the tetramer, but the main function of this region may be to anchor the returning polypeptide chain after the domain-swapped helix M. Flp also contains an additional helix, E, at the beginning of the catalytic domain that is replaced by an extended strand in the prokaryotic recombinase structures. However, this helix is found in human topoisomerase I (46).

In the Flp, Cre, and human topoisomerase I-DNA complexes, the monomer forms a C-shaped clamp around the DNA, with an N-terminal domain bound to the opposite face of the DNA from the catalytic domain. In contrast to the conserved fold of the C-terminal catalytic domain, these N-terminal domains are structurally unrelated to one another. This implies that the catalytic domains of these enzymes share a common ancestor, but that the N-terminal domains have been added in a modular fashion. This idea is supported by the fact that some tyrosine recombinases function with no N-terminal domain (37). It is quite possible that Flp and Cre diverged separately from ancestral topoisomerases, rather than sharing a common recombinase ancestor. The pseudofourfold symmetry of their synaptic complexes (discussed below) may have evolved convergently because the strand-exchange reaction is simplest when the Holliday junction is held in a square planar conformation.

COMPARISON OF Flp AND Cre SYNAPTIC COMPLEXES

Flp and Cre share a reaction pathway and a conserved catalytic domain. From a distance, their complexes with DNA also look quite similar: a pseudofourfold symmetric tetramer of the protein holds the Holliday junction that is the reaction intermediate in a nearly square planar conformation with matching pseudofourfold symmetry. However, the manner in which the complexes are assembled differs in several major aspects.

Both recombinases use cyclic helix swapping among monomers to assemble the tetramer but choose different helices to swap (Fig. 3). In Cre, helix N packs onto a neighboring monomer, whereas in Flp helix M, carrying the catalytic tyrosine, is swapped to the neighbor. The turn between helices M and N crosses between monomers in both enzymes; however, in the Flp case it returns the polypeptide chain back to the original monomer. The fact that Flp has an additional four residues between the conserved catalytic residues HxxR and the active site tyrosine, combined with experiments by the Jayaram lab, had predicted such an arrangement (8, 17, 53).

The relative orientation of the individual catalytic domains within the complexes differs. Relative to Cre, each Flp monomer is farther from the center of the Holliday junction and rotated about the axis of the DNA duplex to which it is bound. This correlates with the longer spacer in the Flp structure (7 bases versus 6 for Cre). It changes the relative orientations of the monomers, such that even without the differences in helix swapping, the protein-protein contacts would vary.

In both systems the N-terminal domains also make important protein-protein contacts, which necessarily differ since these domains are structurally unrelated. The N-terminal domains of Cre pack directly against one another. In Flp, contacts among the N-terminal domains are mediated by the protruding helix that connects the N- and C-terminal domains. This amphipathic helix nestles into a hydrophobic pocket on the N-terminal domain of a neighboring monomer. Since this helix is flanked by long loops, the Flp synaptic complex appears more flexible than that of Cre. This additional flexibility may explain Flp's ability to catalyze recombination of DNA sites with variable spacer lengths.

Finally, although neither complex exhibits perfect fourfold symmetry, the Flp complex deviates much further from true fourfold symmetry than the Cre complex. As discussed below, this implies that larger conformational changes are necessary to accomplish the complete recombination reaction in the Flp case.

THE ACTIVE SITE

Aside from the striking difference in the connectivity of helix M, bearing the catalytic tyrosine, the overall makeup of the Flp active site is quite similar to that of related enzymes. However, the details that probably underlie the control of activity are different. The roles of the conserved residues in the phosphotransfer reaction will not be discussed in detail here, since they have been nicely reviewed elsewhere, and because the exact positioning of these residues in the Flp structure may have been affected by hydrolysis of the phosphotyrosine intermediate, which adds an extra oxygen to the active site. Instead this chapter will focus on control of activity.

As expected from the half-of-the-sites activity of the enzyme, two of the monomers within the Flp complex appear to be in an active conformation, and the other two appear to be inactive (Color Plate 19 [see color insert]). These conformations differ most in the positioning of the nucleophilic tyrosine Y343. For the catalytic centers formed within the active pair of monomers, helix M (donated in *trans* by neighboring inactive monomers) is well ordered, and Y343 is poised to attack the scissile phosphate. In the inactive pair of catalytic centers, helix M is disordered, and although Y343 from one monomer can be located in the electron density, its hydroxyl lies almost 10 Å away from its target phosphorus atom.

The comparison with Cre is particularly interesting here: in the inactive Cre monomers, the displacement of the nucleophilic tyrosine from the scissile phosphate is strikingly similar to that in Flp. However, as discussed below, in Cre, helix M bearing this tyrosine is better ordered in the inactive than in the active catalytic centers.

Of the other conserved active site residues in Flp, R191, H305, and R308 bind the scissile phosphate in both pairs of monomers (Color Plate 19). K223 forms a minor groove hydrogen bond to the adjacent base. In the active pair of monomers, it also contacts the scissile phosphate, whereas in the inactive pair, it is less well ordered, and its tip lies closer to the free 5′ hydroxyl than the phosphate. The equivalent lysines in human topoisomerase I and the active Cre monomers form similar hydrogen bonds in the minor groove (20, 46). Unlike the small difference in the position of this lysine in the two types of Flp active site, in the inactive Cre monomers it is pulled far away from the active site. In Flp, the position of W330 is more varied: in the active pair, it interacts with the free 5′ hydroxyl, but it is farther from the DNA and less well ordered in the inactive pair.

After the phosphotyrosine has been formed, it is attacked by the 5′ hydroxyl of a new incoming DNA strand. This ligation reaction is highly sensitive to the ability of the first 2 to 3 bases of the incoming strand to form Watson-Crick base pairs with their new partner strand (32). Although water is present at approximately 55 M, ligation is vastly favored over hydrolysis (26). These features of the ligation reaction are probably enforced by contacts between the protein and the backbone of the spacer DNA near the active site. The N-terminal domain contacts the outer two phosphates of the spacer on the uncleaved strand. When the incoming strand forms Watson-Crick base pairs with the uncleaved strand, the 5′ hydroxyl is properly oriented to attack the phosphotyrosine, whereas non-Watson-Crick base pairs would distort the backbone geometry and hold the 5′ hydroxyl of a mispaired incoming strand away from the scissile bond. Thus, the enzyme provides a mechanism to localize the 5′ hydroxyl of an appropriate incoming strand but none to localize an attacking water molecule.

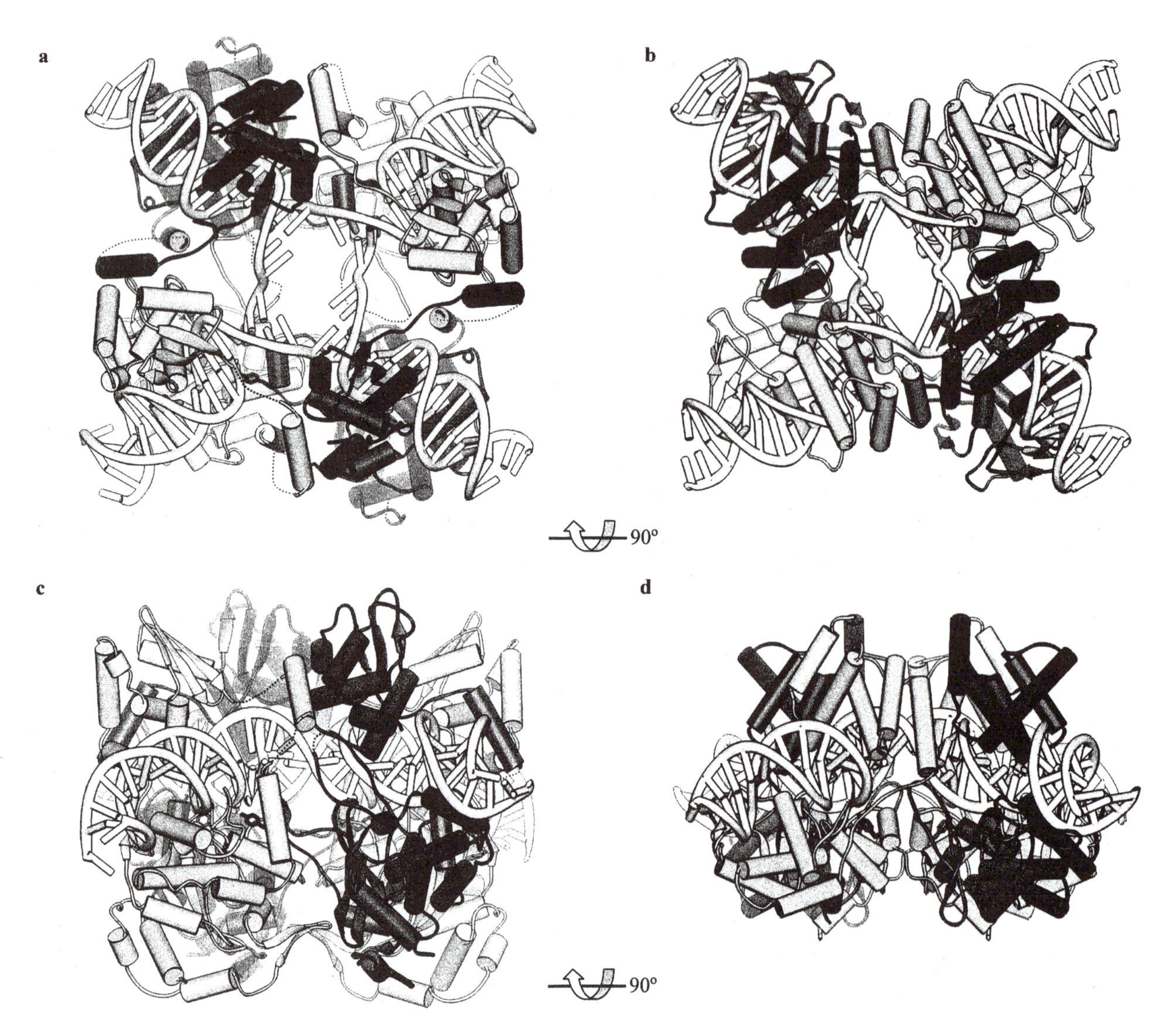

Figure 3. Synaptic complexes of Flp and Cre. (a) Tube-and-arrow representation of the Flp complex. The N-terminal domain is above the DNA, and the catalytic domain, below. Inactive monomers are shaded. The *trans*-donated active site tyrosine is shown in black. The tyrosine donated by one of these monomers is disordered. Dotted lines represent connections not visible in the electron density. (b) Similar representation of the Cre complex (18). The nucleophilic tyrosine is in dark gray but is largely occluded in this view. (c) and (d) The Flp and Cre complexes, rotated such that the catalytic domains are at the bottom of the picture. (e) and (f) Cartoons of the Flp and Cre complexes, as seen from the underside of panels a and b (catalytic domains above the DNA). Helices M and N are shown explicitly. Note that in both cases, the helix M-N segment that traverses a type I interface contains the active tyrosine, regardless of the provenance of that tyrosine.

Contacts between the catalytic domain and the incoming strand may help to prevent hydrolysis and to regulate enzymatic activity. Hydrophobic interaction between Cγ of T224 and the deoxyribose of the first incoming nucleotide may help orient the adjacent K223. The equivalent lysine functions as a general acid during the cleavage reaction of vaccinia virus topoisomerase; thus, it is reasonable to expect that it functions as a general base in the ligation reaction, accepting a proton from the attacking hydroxyl. Rendering its orientation dependent on the presence of an incoming DNA strand would help to prevent hydrolysis of the phosphotyrosine intermediate. This correlates with the fact that the pH optimum for the attack of small nucleophiles such as water is higher than that for ligation in Flp and in the type IB topoisomerase (13, 26, 43). The 3′ phosphate of the first incoming base is hydrogen bonded to S331, which is adjacent to the conserved W330. This also may couple proper assembly of the active site to the presence of an incoming strand. Additionally, S331 is located on the turn between helices L and M that crosses between mono-

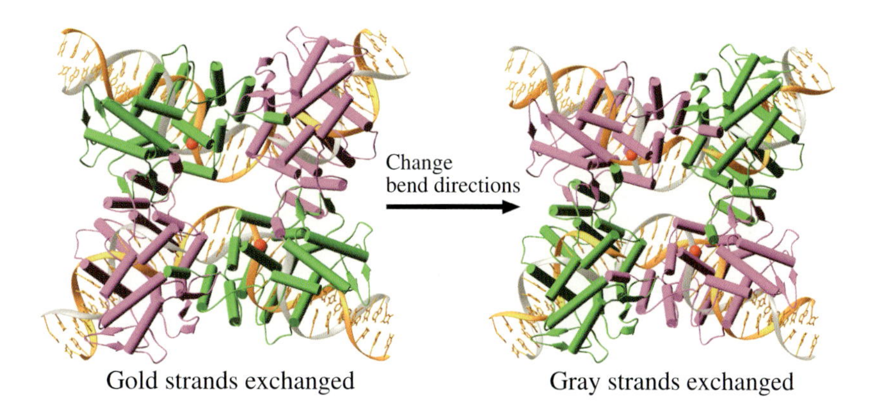

Color Plate 11 (chapter 6). Order of strand exchange. Ribbon-cylinder drawings of the Cre R173K-DNA complex (38) are shown, with phosphates activated for cleavage indicated by red spheres. In the synapse on the left, the gold strands are activated for cleavage by the green recombinase subunits. If the synapse is dissociated, the *loxP* sites are bent in the opposite directions, and the synapse is re-formed, then the resulting new synaptic complex will have gray strands activated for cleavage by the purple subunits.

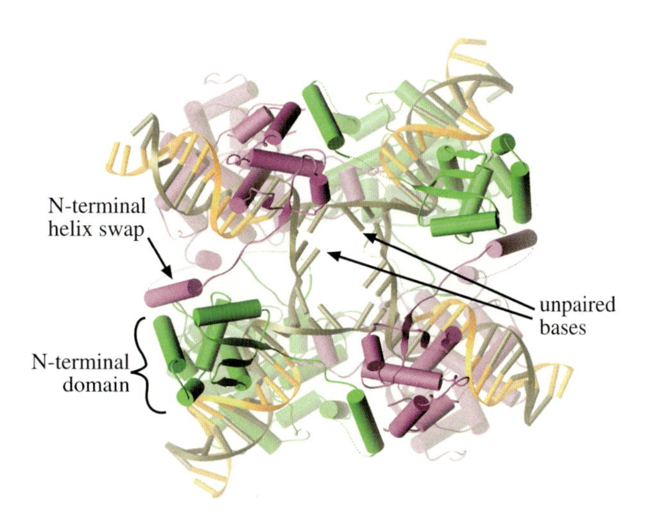

Color Plate 12 (chapter 6). Structure of the Flp recombinase-DNA complex. The DNA is a nicked HJ with an odd number of bases in the crossover region as a result of a base pairing rearrangement following cleavage of a suicide substrate (21). Each strand has an unpaired adenine base at the branch point. Cyclic interactions between recombinase subunits are indicated. The view is from the N-terminal domain side of the complex.

a

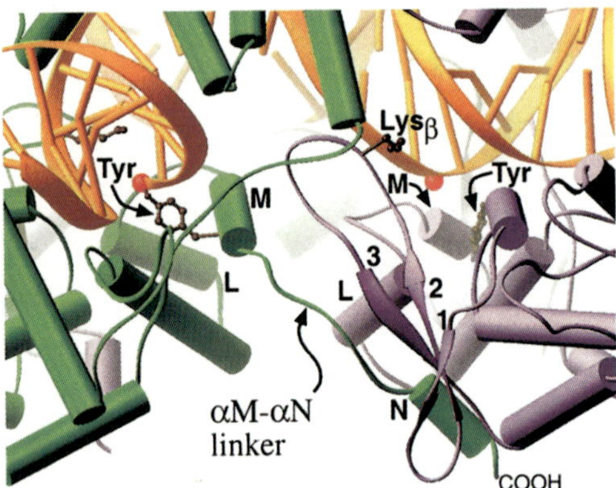

Type I Interface

b

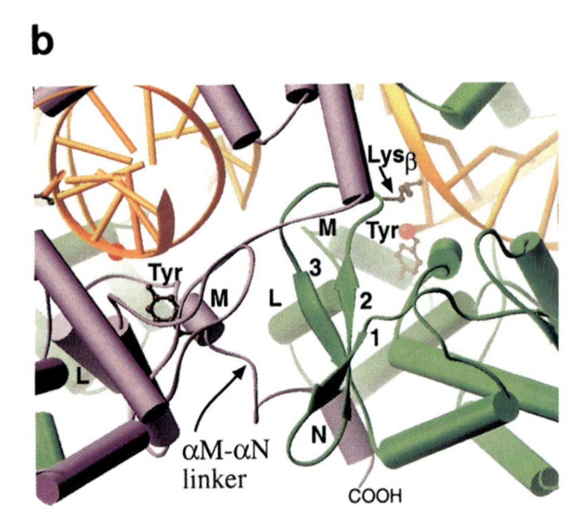

Type II Interface

Color Plate 13 (chapter 6). Interfaces between catalytic domains in the covalent Cre-DNA intermediate complex (39). The green (active) subunit has cleaved the DNA substrate and the purple (inactive) subunit has not cleaved the substrate. (a) Type I interface. Lys-β in the inactive subunit is near (but not in) the minor groove of the DNA substrate. (b) Type II interface. Lys-β in the active subunit has entered the active site. Note the different conformations of the linker peptides connecting helices M and N and the different conformation of the β2-β3 loop in the active and inactive subunits. The type I and type II interfaces shown are those circled in the top middle panel of Fig. 8.

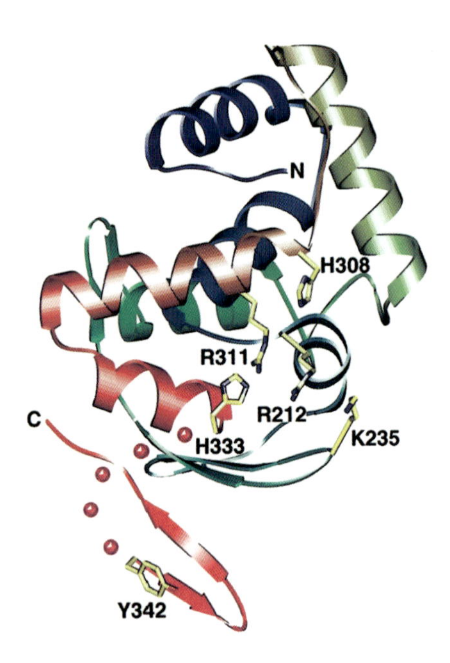

Color Plate 14 (chapter 7). λ Int catalytic domain. The ribbon diagram is colored with a gradient from the NH_2 terminus (blue) to the COOH terminus (red) and shows a seven-helix bundle cradled by a β-sheet and two β-hairpins. The flexible loop (disordered in the structure) connecting the last α-helix with the COOH-terminal β-hairpin is depicted with red balls. Catalytically essential residues R212, K235, H308, R311, and H333 face a small depression in the protein surface that forms the putative DNA-binding surface. The nucleophilic Y342 is physically separated from this region on a flexible loop but can easily be modeled into a proximal orientation. (Adapted from reference 112 with permission; image by Hyock Joo Kwon).

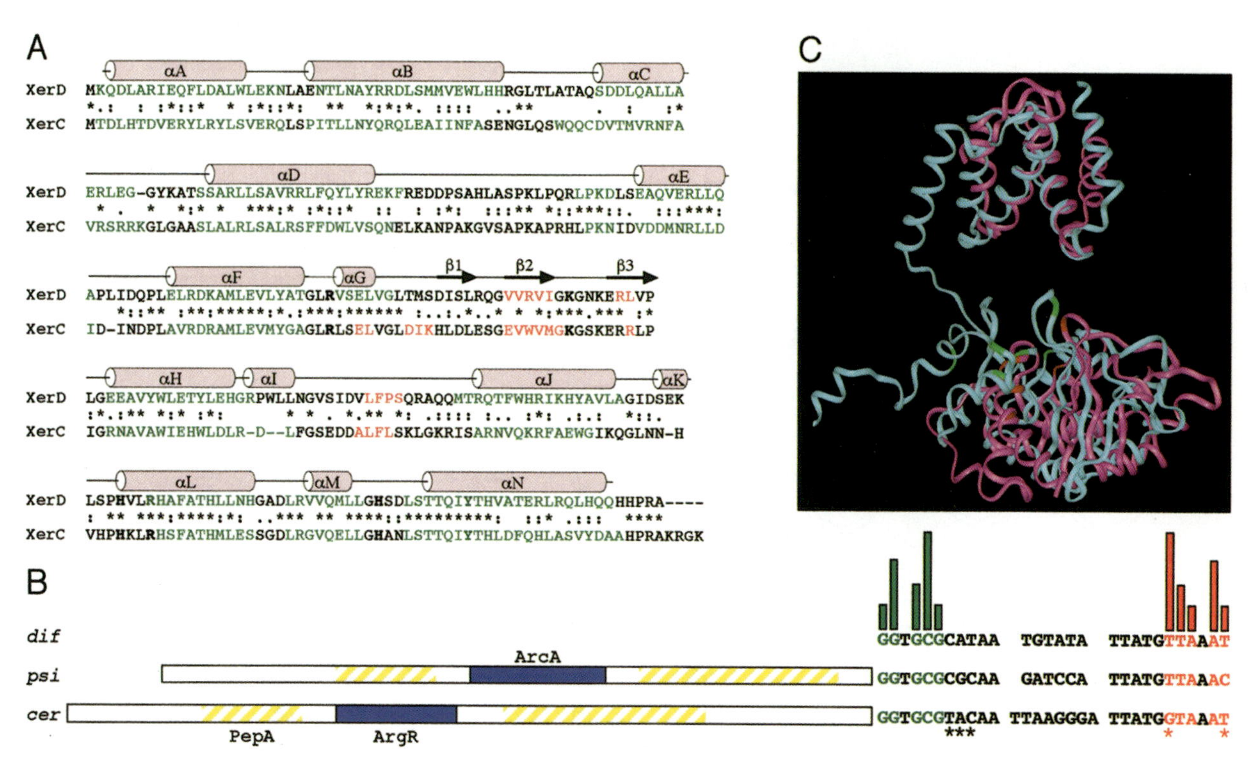

Color Plate 15 (chapter 8). The Xer recombination machine. (A) Sequence comparison of the two Xer recombinases. The primary sequence of XerC and XerD were aligned. Identical residues are indicated by an asterisk, and conservative substitutions are indicated by dots. The conserved catalytic residues of the tyrosine recombinase family are in bold. Green residues were predicted to adopt an α-helix conformation by computer calculation, whereas residues in red were predicted to form β-sheets (PredictProtein PHD software v1.96) (60). Cylinders and arrows represent α-helices and β-sheets, respectively, in XerD, as determined by crystallography. (B) Sequence comparison of the *dif*, *psi*, and *cer* recombination sites. The core sites were aligned with the DNA binding site of XerC on the left side. Green and red residues are thought to be involved in determining the specificity of binding of XerC and XerD, respectively. The size of the bars on top of the sequences indicates their relative importance for binding. Asterisks indicate residues that differ between the binding sites of the three recombination sites. The accessory sequences of *psi* and *cer* are represented by bars, with the binding sites of ArgR and P-ArcA in blue and regions interacting with PepA in yellow diagonal stripes. (C) Three-dimensional structure comparison of Cre and XerD. The two structures are shown as blue and pink ribbons respectively. The positions of the catalytic residues are highlighted in green and red, respectively. The N termini of the proteins (on top) were aligned with respect to each other. Note that despite differences in the structure of the C terminus of the proteins, the active-site residues are remarkably close.

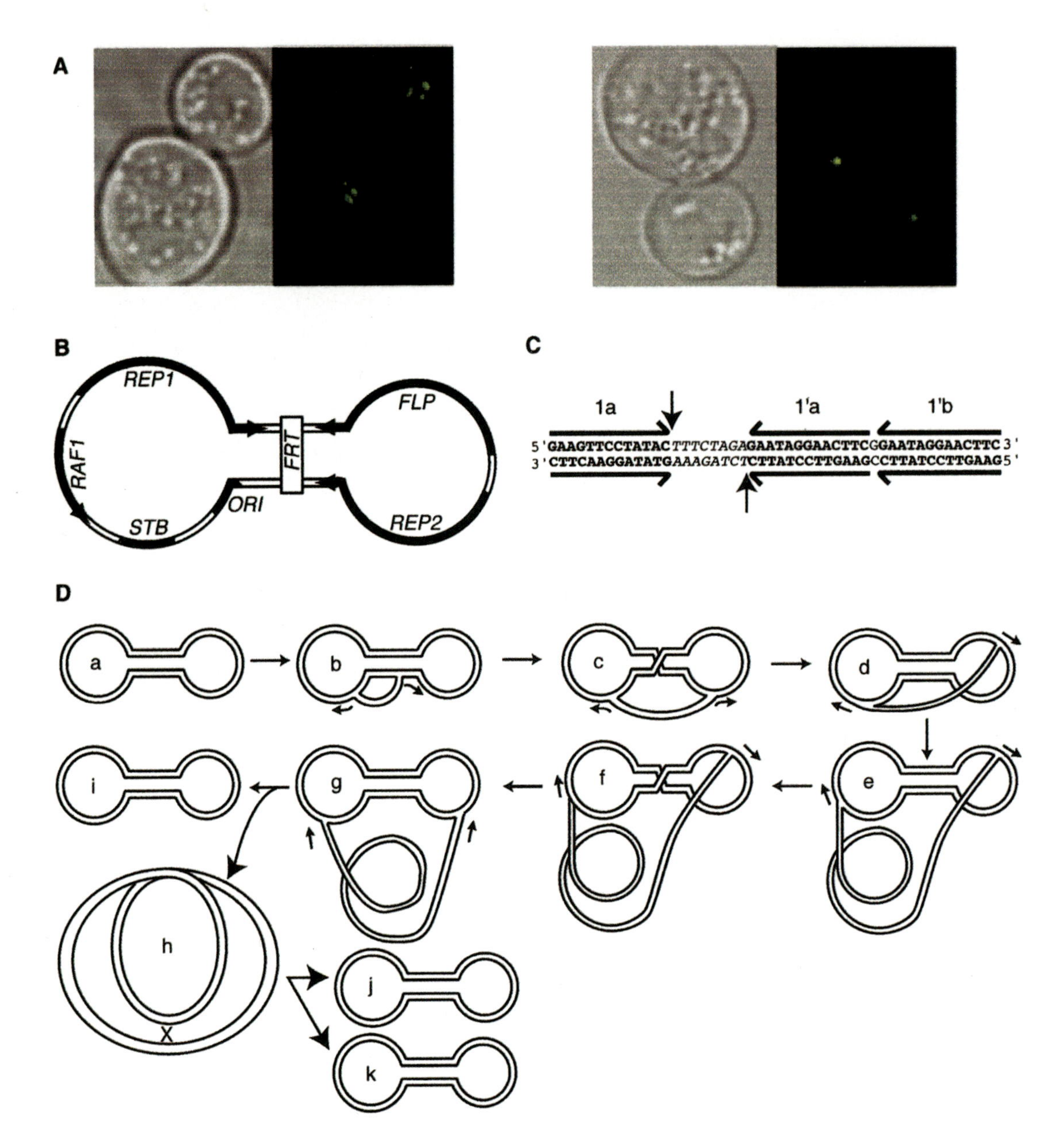

Color Plate 16 (chapter 11). (A, left) A 2μm circle-derived test plasmid carrying multiple copies of the lac operator DNA is visualized in a dividing yeast cell by the associated green fluorescence from the GFP-lac repressor hybrid expressed in the cell (137). (A, right) For comparison, a haploid chromosome, which is also tagged by GFP-lac repressor, is shown. The plasmids are seen most often as a tetrad cluster. (B) Conventionally, the 2μm circle is represented as a dumbbell, with the parallel lines indicating the inverted repeat sequence present within the plasmid genome. The Rep1 and Rep2 proteins along with the *cis*-acting locus *STB* constitute the plasmid-coded components of the partitioning system. The Raf1 protein has not been well characterized, but is a likely positive regulator of *FLP* gene expression. The Flp protein and its target sites (*FRT*) within the inverted repeat constitute the plasmid amplification system. The plasmid replication origin is denoted by *ORI*. (C) In the expanded view of the *FRT* site, the Flp-binding elements are named 1a, 1′a, and 1′b, and their relative orientations are represented by the arrowheads. The sequence of the strand-exchange region (or spacer) is written in italics. The phosphodiester bonds on the top and bottom strands that are broken and joined during recombination are indicated by vertical arrows. The 1′b element is dispensable during recombination; the minimal *FRT* site is constituted by 1a, 1′a, and the spacer segment sandwiched between them. (D) The Futcher model for plasmid amplification is adapted from Broach and Volkert (8). A bidirectional replication event (b) is converted by Flp-mediated recombination into two unidirectional replication forks (c) leading to plasmid amplification (d to f). A second recombination event (f) terminates amplification (g). The replication products h and i are the multimeric and monomeric plasmids, respectively. Resolution of h yields plasmid monomers (j, k, etc.).

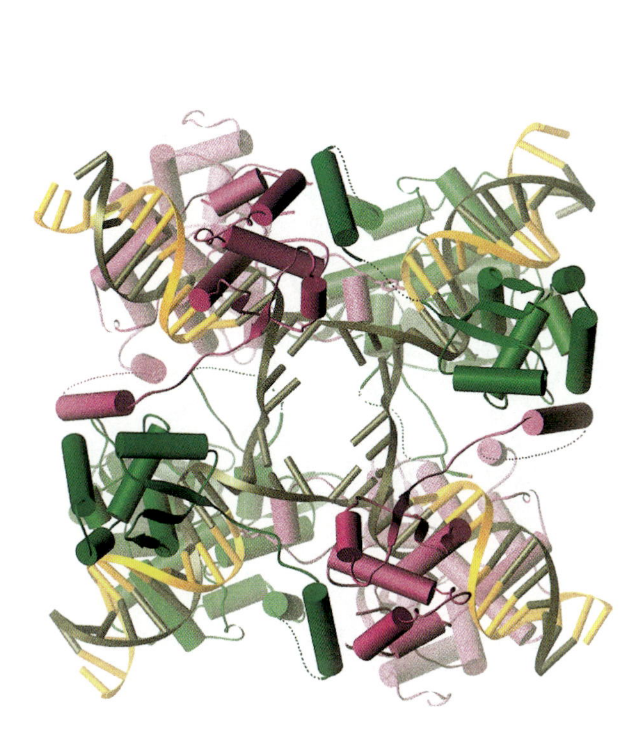

Color Plate 17 (chapter 11). In the structure of a Flp-DNA complex (18), the green Flp monomers are adjacent to the two scissile phosphodiester bonds that are ready for cleavage. The purple Flp monomers are adjacent to the phosphodiesters that are refractory to cleavage in this state. The color scheme matches that used by Guo et al. (47) in their representation of the Cre-DNA structure. The Cre and Flp-DNA structures are quite similar in their overall geometry and protein-DNA interactions, consistent with a similar reaction pathway for the integrase tyrosine recombinase family. A notable exception is that the active-site tyrosines from Cre and Flp are positioned differently so that cleavage occurs in *cis* for Cre and in *trans* for Flp. In the Flp-DNA structure, it is the purple tyrosines that are poised to attack the target phosphates.

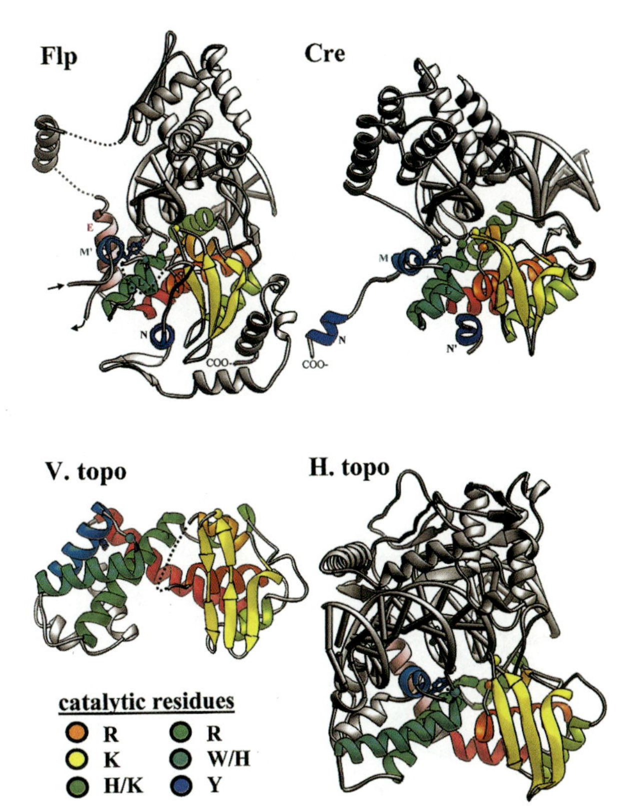

Color Plate 18 (chapter 12). Comparison of related catalytic domains. (Top) Monomers of Flp (11) and Cre (20) are shown, viewed from near the center of the Holliday junction. Secondary-structure elements that are conserved between the two enzymes are colored in rainbow order from helix F (red) to N (blue). The scissile phosphate is shown as a gray sphere, the nucleophilic tyrosine is in blue, and the alpha Carbons of the other conserved active site residues are shown as colored spheres (see key, lower left). The first conserved histidine (green) is replaced by lysine in the topoisomerases, while the second conserved histidine (cyan) is replaced by tryptophan in Flp and Cre. Helices mentioned in the text are labeled. Both subunits are in the active conformation. (Bottom) The catalytic domain of vaccinia virus topoisomerase (V. topo) (12) and the DNA-bound complex of human topoisomerase I (H. topo) (46) are shown in a similar coloring scheme. In the viral topoisomerase, the tyrosine-bearing helix (equivalent to M in Flp and Cre) probably reorients upon DNA binding. The active site tyrosine of the human topoisomerase had been mutated to phenylalanine. An ~75-amino-acid insertion between the helices equivalent to L and M (dotted line) was removed in this structure.

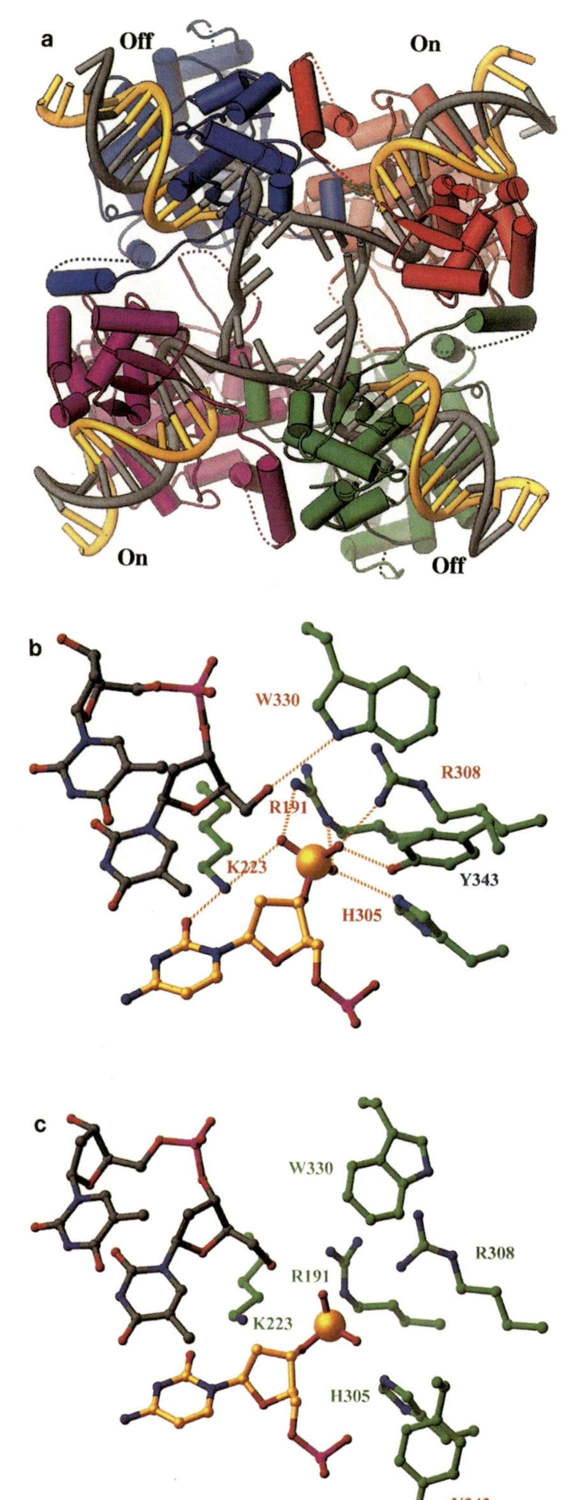

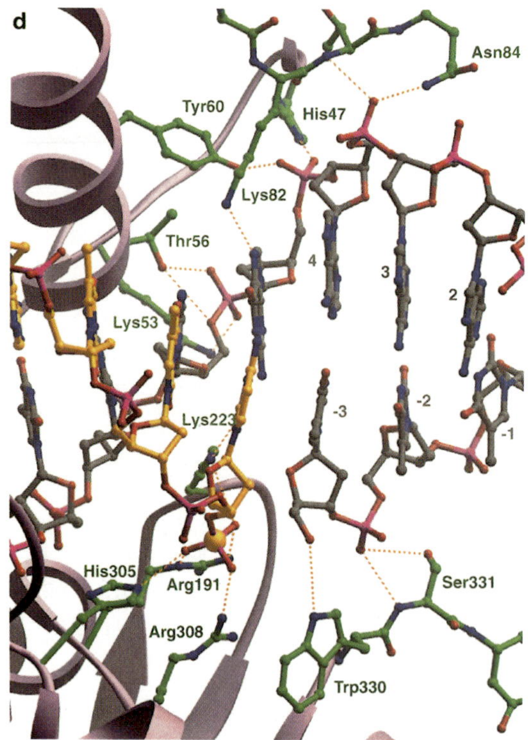

Color Plate 19 (chapter 12). The active site. (a) View of the Flp synaptic complex, colored as in parts b, c, and d of this figure and Color Plate 20. (b) Close-up view of the active catalytic center residing in the monomer that is red in panel a. The nucleophilic tyrosine is provided by the blue monomer and is hydrogen bonded to the scissile phosphate (gold sphere). (c) The inactive catalytic center residing in the monomer that is green in panel a. The hydroxyl of the tyrosine donated by the red monomer is not properly oriented for catalysis. Explicit hydrogen bonds are not shown, as this region is less well ordered in the electron density than that shown in panel b. (d) Protein-DNA contacts in the spacer region. One of the active monomers is shown. The C atoms of protein side-chains, DNA top strand and bottom strand are colored in green, gold, and gray, respectively, and the large gold sphere denotes the scissile phosphate. Hydrogen bonds are shown in dotted lines, and bases are numbered as in Fig. 2a. Y343 from an adjacent monomer is also hydrogen bonded to the scissile phosphate but was removed for clarity. These contacts may ensure that the positioning of the 5′ OH of T-3 in the active site is dependent on the formation of Watson-Crick base pairs in the adjacent spacer. Panels b to d are reprinted from reference 11 with permission from Elsevier Science.

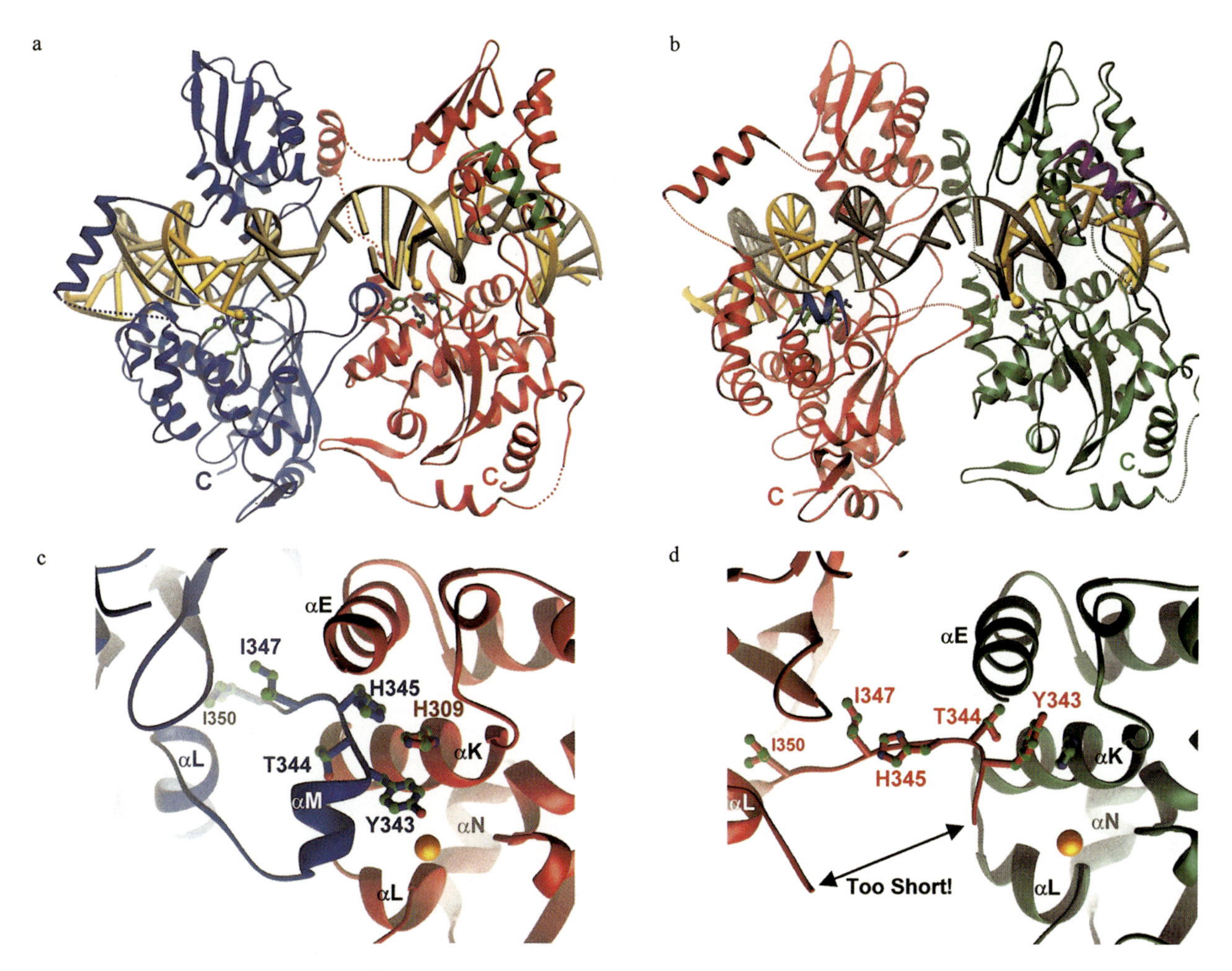

Color Plate 20 (chapter 12). Alternate interfaces within the tetramer. (a) A type I interface viewed from the center of the Holliday junction, with the catalytic domains below the DNA. The coloring scheme is as in Color Plate 19a. This pairing creates a functional active site in the red monomer. Y343, R191, H305, and R308 are shown in green, and the scissile phosphate as gold sphere. Additional *trans* interactions are also shown. (b) A type II interface. The inactive green monomer is in the same orientation as the red monomer in panel a. Y343 of the red monomer can be seen in the green one's active site, but it is not properly positioned for catalysis. The catalytic domains are farther apart in this pairing. (c) Close-up view of the dimer in panel a, viewed from the N-terminal domain. The N-terminal domain and the DNA have been removed for clarity, but the position of the scissile phosphate is marked by a gold sphere. (d) Similar close-up view of the dimer in panel b, with the green monomer in the same orientation as the red one in panel c. Helix M of the red monomer is disordered and may be unraveled in order to accommodate that longer distance that its flanking turns must traverse in this interface. Panels a and b are reprinted from reference 11 with permission from Elsevier Science.

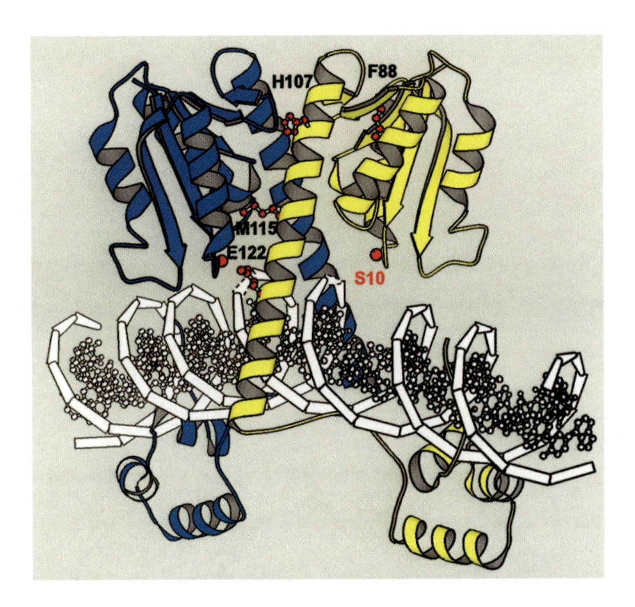

Color Plate 21 (chapter 13). Structural model of the Hin dimer bound to *hixL* (90). The DNA is oriented into the plane of the page, with the dashed segments depicting where the cleavages occur. The DNA binding domains at the bottom are connected to the catalytic and dimerization domains at the top via the long coiled-coil α-helices (α-helix E). The catalytic domains are oriented approximately perpendicular to the DNA axis. The locations of the active-site serine 10 on each subunit are denoted by the red sphere and the side chains of amino acids (Phe 88, His 107, Met 115, and Glu 122), where substitutions generate strong Fis-independent mutants, are shown for one subunit. The model is based on the crystal structures of DNA complexes of the γδ resolvase and the Hin DNA binding domain (41, 61, 291).

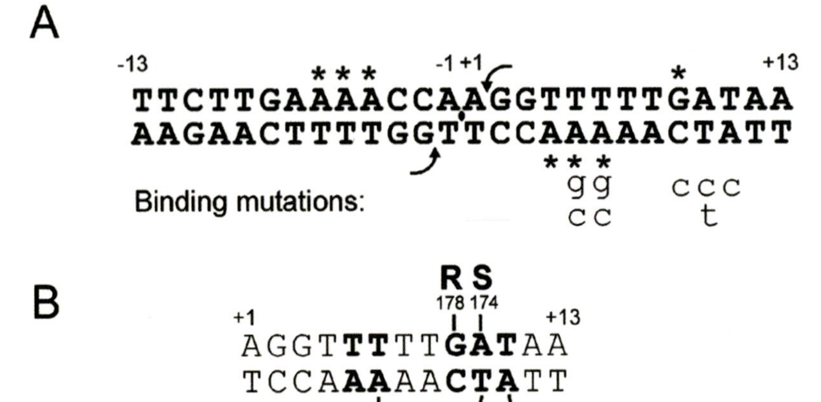

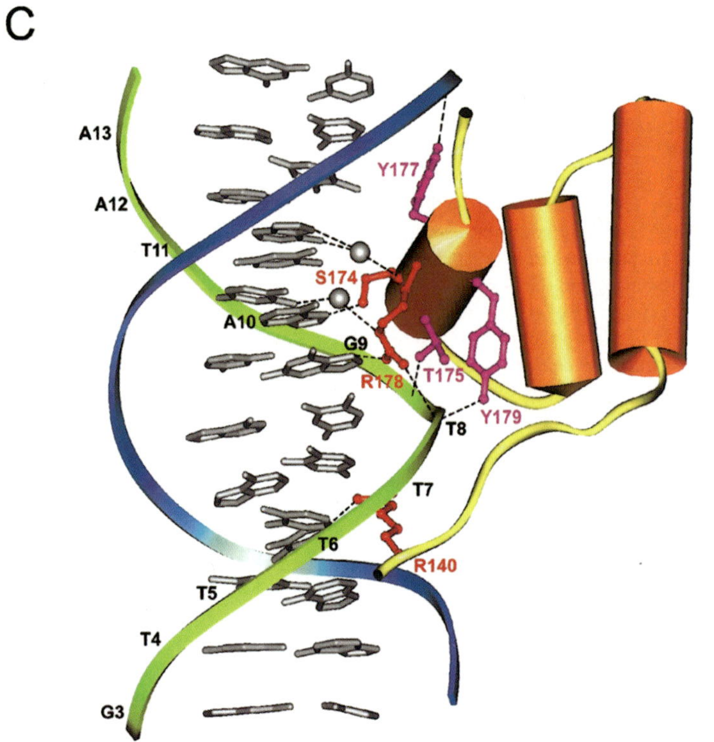

Color Plate 22 (chapter 13). Recognition of the recombination site by Hin. (A) Sequence of the *hixL* Hin binding site. Arrows point to the positions of Hin-induced cleavages. Asterisks denote bases where methylation by dimethyl sulfate (major groove N7 position of G and minor groove N3 position of A) inhibits Hin binding (74). Base pair substitutions that strongly inhibit binding are shown for the right half-site (41, 104, 105). (B) The sequence of the *hixL* half-site within the invertible segment for which crystal structures in a complex with the Hin DNA binding domain have been determined (41, 61). The bases involved in sequence-specific recognition by Hin are in bold and the amino acids which contact these bases shown. The water molecules (w) which are hydrogen bonded to the ε-imino of Arg 178 and the main chain carbonyl oxygen of Ser 174 are important in mediating specificity at base pairs 10 and 11, respectively. Arg 140, along with Gly 139, makes multiple minor groove contacts with the nucleotides at positions 5 and 6. (C) Structure of the Hin DNA binding domain (residues 139–182 shown) bound to the *hixL* half-site, as determined by X-ray crystallography (41, 61). The side chains of amino acids that contact bases (red) and the phosphate backbone (pink) are displayed along with the ordered water molecules (gray spheres) involved in indirect hydrogen bonding.

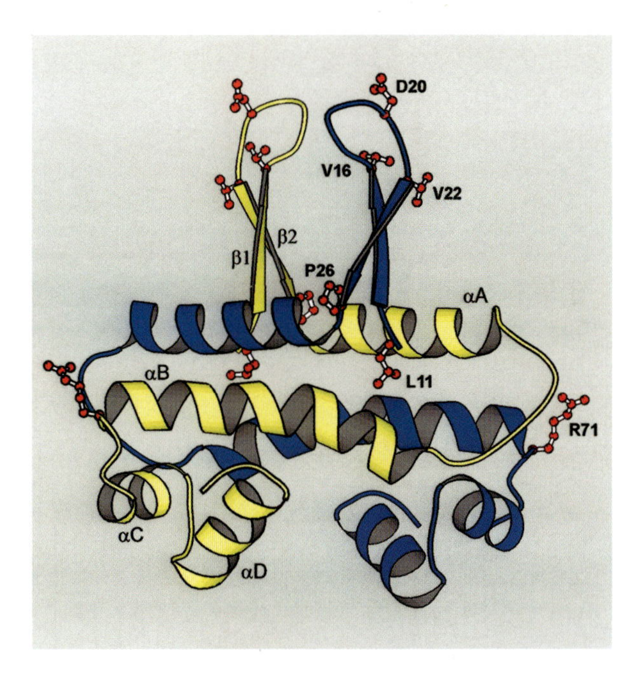

Color Plate 23 (chapter 13). Structure of the Fis protein. The model shown is from the K36E mutant crystal structure (hexagonal form [242]). The polypeptide chain from each subunit of the homodimer begins at Val 10 and ends at the C terminus (Asn 98). The N-terminal β-hairpin arms (β-1 and β-2) are mobile and are oriented somewhat differently in other crystal forms (40). Four α-helices (A to D) make up the core of the protein, with helices C and D constituting a helix-turn-helix DNA binding motif. The side chains of amino acids which are critical for Hin-activation (Val 16, Asp 20, Val 22), the structural integrity of the mobile β-hairpin arm region (Leu 11, Pro 26), and transcriptional activation plus DNA bending (Arg 71) are displayed (22, 187, 211, 242, 292).

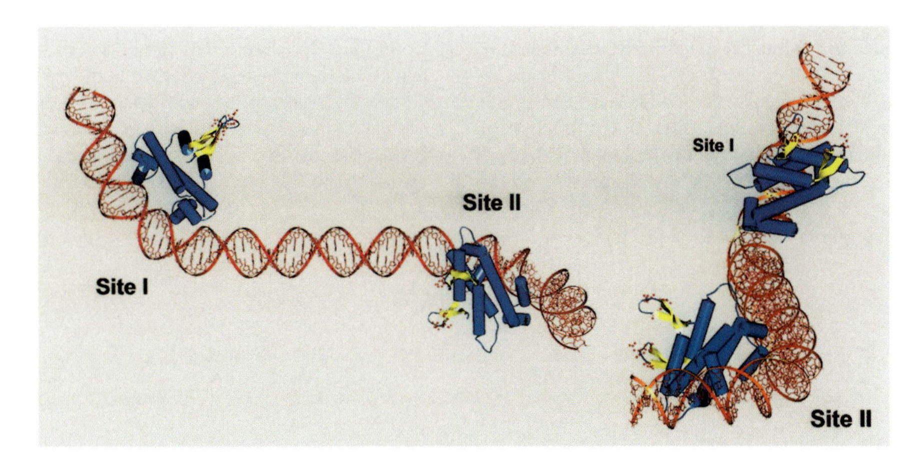

Color Plate 24 (chapter 13). Model of the Fis-bound *hin* enhancer segment (188). An 83-bp DNA segment comprising the *hin* enhancer and surrounding sequence is shown in two views. The *hixL* proximal (site I) and distal (site II) binding sites are labeled, and the Fis protein (blue) is docked onto each binding site based on measurements of Fis-induced bending, the locations and efficiency of site-directed DNA scission induced by different Fis-1,10 phenanthroline chimeras, chemical and nuclease footprinting data, and mutation studies. The DNA twist for the DNA between the Fis binding sites is 32.2°. The β-hairpin arm region of Fis is highlighted in yellow, and the side chains of Fis residues Val 16, Asp 20, and Val 22 that activate Hin are displayed in ball-and-stick representation.

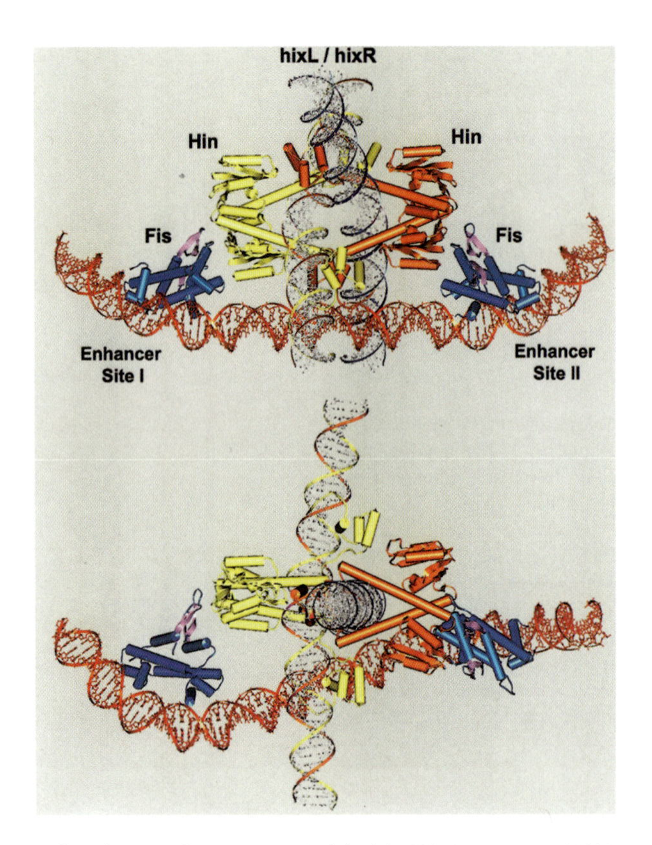

Color Plate 25 (chapter 13). Model of the Hin invertasome (188). The Hin-bound *hixL* and *hixR* complexes (Color Plate 21) are assembled onto the Fis-bound *hin* enhancer (Color Plate 24). In this arrangement, the *hix* DNA strands are located close to each other in a perpendicular orientation. The tips of the mobile β-hairpin arms of the Fis dimers, highlighted in pink, are positioned near the top of the dimer interface in the catalytic domain of Hin. The bottom view is rotated 90° about the *x* axis.

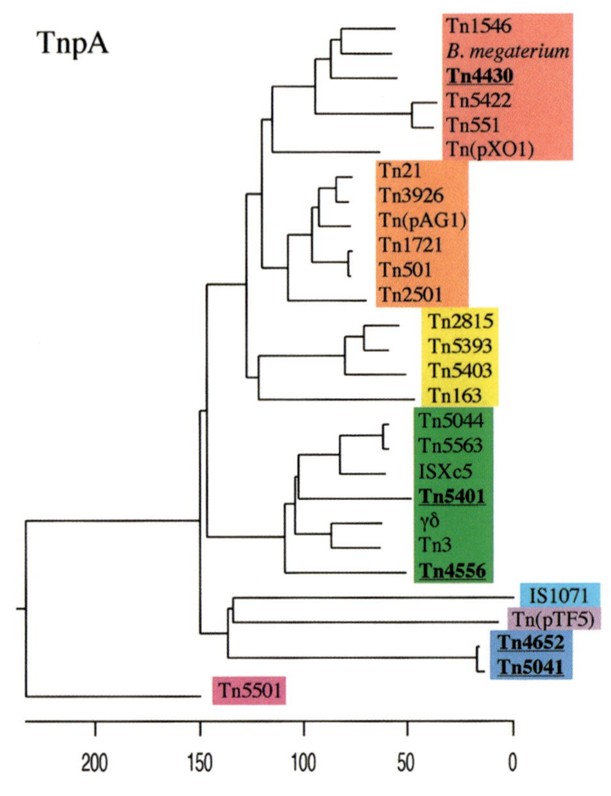

Color Plate 26 (chapter 14). Phylogenetic tree of transposase proteins. The eight clusters described in the text are boxed in different colors. The transposons in boldface, underlined type are those that encode TnpI proteins rather than resolvases. The scale indicates PAM distances (the number of accepted point mutations per 100 residues separating two protein sequences).

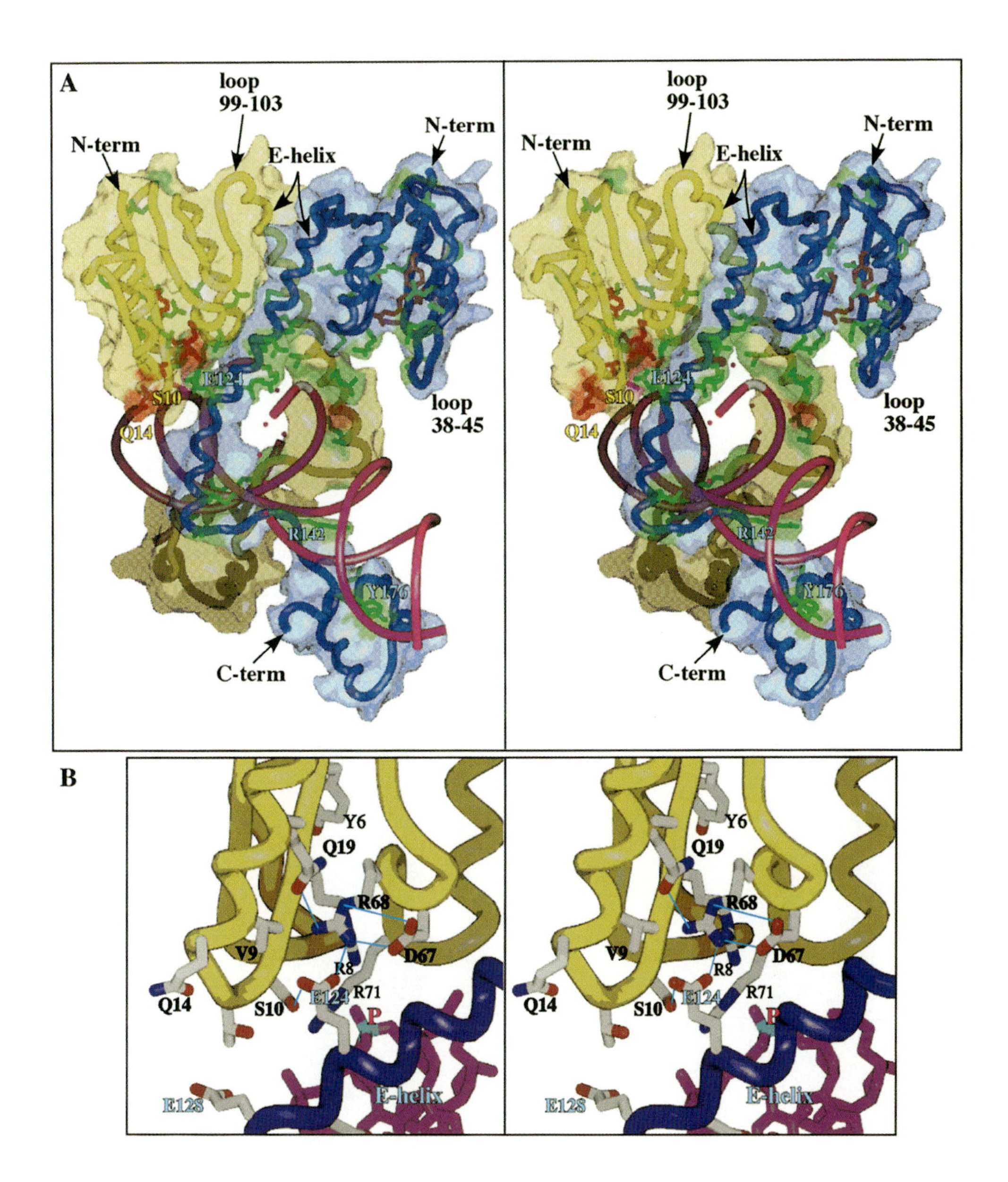

Color Plate 27 (chapter 14). Resolvase three-dimensional structure. (A) Structure of the γδ resolvase dimer bound to site I (139); stereo view rendered in SPOCK (J. A. Christopher, SPOCK: the Structural Properties Observation and Calculation Kit [Program Manual], Center for Macromolecular Design, Texas A&M University, College Station]). The two resolvase monomers are colored yellow and blue. The Ser10 nucleophile is colored magenta; other conserved side chains (and their associated surfaces if exposed) are shown in red (totally conserved) or green (highly conserved). A few landmarks are indicated; individual residues are labeled in yellow or pale blue to correspond with the monomer. The white segment on the DNA worm indicates the position of resolvase cleavage (but note that the readily visible one is destined for cleavage by the active site of the blue subunit, which is both distant and barely visible in this view). For further details, see the text. (B) Active site of γδ resolvase (stereo view). The active site of the yellow subunit is shown, indicating the close approach of Glu124 from the E-helix of the blue subunit and the H-bonding network (cyan lines) that interconnect the active site residues Ser10, Gln19, Asp67, Arg68, and Glu124. Note the distance (~11 Å) between Ser10 and the scissile phosphate (cyan; labeled P).

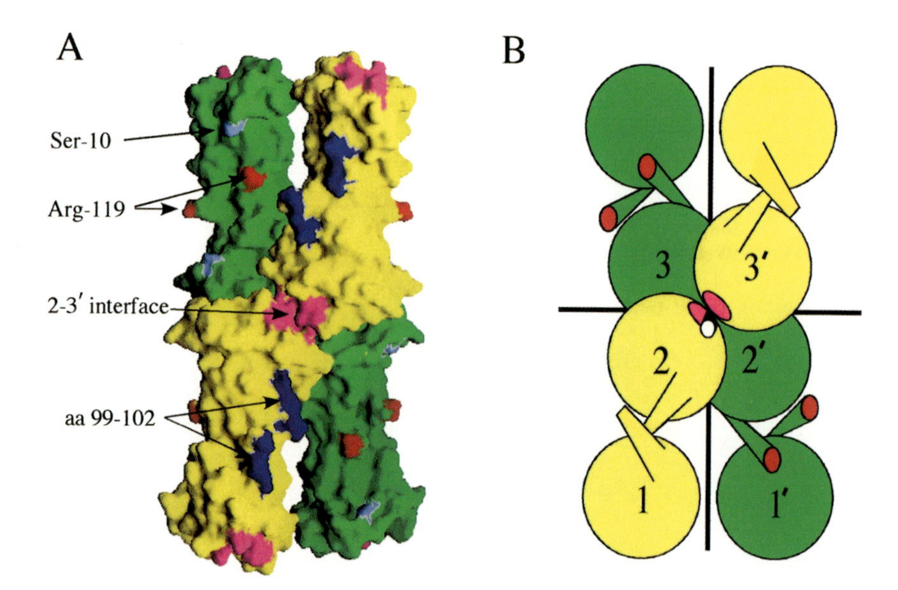

Color Plate 28 (chapter 14). The crystallographic tetramer of dimers of the γδ resolvase catalytic domain (107, 114). (A) Surface representation (GRASP) (96) of four dimers of resolvase; (B) schematic of the same assembly. The structures are for residues 1 to 119 and exclude the portion of resolvase from the middle of the E-helix through the DNA binding domain. The E-helices, represented as the cylinders in panel B, mark the dimer interface. The assembly has 222 (D2) symmetry (three orthogonal twofold axes: the two shown in panel B and the third running through the center of the assembly perpendicular to the plane of the page). The magenta surfaces are residues 2, 32, 54, and 56, which form the 2,3′ interfaces that connect adjacent yellow (and green) dimers. The other colored surface patches are as follows: dark blue, residues 99 to 102; red, residue 119, marking the end of the visible portion of the E-helix; light blue, Ser10.

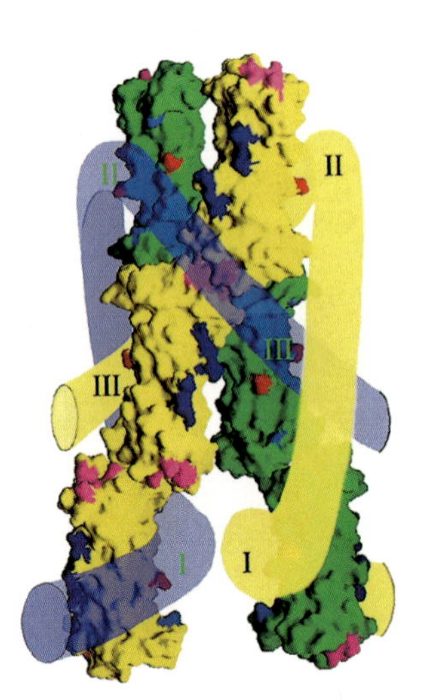

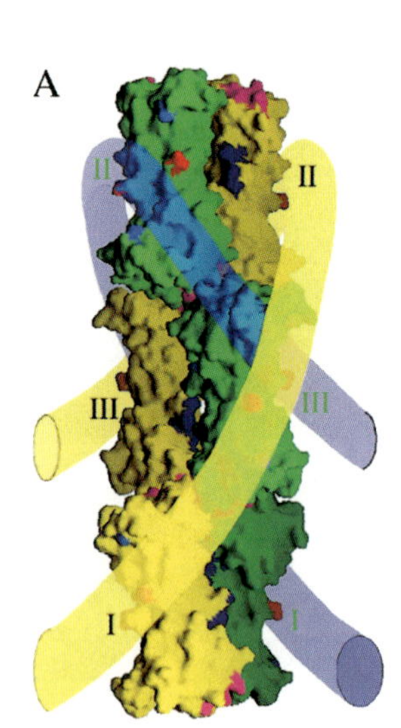

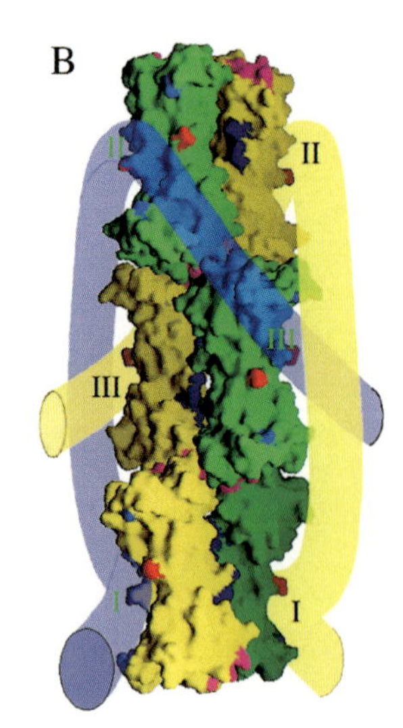

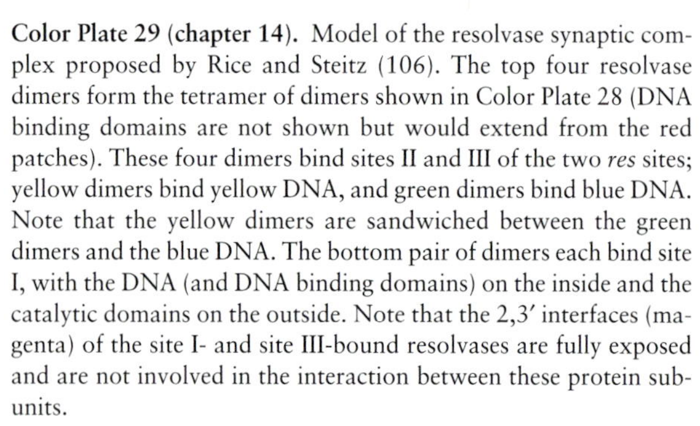

Color Plate 29 (chapter 14). Model of the resolvase synaptic complex proposed by Rice and Steitz (106). The top four resolvase dimers form the tetramer of dimers shown in Color Plate 28 (DNA binding domains are not shown but would extend from the red patches). These four dimers bind sites II and III of the two *res* sites; yellow dimers bind yellow DNA, and green dimers bind blue DNA. Note that the yellow dimers are sandwiched between the green dimers and the blue DNA. The bottom pair of dimers each bind site I, with the DNA (and DNA binding domains) on the inside and the catalytic domains on the outside. Note that the 2,3′ interfaces (magenta) of the site I- and site III-bound resolvases are fully exposed and are not involved in the interaction between these protein subunits.

Color Plate 30 (chapter 14). New models for the resolvase synaptic complex (114a). The models shown in the two panels differ only in the path of the DNA between sites II and I. The three green resolvase dimers interact via the 2′,3′ interface (magenta surface, mostly buried from this perspective), as do the three yellow dimers. Colored surface patches are as in Color Plate 28. Note that all adjacent pairs of yellow and green dimers interact with the blue surface loop (residues 99 to 102) in close contact. The alternative DNA wraps, identical between sites II and III, differ in the following ways: (A) the site I segments cross to form a (+) node, and the resolvase bound to the left half of site I is anticipated to interact with the resolvase at site III (of the same *res* site); (B) the site I segments cross to form a (−) node, and the resolvase bound to the right half of site I is anticipated to interact with the resolvase at site III (of the opposing *res* site). For more details, see the text.

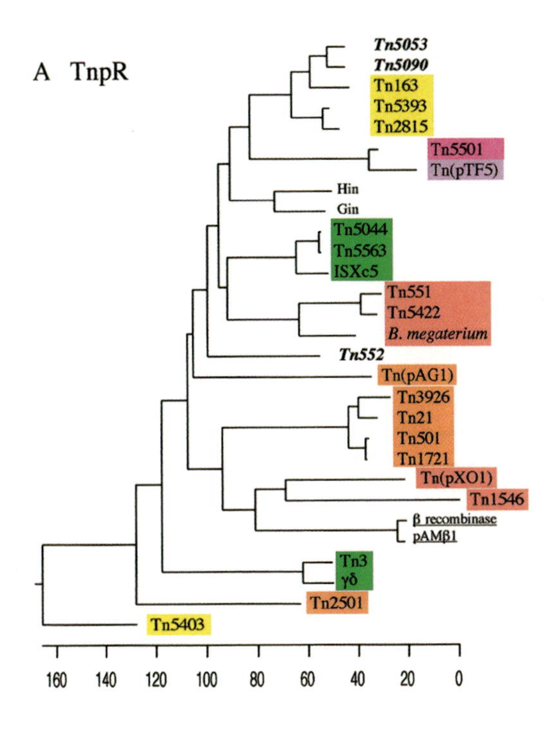

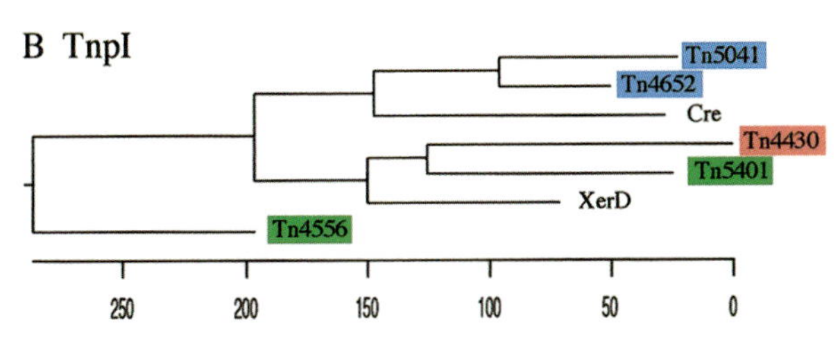

Color Plate 31 (chapter 14). Phylogenetic trees of site-specific recombinases involved in cointegrate resolution. (A) Cointegrate resolvases of the resolvase (serine recombinase) family. The colored boxes are coded as in Color Plate 26 to indicate the different transposase (TnpA) clusters. The uncolored transposons in boldface italic type are the non-Tn*3*-family elements that encode resolvases. Included in this tree are two examples of DNA invertases (Hin and Gin) and two examples of the plasmid-encoded β recombinase group (underlined). (B) Cointegrate resolvases of the integrase-(tyrosine recombinase) family. Colored boxes are as in panel A. The integrase-family recombinases Cre and XerD are included for comparison.

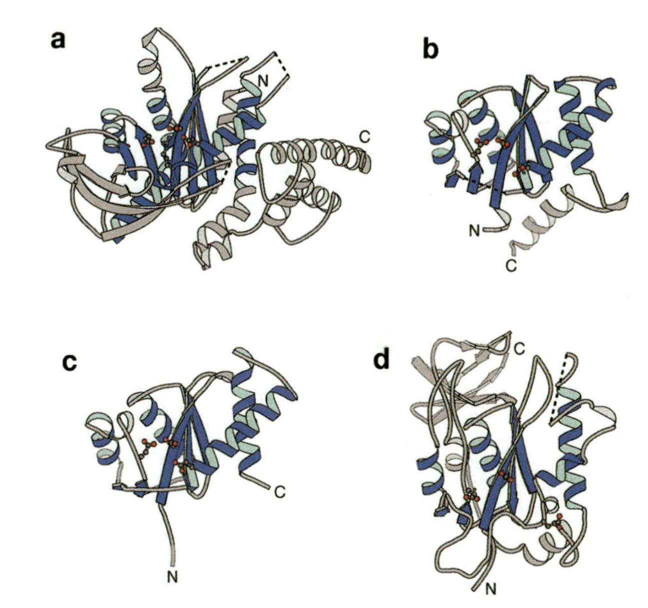

Color Plate 32 (chapter 17). Structural comparison between Tn*5* inhibitor protein (a), HIV type 1 integrase (b), avian sarcoma virus integrase (c), and Mu transposase (d). The structural features common to all of these proteins are colored in blue. For further details see reference 41. Reprinted from the *Journal of Biological Chemistry* (41) with permission of the publisher.

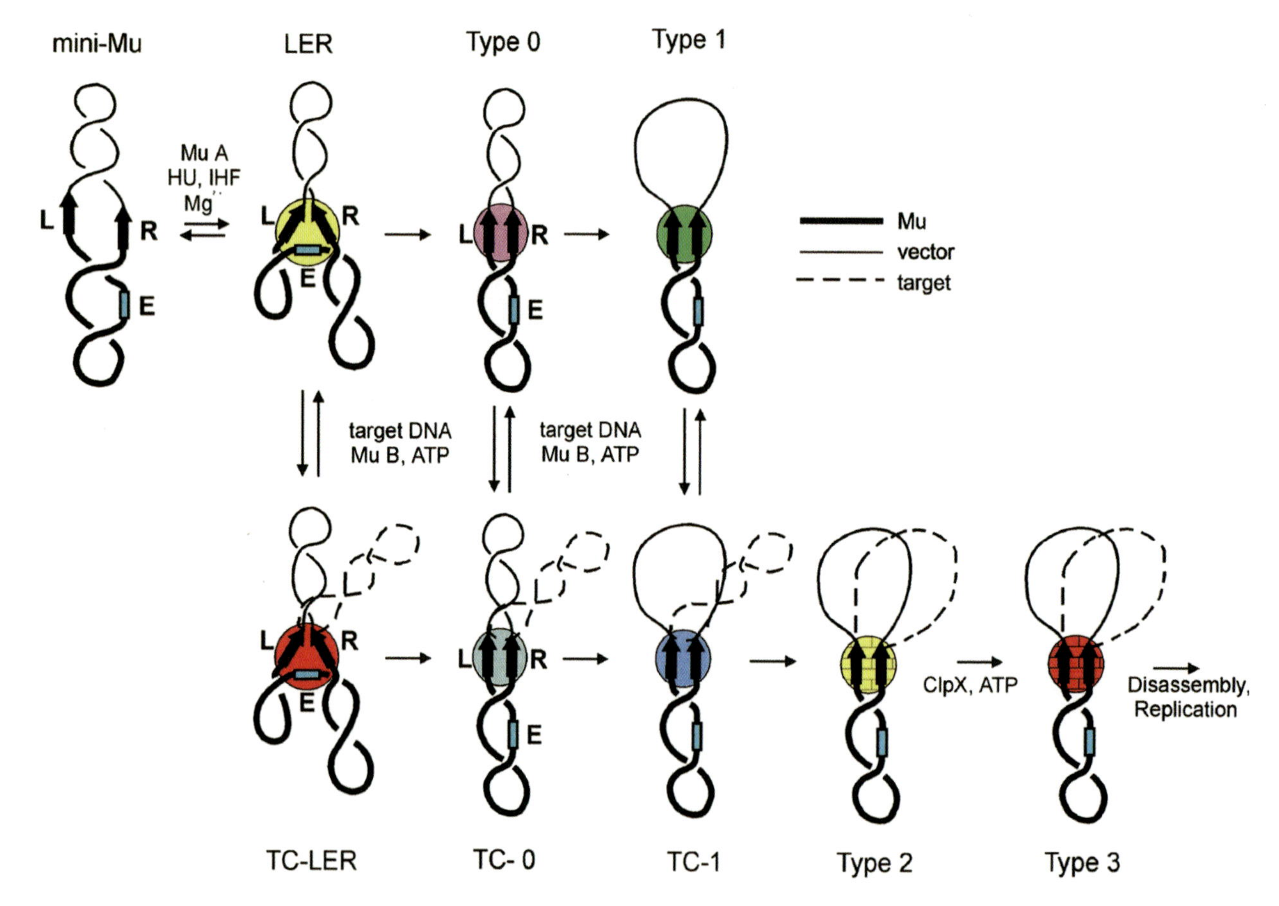

Color Plate 33 (chapter 17). Transpososomes: higher-order nucleoprotein complexes which mediate Mu DNA transposition. Reprinted from *Biochemistry and Cell Biology* (25) with permission of the publisher.

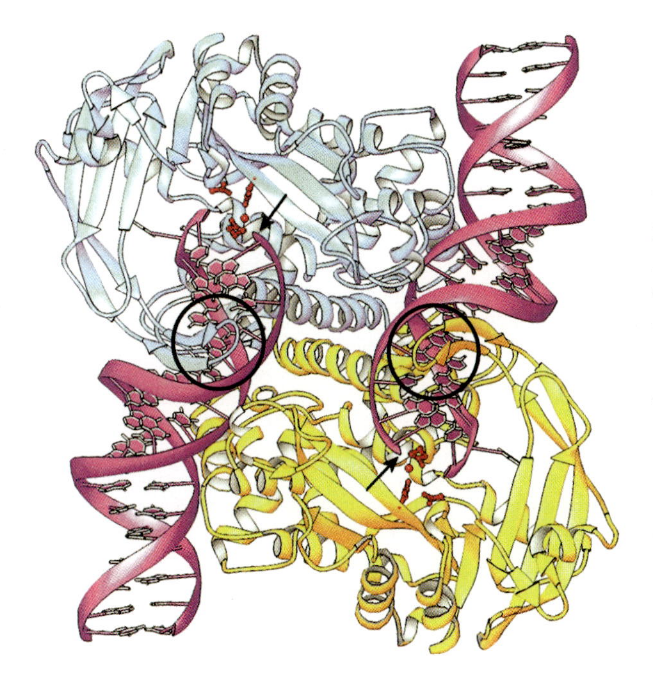

Color Plate 34 (chapter 18). Tn*5* synaptic complex. A ribbon representation of the transposase (Tnp)/DNA dimer in which the two Tnp molecules are blue and yellow, the DNA is purple. The structure demonstrates that each DNA molecule is contacted by both Tnp molecules. We hypothesize that one Tnp binds the DNA initially, and then upon synapsis (dimerization), the DNA is contacted by multiple groups of the other Tnp molecule and the 3′-OH end of the DNA is positioned precisely in the active site of this second Tnp molecule. The active-site DDE residues (D97, D188, and E326) and the Mn^{2+} found in the active site are red, the location of the 3′-OH groups on the DNA are indicated with arrows, and the location of a β hairpin that clamps the DNA is circled. The β hairpin is largely unstructured in the Inh structure. The structure was determined to 2.3-Å resolution. Tnp was the hyperactive triple mutant E54K, M56A, L372P (29). The DNAs were 20-bp duplexes containing the OE sequence (see Fig. 1D) with one additional base pair at the transposon end. The figure is similar to one presented in reference 17.

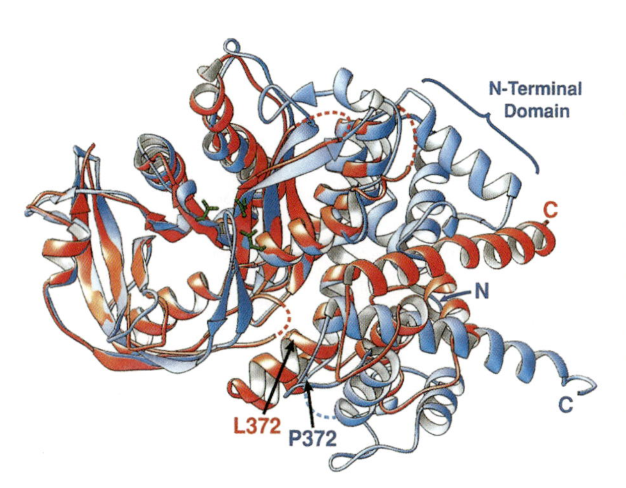

Color Plate 35 (chapter 18). Comparison of Inh and Tnp structures. The Tnp structure (in blue) is from the synaptic complex structure (17; see Color Plate 34). The Inh structure (in red) is from the Inh dimer structure (16). Indicated on the structures are catalytic active-site DDE residues in green and the location of the L372 residue in Inh and the P372 residue in the hyperactive Tnp. The structures superimpose well in most regions except that Inh does not have the N-terminal DNA binding domain, and the C terminus of Tnp has moved substantially either because of the disordering caused by the L372P mutation or because Tnp is in the form of the synaptic complex or both. If the N terminus were on the Inh structure, the N terminus and C terminus would partially occupy the same space. This is a model for the Tnp structure when it is free in solution, and the N-C overlap likely contributes to Tnp inactivity.

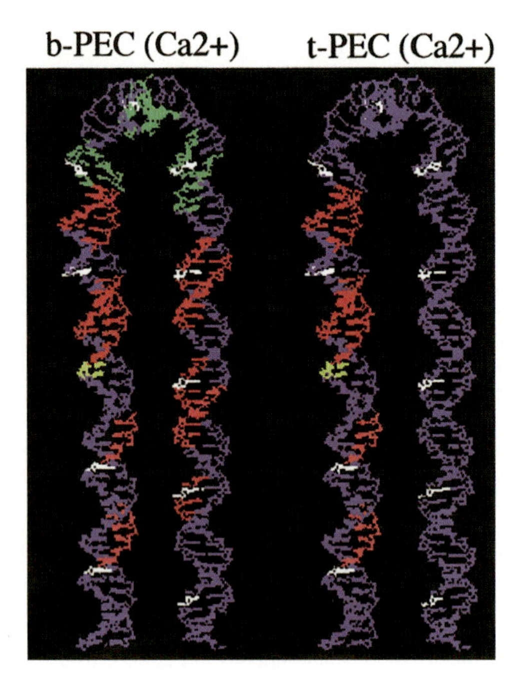

Color Plate 36 (chapter 20). Transposase-DNA contacts in the b-PEC and t-PEC. Red and green, positions at which transposase and IHF, respectively, protect outside-end DNA against attack by hydroxyl radicals; yellow, position of enhanced cleavage by hydroxyl radicals (corresponds with the scissile phosphate on the transferred strand). For point of reference, every tenth residue from the scissile phosphate is shown in white. The U-turn is modeled from the IHF/H′ structure (102). OP-Cu footprinting of the b-PEC reveals a much broader region of hypersensitivity over the transposon-donor junction than does hydroxyl radical footprinting (not shown; Allingham and Haniford, unpublished). This figure was kindly provided by R. Chalmers.

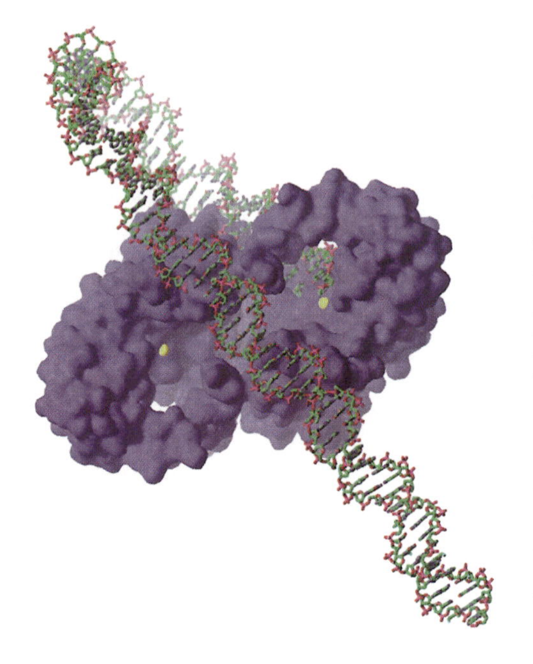

Color Plate 37 (chapter 20). Target capture may be blocked by the position of the transposon end at early stages in the reaction. The Tn*5* transposase dimer, the structure of which was derived from the Tn*5* synaptic complex, is shown interacting with a single transposon end. The position of approximately the first 20 bp is shown as in reference 32. Past this point the DNA helix undergoes a U-turn bend at the IHF site as would be present in the outside end of Tn*10*. The U-turn has been modeled from the IHF/H′ structure as in Color Plate 36. The segment of DNA internal to the IHF site is shown to pass over a deep cleft in which it has been predicted that the Tn*5* synaptic complex binds target DNA (depression between the yellow spheres that represent the 3′-OH termini of the transferred strands). Firing of the molecular spring could be involved in repositioning the transposon DNA so that target DNA can enter the cleft. It is assumed that the organization of the Tn*10* synaptic complex will resemble that of the Tn*5* synaptic complex. This figure was kindly provided by R. Chalmers.

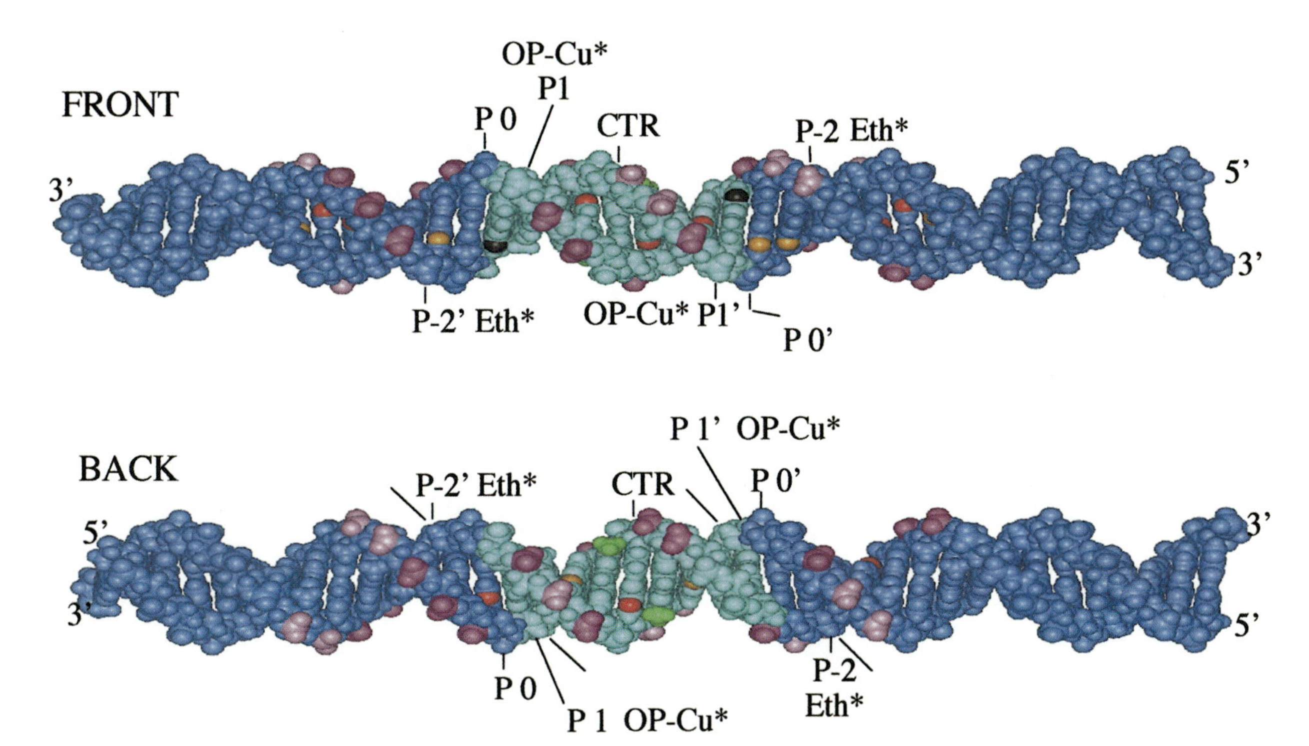

Color Plate 38 (chapter 20). Summary of HisG1 contact probing data. The HisG1 sequence is shown modeled onto B form DNA. The target core is shown in light blue and flanking regions in dark blue (sequence information is provided in Fig. 7). The scissile phosphates are designated P0 and P0′. Sites of OP-Cu protection (pink), OP-Cu protection and ethylation interference (purple), methylation interference (red and orange for strong and weak interference, respectively), methylation enhancement (black), and predicted thymine major groove contacts (green) are shown. Sites of OP-Cu hypersensitivity and ethylation enhancements are denoted by asterisks. Diagonal lines indicate possible positions of DNA kinks through the minor groove that would widen the minor groove and compact the major groove. The "back" view is rotated 180 degrees relative to the "front" view. Adapted from *The Journal of Molecular Biology* (98) with permission of the publisher.

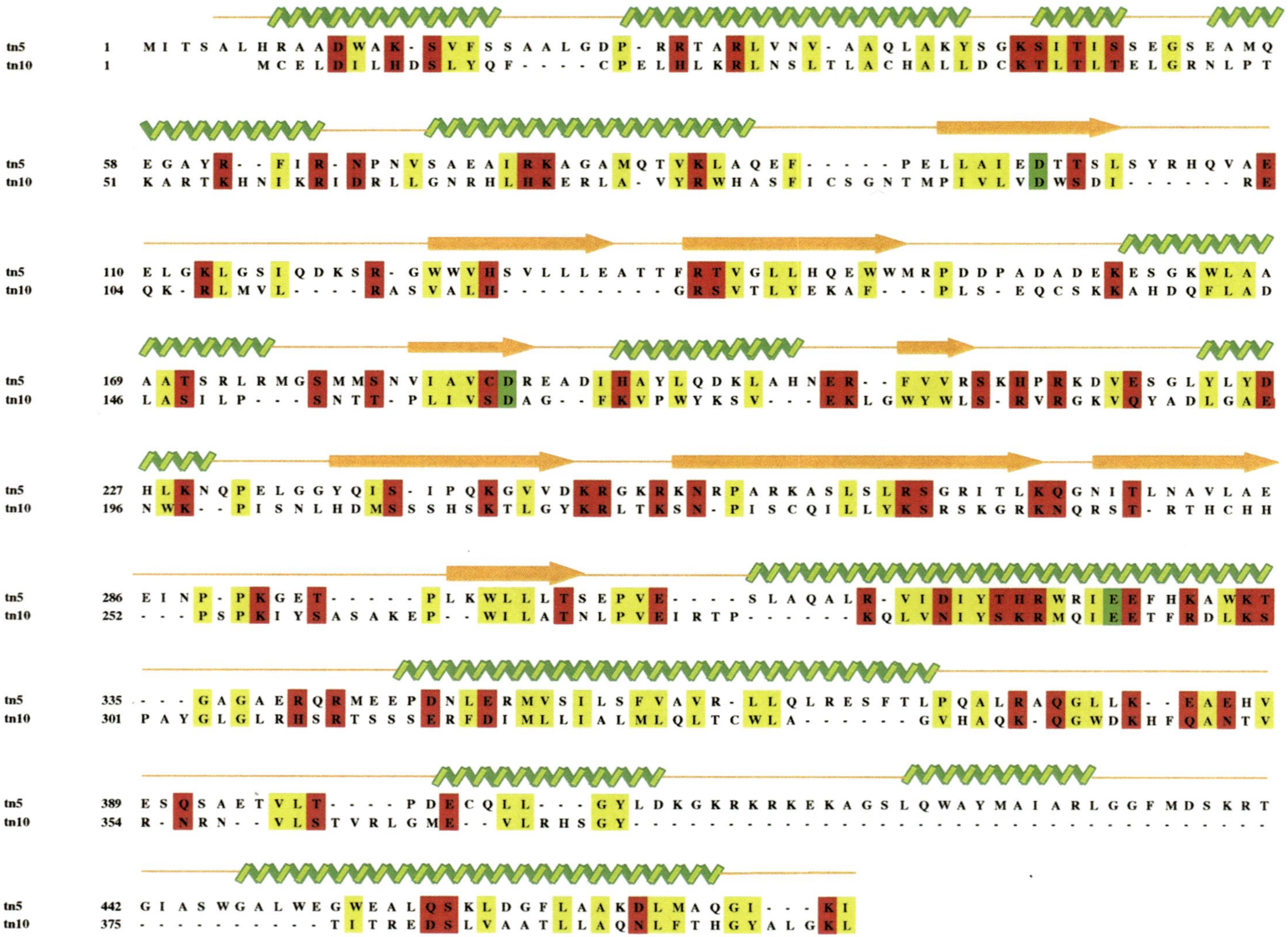

```
tn5    1   M I T S A L H R A A D W A K - S V F S S A A L G D P - R R T A R L V N V - A A Q L A K Y S G K S I T I S S E G S E A M Q
tn10   1             M C E L D I L H D S L Y Q F - - - - C P E L H L K R L N S L T L A C H A L L D C K T L T L T E L G R N L P T

tn5   58   E G A Y R - - F I R - N P N V S A E A I R K A G A M Q T V K L A Q E F - - - - - P E L L A I E D T T S L S Y R H Q V A E
tn10  51   K A R T K H N I K R I D R L L G N R H L H K E R L A - V Y R W H A S F I C S G N T M P I V L V D W S D I - - - - - - R E

tn5  110   E L G K L G S I Q D K S R - G W W V H S V L L L E A T T F R T V G L L H Q E W W M R P D D P A D A D E K E S G K W L A A
tn10 104   Q K - R L M V L - - - - R A S V A L H - - - - - - - - - G R S V T L Y E K A F - - - P L S - E Q C S K K A H D Q F L A D

tn5  169   A A T S R L R M G S M M S N V I A V C D R E A D I H A Y L Q D K L A H N E R - - F V V R S K H P R K D V E S G L Y L Y D
tn10 146   L A S I L P - - - S N T T - P L I V S D A G - - F K V P W Y K S V - - - E K L G W Y W L S - R V R G K V Q Y A D L G A E

tn5  227   H L K N Q P E L G G Y Q I S - I P Q K G V V D K R G K R K N R P A R K A S L S L R S G R I T L K Q G N I T L N A V L A E
tn10 196   N W K - - P I S N L H D M S S S H S K T L G Y K R L T K S N - P I S C Q I L L Y K S R S K G R K N Q R S T - R T H C H H

tn5  286   E I N P - P K G E T - - - - - P L K W L L L T S E P V E - - - - S L A Q A L R - V I D I Y T H R W R I E E F H K A W K T
tn10 252   - - - P S P K I Y S A S A K E P - - W I L A T N L P V E I R T P - - - - - - K Q L V N I Y S K R M Q I E E T F R D L K S

tn5  335   - - - G A G A E R Q R M E E P D N L E R M V S I L S F V A V R - L L Q L R E S F T L P Q A L R A Q G L L K - - E A E H V
tn10 301   P A Y G L G L R H S R T S S S E R F D I M L L I A L M L Q L T C W L A - - - - - - G V H A Q K - Q G W D K H F Q A N T V

tn5  389   E S Q S A E T V L T - - - - P D E C Q L L - - - G Y L D K G K R K R K E K A G S L Q W A Y M A I A R L G G F M D S K R T
tn10 354   R - N R N - - V L S T V R L G M E - - V L R H S G Y - - - - - - - - - - - - - - - - - - - - - - - - - - - - - - - - -

tn5  442   G I A S W G A L W E G W E A L Q S K L D G F L A A K D L M A Q G I - - - K I
tn10 375   - - - - - - - - - - T I T R E D S L V A A T L L A Q N L F T H G Y A L G K L
```

Color Plate 39 (chapter 20). Alignment of Tn*10* and Tn*5* transposase amino acid sequences. Tn*10* and Tn*5* transposase sequences were initially aligned with Clustal W. The primary alignment was used as a guide for further manual alignment that relied on secondary structure predictions for Tn*10* transposase and the crystal structures of the Tn*5* inhibitor (33) and synaptic complex (32). The conserved DDE motif is shown in green, alignment between either charged or polar amino acids in red, and alignment between hydrophobic amino acids in yellow. This figure was contributed by M. Junop.

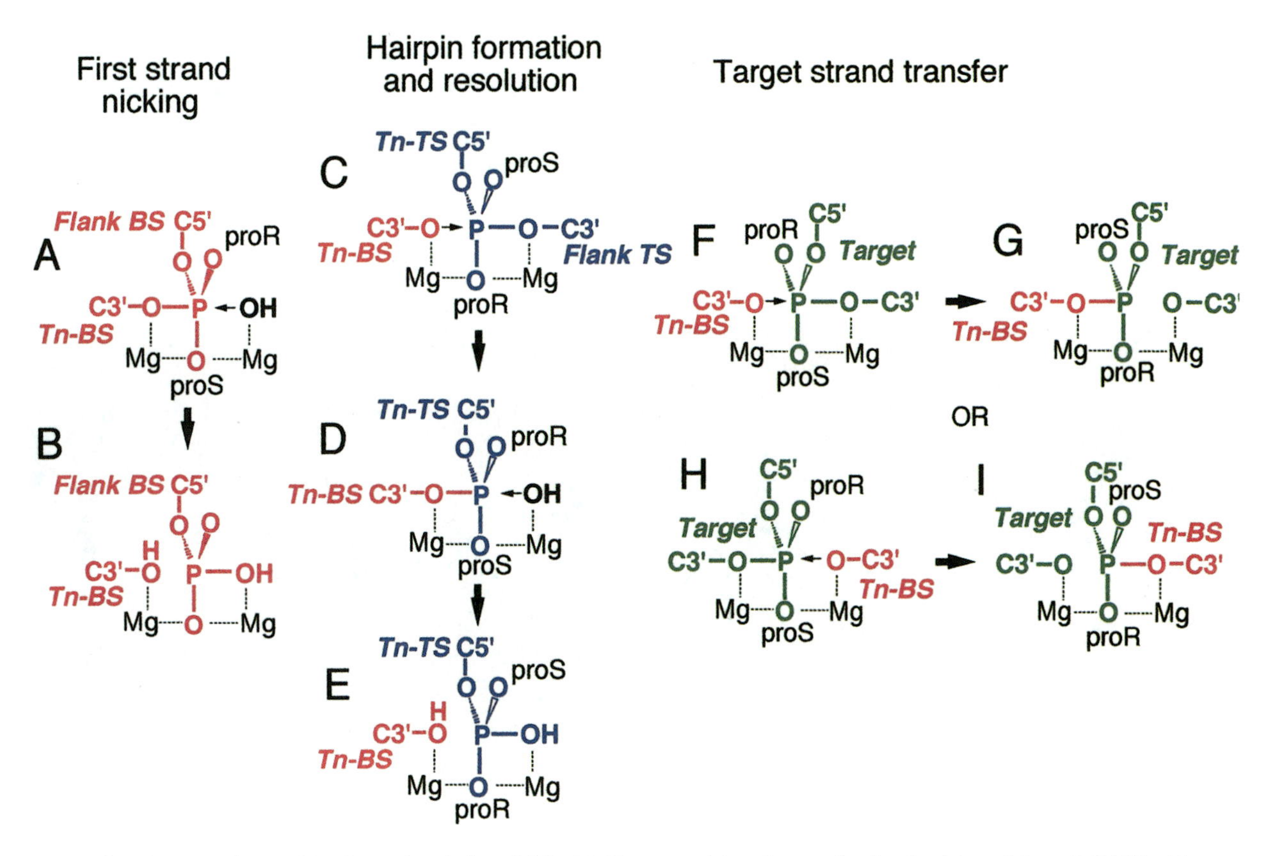

Color Plate 40 (chapter 20). Stereochemical model for Tn*10* transposition. See text for details. Adapted from *Cell* (69) with permission of the publisher.

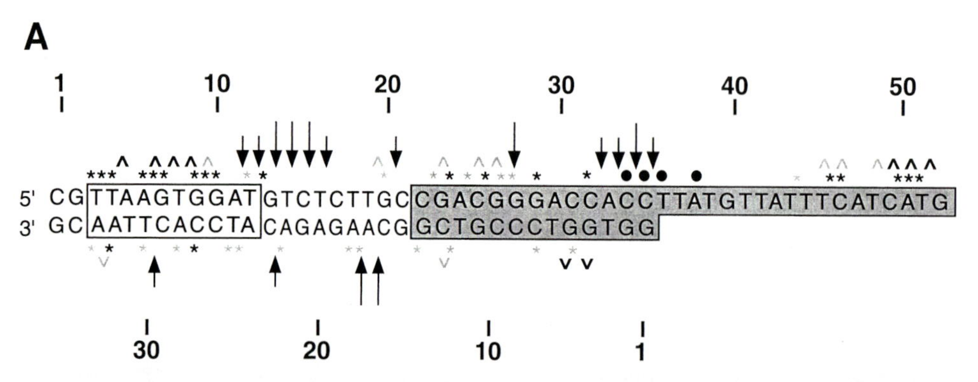

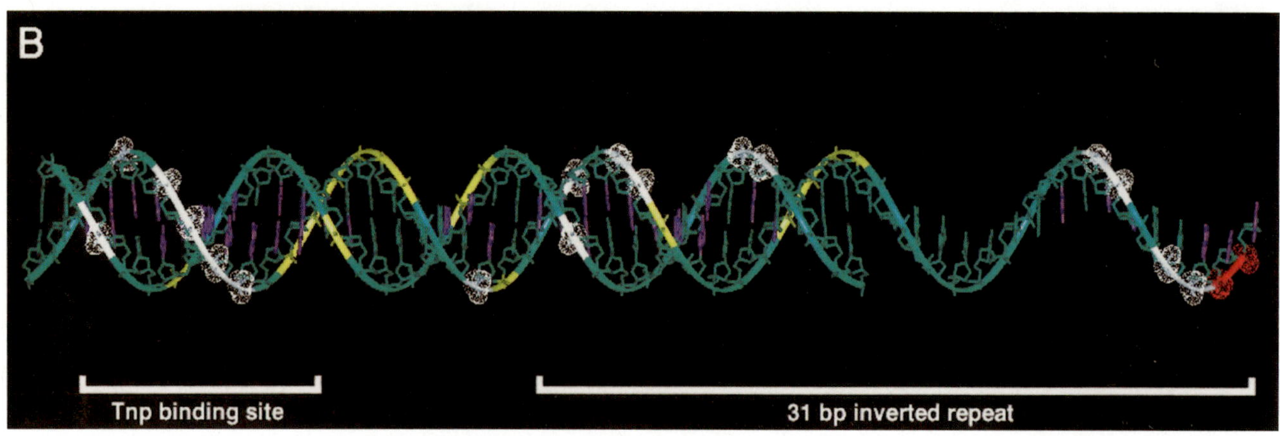

Color Plate 41 (chapter 21). Protein-DNA interactions of P transposase with precleaved oligonucleotide substrates. (A) Summary of ethylation and missing base chemical probing after strand transfer. DNA sequence of 3′ end substrate that was subjected to chemical modification is shown. Asterisks and black circles indicate representation in missing base chemical probing present in reacted population after strand transfer (black asterisks, 0 to 40%; grey asterisks, 40 to 60%; circles, >200%). Carets and arrows indicate representation in phosphate ethylation interference probing in reacted population after strand transfer (black carets, 0 to 40%; grey carets, 40 to 60%; short arrows, 110 to 130%; long arrows, >130%). (B) B-form DNA model of the base and phosphate-specific contacts between transposase and substrate DNA during strand transfer. Both DNA strands are shown in green as ribbons and drawn as B-form DNA, as is the 17-nt single-stranded region. White spheres show phosphates with 10 to 60% normal strand transfer activity when ethylated (both black and grey symbols in panel A); red spheres show phosphates with less than 10% normal strand transfer activity when ethylated; the yellow ribbon shows phosphates with enhanced strand transfer activity when ethylated (short and long arrows in panel A). Purple bases show positions that when removed showed less than 60% strand transfer activity (black and grey asterisks in panel A). Reprinted from reference 11 with permission.

Color Plate 42 (chapter 23). Somatic phenotypes. (a) Somatic excision from the *bz2::Mu1* reporter allele in the triploid aleurone (epidermis) in an active Mutator line. (b) Somatic excision from the *bz2::Mu2* reporter allele in mature anthers in an active Mutator line. (c) Somatic excision from the *bz2::Mu1* reporter allele in the sheath of an adult leaf in an active Mutator line. (d) In an inactive Mutator line, the *a1-mum2* reporter allele shows partial restoration of gene expression resulting in uniform pink coloration; the reporter allele is segregating 1:1 in this ear. The beige kernels are *a1* and the pink kernels are *a1/a1/a1-mum2*.

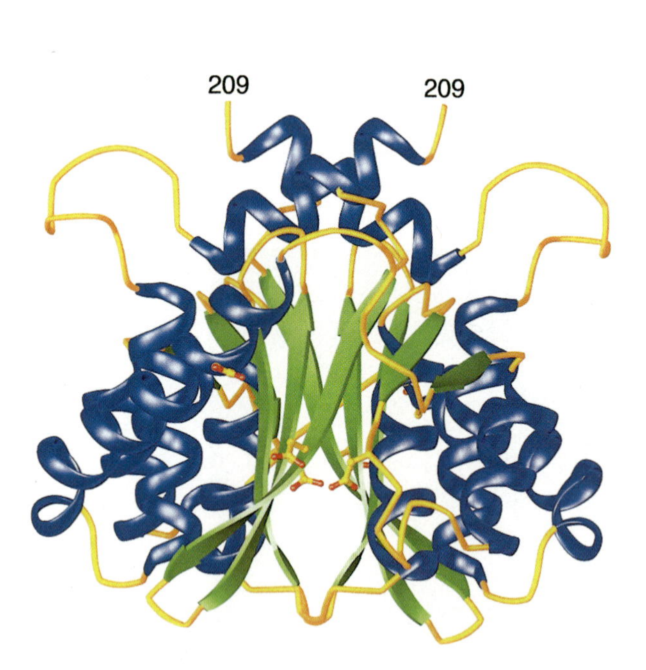

Color Plate 43 (chapter 25). Structure of the catalytic domain of HIV-1 integrase (66) (PDB code 1BIS) represented as ribbon diagram (24) of the dimer. The catalytic residues D64, D116, and E152 are shown as balls and sticks. In this projection, the N terminus (residue 56) is coincident with the α-helix that terminates at residue 209 and is not clearly visible.

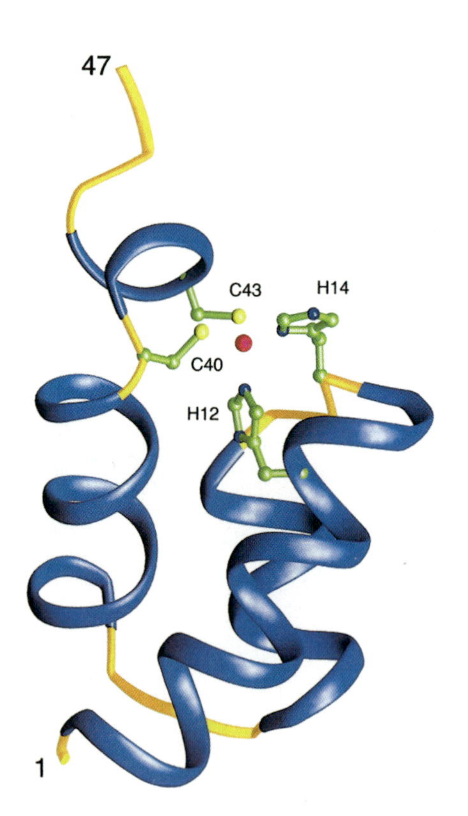

Color Plate 44 (chapter 25). Structure of the N-terminal domain of HIV-1 integrase (23) (PDB code 1WJC) depicted as a ribbon diagram (24) of the monomer, which consists of a bundle of four α-helices. The His and Cys residues of the HHCC motif are shown as balls and sticks. Coordination of zinc, depicted as a magenta sphere, to these residues contributes to stabilization of the structure.

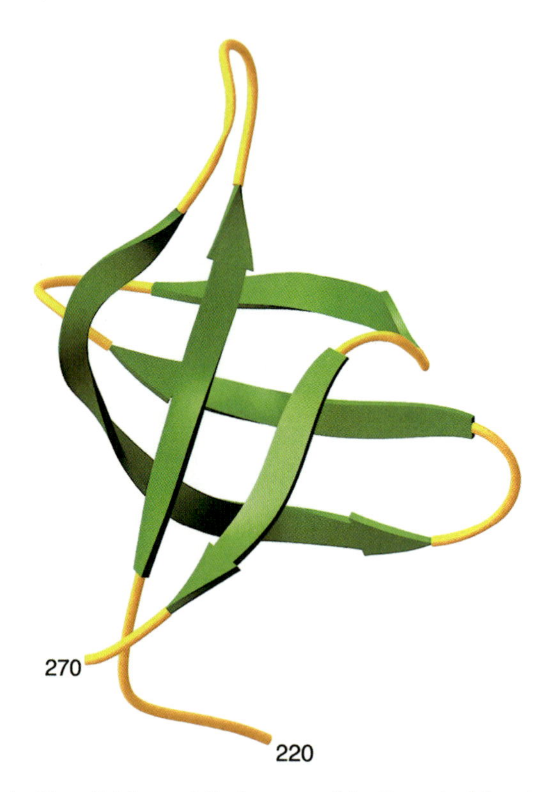

Color Plate 45 (chapter 25). Structure of the C-terminal domain of HIV-1 integrase (105) (PDB code 1IHV) depicted as a ribbon diagram (24) of the monomer.

A. LAGLIDADG: I-*Cre*I

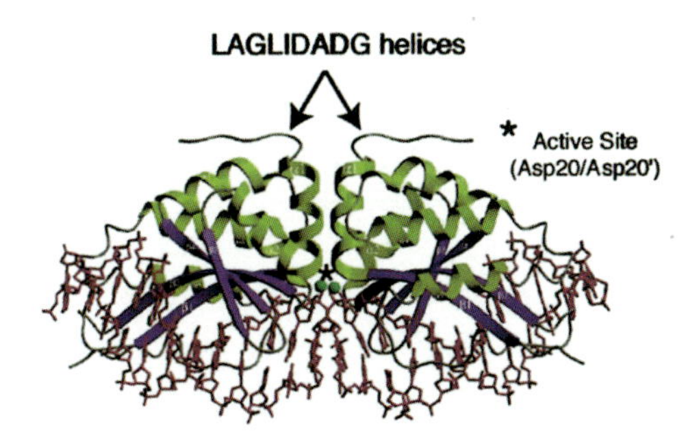

C. H-N-H: Colicin

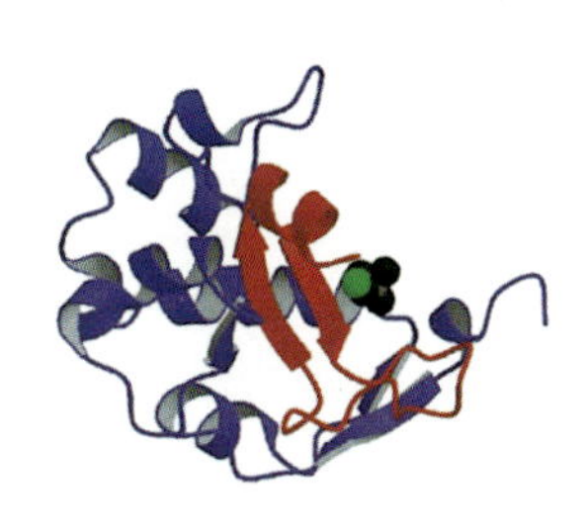

B. GIY-YIG: I-*Tev*I

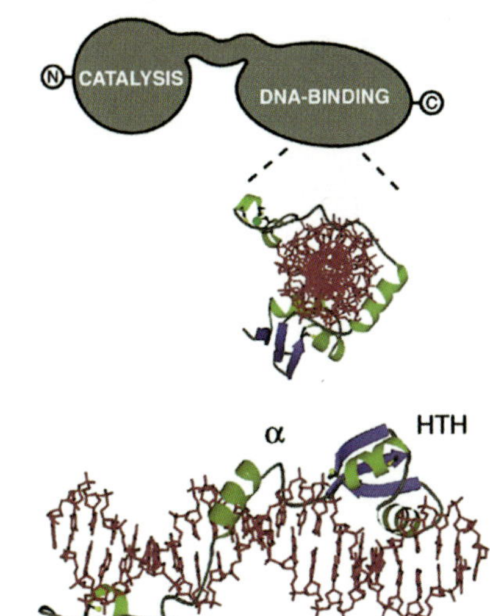

D. His-Cys-Box: I-*Ppo*I

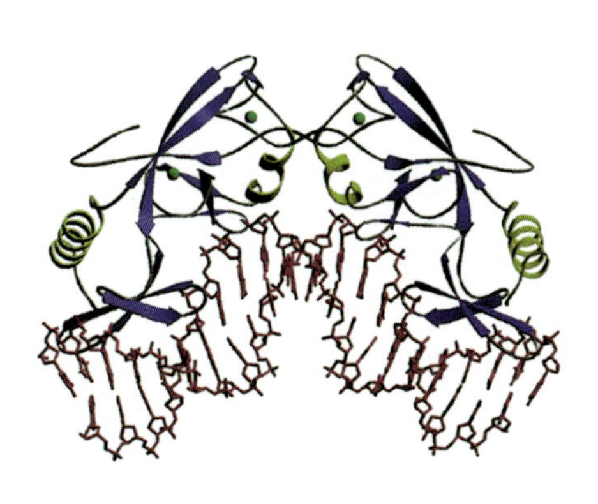

Color Plate 46 (chapter 31). Homing endonuclease structures. Structures representing four families are shown. (A) The crystal structure of I-*Cre*I in complex with its homing site is shown at 3-Å resolution (87). (B) The crystal structure of the DNA-binding domain of I-*Tev*I complexed to the DNA-binding region of the homing site at 2.2-Å resolution (184) is shown. The DNA-binding domain is presented in two views to show the subdomains (Zn, zinc finger; α, minor groove binding helix; HTH, helix-turn-helix) and looking down the DNA helix to show the protein wrapping around the largely undistorted DNA. (C) The crystal structure of the DNase domain of colicin E9 is shown with the H-N-H motif in red, the bound Ni^{2+} ion in green, and the phosphate molecule in black. (D) The crystal structure of I-*Ppo*I is shown complexed to its homing site at 1.8-Å resolution (57).

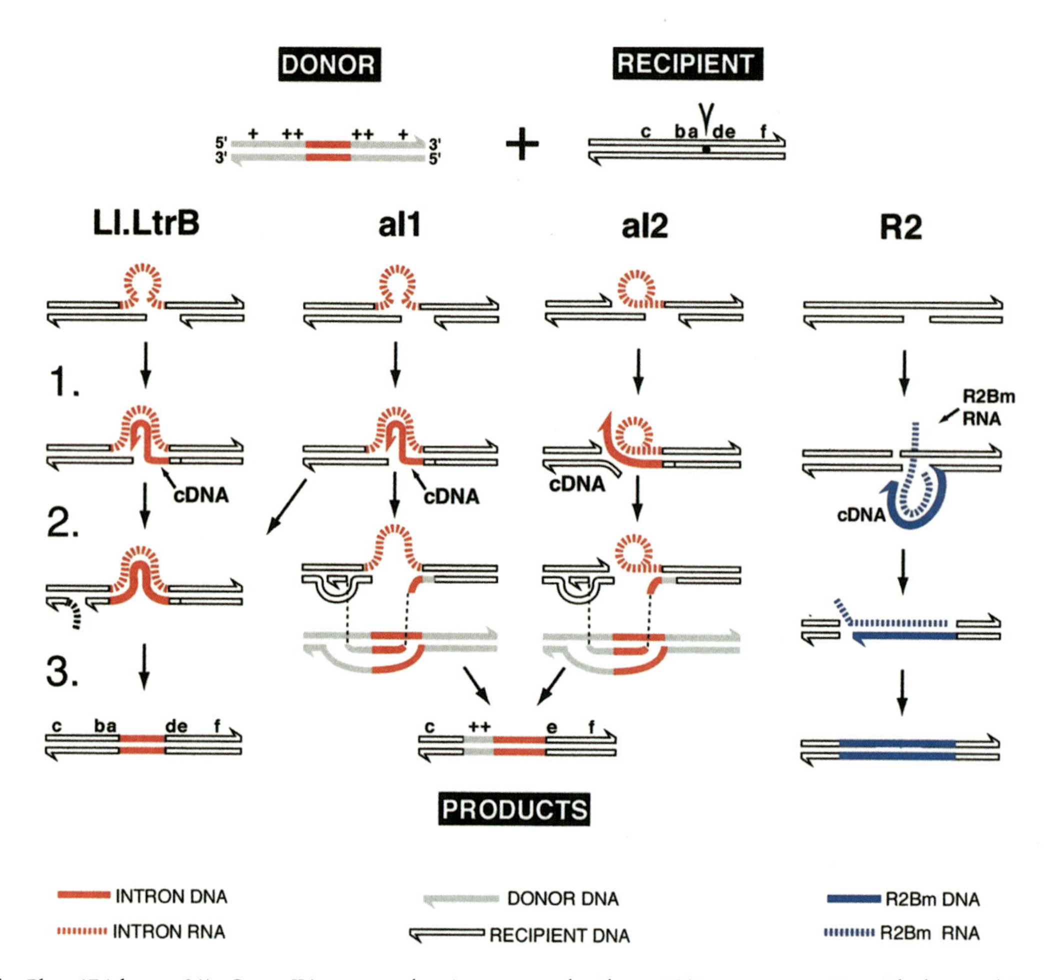

Color Plate 47 (chapter 31). Group II intron retrohoming compared with non-LTR retrotransposition. The bacterial (LI.LtrB) and yeast mitochondrial (aI1 and aI2) retrohoming pathways are shown alongside the retrotransposition pathway of a non-LTR retrotransposon (R2). For group II introns: step 1, cleavage by RNP; step 2, TPRT (cDNA synthesis), and for aI1 and aI2 cDNA undergoes recombination with donor; step 3, RNA displacement or degradation, second-strand synthesis, recombination and/or repair. For R2Bm: step 1, cleavage catalyzed entirely by the protein; step 2, element RNA is template for TPRT; step 3, as for L1.LtrB. Coconversion tracts, marked by plus signs in the products, are indicators of a role for homologous recombination in the retrohoming pathway.

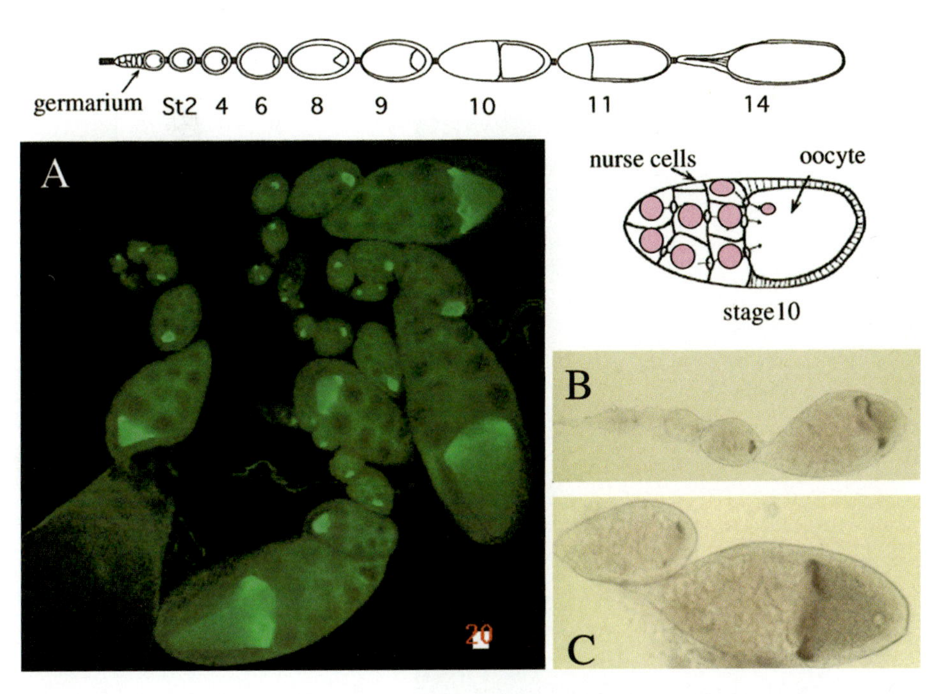

Color Plate 48 (chapter 33). Expression of the protein encoded by ORF1 and of the RNA of the I factor in the ovaries of SF females. The schematic drawings represent one of the 16 ovarioles composing the *Drosophila* ovary and an egg chamber at stage 10 with the oocyte and the nurse cells that compose the germ line (nuclei are in pink). Oogenesis begins in the germarium at the anterior tip of the ovariole where stem cells divide asymmetrically to produce a cystoblast. The cystoblast undergoes divisions to produce 16 cells connected by cytoplasmic bridges. One of them migrates to the posterior region of the cyst and becomes the oocyte, whereas the other 15 develop into polyploid nurse cells. The cyst is covered by a monolayer of somatic follicle cells forming together an egg chamber. The maturation of egg chambers has been divided into 14 morphological stages shown on the scheme. (A) Expression pattern of the ORF1 protein in ovarioles of SF females revealed by immunostaining (the white square is 20 μm). St, stage. (B and C) I factor transcripts revealed by in situ hybridization on ovaries of SF females with use of a DNA probe corresponding to part of ORF1.

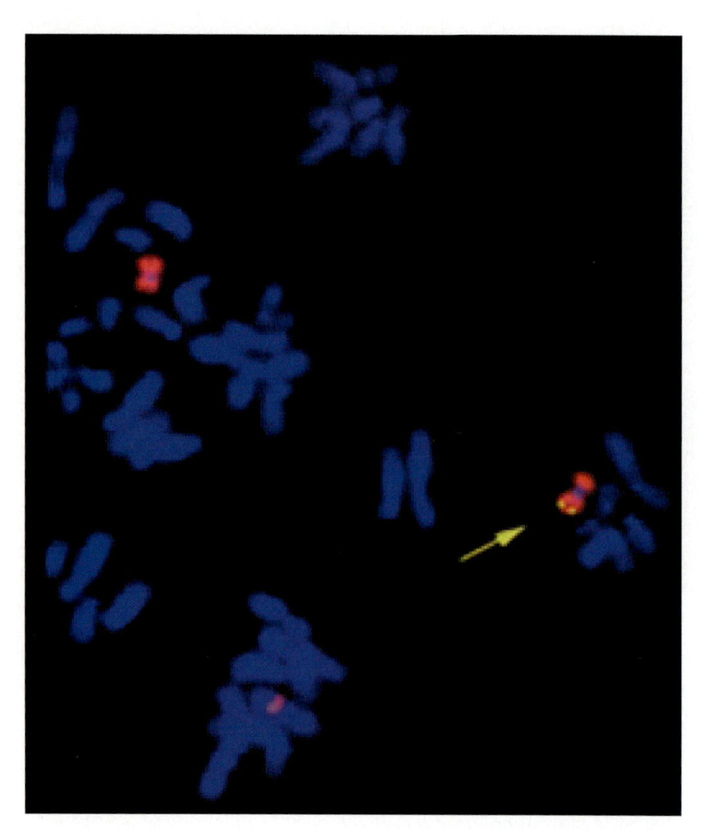

Color Plate 49 (chapter 38). FISH analysis of latently infected IB3 cells. Metaphase chromosomes of IB3 cells (human lung fibroblasts) hybridized with a rhodamine-labeled, chromosome 19-specific probe library (red), a fluorescein-labeled AAV-specific probe (yellow), and counterstained with DAPI (blue). The AAV signal (arrow) localizes to the distal region of chromosome 19q. (Courtesy of Bill Kearns, Johns Hopkins University.)

III. RECOMBINATION VIA DIRECT TRANSESTERIFICATIONS: TRANSPOSITION REACTIONS THAT INVOLVE ONLY DNA INTERMEDIATES

Mobile DNA II
Edited by N. L. Craig et al.

Chapter 15

Insertion Sequences Revisited

MICHAEL CHANDLER AND JACQUES MAHILLON

Michael Chandler • Laboratoire de Microbiologie et Génétique Moléculaires, CNRS, 118 Route de Narbonne, F-31062 Toulouse Cedex, France. **Jacques Mahillon** • Laboratoire de Génétique Microbienne, Université catholique de Louvain, Place Croix du Sud, 2 Bte 12, B-1348 Louvain-la-Neuve, Belgium.

INTRODUCTION

General Scope

The number of identifiable insertion sequences (ISs) continues to increase rapidly. Only approximately 50 had been identified in 1989 (115) compared with more than 500 in 1998 (234) and more than 700 today (see http://www-is.biotoul.fr). This figure does not include the majority of ISs observed in sequenced bacterial genomes.

This chapter is an update of a survey of ISs published in 1998 (234). We have retained the same basic structure: a first section including some key properties of ISs and a second section in which we define and describe the different IS families. Throughout the text we have tried to compare and contrast the different IS families in terms of their transposition mechanism and control of their transposition activity. We have introduced an additional section concerning bacterial genomes and plasmids since a number of genome sequences have become available over the past three years, and a large number of potential ISs have been identified in several of these. We do not pretend that this section is exhaustive. We have also retained a section on eukaryotic insertion sequences. The reader is referred to recent reviews for a detailed description of transposition mechanisms (chapter 2 and references 5, 37, 66, 144, 189, 251, 252, 287, and 379).

ISs can be loosely defined as small (<2.5 kb), generally phenotypically cryptic segments of DNA with a simple genetic organization and capable of inserting at multiple sites in a target DNA molecule. In this definition we voluntarily eliminate several types of mobile genetic elements such as those using RNA as intermediates, the retroviruses, retrotransposons, and retroposons (sections IV and VI and references 37, 108, and 344), DNA elements such as the conjugative transposons (chapter 10 and reference 329), which use a phage λ-type integration and excision mechanism for their translocation, and elements such as bacteriophage Mu (chapter 17 and reference 252), Tn*7* (chapter 19 and reference 67), and transposons of the Tn*554* type (chapter 10 and reference 259), which are large and quite complex. We have, however, included a short section on elements related to transposon Tn*3* since an increasing number of IS-like elements belonging to this group are being identified (chapter 14).

Although they are not considered here in detail, many of these elements share significant functional similarities with ISs and have provided many of the important insights into transposition mechanisms. Reference to these will be included where they may prove useful to understanding the behavior of IS elements.

Nomenclature

Several systems of nomenclature are in operation. One, initiated in 1978 (see reference 208) and centralized until recently by E. Lederberg (Stanford University), attributes a single number to an IS element (e.g., IS*1*). This was adequate when only few ISs were known, but it does not include sufficient information and becomes less transparent with the large numbers of elements known today. A second scheme, which provides some information concerning the source of the element, includes the initials of the bacterial species from which it was isolated (e.g., ISRm*1* for *Rhizobium meliloti*). At present, we are confronted by both types of nomenclature in the literature. Moreover, closely related elements with only few differences at the nucleotide sequence level are in some cases designated isoforms of the parent element and sometimes attributed a specific number. Finally, some elements have been baptized with names that fit none of these rules (e.g., RSalpha-9). Historically certain IS nomenclatures have been reserved for a specific species (e.g., ISRm for *R. meliloti* and ISC for *Sulfolobus*). We have retained these names in view of their common usage. Where appropriate, we had previously simplified certain of the more complicated assignments by adopting the nomenclature that includes the initials of the host bacterial species.

As a general rule, we now suggest the following convention for nomenclature: whenever possible, the first letter of the genus followed by the first two letters of the species and a number should be used (e.g., ISBce*1* for *Bacillus cereus*). Where a conflict is encountered (e.g., *Bacillus cereus* and *Burkholderia cepacia*), the first two letters of both genus and species should be included (e.g., ISBuce*1* for *Burkholderia cepacia*) (M. Chandler and J. Mahillon, Letter, *ASM News* **66**:324, 2000). A list of bacterial species and assigned names can be found at http://www-is.biotoul.fr and is updated intermittently.

GENERAL FEATURES AND PROPERTIES OF IS ELEMENTS

In addition to being small, ISs are genetically compact (Fig. 1). Generally they encode no functions other than those involved in their mobility although members of several families, which include additional genes, are now being identified. They are composed of recombinationally active DNA sequences that define the ends of the element and flank an open reading frame (*orf*) encoding an enzyme, the transposase (Tpase), which recognizes and processes these ends. The Tpase is generally encoded by a single, or sometimes two, *orf* and consumes nearly the entire length of the element.

Terminal IRs

The majority of ISs exhibit short terminal inverted repeat sequences (IR) of between 10 and 40 bp (exceptions are found in the IS*91*, IS*110* families, and the IS*200*/*605* complex; Table 1). To a first approximation, these can often be divided into two functional domains (Fig. 1). Domain I includes the two or three terminal base pairs, and is involved in the cleavages and strand transfer reactions leading to transposition of the element. Domain II is positioned within the IR and is involved in Tpase binding (85, 86, 161, 162, 175, 238, 413). The single terminal Tpase binding sites of ISs are to be contrasted with the multiple and asymmetric protein binding sites observed in the case of bacteriophage Mu (chapter 17 and reference 69), transposons Tn*7* (chapter 19 and reference 67) and probably Tn*552* (320), and the complex En/Spm and Ac elements of maize (chapter 24 and references 121 and 198). It is worth noting that members of the IS*21* family also carry multiple repeated sequences at both ends, which may also represent Tpase binding sites (20, 234). By accommodating different binding patterns at each end, such multiple protein binding sites can provide a functional distinction between the ends either in the assembly or in the activity of the synaptic complex. In addition, indigenous IS promoters are often located partially within the IR sequence upstream of the Tpase gene, by convention IRL. This arrangement may provide a mechanism for autoregulation of Tpase synthesis by Tpase binding. Binding sites for host-specified proteins are also often found within or close to the terminal IRs, and these proteins may play a role in modulating transposition activity or Tpase expression.

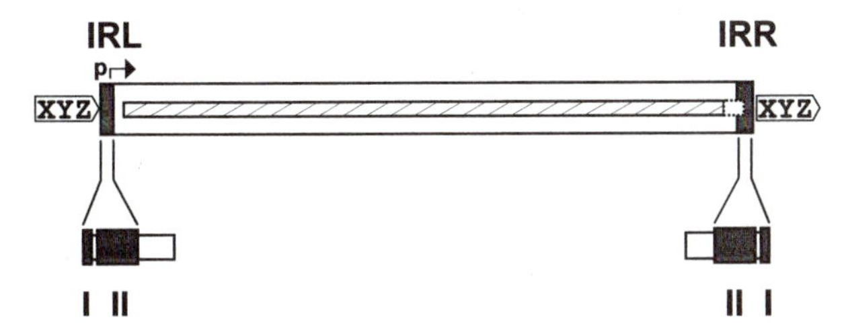

Figure 1. Organization of a typical insertion sequence. The IS is represented as an open box in which the terminal IRs are shown as gray boxes labeled IRL (left inverted repeat) and IRR (right inverted repeat). A single open-reading frame encoding the Tpase is indicated as a hatched box stretching over the entire length of the IS and extending within the IRR sequence. XYZ enclosed in a pointed box flanking the IS represents short directly repeated sequences generated in the target DNA as a consequence of insertion. The Tpase promoter, p, partially localized in IRL, is shown by a horizontal arrow. A typical domain structure (gray boxes) of the IRs is indicated beneath. Domain I represents the terminal base pairs at the very tip of the element whose recognition is required for Tpase-mediated cleavage. Domain II represents the base pairs necessary for sequence-specific recognition and binding by the Tpase.

Domain Structure of Tpases

A general pattern for the functional organization of Tpases appears to be emerging from the limited number that have been analyzed. Many can be divided into topologically distinct structural domains, and although several regions of the protein may contribute to a given function, the isolated domains themselves often exhibit a distinct function. The sequence-specific DNA binding domain and the catalytic domain of the proteins are frequently located principally in the N-terminal and C-terminal regions, respectively (IS*1* [227, 412], IS*30* [351], Mu [206], Tn*3* [232], IS*50* [402], IS*903* [365], IS*911* [291]; for review, see reference 144). One functional interpretation of this arrangement for prokaryotic elements is that it may permit interaction of a nascent protein molecule with its target sequences on the IS, thus coupling expression and activity. This notion is reinforced by the observation that the presence of the C-terminal region of both the IS*50* and IS*10* Tpases appears to reduce binding activity (172, 395), possibly by masking the

Table 1. Major features of prokaryote IS families

Family	Group	Length range[a]	DR[b]	Ends[c]	IR[d]	ORF[e]	Comment(s)[f]
IS*1*		770	9 (8–11)	GGT	Y	2	Phage λ integrase?
IS*3*	IS*2*	1,300–1,350	5	TGA	Y	2	DD(35)E
	IS*3*	1,200–1,300	3 (4)			2	
	IS*51*	1,300–1,400	3 (4)			2	
	IS*150*	1,400–1,550	3–5			2	
	IS*407*	1,200–1,250	4			2	
IS*4*		1,300–1,950	9–12	C (A)	Y	1	DD-E
IS*5*	IS*5*	1,100–1,350	4	GG	Y	1	DD-E
	IS*427*	800–1,000	2–3	Ga/g		(2)	
	IS*903*	1,000–1,100	9	GGC		1	
	IS*1031*	850–950	3	GAG		1	
	ISH*1*	900–1,150	8			1	
	ISL*2*	800–1,100	2–3			1	
IS*6*		750–900	8	GG	Y	1	DD(34)E
IS*21*		1,950–2,500	4 (5, 8)	TG	Y	2	DD-E
IS*30*		1,000–1,250	2–3		Y	1	DD(33)E
IS*66*		2,500–2,700	8	GTA	Y	>3	
IS*91*		1,500–1,850	0		N	1	Rolling circle transposition
IS*110*		1,200–1,550	0		N	1	Site-specific recombinase
IS*200*/IS*605*		700–2,000	0		N	1 (2)	Complex organization
IS*256*		1,300–1,500	8–9	Gg/a	Y	1	DD-E
IS*481*		950–1,100	5–6	TGT (a/g)	Y	1	DD-E
IS*630*		1,100–1,200	2		Y	1	DD-E, eukaryote relatives
IS*982*		1,000	ND	AC	Y	1	DD-E
IS*1380*		1,650	4	Cc/g	Y	1	DD-E
ISAS*1*		1,200–1,350	8	C	Y	1	DD-E?
ISL*3*		1,300–1,550	8	GG	Y	1	DD-E
Tn*3*		>3,000	5	GGGG	Y	1 (2)	DD-E

[a]Length range in base pairs (bp) represents the typical range of each group.
[b]Direct target repeats in base pairs; ND, not determined. Less frequently observed lengths are included in parentheses.
[c]The most conserved nucleotides present at the ends of each IS family are indicated, including some possible variations.
[d]Presence (Y) or absence (N) of terminal inverted repeats.
[e]Numbers in parentheses denote variations.
[f]DDE represents the common acidic triad presumed to be part of the active site of several transposases.

DNA binding domain. This arrangement might favor activity of the protein in *cis,* a property shared by several Tpases (see "Activity in *cis*" below). A similar phenomenon appears to occur with the Tpases of IS*1* (D. Zerbib and M. Chandler, unpublished data) and IS*911* (143, 266). In several cases these domains are assembled into a single protein from consecutive *orf*s by translational frameshifting (see "Programmed translational frameshifting" below).

An additional characteristic of some, if not all, Tpases is their capacity to assemble into multimeric forms essential for their activity (144). This is true of both prokaryotic elements such as bacteriophage Mu (52), IS*50* (75), and IS*911* (142, 143) and of eukaryotic elements such as the retroviruses (5) and the *mariner*-like element, Mos*1* (219, 413a).

DR

Another general feature of IS elements is that, on insertion, most generate short directly repeated sequences (DR) of the target DNA flanking the IS. Attack of each DNA strand at the target site by one of the two transposon ends in a staggered way during insertion provides an explanation for this observation. The length of the DR, between 2 and 14 bp, is characteristic for a given element, and a given element will generally generate a DR of fixed length. However, certain ISs have been shown to generate DRs of atypical length at a low frequency, presumably reflecting small variations in the geometry of the transposition complex (see reference 115 and references therein for further discussion). Note that in certain cases, ISs without DRs can simply result from homologous inter- or intramolecular recombination between two IS elements, each with a different DR sequence, or from the formation of adjacent deletions resulting from duplicative intramolecular transposition (see, for example, references 378 and 393). Three ISs, IS*1549*, IS*1634*, and IS*1630*, have been identified that appear to generate long DRs of quite variable length (45, 284, 387). Two, IS*1549* and IS*1634*, are distantly related to the IS*4* family (see "IS*4* Family" and "Orphans and Emerging Families" below) and one, IS*1630*, belongs to the IS*30* family (see "IS*30* Family" below). The mechanism involved in generating such long DRs is at present unknown.

Control of Neighboring Gene Expression

Many IS elements have been shown to activate the expression of neighboring genes. Initially this was observed for IS*1*, IS*2*, and IS*5* (see reference 115), but other elements such as IS*406* (328), IS*1186* (285), IS*481* (87), IS*928*B (220), ISSg*1* (82), IS*1490* (159), and ISVa*1* (371) have also been shown to exhibit similar properties. Many other examples can be found in the literature. Activation can be due to formation of hybrid promoters when insertion of an IS results in placing an outwardly directed −35 promoter hexamer located in the terminal IRs (see reference 115) at the correct distance from a resident −10 hexamer. Such −35 elements have been observed experimentally in many ISs (IS*1* [293], IS*2* [359], IS*21* [307], IS*30* [70], IS*257* [209], IS*911* [373], IS*982* [220]). Activation can also occur by endogenous transcription "escaping" the IS and traversing the terminal IR (e.g., IS*3* [60], IS*10* [62], IS*481* [87], IS*982* [220]). In the unique case of the *bgl* operon of *Escherichia coli*, where activation can occur by insertion of either IS*1* or IS*5* upstream or downstream of the promoter, it has been suggested that activation involves changes in DNA structure or topology (309, 310, 326).

An inward-directed −10 hexamer has also been detected in the left terminal IR of several elements. When two ends of such an element are juxtaposed, by formation of head-to-tail dimers or of circular copies of the IS, the combination of the IRL −10 hexamer with a −35 hexamer resident in the neighboring IRR can generate relatively strong promoters (IS*21* [307], IS*30* [70], IS*911* [373]). This arrangement can lead to high Tpase expression and consequent increases in transposition activity (see "IS*3* Family," "IS*21* Family," and "IS*30* Family" below and chapter 16).

Control of Transposition Activity

Transposition activity is generally maintained at a low level. An often-cited reason for this is that high activities and the accompanying mutagenic effect of genome rearrangements would be detrimental to the host cell (92). Endogenous transposase promoters, in contrast to those assembled by juxtaposition of −10 and −35 hexamers (see above), are generally weak and many are partially located in the terminal IRs. This would enable their autoregulation by Tpase binding.

While many of the classical mechanisms of controlling gene expression, such as the production of transcriptional repressors (IS*1* [100, 227, 412] and IS*2* [153]) or translational inhibitors (antisense RNA in the case of IS*10* [189]), are known to operate in Tpase expression, several other mechanisms have also been uncovered.

Tpase expression and activity

Impinging transcription. Many ISs have evolved mechanisms that attenuate their activation by impinging transcription following insertion into active host genes. One such mechanism observed with IS*10* and IS*50*, and potentially present in several other ISs, is the sequestering of translation initiation signals in an RNA secondary structure. These ISs carry inverted repeat sequences located close to the left end, which include the ribosome binding site or translation initiation codon for the Tpase gene (see references 76, 192). Transcripts from the resident Tpase promoter include only the distal repeat unit, which is unable to form the secondary structure, while transcripts from neighboring DNA include both repeats.

In several cases, simple transcription across the end of the element has been observed to reduce transposition activity (115, 123). This effect may be the result of disrupting complexes between Tpase and cognate ends.

Programmed translational frameshifting. Translational control of transposition by frameshifting has been demonstrated both for IS*1* (100, 223, 333) and for members of the IS*3* family (chapter 16 and references 57 and 289) but may also occur in several other IS elements (see, for example, IS*5* Family, below). For IS*1* and members of the IS*3* family, the upstream frame appears to carry a DNA recognition domain whereas the downstream frame encodes the catalytic site. The product of the upstream frame alone acts as a modulator of activity, presumably by binding the IR sequences. Frameshifting assembles both domains into a single protein, the Tpase, which directs the cleavages and strand transfer necessary for mobility of the element. The frameshifting frequency is thus critical in determining overall transposition activity.

Although it has yet to be explored in detail, frameshifting could be influenced by host physiology, thus coupling transposition activity to the state of the host cell.

A −1 frameshift involves slippage of the translating ribosome one base upstream and resumption of translation in the alternative phase. This generally occurs at the position of so-called slippery codons in a heptanucleotide sequence of the type Y YYX XXZ in phase 0 (where the bases paired with the anticodon are shown as triplets) and which is read as YYY XXX Z in the shifted −1 (e.g., see references 57, 102, and 119). The sequence A AAA AAG is a common example of this type of heptanucleotide. Ribosomal shifting of this type is stimulated by structures in the mRNA, which tend to impede the progression of the ribosome such as potential ribosome binding sites upstream or secondary structures (stem-loop structures and potential pseudoknots) downstream of the slippery codons (102).

Translation termination. A third potential mechanism derives from the observation that the translation termination codon of Tpase genes of certain elements is located within their IR sequences. Although, to our knowledge, no extensive analysis of the significance of this arrangement has yet been undertaken, it seems possible that it may in some manner couple translation termination, Tpase binding, and transposition activity.

The Tpase gene of several elements does not possess a termination codon. These include IS*240*C, a member of the IS*6* family (Y. Chen and J. Mahillon, unpublished data), two members of the IS*5* family, IS*427* (81) and ISMk*1* (405), and various members of the IS*630* family including IS*870* and ISRf*1* (109). Instead, some of these elements insert into a relatively specific target sequence in which the target DR produced on insertion generates the Tpase termination codon (see IS*630* Family). The relevance of this as a control mechanism has yet to be explored.

Tpase stability. Tpase stability can also contribute to control of transposition activity. The Tpase of IS*903* is sensitive to the *E. coli* Lon protease (86). This sensitivity limits the activity of the Tpase both temporally and spatially and may provide an explanation for the observation that several Tpases function preferentially in *cis* (see below). Indeed mutant IS*903* Tpase derivatives have been isolated that exhibit an increased capacity to function in *trans*. These are more refractory to Lon degradation than the wild-type protein (84). Lon may also be involved in regulating Tn*5* (IS*50*) transposition (17). An observation that might also reflect Tpase instability is the temperature-sensitive nature of IS*1*-mediated adjacent deletions in vivo (303), of Tn*3* transposition (193), and of IS*911* intramolecular recombination both in vivo and in vitro (141). For IS*911*, incubation of the Tpase at 42°C results in an irreversible loss in activity.

Activity in *cis*. Early studies with several transposable elements indicated that transposition activity was more efficient if the Tpase is provided by the element itself or by a Tpase gene located close by on the same DNA molecule. Preferential activity in *cis* reduces the probability that Tpase expression from a given element will activate transposition of related copies elsewhere in the genome. The effect can be of several orders of magnitude and has been observed for a variety of elements including IS*1* (230, 292), IS*10* (255), IS*50* (167), and IS*903* (86, 128). Its magnitude is characteristic for a given IS. This property presumably reflects a preference of the cognate Tpases to bind to transposon ends close to their point of synthesis and is likely to be the result of several phenomena.

For IS*903*, increased stability (86) and expression (84) increase the capacity for Tpase activity in *trans*. Likewise, for IS*10*, mutations that increase translation of the Tpase also decrease the *cis* preference of the enzyme and it has been suggested that *cis* preference is strongly dependent upon the half-life of the Tpase message and the rate at which transcripts are released from their templates (172).

An additional consideration that may promote preferential activity in *cis* is reflected in the N-terminal location of the DNA binding domain in many Tpases. If the N-terminal domain is capable of folding independently of the catalytic domain, this arrangement would permit preferential binding of nascent Tpase polypeptides to neighboring binding sites (see "Domain Structure of Tpases" above). For several Tpases, the N-terminal portion of the protein exhibits a higher affinity for the ends than does the entire Tpase molecule, suggesting that the C-terminal end may in some way mask the DNA binding activity of the N-terminal portion. It remains to be seen whether this is a general property of Tpases.

Host factors

Transposition activity is frequently modulated by host factors. These effects are generally specific for each element. A nonexhaustive list of such factors includes the DNA chaperones (or histone-like proteins) IHF, HU, HNS, and FIS; the replication initiator DnaA; the protein chaperone/proteases ClpX, P, and A; the SOS control protein LexA; and the Dam DNA methylase. Proteins that govern DNA supercoiling in the cell can also influence transposition.

The DNA chaperones may play roles in ensuring the correct three-dimensional architecture in the evolution of various nucleoprotein complexes necessary for productive transposition. They may also be involved in regulating Tpase expression. IHF, HU, HNS, and FIS have all been variously implicated in the case of bacteriophage Mu, in the control of Mu gene expression, or directly in the transposition process (see reference 52 for review). Several elements carry specific binding sites for IHF within, or close to, their terminal IRs. These can lie within (e.g., IS*1* [116], IS*903* [127]) or close to (IS*10* [189]; see chapter 20) the Tpase promoter. IHF appears to influence the nature of IS*10* transposition products by binding to a site 43 bp from one end (53, 321, 342). It also stimulates Tpase binding to the ends of the Tn*3* family member, Tn*1000* or γδ (400). For IS*50*, an element of the same family as IS*10*, both Fis and the replication initiator protein DnaA have been reported to intervene in transposition (311). Finally another "histone-like" protein, HNS, has been reported to stimulate transposition of IS*1* in certain circumstances (341).

Although their mode of action is at present unknown, several other host proteins with otherwise entirely different functions have been implicated in transposition. Acyl carrier protein (ACP) was independently shown to stimulate 3′ end cleavage of Tn*3* by its cognate Tpase (233) and, together with ribosomal protein L29, to greatly increase binding of TnsD (a protein involved in Tn*7* target selection) to the chromosomal insertion site, *att*Tn*7* (338). Moreover ACP and L29 moderately stimulate Tn*7* transposition in vitro while L29 alone has a significant stimulatory effect in vivo (338). The mode of action of these proteins may be similar to that of the accessory proteins PepA and ArgR, which modify the architecture of the synaptic complex in certain XerC/XerD-mediated site-specific recombination reactions (see chapter 8).

Certain factors involved in protein "management," such as ClpX, ClpP, and Lon, have been implicated in transposition. ClpX is essential for bacteriophage Mu growth (250), where it is required for disassembling the Tpase-DNA complex or transpososome in preparation for the assembly of a replication complex (194, 213) and, together with ClpP, also plays a role in proteolysis of the Mu repressor (200, 398). As indicated above, the Lon protease is implicated in proteolysis of the IS*903* Tpase (84, 86). At present, the involvement of these proteins in the transposition of other elements has not been documented.

The third class of host factor includes host cell systems, which act to limit DNA damage and maintain chromosome integrity. Studies with IS*10* (189) and IS*1* (204) have demonstrated that high levels of Tpase in the presence of suitable terminal IRs lead to the induction of the host SOS system. As discussed previously (234), some controversy still exists concerning the role of RecA in Tn*5* (IS*50*) transposition (1, 195, 196, 396) and further investigation is clearly required to understand the apparently incompatible results found in the literature.

The *recBC* genes are also implicated in the behavior of transposons such as Tn*10* and Tn*5* (221, 222) where they affect precise and imprecise excision in a process independent of transposition per se. This is more pronounced with composite transposons in which the component insertion sequences IS*10* and IS*50* are present as inverted repeats, and is stimulated when the transposon is carried by a transfer-proficient conjugative plasmid. It seems probable that such excisions occur by a process involving replication fork slippage (see references 115, 262, 302 for further discussion).

Both DNA polymerase I (322, 358) and DNA gyrase (168, 353) are implicated in the transposition of Tn*5*. While the effect of gyrase may reflect a requirement for optimal levels of supercoiling, the role of PolI remains a matter of speculation. It may be involved in DNA synthesis necessary to repair the single-strand gaps resulting from staggered cleavage of the target and which gives rise to the DRs. DNA gyrase has also been shown to be important in transposition of bacteriophage Mu (277).

Another host function, the Dam DNA methylase, can be important in modulating both Tpase expression and activity. IS*10*, IS*50*, and IS*903* all carry methylation sites (GATC) in the Tpase promoter regions and in each case, promoter activity is increased in a *dam*$^-$ host (316, 407). Additional evidence has been presented that the methylation status of GATC sites within the terminal inverted repeats also modulates the activity of these ends (316). For IS*50*, this can now be understood in terms of steric interference in the Tpase active site, as recently revealed by the determination of the crystal structure of a synaptic complex including its Tpase and a pair of precleaved transposon ends (75). Similar methylation sites have been previously observed in IS*3*, IS*4*, and IS*5*. A survey of the elements included in the IS database has shown that most groups or families contain members that have GATC sites within the first 50 bp of one or both extremities (data not shown). Except for IS*3* itself where strong stimulation of transposition has been observed in a *dam*⁻ host (350), in most of these cases the biological relevance of these sites is unknown.

Reaction Mechanisms

A detailed examination of the reaction mechanisms involved in transposition is outside the scope of this chapter and has been treated in depth elsewhere (e.g., see references 251 and 252 and chapter 2). However, since such mechanisms are pertinent to an understanding of the various behaviors of the IS elements described below, we include here a brief and simplified description.

The process can be divided into several defined steps generally comprising the following: binding of the Tpase to the ends; elaboration of a synaptic complex involving the recombinase, perhaps accessory proteins, and both transposon ends—this step involves either concomitant or subsequent (depending on the element) recruitment of the target DNA, cleavage and strand transfer of the transposon ends into the target; and processing of the strand-transfer complex to a final product. The protein-DNA complexes assembled during this process have been called transpososomes.

Many of the Tpases involved carry a conserved DDE triad of amino acids important in catalysis (see "The DDE motif" below) and the chemistry of cleavage and strand transfer is very similar, if not identical, in most of the limited collection of these transposable elements analyzed to date in detail. They include retroviruses as well as bacteriophage Mu, IS*10*/IS*50*, Tn*7*, Tc*1*/*3*, and IS*911*. The Tpase catalyzes cleavage at the 3′ ends of the element by an attacking nucleophile (generally H_2O) to expose a free 3′OH group (Fig. 2A and B, left-hand panel). The 3′OH in turn acts as a nucleophile in the attack of a 5′ phosphate group in the target DNA in a single-step transesterification reaction (Fig. 2B, right-hand panel). A concerted transfer of both transposon ends to the target site while maintaining the correct strand polarity results in joining each transposon strand to opposite target strands and leaves a 3′OH group on the cleaved target strand. The reaction(s) do not require an external energy source; neither do they appear to involve a covalently linked enzyme-substrate intermediate as do certain site-specific recombination reactions (138). Furthermore, it is worth underlining that since it is the donor strand itself that performs the cleavage-ligation step in the target DNA, no cleaved target molecule is detected in the absence of strand transfer.

It is important to note that not all Tpases are members of the DDE class. These will presumably prove to transpose using alternative reaction mechanisms (see "Other chemistries?" below).

The DDE motif

Over the last few years, it has become clear that many of the enzymes involved in the reactions described above are related and, moreover, are part of a larger family of phosphoryltransferases, which also includes RNaseH and the RuvC "Holliday junction resolvase" (313, 314). An acidic amino acid triad present in all these enzymes is intimately involved in catalysis and its role is presumably in coordinating divalent metal cations (in particular Mg^{2+}) implicated in assisting the various nucleophilic attacking groups and leaving groups during the course of the reaction. For many ISs (the IS*3* and IS*6* families) and the retroviral integrases (IN), this triad is known as the DD(35)E motif and is highly conserved (103, 180, 197) (Fig. 3). In addition, alignments of several Tpases (312) revealed several regions of amino acid conservation, designated N1, N2, and N3 and C1 (237), which encompass the D (N2), D (N3), and E (C1) regions of typical DDE motifs, respectively (Fig. 3). The C1 region is probably the most defined structural element. It appears to be part of an α-helix (see below) and carries additional conserved amino acids, which include a K or R residue approximately 7 amino acids or two helical turns downstream from the E residue (90, 174, 287). Less-well-conserved residues often occur at approximately one helical turn (3 or 4 residues) upstream and downstream E (Fig. 3). In retroviral integrases, the K residue has been shown to interact with the terminal base pairs of the element, presumably contributing to correct positioning of the transposon end in the active site (101, 174). It is remarkable that such a motif can be found in many of the IS families described here. Although this conservation in the primary sequence is lower in certain of the other groups of elements and not all families have been explored in sufficient detail to ensure that the alignments shown in Fig. 3 are biologically relevant, mutagenesis studies with some of these elements (e.g., the Mu, Tn*7*, IS*10*, and Tc*1*/*3* Tpases and the retroviral integrases) clearly underline the importance of these residues. Moreover, structural analysis has shown that these acidic amino acids are arranged close to each other in a similar, three-dimensional manner in other phosphoryltransferases such as RNase H and RuvC (313, 314), which are otherwise unrelated to Tpases.

Major conserved features identified in the structures of the catalytic cores of retroviral integrases, and

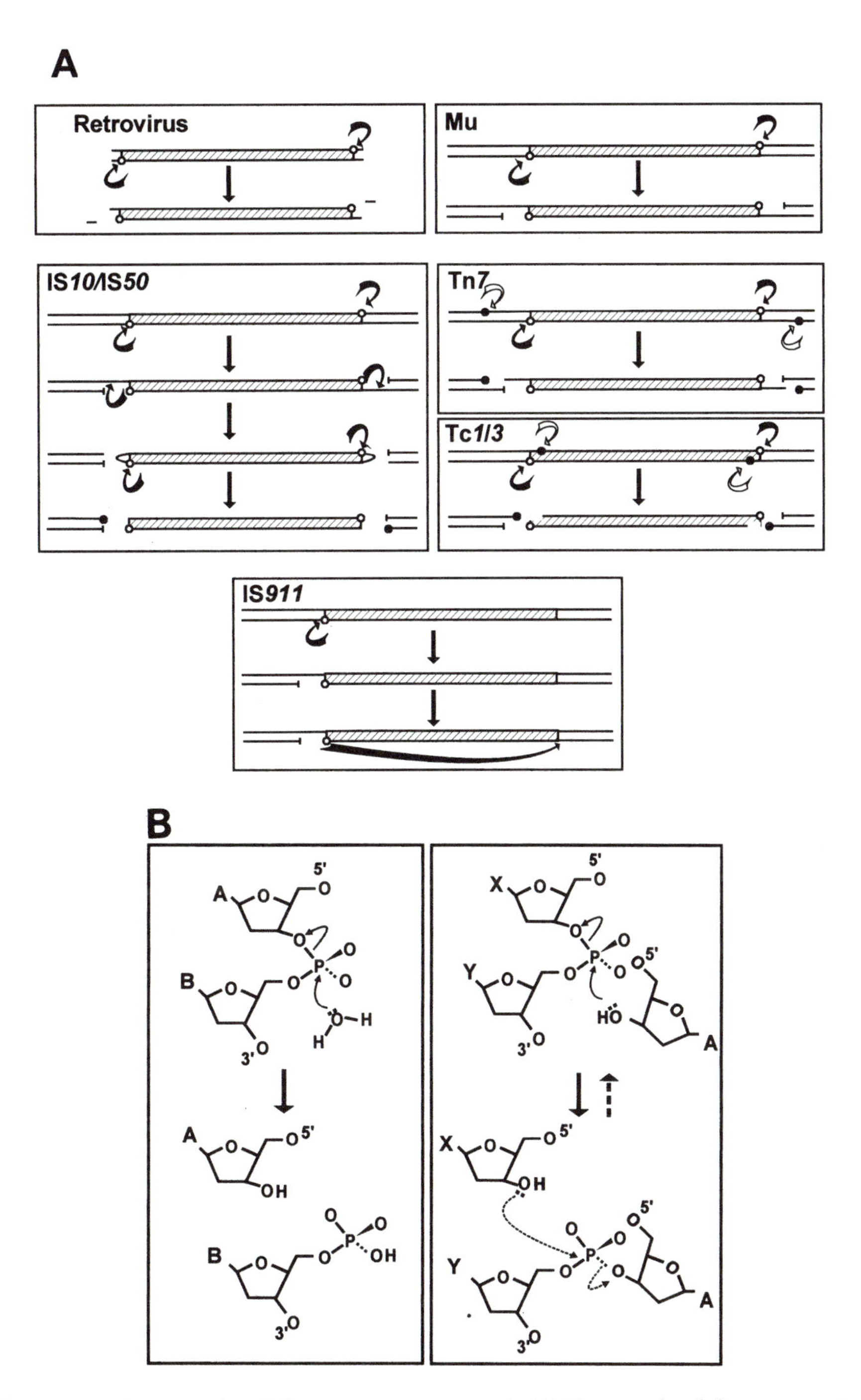

Figure 2. Different types of Tpase-mediated cleavage at transposon ends. (A) Tpase-catalyzed cleavages associated with different transposable elements with DDE Tpases. Transposons are represented by hatched boxes, and flanking donor DNA is represented by black lines. The arrows indicate Tpase-mediated cleavages at the 3′ ends of each element, which give rise to active 3'OH groups shown as open circles and 5′-phosphate groups shown as t-bars. Closed circles indicate 3′OH groups generated in flanking donor DNA. (B) Chemistry of the cleavage and strand transfer events. The left-hand panel shows nucleophilic attack by a water molecule on the transposon phosphate backbone. The nucleotide shown as base A represents the terminal 3′ base of the transposon and that marked B, the neighboring 5′ nucleotide of the vector backbone DNA. Initial attack generates a 3′OH group on the transposon end. The right-hand panel shows a strand transfer event. The 3′OH group at the transposon end acts as a nucleophile in the attack of the target phosphodiester backbone (bases X and Y), joining the 3′ transposon end to a 5′ target end and creating a 3′OH group on the neighboring target base (X). Also shown in this panel as dashed arrows is the "disintegration" reaction in which the 3′OH of the target (X) attacks the newly created phosphodiester bond between the transposon (A) and target (Y) to regenerate the original phosphodiester bond between X and Y.

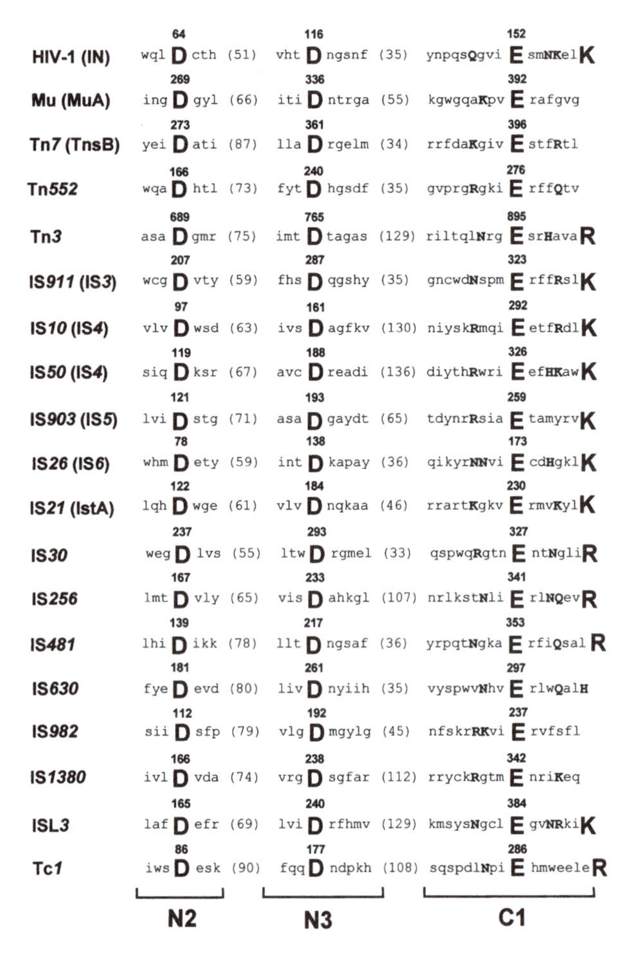

Figure 3. DDE consensus of different families. Individual representative members of each family are shown. Amino acids forming part of the conserved motif are indicated by large bold letters. Uppercase letters indicate conservation within a family and lowercase letters indicate that the particular amino acid is predominant. The numbers in parentheses show the distance in amino acids between the amino acids of the conserved motif. Conservations indicated were derived from previously published alignments or from alignments generated for this chapter. The retroviral integrase alignment is based on reference 287. The overall alignment for the IS*3* family (not shown) is essentially that obtained in reference 287. For IS*21*, see reference 134; *mariner*, see references 90 and 318; IS*630*, see reference 90; IS*4* and IS*5*, see reference 312; IS*256*, see reference 281. N2, N3, and C1 are regions originally defined in the IS*4* family (312).

the Mu and IS*50* Tpases include two β-sheets, each harboring one of the D residues and the long α-helix including the E residue. The α-helix is designated α-4 in HIV and C1 in the Tpases (122, 144, 312). It is one of the most conserved regions in the catalytic core. Mutagenesis and crosslinking studies with IN_{HIV} suggest that it plays a role in positioning both the nucleophile and viral DNA (101, 118, 174). In particular, K159 or K156, which cross-link to the terminal CA dinucleotide, are located on the same side of the α-helix and are strongly conserved in those Tpases shown in Fig. 3. Q148 also lies on the same side and appears to interact with the terminal end of the nonprocessed strand. An amide or basic amino acid is highly conserved at this or the neighboring position and mutation results in severe impairment of catalysis (IN [118], IS*10* [30], IS*50* [chapter 18], IS*903* [365]). Additional convergences in the structure-function relationships of the catalytic domains and their role in target sequestration and positioning will certainly be forthcoming. The recently published structure of a post-cleavage synaptic complex of IS*50* Tpase with its cognate ends, generally supports the model of DNA-protein interactions in the active site (75).

Other chemistries?

Variations and exceptions to this unifying mechanism will certainly emerge. Not all ISs exhibit a well-defined DDE triad. For example, the Tpases of one group of elements, the IS*91* family, show significant similarities with enzymes associated with replicons that use a rolling-circle replication mechanism (chapter 37). Indeed, present evidence (246) suggests that IS*91* has adopted a rolling-circle transposition mechanism similar to that proposed by Galas and Chandler (113). In addition, members of the IS*110* family appear to encode a novel type of site-specific recombinase (211), while active sites for the IS*1* and IS*66* families, IS*200*, and the IS*605* complex (Table 1) have yet to be defined.

Transposition Reactions and Different Types of Gene Rearrangement

While initiation of a transposition reaction proceeds via transfer of the 3′ end of the transposon, the outcome of the reaction is governed by cleavage of the 5′ (nontransferred) end of the element (Fig. 2A). If cleavage of the 5′ end occurs concurrently with cleavage at the 3′ end, the transposon is physically separated from its donor molecule. Strand transfer to a target then results in direct insertion of the element (Fig. 4A, left). If 5′ strand cleavage does not occur concomitantly with 3′ strand transfer, the donor and target molecules become covalently linked (Fig. 4A, right). Subsequent 5′ strand cleavage will separate the element from the donor backbone and will also result in a direct insertion. On the other hand, while 3′ strand transfer joins transposon and target, it leaves a 3′OH in the target DNA at the junction. This can act as a primer for replication of the element and generate cointegrates where donor and target molecules are separated by a single transposon copy at each junction

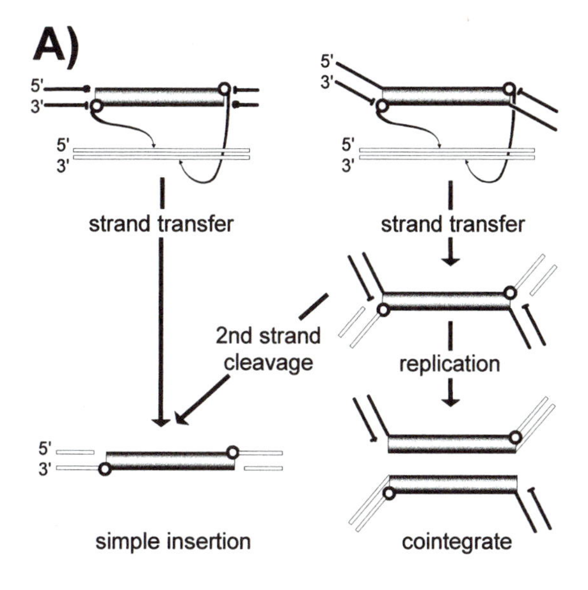

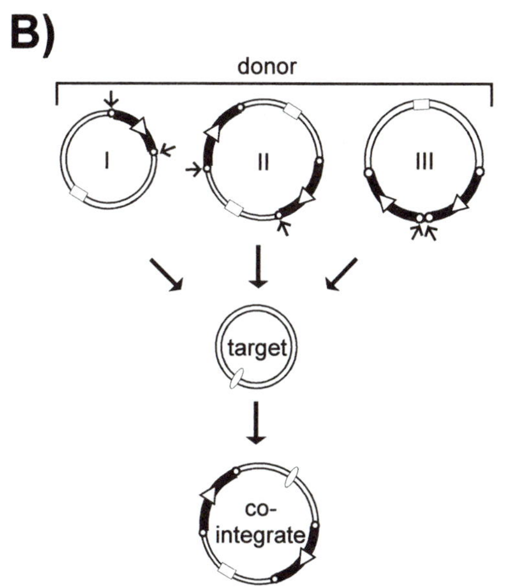

Figure 4. Simple insertions and cointegrate formation. (A) Strand transfer and replication leading to simple insertions and cointegrates. The IS DNA is shown as a shaded box. Liberated transposon 3′OH groups are shown as small shaded circles and those of the donor backbone (bold lines) as filled circles. 5′ phosphates are indicated by a bar. Strand polarity is indicated. Target DNA is shown as unfilled boxes. The left-hand column shows an example of an IS that undergoes double-strand cleavage prior to strand transfer. The right-hand column presents an element that undergoes single-strand cleavage at its ends. After strand transfer, this can evolve into a cointegrate molecule by replication or a simple insertion by second-strand cleavage. (B) Replicative and nonreplicative transposition as mechanisms leading to cointegrates. The figure shows three pathways that generate "cointegrate" molecules by (I) replicative transposition, (II) simple insertion from a dimeric form of the donor molecule, and (III) simple insertion from a donor carrying tandem copies of the transposable element. Transposon DNA is indicated by a heavy line and the terminal repeats by small open circles. The relative orientation is indicated by an open arrowhead. The square and oval symbols represent compatible origins of replication and are included to visually distinguish the different replicons.

(Fig. 4A, right, and 4B). It is, however, important to note that cointegrates identical to those produced by replicative transposition can also be produced by a nonreplicative process from a plasmid dimer (Fig. 4B) (16, 215) or from tandemly repeated copies of an IS element (Fig. 4B) (188, 272, 304, 308, 350, 380).

5′ cleavage can vary from element to element (Fig. 2A) (379). For retroviruses, only the 3′ cleavage occurs, removing 2 bp from the end of the double-strand DNA viral copy. Since no donor backbone is attached to the viral DNA, direct insertion can ensue. During its lytic cycle bacteriophage Mu similarly undergoes only 3′ cleavage of the transferred strand, the donor backbone remains attached, and cointegrate molecules result if replication occurs (Fig. 2A and 4A, right). Both 3′ and 5′ cleavages occur for IS*10*, IS*50*, and Tn*7*, and these elements undergo simple insertion.

Double-strand cleavage at the ends of Tn7 leaves a 5′ 3-bp overhang and involves two proteins, TnsB and TnsA, which cleave the 3′ and 5′ strands, respectively (67). Inactivation of the catalytic domain of TnsA prevents 5′ strand cleavage and results in the formation of branched strand transfer intermediates in vitro and the production of cointegrates in vivo (244). Interestingly, while TnsB is presumed to be structurally similar to typical DDE Tpases (Fig. 3), the structure of TnsA resembles that of a type II restriction endonuclease such as *Fok*I (148) (chapter 19). The use of two distinct strand-specific endonucleases is relatively rare and no IS is at present known to have adopted this type of transposition strategy.

Since only a single Tpase protein is involved in IS*10* and IS*50* transposition, it must catalyze cleavage of both strands. Here, the transferred transposon strand is cleaved first to liberate a 3′OH. However, instead of undergoing strand transfer to a target site, the exposed 3′OH group is directed to attack the *opposite,* complementary, DNA strand (*trans*-strand attack) entirely eliminating flanking donor DNA. This generates a double strand break at the donor ends and a hairpin structure at the transposon end (Fig. 2A), releasing the element from its donor site. In a third step, the bridging phosphodiester bond undergoes hydrolysis to regenerate the terminal 3′OH, which then completes strand transfer to a target DNA molecule. This type of pathway has been adopted in V(D)J recombination (chapter 29 and reference 382), IS*10* (chapter 29 and reference 183), and IS*50* (chapter 18 and reference 25). The geometry of precleaved DNA strands in the active site of the IS*50* synaptic complex shows that the liberated 3′ transposon end is positioned correctly for attacking the opposite strand (75).

Double-strand cleavage has also been demonstrated for the eukaryotic Tc*1*/*3* and *mariner* elements

(Fig. 2A). However, whereas cleavage occurs precisely at the 3′ end of the transferred strand, cleavage at the 5′ ends occurs two bases within the element for Tc*1*/*3* (383), and preferentially 3 bases within the *mariner* elements Himar*1* (203) and Mos*1*. It is interesting to note that for Tc*1*, Tc*3*, Mos*1*, and Himar*1*, 5′ cleavage within the transposon occurs 3′ to a relatively highly conserved CA dinucleotide (5′-CA-GTG for Tc*1*; 5′-CA-GTG for Tc*3*, see [297]; 5′-CCA-GGT or 5′-TCA-GGT for Mos*1*; and 5′-ACA-GGT for Himar*1*). Moreover, early experiments with Tc*1*/*3* indicated that cleavage of the 5′ transposon end may occur before that of the 3′ transferred strand (389). The cognate elements of many "classic" DD35E Tpases are known to terminate with a 5′-CA-3′ dinucleotide on the transferred strand and therefore cleave 3′ to the terminal A. Rather than 5′ cleavage of the nontransferred transposon strand, initial cleavage of the Tc and *mariner* elements could thus be viewed as 3′ cleavage from the vector side. It is tempting to speculate that liberation of the 3′ end of the transferred strand is effected by *trans*-strand attack by the 5′-CA-OH3′. This would generate a hairpin at the donor end, flanks with 2 (Tc) or 3 (*mariner*) bridging nucleotides derived from the transposon in a similar configuration to that observed in V(D)J recombination. It remains to be determined, however, whether it is also capable of correctly catalyzing the predicted hairpin structures.

An interesting variation has been observed in the case of the IS*3* family members IS*2* and IS*911*; a frequent product is a molecule in which only one transposon DNA strand is circularized (Fig. 2A; see "IS*3* Family; IS*911*" below, and chapter 16). This results from a free 3′OH group generated at one transposon end by the Tpase using the *opposite* end as a target. These molecules appear to be processed into transposon circles by "resolving" the complementary strand and the circles can then undergo integration (chapter 16). This type of reaction intermediate generated by sequence-specific recombination between two ends of an element may be generated with several other ISs including members of the IS*21* (21), IS*30* (188), IS*110* (278), IS*256* (224, 295), and ISL*3* (177).

The spectrum of possible DNA rearrangements is probably even larger. A suggestion that certain Tpases may be capable of generating synapses between two ends on different molecules was originally proposed on the basis of the results of a genetic analysis of Tn*5*, (215) and has more recently been demonstrated for IS*10* in vitro (55). Intermolecular reactions of this type can also occur during insertion of IS*911* and provide one element in target site selection (73; C. Loot, C. Turlan, and M. Chandler, unpublished data; see "Target Specificity," below). Similar behavior, as well as the capacity to act on directly repeated IS ends, has been suggested for the IS*1* Tpase in vivo (204). These types of events obviously extend the spectrum of possible DNA rearrangements.

Transposition Immunity

An important property of some transposable elements is that of transposition immunity, in which a target molecule already carrying a copy of an element exhibits a significantly reduced affinity for insertion of a second copy. At present, this phenomenon appears to be limited to the more complex transposons, bacteriophage Mu and Tn*7*, as well as to members of the Tn*3* family. To our knowledge, no insertion sequences have yet been clearly demonstrated to adopt this strategy although some evidence suggests that IS*21* may show immunity (21). It is interesting to note that elements that exhibit this capacity, including IS*21*, show a requirement for ATP. They also tend to carry multiple repeated sequences at each end (20, 21). A priori, this behavior would be inappropriate for elements involved in the formation of compound transposons in which two flanking copies of an IS element mobilize an interstitial DNA segment.

Target Specificity

Where appropriate, insertion patterns of ISs are described in the sections dealing with the individual elements. Insertion specificity has also been treated in detail in a recent review (68). It is perhaps worthwhile, however, to summarize some of the more general issues concerning this aspect of transposition.

Target site selection differs significantly from element to element. Sequence-specific insertion is exhibited to some degree by several elements and varies considerably in stringency. It is strict in the case of one of the two Tn*7* transposition pathways, where insertion occurs exclusively with high efficiency into a unique chromosomal site (*att*Tn*7*) (67), and for IS*91*, which requires a GAAC/CAAG target sequence (247). Insertion sites are less strict, but nevertheless sequence-specific, for members of the IS*630* and *mariner*/Tc families, which both require a TA dinucleotide in the target; for IS*10*, which prefers (but is not restricted to) the symmetric 5′-NGCTNAGCN-3′ heptanucleotide; for IS*50*, which transposes preferentially into 5′-AGNTYWRANCT-3′ (125); for IS*231*, which shows a preference for 5′-GGG(N)$_5$CCC-3′ (137); and for bacteriophage Mu, which shows a preference for 5′-NYG/CRN-3′ (253). In the case of both IS*10* and the Tc*1*/*3* elements, sequences immediately adjacent to the consensus have also been shown to influence target choice (13, 282). A demonstration that IS*10* Tpase directly influences target choice was obtained by isolation of specific Tpase mutants that exhibit distinct alterations in the target sequence (12).

The choice of a target sequence by a given IS can be the result of a subtle combination of several sequence determinants as illustrated by IS*903*. IS*903* generates a 9-bp DR. Although no significant target consensus is obvious to the eye, a statistical analysis suggested that it exhibits a 2-fold symmetry. A synthetic site containing the consensus sequence was shown to be an extraordinarily good target, attracting a very high proportion of all insertions. Mutant target sites confirmed the importance of these features, as well as the importance of the dinucleotide that would occur at the transposon-target junction on insertion. The local DNA context of the target site was also shown to influence its strength (157).

Other elements exhibit regional preferences, for example, GC- or AT-rich DNA segments (IS*186* [335]; IS*1* [112, 249, 411]). Such regional specificity could reflect more global parameters such as local DNA structure. Indeed bent DNA has been evoked as a factor for target choice in retroviral (256), IS*231* (137), IS*4*, and IS*10* integration (136, 294). Recent results with Tn*7* have shown that triple helix DNA strongly targets transposon insertion to a specific end of the triplex-duplex junction (299). This has been interpreted as being the result of recognition of an asymmetric distortion of the target DNA. Other factors that have also been implicated are the degree of supercoiling (IS*50* [218]), replication (Tn*7* [403]; IS*102* [22]), transcription (IS*102* [23]; Tn*7* [78]; Tn*5*/Tn*10* [47]), direction of conjugative transfer (IS*903* [156]), and protein-mediated targeting to, or exclusion from, transcriptional control regions (Mu [390]; yeast Ty*1* [88]; yeast Ty*3* [406]).

A more powerful approach is the use of population-based methods, which provide a statistically more significant picture (256, 296). For phage Mu, not only has it been used to establish the target consensus sequence in vitro (253) but it has also been exploited to investigate occupation of DNA binding sites in vitro (392) and in vivo (391). This technique, known as Mu-printing, relies on the fact that the phage is excluded from insertion in regions of DNA to which regulatory proteins are bound (391). It has also been used to investigate IS*1*-mediated adjacent deletions (378).

Another phenomenon that may reflect insertion-site specificity is the interdigitation of various intact or partial IS elements, which has been noted repeatedly in the literature. Many of these observations are anecdotal and may signal the scars of consecutive but isolated transposition events. Some indication of the statistical significance of this is expected to emerge from the many bacterial genome sequencing projects underway (see "Bacterial Genomes and Plasmids," below). On the other hand, several ISs have been shown to exhibit a true preference for insertion into other elements. A preferred target for IS*231* is the terminal 38 bp of the transposon Tn*4430*, which includes both the sequence-specific and conformational components described above (137), while IS*21* has been reported to show a preference for insertion close to the end of a second copy of the element located in the target plasmid (304), and similar results have been noted for IS*30* (272) and for IS*911* (290; Loot and Chandler, unpublished data; B. Ton-Hoang, unpublished data). In these cases, the site-specific DNA binding properties of the Tpase are presumably implicated. At the mechanistic level, this phenomenon might be related to the capacity of some Tpases to form synaptic complexes with ends located on separate DNA molecules (e.g., reference 55).

THE IS FAMILIES

Occurrence, Variety, and Systematics

The current database (October 2001) contains more than 800 insertion sequences isolated from 86 genera representing 196 bacterial species of both eubacteria and the archaea. This represents 523 individual ISs, of which many have not yet been assigned to a family (Fig. 5). As yet, this includes only a small fraction of the IS sequences identified from the various genome sequencing projects. The majority have not been tested for transposition activity and some may therefore carry inactivating mutations. In many cases, there is also some ambiguity concerning the exact tip of the IS and the length of DRs generated. Where appropriate, these are noted in the text and in the relevant database file at http://www-IS.biotoul.fr. Despite the limitations inherent in the available data, we have been able to include most members of the collection in 19 families based on combinations of the following criteria: (i) similarities in genetic organization (arrangement of open reading frames); (ii) marked identities or similarities in their Tpases (common domains or motifs); (iii) similar features of their ends (terminal IRs); and (iv) fate of the nucleotide sequence of their target sites (generation of a DR of determined length). The general features of these families are shown in

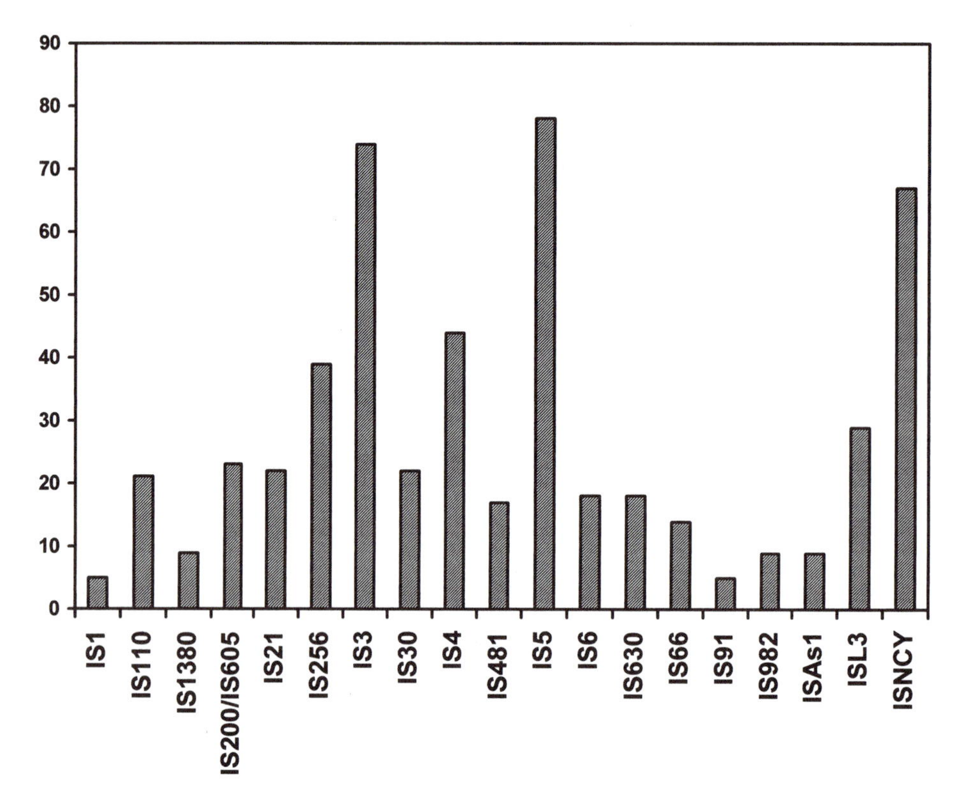

Figure 5. IS distribution among different families. The figure shows the number distribution of the entire IS database into the various IS families. Isoforms are not taken into account.

Table 1 and are presented in greater detail below. It should be noted that two families show some similarities with eukaryote mobile elements: IS*630* is related to the widespread Tc/*mariner* group (90, 283), whereas IS*256* shows a very distant relationship with the plant transposon MuDR (96). We underline here that the classification scheme described below is not rigid and is provided only as a framework. Some families are more coherent and better established than others, and there are numerous uncertainties in several of the attributions of family status. This is particularly true for the IS*200*/IS*605*/IS*607* group. Accumulation of additional sequences has led us to add several families since the appearance of the previous review.

Sequences of the (putative) Tpases were aligned using the PileUp program (Wisconsin Package Version 9.1; Genetics Computer Group [GCG], Madison, Wis. to generate a table of pairwise distances (Distances program, GCG package). These data were converted into dendrograms of relatedness (relationships) by the Growtree program using the UPGMA method (unweighted pair group method using arithmetic averages) (346). We emphasize that we have not explored these in detail and that they do not necessarily represent phylogenetic relationships between the different elements.

Since many IS Tpases carry a DDE motif, this conservation will inevitably be reflected in the alignment results. This background similarity can lead to the erroneous assignment of ISs to a given family. This type of problem is well illustrated by the case of both the IS*4* and IS*5* families. These are rather heterogeneous and can be divided into clear subgroups. Although each subgroup is composed of a core of closely related elements, they can all be related to each other in the family by weak similarities. This is clearly seen in a BLAST search where close members of one group are invariably followed by representatives of other subgroups displaying marginal but significant similarities. No members of other families are found within these searches.

At the other end of the similarity scale, isoforms were defined as elements that show a divergence of less than 5% in the amino acid sequence of their potential proteins or less than 10% in the nucleic acid sequence. This notion of isoforms was helpful when considering similar IS elements from different bacterial species. However, with data accumulating at the present rate from genome sequencing projects, it is becoming less useful, since the number of duplicate ISs in a genome can be considerable. Therefore in the following text we no longer consider isoforms and in the database we have retained only those originating from different species.

Members of most individual families can be distributed across many different eubacterial and archaebacterial genera. Even those IS families that appeared to be restricted to specific bacterial groups, such as IS*1* in the enterobacteria and IS*66* in bacteria of the rhizosphere, have now been observed in phylogenetically distant eubacterial groups (*Synechocystis* and *E. coli*, respectively) (33, 235, 325). An idea of the relative abundance of each family (with the exception of the Tn*3* family) in the database is presented in Fig. 5. The most highly represented are the IS*5* and IS*3* families. The first shows extensive heterogeneity while the second is quite homogenous.

In the following, we first describe the essential features of each IS family. We then review the current state of knowledge about the control of gene expression and transposition mechanism of the better-studied IS sequences within each family. We have not included dendograms for all families but only those that have become apparent or have increased significantly in size since the last survey. Those ISs that are described in separate chapters (IS*10*, IS*50*, IS*91*, IS*911*, and the Tn*3* family) are not treated in detail here.

IS*1* Family

IS*1* was one of the first bacterial insertion sequences to be isolated and characterized (106, 150). The original examples were obtained from an *F′lac-proB* plasmid (IS*1*K: 176) and the multiple drug resistance plasmid R100 (IS*1*R) (270). The nucleotide sequence of many variants of this IS from *Escherichia* and *Shigella* species has been determined. Two examples, IS*1*N and IS*1*H, are significantly different from the others (45 to 47% divergence in nucleotide sequence; 55 to 58% divergence at the protein level) but similar to each other (14 to 19% divergence at the protein level) and might be considered as distinct members of the family (Fig. 6A). Except for IS*1*K(A) and IS*1*R(G), transposition of these elements has not been directly demonstrated experimentally in a controlled way but is implied from the isolation of spontaneous mutants in various genes. IS*1* is a component of several compound transposons such as Tn*9* (225) and Tn*1681* (347) where it is present in direct or inverted orientation flanking a chloramphenicol acetyltransferase and heat-stable toxin gene, respectively.

Although IS*1* was thought to be restricted to the enterobacteria, a distant relative, IS*1*Sa, has recently been detected in the nucleotide sequence of the *Synechocystis* genome (178, 235). There are 3 integral copies of 802 bp and 11 partial or vestigal copies of varying length, all of which can be considered as isoforms. This is illustrated in the dendrogram presented in Fig. 6A based on the relationship between the InsB region (see below). The InsB frame shows about 25 to 30% identity at the protein level with their *E. coli* relatives. However, the InsA frame (see below) and the 22-bp terminal inverted repeats are significantly different (Fig. 6B). Another relative, not included in this figure, has recently been identified in the *Sulfolobus solfataricus* genome (Y18930; coordinates 254390–255563). It carries a single *orf* spanning both *ins*A and *ins*B, is longer (1174 bp), and appears to have long IRs of >50 bp. It is also related to sequences from *Aquifex aeolicus* and *Halobacterium* spp. (B. Berger, personal communication).

Integration of IS*1* is accompanied by duplication of generally 9 bp in the target DNA at the site of insertion (46, 126). DRs of 8, 10, and 14 bp have also been observed (165, 226). The frequency of appearance of these DRs of unusual length is increased by mutations within the Tpase gene (226). The element exhibits a strong preference for insertion into A+T-rich target regions (112, 249, 411).

IS*1* (Fig. 6C) is one of the smallest "autonomous" bacterial insertion sequences isolated so far. It is 768 bp long and includes two approximately 23 bp imperfect inverted repeats (IRL and IRR) located at its ends (176, 270) and two partly overlapping open reading frames (*Ins*A and *ins*B′) located in the 0 and −1 relative translational phases, respectively. The integrity of these frames is essential for transposition (173, 229).

An IS*1* transcript is initiated from a promoter, pIRL, partially located in IRL (228) and is translated to give two products: InsA and InsAB′. The first and more abundant protein is encoded by the *ins*A frame. The small, basic InsA protein binds specifically to the IRs (412, 413) and represses transcription from pIRL (227, 413). It also appears to inhibit transposition, probably by competing with the Tpase for binding to the ends of the element (100). In IS*1*Sa, InsA is longer at its C-terminal end by about 50 amino acids. The second IS*1* protein, InsAB′, is the Tpase of the element. Its production results from a programmed translational frameshift between the *ins*A and *ins*B′ frames. This occurs at a frequency of approximately 1%. The site of frameshifting is an A_6C motif located at the 3′ end of the upstream *ins*A frame (100, 333). A possible A_7 frameshift signal followed by potential secondary structures can also be found in IS*1*Sa.

Natural transposition of IS*1* occurs at a relatively low frequency (approximately 10^{-7} in a standard mating assay). Insertion of an additional A residue within the A_6C motif to yield A_7C or replacing it with GA_2GA_3C fuses the two reading frames leading to constitutive production of the Tpase while eliminating production of InsA. This results in levels of trans-

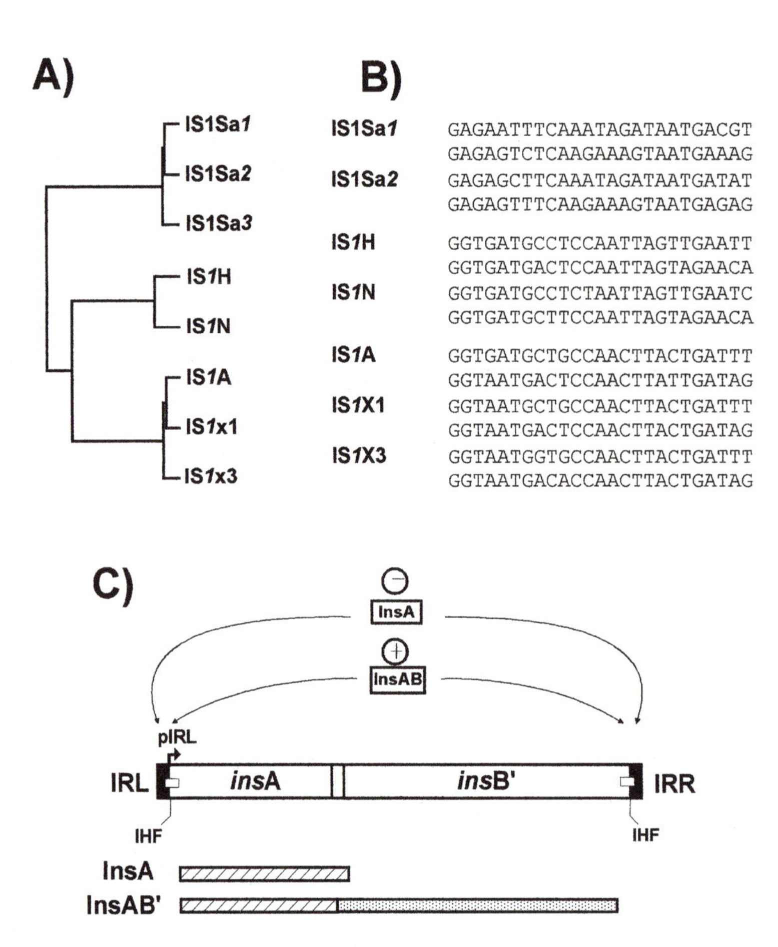

Figure 6. Organization of IS*1*. (A) Dendrogram of the InsB′ reading frames of IS*1* elements from the enterobacteria and IS*1*-like *Synechocystis* elements. (B) Comparison of terminal inverted repeats. (C) Structure of IS*1*. Left (IRL) and right (IRR) inverted terminal repeat are shown as filled boxes. Relative positions of the *ins*A and *ins*B′ reading frames, together with their overlap region, are shown within the open box representing IS*1*. The IS*1* promoter pIRL partially located in IRL is indicated as a small arrow. IHF binding sites located partially within each terminal IR are shown as small open boxes. The InsA protein is represented as a hatched box beneath. The InsA and InsB′ components of the InsAB′ frameshift product are shown as hatched and stippled boxes, respectively. Thin arrows indicate the probable region of action of InsA and InsAB′ proteins. The effect of InsA and InsAB′ on transposition is shown above.

position of between 0.1 and 1% in vivo (99, 100). Indirect evidence has been presented suggesting that a translational restart within the *ins*A frame gives rise to an InsAB′ protein with an N-terminal deletion. It has been proposed that this protein is the true Tpase (241, 242). No independent product of the downstream frame, *ins*B′, alone has been detected. IS*1* transposition activity appears to depend on the ratio InsAB′/InsA (100). Since this is relatively insensitive to the intensity of transcription, this arrangement ensures that IS*1* is not activated by high levels of impinging transcription following insertion into highly expressed genes.

An additional control of Tpase expression may be exercised at the level of transcription termination. Early studies on the organization of IS*1* identified a region at the end of the *ins*A gene, which behaves as a Rho-dependent transcription terminator (160, 293). Premature transcription termination would therefore result in the production of an mRNA lacking the *ins*B frame. The role of this sequence in the control of IS*1* transposition remains to be determined.

IS*1* generates both simple insertions and replicon fusions (cointegrates) composed of two directly repeated copies of the IS, one at each junction between target and donor replicons. The occurrence of stable cointegrates as transposition end products led to the suggestion that transposition of IS*1* can proceed in a replicative manner (114) while simple insertions may occur without replication (27). Thus IS*1* may be capable of both replicative and conservative transposition. More concrete evidence supporting a duplicative transposition pathway was obtained by analyzing the products of intramolecular transposition (378). In addition, these and other studies (331) also detected excised circular copies of the IS*1*-derived transposon and it was suggested that, as in the case of IS*911* (see "IS*3* Family" below), such forms may integrate into a target molecule to give rise to simple insertions. A related type of transposition mechanism was previously proposed for IS*1* transposition (300). Recent experiments have confirmed that such circles in which the IS ends are separated by a spacer of 6 to 9 bp are active in transposition and integrate with high efficiency. Insertion generates a typical target DR and is accompanied by loss of the spacer sequence (341).

High levels of InsAB′ in the presence of suitable IS*1* ends induce the host SOS response, possibly reflecting endonucleolytic activity of the IS*1* Tpase (204). Using this in vivo assay system, originally developed for screening mutants of the IS*10* Tpase (317), it was possible to show that for relatively short derivatives of IS*1*, the level of response depends in a periodic manner on the distance between the ends. The periodicity was found to be about 10 to 11 bp and was also reflected in the transposition activity, suggesting a requirement for correct helical positioning of both ends. Two directly repeated ends were also capable of eliciting the SOS response although they were not capable of giving productive transposition.

The notion that a C-terminal H[200] R[203] Y[231] triad of InsAB′ is catalytically important for IS*1* as it is in integrases of the phage λ family (337) should probably be reevaluated. Although the relative activities of mutants at these positions in IS*1* parallel those of similar mutants in the Flp recombinase (337), this triad is not fully conserved in IS*1*Sa: while the R residue is conserved, the upstream H occurs one residue further upstream and the conserved Y is replaced by an F residue. Moreover, although neither IS*1*N (270) nor IS*1*H (see http://www-IS.biotoul.fr [414]) has been analyzed for transposition activity, in both the H[200] is substituted for an R residue.

IS*3* Family

This large family is widely distributed among eubacteria and is treated in more detail in chapter 16. The G+C content of family members varies from 70% in the mycobacterial examples to 25% in those isolated from *Mycoplasma* species. Despite this enormous variation, they are strikingly similar in many respects and form an extremely coherent and highly related family. Most show a similar G+C content to their host organism. Several family members have been shown to be part of compound transposons. These include IS*3411* flanking genes for citrate utilization (169), IS*4521,* which flanks a heat-stable enterotoxin gene in enterotoxinogenic *E. coli* (154), and IS*1706* (158), which flanks genes of the Clp protease/chaperone family.

Members are characterized by lengths of between 1,200 and 1,550 bp (one exception previously attributed to this family, IS*481,* 1,045 bp, has now been placed in a separate family; see IS*481* Family, below), and terminal IRs in the range of 20 to 40 bp. These repeats are variable but clearly related (Fig. 7B). The majority of the elements terminate with 5′-TG . . . CA-3′ and present an internal block of G/C residues of variable length. IS*3* family members generally have two consecutive and partially overlapping reading frames, *orf*A and *orf*B, in relative translational reading phases 0 and −1, respectively (Fig. 7A). It has been demonstrated in at least three cases (IS*150* [388], IS*3* [332], and IS*911* [288]) that, in addition to the product of the upstream frame, OrfA, a fusion protein, OrfAB, is generated by programmed translational frameshifting (57). However, in contrast to IS*1,* the product of the downstream frame, OrfB, is also detected. The frequency of frameshifting varies from element to element. It is approximately 50% in the case of IS*150* (388) and only 15% for IS*911* (288).

Several members exhibit an organization that does not apparently conform to the generic IS*3* member. In IS*120,* for example, the relationship between the reading phases of the upstream and downstream *orf*s appears to be +1 rather than −1, while in ISNg*1* and ISYe*1* the characteristic motifs of OrfB (see below) are distributed between reading phases. Other members, such as IS*1076,* IS*1138,* IS*1221,* and IS*1141,* exhibit only one long open reading frame. Although these may be true variants, it cannot at present be ruled out that some of these variations are simply due to errors in sequence determination.

Family members from *Mycoplasma* species merit special attention. Not only does the host use a nonuniversal genetic code in which the opal termination codon TGA directs insertion of tryptophan (274), but their genomes are among the smallest bacterial genomes known and extremely rich in AT. To date, five different IS*3* family members have been observed in *Mycoplasma.* Of these, only IS*1138* (and IS*1138*b) has been demonstrated directly to undergo autonomous transposition (26). All exhibit similarly high AT levels, and this unusual base composition could lead to difficulties in sequence determination. It is remarkable that typical IS*3* family characteristics have been maintained in such an "extreme" genetic environment. Nine closely related ISs form a group of isoelements, which have been called IS*1221.* As indicated above, one of these carries a single long reading frame (representing *orf*A plus *orf*B) instead of two consecutive overlapping frames. The others each carry insertions or deletions that destroy either the equivalent

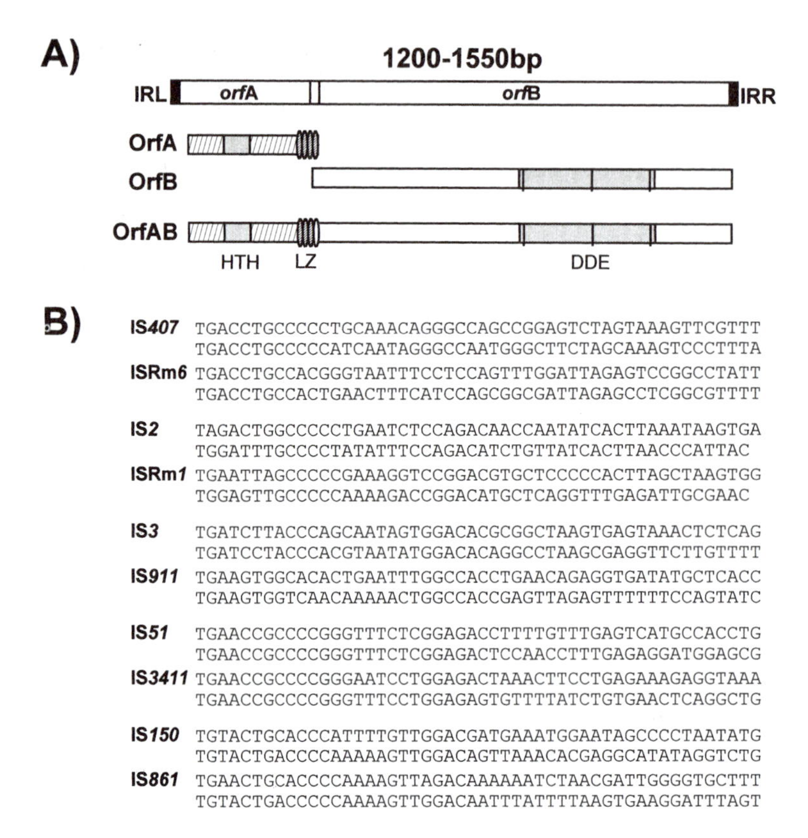

Figure 7. The IS*3* family. (A) General organization of IS*3* family members. The black boxes indicate the left (IRL) and right (IRR) terminal inverted repeats. Transcription probably occurs from a weak promoter located partially in IRL. The two consecutive overlapping open reading frames are indicated (*orf*A and *orf*B) and are arranged in reading phases 0 and −1, respectively. The products of these frames are shown below. OrfA and OrfB are shown as hatched and open boxes, respectively. The position of a potential helix-turn-helix motif (HTH) is shown as a stippled box in OrfA and the DDE catalytic domain as a stippled box in OrfB. A potential leucine zipper (LZ) at the C-terminal end of OrfA and extending into OrfAB is also indicated. Each leucine heptad is indicated by an oval. Those present in the OrfA domain are crosshatched whereas that deriving from the frameshifted product is open. (B) The nucleotide sequence of the terminal IRs of two representative elements of each subgroup is shown.

of *orf*A, *orf*B, or both. Expression studies in *E. coli* indicate that a protein, equivalent to OrfAB, is indeed produced from the long open reading frame of IS*1221*. Interestingly, it appears that a second truncated protein, equivalent to OrfA, may be generated from the *orf*AB frame by translational frameshifting, representing an "inverted" expression pattern to that of the majority of the family members (416). Although this appears not to be a general rule for IS*3* family members originating from *Mycoplasma* hosts, the presence of a similar single-frame arrangement in a second member, IS*1138*, indicates that it might not be rare. Because of the extremely high AT content of these elements, many potential frameshift windows of the A_6G(/C) or A_7 type are expected to occur. Only direct experiment will therefore be able to determine which, if any, of these sequences are used to generate the Tpase or, conversely, an OrfA-like protein.

The predicted primary amino acid sequences of the various OrfA proteins from different members of the family exhibit a relatively strong α helix-turn-α helix motif (Fig. 7A) possibly involved in sequence-specific binding to the terminal IRs of their particular IS (327). The OrfB products on the other hand carry a DD(35)E motif and share additional identities with retroviral integrases and various other Tpases (Fig. 3) (90, 103, 144, 180, 186, 287, 312). Interestingly, many members carry a putative leucine zipper located at the end of OrfA (and sometimes extending into the OrfB region of the OrfAB protein) (34, 381, 416). For IS*911* and IS*2* this has been demonstrated to constitute a multimerization domain (142, 143, 210).

The IS*3* family can be divided into several subgroups (Table 1) defined by deep branching in the alignment of the various OrfB sequences (234). We have designated these the IS*2*, IS*407*, IS*3*, IS*51*, and IS*150* subgroups. Additional members of the family identified subsequently also tend to follow this pattern. One feature that lends biological credence to these subgroups is that they also clearly appear clustered (with some exceptions) in the results of the alignments with the upstream OrfA protein (234). Moreover, there is a strong correlation between the members of each group and the number of base pairs of target DNA duplicated on insertion: for those elements in the IS2 subgroup, insertion invariably leads to a 5-bp DR; for the IS*407* subgroup a 4-bp repeat is observed; while for the other groups a 3-bp repeat is obtained. In the latter cases, some of the elements, e.g., IS*911*, have been shown to occasionally generate 4-bp repeats. This clustering is also exhibited to some extent in the nucleotide sequence of the terminal IRs (Fig. 7B) and is particularly marked in the IS2, IS*51*, and IS*407* subgroups. It can also be observed in the primary sequence details of the putative leucine zipper (data not shown).

Several members carry GATC methylation sites within 50 bp of their ends, which have been shown in one case, IS3, to modulate transposition activity (350). However, this is not a general characteristic of the family, nor is it restricted to any particular subgroup.

Little is known concerning the insertion specificity of members of the family. IS2 exhibits a preference for a region of bacteriophage P1, but the basis of this preference is at present unknown (334). Both IS*911* (290) and IS*150* (399) have been found next to sequences that resemble their IRs and IS*1397* is invariably located within intergenic repeated sequences in *E. coli* (bacterial interspersed mosaic elements or BIMEs [7]).

Finally, an element isolated from the ECOR collection of *E. coli* and closely related to IS*3411* carries a group II intron (104). The implication of this for the regulation of transposition of this element has not been investigated.

The best-characterized members of the IS3 family are IS2, IS3, IS*150*, and IS*911*. These elements generate circular intermediates when supplied with high levels of the fused frame Tpase (214, 332, 399). In the case of IS*911* (374, 375) and IS3 (330), transposon circles can undergo high-frequency integration and appear to be transposition intermediates. These ISs are treated in detail in chapter 16.

IS4 Family

General features

A significant number of new members have been identified over the last three years. Although this family remains extremely heterogenous, members form coherent subgroups or clusters. The choice of retaining these in a single family here was based on the observation that each subgroup is related to others through members with low-level BLAST scores and that no members of other families are found at intervening positions. An overall tree is presented in Fig. 8A. Members of three branches, IS*701*, ISH*27/51*, and IS*942*, originally described as distantly associated with the family, have been removed (see "Orphans and Emerging Families" below).

We have divided the IS4 family into five groups (Fig. 8A). The best defined is the IS*231* cluster, which includes an extensive collection of related but non-isoform IS*231* copies and IS*1151*. It could be extended to include IS*186* and IS4. The second includes IS4Sa and the halobacterial elements ISH*5* and ISH*8*. The three remaining subgroups include IS*10*, IS*50*, and IS*1549*, respectively.

Many members, such as IS*10* and IS*50*, are components of compound transposons (19). A previous alignment of these Tpases (312) revealed several regions of amino acid conservation, N1, N2, and N3 and C1, which encompass the D (N2), D (N3), and E (C1) regions of typical DDE motifs, respectively (Fig. 3). The distance between N3 and C1 in members of the IS4 family is about 100 amino acids compared with 33 to 40 amino acids for Tpases with the canonical DDE motif. In the case of IS*10* and IS*50* (see below), the importance of these regions has received experimental support.

IS*10*, IS*50*, and IS*231*

IS*10*, which forms part of the composite tetracycline resistance transposon Tn*10*, and IS*50*, which forms part of the transposon Tn*5*, are certainly the best-characterized members of this group (Fig. 8C). They have been extensively reviewed (17, 189, 311) and the entire nucleotide sequence of Tn*10* is now available (54, 207). They transpose by a "cut-and-paste" mechanism. The mechanism of transposition of other members of this group is at present unknown.

IS*10* and IS*50* use a similar, if not identical, transposition mechanism. Both are excised via a hairpin intermediate in which the complementary strands at each transposon end are covalently joined (25, 183). However, they exhibit important differences in their organization and control strategies. Although they are treated in detail separately in chapters 18 and 20, it is worthwhile here to compare and contrast some of these differences.

IS*10* carries 22-bp terminal IRs. Those of IS*50* are shorter but the terminal 19 bp are critical for transposition. In both cases they are called outside (OE = IRL) and inside (IE = IRR) ends, which describes their relative position in the Tn*10* and Tn*5*

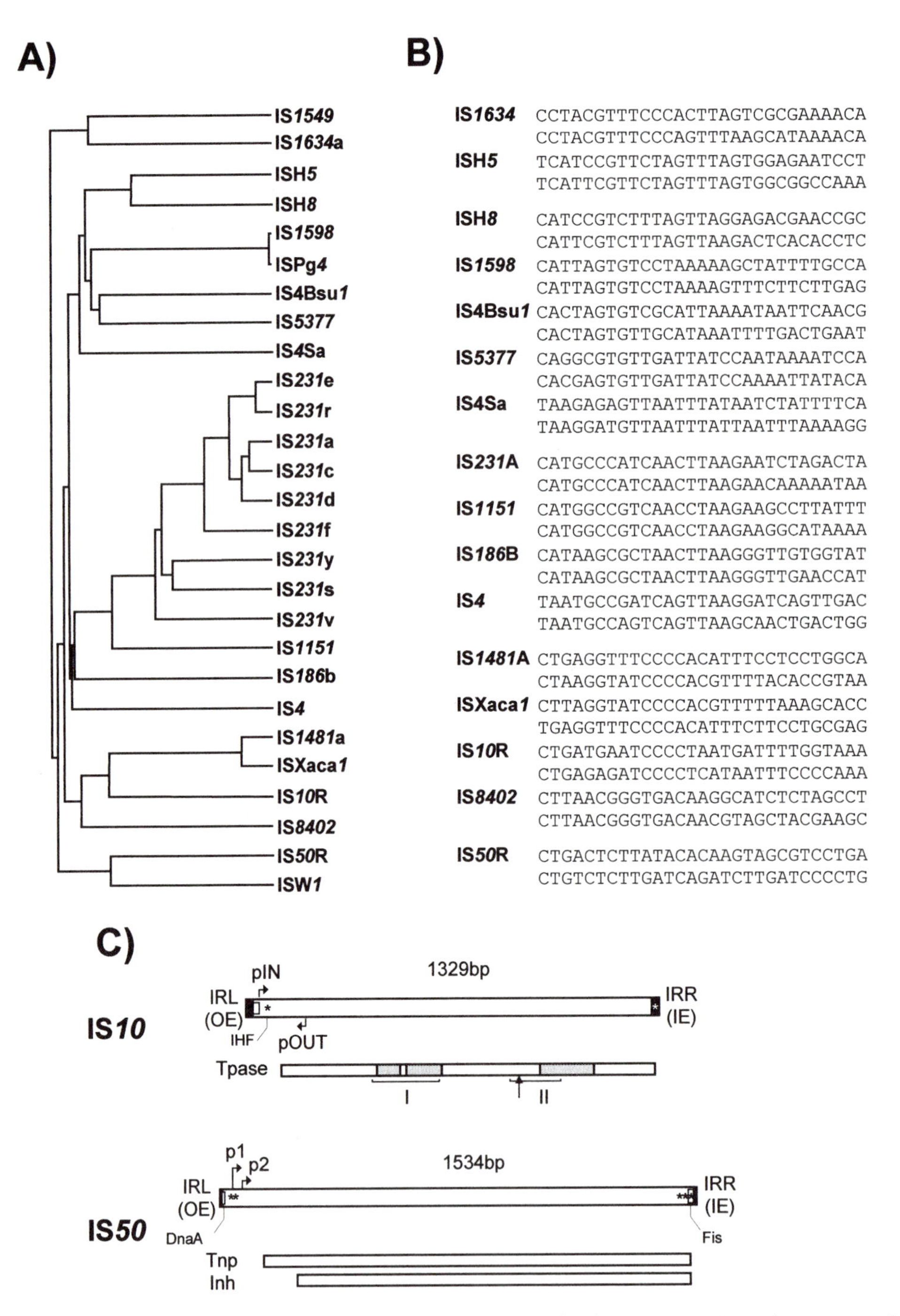

Figure 8. The IS*4* family. (A) Dendrogram of different members of the IS*4* family. (B) Comparison of a representative set of terminal IRs. (C) Organization of IS*10* and IS*50*. IS*10:* The Tpase promoter, pIN, and the anti-RNA promoter, pOUT, are indicated as horizontal arrows. A mechanistically important IHF site is indicated by an open box next to IRL. The Tpase is represented underneath. Stippled boxes indicate the positions of consensus sequence within members of the IS*4* family (from positions 93 to 132, 157 to 187, and 266 to 326). I and II indicate patch I and patch II as defined by mutagenesis (189). The vertical arrow indicates a protease-sensitive site. IS*50:* The promoters for Tpase and inhibitor protein, p1 and p2, are indicated as horizontal arrows. DnaA and Fis binding sites located close to the left and right ends, respectively, are indicated by open boxes.

compound transposons. IS*10* and IS*50* Tpases are expressed from a single long reading frame. The elements exhibit an elaborate ensemble of mechanisms to control their activity and are protected from activation by external transcription by an inverted repeat sequence located close to the left end (see "Impinging transcription" above).

IS*10* contains a Dam methylation site in the −10 hexamer of the Tpase promoter (pIN) (Fig. 8C). Following replication, this site becomes transiently hemimethylated and promoter strength is increased. Dam methylation sites within the Tpase promoter, P1, of IS*50* play a similar role. IS*10* Tpase expression is also controlled in *trans* by a small antisense RNA transcribed from a second, outward-directed promoter located proximal to IRL (pOUT). This complementary RNA interacts with the Tpase mRNA to inhibit ribosome binding. In IS*50,* control in *trans* is exerted by a second protein, the inhibitor protein, Inh. This is translated in the same frame as Tpase, Tnp, but uses an alternative initiation codon and lacks the N-terminal 55 amino acids. It probably employs a separate (and possibly competing) promoter, P2 (Fig. 8C) whose activity is not affected by Dam methylation. Both P1 and P2 are located downstream from the terminal IR (192). It is thought that the inhibitory action of Inh involves the formation of (inactive) heteromultimers between Inh and Tnp.

Dam methylation also exerts control at another level. In addition to the Dam site located in IRL, another site is localized in IRR of IS*10*. Fully methylated sites cause reduced IR activity whereas hemimethylated IRs exhibit increased activity. With this arrangement, both Tpase expression and transposition activity are coupled to replication of the donor molecule. Since transposition of IS*10* is nonreplicative, this ensures that passive replication of the IS occurs before transposition takes place. For IS*50,* transposition activity is reduced by methylation of three consecutive Dam methylation sites located in IE (407) and has been directly attributed to interference with Tnp binding (311).

Binding sites for additional host proteins are observed in both elements. The host IHF protein binds within the left end of IS*10,* interior to the IRL sequence, and subtly influences the nature of transposition products (321, 342). In IS*50,* OE includes a binding site for the host DnaA protein whereas IE carries a binding site for the host protein, Fis. Transposition activity is reduced in a *dna*A host (408) and by the presence of the Fis site (397).

There has been extensive functional analysis of the Tpases (or derivatives) of both elements, by partial proteolysis and mutagenesis (36, 79, 140, 199, 395). In addition, the structure of Inh has been determined (74), as has that of a Tpase complex with the terminal IRs (75). This structure indicates that the two transposon ends are aligned in an antiparallel configuration providing a structural basis for the observation that end cleavage occurs in *trans* (i.e., that Tpase bound at one end catalyzes cleavage of the opposite end; 264). As discussed in the appropriate chapters, these studies are beginning to provide a detailed picture of the transposition mechanism at the chemical level.

The only other member of the IS*4* family that has received some attention is IS*231.* This was isolated flanking a δ-endotoxin crystal protein gene from *B. thuringiensis* (see reference 237), and it has been proposed that it forms part of a composite transposon. IS*231*A is active in *E. coli.* Like IS*10* and IS*50,* a potential ribosome binding site for the Tpase gene would be sequestered in a secondary structure in transcripts originating outside the element. Although most examples carry a single open reading frame, Tpase expression from two elements (IS*231*V and W) may occur by a +1 and +2 frameshift, respectively, but this has yet to be confirmed. Little is known about the transposition mechanism of this element although it exhibits strong target specificity and recent evidence suggests that it transposes by a nonreplicative cut-and-paste mechanism (212).

Recently, several mobile cassettes derived from IS*231* have been identified among *Bacillus cereus* and *B. thuringiensis* strains (J. Mahillon, D. de Palmenaer, and C. Vermeiren, unpublished data). These elements consist of 50 to 80 bp, corresponding to the ends of various iso-IS*231,* flanking genes unrelated to transposition (e.g., *adp,* a D-stereospecific endopeptidase gene). At least in one case, such a cassette (known as MIC*231*) was shown to be trans-complemented by the Tpase of IS*231*A (61).

IS*5* Family

General features

The IS*5* family is also a relatively heterogeneous group and includes sequences from both eubacteria and the archae. The number identified has increased substantially over the past three years. The majority carry only a single *orf.* The major feature that defines this group is the similarities between their putative Tpases (312). In particular this includes the N2, N3, and C1 domains carried by the IS*4* group (Fig. 3). However, Tpases of the IS*5* family exhibit a spacing between the N3 and C1 domains of approximately 40 residues. This group had been divided into six or seven subgroups according to the depth of branching between them (Fig. 9A) (234). Analysis of the largely increased number of members generally confirms

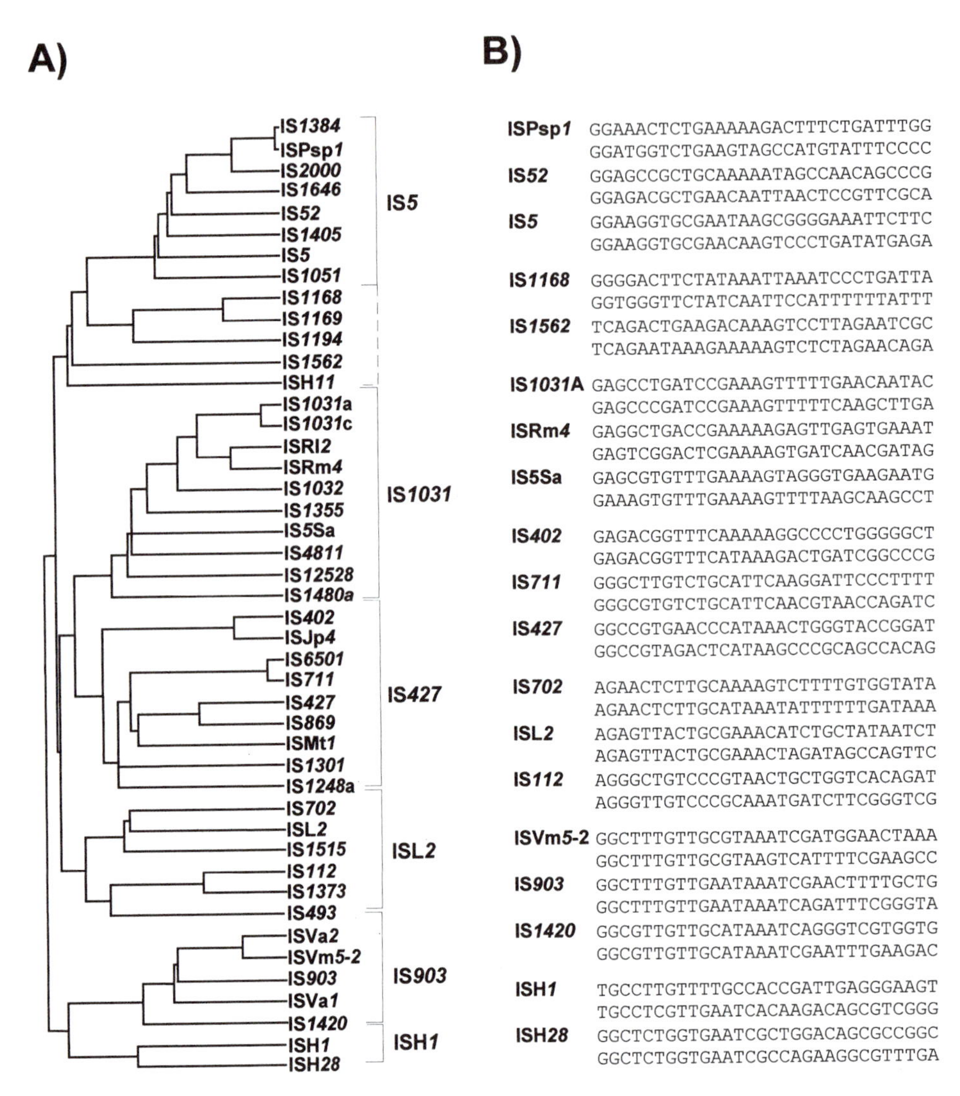

Figure 9. The IS*5* family. (A) Dendrogram of the Tpases of present members of the family showing the different subgroups. (B) Comparison of the terminal IRs of representative members of each subgroup.

these subgroups. Members within each group also generate DRs of similar lengths (IS*5*, 4 bp; ISL2, 2 to 3 bp; IS*1031*, 3 to 4 bp; IS*903*, 8 to 9 bp; and IS*427*, 2 to 3 bp). The IS*903* subgroup appears distinct from the halobacterial ISH*1*, ISH*28*, and ISH*9* elements, which form a cluster. Although included in this figure for convenience, they have been placed in "Orphans and Emerging Families" (see below).

The fact that most members of the IS*427* subgroup exhibit two partially overlapping *orf*s sets them apart. Many of the *orf* pairs occur in a configuration in which the downstream frame is in phase −1 compared with the upstream frame and most have a potential −1 frameshift window. The division is supported by similarities in the terminal IRs (Fig. 9B), by correlation with the length of the target duplication and, to a lesser extent, by the typical length of the entire IS (Table 1). To perform the analysis shown in Fig. 9A, these *orf*s were artificially fused in silico.

Several members of the family are associated with compound transposons. These include IS*903* and IS*602*, which form part of the kanamycin resistance transposons Tn*903* (127) and Tn*602* (354), respectively, and ISVa*1*/ISVa*2*, which form part of a transposon carrying iron transport genes (371).

Several members exhibit GATC sites within their terminal 50 bp. This includes all members of the IS*903* subgroup and many members of the IS*1031* and IS*427* subgroups. IS*903* transposition activity has been shown to be modulated by Dam in vivo (316).

A preferred target sequence, YTAR (often CTAG), is observed for two subgroups, IS*5* and IS*427*, and for two members of the ISL2 group (IS*112* and IS*1373*), in which either all four base pairs or the central TA are duplicated on insertion. It is important to underline that, in many cases, the sequence of the original target site before insertion is not available. This can introduce ambiguities not only in estimating the number of duplicated target base pairs but also in defining the IRs. It is particularly important in several cases where the target repeat is symmetrical (e.g., CTAG) and where it is impossible to distinguish whether the element duplicates 2 or 4 bp and therefore to determine the exact ends of the element. Alignment of the ends of these elements in subgroups has permitted a number of ambiguities to be resolved. Members of the ISL2 group which generate 3-bp DRs exhibit a preference for ANT, while those from the IS*1031* group (which generate exclusively a 3-bp DR) exhibit a preference for insertion sites with the sequence TNA. Neither the small ISH*1* group (8-bp DRs) nor the IS*903* group (9-bp DRs) exhibits marked target specificity (see "IS*903*" and "Target Specificity").

IS*903*

The ends of IS*903* carry IRs of 18 bp, which exhibit the typical two-domain organization (Fig. 1) (85). Tpase has been shown to bind specifically to the ends using a region located in the amino-terminal portion of the protein (83, 365). In addition, a region possibly involved in the formation of higher-order multimers has been identified and residues probably involved in catalysis have been pinpointed among the conserved residues in the catalytic DDE domain (365). An elegant genetic analysis provided strong evidence that IS*903* not only is capable of undergoing direct insertion but can also generate adjacent deletions in a duplicative manner (393). Moreover, point mutations in the terminal base pair of the IRs decrease overall transposition frequency but increase the frequency of cointegrate formation (363). Similarly, mutation of the first nucleotide flanking an IR also influences the level of cointegrate formation (364). The level of cointegrate formation can also be increased by mutation of the Tpase. The molecular nature of these effects requires further investigation.

As discussed above (see "Target Specificity"), a two-fold symmetric consensus target sequence has been elaborated for IS*903* and is highly efficient (157).

IS6 Family

This group of elements was named after the directly repeated insertion sequences in transposon Tn6 (18). Historically, several of these elements were attributed individual IS numbers and have subsequently been determined to be identical. Many have been found as part of compound transposons. The putative Tpases are very closely related and show identity levels ranging from 40 to 94%. Note that one isolate however, IS*15*, corresponds to an insertion of one iso-IS6 (IS*15*Δ) into another (377). All carry short related (15 to 20 bp) terminal IRs (Fig. 10A) and generally create 8-bp DRs. No marked target selectivity has been observed (254). Interestingly, sequences identical to the *Mycobacterium fortuitum* IS*6100* have been identified as part of a plasmid-associated catabolic transposon carrying genes for nylon degradation in *Arthrobacter* spp. (179), from the *Pseudomonas aeruginosa* R1033 plasmid (135), and within the *Xanthomonas campestris* transposon Tn*5393*b (356). Similar copies have also been reported in *Salmonella enterica* serovar Typhimurium (35) and on plasmid pACM1 from *Klebsiella oxytoca* (AF107205). This suggests a relatively recent horizontal dissemination.

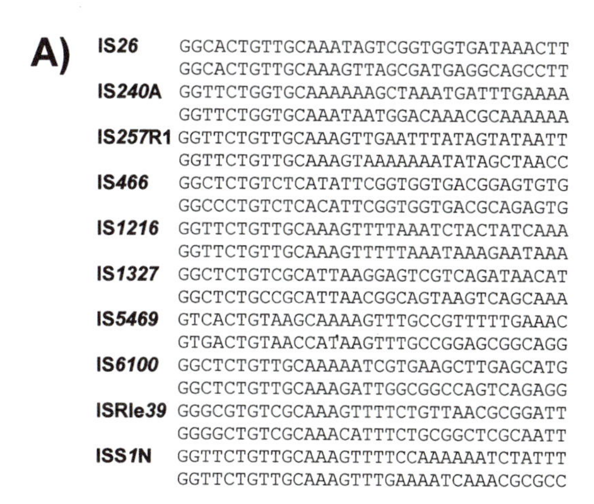

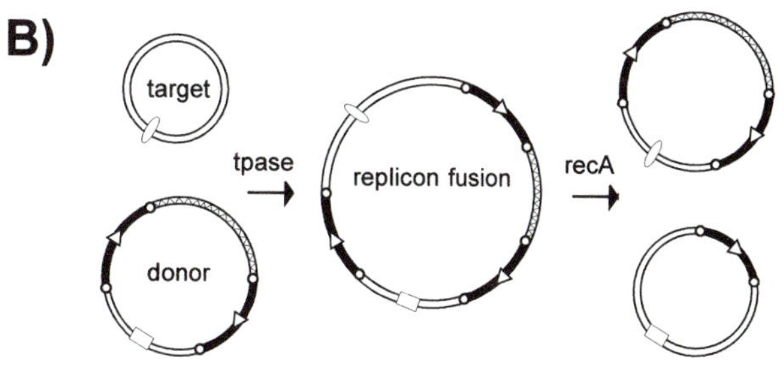

Figure 10. The IS6 family. (A) Terminal inverted repeats. (B) Transposition mechanism. A target plasmid is distinguished by an open oval representing the origin of replication. The transposon carried by the donor plasmid is composed of two copies of the IS (heavy double lines terminated by small circles) in direct relative orientation (indicated by the open arrowhead) flanking an interstitial DNA segment (shown as a zigzag). The donor plasmid is distinguished by an open rectangle representing its origin of replication. Tpase-mediated replicon fusion of the two molecules generates a third copy of the IS in the same orientation as the original pair (open arrowhead). Homologous recombination using the *recA* system between any two copies can, in principle, occur. This will either regenerate the donor plasmid leaving a single IS copy in the target, delete the transposon, or transfer the transposon to the target (as shown) leaving a single copy of the IS in the donor molecule.

In the case of IS*26*, an open reading frame is transcribed from a promoter located within the first 82 bp of the left end. It stretches across almost the entire IS and is required for transposition activity (254), and the predicted amino acid sequence of the corresponding protein exhibits a strong DDE motif (Fig. 3). Translation products of this frame have been demonstrated for IS*240* (80) and IS*26* (K. Vogele and M. Chandler, unpublished data). Little is known concerning the control of Tpase expression although transposition activity of IS*6100* in *Streptomyces lividans* (345) and IS*26* in *E. coli* (Vogele and Chandler, unpublished) is significantly increased when the element is placed downstream from a strong promoter. Insertion of one member, IS*257*, can result in activation of a neighboring gene using both a hybrid promoter and an indigenous promoter (343).

Where analyzed, members of the IS*6* family give rise exclusively to replicon fusions (cointegrates) in which the donor and target replicons are separated by two directly repeated IS copies (e.g., IS*15*Δ, IS*26*, IS*257*, and IS*1936*). Transposition of these elements is therefore presumably accompanied by replication. Following cointegration, a resolution step is required to separate donor and target replicons and to transfer the transposon to the target replicon. Recombination between directly repeated ISs necessary for this separation occurs by homologous recombination and requires a recombination proficient host (Fig. 10B).

IS*21* Family

General features

The members of this homogeneous family have lengths of between 2 and 2.5 kb and are therefore among the largest bacterial IS elements (Fig. 11). They carry related terminal IRs whose lengths may vary between 11 (IS*21*) and 50 (IS*5376*) bp and generally terminate in the dinucleotide 5′-CA-3′. Several members were observed to carry multiple repeated sequences at their ends including part of the terminal IRs (Fig. 11B) (234). These may represent Tpase binding sites. IS*21* itself also carries a degenerate set of such sites (20, 21), which were not previously identified. Insertion of these elements results in a DR of 4 or, more frequently, 5 bp while two members (IS*53* and IS*408*) may generate a DR of 8 bp. They exhibit two consecutive open reading frames: a long upstream frame designated *ist*A and a shorter downstream frame, *ist*B (Fig. 11A). The putative IstA and IstB proteins carry several blocks of highly conserved residues. The *ist*B frame may be located in a relative reading phase of −1 (e.g., IS*21*, IS*5376*) or +1 (e.g., IS*232*, IS*1326*) compared with *ist*A. It can be slightly separated from *ist*A (17 bp, IS*408*) or overlap for one (IS*21*) or several (IS*232*, IS*5376*, IS*1326*) base pairs and is generally preceded by a potential ribosome binding site. The arrangement of the two reading frames suggests that translational coupling could occur (Fig. 11A). The *ist*A reading frame carries a potential helix-turn-helix motif. It also carries a motif related to the widespread integrase DDE signature. Although there is no general conservation of the K/R residue expected seven residues downstream from the final E (Fig. 3), there is good conservation of a basic amino acid at either position 7 or 8 (20, 21). The *ist*B frame carries a relatively well-conserved potential nucleotide triphosphate binding domain (348). Sequence alignments of IstA and of IstB give very similar relationships (234).

Members of the IS*21* family are widespread. In addition to the distinct examples restricted to individual species, several individual ISs are themselves quite widely distributed. Isoforms of IS*21*, several of which are incomplete, have been found in RP4 (IS*8* [117]), in *S. sonnei,* and various clinical isolates of *E. coli* (243), in the EHEC plasmid p0157 (43), and in the *Y. pestis* virulence plasmid pCD1 (IS*640* [152]; IS*21*p [279]). Another family member, IS*100,* is found in several *Yersinia* plasmids, including pCD1 (107, 152, 279, 286), flanking a high pathogenicity island in *Y. pestis* and *Y. pseudotuberculosis* (40) and in the uropathogenic *E. coli* J96 (357).

IS*21*

IS*21* has been studied in some detail. It was discovered in *Pseudomonas aeruginosa* as a constituent of plasmid R68 (315). This conjugative plasmid carries a single copy of IS*21* and is inefficient in chromosome transfer. A derivative of R68, R68.45, exhibiting a marked increase in transfer of chromosomal genes, was isolated (315) and proved to carry two directly repeated copies of IS*21* separated by 3 bp (307). Tandem duplication can be envisaged to occur by various routes. It may be favored by the marked tendency of IS*21* to insert in, or close to, an IS*21* end (304). It is more likely, however, that it occurs by dimerization of the donor plasmid (Fig. 4B, II) and deletion of one intervening plasmid sequence located between the two IS copies (Fig. 4B, III). This type of mechanism has been demonstrated experimentally both for IS*30* (188) and for IS*911* (chapter 16 and reference 380). The tandemly repeated copies of IS*21* promote insertion of the entire plasmid in a transposition event involving the abutted terminal IRs and resulting in loss of the 3 bp located between the two ends in the parental R68.45 plasmid. The resulting replicon fusion thus has the appearance of a true coin-

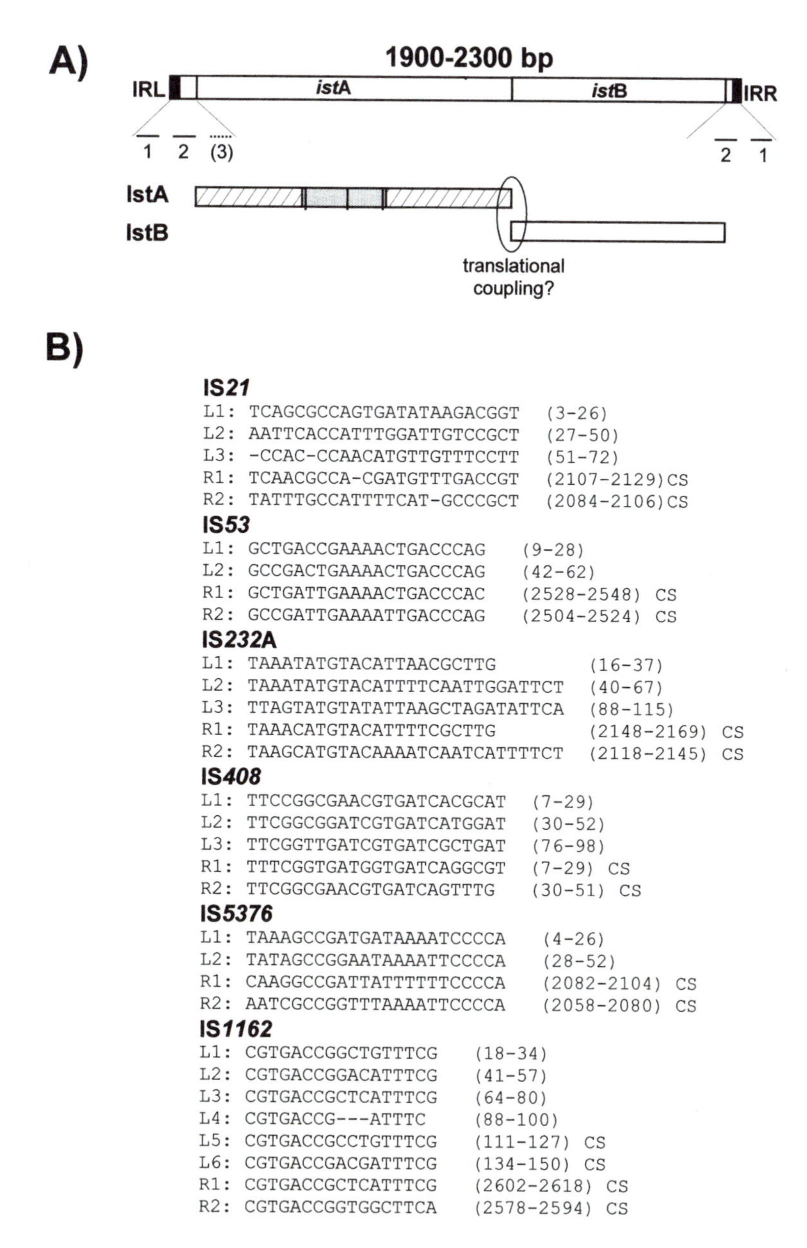

Figure 11. The IS*21* family. (A) General organization. Terminal inverted repeats IRL and IRR are shown as filled boxes. The position of the *ist*A and *ist*B reading frames is also shown. The horizontal lines below show the relative positions of the multiply repeated elements whose sequence is presented in B. IstA (hatched box) together with the potential "DDE" motif (stippled box) and IstB (open box) are indicated below. The possibility of translational coupling between the two reading frames is indicated. (B) The nucleotide sequence of the multiple terminal repeats and their coordinates are presented. CS, complementary strand. L1, L2, L3, and R1, R2, indicate internal repeated sequences at the left and right ends, respectively.

tegrate (i.e., with a single directly repeated copy of the IS at each junction) although transposition does not involve duplication of the insertion sequence during the translocation event. Both IstA and IstB are required for integration (305) and production of a second IstA derivative (from an alternative translational initiation codon eight codons downstream of that of IstA itself) greatly increases the proportion of replicon fusions (306) for reasons that are at present unknown. Integration is optimal when the distance separating the two ends is 4 bp. It is efficient with a 2- or 3-bp separation but inefficient with smaller or larger intervening sequences (324). Like retroviruses, and several other transposable elements, IS*21* terminates with a 5′-CA-3′. Mutation of the terminal A abolishes replicon fusion (134).

The duplication of IS*21* in R68.45 not only creates an efficient integrative donor structure by providing two closely spaced and correctly oriented IRs, but also generates a strong promoter (as measured with a *galK* reporter gene) in which a −35 box, located in the IRR of the upstream element, is combined with a −10 box located in the abutted IRL of the downstream copy. This would direct high levels of production of the IS*21* transposition proteins (307) in a similar manner to the junction promoter created by circularization of IS*911* and IS*2* (chapter 16). Insertion of IS*21* downstream of a strong promoter results in the efficient expression of transposition proteins. Thus IS*21* appears not to have adopted a mechanism that protects it against high levels of impinging transcription. It is presumably the necessity for a transposition substrate carrying two abutted ends and the low frequency of formation of such molecules that acts as a control on transposition activity. More recently, a low level of circular IS*21* copies has been detected in preparations of IS*21*-carrying plasmid DNA (cited in reference 21), raising the possibility that IS*21*, like IS*3* family members, may also transpose via a circular intermediate.

It has been demonstrated in vitro, using *E. coli* cell extracts enriched for IstA, that this protein is responsible for the 3′ end cleavage of IS*21* (305). A substrate carrying two abutted IS*21* ends separated by 3 bp is cleaved specifically at both 3′ ends in an IstA-dependent way.

IS*30* Family

General features

Members of this family have lengths of between 1027 bp (IS*1070*) and 1221 bp (IS*30*) with a single *orf* of between 293 and 383 codons spanning almost the entire length. The termination codon is generally very close to the right end. Alignment of the putative Tpases reveals a well-conserved DD(33)E motif. An open reading frame found originally in a *Spiroplasma citri* phage (ϕSc*1*) but since identified also in the *Spiroplasma citri* genome (see "Bacterial Genomes and Plasmids") is distantly related to the family. Terminal IRs are in the range of 20 to 30 bp and are somewhat heterogenous, although some homologies are apparent (Fig. 12). The tips of the elements, as defined by the various authors, show significant sequence variation. Where it has been determined, 2-bp or 3-bp DRs are generated on insertion. However, one member of this family, IS*1630* (1377 bp) from *Mycoplasma fermentans,* has been reported to be flanked by long but variable DRs of between 19 and 26 bp (45). Various IS*30* family members also have been identified as part of composite transposons (e.g., reference 39).

IS*30*

Of the different members of this family, IS*30* is the best characterized. It shows pronounced insertion

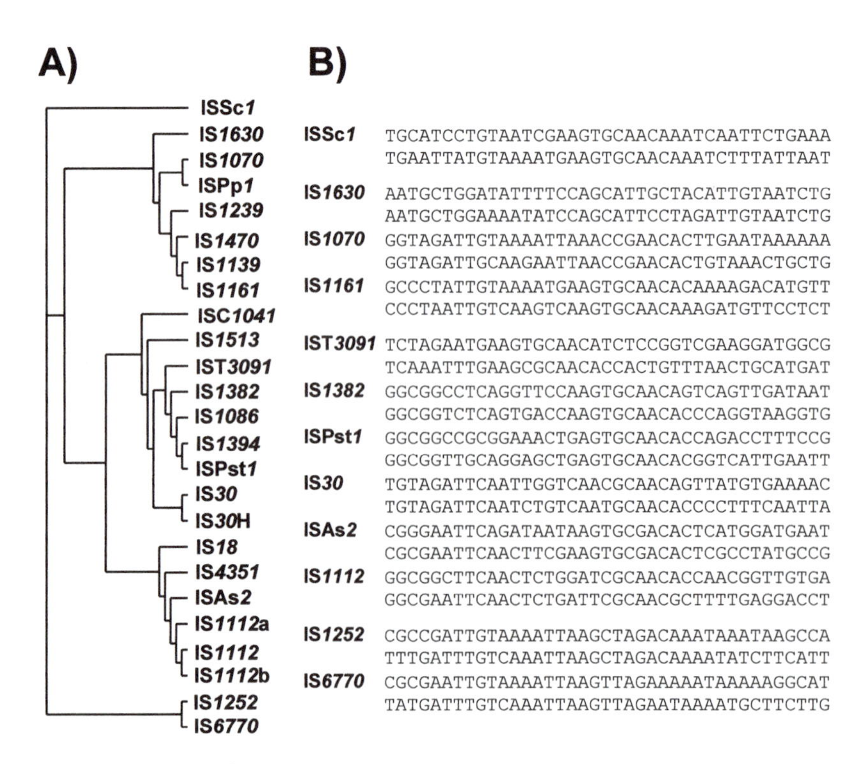

Figure 12. The IS*30* family. (A) Dendrogram based on Tpase alignments. (B) Terminal inverted repeats.

specificity and generates DRs of 2 bp (49). It terminates with the dinucleotide 5′-CA-3′ (72) although many other members of the family do not. The long *orf* is preceded by a relatively weak promoter (p30A), with its −35 hexamer within IRL, capable of driving Tpase expression (71). It also contains a weak internal transcription terminator. Premature termination of transcription at this site could lead to the formation of Tpase molecules truncated at their C-terminal end. Since the specific DNA binding domain that recognizes the terminal IRs is located in the N-terminal third of the protein (351), the potential truncated derivatives may lack the catalytic domain and regulate transposition activity by competition or interaction with a full-length Tpase. This arrangement may play the same role at the transcriptional level as does frameshifting at the translational level for certain other elements. A second weak promoter (p30C) has been located on the opposite strand upstream of a short *orf* (71). The small 150-nucleotide transcript driven from this promoter is contained entirely within the Tpase transcript and acts as an antisense control by reducing translation of the Tpase mRNA (6).

As in the case of IS*21* and IS*911*, the assembly of abutted IRL and IRR copies creates a strong promoter. In a tandem dimer of the element, this may result in increased expression of the Tpase from the downstream copy (70). The IRL-IRR junction is also recombinationally unstable. It promotes transposition reactions at high frequency in vivo (273), presumably in a similar way to the equivalent junction of IS*21* described above. The activity of the junction appears maximal when the spacer between the abutted ends is 2 bp (the length of target duplications introduced when IS*30* inserts). Lower activities are obtained with 1- and 3-bp spacers while little or no activity is observed for 0 bp or >3 bp (T. Naas and W. Arber, personal communication). Strong evidence has recently been presented indicating that donor plasmid dimerization is followed by interaction of one appropriately oriented end from each IS copy to form an abutted IS intermediate (Fig. 4B). The IS junction in this intermediate is highly active in transposition and plays an important role in transposition of IS*30* (188).

IS*66* Family

Members of this family were originally found only in agrobacteria and rhizobia. One member, ISEc*8*, has now been identified in *E. coli* EDL933 adjacent to the locus of the enterocyte effacement (LEE) pathogenicity island (325). Members are between 2.5 and 2.8 kb in length and have very similar terminal IRs of between 15 and 27 bp (Fig. 13B). These elements are generally flanked by 8-bp DRs. Each contains a number of *orf*s, which vary between elements due to slight variations in the nucleotide sequences (either real or due to sequencing errors). We originally used IS*866* as a reference, but recent sequence data obtained from ISRm*14* appears to be more parsimonious since it carries the most representative set of open reading frames (325). ISRm*14* includes three open reading frames on one strand and two on the other. The relationship between some of the members is presented in Fig. 13A. ISRm*14* and ISRmA*3* can be considered as iso-elements although *orf3* of the latter carries a frameshift mutation that introduces a termination codon. A similar situation occurs for IS*866*. In most cases the termination codon of *orf*1 overlaps the initiation codon of *orf*2, raising the possibility that translational coupling is involved in expression of *orf*2. (For a more extensive analysis, see reference 138a.)

An interesting observation concerning this family is the presence of a group II intron maturase signature (191), although its significance is at present unknown. The relationship of the IS*66* family *orf*s to other Tpases is limited (325) and no studies have addressed the transposition mechanism.

IS*91* Family

General features

This family shows important features that distinguish it from many other IS families: there are no significant terminal IRs, no DRs are generated on insertion, and the Tpase is not a member of the DDE class (Table 1). Another notable feature of several members is that while their 3′ ends are relatively constant, there is significant variation in their lengths at the 5′ end. This is related to their particular rolling-circle transposition mechanism (chapter 37). Two members, IS*92*L and R, form part of a composite transposon carrying α-hemolysin synthesis and export genes (410). One, IS*1294*, was discovered on the pUB2380 plasmid (64). Comparison with other members of the family suggests that its Tpase could be extended from 312 to 389 amino acids. This modification would increase the identity to 62% with the Tpase of IS*801*, another member of the family. The putative IS*91* Tpase exhibits 36% identity with that of IS*801*. Recently, related transposons called helitrons have been observed in eukaryotes (178a).

IS*91*

IS*91* (chapter 37) carries very short (7- or 8-bp) imperfect terminal IRs, inserts into a rather specific

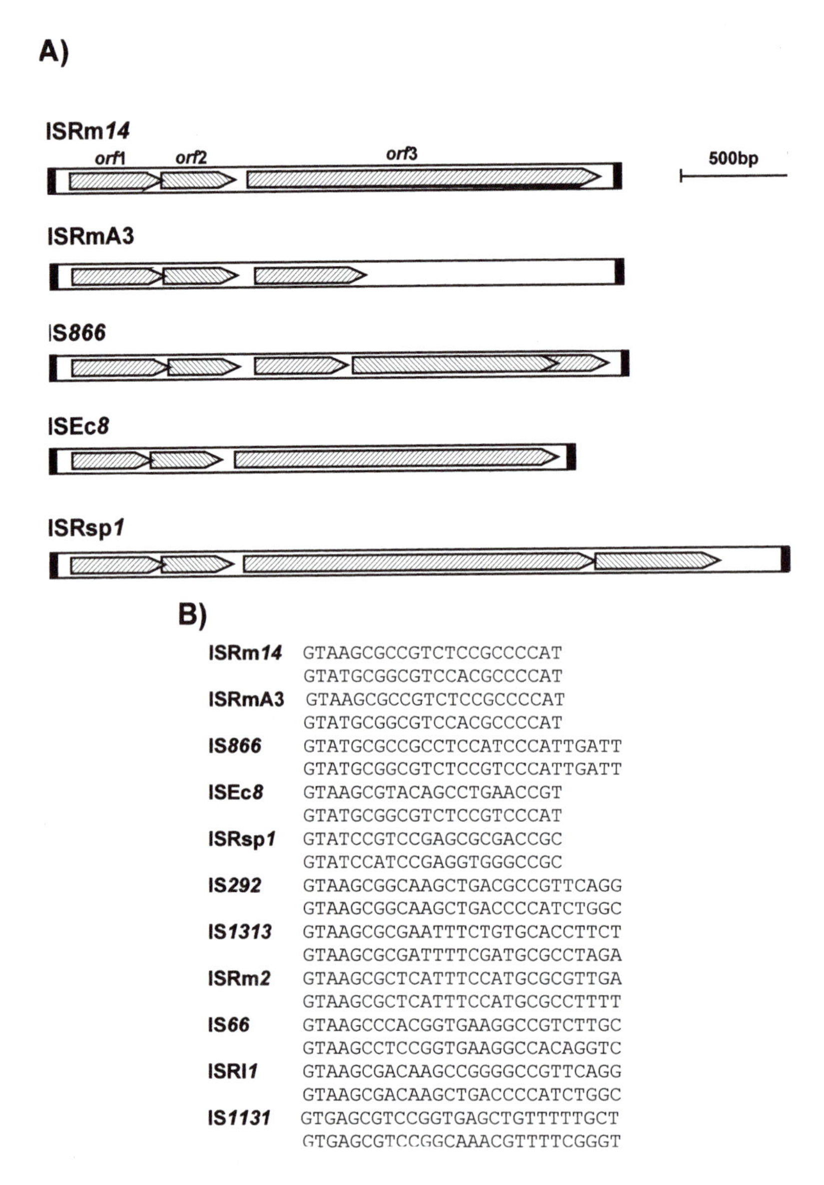

Figure 13. The IS66 family. (A) Organization of IS866. A "best guess" diagram of the open reading frames is shown. All are transcribed from left to right. The difference in shading is simply to facilitate their distinction. Terminal IRs are shown as black boxes. (B) Terminal inverted repeats.

tetranucleotide sequence (GAAC and CAAG), and does not generate DRs upon insertion (247). An alignment of the ends of IS*91* with those of IS*801* and IS*1294* shows highly conserved regions at both the left and right ends and suggests that they also use the same target sequence (248). They carry a long open reading frame whose potential products share significant similarities and are related to a family of replication proteins of bacteriophages and small Gram-positive plasmids that propagate by a rolling circle replication mechanism (166, 248). The similarity is localized to a consensus of four motifs. These Tpases are more related to the φX174 gpA protein than to the plasmid Rep protein family since they carry two tyrosine residues in motif III. In the case of φX174, these are directly involved in catalyzing strand cleavage and transfer (139).

IS*91* appears to transpose by a polarized rolling-circle replicative mechanism (chapter 37 and reference 246) similar to that proposed by Galas and Chandler (113). The right (but not the left) terminal IR and its associated target sequence are essential for transposition. Deletion of IRL results in one-ended transposition, often with a constant end defined by IRR and a variable end defined by a copy of the target consensus located in the donor plasmid. There is a high level of multiple tandem insertions as though the "transposition/replication" apparatus recognizes the donor "target" consensus inefficiently and continues a second round (246). A second member of this family, IS*1294*, appears to promote natural one-ended transposition (64).

IS*110* Family

Like members of the IS*91* family, members of this family show important differences with most of the other IS families. They possess no or very small (<7 bp) IRs and do not generally create DRs although IS*492* might generate 5-bp repeats (9). Little overall similarity can be detected between the ends. A single long, relatively well-conserved *orf* is present and shows some clusters of conservation within the N- and C-terminal portions. These elements share a highly conserved tetrad motif with reverse transcriptase (147) whose significance is at present unknown. None of the signatures typical of rolling circle replication functions found in the Tpases of IS*91*-related elements are detectable in the predicted protein. The inclusion of two additional elements, the *Streptomyces coelicolor* minicircle, IS*117* (145), and a *Moraxella bovis* (and *Moraxella lacunata*) DNA inversion system determining pilus synthesis (Piv) is based on significant similarities in their putative recombinase proteins (211). This family of recombinases is not related to the site-specific recombinases of the λ integrase or resolvase/invertase families. Preliminary modeling studies raise the interesting possibility that this site-specific invertase/Tpase is a phosphoryltransferase, similar to the DDE Tpases, rather than an enzyme whose activity necessitates the formation of phosphotyrosine or phosphoserine enzyme substrate intermediates (370).

Analysis of IS*1383* (205) suggested that the family is divided into two distinct subgroups and that these should be classified as separate families. Alignment of the potential Tpases of presently available members defines two clearly distinguishable groups, IS*1111* and IS*110*. The IS*1111* group includes IS*1111*, IS*1328*, IS*1492*, IS*1533*, IS*5780*, and IS*43211* (Fig. 14), while the IS*110* group contains the remaining members. The two groups have not been separated here into individual families since they are clearly linked by BLAST analysis.

Little information is available concerning the mechanism of transposition of these elements. However, the *S. coelicolor* IS*117* was initially demonstrated in a circular form and integrates at a frequency two orders of magnitude higher than when cloned as a "linear" copy (146). In studies with IS*492* (278) and IS*900* (J. McFadden, personal communication), DNA fragments carrying abutted IS ends have been detected by PCR analysis in vivo and the structures confirmed

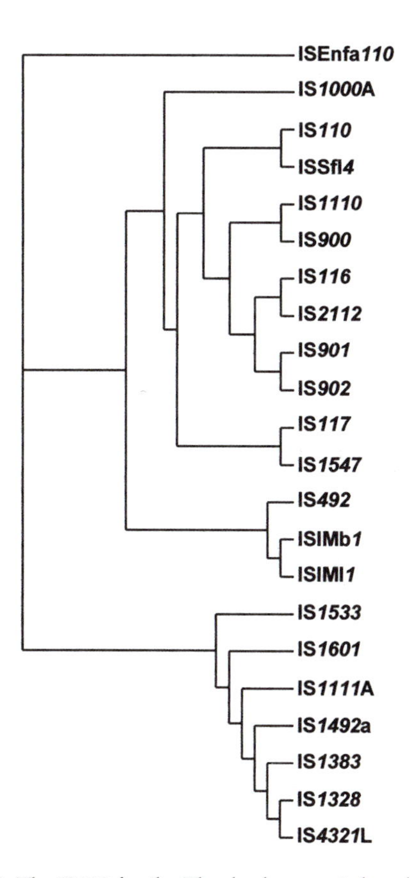

Figure 14. The IS*110* family. The dendrogram is based on Tpase alignments.

by nucleotide sequencing. Their appearance is dependent on an intact Tpase gene. In the case of IS*492* the size of the species carrying the abutted ends is consistent with a circular form of the element and its sequence shows that the ends are separated by 5 bp matching one of the DR copies present in the original insertion. The nucleotide sequence of five IS*492* insertion sites showed that they were remarkably similar with a conserved 5-bp flanking sequence: 5′-CTTGT3-′. This could be extended on one side to 5′-CTTGTTA3-′. IS*492* circle formation also occurs in *E. coli*. Using this system it was shown that excision of the element is Tpase-dependent with reconstitution of the donor molecule. Excision required between 5 and 10 bp of DNA flanking the insertion. Moreover, the junction of IS*492*, like several other ISs, appears to form a strong promoter in *E. coli* (278). For both IS*117* (146) and IS*900* (J. McFadden, personal communication) the target sequences exhibit some similarities to the circle junction, suggesting that insertion may occur by a site-specific recombination.

Finally, although few data are available concerning enzymatic activities of the putative Tpases of this family of elements, IS*900* Tpase was detected by immunological methods in the *Mycobacterium paratuberculosis* host (369) and IS*492* Tpase has been purified and exhibits DNA cleavage activity specific for the ends of the element (D. Perkins-Balding and A. C. Glasgow, personal communication).

IS*256* Family

IS*256* was originally isolated as a component of the compound transposon Tn*4001* (224). Members carry related inverted terminal repeats of between 24 and 41 bp (Fig. 15), and most generate 8-bp DRs, but 9-bp duplications have been observed for some members. Interestingly they also appear to carry internal IRs close to the ends (130, 281). A single long *orf* carrying a potential DDE motif with a spacing of 112 residues between the second D and E residues, together with a correctly placed K/R residue (Fig. 3), is present in most members. The closely related IST2 (*Thiobacillus ferrooxidans*) and IS*6120* (*Mycobacterium smegmatis*) appear to encode two *orfs* (131) and may represent members of a distinct subgroup together with IS*1166*, IS*1413*, and IS*1490* (234). However, the intervening putative amino acid sequence contains residues conserved in the single *orf* of other members of the family. For example, in the case of IST2, a single change of phase extends the N-terminal region of the longer *orf* by 140 amino acids containing additional identities with other family members. It has been suggested that translational frameshifting may occur to generate a single transframe Tpase (131). This subgroup also includes most cases of 9-bp DRs.

IS*1249* shows similarity in a sub-domain of the putative Tpase and also has somewhat related terminal IRs. The other members form three-deep branching groups, each showing between 30 and 40% simi-

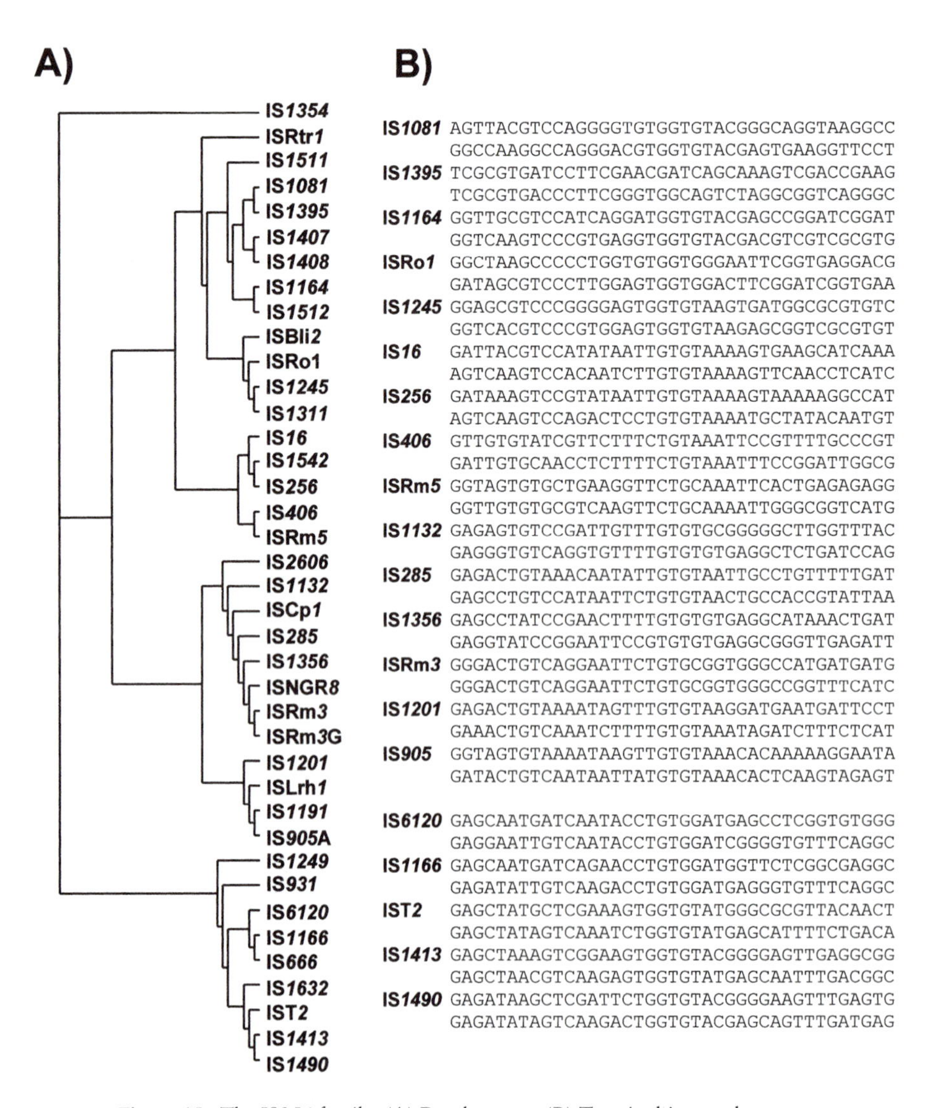

Figure 15. The IS*256* family. (A) Dendrogram. (B) Terminal inverted repeats.

larity. No striking statistical conservation in target sequences can be deduced from the limited number of insertions analyzed. Some evidence suggests that IS*256* generates transposon circles (224). This has been confirmed and analyzed in more detail. The transposon circles appear to resemble those generated by IS*3* family members (295; P. Germon and J. R. Scott, personal communication).

It should also be noted that the putative MurA gene product of the autonomous mutator element of *Zea mays,* MuDR (see "Eukaryotic ISs"), exhibits some weak similarities to the Tpases of this group (96). MuDR generates 9-bp target repeats.

IS*481* Family

This recently defined family contains 13 distinct members (Fig. 16). The founder member, IS*481* from *Bordetella pertussis* (245), was originally placed in the IS*3* family in view of strong similarities in their Tpases. It is, however, significantly shorter (1045 bp) than IS*3* family members and carries only one *orf.* This does not appear to be a deletion derivative of an IS*3*-family member since similar ISs have been identified from various bacteria, including *Arthrobacter nicotinovorans* (ISAni*1*), *Clavibacter michiganensis* (IS*1121*), *Corynebacterium* spp. (ISCgl*1* and IS*5564* [362]), *Streptomyces coelicolor* (ISSco*1* [accession number AF099014]), and from the sequenced genome of *Archaeoglobus fulgidus* (P. Siguier, unpublished data). This argues for the creation of a separate family. One interesting feature of this family is that it includes two closely related members, ISCgl*1* and IS*5564*, from *Corynebacterium,* which have incorporated a chloramphenicol resistance gene. Most members are approximately 1 kb in length and are flanked by CTAG DRs. However, since this sequence is palindromic, it could be part of the IRs.

IS*630* Family

This family shows significant similarities with eukaryote "insertion sequences" such as Tc*1* and Tc*3* of the nematode *Caenorhabditis elegans* and the *mariner* family (90, 283) (Fig. 3). Members are between 1100 and 1200 bp in length and have a high target specificity. Since the conserved target sequence appears duplicated at each end of the element, it is formally possible that one or other copy is an integral part of a terminal IR of the IS. In the absence of the sequence of the target site before insertion, this arrangement therefore introduces an ambiguity not only in determining the exact tips of the IS but also in assessing the number of duplicated target base pairs (see also "IS*5* Family" above).

Detailed studies concerning the target DNA sequence have been carried out in the case of IS*630* (366). The target sequence was determined before and after insertion, and the results suggested that insertion generated a duplication of an invariant target TA di-

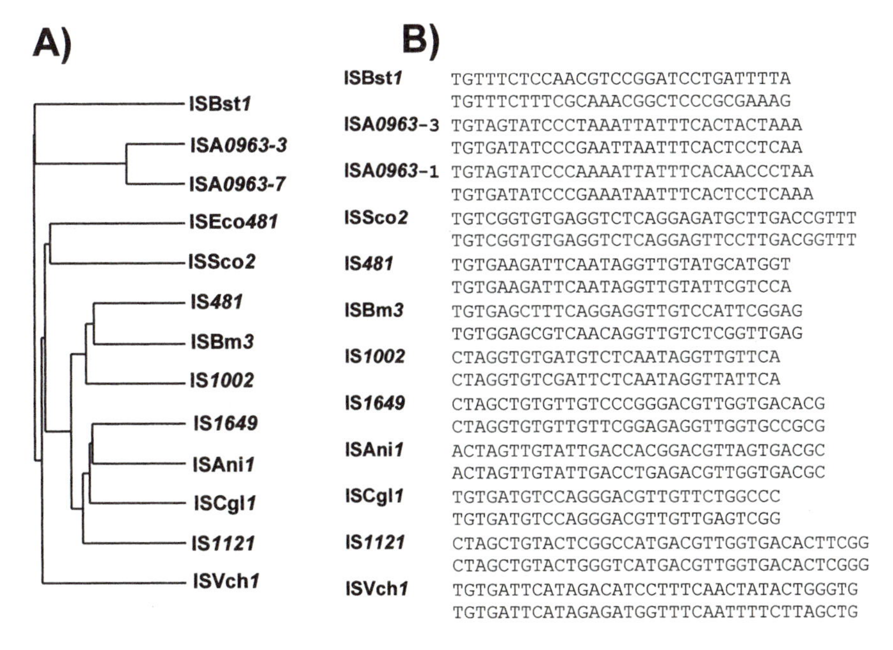

Figure 16. The IS*481* family. (A) Dendrogram. (B) Terminal inverted repeats.

nucleotide. That this dinucleotide does not form an integral part of the IS was investigated by site-specific mutagenesis of the transposon donor to eliminate the terminal TA. Transposition of the IS from the mutated donor molecule resulted in insertions that all exhibited a flanking TA direct repeat. This clearly demonstrates that insertion results in the duplication of the central TA dinucleotide (366). Further analysis demonstrated that IS*630* exhibits a strong preference for a 5′-CTAG-3′ target sequence (367). Moreover, point mutation of the CTAG target sites reduced or eliminated their attractiveness as insertion hotspots.

None of the other insertion sequences of this family have been analyzed in sufficient detail to determine unambiguously the exact tips of the element and the number of target base pairs duplicated. All known insertions of these elements are consistent with a TA duplication (Fig. 17B). In two insertions of the *Agrobacterium vitis* element IS*870* it was possible to determine the target sequence prior to insertion. This was found to carry a single CTAG copy (109), an observation that remains consistent with a simple TA dinucleotide target duplication. For coherence we have therefore assumed that all members of this family generate an identical (TA) target duplication and we have therefore defined the tips of the elements accordingly (Fig. 17).

While IS*895* and ISRm*2011-2* appear to contain two consecutive open reading frames (2, 201), the other members of the family all carry a single long reading frame. Three elements (IS*870,* ISAr*1,* and ISRf*1*) show more than 70% identity, two (IS*1066* and ISRj*1*) show about 50% identity with each other and 40% with the other members. This is reflected in the similarity of their IRs (Fig. 17). It is interesting to note that the three most closely related on the basis of Tpase similarities (IS*870,* ISRf*1,* and ISAr*1*) appear not to carry translation termination codons for the Tpase gene. However, insertion into the specific target site, CTAG, with concomitant duplication (of either 2 or 4 base pairs) generates a TAG termination codon in phase with the Tpase gene. Tc*1* from *C. elegans* is quite distant from this subgroup but is grouped with other bacterial members of this family (Fig. 17). One member, ISRm*2011-2* from *Rhizobium meliloti,* also carries a group II intron, which appears to be active in vivo (239).

A recent analysis of the *Streptococcus pneumoniae* genome revealed several hundred copies of short imperfect palindromic sequences (RUPs) whose ends show strong homology to those of a full-length putative IS*630*-related element also present in the genome (268). Comparison of empty and full sites from different *S. pneumoniae* strains indicated that, like most

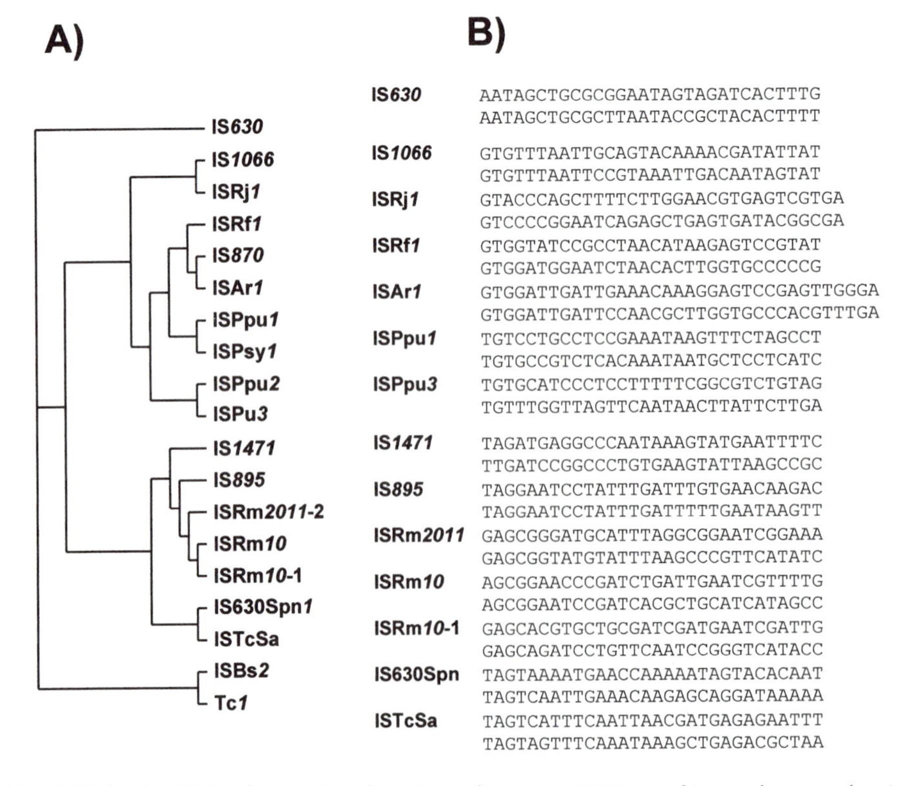

Figure 17. The IS*630* family. (A) Dendrogram based on Tpase alignments. (B) Terminal inverted repeats showing the 3′-TA-5′ target dinucleotide duplicated following insertion.

full-length members of the family, the RUPs are flanked by a TA dinucleotide target repeat. Although transposition of RUP sequences has yet to be demonstrated, they are structurally similar to the IS*231*-related MIC231 elements in *B. thuringiensis* (see "IS*4* Family" above) and to several eukaryote systems where many truncated copies of an element may exist together with a full-length functional copy that is capable of complementing the nonautonomous copies to drive their transposition.

Very little is known concerning the transposition properties of these elements. In the case of IS*630*, while direct insertions can be observed at reasonable frequencies, no cointegrates could be detected (366).

IS*982* Family

Several additional members of this family have recently been identified, and we have therefore included an updated dendrogram (Fig. 18A). Family members are between 962 and 1155 bp in length and carry similar terminal IRs of between 18 and 35 bp (Fig. 18B). Their termini are highly conserved and generally start with 5′-ACCC (Fig. 18). As for several other IS elements, IS*982*B from *Lactococcus lactis* has been shown to provide a −35 hexamer in its right terminal IR, which is capable of forming a hybrid promoter with a resident −10 sequence located at the target site (220). ISSa*4* from *Streptococcus agalactiae* generates 7 bp (or perhaps 9 bp) flanking DRs (349), and ISCce*1* from *Clostridium cellulovorans* is flanked by 7-bp repeats (361). Family members encode a single open reading frame capable of specifying a protein of between 271 and 313 amino acids. A possible DDE motif is also apparent (Fig. 3) but it does not include a convincing conserved downstream K/R residue. Little is known about the transposition of these elements.

IS*1380* Family

Members of this group (Fig. 19) are between 1665 bp (IS*1380*) and 2071 bp (ISBj*1*) in length and each carries a single long open reading frame that includes a potential DDE motif (Fig. 3). Although the founding family member IS*1380* does not itself carry the conserved K/R downstream from the final E, all other members of this family do. The C1 motif is relatively well-conserved. IS*1380* is present in extremely high copy number (approximately 100) in its host *Acetobacter pasteurianus* (360). In addition, ISBj*1* also carries a 112-bp duplication in its right end. The terminal IRs are related (Fig. 19) and are approximately 15 bp long. Insertion appears to generate a target duplication of 4 bp although a member from *S. pneumoniae* (AF154045) is flanked by a 5-bp DR. Note that the external sequence of ISBj*1* is also consistent with a 4-bp duplication.

ISAs*1* Family

Members of this family have lengths of 1207 (ISPg*1*) to 1326 bp (IS*1358*) and generally carry ter-

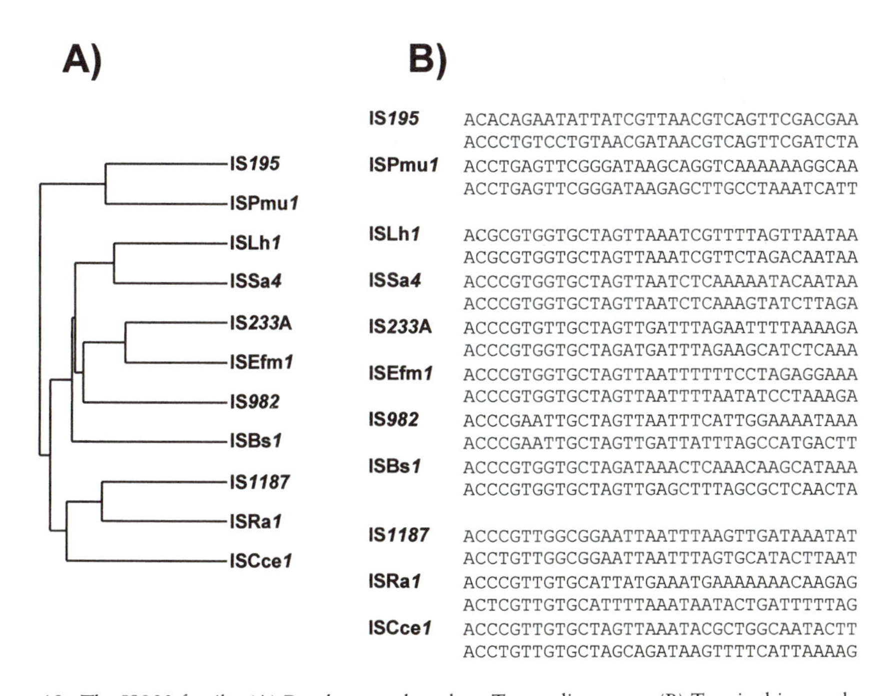

Figure 18. The IS*982* family. (A) Dendrogram based on Tpase alignments. (B) Terminal inverted repeats.

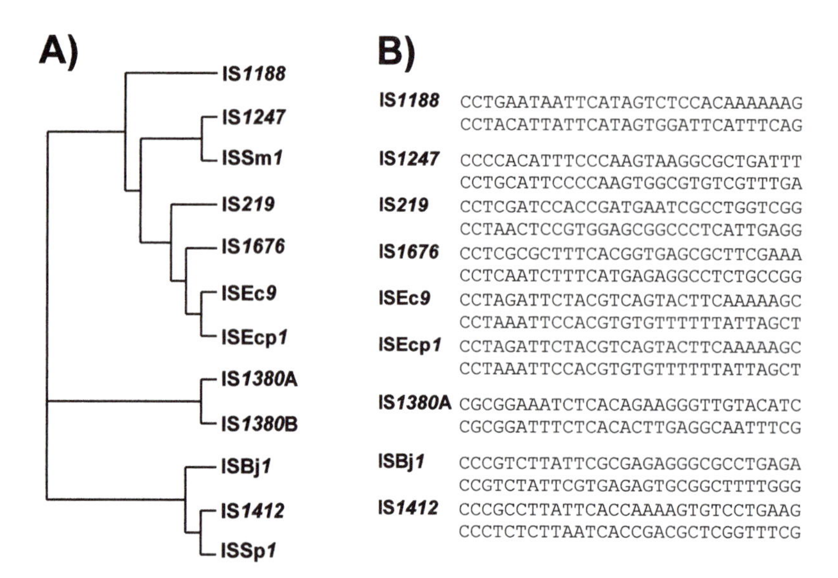

Figure 19. The IS*1380* family. (A) Dendrogram based on Tpase alignments. (B) Terminal inverted repeats.

minal IRs of between 14 and 22 bp (Fig. 20) (although several of the IRs remain to be defined). A single *orf* of between 294 and 376 amino acids occupies almost the entire length and shows between 26 and 50% identity. There are several conserved D and E residues but no clear C1 domain can be detected. The putative Tpase of IS*1358* has been visualized using a phage T7 promoter-driven gene (355) and that of ISAs*1* has been detected in *E. coli* minicells (132).

The family also includes the "H-repeats," which form part of several so-called Rhs (rearrangement hot spot) elements (149, 414). *E. coli* K-12 contains five large repetitions of this type (RhsA to RhsE, scattered around the chromosome with lengths of between 3.7 and 9.6 kb). They represent nearly 1% of the chromosome and provide homology for RecA-dependent rearrangements. These elements are present in many but not all wild-type isolates of *E. coli.* The most prominent Rhs component is a giant core *orf* whose features are suggestive of a cell wall surface ligand-binding protein. Each Rhs element also contains another repeated sequence, the H-rpt element (Hinc-repeat), displaying features of typical insertion sequences, although no transposition activity has yet been detected. For the sake of clarity, the H-rpt DNA sequences B (RhsB), C1 to C3 (RhsC), E (RhsE), and min.*5* (415) as well as H-rptF (414) were renamed ISEc*1* to ISEc*7*, respectively (234). Little is known about the transposition properties of this family of elements. However, recent experiments with the *Vibrio cholerae* element IS*1358* have demonstrated that insertion generates 10-bp DRs and that, in *E. coli,*

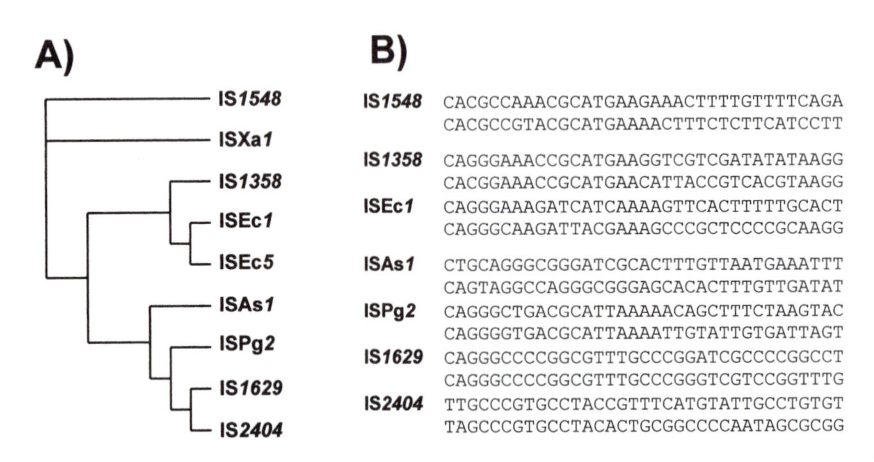

Figure 20. The ISAs*1* family. (A) Dendrogram based on Tpase alignments. (B) Terminal inverted repeats.

it undergoes simple insertion into a target plasmid, pOX38 (94).

ISL*3* Family

Except for ISCx*1*, for which only a partial sequence is available, and IS*1069*, which is significantly larger, members range in size from 1186 bp (ISSg*1*) to 1553 bp (IS*1165*). They carry IRs of between 15 and 39 bp, which are closely related (Fig. 21) and generate a DR of 8 bp. One long open reading frame is present, which gives potential proteins of about 400 to 440 amino acids showing good alignment, particularly in the C-terminal half. IS*1096* harbors two *orf*s. The upstream *orf* exhibits similarities to the ISL*3* family Tpases. The second, *tnp*R, is related to *orf*s from *Agrobacterium rhizogenes* and *Rhizobium* sp. plasmids. In IS*1167*, the reading frame appears to be distributed between two consecutive *orf*s with a potential for translational coupling suggested by overlapping initiation and termination codons (ATGA). The elements fall into three deeply branching groups with about 35%, 30%, and 25 to 60% identities (Fig. 21). Small sequences (130 to 340 bp) related to the IS*1167* IRs have been detected in *Streptococcus sanguis, S. pneumoniae,* and *S. agalactiae* (3, 417). No obvious target sequence specificity has yet been observed although there is some suggestion that there may be a preference for AT-rich regions.

Transposition of most of these elements has been demonstrated, but no analysis of their transposition mechanism has yet appeared. Some evidence has been obtained suggesting that IS*1411* from *P. putida* forms a circular species with abutted IRs separated by 5 bp (177). In one case, IS*31831*, derivatives of which have been used in mutagenesis, the majority of insertions were found to be simple and only rare clones carried accompanying vector DNA (386). DNA species with a size expected for an excised transposon have also been observed (385).

Tn*3* Family

The members of this family are class II transposons (chapter 14 and reference 19). This represents a highly homogeneous group in terms of their transpo-

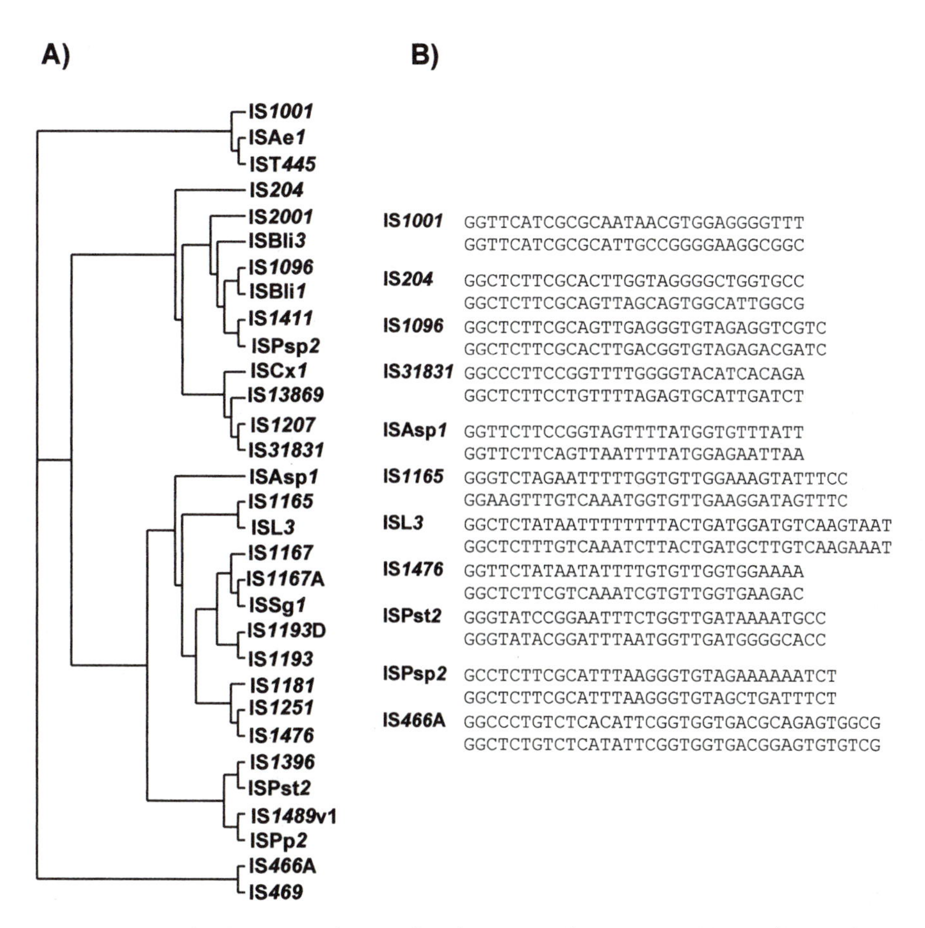

Figure 21. The ISL*3* family. (A) Dendrogram based on Tpase alignments. (B) Terminal inverted repeats.

sition enzymes and terminal IRs. The latter are generally 38 bp long, start with the sequence GGGG, and terminate internally with TAAG. Their Tpases also carry a DDE motif (Fig. 3) (409). Many are complicated in structure and include multiple antibiotic resistance genes carried by another type of transposable element, the integron (chapter 9 and reference 301). Although a complete survey is out of the scope of this chapter, they are included here because several are small and might qualify as insertion sequences (e.g., Tn*1000* [γδ] and IS*101* [340]; IS*1071* [89, 263]; ISXc4/ISXc*5* [217]).

Tn*3* family members transpose using a replicative mechanism. This involves formation of cointegrates promoted by the Tpase, TnpA, which are then resolved by site-specific recombination between the two directly repeated transposon copies. Recombination is mediated by the product of a second gene encoding a site-specific recombinase of the resolvase (TnpR) or phage integrase (TnpI) family. Some of these elements lack phenotypically detectable traits, while others do not appear to carry a site-specific recombinase. Tn*1000* (γδ) carries both *tnp*A and *tnp*R together with a cryptic gene whose function is unknown and Tn*5401* carries *tnp*A and *tnp*I together with a putative poison/antidote system (129). Tn*4430* from *B. thuringiensis* represents a simpler organization and carries only *tnp*A and *tnp*I (236). IS*101*, from plasmid pSC101, carries IRs but no functional Tpase gene, and IS*1071*, from *Alcaligenes*, carries the IRs together with a functional Tpase gene. These represent the simplest type of Tn*3* family member.

The resolution reaction has been studied in exquisite detail for Tn*3* and Tn*1000* (chapter 14). In vitro resolution systems have also been developed for Tn*4430* (B. Hallet, personal communication). However, for members not bearing the site-specific recombination module, homologous recombination can substitute for the highly efficient site-specific recombination reaction, as is the case for members of the IS6 family (see "IS6 Family," above).

The transposition reaction is less well characterized. Tpase binding to the terminal repeats has been described for Tn*3* (163, 232) and γδ (400, 401) and Tpase-induced cleavage of the ends has been observed. In the case of Tn*3*, this is stimulated by the presence of acyl carrier protein (ACP, 233). Members of this family exhibit transposition immunity. An in vitro transposition system has also been developed (164, 231).

IS*200* and the IS*605* Complex

IS*200*

IS*200* was originally identified in *S. enterica* serovar Typhimurium (202), but related elements have since been found in a variety of bacterial species, including certain shigellae (120), clostridia (38), and *S. pneumoniae* (J.-P. Claverys, personal communication). The most recent sequence determination (24) indicates that IS*200* is 707 bp long with a single *orf* preceded by several sequences capable of forming hairpin structures (Fig. 22). One of these, identified within 20 bp of the end, is thought to act as a transcriptional terminator and to prevent impinging transcription from activating the element (202), while another may sequester the ribosome binding site of the Tpase gene (24). IS*200* does not carry terminal IRs. Despite much effort, few cases of IS*200* transposition or other forms of rearrangement have been documented (133). In the few examples of IS*200* insertions in the *S. enterica* serovar Typhimurium genome where the sequence of the unoccupied target site is also known, it seems clear that transposition resulted from a simple insertion of IS*200* alone (24, 44).

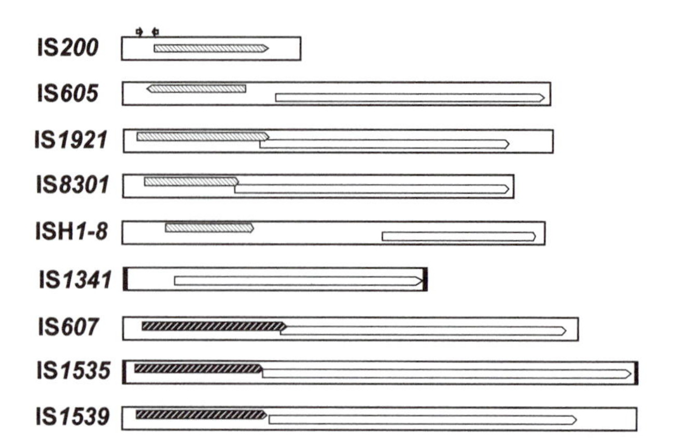

Figure 22. The IS*200* complex. The figure shows the organization of IS*200* (top) with short inverted repeats (open arrows) at the left end and the relative position of the potential open reading frame (hatched box). Selected examples of IS*605* and IS*607* relatives are also included. In all cases the *orf*B frames (unfilled boxes) show clear similarities. The upstream *orf*A frames are similar to that of IS*200* for members of the IS*605* group (thin crosshatching). For the IS*607* group *orf*A (heavy hatching) resemble each other but are not related to those of the IS*605* group. The relative localization of the two frames is indicated with either a significant overlapping region or a one-base overlap, suggesting translational coupling or no overlap at all. Some isolated members carry short IRs. These are indicated by filled boxes.

IS*605*

Although IS*200* has been described as an autonomous element, it is often associated with other *orf*s in the form of a composite element (Fig. 22). This was first observed with IS*605* from *Helicobacter pylori* (184, 372) where it was called *orf*A and is associated with a second divergently transcribed *orf*, *orf*B. OrfB is related to the putative Tpase of IS*1341*. It is also related to the *gip*A gene of Gifsy-1 phage of *S. enterica* serovar Typhimurium, which, since it affects the survival of its host in Peyer's patches, has been implicated in virulence functions (352). IS*605* is found in more than one-third of nearly 300 independent *H. pylori* isolates (184). It is repeated several times in the genome of various strains and both reading frames are always associated (184, 372). That IS*605* is a transposable unit was demonstrated in a mating-out assay in *E. coli* using a native copy to supply transposition functions and a derivative in which *orf*A and *orf*B had been partially replaced by a chloramphenicol resistance gene (184). Moreover, analysis of the metronidazole-resistant strain NCTC11637 indicated that resistance was due to interruption of the nitroreductase gene by a mini-IS*605* and an adjacent deletion (77). The mini-IS*605* was similar to that found at one end of the *cag* pathogenicity island in strain N11638 (51). It includes 41 bp from the left and 33 bp from the right end of IS*605* and is present in multiple copies in strains 26695 and J99 (77). Comparison of occupied and unoccupied sites in various *H. pylori* strains and in the pOX38 plasmid used as a target in the *E. coli* mating-out assay indicated that IS*605* inserts with high sequence specificity 3′ to the sequences 5′-TTTAA-3′ or 5′-TTTAAC-3′ such that the left (IS*200* homolog) end is proximal to the target sequence. A second related sequence, IS*606*, was also identified in *H. pylori*. This was present in one-third of nearly 40 strains, many of which did not carry IS*605*. The potential products of *orf*A and *orf*B reading frames share 25% and 28% identity, respectively, to those of IS*605* itself.

IS*605* is not restricted to *H. pylori* (Fig. 22). A significant number of IS*200*-like isolates from many other species have been found adjacent to *orf*s resembling *orf*B (Table 2). However, the genes are not always found in a divergent orientation. In *Deinococcus radiodurans* (IS*8301*), *Dichelobacter nodosus* (IS*1253* [28]), and *Thermotoga maritima*, for example, they occur in the same orientation and may overlap. A similar arrangement, without overlap, is found in the *Halobacterium* element ISH*1-8* in which *orf*B is related to the putative Tpase of IS*891*. Apparently isolated copies of OrfB-related *orf*s have been reported. This is the case for IS*891*, which has been reported to transpose (8) and IS*1136* (91) and IS*1341* (258), which have not.

Copies of *orf*B-related reading frames have also been found associated with alternative *orf*A frames (Fig. 22). IS*607* identified in a clone obtained from a gastritis strain of *H. pylori* is present in about one-fifth of *H. pylori* strains tested (185). It is composed of two consecutive *orf*s with a nine-codon overlap. The downstream frame is related to *orf*B while the upstream frame is related to potential "resolvase" genes associated with other ISs. IS*607* was shown to transpose at a low frequency in *E. coli* and requires *orf*A but not *orf*B. Comparison of occupied and unoccupied sites in *H. pylori* and in the *E. coli* system indicated that IS*607* terminates with the dinucleotide 5′-GC-3′, carries rather imperfect terminal inverted repeats of approximately 45 bp, and inserts between two adjacent G nucleotides without generating DRs (185).

Another element, IS*1535*, was previously identified as a repeated sequence in the *M. tuberculosis* H37Rv genome where is was proposed to constitute a new IS family (124). IS*1535* (together with IS*1537*, IS*1538*, and IS*1539*) exhibits ends very similar to those of IS*607* and, like IS*607*, is flanked by a G nucleotide in the target site. It also carries two consecutive *orf*s. The downstream *orf*B shows significant similarity with that of IS*605* and IS*607* and the upstream *orf*A shows similarity with *orf*A of IS*607*.

Orphans and Emerging Families

A number of individual and small groups of ISs remain that have yet to be assigned to families. Some of these have been removed from established families after reanalysis in cases where their relationship was found to be quite distant. In other cases there are simply not enough examples identified and it would be premature to designate them to new families. Some, such as IS*1066*, are quite numerous and widespread. Others appear simply as orphan ISs.

Three groups originally assigned to the heterogeneous IS4 family have now been separated because they form outgroups with only marginal similarity to other IS4 family members. One of these, IS*942*, also includes IS*612*–IS*615* from *Bacteroides fragilis*. Two identical ISs carried by plasmids pST01 from *E. coli* (AJ242809) and pTKH11 from *Klebsiella oxytoca* (404) are related to this group and more distantly related to the IS*1380* family. Members of this group are closely related, as judged both by their putative Tpases and by their terminal IRs. The second is the IS*701* group (IS*701*, IS*1357*, IS*1374*, IS*1452*, and IS*1522*), and the third comprises the halobacterial ISH2 and ISH*51*. Another group, which may well prove to constitute a new family, is the ISRm*12* cluster, which includes ISRm*12*, ISRm*13*, IS*1478*, and ISXc6 (http://www-is.biotoul.fr).

BACTERIAL GENOMES AND PLASMIDS

At the time of this writing, the entire nucleotide sequences of more than 50 eubacterial and archaeal

Table 2. Occurrence of IS elements among sequenced bacterial chromosomes[a]

Host	Chromosome size (bp)	IS information			Accession no.
		IS family	No. of members	Copy no.[b]	
Aeropyrum pernix K1	1,669,695	No IS cited			AP000058–AP000064
Aquifex aeolicus VF5[c]	1,551,335	No IS cited			AE000657
Archaeoglobus fulgidus VC-16[c]	2,178,400	IS6	1	2	AE000782
		IS*481*	4	7	
		IS*630*	1	3	
		ISNCY	2	7	
Bacillus halodurans C-125[c]	4,202,353	IS*3*	1	5	NC002570
		IS*4*	2	4, 1p	
		IS*21*	2	3, 1p	
		IS*30*	1	3	
		IS*110*	1	7	
		IS*2001*	1	7	
		IS*605*			
		IS*256*	1	6	
		IS*630*	2	2	
		IS*1380*	1	1	
		ISL*3*	2	42, 4p	
		ISNCY	4	19, 2p	
Bacillus subtillis 168	4,214,814	No IS cited			Z99104–Z99124
Borrelia burgdorferi B31	910,725	IS*605*	1	1	AE00783–AE00794
Buchnera sp. strain APS	640,681	No IS cited			AP001118, AP001119
Campylobacter jejuni NCTC11168	1,641,481	IS*200*	1	1	AL111168
Caulobacter crescentus CB15	4,016,947	IS*3*	1	4	NC002696
		IS*5*	2	8	
		IS6	1	1, 1p	
		IS*30*	1	1	
		IS*110*	2	6	
		IS*481*	1	3	
Chlamydia muridarum	1,069,412	No IS cited			NC002178
Chlamydia trachomatis D/UW-3/CX	1,042,519	No IS cited			AE001273
Chlamydia trachomatis MOPN	1,069,412	No IS cited			NC000117
Chlamydophila pneumoniae AR39	1,229,858	No IS cited			NC002179
Chlamydophila pneumoniae CWL029	1,230,230	No IS cited			NC000922
Chlamydophila pneumoniae J138	1,228,267	No IS cited			NC002491
Clostridium acetobutylicum	3,900,000	IS*200*	1	1	AE001437
		ISNCY	2	2	
Deinococcus radiodurans R1[c]					
Chromosome 1	2,648,638	IS*4*	3	15, 6p	AE000513
		IS*5*	1	1	
		IS*200*	1	1, 1p	
		IS*605*	1	7	
		IS*630*	3	6	
		ISNCY	1	3	
Chromosome 2	412,348	IS*4*	1	3, 2p	AE001825
		IS*481*	1	1, 1P	
Enterococcus faecalis V583	3,218,030	IS*3*	3	5, 4p	
		IS6	1	10	
		IS*110*	1	1	
		IS*256*	3	18	
		IS*605*	1	1	
		ISL*3*	1	1, 1p	
		ISNCY	1	1, 1p	
Escherichia coli K-12 MG1655	4,639,221	IS*1*	1	7	U00096
		IS*3*	5	16	
		IS*4*	2	4	
		IS*5*	1	11	
		IS*30*	1	4	

Continued on following page

Table 2. *Continued*

Host	Chromosome size (bp)	IS information			Accession no.
		IS family	No. of members	Copy no.[b]	
Escherichia coli O157:H7 EDL933	5,528,970	IS1	1	2	NC002655
		IS3	3	39, 13p	
		IS4	1	1, 1p	
		IS30	1	1	
		IS605	1	4	
		IS630	1	2, 2p	
Escherichia coli O157:H7 VT2-Sakai	5,498,450	IS1	1	4, 2p	NC002695
		IS3	5	44, 6p	
		IS4	1	1, 1p	
		IS30	1	4, 3p	
		IS110	1	1, 1p	
		IS256	1	4, 3p	
		IS605	1	3	
		ISNCY	1	1, 1p	
Halobacterium sp. strain NRC-1[c]	2,014,239	IS4	3	14	NC002607
		IS5	3	5, 2p	
		IS605	1	3, 2p	
		ISNCY	3	4	
Haemophilus influenzae Rd KW20	1,830,138	ISNCY	1	3	L42023
Helicobacter pylori 26695[c]	1,667,867	IS605	1	13, 8p	AE000511
		IS606	1	4, 2p	
Helicobacter pylori J99[c]	1,643,831	IS605	1	5, 5p	AE001439
		IS606	1	5, 4p	
Lactococcus lactis IL1403	2,365,589	IS3	4	23, 1p	NC002662
		IS30	1	15	
		IS256	1	1	
Mesorhizobium loti MAFF303099[d]	7,036,074	IS3	?[e]	14, 7p	NC002678
		IS4	1	4, 2p	
pMLa	351,911	IS5	?	19, 16p	NC002679
pMLb	208,315	IS6	?	7, 2p	NC002682
		IS21	?	14, 12p	
		IS66	1	1, 1p	
		IS91	1	4, 3p	
		IS110	?	31, 13p	
		IS481	1	4, 2p	
		IS256	?	7, 7p	
		IS630	?	7, 7p	
		ISL3	1	3, 4p	
		Tn3	1	2	
Methanobacterium thermautotrophicum DH	1,751,377	No IS cited			AE000666
Methanococcus jannaschii	1,664,970	IS6	1	11, 9p	L77117
		IS605	2	2	
		IS982	1	2p	
Mycobacterium tuberculosis CDC1551	4,403,836	IS3	3	6, 1p	NC002755
		IS5	2	4	
		IS110	4	8	
		IS256	3	8, 1p	
		IS481	1	1	
		IS605	1	1, 1p	
		IS607	6	7, 1p	
		ISL3	4	7, 1p	
		ISNCY	2	2	

Continued on following page

Table 2. *Continued*

Host	Chromosome size (bp)	IS information			Accession no.
		IS family	No. of members	Copy no.[b]	
Mycobacterium tuberculosis H37Rv	4,411,529	IS*3*	4	21, 1p	AL123456
		IS*5*	1	2	
		IS*21*	3	3	
		IS*30*	1	1	
		IS*110*	3	5	
		IS*256*	4	9, 1p	
		IS*607*	7	7, 1p	
		ISL*3*	4	6, 4p	
		ISNCY	2	4	
Mycobacterium leprae TN	3,268,203	IS*3*	1	3, 3p	NC002677
		IS*110*	1	1	
		IS*256*	2	5, 4p	
		IS*607*	1	1	
		ISNCY	?	7, 7p	
Mycoplasma genitalium G37	580,074	No IS cited			L43967
Mycoplasma pneumoniae M129	816,394	No IS cited			U00089
Mycoplasma pulmonis UAB CTIP	963,879	IS*3*	1	3, 3p	NC002771
		IS*30*	2	4, 4p	
Neisseria meningitidis Z2491	2,184,406	IS*3*	1	1	AL157952–AL157959
		IS*5*	1	26	
		IS*30*	1	8	
		IS*481*	1	1	
		IS*630*	2	3	
		IS*1016*	3	23	
Neisseria meningiditis MC58*	2,272,351	IS*3*	1	2	AE002098
		IS*5*	1	17	
		IS*30*	1	14	
		IS*630*	1	2	
		IS*1016*	1	16	
Pasteurella multocida PM70	2,257,487	No IS cited			NC002663
Pseudomonas aeruginosa	6,264,403	IS*3*	2	5, 1p	AE004440–AE004968
		IS*110*	1	6	
Pyrococcus abyssi Orsay	1,765,118	IS*605*	2	2	AL096836
Pyrococcus horikoshii OT3	1,738,505	No IS cited			AP000001–AP000007
Rickettsia prowazekii Madrid E	1,111,523	IS*256*	1	1	AJ235269
Staphylococcus aureus N315	2,813,641	IS*3*	?	14, 14p	NC002745
		IS*6*	1	2	
		IS*21*	1	7, 7p	
		IS*630*	1	6, 6p	
		ISL*3*	1	8	
		ISNCY	1	8, 7p	
Staphylococcus aureus Mu50	2,878,040	IS*3*	?	9, 9p	NC002758
		IS*6*	1	2, 1	
		IS*21*	1	6, 6p	
		IS*256*	1	1	
		IS*630*	1	2, 2p	
		ISL*3*	1	10	
		ISNCY	1	6, 6p	
Streptococcus pyogenes M1	1,852,441	IS*3*	?	11, 7p	NC002737
		IS*5*	2	2	
		IS*30*	1	2, 2p	
		IS*110*	1	1, 1p	
		IS*256*	1	2, 2p	
		ISAs*1*	1	4	

Continued on following page

Table 2. *Continued*

Host	Chromosome size (bp)	IS information			Accession no.
		IS family	No. of members	Copy no.[b]	
Sulfolobus solfataricus P2	2,992,245	IS*1*	1	7	NC002754
		IS*4*	4	61	
		IS*5*	1	14	
		IS*110*	3	29	
		IS*256*	1	1	
		IS*605*	2	25	
		IS*630*	2	19	
		ISNCY	6	35	
Synechocystis sp. strain PCC6803[c]	3,573,470	IS*1*	1	14, 11p	AB001339
		IS*3*	1	1p	
		IS*4*	2	22, 13p	
		IS*5*	2	32, 22p	
		IS*256*	1	1, 1p	
		IS*605*	1	1, 1p	
		IS*630*	1	26, 8p	
Thermoplasma acidophilum	1,564,905	IS*5*	1	1	AL139299
		IS*256*	1	2	
		IS*481*	1	1	
		IS*605*	1	1	
		IS*607*	1	2	
Thermoplasma volcanium GSS1	1,584,804	IS*3*	?	5, 5p	NC002689
		IS*4*	?	4, 4p	
		IS*5*	?	4, 4p	
		IS*200*	?	3, 2p	
		IS*256*	?	4, 4p	
		IS*481*	1	1, 1p	
		IS*630*	1	1, 1p	
		ISL*3*	?	4, 4p	
		ISNCY	?	3, 3p	
Thermotoga maritima	1,860,725	IS*605*	1	8	AE000512
Treponema pallidum subsp. *pallidum*	1,138,006	No IS cited			AE000520
Ureaplasma urealyticum	751,719	IS*3*	1	1, 1p	AE002100–AE002158
		IS*256*	1	1, 1p	
Vibrio cholerae El Tor N16961[d]		IS*3*	1	7	AE004093–AE004343
Chromosome 1	2,691,146	IS*5*	1	2	
Chromosome 2	1,072,313	IS*200*	1	5	
		IS*481*	1	1	
		ISAs*1*	1	1	
Xylella fastidiosa	2,679,306	IS*607*	1	2	AE003855–AE004083

[a]A preliminary assessment of IS distribution in available bacterial genomes. In the majority of cases, the IS elements indicated were located from the available annotations of the transposase. The correct family assignment for ISs identified by their annotations was confirmed, and those whose annotations did not assign them to a family were analyzed and identified. In the case of the IS*200*/IS*605*/IS*607* complex, IS*200* indicates that only *orf*A is present.
[b]The total number of copies is indicated, followed by the number of partial (p) copies, where known. Not all genomes have yet been analyzed at this level.
[c]Genomes we have analyzed in detail.
[d]The distribution of ISs between replicons has not been determined here.
[e]We have not determined the number of distinct members.

genomes are publicly available (Table 2) together with a number of large plasmids (Table 3). These provide an important source of both known and previously unknown potential ISs. The level at which each genome has been analyzed for IS content has been variable, and it is probably premature to draw general conclusions concerning the distribution of different IS families between genera and species, their association with particular groups of microbes, their partition between chromosomes and plasmids, or their arrangement on chromosomes. Their number and diversity appear to vary greatly between different species and genera, although this may be partially due to the inability of the homology searches to identify all Tpase

Table 3. Occurrence of IS elements among bacterial plasmids

Plasmid name	Plasmid size (bp)	Plasmid type	Bacterial host	IS family	Members, copy no.[b]	Accession no.
PTi-sakura[a]	206,479	Circular	*Agrobacterium tumefaciens*	IS*21*	1, 1	AB016260
				IS*110*	1, 1	
				IS*3*	1, 1	
Octopine-type Ti[a]	194,140	Circular	*Agrobacterium tumefaciens*	IS*110*	1, 1	AF242881
				IS*66*	3, 3p	
				IS*3*	2, 2	
				IS*5*	1, 1	
				ISNCY	2, 2	
ece1	39,456	Circular	*Aquifex aeolicus*	No IS		AE000667
pXO1	181,654	Circular	*Bacillus anthracis*	IS*3*	1, 2	AF065404
				IS*4*	1, 3	
				IS*5*	1, 1	
pXO2[a]	96,231	Circular	*Bacillus anthracis*	IS*4*	2, 2	AF188935
cp9	9,386	Circular	*Borrelia burgdorferi*	No IS		AE000791
cp26	26,498	Circular	*Borrelia burgdorferi*	No IS		AE000792
lp17	16,823	Linear	*Borrelia burgdorferi*	IS*605*	2, 2	AE000793
lp25	24,177	Linear	*Borrelia burgdorferi*	No IS		AE000785
lp28-1	26,921	Linear	*Borrelia burgdorferi*	No IS		AE000794
lp28-4	27,323	Linear	*Borrelia burgdorferi*	No IS		AE000789
lp28-3	28,601	Linear	*Borrelia burgdorferi*	IS*605*	1, 1	AE000784
lp28-2	29,766	Linear	*Borrelia burgdorferi*	IS*605*	1, 1	AE000786
lp36	36,849	Linear	*Borrelia burgdorferi*	IS*605*	1, 1	AE000788
lp38	38,829	Linear	*Borrelia burgdorferi*	IS*605*	1, 1	AE000787
lp54	53,561	Linear	*Borrelia burgdorferi*	No IS		AE000790
pSOL1[a]	192,060	Circular	*Clostridium acetobutylicum*	Tn*3*	1, 1	AE001438
MP1	177,466	Circular	*Deinococcus radiodurans*	IS*3*	1, 1	AE001826
				IS*4*	1, 7	
				IS*4*	1, 6	
				IS*630*	1, 4	
CP1	45,704	Circular	*Deinococcus radiodurans*	IS*4*	1, 3	AE001827
				IS*5*	1, 1	
				IS*630*	1, 1	
				ISNCY	2, 2	
R751	53,339	Circular	*Enterobacter aerogenes*	IS*110*	2, 2	U67194
pO157	92,077	Circular	*Escherichia coli*	IS*1*	1, 1p	AF074613
				IS*3*	4, 7 (5p)	
				IS*21*	1, 1p	
				IS*91*	1, 1p	
pB171	68,817	Circular	*Escherichia coli*	IS*1*	2, 2	AB024946
				IS*3*	1, 2	
				IS*4*	1, 1	
				IS*21*	1, 2	
				IS*30*	1, 1	
pNRC100	191,346	Circular	*Halobacterium* sp. strain NRC-1	IS*4*	4, 20	AF016485
				IS*5*	1, 2	
				ISNCY	3, 5	
pMRC01	60,232	Circular	*Lactococcus lactis*	IS*6*	1, 3 (1p)	AE001272
				IS*30*	1, 1	
pFZ1	11,014	Circular	*Methanobacterium thermoformicicum*	No IS cited		X68367
pFV1	13,514	Circular	*Methanobacterium thermoformicicum*	No IS cited		X68366
p16	16,550	Circular	*Methanococcus jannaschii*	No IS cited		L77119
p58	58,407	Circular	*Methanococcus jannaschii*	No IS cited		L77118

Continued on following page

Table 3. *Continued*

Plasmid name	Plasmid size (bp)	Plasmid type	Bacterial host	IS family	Members, copy no.[b]	Accession no.
pNGR234a	536,165	Circular	*Rhizobium* sp. strain NGR234	IS*3*	1, 3	AE00064 to AE00109
				IS*4*	1, 1	
				IS*5*	1, 1	
				IS*21*	1, 8 (1p)	
				IS*91*	1, 1	
				IS*110*	1, 6 (1p)	
				ISNCY	2, 5	
R27	180,461	Circular	*Salmonella typhi*	IS*1*	1, 3	AF250878
				IS*3*	1, 1	
				IS*4*	1, 2	
				IS*30*	1, 1	
				ISNCY	1, 1	
pNL1	184,457	Circular	*Sphingomonas aromaticivorans*	IS*630*	2, 2	AF079317
pSCL	11,696	Linear	*Streptomyces clavuligerus*	No IS cited		X54107
pJV1	11,143	Circular	*Streptomyces phaeochromogenes*	No IS cited		U23762
pYVe227	69,673	Circular	*Yersinia enterocolitica*	Tn*3*	1, 1	AF102990
pPCP1	9,610	Circular	*Yersinia pestis*	IS*21*	1, 1	AF053945
pCD1	70,504	Circular	*Yersinia pestis*	IS*3*	4, 4	AF053946
				IS*6*	1, 1	
				IS*21*	2, 3 (1p)	
				IS*256*	2, 4 (2p)	
				IS*605*	1, 2	
pMT1	100,984	Circular	*Yersinia pestis*	IS*3*	1, 1p	AF053947
				IS*21*	1, 2	
				IS*110*	1, 1	
				IS*256*	1, 1	
				IS*605*	1, 1	

[a] Elements analyzed by us; *p*, number of partial copies.

genes. The presence of ISs may therefore be underestimated using present protocols. For example, we have recently identified ISs in two apparently IS-free organisms, *Rickettsia prowazekii* (IS*256* family) and *Pyrococcus abyssi* (IS*605* complex), and have identified several ISs not previously annotated in the *Synechocystis* genome (IS*1* family) (235). In a preliminary analysis (not included in Table 2), we have also detected remnants of ISs in *Aeropyrum pernix* K1 (IS*21*, IS*481*, and IS*110*), *Buchnera* sp. strain APS (IS*21*, IS*3*, and IS*481*), *Chlamydia muridarum*, and *Chlamydia trachomatis* (IS*21*, IS*3*, and IS*481*).

The role of ISs in assembling clusters of genes with specialized functions, such as antibacterial activities, virulence or symbiotic functions, or forming new catabolic pathways, is often best illustrated by bacterial plasmids. This is clearly a result of their capacity for horizontal transmission, which facilitates the sequestration of these accessory functions. Bacterial plasmids can show extremely complex patterns of ISs. The *Rhizobium* symbiotic plasmid pNRG234a, the *Shigella* virulence plasmid pWR100 (40a, 384a), and the *Halobacterium* plasmid pNRC100 are good examples of IS-driven complexity. Those bacteria in which no IS has yet been detected are shown in Tables 2 and 3 but are not included in the text below. Bacteria whose genome sequences are known but not yet published are included in the text but not in the tables.

Eubacteria

Agrobacterium

Although the sequence of the *Agrobacterium tumefaciens* genome is not available, that of two Ti plasmids from *Agrobacterium* species is and has revealed several ISs (Table 3). Moreover, extensive mapping data from other Ti plasmids have identified many other ISs and have provided an idea of the role of ISs in the assembly of these plasmids (109, 207).

Bacillus

No IS elements have been identified in the genome of a common laboratory strain of *Bacillus subti-*

lis, strain 168. However, a member of the IS*4* family, IS4Bsu*1,* has now been identified in the industrial strain NAF4 (261). Initial examination of the unassembled *Bacillus stearothermophilus* genome showed that it carries a variety of IS elements. These include members of the IS*3,* IS*4,* IS*21,* IS*110,* IS*982,* and ISL*3* families. This genome also carries a relatively high number of IS*605* family members (10 copies). Another member of this genus, *Bacillus halodurans,* carries a very large number of elements, in particular over 40 copies of an ISL*3* family member. Although annotated versions of the *Bacillus anthracis* and *B. cereus* genomes were not available, that of two *B. anthracis* virulence plasmids, pXO1 and pXO2, was. They carry members of the IS*3,* IS*4,* and IS*5* families (Table 3). Furthermore, the toxin genes carried by pXO1 were found to be flanked by ISs (271).

Borrelia burgdorferi

Strain B31 of *B. burgdorferi* is at present the most complex eubacterial genome known. It contains a single linear chromosome and 21 plasmids (9 circular and 12 linear [49, 115]). Seven potential members of the IS*605*/IS*200* complex have been identified in this organism. All but one are plasmid borne. In addition, traces of members of the IS*5,* IS*21,* IS*481,* IS*605,* IS*630,* and ISL*3* families are also evident (not included in Tables 2 and 3).

Campylobacter jejuni

Only a single IS, an IS*200*/IS*605 orf*A homologue, has been reported for strain NCTC 11168.

Caulobacter crescentus

C. crescentus CB15 carries eight distinct ISs belonging to six different families.

Clostridium

The preliminary annotated version of the *Clostridium acetobutylicum* genome identifies a single IS related to ISH*4* (from the halobacterial plasmid pNRC100) and one related to IS*200*/IS*605 orf*A (see also references 38, 39, 65). In addition, the sequence of a large plasmid, pSOL1, is also available. It appears to carry one copy of a Tn*3*-related element.

Chlamydia and *Chlamydophila*

It is certainly worth noting that not a single IS has been reported in any of the seven genomes of this group which have been sequenced. However, a preliminary analysis has indicated the presence of several vestigial elements (data not shown). The absence of complete IS copies in this entire group is surprising and merits further investigation.

Deinococcus radiodurans

Strain R1 carries a multiplicity of ISs. Chromosome 1 carries several copies of an orphan IS very distantly related to IS*4,* four copies of IS*2621,* with two divergent *orf*s (one of which is also distantly related to the IS*4* family), and members of the IS*5,* IS*200,* IS*605,* and IS*630* families. It also contains a large number of partial IS sequences. Chromosome 2 carries a second orphan IS and an IS*2621* copy. The megaplasmid MP1 includes seven copies of IS*2621,* six copies of another element also distantly related to IS*4,* one IS*3* family member, and six copies of an IS*630* family member, while the small plasmid CP1 (Table 3) carries two orphans, one IS*5,* one IS*630,* and three copies of an IS*4* family member (O. Delfour, unpublished data).

Enterococcus faecalis

Strain V583 carries a limited number of ISs from at least six families. These include 18 copies of three members of the IS*256* family.

E. coli and other gram-negative enteric bacteria

In the case of probably the best-studied eubacterium, *E. coli* K12, about 42 IS copies have been located on the 4.64-Mb genome of strain MG1655. These, together with other "mobile elements" such as bacteriophage, represent nearly 2% of the genome and include members of some, but not all, of the major IS families: IS*1,* IS*3,* IS*4,* IS*5,* and IS*30* distributed over the entire genome (29). Two strains of *E. coli* O157:H7 (EDL933 and VT2-sakai) also carry numerous ISs. In both cases, a preponderance of IS*3* family members has been observed. Nondomesticated *E. coli* strains, such as different members of the ECOR collection (323) and a collection of clones isolated from old stabs of laboratory strains (260), can exhibit significant variations in IS copy number and distribution. While strains of the phylogenetically related *Shigella* species often exhibit extremely high numbers of ISs (267, 282), neither the number of intact copies nor their partition between chromosomal- and plasmid-borne copies is known.

ISs were first identified in plasmids from enteric bacteria before they were detected in chromosomes (41). Plasmid F, whose entire sequence is now available (accession no. AP001918), was one of the first found to carry IS copies. These are now known as IS*2*, IS*3*, and γδ and are clustered in a small region of the 94.5-kb plasmid. They are important for F integration into the host chromosome. An F-like virulence plasmid, pO157, from *E. coli* O157:H7 has also been determined to carry IS elements, several of which delineate boundaries of virulence segments (43).

Extensive and careful analysis of the sequence of another enteric virulence plasmid, the *Shigella flexneri* pathogenicity plasmid pWR100, has revealed a complex pattern. Here, approximately 80 kb of the 200-kb molecule is composed of material homologous to a set of 12 "standard" known enteric ISs. However, there are few intact IS sequences, and the majority consists of short regions dispersed over the entire length of the plasmid, generally interdigitated with fragments of other ISs. This represents a minimal estimate since other structures resembling ISs have also been detected. The mosaic pattern of ISs presumably reflects the scars of multiple rearrangements which have occurred in plasmid assembly (40a, 384a).

Haemophilus influenzae

The *H. influenzae* Rd genome, the first to be entirely sequenced, appears to be relatively poor in ISs. It is reported to carry five loci with similarities to Tpases. Three of these are copies of IS*1016*-V6, an IS which has yet to be allocated a family but which has a relative in *Neisseria meningitidis* (below).

H. pylori

Strain 26695 carries five copies of the complex IS*605* element and eight partial copies, of which at least two are *orf*A (IS200). It also carries two complete and two partial copies of the related IS*606*. These elements were first identified thanks to the *H. pylori* sequencing project. On the other hand, strain J99 does not carry a single complete IS*605*, although five partial copies are present. It also carries a single complete copy of IS*606* and four partial copies (4). No other ISs have been identified in these genomes. A second type of IS, IS*607*, which includes *orf*B but differs in the *orf*A frame, has been observed in many *H. pylori* strains but not in the two sequenced genomes (185). No other IS family members have been detected.

Lactococcus

The genome sequence of *Lactococcus lactis* subsp. *lactis* strain IL 1403 has been determined and shown to contain 42 copies of 6 different ISs (32), including 23 IS*3* family members and 15 IS*30* family members. IS*1077* is apparently associated with one or two copies of the IS*3* family member IS*904*; all are located in a 1.2-Mb segment of the 2.35-Mb genome. Another 1.5-Mb segment carries the 15 copies of IS*983*. IS*982*, present in 10 copies, has an apparently random distribution (31). This distribution may reflect acquisition of different chromosome segments at different times in the evolutionary history of the IL1403 strain. It is interesting to note that the bacteriocin-producing *Lactococcus* plasmid pMRC01 is partitioned into three approximately 20-kb functional domains separated by ISs (93).

Mycobacterium

M. tuberculosis H37Rv carries 56 *orf*s with homology to Tpases (63). In addition to previously described elements such as IS*6110* (16 copies, all identical to the IS*3* family member IS*987*), IS*1081* (6 copies; IS*256* family), and IS*1547*, members of the IS*5*, IS*21*, IS*30*, IS*110*, IS*256*, and ISL*3* families together with 6 ISs of the proposed IS*1535* family (see IS*200* and the "IS*605* Complex" above) were also identified from the complete genome sequence (124). The only striking aspect of the distribution of these elements on the chromosome is that a 0.6-Mb segment located in the region of the replication origin of the 4.4-Mb genome is devoid of ISs (124). The significance of this is at present unclear. A significantly lower number of ISs have been identified in the *M. tuberculosis* CDC1551 strain. The smaller *Mycobacterium leprae* TN genome also appears limited in the number of different IS families (4) present.

Mollicutes

No IS elements have been identified in the two sequenced *Mycoplasma* genomes, *M. genitalium* G-37 and *M. pneumoniae* M129, whose chromosomes are only 580,074 and 816,394 bp long, respectively. On the other hand, ISs have been identified in several other mycoplasmas, sometimes in high copy number. Using Southern analysis, it has been estimated that *M. mycoides* subsp. *mycoides* small colony type may carry up to 20 copies of IS*1296* and 30 copies of IS*1634* (387), although it is not clear how many of these are entire copies. Multiple IS copies have been identified in *M. pulmonis* (e.g., IS*1138* [26]), *M. hyorhinis* and *M. hypopneumoniae* (IS*1221* [105, 416]), *M. incognitus* (ISMi*1* [155]), and *M. fermentans* (IS*1630* [45]). The completed genome sequence of *M. pulmonis* revealed three complete copies of IS*1138* and five fragments. Four of the fragments and a single complete copy were located in the replication termination region as a duplication (56). Another mollicute, *Ureaplasma urealyticum,* appears to carry single copies of an IS*3* and an IS*256* family member.

The complete genome sequence of the plant pathogen *Spiroplasma citri* also exhibits a variety of IS elements. These include nine *orfs* highly homologous to the Tpase of the IS*30* family element ISSc*1* (found in the *Spiroplasma* phage SpV1) and six *orfs* with similarities to IS*3* family members (F. Laiqret, personal communication).

Neisseria

The genome sequences of two *N. meningitidis* strains, Z2491 belonging to serogroup A and MC58 belonging to serogroup B, reveal a relatively high number of ISs. The 2,148.5-kb genome of *N. meningitidis* Z2491 has been reported to carry 43 complete and partial copies of ISs related to IS*1106* and IS*1655* (276). They are generally located within or near one of the many arrays of repeat elements distributed around the genome. The 2,272-kb genome of *N. meningitidis* MC58 has been reported to contain 22 intact ISs and 29 vestigial sequences (368). These include representatives of the IS*3,* IS*5,* IS*30,* and IS*110* families.

Pseudomonas and *Ralstonia*

The genome of *P. aeruginosa* PAO1 is more than 10-fold larger than that of the smallest genome so far analyzed (*M. genitalium* G37 [Table 2]). However, only 11 chromosomal ISs have been reported. These appear limited to members of the IS*3* and IS*110* families (Table 2). It is worth noting that many ISs have been identified in different strains of *P. aeruginosa,* but these appear to be associated with plasmids or compound transposons. The *Ralstonia solanacedrum* genome (not included in Table 2) is composed of a 3.7-Mb chromosome and a smaller replicon, the megaplasmid, of 2.08 Mb. Both carry a preponderance of IS*3* (17 and 13 copies, respectively), IS*4* (4 and 11 copies), and IS*5* (16 and 23 copies) family members, with few copies of IS*21,* IS*256,* IS*630,* and ISL*3* family members (P. Siguier, personal communication).

Rhizobia

The single chromosome of *Mesorhizobium loti* MAFF303099 (previously *Rhizobium loti*) together with its two resident plasmids carries a large number of ISs. Of particular note is the very high level (31 copies) of IS*110* family members in this strain, but members of many other families are present. A similar distribution of ISs with some important differences can be seen in the genome sequence of *Sinorhizobium meliloti* (previously *Rhizobium meliloti*) (http://sequence.toulouse.inra.fr/meliloti.html). Here, the number of IS families represented is somewhat lower, and the number of IS*110* members is significantly lower. The sequence of an additional large (0.536-Mb) symbiotic plasmid, pNRG234a, from *Rhizobium* sp. strain NGR234 has been reported (111). Thirty-three *orfs* with similarities to known IS Tpases were identified. These correspond to at least 26 IS-like sequences representing 18% of the entire plasmid and include 8 members of the IS*21* family together with members of the IS*3,* IS*4,* IS*5,* IS*91,* and IS*110* families. Some are grouped in clusters separating functionally related genes, such as those involved in amino acid catabolism, nitrogen fixation, and replication and partition functions.

Rickettsia

Although no IS element was reported in the *Rickettsia prowazekii* Madrid E genome, subsequent analysis has revealed a single copy of an IS*256* member (235).

Staphylococcus

The three strains of *Staphylococcus aureus* (COL, N315, and Mu50) all carry moderate numbers of ISs. None of the strains appear to include members of the widespread IS*4* and IS*5* families. *S. aureus* COL was not included in Table 2 because no accession number was available at the time of this writing.

Streptococcus

The genome of *Streptococcus pyogenes* has been assembled. It carries members of the IS*5,* IS*30,* IS*110,* and IS*256* families, together with a single copy of an ISAs*1* family member. The genome carries a preponderance of IS*3* family members (11 copies). Examination of available data for the unassembled genome of *Streptococcus mutans* indicates that it carries members of the IS*3* (six copies), IS*4* (one copy), IS*605* (one copy), and ISL*3* (two copies) families.

Synechocystis

The 3.57-Mb genome of the cyanobacterium *Synechocystis* sp. strain PCC6803 exhibits nearly 100 putative *orfs* with similarities to Tpases (178). However, several of these abut each other and, on closer examination, appear to be part of the same Tpase gene. A preliminary survey has indicated that they may be resolved into about 70 elements which include 26 copies (of which 19 appear to be entire) of the previously described *Synechocystis* ISTcSa (50), related to IS*630* and the Tc/*mariner* family of eukaryotic elements; 22 copies of IS*4* family members with similarities to two previously described *Synechocystis* elements, IS*701* and IS*4*Sa, with about 7 complete copies of IS*4*Sa; 49 (including 33 partial) copies of IS*5* family members, of which 25 are similar to IS*5*Sa (6 complete copies); and 1 copy of each of IS*3*, IS*605*, and IS*256* family members (our unpublished results). Interestingly, as mentioned earlier (see IS*1* Family), we have uncovered three copies of an element of the IS*1* family, together with several partial copies, none of which were predicted from the listed *orf*s. This underlines the necessity for more in-depth analysis of IS sequences in many of the microbial genome sequences.

Thermotoga maritima

Interestingly, the only ISs so far detected in the genome of *T. maritima* MSB8 are members of the IS*605* family.

Vibrio cholerae El Tor N16961

The two chromosomes of *V. cholerae* include several copies of IS*3* family members and single copies of ISAs*1* and IS*481* family members. Although it also carries several copies of the IS*200*/IS*605*-related IS*1004*, there is no associated *orf*B (typical of the IS*605* complex).

Xylella fastidiosa

Two entire copies of an IS*607* element have been detected in this plant pathogen.

Yersinia pestis

The sequences of three *Y. pestis* plasmids have been determined: plasmid pCD1 (70,504 bp) (152, 279), which carries members of the IS*3*, IS*6*, IS*21*, IS*256*, and IS*605* families; pMT1 (100,984 bp) (216), which carries members of the IS*3*, IS*21*, IS*110*, IS*256*, and IS*605* families; and pPCP1 (961 kb; accession no. AL109969), which carries an IS*21* family member.

Archaea

Archaeoglobus fulgidus VC-10 (DSM4304)

Strain VC-10 has been reported to carry about 20 IS copies (199). Seven of these are members of the IS*481* family, two are members of the IS6 family, and three are members of the IS*630* family.

Halobacteria

The halobacteria have been shown to carry large numbers of IS-like repeated sequences (59), although many of these are located on plasmids (265, 280) (Table 3). The 2-Mb chromosome of *Halobacterium* sp. Strain NRC-1 harbors a moderate number of copies (14) of three different IS*4* family members and includes examples of the IS*5* family, IS*605*, and several as yet unclassified ISs.

The nucleotide sequences of two halobacterial plasmids have also been determined. Plasmid pNRC100 carries nearly 30 elements resembling ISs. These are distributed over the entire molecule and include eight different groups or families (Table 3), some of which are related to previously described archaeal ISs. They represent 35.8 kb or nearly 20% of the 0.19-Mb plasmid (265), and their localization with respect to the multiple potential replication and partition genes clearly indicates that they have been important in the assembly of pNRC100. It will be interesting to determine whether, like pWR100, this plasmid is also peppered with IS fragments.

Methanococcus jannaschii (DSM 2661)

Fifteen ISs were identified in *M. jannaschii*, several of which are related to IS*240*, an IS6 family member from *B. thuringiensis* (42).

Pyrococcus

The sequences of three *Pyrococcus* genomes have been determined. These are *Pyrococcus horikoshii*, *P. abyssi* (Table 2), and *Pyrococcus furiosus*. No IS had been reported for any of these, although we have recently detected two copies of an IS*605*-related element in *P. abyssi* (235), and 22 *orfs* related to Tpases of the IS6 family have been detected in the genome of *P. furiosus* (Y. Zivanovic, personal communication).

Sulfolobus

The genome of the crenarchaeote *Sulfolobus solfataricus* carries a large number of ISs. This was evident before the genome sequence became available.

Seven elements had been observed in a 0.156-Mb segment of the genome (336), and four, of which only one (ISC*1217*) corresponded to a previously known element, were identified during mutational studies (240). *S. solfataricus* also harbors 25 elements related to IS*605*, 61 IS*4* family members, and 29 IS*110* family members. A relative of IS*1* has also recently been detected in this genome (Berger, personal communication). At least one IS was also identified from a *Sulfolobus* sp. conjugative plasmid pNOB8 (339). The genomes of the *Crenarchaeota* are treated in detail by Reder et al. (see chapter 46).

Thermoplasma

A limited number of IS-like elements have been identified in an initial scan of the genome of the euryarchaeote *Thermoplasma acidophilum*. These include two members of the IS*256* family (related to IS*905*), an IS*5* family member (related to IS*1186*), an IS*481*-related sequence, one copy related to IS*605*/IS*606*, and two IS*607*-related elements. The *T. volcanium* genome includes members of IS*4*, IS*5*, IS*21*, IS*256*, IS*481*, IS*605*, IS*630*, and ISL*3*, with a total of approximately 29 full-length and partial copies.

EUKARYOTIC ISs

In addition to the large number of bacterial IS elements, a significant number of eukaryote elements have also been documented. The largest class described, and arguably the best-characterized, is the *mariner*/Tc family originally observed in *Drosophila mauritiana* (171) and *C. elegans* (97). These elements are structurally the closest to bacterial ISs. A second type of element that has also received extensive attention is the P element of *D. melanogaster*. A distinguishing character of this element is that Tpase binds to a subterminal region rather than to the terminal IRs. This feature is also a characteristic of two other groups of eukaryotic elements whose lengths exceed 4 kb: the CACTA family (restricted at present to plants) and the Ac family (with members in both plants and insects).

More recently, another class of eukaryotic elements, the helitrons, has been identified. These elements are similar to IS*91* (see above and chapter 37) and may, like IS*91*, transpose using a replicative rolling circle mechanism (178a). Two characteristics of several eukaryotic elements, which were thought to distinguish them from those in eubacteria and archaebacteria are the presence of large numbers of defective copies and the facility with which an intact copy in the same genome can provide functions in *trans* for their mobilization (for reviews, see references 15, 98, and 198). However, studies of certain eubacterial (61, 268) archaebacterial elements (151) and the accumulating data from bacterial genome projects suggest that multiple copies of inactive but potentially transactivated elements may not be restricted to eukaryotes.

The *mariner*/Tc Family

The *mariner* family (chapter 22 and reference 283) includes the Tc elements originally detected in *C. elegans* (97). These have been the subject of extensive analysis. Alignment of their Tpases (90) and their target specificity suggest that they are related to the IS*630* family of bacterial elements. Tc*1*/*3* and *mariner* show significant similarities. Moreover, the insertion specificity of both Tc*1* and Tc*3* is strict. Analysis of 204 Tc*1* and 166 Tc*3* insertions indicated that all occurred into a TA dinucleotide that was duplicated during the insertion event (384), a target choice that is identical to that of IS*630*.

Both Tc*1* and the related Tc*3* element carry a single Tpase gene, *tc1*A and *tc3*A, respectively, each interrupted by one intron at a different position. Tc*1* is 1610 bp long and carries terminal IR sequences of 54 bp, while Tc*3* is 2335 bp long with 462 bp IRs. The reason for this large difference in IR length is unknown. Tc1A and Tc3A also carry the DD(35)E motif (90) (Fig. 3). Like several bacterial Tpases, a sequence-specific DNA binding domain is located at the N-terminal end of both Tpases. It is followed by a region that overlaps the DD(35)E region and binds DNA nonspecifically. Also in the case of Tc1A, band shift assays, footprinting, and methylation interference studies have demonstrated a binding site between nucleotides 5 and 26 of the IR. Nucleotides 1 to 4 are not necessary for binding (390). This observation is consistent with the notion that the Tc*1* IR sequences are organized into a two-domain structure similar to that of several bacterial elements (Fig. 1). The binding site can be further subdivided into two distinct regions: one, closer to the tip, is contacted by the specific N-terminal Tpase domain, and the other is contacted by the nonspecific DNA binding domain (389). A leucine zipper motif has also been detected in the Tpases of certain members of this group (170). As in the IS*3* family, this may represent a multimerization domain although recent structural studies with a truncated Tpase derivative indicated that this region did not assume a coiled-coil motif (95).

Transposon circles have been detected in vivo (97, 298, 319), but their role as possible intermediates in transposition is not entirely clear. Tc*1*/*3* appear to transpose using a cut-and-paste mechanism. Induction of Tc3A in vivo results in excision of Tc*3* as a linear form. One interesting characteristic of these molecules is that while the cleavage at the 3′ ends of the element occurs precisely between the transposon and its flanking sequences, 5′ cleavage occurs two nucleotides within the transposon (383). The result of these cleavages is that two nucleotides of each transposon end remain behind in the donor site leaving a "footprint": TACATA (where the flanking TA dinucleotide is that generated by duplication during the original integration) (Fig. 2A).

The P Element

The P element of *D. melanogaster* (chapter 21) was discovered by its role in the syndrome known as hybrid dysgenesis (see, for example, reference 187). This syndrome is characterized by a loss of fertility in a cross between males that carry an autonomous P element and females that do not and is due to high-level P transposition in germline cells. The element has been used extensively in mutagenesis of its host organism using specifically tailored P derivatives. Like Tc*1*/*3* and *mariner,* it transposes by a cut-and-paste mechanism, and together with these, it is one of the best characterized eukaryotic elements. It is 2.9 kb long, generates 8-bp target repeats on insertion, and specifies an 87-kD Tpase produced from a single *orf,* which carries three introns. Transposition requires approximately 150 bp at each end including 31-bp terminal IRs (257). Tpase does not bind to these, however, but instead recognizes neighboring internal sites (181).

In addition to Tpase binding, two other characteristics distinguish P from the elements discussed above. Studies using an in vitro system have shown that the Tpase requires GTP as a cofactor although nonhydrolyzable analogues are also effective (182). The second characteristic concerns the nature of the double-strand Tpase-induced cleavages at each end. While 3′ cleavage occurs at the terminal nucleotide, 5′ cleavage occurs 17 bp within the element (10). This position within the terminal IRs is directly adjacent to the host factor binding site. The in vitro system has also revealed that both left and right ends are required for cleavage activity, suggesting that prior formation of a synaptic complex is required.

Other Families

CACTA family

Named according to the nucleotide sequence of their highly conserved terminal IRs, this group is currently restricted to plants and includes En/Spm and some Tam elements (Tam*1, 2, 4, 7, 8, 9*) (chapter 24) (121). Full-length members are large (>4 kb), carry nearly identical terminal IRs of 13 bp, and generate target duplications of 3 bp. Transposition requires both the left ~180 bp and the right ~300 bp. The En element produces two proteins by relatively complex differential splicing: TNPA (67 kDa) and TNPD (131 kDa). Both are required for transposition. TNPA binds a reiterated subterminal 12-bp motif but not the terminal IRs. The equivalents of TNPA from different members of this group do not appear to exhibit significant identities in primary sequence. This would be consistent with a role in binding to the element-specific reiterative repeats. Both DNA binding and multimerization domains have been localized for TNPA of En (376). On the other hand, the second and larger protein species show significant identity (45% between TNP2 of Tam*1* and TNPD) and for this reason it has been proposed that these bind the related terminal IRs of the elements and catalyze cleavage.

Ac

The organization of the Ac element of maize (198) is similar to that of Tam*3*, P, and hobo. It is 4565 bp long and carries 11-bp imperfect terminal IRs and numerous reiterated subterminal 6-bp motifs. Insertion results in an 8-bp target duplication. A single *orf* with four introns specifies a 112 kD Tpase. Approximately 240 bp are required at each end for efficient transposition and the frequency decreases as this is reduced to ~100 bp. Both left and right ends are required. Like that of P and En, the Tpase binds to the subterminal repeats probably in a cooperative manner using a domain located in the C-terminal end of the protein. It binds with significantly lower affinity to the terminal IRs using a domain located in the N-terminal end (11). The C-terminal 600 amino acids show homology with the putative Tpases of Tam*3* and hobo, family members from plant and insect species, respectively.

MuDR

This transposon, the mutator element of maize (15), is highly active and has been used extensively to "tag" various maize mutants. The autonomous element is approximately 5 kb long, carries ~200-bp terminal IRs, and generates 9-bp target repeats on insertion. It encodes two open reading frames transcribed in a convergent manner from promoters located in the terminal IRs and capable of specifying proteins of 207 (MURB) and 823 (MURA) amino acids. MURA is required for transposition, shows some similarities with the Tpases of the IS*256* family

(96), and is also capable of binding specifically to the terminal IRs of MuDR (14). Little is known about transposition of this family of elements at the molecular level (chapter 23).

CONCLUSIONS AND PERSPECTIVES

One of the major objectives of this chapter was to update information concerning bacterial insertion sequences that appeared in a previous review (234). Several additional family groups have been added or are emerging. It is clear, however, that this scheme will need further updating as additional ISs are isolated and characterized. Further examples are beginning to accumulate at an increasing rate as a result of efforts in bacterial genome sequencing.

Structuring the known ISs into 19 families underscores the immense diversity of these elements. However, one character exhibited by many of the IS families, which include the majority of known IS elements, is a Tpase that carries a real or potential DDE motif. This amino acid triad has been demonstrated to be mechanistically important in retroviral, phage Mu, Tn7, IS*10*, and Tc*1/3* transposition. It will therefore be of considerable interest to confirm the relevance of this catalytic triad in at least some additional families and to determine the spatial relationship of the residues within the three-dimensional structures. The published structure of a synaptic complex between two precleaved IS*50* ends and a Tpase dimer (chapter 18) has largely validated the active site interactions proposed from studies of the isolated catalytic core of several transposases (144). On the other hand, although the DDE motif appears to be a common theme in many of the elements described here, implying a shared chemistry, the detailed ways in which elements of each family assemble the individual nucleoprotein complexes for cleavage and strand transfer reactions can vary and result in important differences in overall mechanism. Thus certain elements with DDE chemistry generate replicon fusions exclusively (IS*6*, Tn*3*, or replicating phage Mu) while others are excised from their donor molecule prior to forming a synaptic complex with the target molecule (IS*10*, IS*50*, or IS*911*) and yet others appear to require the presence of a target site prior to donor strand cleavage (Tn7). Details of second-strand cleavage also vary (379). Further studies on selected members of the DDE superfamily will undoubtedly uncover additional combinations of reaction steps and extend the known range of DNA transactions that contribute to genome plasticity.

While members of the DDE family represent the majority of known IS elements, a significant fraction do not. These families include IS*1*, IS*91*, and IS*110*, which are at present being actively analyzed, and IS*1380* and ISAs*1*, which to our knowledge are not. From a mechanistic and evolutionary point of view, it is important to understand how these elements may be related to bacteriophage λ integration (IS*1*), single-strand phage and plasmid rolling-circle replication (IS*91*), and site-specific inversion systems (IS*110*).

Several new IS groups have emerged recently. One of the largest, composed principally of examples derived from the genome sequencing projects, is the IS*200*/IS*605* complex. These have yet received little attention (184, 185). In view of their wide distribution and the implication of an *orf*B homologue in survival of *S. enterica* serovar Typhimurium in Peyer's patches (352), a more detailed study of their phylogenetic relationships and transposition mechanism is required.

Finally, we have included here a short synopsis of the occurrence of members of different IS families in the available bacterial genome sequences. These data are far from complete but serve to illustrate the pervasivness of these short and compact segments of mobile DNA.

Acknowledgments. We acknowledge the cooperation of many colleagues in sharing unpublished results, including D. Berg (IS*605*/*607*), K. Derbyshire (IS*903*), B. Berger and D. Haas (IS*21*), D. Perkins-Baldwin, and A. C. Glasgow (IS*492*), S. D. Ehrlich, and A. Bolotin (*Lactococcus lactis* genome), F. Laigret (*Spiroplasma citri* genome), J. Frey (*Mycoplasma* genomes), and Y. Zivanovic (*Pyrococcus furiosus* genome). A. Gaeckle and F. Rodriguez designed and installed our website, J. Perochon and P. Azema developed and maintain the site, P. Seguier and O. Delfour maintain the database, M. Boschet tamed the bibliography, and J. Doris performed some of the Tpase alignments. R. Alazard, G. Duval-Valentin, L. Haren, C. Léonard, C. Loot, C. Normand, P. Rousseau, B. Ton-Hoang, C. Turlan, and C. Vermeiren provided invaluable suggestions during the course of this work.

Work in the Chandler laboratory was supported by the CNRS and grants from the Région Midi-Pyrenées, the Association pour la Recherche contre le Cancer, the Physique et Chimie du Vivant program of the CNRS, and the Microbiology and Infectious Disease Program (French Ministry of Education and Research). Work in the Mahillon laboratory was supported by grants from the FNRS, the Université catholique de Louvain, and the EU MECBAD network.

REFERENCES

1. **Ahmed, A.** 1986. Evidence for replicative transposition of Tn*5* and Tn*9*. *J. Mol. Biol.* **191:**75–84.
2. **Alam, J., J. M. Vrba, Y. Cai, J. A. Martin, L. J. Weislo, and S. E. Curtis.** 1991. Characterization of the IS*895* family of insertion sequences from the cyanobacterium *Anabaena* sp. strain PCC 7120. *J. Bacteriol.* **173:**5778–5783.
3. **Alloing, G., M. C. Trombe, and J. P. Claverys.** 1990. The *ami* locus of the gram-positive bacterium *Streptococcus pneumoniae* is similar to binding protein-dependent transport operons of gram-negative bacteria. *Mol. Microbiol.* **4:**633–644.
4. **Alm, R. A., and T. J. Trust.** 1999. Analysis of the genetic

diversity of *Helicobacter pylori:* the tale of two genomes. *J. Mol. Med.* 77:834–846.

5. **Andrake, M. D., and A. M. Skalka.** 1996. Retroviral integrase, putting the pieces together. *J. Biol. Chem.* **271:** 19633–19636.
6. **Arini, A., M. P. Keller, and W. Arber.** 1997. An antisense RNA in IS*30* regulates the translational expression of the transposase. *Biol. Chem.* **378:**1421–1431.
7. **Bachellier, S., J. M. Clement, M. Hofnung, and E. Gilson.** 1997. Bacterial interspersed mosaic elements (BIMEs) are a major source of sequence polymorphism in *Escherichia coli* intergenic regions including specific associations with a new insertion sequence. *Genetics* **145:**551–562.
8. **Bancroft, I., and C. P. Wolk.** 1989. Characterization of an insertion sequence (IS*891*) of novel structure from the cyanobacterium *Anabaena* sp. strain M-131. *J. Bacteriol.* **171:** 5949–5954.
9. **Bartlett, D. H., and M. Silverman.** 1989. Nucleotide sequence of IS*492,* a novel insertion sequence causing variation in extracellular polysaccharide production in the marine bacterium *Pseudomonas atlantica. J. Bacteriol.* **171:**1763–1766.
10. **Beall, E. L., and D. C. Rio.** 1997. Drosophila P-element transposase is a novel site-specific endonuclease. *Genes Dev.* **11:** 2137–2151.
11. **Becker, H. A., and R. Kunze.** 1997. Maize Activator transposase has a bipartite DNA binding domain that recognizes subterminal sequences and the terminal inverted repeats. *Mol. Gen. Genet.* **254:**219–230.
12. **Bender, J., and N. Kleckner.** 1992. IS*10* transposase mutations that specifically alter target site recognition. *EMBO J.* **11:**741–750.
13. **Bender, J., and N. Kleckner.** 1992. Tn*10* insertion specificity is strongly dependent upon sequences immediately adjacent to the target-site consensus sequence. *Proc. Natl. Acad. Sci. USA* **89:**7996–8000.
14. **Benito, M. I., and V. Walbot.** 1997. Characterization of the maize Mutator transposable element MURA transposase as a DNA-binding protein. *Mol. Cell. Biol.* **17:**5165–5175.
15. **Bennetzen, J. L.** 1996. The Mutator transposable element system of maize. *Curr. Top. Microbiol. Immunol.* **204:** 195–229.
16. **Berg, D. E.** 1983. Structural requirement for IS*50*-mediated gene transposition. *Proc. Natl. Acad. Sci. USA* **80:**792–796.
17. **Berg, D. E.** 1989. Transposon Tn*5,* p.185–210. *In* D. E. Berg and M. M. Howe (ed.), *Mobile DNA.* American Society for Microbiology, Washington, D.C.
18. **Berg, D. E., J. Davies, B. Allet, and J. D. Rochaix.** 1975. Transposition of R factor genes to bacteriophage lambda. *Proc. Natl. Acad. Sci. USA* **72:**3628–3632.
19. **Berg, D. E., and M. M. Howe (ed.).** 1989. *Mobile DNA.* American Society for Microbiology, Washington, D.C.
20. **Berger, B.** 2000. La séquence d'insertion IS*21:* la famille de cet élément bactérien, sa spécificité d'insertion et son utilisation pour une méthode de linker insertion mutagenesis in vitro. Ph.D. thesis. Université de Lausanne, Lausanne, Switzerland.
21. **Berger, B., and D. Haas.** 2001. Transposase and cointegrase: specialised transposition proteins of the bacterial insertion sequence IS*21. Cell. Mol. Life Sci.* **58:**403–419.
22. **Bernardi, F., and A. Bernardi.** 1987. Role of replication in IS*102*-mediated deletion formation. *Mol. Gen. Genet.* **209:** 453–457.
23. **Bernardi, F., and A. Bernardi.** 1988. Transcription of the target is required for IS*102* mediated deletions. *Mol. Gen. Genet.* **212:**265–270.
24. **Beuzon, C. R., and J. Casadesus.** 1997. Conserved structure of IS*200* elements in *Salmonella. Nucleic Acids Res.* **25:** 1355–1361.
25. **Bhasin, A., I. Y. Goryshin, and W. S. Reznikoff.** 1999. Hairpin formation in Tn*5* transposition. *J. Biol. Chem.* **274:** 37021–37029.
26. **Bhugra, B., and K. Dybvig.** 1993. Identification and characterization of IS*1138,* a transposable element from *Mycoplasma pulmonis* that belongs to the IS*3* family. *Mol. Microbiol.* **7:**577–584.
27. **Biel, S. W., and D. E. Berg.** 1984. Mechanism of IS*1* transposition in *E. coli:* choice between simple insertion and cointegration. *Genetics* **108:**319–330.
28. **Billington, S. J., M. Sinistaj, B. F. Cheetham, A. Ayres, E. K. Moses, M. E. Katz, and J. I. Rood.** 1996. Identification of a native *Dichelobacter nodosus* plasmid and implications for the evolution of the vap regions. *Gene* **172:**111–116.
29. **Blattner, F. R., G. Plunkett, C. A. Bloch, N. T. Perna, V. Burland, M. Riley, J. Collado-Vides, J. D. Glasner, C. K. Rode, G. F. Mayhew, J. Gregor, N. W. Davis, H. A. Kirkpatrick, M. A. Goeden, D. J. Rose, B. Mau, and Y. Shao.** 1997. The complete genome sequence of *Escherichia coli* K-12. *Science* **277:**1453–1474.
30. **Bolland, S., and N. Kleckner.** 1996. The three chemical steps of Tn*10*/IS*10* transposition involve repeated utilization of a single active site. *Cell* **84:**223–233.
31. **Bolotin, A., P. Winker, S. Mauger, O. Jaillon, K. Malarme, J. Weissenbach, S. D. Ehrlich, and A. Sorokin.** 2001. The complete genome sequence of the lactic acid bacterium Lactococcus lactis ssp. IL1403. *Genome Sci.* **11:**731–753.
32. **Bolotin, A., S. Mauger, K. Malarme, S. D. Ehrlich, and A. Sorokin.** 1999. Low-redundancy sequencing of the entire *Lactococcus lactis* IL*1403* genome. *Antonie Leeuwenhoek* **76:**27–76.
33. **Bortolini, M. R., L. R. Trabulsi, R. Keller, G. Frankel, and V. Sperandio.** 1999. Lack of expression of bundle-forming pili in some clinical isolates of enteropathogenic *Escherichia coli* (EPEC) is due to a conserved large deletion in the bfp operon. *FEMS Microbiol. Lett.* **179:**169–174.
34. **Boursaux-Eude, C., I. Saint Girons, and R. Zuerner.** 1995. IS*1500,* an IS*3*-like element from *Leptospira interrogans. Microbiology* **141:**2165–2173.
35. **Boyd, D. A., G. A. Peters, L. Ng, and M. R. Mulvey.** 2000. Partial characterization of a genomic island associated with the multidrug resistance region of *Salmonella enterica* typhymurium DT104. *FEMS Microbiol. Lett.* **189:**285–291.
36. **Braam, L. A., and W. S. Reznikoff.** 1998. Functional characterization of the Tn*5* transposase by limited proteolysis. *J. Biol. Chem.* **273:**10908–10913.
37. **Brown, P. O.** 1997. Integration, p. 161–203. *In* J. M. Coffin, S. H. Hughes, and H. E. Varmus (ed.), *Retroviruses.* Cold Spring Harbor Laboratory Press, Plainview, N.Y.
38. **Brynestad, S., L. A. Iwanejko, G. S. Stewart, and P. E. Granum.** 1994. A complex array of Hpr consensus DNA recognition sequences proximal to the enterotoxin gene in *Clostridium perfringens* type A. *Microbiology* **140:**97–104.
39. **Brynestad, S., B. Synstad, and P. E. Granum.** 1997. The *Clostridium perfringens* enterotoxin gene is on a transposable element in type A human food poisoning strains. *Microbiology* **143:**2109–2115.
40. **Buchrieser, C., R. Brosch, S. Bach, A. Guiyoule, and E. Carniel.** 1998. The high-pathogenicity island of *Yersinia pseudotuberculosis* can be inserted into any of the three chromosomal asn tRNA genes. *Mol. Microbiol.* **30:**965–978.

40a. **Buchreiser, C., P. Glaser, C. Rusniok, H. Nedjari, H. d'Hauteville, F. Kunst, P. Sansonetti, and C. Parsot.** 2000.

The virulence plasmid pWR100 and the repertoire of proteins secreted by the type III secretion apparatus of Shigella flexneri. *Mol. Microbiol.* **38**:760–771.
41. **Bukhari, A. I., J. A. Shapiro, and S. L. Adhya.** 1977. *DNA Insertion Elements, Plasmids, and Episomes.* Cold Spring Harbor Laboratory, Cold Spring Harbor, N.Y.
42. **Bult, C. J., O. White, G. J. Olsen, L. Zhou, R. D. Fleischmann, G. G. Sutton, J. A. Blake, L. M. FitzGerald, R. A. Clayton, J. D. Gocayne, A. R. Kerlavage, B. A. Dougherty, J. F. Tomb, M. D. Adams, C. I. Reich, R. Overbeek, E. F. Kirkness, K. G. Weinstock, J. M. Merrick, A. Glodek, J. L. Scott, N. M. Geoghagen, and J. C. Venter.** 1996. Complete genome sequence of the methanogenic archaeon, *Methanococcus jannaschii. Science* **273**:1058–1073.
43. **Burland, V., Y. Shao, N. T. Perna, G. Plunkett, H. J. Sofia, and F. R. Blattner.** 1998. The complete DNA sequence and analysis of the large virulence plasmid of *Escherichia coli* O157:H7. *Nucleic Acids Res.* **26**:4196–4204.
44. **Burnens, A. P., J. Stanley, R. Sack, P. Hunziker, I. Brodard, and J. Nicolet.** 1997. The flagellin N-methylase gene fliB and an adjacent serovar-specific IS*200* element in *Salmonella typhimurium. Microbiology* **143**:1539–1547.
45. **Calcutt, M. J., J. L. Lavrrar, and K. S. Wise.** 1999. IS*1630* of *Mycoplasma fermentans,* a novel IS*30*-type insertion element that targets and duplicates inverted repeats of variable length and sequence during insertion. *J. Bacteriol.* **181**:7597–7607.
46. **Calos, M. P., L. Johnsrud, and J. H. Miller.** 1978. DNA sequence at the integration sites of the insertion element IS*1*. *Cell* **13**:411–418.
47. **Casadesus, J., and J. R. Roth.** 1989. Transcriptional occlusion of transposon targets. *Mol. Gen. Genet.* **216**:204–209.
48. **Casjens, S., N. Palmer, R. van Vugt, W. M. Huang, B. Stevenson, P. Rosa, R. Lathigra, G. Sutton, J. Peterson, R. J. Dodson, D. Haft, E. Hickey, M. Gwinn, O. White, and C. M. Fraser.** 2000. A bacterial genome in flux: the twelve linear and nine circular extrachromosomal DNAs in an infectious isolate of the Lyme disease spirochete *Borrelia burgdorferi. Mol. Microbiol.* **35**:490–516.
49. **Caspers, P., B. Dalrymple, S. Iida, and W. Arber.** 1984. IS*30*, a new insertion sequence of *Escherichia coli* K12. *Mol. Gen. Genet.* **196**:68–73.
50. **Cassier-Chauvat, C., M. Poncelet, and F. Chauvat.** 1997. Three insertion sequences from the cyanobacterium *Synechocystis* PCC6803 support the occurrence of horizontal DNA transfer among bacteria. *Gene* **195**:257–266.
51. **Censini, S., C. Lange, Z. Xiang, J. E. Crabtree, P. Ghiara, M. Borodovsky, R. Rappuoli, and A. Covacci.** 1996. cag, a pathogenicity island of *Helicobacter pylori,* encodes type I-specific and disease-associated virulence factors. *Proc. Natl. Acad. Sci. USA* **93**:14648–14653.
52. **Chaconas, G., B. D. Lavoie, and M. A. Watson.** 1996. DNA transposition: jumping gene machine, some assembly required. *Curr. Biol.* **6**:817–820.
53. **Chalmers, R., A. Guhathakurta, H. Benjamin, and N. Kleckner.** 1998. IHF modulation of Tn*10* transposition: sensory transduction of supercoiling status via a proposed protein/DNA molecular spring. *Cell* **93**:897–908.
54. **Chalmers, R., S. Sewitz, K. Lipkow, and P. Crellin.** 2000. Complete nucleotide sequence of Tn*10*. *J. Bacteriol.* **182**:2970–2972.
55. **Chalmers, R. M., and N. Kleckner.** 1996. IS*10*/Tn*10* transposition efficiently accommodates diverse transposon end configurations. *EMBO J.* **15**:5112–5122.
56. **Chambaud, I., R. Heilig, S. Ferris, V. Barbe, D. Samson, F. Galisson, I. Moszer, K. Dybvig, H. Wroblenski, A. Viari, E. P. Rocha, and A. Blanchard.** 2001. The complete genome sequence of the murine respiratory pathogen Mycoplasma pulmonis. *Nucleic Acids Res.* **29**:2145–2153.
57. **Chandler, M., and O. Fayet.** 1993. Translational frameshifting in the control of transposition in bacteria. *Mol. Microbiol.* **7**:497–503.
58. Reference deleted.
59. **Charlebois, R. L., and W. F. Doolittle.** 1989. Transposable elements and genome structure in halobacteria, p. 297–307. *In* D. E. Berg and M. M. Howe (ed.), *Mobile DNA.* American Society for Microbiology, Washington, D.C.
60. **Charlier, D., J. Piette, and N. Glansdorff.** 1982. IS*3* can function as a mobile promoter in *Escherichia coli. Nucleic Acids Res.* **10**:5935–5948.
61. **Chen, Y., P. Braathen, C. Leonard, and J. Mahillon.** 1999. MIC231, a naturally occurring mobile insertion cassette from *Bacillus cereus. Mol. Microbiol.* **32**:657–668.
62. **Clampi, M. S., M. B. Schmid, and J. R. Roth.** 1982. Transposon Tn*10* provides a promoter for transcription of adjacent sequences. *Proc. Natl. Acad. Sci. USA* **79**:5016–5020.
63. **Cole, S. T., R. Brosch, J. Parkhill, T. Garnier, C. Churcher, D. Harris, S. V. Gordon, K. Eiglmeier, S. Gas, C. E. Barry, F. Tekaia, K. Badcock, D. Basham, D. Brown, T. Chillingworth, R. Connor, R. Davies, K. Devlin, T. Feltwell, S. Gentles, N. Hamlin, S. Holroyd, T. Hornsby, K. Jagels, and B. G. Barrell.** 1998. Deciphering the biology of *Mycobacterium tuberculosis* from the complete genome sequence. *Nature* **393**:537–544. (Erratum, **396**:190.)
64. **Comanducci, A., H. M. Dodd, and P. M. Bennett.** 1989. pUB2380: An R plasmid encoding a unique, natural one-ended transposition system, p. 305–311. *In* L. O. Butler, C. Harwood, and B. E. B. Moseley (ed.), *Genetic Transformation and Expression.* Intercept, Andover, Md.
65. **Cornillot, E., B. Saint-Joanis, G. Daube, S. Katayama, P. E. Granum, B. Canard, and S. T. Cole.** 1995. The enterotoxin gene (cpe) of *Clostridium perfringens* can be chromosomal or plasmid-borne. *Mol. Microbiol.* **15**:639–647.
66. **Craig, N. L.** 1995. Unity in transposition reactions. *Science* **270**:253–254.
67. **Craig, N. L.** 1996. Transposon Tn*7*. *Curr. Top. Microbiol. Immunol.* **204**:27–48.
68. **Craig, N. L.** 1997. Target site selection in transposition. *Annu. Rev. Biochem.* **66**:437–474.
69. **Craigie, R., M. Mizuuchi, and K. Mizuuchi.** 1984. Site-specific recognition of the bacteriophage Mu ends by the Mu A protein. *Cell* **39**:387–394.
70. **Dalrymple, B.** 1987. Novel rearrangements of IS*30* carrying plasmids leading to the reactivation of gene expression. *Mol. Gen. Genet.* **207**:413–420.
71. **Dalrymple, B., and W. Arber.** 1985. Promotion of RNA transcription on the insertion element IS*30* of *Escherichia coli* K12. *EMBO J.* **4**:2687–2693.
72. **Dalrymple, B., P. Caspers, and W. Arber.** 1984. Nucleotide sequence of the prokaryotic mobile genetic element IS*30*. *EMBO J.* **3**:2145–2149.
73. **Danilevich, V. N., and D. A. Kostiuchenko.** 1985. Immunity to repeated transposition of the insertion sequence IS*21*. *Mol. Biol. (Engl. Transl. Mol. Biol. Mosc.)* **19**:1242–1250.
74. **Davies, D. R., L. M. Braam, W. S. Reznikoff, and I. Rayment.** 1999. The three-dimensional structure of a Tn*5* transposase-related protein determined to 2.9-A resolution. *J. Biol. Chem.* **274**:11904–11913.
75. **Davies, D. R., I. Y. Goryshin, W. S. Reznikoff, and I. Rayment.** 2000. Three-dimensional structure of the Tn*5* synaptic complex transposition intermediate. *Science* **289**:77–85.

76. **Davis, M. A., R. W. Simons, and N. Kleckner.** 1985. Tn*10* protects itself at two levels from fortuitous activation by external promoters. *Cell* **43:**379–387.
77. **Debets-Ossenkopp, Y. J., R. G. Pot, D. J. van Westerloo, A. Goodwin, C. M. Vanderbroucke-Grauls, D. E. Berg. P. S. Hoffman, and J. G. Kusters.** 1999. Insertion of mini-IS*605* and deletion of adjacent sequences in the nitroreductase (rdxA) gene cause metronidazole resistance in *Helicobacter pylori* NCTC 11637. *Antimicrob. Agents Chemother.* **43:** 2657–2662.
78. **DeBoy, R. T., and N. L. Craig.** 2000. Target site selection by Tn7: attTn7 transcription and target activity. *J. Bacteriol.* **182:**3310–3313.
79. **de la Cruz, N. B., M. D. Weinreich, T. W. Wiegand, M. P. Krebs, and W. S. Reznikoff.** 1993. Characterization of the Tn*5* transposase and inhibitor proteins: a model for the inhibition of transposition. *J. Bacteriol.* **175:**6932–6938.
80. **Delecluse, A., C. Bourgouin, A. Klier, and G. Rapoport.** 1989. Nucleotide sequence and characterization of a new insertion element, IS*240*, from *Bacillus thuringiensis* israelensis. *Plasmid* **21:**71–78.
81. **De Meirsman, C., C. Van Soom, C. Verreth, A. Van Gool, and J. Vanderleyden.** 1990. Nucleotide sequence analysis of IS*427* and its target sites in *Agrobacterium tumefaciens* T37. *Plasmid* **24:**227–234.
82. **Demuth, D. R., Y. Duan, H. F. Jenkinson, R. McNab, S. Gil, and R. J. Lamont.** 1997. Interruption of the *Streptococcus gordonii* M5 sspA/sspB intergenic region by an insertion sequence related to IS*1167* of *Streptococcus pneumoniae. Microbiology* **143:**2047–2055.
83. **Derbyshire, K. M., and N. D. Grindley.** 1992. Binding of the IS*903* transposase to its inverted repeat in vitro. *EMBO J.* **11:**3449–3455.
84. **Derbyshire, K. M., and N. D. Grindley.** 1996. *cis* preference of the IS*903* transposase is mediated by a combination of transposase instability and inefficient translation. *Mol. Microbiol.* **21:**1261–1272.
85. **Derbyshire, K. M., L. Hwang, and N. D. Grindley.** 1987. Genetic analysis of the interaction of the insertion sequence IS*903* transposase with its terminal inverted repeats. *Proc. Natl. Acad. Sci. USA* **84:**8049–8053.
86. **Derbyshire, K. M., M. Kramer, and N. D. Grindley.** 1990. Role of instability in the *cis* action of the insertion sequence IS*903* transposase. *Proc. Natl. Acad. Sci. USA* **87:** 4048–4052.
87. **DeShazer, D., G. E. Wood, and R. L. Friedman.** 1994. Molecular characterization of catalase from *Bordetella pertussis:* identification of the *katA* promoter in an upstream insertion sequence. *Mol. Microbiol.* **14:**123–130.
88. **Devine, S. E., and J. D. Boeke.** 1996. Integration of the yeast retrotransposon Ty*1* is targeted to regions upstream of genes transcribed by RNA polymerase III. *Genes Dev.* **10:**620–633.
89. **Di, G. D., M. Peel, F. Fava, and R. C. Wyndham.** 1998. Structures of homologous composite transposons carrying *cbaABC* genes from Europe and North America. *Appl. Environ. Microbiol.* **64:**1940–1946.
90. **Doak, T. G., F. P. Doerder, C. L. Jahn, and G. Herrick.** 1994. A proposed superfamily of transposase genes: transposon-like elements in ciliated protozoa and a common "D35E" motif. *Proc. Natl. Acad. Sci. USA* **91:**942–946.
91. **Donadio, S., and M. J. Staver.** 1993. IS*1136*, an insertion element in the erythromycin gene cluster of *Saccharopolyspora erythraea. Gene* **126:**147–151.
92. **Doolittle, W. F., T. B. Kirkwood, and M. A. Dempster.** 1984. Selfish DNAs with self-restraint. *Nature* **307:**501–502.
93. **Dougherty, B. A., C. Hill, J. F. Weidman, D. R. Richardson, J. C. Venter, and R. P. Ross.** 1998. Sequence and analysis of the 60 kb conjugative, bacteriocin-producing plasmid pMRC01 from *Lactococcus lactis* DPC3147. *Mol. Microbiol.* **29:**1029–1038.
94. **Dumontier, S., P. Trieu-Cuot, and P. Berche.** 1998. Structural and functional characterization of IS*1358* from *Vibrio cholerae. J. Bacteriol.* **180:**6101–6106.
95. **Eijkelenboom, A. P., F. M. van den Ent, A. Vos, J. F. Doreleijers, K. Hard, T. D. Tullius, R. H. Plasterk, R. Kaptein, and R. Boelens.** 1997. The solution structure of the amino-terminal HHCC domain of HIV-2 integrase: a three-helix bundle stabilized by zinc. *Curr. Biol.* **7:**739–746.
96. **Eisen, J. A., M. I. Benito, and V. Walbot.** 1994. Sequence similarity of putative transposases links the maize Mutator autonomous element and a group of bacterial insertion sequences. *Nucleic Acids Res.* **22:**2634–2636.
97. **Emmons, S. W., L. Yesner, K. S. Ruan, and D. Katzenberg.** 1983. Evidence for a transposon in Caenorhabditis elegans. *Cell* **32:**55–65.
98. **Engels, W. R.** 1996. P elements in Drosophila. *Curr. Top. Microbiol. Immunol.* **204:**103–123.
99. **Escoubas, J. M., D. Lane, and M. Chandler.** 1994. Is the IS*1* transposase, InsAB′, the only IS*1*-encoded protein required for efficient transposition? *J. Bacteriol.* **176:**5864–5867.
100. **Escoubas, J. M., M. F. Prere, O. Fayet, I. Salvignol, D. Galas, D. Zerbib, and M. Chandler.** 1991. Translational control of transposition activity of the bacterial insertion sequence IS*1*. *EMBO J.* **10:**705–712.
101. **Esposito, D., and R. Craigie.** 1998. Sequence specificity of viral end DNA binding by HIV-1 integrase reveals critical regions for protein-DNA interaction. *EMBO J.* **17:** 5832–5843.
102. **Farabaugh, P. J.** 1997. *Programmed Alternative Reading of the Genetic Code.* R. G. Landes Company, Austin, Tex.
103. **Fayet, O., P. Ramond, P. Polard, M. F. Prere, and M. Chandler.** 1990. Functional similarities between retroviruses and the IS*3* family of bacterial insertion sequences? *Mol. Microbiol.* **4:**1771–1777.
104. **Ferat, J. L., M. Le Gouar, and F. Michel.** 1994. Multiple group II self-splicing introns in mobile DNA from *Escherichia coli. C. R. Acad. Sci. III* **317:**141–148.
105. **Ferrell, R. V., M. B. Heidari, K. S. Wise, and M. A. McIntosh.** 1989. A *Mycoplasma* genetic element resembling prokaryotic insertion sequences. *Mol. Microbiol.* **3:**957–967.
106. **Fiandt, M., W. Szybalski, and M. H. Malamy.** 1972. Polar mutations in *lac, gal* and phage lambda consist of a few IS-DNA sequences inserted with either orientation. *Mol. Gen. Genet.* **119:**223–231.
107. **Filippov, A. A., P. V. Oleinikov, V. L. Motin, O. A. Protsenko, and G. B. Smirnov.** 1995. Sequencing of two *Yersinia pestis* IS elements, IS*285* and IS*100*. *Contrib. Microbiol. Immunol.* **13:**306–309.
108. **Finnegan, D. J.** 1997. Transposable elements: how non-LTR retrotransposons do it. *Curr. Biol.* **7:**R245–R248.
109. **Fournier, P., F. Paulus, and L. Otten.** 1993. IS*870* requires a 5′-CTAG-3′ target sequence to generate the stop codon for its large ORF1. *J. Bacteriol.* **175:**3151–3160.
110. **Fraser, C. M., S. Casjens, W. M. Huang, G. G. Sutton, R. Clayton, R. Lathigra, O. White, K. A. Ketchum, R. Dodson, E. K. Hickey, M. Gwinn, B. Dougherty, J. F. Tomb, R. D. Fleischmann, D. Richardson, J. Peterson, A. R. Kerlavage, J. Quackenbush, S. Salzberg, M. Hanson, V. R. Van, N. Palmer, M. D. Adams, J. Gocayne, and J. C. Venter.** 1997. Genomic sequence of a Lyme disease spirochaete, *Borrelia burgdorferi. Nature* **390:**580–586.

111. **Freiberg, C., R. Fellay, A. Bairoch, W. J. Broughton, A. Rosenthal, and X. Perret.** 1997. Molecular basis of symbiosis between *Rhizobium* and legumes. *Nature* **387:**394–401.
112. **Galas, D. J., M. P. Calos, and J. H. Miller.** 1980. Sequence analysis of Tn*9* insertions in the *lacZ* gene. *J. Mol. Biol.* **144:** 19–41.
113. **Galas, D. J., and M. Chandler.** 1981. On the molecular mechanisms of transposition. *Proc. Natl. Acad. Sci. USA* **78:** 4858–4862.
114. **Galas, D. J., and M. Chandler.** 1982. Structure and stability of Tn*9*-mediated cointegrates. Evidence for two pathways of transposition. *J. Mol. Biol.* **154:**245–272.
115. **Galas, D. J., and M. Chandler.** 1989. Bacterial insertion sequences, p. 109–162. *In* D. E. Berg and M. M. Howe (ed.), *Mobile DNA*. American Society for Microbiology, Washington, D.C.
116. **Gamas, P., D. Galas, and M. Chandler.** 1985. DNA sequence at the end of IS*1* required for transposition. *Nature* **317:** 458–460.
117. **Gerlitz, M., O. Hrabak, and H. Schwab.** 1990. Partitioning of broad-host-range plasmid RP*4* is a complex system involving site-specific recombination. *J. Bacteriol.* **172:**6194–6203.
118. **Gerton, J. L., S. Ohgi, M. Olsen, J. DeRisi, and P. O. Brown.** 1998. Effects of mutations in residues near the active site of human immunodeficiency virus type 1 integrase on specific enzyme-substrate interactions. *J. Virol.* **72:**5046–5055.
119. **Gesteland, R. F., and J. F. Atkins.** 1996. Recoding: dynamic reprogramming of translation. *Annu. Rev. Biochem.* **65:** 741–768.
120. **Gibert, I., J. Barbe, and J. Casadesus.** 1990. Distribution of insertion sequence IS*200* in *Salmonella* and *Shigella*. *J. Gen. Microbiol.* **136:**2555–2560.
121. **Gierl, A.** 1996. The En/Spm transposable element of maize. *Curr. Top. Microbiol. Immunol.* **204:**145–159.
122. **Goldgur, Y., F. Dyda, A. B. Hickman, T. M. Jenkins, R. Craigie, and D. R. Davies.** 1998. Three new structures of the core domain of HIV-1 integrase: An active site that binds magnesium. *Proc. Natl. Acad. Sci. USA* **95:**9150–9154.
123. **Goosen, N., and P. van de Putte.** 1986. Role of Ner protein in bacteriophage Mu transposition. *J. Bacteriol.* **167:**503–507.
124. **Gordon, S. V., B. Heym, J. Parkhill, B. Barrell, and S. T. Cole.** 1999. New insertion sequences and a novel repeated sequence in the genome of *Mycobacterium tuberculosis* H37Rv. *Microbiology* **145:**881–892.
125. **Goryshin, I. Y., J. A. Miller, Y. V. Kil, V. A. Lanzov, and W. S. Reznikoff.** 1998. Tn*5*/IS*50* target recognition. *Proc. Natl. Acad. Sci. USA* **95:**10716–10721.
126. **Grindley, N. D.** 1978. IS*1* insertion generates duplication of a nine base pair sequence at its target site. *Cell* **13:**419–426.
127. **Grindley, N. D., and C. M. Joyce.** 1980. Genetic and DNA sequence analysis of the kanamycin resistance transposon Tn*903*. *Proc. Natl. Acad. Sci. USA* **77:**7176–7180.
128. **Grindley, N. D., and C. M. Joyce.** 1981. Analysis of the structure and function of the kanamycin-resistance transposon Tn*903*. *Cold Spring Harbor Symp. Quant. Biol.* **45**(1): 125–133.
129. **Gronlund, H., and K. Gerdes.** 1999. Toxin-antitoxin systems homologous with relBE of *Escherichia coli* plasmid P307 are ubiquitous in prokaryotes. *J. Mol. Biol.* **285:**1401–1415.
130. **Guedon, G., F. Bourgoin, M. Pebay, Y. Roussel, C. Colmin, J. M. Simonet, and B. Decaris.** 1995. Characterization and distribution of two insertion sequences, IS*1191* and iso-IS*981*, in *Streptococcus thermophilus:* does intergeneric transfer of insertion sequences occur in lactic acid bacteria co-cultures? *Mol. Microbiol.* **16:**69–78.
131. **Guilhot, C., B. Gicquel, J. Davies, and C. Martin.** 1992. Isolation and analysis of IS*6120*, a new insertion sequence from *Mycobacterium smegmatis*. *Mol. Microbiol.* **6:**107–113.
132. **Gustafson, C. E., S. Chu, and T. J. Trust.** 1994. Mutagenesis of the paracrystalline surface protein array of *Aeromonas salmonicida* by endogenous insertion elements. *J. Mol. Biol.* **237:**452–463.
133. **Haack, K. R.** 1995. The activity of IS*200* in *Salmonella typhimurium*. Ph.D. thesis. University of Utah, Salt Lake City.
134. **Haas, D., B. Berger, S. Schmid, T. Seitz, and C. Reimmann.** 1996. Insertion sequence IS*21:* related insertion sequence elements, transpositional mechanisms, and application to linker insertion mutagenesis, p. 238–249. *In* T. Nakazawa, K. Furukawa, D. Haas, and S. Silver (ed.), *Molecular Biology of Pseudomonads*. ASM Press, Washington, D.C.
135. **Hall, R. M., H. J. Brown, D. E. Brookes, and H. W. Stokes.** 1994. Integrons found in different locations have identical 5′ ends but variable 3′ ends. *J. Bacteriol.* **176:**6286–6294.
136. **Hallet, B.** 1993. Transposition et mecanismes de specificite de cible d'IS*231*A, une sequence d'insertion de *Bacillus thuringiensis*. Ph.D. thesis. Université catholique de Louvain, Louvain la Neuve, Belgium.
137. **Hallet, B., R. Rezsohazy, J. Mahillon, and J. Delcour.** 1994. IS*231*A insertion specificity: consensus sequence and DNA bending at the target site. *Mol. Microbiol.* **14:**131–139.
138. **Hallet, B., and D. J. Sherratt.** 1997. Transposition and site-specific recombination: adapting DNA cut-and-paste mechanisms to a variety of genetic rearrangements. *FEMS Microbiol. Rev.* **21:**157–178.
138a. **Han, C.-G., Y. Shiga, T. Tobe, C. Sasakawa, and E. Ohtsubo.** 2001. Structural and functional characterization of IS*679* and IS*66*-family elements. *J. Bacteriol.* **183:** 4296–4304.
139. **Hanai, R., and J. C. Wang.** 1993. The mechanism of sequence-specific DNA cleavage and strand transfer by phi X174 gene A* protein. *J. Biol. Chem.* **268:**23830–23836.
140. **Haniford, D. B., A. R. Chelouche, and N. Kleckner.** 1989. A specific class of IS*10* transposase mutants are blocked for target site interactions and promote formation of an excised transposon fragment. *Cell* **59:**385–394.
141. **Haren, L., M. Betermier, P. Polard, and M. Chandler.** 1997. IS*911*-mediated intramolecular transposition is naturally temperature sensitive. *Mol. Microbiol.* **25:**531–540.
142. **Haren, L., C. Normand, P. Polard, R. Alazard, and M. Chandler.** 2000. IS*911* transposition is regulated by protein-protein interactions via a leucine zipper motif. *J. Mol. Biol.* **296:** 757–768.
143. **Haren, L., P. Polard, B. Ton-Hoang, and M. Chandler.** 1998. Multiple oligomerisation domains in the IS*911* transposase: A leucine zipper motif is essential for activity. *J. Mol. Biol.* **283:**29–41.
144. **Haren, L., B. Ton-Hoang, and M. Chandler.** 1999. Integrating DNA: transposases and retroviral integrases. *Annu. Rev. Microbiol.* **53:**245–281.
145. **Henderson, D. J., D. F. Brolle, T. Kieser, R. E. Melton, and D. A. Hopwood.** 1990. Transposition of IS*117* (the *Streptomyces coelicolor* A 3 (2) mini-circle) to and from a cloned target site and into secondary chromosomal sites. *Mol. Gen. Genet.* **224:**65–71.
146. **Henderson, D. J., D. J. Lydiate, and D. A. Hopwood.** 1989. Structural and functional analysis of the mini-circle, a transposable element of *Streptomyces coelicolor* A3(2). *Mol. Microbiol.* **3:**1307–1318.
147. **Hernandez Perez, M., N. G. Fomukong, T. Hellyer, I. N.**

Brown, and J. W. Dale. 1994. Characterization of IS*1110*, a highly mobile genetic element from *Mycobacterium avium*. *Mol. Microbiol.* **12:**717–724.

148. **Hickman, A. B., Y. Li, S. V. Mathew, E. W. May, N. L. Craig, and F. Dyda.** 2000. Unexpected structural diversity in DNA recombination: the restriction endonuclease connection. *Mol. Cell* **5:**1025–1034.
149. **Hill, C. W., C. H. Sandt, and D. A. Vlazny.** 1994. Rhs elements of *Escherichia coli:* a family of genetic composites each encoding a large mosaic protein. *Mol. Microbiol.* **12:** 865–871.
150. **Hirsch, H. J., P. Starlinger, and P. Brachet.** 1972. Two kinds of insertions in bacterial genes. *Mol. Gen. Genet.* **119:** 191–206.
151. **Hofman, J. D., L. C. Schalkwyk, and W. F. Doolittle.** 1986. ISH*51:* a large, degenerate family of insertion sequence-like elements in the genome of the archaebacterium, *Halobacterium volcanii. Nucleic Acids Res.* **14:**6983–7000.
152. **Hu, P., J. Elliott, P. McCready, E. Skowronski, J. Garnes, A. Kobayashi, R. R. Brubaker, and E. Garcia.** 1998. Structural organization of virulence-associated plasmids of *Yersinia pestis. J. Bacteriol.* **180:**5192–5202.
153. **Hu, S. T., J. H. Hwang, L. C. Lee, C. H. Lee, P. L. Li, and Y. C. Hsieh.** 1994. Functional analysis of the 14 kDa protein of insertion sequence 2. *J. Mol. Biol.* **236:**503–513.
154. **Hu, S. T., and C. H. Lee.** 1988. Characterization of the transposon carrying the STII gene of enterotoxigenic *Escherichia coli. Mol. Gen. Genet.* **214:**490–495.
155. **Hu, W. S., R. Y. Wang, R. S. Liou, J. W. Shih, and S. C. Lo.** 1990. Identification of an insertion-sequence-like genetic element in the newly recognized human pathogen *Mycoplasma incognitus. Gene* **93:**67–72.
156. **Hu, W. Y., and K. M. Derbyshire.** 1998. Target choice and orientation preference of the insertion sequence IS*903*. *J. Bacteriol.* **180:**3039–3048.
157. **Hu, W.-Y., W. Thompson, C. E. Lawrence, and K. M. Derbyshire.** 2001. Anatomy of a preferred target site for the bacterial insertion sequence IS903. *J. Mol. Biol.* **306:**403–416.
158. **Huang, D. C., M. Novel, and G. Novel.** 1991. A transposon-like element on the lactose plasmid of *Lactococcus lactis* subsp. *lactis* Z270. *FEMS Microbiol. Lett.* **61:**101–106.
159. **Hubner, A., and W. Hendrickson.** 1997. A fusion promoter created by a new insertion sequence, IS*1490*, activates transcription of 2,4,5-trichlorophenoxyacetic acid catabolic genes in *Burkholderia cepacia* AC1100. *J. Bacteriol.* **179:** 2717–2723.
160. **Hubner, P., S. Iida, and W. Arber.** 1987. A transcriptional terminator sequence in the prokaryotic transposable element IS*1*. *Mol. Gen. Genet.* **206:**485–490.
161. **Huisman, O., P. R. Errada, L. Signon, and N. Kleckner.** 1989. Mutational analysis of IS*10*'s outside end. *EMBO J.* **8:** 2101–2109.
162. **Ichikawa, H., K. Ikeda, J. Amemura, and E. Ohtsubo.** 1990. Two domains in the terminal inverted-repeat sequence of transposon Tn*3*. *Gene* **86:**11–17.
163. **Ichikawa, H., K. Ikeda, W. L. Wishart, and E. Ohtsubo.** 1987. Specific binding of transposase to terminal inverted repeats of transposable element Tn*3*. *Proc. Natl. Acad. Sci. USA* **84:** 8220–8224.
164. **Ichikawa, H., and E. Ohtsubo.** 1990. In vitro transposition of transposon Tn*3*. *J. Biol. Chem.* **265:**18829–18832.
165. **Iida, S., R. Hiestand-Nauer, and W. Arber.** 1985. Transposable element IS*1* intrinsically generates target duplications of variable length. *Proc. Natl. Acad. Sci. USA* **82:**839–843.
166. **Ilyina, T. V., and E. V. Koonin.** 1992. Conserved sequence motifs in the initiator proteins for rolling circle DNA replication encoded by diverse replicons from eubacteria, eucaryotes and archaebacteria. *Nucleic Acids Res.* **20:**3279–3285.
167. **Isberg, R. R., A. L. Lazaar, and M. Syvanen.** 1982. Regulation of Tn*5* by the right-repeat proteins: control at the level of the transposition reaction? *Cell* **30:**883–892.
168. **Isberg, R. R., and M. Syvanen.** 1982. DNA gyrase is a host factor required for transposition of Tn*5*. *Cell* **30:**9–18.
169. **Ishiguro, N., and G. Sato.** 1988. Nucleotide sequence of insertion sequence IS*3411*, which flanks the citrate utilization determinant of transposon Tn*3411*. *J. Bacteriol.* **170:** 1902–1906.
170. **Ivics, Z., Z. Izsvak, A. Minter, and P. B. Hackett.** 1996. Identification of functional domains and evolution of Tc*1*-like transposable elements. *Proc. Natl. Acad. Sci. USA* **93:** 5008–5013.
171. **Jacobson, J. W., M. M. Medhora, and D. L. Hartl.** 1986. Molecular structure of a somatically unstable transposable element in Drosophila. *Proc. Natl. Acad. Sci. USA* **83:** 8684–8688.
172. **Jain, C., and N. Kleckner.** 1993. Preferential *cis* action of IS*10* transposase depends upon its mode of synthesis. *Mol. Microbiol.* **9:**249–260.
173. **Jakowec, M., P. Prentki, M. Chandler, and D. J. Galas.** 1988. Mutational analysis of the open reading frames in the transposable element IS*1*. *Genetics* **120:**47–55. (Erratum, **121:** 393, 1989.)
174. **Jenkins, T. M., D. Esposito, A. Engelman, and R. Craigie.** 1997. Critical contacts between HIV-1 integrase and viral DNA identified by structure-based analysis and photo-cross-linking. *EMBO J.* **16:**6849–6859.
175. **Johnson, R. C., and W. S. Reznikoff.** 1983. DNA sequences at the ends of transposon Tn*5* required for transposition. *Nature* **304:**280–282.
176. **Johnsrud, L.** 1979. DNA sequence of the transposable element IS*1*. *Mol. Gen. Genet.* **169:**213–218.
177. **Kallastu, A., R. Horak, and M. Kivisaar.** 1998. Identification and characterization of IS*1411*, a new insertion sequence which causes transcriptional activation of the phenol degradation genes in *Pseudomonas putida. J. Bacteriol.* **180:** 5306–5312.
178. **Kaneko, T., S. Sato, H. Kotani, A. Tanaka, E. Asamizu, Y. Nakamura, N. Miyajima, M. Hirosawa, M. Sugiura, S. Sasamoto, T. Kimura, T. Hosouchi, A. Matsuno, A. Muraki, N. Nakazaki, K. Naruo, S. Okumura, S. Shimpo, C. Takeuchi, T. Wada, A. Watanabe, M. Yamada, M. Yasuda, and S. Tabata.** 1996. Sequence analysis of the genome of the unicellular cyanobacterium *Synechocystis* sp. strain PCC6803. II. Sequence determination of the entire genome and assignment of potential protein-coding regions. *DNA Res.* **3:**109–136.

178a. **Kapitonov, V. V., and J. Jurka.** 2001. Rolling circle transposons in eukaryotes. *Proc. Natl. Acad. Sci. USA* **98:** 8714–8719.

179. **Kato, K., K. Ohtsuki, H. Mitsuda, T. Yomo, S. Negoro, and I. Urabe.** 1994. Insertion sequence IS*6100* on plasmid pOAD2, which degrades nylon oligomers. *J. Bacteriol.* **176:** 1197–1200.
180. **Katzman, M., J. P. Mack, A. M. Skalka, and J. Leis.** 1991. A covalent complex between retroviral integrase and nicked substrate DNA. *Proc. Natl. Acad. Sci. USA* **88:**4695–4699.
181. **Kaufman, P. D., R. F. Doll, and D. C. Rio.** 1989. Drosophila P element transposase recognizes internal P element DNA sequences. *Cell* **59:**359–371.
182. **Kaufman, P. D., and D. C. Rio.** 1992. P element transposition in vitro proceeds by a cut-and-paste mechanism and uses GTP as a cofactor. *Cell* **69:**27–39.

183. **Kennedy, A. K., A. Guhathakurta, N. Kleckner, and D. B. Haniford.** 1998. Tn*10* transposition via a DNA hairpin intermediate. *Cell* **95:**125–134.
184. **Kersulyte, D., N. S. Akopyants, S. W. Clifton, B. A. Roe, and D. E. Berg.** 1998. Novel sequence organization and insertion specificity of IS*605* and IS*606:* chimaeric transposable elements of *Helicobacter pylori. Gene* **223:**175–186.
185. **Kersulyte, D., A. K. Mukhopadhyay, M. Shirai, T. Nakazawa, and D. E. Berg.** 2000. Functional organization and insertion specificity of IS*607,* a chimeric element of *Helicobacter pylori. J. Bacteriol.* **182:**5300–5308.
186. **Khan, E., J. P. Mack, R. A. Katz, J. Kulkosky, and A. M. Skalka.** 1991. Retroviral integrase domains: DNA binding and the recognition of LTR sequences. *Nucleic Acids Res.* **19:**851–860. (Erratum, **19:**1358.)
187. **Kidwell, M. G.** 1984. Hybrid dysgenesis in *Drosophila melanogaster:* partial sterility associated with embryo lethality in the P-M system. *Genet. Res.* **44:**11–28.
188. **Kiss, J., and F. Olasz.** 1999. Formation and transposition of the covalently closed IS*30* circle: the relation between tandem dimers and monomeric circles. *Mol. Microbiol.* **34:**37–52.
189. **Kleckner, N., R. M. Chalmers, D. Kwon, J. Sakai, and S. Bolland.** 1996. Tn*10* and IS*10* transposition and chromosome rearrangements: mechanisms and regulation in vivo and in vitro, p. 49–82. *In* H. Saedler and A. Gierl (ed.), *Transposable Elements,* 204th ed. Springer, Heidelberg, Germany.
190. **Klenk, H. P., R. A. Clayton, J. F. Tomb, O. White, K. E. Nelson, K. A. Ketchum, R. J. Dodson, M. Gwinn, E. K. Hickey, J. D. Peterson, D. L. Richardson, A. R. Kerlavage, D. E. Graham, N. C. Kyrpides, R. D. Fleischmann, J. Quackenbush, N. H. Lee, G. G. Sutton, S. Gill, E. F. Kirkness, B. A. Dougherty, K. McKenney, M. D. Adams, B. Loftus, and J. C. Venter.** 1997. The complete genome sequence of the hyperthermophilic, sulphate-reducing archaeon *Archaeoglobus fulgidus. Nature* **390:**364–370. (Erratum, **394:**101, 1998.)
191. **Knoop, V., and A. Brennicke.** 1994. Evidence for a group II intron in *Escherichia coli* inserted into a highly conserved reading frame associated with mobile DNA sequences. *Nucleic Acids Res.* **22:**1167–1171.
192. **Krebs, M. P., and W. S. Reznikoff.** 1986. Transcriptional and translational initiation sites of IS*50.* Control of transposase and inhibitor expression. *J. Mol. Biol.* **192:**781–791.
193. **Kretschmer, P. J., and S. N. Cohen.** 1979. Effect of temperature on translocation frequency of the Tn*3* element. *J. Bacteriol.* **139:**515–519.
194. **Kruklitis, R., D. J. Welty, and H. Nakai.** 1996. ClpX protein of *Escherichia coli* activates bacteriophage Mu transposase in the strand transfer complex for initiation of Mu DNA synthesis. *EMBO J.* **15:**935–944.
195. **Kuan, C. T., S. K. Liu, and I. Tessman.** 1991. Excision and transposition of Tn*5* as an SOS activity in *Escherichia coli. Genetics* **128:**45–57.
196. **Kuan, C. T., and I. Tessman.** 1991. LexA protein of *Escherichia coli* represses expression of the Tn*5* transposase gene. *J. Bacteriol.* **173:**6406–6410.
197. **Kulkosky, J., K. S. Jones, R. A. Katz, J. P. Mack, and A. M. Skalka.** 1992. Residues critical for retroviral integrative recombination in a region that is highly conserved among retroviral/retrotransposon integrases and bacterial insertion sequence transposases. *Mol. Cell. Biol.* **12:**2331–2338.
198. **Kunze, R.** 1996. The maize transposable element activator (Ac). *Curr. Top. Microbiol. Immunol.* **204:**161–194.
199. **Kwon, D., R. M. Chalmers, and N. Kleckner.** 1995. Structural domains of IS*10* transposase and reconstitution of transposition activity from proteolytic fragments lacking an interdomain linker. *Proc. Natl. Acad. Sci. USA* **92:**8234–8238.
200. **Laachouch, J. E., L. Desmet, V. Geuskens, R. Grimaud, and A. Toussaint.** 1996. Bacteriophage Mu repressor as a target for the *Escherichia coli* ATP-dependent Clp protease. *EMBO J.* **15:**437–444.
201. **Labes, G., and R. Simon.** 1990. Isolation of DNA insertion elements from *Rhizobium meliloti* which are able to promote transcription of adjacent genes. *Plasmid* **24:**235–239.
202. **Lam, S., and J. R. Roth.** 1983. IS*200:* a Salmonella-specific insertion sequence. *Cell* **34:**951–960.
203. **Lampe, D. J., M. E. Churchill, and H. M. Robertson.** 1996. A purified *mariner* transposase is sufficient to mediate transposition in vitro. *EMBO J.* **15:**5470–5479.
204. **Lane, D., J. Cavaille, and M. Chandler.** 1994. Induction of the SOS response by IS*1* transposase. *J. Mol. Biol.* **242:**339–350.
205. **Lauf, U., C. Muller, and H. Herrmann.** 1999. Identification and characterisation of IS*1383,* a new insertion sequence isolated from *Pseudomonas putida* strain H. *FEMS Microbiol. Lett.* **170:**407–412.
206. **Lavoie, B. D., and G. Chaconas.** 1996. Transposition of phage Mu DNA. *Curr. Top. Microbiol. Immunol.* **204:** 83–102.
207. **Lawley, T. D., V. Burland, and D. E. Taylor.** 2000. Analysis of the complete nucleotide sequence of the tetracycline-resistance transposon Tn*10. Plasmid* **43:**235–239.
208. **Lederberg, E. M.** 1981. Plasmid reference center registry of transposon (Tn) allocations through July 1981. *Gene* **16:** 59–61.
209. **Leelaporn, A., N. Firth, M. E. Byrne, E. Roper, and R. A. Skurray.** 1994. Possible role of insertion sequence IS*257* in dissemination and expression of high- and low-level trimethoprim resistance in staphylococci. *Antimicrob. Agents Chemother.* **38:**2238–2244.
210. **Lei, G. S., and S. T. Hu.** 1997. Functional domains of the InsA protein of IS2. *J. Bacteriol.* **179:**6238–6243.
211. **Lenich, A. G., and A. C. Glasgow.** 1994. Amino acid sequence homology between Piv, an essential protein in site-specific DNA inversion in *Moraxella lacunata,* and transposases of an unusual family of insertion elements. *J. Bacteriol.* **176:** 4160–4164.
212. **Léonard, C., and J. Mahillon.** 1998. IS*231*A transposition: conservative versus replicative pathway. *Res. Microbiol.* **149:** 549–555.
213. **Levchenko, I., L. Luo, and T. A. Baker.** 1995. Disassembly of the Mu transposase tetramer by the ClpX chaperone. *Genes Dev.* **9:**2399–2408.
214. **Lewis, L. A., and N. D. Grindley.** 1997. Two abundant intramolecular transposition products, resulting from reactions initiated at a single end, suggest that IS2 transposes by an unconventional pathway. *Mol. Microbiol.* **25:**517–529.
215. **Lichens-Park, A., and M. Syvanen.** 1988. Cointegrate formation by IS*50* requires multiple donor molecules. *Mol. Gen. Genet.* **211:**244–251.
216. **Lindler, L. E., G. V. Plano, V. Burland, G. F. Mayhew, and F. R. Blattner.** 1998. Complete DNA sequence and detailed analysis of the *Yersinia pestis* KIM5 plasmid encoding murine toxin and capsular antigen. *Infect. Immun.* **66:**5731–5742.
217. **Liu, C. C., H. R. Wang, H. C. Chou, W. T. Chang, and J. Tu.** 1992. Analysis of the genes and gene products of Xanthomonas transposable elements ISXc5 and ISXc4. *Gene* **120:** 99–103.
218. **Lodge, J. K., and D. E. Berg.** 1990. Mutations that affect Tn*5*

insertion into pBR322: importance of local DNA supercoiling. *J. Bacteriol.* **172:**5956–5960.
219. **Lohe, A. R., D. T. Sullivan, and D. L. Hartl.** 1996. Subunit interactions in the mariner transposase. *Genetics* **144:** 1087–1095.
220. **Lopez de Felipe, F., C. Magni, D. de Mendoza, and P. Lopez.** 1996. Transcriptional activation of the citrate permease P gene of *Lactococcus lactis* biovar diacetylactis by an insertion sequence-like element present in plasmid pCIT264. *Mol. Gen. Genet.* **250:**428–436.
221. **Lundblad, V., and N. Kleckner.** 1985. Mismatch repair mutations of *Escherichia coli* K12 enhance transposon excision. *Genetics* **109:**3–19.
222. **Lundblad, V., A. F. Taylor, G. R. Smith, and N. Kleckner.** 1984. Unusual alleles of recB and recC stimulate excision of inverted repeat transposons Tn*10* and Tn*5*. *Proc. Natl. Acad. Sci. USA* **81:**824–828.
223. **Luthi, K., M. Moser, J. Ryser, and H. Weber.** 1990. Evidence for a role of translational frameshifting in the expression of transposition activity of the bacterial insertion element IS*1*. *Gene* **88:**15–20.
224. **Lyon, B. R., M. T. Gillespie, and R. A. Skurray.** 1987. Detection and characterization of IS*256*, an insertion sequence in *Staphylococcus aureus*. *J. Gen. Microbiol.* **133:**3031–3038.
225. **MacHattie, L. A., and J. B. Jackowski.** 1977. Physical structure and deletion effects of the chloramphenicol resistance element Tn*9* in phage lambda, p. 219–228. *In* A. I. Bukhari, J. A. Shapiro, and S. L. Adhya (ed.), *DNA Insertion Elements, Plasmids, and Episomes.* Cold Spring Harbor Laboratory, Cold Spring Harbor, N.Y.
226. **Machida, C., and Y. Machida.** 1987. Base substitutions in transposable element IS*1* cause DNA duplication of variable length at the target site for plasmid co-integration. *EMBO J.* **6:**1799–1803.
227. **Machida, C., and Y. Machida.** 1989. Regulation of IS*1* transposition by the *insA* gene product. *J. Mol. Biol.* **208:**567–574.
228. **Machida, C., Y. Machida, and E. Ohtsubo.** 1984. Both inverted repeat sequences located at the ends of IS*1* provide promoter functions. *J. Mol. Biol.* **177:**247–267.
229. **Machida, Y., C. Machida, and E. Ohtsubo.** 1984. Insertion element IS*1* encodes two structural genes required for its transposition. *J. Mol. Biol.* **177:**229–245.
230. **Machida, Y., C. Machida, H. Ohtsubo, and E. Ohtsubo.** 1982. Factors determining frequency of plasmid cointegration mediated by insertion sequence IS*1*. *Proc. Natl. Acad. Sci. USA* **79:**277–281.
231. **Maekawa, T., K. Yanagihara, and E. Ohtsubo.** 1996. A cell-free system of Tn*3* transposition and transposition immunity. *Genes Cells* **1:**1007–1016.
232. **Maekawa, T., J. Amemura-Maekawa, and E. Ohtsubo.** 1993. DNA binding domains in Tn*3* transposase. *Mol. Gen. Genet.* **236:**267–274.
233. **Maekawa, T., K. Yanagihara, and E. Ohtsubo.** 1996. Specific nicking at the 3′ ends of the terminal inverted repeat sequences in transposon Tn*3* by transposase and an *Escherichia coli* protein ACP. *Genes Cells* **1:**1017–1030.
234. **Mahillon, J., and M. Chandler.** 1998. Insertion sequences. *Microbiol. Mol. Biol. Rev.* **62:**725–774.
235. **Mahillon, J., C. Leonard, and M. Chandler.** 1999. IS elements as constituents of bacterial genomes. *Res. Microbiol.* **150:** 675–687.
236. **Mahillon, J., and D. Lereclus.** 1988. Structural and functional analysis of Tn*4430*: identification of an integrase-like protein involved in the co-integrate-resolution process. *EMBO J.* **7:** 1515–1526.
237. **Mahillon, J., R. Rezsohazy, B. Hallet, and J. Delcour.** 1994. IS*231* and other *Bacillus thuringiensis* transposable elements: a review. *Genetica* **93:**13–26.
238. **Makris, J. C., P. L. Nordmann, and W. S. Reznikoff.** 1988. Mutational analysis of insertion sequence 50 (IS*50*) and transposon 5 (Tn*5*) ends. *Proc. Natl. Acad. Sci. USA* **85:** 2224–2228.
239. **Martinez-Abarea, F., S. Zekri, and N. Toro.** 1998. Characterization and splicing in vivo of a *Sinorhizobium meliloti* group II intron associated with particular insertion sequences of the IS*630*-Tc*1*/IS*3* retroposon superfamily. *Mol. Microbiol.* **28:** 1295–1306.
240. **Martusewitsch, E., C. W. Sensen, and C. Schleper.** 2000. High spontaneous mutation rate in the hyperthermophilic archaeon *Sulfolobus solfataricus* is mediated by transposable elements. *J. Bacteriol.* **182:**2574–2581.
241. **Matsutani, S.** 1994. Genetic evidence for IS*1* transposition regulated by InsA and the delta InsA-B′-InsB species, which is generated by translation from two alternative internal initiation sites and frameshifting. *J. Mol. Biol.* **240:**52–65.
242. **Matsutani, S.** 1997. Genetic analyses of the interactions of the IS*1*-encoded proteins with the left end of IS*1* and its insertion hotspot. *J. Mol. Biol.* **267:**548–560.
243. **Matsutani, S., and E. Ohtsubo.** 1993. Distribution of the *Shigella sonnei* insertion elements in Enterobacteriaceae. *Gene* **127:**111–115.
244. **May, E. W., and N. L. Craig.** 1996. Switching from cut-and-paste to replicative Tn*7* transposition. *Science* **272:**401–404.
245. **McPheat, W. L., and T. McNally.** 1987. Isolation of a repeated DNA sequence from *Bordetella pertussis*. *J. Gen. Microbiol.* **133** (Pt. 2)**:**323–330.
246. **Mendiola, M. V., I. Bernales, and F. de la Cruz.** 1994. Differential roles of the transposon termini in IS*91* transposition. *Proc. Natl. Acad. Sci. USA* **91:**1922–1926.
247. **Mendiola, M. V., and F. de la Cruz.** 1989. Specificity of insertion of IS*91*, an insertion sequence present in alpha-haemolysin plasmids of *Escherichia coli*. *Mol. Microbiol.* **3:**979–984.
248. **Mendiola, M. V., and F. de la Cruz.** 1992. IS*91* transposase is related to the rolling-circle-type replication proteins of the pUB110 family of plasmids. *Nucleic Acids Res.* **20:**3521.
249. **Meyer, J., S. Iida, and W. Arber.** 1980. Does the insertion element IS*1* transpose preferentially into A + T-rich DNA segments? *Mol. Gen. Genet.* **178:**471–473.
250. **Mhammedi-Alaoui, A., M. Pato, M. J. Gama, and A. Toussaint.** 1994. A new component of bacteriophage Mu replicative transposition machinery: the *Escherichia coli* ClpX protein. *Mol. Microbiol.* **11:**1109–1116.
251. **Mizuuchi, K.** 1992. Polynucleotidyl transfer reactions in transpositional DNA recombination. *J. Biol. Chem.* **267:** 21273–21276.
252. **Mizuuchi, K.** 1992. Transpositional recombination: mechanistic insights from studies of mu and other elements. *Annu. Rev. Biochem.* **61:**1011–1051.
253. **Mizuuchi, M., and K. Mizuuchi.** 1993. Target site selection in transposition of phage Mu. *Cold Spring Harbor Symp. Quant. Biol.* **5884:**515–523.
254. **Mollet, B., S. Iida, and W. Arber.** 1985. Gene organization and target specificity of the prokaryotic mobile genetic element IS*26*. *Mol. Gen. Genet.* **201:**198–203.
255. **Morisato, D., J. C. Way, H. J. Kim, and N. Kleckner.** 1983. Tn*10* transposase acts preferentially on nearby transposon ends in vivo. *Cell* **32:**799–807.
256. **Muller, H. P., and H. E. Varmus.** 1994. DNA bending creates favored sites for retroviral integration: an explanation for

preferred insertion sites in nucleosomes. *EMBO J.* **13:** 4704–4714.
257. **Mullins, M. C., D. C. Rio, and G. M. Rubin.** 1989. *cis*-acting DNA sequence requirements for P-element transposition. *Genes Dev.* **3:**729–738.
258. **Murai, N., H. Kamata, Y. Nagashima, H. Yagisawa, and H. Hirata.** 1995. A novel insertion sequence (IS)-like element of the thermophilic bacterium PS3 promotes expression of the alanine carrier protein-encoding gene. *Gene* **163:**103–107.
259. **Murphy, E.** 1989. Transposable elements in gram-positive bacteria, p. 269–288. *In* D. E. Berg and M. M. Howe (ed.), *Mobile DNA*. American Society for Microbiology, Washington, D.C.
260. **Naas, T., M. Blot, W. M. Fitch, and W. Arber.** 1994. Insertion sequence-related genetic variation in resting *Escherichia coli* K-12. *Genetics* **136:**721–730.
261. **Nagai, T., L.-S. Phan Tran, Y. Inatsu, and Y. Itoh.** 2000. A new IS*4* family insertion sequence, IS*4Bsu1*, responsible for genetic instability of poly-γ-glutamic acid production in *Bacillus subtilis. J. Bacteriol.* **182:**2387–2392.
262. **Nagel, R., and A. Chan.** 2000. Enhanced Tn*10* and mini-Tn*10* precise excision in DNA replication mutants of *Escherichia coli* K12. *Mutat. Res.* **459:**275–284.
263. **Nakatsu, C., J. Ng, R. Singh, N. Straus, and C. Wyndham.** 1991. Chlorobenzoate catabolic transposon Tn*5271* is a composite class 1 element with flanking class II insertion sequences. *Proc. Natl. Acad. Sci. USA* **88:**8312–8316.
264. **Naumann, T. A., and W. S. Reznikoff.** 2000. Trans catalysis in Tn*5* transposition. *Proc. Natl. Acad. Sci. USA* **97:** 8944–8949.
265. **Ng, W. V., S. A. Ciufo, T. M. Smith, R. E. Bumgarner, D. Baskin, J. Faust, B. Hall, C. Loretz, J. Seto, J. Slagel, L. Hood, and S. DasSarma.** 1998. Snapshot of a large dynamic replicon in a halophilic archaeon: megaplasmid or minichromosome? *Genome Res.* **8:**1131–1141.
266. **Normand, C., G. Duval-Valentin, L. Haren, and M. Chandler.** 2001. The terminal inverted repeats of IS911: requirement for synaptic complex assembly and activity. *J. Mol. Biol.* **308:** 853–871.
267. **Nyman, K., K. Nakamura, H. Ohtsubo, and E. Ohtsubo.** 1981. Distribution of the insertion sequence IS*1* in gram-negative bacteria. *Nature* **289:**609–612.
268. **Oggioni, M. R., and J.-P. Claverys.** 1999. Repeated extragenic sequences in procaryotic genomes: a proposal for the origin and dynamics of the RUP element in *Streptococcus pneumoniae. Microbiology* **145:**2647–2653.
269. **Ohtsubo, H., K. Nyman, W. Doroszkiewicz, and E. Ohtsubo.** 1981. Multiple copies of iso-insertion sequences of IS*1* in *Shigella dysenteriae* chromosome. *Nature* **292:**640–643.
270. **Ohtsubo, H., and E. Ohtsubo.** 1978. Nucleotide sequence of an insertion element, IS*1*. *Proc. Natl. Acad. Sci. USA* **75:** 615–619.
271. **Okinaka, R., K. Cloud, O. Hampton, A. Hoffmaster, K. Hill, P. Keim, T. Koehler, G. Lamke, S. Kumano, J. Mahillon, D. Manter, Y. Martinez, D. Ricke, R. Svensson, and P. Jackson.** 1999. Sequence, assembly and analysis of pX01 and pX02. *J. Appl. Microbiol.* **87:**261–262.
272. **Olasz, F., T. Farkas, J. Kiss, A. Arini, and W. Arber.** 1997. Terminal inverted repeats of insertion sequence IS*30* serve as targets for transposition. *J. Bacteriol.* **179:**7551–7558.
273. **Olasz, F., R. Stalder, and W. Arber.** 1993. Formation of the tandem repeat (IS*30*)2 and its role in IS*30*-mediated transpositional DNA rearrangements. *Mol. Gen. Genet.* **239:** 177–187.
274. **Osawa, S., T. H. Jukes, K. Watanabe, and A. Muto.** 1992. Recent evidence for evolution of the genetic code. *Microbiol. Rev.* **56:**229–264.
275. **Otten, L., J. Canaday, J. C. Gerard, P. Fournier, P. Crouzet, and F. Paulus.** 1992. Evolution of agrobacteria and their Ti plasmids—a review. *Mol. Plant-Microbe Interact.* **5:** 279–287.
276. **Parkhill, J., M. Achtman, K. D. James, S. D. Bentley, C. Churcher, S. R. Klee, G. Morelli, D. Basham, D. Brown, T. Chillingworth, R. M. Davies, P. Davis, K. Devlin, T. Feltwell, N. Hamlin, S. Holroyd, K. Jagels, S. Leather, S. Moule, K. Mungall, M. A. Quail, M. A. Rajandream, K. M. Rutherford, M. Simmonds, J. Skelton, S. Whitehead, B. G. Spratt, and B. G. Barrell.** 2000. Complete DNA sequence of a serogroup A strain of *Neisseria meningitidis* Z2491. *Nature* **404:** 502–506.
277. **Pato, M. L., and M. Banerjee.** 1996. The Mu strong gyrase-binding site promotes efficient synapsis of the prophage termini. *Mol. Microbiol.* **22:**283–292.
278. **Perkins-Balding, D., G. Duval-Valentin, and A. C. Glasgow.** 1999. Excision of IS*492* requires flanking target sequences and results in circle formation in *Pseudoalteromonas atlantica. J. Bacteriol.* **181:**4937–4948.
279. **Perry, R. D., S. C. Straley, J. D. Fetherston, D. J. Rose, J. Gregor, and F. R. Blattner.** 1998. DNA sequencing and analysis of the low-Ca^{2+}-response plasmid pCD1 of *Yersinia pestis* KIM5. *Infect. Immun.* **66:**4611–4623.
280. **Pfeifer, F., and U. Blaseio.** 1990. Transposition burst of the ISH27 insertion element family in *Halobacterium halobium. Nucleic Acids Res.* **18:**6921–6925.
281. **Picardeau, M., T. J. Bull, and V. Vincent.** 1997. Identification and characterization of IS-like elements in *Mycobacterium gordonae. FEMS Microbiol. Lett.* **154:**95–102.
282. **Plasterk, R. H.** 1996. The Tc1/*mariner* transposon family. *Curr. Top. Microbiol. Immunol.* **204:**125–143.
283. **Plasterk, R. H., Z. Izsvak, and Z. Ivics.** 1999. Resident aliens: the Tc1/*mariner* superfamily of transposable elements. *Trends Genet.* **15:**326–332.
284. **Plikaytis, B. B., J. T. Crawford, and T. M. Shinnick.** 1998. IS*1549* from *Mycobacterium smegmatis* forms long direct repeats upon insertion. *J. Bacteriol.* **180:**1037–1043.
285. **Podglajen, I., J. Breuil, and E. Collatz.** 1994. Insertion of a novel DNA sequence, IS*1186*, upstream of the silent carbapenemase gene cfiA, promotes expression of carbapenem resistance in clinical isolates of *Bacteroides fragilis. Mol. Microbiol.* **12:**105–114.
286. **Podladchikova, O. N., G. G. Dikhanov, A. V. Rakin, and J. Heesemann.** 1994. Nucleotide sequence and structural organization of *Yersinia pestis* insertion sequence IS*100. FEMS Microbiol. Lett.* **121:**269–274.
287. **Polard, P., and M. Chandler.** 1995. Bacterial transposases and retroviral integrases. *Mol. Microbiol.* **15:**13–23.
288. **Polard, P., M. F. Prere, M. Chandler, and O. Fayet.** 1991. Programmed translational frameshifting and initiation at an AUU codon in gene expression of bacterial insertion sequence IS*911*. *J. Mol. Biol.* **222:**465–477.
289. **Polard, P., M. F. Prere, O. Fayet, and M. Chandler.** 1992. Transposase-induced excision and circularization of the bacterial insertion sequence IS*911*. *EMBO J.* **11:**5079–5090.
290. **Polard, P., L. Seroude, O. Fayet, M. F. Prere, and M. Chandler.** 1994. One-ended insertion of IS*911*. *J. Bacteriol.* **176:** 1192–1196.
291. **Polard, P., B. Ton-Hoang, L. Haren, M. Betermier, R. Walczak, and M. Chandler.** 1996. IS*911*-mediated transpositional recombination in vitro. *J. Mol. Biol.* **264:**68–81.
292. **Prentki, P., M. H. Pham, P. Gamas, M. Chandler, and D. J.**

Galas. 1987. Artificial transposable elements in the study of the ends of IS*1*. *Gene* **61**:91–101.

293. **Prentki, P., B. Teter, M. Chandler, and D. J. Galas.** 1986. Functional promoters created by the insertion of transposable element IS*1*. *J. Mol. Biol.* **191**:383–393.
294. **Pribil, P. A., and D. Haniford.** 2000. Substrate recognition and induced DNA deformation by transposase at the target-capture stage of Tn*10* transposition. *J. Mol. Biol.* **303:** 145–159.
295. **Prudhomme, M., C. Turlan, J.-P. Claverys, and M. Chandler.** 2002. Diversity of Tn*4001* transposition products: the flanking IS*256* elements can form tandem dimers and IS circles. *J. Bacteriol.* **184**:433–443.
296. **Pryciak, P. M., and H. E. Varmus.** 1992. Nucleosomes, DNA-binding proteins, and DNA sequence modulate retroviral integration target site selection. *Cell* **69**:769–780.
297. **Radice, A. D., B. Bugaj, D. H. Fitch, and S. W. Emmons.** 1994. Widespread occurrence of the Tc*1* transposon family: Tc*1*-like transposons from teleost fish. *Mol. Gen. Genet.* **244:** 606–612.
298. **Radice, A. D., and S. W. Emmons.** 1993. Extrachromosomal circular copies of the transposon Tc*1*. *Nucleic Acids Res.* **21:** 2663–2667.
299. **Rao, J. E., P. S. Miller, and N. L. Craig.** 2000. Recognition of triple-helical DNA structures by transposon Tn7. *Proc. Natl. Acad. Sci. USA* **97**:3936–3941.
300. **Read, H. A., S. S. Das, and S. R. Jaskunas.** 1980. Fate of donor insertion sequence IS*1* during transposition. *Proc. Natl. Acad. Sci. USA* **77**:2514–2518.
301. **Recchia, G. D., and R. M. Hall.** 1997. Origins of the mobile gene cassettes found in integrons. *Trends Microbiol.* **5:** 389–394.
302. **Reddy, M., and J. Gowrishankar.** 2000. Characterization of the *uup* locus and its role in transposon excisions and tandem repeat deletions in *Escherichia coli*. *J. Bacteriol.* **182:** 1978–1986.
303. **Reif, H. J., and H. Saedler.** 1974. IS*1* is involved in deletion formation in the *gal* region of *Escherichia coli* K12. *Mol. Gen. Genet.* **137**:17–28.
304. **Reimmann, C., and D. Haas.** 1987. Mode of replicon fusion mediated by the duplicated insertion sequence IS*21* in *Escherichia coli*. *Genetics* **115**:619–625.
305. **Reimmann, C., and D. Haas.** 1990. The istA gene of insertion sequence IS*21* is essential for cleavage at the inner 3′ ends of tandemly repeated IS*21* elements in vitro. *EMBO J.* **9:** 4055–4063.
306. **Reimmann, C., and D. Haas.** 1993. Mobilization of chromosomes and nonconjugative plasmids by cointegrative mechanisms, p. 137–188. *In* D. B. Clewell (ed.), *Bacterial Conjugation*. Plenum Press, New York, N.Y.
307. **Reimmann, C., R. Moore, S. Little, A. Savioz, N. S. Willetts, and D. Haas.** 1989. Genetic structure, function and regulation of the transposable element IS*21*. *Mol. Gen. Genet.* **215:** 416–424.
308. **Reimmann, C., M. Rella, and D. Haas.** 1988. Integration of replication-defective R68.45-like plasmids into the Pseudomonas aeruginosa chromosome. *J. Gen. Microbiol.* **134:** 1515–1523.
309. **Reynolds, A. E., J. Felton, and A. Wright.** 1981. Insertion of DNA activates the cryptic bgl operon in E. coli K12. *Nature* **293**:625–629.
310. **Reynolds, A. E., S. Mahadevan, S. F. LeGrice, and A. Wright.** 1986. Enhancement of bacterial gene expression by insertion elements or by mutation in a CAP-cAMP binding site. *J. Mol. Biol.* **191**:85–95.
311. **Reznikoff, W. S.** 1993. The Tn*5* transposon. *Annu. Rev. Microbiol.* **47**:945–963.
312. **Rezsohazy, R., B. Hallet, J. Delcour, and J. Mahillon.** 1993. The IS*4* family of insertion sequences: evidence for a conserved transposase motif. *Mol. Microbiol.* **9**:1283–1295.
313. **Rice, P., R. Craigie, and D. R. Davies.** 1996. Retroviral integrases and their cousins. *Curr. Opin. Struct. Biol.* **6**:76–83.
314. **Rice, P., and K. Mizuuchi.** 1995. Structure of the bacteriophage Mu transposase core: a common structural motif for DNA transposition and retroviral integration. *Cell* **82:** 209–220.
315. **Riess, G., B. W. Holloway, and A. Puhler.** 1980. R68.45, a plasmid with chromosome mobilizing ability (Cma) carries a tandem duplication. *Genet. Res.* **36**:99–109.
316. **Roberts, D., B. C. Hoopes, W. R. McClure, and N. Kleckner.** 1985. IS10 transposition is regulated by DNA adenine methylation. *Cell* **43**:117–130.
317. **Roberts, D., and N. Kleckner.** 1988. Tn10 transposition promotes RecA-dependent induction of a lambda prophage. *Proc. Natl. Acad. Sci. USA* **85**:6037–6041.
318. **Robertson, H. M., and D. J. Lampe.** 1995. Recent horizontal transfer of a mariner transposable element among and between Diptera and Neuroptera. *Mol. Biol. Evol.* **12**:850–862.
319. **Rose, A. M., and T. P. Snutch.** 1984. Isolation of the closed circular form of the transposable element Tc1 in Caenorhabditis elegans. *Nature* **311**:485–486.
320. **Rowland, S. J., D. J. Sherratt, W. M. Stark, and M. R. Boocock.** 1995. Tn*552* transposase purification and in vitro activities. *EMBO J.* **14**:196–205.
321. **Sakai, J. S., N. Kleckner, X. Yang, and A. Guhathakurta.** 2000. Tn10 transpososome assembly involves a folded intermediate that must be unfolded for target capture and strand transfer. *EMBO J.* **19**:776–785.
322. **Sasakawa, C., Y. Uno, and M. Yoshikawa.** 1981. The requirement for both DNA polymerase and 5′ to 3′ exonuclease activities of DNA polymerase I during Tn*5* transposition. *Mol. Gen. Genet.* **182**:19–24.
323. **Sawyer, S. A., D. E. Dykhuizen, R. F. DuBose, L. Green, T. Mutangadura-Mhlanga, D. F. Wolczyk, and D. L. Hartl.** 1987. Distribution and abundance of insertion sequences among natural isolates of Escherichia coli. *Genetics* **115:** 51–63.
324. **Schmid, S., T. Seitz, and D. Haas.** 1998. Cointegrase, a naturally occurring, truncated form of IS21 transposase, catalyzes replicon fusion rather than simple insertion of IS*21*. *J. Mol. Biol.* **282**:571–583.
325. **Schneiker, S., B. Kosier, A. Puhler, and W. Selbitschka.** 1999. The Sinorhizobium meliloti insertion sequence (IS) element ISRm14 is related to a previously unrecognized IS element located adjacent to the Escherichia coli locus of enterocyte effacement (LEE) pathogenicity island. *Curr. Microbiol.* **39:** 274–281.
326. **Schnetz, K., and B. Rak.** 1992. IS*5*: a mobile enhancer of transcription in *Escherichia coli*. *Proc. Natl. Acad. Sci. USA* **89**:1244–1248.
327. **Schwartz, E., M. Kroger, and B. Rak.** 1988. IS150: distribution, nucleotide sequence and phylogenetic relationships of a new E. coli insertion element. *Nucleic Acids Res.* **16:** 6789–6802.
328. **Scordilis, G. E., H. Ree, and T. G. Lessie.** 1987. Identification of transposable elements which activate gene expression in *Pseudomonas cepacia*. *J. Bacteriol.* **169**:8–13.
329. **Scott, J. R., and G. G. Churchward.** 1995. Conjugative transposition. *Annu. Rev. Microbiol.* **49**:367–397.
330. **Sekine, Y., K. Aihara, and E. Ohtsubo.** 1999. Linearization

and transposition of circular molecules of insertion sequence IS3. *J. Mol. Biol.* **294:**21–34.

331. **Sekine, Y., N. Eisaki, K. Kobayashi, and E. Ohtsubo.** 1997. Isolation and characterization of IS1 circles. *Gene* **191:** 183–190.
332. **Sekine, Y., N. Eisaki, and E. Ohtsubo.** 1994. Translational control in production of transposase and in transposition of insertion sequence IS3. *J. Mol. Biol.* **235:**1406–1420.
333. **Sekine, Y., and E. Ohtsubo.** 1989. Frameshifting is required for production of the transposase encoded by insertion sequence 1. *Proc. Natl. Acad. Sci. USA* **86:**4609–4613.
334. **Sengstag, C., and W. Arber.** 1987. A cloned DNA fragment from bacteriophage P1 enhances IS2 insertion. *Mol. Gen. Genet.* **206:**344–351.
335. **Sengstag, C., S. Iida, R. Hiestand-Nauer, and W. Arber.** 1986. Terminal inverted repeats of prokaryotic transposable element IS186 which can generate duplications of variable length at an identical target sequence. *Gene* **49:**153–156.
336. **Sensen, C. W., H. P. Klenk, R. K. Singh, G. Allard, C. C. Chan, Q. Y. Liu, S. L. Penny, F. Young, M. E. Schenk, T. Gaasterland, W. F. Doolittle, M. A. Ragan, and R. L. Charlebois.** 1996. Organizational characteristics and information content of an archaeal genome: 156 kb of sequence from Sulfolobus solfataricus P2. *Mol. Microbiol.* **22:**175–191.
337. **Serre, M. C., C. Turlan, M. Bortolin, and M. Chandler.** 1995. Mutagenesis of the IS*1* transposase: importance of a His-Arg-Tyr triad for activity. *J. Bacteriol.* **177:**5070–5077.
338. **Sharpe, P. L., and N. L. Craig.** 1998. Host proteins can stimulate Tn7 transposition: a novel role for the ribosomal protein L29 and the acyl carrier protein. *EMBO J.* **17:**5822–5831.
339. **She, Q., H. Phan, R. A. Garrett, S. V. Albers, K. M. Stedman, and W. Zillig.** 1998. Genetic profile of pNOB8 from Sulfolobus: the first conjugative plasmid from an archaeon. *Extremophiles* **2:**417–425.
340. **Sherratt, D.** 1989. Tn*3* and related transposable elements: site-specific recombination and transposition, p. 163–184. *In* D. E. Berg and M. M. Howe (ed.), *Mobile DNA*. American Society for Microbiology, Washington, D.C.
341. **Shiga, Y., Y. Sekine, and E. Ohtsubo.** 1999. Transposition of IS1 circles. *Genes Cells* **4:**551–561.
342. **Signon, L., and N. Kleckner.** 1995. Negative and positive regulation of Tn10/IS10-promoted recombination by IHF: two distinguishable processes inhibit transposition off of multicopy plasmid replicons and activate chromosomal events that favor evolution of new transposons. *Genes Dev.* **9:** 1123–1136.
343. **Simpson, A. E., R. A. Skurray, and N. Firth.** 2000. An IS*257*-derived hybrid promoter directs transcription of a *tetA*(K) tetracycline resistance gene in the *Staphylococcus aureus* chromosomal *mec* region. *J. Bacteriol.* **182:**3345–3352.
344. **Skalka, A. M.** 1993. Retroviral DNA integration: lessons for transposon shuffling. *Gene* **135:**175–182.
345. **Smith, B., and P. Dyson.** 1995. Inducible transposition in *Streptomyces lividans* of insertion sequence IS*6100* from *Mycobacterium fortuitum*. *Mol. Microbiol.* **18:**933–941.
346. **Sneath, P. H. A., and R. R. Sokal.** 1973. *Numerical Taxonomy. The Principles and Practice of Numerical Classification.* W. H. Freeman and Company, San Francisco, Calif.
347. **So, M., F. Heffron, and B. J. McCarthy.** 1979. The *Escherichia coli* gene encoding heat stable toxin is a bacterial transposon flanked by inverted repeats of IS*1*. *Nature* **277:** 453–456.
348. **Solinas, F., A. M. Marconi, M. Ruzzi, and E. Zennaro.** 1995. Characterization and sequence of a novel insertion sequence, IS*1162*, from *Pseudomonas fluorescens*. *Gene* **155:**77–82.
349. **Spellerberg, B., S. Martin, C. Franken, R. Berner, and R. Lutticken.** 2000. Identification of a novel insertion sequence element in *Streptococcus agalactiae*. *Gene* **241:**51–56.
350. **Spielmann-Ryser, J., M. Moser, P. Kast, and H. Weber.** 1991. Factors determining the frequency of plasmid cointegrate formation mediated by insertion sequence IS*3* from *Escherichia coli*. *Mol. Gen. Genet.* **226:**441–448.
351. **Stalder, R., P. Caspers, F. Olasz, and W. Arber.** 1990. The N-terminal domain of the insertion sequence 30 transposase interacts specifically with the terminal inverted repeats of the element. *J. Biol. Chem.* **265:**3757–3762.
352. **Stanley, T. L., C. D. Ellermeier, and J. M. Slauch.** 2000. Tissue-specific gene expression identifies a gene in the lysogenic phage gifsy-1 that affects *Salmonella enterica* serovar Typhimurium survival in Peyer's patches. *J. Bacteriol.* **182:** 4406–4413.
353. **Sternglanz, R., S. DiNardo, K. A. Voelkel, Y. Nishimura, Y. Hirota, K. Becherer, L. Zumstein, and J. C. Wang.** 1981. Mutations in the gene coding for *Escherichia coli* DNA topoisomerase I affect transcription and transposition. *Proc. Natl. Acad. Sci. USA* **78:**2747–2751.
354. **Stibitz, S., and J. E. Davies.** 1987. Tn*602:* a naturally occurring relative of Tn*903* with direct repeats. *Plasmid* **17:** 202–209.
355. **Stroeher, U. H., K. E. Jedani, B. K. Dredge, R. Morona, M. H. Brown, L. E. Karageorgos, M. J. Albert, and P. A. Manning.** 1995. Genetic rearrangements in the rfb regions of *Vibrio cholerae* O1 and O139. *Proc. Natl. Acad. Sci. USA* **92:** 10374–10378.
356. **Sundin, G. W., D. E. Monks, and C. L. Bender.** 1995. Distribution of the streptomycin-resistance transposon Tn*5393* among phylloplane and soil bacteria from managed agricultural habitats. *Can. J. Microbiol.* **41:**792–799.
357. **Swenson, D. L., N. O. Bukanov, D. E. Berg, and R. A. Welch.** 1996. Two pathogenicity islands in uropathogenic *Escherichia coli* J96: cosmid cloning and sample sequencing. *Infect. Immun.* **64:**3736–3743.
358. **Syvanen, M., J. D. Hopkins, and M. Clements.** 1982. A new class of mutants in DNA polymerase I that affects gene transposition. *J. Mol. Biol.* **158:**203–212.
359. **Szeverenyi, I., T. Bodoky, and F. Olasz.** 1996. Isolation, characterization and transposition of an (IS2)2 intermediate. *Mol. Gen. Genet.* **251:**281–289.
360. **Takemura, H., S. Horinouchi, and T. Beppu.** 1991. Novel insertion sequence IS*1380* from *Acetobacter pasteurianus* is involved in loss of ethanol-oxidizing ability. *J. Bacteriol.* **173:** 7070–7076.
361. **Tamaru, Y., and R. H. Doi.** 2000. The *engL* gene cluster of *Clostridium cellulovorans* contains a gene for cellulosomal ManA. *J. Bacteriol.* **182:**244–247.
362. **Tauch, A., Z. Zheng, A. Puhler, and J. Kalinowski.** 1998. *Corynebacterium striatum* chloramphenicol resistance transposon Tn*5564:* genetic organization and transposition in *Corynebacterium glutamicum*. *Plasmid* **40:**126–139.
363. **Tavakoli, N. P., and K. M. Derbyshire.** 1999. IS*903* transposase mutants that suppress defective inverted repeats. *Mol. Microbiol.* **31:**1183–1195.
364. **Tavakoli, N. P., and K. M. Derbyshire.** 2001. Tipping the balance between replicative and simple transposition. *EMBO J.* **20:**2923–2930.
365. **Tavakoli, N. P., J. DeVost, and K. M. Derbyshire.** 1997. Defining functional regions of the IS*903* transposase. *J. Mol. Biol.* **274:**491–504.
366. **Tenzen, T., S. Matsutani, and E. Ohtsubo.** 1990. Site-specific transposition of insertion sequence IS*630*. *J. Bacteriol.* **172:** 3830–3836.
367. **Tenzen, T., and E. Ohtsubo.** 1991. Preferential transposition

of an IS*630*-associated composite transposon to TA in the 5′-CTAG-3′ sequence. *J. Bacteriol.* **173:**6207–6212.

368. **Tettelin, H., N. J. Saunders, J. Heidelberg, A. C. Jeffries, K. E. Nelson, J. A. Eisen, K. A. Ketchum, D. W. Hood, J. F. Peden, R. J. Dodson, W. C. Nelson, M. L. Gwinn, R. DeBoy, J. D. Peterson, E. K. Hickey, D. H. Haft, S. L. Salzberg, O. White, R. D. Fleischmann, B. A. Dougherty, T. Mason, A. Ciecko, D. S. Parksey, E. Blair, H. Cittone, E. B. Clark, M. D. Cotton, T. R. Utterback, H. Khouri, H. Qin, J. Vamathevan, J. Gill, V. Scarlato, V. Masignani, M. Pizza, G. Grandi, L. Sun, H. O. Smith, C. M. Fraser, E. R. Moxon, R. Rappuoli, and J. C. Venter.** 2000. Complete genome sequence of *Neisseria meningitidis* serogroup B strain MC58. *Science* **287:**1809–1815.
369. **Tizard, M. L., M. T. Moss, J. D. Sanderson, B. M. Austen, and J. Hermon-Taylor.** 1992. p. 43, the protein product of the atypical insertion sequence IS*900*, is expressed in *Mycobacterium paratuberculosis*. *J. Gen. Microbiol.* **138:** 1729–1736.
370. **Tobiason, D. M., J. M. Buchner, W. H. Thiel, K. M. Gernert, and A. C. Glasgow Karls.** 2001. Conserved amino acid motifs from the novel Piv/MooV family of DNA transposases and site-specific invertases are required for catalysis of DNA inversion by Piv. *Mol. Microbiol.* **39:**641–651.
371. **Tolmasky, M. E., and J. H. Crosa.** 1995. Iron transport genes of the pJM1-mediated iron uptake system of *Vibrio anguillarum* are included in a transposonlike structure. *Plasmid* **33:** 180–190.
372. **Tomb, J. F., O. White, A. R. Kerlavage, R. A. Clayton, G. G. Sutton, R. D. Fleischmann, K. A. Ketchum, H. P. Klenk, S. Gill, B. A. Dougherty, K. Nelson, J. Quackenbush, L. Zhou, E. F. Kirkness, S. Peterson, B. Loftus, D. Richardson, R. Dodson, H. G. Khalak, A. Glodek, K. McKenney, L. M. Fitzgerald, N. Lee, M. D. Adams, and J. C. Venter.** 1997. The complete genome sequence of the gastric pathogen *Helicobacter pylori*. *Nature* **388:**539–547.
373. **Ton-Hoang, B., M. Betermier, P. Polard, and M. Chandler.** 1997. Assembly of a strong promoter following IS*911* circularization and the role of circles in transposition. *EMBO J.* **16:**3357–3371.
374. **Ton-Hoang, B., P. Polard, and M. Chandler.** 1998. Efficient transposition of IS*911* circles in vitro. *EMBO J.* **17:** 1169–1181.
375. **Ton-Hoang, B., P. Polard, L. Haren, C. Turlan, and M. Chandler.** 1999. IS*911* transposon circles give rise to linear forms that can undergo integration in vitro. *Mol. Microbiol.* **32:**617–627.
376. **Trentmann, S. M., H. Saedler, and A. Gierl.** 1993. The transposable element En/Spm-encoded TNPA protein contains a DNA binding and a dimerization domain. *Mol. Gen. Genet.* **238:**201–208.
377. **Trieu-Cuot, P., and P. Courvalin.** 1984. Nucleotide sequence of the transposable element IS*15*. *Gene* **30:**113–120.
378. **Turlan, C., and M. Chandler.** 1995. IS*1*-mediated intramolecular rearrangements: formation of excised transposon circles and replicative deletions. *EMBO J.* **14:**5410–5421.
379. **Turlan, C., and M. Chandler.** 2000. Playing second fiddle: second-strand processing and liberation of transposable elements from donor DNA. *Trends Microbiol.* **8:**268–274.
380. **Turlan, C., B. Ton-Hoang, and M. Chandler.** 2000. The role of tandem IS dimers in IS*911* transposition. *Mol. Microbiol.* **35:**1312–1325.
381. **Utsumi, R., M. Ikeda, T. Horie, M. Yamamoto, A. Ichihara, Y. Taniguchi, R. Hashimoto, H. Tanabe, K. Obata, and M. Noda.** 1995. Isolation and characterization of the IS*3*-like element from *Thermus aquaticus*. *Biosci. Biotechnol. Biochem.* **59:**1707–1711.
382. **van Gent, D. C., K. Mizuuchi, and M. Gellert.** 1996. Similarities between initiation of V(D)J recombination and retroviral integration. *Science* **271:**1592–1594.
383. **van Luenen, H. G., S. D. Colloms, and R. H. Plasterk.** 1994. The mechanism of transposition of Tc*3* in C. elegans. *Cell* **7984:**293–301.
384. **van Luenen, H. G., and R. H. Plasterk.** 1994. Target site choice of the related transposable elements Tc1 and Tc3 of *Caenorhabditis elegans*. *Nucleic Acids Res.* **22:**262–269.

384a. **Venkatesan, M. M., M. B. Goldberg, D. J. Rose, E. J. Grotbeck, V. Burland, and F. Blattner.** 2001. Complete DNA sequence and analysis of the large virulence plasmid of *Shigella flexneri*. *Infect. Immun.* **69:**3271–3285.

385. **Vertes, A., Y. Asai, M. Inui, M. Kobayashi, and H. Yukawa.** 1995. The corynebacterial insertion sequence IS*31831* promotes formation of an excised transposon fragment. *Biotechnol. Lett.* **17:**1143–1148.
386. **Vertes, A. A., M. Inui, M. Kobayashi, Y. Kurusu, and H. Yukawa.** 1994. Isolation and characterization of IS*31831*, a transposable element from *Corynebacterium glutamicum*. *Mol. Microbiol.* **11:**739–746.
387. **Vilei, E. M., J. Nicolet, and J. Frey.** 1999. IS*1634*, a novel insertion element creating long, variable-length direct repeats which is specific for *Mycoplasma mycoides* subsp. *mycoides* small-colony type. *J. Bacteriol.* **181:**1319–1323.
388. **Vogele, K., E. Schwartz, C. Welz, E. Schiltz, and B. Rak.** 1991. High-level ribosomal frameshifting directs the synthesis of IS*150* gene products. *Nucleic Acids Res.* **19:** 4377–4385.
389. **Vos, J. C., and R. H. Plasterk.** 1994. Tc*1* transposase of *Caenorhabditis elegans* is an endonuclease with a bipartite DNA binding domain. *EMBO J.* **13:**6125–6132.
390. **Vos, J. C., H. G. van Luenen, and R. H. Plasterk.** 1993. Characterization of the *Caenorhabditis elegans* Tc*1* transposase in vivo and in vitro. *Genes Dev.* **7:**1244–1253.
391. **Wang, X., and N. P. Higgins.** 1994. 'Muprints' of the lac operon demonstrate physiological control over the randomness of in vivo transposition. *Mol. Microbiol.* **12:**665–677.
392. **Wei, S. Q., K. Mizuuchi, and R. Craigie.** 1998. Footprints on the viral DNA ends in moloney murine leukemia virus preintegration complexes reflect a specific association with integrase. *Proc. Natl. Acad. Sci. USA* **95:**10535–10540.
393. **Weinert, T. A., N. A. Schaus, and N. D. Grindley.** 1983. Insertion sequence duplication in transpositional recombination. *Science* **222:**755–765.
394. **Weinreich, M. D., A. Gasch, and W. S. Reznikoff.** 1994. Evidence that the cis preference of the Tn5 transposase is caused by nonproductive multimerization. *Genes Dev.* **8:** 2363–2374.
395. **Weinreich, M. D., L. Mahnke-Braam, and W. S. Reznikoff.** 1994. A functional analysis of the Tn5 transposase. Identification of domains required for DNA binding and multimerization. *J. Mol. Biol.* **241:**166–177.
396. **Weinreich, M. D., J. C. Makris, and W. S. Reznikoff.** 1991. Induction of the SOS response in *Escherichia coli* inhibits Tn*5* and IS*50* transposition. *J. Bacteriol.* **173:**6910–6918.
397. **Weinreich, M. D., and W. S. Reznikoff.** 1992. Fis plays a role in Tn*5* and IS*50* transposition. *J. Bacteriol.* **174:**4530–4537.
398. **Welty, D. J., J. M. Jones, and H. Nakai.** 1997. Communication of ClpXP Protease Hypersensitivity to Bacteriophage Mu Repressor Isoforms. *J. Mol. Biol.* **272:**31–41.
399. **Welz, C.** 1993. Functionelle Analyse des Bakteriellen Insertion elements IS*150*. Ph.D. thesis. Albert-Ludwigs-Universität Freiburg, Freiburg, Germany.

400. **Wiater, L. A., and N. D. Grindley.** 1988. Gamma delta transposase and integration host factor bind cooperatively at both ends of gamma delta. *EMBO J.* **7:**1907–1911.
401. **Wiater, L. A., and N. D. Grindley.** 1991. Gamma delta transposase. Purification and analysis of its interaction with a transposon end. *J. Biol. Chem.* **266:**1841–1849.
402. **Wiegand, T. W., and W. S. Reznikoff.** 1994. Interaction of Tn*5* transposase with the transposon termini. *J. Mol. Biol.* **235:**486–495.
403. **Wolkow, C. A., R. T. DeBoy, and N. L. Craig.** 1996. Conjugating plasmids are preferred targets for Tn7. *Genes Dev.* **10:** 2145–2157.
404. **Wu, S. W., K. Dornbusch, and G. Kronvall.** 1999. Genetic characterization of resistance to extended-spectrum β-lactams in *Klebsiella oxytoca* isolates recovered from patients with septicemia at hospitals in the Stockholm area. *Antimicrob. Agents Chemother.* **43:**1294–1297.
405. **Yang, M., B. C. Ross, and B. Dwyer.** 1993. Identification of an insertion sequence-like element in a subspecies of *Mycobacterium kansasii. J. Clin. Microbiol.* **31:**2074–2079.
406. **Yieh, L., G. Kassavetis, E. P. Geiduschek, and S. B. Sandmeyer.** 2000. The Brf and TBP subunits of the RNA polymerase III transcription factor IIIB mediate position-specific integration of the gypsy-like element, Ty3. *J. Biol. Chem.* **275:** 29800–29807.
407. **Yin, J. C., M. P. Krebs, and W. S. Reznikoff.** 1988. Effect of dam methylation on Tn*5* transposition. *J. Mol. Biol.* **199:** 35–45.
408. **Yin, J. C., and W. S. Reznikoff.** 1987. *dnaA*, an essential host gene, and Tn*5* transposition. *J. Bacteriol.* **169:**4637–4645.
409. **Yurieva, O., and V. Nikiforov.** 1996. Catalytic center quest: comparison of transposases belonging to the Tn3 family reveals an invariant triad of acidic amino acid residues. *Biochem. Mol. Biol. Int.* **38:**15–20.
410. **Zabala, J. C., J. M. Garcia-Lobo, E. Diaz-Aroca, F. de la Cruz, and J. M. Ortiz.** 1984. *Escherichia coli* alpha-haemolysin synthesis and export genes are flanked by a direct repetition of IS*91*-like elements. *Mol. Gen. Genet.* **197:**90–97.
411. **Zerbib, D., P. Gamas, M. Chandler, P. Prentki, S. Bass, and D. Galas.** 1985. Specificity of insertion of IS1. *J. Mol. Biol.* **185:**517–524.
412. **Zerbib, D., P. Polard, J. M. Escoubas, D. Galas, and M. Chandler.** 1990. The regulatory role of the IS*1*-encoded InsA protein in transposition. *Mol. Microbiol.* **4:**471–477.
413. **Zerbib, D., P. Prentki, P. Gamas, E. Freund, D. J. Galas, and M. Chandler.** 1990. Functional organization of the ends of IS1: specific binding site for an IS1-encoded protein. *Mol. Microbiol.* **4:**1477–1486.
413a. **Zhang, L., A. Dawson, and D. J. Finnegan.** 2001. DNA-binding activity and subunit interaction of the mariner transposase. *Nucleic Acids Res.* **29:**3566–3575.
414. **Zhao, S., and C. W. Hill.** 1995. Reshuffling of Rhs components to create a new element. *J. Bacteriol.* **177:**1393–1398.
415. **Zhao, S., C. H. Sandt, G. Feulner, D. A. Vlazny, J. A. Gray, and C. W. Hill.** 1993. Rhs elements of *Escherichia coli* K-12: complex composites of shared and unique components that have different evolutionary histories. *J. Bacteriol.* **175:** 2799–2808.
416. **Zheng, J., and M. A. McIntosh.** 1995. Characterization of IS*1221* from *Mycoplasma hyorhinis:* expression of its putative transposase in *Escherichia coli* incorporates a ribosomal frameshift mechanism. *Mol. Microbiol.* **16:**669–685.
417. **Zhou, L., F. M. Hui, and D. A. Morrison.** 1995. Characterization of IS*1167*, a new insertion sequence in *Streptococcus pneumoniae. Plasmid* **33:**127–138.

Mobile DNA II
Edited by N. L. Craig et al.

Chapter 16

Transposition of IS*911*

P. ROUSSEAU, C. NORMAND, C. LOOT, C. TURLAN, R. ALAZARD, G. DUVAL-VALENTIN, AND M. CHANDLER

INTRODUCTION

IS*911* is a member of the IS*3* family of bacterial insertion sequences first defined by Rak and coworkers (58). At present, more than 90 elements have been reported, of which more than 60 represent distinct members. These are distributed over 26 genera and 51 species of eubacteria (http://www-is.biotoul.fr). No examples have yet been described from the archaea (see chapter 15 and reference 34). It is highly homogenous in spite of exhibiting a wide range of G+C contents generally similar to that of their host strains, and of their presence in hosts such as *Mycoplasma* with a nonuniversal genetic code (39). The family can be divided into several subgroups depending on features such as the number of target base pairs duplicated on insertion, similarities between the terminal inverted repeats, and the degree of sequence similarity between transposases. Although there are likely to be differences in the details of their transposition pathways and in the way transposition is regulated, different members of this family probably use a similar transposition mechanism. This review is centered on studies of IS*911* as a model system for studying the large IS*3* group. However, important insights and some intriguing differences have been revealed from other members and these will be described where appropriate.

P. Rousseau, C. Normand, C. Loot, C. Turlan, R. Alazard, G. Duval-Valentin, and M. Chandler • Laboratoire de Microbiologie et Génétique Moléculaires, CNRS, 118 Route de Narbonne, F-31062 Toulouse Cedex, France.

ISOLATION AND ORGANIZATION

Isolation and Distribution

IS*911* was initially isolated from *Shigella dysenteriae* as an insertion into the cI gene of bacteriophage λ. This resulted in a phenotypically virulent phage unable to lysogenize a naïve host but subject to repression following superinfection of an *Escherichia coli* strain lysogenic for wild type λ (48). Restriction and nucleotide sequence analysis showed that the cI gene was interrupted by a 1,250-bp DNA segment exhibiting the characteristics of an insertion sequence related to previously identified ISs of the IS*3* family. Southern blots of genomic DNA using an internal IS*911* fragment as a probe showed that the element was present in multiple copies in the original host strain and in type strains of other *Shigella* species. Two vestigial copies were also detected in the chromosome of *E. coli* K12 and were located at map positions 5.8′ (280 to 285 kb) and 97.1′ (4580 to 4585 kb). Interestingly, both are interrupted by a copy of IS*30* (3; M.-F. Prère, O. Fayet, and M. Chandler, unpublished results). One of these carries functional terminal inverted repeats and could be activated by supplying IS*911* transposase in *trans* (16). Subsequent studies identified entire or truncated copies of IS*911* in several *E. coli* virulence plasmids (pO157H [56]; pB171 [65]), in pathogenicity islands of uropathogenic *E. coli* (63), in various other clinical isolates of *E. coli* and other enterobacteria such as *Klebsiella oxytoca* (Prère et al., unpublished), and in bacteria of the phylloplane such as *Pantoea* (*Enterobacter*) *agglomerans* (57).

General Organization

IS*911* (Fig. 1) is 1,250 bp long. It is bordered by imperfect 36-bp terminal inverted repeats (IRL and IRR) and was shown to generate a 3-bp (or occasionally 4-bp) target duplication on insertion (48). The IRs of different members of the family are quite similar. Most terminate in a 5′CA3′ dinucleotide and many (but not IS*911*) exhibit a short GC-rich stretch at about 8 bp into the IR (34). Additional conserved bases are also apparent. Like most other members of the IS*3* family, IS*911* carries two consecutive and partially overlapping open reading frames, *orfA* and *orfB*. These are under control of a weak promoter: p_{IRL} (partially located in IRL). The 5′ end of *orfB* overlaps the 3′ end of *orfA* and occurs in reading phase −1 relative to *orfA*. Transcription initiates at G_{54} (66). A short inverted repeat sequence is located between coordinates 31 and 40 and 46 and 52 and includes the −10 hexamer of p_{IRL}. Although the role of this sequence in IS*911* has not been investigated,

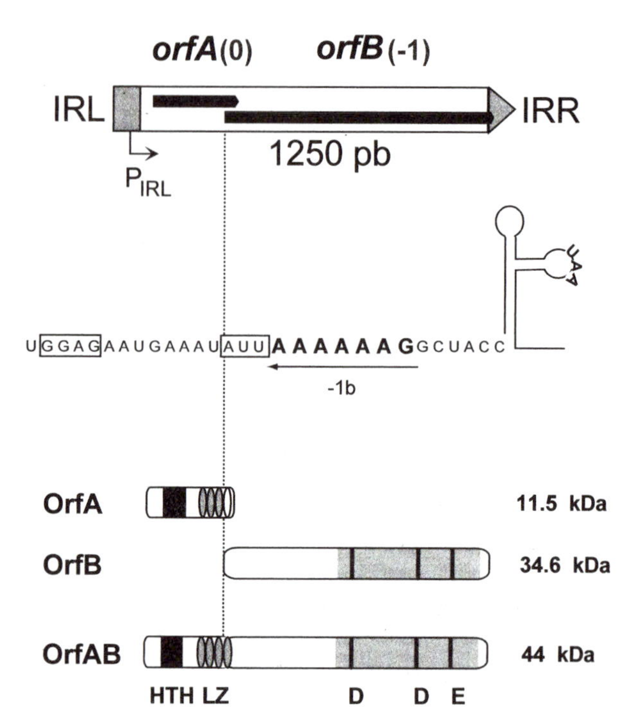

Figure 1. Organization of IS*911*. A cartoon of the IS is presented at the top of the figure. This includes the overlapping reading frames *orfA* and *orfB* (black boxes) together with their relative reading phases (0 and −1). The point of translational frameshifting is shown as a dotted vertical line. The 36 bp left and right terminal inverted repeats are shown as a heavy gray box and triangle, respectively. The endogenous promoter p_{IRL} is also indicated. The RNA sequence encompassing the frameshift window is shown below. The potential *orfB* ribosome binding site is boxed, as is its AUU initiation codon. The ribosome engages the A_6G slippery codons and slips one nucleotide to the left, as shown by the arrow. The structured downstream region is shown to the right and the *orfA* termination codon is indicated. The bottom of the figure shows the organization of IS*911* proteins together with their molecular mass. The α helix-turn-α helix motif (HTH) is shown as a black square. Individual heptads of the leucine zipper (LZ) are represented by four ellipses. The unfilled ellipse shown in OrfA represents the heptad, which differs between OrfA and OrfAB. The catalytic domain is shown as a gray box including the DDE signature (black upright lines).

in the related IS2 element a similar sequence appears to function as a binding site for the OrfA protein, which represses promoter activity (21).

Expression of IS*911*-Encoded Proteins

As illustrated in Fig. 1, IS*911* expresses three proteins: OrfA, OrfB, and the transposase of the element, OrfAB (44).

OrfA and OrfB expression

The 101 codon OrfA initiates at an ATG, terminates with TAA, and has a predicted molecular mass of 11.5 kDa, corresponding well to that estimated from SDS-PAGE. Inspection of the sequence of OrfA located a potential α helix-turn-α helix (HTH) motif probably involved in sequence-specific binding to the terminal IRs of the element and a C-terminal leucine zipper (LZ) motif involved in protein multimerization (see below; 18, 44, 48). Most IS*3* family members exhibit a similarly placed HTH signature (see for example: 13, 48) while the LZ motif has been previously observed in other members of the family (4, 29, 33, 71, 76) and appears to be conserved in the majority of known members (17).

OrfB is 299 residues long and has a predicted molecular mass of 34.6 kDa. Its TAA termination codon lies just within IRR. The initiation codon of OrfB is AUU. This is an unusual initiation codon restricted in *E. coli* to the *inf*C gene, which encodes the translation initiation factor IF3 (54) (Fig. 1). The sequence located immediately upstream also shows some similarity to the equivalent region of the IF3 message (44) and carries a potential ribosome binding site. Studies using $T7_{p\phi10}$ or p_{tac} to drive transcription indicated that initiation at the AUU of OrfB occurs at low levels (44, 51). Initial inspection of the OrfB amino acid sequence showed that it shares significant similarities with retroviral integrases (12, 24, 27). In particular, such comparisons defined the highly conserved amino acid triad DDE common to all IS*3* family members and to many of this type of phosphoryltransferase enzymes. This constitutes part of the active site (for reviews, see chapters 2 and 15 and references 9, 43, 52, and 53). OrfB carries neither the HTH nor the LZ motif.

IS*3* and IS*150* also produce proteins equivalent to OrfB. For IS*3*, OrfB expression is translationally coupled to that of OrfA by virtue of overlapping OrfA termination and OrfB initiation codons. This is modulated by a potential downstream pseudoknot (60). Translation initiation of OrfB occurs in phase with *orfB* for IS*911* while for IS*150*, the OrfB initiation codon is out of phase and expression requires a −1 frameshift after initiation (73, see below). No OrfB product has yet been detected for IS*2* (22).

Transposase OrfAB as a fusion protein

OrfAB, the transposase of the element, is produced in an unusual way. It is assembled from *orfA* and *orfB* by a programmed −1 translational frameshift occurring near the 3′ end of *orfA*. This was first demonstrated for the related IS*150* (73). Frameshifting generates a protein that combines the HTH motif of *orfA*, an LZ motif, and the DD(35)E catalytic domain of *orfB* (44). OrfAB (382 amino acids) shares its 86 N-terminal amino acids with OrfA (100 amino acids) and its 296 C-terminal amino acids with OrfB (299 amino acids). Rephasing of the ribosome to generate OrfAB occurs on a group of "slippery codons" with a frequency of about 15% (measured using systems driven by two different promoters; $T7p_{\phi10}$: 44; p_{tac}: 51). OrfA is therefore normally expressed at significantly higher levels than OrfAB. Frameshifting provides an ingenious way of combining different functional domains into different proteins. The frameshifting mechanism adopted by IS*911* is similar to that used in some retroviruses to generate the Pol-Gag "polyprotein" (23). The relevant sequences involved in frameshifting in IS*911* are shown in Fig. 1. The group of slippery codons is A AAA AAG and is directly preceded by the AUU OrfB initiation codon. Since *E. coli* does not encode a tRNA with an anti-codon for AAG, this is decoded as lysine (AAA) without pairing at the wobble position (68a). The presence of the upstream RBS and the downstream secondary structure has been shown experimentally to stimulate rephasing of the ribosome in a −1 direction to decode the sequence as AAA AAA G (51; Prère and Fayet, personal communication). This secondary structure has received experimental support (51). Different groups of codons have been observed to allow rephasing of the ribosome (11, 75), and although the most common motif is A_6G, different members of the IS*3* family carry a variety of these (e.g., A_3G for IS*3*) (60; for a review, see reference 11).

Experimentally, the OrfAB protein can be produced constitutively (without the necessity of frameshifting) simply by insertion of an additional base pair within the A_6G frameshift window. This fuses *orfA* and *orfB* and eliminates OrfA expression by placing the normal TAA termination codon out of phase. Addition of an A residue (to generate A_7G) does not eliminate frameshifting, which can then occur on the A_7 sequence. In practice, to circumvent this, the sequence ACT CAA AAA AG has been used (44). Here the AUU initiation codon of OrfB is eliminated by mutation to ACT and an additional C residue has been inserted before the A stretch. This eliminates frameshifting but does not change the primary amino acid sequence of OrfAB (44).

Two similarly located partially overlapping reading frames in IS*3* and IS*150* also produce three proteins (60, 73). The transposase, OrfAB, like that of IS*911*, is a fusion product of the two *orfs* generated by a −1 translational frameshift (60, 73). For IS*3*, frameshifting is also stimulated by a presumed pseudoknot involving the downstream secondary structure (60). Two similarly arranged *orfs* occur in IS*2* (22, 30) and have been shown to encode OrfA and OrfAB equivalents (22). This organization is observed in a number of other, less-well-characterized, elements (including IS*51*, IS*222*, IS*600*, IS*1133*, IS*1222*, and IS*3411*; Prère and Fayet, personal communication; see also reference 11). The frequency of frameshifting can be quite variable from element to element (15% for IS*911*: 44, 51; 50% for IS*150*: 73; 6% for IS*3*: 60).

THE TRANSPOSITION PATHWAY: AN OVERVIEW

Our present understanding of the steps in IS*911* transposition is shown in Fig. 2 (42, 47, 66, 67). Like all known reactions catalyzed by the DDE class of transposases (chapters 1, 2, 18, and 20 and references 19, 35, and 69), the chemical steps of IS*911* transposition involve single-strand cleavage at the transposon end and strand transfer to a target. When the plasmid content of *E. coli* carrying transposase and transposon donor plasmids was analyzed, two additional species were observed. Restriction mapping, DNA sequence determination, and electron microscopy showed that one is a covalently closed transposon circle (45). The other species is a molecule in which only one of the two transposon strands is circularized to form a figure-eight structure (42). The overall pathway can be viewed as a series of well-defined steps in which OrfAB catalyzes strand processing and OrfA stimulates the final integration step:

- Synaptic complex A: The initial step must involve recognition of the terminal IRs by the transposase and their assembly into a synaptic complex where they are correctly positioned for subsequent chemical steps.
- Cleavage: Once assembled into a synaptic complex, one transposon strand at one or the other end is cleaved, probably using water as a nucleophile, to generate a free 3′OH (for review, 19).
- Strand transfer leading to figure-eight formation: The liberated 3′OH then acts as a nucleophile, which directs strand transfer to the same strand 3 bases 5′ to the other end of the element. This generates a molecule in which a single transposon strand is circularized, giving rise to a figure-eight structure where the terminal inverted repeats of the transposon are joined by a single-stranded bridge and separated by three bases derived from flanking DNA. The reaction can be viewed as a one-ended site-specific transposition event (Fig. 2).

Note that this strand transfer product, like those of other DDE elements (7, 74), can undergo "disintegration." This can be considered as a reversal of the strand-transfer reaction where nucleophilic attack of the newly formed phosphate bond by the free 3′OH on the target generated by strand transfer releases the transferred strand.

These three initial steps can be accomplished by OrfAB alone.

- Figure-eight conversion to a circular form: Kinetic data suggest that the figure-eight then gives rise to a circular transposon form, which carries terminal abutted IRs separated by three base pairs of DNA flanking the original insertion. This step may involve host factors.
- Promoter assembly: A remarkable consequence of the formation of the circular transposon is to assemble a new promoter, p_{junc}, in which a −35 hexamer is contributed by IRR and a −10 hexamer by IRL (Fig. 2). p_{junc} is significantly stronger than the indigenous promoter p_{IRL}, is correctly placed to drive a high level of transposase synthesis, and plays an active role in vivo in controlling IS*911* transposition. Integration (Fig. 2) results in the disassembly of p_{junc}.
- Synaptic complex B and integration: The formation of a circle junction brings both transposon ends together in an inverted orientation. This active junction must then participate in a second type of synaptic complex with the target DNA. Two single-strand cleavages, one at each abutted IR, would linearize the transposon circle, permitting the two liberated 3′OH groups to direct coordinated strand transfer. The final step requires OrfAB but is greatly stimulated by OrfA. Note that it is not known whether target capture occurs before or after cleavage of the circle junction.

However, even though there is strong evidence for this series of steps, they may not constitute the exclusive transposition pathway (60).

IS*911* PROTEINS: FUNCTIONAL ORGANIZATION

Before describing in detail each of the steps outlined above, it is important to appreciate the organization and inter-relationship of the IS*911* proteins involved.

OrfAB

The 382-residue OrfAB has a theoretical isoelectric point (pI = 10.5) consistent with its capacity to bind DNA. In the absence of structural data, com-

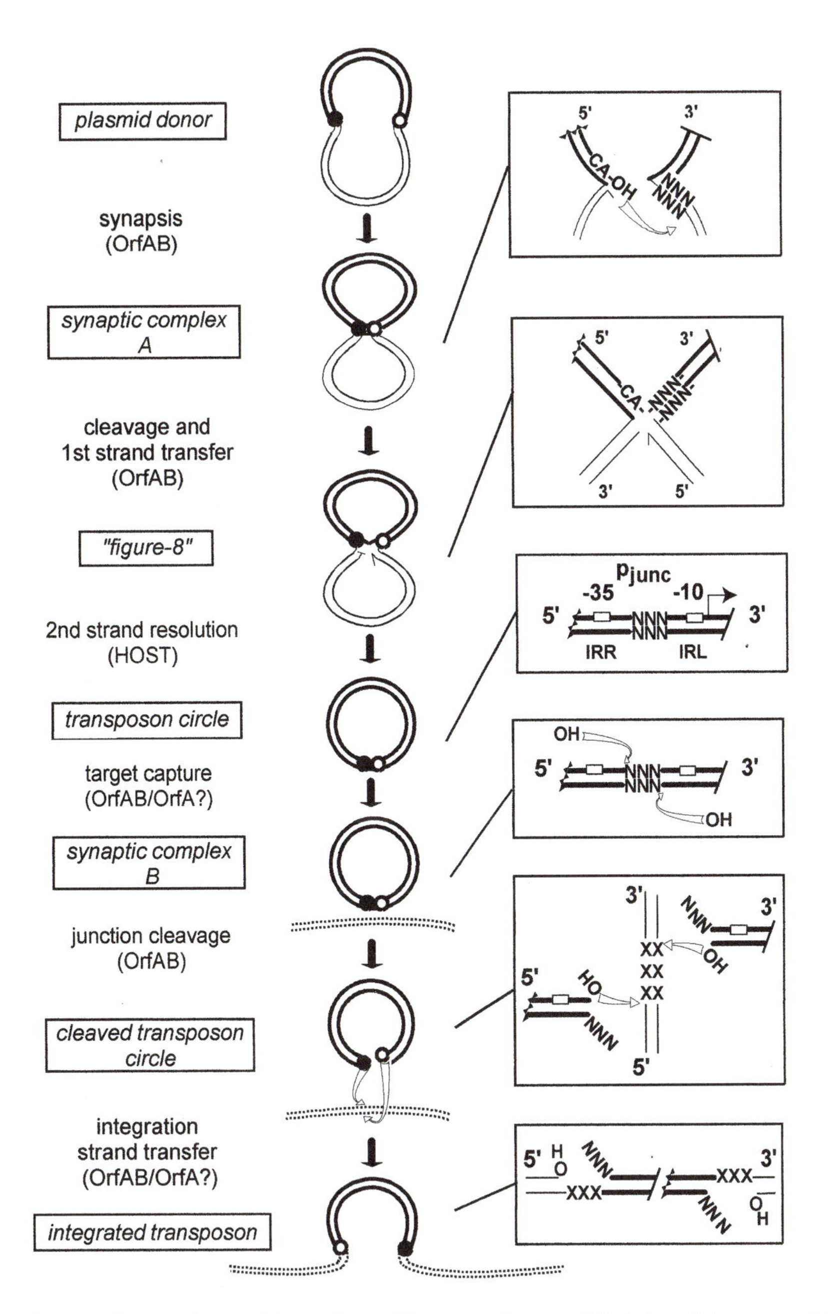

Figure 2. Steps in the overall IS*911* transposition pathway. The center shows a global view of the transposition process. The transposon is shown in bold, donor backbone sequences as fine lines, and target DNA as dotted lines. The small filled and unfilled circles represent IRL or IRR. The panels on the right of the figure show a finer view of the various steps. They illustrate (descending): cleavage at the terminal 5′CA3′ to generate a 3′OH and attack 3 bases (NNN) 5′ of the opposite (target) end. The junction of the figure eight with a free 3′OH in the vector strand is indicated (half arrow), together with the IRR-IRL circle junction and the associated promoter p_{junc} with −35 and −10 regions (unfilled boxes), the position of nucleophilic attack on both strands of the junction (arrows), target attack by the two liberated 3′OH groups on either side of the prospective 3-bp flanking target repeat (arrows), and the final insertion product ready to undergo repair.

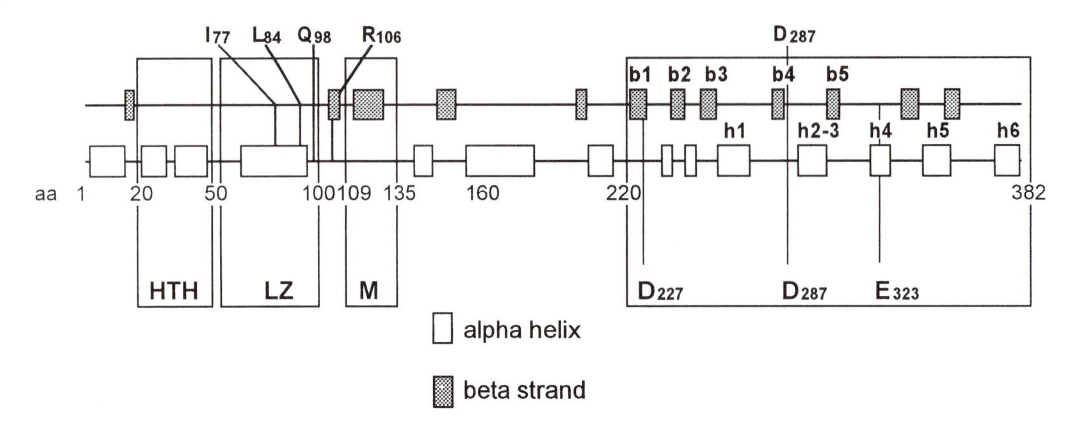

Figure 3. Predicted secondary structures. The map presents predicted β strands (filled boxes) and α helices (unfilled boxes). These are numbered b1 to b5 and h1 to h5, respectively, in order of those found in the crystal structure of the catalytic domain of HIV and ASV integrases. The positions of the mutations referred to in the text are indicated above. The relative positions of the different domains of the protein are boxed and indicated below.

puter analysis has identified potential secondary structure elements in the primary sequence (Fig. 3). OrfAB can be divided into three regions: an N-terminal domain carrying an HTH motif, a C-terminal domain homologous to the "DDE" catalytic domain of integrases and other transposases, and an intervening region composed of several beta-strands and two long alpha-helices, the first of which carries a canonical LZ motif.

The potential alpha-helix–turn–alpha-helix domain

Secondary structure predictions suggest that the potential HTH DNA binding motif is located in the first 50 amino acids of OrfAB, between residues 25 and 45 (Fig. 3). It shows significant similarities with the HTH of CysB, a controlling element of the cysteine regulon of *Haemophilus influenzae* and a member of the LysR family of transcriptional regulators (55). This localization of a potential sequence-specific DNA binding domain in the N-terminal part of OrfAB is supported by the observation that a truncated form of OrfAB, OrfAB[1–135], containing amino acids 1–135, is able to bind IS*911* IRs (15). A similar N-terminal localization of a sequence-specific DNA binding domain has also been observed for IS2 (29).

The DDE catalytic domain

The catalytic domain containing the DDE motif lies between residues 220 and 382. The arrangement of predicted secondary structure elements in this region (Fig. 3) suggests that the structure of this domain is remarkably similar to those of HIV and ASV integrases for which structural data are available (5, 10). This includes five characteristic beta-strands and six alpha-helices. The position of the three catalytic residues D, D, and E on this secondary structure is also conserved. The OrfAB mutant in this domain is inactive in catalysis in vivo (D287G [15]) and the C-terminal region of OrfAB alone is capable of promoting disintegration (47). Furthermore, as appears to be the case for other catalytic core domains of this type, this region is able to promote homomultimerization of OrfAB in vivo (see next section). Characterization of catalytic domains of this family is treated in detail elsewhere (for reviews, see references 35, 53, and 68).

The multimerization domains

The region of OrfAB located between the HTH and DDE domains carries a canonical leucine zipper motif extending between residues 63 and 95 (18). This is composed of four heptads, each with a correctly spaced leucine residue (or in one case, a permitted isoleucine) at every seventh position. Each heptad repeat is expected to form two alpha-helical turns. Two such LZ structures generally align in a parallel configuration to produce dimers (1, 31). The presence of an LZ structural motif is supported by the results of computer analysis using COIL and PERCOIL (for review: 31). This indicates that the sequence between P62 and P96, corresponding exactly to the putative LZ sequence, has a strong probability of folding into a coiled-coil structure (P = 0.95 to 1) (Fig. 3) (18). The role of this region of the protein in multimerization has been confirmed experimentally using a series of deletion derivatives and site-directed point mutations. Coimmunoprecipitation experiments, in which two derivative proteins (one tagged at its C-terminal end with the HA antigen) were expressed in the same cell, demonstrated that the region identified as an LZ was sufficient to promote multimerization (18). Similar conclusions were obtained in an in vivo genetic assay based on gene fusions with phage λ repressor (17). The stoichiometry of the multimers remains to be determined. In addition, introduction of the L84P mutation in the fourth heptad eliminated LZ-mediated multimerization as judged by both the physical and genetic assays. This mutation and that of another key LZ residue, I77P, also eliminate transposase activity in vivo and in vitro (18).

Not only did the coimmunoprecipitation experiments provide evidence for the multimerization capacity of the LZ, they also defined two other regions of the protein capable of promoting multimerization: the region M, located directly downstream from the LZ motif, included between residues 109 to 135, and the region carrying the catalytic core, the DDE motif, between residues 220 and 382 (18).

The transposase of IS*911* thus contains at least three domains that presumably work in collaboration to promote multimerization: the LZ motif, a central 109–135 region (M), and the DDE domain (Fig. 3). The functional role of these multimerization domains will be discussed below.

Temperature sensitivity

Formation of the first IS*911* transposition intermediate (figure-eight molecules) is naturally temperature sensitive (16). Very little figure-eight or transposon-circle product could be observed at 42°C while levels were optimal at 30°C in vivo. Activity at 30°C was found to be sufficiently high to detect circular copies of a single defective chromosomal IS*911* copy. This temperature sensitive phenotype is due directly to OrfAB activity. Indeed, it occurs in vitro with purified OrfAB and, moreover, can be overcome by mutation of the protein. Two temperature-resistant point mutants, Q98R and R106S, were obtained. Both were located between the LZ motif and the second multimerization region, M. The detailed effects of these mutations are at present unknown. This inter-domain region may be important for the correct assembly of the transposase complex.

OrfA

The first 86 residues of the 100-residue protein OrfA, produced from the upstream reading frame, are shared with OrfAB. OrfA thus shares its HTH and most of the LZ motif with OrfAB. The frameshift event that generates OrfAB occurs toward the C-terminal end of OrfA and consequently, the last 14 residues of OrfA are different from those of OrfAB. Frameshifting occurs within the third heptad of the LZ (Fig. 1). Nevertheless, despite this divergence in primary sequence, a leucine residue is conserved in the correct position downstream in both proteins and completes the LZ motif. This sequence divergence might provide the basis for distinguishing between homo- and heteromultimerization of the two proteins. As in the case of OrfAB, the LZ motif is necessary for multimerization and is essential for OrfA activity (17).

Different roles for OrfA have been proposed for different IS*3* family members: stimulation of transposon circle integration for IS*911* (68); inhibition of transposition for IS*3* (62); and repression of transposase expression for IS*2* (21). These are discussed below (see "OrfA Function in Other IS*3* Family Members").

OrfB

OrfB is a 299-residue protein that shares its last 296 residues with OrfAB. It contains the entire catalytic domain, the M region, and the region in which the temperature-resistant mutations are located. OrfB differs from OrfAB by the absence of the N-terminal DNA binding domain (HTH motif) and of the LZ multimerization domain (Fig. 3). This protein is able to promote strand transfer (disintegration; see "The DDE Catalytic Domain" above and reference 47). The function of OrfB in IS*911* is unknown, but in the case of IS*3*, where OrfA has been reported to act as a transposition inhibitor, the protein seems to enhance this inhibitory activity (62).

Additional Protein Species

Although OrfA and OrfAB are the major translation products expressed from IS*911*, additional protein species can be observed. These have been revealed by Western analysis of crude extracts using anti-OrfA polyclonal antibodies. The major product, OrfAB*, corresponds to a truncated derivative thought to contain the N-terminal region of about 150 amino acids. This may be produced by mRNA degradation, premature transcription termination, premature translation termination, or posttranslational proteolysis. It is not clear whether this truncated protein, which is able to bind IRs and may form heteromultimers with the transposase (see "Synaptic Complex A" below), plays a real biological role (17, 18). For IS*3*, this type of product has been demonstrated to act as a transposition inhibitor (59).

ROLE OF IS*911* PROTEINS IN THE TRANSPOSITION PATHWAY

Synaptic Complex A

As a first step in IS*911* transposition, OrfAB must bind and bridge the two IRs. Although full-length OrfAB is extremely inefficient in binding IS*911* ends in vitro, pairing of two IS*911* ends by a form of OrfAB deleted for the C-terminal region, OrfAB[1–149], has been observed in vitro using a gel shift assay (18). The HTH, LZ, and M regions are required for OrfAB-IR interactions, suggesting that multimerization of the protein is important for its DNA binding activity (18).

Incubation of a DNA fragment carrying the IRR sequence with low concentrations of OrfAB[1–149] generated a major specific complex, complex I. At higher concentrations, a second, faster migrating complex, complex II, appeared. Complex I was found to disappear at high protein concentrations with a concomitant increase in the level of complex II. As judged by mixing experiments, complex I carries a pair of IR fragments retained in a protein bridge by OrfAB[1–149] (17). These interactions seem to be highly cooperative since no intermediates in the formation of complex I could be identified. Complex II is thought to result from titration of the IRs by increasing amounts of protein and to contain only one IR fragment bound by OrfAB[1–149] (17). A model for these complexes is presented in Fig. 4.

A third complex (complex III) migrated significantly faster than complexes I and II and was the major species observed with preparations of full-length OrfAB. Analysis of the protein content of complex III failed to detect full-length OrfAB but revealed that it included derivatives truncated for the C-terminal end. These were originally present as low-level contaminating species in the protein preparation (17, 18). This implies that full-length OrfAB binds poorly to IS*911* ends. Complex III could also be generated by mixing OrfAB[1–135] or OrfAB[1–149] with either a shorter OrfAB derivative, OrfAB[1–109], or OrfA itself although neither of the shorter proteins appeared to bind the IRs alone (17). Moreover, no OrfA could be detected in the complex. This suggests that formation of complex III involves interactions between proteins with and without the M region (residues 109 to 135). The LZ motif is certainly involved since mutations L84R or L84P (Fig. 3) abolish or strongly reduce formation of complex III (18). The rapid mobility of complex III may reflect the presence of a lower order OrfAB[1–149] multimer bound to a single end (Fig. 4) (17).

Although these results clearly demonstrate that OrfA modifies the binding properties of OrfAB-[1–149], care should be exercised in extrapolating them to full-length transposase. The presence of the C-terminal region, which modifies the multimerization properties of OrfAB (see "OrfAB" above), may also change the consequences of OrfA interactions. Indeed, in vitro, OrfA appears to have no notable effect on OrfAB-mediated figure-eight formation, which involves formation of synapse I (15) while it strongly modifies full-length OrfAB activity in pro-

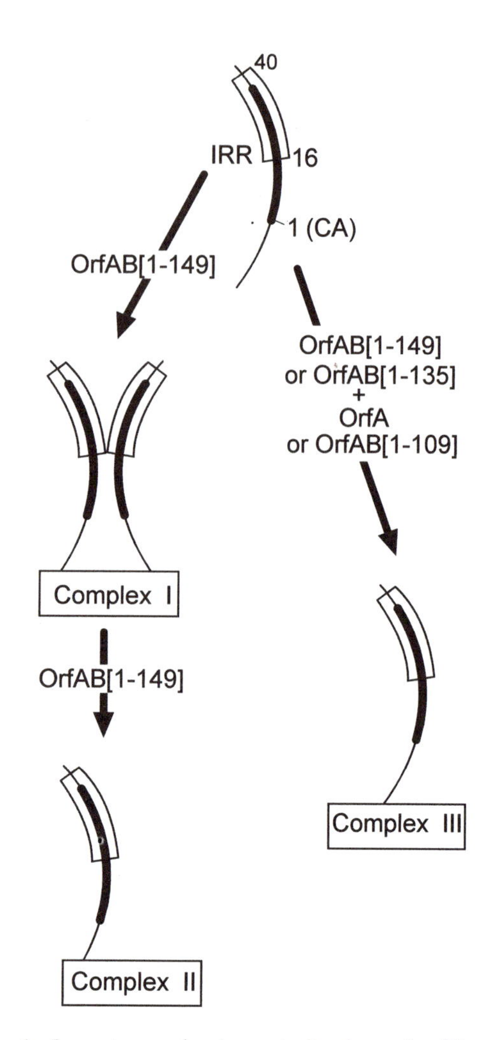

Figure 4. Synaptic complex A or paired-end complex. The scheme represents a model for complexes between the terminal inverted repeats and the truncated OrfAB derivatives. IRR is shown as a heavy black line with the terminal CA base pair labeled. The unfilled box shows the extent of protection by OrfAB[1–149] against DNase I attack in each isolated complex. Note that protection extends beyond the internal boundary of the IR. Complex I is shown as two paired ends retained by an unknown number of OrfAB[1–149] monomers. Complex II derives from complex I by titration of the DNA with high concentrations of protein. Complex III is shown to be generated by mixing a derivative with an M region (OrfAB[1–135], OrfAB[1–149]), and one without (OrfA, OrfAB[1–109]). The resulting uniquely bound OrfAB[1–135] or OrfAB[1–149] is shown either to change stoichiometry or configuration.

moting integration through synapse II (see "Synaptic Complex B and Integration" below).

All three complexes showed similar protection patterns when analyzed by DNase I footprinting: OrfAB[1–149] protects the interior of both IRs. On IRR, protection extends between bases 40 and 16, leaving the terminal 15 base pairs accessible (Fig. 4) (36). This accessible region is presumably contacted by the C-terminal part of OrfAB, which contains the catalytic core. The extent of the binding site on IRR has been confirmed by analysis of the capacity of ends carrying consecutive deletions to bind OrfAB[1–149]. Little difference could be observed between the protection pattern of IRR and IRL (36).

Cleavage and Strand Transfer

While the initial strand cleavage event has not been characterized in detail, the molecules that result from the subsequent strand transfer have been analyzed. First, cleavage at one end liberates a 3′OH, which acts as a nucleophile to attack the same strand 3 bases 5′ to the other end of the element (Fig. 2). The resulting molecules were observed in plasmid preparations from overnight cultures of *E. coli* carrying pairs of plasmids that provide the transposase and an IS*911*-based transposon. Their structure was confirmed by PCR, sequence analysis, and by electron microscopy (42). They are in the form of a figure eight in which a single transposon strand is circularized to create a single-stranded bridge between the terminal inverted repeats (Fig. 2). Mapping the backbone vector strand interruption by primer extension demonstrated that the transposase cleaves only one strand and that the resulting nick on the vector is located exactly at the 3′ end of the transposon (42). The subsequent strand transfer reaction leaves a 3′OH at the other end of the broken vector strand.

The single-ended nature of this recombination reaction was established as follows. The terminal TG dinucleotide is highly conserved in all members of the IS*3* family. Mutation of the dinucleotide at both IS*911* ends eliminates formation of figure-eight molecules in vivo (42) and in vitro (47), whereas mutation at one or the other end does not. Using a derivative in which the sequence of one of the flanking 3-bp direct target repeats was changed, it was observed that the interstitial 3 nucleotide single-strand bridge derived from the mutated end. This shows that the attacking 3′OH is located exclusively at the wild-type end.

The strand-transfer reaction is highly specific and occurs 3 bp 5′ from the targeted end. The nature of this specificity has been investigated using a set of consecutive deletions from the terminal CA of the targeted ends as in vivo substrates for figure-eight and circle formation. The results showed that the target phosphodiester bond is always at the same distance from the transposase binding site in the target end. This implies that the position of the target DNA in the synapse is a major factor in determining the position of nucleophilic attack in strand transfer (36).

Formation of the figure-eight species has been reproduced in vitro with an 85% pure preparation of OrfAB (47). The reaction requires Mn^{2+} or Mg^{2+} but is more efficient in a buffer containing Mn^{2+}, a property typical of DDE transposases (72). No effect could be detected when OrfA was added to the in vitro reaction. IS*911* transposase is unable to catalyze double-strand cleavage as shown by the absence of linear products of either the transposon or the donor plasmid in the reaction.

Similar figure-eight molecules have been identified from IS*2* (30) and have been postulated for IS*3* (59).

Figure-Eight Conversion to Transposon Circles

A circular transposon is the other major species observed in cultures of *E. coli* carrying the transposase and transposon donor plasmids. It carries an IRL-IRR junction in which the abutted ends are separated by one copy of the three flanking target base pairs duplicated in the original insertion (45). Other members of the IS*3* family also generate transposon circles (28, 30, 59, 60). IS*3*, like IS*911*, provokes a 3-bp flanking target duplication and carries a 3-bp intervening sequence, but IS*2*, which provokes a 5-bp target duplication, carries a single base pair separating the ends. A correlation between the number of interstitial base pairs and the length of the target duplication is therefore not a strict rule.

As in the case of the figure-eight molecules, when one transposon end is inactivated with a terminal dinucleotide mutation, the 3-bp interstitial sequence of the covalent transposon circle derives from DNA flanking the inactive end (45). This polarized reaction occurs naturally for IS*2*. The right end of IS*2* is defined by the conserved 5′-CA-3′ dinucleotide while the left end carries a nonfunctional terminal 5′-TA-3′. Here, the single interstitial base pair is derived from the base neighboring IRL (30).

Kinetic experiments in vivo suggested that the figure-eight form is the precursor of transposon circles (42; G. Duval-Valentin, unpublished data). The figure eight was found to accumulate after transposase induction and to disappear concomitantly with increasing levels of circles following inhibition of protein expression by treatment with chloramphenicol (42) or heat inactivation of the transposase (G. Duval-Valentin, unpublished data). However, while figure-eight molecules can be readily formed in vitro in a purified system, it has not yet been possible to generate transposon circles. This suggests that host factors may be implicated in "resolution" of the second strand and in the conversion of figure-eight molecules to circles. Resolution does not require OrfAB. Not only does conversion occur efficiently in vivo at 42°C under conditions where OrfAB is inactive (G. Duval-Valentin, unpublished data), but figure-eight molecules preformed in vitro from a dimeric plasmid donor can be converted to the equivalent resolved circle following transformation of a naïve host strain in the absence of OrfAB (70).

In principle, conversion to circles could occur using host recombination/repair or replication pathways (Fig. 5A) (47). The junction of the bridged figure-eight molecule resembles a Holliday recombination intermediate except for a single-strand interruption at the crossover point (Fig. 2) (42). The second strand could simply be resolved by consecutive host-mediated cleavage and repair (Fig. 5A, top). A second possibility, suggested by the presence of a 3′OH group on the donor backbone, is that this could serve as a primer for leading strand replication. Coupled with lagging strand synthesis on the circularized strand, this would liberate the original donor plasmid and the transposon as a circular double-strand molecule (Fig. 5A, bottom).

Assembly of a Strong Promoter: p_{junc}

The juxtaposition of IRL and IRR in the transposon circle generates a strong promoter, p_{junc}, which is ideally placed to drive OrfA and OrfAB expression.

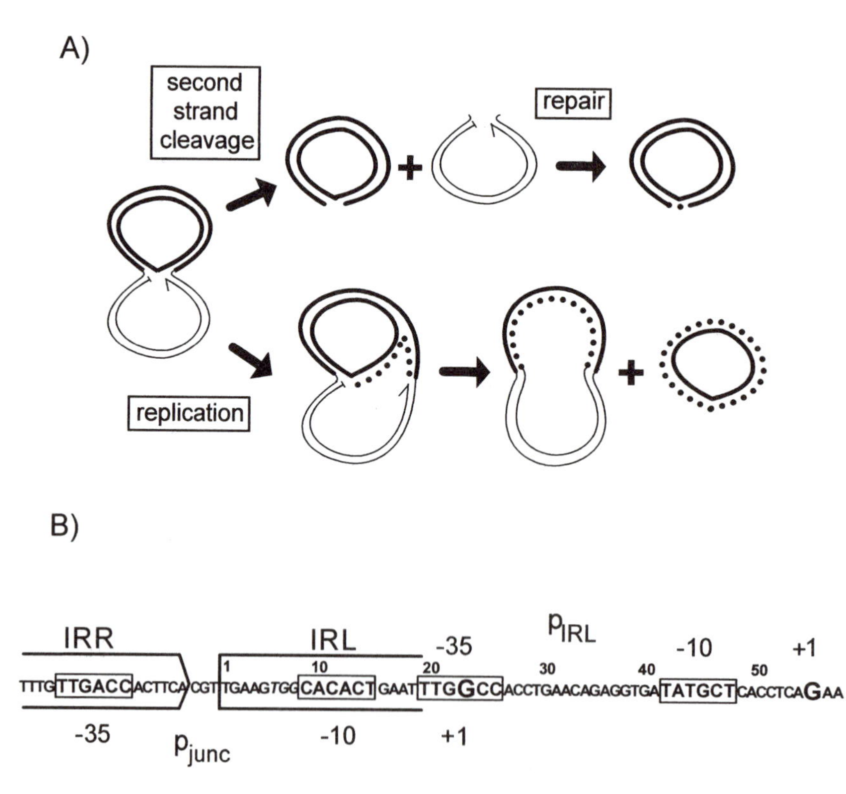

Figure 5. Conversion of the figure-eight form to a circle and junction promoter assembly. (A) Resolution of the figure-eight form into a transposon circle may occur by resolution and repair (top pathway) or replication (bottom pathway). Transposon DNA is indicated in bold, donor backbone DNA by a thin line, and newly replicated DNA by a dotted line. Second-strand cleavage (top) would yield a gapped open circular form of the transposon, which could subsequently be repaired. The fate of the donor plasmid backbone is not determined. The pathway represented in the lower panels invokes replication using the 3′OH group (half arrow) located on the vector backbone at the junction with the transposon as a primer. (B) Schematic representation of the junction region. The proposed −35 and −10 regions are boxed, and the corresponding nucleotides are indicated in large script. The TGN sequence, also important in promoter activity (in this case TGG), is shown in italics. Those indicated below show p_{junc} while those above are p_{IRL}. The nucleotides G, in bold and marked +1, locate the transcriptional start for p_{junc} at nucleotide 22 and for p_{IRL} at 56. The relative positions of IRR and IRL are also shown. Nucleotide coordinates are shown above the sequence initiating at the terminal T of IRL.

This was initially observed in studies with an artificial IS*911*-derived transposon carrying an *orfA-lacZ* translational fusion between copies of IRL and IRR. Whereas p_{IRL} was not sufficiently strong to generate a Lac^+ phenotype on MacConkey lactose plates, the lac^- host exhibited a strong Lac^+ phenotype in the presence of transposon circles produced when OrfAB was provided in *trans* (66). Inspection of the nucleotide sequence of the circle junction revealed a potential −35 hexamer in IRR oriented outwards and a potential −10 within IRL oriented inwards (Fig. 2 and 5B). The −35 hexamer (TTGACC) is a rather good match to the consensus TTGACA although the −10 sequence (CACACT) is somewhat distant from the TATAAT consensus. In addition, a conserved TGN sequence known to enhance promoter activity (6) is located immediately upstream of the −10 sequence. Formation of the circle junction in which IRL and IRR are separated by 3-bp places these two elements with the correct spacing of 17 bp (20). Localization of the point of transcription initiation using reverse transcriptase indicated that it occurs at G_{22} rather than at G_{54}, characteristic of transcription from p_{IRL} (66). The activity of the junction promoter, p_{junc}, was approximately 45-fold higher than that of the indigenous promoter, p_{IRL}, as judged by β-galactosidase expression from *orfA-lacZ* fusions. Moreover, p_{junc} was found to generate a 2.5-fold higher level of β-galactosidase than the strong p_{lacUV5} promoter under standard assay conditions (9a).

Assembly of the p_{junc} promoter following circularization could reflect an unusual mechanism for controlling transposition activity. In this scheme, formation of the figure-eight and circular transposon forms would depend on transposase expression from the weak promoter p_{IRL}. Assembly of p_{junc} in the rare circles produced would be expected to significantly boost transposase expression. High expression of OrfA and OrfAB would facilitate circle integration (Fig. 2). This would lead to disassembly of p_{junc} and thus to a lower level of transposition proteins. This arrangement, resulting in a transiently high level of transposase synthesis, would act as a powerful feedback regulatory circuit. Experiments using a mutated IS*911* with an inactive p_{junc} have confirmed that the junction promoter is important for efficient transposition (9a).

Despite its regulatory potential, not all IS*3* family members appear to possess this type of promoter assembly mechanism. Quantitative assessment of known IRL-IRR junctions from other members of the family in a standardized *lacZ* translational fusion vector indicated that IS2 (see also 64) and IS*600* can drive high levels of β-galactosidase synthesis while IS*3* and IS*987* do not (9a). This type of junction promoter activity has been detected in members of several other families (IS*21* [50], IS*492* [40], and IS*30* [8]).

Synaptic Complex B and Integration

Circles are active intermediates in IS*911* transposition. Plasmids carrying the IRR-IRL junction were found to be active in transposition in vivo when provided with OrfAB. This activity could be increased significantly when OrfA was supplied together with OrfAB from a plasmid carrying a wild-type configuration of IS*911* genes (67). Similar results were obtained in vitro (68) where it was further shown that coordinated insertion of both transposon ends of a transposon circle into a plasmid target required both OrfA and OrfAB. Indeed, in the absence of OrfA in vitro, the major products were those in which a single end had been transferred to the target (67). Moreover, the efficiency of this reaction could be improved if OrfA and OrfAB were preincubated together. This implies that protein-protein interactions between OrfA and OrfAB are involved in intermolecular transposition.

The junction was also found to generate high levels of adjacent deletions in vivo and in vitro. Topological analysis of intramolecular products obtained in vitro from a purified transposon circle in the absence of a target DNA showed that although a low level of catenanes and knots was detected, the majority products were uncatenated or unknotted deletions. Catenanes and knots are the result of trapping even and odd numbers of nodes in a supercoiled DNA molecule between the intramolecular target site and the IRR-IRL junction. Their appearance implies that the two sites encounter each other by random collision. On the other hand, the majority of products are generated by exclusion of intervening nodes, a result which suggests that the transposase may track along the DNA to encounter the final target site (67).

Integration requires the cleavage of both transposon strands at the IRL-IRR junction and their coordinated transfer into the target DNA. This process could occur in a stepwise manner, in which first one and then the second end in the circle junction is cleaved and transferred. Another possible pathway would require two concerted single-strand cleavages followed by concerted strand transfer at both cleaved ends. The latter possibility implies the transitory formation of a linear IS*911* form from the transposon circle. Studies in vivo and in vitro have demonstrated that OrfAB cleaves each strand of the IRR-IRL junction at the 3′ terminal phosphodiester bond, leaving a 3 base 5′ overhang and a recessed 3′OH. Mutation of the terminal dinucleotide at one or both ends of the junction significantly reduces cleavage of the mutated end in vivo and in vitro, and the same single- and double-end mutations strongly reduce the transposition frequency in vivo (68). Moreover, although inefficient, integration of a precleaved linear IS*911* derivative can be accomplished in vitro. Cleavage and integration do not therefore appear to require supercoiling in the donor molecule. Integration of a linear molecule carrying the IRR-IRL junction occurs efficiently in vitro (68), confirming that donor supercoiling is not required. However, the transposition frequency is sensitive to the length of the DNA separating the ends in the junction. It is optimal at 3 bp and efficient with both 2 and 4 bp, but falls off significantly when the spacing is increased to 9 or 13 bp (68).

Like that of IS*911*, the transposase of IS*3* can generate a linear IS copy with a 3-base 5′ overhang (59). It was proposed that the transposase may also have some capacity to cleave the complementary strand since linear forms of IS*3* appeared to be directly excised from donor molecule by double-strand cleavage (61). However, the majority of linear forms have since been demonstrated to result from cleavage of IS*3* circles which, like those of IS*911*, are also efficient in integration (59).

These results indicate that the transposition pathways of IS*2* (30) and IS*3* (59) are largely similar to that of IS*911*.

OrfA Function in Other IS*3* Family Members

Although the OrfAB fusion protein is considered to be the transposase for the IS*3* family, the stimulatory role observed for IS*911* OrfA is not the general rule. For IS*2*, OrfA (InsA) was reported to reduce transposition in vivo by a factor of 6 (21). This was ascribed to the capacity of the protein to bind IRL and repress the indigenous p_{IRL} promoter (21). It has also been proposed that OrfA and OrfAB (InsAB′) compete for IRL binding (22). Indeed these authors have provided evidence that OrfAB binds to both IRL and IRR and that the binding site overlaps that of OrfA on IRL. It has not been ruled out, however, that as in the case of IS*911*, contaminating C-terminal truncated derivatives of OrfAB are responsible for the observed binding. The functional domains of OrfA from IS*2* have been explored (29). The protein is capable of forming multimers as judged both by a yeast two-hybrid system and by glutaraldehyde crosslinking. Deletion analysis has shown that a region located between amino acids 58 and 105, which includes a leucine zipper motif (LZ), is necessary for multimerization. Furthermore, since deletion of this potential LZ motif also abolishes DNA binding, multimerization seems to be necessary for IRL binding. Like that of IS*2*, the OrfA protein of IS*3* appears to be a transposition inhibitor (62). However, this protein does not affect gene expression of the element and it is proposed that OrfA competes with the transposase for IR binding (62). Although no extensive studies on the equivalent IS*150* proteins have been published, reduction in frameshift frequency by about 80% of the wild-type level (itself, about 30%) did not affect the frequency of direct insertion but reduced the frequency of formation of cointegrates by a factor of 10^2 (73).

In conclusion, by IS*2*, IS*3*, IS*150*, and IS*911* probably share the same transposition pathway but may be differently regulated by OrfA and possibly by OrfB. Further analysis is required to explore these results.

INTER-IS RECOMBINATION

The transposition pathway outlined above clearly provides a model for the generation of the majority product of IS*911* transposition: a simple insertion of the element into a target molecule (Fig. 2). However, it does not provide an explanation for the formation of cointegrate structures. In early studies with IS*911*, a small but significant number of cointegrate products composed of fused donor and recipient molecules separated by directly repeated copies of the IS were observed (48). This type of structure represents about 5% of detectable transposition products (41) and cannot easily be accommodated by the model.

Involvement of Plasmid Dimers

It was suggested some time ago (2) that cointegrate molecules could be generated by transposition from a dimer donor plasmid where two directly repeated copies of the donor backbone plasmid are separated by two directly repeated copies of the IS. The original proposal concerned transposons such as Tn*5* and Tn*10* (chapters 18 and 20), which transpose using a "cut-and-paste" mechanism where the transposon is excised from its donor DNA backbone prior to insertion. Transposition of one copy of the backbone flanked by two copies of the IS would occur if an appropriate single end of each element is used. Insertion of this transposon into a target replicon would generate a replicon fusion.

This type of transposition event can also explain the formation of cointegrates by IS*911* from a dimeric plasmid donor (Fig. 6A). Under this scheme, cleavage and strand transfer would occur between an IRL of one IS copy and an IRR of the other to generate the

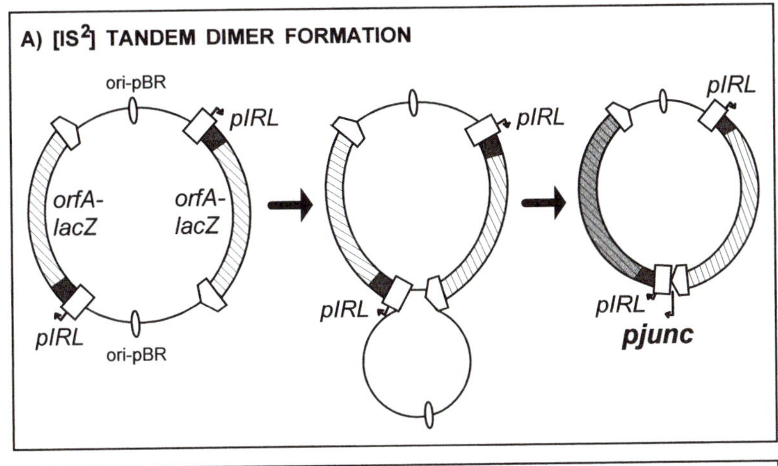

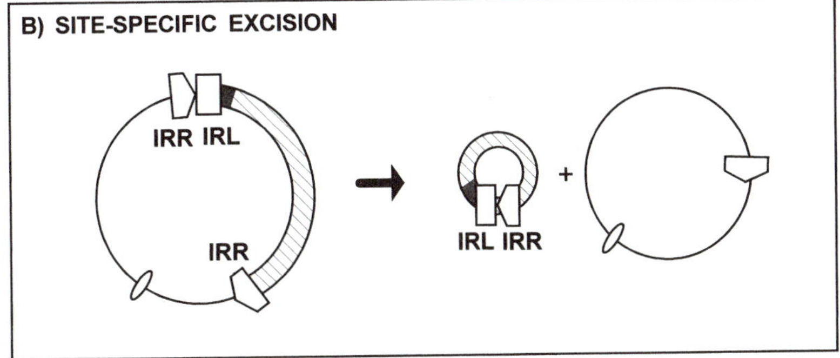

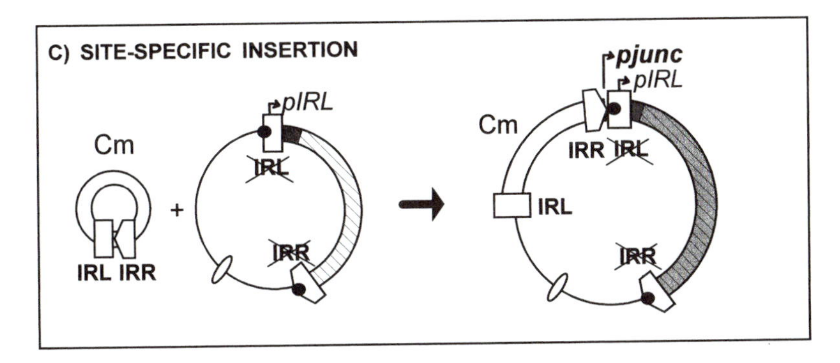

Figure 6. $[IS]^2$ formation from plasmid dimers, targeted insertion, and excision. (A) $[IS]^2$ formation from plasmid dimers. A plasmid dimer is presented on the left. The transposon containing an *orfA-lacZ* translational fusion is shown as a cross-hatched box. Open square and pointed boxes indicate the left and right terminal inverted repeats of IS*911*. The resident promoter, p_{IRL}, partially located in IRL, is also shown. The open oval represents the plasmid origin of replication. Activation of the *orfA-lacZ* fusion by formation of an adjacent p_{junc} is indicated by the gray boxes. Reaction with OrfAB in vitro generates an inter-IS figure-eight form (center), which is converted into a form carrying a tandem IS dimer ($[IS]^2$) following transformation into a suitable *recA* host strain. The IRR-IRL junction formed in this way carries the strong p_{junc} promoter, which drives expression of the *orfA-lacZ* translational fusion. (B) Targeted excision. The donor plasmid carries copies of IRR (pointed box) and IRL (square box) in the configuration of an active junction (IRR-IRL) with an additional appropriately oriented IRR. The transposon is indicated by a cross-hatched box. (C) Targeted insertion. A transposon circle carrying a chloramphenicol resistance gene (Cm) is shown, together with a target plasmid carrying an inactivated transposon including an *orfA-lacZ* fusion (cross-hatched box). The small black circles on IRR and IRL indicate the presence of a mutation of the terminal CA dinucleotide. Activation of the *orfA-lacZ* fusion by insertion with formation of the junction promoter is indicated by a gray box.

equivalent of a figure-eight molecule (inter-IS recombination [Fig. 6A and 7]) that could be resolved in the same way as the simple figure-eight (intra-IS recombination [Fig. 7]). Instead of giving rise to a circular IS copy, resolution of the inter-IS figure-eight would generate a molecule carrying a single copy of the donor backbone together with a head-to-tail tandem IS ($[IS]^2$). Like the transposon circle, this would

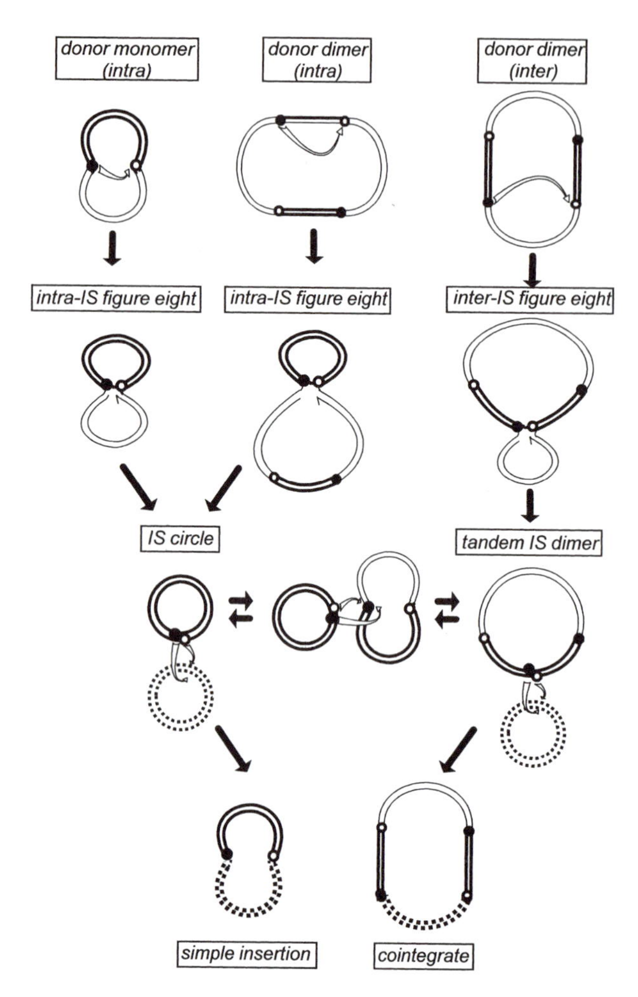

Figure 7. Recombination pathways in IS*911* transposition. Transposon DNA is indicated by bold lines, donor plasmid DNA is indicated by thin lines, and target DNA is indicated by dotted lines. Left and right transposon ends are arbitrarily indicated by small filled and open circles. Half-arrows within the figure-eight forms indicate 3′OH ends, which may act as replication primers in the formation of circular forms from the figure-eight intermediates. The left part of the figure represents a monomeric plasmid generating a simple figure-eight form, which evolves into an IS circle and then undergoes simple insertion into a target DNA molecule. The central part of the figure shows a plasmid dimer, which undergoes intra-IS recombination to generate a figure-eight molecule, which then evolves into an IS circle and likewise undergoes simple insertion. The right part of the figure shows the same dimeric plasmid undergoing inter-IS recombination and evolution of the corresponding figure-eight molecule into a plasmid carrying a tandem IS dimer. Integration of this structure results in the formation of a cointegrate. A second pathway between IS dimers and IS monomeric circles is shown, which involves targeted integration into a monomeric plasmid to generate the $[IS]^2$ copy and excision from the $[IS]^2$ to generate the IS circle and a plasmid monomer.

carry an active IRL-IRR junction and assemble p_{junc}. Integration using this junction would then generate a cointegrate.

Tandem IS structures have been observed for other ISs such as IS*30* (38), IS*21* (50), and IS*256* (32, 49), as well as the IS*3* family member, IS*2* (64). $[IS]^2$ copies may prove to be widespread as transposition intermediates. One implication of this is that the nature of the plasmid donor plays an important role in determining the nature of the transposition products.

$[IS]^2$ Formation from Figure-Eight Structures

The authenticity of this pathway has been examined in some detail. A dimeric plasmid in which the pair of ISs carried the *orfA-lacZ* translational fusion between two IS*911* ends was used in a standard in vitro reaction with OrfAB to generate both inter- and intra-IS figure-eight molecules (Fig. 7). In a typical experiment, where about 20% of the dimer substrate is converted to figure-eight molecules, about equal quantities of inter- and intra-IS recombination products were observed by gel electrophoresis. Following transformation with this deproteinized reaction mixture, inter-IS figure-eight molecules were resolved into an $[IS911]^2$ structure in the absence of transposase (70). Such molecules generate p_{junc} and activate an *orfA-lacZ* reporter gene rendering the host lac$^+$ (Fig. 6A). The notion that $[IS]^2$ forms are generated in the same way as IS circles is reinforced by the observation that the abutting IRs in $[IS2]^2$ molecules are separated by a single base pair (64) as they are in IS circles (30) rather than by the 3 bp generated as a direct target repeat following insertion.

Although perfectly stable in the absence of IS*911* proteins, propagation of the $[IS911]^2$ form in the presence of OrfAB resulted in the appearance of many deletion derivatives. Two of the major products were found to be a transposon circle and a monomeric plasmid donor in which one of the two IS copies had been excised. Similar products were obtained, in vitro and in vivo, using a plasmid carrying three IS*911* ends: two flanking the *orfA-lacZ* fusion and the third, a copy of IRR, forming an active junction with the IRL copy (Fig. 6B) (70). The ability to generate transposon circles in vitro is in strict contrast to the results obtained with monomeric plasmids (see above; 47). Circle formation using the IRR-IRL junction was proposed to occur by an intramolecular "integration" pathway where a concerted integration of IRR and IRL is targeted to the isolated IRR. This was confirmed by use of various combinations of mutations of the terminal 5′CA3′ dinucleotide and sequencing of the in vitro products (70). These results therefore demonstrate that transposition reactions can be targeted to a sequence resembling a transposon end by the OrfAB transposase.

[IS][2] Formation from Targeted Circle Insertions

The targeted intramolecular insertion of IRR-IRL into a transposon end could have important implications on the sequence specificity of intermolecular insertion. It may provide an explanation for the formation of the initial isolate of IS*911*. The sequence of the target λ cI gene flanking IRR in this isolate shows significant similarities with the IRs (30% identity with IRR and 47% with IRL). This sequence, inverted with respect to IRR, carries a conserved 5′-CA-3′ correctly oriented and spaced. It effectively forms an active junction with IRR and can be used as a surrogate end during transposition (45, 46).

Targeted insertion of transposon circles into a transposon end has been examined in an in vitro transposition reaction. This used a transposon circle carrying a chloramphenicol resistance gene and a target plasmid carrying an IS*911* derivative with an *orfA-lacZ* fusion where both ends had been inactivated by mutation of the conserved terminal dinucleotide. Insertion was monitored by transformation and selection for Cm^r. Site-specific integration was examined by formation of the p_{junc} promoter by insertion of the transposon circle at the resident IRL in the target plasmid (Fig. 6C and 7). This occurred at detectable frequencies in the presence of OrfAB and OrfA (B. Ton-Hoang, personal communication) and at significantly higher frequencies in the presence of OrfAB alone (>8% [C. Loot, unpublished data]). This is therefore an alternative route by which [IS][2] derivatives can be formed and could also give rise to cointegrate molecules. Insertions targeted to real or surrogate IR sequences have been observed with other IS elements such as IS*30* (37) and may occur by a related mechanism (25).

CONCLUSIONS AND PERSPECTIVES

Many transposable elements are separated from their flanking donor DNA prior to insertion into a target site. This behavior, termed cut-and-paste, requires processing of both strands at each end of the element. In the case of elements encoding a DDE transposase, cleavage of the first strand is probably similar (see chapter 2). On the other hand, different strategies have been adopted for resolution of the second, complementary, strand to liberate the transposon (see 69). These strategies include the use of two distinct endonucleases, each specific for one of the two DNA strands, as is the case for Tn7 (chapter 19); attack of the second strand by the terminal 3′OH group generated on the first strand, as is the case for IS*10* and IS*50* (see chapters 18 and 20); or prior generation of a linear substrate free from flanking donor DNA, as is the case for retroviruses (see chapters 25 to 27). The studies of IS*911* have shown that this element and other members of the IS*3* family have chosen a different route. Here, a liberated 3′OH at one transposon end attacks the same strand at the opposite end to generate a figure-eight form. Resolution of the second strand occurs by a mechanism independent of transposase to generate a transposon circle. Available evidence suggests that this mechanism is widespread and may have been adopted by members of several other IS families with DDE transposases including the IS*21*, IS*30*, and IS*256* families (14, 25, 32, 49).

Site-specific targeting involved in generating the figure eight is not limited to intra-IS reactions (involving interaction of an IRR and IRL from the same IS). Inter-IS reactions can clearly extend the recombination repertoire. This is illustrated in Fig. 7 where such interactions can lead to formation of tandem IS dimers either by resolution of an inter-IS figure eight formed with a dimeric donor plasmid or by inter-IS targeting between a transposon circle and a monomeric donor plasmid. Both pathways generate an active IRR-IRL junction capable of integrating into a target molecule to generate a cointegrate.

Several important questions concerning the mechanism of IS*911* transposition remain to be resolved. These include the stoichiometry of the paired end complexes with OrfAB[1–149] and their relationship with those presumably formed with full-length OrfAB; a detailed structural analysis of the proteins; and a definition of the mechanism involved in evolution from figure eight to transposon circle. Additional studies are also required to define the roles of OrfA and OrfB in IS*911* and for other members of the family.

Acknowledgments. We thank past members of the Mobile Genetic Elements Group (M. Bétermier, O. Fayet, L. Haren, P. Polard, M.-F. Prère, and B. Ton-Hoang) for their contributions to the study of IS*911*, B. Cointin for technical assistance, and O. Fayet for reading the manuscript.

This work benefited from grants from the CNRS (UPR9007) and the Programme Physique et Chimie du vivant, the Association pour la Recherche contre le Cancer (6406), the Région Midi-Pyrenées, and the MENRST (Programme microbiologie).

REFERENCES

1. **Baxevanis, A. D., and C. R. Vinson.** 1993. Interactions of coiled coils in transcription factors: where is the specificity? *Curr. Opin. Genet. Dev.* **3**:278–285.
2. **Berg, D. E.** 1983. Structural requirement for IS*50*-mediated gene transposition. *Proc. Natl. Acad. Sci. USA* **80**:792–796.
3. **Blattner, F. R., G. Plunkett, C. A. Bloch, N. T. Perna, V. Burland, M. Riley, J. Collado-Vides, J. D. Glasner, C. K. Rode, G. F. Mayhew, J. Gregor, N. W. Davis, H. A. Kirkpatrick, M. A. Goeden, D. J. Rose, B. Mau, and Y. Shao.** 1997. The complete genome sequence of *Escherichia coli* K-12. *Science* **277**: 1453–1474.

4. **Boursaux-Eude, C., I. Saint Girons, and R. Zuerner.** 1995. IS1500, an IS3-like element from Leptospira interrogans. *Microbiology* **141:**2165–2173.
5. **Bujacz, G., M. Jaskolski, J. Alexandratos, A. Wlodawer, G. Merkel, R. A. Katz, and A. M. Skalka.** 1995. High-resolution structure of the catalytic domain of avian sarcoma virus integrase. *J. Mol. Biol.* **253:**333–346.
6. **Burrus, V., Y. Roussel, B. Decaris, and G. Guedon.** 2000. Characterization of a novel integrative element, ICESt1, in the lactic acid bacterium *Streptococcus thermophilus*. *Appl. Environ. Microbiol.* **66:**1749–1753.
7. **Chow, S. A., K. A. Vincent, V. Ellison, and P. O. Brown.** 1992. Reversal of integration and DNA splicing mediated by integrase of human immunodeficiency virus. *Science* **255:**723–726.
8. **Dalrymple, B.** 1987. Novel rearrangements of IS30 carrying plasmids leading to the reactivation of gene expression. *Mol. Gen. Genet.* **207:**413–420.
9. **Doak, T. G., F. P. Doerder, C. L. Jahn, and G. Herrick.** 1994. A proposed superfamily of transposase genes: transposon-like elements in ciliated protozoa and a common "D35E" motif. *Proc. Natl. Acad. Sci. USA* **91:**942–946.

9a. **Duval-Valentin, G., C. Normand, V. Khemici, B. Marty, and M. Chandler.** 2001. Transient promoter formation: a new feedback mechanism for regulation of IS911 transposition. *EMBO J.* **20:**5802–5811.

10. **Dyda, F., A. B. Hickman, T. M. Jenkins, A. Engelman, R. Craigie, and D. R. Davies.** 1994. Crystal structure of the catalytic domain of HIV-1 integrase: similarity to other polynucleotidyl transferases. *Science* **266:**1981–1986.
11. **Farabaugh, P. J.** 1997. *Programmed Alternative Reading of the Genetic Code.* R. G. Landes Company, Austin, Tex.
12. **Fayet, O., P. Ramond, P. Polard, M. F. Prere, and M. Chandler.** 1990. Functional similarities between retroviruses and the IS3 family of bacterial insertion sequences? *Mol. Microbiol.* **4:**1771–1777.
13. **Fu, R., and G. Voordouw.** 1998. IS*D1,* an insertion element from the sulfate-reducing bacterium *Desulfovibrio vulgaris* Hildenborough: structure, transposition, and distribution. *Appl. Environ. Microbiol.* **64:**53–61.
14. **Haas, D., B. Berger, S. Schmid, T. Seitz, and C. Reimmann.** 1996. Insertion sequence IS*21:* related insertion sequence elements, transpositional mechanisms, and application to linker insertion mutagenesis, p. 238–249. *In* T. Nakazawa, K. Furukawa, D. Haas, and S. Silver (ed.), *Molecular Biology of Pseudomonads.* ASM Press, Washington, D.C.
15. **Haren, L.** 1998. *Relations structure/fonction des facteurs protéiques exprimés par l'elément transposable bacterien IS911.* Ph.D. thesis. Université Paul Sabatier, Toulouse, France.
16. **Haren, L., M. Betermier, P. Polard, and M. Chandler.** 1997. IS*911*-mediated intramolecular transposition is naturally temperature sensitive. *Mol. Microbiol.* **25:**531–540.
17. **Haren, L., C. Normand, P. Polard, R. Alazard, and M. Chandler.** 2000. IS*911* transposition is regulated by protein-protein interactions via a leucine zipper motif. *J. Mol. Biol.* **296:**757–768.
18. **Haren, L., P. Polard, B. Ton-Hoang, and M. Chandler.** 1998. Multiple oligomerisation domains in the IS*911* transposase: a leucine zipper motif is essential for activity. *J. Mol. Biol.* **283:**29–41.
19. **Haren, L., B. Ton-Hoang, and M. Chandler.** 1999. Integrating DNA: transposases and retroviral integrases. *Annu. Rev. Microbiol.* **53:**245–281.
20. **Harley, C. B., and R. P. Reynolds.** 1987. Analysis of *E. coli* promoter sequences. *Nucleic Acids Res.* **15:**2343–2361.
21. **Hu, S. T., J. H. Hwang, L. C. Lee, C. H. Lee, P. L. Li, and Y. C. Hsieh.** 1994. Functional analysis of the 14 kDa protein of insertion sequence 2. *J. Mol. Biol.* **236:**503–513.
22. **Hu, S. T., L. C. Lee, and G. S. Lei.** 1996. Detection of an IS2-encoded 46-kilodalton protein capable of binding terminal repeats of IS2. *J. Bacteriol.* **178:**5652–5659.
23. **Jacks, T.** 1990. Translational suppression in gene expression in retroviruses and retrotransposons. *Curr. Top. Microbiol. Immunol.* **157:**93–124.
24. **Khan, E., J. P. Mack, R. A. Katz, J. Kulkosky, and A. M. Skalka.** 1991. Retroviral integrase domains: DNA binding and the recognition of LTR sequences. *Nucleic Acids Res.* **19:**851–860. (Erratum, **19:**1358.)
25. **Kiss, J., and F. Olasz.** 1999. Formation and transposition of the covalently closed IS*30* circle: the relation between tandem dimers and monomeric circles. *Mol. Microbiol.* **34:**37–52.
26. **Kleckner, N.** 1979. DNA sequence analysis of Tn10 insertions: origin and role of 9 bp flanking repetitions during Tn10 translocation. *Cell* **16:**711–720.
27. **Kulkosky, J., K. S. Jones, R. A. Katz, J. P. Mack, and A. M. Skalka.** 1992. Residues critical for retroviral integrative recombination in a region that is highly conserved among retroviral/retrotransposon integrases and bacterial insertion sequence transposases. *Mol. Cell Biol.* **12:**2331–2338.
28. **Kusumoto, M., Y. Nishiya, and Y. Kawamura.** 2000. Reactivation of insertionally inactivated Shiga toxin 2 genes of *Escherichia coli* O157:H7 caused by nonreplicative transposition of the insertion sequence. *Appl. Environ. Microbiol.* **66:**1133–1138.
29. **Lei, G. S., and S. T. Hu.** 1997. Functional domains of the InsA protein of IS2. *J. Bacteriol.* **179:**6238–6243.
30. **Lewis, L. A., and N. D. Grindley.** 1997. Two abundant intramolecular transposition products, resulting from reactions initiated at a single end, suggest that IS2 transposes by an unconventional pathway. *Mol. Microbiol.* **25:**517–529.
31. **Lupas, A.** 1996. Coiled coils: new structures and new functions. *Trends. Biochem. Sci.* **21:**375–382.
32. **Lyon, B. R., M. T. Gillespie, and R. A. Skurray.** 1987. Detection and characterization of IS256, an insertion sequence in *Staphylococcus aureus*. *J. Gen. Microbiol.* **133:**3031–3038.
33. **Macrina, F. L., K. R. Jones, and P. Laloi.** 1996. Characterization of IS*199* from *Streptococcus mutans* V403. *Plasmid* **36:**9–18.
34. **Mahillon, J., and M. Chandler.** 1998. Insertion sequences. *Microbiol. Mol. Biol. Rev.* **62:**725–774.
35. **Mizuuchi, K.** 1997. Polynucleotidyl transfer reactions in site-specific DNA recombination. *Genes Cells* **2:**1–12.
36. **Normand, C., G. Duval-Valentin, L. Haren, and M. Chandler.** 2001. The terminal inverted repeats of IS911: requirement for synaptic complex assembly and activity. *J. Mol. Biol.* **308:**853–871.
37. **Olasz, F., T. Farkas, J. Kiss, A. Arini, and W. Arber.** 1997. Terminal inverted repeats of insertion sequence IS*30* serve as targets for transposition. *J. Bacteriol.* **179:**7551–7558.
38. **Olasz, F., R. Stalder, and W. Arber.** 1993. Formation of the tandem repeat (IS30)2 and its role in IS*30*-mediated transpositional DNA rearrangements. *Mol. Gen. Genet.* **239:**177–187.
39. **Osawa, S., T. H. Jukes, K. Watanabe, and A. Muto.** 1992. Recent evidence for evolution of the genetic code. *Microbiol. Rev.* **56:**229–264.
40. **Perkins-Balding, D., G. Duval-Valentin, and A. C. Glasgow.** 1999. Excision of IS*492* requires flanking target sequences and results in circle formation in *Pseudoalteromonas atlantica*. *J. Bacteriol.* **181:**4937–4948.
41. **Polard, P.** 1993. Etude de l'Elément Transposable Bacterien IS*911:* mise en evidence et charactérisation des facteurs

necessaires a sa transposition. Ph.D. thesis. Université Paul Sabatier, Toulouse, France.
42. **Polard, P., and M. Chandler.** 1995. An in vivo transposase-catalyzed single-stranded DNA circularization reaction. *Genes Dev.* **9:**2846–2858.
43. **Polard, P., and M. Chandler.** 1995. Bacterial transposases and retroviral integrases. *Mol. Microbiol.* **15:**13–23.
44. **Polard, P., M.-F. Prère, M. Chandler, and O. Fayet.** 1991. Programmed translational frameshifting and initiation at an AUU codon in gene expression of bacterial insertion sequence IS911. *J. Mol. Biol.* **222:**465–477.
45. **Polard, P., M.-F., Prère, O. Fayet, and M. Chandler.** 1992. Transposase-induced excision and circularization of the bacterial insertion sequence IS911. *EMBO J.* **11:**5079–5090.
46. **Polard, P., L. Seroude, O. Fayet, M.-F. Prère, and M. Chandler.** 1994. One-ended insertion of IS*911*. *J. Bacteriol.* **176:** 1192–1196.
47. **Polard, P., B. Ton-Hoang, L. Haren, M. Betermier, R. Walczak, and M. Chandler.** 1996. IS*911*-mediated transpositional recombination in vitro. *J. Mol. Biol.* **264:**68–81.
48. **Prère, M.-F., M. Chandler, and O. Fayet.** 1990. Transposition in *Shigella dysenteriae:* isolation and analysis of IS*911*, a new member of the IS*3* group of insertion sequences. *J. Bacteriol.* **172:**4090–4099.
49. **Prudhomme, M., C. Turlan, J.-P. Claverys, and M. Chandler.** 2002. Diversity of Tn*4001* transposition products: the flanking IS*256* elements can form tandem dimers and IS circles. *J. Bacteriol.* **184:**433–443.
50. **Reimmann, C., R. Moore, S. Little, A. Savioz, N. S. Willetts, and D. Haas.** 1989. Genetic structure, function and regulation of the transposable element IS*21*. *Mol. Gen. Genet.* **215:** 416–424.
51. **Rettberg, C. C., M. F. Prere, R. F. Gesteland, J. F. Atkins, and O. Fayet.** 1999. A three-way junction and constituent stem-loops as the stimulator for programmed −1 frameshifting in bacterial insertion sequence IS*911*. *J. Mol. Biol.* **286:** 1365–1378.
52. **Rezsohazy, R., B. Hallet, J. Delcour, and J. Mahillon.** 1993. The IS*4* family of insertion sequences: evidence for a conserved transposase motif. *Mol. Microbiol.* **9:**1283–1295.
53. **Rice, P., R. Craigie, and D. R. Davies.** 1996. Retroviral integrases and their cousins. *Curr. Opin. Struct. Biol.* **6:**76–83.
54. **Sacerdot, C., G. Fayat, P. Dessen, M. Springer, J. A. Plumbridge, M. Grunberg-Manago, and S. Blanquet.** 1982. Sequence of a 1.26-kb DNA fragment containing the structural gene for *E. coli* initiation factor IF3: presence of an AUU initiator codon. *EMBO J.* **1:**311–315.
55. **Schell, M. A.** 1993. Molecular biology of the lysR family of transcriptional regulators. *Annu. Rev. Microbiol.* **47:** 597–626.
56. **Schmidt, H., B. Henkel, and H. Karch.** 1997. A gene cluster closely related to type II secretion pathway operons of gram-negative bacteria is located on the large plasmid of enterohemorrhagic *Escherichia coli* O157 strains. *FEMS Microbiol. Lett.* **148:**265–272.
57. **Schnabel, E. L., and A. L. Jones.** 1999. Distribution of tetracycline resistance genes and transposons among phylloplane bacteria in Michigan apple orchards. *Appl. Environ. Microbiol.* **65:**4898–4907.
58. **Schwartz, E., M. Kroger, and B. Rak.** 1988. IS150: distribution, nucleotide sequence and phylogenetic relationships of a new E. coli insertion element. *Nucleic Acids Res.* **16:** 6789–6802.
59. **Sekine, Y., K. Aihara, and E. Ohtsubo.** 1999. Linearization and transposition of circular molecules of insertion sequence IS3. *J. Mol. Biol.* **294:**21–34.
60. **Sekine, Y., N. Eisaki, and E. Ohtsubo.** 1994. Translational control in production of transposase and in transposition of insertion sequence IS3. *J. Mol. Biol.* **235:**1406–1420.
61. **Sekine, Y., N. Eisaki, and E. Ohtsubo.** 1996. Identification and characterization of the linear IS3 molecules generated by staggered breaks. *J. Biol. Chem.* **271:**197–202.
62. **Sekine, Y., K. Izumi, T. Mizuno, and E. Ohtsubo.** 1997. Inhibition of transpositional recombination by OrfA and OrfB proteins encoded by insertion sequence IS3. *Genes Cells* **2:** 547–557.
63. **Swenson, D. L., N. O. Bukanov, D. E. Berg, and R. A. Welch.** 1996. Two pathogenicity islands in uropathogenic *Escherichia coli* J96: cosmid cloning and sample sequencing. *Infect. Immun.* **64:**3736–3743.
64. **Szeverenyi, I., T. Bodoky, and F. Olasz.** 1996. Isolation, characterization and transposition of an (IS2)2 intermediate. *Mol. Gen. Genet.* **251:**281–289.
65. **Tobe, T., T. Hayashi, C. G. Han, G. K. Schoolnik, E. Ohtsubo, and C. Sasakawa.** 1999. Complete DNA sequence and structural analysis of the enteropathogenic *Escherichia coli* adherence factor plasmid. *Infect. Immun.* **67:**5455–5462.
66. **Ton-Hoang, B., M. Betermier, P. Polard, and M. Chandler.** 1997. Assembly of a strong promoter following IS911 circularization and the role of circles in transposition. *EMBO J.* **16:** 3357–3371.
67. **Ton-Hoang, B., P. Polard, and M. Chandler.** 1998. Efficient transposition of IS911 circles in vitro. *EMBO J.* **17:** 1169–1181.
68. **Ton-Hoang, B., P. Polard, L. Haren, C. Turlan, and M. Chandler.** 1999. IS911 transposon circles give rise to linear forms that can undergo integration in vitro. *Mol. Microbiol.* **32:** 617–627.
68a. **Tsuchihashi, Z., and P. O. Brown.** 1992. Sequence requirements for efficient translational frameshifting in the Escherichia coli dnaX gene and the role of an unstable interaction between tRNA(Lys) and an AAG lysine codon. *Genes Dev.* **6:** 511–519.
69. **Turlan, C., and M. Chandler.** 2000. Playing second fiddle: second-strand processing and liberation of transposable elements from donor DNA. *Trends Microbiol.* **8:**268–274.
70. **Turlan, C., B. Ton-Hoang, and M. Chandler.** 2000. The role of tandem IS dimers in IS911 transposition. *Mol. Microbiol.* **35:**1312–1325.
71. **Utsumi, R., M. Ikeda, T. Horie, M. Yamamoto, A. Ichihara, Y. Taniguchi, R. Hashimoto, H. Tanabe, K. Obata, and M. Noda.** 1995. Isolation and characterization of the IS3-like element from Thermus aquaticus. *Biosci. Biotechnol. Biochem.* **59:**1707–1711.
72. **van Gent, D. C., A. A. Groeneger, and R. H. Plasterk.** 1992. Mutational analysis of the integrase protein of human immunodeficiency virus type 2. *Proc. Natl. Acad. Sci. USA* **89:** 9598–9602.
73. **Vogele, K., E. Schwartz, C. Welz, E. Schiltz, and B. Rak.** 1991. High-level ribosomal frameshifting directs the synthesis of IS150 gene products. *Nucleic Acids Res.* **19:**4377–4385.
74. **Vos, J. C., and R. H. Plasterk.** 1994. Tc1 transposase of Caenorhabditis elegans is an endonuclease with a bipartite DNA binding domain. *EMBO J.* **13:**6125–6132.
75. **Weiss, R. B., D. M. Dunn, M. Shuh, J. F. Atkins, and R. F. Gesteland.** 1989. E. coli ribosomes re-phase on retroviral frameshift signals at rates ranging from 2 to 50 percent. *New Biol.* **1:**159–169.
76. **Zheng, J., and M. A. McIntosh.** 1995. Characterization of IS1221 from Mycoplasma hyorhinis: expression of its putative transposase in Escherichia coli incorporates a ribosomal frameshift mechanism. *Mol. Microbiol.* **16:**669–685.

Mobile DNA II
Edited by N. L. Craig et al.

Chapter 17

Transposition of Phage Mu DNA

GEORGE CHACONAS AND RASIKA M. HARSHEY

George Chaconas • Department of Biochemistry, University of Western Ontario, London, Ontario N6A 5C1, Canada. **Rasika M. Harshey** • Section of Molecular Genetics and Microbiology and Institute of Cellular and Molecular Biology, University of Texas at Austin, Austin, TX 78712.

INTRODUCTION

Mu is a temperate phage which derives its name (mutator) from its ability to integrate into numerous sites in the *Escherichia coli* genome and cause mutations by insertional inactivation (162). The importance of Mu as a transposable element was recognized very early by Ahmad Bukhari at Cold Spring Harbor and by Arianne Toussaint in Brussels (see reference 161). This realization, combined with the highly efficient amplification of Mu as a virus, was instrumental not only in rapidly advancing our understanding of the many DNA rearrangements associated with transposition (see reference 21), but also in leading to the development of the first in vitro transposition system by Kiyoshi Mizuuchi (109).

Two pathways for Mu transposition are generally distinguished in vivo (see references 26, 66, and 133). Upon infection of a sensitive host, the linear virion DNA is converted into a noncovalently closed circular form with the help of an injected protein (51, 68), which likely protects the ends from nucleases (see reference 133). Infecting DNA is not replicated prior to integration; this process is also referred to as conservative integration (5, 67, 100). However, in the ensuing lytic cycle the initial conservative event is immediately followed by replication (67), where Mu DNA is amplified over 100-fold to produce "cointegrates" (28). It is unclear whether the distinction between "conservative" and "replicative" reflects a true duality in mechanism (double-strand versus single-strand donor cleavage, respectively) or whether these terms apply to two successive snapshots along a single pathway for Mu transposition during an infectious cycle; that is, the integration of infecting Mu DNA could occur through the cointegrate pathway. In the case of a single-transposition pathway, the minority of events that lead to stable lysogeny (see reference 66) might conceivably arise from repair of the strand transfer intermediate to generate a simple insertion as has been observed in vitro (109) (see below).

Of the several models for replicative transposition, the one proposed first by Jim Shapiro (8, 152) was shown to be true for Mu in vitro, lending the term "Shapiro intermediate" to the branched-strand transfer product of this cointegrate pathway (see reference 112). As more transposable elements are studied in vitro, the cointegrate mode of transposition (also deduced from in vivo studies for the Tn*3* family of transposons [54]) appears to be an exception rather than a rule. For example, all prokaryotic elements studied in vitro thus far (Tn*5*, Tn*7*, Tn*10*, IS*911*) transpose by a conservative cut-and-paste mechanism which includes hairpin intermediates in the case of Tn*10* and Tn*5* (see chapters in this book). The cut-and-paste mechanism appears to be very popular in eukaryotic organisms as well (see references 65 and 101). Eukaryotic elements that couple integration to replication (not necessarily in that order) are retroviruses, retrotransposons, and group II introns (see chapters in this book).

Mu DNA TRANSPOSITION IN VITRO

Our knowledge of the DNA transposition process in Mu was catapulted forward through in vitro studies which provided a paradigm for the study of other moveable elements (for earlier reviews, see references 25, 29, 30, 63, 93, 110–112, 114, 133, and 161). Although the Mu replicative transposition process has now been studied in vitro for nearly 2 decades, conservative integration of infecting Mu DNA, if indeed it proceeds through a cut-and-paste mechanism (see above), has not been observed in vitro.

Figure 1 depicts the replicative transposition reaction using a mini-Mu donor molecule and a small circular target DNA. Early in vitro studies in replication-competent *E. coli* extracts resulted in a cointegrate product (109). Elimination of polyvinyl alcohol from the reactions (polyvinyl alcohol is required for the DNA replication process) resulted in the accumulation of Shapiro-type intermediates, or θ structures, allowing biochemical and electron microscopic characterization (36, 107). Shortly thereafter, it became possible to perform the cleavage and strand transfer steps of Mu DNA transposition with purified proteins (35), resulting in a more detailed dissection of this process.

The Mu cleavage and strand transfer reactions differ from the breakage and reunion reactions promoted by site-specific recombinases and other topoisomerases. The latter enzymes nick DNA via a nucleophilic attack from the hydroxyl group of an active site serine or tyrosine (see reference 110). The resulting covalent protein-DNA intermediate conserves the energy of the broken phosphodiester bond, and this energy is subsequently used for the strand transfer step. In Mu transposition, however, donor cleavage occurs through a direct nucleophilic attack by water at the Mu-host junction to liberate free 3′ OH groups at the Mu ends (115). In the subsequent one-step transesterification reaction, these 3′ OH groups then function as nucleophiles in a direct in-line attack on the target DNA (113).

Pioneering experiments designed by Kiyoshi Mizuuchi to sort out the Mu reaction chemistry have become the "gold standard" in the transposition field (see references 80, 110, and 115 and other chapters in this book). The essence of the methodology has

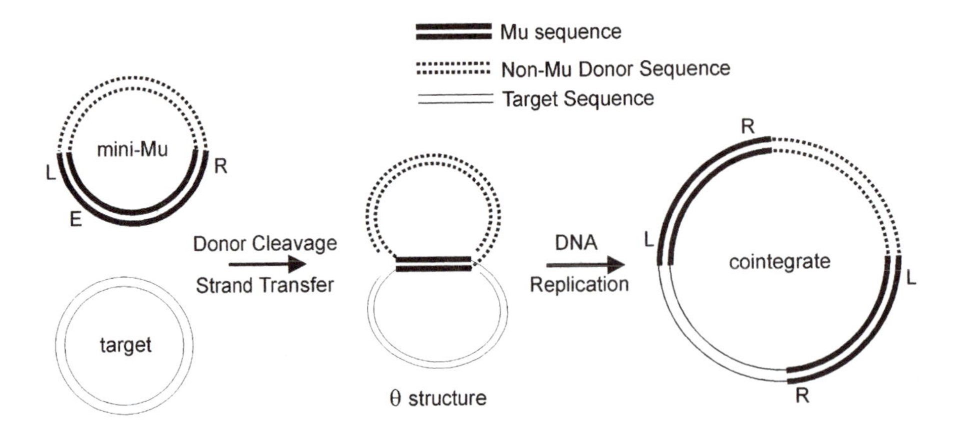

Figure 1. Replicative transposition of Mu DNA in vitro. L, R, and E represent the Mu left end, right end, and transpositional enhancer, respectively. Reprinted from *Biochemistry and Cell Biology* (25) with permission of the publisher.

been to determine the stereochemistry of a reaction step by using chiral phosphorothioates. An inversion of chirality is expected for single-step reactions, while two-step reactions (those proceeding through a protein-DNA intermediate) would invert chirality twice, resulting in the same stereochemistry as the starting product. The first experiments were performed for the strand transfer step in the Mu reaction (113). The Mu donor cleavage step was refractory to analysis in this manner because of the lack of a chiral center. The methodology was, however, amenable to analysis of the DNA cleavage step using human immunodeficiency virus (HIV) integrase, where a chiral center was obtainable (46), and has since been used in a variety of systems. Both cleavage and strand transfer in the HIV system displayed an inversion of chirality, again indicative of a single-step reaction in both cases, without a covalent protein-DNA intermediate.

More recently, Mizuuchi has solved the problem of analyzing the stereochemistry of the Mu cleavage reaction by developing a new approach (115). A phosphorothioate of known chirality is introduced at the scisslie bond, and the cleavage step is performed in the presence of $H_2{}^{18}O$ to label the cleavage product. The stereospecificity of the cleavage step is then deduced by following the ^{18}O label during a subsequent ligation step. Because of the stereochemical selectivity of T4 DNA ligase, the ^{18}O label will be either lost or retained, depending upon the chirality of the substrate. Results from these experiments directly demonstrated an inversion of chirality in the Mu donor cleavage step.

DNA SITES

Ends

The sophistication of Mu as a transposable element is reflected in the number and intricate arrangement of DNA sites required for transposition. The simple inverted repeats found at the ends of many transposons are replaced in Mu by three Mu A (transposase) binding sites at each end (39, 56, 187), also referred to as *att* or end sites (shown in Fig. 2). The arrangement of the *att* sites differs at both ends, with the three left *att* sites all in the same orientation but with an ~80-bp space between L1 and L2 and a smaller space separating L2 and L3; L2 is a weaker site, showing less protection by Mu A, and has been termed a half-site (187). In contrast, the right end has three Mu A binding sites abutting each other; the R3 site is in the opposite orientation of R1 and R2. The reason for this complex arrangement of transposase binding sites is not yet fully appreciated but is in part related to the presence of the *pac* site which resides between L1 and L2 and is required for the headful packaging of Mu phage DNA (57, 64) and could in part be related to elaborate *att*-enhancer interactions

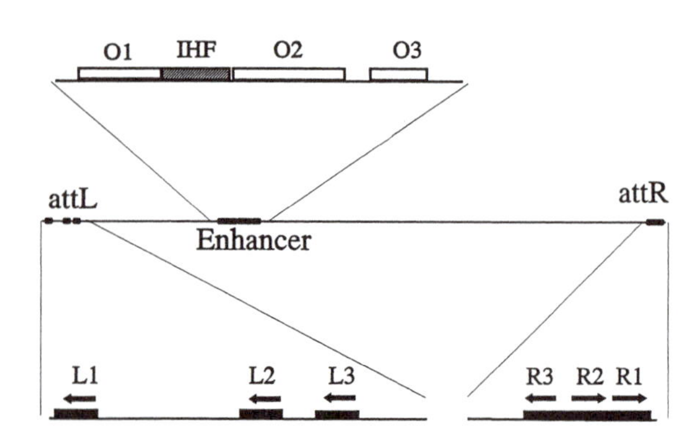

Figure 2. Regions of Mu DNA required for the strand transfer reaction. The substructures of the three required regions are shown in the enlargements. Reprinted from the *Journal of Biological Chemistry* (7) with permission of the publisher.

needed for achieving a correct Mu synapse (see below).

The binding affinity of Mu A for each of the six *att* sites varies (39), as does the relative importance of the sites for transposition (7, 58, 94). The core sites L1, R1, and R2 are essential for the in vitro reaction; however, any one of the accessory sites (L2, L3, and R3) can be eliminated without severe effects. Mu A has a higher affinity for the end sites than the enhancer sites (see below) (39, 97). A complex circuit of interactions exists between the ends and enhancer, and it appears that all six Mu end binding sites are involved in enhancer interactions (see below) (7, 93, 116).

The Enhancer

The transpositional enhancer (97, 119, 160), also referred to as the internal activating sequence, is another accoutrement in the arsenal of the complicated Mu system (Fig. 2). The enhancer consists of three large sites, O1 to O3, originally described as the operator region (see reference 52). A binding site for *E. coli* integration host factor (IHF) lies between O1 and O2 (71). This region is also a hub of transcriptional regulatory activity on the Mu genome, involved in regulating divergent promoters for transposition and repressor functions (71, 164, 165).

Mu (and the related phage D108 [43]) is the only known transposon to carry an enhancer element, which is located about 1 kb from the left end but functions in a distance-independent manner. A functional enhancer in vitro requires O1, the intervening IHF site, and only part of O2 (160); O3 is expendable for enhancer activity, although its presence promotes optimal reaction in the absence of IHF (73). The presence of the enhancer increases the efficiency of transposition over two orders of magnitude in vivo (97) and accelerates the initial rate of transposition in vitro by a similar amount (160). The enhancer element plays an essential role in assembly of the transpososome at the two ends but is not required for the actual chemical steps of the reaction (117, 156). Although the enhancer normally acts in *cis,* it can be active in *trans* when supplied at a 40-fold molar excess near the end of an unlinked linear DNA molecule in the presence of IHF (156).

Enhancers control the rate and directionality of site-specific recombination reactions by promoting the assembly of topologically correct, recombination-competent structures only when the target sites are arranged in a defined orientation in negatively supercoiled substrates (33, 55, 78, 79, 154). In addition to their structural role, enhancers can also modulate the activity of the recombinase by promoting specific protein-protein interactions (104). The enhancer plays similar roles in Mu transposition (see "Regulatory Role of Transpososomes," below) (73, 173, 183), and interactions have been deduced between both sides of the enhancer and both Mu ends (7). Recent elegant experiments using Mu A protein variants with altered enhancer binding specificity (73) have definitively established crisscrossed interactions between O1 and R1 and between O2 and L1 as previously proposed (R. G. Allison and G. Chaconas, unpublished data cited in reference 93).

The SGS

An essential site in the center of the Mu genome is worthy of mention here, even though it is not required for transposition in vitro. The Mu strong gyrase site (SGS) is required for Mu replication in vivo and is believed to function by influencing efficient synapsis of the Mu ends (134–138). The SGS is thought to act by localizing the 37-kb Mu prophage DNA into a single loop of plectonemically supercoiled DNA upon binding of DNA gyrase to the site. The SGS, predicted to reside at the turnaround of the loop, would constrain the two Mu ends to the same supercoiled domain to facilitate synapsis. The SGS is not needed for in vitro reactions, likely because the distance between the ends is small on plasmid substrates, and there is no possibility of the two Mu ends residing in different supercoiled domains as could occur in the host chromosome.

Target DNA

Although Mu is fastidious with respect to its end and enhancer sites, it is relatively flexible about its target sites. Mu insertions can be recovered in virtually any gene of interest, and a target site consensus NY(G/C)RN has been determined (120). Nonetheless, hot spots for Mu integration on small plasmids have been characterized both in vivo (24) and in vitro (120). In vitro, the observed hot spots are associated with DNA sites that have a high affinity for Mu B, which recruits the target DNA. Interestingly, changing some flanking host sequences directly adjacent to the Mu ends, which constitute a target site in the previous round of transposition, results in substrates poorly reactive in the donor cleavage reaction (179); the inhibition is at the cleavage step, and formation of stable synaptic complexes (see below) is unaffected. The inhibitory flanking sequences may be refractory to the necessary DNA deformations, which are believed to be involved in donor cleavage (94, 149, 172).

Both in vitro and in vivo experiments suggest that an appropriate target for Mu integration must lack Mu A binding sites (1, 4, 40, 102). Binding of Mu A to a potential target with Mu A binding sites confers target or transposition immunity, which may protect an entire supercoiled domain of the bacterial chromosome from Mu insertion. The mechanism of target immunity is described in the section on Mu B protein below. Target selection in vivo is influenced by additional factors, and Mu integration appears to be directed to the regulatory regions around nonexpressed genes (170), likely minimizing the damaging effects of integration.

DNA SUPERCOILING

The Mu DNA transposition reaction in vitro has a strict requirement for negative DNA supercoiling in the donor substrate under standard reaction conditions (109). DNA supercoiling is not required for the chemical steps of the reaction but is crucial for the assembly of Mu transposition complexes (38, 117, 155, 171, 172).

The conformational and torsional effects of supercoiling (167, 168) are likely to be manifested at different steps of the Mu assembly pathway (78, 131). First, supercoiling improves the binding affinity of Mu A to end binding sites (88) and of HU to the L1-L2 spacer (83). In the case of HU, stable binding is actually dependent on supercoiling. Conformational effects of supercoiling can also promote the formation of higher-order complexes which contain wrapped DNA with negative writhe and/or trapped negative superhelical nodes. At present the topology of the Mu transpososomes remains to be established, but they likely contain some elements of negative supercoiling (37, 156). In addition, negative supercoiling in the donor DNA would favor the alignment of the two Mu ends in their proper orientation during synapsis, providing a "topological filter" (see reference 49). Elegant experiments by Craigie and Mizuuchi (37) using donor DNA comprised of multiply catenated or knotted DNA molecules have shown that proper interwrapping of the two Mu ends to give a fixed geometrical configuration is an essential feature of the reaction. Only Mu DNA ends provided with the correct geometry were utilized for productive synapsis even if they were on separate but catenated molecules or if they were present as direct repeats on a knotted DNA molecule. Thus, DNA supercoiling favors synapsis of correctly oriented Mu ends but hinders that of Mu ends in the wrong orientation.

The conformational effects of DNA supercoiling are also known to dramatically increase the local concentration of DNA sites on a given molecule by a factor of about 100 (166, 169). This would be expected to impart an impressive kinetic stimulation to a reaction requiring the binding and interaction of multiple proteins at multiple DNA sites. A demonstration of this effect is illustrated by the kinetic properties of the Mu reaction when the Mu enhancer is provided in *trans*. A 40-fold molar excess of the Mu enhancer in *trans* resulted in a reaction rate 25 times slower than that with the enhancer in *cis* on a supercoiled substrate (156).

The torsional effects of DNA supercoiling are likely to be expressed at both the enhancer and the Mu ends. At the enhancer, supercoiling must favor DNA bending. This has been deduced from the observation that the level of DNA supercoiling required for optimal reaction in vitro can be cut in half if IHF is included in the reaction (157). IHF binds at the enhancer between O1 and O2, inducing a sharp bend (71, 157, 160). The same, or a similar, DNA deformation can apparently be facilitated by high levels of DNA supercoiling in the absence of IHF (157).

At the *att* ends, torsional effects are exhibited at two places. The first is DNA bending between L1 and L2 to facilitate stable binding of HU in this region (83). The supercoiling dependence of this process is very similar to that of the overall reaction (in the presence of IHF), suggesting that it may consume a significant portion of the free energy of supercoiling. DNA bending in the region of the Mu-host junction (94) and helix destabilization in the flanking host DNA (149, 172) found upon stable synapsis are also likely to be driven by the free energy stored in the superhelical substrate (171, 172). Moreover, protein conformational changes in Mu A which occur upon stable complex formation are also possible candidates for transitions driven by the free energy of DNA supercoiling.

Finally, although DNA supercoiling is not required for the transesterification step, supercoiled substrates are superior targets, and target capture (TC) complexes can only be stabilized for observation when a superhelical target molecule is used (122). Both the conformational and torsional effects of DNA supercoiling may influence target capture.

PROTEINS

Mu A

The chemical steps of transposition are catalyzed by the transposase, Mu A, which binds primarily at the Mu ends. Mu A expression is negatively regulated in vivo by the lysogenic repressor c, which binds to an adjacent operator region on the Mu genome (see reference 52). As described above, the operator is none other than the enhancer element required for transposition. The N termini of Mu A and Mu c share extensive homology (70, 121) and, accordingly, share binding sites (39, 97, 119). It stands to reason, therefore, that Mu A and Mu c would battle for control at several levels in vivo. Indeed, Mu c blocks Mu transposition in vitro (119). The activity of Mu A is unstable in vivo (17, 48, 139), possibly related to the observed degradation of Mu A monomers by ClpXP (*E. coli* molecular chaperone ClpX complexed with its protease ClpP) in vitro (98). ClpXP degrades Mu c both in vitro and in vivo (50, 77, 176) and thus plays an important role in the lytic cycle of the phage.

Mu A protein is 663 amino acids (aa) long (75 kDa) (70) and can be divided into three major globular domains to which various functions have been mapped (126) (Fig. 3) and for which partial structures are available. The N-terminal domain Iα interacts with the operator-enhancer (97, 119) through a winged-helix DNA-binding motif (31a, 31b). Domain Iβγ interacts with each *att* sequence (81, 97, 126, 130, 187) using separate helix-turn-helix DNA binding motifs in Iβ and Iγ, each of which recognizes half of the Mu A consensus sequence (32, 150). Individual Mu A binding sites are contacted by monomeric Mu A on one face of the helix, over two major and an intervening minor groove (187), accompanied by DNA bending (2, 42, 88).

The central Mu A domain contains a triad of catalytic acidic residues known as the Asp-Asp-Glu (or DDE) motif at positions 269, 336, and 392, respectively (12, 82, 85). These residues are thought to coordinate divalent metal ions implicated in assisting the various nucleophilic attacking groups during the cleavage and strand transfer reactions. The normal cation for the Mu reaction appears to be Mg^{2+}; however, various other cations, including Ca^{2+} (117), Mn^{2+} (12, 82), Zn^{2+}, and Co^{2+} (172), can substitute for Mg^{2+} at different points along the reaction pathway. The DDE motif is phylogenetically conserved in transposases from both prokaryotic and eukaryotic transposable elements (65, 101). While the DDE domain of the HIV integrase is active in a disintegration assay (an apparent reversal of integration) (22, 31), no catalytic activity has been observed for the isolated domain II of Mu A; this domain can, however, allosterically activate the Mu A tetramer for intramolecular strand transfer (180).

A crystal structure of domain II (144) has revealed a remarkable similarity in the overall topology of this domain to DDE domains from the HIV (44), avian sarcoma virus (19, 20), and other retroviral integrases (see chapter 25) as well as to the Tn*5* transposase-related inhibitor protein (41), in spite of little or no similarity in their primary sequence (Color Plate 32 [see color insert]). Mu A and these other transposition proteins are also structurally related to RNase H and the RuvC resolvase and belong to a structurally related superfamily of polynucleotidyl transferases (see references 143 and 186).

The structure of domain II also shows a C-terminal subdomain (IIβ) with a large positively charged surface which may account for the observed DNA binding activity of domain II (126). Domain IIβ comprises a distinct complementation group from domain IIα and may play a role in complex assembly, structural transitions, and binding of the donor DNA flanking the Mu cleavage site (85, 129). Mutations in domains IIβ and IIIα are phenotypically similar and the two structural domains may constitute a single functional domain (Fig. 3).

Domain III of Mu A is functionally divisible into two regions. A 26-aa peptide in IIIα has a nonspecific

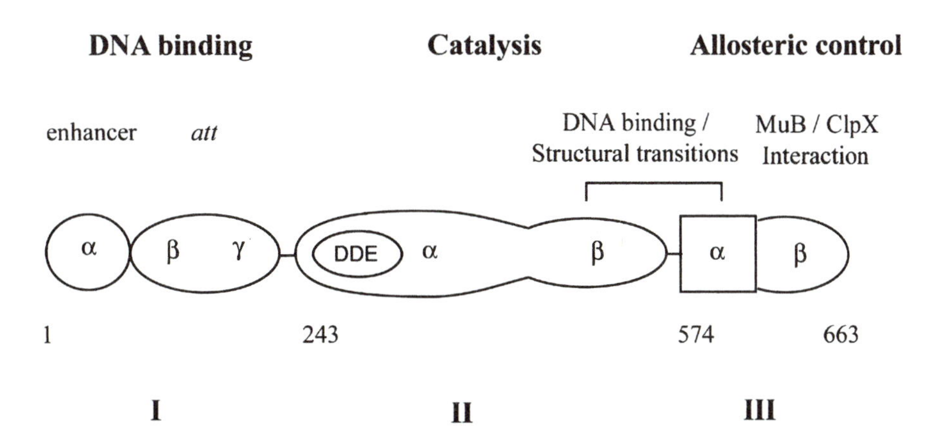

Figure 3. Domain structure of Mu A. On the basis of limited proteolysis, three domains (I to III) were assigned to Mu A protein, whose general functions are indicated. Amino acid numbers corresponding to the amino terminus of each major domain are shown beneath the structure. DDE refers to residues in the catalytic triad. See the text for details.

DNA-binding and nuclease activity associated with positively charged residues RRRKQ (181); similar basic motifs occur in other transposases and integrases. The phenotypes of mutations in domain IIIα of Mu A suggest that this region is involved in contacts with the Mu-host junction region and may be involved in activation of the substrate for cleavage. As noted above, residues in domain IIIα, along with residues in IIβ, are implicated in assembly of the transpososome and in structural or catalytic transitions as the complex passes from assembly to cleavage to strand transfer (85, 123, 129).

The C-terminal end of domain III (IIIβ) interacts with Mu B protein (14, 69, 96, 178), with the last 10 residues encoding overlapping recognition for the disassembly factor ClpX as well (99). As discussed below, Mu B modulates the activity of Mu A at several stages and supplies Mu A with captured target DNA for the integration of Mu ends.

Mu B

Mu B protein (along with the Mu enhancer) is responsible for the phenomenal efficiency of Mu as a transposable element. Mu B stimulates the activity of Mu A over 2 orders of magnitude in vivo (see reference 26), likely due to its ability to stimulate catalytic activities of Mu A (14, 158, 159) as well as to its ability to deliver target DNA to the transposition complex (103, 122).

Mu B is a 312-aa (106) ATP-dependent DNA binding protein (96, 103), whose weak intrinsic ATPase activity is stimulated by DNA and Mu A (3). It can be divided into two domains, with the larger 25-kDa N-terminal domain showing both nonspecific DNA binding and ATP binding (163), consistent with a putative HTH motif at residues 19 to 40 (106) and homology with nucleotide binding folds at residues 86 to 115 and 156 to 185 (26). Little is known about the way in which Mu B binds to DNA. Modification of the protein with the cysteine-modifying agent *N*-ethylmaleimide (NEM) interferes with DNA binding without interfering with stimulation of strand transfer by Mu A (14, 158), an effect that can also be mimicked by alteration of a single cysteine residue in the ATP fold at position 99 (108). NEM modification does not block ATPase activity but does alter the dependence upon Mu A and DNA (158).

Site-directed mutations in the nucleotide binding loops of Mu B produced two classes of ATPase mutants: those defective in target binding but able to stimulate strand transfer by Mu A at high ATP concentrations (similar to NEM-modified Mu B) and those that could bind to the target as well as stimulate Mu A but not allow use of the Mu B-bound DNA as target (182). Because the Mu B-ADP complex can support strand transfer but not target binding, ATP hydrolysis has been proposed to "hand off" bound DNA to Mu A for use as target for strand transfer (182). However, nonhydrolyzable analogs of ATP support intermolecular transposition (4, 163), although it is not clear whether they do so by a mechanism analogous to the ATP hydrolysis pathway. Wild-type Mu B can deliver target DNA to the Mu A complex at any stage of the reaction (see below and reference 122).

Not much is known about the active form of Mu B or how it interacts with Mu A to suppress a variety of defects that compromise transpososome assembly (116, 129, 158, 159, 179). The Mu A-Mu B interaction is also responsible for a phenomenon called target immunity, defined as the capacity of one copy of a transposable element in a target DNA to substantially reduce the frequency of subsequent insertions elsewhere in the same target of another copy of the element, thus preventing genocide as well as suicide. The molecular basis underlying this phenomenon is the ability of *att*-bound Mu A to stimulate hydrolysis of ATP by Mu B, followed by dissociation of Mu B from the potential target DNA (1, 4). Thus, Mu B preferentially distributes itself to non-Mu DNA, capturing it as a target for Mu integration. How the balance between the two different outcomes, transposition and immunity, is achieved by similar protein-protein and protein-DNA interactions is not yet understood but may well depend on the active forms of the proteins involved in the two processes.

Mu B is required for Mu replication in vivo (see reference 26). Deletion of 18 aa from its C terminus allows efficient integration of infecting Mu DNA but blocks replicative transposition (27). Since Mu B is eventually displaced from the strand transfer complex by ClpX in preparation for DNA replication (125), the C-terminal truncation in Mu B either may affect this displacement or may directly or indirectly affect some other step in replication.

HU

E. coli HU is a small, basic, sequence-independent DNA-binding and bending protein (for reviews, see references 18, 59, and 132) that affects an early step in Mu transposition (35, 38, 155, 173) and is incorporated into Mu transpososomes (90). By coupling HU to a DNA cleaving reagent, iron-EDTA, the first "footprint" of the protein was observed (92). Binding of a single HU dimer mapped to the spacer region between *att* sites L1 and L2, producing a cleavage pattern suggesting a dramatic 155° DNA bend (95). HU is thus thought to play a role in closing the gap between the L1 and L2 sites by wrapping the DNA between them, a process facilitated by DNA supercoiling even in the absence of Mu A (83). Modulation of the L1-L2 architecture is likely an essential step in the formation of the left-end, enhancer and right-end complex (LER) (173), the first characterized nucleoprotein intermediate in the transposition pathway (Color Plate 33 [see color insert]). Consistent with this role, HU specifically promotes assembly of Mu A complexes on the left Mu end under dimethyl sulfoxide (DMSO) assay conditions (13, 128). HU has also been shown to bind with high affinity to assembled transpososomes depleted of HU (91). Since HU can bind with high affinity to altered DNA structures (23, 140, 141), it is speculated that an additional role for HU may be to stabilize an altered DNA conformation detected near the Mu ends in the assembled complex before and after cleavage (94, 172). HU is not required for the strand transfer of artificially precleaved ends (38).

HU mutants defective in both (but not single) subunits of this heterodimeric protein do not support Mu transposition in vivo (72). Mutations in another protein, H-NS, which, like HU, is abundant and is implicated in organizing the bacterial chromosome, have the opposite effect of elevating Mu transposition; H-NS is thought to act by stabilizing the Mu repressor (47).

IHF

With fully supercoiled mini-Mu substrates ($\sigma = -0.06$), HU is the only host factor required for transpososome assembly. However, at lower levels of donor DNA supercoiling ($\sigma = -0.025$) that approximate those found in vivo, *E. coli* IHF (see reference 132) is required for efficient reaction to occur (157). IHF binds stoichiometrically at the enhancer (160) and likely plays an architechtural role by bending the DNA sharply (145, 146) between the O1 and O2 sites of the enhancer (71). IHF is absolutely required for a linear enhancer to function in *trans* when Mu ends are present on supercoiled DNA (156).

IHF does not appear to be required for Mu transposition in vivo; although Mu will not grow on IHF mutants, the inhibition can be overcome if Mu A and B proteins are supplied from a plasmid (see reference 133). Nonetheless, IHF is a potent stimulator of the reaction under a variety of conditions and is probably normally present in the enhancer region because of its essential role in transcriptional regulation.

Other Host Proteins

Besides the roles of HU, IHF, and the ClpXP proteins discussed above, most *E. coli* replication proteins are required for Mu DNA replication (see below and reference 75). While a requirement for DNA gyrase is consistent with the need for a negatively supercoiled substrate for transposition, gyrase is also implicated in acting at the central SGS site to organize DNA topology and promote Mu end synapsis in vivo (see "DNA Supercoiling" and "The SGS" above).

REGULATORY ROLE OF TRANSPOSOSOMES

The Mu strand transfer reaction occurs in the context of higher-order nucleoprotein complexes called transpososomes (38, 155), the majority of which have been defined by the Chaconas laboratory. Higher-order complexes are ones which have multiple proteins bound to multiple sites and demonstrate a high level of cooperativity in their assembly. They also usually involve precisely positioned DNA bends and often require DNA supercoiling for assembly. Such complexes play a role in a variety of high-precision DNA metabolic events (45) and have been found in an assortment of DNA transposition and site-specific recombination systems (62). A higher-order complex can impose complex regulatory properties on the reaction at several levels and can provide a strict safeguard against transposition in the wrong place at the wrong time (34). In the case of Mu phage this is a particularly important feature, since the transposition machinery is incredibly efficient and able to perform about 100 events per cell in less than an hour. Any slipups giving rise to a partial or complete reaction at the wrong time would needlessly kill the cell without achieving the "selfish" task of making more copies of itself. The use of transpososomes allows for the assembly of a carefully honed and highly active transposition machine from a collection of inert components.

The enhancer plays many roles in regulation of assembly. First, enhancer (operator) occupancy by the Mu repressor c would block assembly, serving as a checkpoint for gauging the physiological state of the cell. Second, multiple end-enhancer interactions in the LER must ensure a highly precise transpososome architecture (7, 73). Finally, the bipartite association of a catalytically inert Mu A monomer with an enhancer and an att site in the LER unlocks its catalytic potential while simultaneously promoting its association with additional Mu A monomers for tetramerization. Deletions that remove 80 and 84 aa from the N terminus of Mu A confer partial enhancer independence (183). Thus, at least part of the negative controls operating on the monomer to keep it catalytically silent must be contributed by the enhancer binding region of Mu A. Assembly is choreographed in a manner designed to coordinate strand cleavage at the Mu ends and to promote activation of Mu A monomers such that active sites are assembled only at the correct subunit interfaces (those at L1 and R1) within the transpososome (127). In this context, the enhancer has been proposed to be a DNA chaperone that channels monomers into a native complex along one pathway or rescues complexes trapped in a kinetic pit via alternate pathways (183).

Two other important safeguards have been built into the Mu system. The first is "domain sharing" or an "interwoven architecture," in which domains from different Mu A monomers assemble to build a functional active site (6, 85, 127, 184). The second is catalysis in *trans,* whereby the catalytic residues on a Mu A monomer bound at the left end promote the reaction chemistry at the right end and vice versa (6, 127, 148). These features ensure that an active site is not assembled until multimerization of Mu A occurs and that donor cleavage cannot occur until the two Mu ends have synapsed in a stable complex. Another layer of regulatory complexity is the concerted cleavage and strand transfer at the two Mu ends through cooperative transitions in the transpososome. Thus, substitution of an incorrect nucleotide at only one end inhibits assembly of the transpososome (159, 173), and cleavage of only one end suppresses strand transfer, thus aborting partial reactions (127, 177).

Potential regulatory roles in transpososome assembly may also be imagined for both the *E. coli* IHF and HU proteins. As noted in a previous section, IHF is not an essential component for transposition in vivo or in vitro, yet it can have a dramatic stimulatory role on the in vitro reaction under in vivo levels of DNA supercoiling (157). Since IHF is required for expression of the early promoter from which Mu A and B are transcribed (84), it would normally be present at the enhancer during Mu transposition in vivo. It is therefore in the right place at the right time to be a player on the transposition team, and its level of participation, coupled with the superhelical state of the DNA, could certainly be a modulating factor of reaction efficiency. HU on the other hand is essential for the reaction. HU binding to the L1-L2 spacer region is supercoiling dependent (83). The variation of DNA supercoiling and HU levels with growth phase makes these parameters attractive candidates for modulators of transpososome assembly in response to growth phase and environmental stimuli.

Finally, stable transpososomes kinetically and thermodynamically favor completion of the reaction. All of the protein components and DNA sites do not need to meet in one colossal and productive collision. Instead, sequential assembly occurs and favors completion of the reaction. Moreover, the activation energies required for each structural transition are likely far more easily surmountable than if there were a single assembly and reaction step with a much higher energy barrier.

TRANSPOSOSOME FEATURES

The structural and functional core of Mu transpososomes is a stable, functionally active Mu A tetramer (13, 94, 117). Events that lead to tetramerization likely occur in the following sequence. Catalytically inactive Mu A monomers are recruited to the Mu ends. DNA branching and/or slithering leads to interactions of the end-bound monomers (assisted by HU and IHF proteins) with the enhancer to form the unstable three-site complex, the LER. DNA-protein and protein-protein interactions within the LER lead to protein conformational changes that favor oligomerization of Mu A. In the first stable intermediate, the type 0 intermediate, Mu A has attained its active tetrameric configuration and the Mu ends are destabilized in readiness for the transposition chemistry. Cleavage of the Mu ends leads to formation of the type 1 transpososome. Strand transfer of the Mu ends into a target DNA completes transposition to produce the type 2 complex, the stablest of the tetrameric Mu A complexes. In preparation for replication, there ensues a gradual destabilization of the type 2 complex, leading to the type 3 complex, with subsequent total removal of Mu A followed by assembly of a replication complex.

The LER

As shown in Color Plate 33, the earliest characterized complex is the LER, a three-site synaptic complex containing the Mu left and right ends and the transpositional enhancer (173). The LER is the gatekeeper for the assembly of the more stable transpososomes. It is a transient complex, present in maximal amounts 10 s after the start of the reaction, and can only be detected after glutaradehyde cross-linking. The LER can be accumulated using donor substrates carrying terminal base pair mutations. These substrates cannot be processed beyond the LER stage. Similarly, the Mu A mutant T585D, which cannot proceed beyond the LER stage, is very useful for stockpiling the intermediate (123). Experiments with the mutants noted above indicate that the active site of Mu A has not yet engaged the Mu-host junction region in this complex. Thus, terminal base pair recognition is essential for proceeding to the next stage of the reaction. The number of Mu A molecules in the LER has been difficult to address because of the propensity of Mu A to aggregate under conditions which are needed to stabilize the LER. The topological properties of the complex also remain to be characterized, as do the full set of end-enhancer interactions which govern its assembly. Finally, whether there is a single defined pathway for LER assembly or whether either of the observed two-site complexes (LR or ER) (173) can serve as precursors for the LER remains to be determined.

The Type 0 Complex

The LER is converted into the type 0 or stable synaptic complex (117), the formation of which is the the rate-limiting step of the cleavage reaction (172). The type 0 complex is normally short-lived in the presence of Mg^{2+} due to rapid processing through the reaction sequence (117). It can, however, be accumulated when Ca^{2+} is used as the divalent metal ion or when Mu A catalytic mutants are used (12, 82). The type 0 complex is characterized by a stable Mu A tetramer which forms the structural and functional core of the subsequent complexes (94, 117). The enhancer is not stably associated with the type 0 transpososome. The reaction chemistry has not yet occurred in this complex, but the active site has engaged the region around the scissile phosphate. DNA footprinting studies have shown that protection of the Mu ends in the type 0 complex is more extensive than on linear DNA or the LER, and the protection now extends beyond the normal L1 and R1 footprints to cover the Mu-host junction as well as an additional 10 to 13 bp into the host (94, 117, 118). In addition, flanking DNA adjacent to the cleavage site is destabilized upon formation of the type 0 complex (149, 172). The active site is poised for action with the L1-bound subunit ready to promote catalysis at R1 and the R1-bound subunit poised for catalysis at L1. These two subunits will perform both chemical steps of the reaction in *trans* (6, 127, 148).

The Type 1 Complex

The type 0 transpososome is transformed into the type 1 complex, or cleaved donor complex, as a result of DNA cleavage at both Mu ends in the presence of Mg^{2+}. The type 1 complex shows greater stability than the type 0 complex but displays a similar footprint, with the L1, R1, and R2 sites stably protected by the Mu A tetramer (88, 94, 118), as well as sequences beyond the Mu-host junction (94, 118). The great strength of the noncovalent interactions holding the type 1 complex together is evident from the ability of the complex to withstand the torsional strain in the Mu DNA domain which remains supercoiled. The vector domain is relaxed due to free rotation of the DNA around the nicks (155). An altered structure in the vicinity of the Mu-host junction is evident from an enhanced reactivity to hydroxyl radicals (94).

It is important to note that although the stable Mu A tetramer is the functional and structural core of the Mu transpososome, the complexes also contain loosely associated Mu A, which can be removed by increasing the ionic strength (94) or by challenge with heparin (88) or Mu A binding sites (118). The Mu A tetramer in the type 1 complex cannot promote strand transfer in the absence of Mu B unless additional Mu A monomers are present to activate the strand transfer activity (180).

TC Complexes

The TC complexes are formed when the Mu B protein recruits a target molecule to the transposition complex (122). The target DNA is unreacted in this complex and is held in place by noncovalent interactions. The formation of a TC complex, which rapidly undergoes transesterification, is the rate-limiting step in the overall strand transfer reaction. A particularly interesting feature of the Mu TC complexes is that target DNA can be recruited to the Mu transpososome at several points along the reaction pathway. Other transposons, such as Tn7 (9, 10) or Tn*10* (147), can only capture a target molecule at a defined point, before or after donor cleavage, respectively. The Mu TC complexes provide alternate pathways to strand transfer, which may help to maximize transposition potential. They may also be particularly important for the efficient completion of strand transfer as part of the Mu immunity surveillance system. As described earlier, target immunity ensures that a Mu DNA molecule does not integrate into itself or another Mu DNA molecule. If an immune target is recruited to the transpososome at an early stage of the reaction, such as the LER, and then is evicted, a new target may still be brought in regardless of whether the complex has advanced to the type 0 or type 1 stage. This flexibility may be important particularly duri ng later stages of the lytic cycle when Mu DNA represents a very substantial portion of the total DNA in the cell and capturing a nonimmune target may require several tries.

The Type 2 Complex

The type 2 complex, or strand transfer complex (STC), is the culmination of the strand transfer reaction. The transesterification reaction converts the TC-1 into the type 2 complex and covalently links the 3′ ends of Mu to the target DNA. This transpososome contains the θ structure or Shapiro-type intermediate depicted in Fig. 1 and is the most stable of all the complexes. The type 2 complex also contains the Mu B protein (90) and shows considerable resistance to either heat or urea (155). For this reason, and because the 3′ host DNA ends which will serve as replication primers are hidden within the transpososome, the type 2 complex must be actively disassembled for the subsequent DNA replication process, which will convert the strand transferred product into a cointegrate.

The Type 3 Complex

Modification of the type 2 complex, leading to its eventual total disassembly, begins with interaction of the transpososome with the ClpX protein in the presence of ATP (77, 87, 98, 99). ClpX interacts with a region of Mu A near the C terminus (99), destabilizing Mu A in the type 2 complex (77, 87) and resulting in conversion into the type 3 complex or STC2. Mu A is still present in this complex and is subsequently removed through the action of one or more proteins present in the partially purified host fraction Mu replication factor α_2 (87). Details of the replication process are summarized below. An essential role for the ClpX protein in the Mu DNA replication process in vivo has also been shown (105).

DNA REPLICATION

The last several years have seen a number of important advances in our understanding of the Mu DNA replication process. This has resulted primarily from studies of Mu DNA replication in vitro in the Nakai laboratory (74–77, 86, 87, 125). Transpososome disassembly results in the formation of the prereplisome (STC3), which lacks Mu A but which contains host protein component(s) which cover the leading-strand primer. The host PriA protein then loads onto the lagging strand of the template at the fork (previously created by strand transfer to a target DNA molecule). This is followed by addition of PriB, PriC, and DnaT to the PriA-DNA complex, resulting in a PriABC-DnaT complex. The host DnaB-DnaC complex then jumps on, and the 3′-to-5′ helicase activity of the PriA protein creates a landing site for DnaB by unwinding the lagging strand to which PriA is bound. The resulting preprimosome has DnaB loaded on the lagging strand and DnaC ejected from the complex. Formation of the replisome by addition of DNA Poll III holoenzyme then initiates the elongation phase of the reaction. The in vitro replication reaction using this ϕX174-type primosome system for initiation shows a clear bias for initiation at the Mu left end as previously noted for an intact Mu prophage (142).

Another interesting feature of the Mu DNA replication reaction in vitro using host cell extracts, which has not been correlated with previous in vivo data, is the amplification of Mu DNA through the use of a rolling-circle replication mechanism (124). Whether this has been previously overlooked or undetectable in in vivo studies or whether it only occurs in host extracts in vitro remains to be established. A rolling-circle mechanism for Mu DNA replication during lytic development would be an interesting twist for an exotic phage such as Mu and would allow for processive generation of multiple copies of Mu DNA from a single initiation event.

THE IN VITRO SYSTEM WITH OLIGONUCLEOTIDE SUBSTRATES

Transposition reactions are generally performed at pH 7.5 to 8, buffered with Tris or HEPES, 10 mM Mg^{2+}, and 150 mM salt. DNA supercoiling in the mini-Mu donor substrate, correctly oriented L, R, and E sequences, and HU protein are absolutely essential for reaction under these conditions. Inclusion of either 15% glycerol, 15% DMSO, or a combination of the two relaxes the need for DNA supercoiling, for a correct orientation of the L and R ends, for the enhancer, for HU, for the enhancer-binding domain Iα, and for the Mu B-interacting domain IIIβ of Mu A (119). The altered reaction conditions likely influence both Mu A and DNA conformations. Although tetrameric Mu A can be assembled on any single *att* site, the chemistry of the reaction occurs faithfully at the normal phosphodiester bonds; however, the strand transfer reaction is preferentially intramolecular (13, 149).

Under the altered conditions divalent cations were shown not to be required for assembly of paired-end complexes if the R1 site was precleaved, if the flanking DNA outside the Mu sequences was very short, or if the two flanking strands were not complementary (149). Linear precleaved *attR* substrates assembled stable complexes even in the absence of DMSO and all other cofactors (38, 128). The stability of the transpososomes on oligonucleotide substrates containing R1-R2 was critically dependent on the presence of a single flanking nucleotide on the noncut strand, while cleavage additionally required at least two flanking nucleotides on the strand to be cleaved (149). Although cleavage required Mg^{2+}, strand transfer took place in the presence of Ca^{2+} ions. Bearing in mind that normal assembly requirements have been bypassed, a minimal system consisting of synthetic oligonucleotides (generally R1-R2 sites) and Mu A protein nonetheless allows greater flexibility in dissection of the transposition reaction.

SUBUNIT MIXING EXPERIMENTS: DISCREPANCIES AND AGREEMENTS

Cleavage in *trans* and Two- versus Four-Subunit Models

Oligonucleotide substrates were initially used to show that Mu A subunits catalyze recombination in *trans* (6, 148); supercoiled substrates also showed *trans* recombination when Mu A subunits with altered *att* DNA binding specificity were directed to occupy specific *att* sites (127). However, differences in results between these two substrates have been reported specifically for the cleavage reaction by the Harshey group. Thus, when DDE^{+} subunits were loaded at R2, cleavage at R1 was observed on oligonucleotide substrates but not on supercoiled substrates; the strand transfer results were consistent between the two substrates (127, 130, 185). The discrepancy in the cleavage reaction on topologically different substrates is presently unresolved but could arise from alternate catalytic configurations of Mu A under the different reaction conditions. There is overall agreement, however, that only two subunits, acting in *trans*, carry out complete transposition on both supercoiled (127) and oligonucleotide substrates (177). Earlier conclusions (based on supercoiled substrates) that complete transposition requires two subunits for cleavage and four for strand transfer (15) or four subunits for both cleavage and strand transfer (11) were based on mixing catalytically or assembly-defective Mu A subunits and analyzing the yield of both cleaved and strand transfer complexes as well as the ratio of the subunits recovered from them at different initial ratios of the two proteins. In light of the strong inhibition of strand transfer observed on both supercoiled substrates (127) and on oligonucleotide substrates (177) when only one end was cleaved, the previous discrepancies between the two and four subunit models are now explained.

Domain Sharing or an Interwoven Architecture

Subunit mixing experiments employing deletion and point mutants of Mu A, as well as hybrid proteins in which domain Iα is substituted with that from the Mu A-related transposase D108 A, have been particularly useful in dissecting the function of individual domains. A model for shared active sites such as that seen for the Flp recombinase (53) was proposed based on complementation for cleavage obtained by mixing a DDE^{-} point mutant with a domain IIIα deletion mutant of Mu A (181, 184). The inability of this mixture to complete strand transfer after cleavage on supercoiled substrates but not on precleaved linear substrates led to the idea of "reciprocal domain sharing" (184). While the contribution of domain IIIα to the active site still remains to be determined, it is apparent that the inability of the mixed complexes to complete strand transfer on supercoiled substrates was due to the inhibition of strand transfer of singly cleaved ends discussed above (127). The inconsistency between experiments showing that a domain IIIα deletion mutant mixed with a DDE^{-} mutant can function on precleaved linear substrates (184) or on oligonucleotide substrates (6, 185) and those showing that this deletion mutant does not support strand transfer of precleaved ends when mixed with wild-type Mu A (15) is not yet resolved. It is possible that DDE^{-} mutants are more permissive (or wild-type Mu A is more "recalcitrant") in interacting with partner subunits. Differences between these proteins in their intersubunit cooperativity have been observed in other experiments (73).

How Many Mu A Subunits Need To Interact with the Enhancer?

Mixtures of DDE^{-} and domain Iα deletion mutants of Mu A allow cleavage on supercoiled substrates, suggesting that not all four monomers need to interact with the enhancer (183, 184). At least one and possibly two Iα^{-} monomers were shown to be incorporated into the mixed tetramer (184). However, mixtures of a domain Iα deletion mutant with wild-type Mu A failed to assemble functional complexes (116). This discrepancy could either be due to differences in the specific domain Iα deletion mutant employed in the two experiments or (more likely) to the permissiveness of the DDE^{-} mutant but not wild-type Mu A to pair with the Iα deletion. It is therefore possible that all four subunits of the transpososome tetramer must interact with the enhancer in the wild-type complex.

How Many Mu A Subunits Need To Interact with the Mu B Protein?

Mixtures of DDE^{-} and domain IIIβ deletion mutants have been reported to support single-end cleavage and intermolecular strand transfer on supercoiled substrates, with recovery of the two proteins in equimolar amounts in the STCs (116). Similar results were observed in experiments pairing a DDE^{-}-IIIβ double mutant with wild-type Mu A. These experiments have been interpreted as suggesting that all four subunits need not interact with Mu B and that the position of the Mu B-interacting subunits is not unique (116).

USES OF Mu

Mu was the first transposable element used for genetic analysis in bacteria, its high transposition frequency and relatively low specificity giving it a distinct advantage (see reference 16). A variety of Mu derivatives with and without selectable markers are available, including ones that can be used as probes for transcription and translation, as cloning vehicles, as mobile sources of transcriptional promoters, and as movable primers for DNA sequencing studies (see reference 16). A variation of the minimal transposition system described above, which uses two short R ends on a single linear precut DNA fragment with selectable markers in between, has recently been used to generate mini-Mu insertions in plasmids in vitro (61) and for subsequent DNA sequencing studies (60). This system is now available commercially.

Mu transposition has also been used in creative ways to analyze various cellular and biochemical parameters. For example, β-galactosidase activity of Mud*lac* transposons has been used as a tool for vital staining to visualize clonal and nonclonal patterns of organization in bacterial colonies (153). Formation of Mu-mediated *araB-lacZ* coding sequence fusions has been used to illustrate physiological regulation under specific conditions, providing an example of adaptive mutation (89, 151). A method called Mu-printing, a PCR technique that uses one primer within the Mu R end and the other within the chromosomal region of interest to identify patterns of transposition, has been used to demonstrate physiological control over Mu transposition in vivo (170). Finally, a technique dubbed Mu-mediated PCR has been used to obtain a footprint of retroviral preintegration complexes that were available in only limited amounts (174, 175). Mu A protein was used to insert oligonucleotides corresponding to the ends of Mu DNA into the DNA to be footprinted. Subsequent to the Mu reaction in the presence or absence of the protein to be footprinted, the distribution of insertions in the target DNA was analyzed in a manner similar to that of ligation-mediated PCR, using one primer specific to the region of interest and a second primer specific to the Mu end.

SUMMARY

Mu is the most complex of known transposable elements, likely because of its lifestyle as a temperate phage and its use of transposition as a mechanism for genome amplification during the viral lytic cycle. The complexity of Mu is reflected in multiple binding sites for the transposase protein Mu A at its ends, involvement of a distant enhancer element, requirement for a second protein Mu B, and recruitment of two host proteins that modulate reaction efficiency in response to DNA supercoiling.

Mu transposes by a replicative cointegrate mechanism both in vivo and in vitro. There is no evidence to date of an alternate cut-and-paste mode of Mu transposition; however, this alternative cannot yet be ruled out. Although the integration of infecting Mu DNA occurs without prior DNA replication, this could occur via a cointegrate pathway followed by repair of the θ intermediate in the minority of events that give rise to stable lysogens.

The mechanism of phosphoryl transfer in Mu transposition involves successive nucleophilic attacks first by OH groups of water on specific phosphodiester bonds at Mu ends and then by in-line attack by the newly liberated DNA 3′ OH groups on target phosphodiester bonds spaced 5 bp apart. A single active site of one transposase protomer catalyzes the successive reaction chemistry. This phosphoryl transfer mechanism is common to HIV and Tn*10* transposition. Crystal structures of the catalytic core domains of Mu transposase and HIV integrase reveal a remarkable similarity in overall topology in spite of little or no similarity in their primary sequence. The DDE catalytic motif in Mu A is common to a large number of transposases.

Mu transposition occurs in the context of higher-order nucleoprotein complexes called transpososomes, which regulate the reaction at many different levels. Transpososomes have now been observed in several other elements studied in vitro (Tn*10* and Tn*5*), and the structure of a dimeric Tn*5* synaptic complex has now been solved (see chapter 18). A series of these complexes in Mu have been identified at different stages of the reaction pathway. The structural and functional core of Mu transpososomes is a stable Mu A tetramer. Only two subunits within this tetramer act in *trans* to carry out complete transposition. The function of the other two subunits is presently unknown, but their presence is essential for structural and functional integrity of the transpososome. Completion of strand transfer is followed by gradual destabilization of the transpososome in preparation for DNA replication.

Mu has served as a paradigm and a guiding light for many of the advances in the transposition field in the last 3 decades. While much has been learned, more remains to be understood. The next frontier will be to determine the structure of the transpososomes at each stage of the reaction pathway in order to understand the contribution of each component to the remarkable success of Mu as a transposable element.

Acknowledgments. G.C. acknowledges the support of a fellowship from the John Simon Guggenheim Memorial Foundation and an MRC Distinguished Scientist Award and operating grant from the Medical Research Council of Canada. R.M.H. acknowledges support from NIH grant GM 33247 and Robert Welch Foundation grant F-1531.

We thank Bill Reznikoff for permission to reproduce his previously published figure (Color Plate 32). R.M.H. also thanks Ki-Hoon Yoon for help with illustrations.

ADDENDUM IN PROOF

Since submission of this chapter, the following papers related to Mu DNA transposition have been published or accepted for publication: C. J. Coros and G. Chaconas, *J. Mol. Biol.* **310:** 299–309, 2001; L. H. Hung, G. Chaconas, and G. S. Shaw, *EMBO J.* **19:**5625–5634, 2000; H. Jiang and R. M. Harshey, *J. Biol. Chem.* **276:**4373–4381, 2001; J. M. Jones and H. Nakai, *Mol. Microbiol.* **36:**519–527, 2000; A. K. Kennedy, D. B. Haniford, and K. Mizuuchi, *Cell* **101:**295–305, 2000; I. Lee and R. M. Harshey, *J. Mol. Biol.*, in press; D. Manna, X. Wang, and N. P. Higgins, *J. Bacteriol.* **183:**3328–3335, 2001; S. Mariconda, S. Y. Namgoong, K. H. Yoon, H. Jiang, and R. M. Harshey, *J. Biosci.* **25:**347–360, 2000; H. Nakai, V. Doseeva, and J. M. Jones, *Proc. Natl. Acad. Sci. USA* **98:**8247–8254, 2001; L. Paolozzi, G. Fabozzi, and P. Ghelardini, *Microbiology* **146:**591–598, 2000; M. L. Pato and M. Banerjee, *Mol. Microbiol.* **37:**800–810, 2000; S. S. Rai, D. O'Handley, and H. Nakai, *J. Mol. Biol.* **312:**311–322, 2001; L. A. Roldan and T. A. Baker, *Mol. Microbiol.* **40:**141–155, 2001; F. H. Schagen, H. J. Rademaker, S. J. Cramer, H. van Ormondt, A. J. van der Eb, P. van de Putte, and R. C. Hoeben, *Nucleic Acids Res.* **28:**E104, 2000; K. E. Scheirer and N. P. Higgins, *Biochimie* **83:**155–159, 2001; H. Vilen, S. Eerikainen, J. Tornberg, M. S. Airaksinen, and H. Savilahti, *Transgenic Res.* **10:**69–80, 2001; and J. M. Wojciak, J. Iwahara, and R. T. Clubb, *Nat. Struct. Biol.* **8:**84–90, 2001.

REFERENCES

1. **Adzuma, K., and K. Mizuuchi.** 1989. Interaction of proteins located at a distance along DNA: mechanism of target immunity in the Mu DNA strand-transfer reaction. *Cell* **57:**41–47.
2. **Adzuma, K., and K. Mizuuchi.** 1988. MuA protein-induced bending of the Mu end DNA, p. 97–104. *In* W. K. Olson, M. H. Sarma, R. H. Sarma, and M. Sundaralingam (ed.), *Structure and Expression*, vol. 3. Adenine Press, New York, N.Y.
3. **Adzuma, K., and K. Mizuuchi.** 1991. Steady-state kinetic analysis of ATP hydrolysis by the B protein of bacteriophage Mu. Involvement of protein oligomerization in the ATPase cycle. *J. Biol. Chem.* **266:**6159–6167.
4. **Adzuma, K., and K. Mizuuchi.** 1988. Target immunity of Mu transposition reflects a differential distribution of Mu B protein. *Cell* **53:**257–266.
5. **Akroyd, J. E., and N. Symonds.** 1983. Evidence for a conservative pathway of transposition of bacteriophage Mu. *Nature* **303:**84–86.
6. **Aldaz, H., E. Schuster, and T. A. Baker.** 1996. The interwoven architecture of the Mu transposase couples DNA synapsis to catalysis. *Cell* **85:**257–269.
7. **Allison, R. G., and G. Chaconas.** 1992. Role of the A protein-binding sites in the *in vitro* transposition of Mu DNA. A complex circuit of interactions involving the Mu ends and the transpositional enhancer. *J. Biol. Chem.* **267:**19963–19970.
8. **Arthur, A., and D. Sherratt.** 1979. Dissection of the transposition process: a transposon-encoded site-specific recombination system. *Mol. Gen. Genet.* **175:**267–274.
9. **Bainton, R., P. Gamas, and N. L. Craig.** 1991. Tn7 transposition *in vitro* proceeds through an excised transposon intermediate generated by staggered breaks in DNA. *Cell* **65:** 805–816.
10. **Bainton, R. J., K. M. Kubo, J. N. Feng, and N. L. Craig.** 1993. Tn7 transposition: target DNA recognition is mediated by multiple Tn7-encoded proteins in a purified *in vitro* system. *Cell* **72:**931–943.
11. **Baker, T. A., E. Kremenstova, and L. Luo.** 1994. Complete transposition requires four active monomers in the Mu transposase tetramer. *Genes Dev.* **8:**2416–2428.
12. **Baker, T. A., and L. Luo.** 1994. Identification of residues in the Mu transposase essential for catalysis. *Proc. Natl. Acad. Sci. USA* **91:**6654–6658.
13. **Baker, T. A., and K. Mizuuchi.** 1992. DNA-promoted assembly of the active tetramer of the Mu transposase. *Genes Dev.* **6:**2221–2232.
14. **Baker, T. A., M. Mizuuchi, and K. Mizuuchi.** 1991. MuB protein allosterically activates strand transfer by the transposase of phage Mu. *Cell* **65:**1003–1013.
15. **Baker, T. A., M. Mizuuchi, H. Savilahti, and K. Mizuuchi.** 1993. Division of labor among monomers within the Mu transposase tetramer. *Cell* **74:**723–733.
16. **Berg, C. M., D. E. Berg, and E. A. Groisman.** 1989. Transposable elements and genetic engineering of bacteria, p. 879–925. *In* D. E. Berg and M. M. Howe (ed.), *Mobile DNA.* American Society for Microbiology, Washington, D.C.
17. **Betermier, M., R. Alazard, F. Ragueh, E. Roulet, A. Toussaint, and M. Chandler.** 1987. Phage Mu transposase: deletion of the carboxy-terminal end does not abolish DNA-binding activity. *Mol. Gen. Genet.* **210:**77–85.
18. **Bianchi, M. E.** 1994. Prokaryotic HU and eukaryotic HMG1: a kinked relationship. *Mol. Microbiol.* **14:**1–5.
19. **Bujacz, G., M. Jaskolski, J. Alexandratos, A. Wlodawer, G. Merkel, R. A. Katz, and A. M. Skalka.** 1996. The catalytic domain of avian sarcoma virus integrase-conformation of the active-site residues in the presence of divalent cations. *Structure* **4:**89–96.
20. **Bujacz, G., M. Jaskolski, J. Alexandratos, A. Wlodawer, G. Merkel, R. A. Katz, and A. M. Skalka.** 1995. High-resolution structure of the catalytic domain of avian sarcoma virus integrase. *J. Mol. Biol.* **253:**333–346.
21. **Bukhari, A. I., J. A. Shapiro, and S. L. Adhya.** 1977. *DNA Insertion Elements, Plasmids and Episomes.* Cold Spring Harbor Laboratory, Cold Spring Harbor, N.Y.
22. **Bushman, F. D., A. Engelman, I. Palmer, P. Wingfield, and R. Craigie.** 1993. Domains of the integrase protein of human immunodeficiency virus type 1 responsible for polynucleotidyl transfer and zinc binding. *Proc. Natl. Acad. Sci. USA* **90:** 3428–3432.
23. **Castaing, B., C. Zelwer, J. Laval, and S. Boiteux.** 1995. HU protein of *Escherichia coli* binds specifically to DNA that contains single-strand breaks or gaps. *J. Biol. Chem.* **270:** 10291–10296.
24. **Castilho, B. A., and M. J. Casadaban.** 1991. Specificity of mini-Mu bacteriophage insertions in a small plasmid. *J. Bacteriol.* **173:**1339–1343.
25. **Chaconas, G.** 1999. Studies on a "jumping gene machine":

higher-order nucleoprotein complexes in Mu DNA transposition. *Biochem. Cell Biol.* **77:**487–491.
26. **Chaconas, G.** 1987. Transposition of bacteriophage Mu DNA *in vivo,* p. 137–158. *In* N. Symonds, A. Toussaint, P. Van de Putte, and M. M. Howe (ed.), *Phage Mu.* Cold Spring Harbor Laboratory, Cold Spring Harbor, N.Y.
27. **Chaconas, G., E. B. Giddens, J. L. Miller, and G. Gloor.** 1985. A truncated form of the bacteriophage Mu B protein promotes conservative integration, but not replicative transposition, of Mu DNA. *Cell* **41:**857–865.
28. **Chaconas, G., R. M. Harshey, N. Sarvetnick, and A. I. Bukhari.** 1981. Predominant end-products of prophage Mu DNA transposition during the lytic cycle are replicon fusions. *J. Mol. Biol.* **150:**341–359.
29. **Chaconas, G., B. D. Lavoie, and M. A. Watson.** 1996. DNA transposition: jumping gene machine, some assembly required. *Curr. Biol.* **6:**817–820.
30. **Chaconas, G., and M. G. Surette.** 1988. Mechanism of Mu DNA transposition. *BioEssays* **9:**205–208.
31. **Chow, S. A., K. A. Vincent, V. Ellison, and P. O. Brown.** 1992. Reversal of integration and DNA splicing mediated by integrase of human immunodeficiency virus. *Science* **255:** 723–726.
31a. **Clubb, R. T., M. Mizuuchi, J. R. Huth, J. G. Omichinski, H. Savilahti, K. Mizuuchi, G. M. Clore, and A. M. Gronenborn.** 1996. The wing of the enhancer-binding domain of phage Mu transposase is flexible and is essential for efficient transposition. *Proc. Natl. Acad. Sci. USA* **93:**1146–1150.
31b. **Clubb, R. T., J. G. Omichinski, H. Savilahti, K. Mizuuchi, A. M. Gronenborn, and G. M. Clore.** 1994. A novel class of winged helix-turn-helix protein: the DNA binding domain of Mu transposase. *Structure* **2:**1041–1048.
32. **Clubb, R. T., S. Schumacher, K. Mizuuchi, A. M. Gronenborn, and G. M. Clore.** 1997. Solution structure of the Iγ subdomain of the Mu end DNA-binding domain of phage Mu transposase. *J. Mol. Biol.* **273:**19–25.
33. **Colloms, S. D., J. Bath, and D. J. Sherratt.** 1997. Topological selectivity in Xer site-specific recombination. *Cell* **88:** 855–864.
34. **Craigie, R.** 1996. Quality control in Mu DNA transposition. *Cell* **85:**137–140.
35. **Craigie, R., D. J. Arndt-Jovin, and K. Mizuuchi.** 1985. A defined system for the DNA strand-transfer reaction at the initiation of bacteriophage Mu transposition: protein and DNA substrate requirements. *Proc. Natl. Acad. Sci. USA* **82:** 7570–7574.
36. **Craigie, R., and K. Mizuuchi.** 1985. Mechanism of transposition of bacteriophage Mu: structure of a transposition intermediate. *Cell* **41:**867–876.
37. **Craigie, R., and K. Mizuuchi.** 1986. Role of DNA topology in Mu transposition: mechanism of sensing the relative orientation of two DNA segments. *Cell* **45:**793–800.
38. **Craigie, R., and K. Mizuuchi.** 1987. Transposition of Mu DNA: joining of Mu to target DNA can be uncoupled from cleavage at the ends of Mu. *Cell* **51:**493–501.
39. **Craigie, R., M. Mizuuchi, and K. Mizuuchi.** 1984. Site-specific recognition of the bacteriophage Mu ends by the Mu A protein. *Cell* **39:**387–394.
40. **Darzins, A., N. E. Kent, M. S. Buckwalter, and M. J. Casadaban.** 1988. Bacteriophage Mu sites required for transposition immunity. *Proc. Natl. Acad. Sci. USA* **85:**6826–6830.
41. **Davies, D. R., L. M. Braam, W. S. Reznikoff, and I. Rayment.** 1999. The three-dimensional structure of a Tn5 transposase-related protein determined to 2.9-A resolution. *J. Biol. Chem.* **274:**11904–11913.
42. **Ding, Z. M., R. M. Harshey, and L. H. Hurley.** 1993. (+)-CC-1065 as a structural probe of Mu transposase-induced bending of DNA: overcoming limitations of hydroxyl-radical footprinting. *Nucleic Acids Res.* **21:**4281–4287. (Erratum, **22:**256, 1994.)
43. **DuBow, M. S.** 1987. Transposable Mu-like phages, p. 201–213. *In* N. Symonds, A. Toussaint, P. Van de Putte, and M. M. Howe (ed.), *Phage Mu.* Cold Spring Harbor Laboratory, Cold Spring Harbor, N.Y.
44. **Dyda, F., A. B. Hickman, T. M. Jenkins, A. Engelman, R. Craigie, and D. R. Davies.** 1994. Crystal structure of the catalytic domain of HIV-1 integrase: similarity to other polynucleotidyl transferases. *Science* **266:**1981–1986.
45. **Echols, H.** 1990. Nucleoprotein structures initiating DNA replication, transcription, and site-specific recombination. *J. Biol. Chem.* **265:**14697–14700.
46. **Engelman, A., K. Mizuuchi, and R. Craigie.** 1991. HIV-1 DNA integration: mechanism of viral DNA cleavage and DNA strand transfer. *Cell* **67:**1211–1221.
47. **Falconi, M., V. McGovern, C. Gualerzi, D. Hillyard, and N. P. Higgins.** 1991. Mutations altering chromosomal protein H-NS induce mini-Mu transposition. *New Biol.* **3:**615–625.
48. **Gama, M. J., A. Toussaint, and M. L. Pato.** 1990. Instability of bacteriophage Mu transposase and the role of host Hfl protein. *Mol. Microbiol.* **4:**1891–1897.
49. **Gellert, M., and H. Nash.** 1987. Communication between segments of DNA during site-specific recombination. *Nature* **325:**401–404.
50. **Geuskens, V., A. Mhammedi-Alaoui, L. Desmet, and A. Toussaint.** 1992. Virulence in bacteriophage Mu: a case of trans-dominant proteolysis by the *Escherichia coli* Clp serine protease. *EMBO J.* **11:**5121–5127.
51. **Gloor, G., and G. Chaconas.** 1986. The bacteriophage Mu N gene encodes the 64-kDa virion protein which is injected with, and circularizes, infecting Mu DNA. *J. Biol. Chem.* **261:** 16682–16688.
52. **Goosen, N., and P. van de Putte.** 1987. Regulation of transcription, p. 41–52. *In* N. Symonds, A. Toussaint, P. Van de Putte, and M. M. Howe (ed.), *Phage Mu.* Cold Spring Harbor Laboratory, Cold Spring Harbor, N.Y.
53. **Grainge, I., and M. Jayaram.** 1999. The integrase family of recombinase: organization and function of the active site. *Mol. Microbiol.* **33:**449–456.
54. **Grindley, N. D.** 1983. Transposition of Tn3 and related transposons. *Cell* **32:**3–5.
55. **Grindley, N. D. F.** 1994. Resolvase-mediated site-specific recombination, p. 236–267. *In* F. Eckstein and D. M. J. Lilley (ed.), *Nucleic Acids and Molecular Biology,* vol. 8. Springer-Verlag, Berlin, Germany.
56. **Groenen, M. A., and P. van de Putte.** 1986. Analysis of the ends of bacteriophage Mu using site-directed mutagenesis. *J. Mol. Biol.* **189:**597–602.
57. **Groenen, M. A., and P. van de Putte.** 1985. Mapping of a site for packaging of bacteriophage Mu DNA. *Virology* **144:** 520–522.
58. **Groenen, M. A., and P. van de Putte.** 1987. The requirements for a high level of transposition of bacteriophage Mu. *J. Cell Sci. Suppl.* **7:**41–50.
59. **Grosschedl, R.** 1995. Higher-order nucleoprotein complexes in transcription: analogies with site-specific recombination. *Curr. Opin. Cell Biol.* **7:**362–370.
60. **Haapa, S., S. Suomalainen, S. Eerikainen, M. Airaksinen, L. Paulin, and H. Savilahti.** 1999. An efficient DNA sequencing strategy based on the bacteriophage Mu *in vitro* DNA transposition reaction. *Genome Res.* **9:**308–315.
61. **Haapa, S., S. Taira, E. Heikkinen, and H. Savilahti.** 1999. An efficient and accurate integration of mini-Mu transposons

in vitro: a general methodology for functional genetic analysis and molecular biology applications. *Nucleic Acids Res.* **27:** 2777–2784.

62. **Hallet, B., and D. J. Sherratt.** 1997. Transposition and site-specific recombination: adapting DNA cut-and-paste mechanisms to a variety of genetic rearrangements. *FEMS Microbiol. Rev.* **21:**157–178.
63. **Haniford, D. B., and G. Chaconas.** 1992. Mechanistic aspects of DNA transposition. *Curr. Opin. Genet. Dev.* **2:**698–704.
64. **Harel, J., L. Duplessis, J. S. Kahn, and M. S. DuBow.** 1990. The cis-acting DNA sequences required in vivo for bacteriophage Mu helper-mediated transposition and packaging. *Arch. Microbiol.* **154:**67–72.
65. **Haren, L., B. Ton-Hoang, and M. Chandler.** 1999. Integrating DNA: transposases and retroviral integrases. *Annu. Rev. Microbiol.* **53:**245–281.
66. **Harshey, R. M.** 1987. Integration of infecting Mu DNA, p. 111–135. *In* N. Symonds, A. Toussaint, P. Van de Putte, and M. M. Howe (ed.), *Phage Mu.* Cold Spring Harbor Laboratory, Cold Spring Harbor, N.Y.
67. **Harshey, R. M.** 1984. Transposition without duplication of infecting bacteriophage Mu DNA. *Nature* **311:**580–581.
68. **Harshey, R. M., and A. I. Bukharl.** 1983. Infecting bacteriophage Mu DNA forms a circular DNA-protein complex. *J. Mol. Biol.* **167:**427–441.
69. **Harshey, R. M., and S. D. Cuneo.** 1986. Carboxyl-terminal mutants of phage Mu transposase. *J. Genet.* **65:**159–174.
70. **Harshey, R. M., E. D. Getzoff, D. L. Baldwin, J. L. Miller, and G. Chaconas.** 1985. Primary structure of phage Mu transposase: homology to Mu repressor. *Proc. Natl. Acad. Sci. USA* **82:**7676–7680.
71. **Higgins, N. P., D. A. Collier, M. W. Kilpatrick, and H. M. Krause.** 1989. Supercoiling and integration host factor change the DNA conformation and alter the flow of convergent transcription in phage Mu. *J. Biol. Chem.* **264:** 3035–3042.
72. **Huisman, O., M. Faelen, D. Girard, A. Jaffe, A. Toussaint, and J. Rouvlere-Yaniv.** 1989. Multiple defects in *Escherichia coli* mutants lacking HU protein. *J. Bacteriol.* **171:** 3704–3712.
73. **Jiang, H., J. Y. Yang, and R. M. Harshey.** 1999. Criss-crossed interactions between the enhancer and the att sites of phage Mu during DNA transposition. *EMBO J.* **18:**3845–3855.
74. **Jones, J. M., and H. Nakai.** 1999. Duplex opening by primosome protein PriA for replisome assembly on a recombination intermediate. *J. Mol. Biol.* **289:**503–516.
75. **Jones, J. M., and H. Nakai.** 1997. The ϕX174-type primosome promotes replisome assembly at the site of recombination in bacteriophage Mu transposition. *EMBO J.* **16:** 6886–6895.
76. **Jones, J. M., and H. Nakai.** 2000. PriA and phage T4 gp59: factors that promote DNA replication on forked DNA substrates. *Mol. Microbiol.* **36:**519–527.
77. **Jones, J. M., D. J. Welty, and H. Nakai.** 1998. Versatile action of Escherichia coli ClpXP as protease or molecular chaperone for bacteriophage Mu transposition. *J. Biol. Chem.* **273:** 459–465.
78. **Kanaar, R., and N. R. Cozzarelli.** 1992. Roles of supercoiled DNA structure in DNA transactions. *Curr. Opin. Struct. Biol.* **2:**369–379.
79. **Kanaar, R., A. Klippel, E. Shekhtman, J. M. Dungan, R. Kahmann, and N. R. Cozzarelli.** 1990. Processive recombination by the phage Mu Gin system: implications for the mechanisms of DNA strand exchange, DNA site alignment, and enhancer action. *Cell* **62:**353–366.
80. **Kennedy, A. K., D. B. Haniford, and K. Mizuuchi.** 2000. Single active site catalysis of the successive phosphoryl transfer steps by DNA transposases: insights from phosphorothioate stereoselectivity. *Cell* **101:**295–305.
81. **Kim, K., and R. M. Harshey.** 1995. Mutational analysis of the att DNA-binding domain of phage Mu transposase. *Nucleic Acids Res.* **23:**3937–3943.
82. **Kim, K., S. Y. Namgoong, M. Jayaram, and R. M. Harshey.** 1995. Step-arrest mutants of phage Mu transposase. Implications in DNA-protein assembly, Mu end cleavage, and strand transfer. *J. Biol. Chem.* **270:**1472–1479.
83. **Kobryn, K., B. D. Lavoie, and G. Chaconas.** 1999. Supercoiling-dependent site-specific binding of HU to naked Mu DNA. *J. Mol. Biol.* **289:**777–784.
84. **Krause, H. M., M. R. Rothwell, and N. P. Higgins.** 1983. The early promoter of bacteriophage Mu: definition of the site of transcript initiation. *Nucleic Acids Res.* **11:** 5483–5495.
85. **Krementsova, E., M. J. Giffin, D. Pincus, and T. A. Baker.** 1998. Mutational analysis of the Mu transposase. Contributions of two distinct regions of domain II to recombination. *J. Biol. Chem.* **273:**31358–31365.
86. **Kruklitis, R., and H. Nakai.** 1994. Participation of the bacteriophage Mu A protein and host factors in the initiation of Mu DNA synthesis *in vitro.* *J. Biol. Chem.* **269:** 16469–16477.
87. **Kruklitis, R., D. J. Welty, and H. Nakai.** 1996. Clpx protein of *Escherichia coli* activates bacteriophage Mu transposase in the strand transfer complex for initiation of Mu DNA synthesis. *EMBO J.* **15:**935–944.
88. **Kuo, C. F., A. H. Zou, M. Jayaram, E. Getzoff, and R. Harshey.** 1991. DNA-protein complexes during attachment-site synapsis in Mu DNA transposition. *EMBO J.* **10:**1585–1591.
89. **Lamrani, S., C. Ranquet, M. J. Gama, H. Nakai, J. A. Shapiro, A. Toussaint, and G. Maenhaut-Michel.** 1999. Starvation-induced Mucts62-mediated coding sequence fusion: a role for ClpXP, Lon, RpoS and Crp. *Mol. Microbiol.* **32:** 327–343.
90. **Lavoie, B. D., and G. Chaconas.** 1990. Immunoelectron microscopic analysis of the A, B, and HU protein content of bacteriophage Mu transpososomes. *J. Biol. Chem.* **265:** 1623–1627.
91. **Lavoie, B. D., and G. Chaconas.** 1994. A second high affinity HU binding site in the phage Mu transpososome. *J. Biol. Chem.* **269:**15571–15576.
92. **Lavoie, B. D., and G. Chaconas.** 1993. Site-specific HU binding in the Mu transpososome: conversion of a sequence-independent DNA-binding protein into a chemical nuclease. *Genes Dev.* **7:**2510–2519.
93. **Lavoie, B. D., and G. Chaconas.** 1995. Transposition of phage Mu DNA. *Curr. Top. Microbiol. Immunol.* **204:** 83–99.
94. **Lavoie, B. D., B. S. Chan, R. G. Allison, and G. Chaconas.** 1991. Structural aspects of a higher order nucleoprotein complex: induction of an altered DNA structure at the Mu-host junction of the Mu type 1 transpososome. *EMBO J.* **10:** 3051–3059.
95. **Lavoie, B. D., G. S. Shaw, A. Millner, and G. Chaconas.** 1996. Anatomy of a flexer-DNA complex inside a higher-order transposition intermediate. *Cell* **85:**761–771.
96. **Leung, P. C., and R. M. Harshey.** 1991. Two mutations of phage Mu transposase that affect strand transfer or interactions with B protein lie in distinct polypeptide domains. *J. Mol. Biol.* **219:**189–199.
97. **Leung, P. C., D. B. Teplow, and R. M. Harshey.** 1989. Interaction of distinct domains in Mu transposase with Mu DNA

ends and an internal transpositional enhancer. *Nature* **338:** 656–658.

98. **Levchenko, I., L. Luo, and T. A. Baker.** 1995. Disassembly of the Mu transposase tetramer by the Clpx chaperone. *Genes Dev.* **9:**2399–2408.
99. **Levchenko, I., M. Yamauchi, and T. A. Baker.** 1997. ClpX and MuB interact with overlapping regions of Mu transposase: implications for control of the transposition pathway. *Genes Dev.* **11:**1561–1572.
100. **Liebart, J. C., P. Ghelardini, and L. Paolozzi.** 1982. Conservative integration of bacteriophage Mu DNA into pBR322 plasmid. *Proc. Natl. Acad. Sci. USA* **79:**4362–4366.
101. **Mahillon, J., and M. Chandler.** 1998. Insertion sequences. *Microbiol. Mol. Biol. Rev.* **62:** 725–774.
102. **Manna, D., and N. P. Higgins.** 1999. Phage Mu transposition immunity reflects supercoil domain structure of the chromosome. *Mol. Microbiol.* **32:**595–606.
103. **Maxwell, A., R. Craigie, and K. Mizuuchi.** 1987. B protein of bacteriophage mu is an ATPase that preferentially stimulates intermolecular DNA strand transfer. *Proc. Natl. Acad. Sci. USA* **84:**699–703.
104. **Merickel, S. K., M. J. Haykinson, and R. C. Johnson.** 1998. Communication between Hin recombinase and Fis regulatory subunits during coordinate activation of Hin-catalyzed site-specific DNA inversion. *Genes Dev.* **12:**2803–2816.
105. **Mhammedi-Alaoui, A., M. Pato, M. J. Gama, and A. Toussaint.** 1994. A new component of bacteriophage Mu replicative transposition machinery: the Escherichia coli ClpX protein. *Mol. Microbiol.* **11:**1109–1116.
106. **Miller, J. L., S. K. Anderson, D. J. Fujita, G. Chaconas, D. L. Baldwin, and R. M. Harshey.** 1984. The nucleotide sequence of the B gene of bacteriophage Mu. *Nucleic Acids Res.* **12:**8627–8638.
107. **Miller, J. L., and G. Chaconas.** 1986. Electron microscopic analysis of in vitro transposition intermediates of bacteriophage Mu DNA. *Gene* **48:**101–108.
108. **Millner, A., and G. Chaconas.** 1998. Disruption of target DNA binding in Mu DNA transposition by alteration of position 99 in the Mu B protein. *J. Mol. Biol.* **275:**233–243.
109. **Mizuuchi, K.** 1983. *In vitro* transposition of bacteriophage Mu: a biochemical approach to a novel replication reaction. *Cell* **35:**785–794.
110. **Mizuuchi, K.** 1997. Polynucleotidyl transfer reactions in site-specific DNA recombination. *Genes Cells* **2:**1–12.
111. **Mizuuchi, K.** 1992. Polynucleotidyl transfer reactions in transpositional DNA recombination. *J. Biol. Chem.* **267:** 21273–21276.
112. **Mizuuchi, K.** 1992. Transpositional recombination: mechanistic insights from studies of Mu and other elements. *Annu. Rev. Biochem.* **61:**1011–1051.
113. **Mizuuchi, K., and K. Adzuma.** 1991. Inversion of the phosphate chirality at the target site of Mu DNA strand transfer: evidence for a one-step transesterification mechanism. *Cell* **66:**129–140.
114. **Mizuuchi, K., and R. Craigie.** 1986. Mechanism of bacteriophage Mu transposition. *Annu. Rev. Genet.* **20:**385–429.
115. **Mizuuchi, K., T. J. Nobbs, S. E. Halford, K. Adzuma, and J. Qin.** 1999. A new method for determining the stereochemistry of DNA cleavage reactions: application to the SfiI and HpaII restriction endonucleases and to the MuA transposase. *Biochemistry* **38:**4640–4648.
116. **Mizuuchi, M., T. A. Baker, and K. Mizuuchi.** 1995. Assembly of phage Mu transpososomes: cooperative transitions assisted by protein and DNA scaffolds. *Cell* **83:**375–385.
117. **Mizuuchi, M., T. A. Baker, and K. Mizuuchi.** 1992. Assembly of the active form of the transposase-Mu DNA complex: a critical control point in Mu transposition. *Cell* **70:**303–311.
118. **Mizuuchi, M., T. A. Baker, and K. Mizuuchi.** 1991. DNase protection analysis of the stable synaptic complexes involved in Mu transposition. *Proc. Natl. Acad. Sci. USA* **88:** 9031–9035.
119. **Mizuuchi, M., and K. Mizuuchi.** 1989. Efficient Mu transposition requires interaction of transposase with a DNA sequence at the Mu operator: implications for regulation. *Cell* **58:**399–408.
120. **Mizuuchi, M., and K. Mizuuchi.** 1993. Target site selection in transposition of phage Mu. *Cold Spring Harbor Symp. Quant. Biol.* **58:**515–523.
121. **Mizuuchi, M., R. A. Weisberg, and K. Mizuuchi.** 1986. DNA sequence of the control region of phage D108: the N-terminal amino acid sequences of repressor and transposase are similar both in phage D108 and in its relative, phage Mu. *Nucleic Acids Res.* **14:**3813–3825.
122. **Naigamwalla, D. Z., and G. Chaconas.** 1997. A new set of Mu DNA transposition intermediates: alternate pathways of target capture preceding strand transfer. *EMBO J.* **16:** 5227–5234.
123. **Naigamwalla, D. Z., C. J. Coros, Z. Wu, and G. Chaconas.** 1998. Mutations in domain IIIα of the Mu transposase: evidence suggesting an active site component which interacts with the Mu-host junction. *J. Mol. Biol.* **282:**265–274.
124. **Nakai, H.** 1993. Amplification of bacteriophage Mu DNA by rolling circle DNA replication *in vitro*. *J. Biol. Chem.* **268:** 23997–24004.
125. **Nakai, H., and R. Kruklitis.** 1995. Disassembly of the bacteriophage Mu transposase for the initiation of Mu DNA replication. *J. Biol. Chem.* **270:**19591–19598.
126. **Nakayama, C., D. B. Teplow, and R. M. Harshey.** 1987. Structural domains in phage Mu transposase: identification of the site-specific DNA-binding domain. *Proc. Natl. Acad. Sci. USA* **84:**1809–1813.
127. **Namgoong, S. Y., and R. M. Harshey.** 1998. The same two monomers within a MuA tetramer provide the DDE domains for the strand cleavage and strand transfer steps of transposition. *EMBO J.* **17:**3775–3785.
128. **Namgoong, S. Y., M. Jayaram, K. Kim, and R. M. Harshey.** 1994. DNA-protein cooperativity in the assembly and stabilization of Mu strand transfer complex. Relevance of DNA phasing and att site cleavage. *J. Mol. Biol.* **238:**514–527.
129. **Namgoong, S. Y., K. Kim, P. Saxena, J. Y. Yang, M. Jayaram, D. P. Giedroc, and R. M. Harshey.** 1998. Mutational analysis of domain IIβ of bacteriophage Mu transposase: domains IIα and IIβ belong to different catalytic complementation groups. *J. Mol. Biol.* **275:**221–232.
130. **Namgoong, S. Y., S. Sankaralingam, and R. M. Harshey.** 1998. Altering the DNA-binding specificity of Mu transposase *in vitro*. *Nucleic Acids Res.* **26:**3521–3527.
131. **Nash, H. A.** 1990. Bending and supercoiling of DNA at the attachment site of bacteriophage lambda. *Trends Biochem. Sci.* **15:**222–227.
132. **Nash, H. A.** 1996. The HU and IHF proteins: accessory factors for complex protein-DNA assemblies, p. 149–179. *In* E. C. C. Lin and A. S. Lynch (ed.), *Regulation of Gene Expression in* Escherichia coli. R. G. Landes Company, Austin, Tex.
133. **Pato, M.** 1989. Bacteriophage Mu, p. 23–52. *In* D. E. Berg and M. M. Howe (ed.), *Mobile DNA*. American Society for Microbiology, Washington D.C.
134. **Pato, M. L.** 1994. Central location of the Mu strong gyrase binding site is obligatory for optimal rates of replicative transposition. *Proc. Natl. Acad. Sci. USA* **91:**7056–7060.
135. **Pato, M. L., and M. Benerjee.** 1996. The Mu strong gyrase-

binding site promotes efficient synapsis of the prophage termini. *Mol. Microbiol.* **22:**283–292.
136. **Pato, M. L., and M. Banerjee.** 1999. Replacement of the bacteriophage Mu strong gyrase site and effect on Mu DNA replication. *J. Bacteriol.* **181:**5783–5789.
137. **Pato, M. L., M. M. Howe, and N. P. Higgins.** 1990. A DNA gyrase-binding site at the center of the bacteriophage Mu genome is required for efficient replicative transposition. *Proc. Natl. Acad. Sci. USA* **87:**8716–8720.
138. **Pato, M. L., M. Karlok, C. Wall, and N. P. Higgins.** 1995. Characterization of Mu prophage lacking the central strong gyrase binding site: localization of the block in replication. *J. Bacteriol.* **177:**5937–5942.
139. **Pato, M. L., and C. Reich.** 1982. Instability of transposase activity: evidence from bacteriophage Mu DNA replication. *Cell* **29:**219–225.
140. **Pinson, V., M. Takahashi, and J. Rouviere-Yaniv.** 1999. Differential binding of the *Escherichia coli* HU, homodimeric forms and heterodimeric form to linear, gapped and cruciform DNA. *J. Mol. Biol.* **287:**485–497.
141. **Pontiggia, A., A. Negri, M. Beltrame, and M. E. Blanchi.** 1993. Protein HU binds specifically to kinked DNA. *Mol. Microbiol.* **7:**343–350.
142. **Reich, C., B. T. Waggoner, and M. L. Pato.** 1984. Synchronization of bacteriophage Mu DNA replicative transposition: analysis of the first round after induction. *EMBO J.* **3:** 1507–1511.
143. **Rice, P., R. Craigie, and D. R. Davies.** 1996. Retroviral integrases and their cousins. *Curr. Opin. Struct. Biol.* **6:**76–83.
144. **Rice, P., and K. Mizuuchi.** 1995. Structure of the bacteriophage Mu transposase core: a common structural motif for DNA transposition and retroviral integration. *Cell* **82:** 209–220.
145. **Rice, P. A.** 1997. Making DNA do a U-turn: IHF and related proteins. *Curr. Opin. Struct. Biol.* **7:**86–93.
146. **Rice, P. A., S. Yang, K. Mizuuchi, and H. A. Nash.** 1996. Crystal structure of an IHF-DNA complex: a protein-induced DNA U-turn. *Cell* **87:**1295–1306.
147. **Sakai, J., and N. Kleckner.** 1997. The Tn10 synaptic complex can capture a target DNA only after transposon excision. *Cell* **89:**205–214.
148. **Savilahti, H., and K. Mizuuchi.** 1996. Mu transpositional recombination-donor DNA cleavage and strand transfer in trans by the Mu transposase. *Cell* **85:**271–280.
149. **Savilahti, H., P. A. Rice, and K. Mizuuchi.** 1995. The Phage Mu transpososome core-DNA requirements for assembly and function. *EMBO J.* **14:**4893–4903.
150. **Schumacher, S., R. T. Clubb, M. Cal, K. Mizuuchi, G. M. Clore, and A. M. Gronenborn.** 1997. Solution structure of the Mu end DNA-binding Iβ subdomain of phage Mu transposase: modular DNA recognition by two tethered domains. *EMBO J.* **16:**7532–7541.
151. **Shapiro, J. A.** 1997. Genome organization, natural genetic engineering and adaptive mutation. *Trends Genet.* **13:** 98–104.
152. **Shapiro, J. A.** 1979. Molecular model for the transposition and replication of bacteriophage Mu and other transposable elements. *Proc. Natl. Acad. Sci. USA* **76:**1933–1937.
153. **Shapiro, J. A.** 1984. The use of Mudlac transposons as tools for vital staining to visualize clonal and non-clonal patterns of organization in bacterial growth on agar surfaces. *J. Gen. Microbiol.* **130:**1169–1181.
154. **Stark, W. M., M. R. Boocock, and D. J. Sherratt.** 1992. Catalysis by site-specific recombinases. *Trends Genet.* **8:**432–439. (Erratum, 9:45, 1993.)
155. **Surette, M. G., S. J. Buch, and G. Chaconas.** 1987. Transpososomes: stable protein-DNA complexes involved in the in vitro transposition of bacteriophage Mu DNA. *Cell* **49:** 253–262.
156. **Surette, M. G., and G. Chaconas.** 1992. The Mu transpositional enhancer can function in trans: requirement of the enhancer for synapsis but not strand cleavage. *Cell* **68:** 1101–1108.
157. **Surette, M. G., and G. Chaconas.** 1989. A protein factor which reduces the negative supercoiling requirement in the Mu DNA strand transfer reaction is Escherichia coli integration host factor. *J. Biol. Chem.* **264:**3028–3034.
158. **Surette, M. G., and G. Chaconas.** 1991. Stimulation of the Mu DNA strand cleavage and intramolecular strand transfer reactions by the Mu B protein is independent of stable binding of the Mu B protein to DNA. *J. Biol. Chem.* **266:** 17306–17313.
159. **Surette, M. G., T. Harkness, and G. Chaconas.** 1991. Stimulation of the Mu A protein-mediated strand cleavage reaction by the Mu B protein, and the requirement of DNA nicking for stable type 1 transpososome formation. *In vitro* transposition characteristics of mini-Mu plasmids carrying terminal base pair mutations. *J. Biol. Chem.* **266:**3118–3124.
160. **Surette, M. G., B. D. Lavoie, and G. Chaconas.** 1989. Action at a distance in Mu DNA transposition: an enhancer-like element is the site of action of supercoiling relief activity by integration host factor (IHF). *EMBO J.* **8:**3483–3489.
161. **Symonds, N., A. Toussaint, P. Van de Putte, and M. M. Howe.** 1987. *Phage Mu.* Cold Spring Harbor Laboratory, Cold Spring Harbor, N.Y.
162. **Taylor, A. L.** 1963. Bacteriophage-induced mutations in *E. coli. Proc. Natl. Acad. Sci. USA* **50:**1043–1051.
163. **Teplow, D. B., C. Nakayama, P. C. Leung, and R. M. Harshey.** 1988. Structure-function relationships in the transposition protein B of bacteriophage Mu. *J. Biol. Chem.* **263:** 10851–10857.
164. **van Ulsen, P., M. Hillebrand, M. Kainz, R. Collard, L. Zullanello, P. van de Putte, R. L. Gourse, and N. Goosen.** 1997. Function of the C-terminal domain of the alpha subunit of Escherichia coli RNA polymerase in basal expression and integration host factor-mediated activation of the early promoter of bacteriophage Mu. *J. Bacteriol.* **179:**530–537.
165. **van Ulsen, P., M. Hillebrand, L. Zulianello, P. van de Putte, and N. Goosen.** 1997. The integration host factor-DNA complex upstream of the early promoter of bacteriophage Mu is functionally symmetric. *J. Bacteriol.* **179:**3073–3075.
166. **Vologodskii, A., and N. R. Cozzarelli.** 1996. Effect of supercoiling on the juxtaposition and relative orientation of DNA sites. *Biophys. J.* **70:**2548–2556.
167. **Vologodskii, A. V., and N. R. Cozzarelli.** 1994. Conformational and thermodynamic properties of supercoiled DNA. *Annu. Rev. Biophys. Biomol. Struct.* **23:**609–643.
168. **Vologodskii, A. V., and N. R. Cozzarelli.** 1993. Monte Carlo analysis of the conformation of DNA catenanes. *J. Mol. Biol.* **232:**1130–1140.
169. **Vologodskii, A. V., S. D. Levene, K. V. Klenin, M. Frank-Kamenetskii, and N. R. Cozzarelli.** 1992. Conformational and thermodynamic properties of supercoiled DNA. *J. Mol. Biol.* **227:**1224–1243.
170. **Wang, X., and N. P. Higgins.** 1994. 'Muprints' of the lac operon demonstrate physiological control over the randomness of in vivo transposition. *Mol. Microbiol.* **12:**665–677.
171. **Wang, Z., and R. M. Harshey.** 1994. Crucial role for DNA supercoiling in Mu transposition: a kinetic study. *Proc. Natl. Acad. Sci. USA* **91:**699–703.
172. **Wang, Z., S.-Y. Namgoong, X. Zhang, and R. M. Harshey.** 1996. Kinetic and structural probing of the pre-cleavage syn-

aptic complex (type 0) formed during phage Mu transposition: action of metal ions and reagents specific to single stranded DNA. *J. Biol. Chem.* **271**:9619–9626.
173. **Watson, M. A., and G. Chaconas.** 1996. Three-site synapsis during Mu DNA transposition: a critical intermediate preceding engagement of the active site. *Cell* **85**:435–445.
174. **Wei, S. Q., K. Mizuuchi, and R. Craigie.** 1998. Footprints on the viral DNA ends in moloney murine leukemia virus preintegration complexes reflect a specific association with integrase. *Proc. Natl. Acad. Sci. USA* **95**:10535–10540.
175. **Wei, S. Q., K. Mizuuchi, and R. Craigie.** 1997. A large nucleoprotein assembly at the ends of the viral DNA mediates retroviral DNA integration. *EMBO J.* **16**:7511–7520.
176. **Welty, D. J., J. M. Jones, and H. Nakai.** 1997. Communication of ClpXP protease hypersensitivity to bacteriophage Mu repressor isoforms. *J. Mol. Biol.* **272**:31–41.
177. **Williams, T. L., E. L. Jackson, A. Carritte, and T. A. Baker.** 1999. Organization and dynamics of the Mu transpososome: recombination by communication between two active sites. *Genes Dev.* **13**:2725–2737.
178. **Wu, Z., and G. Chaconas.** 1994. Characterization of a region in phage Mu transposase that is involved in interaction with the Mu B protein. *J. Biol. Chem.* **269**:28829–28833.
179. **Wu, Z., and G. Chaconas.** 1992. Flanking host sequences can exert an inhibitory effect on the cleavage step of the *in vitro* Mu DNA strand transfer reaction. *J. Biol. Chem.* **267**: 9552–9558.
180. **Wu, Z., and G. Chaconas.** 1997. The Mu transposase tetramer is inactive in unassisted strand transfer: an auto-allosteric effect of Mu A promotes the reaction in the absence of Mu B. *J. Mol. Biol.* **267**:132–141.
181. **Wu, Z., and G. Chaconas.** 1995. A novel DNA binding and nuclease activity in domain III of Mu transposase: evidence for a catalytic region involved in donor cleavage. *EMBO J.* **14**:3835–3843.
182. **Yamauchi, M., and T. A. Baker.** 1998. An ATP-ADP switch in MuB controls progression of the Mu transposition pathway. *EMBO J.* **17**:5509–5518.
183. **Yang, J. Y., M. Jayaram, and R. M. Harshey.** 1995. Enhancer-independent variants of phage Mu transposase: enhancer-specific stimulation of catalytic activity by a partner transposase. *Genes Dev.* **9**:2545–2555.
184. **Yang, J. Y., K. Kim, M. Jayaram, and R. M. Harshey.** 1995. Domain sharing model for active site assembly within the Mu a tetramer during transposition: the enhancer may specify domain contributions. *EMBO J.* **14**:2374–2384.
185. **Yang, J.-Y., M. Jayaram, and R. M. Harshey.** 1996. Positional information within the Mu transposase tetramer: catalytic contributions of individual monomers. *Cell* **85**: 447–455.
186. **Yang, W., and T. A. Steitz.** 1995. Recombining the structures of HIV integrase, RuvC and RNase H. *Structure* **3**:131–134.
187. **Zou, A. H., P. C. Leung, and R. M. Harshey.** 1991. Transposase contacts with Mu DNA ends. *J. Biol. Chem.* **266**: 20476–20482.

Mobile DNA II
Edited by N. L. Craig et al.

Chapter 18

Tn*5* Transposition

WILLIAM S. REZNIKOFF

INTRODUCTION

Tn*5* is a composite transposon within the IS*4* family of mobile elements (62). Tn*5*, Tn*10*, and presumably other IS*4* family elements transpose via a cut-and-paste mechanism. Tn*5* is an important object of study not because it is clinically important (it has only been isolated twice from natural sources [6; J. Davies, personal communication]), but rather because it is an excellent model system for studying cut-and-paste transposition and similar DNA rearrangement processes such as HIV-1 DNA integration. In addition, Tn*5* is a powerful tool for studying and manipulating DNA molecules in vivo and in vitro.

The importance of Tn*5* as an object of basic research results from a long history of genetic studies, more recent biochemical studies made possible by genetic analyses and by new X-ray crystallographic mo-

William S. Reznikoff • Department of Biochemistry, University of Wisconsin, 433 Babcock Dr., Madison, WI 53706.

lecular structure determinations. The genetic studies not only defined various domains critical for Tnp activity, but also resulted in the isolation of the hyperactive mutations that facilitated the biochemical experiments. The molecular structures are more complete than those available for any other transposase or retroviral integrase and include the structure of the full-length Tn*5* transposase bound to the specific transposon end sequences (the so-called synaptic or paired end complex) (Color Plate 34 [see color insert]) (see references 16, 17). This structure provides us with a molecular scaffold with which to analyze different mutants and the biochemical mechanism of transposition. The hyperactive mutations mentioned above also made possible the isolation of the Tnp-DNA synaptic complexes used for the X-ray crystallographic studies.

Tn*5* transposition is a useful tool for analyzing and manipulating DNA molecules, because it is relatively simple, efficient, and random; many variations in the basic Tn*5* structure have been constructed; and convenient in vivo, in vitro, and mixed in vitro/in vivo methodologies for performing Tn*5* transposition exist. Uses of Tn*5* as a laboratory tool have been described previously (3, 11, 26, 51, 59, 77, 89; see chapter 3).

In this chapter we first present an overall view of the structure of Tn*5*. Second, we study the mechanism behind the Tn*5* transposition process by using the available genetics, biochemistry, and molecular structures as guides. Third, we discuss the regulation of Tn*5* transposition referring to both transposon functions and host functions. Finally, we discuss the use of the Tn*5* system as a practical tool.

The reader is reminded that several excellent reviews focused on Tn*5* have already been published. For instance, the previous edition of this text contained chapters by Berg (4) and by Berg et al. (3). The current chapter focuses primarily on new material not available in 1989.

Tn*5* STRUCTURE

Tn*5* is a 5.8-kb composite transposon in which three antibiotic resistance determinants are bracketed by two inverted 1.5-kb IS*50* insertion sequences, IS*50*R and IS*50*L (Fig. 1A). Below I present an overview of the internal region containing the antibiotic resistance determinants. Then, I present in more detail the structure of IS*50,* because it is this element that encodes the key functions required for transposition and an important transposition regulatory function as well. The internal region plays no known role in the transposition process.

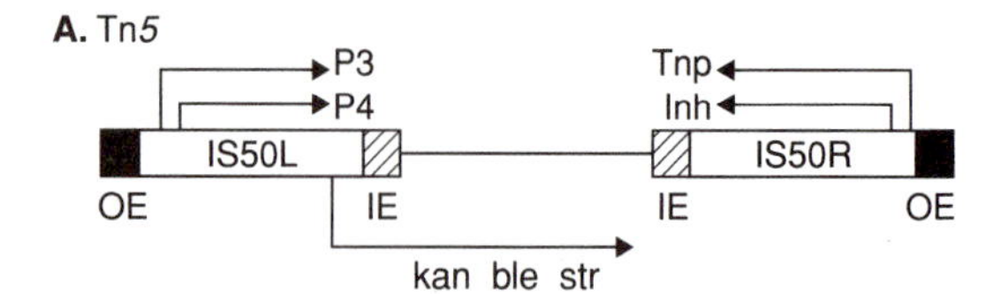

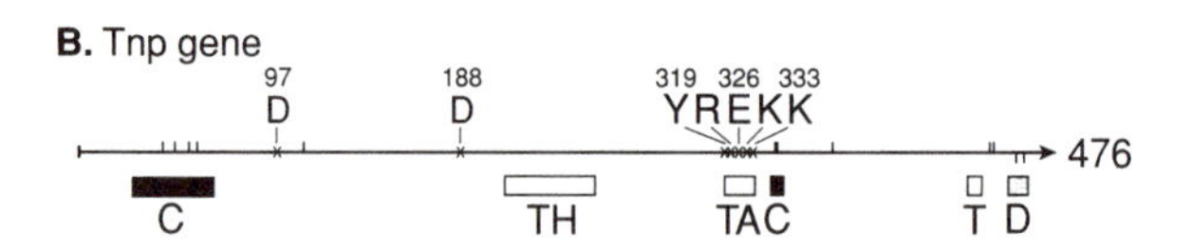

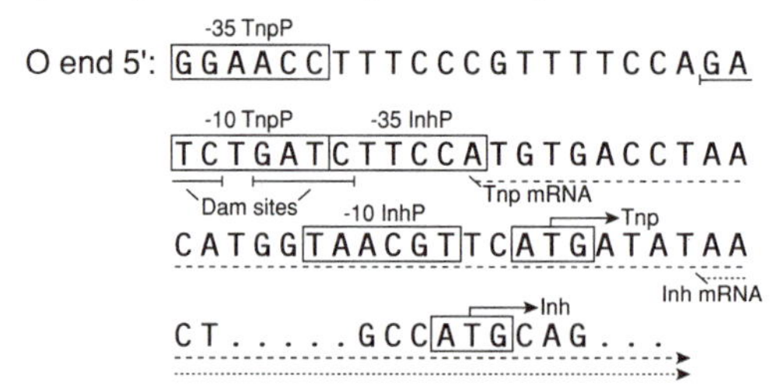

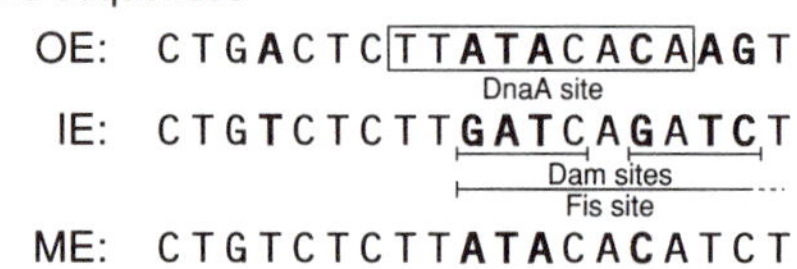

Figure 1. Tn*5* structure. (A) Tn*5*. Two nearly identical insertion sequences (IS*50*L and IS*50*R) bracket three genes that encode resistance to kanamycin, bleomycin, and streptomycin. IS*50*R encodes two proteins that play important roles in Tn*5* transposition: the transposase (Tnp) and the inhibitor of transposition (Inh). Both proteins are read in the same reading frame, but Inh is initiated 56 residues downstream of Tnp. IS*50*L encodes two inactive truncated versions of these proteins (P3 and P4) because of an ochre codon 26 residues from the C terminus. The *cis* sites that define the ends of the transposable elements and at which Tnp acts are 19-bp sequences called OE (outside end) and IE (inside end) (Fig. 1D). For Tn*5* transposition, Tnp acts at two OE sequences (see Fig. 2). For IS*50* transposition, Tnp acts at one OE and one IE. (B) Schematic of Tnp primary structure. The locations of critical active-site residues are indicated with Xs. These include the DDE motif, the YREK signature sequence found in IS*4* family transposases, and K330. K330 is found in retroviral integrases and is similar to R found in other IS*4* family transposases. The black boxes marked with Cs locate domains involved in *cis* OE contacts. The open boxes marked with Ts locate domains involved in *trans* OE contacts; TH is the *trans* hairpin clamp contacts and TA is the *trans* active-site contacts. The shaded box is the location of the Tnp-Tnp dimer interface. Upward facing marks indicate hyperactive mutations. Downward facing marks are defective mutations caused by dimerization failure. (C) Tnp and Inh regulatory elements. The mRNAs that encode Tnp and Inh are programmed by two overlapping promoters. Two Dam methylation sites overlap the Tnp promoter −10 region. (D) End sequences. The nontransferred strands are presented for OE and IE with position 1 (5′) on the left. In addition, a hyperactive end sequence called the mosaic end (ME) is shown. The seven positions at which OE and IE differ are indicated in bold. The bold letters in the ME sequence indicate positions at which it has an OE-specific base in place of an IE base. Also shown are sites specific for host proteins: DnaA, Dam, and Fis.

Tn*5* Antibiotic Resistance Genes

The three antibiotic resistance genes encode resistance to kanamycin, bleomycin, and streptomycin (52). There are three particularly interesting stories related to these genes: the location of the antibiotic resistance promoter, the cryptic nature of the streptomycin resistance gene, and the widespread biotechnology use of the kanamycin resistance gene.

The expression of the antibiotic resistance genes is programmed by a constitutive promoter located within the adjacent IS*50*L sequence (66). This arrangement in effect couples these genes with the transposition elements. The promoter activity is a consequence of a single base pair difference between IS*50*L and IS*50*R that generates a sequence closer to the canonical σ70 promoter sequence (67). One possible interpretation is that the IS*50* element inserted itself between the original promoter and the antibiotic resistance genes, and that the new promoter was generated through an accidental secondary point mutation.

The second interesting feature is that the streptomycin resistance gene is cryptic in *Escherichia coli,* requiring a 2-amino-acid deletion near the C terminus to be active (53). The inactivity of the streptomycin resistance function is presumably caused by some sort of post-translational processing event missing in *E. coli.* Because the native gene is active in other bacteria (*Methylobacterium organophilum* and *Rhizobium* spp.) (52), it is likely that this gene, and presumably Tn*5* itself, did not originate in an *E. coli*-type bacterium.

Finally, the gene that encodes kanamycin resistance has had, since its discovery, a biotechnology history of its own. All studies that use a gene for G418 resistance either use the kanamycin resistance gene originally from Tn*5* or a related gene from Tn*903* (18).

General Overview of IS*50* Structure

All the specific genetic information involved in Tn*5* transposition is embedded in the IS*50* elements (Fig. 1A) (66). IS*50*R encodes two proteins. The transposase (Tnp), which catalyzes transposition, is 476 amino acids long (46, 67). Fig. 1B presents a primary structure schematic of Tnp. As discussed in more detail in the regulation section, Tnp functions primarily in *cis* (35, 43, 67). The second protein (Inh) is a *trans*-active inhibitor of transposition (10, 37, 43). Tnp and Inh are read in the same reading frame but differ because Inh lacks the first 55 amino acids of Tnp (46, 67). The syntheses of the two proteins are programmed by two different overlapping and possibly competitive promoters (Fig. 1C) (46). A host function, Dam methylase, plays an important role in modulating these promoter activities (84). This is related to an interesting regulatory mechanism as described below.

The IS*50* elements are bounded by two 19-bp *cis*-active sites (the outside end [OE] and the inside end [IE]) at which the transposase acts (Fig. 1D) (40, 68). OE and IE differ at 7 of 19 positions but function at roughly the same efficiencies in the presence of wild-type Tnp (57, 90; T. A. Naumann and W. S. Reznikoff, unpublished data). Thus, the recognition of these sequences by Tnp is compromised to accommodate the differences. A19-bp sequence that combines elements of both OE and IE, the so-called mosaic sequence (ME), functions at least 12- to 45-fold better than either OE or IE (90). Moreover, it is also possible to isolate hyperactive Tnp mutations that prefer OE but are impaired for IE recognition and vice versa (57, 91; Naumann and Reznikoff, unpublished). These studies are discussed in more detail later. Finally, the OE and IE sequences are also recognized by host functions leading to host regulation of transposition. These host functions include DnaA (for the OE) (25, 85), and FIS (80) and Dam methylase (68, 84) (for the IE). For a more detailed discussion, see the regulation section.

The two IS*50* elements differ in sequence at a single position. IS*50*R is fully functional for transposition (and its element-encoded regulation). IS*50*L contains an ocher codon that prematurely terminates the two encoded proteins 26 codons early (67). The ocher codon results from the identical change that generates the antibiotic resistance promoter activity mentioned above (66, 67). The resulting truncated versions of Tnp and Inh (P3 and P4) are nonfunctional.

The outline above provides the basic information regarding Tn*5* structure, which is covered in more detail by Berg (4). I present more specific details on the structure of Tnp, Inh, and the end sequences, as necessary, in our discussion of the transposition mechanism and the regulation of transposition. Details of the Tnp and Inh regulatory elements and host function recognition of end sequences are also discussed in the regulation section.

MECHANISM OF Tn*5* TRANSPOSITION

Tn*5* transposes through a complex, multistep, cut-and-paste process as outlined in Fig. 2. The first steps (Tnp binding to the specific 19-bp end sequences followed by dimerization of the Tnp-DNA structure) lead to formation of the synaptic complex. After synaptic complex formation three reactions occur which result in cleavage of the transposon DNA free from the adjacent donor DNA. These reactions require the

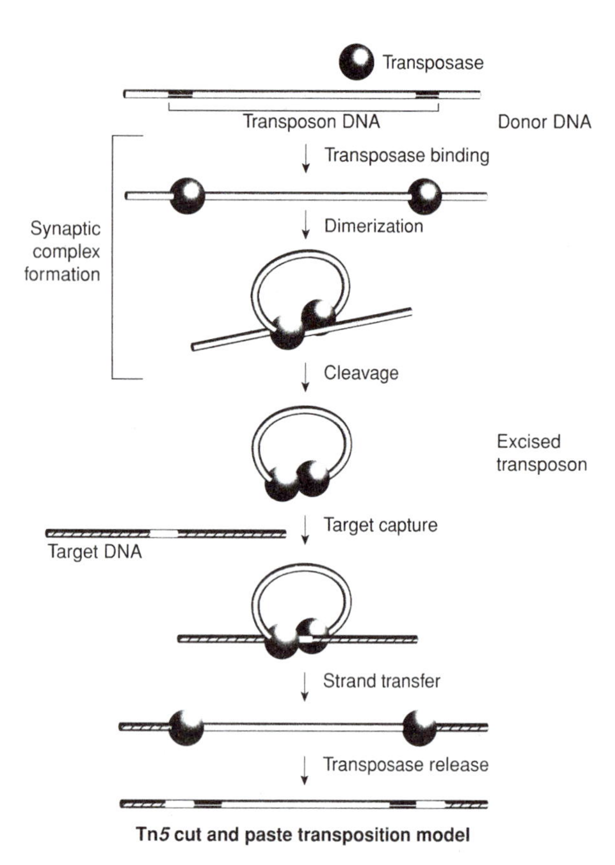

Figure 2. Tn*5* transposes via a cut-and-paste conservative mechanism. Monomeric Tnp binds to the end sequences. Dimerization of Tnp and formation of *trans* Tnp-DNA contacts lead to formation of the synapse. Tnp (in the presence of Mg^{2+}) cleaves the transposon DNA free from the donor backbone DNA through a three-step reaction.

presence of divalent cations such as Mg^{2+}. The released Tnp-transposon synaptic complex then binds to target DNA, and strand transfer occurs. Finally, Tnp is released from the products and cell functions patch repair the duplicated donor sequences. This multistep process for Tn*5* transposition is discussed in brief in a recent review (61) and in more detail below.

For all of the steps up to and including transposon DNA integration, the only necessary macromolecules are Tnp, a transposon DNA defined by two inverted 19-bp end sequences (either OE, IE, or ME), and target DNA (29). This is not to say that no host proteins affect these steps, but rather that, if they do have an impact, they act as modulators. After integration of the Tn*5* DNA, the removal of the transposase from the products and the repair of the DNA joints between the transposon and the target are likely to be performed by host functions.

Synaptic Complex Formation

A key event in transposition is the formation of the synaptic complex consisting of two molecules of the transposase and the two end DNA sequences (7, 9). Formation of the synaptic complex involves multiple steps, and it is also likely to be a disfavored process, thus leading to the biologically important low frequency of transposition. The resulting precleavage synaptic complex is critical because it provides the structure necessary for the catalytic steps. That is, catalysis is performed in *trans*; the subunit that performs catalysis at one end is the subunit that first binds to the other end; thus synapsis must occur before catalysis (57).

Synaptic complex structure

The structure presented in Color Plate 34 represents a synaptic complex after the transposon end DNA sequences have been cleaved free of donor DNA (17). The Tnp in the structure is a hyperactive mutant protein (Tnp E54K, M56A, L372P). The end sequence is the OE. Conformational changes are likely to occur in the synaptic complex after transposon-donor DNA cleavage and thus the structure in Color Plate 34 does not precisely represent the initial synaptic complex. Nonetheless, it certainly provides valuable insights into the structure of the synaptic complex at all stages. The structure is consistent with almost all of the genetic and biochemical evidence that currently exists. Moreover, the catalytic core has an overall structure similar to that found for the retroviral integrases and MuA transposase and has the same DDE active-site motif (14, 16, 24, 63, 82). The DDE residues (aspartate 97, aspartate 188, and glutamate 326) are indicated in red in Color Plate 34. E326 is part of a signature sequence found in IS*4* family transposases (62). This signature sequence includes Y319, R322, E326, and K333. The important DDYREK residues are also shown in the Tnp primary structure representation (Fig. 1B).

The structure of Tnp in the synaptic complex is generally similar to the previously solved Inh structure (16) with three major exceptions (Color Plates 34 and 35 [see color insert]). First, Inh is missing the N terminus with its important DNA binding domain. Second, the sequence between residues 243 and 255 is disordered in the Inh structure but exists as a β hairpin in the Tnp synaptic complex structure. This β hairpin forms a clamp over the end DNA (circled in Color Plate 34). Third, Tnp in the synaptic complex has a disordered region from proline 373 to glutamine 391 that is associated with a relocation of the C terminus so that it no longer occupies overlapping space with the N terminus. The relocation of the C terminus also results in the critical 3′OH ends of the DNAs (indicated with arrows in Color Plate 34) being 41 Å apart versus the 65 Å as would have been seen for the Inh structure. The disordered region probably results from the introduction of a proline at position 372 because of the L372P mutation. There is only one divalent cation (Mn^{2+}, shown as a red ball in Color Plate 34) found in the active site in this structure, but a recent structural analysis of the synaptic complex in the presence of Cd^{2+} places 2 cadmiums in the active site (S. Lovell, I. Y. Goryshin, W. S. Reznikoff, and I. Rayment, unpublished data). The structure is dimeric, symmetrical, and antiparallel. Each DNA molecule makes contact with both Tnp molecules. The simplest interpretation of this arrangement is that one Tnp makes initial (or primary or *cis*) contacts with one end DNA sequence. Then, upon synapsis, the other contacts are formed.

There are three surprising aspects to the structure. The first surprise is the great complexity of the protein-DNA contacts (see also Fig. 3). The second surprise is that protein-DNA contacts and not protein-protein contacts provide most of the interactions stabilizing the complex. The final surprise is the clarity with which the structure correlates with the hairpin-cleavage mechanism to be discussed later. More details of the structure and how it has helped shape our understanding of Tn*5* transposition are described in the next sections.

The synaptic complexes for other mobile genetic elements will likely differ from the Tn*5* synaptic complex structure in interesting ways. For instance, the related transposase from Tn*10* lacks the C-terminal α-helix that makes the critical protein-protein dimer contact for the Tn*5* synaptic complex (13, 73). How Tn*10* Tnp accomplishes the necessary protein-protein interaction is not known. Nonetheless, the Tn*5* synaptic complex structure should help in the analyses of these other systems.

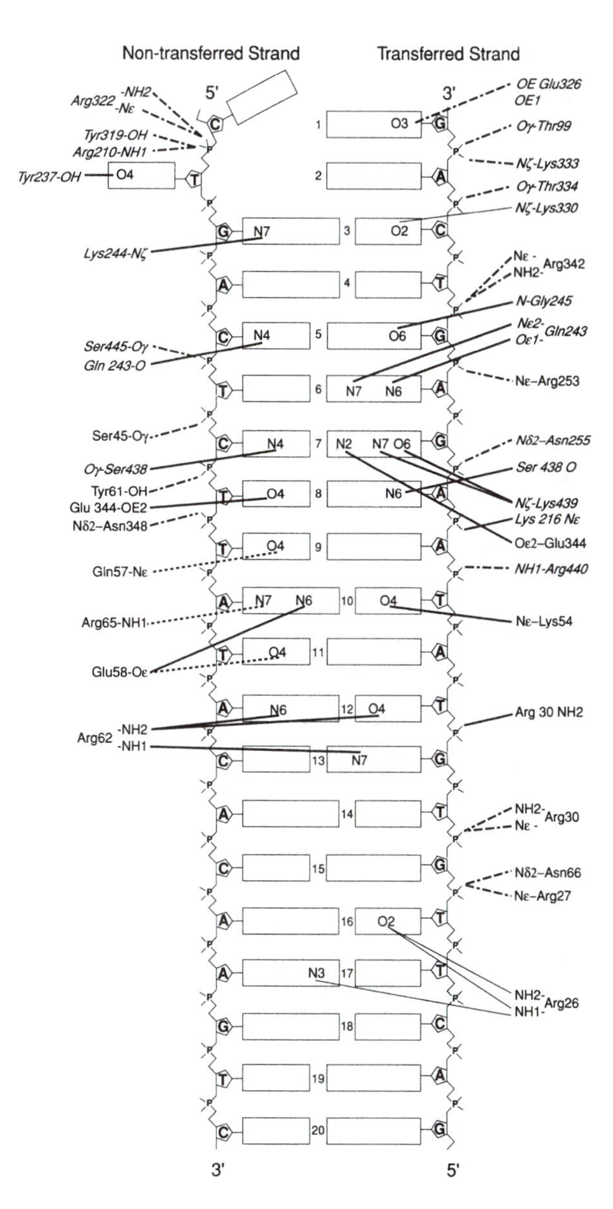

Figure 3. Tnp-OE contacts. The structure data (17) were used to generate a model for the Tnp-OE contacts. Residues in roman type represent *cis* contacts, and residues in italics represent *trans* contacts. Thick lines represent base-specific contacts in the major groove. Thin lines represent base-specific contacts in the minor groove. Dashed lines represent water-mediated interactions with bases. Phosphate interactions are also shown. This figure is based on the structure analysis by Douglas Davies and Scott Lovell.

Synaptic complex formation is a complicated process

I present the formation of the synaptic complex as a multistep reaction starting first with the two transposase monomers binding to the two end sequences, proceeding through protein dimerization via the C-terminal–C-terminal interaction and ending with the formation of the multiple *trans*-protein-DNA contacts. Although this is a conceptually pleasing model, the order of the steps and the number of steps have not been determined.

Three lines of evidence support the idea that synaptic complex formation is a very complex process. First, biochemical studies demonstrate that Tnp domains presumed to be critical for the two subreactions (the N terminus for the primary DNA-binding event, and the C terminus for dimerization) inhibit each other's activities. Gel-retardation studies have been used to show that the C terminus inhibits the Tnp::OE binding reaction. No monomeric Tnp::OE complexes have been detected when full-length Tnp is used. However, monomeric complexes can easily be formed when C-terminal deletion variants are used (i.e., Δ369 Tnp missing 108 C-terminal residues forms monomeric complexes with OE containing DNA) (79, 87). Likewise, Tnp dimerization has not been detected by gel-exclusion chromatography or cross-linking unless the N terminus is removed (12). These observations are important mechanistically because they suggest that the full-length transposase must undergo a conformational change to relieve the C terminus-N terminus inhibition before either DNA binding or dimerization.

Douglas Davies has studied the model of the N-terminal–C-terminal inhibition by superimposing the structure of Inh with Tnp (as exemplified by the E54K, L372P Tnp found in the synaptic complex). The bulk of Tnp matches the Inh structure except for the N-terminal DNA binding domain and the C-terminal domain reorientation (Color Plate 35). In this representation, and from all different perspectives, portions of the N-terminal DNA binding domain of Tnp and the Inh C-terminal domain occupy overlapping space. This simple structural analysis supports the biochemical observations that for free wild-type Tnp these two domains inhibit each other's activities. Thus the N-terminus–C-terminus clash has to be relieved before synapsis by a realignment of the C terminus. This model also explains the hyperactive phenotype of the leucine to proline mutation at position 372, because it disorients a critical α-helix and would facilitate this realignment (17). As described in more detail in the regulation section, this C-terminus–N-terminus inhibition is likely to be an important mechanism for down-regulating the frequency of transposition and partially explains the *cis*-active nature of Tnp.

A second body of evidence that indicates synapsis is a complex process is the many protein-DNA contacts revealed by the crystal structure (Color Plate 34). At the very least, if DNA binding is a first step, it must involve only a portion of the overall protein-DNA contacts that eventually are formed because each Tnp monomer interacts with both end DNA sequences (Figs. 1B and 3). Therefore, some DNA contacts can only be formed in the context of dimerization. In fact, *trans*-protein-DNA contacts make more contributions to stabilizing the eventual synaptic complex than the protein-protein contacts and, as described later, one group of *trans*-protein-DNA contacts requires a protein conformational change.

Once synaptic complexes are formed, however, they are extremely stable. In vitro formed complexes can be stored for extended periods at low temperatures (26).

Initial end sequence binding

The transposase-end sequence binding interaction is extremely complex especially in terms of the protein partner. A key feature of the "primary" or "initial" binding interaction (the recognition of end sequence positions 6 to 19 by the N-terminal sequences of the transposase) was suggested by genetic studies that involved the isolation of null, hyperactive, and change-of-specificity transposase mutations specifically related to the DNA-binding step (Fig. 1B) (57, 57a, 79, 91). Mutations that enhance both OE- and IE-mediated transposition and enhance Tnp binding to both OE and IE containing DNA are located at residues 41 (Y to H), 47 (T to P), and 54 (E to V). An additional change at position 54 (E to K) changes the specificity of end sequence binding, greatly enhancing OE binding but decreasing IE binding. Another change of specificity mutation, enhancing IE recognition at the expense of OE, has been located at position 58 (E to V). In addition, other mutations in the N terminus abolish end DNA binding. These data all suggest that a critical component of the end DNA-binding activity resides in the N terminus. An examination of the cocrystal structure shows the location of critical N-terminus residues vis-à-vis the 19-bp OE sequence (17) (Color Plate 34 and Fig. 3). These initial binding contacts are located at residues 26, 27, 30, 45, 54, 57, 58, 61, 62, 65, and 66.

In addition to the residues in the N-terminal DNA binding domain, another short domain provides contacts to the end DNA in *cis*. These contacts are to residues 342, 344, and 348 (Figs. 1B and 3). We do not know whether these contacts are made during the initial end sequence binding reaction or later during synapsis, although an E344K change does introduce a 20-fold bias for the IE relative to the OE (57a).

The initial Tnp-DNA binding reaction is likely to involve a subset of the OE/IE 19 bp, specifically positions 6 to 19. These observations for Tn*5* are generally consistent with those inferred from genetic studies with Tn*10* and Tn*903* (21, 34). Mutations at most positions (with the likely exception of position 4) are deleterious to the overall transposition reaction (3, 49, 58, 90). However, these in vivo effects cannot be deconvoluted into individual process steps without detailed biochemistry (which has only been performed on a few IE sequence mutations). Even the biochemical results would not determine whether a particular mutation inhibits transposition by generating an interfering group, or indirectly by generating DNA microconformational changes, or directly by removing a specific contact. Four alternative approaches have been used to study the nucleotide pairs involved in the primary Tnp-DNA binding step. These include performing a so-called missing nucleoside experiment that examines whether the removal of a nucleoside blocks formation of the Tnp::DNA monomer complex (9, 39), performing a methylation inhibition experiment that examines whether methylation of guanines in the major groove prevents binding (39), performing an OH radical protection experiment that studies which sugar residues are protected from OH attack by prebinding Tnp (9), or examining the cocrystal structure for nucleosides specifically contacting the N-terminal DNA binding domain (17).

The missing nucleoside experiment was performed by forming monomeric complexes of Δ369 Tnp with ME containing DNA that had on the average no more than one missing nucleoside. Electrophoresis was used to separate bound from free DNAs, and then each population was examined for the species present. Although the two DNA strands gave slightly different results, in general, removal of nucleosides at positions 6 to 19 inhibits binding (9). Similar results have been reported for the more complex Tnp + Inh heterodimer binding reaction to OE (39) and are generally consistent with the *cis* protein-OE contacts observed in the synaptic complex structure (Color Plate 34 and Fig. 3) (17).

The G methylation-inhibition experiment was accomplished by pretreating OE DNA with dimethyl sulfate and then isolating Tnp + Inh-OE heterodimer complexes and examining the bound and the free DNA for the presence of modifications. Methylation of the Gs in the major groove at positions 7, 13, and 15 inhibit complex formation (39). The binding inhibition by methylation in the major groove at position 13 is obviously consistent with the structure because there is a major groove contact between the N-terminal DNA binding domain and the G at position 13 (Fig. 3). The other methylation inhibition results cannot at the moment be easily explained from the synaptic complex structure.

The OH radical protection experiment provided results similar to the missing nucleoside inhibition experiment (9); that is, the C-terminal deletion variant of Tnp, when bound to ME containing DNA, protected a sequence from positions 6 to 19. In the protection experiment, however, the protection was not continuous. A region between positions 10 and 15 was not as well protected, which presumably means that Tnp was not in as close a contact with the sugar-phosphate backbone at these positions. These results are consistent with the crystallographic data discussed below.

The cocrystal structure of the synaptic complex supports the conclusion that OE positions 6 to 17 are important for the N-terminal OE recognition reaction. The detailed interactions are schematically presented in Fig. 3. Specifically, the N-terminal DNA binding domain (thought to define the primary or initial or *cis* DNA contact) makes base contacts at positions 9, 10, 11, 12, 13, 16, and 17, and phosphate contacts at positions 6–7, 7–8, 8–9, 14–15, and 15–16 (17).

The Tnp binding to the end sequences is likely to be a compromised suboptimal reaction. In part, this is because Tnp must have evolved to recognize two quite different sequences, the OE and the IE (Fig. 1D), with comparable affinities. On the basis of this model we isolated four Tnp mutations that enhanced OE recognition and examined them for IE recognition (91). Three mutations enhanced OE and IE recognition, demonstrating that wild-type Tnp is generally impaired in end sequence recognition (this is discussed later as it relates to transposition regulation). One mutation, E54K, manifests enhanced OE binding but diminished IE binding supporting the model above. With use of the reverse approach we have also isolated a multisite mutation (R8C, E54V, E344K, L372Q; with the major effect due to E58K) with the opposite phenotype (IE preferred over OE) (57, 57a).

Tnp binding to DNA fragments containing the OE attached to donor backbone DNA results in DNA bending and/or distortion at or near the end sequence-donor backbone cleavage site. Intuitively this is exactly what one might expect to occur at a site about to be cleaved. The apparent bending has been studied through three types of experiments. First, the mobility of a Tnp::OE complex in polyacrylamide gel electrophoresis experiments had the properties expected of a DNA bent near the cleavage site (87, 88). Second, OH radical footprinting experiments of both the Δ369 Tnp::ME bound complex and the synaptic complex showed hypersensitivity near the cleavage site (9). The hypersensitivity is interpreted to mean that the DNA is distorted in this region. Third, the missing nucleoside experiment showed that DNAs modified near the cleavage site actually bound Δ369 Tnp better as if the flexibility associated with nucleoside removal enhances the bending needed for binding (9, 39).

The cocrystal Tnp::OE synaptic complex structure shows that the nucleotide pairs at positions +1 and +2 are denatured (Color Plate 34 and Fig. 3). This localized denaturation might be related to the −1/+1 bend described above (17). In addition to the bend described above, the cocrystal structure shows a bend of 40° near position 10.

Dimerization of end-bound Tnp monomers

Dimerization of Tnp::end sequence DNA complexes to form the synaptic complex involves protein-protein interactions, *trans* protein-DNA interactions, and conformational changes in the transposase. As mentioned before, it is likely that the *trans* protein-DNA interactions play the dominant role in dimerization. In total, two-thirds of the surface area buried during synaptic complex formation involves protein/DNA contacts (17).

The critical Tnp domain involved in protein dimerization leading to synapsis is located in the C-terminal α-helix (Color Plate 34 and Fig. 1B) (17, 73, 79). As shown by crystallographic analysis of Inh and the synaptic complex (16, 17), these helices cross with a "scissor-like" interface involving residues 458 to 468, with glycine 462 being at the crossover point. Genetic analyses have pointed to the importance of these residues in Inh dimerization, Inh-Tnp dimerization, and, most importantly, synaptic complex formation (12, 73). In particular, amino acid changes at positions 462 and 466 blocked synapsis as seen in gel retardation experiments.

The OH radical protection and missing nucleoside experiments were also used to study protein-DNA contacts involved in synapsis. Both of these studies suggested that the contacts to ME base pairs 6 to 18 in the monomeric Tnp::end DNA sequence complexes were extended down to position 2 on synapsis, suggesting that additional protein-DNA contacts were made (9). This observation is consistent with predictions from genetic studies performed with Tn*10* and Tn*903* (21, 34).

The proposed extension of Tnp-DNA contacts upon synapsis was dramatically confirmed by the synaptic complex cocrystal structure (Color Plate 34 and Fig. 3) (17). Three regions of the transposase, including residues 210 to 255, active-site residues 319 to 334, and residues 438 to 440, have *trans* DNA contacts (Fig. 1B). These contacts are quite extensive, involving bases and phosphate groups at positions 1/2, 2, 2/3, 3, 5, 5/6, 6, 7, 7/8, and 9/10. The contacts to lysines 330 and 333 seem to position the DNA tip in the active site. Note that lysine 333 is part of the YREK signature sequence for the IS*4* transposase family (62) and that lysines are found in a similar position in retroviral integrases (see reference 38). Mutation of lysine 333 severely impairs synaptic complex formation, suggesting that this Tnp-DNA contact plays a role in stabilizing the synaptic complex (Naumann and Reznikoff, unpublished). The lysine contacts were previously suggested by cross-linking studies with the retroviral integrases (38). Thus the role of *trans* protein-DNA complexes in synapsis is likely to be of general importance for transposition.

The contacts involving residues 243, 244, 245, 253, and 255 (forming the β hairpin clamp over the DNA [see circle in Color Plate 34]) may be the first *trans* protein-DNA contacts. Then a series of contacts near position 2 of the end DNA sequence precisely orient the DNA in the active site for catalysis. These contacts involve residues 99, 326, 330, 333, and 334 to the transferred (3′) strand and residues 210, 237, 319, and 322 to the nontransferred strand (see Fig. 3). The contacts to nontransferred strand positions 1 and 2 also likely play an important role in DNA hairpin formation as described subsequently.

Additional *trans*-transposase-end DNA contacts are also formed. These include interactions between residues 216, 243, 245, 255, 438, 439, 440, and 445 and end sequence positions 5 to 9 (see Fig. 3).

The Tnp-OE interaction associated with the dimerization step in synapsis is also likely to be suboptimal. In particular, we note that the mutant end sequence, ME (Fig. 1D), facilitates a 12- to 45-fold enhancement of transposition relative to OE-mediated transposition, yet the mosaic end sequence is slightly impaired in monomer Δ369 Tnp binding (90). ME surprisingly differs from the OE at the following positions: 4, 17, and 18. Positions 17 and 18 are far removed from any *trans* contacts (Fig. 3) and thus should not be critical for synapsis. Previous in vivo experiments had suggested that position 4 is also unimportant (49). However, recent experiments have demonstrated that the critical difference between ME and OE is at position 4 and that this position affects synaptic complex formation (M. Steiniger-White and W. S. Reznikoff, unpublished results).

Transposon-DBB DNA Cleavage

Tn*5* transposase catalyzes transposon donor backbone (DBB) cleavage through a three-step, Mg^{2+} dependent, hairpin mechanism similar to that found for Tn*10* transposase and Rag 1 and 2 mediated V(D)J joining (Fig. 4) (8, 44, 54). Transposon-donor backbone DBB cleavage has been found to occur in *trans* (57). The *trans* nature of the catalysis was confirmed by the *trans* nature of Tnp-DNA contacts seen in the crystal structure (Color Plate 34) (17).

Trans catalysis means that the Tnp monomer that initially binds to one end of the transposon performs the catalytic reactions on the other transposon end. This was determined by transposon-DBB DNA cleavage experiments by using mixed assays with two transposases of different end DNA-binding specificities (57). One Tnp preferentially bound to OE and the other to IE. The transposon DNA was defined by

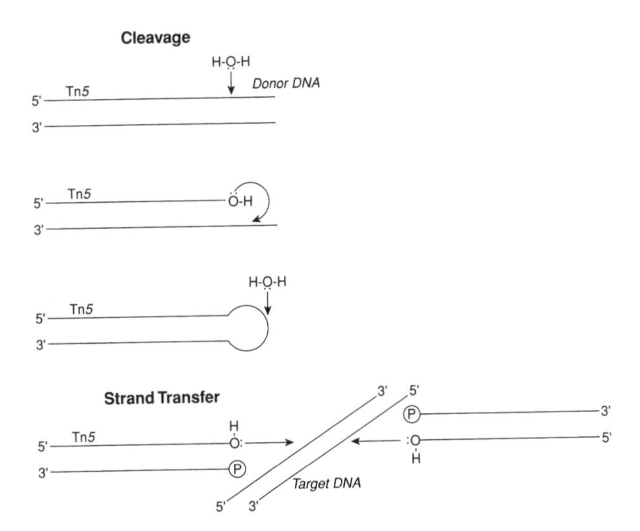

Figure 4. Transposon-DBB cleavage and strand transfer. Tn*5* Tnp catalyzes cleavage through a hairpin intermediate that involves three sequential phosphoryl transfer reactions. In the first step, the transposon-transferred strand-donor DNA junction is nicked. The released 3′-OH of the transferred strand then attacks the opposite strand to form a hairpin intermediate. The hairpin is then nicked. Strand transfer then follows after target capture. The attack of the 3′-OH groups during strand transfer is staggered 9 bp apart.

one OE sequence and one IE sequence as found for IS*50* (Fig. 1A). One Tnp also contained an inactivating mutation in the catalytic active site (E326A). When the OE-specific Tnp was catalytically inactive, cleavage occurred next to the OE. When the IE-specific Tnp was catalytically inactive, cleavage occurred next to the IE.

All three steps in transposon-DNA cleavage have been tracked by denaturing gel electrophoresis of reaction products generated by a synaptic complex that consists of two Tnp molecules and two DNA fragments containing the ME-DBB junction (8). The first reaction involves 3′ strand nicking in which H_2O is a source for the nucleophile (Fig. 4) (the V(D)J joining reverses the polarity in that the 5′ strand relative to the signal sequence is nicked [54]). This first step is thought to be universal in DNA transposition and retroviral insertion (15, 33, 55).

Second, the 3′-OH attacks the 5′ strand to form a DNA hairpin structure, and thus releases the transposon from the donor backbone DNA (8) (Fig. 4). This is seen as the generation of a molecule that on denaturing gels has double the length of the final ME-containing product. This product can snap back on renaturation to form the expected length molecule, and it has the expected hairpin sequence. This hairpin generation reaction is key to the understanding of how a single active site can cleave the two phosphodiester bonds with opposite orientations at the transposon-DBB boundary. This step is missing in replicative transposition as exemplified in Mu and Tn*3* replicative transposition, and also missing in retroviral DNA processing and Tn7 transposition (33).

The Tn*5* synaptic complex cocrystal structure presents a blunt-ended, cleaved DNA molecule (17). Nonetheless, the tip of the DNA has a conformation that is compatible with a hairpin structure in which the two strands are covalently attached to one another. Of particular note are the observations that the two terminal base pairs are denatured and the thymine at position 2 is flipped out to fit into a protein pocket (Color Plate 34 and Fig. 3). At least three nonpaired nucleotides are necessary to form a loop; therefore, the two denatured nucleotide pairs are required for the hairpin formation. The flipped out thymine at position 2 of the 5′ strand seems to bend the strand, making it a suitable target for the 3′-OH attack. It seems that tyrosine 237 makes a hydrogen-bonded contact with the flipped-out thymine at position 2 (Fig. 3) and tryptophan 298 makes a staking contact with this base (not shown). In addition, several contacts with the phosphate between positions 1 and 2 of the nontransferred strand seem to position this strand for the attack of the 3′-OH (notice contacts by residues 210, 319, and 322 in Fig. 3). The resulting structure has the 3′-OH and the 5′-OH ends of the two strands ending within 3.5 Å of each other. Thus, if the transposon-DBB boundary bending or distortion described before is related to the denaturation of the last 2 bp and the rotation of the thymine, it is a key event allowing the hairpin attack.

The final step is the nicking of the DNA hairpin, again using H_2O as a nucleophile (Fig. 4). This step generates the blunt-ended intermediate transposon molecule.

Limited genetic work has been done relevant to the cleavage step for the Tn*5* system. Not surprisingly substitutions of any of the catalytic site DDE residues destroy activity (Naumann and Reznikoff, unpublished). Residues Y319 and R322 are part of the YREK signature motif for IS4 family transposases (Fig. 1B) (62). The crystal structure would predict that both of these residues would interact with the 5′-phosphate of the flipped-out thymine at position 2 and thus might play a role in hairpin formation (17) (Fig. 3). Substitutions of R322 are deleterious for all catalytic steps, whereas substitutions of Y319 are less so (Naumann and Reznikoff, unpublished). No work on the end sequence determinants for cleavage has been performed.

Tn*5* and Tn*10* transposases presumably have very similar catalytic mechanisms, but the catalytic mechanism for Tn*10* transposase has been studied more extensively, especially regarding the stereochemistry of the reaction (45) (see chapter 20).

Target Capture and Strand Transfer

The released transposase-transposon complex next must bind target DNA, and then the strand transfer of the 3′ ends into the target is catalyzed so that two ends go into the target spaced 9 bp apart resulting in a 9-bp duplication of the target sequence (6, 71). Tn*5* has been used for many years as a mutagenic tool so we have a considerable body of evidence that Tn*5* transposition target selection is generally random when considered on a genome-wide basis. The one bias that was recognized was for negative superhelical DNA (36, 48). This same bias has been found for in vitro transposition experiments (H. Yigit and W. S. Reznikoff, unpublished).

However, saturation transposition trials, either in vivo or in vitro, in which the targets are less than 1000 bp long, have indicated that two interesting biases (but not absolute requirements) do exist (28). In general, the target is centered on an A/T, the 9-bp ends have G/C and C/G bps, and adjacent positions are A/T and T/A. More surprisingly independent inserts tend to be found in small clusters with a 5-bp periodicity between inserts being the norm. The simplest way to explain these insert biases is to posit that Tn*5* transposase does bind targets with some modest sequence biases and that, in the presence of excess transposase, it tends to form filaments of Tnp dimers on the target. Within the filament, one presumably random dimer would insert its transposon.

The cocrystal structure is compatible with the general requirements for target capture (17). The 3′-transposon ends (indicated with arrows in Color Plate 34) are spaced 41 Å apart. This is slightly more than that predicted for a 9-bp target on B DNA (35 Å). This difference could reflect Tn*5* Tnp bias towards negatively supercoiled target DNA. The 3′ ends are located within a positively charged cleff appropriate for interacting with the target's phosphate groups.

Strand transfer involves the Mg^{2+}-dependent nucleophilic attack by the 3′ transposon ends on the target's phosphates (Fig. 4). This is a common mechanism for all DNA-transposable elements (15, 33, 55). For Tn*5* we know that the catalysis for this strand transfer is also performed by the active site in *trans* to the primary end DNA binding site (57).

Transposase Release

Tnp needs to disengage from transposition products so that the gap repair of the 9-base single-stranded targets on either end can occur. However, in vitro Tn*5* Tnp remains bound to the transposition products unless they are treated with phenol or detergent (29). Thus, some factor is missing from the in vitro system that would accomplish this event. We have preliminary evidence suggesting two alternative mechanisms. For constructs that end in OE sequences, it is possible that DnaA might play a role because, as described later, it also binds to the OE and Tnp appears to fall off OE in its presence (25, 85; A. Bhasin and W. S. Reznikoff, unpublished data). A second model comes from studies on Tnp proteolysis. Tnp cleaves in vivo after residues 30, 40, and 62 (13). Change of lysine 40 to glutamine has very interesting effects (77a). No proteolysis at that position occurs. The in vitro transposition properties are close to normal. But the in vivo transposition properties are severely compromised. This suggests that residue 40 within the critical end DNA binding domain is recognized by some host factor (a protease?), leading to disengagement of Tnp from the transposition products in vivo.

REGULATION OF Tn*5* TRANSPOSITION

Tn*5* propagates within cells and it is obviously important that the host cells survive for the propagation to be successful. However, transposition itself is likely to be a deleterious event both in terms of the double end breaks left behind in donor DNA and in terms of target genes suffering insertion mutations. Thus, it is not at all surprising that Tn*5* transposition is highly regulated so that it occurs at a very low frequency (a typical transposition assay registers less than one event per 10^5 cells per generation). This regulation takes at least two general forms: maintaining a very low level of transposition activity in general, and tying transposition events that do occur either to DNA replication or to the entry of DNA into the cell. With this as a framework, we discuss Tn*5* transposition regulation from the following perspectives.

1. The transposase and the transposon end sequences are structured to have low levels of activity.

2. The transposase acts primarily in *cis*, on the DNA that is expressing the transposase gene.

3. Tn*5* (IS*50*R) encodes an N-terminal truncated version of the transposase (the inhibitor or Inh) that inhibits transposase activity. Inh functions efficiently in *trans*.

4. Dam DNA methylation down-regulates Tnp expression, up-regulates Inh expression, and inhibits Tnp recognition of the IE. This effect of Dam DNA methylation in conjunction with the strong *cis* bias of Tnp activity likely results in Tn*5* transposition being tied to the passage of the replication fork. Furthermore, Tn*5* DNA that is newly introduced into the cell is frequently hypomethylated and thus should transpose more frequently.

5. Spurious expression of Tn*5* Tnp off of readthrough transcripts originating from outside the element is blocked.

In addition, I discuss the modulation of Tn*5* transposition by several host functions, although it is not clear whether these host functions respond to various cell cycle or environmental influences to alter their modulation of Tn*5* transposition.

Low Transposition Frequencies: Tnp and End Sequences Are Designed That Way

Transposase mutations that enhance the frequency of Tn*5* transposition can be readily isolated (20, 47, 57, 57a, 78, 81, 91). This observation means that the wild-type transposase has a suboptimal structure. The hyperactive Tnp mutations isolated to date fall into four different groups, at least three of which act additively with each other. Various combinations of these hyperactive mutations can raise the transposition frequency more than 10^3 fold (29).

The first group of hyperactive mutations includes the various changes that enhance the end DNA-binding activity of the protein. These include mutations at residues Y41, T47, E54, E58, and E344 (Fig. 1B) (57, 57a, 91). In three cases the enhancement is sequence specific: OE for E54K (91), and IE for E58V and E344K (57, 57a).

The second group of mutations includes E110K and E345K (Fig. 1B) (81). In the cocrystal structure, E110 and E345 face each other with an Mn^{2+} forming a salt bridge (17). Either E110K or E345K would relieve the necessity for having a divalent cation between the residues. It is not clear from the structure why the E110-E345 interaction is important. One guess is that this interaction precisely positions residues 342, 344, and 348 to interact with the end DNA.

The precise residue at position 372 has a large influence. In particular, the L372P mutation gives rise to Tnp with a hyperactive and a *trans*-active phenotype (Fig. 1B and Color Plate 35) (78). L372Q gives rise to a hyperactive phenotype but its effect on *cis* versus *trans* activity is unknown (57, 57a). The L372P substitution breaks an α-helix, and the resulting disordered structure is likely to destabilize the C-terminal-N-terminal inhibitory interaction discussed before and modeled in Color Plate 35 (3). The L372Q change presumably has a similar effect. Thus these changes at residue 372 may facilitate the Tnp conformational change necessary for forming synaptic complexes. The role of the C-terminal-N-terminal interaction in down-regulating Tnp activity and the possible relationship of this interaction to the *cis*-active nature of Tnp are discussed again later.

Two hyperactive mutations are located in the C-terminal α-helix: L449F (20) and E451Q (Fig. 1B) (47). No mechanistic explanation for the hyperactive mutant phenotypes exists for these mutations, but perhaps they enhance the Tnp dimerization reaction mediated by the C-terminal α-helix or perhaps they alter the N-terminal-C-terminal inhibitory interaction.

As mentioned before, the OE and IE sequences are also suboptimal. In an experiment designed to determine how the E54K hyperactive mutant of Tnp discriminates between OE and IE, we made a library of OE/IE mosaic sequences. (We varied the positions that differed between OE and IE in all possible combinations.) The resulting mosaic sequences were analyzed with respect to their effects on the transposition frequency (90). Through this analysis we discovered that one sequence, the so-called mosaic end (ME) (Fig. 1D), gave rise to a 12- to 45-fold enhancement of transposition for wild-type Tnp and E54K hyperactive Tnp. The ME sequence enhances transposition by stimulating synaptic complex formation and not monomer Tnp binding.

Transposase Acts Primarily in *cis*

Tn*5* Tnp (and transposases of some other transposons such as Tn*10* and Tn*903* [22, 56]) has a strong bias to act in *cis* (35, 43). That is, Tn*5* Tnp preferentially mobilizes transposon DNA that encodes the transposase or is located close by on the same DNA that encodes the transposase, but does not efficiently mobilize transposon DNA located on a second replicon or an extensive distance away on the same replicon. *Cis* action by the transposase should contribute to the low frequency of transposition. If the *cis* preference is caused by the transposase only being active transiently, the availability of active Tnp is severely limited. Furthermore, if multiple copies of Tn*5* are present, transposition is restricted to the copy that is in the process of synthesizing Tnp.

Over the years we have presented various proposed explanations for Tnp *cis* activity. Our current data suggest that two related mechanisms are important in the phenomenon. The essence of the two models is that the Tnp N-terminus contains the key DNA-binding determinants involved in the initial Tnp-terminal sequence interaction, and the C-terminus inhibits N-terminus function (described above and in Color Plate 35) and makes the full-length Tnp susceptible to inactivating dimerization with Inh (see below). Thus Tnp would be active if the N-terminus binds to the end DNA sequences before synthesis of the C-terminus while it is still tethered to the mRNA. In such a circumstance the inhibitory C-terminus-N-terminus interaction could not occur, and the time-frame for C-terminus-mediated interaction with Inh is shortened. This generates a *cis*-active Tnp with a short functional half-life. In this scenario, L372P decreases the possibility of the C-terminus-N-terminus interaction and thus enhances the chances for free full-length Tnp to be active.

The Tn5 Inh Inhibits Tnp

A key mechanism for down-regulating Tn*5* transposition involves the synthesis of N-terminal truncated versions of Tnp. Inh (and other N-terminal truncated versions of Tnp mentioned below) lack residues playing a key role in end DNA binding and are thus inactive as transposases. However, their dimerization activity is enhanced (12) (because of the absence of the inhibitory N-terminus) and they can readily dimerize with Tnp to form inactive heterodimers (19). Inh binding to Tnp should presumably block the Tnp-Tnp interaction involved in synapsis. Thus Inh is a potent *trans*-dominant negative mutant version of Tnp (41). In steady-state conditions, a 4-fold excess of Inh is present in the cells (42) and blocking Inh synthesis (for instance, by using a mutation that changes the Inh AUG to GCC) increases the transposition frequency 10-fold (81, 86). In vitro experiments have also demonstrated that Inh is a potent inhibitor of transposition and that this inhibition is blocked by a mutation in Inh's C-terminal dimerization domain (12). Thus, Inh plays a significant role in down-regulating Tn*5* transposition.

There are other N-terminal truncated inhibitory versions of Tnp present in Tn*5*-containing cells. Under steady-state conditions, in vivo proteolysis generates Tnp derivatives cleaved adjacent to residues 30, 40, and 62 (13, 77a). It is likely that all of these truncated products can also function as inhibitors.

Dam Methylation Regulates Competing Tnp and Inh Promoters

Tnp and Inh synthesis, which is affected in opposite ways by the Dam methylase system, is programmed by different mRNAs (46, 84). Thus, the transposition frequency is tied to the DNA replication cycle in which DNA is transiently hemimethylated after passage of the DNA replication fork, and Tn*5* DNAs newly introduced to the cell will likely have a higher transposition frequency than resident Tn*5*s. Dam methylation regulation of transposition has also been shown for Tn*10* (64).

The transcription-controlling elements programming Tnp and Inh mRNAs are shown in Fig. 1C. The start sites are displaced by 31 bp, suggesting that the relevant two RNA polymerase binding events should be competitive (46). In addition, the upstream Tnp promoter sequence contains two Dam methylation sites overlapping the −10 sequence. In wild-type cells, the level of Tnp mRNA falls fourfold and the level of Inh mRNA increases twofold relative to that found for *dam* cells. These switches are qualitatively mirrored in the Tnp and Inh protein synthesis levels and in transposition frequencies. The observed Dam regulation is abolished by site-specific mutations of the Tnp promoter that destroy the Dam sites (84).

A striking example of the biological impact of this transcription regulation can be found in the analysis of Tn*5* protein expression and transposition resulting from the infection of cells by λ::Tn*5* (65). λ DNA is hypomethylated as it is packaged in the virion. Upon infection there is a burst of transposition and high levels of Tnp versus Inh expression. The levels of Tnp drop relative to Inh with time. These are exactly the expected results from the proposed system in which Tnp mRNA synthesis is negatively regulated by Dam DNA methylation, Inh mRNA synthesis is positively affected by the drop in Tnp mRNA synthesis, and the λ DNA is hypomethylated immediately upon infection but becomes methylated after residence in the cell. This strategy offers an elegant way to have elevated Tn*5* transposition immediately upon DNA entry into the cell but lower steady-state levels of transposition after the newly entered DNA establishes residence.

An interesting coupling also occurs between the Dam regulation of Tnp mRNA synthesis and the previously described *cis*-active property of Tnp. In a cell containing multiple copies of Tn*5*, only the copy that has recently been replicated should synthesize elevated levels of Tnp and those Tnp molecules should act primarily on the newly replicated version of Tn*5*. As described below, this coupling should be even more strenuous for IS*50*.

Dam Also Regulates IE-Mediated Transposition

Although Tn*5* transposition is regulated by Dam DNA methylation via its effect on Tnp versus Inh mRNA synthesis (84), IS*50* transposition is regulated by Dam through an additional regulatory strategy. Dam DNA methylation directly modifies the IE sequence that is recognized by Tnp during the transposition of IS*50* (49, 58, 84). The IE sequence contains two Dam methylation sites (Fig. 1D). Transposition of IS*50*-like structures is depressed 1,000-fold by Dam DNA methylation, 100-fold of which is probably related to IE sequence methylation (49, 57a, 58, 84). In vitro Tnp-binding assays also indicate that Dam methylation of IE inhibits Tnp binding (39).

The analysis of the E54K, M56A, L372P Tnp-OE cocrystal structure (Color Plate 34 and Fig. 3) (17) indicates that there are major groove contacts by Tnp residues 54 and 58 to OE bp 10, and residue 62 to OE bp 12 and 13. Dam methylation would put methyl groups in the major groove at bp 11 and 12 of IE, suggesting that a simple steric blocking of major groove interactions in this region may explain the Dam DNA methylation inhibition effect. However, the analysis is more complex. It is necessary to be careful in extrapolating the cocrystal structure data because the DNA in the complex is the OE not the IE. Second, a single residue change in Tnp (a change at residue 58 from E to V) generates a transposase that more than wipes out the Dam DNA methylation inhibition effect, resulting in a transposase that actually prefers methylated IE DNA as a recognition sequence (57, 57a). The observed reverse effect generated by the E58V mutation suggests that the valine can provide a beneficial hydrophobic contact with a methyl group introduced by Dam.

Blocking Expression of Readthrough Transcripts

Many Tn*5* inserts could land in highly expressed genes. This would result in the generation of readthrough transcripts for Tnp and potentially high levels of Tnp synthesis. However, Tnp is not translated from readthrough transcripts (46), because an RNA secondary structure is formed in these readthrough RNAs that blocks the Tnp translation initiation signals (72). This is a common mechanism shared by several other transposable elements (60).

Other Host Factors That Modulate Transposition

Several host factors have been implicated as modulating participants in the Tn*5* transposition process. That is, for the wild-type in vivo system, mutations in several host functions increase or decrease the frequency of transposition. However, for the hyperactive in vitro system in which the E54K, M56A, L372P Tnp is used, these host functions are clearly not required for efficient transposition. These host functions are DNA gyrase, DNA polymerase I, topoisomerase I, DnaA, factor for inversion stimulation (FIS), integration host factor (IHF), and SulA (Table 1). Unfortunately, little work has been accomplished on the role of many of these host functions in Tn*5* transposition since the review of Berg in the first edition of this text (4) so these will not be discussed further here. We briefly discuss DnaA and topoisomerase I.

DnaA and Tn*5* Transposition

DnaA has a binding site within the OE sequence (Fig. 1D) (25, 85). The obvious question is: Does DnaA play a role in Tn*5* transposition? The in vivo evidence is clear that DnaA and its binding site in the OE act as positive elements in Tn*5* transposition, suggesting that DnaA binding to OE stimulates transposition or the recovery of transposition products by at least 10-fold (85). Given the complex protein-DNA contacts present in the transposase-end DNA complexes (Color Plate 34), it is hard to imagine how DnaA and Tnp could bind to OE DNA simultaneously. As mentioned previously, our current idea is that perhaps DnaA helps to disengage Tnp from final transposition products. Preliminary footprinting experiments have suggested that DnaA and Tnp binding

Table 1. Host functions and Tn*5* transposition

Host function	Effect on Tn*5* transposition	Reference(s)
DNA gyrase	Stimulates transposition by generating supercoiled target	36
DNA polymerase I	Stimulates transposition 10-fold by unknown mechanism	69, 75
DNA topoisomerase I	Stimulates transposition by relaxing donor DNA and facilitating target capture	74, 83; Yigit and Reznikoff, unpublished
DnaA	Stimulates transposition 10-fold by unknown mechanism	25, 85
Fis	Stimulates transposition in Dam^+ cells through an unknown mechanism, inhibits IS*50* transposition in Dam^- cells by binding to IE	80
IHF	Stimulates transposition in Dam^- cells through an unknown mechanism	50
SulA	Inhibits transposition through an unknown mechanism	70

to OE are competitive and that addition of DnaA to premade Tnp::OE complexes yields a DnaA::OE footprint and not a Tnp::OE footprint (Bhasin and Reznikoff, unpublished).

Topoisomerase I and Tn*5* Transposition

Early in vivo experiments suggested that topoisomerase I might stimulate Tn*5* transposition (74). The discovery that Tn*5* transposition prefers relaxed donor DNA (36) suggested an indirect model for the topoisomerase I-mediated stimulation. However, recent studies have indicated an additional, more direct role for topoisomerase I that does not involve its DNA relaxation activity. In essence topoisomerase I copurifies with Tnp and directly interacts with Tnp through pull-down experiments (83). Moreover, topoisomerase I stimulates Tn*5* transposition in vitro, particularly when supercoiled target DNA is used, and this in vitro stimulation (and the in vivo stimulation) could occur in the absence of DNA relaxation activity (Yigit and Reznikoff, unpublished). The simplest model available is that topoisomerase I stimulates target capture.

Tn*5* AS A TOOL

Transposition is a powerful tool for the geneticist. In its simplest application it is a means for generating knockout mutations. However, its power is greatly expanded because the internal sequences of a transposon can be modified to contain any desirable genetic information and still function as a transposon if transposase is supplied in *trans*. The Tn*5* system is particularly useful because Tn*5* transposition is random, the Tn*5*-specific requirements are simple (19-bp end sequences and Tnp), a wide variety of Tn*5*-like elements have been constructed, and a simple, robust, commercially available in vitro transposition system exists.

Below, I discuss various genetic options for the DNA to be located between the end sequences and, for some of these options, I highlight some particularly interesting applications. Then, I discuss the advantages of using transposable elements as sequencing tools. Finally, I discuss some unique applications of the Tn*5* in vitro system. General reviews of in vivo transposition systems as tools are available (see reference 3, for instance).

DNA between the Ends

Tn*5* transposition, as well as that of many other transposons, does not require any genetic functions encoded by the sequences between the ends as long as active Tnp is available. The only likely limitations for the transposon's content are length considerations. The ability to form a synaptic complex should be influenced by the length of DNA between the ends in two ways. First, the DNA length must be long enough to form the required loop. The shortest Tn*5*-like transposon to be tried to date is 264 bp long (59). From studies of DNA looping in which the donor backbone was varied (27), we estimate that a Tn*5*-like transposon less than 100 bp in length would not function efficiently. Second, the ends should not be too far apart, because very long intervening DNA would lower the relative concentration of the two ends and inhibit synapsis. This upper limit has not been studied but we have used Tn*5*-like constructs that are 11 kb in length with no apparent reduction in transposition frequencies (89).

Within these rough length constraints, the experimenter is likely to consider incorporating the following elements: a selectable marker, a reporter function, a landmark function, and controlling elements. The selectable marker is obvious and will not be discussed further. Examples of the other elements are discussed below.

Reporter functions

A variety of reporter functions can be included within the body of the transposon. In general, the reporter gene abuts one end sequence and is structured to result in one of two types of fusion. For transcription or operon fusions the reporter gene is preceded by its own translation initiation signals. A typical use of such transcription fusions is to study the transcriptional regulation of the gene suffering the insertion. For translational fusions the reporter gene lacks its normal AUG and can only manifest function when transposition insertion results in a gene fusion in the correct reading frame. An obvious requirement for generating translational fusions is that there be no in-frame nonsense codons between the end of the transposon and the reporter gene open reading frame. Translational fusions can also be used to study gene regulation (transcriptional and translational), but they have other useful applications as well, for instance, studying membrane protein topology (11, 77). I give a brief example of how Tn*5* translational fusion technology has been used to study membrane protein topology.

Within the context of the *E. coli* cell, β-galactosidase functions as a cytoplasmic protein; it cannot function when embedded in the membrane or when localized in the periplasm. Alkaline phosphatase, on the other hand, functions exclusively as a periplasmic protein. With these principles, Beckwith's lab pioneered the use of Tn*5*-*lacZ* and Tn*5*-*phoA* to create random translational fusions within the gene for a membrane protein, and by looking for activity of the β-galactosidase and alkaline phosphatase fusions, they mapped out the cytoplasmic, transmembrane, and periplasmic domains of the protein (11, 77).

Inserting landmarks

The transposon-generated insertions can introduce one of a variety of "landmark" sequences. The simplest example of a landmark is a primer binding site for DNA sequencing. DNA sequencing applications are discussed separately. Because the primer can be synthesized to fit the transposon, no specialized constructions of the transposon are necessary. Other examples of landmark sites include the sequences for site-specific recombination systems (such as a λ att site, an FRT site, or a *lox* site). These sites allow the subsequent incorporation of additional genetic information. Rare restriction sites for subsequent physical mapping purposes can also be incorporated in the transposon.

One of the more interesting landmark insertion techniques involves the insertion of epitopes or protease cleavage sites at more-or-less random locations within a protein (51). An example construct for this technology would involve a transposon that generates translation fusions upon insertion, but the transposon ends have a specific restriction site at their internal borders. After isolation and characterization of the intact in-frame insertion, the entire inside of the transposon can be excised leaving a short in-frame insertion encoding an epitope and/or a protease-sensitive sequence. In essence this is a form of linker-scanning mutagenesis. This technology maps regions of the protein that will allow insertion without impairing activity or can precisely position protease cleavage sites within the protein.

Controlling elements

Transposons have been used to deliver a variety of *cis*-active DNA sequences. These include regulated promoters, transcription termination signals (polar mutations), origins of replication, and origins for DNA transfer.

Transposons as Tools in Genomics: Sequencing Primer Binding Sites and Knockouts in One

The key initial step in DNA-sequencing protocols is to generate a template that has a primer binding site adjacent to the DNA to be sequenced. There are four approaches to this goal: (i) cloning the target DNA into a standard vector containing the primer binding sites, (ii) performing a primer walking experiment in which new primers are generated for each step from the previous sequence determination, (iii) generating nested deletions in which various sections of the target DNA are brought next to a standard primer binding site (see below for a Tn*5* mechanism for generating nested deletions [89]), or (iv) inserting the primer binding sites via transposition.

The concept of using transposons to "sprinkle" primer binding sites at random into a target is not new (for instance see references 1, 59). There are some good reasons for this. First, one obtains "two for the price of one." That is, for each insert one is generating two primer binding sites facing out from the two ends of the transposon. Second, and generally overlooked, with transposon-delivered primer binding sites one is simultaneously creating knockout mutations. One can archive the inserts, and during the functional analysis of the target DNA, one can retrieve the appropriate inserts, reintroduce them into the in vivo host, and then study the functional consequences of the insertion knockout.

The use of transposons as mobile primer binding sites deserves a fresh look because of two technical developments. First, the development of efficient in vitro intermolecular transposition systems (2, 23, 29–32) allows the investigator to bypass in vivo methodologies. Second, the newer generation sequencing protocols/machines are sufficiently sensitive to bypass the need of high-copy number templates. In fact, direct sequencing from chromosomal inserts in bacteria is easily doable.

Tn*5* In Vitro Applications

During the past several years efficient in vitro intermolecular transposition systems have been developed on the basis of several transposable elements. In addition to the Tn*5* system (29) these include those derived from Ty1 (23), Tn*7* (31), Mu (32), Mariner (2), and Tn*502* (30). These offer convenient tools for tagging sequences cloned in various vectors. Intermolecular transposition can also be used to clone target sequences because linearized sequences defined by transposon ends can be used in in vitro systems (indeed such linearized molecules are preferred in the Tn*5* system [29; Yigit and Reznikoff, submitted]). In essence the target sequence is PCR amplified by using primers that contain the required transposon end sequences, and then a standard transposition reaction is performed in the presence of the desired target vector.

Tn*5* in vitro transposition also can be used for the generation of nested deletions and nested inversion/deletions through intramolecular transposition (76, 89) (Fig. 5). This technology involves cloning of the

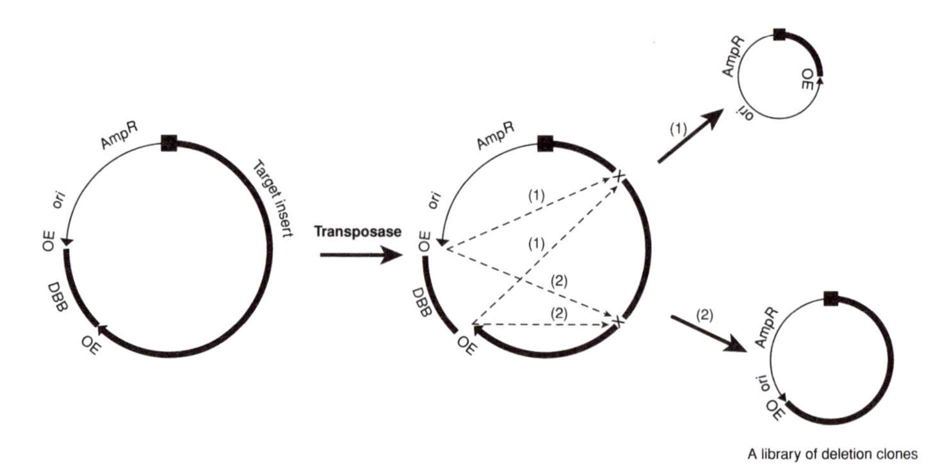

Figure 5. Intramolecular transposition yielding nested deletions. Intramolecular transposition events yield two types of products depending on the precise orientation of the strand transfer reaction: deletion circles or inversion circles. The generation of deletion circles can be used as a tool for generating nested deletion families. The generation of two such deletions is pictured. The generation of inversion circles is not shown. Both types of products are convenient templates for sequencing. In the case pictured, the transposon has been constructed to have a target sequence, an origin of replication, and a selectable marker. Only deletion products from one side (carrying the origin and the selectable marker) are inherited after transformation.

target gene of choice into the body of the transposon and then performing an in vitro transposition reaction under dilute DNA conditions so that the transposon ends likely do not find intermolecular targets. Two types of products result from intramolecular transposition events depending on the orientation of the transposon attack on the target. The first type of product consists of two deletion circles as shown in Fig. 5. By proper design of the transposon body with an origin of replication and a selectable marker, one of these circles can propagate in cells after electroporation. The second class of products consists of inversion circles (not shown in Fig. 5). The formation of deletion circles and the inversion circles is useful for bidirectional sequencing of relatively small (several kilobase) targets and for performing a deletion analy-

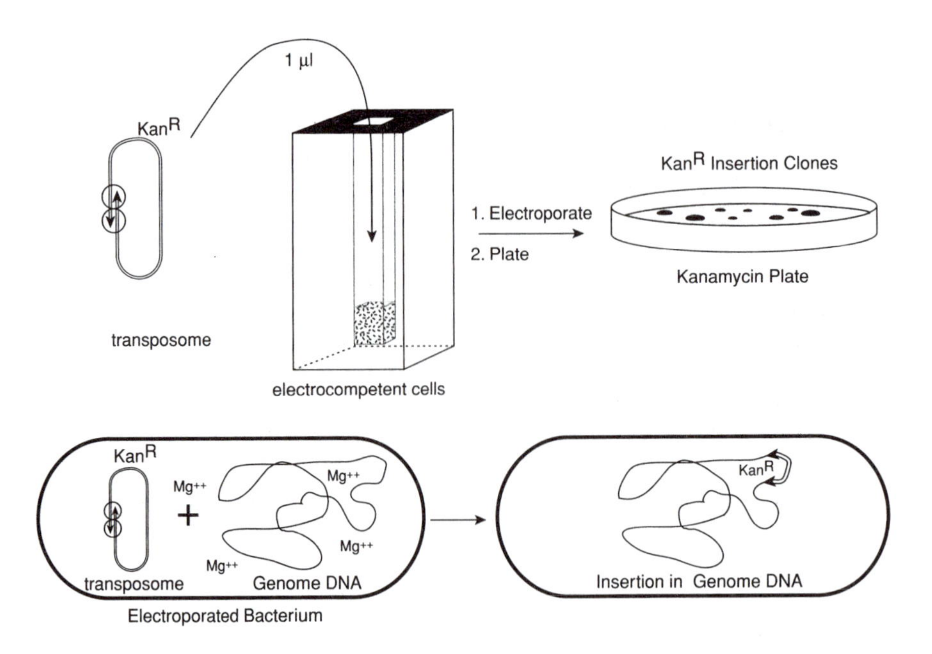

Figure 6. Electroporation of preformed transposon-Tnp complexes. Tnp forms stable synaptic complexes with precleaved transposon DNA in the absence of Mg^{2+}. Electroporation of these complexes into target cells yields transposition events. This technology avoids the need to produce Tnp in the target cells.

sis of a protein or an RNA encoded by the target DNA.

A unique Tn*5* technology exists that allows one to make in vivo inserts without encoding the transposase within the target cells. In essence Tnp::transposon transposition complexes are made in vitro. These can be stored in the absence of Mg^{2+} for several months. The complexes are then electroporated into the cells, and transposition events occur presumably because the cell provides the necessary Mg^{2+} and the target DNA (Fig. 6). This technology has been used successfully in several gram-negative bacterial species (26) and in *Mycobacterium smegmatis* (K. M. Derbyshire, personal communication).

Finally, the existence of two hyperactive Tn*5* transposases with distinct end DNA specificities (Tnp sC7v2.0 to the IE and Tnp EK-LP to the OE [57, 57a, 91]) allows one to perform two sequential transposition events, each of which uses different end sequences. One application of this capability is in the generation of random gene fusion libraries (I. Y. Goryshin, T. A. Naumann, and W. S. Reznikoff, unpublished data). In this application the initial transposable element is defined by two IE sequences with OE sequences adjacent to the ends in inverted orientation. A random insert library into a plasmid containing gene "X" is generated by using IE-specific hyperactive Tnp (Tnp sC7v2.0). This insert library has generated a second class of transposons defined by OE sequences (a library of OE-X fragment-plasmid sequence-X′ fragment-OE transposons). The next transposition event uses the hyperactive OE-specific transposase (Tnp EK-LP) to make inserts into gene Y. The result is a library of Y-X and X′-Y′ fusions, each of which has an IE-OE linker sequence at the point of fusion (Fig. 7).

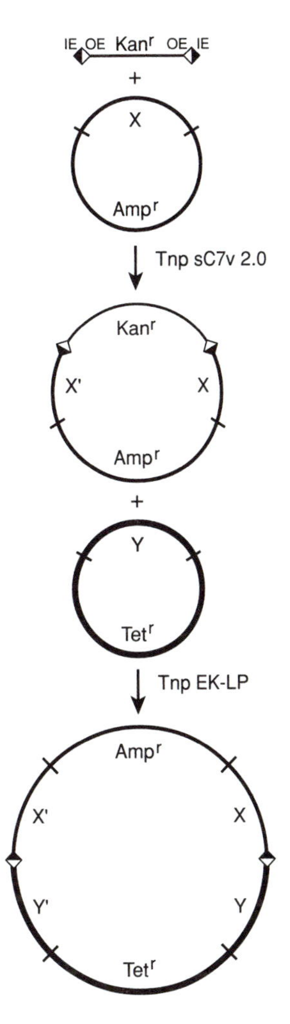

Figure 7. Random gene fusion technology. Two sequential transposition events are used to generate a random library of X-Y and Y-X fusions. The initial transposon is defined at its ends by IE sequences. Inside each IE sequence is an OE sequence in inverted orientation. The first transposition event uses the IE-specific hyperactive transposase sC7v2.0 (57, 57a) to make a library of inserts in gene X. The second transposition event uses the OE-specific hyperactive transposase EK-LP (91) to transpose the X and X′ containing transposons into gene Y.

Beyond Bacteria

The in vitro Tn*5* transposition system has demonstrated that efficient Tn*5* transposition can occur in the absence of host functions (29). This observation suggests that Tn*5* transposition could also occur in the cells of a wide variety of organisms. Indeed, we have demonstrated that electroporation of premade Tn*5* transposition complexes into *Saccharomyces cerevisiae* gives rise to bona fide Tn*5* transposition events (26). Recently we have learned that the same basic technology has been successfully applied to *Drosophila melanogaster* and *Aedes aegypti* (D. O'Brochta, personal communication). The basic approach involved injecting a preformed transposition complex into the insect preblastoderm and finding animals that expressed the Tn*5*-encoded function (GFP) among the progeny. This observation opens up many avenues for the future practical use of Tn*5* as a scientific tool.

CONCLUSION

Tn*5* is a powerful system for studying the molecular basis of DNA transposition. Along with the analysis of other DNA transposons, these studies should yield a detailed molecular picture of DNA transposition. Moreover, because its structure and molecular mechanisms of DNA cleavage and integration have striking similarities to those of retroviral integrases,

Tn*5* Tnp should be a good surrogate for studying HIV-1 integrase. Finally, Tn*5* transposition is a useful tool for studying and manipulating DNA structure.

Acknowledgments. I thank Douglas Davies for developing the synaptic complex, transposase-inhibitor, and DNA contact figures (Color Plates 34 and 35 and Fig. 3), the Biochemistry media facility for the illustrations, and my laboratory for many years of imaginative and productive research. I thank Carol Pfeffer for preparing the manuscript.

The research work in my laboratory was supported by NIH grant GM 50692, NSF grant MCB 84089, the University of Wisconsin University Industrial Relations grant 1680, and a USDA Hatch grant (4184).

REFERENCES

1. **Adachi, T., M. Mizuuchi, E. A. Robinson, E. Appella, M. H. O'Dea, M. Gellert, and K. Mizuuchi.** 1987. DNA sequence of the *E. coli* gyrB gene: application of a new sequencing strategy. *Nucleic Acids Res.* **15:**771–784.
2. **Akerley, B. J., E. J. Rubin, A. Camilli, D. J. Lampe, H. M. Robertson, and J. J. Mekalanos.** 1998. Systematic identification of essential genes by in vitro mariner mutagenesis. *Proc. Natl. Acad. Sci. USA* **95:**8927–8932.
3. **Berg, C. M., D. E. Berg, and E. Groisman.** 1989. Transposable elements and the genetic engineering of bacteria, p. 879–925. *In* D. E. Berg and M. M. Howe (ed.), *Mobile DNA.* American Society for Microbiology, Washington, D.C.
4. **Berg, D. E.** 1989. Transposon Tn*5*, p. 185–210. *In* D. E. Berg and M. M. Howe (ed.), *Mobile DNA.* American Society for Microbiology, Washington, D.C.
5. **Berg, D. E., J. Davies, B. Allet, and J.-D. Rochaix.** 1975. Transposition of R factor genes to bacteriophage λ. *Proc. Natl. Acad. Sci. USA* **72:**3628–3632.
6. **Berg, D. E., L. Johnsrud, L. McDivitt, R. Ramabhadran, and B. J. Hirschel.** 1982. Inverted repeats of Tn*5* are transposable elements. *Proc. Natl. Acad. Sci. USA* **79:**2632–2635.
7. **Bhasin, A.** 2000. Ph.D. thesis. University of Wisconsin, Madison.
8. **Bhasin, A., I. Y. Goryshin, and W. S. Reznikoff.** 1999. Hairpin formation in Tn*5* transposition. *J. Biol. Chem.* **274:** 37021–37029.
9. **Bhasin, A., I. Y. Goryshin, M. Steiniger-White, D. York, and W. S. Reznikoff.** 2000. Characterization of a Tn*5* pre-cleavage synaptic complex. *J. Mol. Biol.* **302:**49–63.
10. **Biek, D., and J. R. Roth.** 1980. Regulation of Tn*5* transposition in *Salmonella typhimurium. Proc. Natl. Acad. Sci. USA* **77:**6047–6051.
11. **Boyd, D., B. Traxler, G. Jander, W. Prinz, and J. Beckwith.** 1993. *Gene Fusion Approaches to Membrane Protein Topology.* Rockefeller University Press, New York, N.Y.
12. **Braam, L. M., I. Y. Goryshin, and W. S. Reznikoff.** 1999. The mechanism of Tn*5* inhibition carboxyl-terminal dimerization. *J. Biol. Chem.* **274:**86–92.
13. **Braam, L. M., and W. S. Reznikoff.** 1998. Functional characterization of the Tn*5* transposase by limited proteolysis. *J. Biol. Chem.* **273:**10908–10913.
14. **Bujacz, G., M. Jaskolski, J. Alexandratos, A. Wlodawer, G. Merkel, R. A. Katz, and A. M. Skalka.** 1995. High-resolution structure of the catalytic domain of avian sarcoma virus integrase. *J. Mol. Biol.* **253:**333–346.
15. **Craig, N.** 1995. Unity in transposition reactions. *Science* **270:** 253–254.
16. **Davies, D. R., L. M. Braam, W. S. Reznikoff, and I. Rayment.** 1999. The three-dimensional structure of a Tn*5* transposase-related protein determined to 2.9Å resolution. *J. Biol. Chem.* **274:**11904–11913.
17. **Davies, D. R., I. Y. Goryshin, W. S. Reznikoff, and I. Rayment.** 2000. Three-dimensional structure on the Tn*5* synaptic complex transposition intermediate. *Science* **289:**77–85.
18. **Davies, J.** 1986. A new look at antibiotic resistance. *FEMS Microbiol. Rev.* **39:**363–371.
19. **de la Cruz, N. B., M. D. Weinreich, T. W. Wiegand, M. P. Krebs, and W. S. Reznikoff.** 1993. Characterization of the Tn*5* transposase and inhibitor proteins: A model for the inhibition of transposition. *J. Bacteriol.* **175:**6932–6938.
20. **DeLong, A., and M. Syvanen.** 1990. Membrane association of the Tnp and Inh proteins of IS*50*R. *J. Bacteriol.* **172:** 5516–5519.
21. **Derbyshire, K. M., L. Hwang, and N. D. F. Grindley.** 1987. Genetic analysis of the interaction of the insertion sequence IS*903* transposase with its terminal inverted repeats. *Proc. Natl. Acad. Sci. USA* **84:**8049–8053.
22. **Derbyshire, K. M., M. Kramer, and N. D. F. Grindley.** 1990. Role of instability in the cis action of the insertion sequence IS*903* transposase. *Proc. Natl. Acad. Sci. USA* **87:**4048–4052.
23. **Devine, S. E., and J. D. Boeke.** 1994. Efficient integration of artifical transposons into plasmid targets in vitro: a useful tool for DNA mapping, sequencing and genetic analysis. *Nucleic Acids Res.* **22:**3765–3772.
24. **Dyda, F., A. B. Hickman, T. M. Jenkins, A. Engelman, R. Craigie, and D. R. Davies.** 1994. Crystal structure of the catalytic domain of HIV-1 integrase: similarity to other polynucleotidyl transferases. *Science* **266:**1981–1986.
25. **Fuller, R. S., B. E. Funnell, and A. Kornberg.** 1984. The *dna*A protein complex with the *E. coli* chromosomal replicative origin (*oriC*) and other DNA sites. *Cell* **38:**889–900.
26. **Goryshin, I. Y., J. Jendrisak, L. M. Hoffman, R. Meis, and W. S. Reznikoff.** 2000. Insertional transposon mutagenesis by electroporation of released Tn*5* transposition complexes. *Nat. Biotechnol.* **18:**97–100.
27. **Goryshin, I. Y., Y. V. Kil, and W. S. Reznikoff.** 1994. DNA length, bending and twisting constraints on IS50 transposition. *Proc. Natl. Acad. Sci. USA* **91:**10834–10838.
28. **Goryshin, I. Y., J. A. Miller, Y. V. Kil, V. A. Lanzov, and W. S. Reznikoff.** 1998. Tn*5*/IS*50* target recognition. *Proc. Natl. Acad. Sci. USA* **95:**10716–10721.
29. **Goryshin, I. Y., and W. S. Reznikoff.** 1998. Tn*5 in vitro* transposition. *J. Biol. Chem.* **273:**7367–7374.
30. **Griffin, T. J., IV, L. Parsons, A. E. Leschziner, J. DeVost, K. M. Derbyshire, and N. D. F. Grindley.** 1999. In vitro transposition of Tn*552*: a tool for DNA sequencing and mutagenesis. *Nucleic Acids Res.* **27:**3859–3865.
31. **Gwinn, M. L., A. E. Stellwagen, N. L. Craig, J. F. Tomb, and H. O. Smith.** 1997. In vitro Tn7 mutagenesis of *Haemophilus influenzae* Rd and characterization of the role of *atpA* in transformation. *J. Bacteriol.* **179:**7315–7320.
32. **Haapa, S., S. Taira, E. Heikkinen, and H. Salvilahti.** 1999. An efficient and accurate integration of mini-Mu transposons *in vitro:* a general methodology for functional genetic analysis and molecular biology applications. *Nucleic Acids Res.* **27:** 2777–2784.
33. **Haren, L., B. Ton-Hoang, and M. Chandler.** 1999. Integrating DNA: transposases and retroviral integrases. *Annu. Rev. Microbiol.* **53:**245–281.
34. **Huisman, O., P. R. Errada, L. Signon, and N. Kleckner.** 1989. Mutational analysis of IS10's outside end. *EMBO J.* **8:** 2101–2109.
35. **Isberg, R., and M. Syvanen.** 1981. Replicon fusion promoted

by the inverted repeats of Tn5. The right repeat is an insertion sequence. *J. Mol. Biol.* **150:**15–32.

36. **Isberg, R., and M. Syvanen.** 1982. DNA gyrase is a host factor required for transposition of Tn5. *Cell* **30:**9–18.
37. **Isberg, R. R., A. L. Lazaar, and M. Syvanen.** 1982. Regulation of Tn5 by the right repeat proteins: control at the level of the transposition reactions? *Cell* **30:**883–892.
38. **Jenkins, T. M., D. Esposito, A. Engelman, and R. Craigie.** 1997. Critical contacts between HIV-1 integrase and viral DNA identified by structure-based analysis and photo-crosslinking. *EMBO J.* **16:**6849–6859.
39. **Jilk, R. A., D. York, and W. S. Reznikoff.** 1996. The organization of the outside end of the transposon Tn5. *J. Bacteriol.* **178:**1671–1679.
40. **Johnson, R. C., and W. S. Reznikoff.** 1983. DNA sequences at the ends of transposon Tn5 required for transposition. *Nature* **304:**280–282.
41. **Johnson, R. C., and W. S. Reznikoff.** 1984. Copy number control of Tn5 transposition. *Genetics* **107:**9–18.
42. **Johnson, R. C., and W. S. Reznikoff.** 1984. The role of the IS50R proteins in the promotion and control of Tn5 transposition. *J. Mol. Biol.* **177:**645–661.
43. **Johnson, R. C., J. C.-P. Yin, and W. S. Reznikoff.** 1982. The control of Tn5 transposition in *Escherichia coli* is mediated by protein from the right repeat. *Cell* **30:**873–882.
44. **Kennedy, A. K., A. Guhathakurta, N. Kleckner, and D. B. Haniford.** 1998. Tn10 transposition via a DNA hairpin intermediate. *Cell* **95:**125–134.
45. **Kennedy, A. K., D. B. Haniford, and K. Mizuuchi.** 2000. Single active site catalysis of the successive phosphoryl transfer steps by DNA transposases: insights from phosphorothioate stereoselectivity. *Cell* **101:**295–305.
46. **Krebs, M. P., and W. S. Reznikoff.** 1986. Transcriptional and translational sites of IS50. Control of transposase and inhibitor expression. *J. Mol. Biol.* **192:**781–791.
47. **Krebs, M. P., and W. S. Reznikoff.** 1988. Use of a Tn5 derivative that creates *lacZ* translational fusions to obtain a transposition mutant. *Gene* **63:**277–285.
48. **Lodge, J. K., and D. E. Berg.** 1990. Mutations that affect Tn5 insertion into pBR322: importance of local DNA supercoiling. *J. Bacteriol.* **172:**5956–5960.
49. **Makris, J. C., P. L. Nordmann, and W. S. Reznikoff.** 1988. Mutational analysis of IS50 and Tn5 ends. *Proc. Natl. Acad. Sci. USA* **85:**2224–2228.
50. **Makris, J. C., P. L. Nordmann, and W. S. Reznikoff.** 1990. The role of IHF in IS50 and Tn5 transposition. *J. Bacteriol.* **172:**1368–1373.
51. **Manoil, C., and B. Traxler.** 2000. Insertion of in-frame sequence tags into proteins using transposons. *Methods* **20:** 55–61.
52. **Mazodier, P., P. Cossart, E. Giraud, and F. Gasser.** 1985. Completion of the nucleotide sequence of the central region of Tn5 confirms the presence of three resistance genes. *Nucleic Acids Res.* **13:**195–205.
53. **Mazodier, P., O. Genilloud, E. Giraud, and F. Gasser.** 1986. Expression of Tn5-encoded syreptomycin resistance in *E. coli. Mol. Gen. Genet.* **204:**404–409.
54. **McBlane, J. F., D. C. van Gent, D. A. Ramsden, C. Romeo, C. A. Cuomo, M. Gellert, and M. A. Oettinger.** 1995. Cleavage at a V(D)J recombination signal requires only RAG1 and RAG2 proteins and occurs in two steps. *Cell* **83:**387–395.
55. **Mizuuchi, K.** 1997. Polynucleotidyl transfer reactions in site-specific DNA recombination. *Genes Cells* **2:**1–12.
56. **Morisato, D., J. C. Way, H.-J. Kim, and N. Kleckner.** 1983. Tn*10* transposase acts preferentially on nearby transposon ends in vivo. *Cell* **32:**799–807.
57. **Naumann, T. A., and W. S. Reznikoff.** 2000. *Trans* catalysis in Tn5 transposition. *Proc. Natl. Acad. Sci. USA* **97:**8944–8949.

57a. **Naumann, T. A., and W. S. Reznikoff.** Tn5 transposase with an altered specificity for transposon ends. *J. Bacteriol.*, in press.

58. **Phadnis, S. H., and D. E. Berg.** 1987. Identification of base pairs in the IS*50* O end needed for IS*50* and Tn5 transposition. *Proc. Natl. Acad. Sci. USA* **84:**9118–9122.
59. **Phadnis, S. H., H. V. Huang, and D. E. Berg.** 1989. Tn5*supF*, a 264-base-pair transposon derived from Tn5 for insertion mutagenesis and sequencing DNAs cloned in phage lambda. *Proc. Natl. Acad. Sci. USA* **86:**5908–5912.
60. **Reznikoff, W. S.** 1993. The Tn5 transposon. *Annu. Rev. Microbiol.* **47:**945–963.
61. **Reznikoff, W. S., A. Bhasin, D. R. Davies, I. Y. Goryshin, T. Naumann, I. Rayment, M. Steiniger-White, and S. S. Twining.** 1999. Tn5: a molecular window on transposition. *Biochem. Biophys. Res. Commun.* **266:**729–734.
62. **Rezsohazy, R., B. Hallet, J. Delcour, and J. Mahillon.** 1993. The IS4 family of insertion sequences: evidence for a conserved transposase motif. *Mol. Microbiol.* **9:**1283–1295.
63. **Rice, P., and K. Mizuuchi.** 1995. Structure of the bacteriophage Mu transposase core: a common structural motif for DNA transposition and retroviral integration. *Cell* **82:** 209–220.
64. **Roberts, D. E., B. C. Hoopes, W. R. McClure, and N. Kleckner.** 1985. IS10 transposition is regulated by DNA adenine methylation. *Cell* **43:**117–130.
65. **Rosetti, O. L., R. Altman, and R. Young.** 1984. Kinetics of Tn5 transposition. *Gene* **32:**91–98.
66. **Rothstein, S. J., R. A. Jorgensen, K. Postle, and W. S. Reznikoff.** 1980. The inverted repeats of Tn5 are functionally different. *Cell* **19:**795–805.
67. **Rothstein, S. J., and W. S. Reznikoff.** 1981. The functional differences in the inverted repeats of Tn5 are caused by a single base pair nonhomology. *Cell* **23:**191–199.
68. **Sasakawa, C., G. F. Carle, and D. E. Berg.** 1983. Sequences essential for transposition at the termini of IS*50*. *Proc. Natl. Acad. Sci. USA* **80:**7293–7297.
69. **Sasakawa, C., Y. Uno, and M. Yoshikawa.** 1981. The requirement for both DNA polymerase and 5′ to 3′ exonuclease activities of DNA polymerase I during Tn5 transposition. *Mol. Gen. Genet.* **182:**19–24.
70. **Sasakawa, C., Y. Uno, and M. Yoshikawa.** 1987. *Ion-sul*A regulatory function affects the efficiency of transposition of Tn5 from λ *b*221 *c*1857 Pam Oam to the chromosome. *Biochem. Biophys. Res. Commun.* **142:**879–884.
71. **Schaller, H.** 1979. The intergenic region and the origins for filamentous phage DNA replication. *Cold Spring Harbor Symp. Quant. Biol.* **43:**401–408.
72. **Schulz, V. P., and W. S. Reznikoff.** 1991. Translation initiation of IS*50*R read through transcripts. *J. Mol. Biol.* **221:**65–80.
73. **Steiniger-White, M., and W. S. Reznikoff.** 2000. The C-terminal alpha helix of Tn5 transposase is required for synaptic complex formation. *J. Biol. Chem.* **275:**23127–23133.
74. **Sternglanz, R., S. DiNardo, K. A. Voelker, Y. Nishimura, Y. Hirota, K. Becherer, L. Zumstein, and J. C. Wang.** 1981. Mutations in the gene coding for *E. coli* DNA topoisomerase I affect transcription and transposition. *Proc. Natl. Acad. Sci. USA* **78:**2747–2751.
75. **Syvanen, M., J. D. Hopkins, and M. Clements.** 1982. A new class of mutants in DNA polymerase I that affects gene transposition. *J. Mol. Biol.* **158:**203–212.
76. **Tomcsanyi, T., C. M. Berg, S. H. Phadnis, and D. E. Berg.** 1990. Intramolecular transposition by a synthetic IS50 (Tn5) derivative. *J. Bacteriol.* **172:**6348–6354.
77. **Traxler, B., D. Boyd, and J. Beckwith.** 1993. The topological

analysis of integral cytoplasmic membrane proteins. *J. Membr. Biol.* **132:**1–11.

77a. **Twining, S. S., I. Y. Goryshin, A. Bhasin, and W. S. Reznikoff.** 2001. Functional characterization of arginine 30, lysine 40 and arginine 62 in Tn*5* transposase. *J. Biol. Chem.* **276:** 23135–23143.

78. **Weinreich, M. D., A. Gasch, and W. S. Reznikoff.** 1994. Evidence that the *cis* preference of the Tn*5* transposase is caused by non-productive multimerization. *Genes Dev.* **8:**2363–2374.

79. **Weinreich, M. D., L. Mahnke-Braam, and W. S. Reznikoff.** 1994. A functional analysis of the Tn5 transposase: identification of domains required for DNA binding and multimerization. *J. Mol. Biol.* **241:**166–177.

80. **Weinreich, M. D., and W. S. Reznikoff.** 1992. Fis plays a role in Tn5 and IS50 transposition. *J. Bacteriol.* **174:**4530–4537.

81. **Wiegand, T. W., and W. S. Reznikoff.** 1992. Characterization of two hypertransposing Tn*5* mutants. *J. Bacteriol.* **174:** 1229–1239.

82. **Yang, Z. N., T. C. Muesser, F. D. Bushman, and C. C. Hyde.** 2000. Crystal structure of an active two-domain derivative of Rous sarcoma virus integrase. *J. Mol. Biol.* **296:**535–548.

83. **Yigit, H., and W. S. Reznikoff.** 1999. *Escherichia coli* DNA topoisomerase I copurifies with Tn*5* transposase, and Tn*5* transposase inhibits topoisomerase I. *J. Bacteriol.* **181:** 3185–3192.

84. **Yin, J. C.-P., M. P. Krebs, and W. S. Reznikoff.** 1988. The effect of *dam* methylation on Tn*5* transposition. *J. Mol. Biol.* **199:**35–46.

85. **Yin, J. C.-P., and W. S. Reznikoff.** 1987. *dnaA*, an essential host gene, and Tn*5* transposition. *J. Bacteriol.* **169:** 4637–4645.

86. **Yin, J. C.-P., and W. S. Reznikoff.** 1988. p2 and the inhibition of Tn*5* transposition. *J. Bacteriol.* **170:**3008–3015.

87. **York, D., and W. S. Reznikoff.** 1996. Purification and biochemical analysis of a momomeric form of Tn5 transposase. *Nucleic Acids Res.* **24:**3790–3796.

88. **York, D., and W. S. Reznikoff.** 1997. DNA-binding and phasing analyses of Tn*5* transposase and a monomereic variant. *Nucleic Acids Res.* **25:**2153–2160.

89. **York, D., J. Welch, I. Y. Goryshin, and W. S. Reznikoff.** 1998. Simple and efficient generation *in vitro* of nested deletions and inversions: Tn*5* intramolecular transposition. *Nucleic Acids Res.* **26:**1927–1933.

90. **Zhou, M., A. Bhasin, and W. S. Reznikoff.** 1998. Molecular genetic analysis of transposase-end sequence recognition: cooperation of three adjacent base pairs in specific interaction with a mutant Tn*5* transposase. *J. Mol. Biol.* **276:**913–925.

91. **Zhou, M., and W. S. Reznikoff.** 1997. Tn*5* transposase mutants that alter DNA binding specificity. *J. Mol. Biol.* **271:** 362–373.

Mobile DNA II
Edited by N. L. Craig et al.

Chapter 19

Tn7

Nancy L. Craig

Nancy L. Craig • Howard Hughes Medical Institute and Department of Molecular Biology and Genetics, 615 PCTB, Johns Hopkins School of Medicine, 725 N. Wolfe St., Baltimore, MD 21205.

INTRODUCTION

The bacterial transposon Tn7 is a sophisticated mobile element that pursues several alternative lifestyles to promote its propagation (11, 13, 24, 25, 27). The ability of Tn7 to use different targeting pathways and to control its transposition according to the target DNA that is present provides Tn7 with these alternative lifestyles (Fig. 1). Most other transposable elements recognize only a single type of target site.

In one pathway, Tn7 inserts at high frequency into a specific site called an attachment site, *attTn7*, that is present in the chromosomes of many bacteria (25, 46, 66). Insertion into this site is nondeleterious to the bacterial host. The ability of Tn7 to join with the bacterial chromosome in a nondeleterious way promotes the survival of the host cell, and hence of Tn7, and thus facilitates the vertical transmission of Tn7 to daughter cells. In another pathway, Tn7 preferentially transposes to conjugating plasmids, i.e., those plasmids that are transferred by mating between bacterial cells (124). This preference for conjugating plasmids and the generally broad host range of such plasmids favors the horizontal dispersal of Tn7 among bacterial populations. The final feature of a potential target DNA that can be sensed by Tn7 is the presence of a preexisting copy of Tn7 in that DNA (4, 52). Tn7 avoids inserting into such DNAs, a process called target immunity or transposition immunity. The utility of this phenomenon is probably to avoid generating plasmids containing multiple copies of Tn7 that could then be substrates for deleterious homologous recombination reactions and also to ensure that Tn7 does not insert within itself, thereby self-destructing.

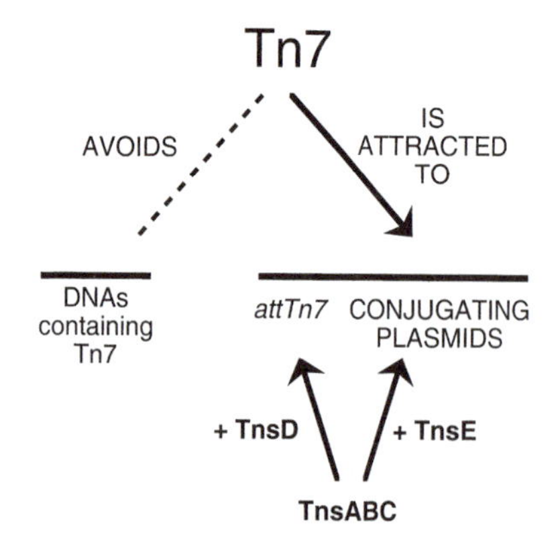

Figure 1. Tn7 transposition is modulated by different aspects of target DNAs. Tn7 is preferentially attracted to *attTn7*, a single specific site in the chromosomes of many bacteria, a lifestyle that promotes vertical transmission of the element. Tn7 is also preferentially attracted to conjugating plasmids, likely because of the unique form of lagging-strand synthesis that occurs upon mating. These different targeting pathways are mediated by different subsets of Tn7-encoded transposition proteins. Preferential insertion on conjugating plasmids provides a lifestyle that promotes dispersal of Tn7 between organisms. Tn7 avoids DNA containing Tn7, thereby avoiding self-destruction.

The ability of Tn7 to recognize two different classes of target sites, one being *attTn7* and the other being non-*attTn7* sites, derives from the fact that Tn7 encodes multiple transposition proteins (TnsA, TnsB, TnsC, TnsD, and TnsE) and that different subsets of these proteins direct Tn7 into different classes of target sites (61, 98, 120) (Fig. 1). Insertion into *attTn7* is mediated by TnsABC+D (8, 98, 120). The TnsABC+E pathway does not recognize particular target sequences; rather this pathway recognizes and preferentially directs insertion into DNA undergoing a distinctive form of lagging-strand DNA synthesis (83, 124). Each transposition pathway thus requires a distinctive target selector, TnsD or TnsE, and a common set of proteins, TnsABC. In addition to acting as target selectors, TnsD and TnsE also act as "activators" of the common TnsABC proteins; only exceedingly low levels of recombination can be detected with just TnsABC. A feature of both the TnsD and TnsE pathways is the recognition of particular DNA structures at the target site (62, 83, 92).

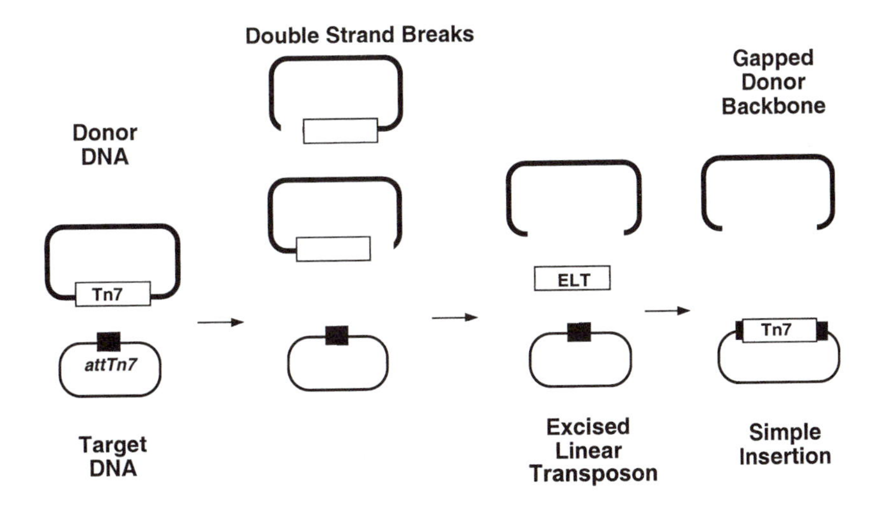

Figure 2. Tn7 translocates via cut-and-paste transposition. The substrate DNAs are a donor DNA containing Tn7 and a target plasmid containing *attTn7*. Recombination initiates by double-strand breaks at either end of the Tn7 element; a second break generates an excised transposon species that is then inserted into *attTn7*. The other product of the reaction is a gapped donor backbone plasmid.

An in vitro Tn7 transposition system in which a mini-Tn7 element translocates from a donor plasmid and inserts into an *attTn7* site on a target plasmid has been reconstituted with purified proteins, allowing dissection of the reaction components and transposition mechanism (7, 8). Recombination in vitro requires TnsA, TnsB, TnsC, and TnsD and the cofactors ATP and Mg^{2+}. Recombination occurs through a cut-and-paste pathway in which the element is excised from the donor site at each end of the element and then joined to the target DNA to form a simple insertion in the target DNA (Fig. 2). Tn7 insertion is accompanied by a 5-bp target sequence duplication.

The roles of the Tns proteins in recombination have also been determined. The Tn7 transposase is distinctive because it is formed by a collaboration between two polypeptides, TnsA and TnsB (18, 68, 71, 102). Other characterized elements have a transposase formed by a single polypeptide. The target-selecting proteins interact with DNA: TnsD binds to *attTn7* (8, 119) and TnsE binds preferentially to DNAs with recessed 3′ ends (83). A very interesting aspect of Tn7 transposition is how transposition is controlled. TnsC controls transposition through interaction with both the transposase bound to the ends of the transposon (8, 68, 109) (Z. Skelding, R. Sarnovsky, and N. L. Craig, unpublished data) and with the target DNA bound by the target-selecting proteins (8, 92, 109). TnsC is an ATPase (111) and an ATP-dependent DNA-binding protein (43), and these activities are apparently central to the control of Tn7 transposition; TnsC-ATP promotes transposition and the ATP state of TnsC is likely determined through its interactions with the target DNA and proteins (108).

THE STRUCTURE OF Tn7

Tn7 was first isolated from a trimethoprim- and streptomycin-resistant bacterium in a calf that was being treated prophylactically with trimethoprim in its feed (11, 13).

Transposition Functions

The Tns proteins

Most mobile DNAs encode only one or two proteins that mediate mobility. By contrast, Tn7 encodes five transposition genes, *tnsA*, *tnsB*, *tnsC*, *tnsD*, and *tnsE* (Fig. 3). The five *tns* genes are clustered and lie in the right half of Tn7; antibiotic resistance genes occupy the left half of Tn7. The *tns* genes are all oriented in the same direction with their 5′ ends toward the right end of Tn7. These genes were identified genetically by analysis of deletion and insertion derivatives of Tn7 (52, 98, 120) and by gene sequencing (41, 82), protein identification (82), and protein purification (TnsA [8], TnsB [5, 114], TnsC [43], TnsD [8], and TnsE [83]).

Relatively little is known about the regulation of *tns* gene expression. It is known that TnsB is a repressor of the transcription of an operon that includes *tnsA* and *TnsB* (73, 120). Obvious unanswered ques-

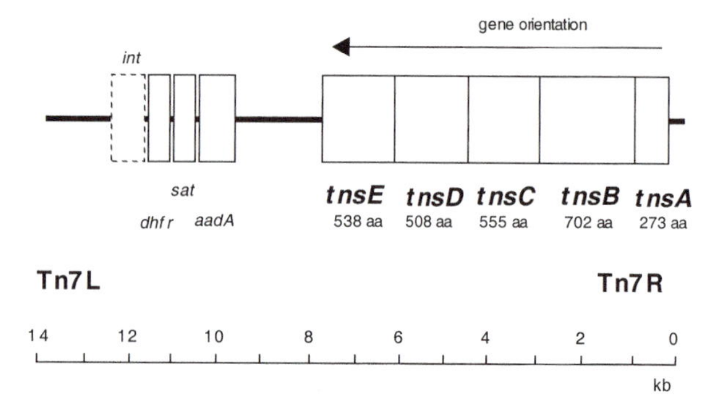

Figure 3. Map of Tn7. The transposition genes (*tnsA, tnsB, tnsC, tnsD,* and *tnsE*) are all carried in the right end of Tn7 with their 5′ termini closest to the right terminus of Tn7. The left end of Tn7 carries several antibiotic resistance genes: *dhfr,* which provides resistance to trimethoprim; *sat,* which provides resistance to streptothricin; and *aadA,* which provides resistance to streptomycin and spectinomycin. These antibiotic resistance cassettes are part of an integron element.

tions include whether the remainder of the *tns* genes are part of this operon and whether certain cellular conditions might influence the expression of the *tns* genes. Understanding the *tns* expression circuitry awaits direct analysis of *tns* gene transcription and translation.

All the Tns proteins likely participate directly in transposition rather than simply providing a regulatory function. TnsA and TnsB together form the transposase that binds specifically to the ends of Tn7 (5, 114) and execute the chemical steps of transposition, the double-strand breaks that cut Tn7 away from the flanking donor DNA and the subsequent joining of the Tn7 ends to the target (18, 71, 102). As described in detail below, TnsB is a member of the retroviral integrase superfamily of recombinases, whereas TnsA resembles a restriction enzyme.

TnsC controls the activity of the transposase, activating the transposase when TnsC and the target-selecting proteins interact with an appropriate target DNA (8, 107, 111). TnsD, which directs Tn7 insertion into *attTn7,* binds specifically to *attTn7* (8). Because TnsE is also a DNA-binding protein (83), it seems likely that the interaction of TnsE with a target DNA will also play a role in transposition. Although there may also be important protein-protein interactions between TnsC and these target proteins, the direct interaction of TnsC itself with the target DNA appears to play a key role in transposition (92).

There are no obvious protein sequence similarities between TnsC, TnsD, and TnsE and other proteins except those analogues encoded by other transposons relatively similar to Tn7.

The Ends of Tn7 Provide DNA Substrates for Transposition

The ends of Tn7 provide the *cis*-acting sequences upon which the TnsAB transposase acts (Fig. 4). A fully functional mini-Tn7 element can be created by flanking a DNA segment with a 165-bp segment from the left end of Tn7 (Tn7L) and a 90-bp segment from the right end of Tn7 (Tn7R) (4, 30). A notable feature of these end segments is that they contain multiple binding sites for TnsB, a sequence-specific DNA-binding protein that is part of the Tn7 transposase (3, 5, 73, 114). The arrangement of these TnsB binding sites is different in each end of Tn7: Tn7L contains these separated TnsB sites, whereas there are four closely juxtaposed sites in Tn7R. A mini-Tn7 element containing less than the complete Tn7 ends can also transpose, although at lower frequency. For example, a mini-Tn7 element that contains a 165-bp Tn7L segment and a 70-bp Tn7R segment that includes only three TnsB sites can transpose, although at a frequency less than that of a mini-Tn7 element with wild-type ends (4). Recombination of a mini-Tn7 element containing a 41-bp Tn7R end can also be detected in vitro (19). The ability of these shorter segments to

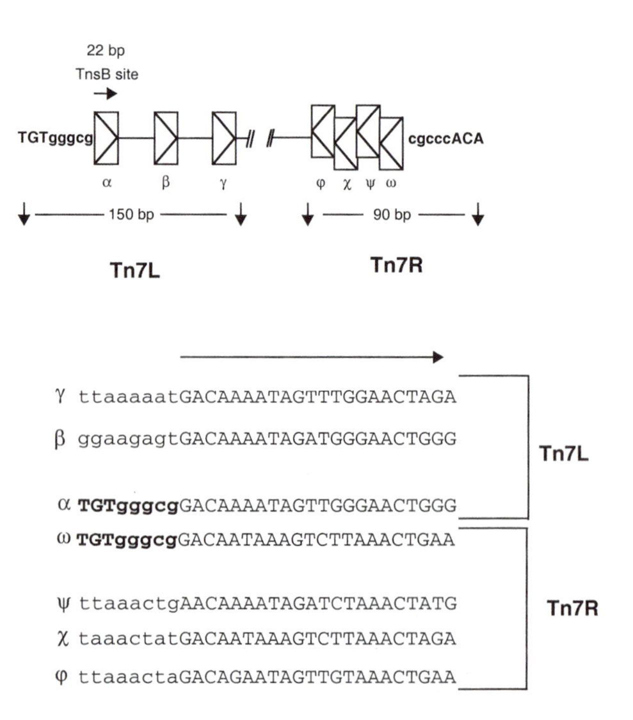

Figure 4. The ends of Tn7. The Tn7L and Tn7R end segments contain multiple copies of the 22-bp TnsB binding sequence as defined by TnsB-footprinting studies. The 30-bp nearly perfect inverted repeats contain a perfect 8-bp terminal inverted repeat (bold) containing a critical 5′-TGT . . . .ACA-3′, a 5-bp spacer region and a TnsB binding site.

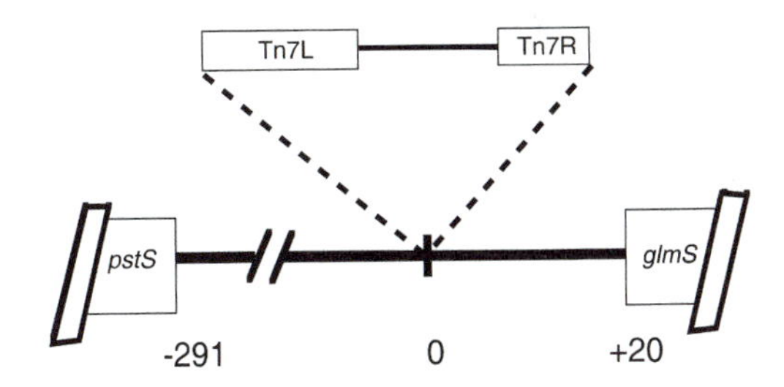

Figure 5. Tn7 insertion into *attTn7*. The specific point of Tn7 insertion lies between *pstS* and *glmS*. Tn7 inserts into *attTn7* in a specific orientation with the right end of Tn7 adjacent to *glmS*. The position designated 0 is the center of the target 5 bp duplicated upon Tn7 insertion, base pairs rightward toward *oriC* are designated + and base pairs leftward are designated −.

transpose is especially useful in using Tn7 to make translational fusions (see below).

Not only are the Tn7L and Tn7R ends structurally distinct, they are also functionally distinct. For example, when Tn7 insertion into *attTn7* occurs, the insertion virtually always occurs so that the Tn7R end is oriented in a particular direction with respect to the target site, i.e., toward *glmS* (7, 47, 67, 72) (Fig. 5). Backward insertion does, however, occur at very low frequency (32, 54). Although the different arrangements of TnsB binding sites on Tn7L and Tn7R must be key in determining this orientation specificity, the mechanism by which orientation selection occurs is not clear. Another indication of the functional asymmetry of the ends of Tn7 is that, whereas mini-Tn7 elements containing two Tn7R ends are active in transposition, elements containing two Tn7L ends are not (4). The 30 terminal bases of each end of Tn7 are very similar, being nearly identical inverted repeats (67) (Fig. 4). Of these 30-bp segments, the extreme termini are 8-bp perfect inverted repeats and the interior 22 bp are highly related TnsB binding sites (5, 67, 73, 114). Mutational analysis of the terminal 8-bp segment has revealed that it is composed of two subsections that have different functions (44; R. Sarnovsky, Z. Skelding, Y. Li, P. Gary, and N. L. Craig, unpublished data). The 3 bp at the extreme termini at each end, 5′-TGT. . . . . ACA-3′, play a critical role in recombination; mutations at these sites block recombination. However, the terminal base pairs, 5′-TGT. . . . . ACA-3′, are not required for transposase binding; the interior TnsB binding sites lack these sequences.

The identities of the remaining positions adjacent to the terminal TnsB binding sites, bp 4 to 8, seem to be less critical for recombination, because changing the identity of these base pairs has only very modest effects on transposition. However, changing the number of base pairs in this region between the TnsB binding site and the 3-terminal nucleotides blocks transposition. An attractive view of the 30-bp terminal inverted repeats is that the internal 22-bp TnsB segment is critical for transposase binding, and the terminal 5′-TGT. . . . . ACA-3′ base pairs facilitate a post-binding cleavage reaction. The intervening 5 bp likely provide a spacer region between the TnsB binding site and the 3-bp terminal cleavage signal.

The most interior of the nonterminal TnsB sites have the highest affinity for TnsB (3, 5). The other TnsB sites on the interior of the Tn7 end segments likely play key roles in "loading" the transposase onto the terminal sites.

Topological Requirements of the Substrate DNAs

The topological requirements of the donor DNA substrate have been examined in the in vitro Tn7 transposition system. Under our standard reaction conditions for TnsABC+D transposition, relaxed DNA containing a mini-Tn7 element seems to be as effective a donor substrate as supercoiled DNA (44). Thus, Tn7 transposition does not involve an intrinsic requirement for DNA supercoiling. Under some alternative reaction conditions, however, supercoiling of the mini-Tn7-containing donor substrate is required for transposition (6, 18). Such a requirement for supercoiling could reflect several roles for supercoiling, perhaps promoting the juxtaposition of the two ends for synapsis or functioning in some postsynapsis step requiring an alteration in DNA structure.

Two Tn7 end segments on a circular donor DNA are required for transposition. Probably a key consequence of the presence of the two ends on a single plasmid is to promote the high local concentration of the ends such that their interaction is facilitated. However, when using a donor substrate that has already undergone the breakage step that cuts the transposon ends away from the flanking donor DNA, two Tn7R ends from two different DNA molecules can collaborate to insert together into a target DNA (44).

Under our standard reaction conditions for in vitro TnsABC+D transposition, relaxed target DNA containing an *attTn7* site appears to be as effective a substrate as supercoiled *attTn7* DNA. However, under other conditions, supercoiled *attTn7* DNA is the favored target (6). This finding is consistent with our view that target DNA structure, in particular the introduction of DNA distortions into *attTn7*, is a key step in using this site as a target (see below).

Tn7 Antibiotic Resistance Genes

Three genes encoding enzymes that provide resistance to several different antibiotics are present in the left half of Tn7 (Fig. 3). Cells containing Tn7 are resistant to antifolates such as trimethoprim and to the aminoglycosides streptothricin, streptomycin, and spectinomycin. The *dhfr* gene encodes a novel dihydrofolate reductase protein that is insensitive to inhibition by trimethoprim (39, 105). The *sat* gene encodes a transacetylase that inactivates streptothricin (113, 115). The *aadA* gene encodes an adenylyltransferase that inactivates streptomycin and spectinomycin (40).

These antibiotic resistance genes appear to be part of an "integron" system in which resistance cassettes can undergo rearrangement through excision and reintegration mediated by an integrase encoded adjacent to the cassettes (50, 113) (see chapter 9). There is an open reading frame (ORF) to the left of the Tn7 *dhfr* gene that is homologous to a lambda phage integrase (113). However, this Tn7 integrase is inactive because of a stop codon that would result in a truncated protein. This ORF is not required for Tn7 transposition (98, 120).

Transposons Tn*1825* and Tn*1826* whose transposition functions are closely related to those of Tn7 contain a *sat* and an *aadA* gene but lack a *dhfr* gene (116). This rearrangement likely occurred via excision of the *dhfr* cassette.

Tn7 TRANSPOSITION PATHWAYS

An in vitro Tn7 transposition system in which a mini-Tn7 element excises from a donor plasmid and inserts into an *attTn7* site on a target plasmid has been reconstituted using purified proteins. Recombination in vitro requires TnsA, TnsB, TnsC, and TnsD and the cofactors ATP and Mg^{2+} (7, 8).

Tn7 Usually Moves by a Cut-and-Paste Mechanism

Tn7 moves by a cut-and-paste reaction: the transposon is disconnected by double-strand breaks at each end of the element from the donor DNA backbone that flanks the element in the donor site; it is then joined to the target DNA to generate the transposition product called a simple insertion (7, 8) (Fig. 6). The other product of this reaction is a gapped donor backbone molecule. Recombination initiates with a double-strand break that can occur at either end of the Tn7 element; this initial break is followed by a second double-strand break to generate an excised transposon species that is an intermediate in recombination.

The double-strand breaks that excise Tn7 from the donor backbone are actually staggered (7) (Fig. 6). Cleavage at the 3′ ends of the transposon occurs precisely at the termini of the transposon, exposing 3′-OHs. However, the cleavages at the 5′ ends are displaced, occurring 3 nucleotides into the flanking donor DNA. Thus, the excised transposon has 3 nucleotides from the flanking donor site at its 5′ termini.

The 3′-OH ends of Tn7 then attack each strand of the target DNA, joining at positions that are staggered by 5 bp on the top and bottom strands of the target DNA. The transposon ends themselves are likely the nucleophiles that directly attack the target DNA. This reaction chemistry is presumed from the similarity of TnsB, which executes the reactions at the 3′ ends of the transposon, to the retroviral integrase superfamily of recombinases, which have been directly shown to perform such one-step transesterification reactions in the target joining step (76) (see chapter 2). Moreover, each end of Tn7 can join independently to the target DNA, that is, one target strand is joined to a Tn7 end, whereas the other target strand remains intact, revealing that recombination does not involve a double-strand break at the target site into which Tn7 is ligated (44).

The newly inserted transposon is flanked by short 5-bp gaps, reflecting the staggered positions of Tn7 end joining on the top and bottom strands on the target DNA (Fig. 6). These gaps are repaired by host functions. However, these gaps are not directly accessible to a polymerase after transpositional attack because of the presence of the Tns proteins on the recombination product (C. Johnson, J. Holder, and N. L. Craig, unpublished observation). When the recombination product bound by Tns proteins is incubated in a cellular extract, repair can occur (Johnson et al., unpublished). Thus, cellular functions must make the gaps accessible and then mediate repair. The 3 nucleotides at the 5′ ends of the transposon that derive from the donor site are removed in this repair process so that the newly inserted Tn7 element is flanked by 5-bp target sequence duplications.

This conservative cut-and-paste recombination pathway is observed in vitro. There is also evidence that cut-and-paste transposition occurs in vivo. When transposition of a heteroduplex version of Tn7 that contains one strand of *lac*$^+$ information and one strand of *lac*$^-$ information occurs from a bacteriophage lambda donor, the progeny of the insertion product contain both *lac*$^+$ and *lac*$^-$ information (M. Lopata, K. Orle, R. Gallagher, and N. L. Craig, unpublished data). The finding of both *lac*$^+$ and *lac*$^-$ information argues that insertion occurred by a nonreplicative mechanism (15, 51).

The 3′ and 5′ Ends of Tn7 Are Cleaved by Different Proteins

A very interesting aspect of Tn7 transposition is that the cleavages at the 3′ and 5′ ends of the transpo-

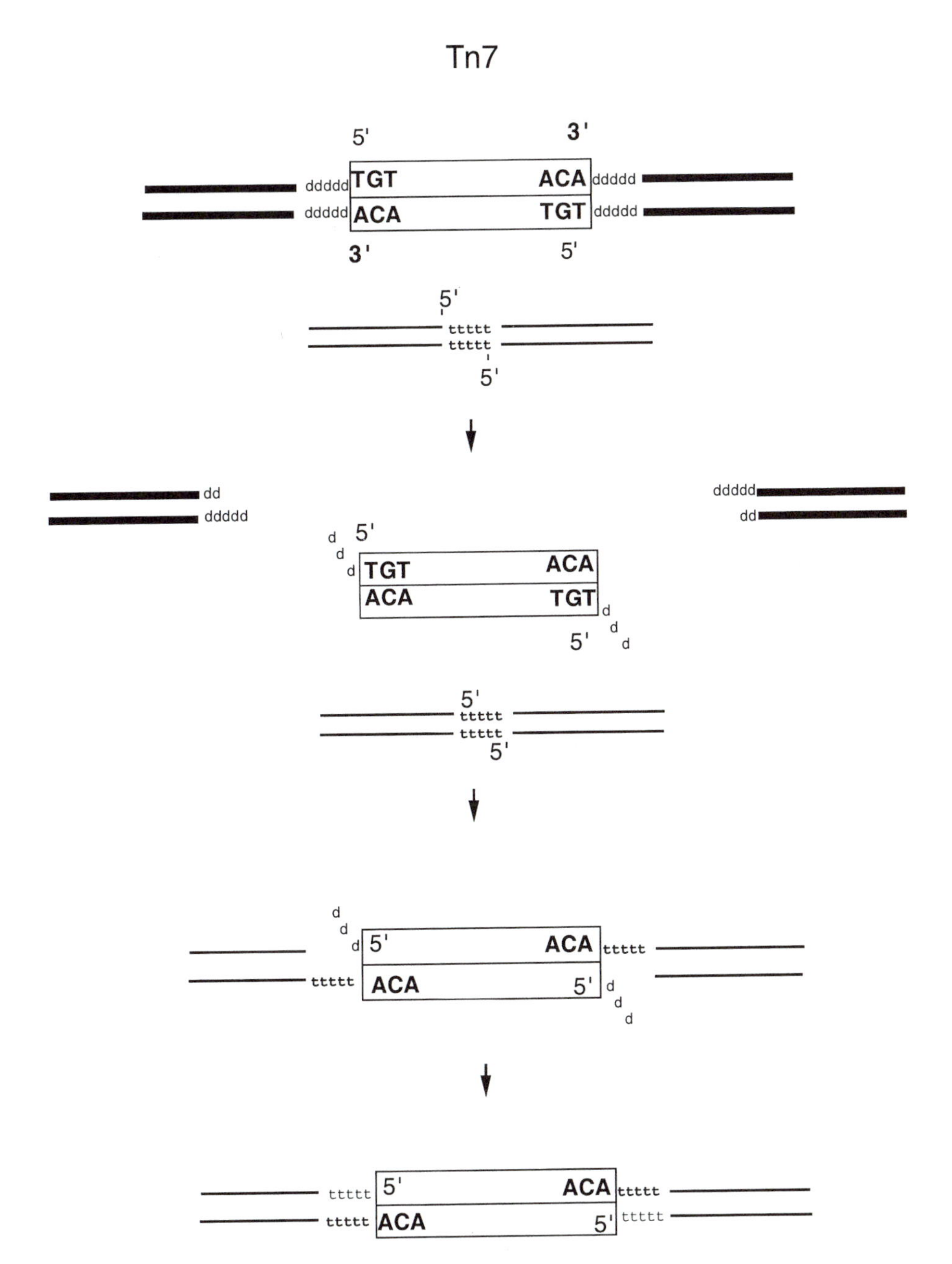

Figure 6. Tn7 transposition chemistry. In the donor site, Tn7 is flanked by a 5-bp duplication of target sequences resulting from the insertion of Tn7 into that site. Tn7 is excised from the donor site by double-strand breaks. Cleavage at the transposon ends is staggered: cleavage occurs precisely at the 3′ ends of the element but occurs within the flanking donor DNA such that 3 nucleotides of flanking donor sequences (indicated by "d") are attached to the 5′ ends of the transposon. The 3′ ends of the transposon attack staggered positions on the target DNA (indicated by "t"), generating a new insertion flanked by short gaps. The donor nucleotides on the 5′ ends of the transposon are removed and the gaps are repaired by host functions.

son are mediated by different activities. This was first suggested by the observation that mutations in the substrate transposon ends change the DNA cleavage patterns at the end of Tn7. For example, mutation of the two-terminal CA-3′ nucleotides blocks cleavage at the 3′ ends of the transposon but does not block cleavage at the 5′ ends of the element (44).

We now know that the Tn7 transposase that executes the chemical steps of transposition is formed by two proteins, TnsA and TnsB (18, 68, 71, 102). These proteins have unique roles but act interdependently; no recombination is observed in the presence of only one subunit of the transposase. TnsB binds specifically to the ends of the transposon (5, 114) and likely

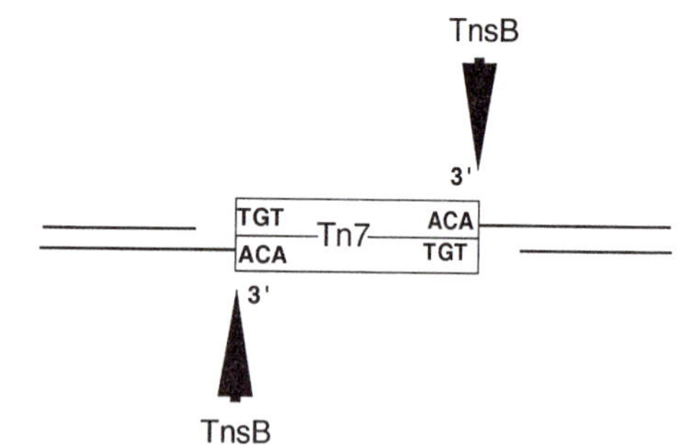

Figure 7. The processing events at the ends of Tn*7* are mediated by two different polypeptides. TnsB mediates the cleavages at the 3′ ends of Tn*7* and the joining of the exposed 3′-OHs to the target DNA. TnsA mediates the cleavages at the 5′ ends of Tn*7*. Transposase activity requires TnsA and TnsB: no activity is seen in the presence of either protein alone.

recruits TnsA, which alone has no observable DNA-binding activity, to the transposon ends. TnsB executes the cleavages that cut the 3′-transposon ends away from the donor DNA and also the joining steps that link these ends to the target DNA (102). TnsA executes the cleavages that disconnect the 5′-Tn*7* ends from the flanking donor DNA (71, 102). Thus, Tn*7* transposition involves double-strand breaks where each DNA strand is cut by a different polypeptide (Fig. 7).

Although the catalytic activity of TnsA is not required in the TnsB-mediated joining of the 3′ ends of Tn*7* to the target DNA, the presence of TnsA is still required (102). Thus, TnsA has two roles in recombination, executing cleavage at the 5′ ends of Tn*7* and also modulating the activity of TnsB. Similarly, TnsB also has multiple roles: binding specifically to the ends of Tn*7* (5, 114), mediating the breakage and joining events at the 3′ ends of Tn*7* (102), and likely recruiting TnsA to the ends of Tn*7* and modulating its activity (102).

Other Cut-and-Paste Transposons Use a Single Polypeptide Transposase To Cleave the 3′ and 5′ Transposon Ends

The strategy of using two different polypeptides that mediate the chemical events at the ends of a transposon is thus far unique to Tn*7* (118). IS*10* (20, 57, 58) and IS*50* (17) use a different series of events catalyzed by a single polypeptide to promote a double-strand break (Fig. 8). The 3′ transposon end is cleaved by a transposase containing only one type of polypeptide to expose a 3′-OH. This 3′-OH then intramolecularly attacks the 5′ transposon strand, disconnecting the transposon from the donor DNA and generating a hairpin transposon end. The hairpin is then opened to give a free 3′-OH on the transposon ends that can attack the target DNA. The chemistry of DNA strand breakage remains to be determined for other elements that use double-strand breaks such as *Drosophila* P elements and the widespread Tc1/ mariner family.

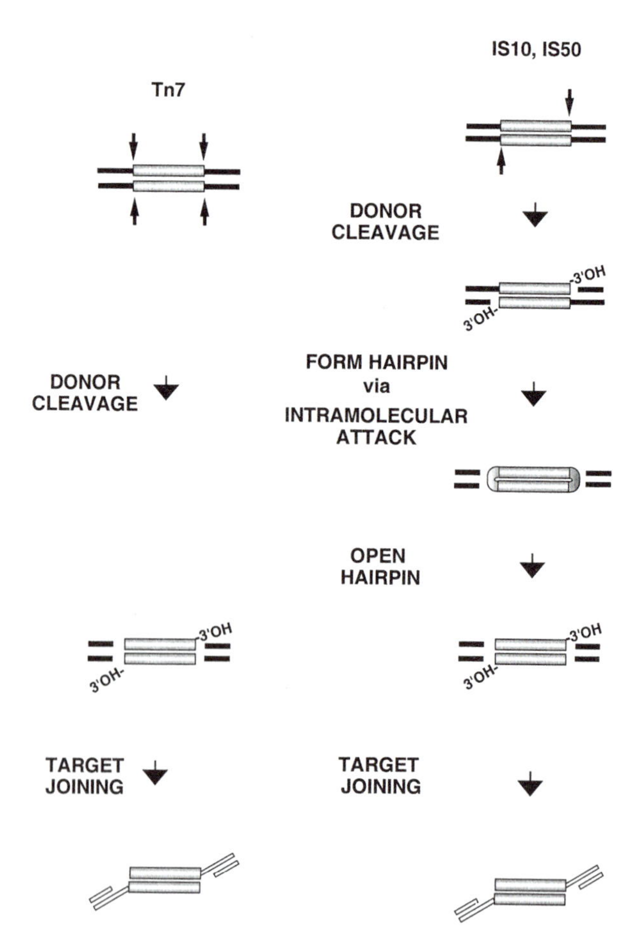

Figure 8. Double-strand breaks at transposon ends can occur by multiple mechanisms. Whereas Tn*7* uses cleavages by two different polypeptides to generate double-strand breaks at the ends of Tn*7*, IS*10* and IS*50* use a one polypeptide and a hairpinning mechanism to generate double-strand breaks. With IS*10* and IS*50*, a nick is introduced at the 3′ end of the transposon. The resulting 3′-OH intramolecularly attacks the 5′ strand of the transposon to form a hairpin on the transposon end; this step also disconnects the transposon from the flanking donor DNA. The transposase then opens the hairpin, again exposing the 3′ end of the transposon which can go on to attack the target DNA.

Similar intramolecular hairpinning reactions via transesterifications promoted by the RAG proteins also underlie V(D)J recombination (see chapter 29).

The Fate of the Gapped Donor Backbone

What is the fate of the gapped donor DNA from which Tn7 is excised? We have found that the gap in the donor chromosome from which Tn7 is excised can be repaired via double-strand break-induced homologous recombination (49) (Fig. 9). Such repair can be visualized by analyzing the fate of a Tn7 donor site in a *lac* gene (*lac*::Tn7) in a cell that also contained a nonoverlapping *lac*$^-$ heteroallele. We found that transposition was accompanied by conversion of the *lac*::Tn7 donor site to *lac*$^+$, indicating that repair using the *lac*$^-$ heteroallele occurred. Such repair of transposon donor sites has also been observed with the P elements of *Drosophila* (38) and the Tc1 element of *Caenorhabditis elegans* (85).

Another possibility is that the gapped chromosome may sometimes be lost from the cell, for example, by degradation. Such loss could often be tolerated because rapidly growing *Escherichia coli* have multiple chromosomes.

In IS*10*, transposition is tightly tied to DNA replication, transposition being triggered after passage of the replication fork (97). Thus, transposition occurs only when there are two copies of the donor chromosome such that if IS*10* excises from one chromosomal copy leading to loss of that chromosome, the other chromosome will remain intact. Key to this regulation are GATC methylation sites in the ends of IS*10* that can affect both substrate activity and transposase expression. A similarly placed GATC site in the end of IS*50* also promotes transposition after replication (34, 117).

There is a GATC site in Tn7R among the TnsB binding sites. However, manipulation of this site in Tn7 has not provided evidence for any linkage of Tn7 transposition to replication of the donor site (31). It should be noted, however, that target site replication can play a key role in TnsABC+E transposition (see below).

Cut-and-paste Tn7 transposition can be biologically replicative

The standard Tn7 transposition reaction observed in vitro is a conservative cut-and-paste one in which the element is excised from the donor site and then inserted into the target site. There are several mechanisms by which this reaction could be effectively replicative in vivo, i.e., spread the transposon to new sites without losing the element from the original donor. One strategy would be for transposition to occur when there are at least two copies of the chromosome (Fig. 10). This will often be true when Tn7 transposes from the *E. coli* chromosome because *attTn7* is located near *oriC*. If insertion occurs into

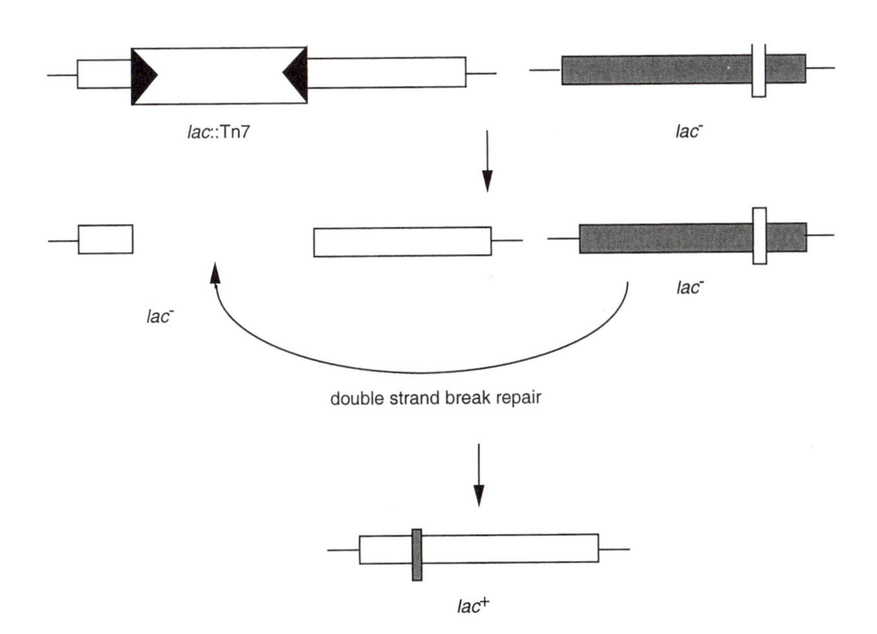

Figure 9. The gapped donor site can be repaired. The gapped donor site that results from transposon excision can be repaired by double-strand break repair using another copy of the donor site. Such repair can be visualized by examination of cells containing *lac*$^-$ heteroalleles: one *lac*::Tn7 allele and the other a *lac*$^-$ heteroallele mutant at a site that does not overlap the Tn7 insertion site. When Tn7 transposition occurs, that is, Tn7 excises from the *lac*::Tn7 site, the resulting gap can use information in the *lac*$^-$ heteroallele to convert the gapped region of the Tn7 donor site to *lac*$^+$.

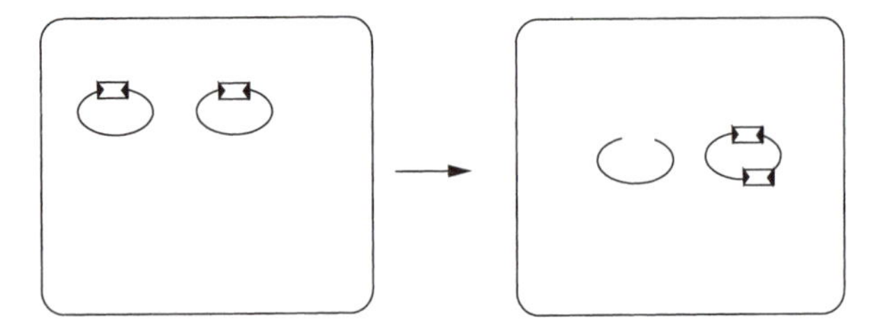

Figure 10. Cut-and-paste transposition can be biologically replicative. The large square is a bacterial cell with two copies of the bacterial chromosome, each containing a transposable element. Transposition of the transposon from one DNA into the other results in one gapped chromosome and one chromosome containing two copies of the element, one at the donor site and one at the new site of insertion. Thus, when cells containing the bacterial chromosome with the two copies of the element are examined, transposition looks replicative, although the element actually moved by cut-and-paste transposition.

the chromosome from which excision has not occurred, that target chromosome will now contain two copies of Tn*7*, one at the original donor site and the other at the new target site.

Switching from Cut-and-Paste to Replicative Transposition

Not all transposable elements move by cut-and-paste transposition. Several bacterial transposons, including phage Mu and Tn*3*, move by replicative transposition, during which a copy of the transposable element is actually generated by the act of transposition (Fig. 11). The products of replicative transposition are cointegrates in which two copies of the transposon link the donor backbone and a target DNA. Tn*7* can switch from an element that moves by cut-and-paste transposition to one that moves by replicative transposition by mutation of the TnsA transposase subunit (71). TnsA mediates cleavage at the 5′ ends of the element; mutation of this protein does not affect cleavage and strand transfer at the 3′ ends of Tn*7* by TnsB, but rather blocks cleavage at the 5′ ends of the element. The product of transposition in vitro using TnsA mutants is a fusion product that covalently joins the donor backbone, the transposon, and the target DNA. The 3′ ends of the transposon are cleaved from the donor DNA and joined to the target DNA, but the donor DNA remains joined to the transposon at its 5′ ends. Such a fusion product is equivalent to the strand transfer intermediate observed in Mu transposition (see chapter 17) (28) and is also a likely intermediate in Tn*3* transposition (see chapter 14).

The Tn*7* fusion reaction not only occurs efficiently in the test tube but also in vivo (71). Cointegrates that result from the replication of the fusion product are the products of Tn*7* transposition in vivo using these TnsA mutants that generate fusion products in vitro. An interesting unanswered question in Tn*7* transposition is what DNA replication machine(s) mediates the repair of the 5-bp gaps that flank the newly inserted transposon in a simple insertion product and process the replication of the entire element to generate a cointegrate.

Thus, a mutation in a transposase component can convert Tn*7* from a cut-and-paste element to a replicative element that generates a cointegrate. An interesting possibility is that cellular conditions may exist that would modify the activity of TnsA so that wild-type Tn*7* generates cointegrates.

Assembly of the Entire Tn7 Transposition Machine Occurs before Strand Cleavage

An interesting feature of the Tn*7* system is that little recombination is observed when any component of the reaction is omitted (8). Thus, no partial transposition reactions occur without the capacity to complete recombination. Most notable is that no cleavage at the ends of the transposon in the donor site is observed unless an *attTn7* target site is present. We interpret these observations to mean that a nucleoprotein

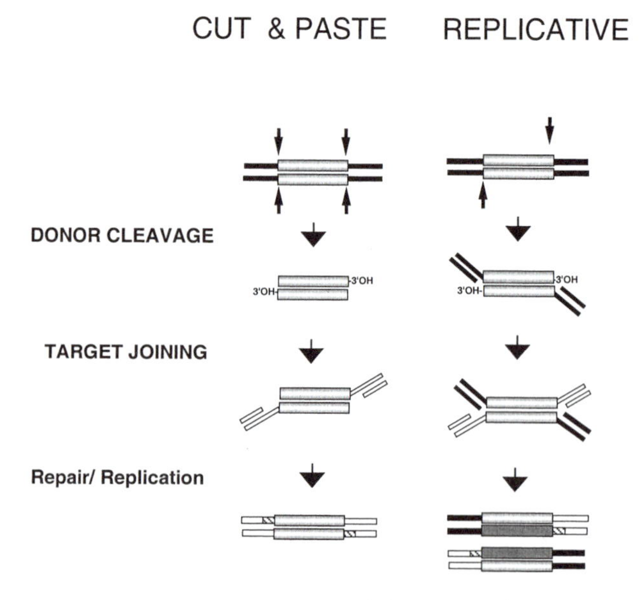

Figure 11. Tn*7* transposition can switch from a cut-and-paste mechanism to a replicative mechanism. The cut-and-paste pathway in which Tn*7* excises from the donor site is shown on the left. On the right in the replicative pathway, nicks occur at the 3′ ends of the transposon. These 3′ ends then attack the target DNA to form the fusion product in which the transposon links the donor and target DNAs. Repair and replication of this fusion product yield the cointegrate that contains two copies of the mobile element.

complex containing the donor and target DNAs and the recombination proteins must be formed before the double-strand breaks at the transposon ends that initiate recombination occur. The requirement for the formation of a nucleoprotein complex containing an appropriate target DNA provides a mechanism by which evaluation of the target DNA can control Tn7 transposition; recombination (i.e., breakage at the transposon ends) does not occur in the absence of an appropriate target DNA.

Through cross-linking studies, we have identified nucleoprotein complexes that are precursors to recombination which contain the Tn7 donor and target DNAs (R. Sarnovsky, Z. Skelding, and N. L. Craig, unpublished data). Formation of these complexes appears to be mediated through interactions between TnsC on the target DNA with TnsA (68, 109) and TnsB on the transposon end DNAs (Skelding et al., unpublished).

The bacteriophage Mu can also likely evaluate its target before initiating transposition (79, 80, 123). However, this strategy of evaluation of the target DNA before cleavage of the transposon from the flanking donor DNA is not universal. With IS*10*, for example, transposon excision must occur before the recognition of the target DNA (101).

TnsAB: THE Tn7 TRANSPOSASE

In contrast to other transposases that are formed by a single protein, the Tn7 transposase is formed by a collaboration between two proteins. One subunit, TnsB, is a member of the retroviral integrase superfamily and the other subunit, TnsA, resembles a type II restriction enzyme. Under our standard reactions conditions, transposase activity is seen only when the transposase is activated by TnsC and other components of the transposition machinery.

Identification of the TnsAB Transposase

Because TnsB is the sequence-specific DNA-binding protein that binds to multiple sites on the ends of the transposon (5, 114), it was expected that this protein would be the transposase. However, no recombination can be observed under any conditions with just TnsB. Several different observations revealed that TnsB and TnsA together form the Tn7 transposase. With just TnsAB, DNA breakage at the ends of Tn7 and intramolecular joining of the resulting 3′-OH ends can occur under atypical reaction conditions, i.e., in the presence of Mn^{2+} and high glycerol (18). The products of these reactions are circularized Tn7 species in which the exposed 3′-OH from one end of Tn7 joins intramolecularly to the 5′ strand at the other end of the element (Fig. 12). Circles that involve double-strand breakage at either end followed by intramolecular joining are observed. The underlying chemistry of this reaction (exposure of 3′-Tn7 ends and joining of these ends to a target DNA) is the same chemistry that underlies the standard intermolecular transposition pathway (7) but, in this case, the target is an intramolecular one. The basis of the permissiveness of these Mn^{2+} and high-glycerol reaction conditions is not known.

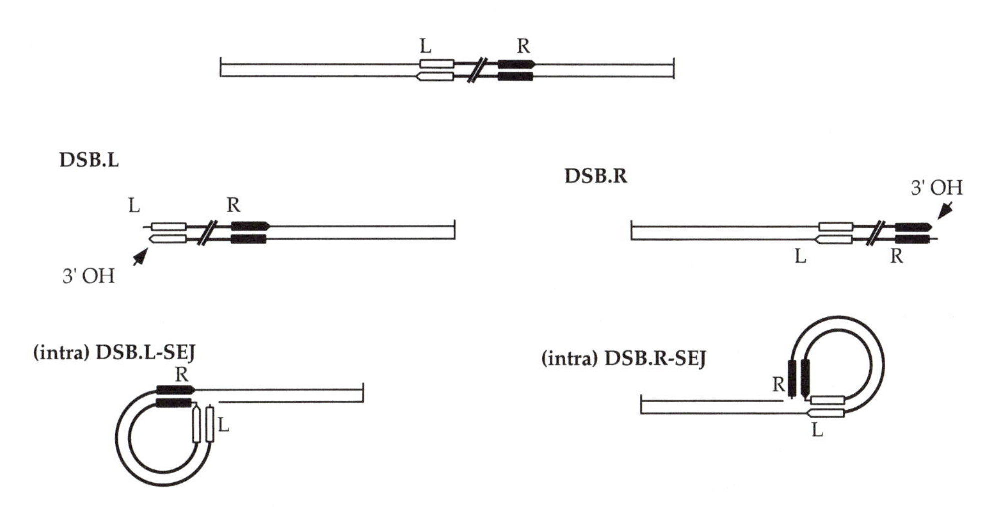

Figure 12. Products of intramolecular Tn7 transposition. The upper line shows a linearized Tn7-containing plasmid DNA before recombination. Recombination initiates by a double-strand break at either the left end (DSB.L) or right end (DSB.R) of Tn7. The 3′-OH ends exposed by these double-strand breaks then attack the 5′ strands at the other ends of the transposon, generating species called "single end joins" (SEJ) that contain circularized versions of Tn7.

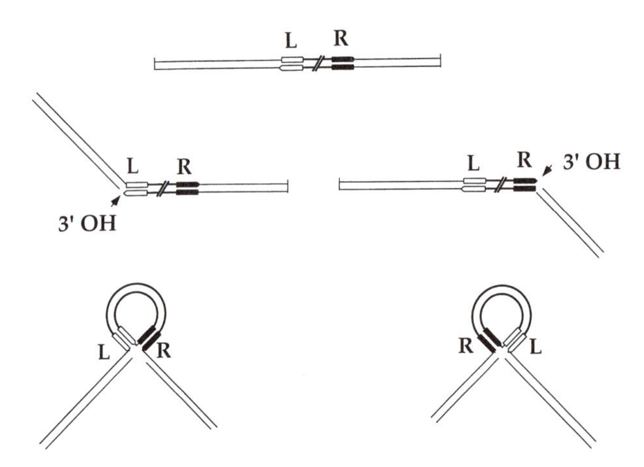

Figure 13. Tn7 "figures of eight." The upper line shows a linearized Tn7-containing plasmid DNA before recombination. When recombination occurs with $TnsA^{mut}+TnsB^{wt}$, where $TnsA^{mut}$ is blocked for cleavage at the 5′ ends of Tn7, only single-strand cleavage occurs at either the left or right 3′ end of Tn7. These exposed 3′ ends then execute an intramolecular attack at the 5′ strand at the other end of the transposon, generating figures of eight.

Another form of intramolecular joining is observed when TnsAB recombination under the Mn^{2+}/high-glycerol reaction conditions is performed with a TnsA mutant that fails to execute cleavage at the 5′ ends of the transposon (Fig. 13). In this case, the recombination product is a "figure of eight" first described for IS*911* (88): single-strand cleavage at one 3′ end of the transposon is followed by the attack of this exposed 3′ end of the same strand at the 5′ end of the molecule.

Another piece of evidence supporting the view that TnsAB forms the transposase is the isolation of TnsA and TnsB gain of function mutations that allow recombination in the presence of only TnsA and TnsB (68) (Fig. 14A). The ability to do recombination with only TnsAB has been observed to date only when certain TnsA and TnsB mutations are combined. It remains to be determined whether this requirement represents simply a threshold effect or some allele-specific interactions.

These gain-of-function TnsB mutations lie in the catalytic center of TnsB (see below). An attractive explanation for the gain-of-function consequences of these mutations is that they reconfigure the active site so that breakage and joining can now occur in a way that usually occurs only in the presence of the activators TnsC and TnsD or TnsE. The gain-of-function TnsA mutations lie in the C-terminal domain of the protein whose function is not understood; thus, the basis of their gain-of-function phenotype is not yet known. One interesting possibility is that they may affect interactions between TnsA and TnsB that are key to transposase activation.

TnsB: a Member of the Retroviral Integrase Superfamily

TnsB mediates recognition of the Tn*7* ends

A key step performed by the transposase is the recognition of the transposon ends. The TnsB subunit of the Tn7 transposase performs this function: TnsB is a sequence-specific DNA-binding protein that interacts with multiple sites within the Tn7 end segments that are required for recombination (3, 5, 114). These TnsB sites have been critical to recombination through deletion analysis of the Tn7 ends (4). The end-specific DNA binding domain of TnsB, which appears to be composed of two subdomains, is present in the amino terminus of TnsB (R. Sarnovsky and N. L. Craig, unpublished data) (Fig. 14A). How these subdomains actually bind to DNA remains to be determined; there are no readily identifiable DNA-binding motifs within TnsB. The actual roles of these multiple subdomains in recombination also remains to be determined.

TnsB mediates synapsis of the Tn*7* ends

A key step in recombination is the synapsis of the two ends of the transposable element by the transposase. This synapsis step ensures that recombination occurs only with a suitable substrate (i.e., a DNA segment flanked by two Tn7 ends). By cross-linking studies, we have established that TnsB alone can mediate synapsis between the ends of Tn7 (Sarnovsky et al., unpublished). Thus, TnsB has the capacity to recognize and juxtapose the two ends of Tn7, activities that must be key to the initiation of recombination.

TnsB is a member of the retroviral integrase superfamily

Inspection of the sequence of TnsB reveals that it contains a central core region with limited, albeit significant sequence similarity to members of the retroviral integrase superfamily (26, 89, 102) (Fig. 14B). This superfamily includes retroviral integrases, integrases of some retroviral-like transposable elements such as yeast Ty1 and *Drosophila* copia, transposases of the Tc1/mariner family, and certain bacterial transposases including the phage Mu transposase MuA, transposases of the IS*3* family and of the IS*4* family which includes IS*50* and IS*10* (36, 63, 87). Essential features of this conserved region are several highly conserved acidic amino acids, the so-called "DDE" motif. Mutation of these conserved amino acids in many transposases of this superfamily including TnsB (102) imposes a catalytic block on recombination.

Crystallographic analysis of this core domain from HIV (35), MuA (95), and IS*50* (29) reveals that the conserved acidic residues cluster within the protein and can interact with Mg^{2+}, an essential cofactor in all transposition reactions that have been analyzed to date. This central domain is presumed to be the active site where the DNA breakage and joining reactions that underlie recombination occur.

Mechanistic studies on the DNA breakage and joining reactions promoted by MuA (58, 78), the HIV integrase (37), and the IS*10* (58) transposase have revealed that the initial breakage reaction at the ends of the transposon occurs through a simple hydrolysis step to expose the 3′-OH end of the element (75). The strand transfer step is thought to occur via direct attacks of the 3′-OH transposon ends on the target DNA, via one-step transesterification reactions that do not involve protein-DNA intermediates. Given the similarity of TnsB to these other transposases and the similarity of the chemistry of all the transposition reactions, it is likely that the breakage and joining actions of TnsB occur by the same direct, one-step transesterification reactions.

An issue that remains to be established for Tn7 and for many other transposable elements is the relative distribution of the 3′ end catalytic steps with regard to the TnsB-like protomers. In IS*10*, several studies of the chemistry of the breakage and joining steps support the view that all steps at each 3′ end of the element are mediated by a single active site in a single protomer of transposase (20, 58). Whether this strategy is followed by Tn7 needs to be determined experimentally.

As noted above, gain-of-function mutations in TnsB have been isolated that in combination with certain mutations in TnsA, allow recombination in the presence of just TnsAB. These TnsB mutations lie within the active-site domain of the protein (Fig. 14A). Indeed, one gain-of-function mutation is M366I, near one of the conserved acidic residues D361; another gain-of-function mutation is A325T. A plausible model for the basis of the phenotypes of these mutants is that these mutations allow some changes in the architecture of the active site in the absence of the usual activators of transposition, thereby allowing the active site to perform recombination that usually occurs only in the presence of the other Tns proteins.

TnsA: a Restriction Enzyme-Like Protein

Although TnsB can recognize the ends of Tn7 in the absence of TnsA (or any other Tns protein), no recombination is detected under any conditions or in the presence of any combination of Tns proteins in the absence of TnsA. The finding that mutations in TnsA specifically blocked cleavage at the 5′ ends of Tn7 suggested that TnsA was involved in these chemical steps of transposition but did not show directly that there was an active site for DNA hydrolysis in TnsA (71, 102). The hypothesis that TnsA was directly involved in 5′ end cleavage was first suggested by evidence that an amino acid in TnsA interacted directly with Mg^{2+}, a likely cofactor in end cleavage (102). The interaction of a particular aspartate, D114, with Mg^{2+}, whose mutation blocked 5′ end cleavage in the presence of Mg^{2+}, was demonstrated by showing that the activity of TnsA for 5′ end cleavage was restored by changing D114 to a cysteine and performing recombination in the presence of Mn^{2+} rather than Mg^{2+}.

Further evidence that TnsA participates directly in recombination comes from the crystal structure of TnsA; TnsA most resembles a type II restriction enzyme (53). The protein contains two domains, an N-terminal domain containing the active site and a C-terminal domain whose function is not yet known. Mutation of the amino acids in TnsA that correspond to the active-site amino acids in *Bam*HI and *Hin*dIII block cleavage at the 5′ ends of Tn7, arguing strongly that this is the active-site region of TnsA. In the TnsA crystal, two Mg^{2+} ions are present in the active site even in the absence of a DNA substrate, a feature distinct from that of the restriction endonucleases. The amino acid indicated to be in contact with Mg^{2+} by the cysteine-Mn^{2+} experiment described above, D114, is indeed in contact with Mg^{2+} in the TnsA crystal (Fig. 14C).

The function of the C-terminal domain of TnsA is unknown. One possibility is that this domain is involved in protein-protein interactions. As noted above, we have isolated gain-of-function TnsA mutations in the C-terminal domain of TnsA that allow recombination in the presence of just TnsA and TnsB. It is not yet clear why alterations in these amino acids result in a gain-of-function phenotype.

The presence of TnsA is also required for the action of TnsB at the target-joining step in which the 3′ ends of the transposon are joined to the target DNA (102). Thus, TnsA is involved both in the chemical steps of end cleavage and in controlling the activity of TnsB.

The sequences of TnsA proteins from several other Tn7-like elements have been determined (55, 81). Although the amino-terminal domain that contains the active-site region is highly conserved among these different TnsA species, the C-terminal regions are highly diverged. An interesting but unanswered question is whether the structures of these variable C-terminal regions are conserved or diverged.

A

DNA-binding domain

catalytic domain

TnsB D D E 702a.a.

A325T

M366I

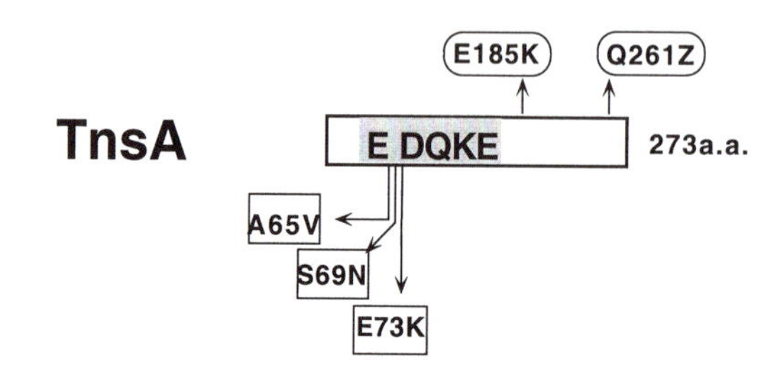

Recombination with $TnsA^{mut}$ + $TnsB^{mut}$

Recombination with $TnsA^{mut}$ + $TnsB^{wt}$ + $TnsC^{wt}$

B

```
                        10        20        30        40        50        60
                ....*....|....*....|....*....|....*....|....*....|....*....|
consensus    1 RASRPNELWQMDFT---PLPVLGKGg-------kKYLLVIVDDFSRFVVAYPLKSKs--- 47
RSV          5 RGLGPLQIWQTDFT---LEPRMAPR---------SWLAVTVDTASSAIVVTQHGRV---- 48
HIV          4 QVDCSPGIWQLDCT---HLE----G---------KVILVAVHVASGYIEAEVIPAEt--- 44
Tn402      172 AVTAPLEQVQIDHT---VIDLIVVDdrdrqpigrPYLTLAIDVFTRCVLGMVVTLEap-- 226
Tn552      155 ESSRPNEIWQADHT---LLDIYILDqkgn--inrPWLTIIMDDYSRAIAGYFISFDa--- 206
Tn5468     276 ETVGPASRYQIDATiadVYLVSRYDrtki--vgrPVLYIVIDVFSRMITGVYVGFEgpsw 333
Tn7 TnsB   262 QALGPGSRYEIDATiadIYLVDHHDrqki--igrPTLYIVIDVFSRMITGFYIGFEnpsy 319
IS3        124 YASGPNQKWAGDIT---YLR--TPEg-------wLYLAVVIDLWSRAVIGWSMSPRm--- 168
               -----------D------------------------------------------------

                        70        80        90       100       110       120
                ....*....|....*....|....*....|....*....|....*....|....*....|
consensus   48 ----SAETVFDLLKAAVERRGGk--------------PKTIHSDNGSEFTSKAFQELLKE 89
RSV         49 ----TSVAVQHHWATAIAVLGR---------------PKAIKTDNGSCFTSKSTREWLAR 89
HIV         45 ----GQETAYFLLK-LAGRWPVk----------------tVHTDNGSNFTSTTVKAACWW 83
Tn402      227 ---sAVSVGLCLVHVACDKRPWleglnvemdwqmsgkPLLLYLDNAAEFKSEALRRGCEQ 283
Tn552      207 ----PNAQNTALTLHQAIWNKNntnwp------vcgiPEKFYTDHGSDFTSHHMEQVAID 256
Tn5468     334 vgamMALSNTATEKVEYCRQFGveigaad--wpcqalPDALLGDRG-EIAGSAIETLINN 390
Tn7 TnsB   320 vvamQAFVNACSDKTAICAQHDieisssd--wpcvglPDVLLADRG-ELMSHQVEALVSS 376
IS3        169 ----TAQLACDALQMALWRRKRp-------------rNVIVHTDRGGQYCSADYQAQLKR 211
               ------------------------------------------D-----------------

                       130       140       150       160       170       180
                ....*....|....*....|....*....|....*....|....*....|....*....|
consensus   90 LGIKHSFSRPYSPQDNGVVERFNRTLKRELRKLLSFS--------------------SL 128
RSV         90 WGIAHTTGIPGNSQGQAMVERANRLLKDRIRVLAEGDgfm---------------kriPT 134
HIV         84 AGIKQEFGIPYNPQSQGVIESMNKELKKIIGQVRDQ----------------------A 120
Tn402      284 HGIRLDYRPLGQPHYGGIVERIIGTAMQMIHDELPGTtfsnpdqrgd--ydsenkaalTL 341
Tn552      257 LKINLMFSKVGVPRGRGKIERFFQTVNQTFLEQLPGYinnndt---------ssdlidFQ 307
Tn5468     391 FQVRVENAAPYRADWKGIVEQRFRLIPARFKAYVAGYvapdfqerga-rdyrldatldID 449
Tn7 TnsB   377 FNVRVESAPPRRGDAKGIVESTFRTLQAEFKSFAPGIvegsrikshgetdyrldaslsVF 436
IS3        212 HNLRGSMSAKGCCYDNACVESFFHSLKVECIHGEHFIs-------------------RE 251
               ------------------E-----------------------------------------

                       190       200       210       220
                ....*....|....*....|....*....|....*....|.
consensus  129 EEWEEALATALYLYNRRRRHSLLGg---------TPAERLA 160
RSV        135 SKQGELLAKAMYALNHKERGENTK----------TPIQKHW 165
HIV        121 EHLKTAVQMAVFIHNHKRKGGIGGy---------SAGERIV 152
Tn402      342 RELERWLTLAVGTYHGSVHNGLLQ----------PPAARWA 372
Tn552      308 NFEEKLRYFLIEDYNQKEHSA-IQs---------TPINRWN 338
Tn5468     450 QFTRIVLYCILYYNNQHFLRDYEKs-----------SDMIA 479
Tn7 TnsB   437 EFTQIILRTILFRNNHLVMDKYDRdadfptdlpsIPVQLWQ 477
IS3        252 IMRATVFNYIECDYNRWRRHSWCGg--------lSPEQFEN 284
               -----------------------------------------
```

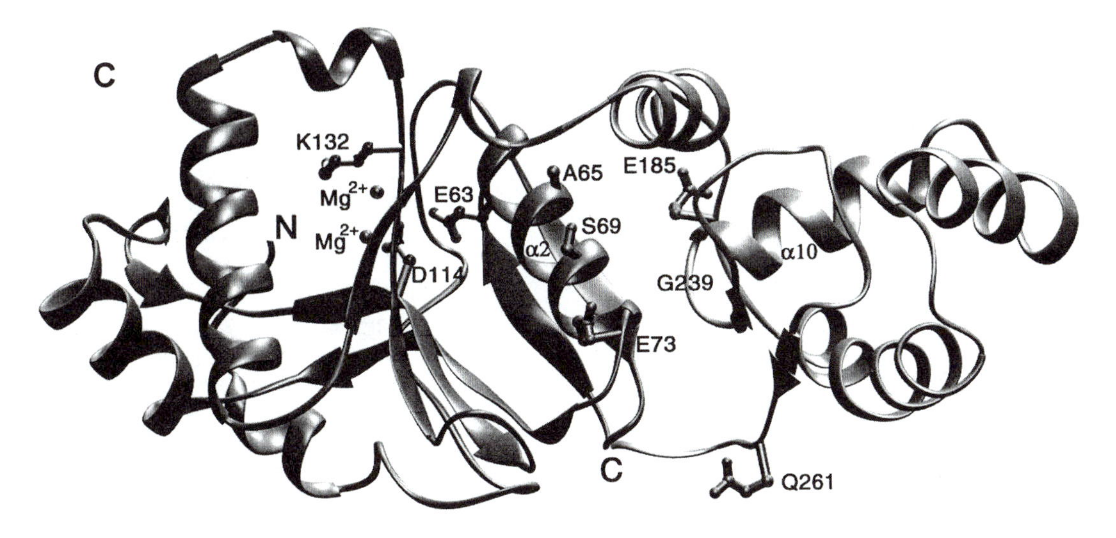

Figure 14. Structures of TnsB and TnsA. (A) TnsB and TnsA are indicated as open bars. The DNA binding domain of TnsB (1 to 241) that specifically recognizes the ends of Tn7 is indicated by a cross-hatched bar. The catalytic amino acids D273, D361, and E396 are indicated in TnsB and E63, D114, Q130, K132, and E149 in TnsA (gray bars). Gain-of-function mutations in TnsA and TnsB that allow recombination under the indicated conditions are shown. (B) Alignment of TnsB with other members of the Retroviral Integrase Superfamily. This alignment of the catalytic domains of these recombinases was generated by the Conserved Domain Database at the National Center for Biotechnology Information. The acidic amino acids in bold are the DDE motif. (C) Gain-of-function mutations in TnsA that allow recombination in the presence of $TnsA^{mut}TnsB^{wt}TnsC^{wt}$ lie on the back of the helix containing the TnsA active-site amino acids E63. Reprinted from reference 68 with permission.

Activation of TnsAB by TnsC

As described in detail below, our working model of Tn7 transposition is that the TnsAB transposase is activated by interaction with TnsC and that the ability of TnsC to activate the transposase is determined by its interaction with an appropriate target DNA and targeting proteins (108). Several observations have provided evidence that both TnsA and TnsB can interact with and be activated by TnsC.

Evidence for TnsA-TnsC interaction

We have determined by several means, including affinity chromatography in vitro (68, 111) and a two-hybrid analysis in vivo (Y. Li, F. Lu, and N. L. Craig, unpublished data), that TnsA and the C terminus of TnsC interact with each other. An attractive model is that this interaction underlies the pathway by which TnsC interacts with and activates TnsA.

We have also isolated gain-of-function TnsA mutants that give increased recombination in the presence of TnsC (68). The TnsA amino acids changed in these mutants lie on one face of a helix that is exposed on one side of TnsA, whereas the other side of this helix supports one of the amino acids of the TnsA active site (Fig. 14C). We speculate that these gain-of-function mutants alter TnsA-TnsC interactions to increase their efficacy, resulting in a reconfiguration of the TnsA active-site region in the absence of the other usual activators of transposition.

Evidence for TnsB-TnsC interaction

There is also evidence that interactions between TnsB and TnsC occur that are key to activation of the transposase. Through cross-linking studies, we have identified nucleoprotein complexes containing the Tn7 donor and target DNAs (Skelding et al., unpublished). These complexes can be formed in the presence of only TnsBC+D when using an *attTn7*-containing target or in the presence of TnsBC*, where TnsC* is a gain-of-function version of TnsC that allows recombination in the absence of TnsD and TnsE. These observations reveal that interactions between TnsB and TnsC must play a key role in the formation of these recombination complexes.

We have also isolated TnsB mutants with C-terminal alterations that fail to form these donor-target complexes and also fail to promote recombination, although they are active in other recombination steps, for example, binding to and synapsing the ends of Tn7 (Z. Skelding and N. L. Craig, unpublished data). Thus, it seems likely that interactions between the C termini of TnsB and TnsC play a key role in forming donor-target nucleoprotein complexes that are important for recombination.

In the presence of TnsA and TnsBC+D, (i.e., conditions under which recombination occurs), formation of the donor-target complexes is greatly increased.

Thus, our working model of Tn7 transposition is that both subunits of the TnsAB transposase are contacted by TnsC to activate the transposase. An interaction also occurs between TnsB and TnsC that has a negative impact on transposition during transposition immunity (see below).

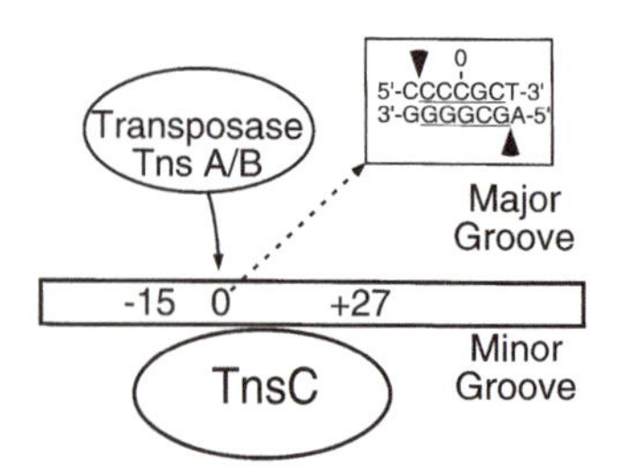

Figure 15. TnsC bound to the target DNA provides a platform for the transposase. The TnsAB transposase bound to the ends of Tn7 is positioned on the target DNA through the interaction with TnsC. The interaction of TnsC with the transposase activates the breakage and joining events that underlie transposition. The boundaries of TnsC interaction with the target DNA are those observed in TnsC-TnsD-*attTn7*. The transposase attacks the target site at 5′-staggered positions on the target DNA.

TnsC: THE MEDIATOR OF TRANSPOSASE-TARGET INTERACTIONS AND THE CONTROLLER OF TRANSPOSASE ACTIVITY

The activity of the Tn7 transposase is modulated by the presence of other Tns proteins and the target DNA. Central to this regulation is TnsC, the protein that can directly communicate with the transposase and with the target proteins and DNAs. TnsC is an ATPase (43, 109) and an ATP-dependent, sequence-independent DNA-binding protein (43). Our working model is that TnsC controls the activity of the transposase (108) by interacting with both subunits of the transposase, TnsA and TnsB. We suggest that the ability of TnsC to activate the transposase is determined by its interactions with ATP and DNA. The distribution of TnsC on DNA plays a key role in target site selectivity.

TnsC Interacts with the Target DNA

The ability of TnsC to bind DNA is intimately related to its ability to select insertion sites and activate transposition. DNAs that are targets for Tn7 insertion are bound by TnsC (8, 62, 92, 106). TnsC is recruited to target certain sites through interactions with target proteins and structures. DNAs that are not targets for Tn7 insertion are not bound by TnsC (106). TnsC appears to form a platform on the target DNA at the point of Tn7 insertion, which can then recruit and activate the transposase bound to the ends of Tn7 (Fig. 15). This disposition of TnsC is suggested by comparison of the patterns of DNA protection in TnsC-TnsD-*attTn7* and TnsD-*attTn7* complexes (62). These protection experiments also revealed that TnsC binds to the minor groove of DNA. The binding of TnsC to the minor groove at the insertion site provides room in the major groove for the transposase and the ends of Tn7, a configuration consistent with the position on the target DNA that must be attacked by the ends of Tn7.

TnsC on the target DNA is thus positioned to interact with the transposase and activate DNA breakage and joining. In this view, TnsC is the component that actually guides the transposase to the target DNA, not the transposase itself. This "indirect" strategy is reminiscent, for example, of that used by yeast Ty3 in which the transposase is positioned on the target site through the interaction with host transcription factors (see chapter 27) (125). By contrast, for other elements such as IS*10* (see chapter 20) (16), transposase itself is the key selector of the target DNA through direct interaction with the insertion site.

TnsC binds preferentially to alternative DNA structures

The position of TnsC on the target DNA, and hence the selection of Tn7 insertion site, can be highly influenced by the target DNA structure. No preference for TnsC binding to any particular DNA structure on duplex DNA has been discernible. However, compared with binding to duplex DNA, TnsC binds preferentially to regions of pyrimidine triplex DNA that are formed by the annealing of pyrimidine oligonucleotides to polypurine tracts (91, 92) (see below). This binding selectivity by TnsC underlies the fact that TnsABC-dependent insertions occur preferentially adjacent to regions of triplex DNA (see below). It is not yet known what features of the triplex DNA attract TnsC.

Alternative DNA structures also play a key role in recruiting TnsC to TnsD-*attTn7* (62) (see below). The binding of TnsD to *attTn7* imposes alterations of DNA structure as detected by a variety of footprinting methods, and the formation of these alterations is key to *attTn7* function.

Activation of the Transposase by TnsC and the Role of ATP

Several different kinds of observations have revealed that TnsC can activate the transposase. Under standard in vitro recombination conditions, the presence of an appropriate target protein and DNA is required in addition to TnsC to provoke transposition (8). However, TnsC mutants have been isolated that stimulate recombination in the absence of any other factor, pointing to TnsC as the protein that directly communicates with the transposase (107, 109) (see below). Moreover, DNA cleavage and intramolecular joining by TnsAB under alternative recombination conditions, high glycerol or high glycerol + Mn^{2+}, can be provoked by TnsC (18, 109).

The ATP state of TnsC seems to be a key feature in its ability to affect the transposase. TnsABC+D recombination requires ATP; no recombination is observed in the presence of ADP (8). We have also found that recombination occurs constitutively in the presence of TnsABC and the nonhydrolyzable ATP analogue adenosine 5′-(β,γ-iminotriphosphate) (AMP-PNP) (7, 8). Moreover, the gain-of-function mutants of TnsC that can perform recombination in the absence of TnsD and TnsE are altered in their interactions with ATP and DNA (109) (see below). Thus, our working model is that when TnsC is ATP bound, it can activate the transposase to execute breakage and joining, but that TnsC cannot activate the transposase when it is bound to ADP or lacks nucleotide (108) (Fig. 16).

What part of TnsC interacts with the transposase? We have found that a truncated TnsC containing only the C-terminal half of the protein can stimulate transposition (109). This TnsC segment notably lacks detectable DNA-binding activity and the ability to interact with ATP.

Controlling the ATP State of TnsC

The ATPase activity of TnsC provides a mechanism for the protein to switch between an active ATP-bound state that can stimulate recombination and an ADP-bound (or lacking nucleotide) state that does not stimulate recombination (109). We hypothesize that the ATP state of TnsC does not change stochastically, but rather that it is influenced by the presence of particular targeting proteins and target DNAs (108) (Fig. 16). As described below, TnsC is recruited to TnsD-*attTn7* to form a target complex that can interact with the transposase bound to the ends of Tn7; formation of this complex requires ATP (8). An attractive hypothesis is that when TnsC is recruited to TnsD-*attTn7*, it forms a TnsC-ATP-TnsD-*attTn7* complex that is the transposase activator. Supporting this hypothesis, we have observed an increase in ATP binding when TnsC is incubated with TnsD and *attTn7* (110).

Tn7's aversion to certain DNAs is also related to TnsC and its ATP state. As described below, Tn7 exhibits target immunity; that is, Tn7 does not insert into DNAs that already contain a copy of Tn7 (106). The establishment of immunity requires ATP hydrolysis. Without hydrolysis, immunity is bypassed and Tn7 insertion into targets containing Tn7 readily occurs. Furthermore, the execution of immunity is accompanied by the removal of TnsC from a Tn7-containing target DNA and its redistribution to a non-Tn7-containing DNA that can serve as a target for Tn7 insertion. We speculate that TnsC is removed from Tn7-containing targets through ATP hydrolysis, which results in the decreased ability of TnsC to bind DNA. As discussed below, the binding of TnsB to the Tn7 copy preexisting in the target DNA is the initiator of immunity, and we have suggested that the interaction of TnsB with TnsC provokes ATP hydrolysis that results in the redistribution of TnsC away from the immune, Tn7-containing target.

We suggest that TnsC functions as a nucleotide-dependent molecular switch in a manner similar to that of GTP-dependent molecular switches such as Ras (21, 108), i.e., both of these proteins are active in their nucleotide triphosphate state and inactive upon nucleotide triphosphate hydrolysis (Fig. 16). Moreover, we are attracted to the view that other components of the Tn7 transposition machinery influence the nucleotide state of TnsC in a manner analogous

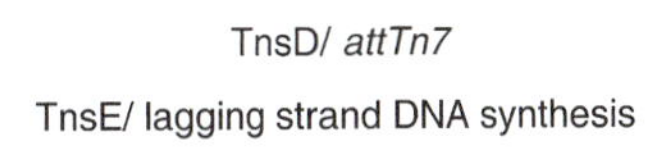

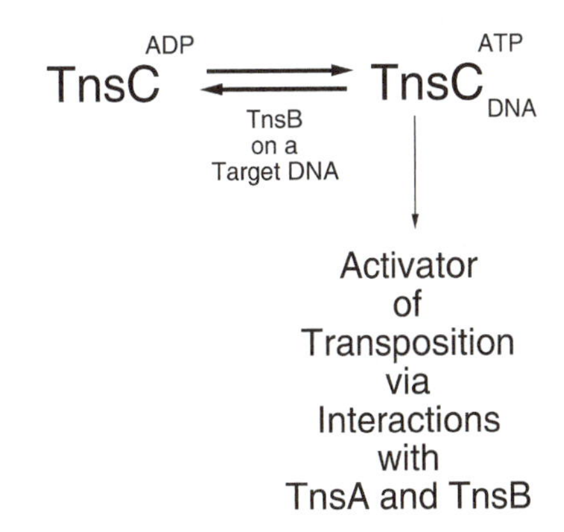

Figure 16. Modulation of TnsC activity. TnsC stimulates the transposase to execute breakage and joining when bound to ATP and the target DNA. The presence of particular targeting proteins and target DNAs determines the ATP and DNA state of TnsC.

to how guanine exchange factors (GEFs) and guanine-activating proteins (GAPs) modulate Ras. For example, TnsD-*attTn7* may promote the formation of TnsC-ATP by facilitating ADP-ATP exchange, i.e., acting like a GEF. Alternatively, TnsB when bound to an immune target DNA may activate TnsC ATP hydrolysis, i.e., act like a GAP, in promoting the removal of TnsC and making that target DNA immune.

TnsC Gain-of-Function Mutants That Provoke Transposition in the Absence of TnsD or TnsE

Support for the view that TnsC plays a critical role in controlling transposase activity has come from the isolation and characterization of TnsC gain-of-function mutants that promote recombination in the presence of TnsAB but in the absence of TnsD and TnsE (107, 109). Mutations at different positions throughout the TnsC polypeptide that result in this phenotype have been isolated (Fig. 17). One particularly interesting mutant is $TnsC^{A225}$ in which the altered amino acids lie just a few amino acids away from one component of the amino acids of the Walker B purine nucleotide utilization motif in TnsC. Analysis of this mutant TnsC protein in vitro has revealed that it has a slower rate of ATP hydrolysis than does wild-type TnsC and binds more tightly to DNA (109). These altered properties of TnsC could account for its gain-of-function phenotype. We speculate that because of this slower rate of ATP hydrolysis, this TnsC mutant spends more time in the active ATP-bound state than does the wild type and is able to bind to DNA in the absence of the usual targeting proteins and DNAs. This mutant is still responsive to target inputs including TnsD, TnsE, and immune targets, consistent with its ability to do some ATP hydrolysis.

Other gain-of-function mutants are insensitive to such target signals and seem to be constitutive in their ability to activate the transposase. In vitro, one of these mutants is no longer sensitive to ADP inhibition of its ability to stimulate TnsAB. Perhaps this mutant protein is now permanently in a configuration that activates TnsAB.

These gain-of-function TnsC mutations are distributed throughout the TnsC polypeptide (Fig. 17). This may suggest that multiple steps are involved in changing TnsC to a form that can activate the TnsAB transposase.

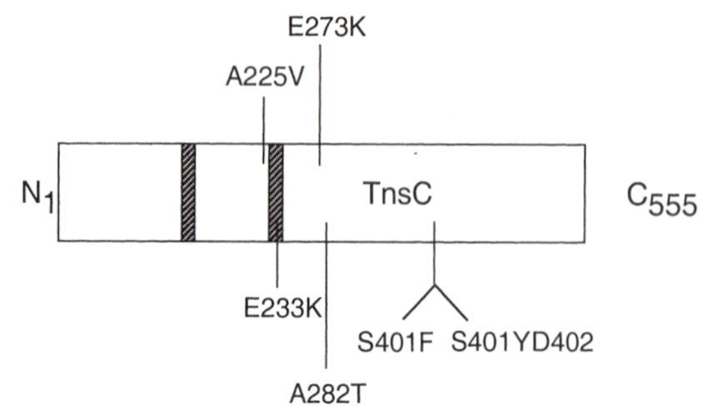

Figure 17. Positions of gain-of-function mutations in TnsC that allow recombination with TnsA + TnsB + $TnsC^{mut}$. The TnsC polypeptide is indicated by the open bar; the positions of the Walker A and B motifs for purine nucleotide utilization are indicated by hatched bars. Class I gain-of-function mutations remain sensitive to target DNA signals; they are sensitive to transposition immunity and are directed to appropriate target sites by TnsD and TnsE. Class II mutants are insensitive to target signals; they promote transposition into immune targets and do not respond to TnsD and TnsE. Reprinted from reference 107 with permission.

TnsABC TRANSPOSITION

Recombination with wild-type Tns proteins requires TnsABC+D or TnsABC+E. Virtually no transposition is observed with only TnsABC in vivo. TnsABC recombination in vitro can be observed only through the use of very sensitive PCR detection methods (19). Nonetheless, the observation that some TnsABC recombination can occur is important, because it allows analysis of target selection by TnsABC in the absence of the usual target selectors TnsD and TnsE.

TnsABC: Transposition to Duplex DNA Targets

TnsABC-dependent transposition in vitro displays relatively little target-site selectivity on duplex DNA targets (19). Although variation in the frequency of insertions at different internucleotide junctions is observed, the selectivity of insertion is sufficiently low that many, many insertions are observed within each relatively short (i.e., 50 to 100-bp) segment, within a duplex plasmid DNA.

The target-site selectivity of TnsABC insertions involving the TnsC gain-of-function mutants has also been examined in vitro (19). Insertions using $TnsABC^{A225V}$, like those obtained with $TnsABC^{wt}$, display relatively little target-site selectivity on duplex DNA (Fig. 18A). Although some sites do seem to be preferred over other sites, it has not been possible from sequence inspection of the insertion sites and nearby DNA to determine why one site might be preferred over another.

The high frequency of $TnsABC^{A225V}$ transposition coupled with the low level of target discrimina-

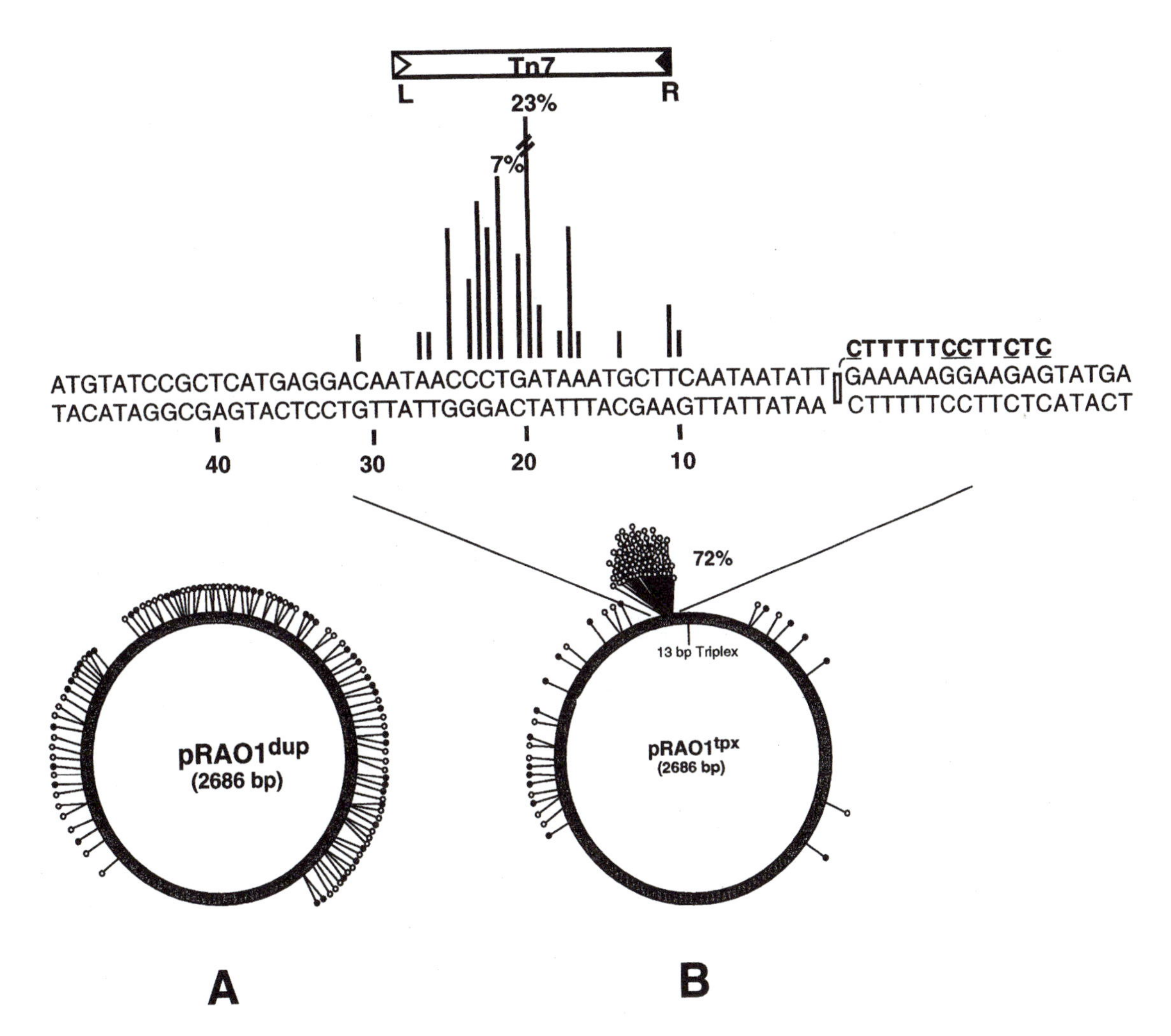

Figure 18. Comparison of patterns of insertions promoted by TnsABC^{A225V} into duplex- and triplex-containing DNA targets. (A) Transposition was performed in vitro using a duplex DNA. The transposition products were recovered by transformation into *E. coli* and the sequences of 100 products determined. The filled circles represent insertions in one orientation with respect to the target backbone and the open circles in the other. The hatched region is the origin-of-replication region. (B) Transposition was performed in vitro into a triplex DNA target formed by annealing a triplex-forming oligonucleotide to the plasmid and psoralen cross-linking. Insertions were recovered and mapped as described above.

tion makes this transposition system a very useful tool for gene and genome analysis (19) (see below).

TnsABC: the Triplex Recognition Pathway

Although Tn7 TnsABC transposition displays relatively low target-site selectivity on duplex target DNA, it does display a profound insertion selectivity for DNA triplexes that are formed by the annealing of a short (15-bp) pyrimidine oligonucleotide to a polypurine tract in the target DNA (91, 92) (Fig. 18B). Tn7 can recognize both natural triplexes (i.e., H-form DNA) and artificial triplexes formed by the annealing of an oligonucleotide to a plasmid. Tn7 is able to recognize only pyrimidine triplexes; no preferential insertion is seen with a triplex structure in which the third strand is formed of purines. Another very interesting aspect of the triplex-targeting pathway is that a profound orientation basis in insertion is also observed; virtually all insertions near the triplex occur so that the right end of Tn7 points toward the triplex region. The pattern of site and orientation specificity is reminiscent of the pattern of TnsABC+D-dependent insertion into *attTn7* where insertions in a particular place in a particular orientation are also observed.

The exact nature of the signal in and around the triplex that is recognized by TnsABC remains to be established. DNA-binding studies have revealed that TnsC is the component of the TnsABC machinery that interacts specifically with the triplex (92). Apparently both the triplex itself and the psoralen cross-link used when making synthetic triplexes by cross-linking a triplex-forming oligonucleotide to a target plasmid can contribute to defining the position of insertion (91). When a triplex-forming oligonucleotide cross-linked at its 5′ end on a plasmid is used as a target, insertions are observed to be clustered in a 10-bp region about 20 bp from the 5′ end of the triplex region. When the triplex is cross-linked to the target plasmid via its 3′ end, a cluster of specific insertions is again observed about 20 bp from the position of the psoralen cross-link. When the oligonucleotide fragment is cross-linked at both termini, specific insertion into the region of the triplex again occurs, demonstrating that the targeting signal recognized by TnsABC is not simply a single-stranded DNA attached to a DNA duplex.

We speculate that TnsABC is attracted by some alteration in DNA structure that is imposed in the triplex region per se that may exist at the junctions of triplex-duplex DNA, and also an alteration at the position of the psroralen cross-link. It has not been possible to mimic the preferential triplex-dependent insertion of Tn7 using DNAs containing short mismatched regions or bent DNAs (90).

Although insertions mediated by both TnsABC^{wt} and TnsABC^{A225V} do preferentially occur at positions of triplex DNA, the relative frequencies of transposition promoted by these different systems on triplex DNA is about the same as their relative frequencies on duplex DNA (92). TnsABC^{wt} insertion occurs at a very low frequency on triplex-containing DNA, whereas TnsABC^{A225V} insertion occurs at a high frequency. Thus, although recognition of a triplex DNA can position the transposition machinery on a target DNA, the presence of a triplex target is not sufficient to activate the TnsABC^{wt} machinery.

TnsABC+D TRANSPOSITION: RECOGNITION OF *attTn7*

With use of TnsABC+D, Tn7 transposes to *attTn7* because TnsD is a sequence-specific DNA-binding protein that binds specifically to *attTn7*. Recruitment of TnsC to the TnsD-*attTn7* complex can then attract the TnsAB transposase, thereby promoting Tn7 insertion into *attTn7* (8).

attTn7 Structure

E. coli attTn7

The specific chromosomal insertion site of Tn7 in *E. coli*, *attTn7*, which is used in the TnsABC+D pathway, is about 14 kb to the left of the origin of replication (*oriC*), at minute 84.3. The actual point of Tn7 insertion in *attTn7* lies about 20 bp downstream of the 3′ end of the *glmS* gene that encodes glutamine synthetase (L-glutamine:D-fructose-6-phosphate aminotransferase) (46, 67, 121) (Fig. 5). *glmS* is the second gene of the *glmUS* operon (86). Both GlmS and GlmU (*N*-acetylglucosamine-1-phosphate uridyltransferase) are involved in synthesis of *N*-acetylglucosamine. The Tn7 insertion point is about 300 bp upstream of the *pstS* gene that encodes a high-affinity phosphate-specific transporter. The insertion of Tn7 into *attTn7* has not been deleterious.

The high frequency with which Tn7 inserts into *attTn7* is notable. For example, analysis of chromosomal *attTn7* occupancy by Tn7 using Southern blotting in the presence of a plasmid source of Tn7 has revealed that somewhere between 1% and 10% of the *attTn7* sites are filled by Tn7 (30). This high frequency of transposition is observed in the absence of any selection for Tn7 insertion.

Insertion of Tn7 into *attTn7* is highly orientation specific, with Tn7R virtually always inserting adjacent to *glmS* (47, 66, 67, 72). However, rare insertions in the opposite orientation have also been observed (32, 54).

The Tn7 insertion point lies within the transcription terminator of *glmS* (47) (Fig. 19). Experiments examining Tn7 insertions into chromosomal *attTn7* have shown that transcription of *attTn7* actually decreases the frequency of Tn7 insertion into *attTn7* (32). This inhibition suggests that there may be a linkage between Tn7 transposition and host-cell metabolism.

The molecular basis of the negative effect of tran-

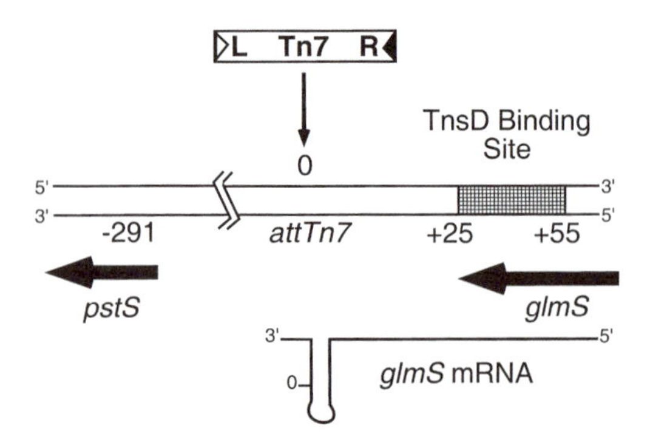

Figure 19. Tn7 insertion into *E. coli attTn7*. Tn7 inserts about 25 bp to the 5′ side of the TnsD binding site. The TnsD binding site lies in the C-terminal region of the *glmS* gene and the actual point of Tn7 insertion lies in the *glmS* transcription terminator. The middle of the 5 bp duplicated upon Tn7 insertion is designated 0; sequences to the right are indicated by plus signs, and sequences to the left are indicated by a minus sign.

scription of *attTn7* on Tn7 insertion into *attTn7* is not understood. This inhibition may reflect a "sweeping off" of the Tn7 targeting proteins from the target DNA by RNA polymerase or may result from some alteration in DNA structure at the point of the insertion that discourages Tn7 insertion.

Other experiments using *attTn7* targets on plasmids revealed no difference in *attTn7* target activity in the presence or absence of transcription across *attTn7* or in the presence of the terminator (47). These observed differences between *attTn7* function in the chromosome versus in a plasmid may have been obtained because of a large excess of targets in the plasmid system or some feature that makes plasmid structure different from that of the chromosome.

attTn7 is a target for Tn7 insertion because TnsD binds to *attTn7* (8). The TnsD binding site lies within the region of *glmS* that encodes the C terminus of GlmS which is highly conserved and actually contains the active-site region (Fig. 19). Thus, Tn7 insertion into a specific chromosomal position depends on recognition of sequences adjacent to the point of insertion rather than directly at the point of insertion. This clever strategy allows Tn7 to exploit recognition of a highly conserved region but to avoid inactivation of that gene through insertion.

attTn7 sites in other bacteria

Site-specific insertion of Tn7 has been observed in the chromosomes of many different bacteria (reviewed in references 24 to 26). This conservation of *attTn7* is likely because the *glmS* gene whose C terminus contains the essential information for *attTn7* target activity is highly conserved because of its central role in cell-wall metabolism.

Although Tn7 insertion does occur downstream of the *glmS* gene in several organisms, the precise spacing of the Tn7 insertion site from the essential *attTn7* sequences in *glmS* can vary. The basis of this variation is not known. By contrast, analyses of hundreds of insertions into *E. coli attTn7* have shown that insertion generally occurs at exactly the same position (47, 72). In *E. coli attTn7,* however, the sequence of the point of insertion can be changed without changing the position of Tn7 insertion (see below).

Pseudo-*attTn7* sites in *E. coli*

When *attTn7* is not available, TnsABC + D insertions still occur but only at low frequency. These insertions occur repeatedly at a few sites (61). Sequence analysis of these sites reveals that they are related in sequence to *attTn7* in the region of the TnsD binding site. Thus, TnsD interaction with these pseudo-*attTn7* sites can also attract Tn7.

Definition of Sequences in *attTn7* Required for *attTn7* Target Activity

What sequences within *attTn7* are actually required to promote the high-frequency, site- and orientation-specific insertion of Tn7? Analysis of various deletion derivatives of *attTn7* has revealed that the required sequence information lies to the right of the specific point of Tn7 insertion, extending from about +27 to +57, 0 being the central position in the 5-bp sequence duplicated upon Tn7 insertion (8, 47, 62, 119). As described in more detail below, this required region corresponds to the sequences required for TnsD binding (Fig. 20). The sequences at the actual point of Tn7 insertion do not determine *attTn7* target activity and can be changed without altering *attTn7* target activity.

Under our standard in vitro recombination reaction conditions, both supercoiled and relaxed *attTn7* DNAs are effective targets for Tn7 insertion. However, under conditions in which recombination is limited, recombination is favored by a supercoiled *attTn7* substrate (6). This observation supports the view that an alteration in *attTn7* DNA structure is important to *attTn7* target activity. An attractive hypothesis is that DNA supercoiling facilitates formation of the distortion imposed by TnsD to activate transposition (see below).

TnsD Binds to *attTn7*

TnsD binds specifically to *attTn7* (8, 62). Inspection of the amino acid sequence of TnsD has not re-

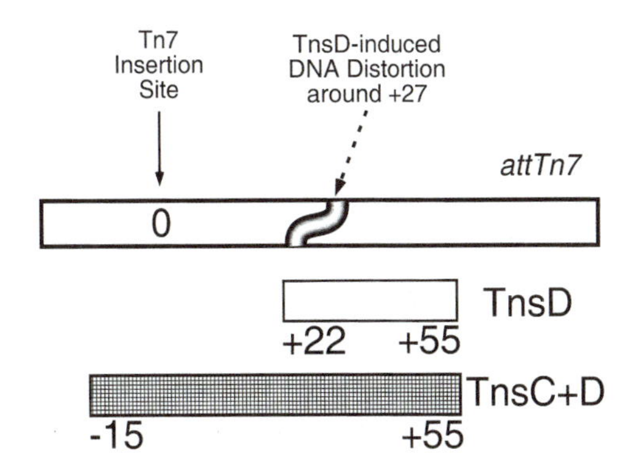

Figure 20. Positions of Tns protein binding on *attTn7*. The positions of TnsD and TnsC-TnsD binding on *attTn7* as evaluated by various footprinting methods are shown. The position of a TnsD-induced distortion in DNA at +27 is also indicated. Deletion of this region of distortion decreases TnsD and TnsCD binding to *attTn7* and *attTn7* target activity.

vealed any DNA-binding motifs or homologies to proteins other than the TnsD homologues of other Tn7-like transposons.

TnsD binding to *attTn7* has been examined by a variety of methods including footprinting with enzymatic and chemical attack and binding interference analysis with chemical attack reagents (8, 62). The picture that has emerged is of a complex binding site that includes a region of core "conventional" protein binding and a region of TnsD-induced distortion 5′ of this region of conventional binding in *attTn7* DNA (Fig. 20).

The region of *attTn7* that interacts with TnsD extends from about +22 to about +57. The region of the core binding lies most distal from the insertion site of Tn7, extending from +30 to about +57. Footprinting analysis using dimethyl sulfate (DMS) shows that TnsD in this region interacts with *attTn7* primarily in the major groove (62). TnsD also imposes a modest bend on *attTn7* in this region (P. Kuduvalli and N. L. Craig, unpublished data).

Several observations have revealed a region of DNA distortion imposed on *attTn7* in the 5′ region of TnsD interaction (62). One indication of this distortion is that hypercleavage of DNA is observed during attack by hydroxyl radicals, potassium permanganate, and 1,5-phenanthroline Cu. Another indication that TnsD binding to *attTn7* involves DNA distortion is that gapping of the *attTn7* site in this region actually stimulates the binding of TnsD to *attTn7*.

The region of TnsD-imposed distortion is essential to *attTn7* target activity (62). When the sequence of the distorted region is changed, TnsD binding to both *attTn7* and *attTn7* target activity is reduced. As discussed below, the region of distortion imposed by TnsD likely plays a critical role in attracting TnsC to *attTn7* and hence in promoting transposition.

The Formation of a TnsC-TnsD-*attTn7* Complex

TnsC binds specifically to the TnsD-*attTn7* complex (8). In this TnsC-TnsD-*attTn7* complex, the positions of protein interaction with DNA as defined by hydroxyl radical footprinting and interference studies now extend from the distal edge of the TnsD binding site at +57 across the insertion site to a position about 15 bp to the left of the insertion site (−15) (8, 62) (Fig. 20). Methylation protection experiments provide additional information about the TnsC-TnsD-*attTn7* complex. Protection in the major groove is observed in the righthand side of the *attTn7* site which contains the specific determinants for TnsD binding. By contrast, the rest of the region displays protection in the minor groove. These findings suggest that in the TnsC-TnsD-*attTn7* complex, TnsD occupies its specific binding site in *attTn7* and then TnsC is recruited to interact with the rest of *attTn7*. Because of the ability of TnsC to bind to triplex DNA which contains alterations in DNA structure (92), and the similarities in the relative positions of Tn7 insertions in the TnsABC-triplex pathway and the TnsABC+D-*attTn7* pathway (Fig. 21), we are attracted to the view that the region of distorted DNA introduced by TnsD is the signal that attracts TnsC to the TnsD-*attTn7* complex (62).

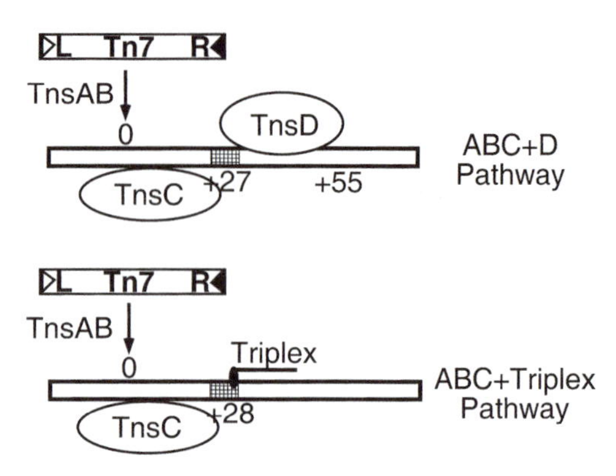

Figure 21. Target DNA structure plays a key role in Tn7 insertion. Similar features of the target DNA and Tn7 insertion during insertion into *attTn7* and adjacent to triplex DNA emphasize the critical role of target DNA structure. Target DNA distortions indicated by cross-hatching in *attTn7* and at triplex DNA have been detected by DNA footprinting. Reprinted from reference 62 with permission.

Our working model is that TnsC interacts with the target DNA in the minor groove. This is especially pleasing in considering how the ends of Tn7 bound by transposase must interact with the actual point of insertion (67). TnsC binding the minor groove could be providing a platform for the interaction of the TnsAB transposase in the major groove. Interaction of the ends of Tn7 with the major groove of the target DNA is compatible with the observed positions of the joining of the ends to the target DNA (Fig. 22).

Also notable in the interference studies of TnsC-TnsD-*attTn7* complexes is that when the region immediately surrounding the point of Tn7 insertion is made more flexible (i.e., gapped by hydroxyl radical treatment), the formation of these complexes increases (62). This observation supports the hypothesis that some DNA deformation must occur around the point of insertion to make this DNA accessible to the recombination machinery.

Why does the Tn7 system use this elaborate targeting strategy in which the protein that activates transposition, TnsC, interacts with the target DNA itself rather than simply being recruited to *attTn7* by interaction with TnsD? One rationale for the DNA-based target signal is that this strategy ensures that

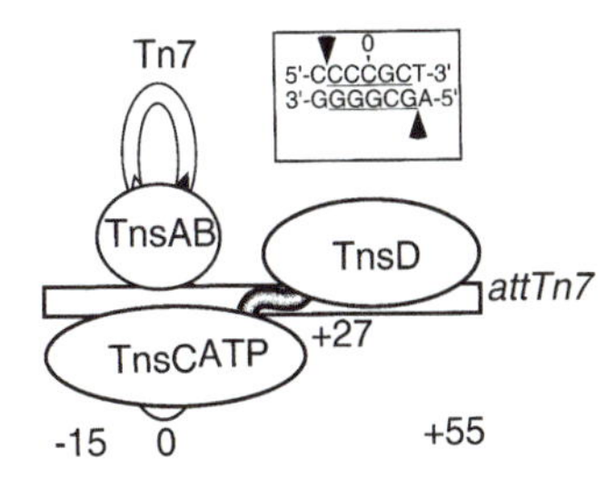

Figure 22. A model for Tn7 insertion at *attTn7*. The binding of TnsD to *attTn7* results in the formation of a TnsC-TnsD-*attTn7* complex. The binding of TnsD results in a DNA distortion (shaded bar at +27) which attracts TnsC. TnsC that is bound in the minor groove provides a platform for the interaction and activation of the TnsAB transposase that is bound on the transposon ends. The transposase and transposon ends must attack the target DNA across the major groove.

recombination will always occur at the actual target DNA itself, not "in solution" where activation by protein-protein interactions between TnsC and TnsD could occur.

Is There a TnsC-TnsD Interaction Essential to Transposition?

The work described above strongly suggests that TnsC recognizes a DNA signal in *attTn7* induced by TnsD. It may also be that there are protein-protein interactions between TnsC and TnsD that are critical to recombination. We have observed a protein-protein interaction between TnsC and TnsD using two-hybrid analysis (R. Mitra and N. L. Craig, unpublished data), but it remains to be established whether this interaction is key to transposition.

Activation of TnsC in TnsC-TnsD-*attTn7* Complexes

A central question in Tn7 transposition is how does the TnsC-TnsD-*attTn7* complex activate transposition? One attractive hypothesis is that the formation of this complex results in the formation of TnsC-ATP at *attTn7,* which, as we have suggested above, is the actual activator of transposition. Support for this view derives from the observation that increased ATP binding is observed in the TnsC-TnsD-*attTn7* complex compared with TnsC alone and that this increase depends on the presence of both TnsD and *attTn7* (110).

Host Proteins in TnsABC+D Transposition

Tn7 insertion into *attTn7* in vitro can occur in the presence of just TnsABC+D (8). However, several lines of evidence suggest that TnsABC+D transposition involves host proteins.

Host proteins can stimulate TnsD binding to *attTn7* and transposition to *attTn7* in vitro

Although TnsD alone can bind to *attTn7,* its binding is stimulated in the presence of a crude extract from a non-Tn7-containing cell (8, 104). Fractionation of the host extract revealed that two *E. coli* proteins, L29, a component of the large subunit of the ribosome, and acyl carrier protein (ACP), which is involved in fatty acid biosynthesis, can stimulate TnsD binding to *attTn7* (104). The presence of these host proteins in a binding reaction leads to a host protein-TnsD-*attTn7* complex of a mobility distinct from that of the TnsD-*attTn7* complex. Moreover, the presence of these host proteins increases the amount of TnsD-*attTn7* complex that is observed at a limiting concentration of TnsD. Although the greatest stimulation of TnsD binding is observed in the presence of both L29 and ACP, each factor alone can stimulate TnsD binding.

The presence of L29 and ACP can also stimulate the amount of transposition that occurs in a TnsABC+D reaction in vitro.

L29 plays a role in TnsABC+D recombination in vivo

The role of L29 in TnsABC+D recombination in vivo was probed by comparing transposition in $L29^+$ and $L29^-$ strains (104). Transposition in the TnsABC+D pathway was decreased substantially in the L29 mutant, whereas no decrease in recombination was observed in TnsABC+E recombination. The significance of this involvement of L29 in Tn7 recombination remains to be determined. It may represent a pathway through which cellular metabolism can influence transposition.

The in vivo role of ACP in Tn7 transposition has not been evaluated because of technical difficulties; ACP is an essential protein.

TnsABC+E TRANSPOSITION: RECOGNITION OF LAGGING-STRAND DNA SYNTHESIS

At low frequency compared with the TnsABC+D pathway, TnsABC+E also promotes Tn7 transposition (61, 98, 120). TnsE-dependent insertion sites are not related in sequence to *attTn7* or to each other (60, 66). However, TnsE-dependent insertion does not occur at random; rather, certain replicons and positions within replicons are preferred positions for TnsE-dependent insertion. A precise definition of the TnsE target signal(s) awaits reconstitution of an in vitro TnsABC+E transposition system.

TnsE Target Sites

TnsE promotes preferential insertion into conjugating DNAs

When TnsABC + E transposition occurs in a cell that contains the bacterial chromosome and a conjugating plasmid, (i.e., a plasmid that can move between cells by mating), insertion occurs preferentially into the conjugating plasmid, although the size of such a plasmid is far less than the size of the *E. coli* chromosome (124). Conjugation is the feature of these plasmids that attracts Tn7; blocking conjugation blocks this preferential TnsE-dependent insertion in the plasmid. Furthermore, making a usually nontransferable plasmid transferable by the addition of an *oriT* sequence that is the initiation site for conjugal transfer in the presence of Tra(nsfer) proteins also makes the plasmid a preferred target for Tn7 insertion.

This ability of TnsE to preferentially target conjugating plasmids is not the result of the bacterial chromosome being inaccessible to Tn7. $TnsABC^{A225V}$ insertions that occur in the absence of TnsE occur into the bacterial chromosome rather than into the plasmid (83, 84, 124). Thus, TnsE directs insertion to conjugating DNAs rather than insertion into the chromosome being precluded. It has also been shown that the preferential insertion of Tn7 into conjugating DNAs occurs in the recipient rather than the donor cells (124).

Why might Tn7 have an insertion pathway that preferentially directs insertions into conjugating plasmids? Such plasmids usually have a broad host range. Thus, the preferential insertion of Tn7 into such plasmids is useful in promoting Tn7 dispersal through bacterial populations.

A very interesting feature of these TnsE-dependent insertions into conjugating plasmids is that although insertions occur at many different sites, all insertions occur in a single orientation with respect to the target plasmid (12, 14, 124). These findings suggest that some directional process associated with conjugation (e.g., the transfer of DNA into the recipient or the lagging-strand DNA replication in the recipient that is required to generate an intact DNA duplex from the incoming single strand) is the actual target for insertion.

TnsE insertion appears to target lagging-strand DNA synthesis

At low frequency compared with the frequency of Tn7 insertion into conjugating DNA, TnsABC + E insertion into the *E. coli* chromosome does occur. The distribution of these chromosomal TnsE-dependent insertions recovered from cells lacking conjugal plasmids is not random. With wild-type TnsE, insertions occur preferentially into the terminus region of *E. coli* DNA replication (84) (see below). When using TnsE mutants that promote transposition at much higher frequency than the wild-type protein, a much broader distribution of insertions is now observed and TnsE-dependent insertions outside the terminus region are detected more readily (83).

Notably, the orientation of the TnsE-dependent insertions outside the terminus is different on either side of *oriC* (Fig. 23). Although these insertions differ in their orientation with respect to the chromosome, they all occur in the same relative orientation with respect to the movement of the replication fork on each side of *oriC,* that is, in each replicore (Fig. 24). These TnsE mutants also still promote insertion in a single orientation into conjugating plasmids.

Comparison of the relative orientations of the TnsE insertions in conjugating DNAs and in the *E. coli* replicores reveals that they all occur in a single orientation with respect to lagging-strand DNA synthesis (Fig. 24). That is, the left and right ends of Tn7 align in the same direction as the 5′ to 3′ orientation of Okazaki fragments. These results suggest that the key determinant of the target signal recognized by TnsE is likely a component of the replication fork that is involved in lagging-strand DNA replication.

Although lagging-strand DNA synthesis seems to be a critical signal for insertion into both conjugating plasmids and into the *E. coli* chromosome, the conju-

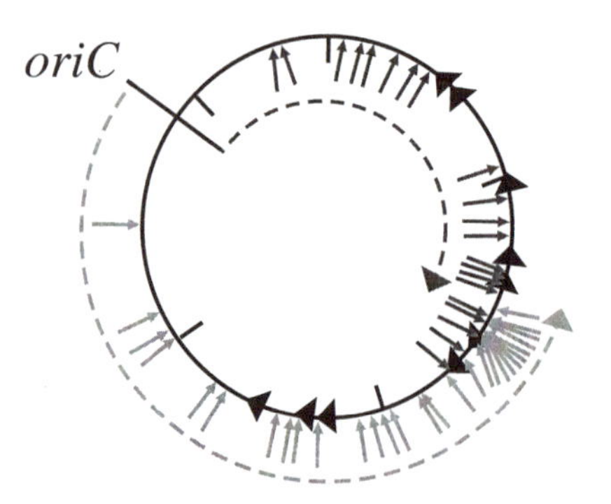

Figure 23. Orientation of TnsABC + E insertions in the *E. coli* chromosome. The position of the origin of replication (*oriC*) and the *ter* sites (triangles) are marked as the two chromosomal replicores (dashed lines). The positions of 50 TnsABC + E insertions (arrows) promoted by TnsE mutants that are more active than wild-type TnsE are also marked; those outside the *E. coli* chromosomal circle lie in one orientation with respect to the chromosome, whereas those inside are inserted in the opposite orientation. Reprinted from reference 83 with permission.

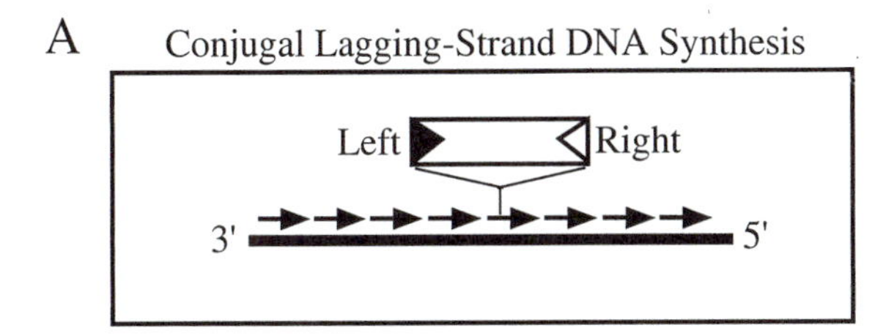

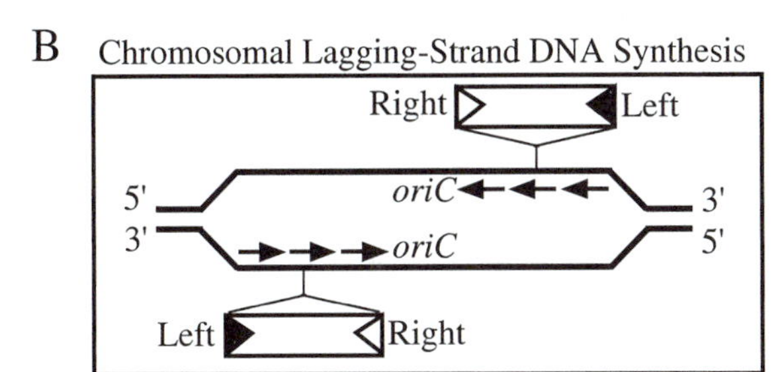

Figure 24. Disposition of TnsABC+E insertions with respect to lagging-strand DNA synthesis in various replicons. (A) Tn7 insertion into conjugating plasmids occurs in a single orientation in recipient cells, independent of the insertion position within the plasmid. Insertion occurs in the orientation indicated with respect to lagging-strand DNA synthesis in the recipient. (B) Tn7 insertions occur in different orientations in each of the chromosomal replicores as indicated; in both replicores, insertion occurs in the same orientation with respect to lagging-strand DNA synthesis.

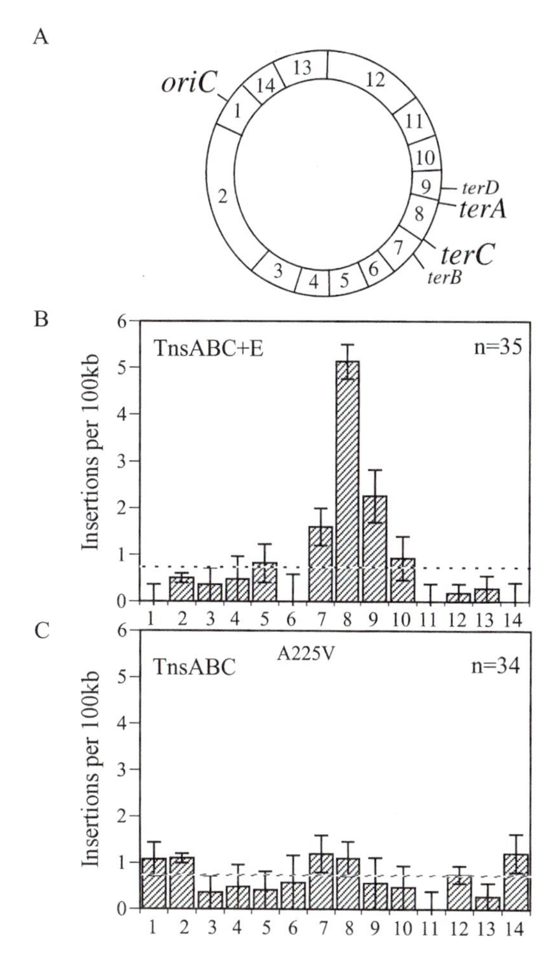

Figure 25. Positions of TnsABC + E insertions in the *E. coli* chromosome. (A) A physical map of the *E. coli* chromosome based on digestion with certain rare cutting restriction enzymes is shown. The positions of the origin of DNA replication (*oriC*) and several of the possible termination sites (*terA* to *terD*) are shown. (B) The distribution of 35 TnsABC+E insertions per 100 kb within the chromosome. The error bar represents the change in insertion frequency with one more or one less insertion in that region. (C) The distribution of 34 TnsABC^{A225V}-mediated insertions that occur in the absence of TnsE.

gating plasmid is still a much "hotter" target per unit length of DNA (124). Thus, there must be some other feature that makes a conjugating plasmid a "hot" target. Perhaps because only lagging-strand synthesis need occur at the first round of plasmid replication, this replication machine is especially accessible to TnsABC + E.

TnsE-dependent insertions occur preferentially into the replication terminus region

With use of TnsABC + E in cells that lack a conjugating plasmid, insertions occur preferentially into the region of termination of chromosomal DNA replication (84) (Fig. 25). In some cases, insertion occurs within just a few hundred base pairs from the *ter* sites (22) that can block the progression of the replication fork. Because of the wide distribution of the many *ter* sites, it is impossible to say from which direction any particular fork to be terminated actually entered the termination region. Thus, no consistent statement can be made about the relative orientation of Tn7 insertion within the terminus region and the direction of movement of the replication fork.

Termination of replication at *ter* sites involves the action of the Tus protein, an antihelicase (22). Although a small but reproducible effect on the frequency of Tn7 TnsABC + E insertion is seen in *tus*$^-$ strains, insertions still generally occur preferentially in the terminus region in this background (84). Thus, Tus is not directly involved in the targeting of TnsABC + E insertions to the terminus region. It is likely, however, that most termination events still occur in the region spanned by the *ter* sites, because the replication forks from each replicore will encounter each other in this region.

One model to account for the preferential insertion of Tn7 in the replication terminus region promoted by TnsABC + E could be that the terminus region is uniquely accessible to the transposition machinery because of the structure of the chromosome itself. However, this explanation is unlikely because insertions are seen throughout the chromosome when transposition is promoted by the TnsABC^{A225V} system (83, 84, 124) (Fig. 25). Thus, TnsE seems to actively promote insertions into the terminus region. An interesting possibility to account for the preferential insertion of Tn7 into the terminus region is that the lagging-strand replication machine is particularly accessible to TnsE in this terminus region.

The TnsE pathway can also preferentially target double-strand breaks

The repair of double-strand breaks in DNA must also involve lagging-strand DNA synthesis. Could such breaks (or some aspect of their processing) be special targets for the TnsABC+E pathway? We found that when double-strand breaks were introduced into the chromosome outside of the terminus region, the frequency of Tn7 transposition increased and the resulting Tn7 insertions occurred in the region of the double-strand break (84) (Fig. 26). We generated double-strand breaks at specific positions in the chromosome by inducing the transposition of excision^{+} insertion^{-} derivatives of Tn*10*. Actual excision of the Tn*10* transposon was required to observe the increase in Tn7 transposition and its targeting to the region of Tn*10* excision.

The observation that TnsABC+E promotes preferential insertion into conjugating DNAs in recipient cells where only lagging-strand DNA synthesis is occurring and that insertion in the conjugating plasmid and in each chromosomal replicore supports the view that lagging-strand DNA synthesis is likely a key feature of a TnsABC+E target signal. The finding that Tn7 insertion is also preferentially targeted to the region of double-strand breaks whose repair must involve lagging-strand DNA synthesis supports this view (23). A more precise definition of target recognition by TnsE awaits the development of an in vitro TnsABC+E transposition system.

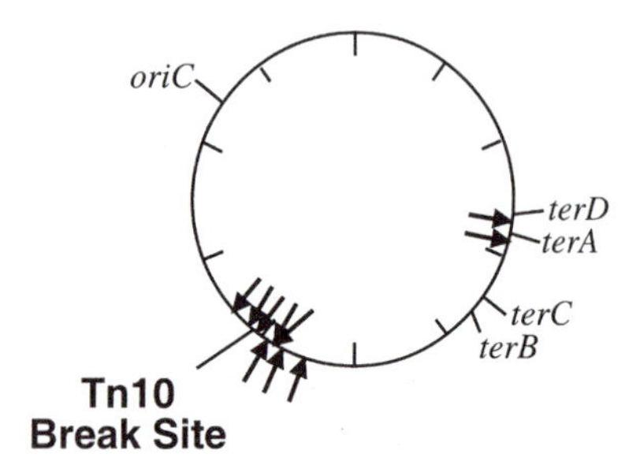

Figure 26. The distribution of Tn7 TnsABC+E insertions in a chromosome containing a double-strand break. A double-strand break was introduced into the bacterial chromosome by Tn*10* excision and the positions of Tn7 insertions determined. Reprinted from reference 84 with permission.

TnsE Binds to DNA

Inspection of the TnsE sequence has not been informative in deciphering the activity of this protein.

Gain-of-function TnsE mutants that promote increased frequencies of TnsABC+E-dependent insertion have been isolated (83). In all these mutants, two amino acids in the C terminus of the protein are changed. (The use of hydroxylamine as a mutagen likely biased the classes of mutants that could be recovered.) When these two mutations are combined, transposition increases about 1,000-fold to a level approaching that of the TnsD pathway.

Although a TnsABC+E in vitro transposition system has not been established, TnsE has been purified as a His-tagged derivative of TnsE and found to have DNA-binding activity (83). Moreover, TnsE mutants that promote increased transposition in vivo have increased DNA binding in vitro, suggesting that this DNA-binding activity is relevant to the activity of TnsE in transposition. These DNA-binding experiments have also shown that the preferred substrate for TnsE binding is a structure containing a recessed 3′-OH end. This is a DNA structure that should be prominent in DNA undergoing lagging-strand DNA synthesis, a structure that is apparently TnsE's preferred target for insertion in vivo.

Further work is needed to probe the connection between the DNA-binding activity of TnsE and its role in target selection. The observation that TnsE, like TnsD, has a DNA-binding activity suggests that these target-selecting proteins act by interacting with the target DNA in both pathways. The fact that TnsD and TnsE interact with different DNAs (TnsD with *attTn7* and TnsE with DNAs with recessed 3′ ends) can account for the different targets selected in the TnsD and TnsE pathways. It will be interesting to see if TnsC is activated by similar mechanisms in both pathways.

AVOIDING SELF: TARGET IMMUNITY

A very interesting aspect of Tn7 biology is that it displays transposition immunity; that is, Tn7 does not insert into DNAs that already contain Tn7 (4, 52). Thus, DNAs containing Tn7 are said to be immune to insertion by a second Tn7. Transposition immunity is *cis*-acting, that is, it is effective on DNA segments that contain Tn7, rather than being a global inhibitor of transposition. The feature of Tn7 in a target DNA that renders that target immune is the presence of binding sites for TnsB, (i.e., the ends of Tn7). Transposition immunity was originally observed with Tn*3* (64) and has been studied extensively with bacteriophage Mu (1, 2, 70, 77, 94).

Over what distances is Tn7 target immunity effective; i.e., when Tn7 is present in a potential target DNA, over what distance in that DNA will the frequency of insertion of a second Tn7 be decreased? We have used the fact that Tn7 inserts into *attTn7* to probe this question in vivo (30). We constructed a series of bacterial chromosomes containing nonmobile Tn7 derivatives at different distances from *attTn7* and then examined the frequency of transposition of a mobile Tn7 from a plasmid to the chromosomal *attTn7* site (Fig. 27). We found that the presence of the nonmobile Tn7 elements in the chromosome decreased the frequency of Tn7 insertion into *attTn7,* (i.e., imposed immunity on *attTn7*). Moreover, the degree of inhibition was inversely related to the distance of the nonmobile Tn7 to *attTn7*. For example, when the ends of Tn7 are located about 8 kb away from an *attTn7* target, a very strong inhibition of insertion into *attTn7* is observed. By contrast, as the Tn7 ends are moved to positions more distal from *attTn7* in the chromosome, less inhibition of insertion into *attTn7* is observed. Even when the nonmobile Tn7 ends and *attTn7* are far apart, >175 kb, some inhibition of insertion is still seen, indicating that chromosomal positions even at these large distances are capable of interacting. In control experiments, a nonmobile copy of Tn7 located at the other side of the bacterial chromosome from *attTn7,* several megabases away, had no effect on *attTn7* insertion, emphasizing the *cis*-acting nature of transposition immunity.

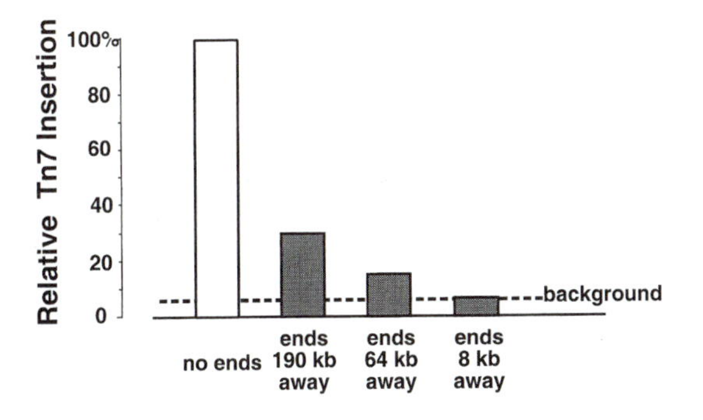

Figure 27. Transposition immunity in the *E. coli* chromosome. The mean fraction of Tn7 insertion from F::Tn7 into *attTn7* in chromosomes containing immobile mini-Tn7 ends at different distances from *attTn7* is expressed relative to Tn7 insertion into chromosomes without a mini-Tn7 element (100% = 2.3% Tn7 occupancy) as assayed by Southern blotting. The contribution of background hybridization to the Tn7 insertion signal in cells without a mini-Tn7 element was determined from assays of cells lacking Tn7 and is displayed as a dashed line. Reprinted from reference 30 with permission.

Long-range interactions in the bacterial chromosome have also been detected with use of other site-specific recombination systems (122).

Transposition immunity can also be observed in the purified Tn7 transposition system (8). Tn7 does not insert into a DNA that already contains a Tn7 end but does insert into a DNA lacking a Tn7 end. The ability to observe immunity in vitro has allowed investigation of its mechanism. The presence of the ends of Tn7, in particular the presence of TnsB binding sites, in a target DNA, is the feature that renders that DNA immune to further Tn7 insertion. The finding that it is the ends of Tn7 that impose transposition immunity is consistent with the finding that, for target immunity to be established, TnsB, the end binding protein, must be incubated with the targeting proteins and potential target DNAs (106). The central role of TnsB in imposing transposition immunity is also supported because we have isolated TnsB mutants that are altered in their ability to impose transposition immunity (Skelding and Craig, unpublished).

We have found that the establishment of immunity on a target DNA containing Tn7 ends is correlated with the loss of TnsC, a key target-selecting protein, from that DNA (106). Moreover, this loss of TnsC from an immune target requires that ATP hydrolysis by TnsC be possible; insertions into a target already containing a copy of Tn7 do occur in the presence of a nonhydrolyzable analogue of ATP (8, 106).

Previous work with the bacteriophage Mu system also established that transposition immunity in this system involves the redistribution of MuB, the ATP-utilizing analogue of TnsC, from a target containing a Mu end to a target without a Mu end (1, 2).

How do the many potential target site on a target DNA communicate with a Tn7 end at some distal position in that target to result in transposition immunity (i.e., to block insertions in that DNA at many positions)? We have found that when a plasmid containing a Tn7 end is catenated to a plasmid containing a Tn7 target site, target immunity is imposed on the catenated target plasmid; that is, no insertions are observed even though there is no covalent connection between the two plasmids (106). This result suggests that the Tn7 end and the potential target site (i.e., a place of TnsC interaction with the target DNA) encounter each other in solution by random collision rather than by some signal that is propagated through DNA. This model would also account for the apparent distance effect of transposition immunity; that is, sites that are closer to each other have stronger effects, whereas sites that are more distal show less immunity. Thus, a target containing Tn7 ends is immune because

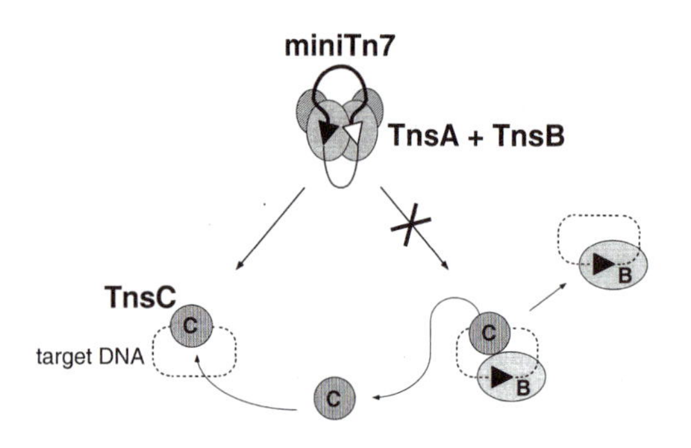

Figure 28. A model for Tn7 transposition immunity. Tn7 insertion into a target molecule results from the interaction of the TnsAB transposase bound to the Tn7 element with TnsC on an appropriate target DNA (or TnsCD-*attTn7* or TnsCE-target DNA). A target DNA containing a Tn7 end is immune to Tn7 insertion because TnsB interacts with that end and with TnsC to result in the removal of TnsC from the target DNA. Reprinted from reference 106 with permission.

the presence of TnsB binding sites on that DNA results in a high local concentration of TnsB, which then likely interacts with TnsC to promote the removal of TnsC from that target DNA (Fig. 28).

Similar experiments with the bacteriophage Mu system also established that the interactions which establish immunity occur in solution rather than being propagated through DNA (1, 2).

A striking feature of transposition immunity is the involvement of ATP in this process (8). We find that transposition immunity can be bypassed (i.e., insertions into Tn7-containing molecules do occur) in transposition reactions that contain a nonhydrolyzable ATP analogue rather than ATP. One reasonable explanation for this result is that ATP hydrolysis is related to the ability of Tn7 to bind DNA, binding being decreased in the presence of ADP compared with ATP (43). Thus, it may be that the interaction of TnsB with TnsC on a target DNA prompts TnsC ATP hydrolysis, thus removing TnsC from that potential target such that no insertions into that target occur (106, 108).

Transposition immunity is often considered and studied as the ability of an element in an intermolecular target DNA such as a plasmid to block the insertion of another copy of the element into that plasmid so that likely deleterious rearrangement of a plasmid containing two elements does not occur. Another function of transposition immunity is likely to block deleterious intramolecular rearrangements of a DNA containing a transposable element.

Tn7 RELATIVES

Tn7 is part of a family of bacterial transposable elements that encode both a transposase and an ATP-utilizing protein that controls the activity of the transposase and often plays a central role in target-site selection (Table 1). Tn7 and bacteriophage Mu (see chapter 17), which uses transposition for the establishment of lysogeny and for the replication of its DNA during lytic growth, are the best studied and understood members of this superfamily. Although there is little protein-protein homology between the Tn7 and Mu transposition proteins, there are considerable functional similarities. In both cases, a retroviral-like integrase executes breakage and joining at the 3′ ends of the mobile element and there is an ATP-utilizing protein that plays a key role in target selection. Tn7 is distinguished by its more elaborate transposase and its use of multiple target-selecting proteins to allow the use of different classes of target sites.

The Role of the Transposase in Determining the Nature of the Transposition Product

The members of the ATP-subunit superfamily of mobile elements vary in the constitution of their transposase and thus in the kind of transposition product they form. The product of Tn7 transposition is a simple insertion because transposition occurs by a cut-and-paste mechanism in which the transposon is disconnected from the donor site by double-strand breaks before its insertion into the target (7) (Fig. 11). Tn7 executes these double-strand breaks by having a heteromeric transposase; one subunit, TnsB, makes cleavages at the 3′ ends of the transposon, whereas the other subunit, TnsA, makes cleavages at the 5′ ends of the transposon (71, 102). We hypothesize that Tn7 relatives containing a TnsAB-like transposase will also move by a cut-and-paste mechanism.

Transposases that have only a TnsB-like subunit will make cointegrate products if they, like Mu, execute breakage at only the 3′ end of the transposon (Fig. 11); however, they may use the IS*10*/IS*50* strategy of hairpinning to make double-strand breaks (Fig. 8). Use of the double-strand break pathway will lead to simple insertions as the direct products of transposition, whereas a single-strand nicking pathway will lead to a fusion product, followed by replication to make a cointegrate. Some transposons also encode a site-specific resolution system so that, although simple insertions are observed, they result from the two-step process of transposition followed by conservative site-specific recombination (see chapter 14).

Table 1. Transposons with ATP-utilizing proteins

Transposon	Retroviral integrase transposase subunit	Restriction endonuclease transposase subunit	ATP-utilizing protein	Targeting protein	Targeting	Primary transposition product	Other determinants[a]
Tn7	TnsB	TnsA	TnsC	TnsD	*attTn7*	Simple insertion	Tp[r], Spec/St[r]
Tn7	TnsB	TnsA	TnsC	TnsE	Conjugal plasmids	Simple insertion	Tp[r], Spec/St[r]
Tn7	TnsB	TnsA[mut b]	TnsC	TnsD	*attTn7*	Cointegrate	Tp[r], Spec/St[r]
Tn7	TnsB	TnsA[mut]	TnsC	TnE	Conjugal plasmids	Cointegrate	Tp[r], Spec/St[r]
Tn*5468*	TnsB	TnsA	TnsC	TnsD	*attTn7*	ND	ND
Tn*5053*	TniA		TniB	TniQ?	*res* sites	Cointegrate[c]	Mer[r]
Tn*402*	TniA		TniB	TniQ?	*res* sites	Cointegrate[c]	Tp[r], Sul[r], qacE
Tn*552*	p480		p271	?	*res* sites	Cointegrate	Tc[r]
Mu	MuA		MuB	MuB	Many sites	Cointegrate/ simple insertion	Bacteriophage genes
IS*21*	IstA		IstB	?		Simple insertion	

[a]Tp, trimethoprim resistance; Spec/Strep, spectinomycin/streptomycin resistance; Mer, mercury resistance; Sul, sulfonamide resistance; qacE, multidrug exporter.
[b]TnsA[mut], TnsA with change in catalytic amino acid, for example, TnsA[D114A].
[c]+, primary transposition product before resolution by element-encoded resolvase.

Tn7 Relatives with a TnsAB Transposase

Two transposons called Tn*1825* and Tn*1826* have been described whose transposition functions are related to Tn7 as tested at the level of restriction mapping and complementation (116). These elements have not been analyzed at the DNA sequence level; thus, the exact relationship among the transposition functions of these elements remains to be determined. Tn*1825*, Tn*1826*, and Tn7 are related in the antibiotic resistance determinants they encode as described above.

A very close relative of Tn7, Tn*5468*, has recently been characterized during a sequence analysis of the *atp* operon region of *Thiobacillus ferrooxidans* (81) (Table 1). The identification of this close Tn7 relative in *T. ferrooxidans* is somewhat surprising because of the very different biological niches occupied by enteric bacteria and this acidophilic, obligately chemolithoautotrophic organism.

In both *E. coli* and *T. ferrooxidans*, the *glmU* and *glmS* genes are downstream of the ATP operon, and Tn7 and Tn*5468* lie in similar chromosomal locations, both being downstream of *glmS* genes. The right ends of Tn7 and Tn*5468* which are adjacent to *glmS* are similar in organization and sequence. Like the right end of Tn7, Tn*5468* contains four closely juxtaposed repeated sequences that are likely binding sites for the Tn*5468* TnsB.

Tn*5468* contains four ORFs clearly corresponding to TnsA, TnsB, TnsC, and TnsD. Although there is an ORF just beyond Tn*5468* TnsD, there is little similarity between it and Tn7 TnsE; thus, it is not clear whether this ORF is a functional part of Tn*5468*. This ORF of unknown function is closely juxtaposed to what is very likely a chromosomal *recG* gene involved in homologous recombination rather than the the *pstS* gene encoding a phosphate transport protein that is downstream of *attTn7* in *E. coli*. The other features of Tn7, several antibiotic resistance determinants and a left end, are not identifiable in Tn*5468*.

The partial sequence of another TnsA gene distinct from the others described above has also been described (55), suggesting that more close relatives of Tn7 remain to be discovered.

Tn7 Relatives with a TnsB-Like Transposase

Two-protein systems

Both bacteriophage Mu and Tn*552* (99, 100) have a TnsB-like transposase and ATP-utilizing protein. These elements generate cointegrate transposition products, although Tn*552* also encodes a resolution system such that simple insertions are observed as recombination products. Tns*552* targets *res* sites, the sites used by conservative site-specific recombination systems to resolve cointegrate transposition products for insertion, but it has no obvious candidate for a specialized targeting protein. Sequence inspection also suggests that IS*21* has a transposase and an ATP-using protein (93).

Three-protein systems

Two other relatively close relatives to Tn*7* are Tn*5053* (59) and Tn*402* (also called Tn*5090*) (89, 103). These elements contain an apparently single polypeptide transposase whose catalytic domain is similar to that of Tn*7* TnsB and a TnsC-like subunit. These elements make cointegrate transposition products. These elements also encode a third protein, TniQ, required for transposition, although the role of this protein has yet to be determined.

An interesting aspect of these elements is that they, like Tn*7* and Tn*552*, have distinctive target-site selection patterns. Both of these elements insert preferentially around *res* sites (56, 74). How this targeting is imposed is not known.

Tn*5053* encodes mercury resistance determinants in addition to its transposition functions, whereas Tn*402* (Tn*5090*) encodes antibiotic resistance determinants. In both cases, these determinants are part of an integron (112) (see chapter 9), a cluster of genes such as antibiotic resistance or heavy metal resistance cassettes whose members can be exchanged by site-specific recombination promoted by an integron-encoded integrase. As noted above, the antibiotic resistance determinants of Tn*7* are also part of an integron, but the Tn*7* integrase is defective because of a chain-terminating mutation.

TRICKS WITH Tn*7*

Transposons have long been useful tools for the manipulation of genes and genomes. We and others have developed several systems for using Tn*7* for the manipulation of genes and genomes.

Exploiting a Tn*7* System That Has Low Target-Site Selectivity

Transposons have been used in many organisms for insertional mutagenesis. For example, the *tns* genes were defined by in vitro transposon mutagenesis of a plasmid containing the *tns* genes and then examining the Tn*7* transposition phenotypes of these insertion mutants (120). Tn*7* itself, however, is generally ill suited for use as an insertional mutagen because of its target-site selectivity. However, the TnsABC^{A225V} system, which uses a TnsC gain-of-function mutant, displays high-efficiency transposition with relatively little target-site selectivity (19). Thus, this Tn*7* system is well-suited for making insertion mutants or introduction of a marker that can be used as a priming site for DNA sequencing to avoid tedious sequencing methods such as subcloning or primer walking (33, 81).

An in vitro Tn*7* system for insertional mutagenesis has been developed in which a mini-Tn*7* element containing convenient restriction sites for introduction of selectable markers is present on a conditionally replicating plasmid that cannot replicate in strains that lack a specific initiator protein for a specialized DNA replication system (19). Thus, after in vitro mutagenesis, only target plasmids containing a mini-Tn*7* insertion will be recovered. Tn*7* derivatives with short-end sequences that are still efficient in transposition have also been constructed which allows the generation of translational fusions.

The efficiency of Tn*7* TnsABC^{A225V} transposition is sufficiently high that in vitro mutagenesis of genomic DNA can be followed by the recovery of insertions introduced into a host genome by homologous recombination (48) (N. Bachman, M. Biery, J. Boeke, and N. L. Craig, unpublished data). This strategy will likely be a powerful tool for organisms that lack well-defined genetic systems and tamed transposons.

Another method to use insertional mutagenesis is at the gene level. We have developed a version of Tn*7* that, after Tn*7* insertion and excision of most of the transposon by restriction digestion and religation, results in the introduction of five amino acid segments into a target protein (19). Such derivatives should be useful in protein and nucleic acid structure-function analysis.

Exploiting Site-Specific Insertion by Tn*7*

In many cases, it is desirable to evaluate the expression of multiple gene variants within the same chromosomal context. Introduction of test genes into *attTn7* provides an easy method for the reproducible introduction of the gene to be evaluated into a specific site in the chromosome (9, 10). Tn*7* vectors that contain mini-Tn*7* elements with restriction sites for easy cloning are available, as are plasmids encoding TnsABC+D. The frequency of Tn*7* insertion into *attTn7* is sufficiently high that bacteria in which transposition into the *attTn7* site has occurred can be isolated by examination of a few bacterial colonies. Moreover, the *tns* genes can be easily removed by putting them on unstable or conditional replicons.

Another use of site-specific insertion by Tn*7* transposition is to transfer a DNA segment from one replicon to another. This method has been exploited by cloning of the gene of interest into a mini-Tn7 element that then transposes to *attTn7* on a baculovirus vector (65, 69), on an artificial bacterial chromosome vector (42), or on an adenovirus vector (96).

CONCLUDING REMARKS

The study of Tn7 has enriched the study of mobile DNA; there are distinct differences between Tn7 and other elements as well as similarities between Tn7 and other elements.

Tn7 moves by a distinctive cut-and-paste mechanism in which two different polypeptides of the transposase mediate the breakage and joining reactions of transposition. There are few other examples of such heteromeric DNA-processing enzymes. Understanding the interplay between TnsA and TnsB will be facilitated by the ability to modify each of these proteins independently. An interesting question is whether there might be physiological conditions under which 5′ end cleavage by TnsA is modulated to switch between simple insertion and cointegrate transposition products.

A very interesting feature of Tn7 is its ability to evaluate and respond to multiple features of a target DNA. Tn7 transposes at high frequency to *attTn7,* at a lower frequency but preferentially to conjugating plasmids, and avoids DNAs already containing Tn7. The capacity to recognize these different target features provides Tn7 with alternative lifestyles to promote its propagation. Insertion into *attTn7* is nondeleterious to its host and thus provides Tn7 with a "safe" site within the cell. Alternatively, the fact that Tn7 can preferentially transpose to conjugating plasmids provides a means for Tn7 to be dispersed through bacterial populations.

TnsC seems to act as the sensor, integrator, and effector of target signals; it interacts with the target DNA and target proteins and can then interact with the transposase to promote transposition. These roles involve the ability of TnsC to bind and hydrolyze ATP and its ability to bind DNA in the presence of ATP. Our working model is that the recruitment of TnsC to an appropriate target site, for example, TnsD bound to *attTn7,* puts TnsC in an ATP state in which it activates transposition. It will be interesting to see if this same paradigm is indeed followed in the TnsABC + E pathway. We also suggest that the ability of Tn7 to avoid certain targets (i.e., those that are immune because they already contain Tn7) involves the removal of TnsC from such targets by interaction of TnsB bound to the Tn7 ends with TnsC and the ATP hydrolysis activity of TnsC. In this model, other components of the Tn7 transposition machinery modulate the activity of a nucleotide-dependent switch protein in a manner reminiscent of how the activity of G proteins are modulated (21, 108). A molecular understanding of this TnsC switch is an important topic for future research.

The isolation of TnsC gain-of-function mutants that allow transposition in the absence of TnsD and TnsE is particularly interesting. Notably, single amino acid changes in TnsC can convert Tn7 from an element dependent on TnsD and TnsE to one that is independent of target selectors. Understanding the molecular basis of these gain-of-function derivatives is likely to provide deep insights into the control of Tn7 transposition. Given that a single amino acid change can convert a TnsABCDE transposon to a TnsABC transposon emphasizes the utility of Tn7's multiple target pathways. It will be interesting to see if TnsABC-like elements exist in nature.

Another interesting aspect of Tn7 biology is the degree to which the structure of the target DNA plays such an intimate role in transposition. A particular target DNA structure (a pyrimidine triplex) can recruit the TnsABC transposition machinery to promote insertion at the triplex region. This system offers the possibility of targeting Tn7 insertion to sites chosen by an investigator rather than those selected naturally by Tn7. The modular nature of the transposition machinery may also be helpful in using other strategies to target Tn7 insertion to particular locations, for example by fusing a specific DNA binding domain to TnsC or a heterologous domain to TnsD. Tn7's strategy of using a structural DNA distortion imposed on *attTn7* by TnsD as the recruitment signal for TnsC also ensures that recombination will occur at *attTn7* rather than in solution.

The study of Tn7 seems sure to continue to provide an interesting system in which to examine the interaction of a transposable element with its host, to provide interesting problems in protein-DNA and protein-protein interactions to study, and to provide useful tools for the study and manipulation of genes and genomes.

REFERENCES

1. **Adzuma, K., and K. Mizuuchi.** 1989. Interaction of proteins located at a distance along DNA: mechanism of target immunity in the Mu DNA strand-transfer reaction. *Cell* 57:41–47.
2. **Adzuma, K., and K. Mizuuchi.** 1988. Target immunity of Mu transposition reflects a differential distribution of MuB protein. *Cell* 53:257–266.
3. **Arciszewska, L. K., and N. L. Craig.** 1991. Interaction of the Tn7-encoded transposition protein TnsB with the ends of the transposon. *Nucleic Acids Res.* **19**:5021–5029.
4. **Arciszewska, L. K., D. Drake, and N. L. Craig.** 1989. Transposon Tn7 *cis*-acting sequences in transposition and transposition immunity. *J. Mol. Biol.* **207**:35–52.

5. **Arciszewska, L. K., R. L. McKown, and N. L. Craig.** 1991. Purification of TnsB, a transposition protein that binds to the ends of Tn7. *J. Biol. Chem.* **266:**21736–21744.
6. **Bainton, R.** 1992. Ph.D. thesis. University of California, San Francisco.
7. **Bainton, R., P. Gamas, and N. L. Craig.** 1991. Tn7 transposition in vitro proceeds through an excised transposon intermediate generated by staggered breaks in DNA. *Cell* **65:** 805–816.
8. **Bainton, R. J., K. M. Kubo, J.-N. Feng, and N. L. Craig.** 1993. Tn7 transposition: target DNA recognition is mediated by multiple Tn7-encoded proteins in a purified in vitro system. *Cell* **72:**931–943.
9. **Bao, Y., D. P. Lies, H. Fu, and G. P. Roberts.** 1991. An improved Tn7-based system for the single-copy insertion of cloned genes into chromosomes of gram-negative bacteria. *Gene* **109:**167–168.
10. **Barry, G.** 1986. Permanent insertion of foreign genes into the chromosome of soil bacteria. *BioTechniques* **4:**446–449.
11. **Barth, P., and N. Datta.** 1977. Two naturally occurring transposons indistinguishable from Tn7. *J. Bacteriol.* **102:** 129–134.
12. **Barth, P., and N. Grinter.** 1977. Map of plasmid RP4 derived by insertion of transposon C. *J. Mol. Biol.* **113:**455–474.
13. **Barth, P. T., N. Datta, R. W. Hedges, and N. J. Grinter.** 1976. Transposition of a deoxyribonucleic acid sequence encoding trimethoprim and streptomycin resistances from R483 to other replicons. *J. Bacteriol.* **125:**800–810.
14. **Barth, P. T., N. J. Grinter, and D. F. Bradley.** 1978. Conjugal transfer system of plasmid RP4: analysis of transposon 7 insertion. *J. Bacteriol.* **133:**43–52.
15. **Bender, J., and N. Kleckner.** 1986. Genetic evidence that Tn10 transposes by a nonreplicative mechanism. *Cell* **45:** 801–815.
16. **Bender, J., and N. Kleckner.** 1992. IS10 transposase mutations that specifically alter target site recognition. *EMBO J.* **11:**741–750.
17. **Bhasin, A., I. Y. Goryshin, and W. S. Reznikoff.** 1999. Hairpin formation in Tn5 transposition. *J. Biol. Chem.* **274:** 37021–37029.
18. **Biery, M., M. Lopata, and N. L. Craig.** 2000. A minimal system for Tn7 transposition: the transposon-encoded proteins TnsA and TnsB can execute DNA breakage and joining reactions that generate circularized Tn7 species. *J. Mol. Biol.* **297:**25–37.
19. **Biery, M. C., F. Steward, A. E. Stellwagen, E. A. Raleigh, and N. L. Craig.** 2000. A simple *in vitro* Tn7-based transposition system with low target site selectivity for genome and gene analysis. *Nucleic Acids Res.* **28:**1067–1077.
20. **Bolland, S., and N. Kleckner.** 1996. The three chemical steps of Tn10/IS10 transposition involve repeated utilization of a single active site. *Cell* **84:**223–233.
21. **Bourne, H. R., D. A. Sanders, and F. McCormick.** 1990. The GTPase superfamily: a conserved switch for diverse cell functions. *Nature* **348:**125–132.
22. **Bussiere, D. E., and D. Bastia.** 1999. Termination of DNA replication of bacterial and plasmid chromosomes. *Mol. Microbiol.* **31:**1611–1618.
23. **Cox, M. M., M. F. Goodman, K. N. Kreuzer, D. J. Sherratt, S. J. Sandler, and K. J. Marians.** 2000. The importance of repairing stalled replication forks. *Nature* **404:**37–41.
24. **Craig, N. L.** 1991. Tn7: a target site-specific transposon. *Mol. Microbiol.* **5:**2569–2573.
25. **Craig, N. L.** 1989. Transposon Tn7, p. 211–225. *In* D. E. Berg and M. M. Howe (ed.), *Mobile DNA*. American Society for Microbiology, Washington, D.C.
26. **Craig, N. L.** 1996. Transposon Tn7. *Curr. Top. Microbiol. Immunol.* **204:**27–48.
27. **Craig, N. L.** 1995. Unity in transposition reactions. *Science* **270:**253–254.
28. **Craigie, R., D. J. Arndt-Jovin, and K. Mizuuchi.** 1985. A defined system for the DNA strand-transfer reaction at the initiation of bacteriophage Mu transposition: protein and DNA substrate requirements. *Proc. Natl. Acad. Sci. USA* **82:** 7570–7574.
29. **Davies, D. R., I. Y. Goryshin, W. S. Reznikoff, and I. Rayment.** 2000. Three-dimensional structure of the Tn5 synaptic complex transposition intermediate. *Science* **289:**77–85.
30. **DeBoy, R., and N. L. Craig.** 1996. Tn7 transposition as a probe of *cis* interactions between widely separated (190 Kilobases Apart) DNA sites in the *Escherichia coli* chromosome. *J. Bacteriol.* **178:**6184–6191.
31. **DeBoy, R. T.** 1997. Ph.D. thesis. Johns Hopkins University, Baltimore, Md.
32. **DeBoy, R. T., and N. L. Craig.** 2000. Target site selection by Tn7: *attTn7* transcription and target activity. *J. Bacteriol.* **182:**3310–3313.
33. **Devine, S. E., and J. D. Boeke.** 1994. Efficient integration of artificial transposons into plasmid targets *in vitro:* a useful tool for DNA mapping, sequencing and genetic analysis. *Nucleic Acids Res.* **22:**3765–3772.
34. **Dodson, K. W., and D. E. Berg.** 1991. A sequence at the inside end of IS50 down regulates transposition. *Plasmid* **25:** 145–148.
35. **Dyda, F., A. B. Hickman, T. M. Jenkins, A. Engelman, R. Craigie, and D. R. Davies.** 1994. Crystal structure of the catalytic domain of HIV-1 integrase: similarity to other polynucleotidyl transferases. *Science* **266:**1981–1986.
36. **Engelman, A., and R. Craigie.** 1992. Identification of conserved amino acid residues critical for human immunodeficiency virus type 1 integrase function *in vitro*. *J. Virol.* **66:** 6361–6369.
37. **Engelman, A., K. Mizuuchi, and R. Craigie.** 1991. HIV-1 DNA integration: mechanism of viral DNA cleavage and DNA strand transfer. *Cell* **67:**1211–1221.
38. **Engels, W. R., D. M. Johnson-Schlitz, W. B. Eggleston, and J. Sved.** 1990. High-frequency P element loss in Drosophila is homolog dependent. *Cell* **62:**515–525.
39. **Fling, M., J. Kopf, and C. Richards.** 1985. Nucleotide sequence of the transposon Tn7 gene encoding an aminoglycoside-modifying enzyme, 3′(9)-O-nucleotidyltransferase. *Nucleic Acids Res.* **13:**7095–7106.
40. **Fling, M., and C. Richards.** 1983. The nucleotide sequence of the trimethoprim-resistant dihydrofolate reductase gene harbored by Tn7. *Nucleic Acids Res.* **11:**5147–5158.
41. **Flores, C., M. I. Qadri, and C. Lichtenstein.** 1990. DNA sequence analysis of five genes, *tnsA, B, C, D* and *E,* required for Tn7 transposition. *Nucleic Acids Res.* **18:**901–911.
42. **Frengen, E., D. Weichenhan, B. Zhao, K. Osoegawa, M. van Geel, and P. J. de Jong.** 1999. A modular, positive selection bacterial artificial chromosome vector with multiple cloning sites. *Genomics* **58:**250–253.
43. **Gamas, P., and N. L. Craig.** 1992. Purification and characterization of TnsC, a Tn7 transposition protein that binds ATP and DNA. *Nucleic Acids Res.* **20:**2525–2532.
44. **Gary, P. A., M. C. Biery, R. J. Bainton, and N. L. Craig.** 1996. Multiple DNA processing reactions underlie Tn7 transposition. *J. Mol. Biol.* **257:**301–316.
45. Reference deleted.
46. **Gay, N. J., V. L. Tybulewicz, and J. E. Walker.** 1986. Insertion of transposon Tn7 into the *Escherichia coli glmS* transcriptional terminator. *Biochem. J.* **234:**111–117.

47. **Gringauz, E., K. Orle, A. Orle, C. S. Waddell, and N. L. Craig.** 1988. Recognition of *Escherichia coli attTn7* by transposon Tn7: lack of specific sequence requirements at the point of Tn7 insertion. *J. Bacteriol.* **170:**2832–2840.
48. **Gwinn, M. L., A. E. Stellwagen, N. L. Craig, J.-F. Tomb, and H. O. Smith.** 1997. In vitro Tn7 mutagenesis of *Haemophilus influenzae* Rd and characterization of the role of *atpA* in transformation. *J. Bacteriol.* **179:**7315–7320.
49. **Hagemann, A. T., and N. L. Craig.** 1993. Tn7 transposition creates a hotspot for homologous recombination at the transposon donor site. *Genetics* **133:**9–16.
50. **Hall, R. M., D. E. Brookes, and H. W. Stokes.** 1991. Site-specific insertion of genes into integrons: role of the 59-base element and determination of the recombination cross-over point. *Mol. Microbiol.* **5:**1941–1959.
51. **Harshey, R. M.** 1984. Transposition without duplication of infecting bacteriophage Mu DNA. *Nature* **311:**580–581.
52. **Hauer, B., and J. A. Shapiro.** 1984. Control of Tn7 transposition. *Mol. Gen. Genet.* **194:**149–158.
53. **Hickman, A. B., L. Li, S. V. Mathew, E. W. May, N. L. Craig, and F. Dyda.** 2000. Unexpected structural diversity in DNA recombination: the restriction endonuclease connection. *Mol. Cell* **5:**1025–1034.
54. **Hughes, O.** 1993. Ph.D. thesis. University of California, San Francisco.
55. **Inoue, C., K. Sugawara, and T. Kusano.** 1991. The *merR* regulatory gene in *Thiobacillus ferrooxidans* is spaced apart from the *mer* structural genes. *Mol. Microbiol.* **5:**2707–2718.
56. **Kamali-Moghaddam, M., and L. Sundstrom.** 2000. Transposon targeting determined by resolvase. *FEMS Microbiol. Lett.* **186:**55–59.
57. **Kennedy, A., A. Guhathakurta, N. Kleckner, and D. B. Haniford.** 1998. Tn10 transposition via a DNA hairpin intermediate. *Cell* **95:**125–134.
58. **Kennedy, A. K., D. B. Haniford, and K. Mizuuchi.** 2000. Single active site catalysis of the successive phosphoryl transfer steps by DNA transposases: insights from phosphorothioate stereoselectivity. *Cell* **101:**295–305.
59. **Kholodii, G. Y., S. Z. Mindlin, I. A. Bass, O. V. Yurieva, S. V. Minakhina, and V. G. Nikiforov.** 1995. Four genes, two ends, and a *res* region are involved in transposition of Tn5053: a paradigm for a novel family of transposons carrying either a *mer* operon or an integron. *Mol. Microbiol.* **17:** 1189–1200.
60. **Kubo, K.** 1992. Ph.D. thesis. University of California, San Francisco.
61. **Kubo, K. M., and N. L. Craig.** 1990. Bacterial transposon Tn7 utilizes two classes of target sites. *J. Bacteriol.* **172:** 2774–2778.
62. **Kuduvalli, P., J. Rao, and N. L. Craig.** 2001. Target DNA structure plays a critical role in Tn7 transposition. *EMBO J.* **20:**924–932.
63. **Kulkosky, J., K. S. Jones, R. S. Katz, J. P. G. Mack, and A. M. Skalka.** 1992. Residues critical for retroviral integrative recombination in a region that is highly conserved among retroviral/retrotransposon integrases and bacterial insertion sequence transposases. *Mol. Cell. Biol.* **12:**2331–2338.
64. **Lee, C.-H., A. Bhagwhat, and F. Heffron.** 1983. Identification of a transposon Tn3 sequence required for transposition immunity. *Proc. Natl. Acad. Sci. USA* **80:**6765–6769.
65. **Leusch, M. S., S. C. Lee, and P. O. Olins.** 1995. A novel host-vector system for direct selection of recombinant baculoviruses (bacmids) in *Escherichia coli. Gene* **160:**191–194.
66. **Lichtenstein, C., and S. Brenner.** 1981. Site-specific properties of Tn7 transposition into the E. coli chromosome. *Mol. Gen. Genet.* **183:**380–387.
67. **Lichtenstein, C., and S. Brenner.** 1982. Unique insertion site of Tn7 in *E. coli* chromosome. *Nature* **297:**601–603.
68. **Lu, F., and N. L. Craig.** 2000. Isolation and characterization of Tn7 transposase gain-of-function mutants: a model for transposase activation. *EMBO J.* **19:**3446–3457.
69. **Luckow, V. A., S. Lee, G. F. Barry, and P. O. Olins.** 1993. Efficient generation of infectious recombinant baculoviruses by site-specific transposon-mediated insertion of foreign genes into a baculovirus genome propagated in *Escherichia coli. J. Virol.* **67:**4566–4579.
70. **Manna, D., and N. P. Higgins.** 1999. Phage Mu transposition immunity reflects supercoil domain structure of the chromosome. *Mol. Microbiol.* **32:**595–606.
71. **May, E. W., and N. L. Craig.** 1996. Switching from cut-and-paste to replicative Tn7 transposition. *Science* **272:**401–404.
72. **McKown, R. L., K. A. Orle, T. Chen, and N. L. Craig.** 1988. Sequence requirements of *Escherichia coli attTn7,* a specific site of transposon Tn7 insertion. *J. Bacteriol.* **170:**352–358.
73. **McKown, R. L., C. S. Waddell, L. A. Arciszewska, and N. L. Craig.** 1987. Identification of a transposon Tn7-dependent DNA-binding activity that recognizes the ends of Tn7. *Proc. Natl. Acad. Sci. USA* **84:**7807–7811.
74. **Minakhina, S., G. Kholodii, S. Mindlin, O. Yurieva, and V. Nikiforov.** 1999. Tn5053 family transposons are *res* site hunters sensing plasmidal res sites occupied by cognate resolvases. *Mol. Microbiol.* **33:**1059–1068.
75. **Mizuuchi, K.** 1997. Polynucleotidyl transfer reactions in site-specific DNA recombination. *Genes Cells* **2:**1–12.
76. **Mizuuchi, K.** 1992. Polynucleotidyl transfer reactions in transpositional DNA recombination. *J. Biol. Chem.* **267:** 21273–21276.
77. **Mizuuchi, K.** 1992. Transpositional recombination: mechanistic insights from studies of Mu and other elements. *Annu. Rev. Biochem.* **61:**1011–1051.
78. **Mizuuchi, K., and K. Adzuma.** 1991. Inversion of the phosphate chirality at the target site of the Mu DNA strand transfer: evidence for a one-step transesterification mechanism. *Cell* **66:**129–140.
79. **Mizuuchi, M., T. A. Baker, and K. M. Mizuuchi.** 1995. Assembly of phage Mu transpososomes: cooperative transitions assisted by protein and DNA scaffolds. *Cell* **83:**375–385.
80. **Naigamwalla, D. Z., and G. Chaconas.** 1997. A new set of Mu DNA transposition intermediates: alternate pathways of target capture preceding strand transfer. *EMBO J.* **16:** 5227–5234.
81. **Oppon, J. C., R. J. Sarnovsky, N. L. Craig, and D. E. Rawlings.** 1998. A Tn7-like transposon is present in the *glmUS* region of the obligately hemoautolithotrophic bacterium *Thiobacillus ferrooxidans. J. Bacteriol.* **180:**3007–3012.
82. **Orle, K. A., and N. L. Craig.** 1991. Identification of transposition proteins encoded by the bacterial transposon Tn7. *Gene* **104:**125–131.
83. **Peters, J. E., and N. L. Craig.** 2001. Tn7 recognizes target structures associated with DNA replication using the DNA binding protein TnsE. *Genes Dev.* **15:**737–747.
84. **Peters, J. E., and N. L. Craig.** 2000. Tn7 transposes proximal to DNA double-strand breaks and into regions where chromosomal DNA replication terminates. *Mol. Cell* **6:**573–582.
85. **Plasterk, R. H. A., and J. T. M. Groenen.** 1992. Targeted alterations of the *Caenorhabditis elegans* genome by transgene-instructed DNA double strand break repair following Tc1 excision. *EMBO J.* **11:**287–290.
86. **Plumbridge, J.** 1995. Co-ordinated regulation of amino sugar biosynthesis and degradation: the MagC repressor acts as both an activator and a repressor for the transcription of

the *glmUS* operon and requires two separated NagC binding sites. *EMBO J.* **14:**3958–3965.

87. **Polard, P., and M. Chandler.** 1995. Bacterial transposases and retroviral integrases. *Mol. Microbiol.* **15:**1–23.
88. **Polard, P., and M. Chandler.** 1995. An *in vivo* transposase-catalyzed single-stranded DNA circularization reaction. *Genes Dev.* **9:**2846–2858.
89. **Radstrom, P., O. Skold, G. Swedberg, F. Flensburg, P. H. Roy, and L. Sundstrom.** 1994. Transposon Tn*5090* of plasmid R751, which carries an integron, is related to Tn7, Mu, and the retroelements. *J. Bacteriol.* **176:**3257–3268.
90. **Rao, J.** 2001. Ph.D. thesis. Johns Hopkins University, Baltimore, Md.
91. **Rao, J. E., and N. L. Craig.** 2001. Selective recognition of pyrimidine motif triplexes by a protein encoded by the bacterial transposon Tn7. *J. Mol. Biol.* **307:**1161–1170.
92. **Rao, J. E., P. S. Miller, and N. L. Craig.** 2000. Recognition of triple-helical DNA structures by transposon Tn7. *Proc. Natl. Acad. Sci. USA* **97:**3936–3941.
93. **Reimmann, C., R. Moore, S. Little, A. Savioz, N. S. Willetts, and D. Haas.** 1989. Genetic structure, function and regulation of the transposable element IS21. *Mol. Gen. Genet.* **215:** 416–424.
94. **Reyes, I., A. Beyou, C. Mignotte-Vieux, and F. Richaud.** 1987. Mini-mu transduction: cis-inhibition of the insertion of mud transposons. *Plasmid* **18:**183–192.
95. **Rice, P., and K. Mizuuchi.** 1995. Structure of the bacteriophage Mu transposase core: a common structural motif for DNA transposition and retroviral integration. *Cell* **82:** 209–220.
96. **Richards, C. A., C. E. Brown, J. P. Cogswell, and M. P. Weiner.** 2000. The admid system: generation of recombinant adenoviruses by Tn7-mediated transposition in *E. coli. BioTechniques* **29:**146–154.
97. **Roberts, D., B. C. Hoopes, W. R. McClure, and N. Kleckner.** 1985. IS10 transposition is regulated by DNA adenine methylation. *Cell* **43:**117–130.
98. **Rogers, M., N. Ekaterinaki, E. Nimmo, and D. Sherratt.** 1986. Analysis of Tn7 transposition. *Mol. Gen. Genet.* **205:** 550–556.
99. **Rowland, S.-J., and K. G. H. Dyke.** 1989. Characterization of the staphylococcal β-lactamase transposon Tn552. *EMBO J.* **8:**2761–2773.
100. **Rowland, S.-J., and K. G. H. Dyke.** 1990. Tn552, a novel transposable element from *Staphylococcus aureus. Mol. Microbiol.* **4:**961–975.
101. **Sakai, J., and N. Kleckner.** 1997. The Tn10 synaptic complex can capture a target DNA only after transposon excision. *Cell* **89:**205–214.
102. **Sarnovsky, R., E. W. May, and N. L. Craig.** 1996. The Tn7 transposase is a heteromeric complex in which DNA breakage and joining activities are distributed between different gene products. *EMBO J.* **15:**6348–6361.
103. **Shapiro, J. A., and P. Sporn.** 1977. Tn402: a new transposable element determining trimethoprim resistance that inserts in bacteriophage lambda. *J. Bacteriol.* **129:**1632–1635.
104. **Sharpe, P., and N. L. Craig.** 1998. Host proteins can stimulate Tn7 transposition: a novel role for the ribosomal protein L29 and the acyl carrier protein. *EMBO J.* **17:**5822–5831.
105. **Simonsen, C., E. Chen, and A. Levinson.** 1983. Identification of the type I trimethoprim-resistant dihydrofolate reductase specified by the *Escherichia coli* R-plasmid 483: comparison with procaryotic and eucaryotic dihydrofolate reductases. *J. Bacteriol.* **155:**1001–1008.
106. **Stellwagen, A., and N. L. Craig.** 1997. Avoiding self: two Tn7-encoded proteins mediate target immunity in Tn7 transposition. *EMBO J.* **16:**6823–6834.
107. **Stellwagen, A., and N. L. Craig.** 1997. Gain-of-function mutations in TnsC, an ATP-dependent transposition protein which activates the bacterial transposon Tn7. *Genetics* **145:** 573–585.
108. **Stellwagen, A., and N. L. Craig.** 1998. Mobile DNA elements: controlling transposition with ATP-dependent molecular switches. *Trends Biochem. Sci.* **23:**486–490.
109. **Stellwagen, A. E.** 2001. Analysis of gain of function mutants of an ATP-dependent regulator of Tn7 transposition. *J. Mol. Biol.* **305:**633–642.
110. **Stellwagen, A. E.** 1997. Ph.D. thesis. University of California, San Francisco.
111. **Stellwagen, A. E., and N. L. Craig.** 2001. Analysis of gain of function mutants of an ATP-dependent regulator of Tn7 transposition. *J. Mol. Biol.* **305:**633–642.
112. **Stokes, H. W.** 1989. A novel family of potentially mobile DNA elements encoding site-specific gene-integration functions: integrons. *Mol. Microbiol.* **3:**1669–1683.
113. **Sundstrom, L., P. H. Roy, and O. Skold.** 1991. Site-specific insertion of three structural gene cassettes in transposon Tn7. *J. Bacteriol.* **173:**3025–3028.
114. **Tang, Y., C. Lichtenstein, and S. Cotterill.** 1991. Purification and characterization of the TnsB protein of Tn7, a transposition protein that binds to the ends of Tn7. *Nucleic Acids Res.* **19:**3395–3402.
115. **Tietze, E., and J. Brevet.** 1991. The trimethoprim resistance transposon Tn7 contains a cryptic streptothricin resistance gene. *Plasmid* **25:**217–220.
116. **Tietze, E., J. Brevet, and H. Tschape.** 1987. Relationships among the streptothricin resistance transposons in Tn1825 and Tn1826 and the trimethoprim resistance transposon Tn7. *Plasmid* **18:**246–249.
117. **Tomcsanyi, T., and D. E. Berg.** 1989. Transposition effect of adenine (Dam) methylation on activity of O end mutants of IS50. *J. Mol. Biol.* **209:**191–193.
118. **Turlan, C., and M. Chandler.** 2000. Playing second fiddle: second-strand processing and liberation of transposable elements from donor DNA. *Trends Microbiol.* **8:**268–274.
119. **Waddell, C. S., and N. L. Craig.** 1989. Tn7 transposition: recognition of the *attTn7* target sequence. *Proc. Natl. Acad. Sci. USA* **86:**3958–3962.
120. **Waddell, C. S., and N. L. Craig.** 1988. Tn7 transposition: two transposition pathways directed by five Tn7-encoded genes. *Genes Dev.* **2:**137–149.
121. **Walker, J. E., N. J. Gay, M. Saraste, and A. N. Eberle.** 1984. DNA sequence around the *Escherichia coli unc* operon. Completion of the sequence of a 17 kilobase segment containing *asnA, oriC, unc, glmS* and *phoS*. *Biochem J.* **224:**799–815.
122. **Wang, X., and N. P. Higgins.** 1994. 'Muprints' of the *lac* operon demonstrate physiological control over the randomness of *in vivo* transposition. *Mol. Microbiol.* **12:**665–677.
123. **Watson, M. A., and G. Chaconas.** 1996. Three-site synapsis during Mu DNA transposition: a critical intermediate preceding engagement of the active site. *Cell* **85:**435–445.
124. **Wolkow, C. A., R. T. DeBoy, and N. L. Craig.** 1996. Conjugating plasmids are preferred targets for Tn7. *Genes Dev.* **10:** 2145–2157.
125. **Yieh, L., G. Kassavetis, E. P. Geiduschek, and S. B. Sandmeyer.** 2000. The brf and TATA-binding protein subunits of the RNA polymerase III transcription factor IIIB mediate position-specific integration of the gypsy-like element, Ty3. *J. Biol. Chem.* **275:**29800–29807.

Mobile DNA II
Edited by N. L. Craig et al.

Chapter 20

Transposon Tn*10*

David B. Haniford

INTRODUCTION AND BIOLOGY

Structure

Tn*10* is a composite bacterial transposon. It is made up of insertion sequences IS*10*-Right and IS*10*-Left, which are present in inverted orientation and flank genes that confer resistance to tetracycline and several other open reading frames (ORFs) (Fig. 1) (24, 41, 49). The present isolate of Tn*10* came from the enteric bacterium *Shigella flexneri,* where it was identified on the drug resistance transfer plasmid R100 (129). The complete sequence of this isolate has been determined very recently. Tn*10* is 9,147 bp in length and includes 9 ORFs, as shown in Fig. 1. Three of these, including jemA, jemB, and jemC, were previously unknown (24). IS*10*-Right and IS*10*-Left are each 1,329 bp in length and are nearly identical copies of each other. However, only IS*10*-Right contains a long uninterrupted ORF (1,206 bp) that encodes the functional transposase protein (Fig. 1). The two ends of each of IS*10*-Right and IS*10*-Left are referred to as "outside" and "inside," as defined by their location in Tn*10*. Both outside and inside end sequences contain the DNA determinants for transposase binding. The outside end also contains a binding site for the bacterial protein integration host factor (IHF).

Biology

Tn*10* and each of its component insertion sequences can transpose independent of each other, al-

David B. Haniford • Department of Biochemistry, University of Western Ontario, London, Ontario N6A 5C1, Canada.

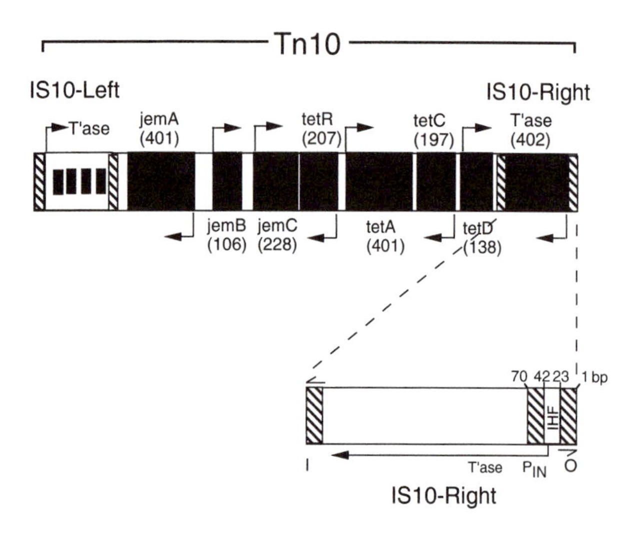

Figure 1. Structure of Tn*10*. The direction of transcription is indicated by arrows. IS*10*-Left encodes a truncated form of transposase that is nonfunctional. Half-arrows indicate nearly perfect inverted repeat sequences at the outside (O) and inside (I) ends of IS*10*-Right. Binding sites for transposase and IHF are indicated by hatched and open rectangles, respectively. IS*10*-Left and IS*10*-Right flank, genes encoding resistance to tetracycline; jemA, a predicted sodium-dependent glutamate permease; jemB, unknown function. jemC has some homology to a family of bacterial transcriptional regulators that repress the arsenic and mercury resistance operons. This figure was adapted from *The Journal of Bacteriology* (24) with permission from the publisher.

though IS*10*-Left transposition requires a source of transposase from either Tn*10* or IS*10*-Right (73). Transposition occurs exclusively by a nonreplicative "cut-and-paste" mechanism (Fig. 2) (10, 50). When the target site for Tn*10*/IS*10* insertion is on a separate

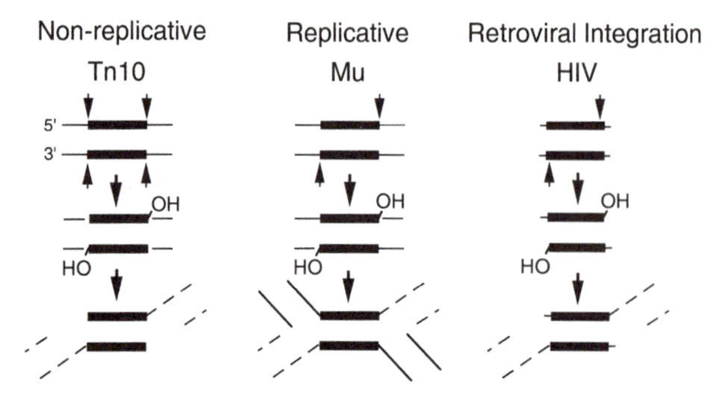

Figure 2. Mechanisms of DNA transposition. In nonreplicative transposition, as performed by Tn*10*, transposon sequences (thick lines) are completely separated from flanking donor sequences (thin lines) at an early stage of the reaction. In contrast, in replicative transposition, as conducted by bacteriophage Mu, the donor DNA remains attached to the transposon subsequent to strand transfer into the target DNA (dashed lines). In retroviral integration cleavage of the viral DNA generated by reverse transcription resembles DNA cleavage in the Mu reaction. Strand transfer in all three systems occurs by the same chemical mechanism.

DNA molecule from that harboring the transposon (i.e., intermolecular transposition), a simple insert is produced. However, if the target site is on the same molecule as the transposon, inversions and deletions occur as alternative reaction products. The diagram in Fig. 3 shows an intermolecular transposition event involving two outside ends and intramolecular transposition events involving two inside ends. The Tn*10* inversion event shown resulting from the intramolecular pathway is of particular importance because each event generates a new composite transposon. In the deletion event chromosomal sequences adjacent to one end of the transposon plus the proximal IS*10* element are lost (106). A single IS*10* element is also capable of generating an "adjacent deletion" (104). In addition to insertion into the same molecule harboring Tn*10*/IS*10*, insertion can also occur into the transposon itself because, unlike other bacterial transposons such as bacteriophage Mu and Tn*7*, Tn*10*/IS*10* does not exhibit transposition immunity (1, 5, 14). Intra-transposon insertion events give rise to unknotted inversion circles, catenated deletion circles, and knotted inversion circles (see Fig. 4 in reference 74). Tn*10*/IS*10* can also appear in a cointegrate, a product that is considered to be a hallmark of replicative transposition (53, 117, 131). Several possible mechanisms for this have previously been discussed (74). More recently, a new mechanism for cointegrate formation via nonreplicative transposition has been proposed on the basis of the observation that Tn*10*/IS*10* ends from two different DNA molecules can be used as substrates for transposase in vitro. In this model the normal steps of the nonreplicative transposition reaction give rise to not only cointegrates but also adjacent deletions and duplicative inversions (Fig. 4) (25).

All the above-mentioned DNA rearrangements mediated by Tn*10*/IS*10* affect the structure of the host genome and have the potential to alter gene expression. A subset of these events can also result in the integration of exogenous DNAs such as plasmids and bacteriophage into the bacterial chromosome.

The next three sections of this review summarize our current understanding of the molecular mechanism of Tn*10*/IS*10* transposition. The last section of this review describes how double-strand cleavages are made in other complex systems including V(D)J recombination. For a detailed summary of the regulation of Tn*10*/IS*10* transposition the reader is referred to references 73 and 74.

MOLECULAR MECHANISM OF Tn*10*/IS*10* TRANSPOSITION

Overview of the Chemical Steps and Comparison with Other Transposition Systems

Tn*10* was the first transposon shown to transpose by a nonreplicative "cut-and-paste" mechanism

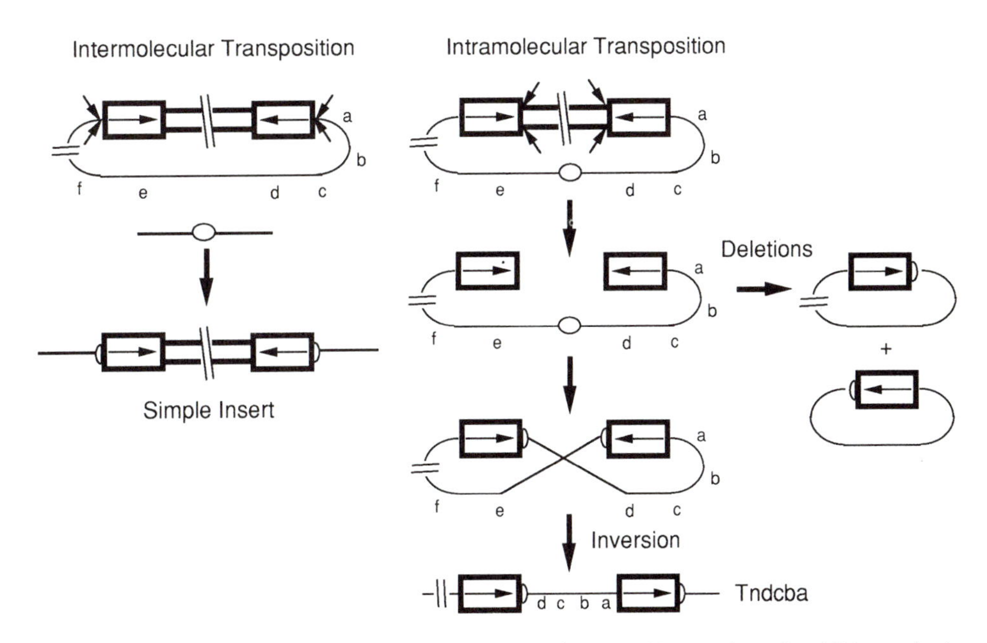

Figure 3. Nonreplicative transposition can generate different types of transposition products. In addition to the formation of simple inserts, Tn*10* can generate both deletions and inversions as alternative reaction products; sites of insertion are indicated by a bubble not attached to a box. Inversions of the type shown generate new composite transposons, e.g., Tn-dcba; adapted from reference 74 with permission of the publisher.

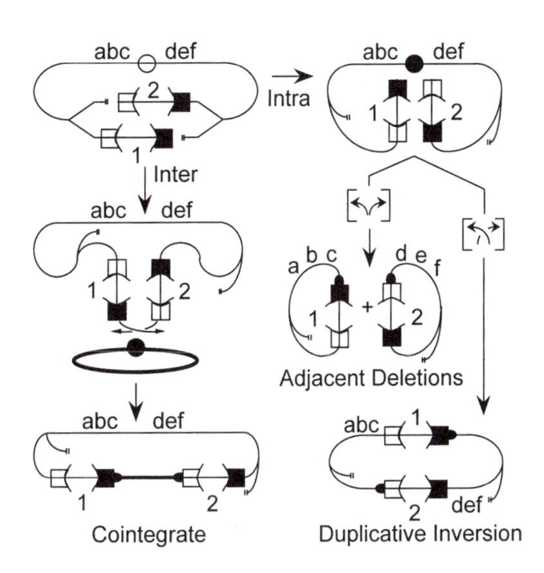

Figure 4. An alternate model for cointegrate and adjacent deletion formation through nonreplicative transposition. Synapsis of transposon ends present on sister chromatids can lead to the formation of products that are characteristic of replicative DNA transposition. If the target for insertion is on a separate DNA molecule from the transposon ends (i.e., intermolecular) a cointegrate can be formed. Alternatively, if the target is on the same molecule, either an adjacent deletion is formed or a duplicative inversion, depending on the particular orientation in which the transposon ends attack the target. Reprinted from *EMBO J.* (25) with permission of the publisher.

(10). In this mode of transposition the transposon DNA is completely separated from the flanking donor DNA at both ends at an early stage of the reaction (50) (Fig. 2). The excised transposon DNA is then inserted into a new target site. Insertion generates nine nucleotide gaps at the site of each new junction, and repair of these gaps by DNA replication and ligation generates a direct repeat (9 bp in length) of the original target sequence (14, 72) (Fig. 3). The separation of transposon and flanking donor DNA at an early stage of the reaction ensures that the theta intermediate characteristic of replicative bacterial transposition cannot form (30, 89, 117) (Fig. 2). This is the main feature that distinguishes these two modes of transposition. All the chemical steps in Tn*10* transposition, with the exception of processing of the 9-bp gaps, are performed by the Tn*10* transposase protein (13, 26).

Although "cut-and-paste" transposition is widespread in both bacteria and eukaryotes, recent work has established that substantially different mechanisms exist for getting to the excised transposon stage (124). Tn*10* generates the excised transposon through three chemical steps (at each transposon end): nicking of the transferred strand (hereafter referred to as first-strand nicking), hairpin formation, and hairpin cleavage or resolution (67). This three-step mechanism is also used by Tn*5* (16) and probably by several plant transposons (see below). In contrast, in Tn*7* transposition the excised transposon intermediate is made simply by a pair of nicks, each of which is performed

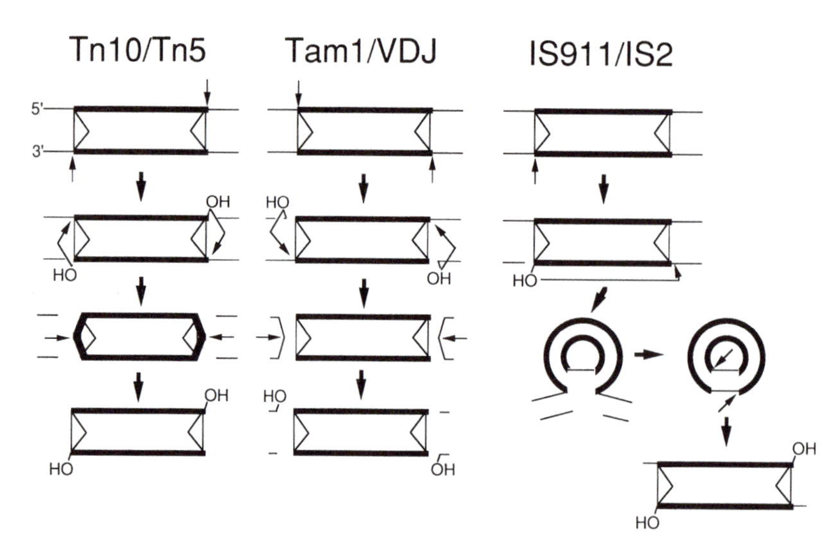

Figure 5. Chemical steps in selected double-strand DNA cleavage reactions. Note that first-strand cleavage reactions in Tn*10*/Tn*5* and Tam1/VDJ recombination are on opposite strands, and this leads to a transposon end hairpin in the former and a donor flank hairpin in the latter. The second chemical step in IS*911*/IS2 transposition resembles that in the reactions above in that it is a transesterification; however, unlike hairpin formation it is an intrastrand event. Adapted from *Cell* (67) with permission of the publisher.

by a different Tn7-encoded protein, TnsA and TnsB (113). Yet another means of progressing to the excised transposon stage is observed for IS*911* and its relatives. A single nick is made by the transposase on the transferred strand at the transposon-donor junction. This is followed by the transfer of the resulting 3′-OH terminus to a phosphate group situated on the same strand but 2–3 bp within the donor DNA adjacent to the other transposon end. The resulting intermediate is processed by an as-yet-undefined mechanism to yield a circular product that is then resolved by the transposase into the excised transposon (81, 97, 116) (Fig. 5).

In all three pathways the 3′-OH termini of the excised transposon are joined to PO_4 groups on separate strands of a target DNA (6, 14, 123). The process of generating the excised transposon fragment is generally referred to as "donor cleavage" whereas the latter step is referred to as "strand transfer." Strand transfer steps of the same chemical polarity follow donor cleavage in replicative transposition as performed by bacteriophage Mu and viral end cleavage as performed in retroviral integration (44, 90). Transposons that undergo donor (or viral) cleavage followed by strand transfer have been referred to as the "classical" transposons (91).

Transposase Interactions with Transposon Ends and Transpososome Formation

Overview

In all the well-studied transposition systems the chemical steps in transposition reactions take place in the context of higher order protein-DNA complexes called "transpososomes." Within the transpososome, the two transposon ends are brought together through a complex series of protein-protein and protein-DNA interactions. The pairing of the transposon ends permits the two ends to be joined to the same short target sequence. In addition, in both Mu and Tn*5* systems, engagement of the active site and the transposon-donor junction requires transpososome assembly (32, 114, 136, 137). The recently solved structure of a Tn*5* transpososome has provided the first detailed view of this phenomenon (32). In general, divalent metal ions such as Ca^{2+}, Mg^{2+}, or Mn^{2+} aid in the assembly-stabilization of transpososomes, but only Mg^{2+} and Mn^{2+} support reaction chemistry (91). One exception to this is the Mu strand transfer reaction where Ca^{2+} can substitute for Mg^{2+} or Mn^{2+} under certain conditions (115).

Tn*10* transposition has been shown to proceed through a defined series of transpososome structures (50, 63, 108, 109). An in vitro reaction system using short DNA fragments containing the outside end of IS*10*-Right, purified transposase, and purified IHF has provided the most detailed picture of the transpososomes formed during the course of the Tn*10* transposition reaction (Fig. 6). Different transpososomes are distinguished by their unique electrophoretic mobilities on native polyacrylamide gels. Initially, three distinct pretarget capture transpososomes were identified, including a precleavage (PEC), single-end break (SEBC), and double-end break (DEBC) transpososome. Both transposon-donor junctions are intact in

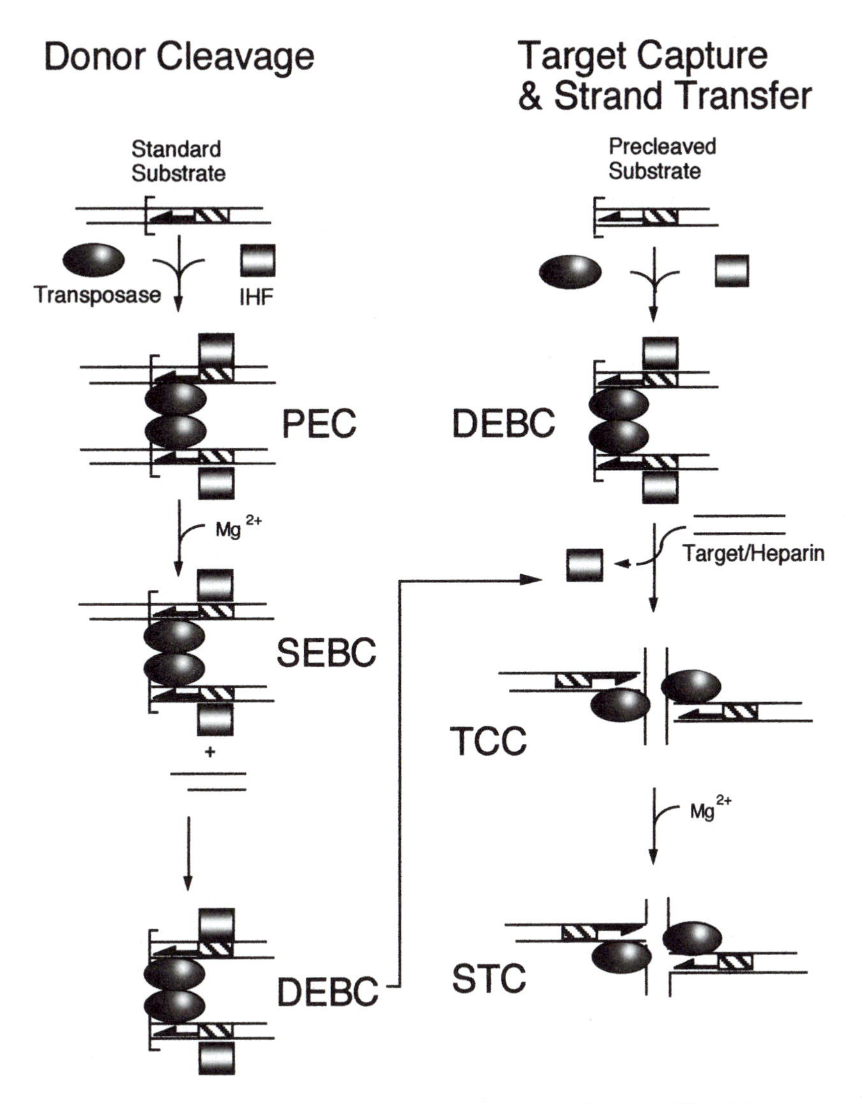

Figure 6. Transpososomes in Tn*10* transposition. In vitro reactions using short, linear outside end fragments provide a convenient means of generating Tn*10* transpososomes. Target capture of and strand transfer into a short linear target fragment are shown. Transpososome assembly does not require the presence of a divalent metal ion but reaction chemistry does. For simplicity transposase is shown bound to only one site on each transposon end fragment and only the equivalent of the b-PEC is shown. Also, the relative arrangement of the two transposon ends within each of the transpososomes is not known. PEC, paired ends complex; SEBC, single-end break complex; DEBC, double-end break complex; TCC, target capture complex; STC, strand transfer complex. Adapted from *The Journal of Molecular Biology* (4) with permission of the publisher.

the PEC. Flanking donor DNA has been completely separated from one transposon end in the SEBC, and flanking donor DNA has been completely removed from both ends in the DEBC. Kinetic analysis of the formation of these species yielded results consistent with transposon excision taking place via a simple irreversible reaction starting at the PEC stage, passing through the SEBC, and culminating with the conversion of SEBC to DEBC (108). The rate of DEBC formation in this system is comparable with that of the appearance of excised transposon fragment in reactions performed with supercoiled substrates both in vitro and in vivo. In reactions with supercoiled substrates the excised transposon is prominent by 20 min, and in the linear fragment system average life spans for PEC and SEBC species are approximately 12 and 15 min, respectively (50, 108). A complex in which transposase is bound to a single end is not readily detected in the Tn*10* system. This is in contrast to the Tn*5* system where early on a "single-end complex" was readily detected but transpososome formation was not (132). In the Tn*10* system, difficulty in detecting a single-end complex could reflect the instability of this complex and/or its rapid conversion to the PEC.

As described in more detail below, target DNA

enters the transpososome relatively late in the reaction (i.e., after DEBC formation) (109). The available evidence suggests the existence of two distinct forms of "target capture complex," one in which transposase engages the target DNA in a sequence-nonspecific binding mode and one in which transposase engages the target DNA in a sequence-specific binding mode (63, 109). Formation of the former transpososome would precede the formation of the latter. In the final step performed by transposase, the sequence-specific target capture complex is converted to a strand transfer complex. The chemical events at the two transposon ends at this stage of the reaction seem to be tightly coupled, unlike in donor cleavage, because a double-end strand transfer complex is the predominant strand transfer species detected (17). This product also has a slightly increased electrophoretic mobility on a native gel relative to the sequence-specific target capture complex, possibly indicating the occurrence of a significant structural transition accompanying the formation of the strand transfer product (63).

Many of the transpososomes formed in Tn*10* transposition are functionally equivalent to transpososomes formed in bacteriophage Mu transposition. For instance, the PEC and type 0 transpososome, the DEBC and type 1 transpososome, and the strand transfer complex and type 2 transpososome are directly comparable (22). Thus, even though these reactions generate different types of transposition products, they pass through similar intermediate stages. Also in both systems the stability of the transpososomes increases as the reaction proceeds (50, 120). This end-product stabilization is likely an important factor in pushing the two types of transposition reactions to completion. One significant difference in the two systems is the number of transposase molecules that make up the core of the transpososomes. For Tn*10* (and also for Tn*5*) two transposase molecules are present, whereas Mu transpososomes contain four molecules of transposase (7, 32, 66, 80).

Organization of protein binding sites in the outside end

The outside end of IS*10* contains binding sites for both transposase protein and the *Escherichia coli* protein IHF (Fig. 1). In contrast, the inside end of IS*10* is missing the IHF site. The primary determinants for transposase binding are in the nearly perfect inverted repeat sequence that spans bp 1 to 23 of the outside and inside ends (58, 130). The IHF site extends from bp 30 to 42. A second region that functions in transposase binding and is internal to the IHF site has recently been identified through footprinting of the PEC and deletion analysis (111) (P. Crellin and R. Chalmers, unpublished results). This region extends from approximately bp 45 to 70 and the contacts are of the nonspecific nature. Mutational studies have further revealed that the terminal inverted repeat region has a bipartite organization, with bp 1 to 3 having a distinct function relative to the rest of the end. It has been inferred that this terminal subdomain contributes sequence-specific information primarily at steps subsequent to the initial interaction between transposase and the outside end. In contrast, the region from bp 6 to 13 is thought to include the primary determinants for the initial interaction between transposase and the outside end (52, 58). A very similar organization exists for the terminal inverted repeat sequences of Tn*5*/IS*50* and Tn*903*/IS*903* (34, 61). From the Tn*5*/IS*50* transpososome structure it is apparent that the bipartite organization of the outside end reflects the formation of contacts between each of the two domains and different functional domains in transposase. One of these domains, the N-terminal binding domain, is likely responsible for the initial binding of transposase to the internal portion of the outside end; the other, the catalytic core domain, is critical for the formation of a stable transpososome (32).

IHF and transpososome assembly

In reactions with supercoiled plasmid substrates IHF addition is not essential for the formation of transposition products unless reactions are performed under suboptimal conditions. For example, at reduced levels of DNA supercoiling, IHF serves as a "supercoiling relief" factor (23). This response is typical of a system in which an IHF-mediated folded protein-DNA complex is a reaction intermediate (103, 121). Consistent with this is the observation that even at high levels of plasmid supercoiling (i.e., $\sigma = -0.6$) the addition of a moderate amount of IHF increases the level of transposition products observed. IHF has a second important effect on the in vitro transposition reaction. At higher IHF concentrations where no additional stimulatory effects on product formation are observed, there is a qualitative change in the strand transfer products formed. This includes a decrease in strand transfer products that are generated by random collision of transposon ends and target DNA (this includes intermolecular and topologically complex intratransposon products) and an increase in one specific product, an unknotted inversion circle. The increase in the level of this one specific product reflects an interaction between transposon ends and target DNA wherein random collision is replaced by an interaction governed by tight geometric and topological constraints. This phenomenon is referred to as channeling (23).

Additional information regarding the precise role of IHF in transpososome assembly has come from experiments conducted with use of the short, linear outside-end fragment assay in which transpososomes are easily detected. The analysis of PEC formation as a function of IHF concentration initially yielded two forms of the PEC that are distinguished by differences in their electrophoretic mobilities. Reactions performed with high amounts of IHF yielded a single form of the PEC, whereas reactions performed with low and intermediate amounts of IHF yielded two forms of the PEC. The PEC formed under "high" IHF conditions has a greater mobility than the PEC that predominates under "low" IHF conditions and on the basis of these relative mobilities the former is referred to as b-PEC (for "bottom" PEC) and the latter as t-PEC (for "top" PEC). Subsequently, it was shown that the two forms of the PEC are interconvertible. The b-PEC can be converted to the t-PEC by titration of IHF out of the complex. This can be accomplished by the addition of a large excess of plasmid DNA or a lesser amount of DNA oligonucleotide containing an IHF binding site. Also, t-PEC generated by this IHF titration could be shifted back to b-PEC by addition of more IHF. The decrease in mobility arising from removal of IHF provides evidence that IHF introduces a significant fold into the DNA of the b-PEC, and for this reason, t-PEC and b-PEC are also referred to as unfolded and folded forms of the PEC, respectively (111).

Further differences in t-PEC and b-PEC have been demonstrated by DNA-footprinting experiments. The t-PEC footprint revealed protection from about bp 2 to 24, thus revealing the lack of occupancy of the IHF site. In contrast the b-PEC footprint revealed a much broader protection pattern (from bp 2 to at least bp 45) that includes the IHF site and indicates IHF occupancy of at least one of the two IHF sites in the PEC. In the b-PEC, contacts are also formed between transposase and a region of the outside end that includes bp 46 to 70 (111; Crellin and Chalmers, unpublished) (Color Plate 36 [see color insert]). Contacts with proximal and distal segments of the end are thought to generate a topologically closed loop in the b-PEC (23).

The idea that the formation of this loop is precisely equivalent to the loading of a molecular spring has been presented. It has also been suggested that in supercoiled substrates transposase could interact with an additional DNA segment, a plectosomal branch point, to establish torsional tension in the spring. This idea comes from the dependence of transposon excision on negative supercoiling under conditions in which IHF is not present. Firing of this spring would coincide with the ejection of IHF from the PEC (23). Consistent with this idea, evidence has been provided that the t-PEC arises from the decay of the b-PEC (111). Possible functions for the firing of such a spring are considered in the next section.

Consequences of IHF release from the b-PEC

The rates of flanking donor DNA release were measured for the t-PEC and b-PEC and only a very small difference was observed. In addition, the amount of conversion of starting substrate to product (i.e., free flanking donor DNA) was very similar at the end point of each reaction. For the b-PEC there is a clear progression from the single-end break complex to the double-end break complex. The appearance of a single-end break species may simply reflect the absence of a mechanism for coordinating the cleavage events at the two ends.

Although the IHF status of a PEC does not seem to have a significant effect on the ability of a PEC to undergo donor cleavage, this is not the case for the conversion of "postcleavage" transpososomes to target capture complex. As noted previously, the addition of a large excess of a supercoiled target plasmid to a standard transpososome assembly reaction results in the titration of IHF off of the b-PEC, giving rise to the t-PEC. Appropriate analysis of such reactions has revealed that the t-PEC readily associates with the target plasmid after donor cleavage is complete, giving rise to a double-end strand transfer product. The effect of IHF on target capture complex formation was measured by titrating IHF into reactions containing postcleavage transpososomes and target DNA. IHF has an inhibitory effect on target capture, indicating that it has the ability to release postcleavage transpososomes from noncovalent target capture complexes (111). Evidence that the presence of IHF in a postcleavage transpososome is inhibitory to target capture also comes from another type of experiment. A postcleavage transpososome assembled under standard conditions could only form a target capture complex with a short piece of DNA containing a Tn*10* insertion hot spot if the transpososome was treated with heparin. At the amount of heparin used to get target capture, IHF was completely titrated off of the uncomplexed transposon end DNA, indicating that IHF would also have been stripped from the postcleavage transpososome (63).

The existence of a form of the PEC in which the transposon ends are tightly constrained (or wrapped) can explain, in part, how IHF channeling occurs. By constraining the transposon ends in a tight wrap the ends would be directed toward a target sequence located at a fixed distance from the terminus of the transposon ends. This distance has been determined to be about 130 bp. This wrap would also provide geometric constraints. In contrast, the transposon ends in the unfolded transpososome would be unconstrained and therefore free to interact with target sites by random collision. However, the available evidence argues that there is a progression from folded to unfolded transpososomes before target capture. For this reason it has been suggested that, subsequent to the ejection of IHF from the transpososome, IHF can rebind to the transposon end and push the reaction down the channeled pathway. The strongest evidence for this is that channeling occurs at an IHF concentration in excess of that required for simple stimulation of transpososome formation. That is, this observation implies that channeling is manifested at a later step than transpososome formation (23). Also, experiments in the linear fragment system have provided evidence that unfolded transpososomes have an increased affinity for IHF relative to naked DNA and this could provide an opportunity for IHF to rebind to the DNA at a reasonable rate so that channeled events could occur at an appreciable frequency relative to unchanneled events (111).

What might be the role of firing a molecular spring in the developing transpososome? One possibility is that the firing of the spring is required to reposition part of the transposon end sequence so that it does not block entry of the target DNA into the transpososome. The diagram in Color Plate 37 (see color insert) shows a model of an IHF-bound Tn*10* end superimposed on the Tn*5* transposase dimer. The model shows that a portion of the Tn*10* end internal to the IHF binding site crosses over the putative target-binding cleft at a right angle. Another possibility is that energy derived from the firing of the spring is used to drive a conformational change in transposase that is required for the binding of the target DNA. For example, a conformational change in transposase might be necessary to allow an exchange of outside-end contacts for contacts with the target DNA.

Donor Cleavage via a Hairpin Intermediate

The two strand cleavage events occur in a specific order

In generating a double-strand break at the transposon-donor junction the transferred strand is always cleaved before the nontransferred strand (Fig. 5). This was shown by using the short, linear fragment assay, by isolating full-length duplex DNA from a cleavage reaction, and then determining by denaturing gel electrophoresis if any of the DNA molecules in this population contained nicks. The product distribution revealed nicking precisely at the transposon-donor junction of only the transferred strand. First-strand nicking is typically a very efficient reaction, with close to 100% of the PEC undergoing this reaction when Mg^{2+} is present. However, this reaction is relatively slow, taking almost 2 h to go to completion with the linear substrate (18).

Evidence for a hairpin intermediate in Tn*10* transposition

In considering the significance of the obligatory order of strand cleavage events, the possibility was raised that a functional interdependency may exist between the two strand cleavages. That is, cleavage of the nontransferred strand could somehow depend on cleavage of the transferred strand (18). This, in fact, is correct because the 3′-OH terminus of the transferred strand serves as the attacking nucleophile in the cleavage of the nontransferred strand. This generates a transposon-end hairpin at the same time as the donor DNA is completely separated from the transposon end (67) (Fig. 5).

A transposon-end hairpin was identified in a kinetic analysis of donor cleavage by using the linear fragment assay. A species with an apparent molecular weight substantially larger than that of any of the other cleavage products was detected in a high-resolution denaturing gel. In addition, formation of this species showed the normal dependencies of a cleavage product, including transpososome assembly and the presence of Mg^{2+} or Mn^{2+}. Several different criteria had to be satisfied before it could be concluded that the species in question was a transposon-end hairpin. Upon isolation from a denaturing gel a hairpin end would have roughly the same mobility on a native gel as the cleaved end species. A hairpin end would run off the diagonal of a two-dimensional gel (first dimension, native conditions, second dimension, denaturing conditions) at the position of the cleaved end in the first dimension. A hairpin end would be sensitive to digestion at the "turn-around" point by single-strand-specific nucleases. All three of these criteria were satisfied, and in addition DNA sequencing of the gel-purified species by using the Maxam-and-Gilbert G reaction yielded a partial DNA sequence consistent with the species being a perfectly self-complementary transposon-end hairpin (67).

Evidence that the hairpin species is an obligatory intermediate in the donor cleavage reaction came in two forms. First, a kinetic analysis revealed that nicked, hairpin, and cleaved end species appeared in appropriate succession, and that hairpin formation and the release of the flanking donor occurred at exactly the same time (Fig. 5). The nicked species is formed first and then disappears with an average life span of approximately 25 to 30 min. The hairpin appears next and then disappears with an average life span of approximately 15 min. The cleaved end species appears last and accumulates during the course of the reaction. Second, the ability to cleave the nontransferred strand was shown to depend on the transferred strand having an exposed 3′-OH terminus. This is expected if the hairpin is an obligatory intermediate, because hairpin formation depends on the nucleophilic attack of this 3′-OH group at the scissile phosphate of the nontransferred strand. Alternatively, if it were possible for nontransferred strand cleavage to take place by a second hydrolytic step instead of hairpin formation, there is no a priori reason for the requirement of the terminal 3′-OH group for this reaction. This was examined by incorporating a dideoxynucleotide (3′-H instead of 3′-OH) at the terminal position of the transferred strand and asking if cleavage of the nontransferred strand could still occur. The experiment clearly showed a dependence of nontransferred strand cleavage on the 3′-OH group. Only a very small amount of cleavage occurred in the reaction with the "dideoxynucleotide substrate" (about 100-fold less than with the isogenic substrate containing a standard nucleotide), and the appearance of this trace amount of cleavage product exhibited delayed kinetics relative to the control reaction. This low level of cleavage can be explained by assuming that when the normal cleavage pathway is blocked nontransferred strand cleavage can occur very slowly and inefficiently by an alternative pathway (67).

Similar approaches have been used to demonstrate that donor cleavage in the Tn*5* system also takes place via a hairpin mechanism. As for Tn*10*, the hairpin is formed in the Tn*5* system at the transposon end (16) (Fig. 5).

Other DNA substrate requirements for Tn*10* hairpin formation

Recent work has shown that hairpin formation, but not first-strand nicking or PEC assembly, is extremely sensitive to the DNA sequence at the tip of the outside end. A mutational analysis of the terminal 4 bp plus the first base pair in the flanking donor DNA was performed, and the mutants were tested for transpososome assembly and the three steps of donor cleavage. Only a subset of bp 1 and 2 mutants exhibited a strong defect in hairpin formation, whereas all the bp 3 mutants exhibited moderate to weak effects on hairpin formation. The extent of the inhibitory effects on hairpin formation was such that the life span of the nicked species was increased at least 30-fold for the strongest mutants and 3- to 4-fold for the weakest mutants. Base pair 4 and −1 mutants were not affected for hairpin formation. None of the mutants examined exhibited defects in either PEC formation or first-strand nicking. These results indicate a special role for the terminal 3 bp in hairpin formation. Specific base pair mismatches, including 1C/1C and 2A/2A, had close to wild-type activities with respect to hairpin formation (J. Allingham and D. Haniford, unpublished results). Rescue of hairpin formation in reactions with substrates containing base pair mismatches in otherwise suboptimal sequences has also been observed in the V(D)J recombination reaction and is consistent with the idea that hairpin formation requires some disruption of base-pairing (31, 99).

An attempt was made to correlate the inhibitory effects of terminal base pair mutations with changes in DNA structure in the PEC. Consistent with the idea that hairpin formation requires some disruption of base-pairing in the transposon-donor junction region, it has been found that this region exhibits hypersensitivity to the DNA-footprinting reagents 1,10-phenanthroline-copper (OP-Cu) (Allingham and Haniford, unpublished) and hydroxyl radicals (Crellin and Chalmers, unpublished) in the wild-type PEC. However, the OP-Cu footprinting did not reveal any significant differences in the PEC footprints derived from wild-type and "hairpin-defective" substrates. This could mean that the hairpin-defective mutations exert their effects after the DNA structure in the junction region has been distorted (and presumably set up for hairpin formation).

The structure of the Tn*5* synaptic complex has provided details of the DNA structure and protein-DNA interactions that are likely relevant to the hairpin formation stage. Nucleotide residues at positions 1 and 2 are not base-paired. In fact, the base at the second position of the nontransferred strand is flipped out of the DNA helix. Sequence-nonspecific interactions between this nucleotide and several active-site residues stabilize this configuration. The close approach of the strand termini that are separated by only 3.5Å is facilitated by a network of interactions, some base-specific, that seem to stabilize a sharp bend in the nontransferred strand. This also permits stacking of the terminal bases (32). The results of the mutational analysis and PEC footprinting in the Tn*10* system are consistent with the idea that the terminal residues in the outside end of Tn*10* undergo a structural transition similar to that observed in the Tn*5* system. In addition, they add to the Tn*5* results by delimiting the area of the end important for hairpin formation and by making it clear that hairpin formation has significantly different dependencies than does first-strand nicking.

Mn^{2+} effects on hairpin formation

Substituting Mn^{2+} for Mg^{2+} has also been shown to extend the life span of the nicked species (i.e., inhibit hairpin formation) (18). This effect is highly dependent on the concentration of Mn^{2+} used. Inhibition of hairpin formation is observed at a relatively high Mn^{2+} concentration (i.e., 5 mM); however, at a lower Mn^{2+} concentration (i.e., 0.18 mM) hairpin formation actually occurs at a faster rate than in reactions performed with 5 mM Mg. In addition, (partial) suppression of the hairpin formation defect caused by terminal base pair mutations is observed at the lower Mn^{2+} concentration (Allingham and Haniford, unpublished). The explanation for the differential effects of "high" and "low" Mn^{2+} is not simply that the former is generally toxic to the reaction because PEC formation and first-strand nicking occurred normally under the "high" Mn^{2+} conditions (18). Alternatively, given the capacity of Mn^{2+} to interact with the bases in DNA in addition to the phosphate backbone, perhaps "high" Mn^{2+} prevents the DNA from adopting the required structure for hairpin formation by affecting the orientation of specific bases that are important for hairpin formation. Such interactions would likely involve Mn^{2+} ions that are not bound to the catalytic core of transposase. It is not known how "low" Mn^{2+} stimulates the wild-type reaction, but it seems reasonable to assume that the partial suppression of the terminal base pair mutations reflects the general stimulatory effect of "low" Mn^{2+}.

Hairpin Cleavage

Hairpin cleavage involves a water-mediated nucleophilic attack on the same phosphate that was attacked by the 3′-OH terminus of the transferred strand in the previous step, hairpin formation. No evidence for opening the hairpin by attack of a neighboring phosphate has been obtained in the Tn*10* system, although a low-frequency "off-center" cleavage event has been reported in the Tn*5* system (16, 67). The average life span of the hairpin has been calculated to be approximately 15 min. Thus, it takes roughly twice as long to form the hairpin as it does to open it. This is consistent with the idea that hairpin formation involves a particularly difficult conformational change in the transpososome.

Hairpin resolution has been detected in a reaction where a preformed hairpin having the same structure as that arising during the transposition reaction was used as a substrate. The preformed hairpin assembled into a transpososome and underwent cleavage at exactly the same site as the "native" hairpin. In addition, the kinetics and overall efficiency of "artificial" versus "native" hairpin resolution were very similar. The ease with which the preformed hairpin substrate can be used at all of the relevant steps of the process provides strong support for a hairpin as a true intermediate in the normal transposition reaction (67).

In general, the base pair mutations that strongly inhibit hairpin formation also negatively affect hairpin resolution (Allingham and Haniford, unpublished). This may be indicative of a common set of contacts governing both steps. In fact, as discussed in more detail below, hairpin opening can be viewed as the simple reverse reaction of hairpin formation and as such hairpin opening may not require any significant movement of DNA strands in the active site relative to hairpin formation.

Target Capture and Strand Transfer

Distinct forms of target capture complex and timing of target capture

Incubation of the DEBC with a supercoiled target DNA leads to a physical association between the two that can be detected in an electrophoretic mobility shift assay. However, only roughly half of the complexes formed are typically competent for strand transfer. This number is significantly less if a divalent metal ion is omitted from the assembly reaction. Furthermore, the SEBC, but not the PEC, is also able to associate with a supercoiled target DNA; however, this association does not lead to strand transfer at significant levels. One conclusion from these results is that a functionally significant interaction between target DNA and transpososome is not made until flanking donor DNA has been removed from both ends of the transposon (109). This is in striking contrast to other transposition systems such as bacteriophage Mu and Tn7 where the entry of the target DNA into the transpososome occurs before donor cleavage (6, 8, 94). As discussed in more detail below an additional requirement for target capture in the Tn*10* system is that the hairpin ends of the excised transposon be opened up.

The observation that some physical associations between DEBC and target DNA are meaningful (from a functional point of view), whereas others are not, is consistent with there being two distinct modes of target binding. Furthermore, Tn*10* exhibits a marked preference for insertion into sites that conform to the following consensus sequence, 5′-GCTNAGC-3′ (48), and this is consistent with the idea that one mode of target capture involves the formation of sequence-specific contacts between transposase and the target DNA. Additional observations strongly support this idea. First, the decay of target capture complexes formed on mixing DEBC, supercoiled target plasmid, and Ca^{2+} seems to be biphasic, with a small population of target capture molecules decaying very rapidly and a larger population decaying much more slowly. When Ca^{2+} was left out of the assembly mixture the percentage of target capture complexes decaying with the faster kinetics increased significantly, which suggested an important role for divalent metal ions in the formation of stable target capture complexes (109). Second, target capture complexes with different relative stabilities have been observed in experiments where the target DNA is a short (25 bp), linear DNA fragment containing a Tn*10* insertion hotspot. Here the 25-bp target DNA is incorporated into a stable target capture complex, whereas excess transposon-end fragment present in the reaction mixture forms an unstable target capture complex. Dissociation of this unstable complex was enhanced by the presence of a divalent metal ion, whereas the stability of the sequence-specific target capture complex increased. In addition, the target capture complex formed with the hotspot fragment was efficiently converted to strand transfer product, whereas almost no strand transfer product was generated from the other target capture complex. Thus, the ability to form a stable target capture complex and undergo strand transfer shows a strong dependence on the sequence of the target DNA (63).

The observations above are consistent with a two-stage model for the assembly of a functional target capture complex. In the first stage the DEBC would bind to a target DNA in a sequence-nonspecific mode, giving rise to a relatively unstable target capture complex. At this point the DEBC would be free to diffuse along the target DNA in the vicinity of its initial contact point. In the second stage sequence-specific DNA binding determinants in transposase would interact with a usable target sequence and this interaction would be responsible for stabilizing the target complex. The mechanism by which a divalent metal ion would help stabilize the sequence-specific target capture complex is not known, although one specific possibility is discussed below.

The rate determining step in strand transfer

There is a significant interval (~1 h) between the formation of the excised transposon and the appearance of strand transfer products both in vivo and in vitro (26, 50, 109). In vitro, the length of this interval can be reduced by preincubating a target DNA with the DEBC before the addition of Mg^{2+}. Part of this enhancement (approximately half) is simply caused by titrating IHF off of the DEBC so that intermolecular target capture can occur (109). This can also be achieved by addition of heparin (63). This step would not contribute to the lag in appearance of the strand transfer product in unstaged reactions where removal of IHF from the transpososomes occurs at the PEC stage of the reaction (111). However, the rest of the enhancement in rate is significant, likely reflecting a slow transition from sequence-nonspecific to sequence-specific target capture. The alternative possibility, that it is the formation of the initial (sequence nonspecific) target capture complex that is slow, is unlikely given the observation that the rate of strand transfer is not significantly influenced by the concentration of target DNA present (109).

Target DNA recognition and distortion by transposase

In addition to residues between the points of strand transfer being important in determining target site selection, residues in the regions immediately flanking the 9-bp "core" contribute significantly to target site selection. This was shown initially by mutational studies on a particular insertion hotspot called HisG1. For instance, a single mutation in one of the consensus base pairs of the target core or two mutations, one in each of the flanking regions, reduced the frequency of target site utilization by approximately 500- to 1000-fold. This study also demonstrated that only the first 5 to 6 bp of the flanking sequence influenced target site utilization (12). These studies have now been complemented by biochemical studies in which contact probing methods were used to define transposase-target DNA interactions within two different sequence-specific target capture complexes, HisG1 and the LacZ* hotspot. Both target capture complexes yielded rather broad protection patterns that extended over approximately 24 bp when probed with the chemical nuclease 1,10-phenanthroline-copper (Fig. 7). This included contacts in the target core and out to approximately 7 to 8 bp in each of the adjacent flanking regions. The patterns of protection were highly symmetric consistent with two symmetrically disposed subunits of transposase being bound. Each of the target footprints also yielded symmetrically disposed sites of OP-Cu hypersensitivity. In each case the hypersensitivity was localized to one phosphodiester bond on each side of the target core. These patterns of OP-Cu hypersensitivity are strongly suggestive of a pair of symmetrically disposed DNA kinks in the target DNA. The positions of the OP-Cu hypersensitivities varied between the two target

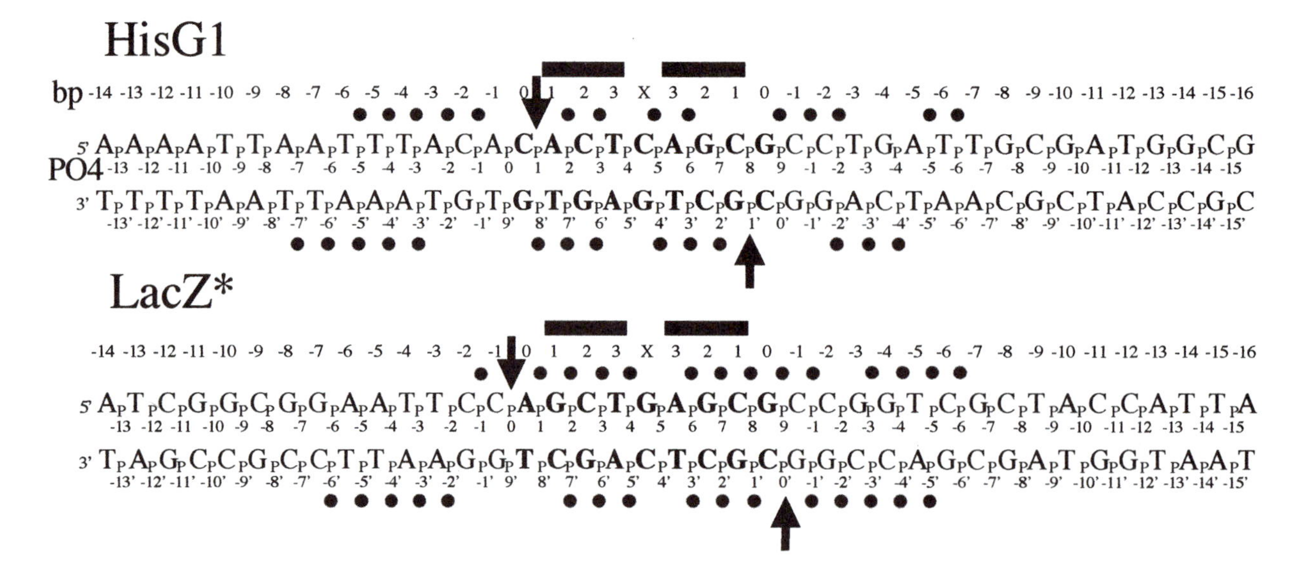

Figure 7. Comparison of OP-Cu footprints for HisG1 and LacZ* Tn*10* insertion hotspots. Positions of OP-Cu protections are indicated by black circles, and OP-Cu hypersensitivities are indicated by arrows. Black rectangles show the positions of the target core half-sites. Residues in the 9-bp target core are indicated in bold. The numbering scheme for both bases and phosphates is shown.

complexes, raising the possibility that the putative DNA kinks are not at exactly the same relative positions in the two target DNAs (98). Presumably, the positioning of the hypersensitive sites (and therefore the putative DNA kinks) is determined by the specific target DNA sequence present.

Dimethyl sulfate (DMS) interference experiments on the HisG1 target capture complex (assembled in the presence of Ca^{2+}) have provided evidence for extensive major and minor groove base contacts on both faces of the target core (Color Plate 38 [see color insert]). These results are consistent with a direct readout mechanism for the recognition of the target core half-sites. Sites of DMS interference were not restricted to the target core because a tight clustering of interference sites was observed in the same parts of the flanking regions that were protected against OP-Cu cleavage (98). Given the absence of sequence conservation in the flanking regions of Tn*10* insertion hotspots it is unlikely that the observed DMS interference is caused by the disruption of base-specific contacts. Instead it is likely that transposase makes extensive backbone contacts in these regions and that DMS modification of bases prevents the backbone from adopting a structure required for optimal contacts with transposase. That is, an indirect readout mechanism is responsible for achieving binding specificity in the flanking regions. This is exactly the situation that has been documented for the interaction between IHF and the H′ IHF binding site (102, 135).

Within the target core the phosphate backbone also seems to contribute significantly to target selection but possibly by a mechanism distinct from that described above for the flanking regions. Phosphorothioate substitutions specifically at either of the two phosphate groups immediately adjacent to the scissile phosphate in both halves of the HisG1 target site significantly decrease the utilization of this site. An even more dramatic inhibition has been observed when the equivalent phosphates in a derivative of HisG1, called MutF, were substituted with phosphorothioate phosphates. MutF differs from HisG1 at four positions, two in each of the flanking regions. Given that these sites of phosphorothioate interference map to the same region as the DNA distortion, it seems likely that the affected phosphates would play an important role in the formation of the DNA distortion, although this has yet to be demonstrated directly. The inhibitory effects of the MutF phosphorothioates on target capture-strand transfer were suppressed by substituting Mn^{2+} for Mg^{2+} (98). This observation raises the possibility that there may be a link between the target DNA distortion and divalent metal ion binding to the target DNA.

Model for target site selection

Although much work remains to be done to understand how the above-discussed pieces of information fit together, the data as they stand suggest the following model for target site selection in Tn*10*

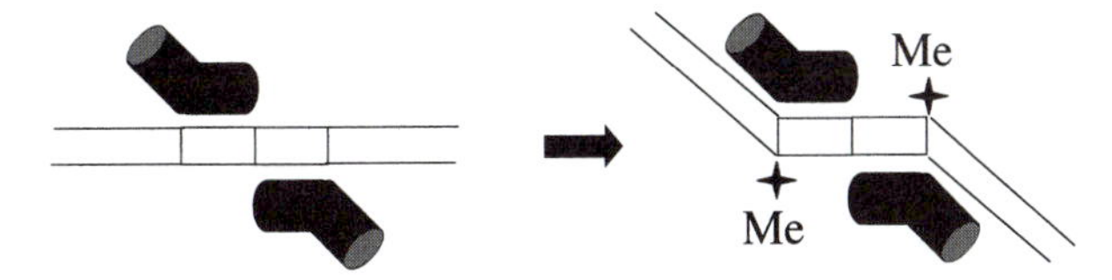

Figure 8. Model for sequence-specific target capture. A transposase dimer is shown binding to the target core of a Tn*10* insertion hotspot (left side). Subsequent to the initial contact with the core, the target DNA is shown to form two kinks, the formation of which allows a second region of transposase to contact flanking DNA (right side). Kink formation may be stabilized by the binding of a divalent metal ion to the DNA at the position of the kink and/or the establishment of the kink may be necessary to set up a high-affinity metal ion binding site. Kink formation may also activate the scissile phosphate for nucleophilic attack. It is also possible that the initial contact with the core alters the conformation of transposase so that it is better able to facilitate kink formation (not shown).

transposition (Fig. 8). After sequence-nonspecific target capture the transposase in the DEBC would use direct sequence readout determinants to bind to an appropriate target core sequence. Upon binding, transposase would then induce a kink in the target DNA at or near each of the core-flanking region junctions. The DNA kink would permit transposase to make extensive contacts in the flanking region, thereby conferring additional stability to the sequence-specific target capture complex, and could also facilitate strand transfer chemistry by activating and/or positioning a specific phosphodiester bond for nucleophilic attack. In this model the sequence of the flanking regions would determine the potential of these regions to contact transposase by influencing the shape of the phosphodiester backbone and possibly the potential of these regions to support a kink in the DNA. In addition, divalent metal ion binding at or near the predicted kink site might favor kink formation by neutralizing unfavorable charge-charge interactions between adjacent phosphate groups.

Outside-end sequence requirements for target capture/strand transfer

In addition to the target DNA sequence, target capture-strand transfer shows a dependence on the sequence at the tip of the transposon ends. In fact, one of the mutations that negatively affects hairpin formation (1A/1T) also has significantly reduced target capture and strand transfer levels. For example, in an experiment with the 1A/1T outside-end substrate present on a plasmid, the mutation (present at both transposon ends) increased the ETF:TnO ratio in vivo by at least 20-fold (52). The excised transposon fragment (ETF) is the full transposon excision product and the TnO is an intramolecular strand transfer product (14, 51). An increase in the ETF:TnO ratio is indicative of a defect in target capture and/or strand transfer. More recently, the short linear fragment system was used to show that the 1A/1T mutation actually inhibits sequence-specific target capture in the presence of Ca^{2+} to a significant degree; no strand transfer product was observed in reactions with Mg^{2+}. One of the mutations at bp 2 that strongly inhibits hairpin formation, 2C/2G, has a much less severe effect on target capture and strand transfer, indicating that the effect is highly localized (Allingham and Haniford, unpublished). A possible explanation for this result is that the terminal base pair mutation alters the position of the transferred and/or nontransferred strand to block the target site binding pocket. As previously noted, this same mutation does not inhibit first-strand nicking, and this could be an indication that it is the position of the nontransferred strand that is being negatively affected.

Altered target site specificity

Substitution of Mn^{2+} for Mg^{2+} has been shown to relax the target sequence requirements for Tn*10* insertion. This was shown in vitro by low-resolution (restriction enzyme) mapping of intramolecular target sites used in the plasmid-based assay in which either Mg^{2+} or Mn^{2+} was present; in the presence of Mn^{2+} a much broader distribution of target sites was used (64). The relaxation of sequence specificity by Mn^{2+} is also a prominent feature of restriction enzyme systems (15). The availability of structural information in several of the restriction enzyme systems has led to some ideas on how this relaxation of specificity by Mn^{2+} can be achieved (71, 134). Given the similarities in catalytic mechanisms in restriction enzyme and transposition systems (93), it is reasonable to predict that a common mechanistic basis for Mn^{2+}-induced relaxed specificity exists in the two systems.

The basic idea coming from the restriction enzyme systems is that sequence-specific DNA binding and catalysis are allosterically coupled. More specifically, it is thought that binding energy derived from sequence-specific DNA binding is used to drive a conformational change in the protein-DNA complex (usually takes the form of a distortion in the DNA structure) that is required for catalysis. This conformational change is thought to be important for correctly positioning the substrate with the various active-site components that are required for reaction chemistry including the divalent metal ion(s). It follows that factors that reduce the extent of the conformational change necessary for substrate-active site alignment would reduce the amount of binding energy required for catalysis and therefore permit the use of suboptimal binding sites (54, 59, 127). Mn^{2+} might accomplish this by one of several different mechanisms. The larger ionic radius of Mn^{2+} relative to Mg^{2+} might reduce the extent of substrate movement required for optimal alignment in the active site. Alternatively, Mn^{2+} binds more tightly to multidentate ligands than does Mg^{2+}, and consequently, it could bind more tightly to a suboptimal binding site (86). This could effectively reduce the amount of movement necessary to properly set up the active site. Given the evidence that sequence-specific target binding in Tn*10* transposition correlates with distortions in the target DNA structure, it is attractive to think that altered target specificity in the Tn*10* system may also correlate with less discriminatory divalent metal ion binding. That is, Mn^{2+} would be able to bind better than Mg^{2+} to a target complex in which the divalent metal ion binding site is not optimally set up. Establishment of an optimal divalent metal ion binding site would itself depend on transposase making appropriate contacts with the target DNA.

STRUCTURE OF Tn*10* TRANSPOSASE

Insights from Partial Proteolysis

The sites in Tn*10* transposase most sensitive to digestion by trypsin are in the interval between amino acids 247 and 256 and amino acid 54 (Fig. 9) (78).

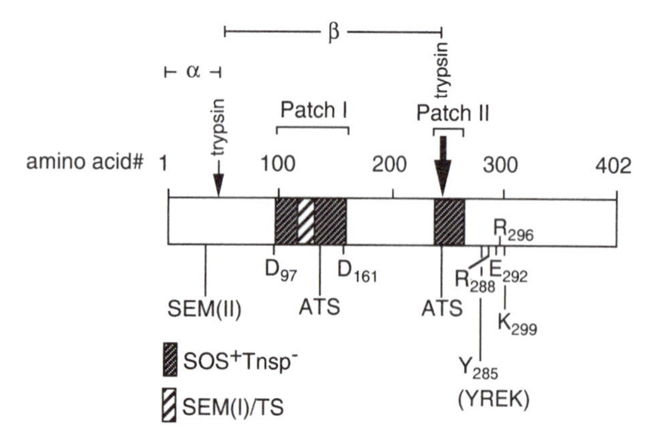

Figure 9. Linear map of Tn*10* transposase showing the relative positions of the major trypsin cleavage sites, the DDE motif, and positions of various classes of mutations. ATS, altered target site specificity mutations; SEM, suppressor of end mutations; TS, transpososome stability mutations. The YREK sequence is part of a highly conserved motif present in the IS*4* family (100) and a basic residue is usually situated between the E and K (60). In Tn*10* transposase this basic residue is R296, which is also the position of a TS mutation. α and β refer to different proteolytic fragments generated by trypsin digestion.

These are likely positions of the most exposed surface loops in the protein and could potentially mark boundaries between distinct structural domains. Many proteolytic fragments generated by trypsin digestion were purified from denaturing gels, refolded, and examined for biochemical activities both on their own and in various combinations. The only biochemical activity observed for any of the individual fragments was a sequence-nonspecific DNA-binding activity, and this was restricted to fragments α and β. However, upon mixing the entire N-terminal fragment (α + β) with the C-terminal fragment, complete DNA transposition was observed at levels similar to that observed in a reaction with the full-length transposase protein (78). This indicates that the assembly of at least the functional domains required to form the PEC requires interaction between the two different fragments.

In reconstitution experiments of the type described above, the region between amino acids 247 and 256 can be left out without seriously affecting the level of transposition activity. This was somewhat unexpected because mutations in this interval can seriously impair the transposition activity. This interval contains a region called Patch II that was defined on the basis of the location of a number of mutations that conferred the SOS^+ $Tnsp^-$ phenotype. These mutants promote Tn*10* excision at roughly wild-type levels (and like wild-type transposase induce the SOS response in *E. coli*); however, they are defective for transposon insertion (51, 105). As discussed in more detail below this defect seems to occur at the level of sequence-specific target capture. It has been suggested that a surface loop contained in this region might have to be repositioned after donor cleavage for the target DNA to enter the active site. It follows that mutations in this surface loop could prevent such a structural transition from happening (78).

Tn*5* transposase has also been subjected to limited proteolysis with trypsin, and the major cleavage sites in Tn*5* transposase match well with those of Tn*10* transposase. The one exception is that there is an additional site of trypsin sensitivity in what turns out to be an extension of the C-terminal region that is not present in Tn*10* transposase (85). An alignment of Tn*10* and Tn*5* transposase sequences is shown in Color Plate 39 (see color insert). Amino acid identity is only about 20%; however, with the exception of the C-terminal extension in Tn*5* there are amino acid similarities spread through the entire length of the two proteins. From the structure of the Tn*5* complex it is apparent that the Tn*5* equivalent of the Tn*10* α domain corresponds to a distinct structural domain that functions in binding the transposon end in the region from bp 5 to 20. However, the major region of proteolytic sensitivity between amino acids 252 and 263 (corresponding to the site between amino acids 247 and 256 in Tn*10*) is not found at a structural domain boundary. Instead, an internal structural domain extending from about amino acid 70 to 370 is apparent in the Tn*5* synaptic complex. In addition to containing a distinct DNA binding domain that makes contacts with the outside end at bp 1 to 7, this structural domain contains the catalytic core of the protein that includes the DDE triad (32). In all the other transposase-integrase proteins for which there are structures available, the residues of the DDE triad are located on a distinct structural domain that is referred to as the catalytic core domain (21, 35, 101). In light of this it is likely that the DDE triad in Tn*10* transposase is also present on a distinct structural domain as opposed to being split into two structural domains as has been proposed previously (78). In addition, sequence homology between Tn*5* and Tn*10* transposase proteins in their extreme N-terminal regions (amino acids 28 to 47 in Tn*10* transposase and amino acids 35 to 54 in Tn*5* transposase) raises the strong possibility that these regions perform similar, if not identical functions (85). Consistent with this is the observation that the α fragment in Tn*10* transposase has DNA-binding activity (78).

The structure of the Tn*5* synaptic complex revealed that each transposon end is bound by domains contributed by separate transposase monomers. In this arrangement the monomer that is tethered to the primary transposase binding determinants at one end provides its active site to the other end so that cleavage occurs in *trans*. This arrangement ensures that reaction chemistry cannot occur until the two transposon ends are paired (32). It is not yet known if "*trans* cleavage" occurs in the Tn*10* system. *Trans* cleavage has also been observed in the Mu transposition system and could turn out to be a general feature of the classical DNA transposition systems (114, 133, 136).

Classes of Tn*10* Transposase Mutants

DDE motif mutants

Members of the bacterial transposase-retroviral integrase family contain a triad of acidic amino acids, the so-called DDE motif, that is a critical part of the active site (37, 77, 84). By analogy to related phosphoryl transfer systems, such as DNA polymerase I of *E. coli*, it has been inferred that the role of the DDE residues in the active site is to bind one or more divalent metal ions that function directly in catalysis (91). In the Tn*10* system, D97, D161, and E292 make up the DDE catalytic triad (17). These residues were identified by isolating mutants that could perform transpososome assembly but not reaction chemistry. In one approach, acidic amino acids in conserved regions of transposase were targeted for mutagenesis and individual mutant proteins were partially purified and screened in vitro for the activities above (17). In another approach, random chemical mutagenesis was used to first generate a pool of dominant negative transposase mutants. These mutants were then screened in vivo for an excision-minus phenotype and the excision-minus proteins were purified and then tested in vitro for their ability to perform transpososome assembly and catalysis (68). Subsequently, evidence was provided that all three of the DDE residues function in divalent metal ion binding (4). This was done by using an approach that previously had been used in both Mu (T. Baker, unpublished results) and Tn7 (113) transposition systems. Each of the DDE motif residues was individually mutated to cysteine. The biochemical activity of each mutant protein was then tested in the presence of either Mg^{2+} or Mn^{2+}. With respect to divalent metal ion binding ligands, mutating Glu or Asp to Cys effectively swaps a sulfhydryl group for a carboxylate group. Although the latter is a strong ligand for both Mg^{2+} and Mn^{2+}, the former is a strong ligand only for Mn^{2+}. The ability of Cys to substitute for either Glu or Asp coupled with the strict dependence of the reaction on Mn^{2+} can therefore provide strong evidence for a divalent metal ion binding role for the residue in question. All three of the "cysteine" mutants (i.e., DC97, DC161, and EC292) were found to be functional with respect to donor cleavage and strand transfer in the presence of Mn^{2+} but not Mg^{2+}, consistent with a role for these three residues in divalent metal ion binding.

One area in which DDE motif mutants differ is in their ability to perform sequence-specific target capture in staged reactions where precleaved ends are provided. For example, the double mutant DA97/DA161 actually stimulates sequence-specific target capture relative to wild-type transposase under conditions where there is not a strict dependence on a divalent metal ion for target capture (i.e., in the presence of an optimal target site) (4). In contrast, the single mutant EA292 has a severe defect in target capture under these same conditions (63). With regard to the EA292 phenotype, one possibility is that E292 is located in the part of the protein that is also responsible for target capture. In the Tn*5* structure the E of the DDE motif is located on an α helix that extends into a cavity that is wide enough to accommodate a DNA duplex and thus represents a potential binding site for the target DNA (32). For Tn*10* transposase a mutation very close to E292, KA299, also causes a defect in target capture (17). This is consistent with the idea that a specific secondary structure element (probably an α helix) that includes E292 and K299 plays an important role in target capture. Residues on this α helix might interact directly with the target DNA. Another possibility that has been raised is that the loss of a divalent metal ion binding ligand (i.e., E292) in the EA292 mutant could destabilize the target capture complex. In this scenario the divalent metal ion bound to E292 would contribute to target capture by interacting with a phosphate group in the target DNA (63).

As noted above the DA97/DA161 double mutant also has interesting target capture properties. Although target capture is enhanced relative to wild type when a divalent metal ion is omitted from the reaction, there is little stimulation of target capture in the presence of a divalent metal ion. By comparison target capture levels for wild-type transposase increase about 6-fold in the presence of a divalent metal ion (4). In the absence of a divalent metal ion the target DNA might approach the D/D residues closely enough to be repelled by the negative charges on the D/D residues and the phosphate oxygen. This would serve to destabilize the target capture complex. However, if the negative charges on the protein were removed by mutating each of the D/D residues to a neutral alanine, as in DA97/DA161, charge repulsion would be eliminated, thereby removing a negative factor in target capture. This could account for the relatively high level of target capture exhibited by DA97/DA161 in the absence of a divalent metal ion.

SOS$^+$Tnsp$^-$ mutants

SOS$^+$Tnsp$^-$ mutants are defective in transposon insertion but functional in donor cleavage. Mutants of this type were originally isolated from a two-part genetic screen in which mutants defective in DNA transposition were subsequently tested for their ability to induce SOS functions. Transposition-minus mutants that retained the ability to induce SOS functions were shown to accumulate an excised transposon fragment, which indicates their competence for donor cleavage and deficiency in some subsequent step in the reaction. A relatively large number (5 to 10%) of randomly identified transposition-minus mutants were in this class. The mutations (which have also been referred to as Exc$^+$Int$^-$) map in clusters in two regions of transposase designated Patch I and Patch II (Fig. 9). Patch I extends from amino acid 100 to 167, although SOS$^+$Tnsp$^-$ mutants are not evenly spread out through this region. In fact, the mutations fall into two clusters, including amino acids 100 to 106 and 134 to 167. As discussed below residues 107 to 127 may correspond to a distinct functional domain involved in binding the transposon-donor junction. Patch II extends from amino acid 243 to 264 (51).

Several SOS$^+$Tnsp$^-$ proteins have been purified and tested for their ability to form a target capture complex with the HisG1 target fragment (including DE100, RQ106, SL135, PS167, RH243, and EK263). All the mutants tested exhibited a significant sequence-specific target capture defect but retained the ability to bind target DNA in a sequence-nonspecific manner (63). One possibility is that a relatively large region of transposase is dedicated to binding the target DNA; accordingly, mutating any one of a large number of amino acid residues within this region results in a target binding defect. This is consistent with the numerous target DNA contacts observed in contact-probing experiments (98). Also, a large number of target DNA contacts are predicted from the numerous basic amino acid residues in the putative target binding cleft in Tn*5* transposase (32).

An alternative mechanism for inhibition of target capture by SOS$^+$Tnsp$^-$ mutants might include blocking the release of flanking donor DNA and/or the nontransferred transposon strand from the active site. Also, as discussed previously, residues in the surface-exposed loop region defined by proteolysis in Patch II may be important in mediating a conformational change that helps reposition the loop region so that it does not inhibit target capture.

Altered target specificity mutants

Mutants of this class exhibit an altered (or more precisely a "relaxed") target specificity (ATS) phenotype. They retain the same preference for base sequence information at the consensus positions of the target core as wild-type transposase; however, they are better able to tolerate deviations from the consensus sequence than wild-type transposase can. This relaxation in target sequence requirements also seems to extend to base pair changes in the flanking regions. Mutants of this type were isolated by using a genetic screen for transposase mutants that had increased transposition frequencies into suboptimal target sites relative to wild-type transposase. Originally, only amino acid substitutions at each of two positions in transposase, amino acids 134 and 249, were found to confer the ATS phenotype (11). Amino acid 134 is in Patch I and amino acid 249 is in the predicted loop region of Patch II (Fig. 9). Subsequently, it was shown that some SOS$^+$Tnsp$^-$ mutants (e.g., RQ106 and RH243) also have an ATS phenotype. This was determined by mapping insertion sites in vivo where it was possible to identify low-frequency transposition events with a papillation assay (62). Furthermore, one particular position 134 allele, CW134, was shown to confer an SOS^+Tnsp^- phenotype (65). These results suggest that SOS^+Tnsp^- and ATS phenotypes result from differentially affecting the same basic function in transposase, that is, target capture. This inference is supported further by the observation that the ATS mutant CY134 is an intragenic suppressor to a variety of different SOS^+Tnsp^- mutants. This suppression can be explained by the CY134 effect on target capture being dominant over the negative effect SOS^+Tnsp^- mutations have on target capture (65).

An intriguing property of the ATS mutant CY134 is that, in contrast to wild type, it forms an extremely stable target capture complex in the absence of a divalent metal ion. This has been shown by competition experiments in which partitioning of the DEBC to different target molecules is monitored after an initial preincubation of the DEBC with one of the two different sized target molecules (P. Pribil and D. Haniford, unpublished results). The model for sequence-specific target capture described above, in which it is proposed that a divalent metal ion helps to stabilize a distorted target DNA structure, could be explained by CY134 having an increased capacity to deform and/or stabilize a deformed target structure. A predisposition to distorting the target DNA could then make CY134 less dependent on making the normal set of contacts with a target site to form a stable-functional target capture complex, thereby accounting for its relaxed specificity.

Hypernicking mutants

Through random in vitro screening of transposase mutants that are defective for transposition, mutants were identified that are defective in the conversion of the initial nicked species to the transposon-end hairpin. Consequently, the product of first-strand nicking accumulates to a much higher level in reactions with these mutants than wild-type reactions—thus the name hypernicking mutants (18). To date six hypernicking mutants have been identified, including AT162, EK263, MI289, PS167, DE100, and YS285 (18) (J. Allingham, A. Dar, and D. Haniford, unpublished results). The first five listed were originally identified as SOS^+Tnsp^- mutants (51). This presents a paradox because SOS^+Tnsp^- mutants all generate the excised transposon intermediate when presented with a supercoiled transposon substrate, yet formation of this product depends on hairpin formation. The answer to this puzzle is likely related to the use of linear substrates for detecting the hypernicking phenotype. Presumably, DNA supercoiling plays an important role in hairpin formation, and in the absence of this factor, certain transposase mutations have exaggerated effects on this stage of the reaction. On the other hand, hairpin formation and strand transfer are both transesterification reactions and require substrate binding to a similar (if not identical) region of the active site (69). Thus, the likelihood of a mutation affecting both steps is expected.

Unlike other members of the group listed above, YS285 is not an SOS^+Tnsp^- mutant. YS285 confers a strong transposition defect, and little or no transposon excision exists in reactions with either supercoiled or linear transposon substrates (51). The tyrosine at amino acid 285 is highly conserved in the IS*4* family of transposase proteins; it is part of the YREK motif (100) (Fig. 9). That a mutation at this residue causes a hypernicking phenotype could be an indication that the hairpin mechanism for donor cleavage is also conserved in the IS*4* family. In support of this the Tn*5* equivalent of Y285, which is Y319 (Color Plate 39), also seems to play a critical role in hairpin formation (32). Y319 in Tn*5* transposase forms a hydrogen bond through its hydroxyl group with the phosphate backbone of the nontransferred strand at the penultimate residue. This interaction is thought to help stabilize the "flipped-out" configuration of the base at this position. However, the YF285 mutant has essentially a wild-type phenotype, even though it is lacking the hydroxyl group required for a hydrogen bond donor (51). This raises the possibility that the primary role of the Y of the YREK motif is to maintain the structural integrity of the active site subsequent to first-strand nicking rather than binding DNA.

Transpososome stability mutants

Transposase mutants that form unstable PECs were identified in a screen for excision-minus mutants. Mutants RH119, SF120, and RQ296 all form PEC reasonably well, but unlike the wild-type PEC these complexes dissociate upon addition of a divalent metal ion (Mg^{2+} or Ca^{2+}) or competitor DNA or high-temperature treatment. This coincides with transposition frequencies that are more than 1000-fold reduced relative to wild type in vivo (66, 68). There is no direct evidence that any of the affected amino acids are involved in binding the transposon ends, but the corresponding residue to R296 in Tn*5* is predicted to be K330 and this residue has been shown to make a minor groove contact with bp 3 of the transferred strand. This is a *trans* contact and accordingly should be important for transpososome stability (32). Also, K156 of HIV integrase, which aligns with R296 of Tn*10* transposase, is predicted to interact with the viral DNA close to the scissile phosphate (60).

R119 and S120 are located in the central region of Patch I where there is a gap with respect to the occurrence of SOS^{+}Tnsp^{-} mutations. Given the different phenotypes conferred by mutations at these positions relative to other positions in Patch I, the central part of Patch I may be a functionally distinct domain. Additional evidence to support this idea has come from the identification of another class of transposase mutants designated as suppressor of end mutations (SEM). These mutants were identified by using a papillation screen in which randomly generated transposase mutants increased the transposition frequency of plasmid substrates containing terminal base pair mutations relative to wild type. One subclass of SEM mutants (class I) includes amino acid substitutions that map to the central portion of Patch I: LF107, VI114, VI121, and AV127. It has been inferred from the activity of these mutants on single- versus double-end substrate mutants that these protein mutants suppress end mutations by influencing a step subsequent to the initial binding of transposase to an end but before the initiation of donor cleavage (110). This, along with the colocalization of mutants for which there is biochemical evidence for effects on PEC stability (i.e., RH119 and SF120), raises the possibility that class I SEM mutants suppress terminal base pair mutations by increasing the stability of the PEC relative to wild-type transposase.

Class II SEM mutants include two alleles of codon 35, DN35 and DK35. DN35 is a slightly (2-fold) hyperactive transposase and, whereas both DN35 and DK35 suppress a variety of terminal base pair mutations at a moderate level (2- to 10-fold), both mutants are extremely strong suppressors of a bp 6 mutation, G-C to T-A, 90- and 35-fold suppression, respectively (110). D35 is in the Nα subdomain of transposase and, as previously mentioned, part of this subdomain, which includes D35, has significant homology to the N-terminal subdomain of Tn*5* transposase (85). This subdomain in Tn*5* transposase functions in binding the outside end and makes numerous contacts in *cis* to residues spanning bp 6 to 20. EK54 in Tn*5* transposase is also a hyperactive transposase mutant. The structure of the Tn*5* synaptic complex shows that K54 forms a base-specific contact with the T at position 10. The hyperactivity of this mutant can thus be rationalized by assuming that replacement of E54 with lysine eliminates a negative factor and contributes a positive factor in PEC assembly. The former would be the loss of an unfavorable charge-charge interaction between E54 and the phosphate backbone and the latter would be the formation of a novel hydrogen bond between K54 and T10 (32). Similarly, DN35/DK35 in Tn*10* transposase might eliminate a negative factor in transposase-end binding by removing an inhibitory negative charge, thus accounting for the general suppression on end mutations and the moderate hyperactivity on wild-type ends. In addition, the large increase in activity on the bp 6 mutation could be an indication that amino acid 35 is close enough to bp 6 to make base-specific contacts. According to this hypothesis, the DN35/DK35 mutations would be exerting their effects primarily at the level of forming a stable initial interaction between transposase and the outside end.

SINGLE-ACTIVE-SITE CATALYSIS

Tn*10* was the first DNA transposition system for which evidence was provided that a single active site is used repeatedly to perform (at each transposon end) the multiple chemical steps in transposition (17). Subsequently, evidence for repeated use of a single active site was obtained in the bacteriophage Mu system (95, 133). This finding was somewhat unexpected because the Mu transpososome consists of four monomers of transposase and therefore potentially four active sites, one for each of the two chemical steps per end. The available evidence in the retroviral integration systems is also consistent with repeated utilization of a single active site (20). Furthermore, structural studies on members of the bacterial transposase-retroviral integrase family of proteins have revealed a remarkably similar active-site structure, implying a common mechanism of active-site function (21, 32, 35, 101). Experiments examining phosphorothioate stereoselectivity in Tn*10*, Mu, and HIV systems have permitted comparison at yet another level for this diverse group of transposable elements, i.e., how the DNA strands are oriented in each of the respective active sites (46, 69, 93).

Evidence for Repeated Use of a Single Active Site in the Tn*10* System

In the course of a complete Tn*10*/IS*10* transposition reaction each of two molecules of transposase present per transpososome performs four phosphoryl transfer steps at each transposon end (Figs. 5 and 6). The available evidence strongly supports the idea that a single active site in each transposase monomer is used repeatedly to catalyze all four steps. Mutations in the catalytic DDE motif block the transitions from PEC to first-strand nicking, first-strand nicking to hairpin formation, hairpin formation to hairpin resolution, and full transposon excision to strand transfer (17, 67). Thus, each of the chemical steps is blocked by any one of three mutations in one functionally related set of amino acids (4). If separate active sites are involved, then there is no a priori reason for a single mutation to block all four steps.

Evidence that the same active site is used repeatedly for the four steps has come from "protein-mixing" experiments. Here reactions containing mixtures of wild-type and catalytic-minus mutants failed to accumulate a singly nicked species at any ratio of wild-type and mutant proteins tested, as would be the case if separate monomers were responsible for nicking each of the two strands. Also, the proportion of strand transfer products formed in experiments with protein mixtures revealed that every doubly cleaved complex underwent double-end strand transfer. This is consistent with the idea that the same monomer that is responsible for donor cleavage at one end is also responsible for the strand transfer of that same end. If different monomers were responsible for donor cleavage and strand transfer steps, then a population of the single-end strand transfer product should have been observed and this was not the case (17). Finally, consistent with the data from the stoichiometric analysis of transpososomes, the distribution of uncleaved, single-end-cleaved, and double-end-cleaved products in mixed reactions was found to best fit a model in which two rather than four transposase molecules are present per transpososome (17, 66).

Stereoselectivity and Chemical Mechanism

Repeated use of a single active site for four consecutive chemical steps requires that the active site can accommodate different DNA substrates. In addition the active site must be able to coordinate substrate engagement in a precise order so that a biologically sensible end product is generated. One of the major challenges to be addressed in the Tn*10* system (and other related systems) is to determine how this coordination is achieved. This problem has been investigated by asking if and how the active site of a transposase (or integrase) discriminates between the two stereoforms of a phosphorothioate (Ps for short) when the Ps replaces a standard phosphate linkage at the scissile position. Discrimination of this type is referred to as "stereoselectivity" and the determination of the stereoselectivity provides information regarding the relative orientation of the substrate in the active site. This is because the two stereoforms of the Ps have different topologies, and as a result, stereoselectivity imposes different constraints on the possible orientations with which the substrate can bind to the active site in a productive manner. Thus, by determining the stereoselectivity at each step of a multistep reaction, information regarding DNA-strand orientation in the active site at the different stages can be obtained. This, along with other information (see below), was used to generate a stereochemical model of the Tn*10* transposition reaction (69). In addition, by comparing the stereoselectivity in different systems, it can be determined whether different active sites are interacting with DNA strands in the same or different orientations and therefore are organized similarly.

The nomenclature used in this type of work is as follows: the two stereoforms of the Ps are referred to as Rp and Sp. For the Rp-Ps a sulfur atom replaces the nonbridging proR oxygen and for the Sp-Ps a sulfur atom replaces the nonbridging proS oxygen atom (36). If catalysis is inhibited by the Rp-Ps but not the Sp-Ps, then the reaction is said to be stereoselective for the Sp-Ps. If the opposite situation is true the reaction is stereoselective for the Rp-Ps.

Stereoselectivity was observed for each of the four chemical steps in Tn*10* transposition. First, strand nicking, hairpin formation, hairpin resolution, and target strand transfer exhibited Rp, Sp, Rp, and Rp stereoselectivities, respectively (69). These stereoselectivities match those of other related reactions. Simple hydrolysis reactions tend to be stereoselective for the Rp-Ps. Examples include donor cleavage in Mu transposition, viral cleavage in HIV integration, restriction enzyme cleavage, and cleavage by the 3′-5′ exonuclease active site of DNA-polymerase I (19, 38, 93). As for Tn*10,* hairpin formation in V(D)J recombination is stereoselective for the Sp-Ps (126). The stereoselectivity of hairpin resolution (Rp) has only been worked out in the Tn*10* system (69). Finally, target strand transfer in Mu transposition and HIV integration exhibits Rp-Ps stereoselectivity (46, 69). These results imply that at comparable stages the respective active sites interact with substrate DNA strands in the same relative orientations. Thus, not only does the catalytic machinery seem to be very similar but the organization of DNA binding sites around the active site is likely very similar in the Tn*10,* Mu, and HIV systems.

Additional information that was required to generate a stereochemical model for Tn*10* transposition included defining the type of reaction chemistry Tn*10* transposase is able to perform. Protein-mediated phosphoryl transfer reactions typically take place by an associative mechanism in which the first step is addition of the nucleophile to give a pentavalent phosphorus intermediate, followed by elimination of the leaving group (75). Associative mechanisms can be broken down further into different classes, some of which can be distinguished by differences in the stereochemical outcomes of the reactions if the scissile phosphate has been substituted with a chiral phosphate such as a Ps. In the in-line mechanism the nucleophile attacks directly opposite from the position where the leaving group exits. This gives rise to a product in which the stereochemical configuration of the Ps has inverted relative to the starting substrate.

However, if two in-line attacks occur as consecutive steps on the same phosphate, then there is retention of the stereochemical configuration in the product. Finally, there is an adjacent attack mechanism where the nucleophile enters on the same side as where the leaving group exits. This also results in retention of stereochemical configuration in the product (40).

Hairpin formation and target strand transfer in Tn*10* transposition were shown by using Ps to proceed with inversion of the stereochemical configuration (69). This is indicative of a one-step in-line attack mechanism where the nucleophile is the 3′-OH group of the transferred strand for both hairpin formation and target strand transfer. Inversion of Ps configuration has also been observed in hairpin formation in V(D)J recombination and target strand transfer in Mu transposition and HIV integration. In all these systems inversion has been demonstrated by treating the respective products with nucleases that have opposite stereoselectivities (i.e., P1 nuclease and snake venom phosphodiesterase) (38, 69, 92, 126). The chemical mechanism of hydrolysis reactions is more difficult to work out relative to transesterification reactions because the product is achiral. The chemical mechanism of each of the two hydrolysis reactions in Tn*10* transposition (i.e., first-strand nicking and hairpin cleavage) has not been defined. However, the reaction analogous to first-strand nicking in Mu transposition has been shown to proceed with inversion and the same situation is likely for viral cleavage in HIV integration (38, 93). This, taken together with the evidence that the Tn*10* active site already uses the one-step in-line attack mechanism for transesterification reactions, makes it very likely that hydrolysis reactions in Tn*10* transposition also take place by the one-step in-line attack mechanism.

A Stereochemical Model for Tn*10* Transposition

Positional information derived from the determination of stereoselectivity and chemical mechanism was used to generate a stereochemical model for the Tn*10* transposition reaction (Color Plate 40 [see color insert]). Construction of this model also required making some well founded assumptions. One assumption was that a common structural basis for the inactivity of specific Ps stereoisomers exists for all the steps. Details of the substrate-active-site structure are not known for any of the chemical steps. However, it was assumed, on the basis of structural details from the 3′-5′ exonuclease domain of DNA polymerase I, that one of the nonbridging oxygen atoms of the scissile phosphate binds to a divalent metal ion in the active site (9). It follows that sulfur substitution at this position would block this interaction and thereby prevent reaction chemistry. This, in fact, is the basis for the Rp stereoselectivity documented in this system (19). It was also assumed as discussed above that all the chemical steps take place by the one-step in-line mechanism.

In the model the active site can be viewed as working in two distinct configurations, a "forward" configuration and a "reverse" configuration. In the forward configuration the attacking nucleophile (water) comes in from the right side and the leaving group (the C3′O) exits on the left side. In the reverse configuration the attacking nucleophile (C3′O) comes in from the left side and the leaving group (C3′O) exits from the right side. In this view, hairpin resolution apparently is essentially a repeat of the first-step, first-strand nicking. The only major difference is that one of the equatorial positions of the triagonal bipyramid is occupied by the C5′O of the bottom strand of the flanking donor DNA in the first step and the C5′O of the top strand in the hairpin resolution step. Hairpin formation involves utilization of the reverse configuration of the active site. It is assumed that the C3′O that was the leaving group after the first step remains bound to the active-site pocket and is in position to attack the top strand of the transposon. The C3′O of the transferred strand has in fact been shown to be bound to a divalent metal ion in the Tn*5* synaptic complex structure (32). Hairpin opening can be viewed as the simple reversal of hairpin formation. This mechanism makes it possible for the single active site to cleave both transposon strands without release of both ends of the bottom strand from the active site.

The stereochemical model for Tn*10* transposition does not depend on a two metal ion active site. Different functional groups in transposase could move into position to act as general acid and general base catalysts at different stages of the reaction. However, this makes for a substantially more complex active site.

Although hairpin resolution is essentially a repeat of first-strand nicking, is the same true for hairpin formation and target strand transfer? The answer to this question appears to be no. This is because the two steps have opposite stereoselectivities (Sp for hairpin formation and Rp for target strand transfer), and this implies that the DNA strands (top strand of transposon-donor DNA for hairpin formation and target DNA for target strand transfer) are oriented differently in the active site. This difference in substrate orientation may provide a mechanism for protecting the strand transfer product from undergoing additional reaction chemistry, either hydrolysis or disintegration. If target strand transfer were simply a repeat of hairpin formation, there would be no a priori reason that prevents at least hydrolysis of the strand transfer product, a step that would be equivalent to hairpin opening. In fact, studies with a preassembled strand transfer product revealed that this substrate does not undergo hydrolysis or disintegration reactions at detectable levels (66).

Another possibility is that the target strand has the same orientation in the active site as the transposon-donor DNA strand but there is a switch in the orientation of the transferred strand (Color Plate 40H). However, it is not clear if it is practical for such a large movement of the DNA to occur in the confines of the active site.

Both configurations of the target capture complex would generate strand transfer products with unique structural features that could explain the absence of additional reactivity. The strand transfer product depicted in Color Plate 40G has a configuration that is not seen at any other stage of the reaction because the 5′ arm of the DNA is projecting out of the page instead of into the page. In the second model the 5′ arm projects into the page but the 3′ end of the transferred strand has moved to the opposite side of the active site. If the leaving group pocket on the left side of the active site cannot accommodate a water nucleophile, then the strand transfer product would fail to undergo hydrolysis.

Although target strand transfer reactions in the HIV, Mu, and Tn*10* systems all exhibit the same Ps stereoselectivity, disintegration is a fairly robust reaction only in the HIV system (27). This observation could be used as an argument against the idea that Rp stereoselectivity at this stage reflects a built-in mechanism for protecting the strand transfer product. However, it remains to be demonstrated that the substrates used to investigate disintegration in retroviral systems are assembled in transposition complexes equivalent to true strand transfer complexes. It is also worth noting that the RAG recombinase performs the equivalent of disintegration (87, 88).

DNA Binding Determinants in the Active Site

From the stereochemical model, predictions for the organization of DNA binding determinants in the active site can be made. The diagram in Fig. 10 summarizes these predictions. It is apparent that some sites are required to accommodate different DNA determinants at different stages of the reaction. For instance, site A would accommodate both the bottom strand of the flanking donor DNA during first-strand nicking and the top strand of the transposon during hairpin formation. Site C would accommodate the top strand of the flanking DNA during hairpin formation and then target DNA. Each of these sites presumably can only accommodate one DNA strand at any given

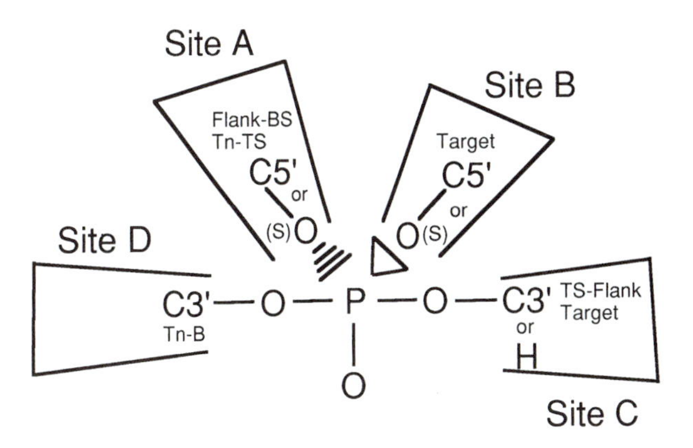

Figure 10. Possible arrangement of DNA strands in the active site as predicted from the stereochemical model shown in Color Plate 40. See text for details.

time. Release of the leaving group from the active site would then be a critical feature of the reaction scheme. It is notable that the timely release of the leaving group seems to be a feature of only the right side of the active site, that is, site C. All the interactions at this site are expected to be of the sequence-nonspecific nature; this could be an important factor in facilitating the release of DNA strands. This includes interactions with the flanking donor DNA and DNA sequences outside of the target core. In contrast, in site D, where it is predicted that there is no release of the leaving group from the active site, the interactions with the DNA are expected to be sequence-specific in nature. This is because the primary determinants for transposase binding to the DNA would be located in this arm of the DNA.

One of the two models for target binding to the active site seems to implicate a new and distinct binding surface in transposase. This is designated site B in Fig. 10. However, the possibility exists that the 5′ arm of the target DNA is actually bound by domain A of the second monomer that is present in the transposome. In this case there would have to be release of the bottom strand flank from this second monomer to accommodate target binding.

Release of the donor flank arm is expected to be a prerequisite for target capture because the flanking region arm of the target DNA is predicted to occupy the same site (site C). In fact, as previously discussed the formation of a stable target capture complex requires prior removal of flanking donor DNA (109). Another prediction is that hairpin opening would be required for target capture. This is because the top strand of the transposon and the scissile phosphate of the hairpin would otherwise be trapped in the active site, thus preventing entry of the target DNA into the same pocket. This possibility has been tested by asking if a transpososome that was assembled using a preformed hairpin substrate could form a stable target capture complex. No target capture was observed with the preformed hairpin substrate unless incubations were conducted in the presence of Mg^{2+} where the hairpin was opened (S. Wardle, K. Mizuuchi, and D. Haniford, unpublished result). Thus, in addition to removal of the flanking DNA, the hairpin must be opened presumably to allow the top-strand arm of the transposon and/or the phosphate group to move out of the active site so that the target DNA can move in.

Divalent Metal Ion Binding to the Active Site

A two-metal ion mechanism

The stereochemical model for Tn*10* transposition predicts the use of two alternate active-site configurations. This requires a change in the chemical properties of the active site at each step. Perhaps the simplest way of achieving this is to use the "two-metal ion mechanism" proposed for alkaline phosphatase and group 1 and 2 self-splicing introns (119). In this mechanism the two divalent metal ions in the active site alternate between roles as general acid and general base catalysts at each step of the reaction. Evidence for the DDE catalytic triad in transposase functioning in divalent metal ion binding has already been discussed. However, all three residues could be involved in coordinating one metal ion or the binding functions may be partitioned in some manner to more than one metal ion.

In an attempt to determine whether the DDE triad of Tn*10* transposase binds more than one divalent metal ion, an experiment was performed in which both Mg^{2+} and Mn^{2+} were included in reactions with active-site mutants DC97 and DC161. The objective was to determine whether the presence of Mg^{2+} would reduce the amount of Mn^{2+} required for catalysis as would be expected if Mn^{2+} was now required to bind to only one instead of two sites. However, addition of Mg^{2+} was not found to lower Mn^{2+} optimum for catalysis (Allingham and Haniford, unpublished).

Furthermore, in the Tn*5* synaptic complex only a single divalent metal ion was observed in the active site. A Mn^{2+} ion was found bound to the C3′O of the transferred strand, D97 and E326; these residues make up the first D and the E of the Tn*5* transposase DDE triad. Surprisingly, the second D (D188) was not found to be a ligand for a divalent metal ion in this structure (32). However, the protein-DNA complex that was crystallized in this study was missing the 5′-PO_4 at the terminus of the nontransferred strand. The 5′-PO_4 could be a critical component for a second divalent metal ion binding site, and its absence could account for the unbound state of D188. One possible argument against this is that the crystallized complex supported strand transfer when it was dissolved and mixed with target DNA. Finally, it is also possible that Tn*10* and Tn*5* transposase proteins simply differ in the number of divalent metal ions they bind. For example, *Eco*RV and *Eco*RI restriction endonucleases have related active-site structures, yet their active sites bind different numbers of divalent metal ions (71, 76, 122, 128, 134).

Changes in divalent metal ion requirements

Mg^{2+} titration experiments have shown differences in the amount of Mg^{2+} required to support reaction chemistry at different stages of Tn*10* transposition. For example, intramolecular strand transfer levels were shown to drop more than 20-fold without a decrease in the levels of transposon excision when the Mg^{2+} concentration was reduced 16.6-fold relative to the standard amount (64). This could be indicative of changes in the affinity of the active site for Mg^{2+} as the reaction progresses. This in turn could reflect changes in the occupancy and positioning of specific DNA strands in the active site and/or changes in the position of different transposase functional groups at different reaction steps. According to the stereochemical model, binding of a Mg^{2+} ion to the right-hand side of the active site would be influenced by the nature of the DNA strand occupying the same side of the active site. Perhaps a small difference in DNA strand position for target versus donor DNA creates a less optimal geometry for Mg^{2+} binding. Alternatively, the active site might be more dynamic than proposed with respect to the movement of functional groups in and out of this part of the protein. Such movements could also alter the geometry and therefore the affinity of a Mg^{2+} binding site in a step-dependent manner.

COMPLEX DOUBLE-STRAND CLEAVAGE REACTIONS IN OTHER SYSTEMS

V(D)J recombination is the process whereby antigen and T-cell receptor genes are assembled from different gene segments in developing B and T cells, respectively. The initial chemical steps in this process are mediated by the RAG recombinase and involve the separation of signal and coding sequences by a double-strand break in the DNA that occurs by a hairpin mechanism (Fig. 5). Evidence for this came initially from the identification of hairpin-coding ends in cells actively undergoing V(D)J recombination and subsequently from in vitro studies using purified RAG1 and RAG2 proteins and model DNA substrates (for recent reviews see references 45, 82, and 96). It is also likely that many plant transposons that transpose by a nonreplicative mechanism use the hairpin mechanism for donor cleavage. Plant transposons, including members of the Ac-Ds family, the Spm family, the Tam family, dLute, and Ascot-1, leave "donor footprints" at their excision sites (28, 29, 39, 83). Donor footprints are short palindromic repeats of the original donor sequence, and their presence can be explained by off-center cleavage of hairpin ends followed by the filling in of those ends that contain a 5′ overhang by a DNA polymerase. However, direct proof of a hairpin mechanism in these systems still awaits the actual detection of a hairpin intermediate. Many other bacterial transposons that transpose by nonreplicative transposition might also use the hairpin mechanism for donor cleavage, although some notable exceptions that have already been discussed include Tn*7* and IS*911*. Other proteins shown to make double-strand breaks in DNA via a hairpin mechanism include the replication initiator protein gpII of the filamentous phage f1(47) and vaccinia virus DNA topoisomerase I (118). Although in these systems double strand break formation via the hairpin mechanism is a side reaction.

V(D)J Recombination

Links to bacterial transposition systems

There is reason to believe that the V(D)J recombination system is closely related to DNA transposition systems. One of the most compelling arguments in support of this is the observation that RAG1/RAG2 can catalyze DNA transposition reactions in vitro (2, 57). This issue is reviewed in reference 107. Additional support for this argument comes more recently from the observations that RAG1 contains a DDE motif (43, 70, 79) and that both RAG1 and Tn*10* transposase possess a 3′ flap endonuclease activity (112).

As shown in Fig. 5, a notable difference between hairpin formation in the V(D)J system relative to bacterial transposition systems (i.e., Tn*10* and Tn*5*) is that in the former the hairpin is formed at the coding end, which is formally equivalent to the donor flank in transposition systems. This is due to first strand nicking on the equivalent of the nontransferred strand of a bacterial transposon.

Multiple pathways for processing DNA ends by the RAG recombinase

As in DNA transposition, the chemical steps mediated by the RAG recombinase take place in the context of a synaptic complex in which recombination signal sequences are paired. The available evidence is consistent with the idea that after hairpin formation the coding ends and the signal ends remain associated in a "cleaved signal complex" (CSC) (3, 55, 56, 125). Different products could then arise depending on how the relevant DNA strands are processed in this complex (see Fig. 3 in reference 42). In one pathway the RAG recombinase or perhaps the Rad50/Mre11/NBS DNA repair complex can open up the coding end hairpins by hydrolysis. The former would be equivalent to the third chemical step in Tn*10*/Tn*5* transposition. In a second pathway, the 3′-OH of the exposed signal end can attack the hairpin, generating an open and shut or hybrid joint. This would be equivalent to the reversal of hairpin formation but by transesterification instead of hydrolysis as occurs in Tn*10*/Tn*5* transposition. In a third pathway, the coding ends (unresolved) dissociate from the active site whereas the signal ends remain bound and could subsequently be joined to a DNA duplex that replaces the coding ends, leading to a signal end transposition event.

It has been proposed that all the steps in the three pathways could be catalyzed by a single active site in RAG1/RAG2 with use of a two-metal ion mechanism for catalysis (42). The model is basically a variation of that originally proposed for Tn*10* transposition to account for the alternation between hydrolysis and transesterification reactions performed by a single active site.

The greater variety of products formed in the V(D)J than in the Tn*10* system could in part be because the active site in RAG1/RAG2 is more flexible with regard to its ability to activate different nucleophiles for attack on one side of the active site. That is, open-and-shut joint formation and signal-end transposition both require that the "right side" of the RAG1/RAG2 active site (in Fig. 3 of reference 42) can activate a C3′O terminus for nucleophilic attack, whereas first-strand nicking and hairpin resolution require activation of a water-nucleophile from the same side. In contrast, it has been proposed that the Tn*10* active site is set up in such a way that only a water-nucleophile can attack from the right side (Color Plate 40). This would exclude the possibility of the C3′O of the flanking DNA opening the hairpin and forming the equivalent of an open-and-shut joint.

In addition, in Tn*10* transposition hairpin resolution must necessarily precede target strand transfer to expose a 3′-OH terminus for joining to the target DNA. This is not the case in V(D)J recombination as the 3′-OH group that is joined to the target DNA in signal-end transposition is exposed immediately after hairpin formation. Thus, if the coding end can readily diffuse out of the cleaved signal complex (and be replaced by a target DNA strand), target strand transfer can occur immediately after hairpin formation. The relative frequency of coding end hairpin resolution events and signal-sequence transposition might then reflect the stability of the cleaved signal-end complex.

Acknowledgments. I thank N. Kleckner for helpful discussions, R. Chalmers for providing unpublished results and helpful discussions, and W. Reznikoff for providing data before publication. I thank members of my own laboratory, J. Allingham and P. Pribil, Simon Wardle, Angie Kennedy, and Barry Stewart, for providing data and illustrations for this review. I also thank R. Chalmers for providing illustrations and M. Junop for providing the alignment between Tn*10* and Tn*5* transposase.

All the work described in this chapter that was performed in my own laboratory was funded by grants from the Medical Research Council of Canada and The Canadian Institutes for Health Research.

REFERENCES

1. **Adzuma, K., and K. Mizuuchi.** 1988. Target immunity of Mu transposition reflects a differential distribution of MuB protein. *Cell* **53:**257–266.
2. **Agrawal, A., Q. M. Eastman, and D. G. Schatz.** 1998. Transposition mediated by RAG1 and RAG2 and its implications for the evolution of the immune system. *Nature* **394:** 744–751.
3. **Agrawal, A., and D. G. Schatz.** 1997. RAG1 and RAG2 form a stable post-cleavage synaptic complex with DNA containing signal ends in V(D)J recombination. *Cell* **89:**43–53.
4. **Allingham, J. S., P. A. Pribil, and D. B. Haniford.** 1999. All three residues of the Tn10 transposase DDE catalytic triad function in divalent metal ion binding. *J. Mol. Biol.* **289:** 1195–1206.
5. **Arciszewska, L. K., D. Drake, and N. L. Craig.** 1989. Transposon Tn7 cis-acting sequences in transposition and transposition immunity. *J. Mol. Biol.* **207:**35–52.
6. **Bainton, R., P. Gamas, and N. L. Craig.** 1991. Tn7 transposition in vitro proceeds through an excised transposon intermediate generated by staggered breaks in DNA. *Cell* **65:** 805–816.
7. **Baker, T. A., and K. Mizuuchi.** 1992. DNA-promoted assembly of the active tetramer of the Mu transposase. *Genes Dev.* **6:**2221–2232.
8. **Baker, T. A., M. Mizuuchi, and K. Mizuuchi.** 1991. MuB protein allosterically activates strand transfer by transposase of phage Mu. *Cell* **65:**1003–1013.
9. **Beese, L. S., and T. A. Steitz.** 1991. Structural basis for the 3′–5′ exonuclease activity of Escherichia coli DNA polymerase I: a two metal ion mechanism. *EMBO J.* **10:**25–33.
10. **Bender, J., and N. Kleckner.** 1986. Genetic evidence that Tn10 transposes by a nonreplicative mechanism. *Cell* **45:** 801–815.
11. **Bender, J., and N. Kleckner.** 1992. IS10 transposase mutations that specifically alter target site recognition. *EMBO J.* **11:**741–750.
12. **Bender, J., and N. Kleckner.** 1992. Tn10 insertion specificity is strongly dependent upon sequences immediately adjacent to the target-site consensus sequence. *Proc. Natl. Acad. Sci. USA* **89:**7996–8000.
13. **Benjamin, H. W., and N. Kleckner.** 1992. Excision of Tn10 from the donor site during transposition occurs by flush double-strand cleavages at the transposon termini. *Proc. Natl. Acad. Sci. USA* **89:**4648–4652.
14. **Benjamin, H. W., and N. Kleckner.** 1989. Intramolecular transposition by Tn10. *Cell* **59:**373–383.
15. **Bennett, S. P., and S. E. Halford.** 1989. Recognition of DNA by type II restriction enzymes. *Curr. Top. Cell. Regul.* **30:** 57–104.
16. **Bhasin, A., I. Y. Goryshin, and W. S. Reznikoff.** 1999. Hairpin formation in Tn5 transposition. *J. Biol. Chem.* **274:** 37021–37029.
17. **Bolland, S., and N. Kleckner.** 1996. The three chemical steps of Tn10/IS10 transposition involve repeated utilization of a single active site. *Cell* **84:**223–233.
18. **Bolland, S., and N. Kleckner.** 1995. The two single-strand cleavages at each end of Tn10 occur in a specific order during transposition. *Proc. Natl. Acad. Sci. USA* **92:**7814–7818.
19. **Brautigam, C. A., and T. A. Steitz.** 1998. Structural principles for the inhibition of the 3′–5′ exonuclease activity of Escherichia coli DNA polymerase I by phosphorothioates. *J. Mol. Biol.* **277:**363–377.
20. **Brown, P. O.** 1997. Integration, p. 161–204. *In* L. M. Coffin, S. H. Hughes, and H. E. Varmus (ed.), *Retroviruses.* Cold Spring Harbor Laboratory Press, Cold Spring Harbor, N.Y.
21. **Bujacz, G., M. Jaskolski, J. Alexandratos, A. Wlodawer, G. Merkel, R. A. Katz, and A. M. Skalka.** 1995. High-resolution structure of the catalytic domain of avian sarcoma virus integrase. *J. Mol. Biol.* **253:**333–346.
22. **Chaconas, G., B. D. Lavoie, and M. A. Watson.** 1996. DNA transposition: jumping gene machine, some assembly required. *Curr. Biol.* **6:**817–820.
23. **Chalmers, R., A. Guhathakurta, H. Benjamin, and N. Kleckner.** 1998. IHF modulation of Tn10 transposition: sensory transduction of supercoiling status via a proposed protein/DNA molecular spring. *Cell* **93:**897–908.
24. **Chalmers, R., S. Sewitz, K. Lipkow, and P. Crellin.** 2000. Complete nucleotide sequence of Tn10. *J. Bacteriol.* **182:** 2970–2972.
25. **Chalmers, R. M., and N. Kleckner.** 1996. IS10/Tn10 transposition efficiently accommodates diverse transposon end configurations. *EMBO J.* **15:**5112–5122.
26. **Chalmers, R. M., and N. Kleckner.** 1994. Tn10/IS10 transposase purification, activation, and in vitro reaction. *J. Biol. Chem.* **269:**8029–8035.
27. **Chow, S. A., K. A. Vincent, V. Ellison, and P. O. Brown.** 1992. Reversal of integration and DNA splicing mediated by integrase of human immunodeficiency virus. *Science* **255:** 723–726.
28. **Coen, E., T. Robbins, J. Almeida, A. Hudson, and R. Carpenter.** 1989. Consequences and mechanisms of transposition in Antirrhinum majus, p. 413–436. *In* D. E. Berg and M. M. Howe (ed.), *Mobile DNA.* American Society for Microbiology, Washington, D.C.
29. **Colot, V., V. Haedens, and J. L. Rossignol.** 1998. Extensive, nonrandom diversity of excision footprints generated by Ds-like transposon Ascot-1 suggests new parallels with V(D)J recombination. *Mol. Cell. Biol.* **18:**4337–4346.
30. **Craigie, R., D. J. Arndt-Jovin, and K. Mizuuchi.** 1985. A defined system for the DNA strand-transfer reaction at the initiation of bacteriophage Mu transposition: protein and DNA substrate requirements. *Proc. Natl. Acad. Sci. USA* **82:** 7570–7574.
31. **Cuomo, C. A., C. L. Mundy, and M. A. Oettinger.** 1996. DNA sequence and structure requirements for cleavage of V(D)J recombination signal sequences. *Mol. Cell. Biol.* **16:** 5683–5690.
32. **Davies, D. R., I. Y. Goryshin, W. S. Reznikoff, and I. Ray-**

ment. 2000. Three-dimensional structure of the Tn5 synaptic complex transposition intermediate. *Science* **289:**77–85.
33. **Davies, D. R., L. Mahnke Braam, W. S. Reznikoff, and I. Rayment.** 1999. The three dimensional structure of a Tn5 transposase-related protein determined to 2.9 A resolution. *J. Biol. Chem.* **274:**11904–11913.
34. **Derbyshire, K. M., L. Hwang, and N. D. F. Grindley.** 1987. Genetic analysis of the interaction of the insertion sequence IS903 transposase with its terminal inverted repeats. *Proc. Natl. Acad. Sci. USA* **84:**8049–8053.
35. **Dyda, F., A. B. Hickman, T. M. Jenkins, A. Engelman, R. Craigie, and D. R. Davies.** 1994. Crystal structure of the catalytic domain of HIV-1 integrase: similarity to other polynucleotidyl transferases. *Science* **266:**1981–1985.
36. **Eckstein, F.** 1985. Nucleoside phosphorothioates. *Annu. Rev. Biochem.* **54:**367–402.
37. **Engelman, A., and R. Craigie.** 1992. Identification of conserved amino acid residues critical for human immunodeficiency virus type 1 integrase function in vitro. *J. Virol.* **66:** 6361–6369.
38. **Engelman, A., K. Mizuuchi, and R. Craigie.** 1991. HIV-1 DNA integration: mechanism of viral DNA cleavage and DNA strand transfer. *Cell* **67:**1211–1221.
39. **Fedoroff, N. V.** 1989. Maize transposable elements, p. 375–411. *In* D. E. Berg and M. M. Howe (ed.), *Mobile DNA*. American Society for Microbiology, Washington, D.C.
40. **Fersht, A.** 1999. Stereochemistry of enzymatic reactions, p. 245–272. *In* M. Russell Julet (ed.), *Structure and Mechanism in Protein Science*. W. H. Freeman and Company, New York, N.Y.
41. **Foster, T. J., M. A. Davis, D. E. Roberts, K. Takeshita, and N. Kleckner.** 1981. Genetic organization of transposon Tn10. *Cell* **23:**201–213.
42. **Fugmann, S. D., A. I. Lee, P. E. Shockett, I. J. Villey, and D. G. Schatz.** 2000. The RAG proteins and V(D)J recombination: complexes, ends, and transposition. *Annu. Rev. Immunol.* **18:**495–527.
43. **Fugmann, S. D., I. J. Villey, L. M. Ptaszek, and D. G. Schatz.** 2000. Identification of two catalytic residues in RAG1 that define a single active site within the RAG1/RAG2 protein complex. *Mol. Cell.* **5:**97–107.
44. **Fujiwara, T., and K. Mizuuchi.** 1988. Retroviral DNA integration: structure of an integration intermediate. *Cell* **54:** 497–504.
45. **Gellert, M.** 1997. Recent advances in understanding V(D)J recombination. *Adv. Immunol.* **64:**39–64.
46. **Gerton, J. L., D. Herschlag, and P. O. Brown.** 1999. Stereospecificity of reactions catalyzed by HIV-1 integrase. *J. Biol. Chem.* **274:**33480–33487.
47. **Greenstein, D., and K. Horiuchi.** 1989. Double-strand cleavage and strand joining by the replication initiator protein of the filamentous phage f1. *J. Biol. Chem.* **264:**12627–12632.
48. **Halling, S. M., and N. Kleckner.** 1982. A symmetrical six-base-pair target site sequence determines Tn10 insertion specificity. *Cell* **28:**155–163.
49. **Halling, S. M., R. W. Simons, J. C. Way, R. B. Walsh, and N. Kleckner.** 1982. DNA sequence organization of IS10-right of Tn10 and comparison with IS10-left. *Proc. Natl. Acad. Sci. USA* **79:**2608–2612.
50. **Haniford, D. B., H. W. Benjamin, and N. Kleckner.** 1991. Kinetic and structural analysis of a cleaved donor intermediate and a strand transfer intermediate in Tn10 transposition. *Cell* **64:**171–179.
51. **Haniford, D. B., A. R. Chelouche, and N. Kleckner.** 1989. A specific class of IS10 transposase mutants are blocked for target site interactions and promote formation of an excised transposon fragment. *Cell* **59:**385–394.
52. **Haniford, D. B., and N. Kleckner.** 1994. Tn10 transposition in vivo: temporal separation of cleavages at the two transposon ends and roles of terminal basepairs subsequent to interaction of ends. *EMBO J.* **13:**3401–3411.
53. **Harayama, S., T. Oguchi, and T. Iino.** 1984. Does Tn10 transpose via the cointegrate molecule? *Mol. Gen. Genet.* **194:**444–450.
54. **Heitman, J.** 1992. How the EcoRI endonuclease recognizes and cleaves DNA. *Bioessays* **14:**445–454.
55. **Hiom, K., and M. Gellert.** 1998. Assembly of a 12/23 paired signal complex: a critical control point in V(D)J recombination. *Mol. Cell.* **1:**1011–1019.
56. **Hiom, K., and M. Gellert.** 1997. A stable RAG1-RAG2-DNA complex that is active in V(D)J recombination. *Cell* **88:** 65–72.
57. **Hiom, K., M. Melek, and M. Gellert.** 1998. DNA transposition by the RAG1 and RAG2 proteins: a possible source of oncogenic translocations. *Cell* **94:**463–470.
58. **Huisman, O., P. R. Errada, L. Signon, and N. Kleckner.** 1989. Mutational analysis of IS10's outside end. *EMBO J.* **8:** 2101–2109.
59. **Jeltsch, A., J. Alves, T. Oelgeschlager, H. Wolfes, G. Maass, and A. Pingoud.** 1993. Mutational analysis of the function of Gln115 in EcoRI restriction endonuclease, a critical amino acid for recognition of the inner thymidine residues in the sequence-GAATTC- and for coupling specific DNA binding to catalysis. *J. Mol. Biol.* **229:**221–234.
60. **Jenkins, T. M., D. Esposito, A. Engelman, and R. Craigie.** 1997. Critical contacts between HIV-1 integrase and viral DNA identified by structure-based analysis and photo-cross-linking. *EMBO J.* **16:**6849–6859.
61. **Jilk, R. A., D. York, and W. S. Reznikoff.** 1996. The organization of the outside end of transposon Tn5. *J. Bacteriol.* **178:** 1671–1679.
62. **Junop, M.** 1997. Ph.D. thesis. University of Western Ontario, London, Ontario, Canada.
63. **Junop, M. S., and D. B. Haniford.** 1997. Factors responsible for target site selection in Tn10 transposition: a role for the DDE motif in target DNA capture. *EMBO J.* **16:**2646–2655.
64. **Junop, M. S., and D. B. Haniford.** 1996. Multiple roles for divalent metal ions in DNA transposition: distinct stages of Tn10 transposition have different Mg^{2+} requirements. *EMBO J.* **15:**2547–2555.
65. **Junop, M. S., D. Hockman, and D. B. Haniford.** 1994. Intragenic suppression of integration-defective IS10 transposase mutants. *Genetics* **137:**343–352.
66. **Kennedy, A. K.** 1999. Ph.D. thesis. University of Western Ontario, London, Ontario, Canada.
67. **Kennedy, A. K., A. Guhathakurta, N. Kleckner, and D. B. Haniford.** 1998. Tn10 transposition via a DNA hairpin intermediate. *Cell* **95:**125–134.
68. **Kennedy, A. K., and D. B. Haniford.** 1996. Isolation and characterization of IS10 transposase separation of function mutants: identification of amino acid residues in transposase that are important for active site function and the stability of transposition intermediates. *J. Mol. Biol.* **256:**533–547.
69. **Kennedy, A. K., D. B. Haniford, and K. Mizuuchi.** 2000. Single active site catalysis of the successive phosphoryl transfer steps by DNA transposases: insights from phosphorothioate stereoselectivity. *Cell* **101:**295–305.
70. **Kim, D. R., Y. Dai, C. L. Mundy, W. Yang, and M. A. Oettinger.** 1999. Mutations of acidic residues in RAG1 define the active site of the V(D)J recombinase. *Genes Dev.* **13:** 3070–3080.

71. **Kim, Y., J. C. Grable, R. Love, P. J. Greene, and J. M. Rosenberg.** 1990. Refinement of EcoRI endonuclease crystal structure: a revised protein chain tracing. *Science* **249:**1307–1309.
72. **Kleckner, N.** 1979. DNA sequence analysis of Tn10 insertions: origin and role of the 9-bp flanking repititions during Tn10 translocation. *Cell* **16:**711–720.
73. **Kleckner, N.** 1989. Transposon Tn*10*, p. 227–268. *In* D. E. Berg and M. M. Howe (ed.), *Mobile DNA*. American Society for Microbiology, Washington, D.C.
74. **Kleckner, N., R. M. Chalmers, D. Kwon, J. Sakai, and S. Bolland.** 1996. Tn10 and IS10 transposition and chromosome rearrangements: mechanism and regulation in vivo and in vitro. *In* H. Saedler and A. Gierl (ed.), *Current Topics in Microbiology and Immunology,* vol. 204. *Transposable Elements.* Springer, Berlin, Germany.
75. **Knowles, J. R.** 1980. Enzyme-catalyzed phosphoryl transfer reactions. *Annu. Rev. Biochem.* **49:**877–919.
76. **Kostrewa, D., and F. K. Winkler.** 1995. Mg^{2+} binding to the active site of EcoRV endonuclease: a crystallographic study of complexes with substrate and product at 2A resolution. *Biochemistry* **34:**683–696.
77. **Kulkosky, J., K. S. Jones, R. A. Katz, J. P. G. Mack, and A. M. Skalka.** 1992. Residues critical for retroviral integrative recombination in a region that is highly conserved among retroviral/retrotransposon integrases and bacterial insertion sequence transposases. *Mol. Cell. Biol.* **12:**2331–2338.
78. **Kwon, D., R. M. Chalmers, and N. Kleckner.** 1995. Structural domains of IS10 transposase and reconstitution of transposition activity from proteolytic fragments lacking an interdomain linker. *Proc. Natl. Acad. Sci. USA* **92:**8234–8238.
79. **Landree, M. A., J. A. Wibbenmeyer, and D. B. Roth.** 1999. Mutational analysis of RAG1 and RAG2 identifies three catalytic amino acids in RAG1 critical for both cleavage steps of V(D)J recombination. *Genes Dev.* **13:**3059–3069.
80. **Lavoie, B. D., B. S. Chan, R. G. Allison, and G. Chaconas.** 1991. Structural aspects of a higher order nucleoprotein complex: induction of an altered DNA structure at the Mu-host junction of the Mu Type 1 transpososome. *EMBO J.* **10:** 3051–3059.
81. **Lewis, L. A., and N. D. F. Grindley.** 1997. Two abundant intramolecular transposition products resulting from reactions initiated at a single end suggest that IS2 transposes by an unconventional pathway. *Mol. Microbiol.* **25:**517–529.
82. **Lewis, S. M.** 1994. The mechanism of V(D)J joining: lessons from molecular, immunological and comparitive analyses. *Adv. Immunol.* **56:**27–150.
83. **Luck, J. E., G. J. Lawrence, E. J. Finnegan, D. A. Jones, and J. G. Ellis.** 1998. A flax transposon identified in two spontaneoius mutant alleles of the L6 rust resistance gene. *Plant J.* **16:**365–369.
84. **Mahillon, J., and M. Chandler.** 1998. Insertion sequences. *Microbiol. Mol. Biol. Rev.* **62:**725–774.
85. **Mahnke Braam, L. A., and W. S. Reznikoff.** 1998. Functional characterization of the Tn5 transposase by limited proteolysis. *J. Biol. Chem.* **273:**10908–10913.
86. **Martell, A. E., and R. M. Smith.** 1977. *Critical Stability Constants,* vol. 3. Plenum, New York, N.Y.
87. **Melek, M., and M. Gellert.** 2000. RAG1/2-mediated resolution of transposition intermediates: two pathways and possible consequences. *Cell* **101:**625–633.
88. **Melek, M., M. Gellert, and D. C. van Gent.** 1998. Rejoining of DNA by the RAG1 and RAG2 proteins. *Science* **280:** 301–303.
89. **Mizuuchi, K.** 1983. In vitro transposition of bacteriophage Mu: a biochemical approach to a novel replication reaction. *Cell* **35:**785–794.
90. **Mizuuchi, K.** 1984. Mechanism of transposition of bacteriophage Mu: polarity of the strand transfer reaction at the initiation of transposition. *Cell* **39:**395–404.
91. **Mizuuchi, K.** 1997. Polynucleotidyl transfer reactions in site-specific DNA recombination. *Genes Cells.* **2:**1–12.
92. **Mizuuchi, K., and K. Adzuma.** 1991. Inversion of the phosphate chirality at the target site of Mu DNA strand transfer: evidence for a one-step transesterification mechanism. *Cell* **66:**129–140.
93. **Mizuuchi, K., T. J. Nobbs, S. E. Halford, K. Adzuma, and J. Qin.** 1999. A new method for determining the stereochemistry of DNA cleavage reactions: application to the SfiI and HpaII restriction endonucleases and to the MuA transposase. *Biochemistry* **38:**4640–4648.
94. **Naigamwalla, D. Z., and G. Chaconas.** 1997. A new set of DNA transposition intermediates: alternate pathways of target capture preceding strand transfer. *EMBO J.* **16:** 5227–5234.
95. **Namgoong, S.-Y., and R. M. Harshey.** 1998. The same two monomers within a MuA tetramer provide the DDE domains for strand cleavage and strand transfer steps of transposition. *Cell* **17:**3775–3785.
96. **Oettinger, M. A.** 1999. V(D)J recombination: on the cutting edge. *Curr. Opin. Cell Biol.* **11:**325–329.
97. **Polard, P., B. Ton-Hoang, L. Haren, M. Betermier, R. Walczak, and M. Chandler.** 1996. IS911-mediated transpositional recombination in vitro. *J. Mol. Biol.* **264:**68–81.
98. **Pribil, P. A., and D. B. Haniford.** 2000. Substrate recognition and induced DNA deformation by transposase at the target capture stage of Tn10 transposition. *J. Mol. Biol.* **303:** 145–159.
99. **Ramsden, D. A., J. F. McBlane, D. C. van Gent, and M. Gellert.** 1996. Distinct DNA sequence and structure requirements for the two steps of V(D)J recombination signal cleavage. *EMBO J.* **15:**3197–3206.
100. **Rezsohazy, R., B. Hallet, J. Delcour, and J. Mahillon.** 1993. The IS4 family of insertion sequences: evidence for a conserved transposase motif. *Mol. Microbiol.* **9:**1283–1295.
101. **Rice, P., and K. Mizuuchi.** 1995. Structure of the bacteriophage. Mu transposase core: a common structural motif for DNA transposition and retroviral integration. *Cell* **82:** 209–220.
102. **Rice, P. A., W. Yang, K. Mizuuchi, and H. A. Nash.** 1996. Crystal structure of an IHF-DNA complex: a protein-induced DNA U-turn. *Cell* **87:**1295–1306.
103. **Richet, E., P. Abcarian, and H. A. Nash.** 1986. The interaction of recombination proteins with supercoiled DNA: defining the role of supercoiling in lambda integrative recombination. *Cell* **46:**1011–1021.
104. **Roberts, D., D. Ascherman, and N. Kleckner.** 1991. IS10 promotes formation of adjacent deletions at low frequency. *Genetics* **128:**37–43.
105. **Roberts, D., and N. Kleckner.** 1988. Tn10 transposition promotes RecA-dependent induction of a lambda prophage. *Proc. Natl. Acad. Sci. USA* **85:**6037–6041.
106. **Ross, D., J. Swan, and N. Kleckner.** 1979. Physical structures of Tn10-promoted deletions and inversions: role of 1400 basepair inverted repetitions. *Cell* **16:**721–731.
107. **Roth, D. B., and N. L. Craig.** 1998. VDJ recombination: a transposase goes to work. *Cell* **94:**411–414.
108. **Sakai, J., R. M. Chalmers, and N. Kleckner.** 1995. Identification and characterization of a pre-cleavage synaptic complex that is an early intermediate in Tn10 transposition. *EMBO J.* **14:**4374–4383.
109. **Sakai, J., and N. Kleckner.** 1997. The Tn10 synaptic complex

can capture a target DNA only after transposon excision. *Cell* **89:**205–214.

110. **Sakai, J., and N. Kleckner.** 1996. Two classes of Tn10 transposase mutants that suppress mutations in the Tn10 terminal inverted repeat. *Genetics* **144:**861–870.
111. **Sakai, J. S., N. Kleckner, X. Yang, and A. Guhathakurta.** 2000. Tn10 transpososome assembly involves a folded intermediate that must be unfolded for target capture and strand transfer. *EMBO J.* **19:**776–785.
112. **Santagata, S., E. Besmer, A. Villa, F. Bozzi, J. S. Allingham, C. Sobacchi, D. B. Haniford, P. Vezzoni, M. C. Nussenzweig, Z. Q. Pan, and P. Cortes.** 1999. The RAG1/RAG2 complex constitute a 3′ flap endonuclease: implications for junctional diversity in V(D)J and transpositional recombination. *Mol. Cell* **4:**935–947.
113. **Sarnovsky, R. J., E. W. May, and N. L. Craig.** 1996. The Tn7 transposase is a heteromeric complex in which DNA breakage and joining activities are distributed between different gene products. *EMBO J.* **15:**6348–6361.
114. **Savilahti, H., and K. Mizuuchi.** 1996. Mu transpositional recombination: donor DNA cleavage and strand transfer in trans by the Mu transposase. *Cell* **85:**271–280.
115. **Savilahti, H., P. A. Rice, and K. Mizuuchi.** 1995. The phage Mu transpososome core: DNA requirements for assembly and function. *EMBO J.* **14:**4893–4903.
116. **Sekine, Y., K. Aihara, and E. Ohtsubo.** 1999. Linearization and transposition of circular molecules of insertion sequence IS3. *J. Mol. Biol.* **294:**21–34.
117. **Shapiro, J. A.** 1979. Molecular model for the transposition and replication of bacteriophage Mu and other transposable elements. *Proc. Natl. Acad. Sci. USA* **76:**1933–1937.
118. **Shuman, S.** 1992. DNA strand transfer reactions catalyzed by vaccinia topoisomerase I. *J. Biol. Chem.* **267:**8620–8627.
119. **Steitz, T. A., and J. A. Steitz.** 1993. A general two-metal-ion mechanism for catalytic RNA. *Proc. Natl. Acad. Sci. USA* **90:**6498–6502.
120. **Surette, M. G., S. J. Buch, and G. Chaconas.** 1987. Transpososomes: stable protein-DNA complexes involved in the in vitro transposition of bacteriophage Mu DNA. *Cell* **49:** 253–262.
121. **Surette, M. G., B. D. Lavoie, and G. Chaconas.** 1989. Action at a distance in Mu DNA transposition: an enhancer-like element is the site of action of supercoiling relief activity by integration host factor (IHF). *EMBO J.* **8:**3483–3489.
122. **Thielking, V., U. Selent, E. Kohler, H. Wolfes, U. Pieper, R. Geiger, U. C, F. K. Winkler, and A. Pingoud.** 1991. Site-directed mutagenesis studies with EcoRV restriction endonuclease to identify regions involved in recognition and catalysis. *Biochemistry* **30:**6416–6422.
123. **Ton-Hoang, B., P. Polard, L. Haren, C. Turlan, and M. Chandler.** 1999. IS911 transposon circles give rise to linear forms that can undergo integration in vitro. *Mol. Microbiol.* **32:**617–627.
124. **Turlan, C., and M. Chandler.** 2000. Playing second fiddle: second-strand processing and liberation of transposable elements from donor DNA. *Trends Microbiol.* **8:**268–274.
125. **van Gent, D. C., K. Hiom, T. T. Paull, and M. Gellert.** 1997. Stimulation of V(D)J cleavage by high mobility group proteins. *EMBO J.* **16:**2665–2670.
126. **van Gent, D. C., K. Mizuuchi, and M. Gellert.** 1996. Similarities between initiation of V(D)J recombination and retroviral integration. *Science* **271:**1592–1594.
127. **Vermote, C. L. M., and S. E. Halford.** 1992. EcoRV restriction endonuclease: communication between catalytic metal ions and DNA recognition. *Biochemistry* **31:**6082–6089.
128. **Vipond, I. B., G. S. Baldwin, and S. E. Halford.** 1995. Divalent metal ions at the active sites of the EcoRV and EcoRI restriction endonucleases. *Biochemistry* **34:**697–704.
129. **Watanabe, T., C. Furuse, and S. Sakaizumi.** 1968. Transduction of various R factors by phage P1 in Escherichia coli and by phage P22 in Salmonella typhimurium. *J. Bacteriol.* **96:** 1791–1795.
130. **Way, J. C., and N. Kleckner.** 1984. Essential sites at transposon Tn10 termini. *Proc. Natl. Acad. Sci. USA* **81:**3452–3456.
131. **Weinert, T. W., K. Derbyshire, F. M. Highson, and N. D. F. Grindley.** 1984. Replicative and conservative transpositional recombination of insertion sequences. *Cold Spring Harbor Symp. Quant. Biol.* **49:**251–260.
132. **Weinrich, M. D., L. Mahnke-Braam, and W. S. Reznikoff.** 1993. A functional analysis of the Tn5 transposase: identification of domains required for DNA binding and multimerization. *J. Mol. Biol.* **241:**166–177.
133. **Williams, T. L., E. L. Jackson, A. Carritte, and T. A. Baker.** 1999. Organization and dynamics of the Mu transpososome: recombination by communication between two active sites. *Genes Dev.* **13:**2725–2737.
134. **Winkler, F. K., D. W. Banner, C. Oefner, D. Tsernoglou, R. S. Brown, S. P. Heathman, R. K. Bryan, P. D. Martin, K. Petratos, and K. S. Wilson.** 1993. The crystal structure of EcoRV endonuclease and of its complexes with cognate and non-cognate DNA fragments. *EMBO J.* **12:**1781–1795.
135. **Yang, C.-C., and H. A. Nash.** 1989. The interaction of E. coli IHF protein with its specific binding sites. *Cell* **57:**869–880.
136. **Yang, J.-Y., M. Jayaram, and R. M. Harshey.** 1996. Positional information within the Mu transposase tetramer: catalytic contributions of individual monomers. *Cell* **85:** 447–455.
137. **Yang, J.-Y., K. Kim, M. Jayaram, and R. M. Harshey.** 1995. A domain sharing model for active site assembly within the MuA tetramer during transposition: the enhancer may specify domain contributions. *EMBO J.* **14:**2374–2384.

Mobile DNA II
Edited by N. L. Craig et al.

Chapter 21

P Transposable Elements in *Drosophila melanogaster*

DONALD C. RIO

Donald C. Rio • Department of Molecular and Cell Biology, University of California, Berkeley, Berkeley, CA 94720-3204.

INTRODUCTION

P transposable elements are one of the best studied eukaryotic mobile DNA elements in metazoans. These elements were initially discovered in the late 1960s because they cause a syndrome of genetic traits termed hybrid dysgenesis (82). Hybrid dysgenesis is the term used to describe a collection of symptoms including high rates of sterility, mutation induction, male recombination (which does not normally occur in *Drosophila melanogaster*), and chromosomal abnormalities and rearrangements (48, 109–111). The effects of hybrid dysgenesis are usually only observed in the germ lines, but not in the soma, of progeny from crosses in which males carrying P transposable elements (termed P or paternally contributing strains) are mated to females that lack autonomously mobile P elements (termed M or maternally contributing strains) (Fig. 1). P strains are typically from natural populations, whereas M strains are normal laboratory stocks. None of the deleterious effects of hybrid dysgenesis are observed in the reciprocal cross of M males × P females or in M × M or P × P crosses. Both the underlying basis for this reciprocal cross effect and the germ line restriction of hybrid dysgenesis have been of considerable interest to those studying the regulation of P element transposition.

The unstable nature and revertability of the mutations caused by hybrid dysgenic crosses first suggested that they might be caused by mobile element insertions. Indeed, a detailed molecular analysis of hybrid dysgenesis-induced mutations at the *white* (w^+) locus allowed the isolation and molecular cloning of P transposable elements (16, 178). The characterization of P elements rapidly led to the development of their use as vectors for efficient germ line gene transfer in *Drosophila* (180, 199). More recently, these elements have found additional and critically important uses as the molecular genetics of *Drosophila* has evolved (54, 179). This utility is likely to increase because the complete genomic DNA sequence of *Drosophila* is now available (141, 181) and because of a recent report of homologous gene targeting (168).

This review focuses on the more recent developments in understanding the mechanism and specificity of P element transposition through detailed biochemical and genetic experiments. It also discusses studies aimed at providing insights into how P element transposition is controlled, how P elements rely on their host cells to provide the functions necessary to aid their successful mobility, and how the damage done to the genomes in which they reside is limited. For reviews about other aspects of P element biology, more historical perspectives, or the invasion of P elements into *Drosophila*, see earlier review articles (50, 52, 55, 106, 107, 159, 160).

P TRANSPOSABLE ELEMENTS IN *DROSOPHILA*

Structure and Protein Products of P Transposable Elements

The initial isolation and DNA sequence analysis of the P elements inserted at the *white* locus following

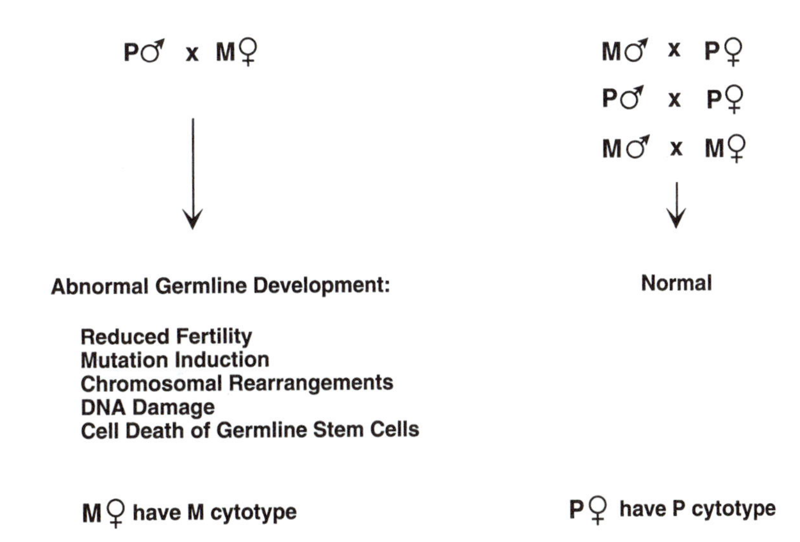

Figure 1. The genetics and symptoms of hybrid dysgenesis. The reciprocal crosses of hybrid dysgenesis are shown. Only when P strain males are mated to M strain females is germ line development abnormal, because of high rates of P element transposition. Progeny from reciprocal M male by P female, P × P or M × M crosses are normal. M females give rise to eggs with a state permissive for P element transposition (M cytotype), whereas P females give rise to eggs with a state restrictive for P element transposition (P cytotype).

dysgenic crosses, and subsequently from P strain genomic DNA libraries, showed that the P elements could vary in size in a typical P strain from 0.5 to 2.9 kb (146). Additional studies showed that P strains carried 50 to 60 P element insertions and that approximately one-third were full-length (2.9 kb), whereas the remainder were internally deleted in different ways (145). All the P elements analyzed to date have a similar structure that includes 31-bp terminal inverted repeats and internal inverted repeats of 11 bp located about 100 bp from the ends (Fig. 2). Between these two repeats, but distant from the site of DNA cleavage, are sites to which the P transposase protein binds (see below; Fig. 2A and B). A series of in vivo studies showed that the 31-bp inverted repeats are required for transposition, that the 11-bp inverted repeats, although not absolutely required, act as transpositional enhancers, and additionally that two different P element ends (a 5′ and 3′ end) must be paired for transposition to occur (94, 140, 180). This pairing of nonequivalent termini during P element transposition is mechanistically similar to the 12/23 rule for pairing the two distinct types of recombination sites during V(D)J recombination in the immune system of vertebrates (154). The spacing between the terminal 31-bp inverted repeats and the transposase binding sites is 21 bp at the 5′ end and 9 bp at the 3′ end (10) similar to the 12- and 23-bp spacers found separating the conserved heptamer and nonamer elements in the V(D)J signals (154). All the P elements that were analyzed initially (146) and subsequently, including a recent set of 2,266 insertions from the Berkeley *Drosophila* Genome Project (BDGP [http://www.fruitfly.org]) showed that an 8-bp duplication of target DNA occurs when P elements integrate (127). Informatics analysis on these data showed that the insertion sites tended to be GC-rich and, in fact, that there was a symmetric palindromic pattern of 14 bp centered on

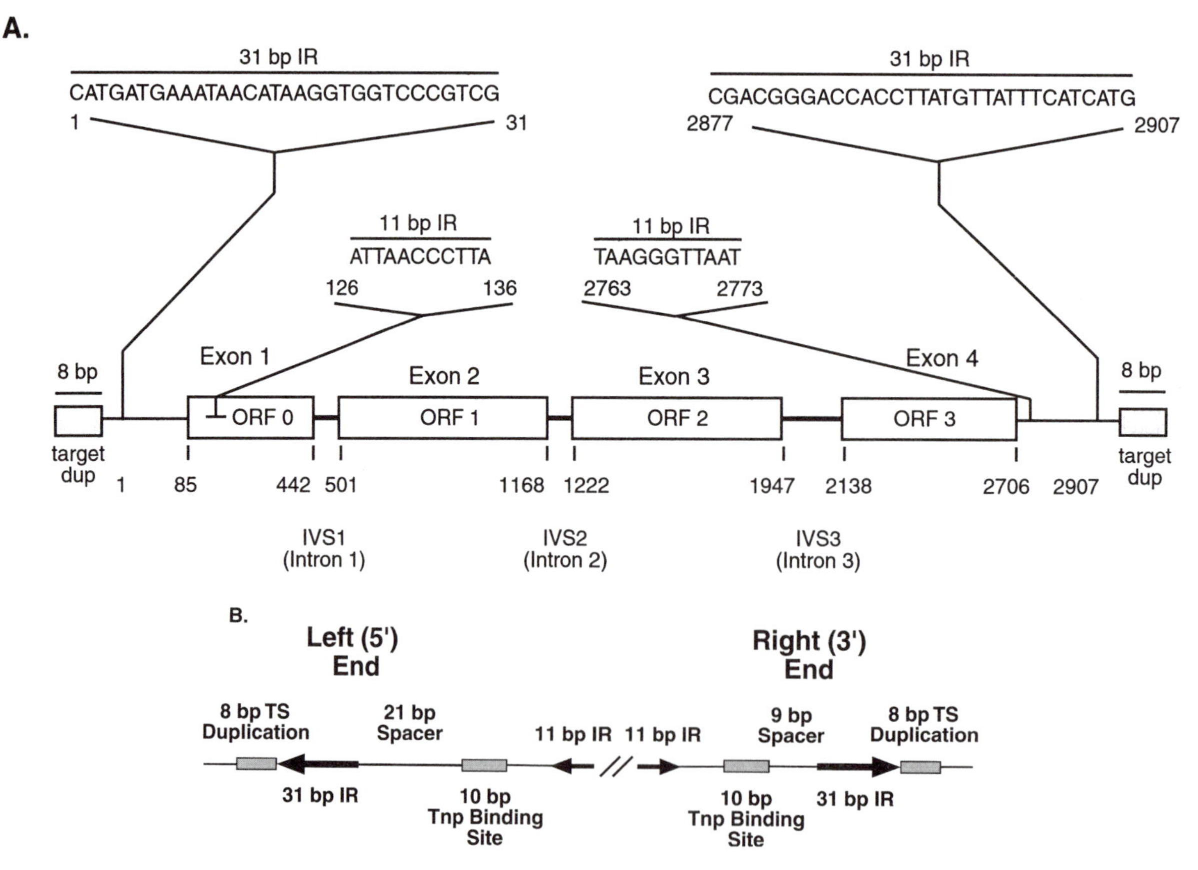

Figure 2. Features of the complete 2.9-kb P element. (A) Sequence features of the 2.9-kb P element. The four coding exons (ORF0, 1, 2, and 3) are indicated by boxes with nucleotide numbers shown. The positions of the three introns (IVS1, 2, 3) are indicated below. The DNA sequences of the terminal inverted 31-bp repeats and the internal 11-bp inverted repeats are shown, with corresponding nucleotide numbers shown above. The 8-bp duplications (dup) of target site DNA are shown by boxes at the ends of the element. (B) *cis*-acting elements of P element transposition. Key sequence features of the left (5′) and right (3′) end are indicated. The terminal 31-bp inverted repeats and the 11-bp internal inverted repeats are indicated by arrows. The transposase binding sites and 8-bp target site duplications are indicated by boxes. The distinct spacer lengths between the 31-bp repeats and the transposase binding sites, 21 bp at the 5′ end and 9 bp at the 3′ end, are indicated above.

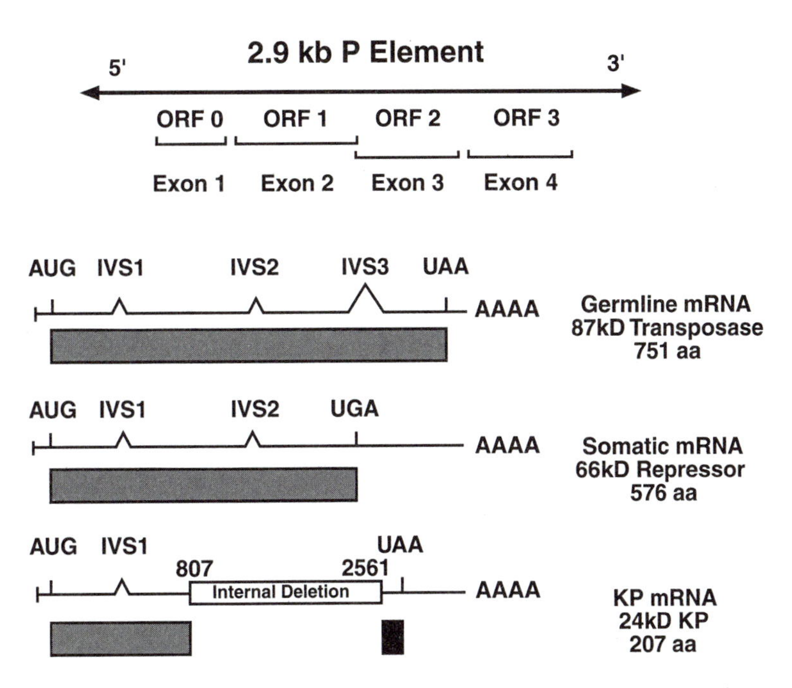

Figure 3. P element mRNAs and proteins. The 2.9-kb P element and four exons (ORF0, 1, 2, and 3) are shown at the top. The germ line mRNA, in which all three introns are removed, encodes the 87-kDa transposase mRNA. The somatic mRNA, from which only the first two introns are removed (and which is also expressed in germ line and somatic cells), encodes the 66-kDa repressor mRNA. Shown at the bottom is a KP element, which contains an internal deletion. This truncated element encodes a 24-kDa repressor protein.

the 8-bp target duplication found at the insertion site (127). This pattern would make sense if the recognition of target site DNA required binding of two (or an even number of) subunits of the transposase protein in a synaptic complex juxtaposing the two P element ends with the insertion sites in the target DNA.

The largest active, complete P element analyzed was 2,907 bp in length and contained four noncontiguous open reading frames (146) (Fig. 2A and 3). Subsequent molecular genetic and biochemical studies showed that all four open reading frames or exons were required for P element mobility and encode the 87-kDa transposase protein (94, 161). These open reading frames are joined together via pre-mRNA splicing (121). In general, the expression of transposase is restricted to germ line cells because splicing of the P element third intron (IVS3) only occurs in the germ line. In addition, it was shown that in somatic cells (and also to a large extent in the germ line) the third intron (IVS3) is retained, producing a functional mRNA that is exported to the cytoplasm and which encodes a 66-kDa protein (161). The 66-kDa protein functions as a repressor of transposition and has been termed a type I repressor (67, 137, 163) (Fig. 3; see the section on the 66-kDa/type I family of repressor proteins, below). In addition, some truncated proteins produced from the smaller, internally deleted P elements can act as repressors of transposition and are called type II repressors (67) (Fig. 3; see the section on the KP/type II family of repressor proteins below). A well-characterized protein of this type II class is the KP protein, identified as a prevalent internally deleted P element found in many naturally occurring populations (18, 87, 194).

In summary, P element transposition in vivo requires about 150 bp of DNA in *cis* at each transposon end, including 31-bp terminal inverted repeats and the transposase binding sites. Upon insertion P elements create an 8-bp duplication of target DNA, and the elements can encode a transposase, as well as repressors of transposition.

Hybrid Dysgenesis and the Control of P Element Transposition

One of the most interesting aspects of hybrid dysgenesis has been the nature of the reciprocal cross effect. The ability of eggs derived from P strain females to repress P element transposition is caused by the presence of a state termed P cytotype that is restrictive for hybrid dysgenesis (48, 49, 108, 109). By contrast, eggs derived from M strain females are permissive for P element transposition and are said to possess the M cytotype. The patterns of inheritance of the P cytotype state are linked to the genotypes (P or M) of the parents in these crosses, yet exhibit both maternal and zygotic components (48, 49, 108, 111, 136, 155). If a mother or great grandmother had P cytotype, then it is probable that the progeny from dysgenic crosses will display P cytotype control. This pattern of inheritance is brought about by a combination of inheritance of P elements and a transfer of components from the females during oogenesis (see "P Cytotype versus Pre-P Cytotype" below). Genetic studies have indicated that although the maternal effect of P cytotype is defined by the maternal cytoplasm and is required for an inhibitory effect, the multigenerational inheritance of the P cytotype state requires that P element DNA be transmitted, either maternally or zygotically, along with maternally derived P strain cytoplasm (79, 169, 171) (see below). Studies using recombinant P elements that mimic certain aspects of P cytotype (67, 136, 137, 163) showed that the 66-kDa protein was a repressor of transposition (137). These and other experiments demonstrated that the genomic position of the repressor elements has a dramatic influence on the potency and maternal effect of the repressive effects (66, 136, 194). It is typically the larger, full-length or near full-length, P elements, but not the much smaller internally deleted ones, that exhibit the germ line maternal effect characteristic of P cytotype (66, 126, 169, 171). Finally, one aspect of P cytotype involves transcriptional repression (125), although the maternal effect is restricted to germ line cells (126). One point regarding the genetics of cytotype control is that it is sometimes difficult to compare effects from different studies because, in addition to the classic test for hybrid gonadal dysgenic (GD) sterility, there are many other commonly used assays for P cytotype repressive effects that can be observed either in the germ line, soma, or both (Table 1). In some cases, repressive effects seen in one assay or with one particular repressor P element are not observed in other cytotype tests. In fact, differences can be observed with the same strains in different assays for P cytotype repression. This strain and assay variability, in addition to the possibility for both transcriptional and posttranscriptional control, have contributed to the difficulties in obtaining a clear understanding P cytotype control.

Some of the most interesting and informative studies regarding the genetics of hybrid dysgenesis have involved the use of a naturally occurring P cytotype repressor strain termed *Lk-P*(1A) (169) (see "P Cytotype versus Pre-P Cytotype" below). Observations regarding the ability of P strain females to block transposition from P strain sperm are akin to both maternal effect and inheritance, but other aspects, such as rapid switching or the stochastic nature of P cytotype, are not explained by these classical definitions (49, 50, 103–105). In particular, a state known as pre-P cytotype can be observed in crosses using *Lk-P*(1A) that gives a single generation of repression, in the absence of the inheritance of P element DNA (171). However, repression of P element mobility also requires the zygotic presence of P element DNA to observe the repressive effects of P cytotype (79, 171). Previous biochemical studies have shown that P strain females can deposit and transmit the 66-kDa repressor protein (and presumably mRNA) in the oocyte cytoplasm, which may play a role as a cytoplasmic determinant of P cytotype. However, a good correlation does not exist between levels of the 66-kDa repressor protein expressed during oogenesis and the ability to transmit P cytotype maternally (136, 137). Other studies have suggested the transmission of an imprinted chromatin state along with the maternal genome as the maternal component of P cytotype (165, 166, 170, 172) (see below), whereas some findings might be consistent with an antisense or double-stranded RNA-mediated maternal transmission of the P cytotype state (see "Possible Control of P Element Transposition by Antisense RNA" below).

Studies on the *Lk-P*(1A) strain have shown that the two complete P elements present in this strain are inserted into the subtelomeric telomere-associated sequences, or *TAS* elements, at the tip of the X-chromosome (165, 170). The *TAS* repeats are 0.8- to 1.8-kb repetitive sequence elements consisting of shorter internal repeat motifs that are often found near the

Table 1. Genetic assays for P cytotype repression

Assay	Tissue(s)	Reference(s)
singed-weak (sn^W) test	Germ line	51, 94, 199
Gonadal dysgenic sterility	Germ line	105
singed female sterility (cytotype-dependent alleles)	Germ line	163
singed-weak (sn^W) bristle mosaics	Soma	121
P[w⁺] white gene excision eye color mosaics	Soma	121
P[w⁺] white gene transposition eye color mosaics	Soma	121
Modified *P[w⁺] white* gene expression	Soma	37
Suppression of Δ2–3 × Birm2 lethality	Soma	56
singed bristle phenotype (cytotype-dependent alleles)	Soma	163
vestigial wing phenotype (cytotype-dependent alleles)	Soma	163, 164
P[lacZ] gene enhancer trap	Soma and/or germ line	125, 126
β-Galactosidase expression	Germ line	
Pre-P cytotype	Germ line	171
trans-Silencing	Germ line	165, 166
Combination effect	Germ line	172

ends of all *Drosophila* chromosomes (95). Mutations in the heterochromatin protein 1 (HP-1) or *Su(var)205* gene reduce the ability of the *Lk-P*(1A) P elements to mediate P cytotype, suggesting a role for subtelomeric heterochromatin in the cytotype effect of this strain (170). Additionally, recombinant enhancer trap P elements inserted at the 1A(X) or 100F(3R) sites can also cause a repressive cytotype effect like the *Lk-P*(1A) P elements or can act in combination with other P elements to produce a P cytotype effect (172). *Lk-P*(1A) can also cause a transcriptional repression in the germ line of heterologous promoters carried in P element vectors inserted into euchromatin (165, 166). This repressive *trans*-silencing effect again suggests that the heterochromatic nature of telomeric P element insertions can be transmitted to other P elements at euchromatic sites (see section on *trans*-silencing, below). The reduction of transcription in the female germ line also alters the ratio of IVS3 splicing, so that under low-transcription conditions a higher percentage of unspliced, 66-kDa repressor-encoding mRNA is produced (166). This and previous genetic observations suggest that either a direct or indirect effect on IVS3 splicing in the germ line may also contribute to increased repressor synthesis and to a reduction in the levels of the transposase protein (126, 145). This type of effect may fit with the notion of a feedback autoregulatory mechanism for maintenance of P cytotype control that involves increasing repressor synthesis or activity (146).

P Elements as Tools for *Drosophila* Genetics

P-element-mediated germ line transformation

One of the most important use of P elements since their identification has been P-element-mediated germ line transformation. The method makes use of the fact that P elements usually only transpose in germ line cells and that a P element carrying a foreign gene can be mobilized in *trans* by using a source of transposase (180, 199) (Fig. 4). Two plasmid DNAs, one carrying the foreign DNA within an internally deleted P element and the second one carrying the transposase gene but lacking P element ends, are coinjected into preblastoderm embryos at the posterior pole where the presumptive germ line cells (pole cells) form (G_0 generation). The recombinant P element is mobilized from the injected plasmid to *Drosophila* chromosomes in the pole cells. These germ line events are then assayed in the next generation after mating (G_1 generation) and transposition is observed in the progeny of this cross (G_2 generation) (Fig. 4). These integrations are stable in the absence of a transposase source, but can be mobilized again by using either injection of a transposase-producing DNA or by crossing to strains that carry a stable source of transposase (78, 164). This procedure is now widely used in many laboratories and typically results in ~20% to 30% of the G_1 progeny carrying the foreign DNA and becoming stably transformed (198).

Mutagenesis and gene tagging

The original experiments describing the molecular basis for hybrid dysgenesis and the cloning of P element DNA used dysgenesis-induced mutations at the *white* locus to isolate the first P elements (16, 178). This was the first example of the use of P elements in directed mutagenesis experiments, but transposon tagging had been used previously with retrotransposons to isolate the *white* locus (17).

Subsequently, two significant improvements have occurred in the use of P elements for mutagenesis experiments. First, the generation and use of single, genetically marked transposons simplified the identification and recovery of induced mutations (38). Second, the identification of single P elements carrying the transposase gene at a defined chromosomal location allowed the ability to induce transposition events by genetic crosses rather than having to inject DNA into embryos (164). Thus, after mobilization the transposase source can easily be removed by genetic crosses and the new insertions isolated in a stable state. This type of single P element mutagenesis has allowed the identification of several thousand lethal P element insertions in distinct genes and has played an important role in the BDGP (http://www.fruitfly.org) (200, 201) (see "P Elements and the *Drosophila* Genome Project" below).

Two additional findings have allowed the extension of single P element mutagenesis to many additional genes, even when the phenotype of a gene disruption is not known beforehand. First, in generals, P elements can transpose preferentially in *cis,* within about 50 to 100 kb from the initial location (213, 223). Thus, even if a P element is not very close to a gene of interest, often there will be one within the local cytogenetic interval (<100 kb) which might be useful for so-called "local hopping." This important finding coupled with the progress of the BDGP on the cytogenetic mapping of single P element insertions throughout the *Drosophila* genome greatly improves the likelihood that one can obtain P element insertions in a gene of interest. Second, insertions of P elements into a target locus can be detected with PCR or plasmid rescue and DNA hybridization (5, 74, 92). This greatly simplifies the identification of new insertions because it can be done with use of pools of potential insertion strains (43, 75, 185). Because molec-

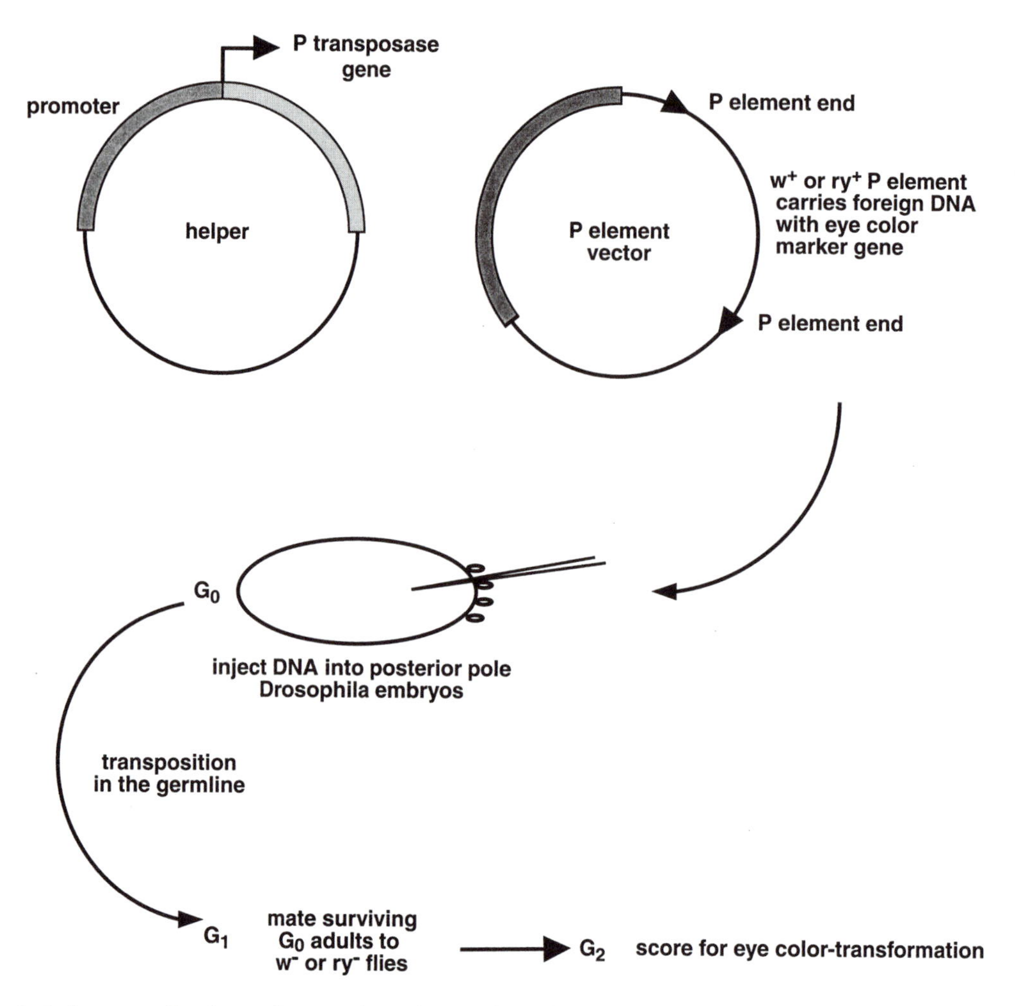

Figure 4. P element-mediated germ line transformation. Outline of the method for germ line transformation of *Drosophila* with P element vectors. Two plasmids, one encoding the P element transposase protein but lacking P element ends and the second plasmid carrying a foreign DNA segment and an eye color marker gene (w^+ or ry^+) within P element ends, are injected into the posterior pole of preblastoderm embryos. The transposase plasmid enters nuclei of presumptive germ line cells, is expressed, and leads to transposition of the P element from the second plasmid into *Drosophila* germ line chromosomes. After development of the injected embryos (G_0 generation), the surviving adults are mated to w^- or ry^- flies (G_1 generation), and the progeny from this cross (G_2 generation) are scored for restoration of wild-type eye color. The transformation frequency is typically ~20% of the fertile G_0 adults.

ular techniques are used, heterozygotes can be analyzed with no prior knowledge of the gene disruption phenotype.

Finally, often when P elements excise there can be a subsequent deletion of DNA flanking the donor DNA site (57, 66, 144). This is thought to result from aberrant DNA repair before nonhomologous end joining or gap repair (see sections on genetic studies of P element transposition and gap repair and on DNA repair pathways, below). P elements often insert into the 5′ region of genes near the promoter and cause either no or mild hypomorphic phenotypes (201). Thus, the phenotypic consequences of these P element insertions can be made more severe by screening for large deletions induced by imprecise excision (200).

Enhancer trapping, GAL4 regulation, and the EP misexpression elements

Based on the use of the *Escherichia coli* β-galactosidase (*lacZ*) gene as a useful histochemical marker in bacterial and yeast genetics, a method was developed called "enhancer trapping" by using the weak P element transposase promoter fused to the *lacZ* gene (147). Upon mobilization of these P-*lacZ* P elements into the *Drosophila* genome it was found that the expression pattern of *lacZ* often corresponded to that of a nearby gene in tissue or temporal specificity because of adjacent enhancers acting on the weak P element promoter. This initial report stimulated several subsequent large-scale screens that identified novel genes and also provided useful markers for specific cell or tissue types in development (13, 15, 218). In addition, the use of specific *lacZ* enhancer trap strains to produce phenotypically marked tissues has allowed genetic screens to be performed that disrupt normal cell behavior, function, development, or migration (73). Finally, *lacZ* expression can be used as a marker to purify discrete cell populations from enhancer trap strains with fluorescence-activated cell sorting (FACS) (115).

The manipulation of gene expression can be used to create new phenotypes or restrict expression of certain gene products to specified cells, tissues, or developmental stages. This became possible with *Drosophila* when it was realized that the yeast activator GAL4 could function in *Drosophila* (24). A two-component system was developed consisting of a P element carrying the GAL4 activator protein, controlled by a specific promoter or enhancer to restrict GAL4 expression, and a second P element carrying the gene to be expressed, placed under the control of a minimal promoter with adjacent GAL4 operator binding sites. This system has been used extensively to express or misexpress specific genes in certain somatic cell types (23). Recently, by modifying the core promoter, the GAL4 system has been adapted for expression in the female germ line during oogenesis and in the early embryo (174).

A new method of gene discovery combines enhancer activation by the GAL4 system with P element insertional mutagenesis. This method, termed modular misexpression, uses a P element with a GAL4-dependent enhancer pointed outward from the P element, called EP (173). Upon mobilization and insertion, genes adjacent to the EP element can be activated by crossing the EP strain to a GAL4 driver strain expressing the activator in a certain cell type. This has been done on a large scale for systematic gain-of-function genetics (175). Another system, similar in concept but which does not use GAL4, was also developed with an eye-specific enhancer to activate genes adjacent to P element insertions ectopically in the eye, screening for altered eye morphology (77). It has also been possible to generate new GAL4 driver lines using P element gene replacement (see "Gene Replacement" below) to copy a GAL4 gene from one transgene to another, to increase the number of GAL4 driver lines with specific expression patterns (186).

Uses of yeast FLP recombinase in *Drosophila*

With the use of P element transformation to introduce the yeast site-specific Flip recombinase (FLP) and the Flip recombinase target (FRT) recombination sites into the *Drosophila* genome, it was shown that the FLP recombinase could catalyze DNA strand exchange between FRT sites (70). Subsequently, it was shown that strand exchange could occur between FRT sites on different chromosomes or sister chromatids (68, 69). After these initial reports, and with the introduction of FRT sites near the centromeres of the major chromosome arms, it became possible to efficiently generate somatic genetic mosaics in embryonic or adult tissues (33, 44). An extension of this method was to use FLP recombination to generate homozygous mutations for genetic screens (34, 211, 220, 221). In combination with the use of the dominant female sterile (DFS) method using the X-linked $ovoD^1$ mutation or autosomal transgenes carrying the $ovoD^1$ gene, the FLP-FRT system can be used to generate homozygous mutant germ line clones which is useful for examining the phenotypes of certain zygotic lethal mutations (32, 34). Finally, a method to induce gene expression in somatic cell clones, termed the FLP-out method, is useful for examining cell autonomy of signaling pathways (7, 205). Here, a transgene is created with a promoter and a gene of interest separated by FRT sites flanking a transcription terminator and a cell autonomous marker gene. Upon induction of FLP recombination, the terminator and the marker gene are deleted, resulting in activation of the gene of interest in marked clones of cells. Thus, the combination of P element transformation with FLP recombination has led to the development of novel methods for *Drosophila* molecular genetics.

Homologous gene targeting

One of the missing genetic tools in *Drosophila* molecular genetics, relative to other model organisms such as yeast or mice, has been gene targeting by homologous recombination (168). In other systems, linear DNA ends are recombinogenic. Indeed, DNA breaks induced by P element transposase can be repaired via a homology-dependent pathway (see sections on the P element transposase and on P element transposition and gap repair, below).

Attempts to obtain homologous targeting in *Drosophila* have focused on the use of the yeast mitochondrial intron-encoded endonuclease I-SceI to generate DNA breaks at specific sites. I-SceI cleaves DNA at an 18-bp recognition site, leaving a 4-nucleotide (nt) 5′ extension. The recognition site size allows cleavage very infrequently, even in large metazoan genomic DNA. This enzyme has been used successfully to introduce specific DNA double-strand breaks in mammalian genomes to study homologous recombination (89) and is active in *Drosophila* (12). Very recently, with use of a combination of P-element-mediated transformation, FLP-FRT recombination, and I-SceI cleavage, a transgenic system of homologous gene targeting to the *yellow* locus was reported (168). This system, which generated extrachromosomal circles by FLP recombination which were then opened with I-SceI to generate linear recombinogenic DNA ends in the female germ line, resulted in bona fide homologous gene-targeting events at ~1 of 500 progeny. This method also generated several other kinds of aberrant and unexpected recombination products. This system should find new and wide applicability with the recent completion of the *Drosophila* genome sequence (1).

Gene replacement

On the basis of studies of one of the double-strand DNA break repair pathways used after P element excision (57) (see on P element transposition and gap repair and on accessory factors, below), it was demonstrated that this gap repair process could use ectopic templates to copy new information near the site of a P element-induced break (66, 90, 120). This repair was reasonably efficient, could be observed at several different loci (119), and exhibited a strong *cis*-preference (58). Conversion tracts were ~1.4 kb, required some short homology at the break to search for a repair template, and at a low frequency could use injected nonchromosomal DNA as a template (46, 66, 100, 101). This method has been particularly useful for large genes that cannot be subjected to normal P element-mediated transformation studies (120).

Recombination methods

It has long been known that recombination is induced at the sites of P elements during hybrid dysgenesis (50, 52, 206, 208). Although this occurs in both males and females, it is more evident in males, which usually exhibit ~1000-fold less meiotic recombination than females. The ability to stimulate recombination at P element sites has had several useful applications in *Drosophila* genetics. One report used recombination between two marked P elements, either in *cis* or *trans*, to generate "designer deletions" with appropriately marked chromosomes (39). Studies on P element-induced male recombination have outlined several pathways for the rearrangements to proceed, one of which, termed hybrid end insertion (HEI), can lead to chromosomal deletions (71, 152, 153, 207, 209). Finally, a recent application of male recombination was used in conjunction with the cytogenetically mapped P element insertion collection from the BDGP (http://www.fruitfly.org) to localize ethyl methanesulfonate-induced gene mutations between two P elements (31, 210). Given the available *Drosophila* genome sequence (1, 177), the bacterial artificial chromosome (BAC) physical map (85), and the ability to use PCR to identify flanking DNA adjacent to a P element insertion (43, 75, 185), the exact boundaries of the chromosomal interval containing a gene of interest can now be precisely determined.

P Elements and the *Drosophila* Genome Project

With the development of efficient single P element mutagenesis strategies (38), many laboratories have undertaken large-scale P element insertional mutation screens (summarized in reference 201). The Berkeley *Drosophila* Genome Project (www.fruitfly.org) has been involved in collecting and further characterizing these lethal P element insertion strains (4, 200). The cytogenetic mapping and further analysis of these P elements have played critical roles in generating sequence tagged sites (STS) useful in the generation of high-resolution P1 and BAC-based physical maps (85). In addition, it is estimated that there are about 3,600 vital genes in *Drosophila* and that the lethal P element insertion collection inactivates >25% (1,045) of these genes (200). Recent completion of the *Drosophila* genome sequence allowed an estimate of 13,601 predicted genes (1, 177, 181), in good agreement with the previous estimate of ~12,000 (133). Regarding the use of P elements as insertional mutagens, it was estimated that about two-thirds of the genes in *Drosophila* do not mutate to give a phenotype even though P elements often insert within 100 to 200 bp upstream of the transcription start site (200). However, from an analysis of a 2.8-Mb region of the *Drosophila* genome around the alcohol dehydrogenase (*ADH*) gene (4), it is known that by mapping P element insertions in this region on the genomic DNA with inverse PCR methods to generate DNA sequence flanking the insertion, regardless of whether the insertions cause a phenotype, that a significant increase in new gene-mutation links can be made on the basis of DNA sequence database searching (1, 177, 200). In fact, the BDGP gene disruption project is going to be scaled up with the EP misexpression P element (see section on enhancer trapping, above), with an estimate that analysis of ~50,000 new random P element insertions should allow tagging of ~70% of the open reading frames that would allow for either misexpression or deletion of the nearby genes (1, 177, 200). Thus, P elements will continue to play an important role in the post-genome sequence era of the *Drosophila* genome project.

MECHANISM OF P ELEMENT TRANSPOSITION

cis-Acting DNA Sites Involved in P Element Transposition

Initial DNA sequencing studies of P elements showed that an 8-bp duplication of target site DNA is created upon insertion and that perfect inverted repeats of 31 bp are present at the termini (146). Two initial tests with P element-mediated transformation showed that the 31-bp inverted repeats were required for transposition (94, 199). A very extensive set of mutagenesis studies subsequently showed that both the terminal 31-bp inverted repeats and the internal transposase binding sites are also required for transposition in vivo (140). Furthermore, internal 11-bp inverted repeats located at ~120 bp from each transposon end seemed to function as transpositional enhancer elements (140). In addition, an artificial P element carrying two 5′ ends in inverted orientation was not active for transposition in vivo (140), demonstrating that the P element transposase protein can distinguish each transposon end and requires appropriate pairing of the 5′ and 3′ ends for transposition. However, a clear understanding of how the *cis*-acting P element sequences function in transposition has required the development of in vitro systems to study the different steps of the transposition reaction (see sections on donor DNA cleavage and strand transfer and disintegration, below).

The P Element Transposase Protein

Initial studies of P elements showed that they encode a *trans*-acting protein product that is required for P element mobility (51, 94, 199). Biochemical studies, genetic experiments, and sequence comparisons of P elements from other *Drosophila* species have provided insights into the functional domains of the protein (Fig. 5). A C_2HC putative zinc knuckle motif is located at the amino terminus of the protein that, in conjunction with an adjacent basic region, constitutes the site-specific DNA binding domain (123, 124). This region is highly conserved in the sequences of P elements from other *Drosophila* species (134). A leucine zipper coiled-coil motif located near the DNA binding region allows protein dimerization, although this is not essential for high-affinity site-specific DNA-protein recognition (124). The KP repressor protein contains an additional dimerization region C-terminal to the leucine zipper motif (124), both of which are present in transposase. A putative regulatory domain is also located in the amino-terminal region and contains potential sites for phosphorylation by the DNA repair-checkpoint phosphatidyl inositol-3-phosphate (PI_3)-related protein kinases DNA-PK, ATM (ataxia telangiectasia mutated), and ATR (ATM related) (S/TQ; QS/T) (47, 113, 196, 197). Alteration of these potential phosphorylation sites by mutagenesis to alanine resulted in both increased and decreased transposase activity in vivo; and in vitro (E. L. Bell, M. Mul, M. Mahoney, W. L. Wang, D. C. Rio, unpublished results). Consistent with these findings are the observations that the transposase protein is phosphorylated on serine, threonine, and tyrosine in vivo; that the bacterially produced transposase protein binds DNA but lacks catalytic activity; and that treatment of native transposase with phosphatase reduces its activity (Beall et al., unpublished). Candidate protein kinases for mediating transposase phosphorylation are either an ATM-kinase family member and/or the cyclin-dependent kinases, such as cdc2/cyclin B (Fig. 6). Al-

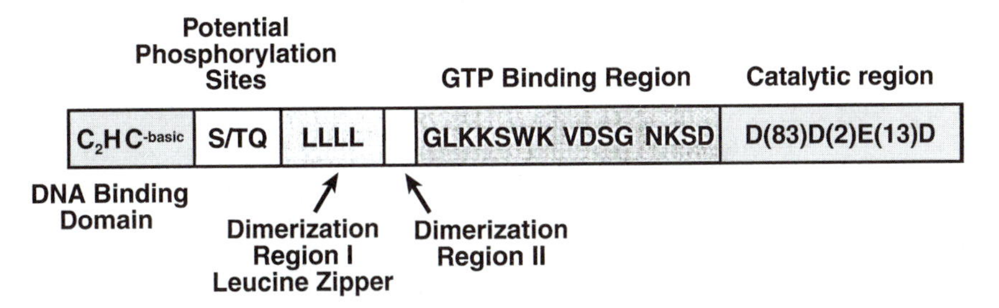

Figure 5. Features of the P element transposase protein. The N-terminal region contains a C_2HC motif and basic region involved in site-specific DNA binding. Adjacent to this region are consensus sites for phosphorylation by the ATM family of DNA repair-checkpoint PI_3-related protein kinases. These sites, when mutated, affect transposase activity in vivo (Beall et al., unpublished). There are two dimerization regions adjacent to the N-terminal DNA binding domain: dimerization region I is a leucine zipper motif and dimerization region II is C-terminal to the leucine zipper but does not resemble any known motif. The central part of the protein contains a GTP binding region, with some sequence motifs found in the GTPase superfamily (139). The C-terminal region is highly acidic and contains four acidic residues (D444, D528, E531, and D545) that, when mutated, block transposase activity in vivo and in vitro (Ahrens et al., unpublished).

```
  1   MKYCKFCCKAVTGVKLIHVPKCAIKRKLWEQSLGCS 36
 37   LGENSQICDTHFNDSQWKAAPAKGQTFKRRRLNADA 72
 73   VPSKVIEPEPEKIKEGYTSGSTQTESCSLFNENKSL 108
109   REKIRTLEYEMRRLEQQLRESQQLEESLRKIFTDTQ 144
145   IRILKNGGQRATFNSDDISTAICLHTAGPRAYNHLY 180
181   KKGFPLPSRTTLYRWLSDVDIKRGCLDVVIDLMDSD 216
217   GVDDADKLCVLAFDEMKVAAAFEYDSSADIVYEPSD 252
253   YVQLAIVRGLKKSWKQPVFFDFNTRMDPDTLNNILR 288
289   KLHRKGYLVVAIVSDLGTGNQKLWTELGISESKTWF 324
325   SHPADDHLKIFVFSDIPHLIKLVRNHYVDSGLTING 360
361   KKLTKKTIQEALHLCNKSDLSILFKINENHINVRSL 396
397   AKQKVKLATQLFSNTTASSIRRCYSLGYDIENATET 432
433   ADFFKLMNDWFDIFNSKLSTSNCIECSQPYGKQLDI 468
469   QPDILNRMSEIMRTGILDKPKRLPFQKGIIVNNASL 504
505   DGLYKYLQENFSMQYILTSRLNQDIVEHFFGSMRSR 540
541   GGQFDHPIPLQFKYRLRKYIIARNTEMLRNSGNIEE 576
                                GMTNLKECVNKNVIP
577   DNSESWLNLDFSSKENENKSKDDEPVDDEPVDEMLS 612
613   NIDFTEMDELTEDAMEYIAGYVIKKLRISDKVKENL 648
649   TFTYVDEVSHGGLIKPSEKFQEKLKELECIFLHYTN 684
685   NNNFEITNNVKDKLILAARNVDVDKQVKSFYFKIRI 720
721   YFRIKYFNKKIEIKNQKQKLIGNSKLLKIKL      751
```

Figure 6. Amino acid sequences of the P element transposase and 66-kDa repressor protein. Single letter code for the 751-amino-acid and 576-amino-acid transposase and 66-kDa proteins. The alternative C terminus of the 66-kDa protein (GMTNLKECVNK-VIP) encoded by sequences from IVS3 is indicated below the corresponding transposase residues (amino acids 562 to 576). The C_2HC DNA binding motif is indicated by boxed shadowed letters. The leucine zipper motif is indicated by boxed, outlined L residues. The three GTP binding region homologies are indicated by white letters on a black background. Potential phosphorylation sites for the ATM family of DNA repair-checkpoint PI_3 protein kinases (S/TQ or QS/T) are indicated by white letters on a black background. Potential phosphorylation sites for cdc2 (TPHL and TPLQ) are indicated by highlighting and underlining. Acidic residues at the C terminus are underlined, and the four acidic residues that, when mutated, inactivate the protein for catalysis in vivo and in vitro are boxed and highlighted.

though *Drosophila* has both Ku p70 and Ku p80, recent sequence analysis of several model organism genomes suggests that no direct ortholog of the DNA-PK catalytic subunit, DNA-PKcs, exists in *Drosophila, Saccharomyces cerevisiae,* or *Caenorhabditis elegans* (1, 63, 177, 181). Analysis of the available *Drosophila* genomic sequence indicates that at least two PI_3-related protein kinase family members exist in *Drosophila:* the previously identified *mei-41* gene and another previously unidentified ATM-like kinase gene (181). In mammals, the homologs of both these kinases (ATR and ATM) have substrate specificities similar to that of DNA-PK (113).

The unusual requirement of GTP as a cofactor for P element transposition (99) suggested that the transposase would bind GTP. Indeed, the central region of the transposase protein contains several sequence motifs found in the GTPase protein superfamily (19, 20) (Fig. 6 and 7). Mutation of some of the key conserved residues abolished GTP binding in vitro and transposase activity in vivo (139). Moreover, alteration of the most conserved NKXD guanine recognition motif to the NKXN xanthine recognition motif (D379N) altered the purine nucleotide requirement for transposase activity from guanosine to xanthosine triphosphate, both in vitro and in vivo, indicating that purine nucleotide cofactor binding is required for transposase activity (139) (Fig. 7).

The carboxyl-terminal region of the transposase protein, including exon 4 (ORF3), contains a high proportion of acidic amino acid residues (Fig. 6). The P element transposase belongs mechanistically to a superfamily of polynucleotidyltransferases, including the transposases from bacteriophage Mu and other bacterial mobile elements, the retroviral integrases, the Holliday junction nuclease RuvC, the RNase H family (72, 150, 158), and the RAG1 V(D)J recombinase (62, 112, 118). These enzymes generally use metal ion-mediated catalysis of phosphodiester bond hydrolysis and formation, where acidic amino acids in the protein active site serve to coordinate a divalent metal ion, usually magnesium (59, 158). Because the P element-encoded 66-kDa repressor protein lacks catalytic activity and because of the high proportion of acidic amino acid residues in the C-terminal part of the protein, it seemed likely that acidic residues found in this region might constitute the catalytic domain of the protein. However, sequence (T. A. Baker, personal communication) and structure-based (P. Rice, personal communication) alignments to other polynucleotidyltransferase superfamily members (150) revealed little or no sequence alignment that could be used to implicate specific transposase residues as corresponding to the conserved motifs of acidic residues found in other transposase or integrase proteins. Nonetheless, random mutagenesis of many of the acidic residues in this region revealed four residues that, when changed to alanine, markedly reduced transposase activity in vivo and in vitro (D444, D528, E531, and D628) (Fig. 6) (K. Ahrens, Y. Mul, W. L. Wang, B. Weinert, and D. C. Rio, unpublished results). Thus, although the P element transposase protein seems to use a catalytic strategy similar to that observed for other members of the polynucleotidyltransferase superfamily, the primary amino acid sequence does not obviously reveal the identity of the catalytic residues. Even though the Mu transposase and the HIV integrase catalytic domains are only 18% identical in amino acid sequence, structural studies revealed a remarkable overall similarity in the folded catalytic domains (72, 158). This lack of sequence identity (150), and the finding of four rather than three acidic amino acid residues required for activity, may suggest that the P element transposase evolved independently from the other members of this polynucleotidyltransferase protein superfamily, possibly

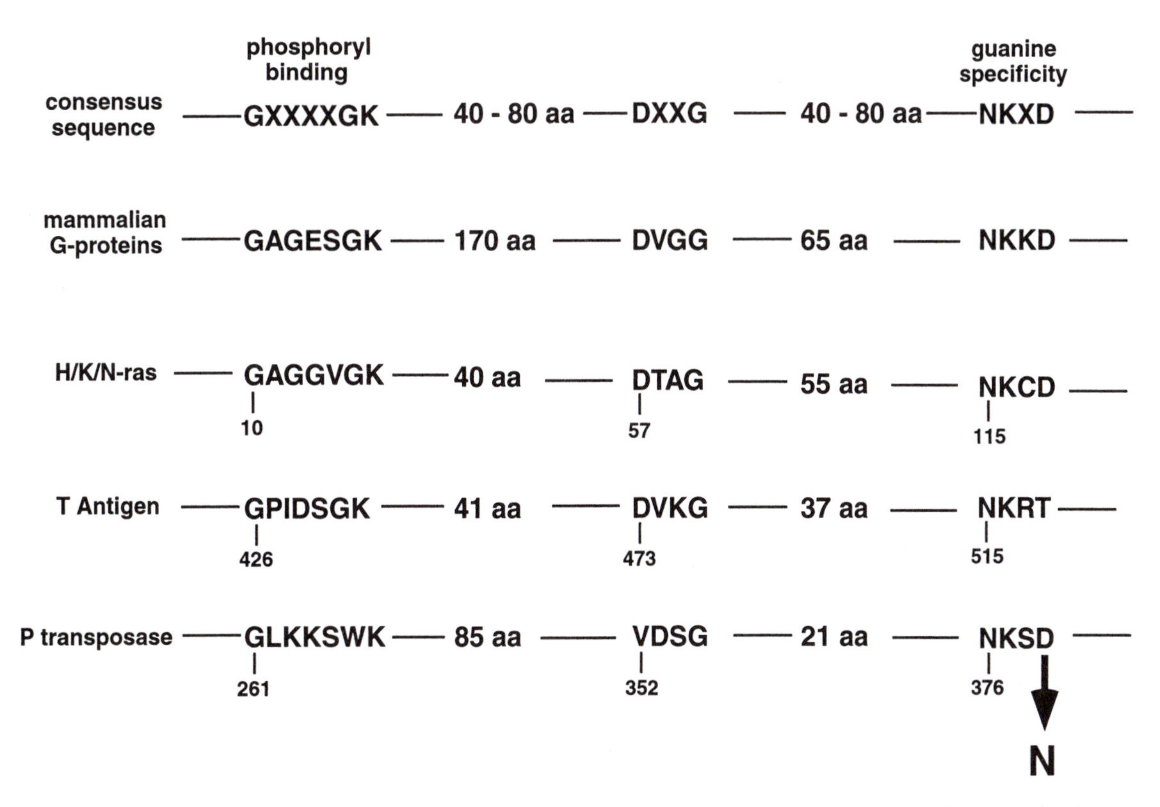

Figure 7. Similarities between P element transposase and GTPase superfamily members. Alignments of regions of P element transposase that bear some resemblance to known G proteins. The conserved motifs for phosphoryl binding and guanine specificity are indicated at the top. Amino acid numbers are given below for ras, T antigen, and P element transposase. The residue D379 that when changed to N switched the specificity from guanosine to xanthosine in P element transposase is indicated at the bottom. Reprinted from reference 139 with permission.

from a metal binding site similar to those found in the nucleic acid polymerase superfamily (25, 91) or those found in restriction or other endonucleases (81).

Purification and Characterization of P Element Transposase

P elements contain four exons (ORF0, 1, 2, and 3 [Fig. 2A and 3]) that are all required for transposition (94) and which are linked to generate an mRNA lacking all three introns that encoded an 87-kDa transposase protein (161) (Fig. 3). A system for purifying the P element transposase protein from *Drosophila* tissue culture cell nuclear extracts was developed by using a combination of conventional and affinity chromotography methods (97). DNase I footprinting experiments showed that the 87-kDa transposase protein binds to sites near each end of the P element, located between the terminal 31-bp and internal 11-bp inverted repeats (97) (Fig. 2B and 8). Surprisingly, the transposase binding sites do not overlap the terminal 31-bp inverted repeats that are required for transposition in vivo (94, 140) (Fig. 2B). The binding site for transposase at the 5′ P element end overlaps the TATA box of the minimal P element promoter, suggesting that transposition and transcription of the P element might be mutually exclusive (97, 98).

These studies showed that P element transposase, unlike most transposase proteins encoded by other transposons, binds site-specifically to internal sites on P element DNA (97). After initial recognition of these binding sites by transposase, an elaborate assembly (or synapsis) must occur to bring the transposase active sites in position relative to the sites of DNA cleavage within the 31-bp terminal inverted repeats for transposition to occur (Fig. 2B).

Development of an In Vitro Reaction for P Element Transposition

The initial purification and characterization of the P element transposase protein led to the development of an in vitro assay system to study the different steps of the transposition reaction: DNA cleavage and integration into a new target site (99). This system relied on the use of a genetic selection in bacteria to detect transposition events that occurred in vitro be-

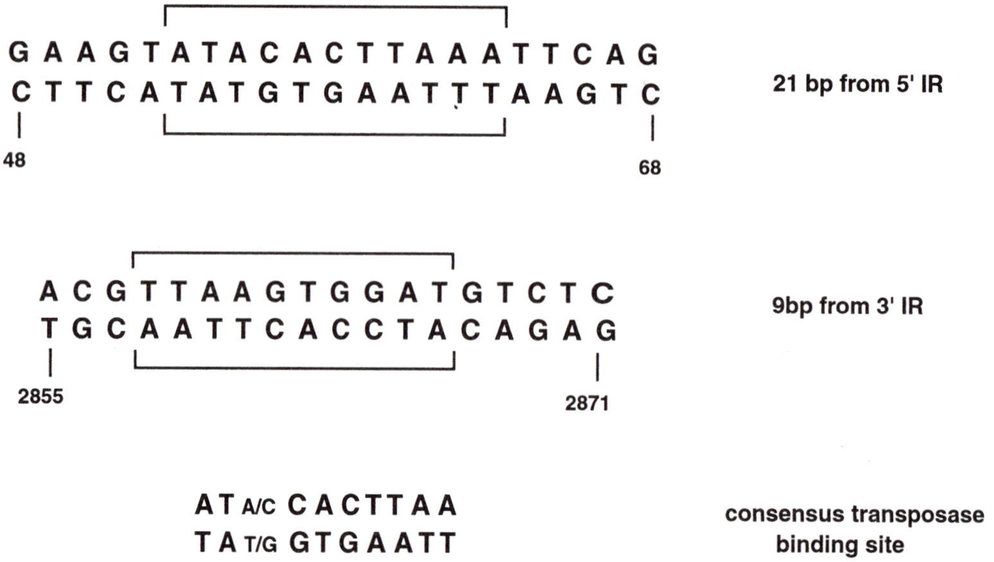

G A A G T A T A C A C T T A A A T T C A G
C T T C A T A T G T G A A T T T A A G T C
48 68
21 bp from 5' IR

A C G T T A A G T G G A T G T C T C
T G C A A T T C A C C T A C A G A G
2855 2871
9bp from 3' IR

AT A/C C A C T T A A
T A T/G G T G A A T T
consensus transposase binding site

Figure 8. Binding sites for transposase on P element DNA. DNA sequences from the 5′ end (nt 48 to 68) and from the 3′ end (nt 2855 to 2871) that are protected from DNase I cleavage by P element transposase. Distance of the beginning of the 10-bp core sequence from the corresponding 31-bp terminal repeat is indicated at the right. A consensus site derived by comparing the two protected regions is shown at the bottom.

tween two different plasmid substrates. The donor plasmid DNA had a P element carrying the bacterial tetracycline resistance gene and was unable to replicate in normal *E. coli* strains. The target plasmid DNA was a standard ampicillin-resistant plasmid. Transposition events were scored after transformation of the reaction products into *E. coli* and selection for tetracycline- and ampicillin-resistant colonies. Similar assays in bacteria had been used to detect V(D)J recombination products recovered from mammalian cells (80).

Analysis of the DNA reaction products by restriction endonuclease cleavage and sequence analysis showed that they corresponded to transfer of the P element from one plasmid to the other. The reaction also required wild-type transposase DNA binding sites on the donor P element. Initially, a mixture of cofactors including ribo- and deoxyribonucleoside triphosphates was tested and systematic omission showed that GTP was the active cofactor for transposition (99). Use of nonhydrolyzable GTP analogs (GTP-γ-S, β, γ-imidoguanosine 5′-triphosphate [GMP-PNP], and β,γ-methyleneguanosine 5′-triphosphate [GMP-PCP]) showed levels of activity equivalent to those observed with normal GTP, indicating that hydrolysis of the high-energy phosphoryl bond was not required for activity. This suggests that GTP may play a structural or allosteric role, in much the same way that GTP modulates the activities of other GTP-binding proteins, such as ras or mammalian Gα subunits (19, 219). Although ATP has been a cofactor for the MuB protein and in Tn7 transposition (203, 204, 222), the GTP requirement for the P element transposase protein is unique among this class of polynucleotidyltransferase proteins.

Additional studies with modified substrate DNAs showed that linear precleaved donor DNA could be used as a substrate, indicating that supercoiling of the donor DNA is not required for transposition. Moreover, 3′-hydroxyl groups at the P element ends were required for activity, indicating that, like other transposition systems, a 3′-hydroxyl group is used as a nucleophile during strand transfer (99). Taken together with the finding that deoxy- or dideoxynucleoside triphosphates neither were required for nor inhibited the transposition reaction, these studies confirmed biochemically the previous genetic experiments suggesting that P element transposition proceeds via a cut-and-paste mechanism (Fig. 9).

Donor DNA Cleavage and P Element Excision

Although the initial description of the in vitro P element transposition reaction relied on the use of a genetic assay in bacteria to detect transposition, analysis of other aspects of the reaction required development of physical assays for the detection and analysis of reaction products and intermediates. In subsequent studies, with use of both radiolabeled and unlabeled P element DNAs, an excised P element transposon

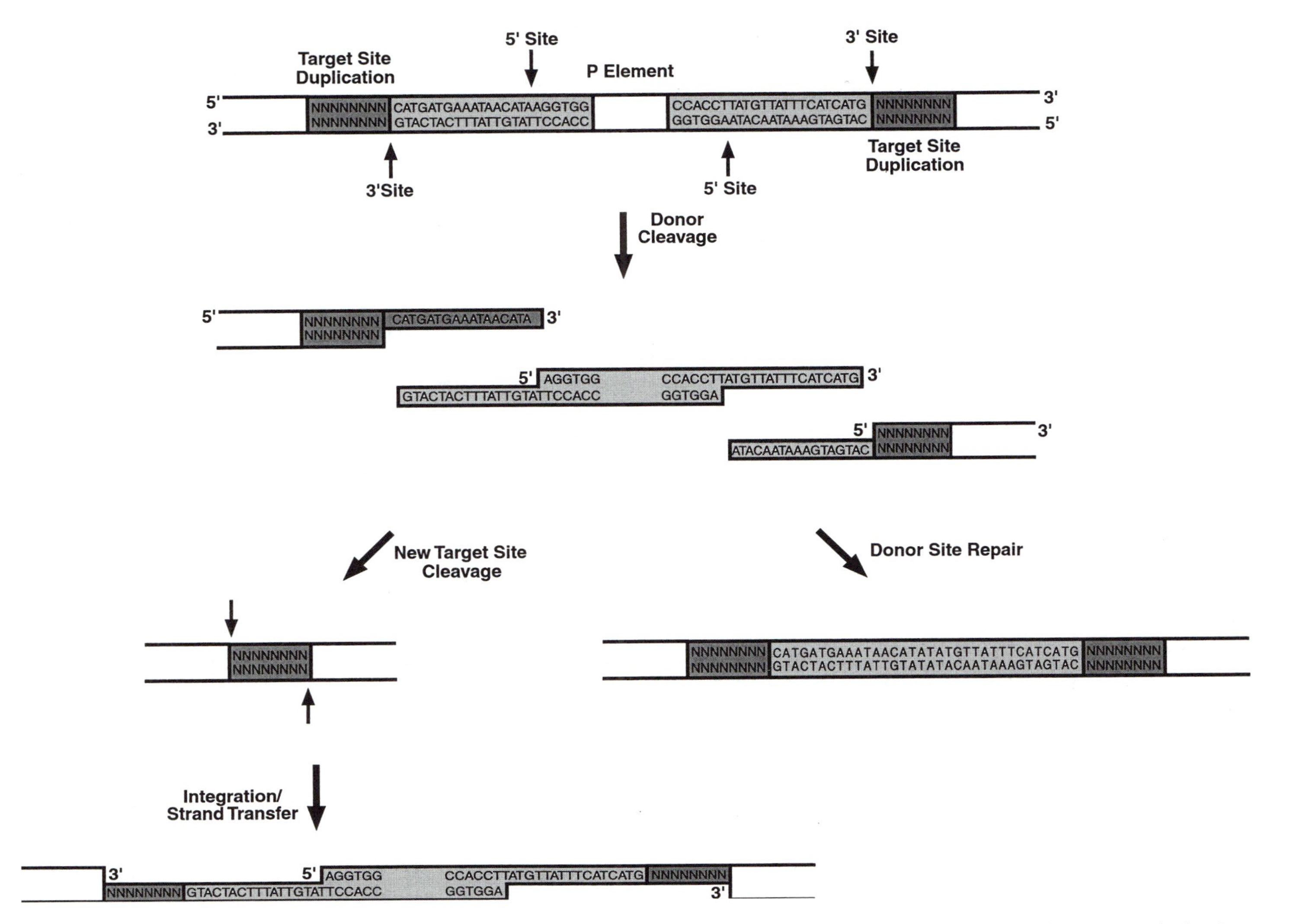

Figure 9. Pathway of DNA cleavage and joining during P element transposition. Shown at the top is the P element donor site with the target site duplications, 31-bp inverted repeats, and the 5′ and 3′ cleavage sites indicated. In the first step of transposition, donor cleavage, both ends of the P element are cleaved. The novel cleavage results in a 17-nt single-strand extension on the P element ends and leaves 17 nt from each P element inverted repeat attached to the donor DNA cleavage site. Once transposon excision occurs, the donor site can be repaired via an NHEJ pathway (shown to the right) or via an SDSA gap repair pathway (not shown) (142). The excised P element then selects a target site and, upon strand transfer, integrates into the donor site, generating a gapped intermediate, which upon repair completes integration, creating an 8-bp duplication of target DNA (bottom). Reprinted from reference 10 with permission.

fragment was detected, directly confirming that transposition occurs through double-strand DNA breaks at the transposon termini (10). A series of mapping experiments using both primer extension and ligation-mediated PCR (LM-PCR) showed that P elements are excised from the donor DNA site as the product of an unexpected set of cleavages at the transposon ends (10). As expected, the 3′ ends of the transposon DNA were cleaved at the junction with the 8-bp target site duplication (Fig. 9). This cleavage yields a 3′-hydroxyl group that functions as the nucleophile in the strand transfer reaction. However, the 5′ ends were cleaved 17 nt into the 31-bp inverted repeats generating novel 17-nt 3′ extensions on both the excised transposon and flanking donor DNAs. The transposase-mediated cleavage sites determined in vitro are consistent with previously characterized in vivo P element excision products, in which 16 or 17 nt derived from each P element end were present at the chromosomal donor DNA site after nonhomologous end joining (NHEJ) repair (9, 202). The generation of this unique 3′ extension may facilitate NHEJ repair after P element excision because this type of structure (a single strand-double strand DNA junction) is bound efficiently by the Ku p70/Ku p80 complex (47, 197). The 17-nt 3′ extension would also facilitate the synthesis-dependent strand annealing (SDSA) gap repair process because these extended 3′ ends on the donor DNA may aid the homology search, facilitate resection, and/or function in the priming of DNA synthesis to copy the template (61, 142) (see Fig. 11). Finally, the unusual nature of the cleavage products could be involved in promoting the irreversibility of the excision reaction. Because the 3′ extension on the P element cleavage product is used by the transposase active site for integration (see next section), a conformational rearrangement probably occurs between the initial cleavage of the duplex donor DNA and the integration of the single-stranded P element termini into a target DNA after excision. An alteration in the active-site geometry after the cleavage reaction could serve to orient the newly generated 3′ ends for strand transfer which act as the nucleophile instead of water in the second stage of the reaction. A similar strategy seems to operate to separate the two steps in intron excision by the *Tetrahymena* ribozyme (28) and in pre-mRNA splicing by the spliceosome (138).

Strand Transfer and Disintegration by P Element Transposase

The ability of the P element transposase protein to use linear DNA as a substrate (99) suggested that it would be possible to use precleaved DNA to separate the first (donor cleavage) and second (integration or strand transfer) steps of the transposition reaction (Fig. 9). Indeed, by using PCR-generated precleaved P element donor DNAs, it was shown that strand transfer could be separated from donor cleavage in vitro (99).

On the basis of the finding that donor DNA cleavage occurs with an unusual staggered 3′ extension (see preceding section), the DNA substrate requirements for strand transfer in vitro were investigated with synthetic oligonucleotide substrates. This approach has been very fruitful for the study of Mu transposase, retroviral integrases, and the RAG1/RAG2 V(D)J recombinase (35, 59, 154, 182). Two types of studies were conducted. First, wild-type or mutant oligonucleotide substrates carrying the 31-bp inverted repeat and transposase were investigated (11). These studies showed that the 17-nt single-stranded 3′ extension is critical for strand transfer because a decrease in the length of the exposed single-stranded DNA extension caused a drastic reduction in activity (11). Second, with use of chemically modified oligonucleotide substrates, with either purine or pyrimidine bases missing or with backbone phosphate groups ethylated, a series of studies were conducted to probe the protein-DNA contacts that were critical for strand transfer activity by P element transposase (11) (Color Plate 41 [see color insert]). These experiments demonstrate that critical contacts between transposase and both the double- and single-stranded regions of the substrate DNA are necessary for strand transfer activity in vitro (11). In addition, there were sites in which DNA modification actually stimulates strand transfer (Color Plate 41A), indicating that distortion of the substrate DNA may facilitate the chemistry of strand transfer. Similar stimulatory effects of distorted DNA substrates have been observed with Mu transposase (182) and HIV integrase (184). Thus, although the initial donor DNA cleavage reaction occurs on duplex DNA, the strand transfer reaction uses a completely different substrate in which the single-stranded region at the P element ends are critical for activity, indicating that there must be significant active-site flexibility in the P element transposase protein during transposition.

Retroviral integrase proteins catalyze a reaction called "disintegration" (35). With use of oligonucleotide substrates that resemble the strand transfer intermediate in retroviral integration, it was shown that integrase cleaves the transferred viral DNA strand with concomitant rejoining of the target DNA strand (35). This reversal is expected because the transesterification reactions of strand transfer should be reversible. In P element transposase, with use of oligonucleotide substrates that mimic a strand transfer intermediate carrying a precleaved P element end joined to target DNA, the purified transposase protein performs disintegration in a fashion similar to the retroviral integrases (11). In fact, two other eukaryotic recombinases, the *C. elegans* Tc1 transposase and the V(D)J RAG1/2 recombinase, also perform disintegration in vitro (132, 215). By contrast, the prokaryotic Tn*10* and Mu transposase proteins do not perform disintegration. The ability of the eukaryotic proteins to perform disintegration may be significant in genomic surveillance to prevent chromosomal translocations and other rearrangements that might follow aberrant transposition events (132). This might be a significant difference between the prokaryotic and eukaryotic recombinases.

Target Site Selection in P Element Transposition

The initial DNA sequence analysis of several cloned P element insertions revealed that 8-bp duplications of the target were found flanking all the P elements analyzed. Comparisons of these target site sequences revealed a general high GC base composition in the 8-bp sites, with the consensus sequence being 5′-GTCCGGAC-3′ (146). A more recent study by the BDGP (http://www.fruitfly.org) of 2,266 nonselected P element insertion site sequences showed that the target site is not only GC-rich, but also actually consists of a 14-bp palindromic sequence surrounding the integration site (127). This observation is consistent with a model in which the target site is recognized by two transposase subunits, with each subunit positioning a cleaved P element terminus for the transesterification reaction of strand transfer into a correctly positioned target DNA molecule.

More globally, many extensive genetic studies have revealed that P element transposition occurs preferentially in *cis*, onto the same chromosome, generally within 100 kb of the starting element (213, 223). These studies also showed that there is a higher transposition frequency in the female versus the male germ line (223). The basis for this difference is not known, but the discovery of preferential transposition in *cis* suggested a more efficient way to target P element insertions to a given region by using this "local transposition" method (223). Given the progress on cytogenetic mapping of P element insertions by the BDGP, most genes should have a P element insertion located roughly within a 100-kb interval of most genes (85, 177). However, the recent report of homologous gene targeting in *Drosophila* may eliminate the need to do local P element hopping (168). A *cis* preference is also seen for the repair of P-induced DNA breaks from ectopic chromosomal templates (58). Whether these two distinct *cis*-effects are related in any way is not known. In addition, P elements can insert into heterochromatic regions, which would be less deleterious for the host cell (223). Heterochromatic insertions may allow transposable elements to play a role in genome evolution and alter the number and arrangements of repetitive DNA sequences (212). Several studies have shown that there are also "hot spots" for P element insertion, such as the *singed* locus on the *Drosophila* X chromosome (167). However, examination of the DNA sequences in the region surrounding the *singed* P element insertion sites has not revealed any enlightening feature(s) unique to the most frequently used target sites (149, 167). Studies on the specificity of P element insertion have shown that whereas the specific structure and content of the recombinant P element may influence the rate of transposition, the hot spots seem to be the same regardless of the initial genomic location of the element (14). One report demonstrated that inclusion of the upstream regulatory region of the *engrailed* gene in a P element allowed the transgenes carrying that element to preferentially insert near the chromosomal *engrailed* locus (96). How those sequences led to the alteration of P element insertion specificity is not known. Finally, the extensive analysis of P element insertion sites by the BDGP, as well as several observations from individual *Drosophila* laboratories, indicates that when P element insertions occur in genes, they are found in the region around, and generally upstream from, the transcriptional initiation site (127, 200, 201). The presumption is that this insertion site specificity is caused by localized chromatin remodeling within promoter regions, indirectly allowing access of the P element transposase protein to the DNA. Thus, P elements transpose preferentially in *cis*, generally insert near the 5′ ends of genes, and can use certain genomic sites preferentially as hot spots for insertion.

A Role for GTP as a Cofactor in P Element Transposition

One of the most novel and surprising findings made during the initial studies of P element transposition in vitro was that the transposase protein uses GTP as a cofactor (99). The fact that nonhydrolyzable GTP analogs support cleavage and strand transfer suggests that the nucleotide cofactor acts to trigger an active conformation to assemble a stable synaptic complex before catalysis, i.e., GTP is an allosteric effector of the transposase protein. This idea is interesting in light of the fact that topoisomerases use ATP as a cofactor for activity and subsequently hydrolyze ATP to recycle the protein for another round of catalysis (216). Although it is known that the P element transposase binds GTP and that GTP activates the protein in vitro and in vivo (99, 139), it is not known if transposase hydrolyzes GTP to recycle for a new round of transposition. Such a switch is used by the GTPase superfamily, in which the GTP-bound state is active and the GDP-bound state is not (19, 219). However, the nucleotide-bound state of P transposase purified from *Drosophila* cells is unknown. This idea of GTP/GDP-modulated assembly is also found with the GTP-binding proteins tubulin and dynamin (45, 183). It is also possible that the GTP-dependent assembly of an active transposition complex could activate an intrinsic GTPase activity, much like the bacterial proteins FtsY and Ffh (151). Indeed, in other transposition systems, namely, Mu and Tn7, accessory factors rather than the recombinase seem to use ATP hydrolysis and their ATP- versus ADP-bound states as switches to control their reaction pathways (203, 204, 222). It will be interesting to investigate whether transposase hydrolyzes GTP and how GTP and GDP affect the conformational state and activity of P element transposase protein.

Biochemical Studies of the P Element KP Repressor Protein

One interesting aspect of P element transposition is how transposition is regulated by repressor proteins. Truncated forms of the transposase protein, such as the 66-kDa or KP proteins (Fig. 3), act as inhibitors of transposition (137, 159, 160, 163). A series of biochemical studies on the KP repressor protein showed that it carries the site-specific DNA binding domain and the leucine zipper dimerization region of the transposase protein, as well as an additional C-terminal dimerization region (Fig. 5). The KP repressor protein inhibits transposase activity by directly binding to the transposase DNA binding sites and blocking the P element DNA cleavage and strand transfer reactions in vitro (123). Mutant KP proteins that lack DNA-binding activity do not inhibit transposition in vitro (123). Moreover, repression by the KP protein in vitro did not require the leucine zipper motif (123), in contrast to a study in vivo in which leucine zipper mutations impaired KP-mediated repression (3). It is possible that both the DNA binding and leucine zipper regions of the KP repressor contribute to transpositional repression.

Genetic Studies of P Element Transposition and a Gap Repair Mechanism

The mechanism of P element transposition was first suggested by an elegant series of genetic experiments examining P element excision events in the presence or absence of an homologous chromosome (57). These studies showed that if a normal copy of the *white* locus was present in females that underwent excision of a P element inserted in the *white* locus on the homologous chromosome, then the frequency of "precise" P element excision was elevated 100-fold compared with the excision frequency for the same P element in the hemizygous state in males. These results were interpreted to mean that excision of a P element resulted in a double-strand DNA break, which was then repaired via a gap repair process with the homologous chromosome as a template (57) (Fig. 10). These findings indicated that P elements are mobilized via a cut-and-paste mechanism, similar to that used by the bacterial transposons Tn*10* and Tn7 (40). These experiments also clarified the ambiguities of previous studies regarding whether P element transposition was conservative or replicative.

The concept of gap repair led to a test of the idea that ectopic transgenes could be used as templates for the repair events and thus be used as an effective gene replacement strategy (66). By using restriction site

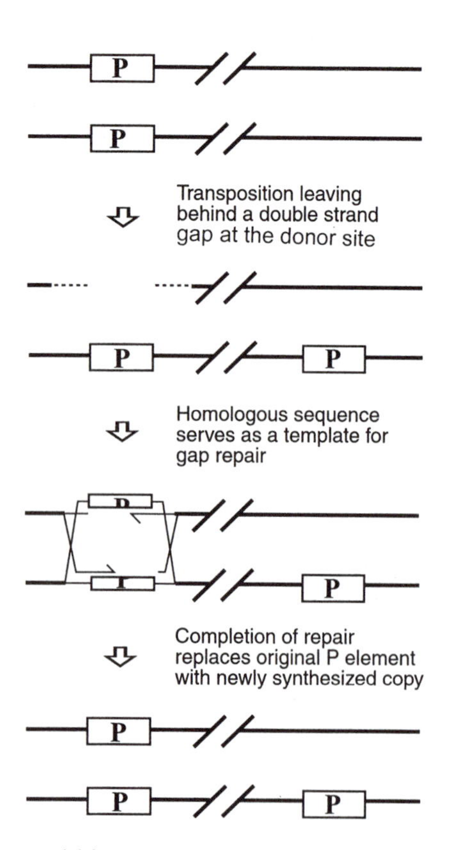

Figure 10. Model for gap repair after P element excision. Homologous chromosomes or sister chromatids, which after undergoing P element excision leave a double-strand gap at the donor site. The homologous sequence then serves as a template for SDSA synthesis (61, 142). Completion of repair replaces the original P element with a newly synthesized copy. If this gap repair process were incomplete, internal deletions of the P element would result.

polymorphisms in the ectopic transgenic template, it was demonstrated that the tract lengths of the gap repair process average ~1.4 kb (66, 142). Furthermore, by comparing different ectopic templates, a dramatic preference of template use in *cis*, on the same chromosome, was observed (58). More recent studies have shown that ~30 bp of homology are required for efficient template use and gap repair (46, 101) and that nonchromosomal plasmid sequences can be used as ectopic templates with lower efficiency (100).

Analysis of the ectopic copying of foreign P element templates showed that incomplete copying across the template from both sides can occur occasionally. Repair by annealing of short regions of homology results in junctions with short direct repeats (36, 41, 52, 93). Incomplete copying can also explain how internally deleted P elements are generated in natural P strains. The generation of internally deleted products is consistent with a model for the gap repair process occurring via a synthesis-dependent strand annealing (SDSA) mechanism first described for replication-linked recombination in bacteriophage T4 (61) (Fig. 11). This model is distinct from the predominant double-strand break repair process in yeast, in which gene conversion events are associated with the crossing over of flanking genetic markers (64, 148). These studies on the mechanism of P element transposition also suggest a way for P elements to increase in copy number because the gap repair process restores a new copy of the P element at the donor site (Fig. 10).

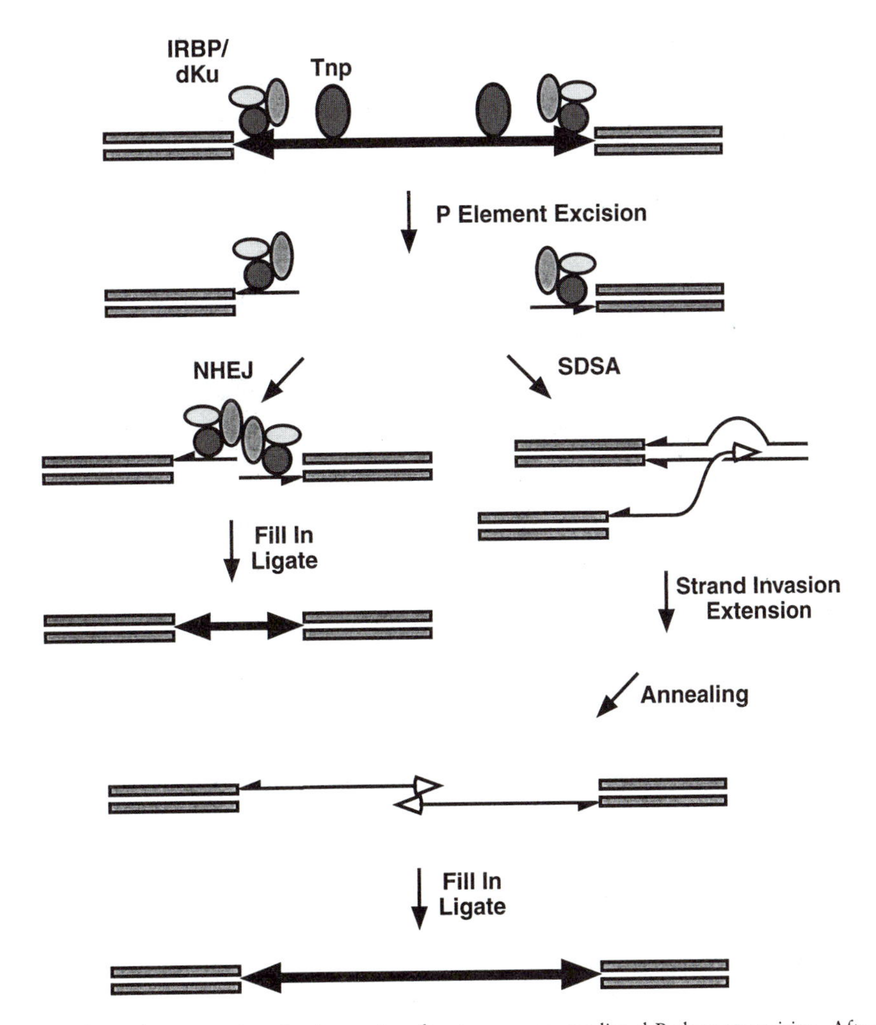

Figure 11. DNA repair pathways used at the donor site after transposase-mediated P element excision. After cleavage of P element DNA by transposase, the *Drosophila* Ku subunits may associate with the single-strand extensions. These cleaved termini can undergo two fates. One involves NHEJ where the cleaved termini are brought together and after DNA polymerase, ligase, and possibly nuclease action rejoining occurs with the P element sequences and target site duplications present (9). The second pathway involves gap repair via the SDSA pathway. Either with or without resection, these single-stranded 3′ extensions find homologous DNA sequences to use as templates for DNA repair synthesis. Copying can occur from either P element or flanking *Drosophila* genomic sequences as templates (54).

Accessory Factors and DNA Repair Pathways Involved in P Element Transposition

With use of extracts prepared from flies lacking P elements (M strains) and DNase I footprint protection assays, a DNA-binding activity was found that binds to the first 16 nt of the P element terminal 31-bp inverted repeats adjacent to the P element cleavage sites (162). Biochemical fractionation showed that the activity [inverted repeat binding protein (IRBP)] copurified with the *Drosophila* homolog of the DNA repair protein Ku p70 (8). The Ku p70/p80 heterodimeric complex is involved in the NHEJ pathway of DNA repair (36, 47) and seems to function in telomere homeostasis in yeast (41, 197). Simultaneously, the *Drosophila* Ku p70/p80 heterodimer was purified as a DNA-binding factor on the yolk protein genes, and the two subunits were called YPF-1 and YPF-2 (88). Cytological mapping indicated that Dm Ku p70 mapped to cytological position 86E2-3 on the right arm of chromosome III (8). This region contains a DNA repair mutant, *mus309,* isolated in a screen for mutants that were sensitive to both X-rays and methyl methane sulfonate (MMS, an X-ray mimetic chemical) (21, 22). A series of molecular genetic studies suggested that *mus309* corresponds to Dm Ku p70 (9). However, recent high-resolution mapping indicates that *mus309* corresponds to the *Drosophila* homolog of the Bloom's syndrome DNA helicase (BLM) (116). Nonetheless, Ku p70 transgenes complement the *mus309* phenotypes, suggesting a role for both Ku p70 and the Bloom's helicase (116) in double-strand break repair. *Mus309* mutants are defective for repair of P element transposase-induced DNA double-strand breaks, as shown by a loss of DNA sequences adjacent to the donor DNA cleavage site after P element excision (9). Similarly, defects in repair were observed in a genetic assay for chromosome loss after P element-induced breaks in strains mutant for *mus309* (9), the PI_3 kinase checkpoint gene *mei-41,* and the mutagen-sensitive *mus101* gene (6). The extensive deletion of DNA sequences flanking the break site after P element excision in *mus309* mutants is analogous to that observed after V(D)J recombination in X-ray-sensitive mutant mammalian cell lines lacking Ku p70 (217). Thus, the Ku p70/p80 heterodimer, which is conserved from yeast to flies to mammals (41, 47), seems to play a similar role in the NHEJ DNA repair pathway (36, 93, 197) in *Drosophila* and is involved in at least one repair pathway after P element transposase-induced double-strand DNA breaks (Fig. 11).

Biochemical studies of the recombinant Dm Ku p70 alone or the Dm Ku p70/p80 heterodimer have shown that these recombinant proteins do not footprint to the P element inverted repeats (Beall et al., unpublished). It is possible that a protein of much lower abundance, unrelated to Dm Ku and missed in the original purification, is responsible for the IRBP footprint. However, there have been conflicting reports regarding the DNA binding specificity of the mammalian Ku heterodimer; some report only binding to DNA gaps or ends, whereas others report binding to specific DNA sites (47, 197). It is possible that the Dm Ku proteins might interact with another, as yet unidentified, protein to target the Ku heterodimer to the P element inverted repeats. Alternatively, it is possible that the native and recombinant Dm Ku p70/p80 proteins are somehow modified differently, which affects site-specific DNA binding. For instance, the Ku subunits in mammals are phosphorylated by the DNA-dependent protein kinase (DNA-PK) (47). Molecular genetic studies of an extensive set of P element excisions at the vestigial locus have shown that many chromosomal P element excision products seem to be repaired by NHEJ (202). Most of these repair products can be interpreted as simple rejoining of cleaved P element donor DNA ends with 0, 1, or 2 bp of homology (Fig. 9 and 11). These data support the idea that one of the predominant pathways for repair after P element excision is NHEJ, as in mammalian cells after V(D)J recombination (154). In addition, older studies using plasmid-based P element excision, and a more recent study using PCR-based analysis in the germ line and soma, reached similar conclusions (11, 65, 144). In fact, the PCR analysis of germ line versus soma repair products showed interesting tissue-specific differences (longer P element inverted repeat sequences at the repair junctions in the germ line versus soma) which correlate with the expression level of the Dm Ku p70 protein, which is high in the female germ line and much lower in somatic cells (65). Recent genetic studies of *Drosophila RAD54* and *mus309* double mutants suggest that, as in yeast and mammals, both NHEJ and homologous recombinational repair act synergistically in the repair of double-strand DNA breaks (93, 114).

A second pathway for repair of P element-induced double-strand DNA breaks is SDSA, in which cleaved P element donor DNA ends, whether processed by resection or not, can be used to search for homologous DNA sequences that are then used as a template for repair DNA synthesis (54, 57, 142) (Fig. 10 and 11) (see preceding section). After extension DNA synthesis, the two DNA strands can anneal at short stretches of homology to promote repair DNA synthesis and subsequent ligation (similar to NHEJ). If this copying occurred from a sister chromatid or homologous chromosome carrying a P element and repair synthesis was interrupted, this could give rise to the internally deleted P elements observed in naturally occurring P strains (54, 146). This "copying" mechanism can also explain the rapid spread of P elements in wild populations, because P elements are thought to have been introduced into *Drosophila* through horizontal transfer from parasitic mites sometime within the past century (53, 55, 86, 106, 107) (Fig. 10).

The availability of the complete *Drosophila* genome sequence (1, 177), the existing DNA repair mutants (21, 22), and facile assays for DNA repair after P-element-induced strand breaks (6) provides new avenues for investigation into the relative roles of the NHEJ and SDSA repair pathways in P element mobility, in the maintenance of genomic integrity (176), and in how the proteins involved may affect the activity of the P element transposition machinery.

REGULATION OF P ELEMENT TRANSPOSITION

P Cytotype

Initial studies on the reciprocal cross effect in hybrid dysgenic crosses led to the description of two regulatory states based on genetic crosses between P and M strains. M strains were said to have an M cytotype state, which was permissive for P element mobility, whereas P strains were said to have P cytotype, a state that was restrictive for P element movement (108, 110). These reciprocal cross experiments followed the repressive effect of P strain females for several generations (48, 49, 108). The segregation pattern of P cytotype exhibited some aspects of strict maternal inheritance in that the cytotype of the great grandmother played a role in determining whether subsequent progeny displayed M or P cytotype. These studies also suggested that P strain-derived chromosomes were required for P cytotype and for the stable inheritance of P cytotype (49). Thus, a combination of genotype and maternal cytoplasm seemed to determine M versus P cytotype and an analogy was made to zygotic induction by bacteriophage (146). However, all these studies were complicated by the heterogeneity between natural P strains and even within any given strain because natural P strains typically carry 30 to 50 different P element insertions (reviewed in references 50 and 52). As more studies using natural P strains were done, outbreeding was used to identify single P element insertions derived from natural P strains that still exhibited P cytotype (67, 155, 169). These studies have led to a clearer picture of P cytotype control and the interplay of P elements with their host.

P Cytotype versus Pre-P Cytotype

The study of one repressor-producing P strain derived from a wild population, called *Lk-P*(1A), has been particularly informative regarding the mechanism and genetics of P cytotype regulation. *Lk-P*(1A) contains two full-length P elements at the tip of the X chromosome at cytological position 1A (169). Because of the telomeric location of these two elements, it was expected that they would be inserted into repetitive DNA. Indeed, DNA sequence analysis showed that both P elements, which lie in inverted orientation separated by ~5 kb, are integrated into subtelomeric heterochromatic repeat sequences known as *TAS* repeats (165, 170). These repeats of ~0.8 to 1.8 kb are often found at *Drosophila* telomeres examined to date, including those on the minichromosome *(Dp)1187,* where the *TAS* repeats have been hot spots for P element insertion (95). *Lk-P*(1A) exhibits a very strong P cytotype repressive effect characteristic of the regulatory properties observed with normal P strains, and although the two P elements at 1A seem to be full length, the strain produces very low levels of transposase (126, 171). This fact allows the repressor elements at 1A to be used in genetic crosses as a stable genetic marker conferring P cytotype, although excision of one of the two P elements occasionally occurs with concordant loss of P cytotype strength (170).

The initial studies on *Lk-P*(1A) used either of two genetic tests for P element cytotype control, GD sterility or the *singed-weak* (sn^w) hypermutability assays (Table 1) (169). *Lk-P*(1A) displayed a P cytotype repression effect equal to that observed with several natural P strains carrying 20 to 30 times the number of P elements (126, 169). The repressive properties of *Lk-P*(1A) were strong in germ line tissues and were transmitted maternally, but repression was observed only weakly in somatic tissues (169). By contrast, a natural P strain, such as Harwich, displays the repressive properties of P cytotype strongly in both somatic and germ line tissues (169).

The stability of the two *Lk-P*(1A) P elements permitted the use of this strain in genetic tests for the maternal effect and inheritance of the P cytotype. Maternally derived cytoplasm from a P strain female can confer repression of P element mobility to offspring for a single generation (maternal effect), whereas the presence of maternally derived P elements is required for inheritance of P cytotype by future generations (maternal inheritance). However, in a case where cytoplasm from a P strain female is inherited without chromosomal P elements, P cytotype can be maintained when chromosomal P elements are inherited paternally (170–172). Thus, the eggs derived from P strain females carry an extrachromosomal determinant termed pre-P cytotype, which is not functional in the absence of either maternally or paternally provided chromosomal P elements (171). These experiments on pre-P cytotype showed a clear separation of the maternal and zygotic components of P cytotype and clearly demonstrated the interplay between P female egg cytoplasm and the P elements entering the zygote either with the sperm or from the egg (171). The nature of this cytoplasmic determinant is not known, but does not seem to correlate with the level of the 66-kDa repressor protein produced during oogenesis (136). It may involve maternal genomic imprinting (see on *trans*-silencing, below) or perhaps inheritance of antisense or duplex RNA leading to RNA interference (193) (see section on possible control by antisense RNA, below).

Transcriptional Control of P Elements by P Cytotype

Studies of P cytotype control with use of natural P strains, as well as *Lk-P*(1A), showed that P[*lacZ*] enhancer trap elements are transcriptionally repressed in both germ line and somatic tissues in a P cytotype-dependent manner (Fig. 12) (Table 1) (125). Although transcriptional repression of germ line-expressed P[*lacZ*] enhancer trap elements by P cytotype exhibited a maternal effect, transcriptional repression of P[*lacZ*] elements in somatic cells occurred zygotically and did not exhibit the maternal effect characteristic of P cytotype. This study, and the observation that full-length P elements carrying the third intron (IVS3) are required for strong P cytotype, led to a model for the mechanism of P cytotype inheritance in which transcriptional repression of the P element promoter in the germ line resulted indirectly in reduced splicing of IVS3 (126). This reduction in splicing would lead to increased production of the 66-kDa repressor protein relative to transposase. This repressor protein would be transmitted to the unfertilized egg and be present upon fertilization to inhibit transposition of P elements coming into the zygote with the sperm DNA (126). In fact, it has been shown directly that the 66-kDa P element protein acts as a repressor of transposition (137, 163) and is deposited in P strain oocytes before fertilization (136, 137). This repressive effect may result from post-transcriptional effects on transposase activity (Fig. 12) (see "Posttranscriptional Control by P Cytotype" below). However, the fact that the truncated N-terminal KP and 66-kDa repressor proteins (Fig. 3) bind to a site that overlaps the TATA box of the P element promoter suggests that direct binding of P element repressor proteins may be responsible for the observed transcriptional repression in the assays using P[*lacZ*] fusion genes described above (124; C. C. Lee and D. C. Rio, unpublished results). In fact, purified transposase protein bound to its recognition site at the P element 5′ end repressed transcription from the P element promoter in vitro by blocking the TFIID/TPB-TATA box interaction (98). Because both the 66-kDa and KP repressor proteins share the C_2HC N-terminal transposase DNA binding domain, it seems plausible that the truncated repressors might act by a similar mechanism (Fig. 12).

P strains can also exhibit a transcriptional effect called *trans*-silencing (see section on *trans*-silencing below) in which heterologous promoters carried in P element vectors undergo germ line-specific transcriptional silencing in P cytotype genetic backgrounds (166). *Trans*-silencing is independent of the 66-kDa repressor protein and should be contrasted with sce-

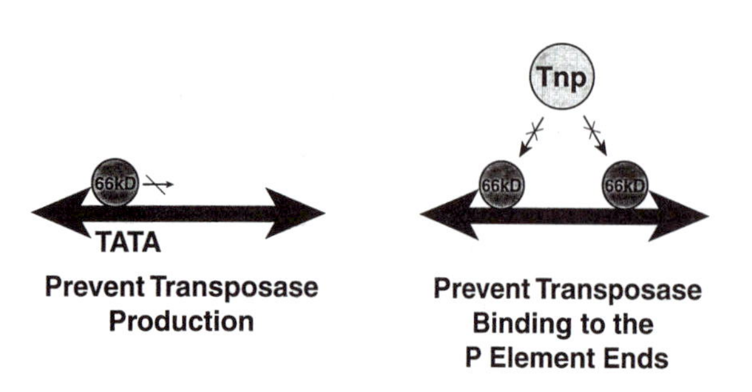

Figure 12. Molecular mechanisms for repression of P element transposition. Two effects of repressor proteins can affect repression: transcriptional (left) and posttranscriptional (right). (Left) The binding of repressor to a transposase site at the 5′ P element end blocks transcription from the P element promoter, reducing transposase mRNA synthesis. (Right) The binding of repressor to the transposase sites occludes transposase-DNA interactions required to initiate P element transposition.

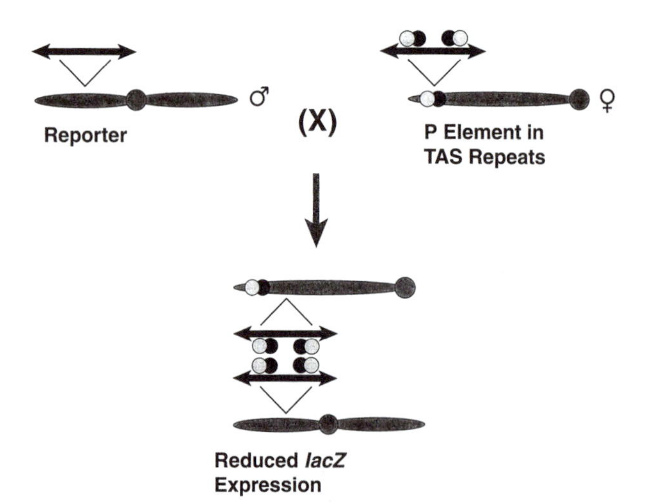

Figure 13. Model for P element *trans*-silencing. Telomeric P element insertions in *TAS* repeats interact with heterochromatin proteins. After mating, the euchromatic *lacZ* reporter P element transgene can form a paired complex with the P element at the telomere (bottom). This association results in transcriptional repression of the reporter, hence the term "*trans*-silencing."

narios in which P element repressor proteins could interact directly with P element DNA sites to block transcription (Fig. 12). In *trans*-silencing, the assembly of heterochromatin proteins with the telomeric P elements may transmit their repressive effects to euchromatic P element insertions via pairing that depends on P element sequence homology (Fig. 13). The transcriptional effects of *trans*-silencing could also affect the ratio of P element IVS3 splicing and thereby modulate the relative levels of the 66-kDa repressor and transposase proteins (Fig. 14) (see section on *trans*-silencing below) (126, 166).

A Role for Inheritance of P Element DNA in P Cytotype

The genetics of P cytotype argued for both extrachromosomal and chromosomal components for the multigenerational inheritance of the P cytotype state. Models for the maintenance of P cytotype argued that some aspect of the repression mechanism would require an autoregulatory feedback loop. Studies with *Lk-P*(1A) clearly demonstrated a role for full-length P elements in the maintenance and inheritance of P cytotype (169–171). The pivotal role of full-length P elements, which are capable of modulating the ratio of repressor to transposase via splicing of the third intron (IVS3), also provides a mechanism for feedback control of repressor synthesis so that in P cytotype the 66-kDa protein would be synthesized preferentially (Fig. 14) (126, 145, 166). Recent studies have revealed important spatial and temporal links between the eukaryotic transcriptional and RNA-processing machineries (83). Moreover, the physical inheritance of P element DNA explains the multigenerational nature of cytotype regulation and why ma-

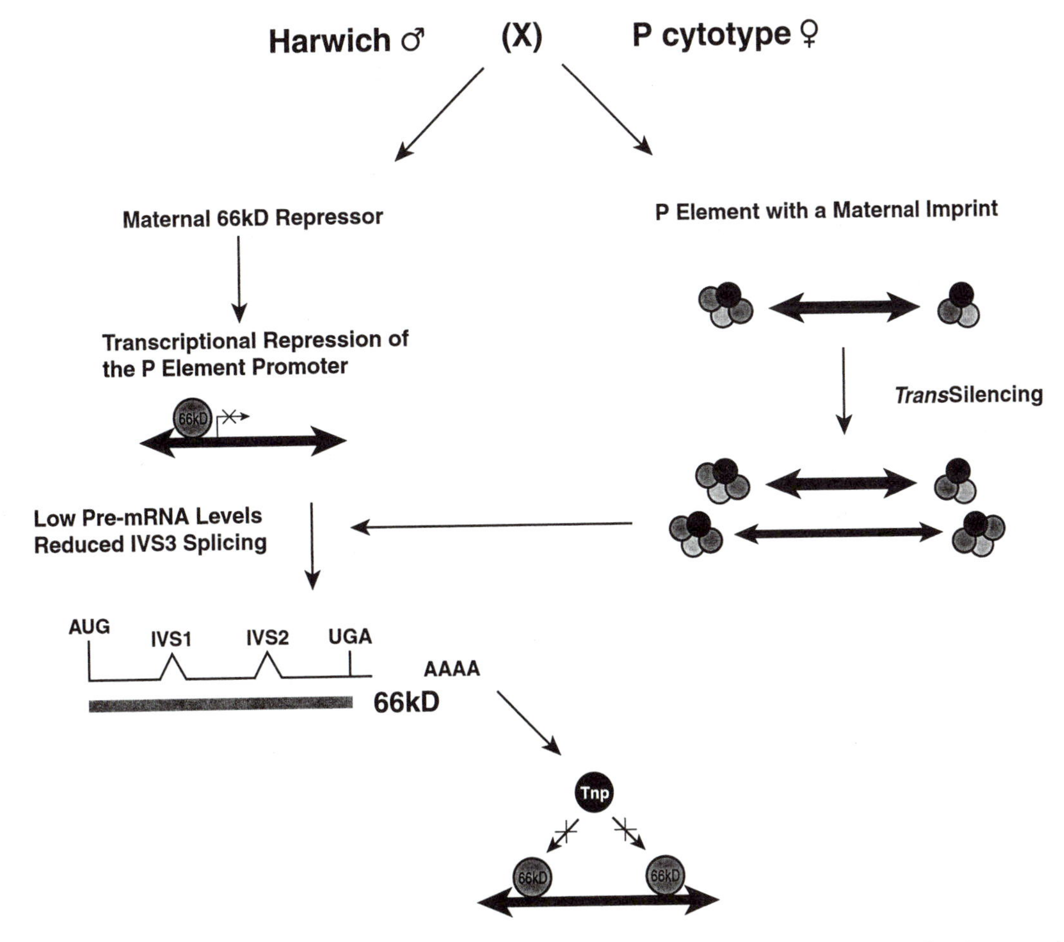

Figure 14. Model for P cytotype regulation. Maternal P cytotype provides several components to control P element activity. Maternal deposition of repressor protein or RNA leads to reduction of P element pre-mRNA synthesis, altering the ratio of repressor and transposase mRNAs, resulting in an autoregulatory loop (126, 145, 166). The maternal genome contains some P elements that acquire a maternal imprint of heterochromatin proteins (clustered grey circles) that, in the zygote, can lead to *trans*-silencing of certain P element insertions. Production and binding of the repressor to P element DNA can inhibit transposase RNA synthesis (left) and block transposase binding to P element DNA (right).

ternal cytoplasm alone is not sufficient for the multigenerational inheritance of the P cytotype state (171). Finally, several studies have suggested that the presence of P element DNA might act to titrate the transposase and disperse it to below the level necessary for transposition (156, 192).

The fact that P cytotype repression by *Lk-P*(1A) is abolished in the heterochromatin protein-1 (HP-1) or *Su(var)*205 mutant (170) suggests that perhaps a chromatin state, rather than simple maternal deposition of P element repressor protein or mRNA, might be maternally transmitted from P strain females. This idea would be similar to parental genomic imprinting in mammals. However, the connection between chromatin structure, P cytotype, and the P element repressor proteins is not clear at present. The *Lk-P*(1A) repressive effect can also be observed on euchromatic P elements in the *trans*-silencing assay (see section on *trans*-silencing, below) which is also consistent with the notion of transmission of a chromatin-based effect (Fig. 14) (166).

The 66-kDa/Type I Family of Repressor Proteins

Initial mapping studies of P element mRNAs showed that the full-length P elements encode a 66-kDa protein in somatic cells when the third intron (IVS3) is retained (121, 161). This protein is encoded by the first three P element exons but lacks sequences found at the C-terminus of transposase (Fig. 3). Because P elements were known to regulate as well as stimulate P element transposition, the 66-kDa protein was a candidate for a P element-encoded repressor protein. Studies of the regulatory effects of P cytotype and the P element-encoded proteins have used several different genetic assays (Table 1), which can differ in their tissue specificity and whether or not they directly assay P element mobility. Two approaches have been used to study the P element-encoded repressor proteins. First, different genomic insertions of in vitro-modified elements expressing a single type of repressor protein have been assayed for the inhibition of P element activity. Second, genetic recombination and outcrossing have been used to identify and isolate P element insertions from natural P strains that show the repressive effects of P cytotype.

The initial studies of the 66-kDa repressor hypothesis used the original four frameshift mutants that defined the transposase coding sequences (94) to ask if these mutations could affect P repressor activity. A recombinant P element containing a frameshift in exon4 (ORF3), called P[*ry*$^+$;*Sal*](89D) could repress P element transposition in both the germ line and soma (163). However, the repressive effects of the modified P[*ry*$^+$;*Sal*](89D) element did not display all aspects of P cytotype. The P[*ry*$^+$;*Sal*](89D) element did not exhibit the reciprocal cross effect characteristic of P cytotype, i.e., it caused repression when inherited from either males or females. The repressive effect was only observed when a single transposase-producing P element was used, but not from the multiple elements present in natural P strains, like π_2. A second study definitively identified the 66-kDa protein as a P element repressor with use of modified P elements engineered to only encode the shorter 66-kDa protein and not the transposase protein (137). These studies showed that the 66-kDa protein-producing P[*ry*$^+$;*66kD*] P elements could inhibit transposase-induced mobility of a P[*w*$^+$] transposon in the somatic cells of the eye (Table 1) and the germ line mobility of the two small P elements at the singed bristle locus (the *singed-weak*, *sn*^{w}, allele) (Table 1). Thus, the P[*ry*$^+$;*66kD*] elements could block transposase activity. As observed with the P[*ry*$^+$;*Sal*](89D) element, repression by the P[*ry*$^+$;*66kD*] elements did not exhibit a maternal effect characteristic of true P cytotype (137). However, upon mobilization of P[*ry*$^+$;*66kD*] P elements to new genomic locations and testing these new insertion sites for maternal repression, one strain (no. 68) was found that exhibited maternal-effect repression (136). A subsequent study showed that a series of truncated C-terminal derivatives similar to the 66-kDa protein could act as repressors (67).

A second line of investigation into the nature of P element repressors involved outcrossing natural P strains to identify loci capable of causing P cytotype repression. The first study of this type isolated a naturally occurring repressor P element lacking exon 4 (ORF3) and which was capable of encoding a 66-kDa-like protein (143). Subsequent studies with natural P strains have identified repressor elements capable of encoding proteins similar to the 66-kDa protein (67). A special type of strain identified from a natural population is the *Lk-P*(1A) strain (see sections on P cytotype and transcriptional control above and section on *trans*-silencing below).

Most of the engineered 66-kDa P elements and some natural repressor elements do not exhibit the maternal component of P cytotype. The maternal effect of P cytotype was proposed to involve deposition of the 66-kDa repressor protein (or mRNA) into oocytes derived from P strain females and that the presence of the repressor in P strain eggs could be responsible for the maternal inheritance of P cytotype (50, 146). Although the 66-kDa repressor was present in oocytes from the P strain π_2, it was significantly reduced in P[*ry*$^+$;*66kD*] oocytes (137). Because P[*ry*$^+$;*Sal*] (89D) and the initially tested P[*ry*$^+$;*66kD*] P elements did not show maternal-effect repression, it was also proposed that the insertion sites or genomic locations of the repressor elements might determine maternal repression. Indeed, a study that mobilized the P[ry^+;*66kD*] elements to new genomic locations identified a novel insertion that showed maternal repression of transposition (136). However, the levels of the 66-kDa repressor protein in oocytes did not correlate with the maternal repression and additional studies showed that maternal expression of the 66-kDa protein was not sufficient for the maternal component of P cytotype (136). These studies did demonstrate that genomic position can be a major determinant of maternal P cytotype (67, 194) and helped to explain why many strains that contain full-length P elements do not exhibit P cytotype. One insight into the role of genomic position came from studies of the *Lk-P*(1A) strain, which has two P elements inserted at the X chromosome telomere (see section on *trans*-silencing below). In conclusion, although the 66-kDa protein can clearly act as a repressor of transposition, probably by a posttranscriptional effect (Fig. 12) (see "Posttranscriptional Control by P Cytotype" below), it does not seem to be the "maternal component" that is inherited through the female germ line to cause P cytotype.

The KP/Type II Family of Repressor Proteins

The P elements present in natural P strains consist predominantly of internally deleted nonautonomous P elements (145, 146). This situation is similar to the Ac-Ds family of elements initially described in maize (131). In many cases, these small internally deleted elements can encode repressor proteins, which have also been found in so-called M′ or Q strains. M′ strains have P element DNA, but are not capable of either inducing or repressing P element mobility, whereas Q strains cannot induce, but can repress, P element mobility (103, 105). M′ or Q strains contain P element DNA, can inhibit transposition to varying degrees, but cannot induce hybrid dysgenesis because they lack a sufficient number of full-length P elements to produce transposase (52). The first, and most abundant, small repressor element described was the KP element, which contains an internal deletion of 1.7 kb from nucleotides 808 to 2,560 (18, 87). This element can encode a 207-amino-acid protein containing the DNA binding and dimerization domains shared with the transposase and 66-kDa proteins (Fig. 3 and 5). The KP element is transcribed in M′ strains to encode a 0.8-kb mRNA, although the protein has not been observed. Some M′ strains may contain up to 30 copies of the KP element, and DNA sequence comparisons of different KP elements from distinct strains have shown high conservation. Thus, in nature, it may be that the KP element in M′ strains is under some type of positive selection as a particularly effective or stable repressor. Genetic studies of the isolated KP elements have shown that repression of transposition is weak and is predominantly chromosomally based, requiring inheritance of the repressor element (67, 155, 194). However, in assays for repressor activity using single transposase-producing elements, a very weak maternal effect could be observed with some single, genetically isolated elements, again reinforcing the role that genomic position of repressor elements plays in maternal P cytotype (79, 155, 194).

Studies of naturally occurring repressor elements have also identified other internally deleted elements, including the D50 and SP elements (155). Both elements, as well as KP elements, were isolated in genetic crosses using the inbred M′ strain Sexi (155). The D50 element is very similar to the deleted KP element, encodes a 204-amino-acid protein, and also contains an insertion of 21 bp of non-P element DNA at the deletion junction (155). In contrast, the SP element contains a more extensive internal deletion from nucleotides 186 to 2,577 and thus could only encode a small 14-amino-acid peptide (155). Both the D50 and SP elements, like the KP element, are weak repressors of P element mobility (155). Once again, it was made clear from these studies that genomic position of the repressor elements, as well as their structure, can influence the repressive properties of individual elements.

Although the KP/type II repressor family can possess many of the same properties of repression as the larger type I repressors, the type II repressors do not, in general, show the maternal effect or inheritance of repression typically characteristic of natural P strains (67, 169). There are also sometimes distinctions between type I and type II repressors regarding their somatic versus germ line transpositional repressive activities (67, 155). However, in some cases these differences may be caused by the different assays used to recover or test the effectiveness of the repressor elements (Table 1). In some cases with certain strain backgrounds, the reduced effectiveness of the short KP/type II repressors may be compensated for by increased copy number of these internally deleted P elements (87, 145, 194).

Posttranscriptional Control by P Cytotype

P strains, such as *Lk-P*(1A), can repress transcription from P[*lacZ*] enhancer trap P elements (125) (see "Transcriptional Control of P Elements by P Cytotype" above). In contrast, modified P elements encoding either the 66-kDa or KP repressors exhibited only weak repression of P[*lacZ*] transcription in the soma (126, 136) and exhibited no repression of

P[*lacZ*] transcription in the germ line (136). However, P[ry^+;*66kD*] elements did show a repressive effect in the sn^w/sn^{x2} cytotype-dependent allele assay (136) (Table 1). The effect of P cytotype on the *singed-weak* allele appears to be transcriptional (149, 165). At least some of the repressive effects of P cytotype on transcription may involve a chromatin-based effect (see "Transcriptional Control of P Elements by P Cytotype" and "*trans*-Silencing and the Role of Telomeric P Elements in P Cytotype Control" above) (Fig. 14) in addition to direct binding of repressor proteins to DNA (Fig. 12).

The function of the 66-kDa and KP repressor proteins may involve posttranscriptional effects on transposase activity. Because these truncated repressor proteins possess both the DNA binding and dimerization domains of transposase (Fig. 3 and 5), it was proposed that the repressor proteins might either compete with transposase for binding to sites on P element DNA or, alternatively, block oligomerization of the transposase protein (Fig. 12). It is known that the KP and 66-kDa proteins bind to the same sites as the transposase (124; Lee and Rio, unpublished) and that mutant KP proteins that cannot bind DNA fail to block transposase activity in vitro (123) (see section on the P element KP repressor protein). In addition, in vivo studies on the KP leucine zipper region indicated that this region is important for repressor activity, supporting the idea that protein-protein interactions between repressor and transposase block activity or that the leucine zipper might promote DNA binding by the repressor protein (3). Thus, both mechanisms of posttranscriptional control by the P element repressor proteins may be operating to modulate transposase activity in P cytotype.

Possible Control of P Element Transposition by Antisense RNA

Several observations suggest that genomic transcription from promoters flanking P element insertions could lead to an inhibitory effect on transposition. Recent studies on transposon suppression in the *C. elegans* germ line, cosuppression in plants and *Drosophila,* and the phenomenon of RNA interference (RNAi) in *C. elegans* and *Drosophila* (60, 187) suggest that the generation of double-stranded P element RNA might serve to block transposase expression in P strains, possibly by causing the targeted degradation of the P element mRNA. One finding of a naturally occurring repressor P element called SP which encoded a short 14-amino-acid peptide led to the suggestion that perhaps the SP insertion caused repression via antisense (or possible duplex) RNA inhibition (155). An extension of these initial findings used antisense P element transgenes to show that at some insertion sites a repressive effect could be observed (193). In light of the new findings regarding a double-stranded RNA surveillance system operating to modulate transposons in the *C. elegans* germ line (102), it seems reasonable that duplex RNA could provide a means to modulate transposase production in certain situations.

trans-Silencing and the Role of Telomeric P Elements in P Cytotype Control

During studies of P cytotype in the germ line using P element transgenes carrying germ line-specific promoters fused to *lacZ,* the phenomenon of *trans*-silencing was observed (166). Euchromatic P elements carrying germ line-specific *hsp83* or *vasa* promoter-*lacZ* gene fusions were transcriptionally silenced by the natural P strain π_2 or the *Lk-P*(1A) strain (Fig. 13). Silencing of the *lacZ* reporter transgenes showed the maternal effect characteristic of P cytotype (166) and could also be observed with recombinant P elements inserted in the telomeric *TAS* repeats at the tip of the X-chromosome (1A) or on the right arm of chromosome 3 (100F) (165). This transcriptional repression was independent of the genomic position of the target element in euchromatin, and it was suggested that it might involve pairing of P element sequences to "spread" the heterochromatic structure of the telomeric *TAS* repeats to the euchromatic P elements (Fig. 13) (166). The telomeric recombinant P element transgenes at 1A and 100F also displayed a so-called "combination effect" whereby the telomeric P elements repressed gonadal dysgenic (GD) sterility when in combination with natural regulatory P elements (172). This effect was dependent on the HP-1/*Su(var)205* protein, just like the P cytotype effects of the *Lk-P*(1A) strain (172). Thus, the telomeric P element insertions may allow the transcriptional inhibitory effects of subtelomeric heterochromatin to be transmitted to euchromatic P element insertions to inhibit their expression.

Control of the Tissue Specificity of P Element Transposition by Alternative RNA Splicing

One of the most startling findings regarding the regulation of P element transposition was the discovery that alternative pre-mRNA splicing restricted expression of the P element transposase, and hence the entire syndrome of traits associated with hybrid dysgenesis, to germ line cells (48, 49, 110). Although most genes in metazoans, including *Drosophila,* display alternative splicing and the basic biochemical mechanism of pre-mRNA intron removal by the spliceosome is conserved in eukaryotic cells, very little is known about how splice site selection occurs in complex nuclear pre-mRNAs (29, 84, 157). The example of the tissue-specific splicing of P element IVS3 has served as a useful model system to investigate how alternative RNA splicing patterns are generated in distinct cell or tissue types (29, 128, 195).

Two lines of investigation of IVS3 splicing led to the conclusion that at least one aspect of this control involves an inhibition of IVS3 splicing in somatic cells. First, using a molecular genetic approach it was shown that mutations in the 5′ exon, upstream from IVS3, caused an activation of IVS3 splicing in somatic cells (30, 122). Here, a 240-bp DNA fragment was fused to the *E. coli lacZ* gene and expressed by using the hsp70 promoter. The wild-type fusion gene, as well as many altered derivatives, exhibited the normal tissue specificity of IVS3 splicing (30). However, specific sequence changes in the 5′ exon activated IVS3 splicing in somatic cells (30). Moreover, these studies showed that when IVS3 splicing was activated somatically, it could be spliced in two different ways: accurate removal of IVS3 and leading to transposase production and a second double splicing event removing two smaller introns and leaving a microexon, all from within IVS3 (30) (Fig. 15A and B). This pattern of splicing was also observed in hypomorphic mutants of the hrp48 gene, in which IVS3 splicing was activated in somatic cells (76). Second, a set of biochemical experiments using mammalian and somatic *Drosophila* cell splicing extracts, showed that while IVS3 splicing occurred in mammalian cell splicing extracts, it was not observed in the *Drosophila* somatic cell extracts (191). Titration of 5′ exon RNA into the somatic cell extract activated IVS3 splicing, suggesting that the action of *trans*-acting factors led to the inhibitory effect observed (189, 191). A series of UV cross-linking and binding and competition experiments identified several proteins of 97 kDa, 65 kDa, 50 kDa, and 45 kDa that seemed to interact with the IVS3 5′ exon (191). Both IVS3 and the 5′ exon had several 5′ splice site-like sequences, and it was thought that these pseudo-5′ splice sites might play a role in regulating IVS3 splicing (189, 191) (Table 2; Fig. 15B). Another study with mammalian splicing extracts showed that mutations in the 5′ exon known to activate IVS3 splicing in *Drosophila* also activated IVS3 splicing in vitro, suggesting the possible conservation of components (214). Indeed, subsequent studies using in vitro splicing assays with *Drosophila* extracts showed that IVS3 splicing was activated by mutations in the 5′ exon (189). RNA-protein interaction studies defined two elements in the 5′ exon termed F1 and F2, both of which bear sequence identity to 5′ splice sites and contain a reiterated sequence motif (AGNUUAAG) (189) (Table 2; Fig. 15C). Mutations in the F1 and F2 sites that activate splicing in vitro inhibit RNA-protein complex formation in nuclear extracts (189) (Fig. 15C). The F1 site binds U1 snRNP (189) and the F2 site binds hrp48 (190) (Fig. 16), a *Drosophila* hnRNP protein similar to mammalian hnRNP A1 (129). hnRNP A1 causes use of distal (upstream) 5′ splice sites in vitro (130), and binding site selection data indicate it binds to sites resembling 5′ splice sites (26). In extract mixing experiments where *Drosophila* somatic cell extract was mixed with mammalian splicing extract, IVS3 splicing was inhibited and there was a concomitant reduction of U1 snRNP binding to the accurate IVS3 5′ splice site (189) (Fig. 16). This observation makes sense in that the U1-5′ splice site interac-

Table 2. Sequences of 5′ splice sites

Splice site	Sequence	Match	Location
Consensus	C A AG/GU AGU A G		
P IVS3 Accurate 5′ splice site	UAG/GUAUGA	6/9	nt 1945–1953
fushi tarazu	CAG/GUAGGC	7/9	
P IVS3 pseudo-5′ splice sites			
Upstream			
1	UAG/GUUAAG	5/9	nt 1925–1933 (F2)
2	UAA/GUAUAG	4/9	nt 1919–1927 (F1)
3	GAG/GUGGAC	5/9	nt 1882–1890
Downstream			
1	AAU/GUAAUU	7/9	nt 1980–1988 (D1)
2	AAU/GUAGGU	7/9	nt 2046–2054 (D2)
3	UAG/GUAGUU	6/9	nt 2050–2058 (D3)
4	AUU/GUGUUU	6/9	nt 2070–2078 (D4)

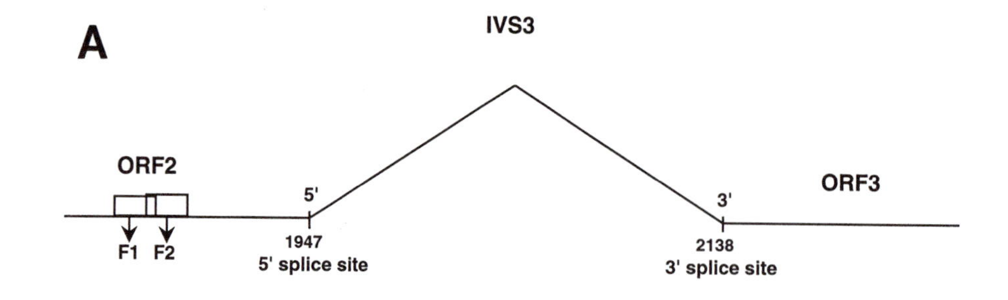

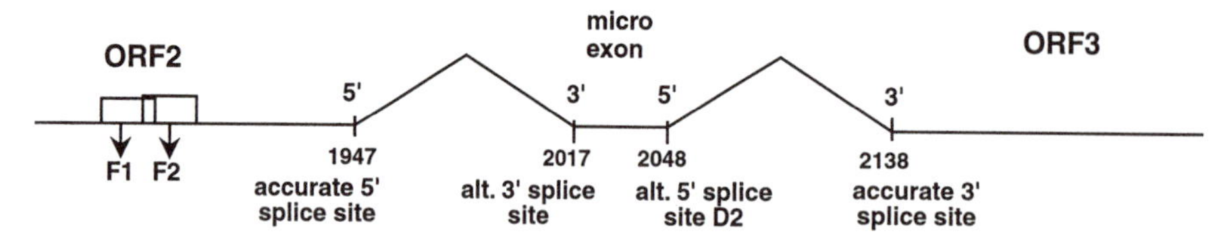

B

UUGUGGAGCAUUUUUUUGGCAGCAUGCGAUCGAGAGGUGGACA

AUUCGACCAUCCCACUCCACUGCAGUUUAAGUAUAGGUUAAG

Accurate IVS3 5' SS　F1　F2

AAAAUAUAUAAUAGGUAUGACAAAUUUAAAAGAAUGCGUAA

1947　alt 3' SS

ACAAAAAUGUAAUUCCAUGAUUUAUAAUUGUUUAAUGUUUAG

D1　alt 5' SS

CUAUAUGUUUCAGGAAAGUUUCAGUUGAGAAUGUAGGUAGUU

D2　D3

AUGUGCUGUCUAUUGUGUUUUGUCUUUUAUCUGUUUCUUUUCA

BP　D4　Accurate IVS3 3' SS

UUUUAUUAUUUAAUCAUUAUCCUUUUGCUUAUCCAGCCAGGA

2138

AUACAGAAAUGUUAAGAAAUUCGGGAAAUAUCGAAGAGGAC

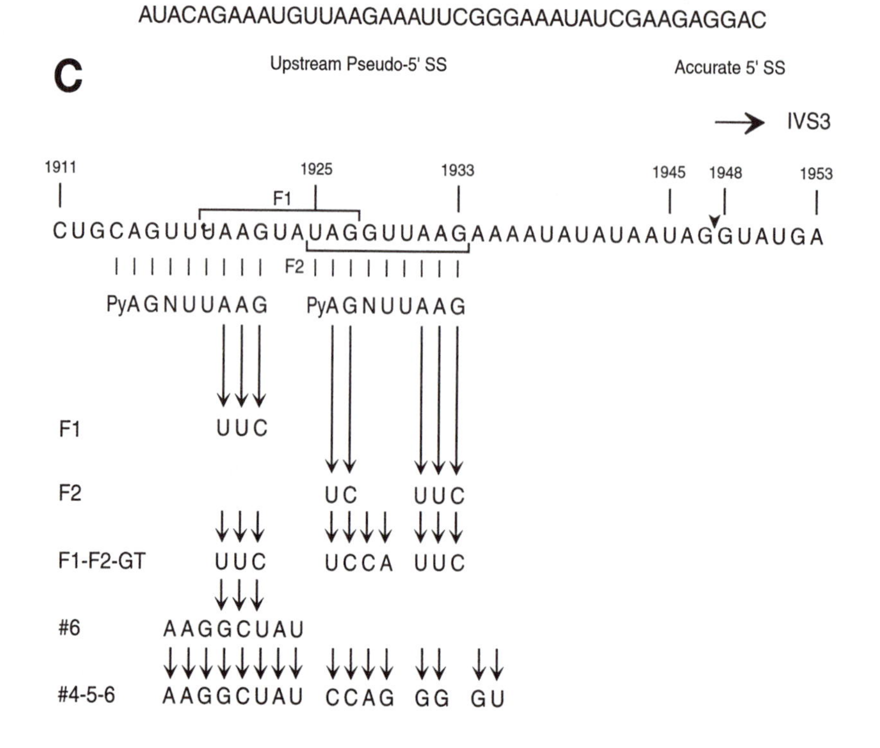

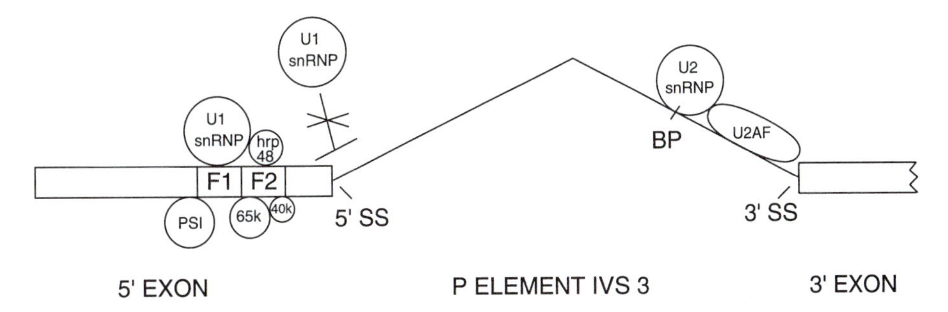

Figure 16. Model for somatic inhibition of IVS splicing. U1 snRNP usually interacts with the IVS3 5′ splice site during the early steps of intron recognition and spliceosome assembly. In somatic cells (and in vitro) this site is blocked (189). Mutations in the upstream negative regulatory element lead to activation of IVS3 splicing in vivo (30) and in vitro (189, 214). The F1 site binds U1 snRNP (189) and the F2 site binds the hnRNP protein, hrp48 (190). An RNA-binding protein containing four KH domains that is expressed highly in somatic cells, called PSI, has also been implicated in IVS3 splicing control (188). Several other uncharacterized proteins from *Drosophila* extracts have been identified by UV cross-linking to the IVS3 5′ exon (191).

tion is an early event in spliceosome assembly (84, 138) and the block to IVS3 splicing in vitro occurs before 5′ exon cleavage and lariat formation (191). These results led to a model in which RNA-protein and snRNP interactions in the 5′ exon cause an inhibition of IVS3 splicing by blocking access of U1 snRNP to the normal IVS3 5′ splice site (189) (Fig. 16). A series of RNA-protein complex formation assays were consistent with the binding of multiple proteins to the 5′ exon regulatory element in a complex that might alter the distribution of U1 snRNP on the P element pre-mRNA and prevent U1 snRNP binding to the IVS3 5′ splice site (189).

Subsequent biochemical studies have been aimed at identifying additional proteins that interact with the IVS3 5′ exon negative regulatory RNA element. By using RNA binding assays both the 97-kDa and 50-kDa proteins were identified as P element somatic inhibitor (PSI) and hrp48, respectively. PSI contains four KH-type RNA binding motifs and a C-terminal repeat of ~100 amino acids directly repeated twice (188). Conserved residues in these repeat motifs are involved in direct interaction between PSI and the U1 snRNP 70k protein (117). A similar structure is found in a mammalian alternative splicing factor called KSRP and a related family of proteins called FBPs (135). PSI interacts with the IVS3 5′ exon RNA, but not with heterologous RNAs of similar length and base composition (188). It is also expressed highly in somatic cells, but at low or undetectable levels in the female germ line (188). Furthermore, addition of anti-PSI antibodies to in vitro splicing reactions resulted in activation of IVS3 splicing and inhibition was restored upon addition of recombinant PSI protein (190). Thus, PSI seems to be expressed highly in the soma, but not the germ line, and in vitro seems to play a role in IVS3 splicing inhibition (188, 190). The hrp48 protein specifically binds to the F2 element in the 5′ exon (190) and is expressed in both the germ line and soma (188). This protein contains two N-terminal RNP-CS type RNA binding domains and a C-terminal glycine-rich domain (129). This structure is characteristic for this class of hnRNP proteins, termed 2XRBD-GLY, of which mammalian hnRNP A1 is a member (27). The expression pattern of hrp48 in both germ line and soma suggests that hrp48 might also play a role in the inhibition of IVS3 splicing in the germ line, where IVS3 is known to be spliced inefficiently (166).

Molecular genetic studies have addressed the

Figure 15. Alternative splicing of the P element third intron (IVS3). (A) Diagram of two IVS3 splicing patterns observed in vivo. The 5′ exon (ORF2), 3′ exon (ORF3), and intron (IVS3) elements are shown schematically. The negative regulatory elements in the 5′ exon (F1 and F2) are indicated. The accurate 5′ (1947) and 3′ (2138) splice sites are shown. The minor alternative D2 5′ (2048) splice site and alternative 3′ splice site (2017) with the flanking microexon is indicated below (30, 176). (B) RNA sequence of the IVS3 region in the P element. The accurate 5′ and 3′ splice sites are indicated, as is the intron branchpoint (BP). Also underlined are the internal (D1, D2, D3, and D4) and upstream (F1 and F2) 5′ splice site-like (pseudo) sequences (Table 2) (189). (C) The 5′ exon negative regulatory element and mutations that activate IVS3 splicing. Sequence of the P element 5′ exon. The F1 and F2 sites (labeled brackets) and the accurate 5′ splice site (arrowhead) are shown. Below is shown a reiterated sequence (PyAGNUUAAG) present overlapping the F1 and F2 sites. Below are shown 5′ exon mutations that activate IVS3 splicing in vitro (30, 122) or in vivo (189, 214).

possible roles of PSI and hrp48 in IVS3 splicing control in vivo. First, ectopic expression of PSI in germ line cells caused a modest reduction in IVS3 splicing (2). Antisense hammerhead ribozyme targeting of PSI mRNA in somatic cells resulted in activation of IVS3 splicing in the soma (2). Both experiments suggest a role for PSI in the reduction of IVS3 splicing. Reduction of hrp48 levels in somatic cells by using hypomorphic P element insertion alleles caused a small activation of IVS3 splicing in the soma (76). These experiments also showed that hrp48 was encoded by an essential gene, and therefore must have additional functions involving cellular mRNAs. Future studies will identify other P element 5′ exon RNA binding factors, the interaction of these proteins with spliceosomal and other RNA processing components and how these gene products might affect processing of cellular RNAs.

Whereas the basic mechanism of pre-mRNA splicing is well understood, how differential splicing is controlled is still unclear. Both exon and intron sequences have been implicated in both positive and negative control of intron splicing (128, 157, 195). Thus, further study of the P element system as a model for the negative control of 5′ splice site selection should provide useful information regarding the control mechanisms for alternative splice site selection.

SUMMARY AND CONCLUSION

P elements in *Drosophila* have become one of the best studied eukaryotic mobile elements. This understanding has transformed them from being known as the causative agents of hybrid dysgenesis to one of the most powerful tools of the *Drosophila* geneticist. Moreover, understanding of the mechanism of P element transposition has led to studies of how these elements can use the *Drosophila* DNA repair machinery to facilitate their movement from one genomic location to another. It will be extremely important and interesting for the future to understand how P element mobility might be linked to the cell cycle and DNA repair checkpoints and how P element transposition might be influenced by extracellular stimuli (42, 131). Although P elements have not been observed to move in species other than *Drosophila* or very closely related species, further studies into the interplay between these elements and the *Drosophila* DNA repair and replication machinery will eventually lead to this goal and provide important insights into the mechanisms of cellular DNA repair and recombination.

Because P elements use differential RNA splicing to restrict their mobility to germ line cells, understanding the mechanistic basis for this block is in general more important. In fact, although most genes in metazoans are differentially spliced (138) and 15% to 20% of known disease gene mutations affect splicing (84), our ability to predict splicing patterns from metazoan genomic sequences is limited. Thus, P elements should continue to reveal how the packaging of pre-mRNA in the nucleus modulates the splicing and transport of differentially processed transcripts. Thus, P elements should continue to be a good model for the study of RNA processing in eukaryotes.

Finally, although studies of the bizarre phenomenon of hybrid dygenesis were responsible for the isolation of P transposable elements, their utility has revolutionized the field of *Drosophila* genetics and caused a large influx of investigators into this field. One could not have imagined the numerous and far-reaching insights regarding many different biological processes that have come from the use of this small piece of DNA. Because of their critical role in *Drosophila* molecular genetics, P elements will continue to provide important biological information in the post-genome sequence era of *Drosophila* biology.

Acknowledgments. I thank Gerry Rubin and Allan Spradling for introducing me to P transposable elements and for many interesting and productive interactions during the past 20 years. I thank Tom Cech, Bob Tjian, and Mike Botchan for sparking my interest in nucleic acid chemistry, protein biochemistry, and transposable elements. I thank all my colleagues who have contributed to our understanding of P elements and to the larger community of investigators studying the mechanisms and regulation of transposable elements, particularly Tania Baker, Nancy Craig, Marty Gellert, Nancy Kleckner, and Kyoshi Mizuuchi. I thank Eileen Beall, Kathy Collins, and Siobhan Roche for their comments on the manuscript, and Danielle Colitti for help with the figures and manuscript.

Finally, I express my gratitude for continued interest and funding from the NIH.

REFERENCES

1. Adams, M. D., S. E. Celniker, R. A. Holt, C. A. Evans, J. D. Gocayne, P. G. Amanatides, S. E. Scherer, P. W. Li, R. A. Hoskins, R. F. Galle, R. A. George, S. E. Lewis, S. Richards, M. Ashburner, S. N. Henderson, G. G. Sutton, J. R. Wortman, M. D. Yandell, Q. Zhang, L. X. Chen, R. C. Brandon, Y. H. Rogers, R. G. Blazej, M. Champe, B. D. Pfeiffer, K. H. Wan, C. Doyle, E. G. Baxter, G. Helt, C. R. Nelson, G. L. Gabor Miklos, J. F. Abril, A. Agbayani, H. J. An, C. Andrews-Pfannkoch, D. Baldwin, R. M. Ballew, A. Basu, J. Baxendale, L. Bayraktaroglu, E. M. Beasley, K. Y. Beeson, P. V. Benos, B. P. Berman, D. Bhandari, S. Bolshakov, D. Borkova, M. R. Botchan, J. Bouck, P. Brokstein, P. Brottier, K. C. Burtis, D. A. Busam, H. Butler, E. Cadieu, A. Center, I. Chandra, J. M. Cherry, S. Cawley, C. Dahlke, L. B. Davenport, P. Davies, B. de Pablos, A. Delcher, Z. Deng, A. D. Mays, I. Dew, S. M. Dietz, K. Dodson, L. E. Doup, M. Downes, S. Dugan-Rocha, B. C. Dunkov, P. Dunn, K. J. Durbin, C. C. Evangelista, C. Ferraz, S. Ferriera, W. Fleischmann, C. Fosler, A. E. Gabrielian, N. S. Garg, W. M. Gelbart, K. Glasser, A. Glodek,

F. Gong, J. H. Gorrell, Z. Gu, P. Guan, M. Harris, N. L. Harris, D. Harvey, T. J. Heiman, J. R. Hernandez, J. Houck, D. Hostin, K. A. Houston, T. J. Howland, M. H. Wei, C. Ibegwam, M. Jalali, F. Kalush, G. H. Karpen, Z. Ke, J. A. Kennison, K. A. Ketchum, B. E. Kimmel, C. D. Kodira, C. Kraft, S. Kravitz, D. Kulp, Z. Lai, P. Lasko, Y. Lei, A. A. Levitsky, J. Li, Z. Li, Y. Liang, X. Lin, X. Liu, B. Mattei, T. C. McIntosh, M. P. McLeod, D. McPherson, G. Merkulov, N. V. Milshina, C. Mobarry, J. Morris, A. Moshrefi, S. M. Mount, M. Moy, B. Murphy, L. Murphy, D. M. Muzny, D. L. Nelson, D. R. Nelson, K. A. Nelson, K. Nixon, D. R. Nusskern, J. M. Pacleb, M. Palazzolo, G. S. Pittman, S. Pan, J. Pollard, V. Puri, M. G. Reese, K. Reinert, K. Remington, R. D. Saunders, F. Scheeler, H. Shen, B. C. Shue, I. Siden-Kiamos, M. Simpson, M. P. Skupski, T. Smith, E. Spier, A. C. Spradling, M. Stapleton, R. Strong, E. Sun, R. Svirskas, C. Tector, R. Turner, E. Venter, A. H. Wang, X. Wang, Z. Y. Wang, D. A. Wassarman, G. M. Weinstock, J. Weissenbach, S. M. Williams, T. Woodage, K. C. Worley, D. Wu, S. Yang, Q. A. Yao, J. Ye, R. F. Yeh, J. S. Zaveri, M. Zhan, G. Zhang, Q. Zhao, L. Zheng, X. H. Zheng, F. N. Zhong, W. Zhong, X. Zhou, S. Zhu, X. Zhu, H. O. Smith, R. A. Gibbs, E. W. Myers, G. M. Rubin, and J. C. Venter.** 2000. The genome sequence of *Drosophila melanogaster. Science* **287:** 2185–2195.

2. **Adams, M. D., R. S. Tarng, and D. C. Rio.** 1997. The alternative splicing factor PSI regulates P-element third intron splicing in vivo. *Genes Dev.* **11:**129–138.
3. **Andrews, J. D., and G. B. Gloor.** 1995. A role for the KP leucine zipper in regulating P element transposition in *Drosophila melanogaster. Genetics* **141:**587–594.
4. **Ashburner, M., S. Misra, J. Roote, S. E. Lewis, R. Blazej, T. Davis, C. Doyle, R. Galle, and N. Harris.** 1999. An exploration of the sequence of a 2.9-Mb region of the genome of *Drosophila melanogaster. Genetics* **153:**179–219.
5. **Ballinger, D. G., and S. Benzer.** 1989. Targeted gene mutations in Drosophila. *Proc. Natl. Acad. Sci. USA* **86:** 9402–9406.
6. **Banga, S. S., A. Velazquez, and J. B. Boyd.** 1991. P transposition in Drosophila provides a new tool for analyzing postreplication repair and double-strand break repair. *Mutat. Res.* **255:**79–88.
7. **Basler, K., and G. Struhl.** 1994. Compartment boundaries and the control of Drosophila limb pattern by hedgehog protein. *Nature* **368:**208–214.
8. **Beall, E. L., A. Admon, and D. C. Rio.** 1994. A Drosophila protein homologous to the human p70 Ku autoimmune antigen interacts with the P transposable element inverted repeats. *Proc. Natl. Acad. Sci. USA* **91:**12681–12685.
9. **Beall, E. L., and D. C. Rio.** 1996. Drosophila IRBP/Ku p70 corresponds to the mutagen-sensitive mus309 gene and is involved in P-element excision in vivo. *Genes Dev.* **10:** 921–933.
10. **Beall, E. L., and D. C. Rio.** 1997. Drosophila P-element transposase is a novel site-specific endonuclease. *Genes Dev.* **11:** 2137–2151.
11. **Beall, E. L., and D. C. Rio.** 1998. Transposase makes critical contacts with, and is stimulated by, single-stranded DNA at the P element termini in vitro. *EMBO J.* **17:**2122–2136.
12. **Bellaiche, Y., V. Mogila, and N. Perrimon.** 1999. I-SceI endonuclease, a new tool for studying DNA double-strand break repair mechanisms in Drosophila. *Genetics* **152:**1037–1044.
13. **Bellen, H. J., C. J. O'Kane, C. Wilson, U. Grossniklaus, R. K. Pearson, and W. J. Gehring.** 1989. P-element-mediated enhancer detection: a vesatile method to study development in Drosophila. *Genes Dev.* **3:**1288–1300.
14. **Berg, C. A., and A. C. Spradling.** 1991. Studies on the rate and site-specificity of P element transposition. *Genetics* **127:** 515–524.
15. **Bier, E., H. Vaessin, S. Shepherd, K. Lee, K. McCall, S. Barbel, L. Ackerman, R. Carretto, T. Uemura, E. Grell, et al.** 1989. Searching for pattern and mutation in the Drosophila genome with a P-lacZ vector. *Genes Dev.* **3:**1273–1287.
16. **Bingham, P. M., M. G. Kidwell, and G. M. Rubin.** 1982. The molecular basis of P-M hybrid dysgenesis: the role of the P element, a P strain-specific transposon family. *Cell* **29:** 995–1004.
17. **Bingham, P. M., R. Levis, and G. M. Rubin.** 1981. Cloning of DNA sequences from the white locus of D. melanogaster by a novel and general method. *Cell* **25:**693–704.
18. **Black, D. M., M. S. Jackson, M. G. Kidwell, and G. A. Dover.** 1987. KP elements repress P-induced hybrid dysgenesis in Drosophila melanogaster. *EMBO J.* **6:**4125–4135.
19. **Bourne, H. R., D. A. Sanders, and F. McCormick.** 1990. The GTPase superfamily: a conserved switch for diverse cell functions. *Nature* **348:**125–132.
20. **Bourne, H. R., D. A. Sanders, and F. McCormick.** 1991. The GTPase superfamily: conserved structure and molecular mechanism. *Nature* **349:**117–127.
21. **Boyd, J. B., M. D. Golino, K. E. Shaw, C. J. Osgood, and M. M. Green.** 1981. Third-chromosome mutagen-sensitive mutants of Drosophila melanogaster. *Genetics* **97:**607–623.
22. **Boyd, J. B., and R. B. Setlow.** 1976. Characterization of post-replication repair in mutagen-sensitive strains of Drosophila melanogaster. *Genetics* **84:**507–526.
23. **Brand, A. H., A. S. Manoukian, and N. Perrimon.** 1994. Ectopic expression in Drosophila. *Methods Cell Biol.* **44:** 635–654.
24. **Brand, A. H., and N. Perrimon.** 1993. Targeted gene expression as a means of altering cell fates and generating dominant phenotypes. *Development* **118:**401–415.
25. **Brautigam, C. A., and T. A. Steitz.** 1998. Structural and functional insights provided by crystal structures of DNA polymerases and their substrate complexes. *Curr. Opin. Struct. Biol.* **8:**54–63.
26. **Burd, C. G., and G. Dreyfuss.** 1994. RNA binding specificity of hnRNP A1: significance of hnRNP A1 high-affinity binding sites in pre-mRNA splicing. *EMBO J.* **13:**1197–1204.
27. **Burd, C. G., and G. Dreyfuss.** 1994. Conserved structures and diversity of functions of RNA-binding proteins. *Science* **265:**615–621.
28. **Cech, T. R.** 1990. Self-splicing of group I introns. *Annu. Rev. Biochem.* **59:**543–568.
29. **Chabot, B.** 1996. Directing alternative splicing: cast and scenarios. *Trends Genet.* **12:**472–478.
30. **Chain, A. C., S. Zollman, J. C. Tseng, and F. A. Laski.** 1991. Identification of a cis-acting sequence required for germ line-specific splicing of the P element ORF2-ORF3 intron. *Mol. Cell. Biol.* **11:**1538–1546.
31. **Chen, B., T. Chu, E. Harms, J. P. Gergen, and S. Strickland.** 1998. Mapping of Drosophila mutations using site-specific male recombination. *Genetics* **149:**157–163.
32. **Chou, T. B., E. Noll, and N. Perrimon.** 1993. Autosomal P[ovoD1] dominant female-sterile insertions in Drosophila and their use in generating germ-line chimeras. *Development* **119:**1359–1369.
33. **Chou, T. B., and N. Perrimon.** 1992. Use of a yeast site-specific recombinase to produce female germline chimeras in Drosophila. *Genetics* **131:**643–653.
34. **Chou, T. B., and N. Perrimon.** 1996. The autosomal FLP-DFS technique for generating germline mosaics in Drosophila melanogaster. *Genetics* **144:**1673–1679.

35. **Chow, S. A., K. A. Vincent, V. Ellison, and P. O. Brown.** 1992. Reversal of integration and DNA splicing mediated by integrase of human immunodeficiency virus. *Science* **255:** 723–726.
36. **Chu, G.** 1997. Double strand break repair. *J. Biol. Chem.* **272:**24097–24100.
37. **Coen, D.** 1990. P element regulatory products enhance zeste repression of a P[white duplicated] transgene in Drosophila melanogaster. *Genetics* **126:**949–960.
38. **Cooley, L., R. Kelley, and A. Spradling.** 1988. Insertional mutagenesis of the Drosophila genome with single P elements. *Science* **239:**1121–1128.
39. **Cooley, L., D. Thompson, and A. C. Spradling.** 1990. Constructing deletions with defined endpoints in Drosophila. *Proc. Natl. Acad. Sci. USA* **87:**3170–3173.
40. **Craig, N. L.** 1995. Unity in transposition reactions. *Science* **270:**253–254.
41. **Critchlow, S. E., and S. P. Jackson.** 1998. DNA end-joining: from yeast to man. *Trends Biochem. Sci.* **23:**394–398.
42. **Curcio, M. J., and D. J. Garfinkel.** 1999. New lines of host defense: inhibition of Ty1 retrotransposition by Fus3p and NER/TFIIH. *Trends Genet.* **15:**43–45.
43. **Dalby, B., A. J. Pereira, and L. S. Goldstein.** 1995. An inverse PCR screen for the detection of P element insertions in cloned genomic intervals in Drosophila melanogaster. *Genetics* **139:** 757–766.
44. **Dang, D. T., and N. Perrimon.** 1992. Use of a yeast site-specific recombinase to generate embryonic mosaics in Drosophila. *Dev. Genet* **13:**367–375.
45. **Desai, A., and T. J. Mitchison.** 1997. Microtubule polymerization dynamics. *Annu. Rev. Cell. Dev. Biol.* **13:**83–117.
46. **Dray, T., and G. B. Gloor.** 1997. Homology requirements for targeting heterologous sequences during P-induced gap repair in Drosophila melanogaster. *Genetics* **147:**689–699.
47. **Dynan, W. S., and S. Yoo.** 1998. Interaction of Ku protein and DNA-dependent protein kinase catalytic subunit with nucleic acids. *Nucleic Acids Res.* **26:**1551–1559.
48. **Engels, W. R.** 1979. Hybrid dysgenesis in Drosophila melanogaster: rules of inheritance of female sterility. *Genet. Res.* **33:** 219–236.
49. **Engels, W. R.** 1979. Extrachromosomal control of mutability in Drosophila melanogaster. *Proc. Natl. Acad. Sci. USA* **76:** 4011–4015.
50. **Engels, W. R.** 1983. The P family of transposable elements in Drosophila. *Annu. Rev. Genet.* **17:**315–344.
51. **Engels, W. R.** 1984. A trans-acting product needed for P factor transposition in Drosophila. *Science* **226:**1194–1196.
52. **Engels, W. R.** 1989. P elements in *Drosophila melanogaster*, p. 437–484. *In* D. E. Berg and M. M. Howe (ed.), *Mobile DNA*. American Society for Microbiology, Washington, D.C.
53. **Engels, W. R.** 1992. The origin of P elements in Drosophila melanogaster. *Bioessays* **14:**681–686.
54. **Engels, W. R.** 1996. P elements in Drosophila. *Curr. Top. Microbiol. Immunol.* **204:**103–123.
55. **Engels, W. R.** 1997. Invasions of P elements. *Genetics* **145:** 11–15.
56. **Engels, W. R., W. K. Benz, C. R. Preston, P. L. Graham, R. W. Phillis, and H. M. Robertson.** 1987. Somatic effects of P element activity in Drosophila melanogaster: pupal lethality. *Genetics* **117:**745–757.
57. **Engels, W. R., D. M. Johnson-Schlitz, W. B. Eggleston, and J. Sved.** 1990. High-frequency P element loss in Drosophila is homolog-dependent. *Cell* **62:**515–525.
58. **Engels, W. R., C. R. Preston, and D. M. Johnson-Schlitz.** 1994. Long-range cis preference in DNA homology search over the length of a Drosophila chromosome. *Science* **263:** 1623–1625.
59. **Esposito, D., and R. Craigie.** 1999. HIV integrase structure and function. *Adv. Virus Res.* **52:**319–333.
60. **Fire, A.** 1999. RNA-triggered gene silencing. *Trends Genet.* **15:**358–363.
61. **Formosa, T., and B. M. Alberts.** 1986. DNA synthesis dependent on genetic recombination: characterization of a reaction catalyzed by purified bacteriophage T4 proteins. *Cell* **47:** 793–806.
62. **Fugmann, S. D., I. J. Villey, L. M. Ptaszek, and D. G. Schatz.** 2000. Identification of two catalytic residues in RAG1 that define a single active site within the RAG1/RAG2 protein complex. *Mol. Cell* **5:**97–107.
63. **Gell, D., and S. P. Jackson.** 1999. Mapping of protein-protein interactions within the DNA-dependent protein kinase complex. *Nucleic Acids Res.* **27:**3494–3502.
64. **Gloor, G. B., and D. H. Lankenau.** 1998. Gene conversion in mitotically dividing cells: a view from Drosophila. *Trends Genet.* **14:**43–46.
65. **Gloor, G. B., J. Moretti, J. Mouyal, and K. J. Keeler.** 2000. Distinct P element excision products in somatic and germline cells of Drosophila. *Genetics* **155:**1821–1830.
66. **Gloor, G. B., N. A. Nassif, D. M. Johnson-Schlitz, C. R. Preston, and W. R. Engels.** 1991. Targeted gene replacement in Drosophila via P element-induced gap repair. *Science* **253:** 1110–1117.
67. **Gloor, G. B., C. R. Preston, D. M. Johnson-Schlitz, N. A. Nassif, R. W. Phillis, W. K. Benz, H. M. Robertson, and W. R. Engels.** 1993. Type I repressors of P element mobility. *Genetics* **135:**81–95.
68. **Golic, K. G.** 1991. Site-specific recombination between homologous chromosomes in Drosophila. *Science* **252:** 958–961.
69. **Golic, K. G., and M. M. Golic.** 1996. Engineering the Drosophila genome: chromosome rearrangements by design. *Genetics* **144:**1693–1711.
70. **Golic, K. G., and S. Lindquist.** 1989. The FLP recombinase of yeast catalyzes site-specific recombination in the Drosophila genome. *Cell* **59:**499–509.
71. **Gray, Y. H. M., M. M. Tanaka, and J. A. Sved.** 1996. P element-induced recombination in Drosophila melanogaster: hybrid element insertion. *Genetics* **144:**1601–1610.
72. **Grindley, N. D., and A. E. Leschziner.** 1995. DNA transposition: from a black box to a color monitor. *Cell* **83:** 1063–1066.
73. **Grossniklaus, U., H. J. Bellen, C. Wilson, and W. J. Gehring.** 1989. P-element-mediated enhancer detection applied to the study of oogenesis in Drosophila. *Genes Dev.* **107:**189–200.
74. **Hamilton, B. A., M. J. Palazzolo, J. H. Chang, K. VijayRaghavan, C. A. Mayeda, M. A. Whitney, and E. M. Meyerowitz.** 1991. Large scale screen for transposon insertions into cloned genes. *Proc. Natl. Acad. Sci. USA* **88:**2731–2735.
75. **Hamilton, B. A., and K. Zinn.** 1994. From clone to mutant gene. *Methods Cell Biol.* **44:**81–94.
76. **Hammond, L. E., D. Z. Rudner, R. Kanaar, and D. C. Rio.** 1997. Mutations in the hrp48 gene, which encodes a Drosophila heterogeneous nuclear ribonucleoprotein particle protein, cause lethality and developmental defects and affect P-element third-intron splicing in vivo. *Mol. Cell Biol.* **17:** 7260–7267.
77. **Hay, B. A., R. Maile, and G. M. Rubin.** 1997. P element insertion-dependent gene activation in the Drosophila eye. *Proc. Natl. Acad. Sci. USA* **94:**5195–5200.
78. **Hazelrigg, T., R. Levis, and G. M. Rubin.** 1984. Transforma-

tion of white locus DNA in drosophila: dosage compensation, zeste interaction, and position effects. *Cell* **36:**469–481.

79. **Heath, E. M., and M. J. Simmons.** 1991. Genetic and molecular analysis of repression in the P-M system of hybrid dysgenesis in Drosophila melanogaster. *Genet. Res.* **57:**213–226.
80. **Hesse, J. E., M. R. Lieber, M. Gellert, and K. Mizuuchi.** 1987. Extrachromosomal DNA substrates in pre-B cells undergo inversion or deletion at immunoglobulin V-(D)-J joining signals. *Cell* **49:**775–783.
81. **Hickman, A. B., Y. Li, S. V. Mathew, E. W. May, N. L. Craig, and F. Dyda.** 2000. Unexpected structural diversity in DNA recombination: the restriction endonuclease connection. *Mol. Cell* **5:**10025–11034.
82. **Hiraizumi, Y.** 1971. Spontaneous recombination in Drosophila melanogaster males. *Proc. Natl. Acad. Sci. USA* **68:** 268–270.
83. **Hirose, Y., and J. L. Manley.** 2000. RNA polymerase II and the integration of nuclear events. *Genes Dev.* **14:**1415–1429.
84. **Horowitz, D. S., and A. R. Krainer.** 1994. Mechanisms for selecting 5′ splice sites in mammalian pre-mRNA splicing. *Trends Genet.* **10:**100–106.
85. **Hoskins, R. A., C. R. Nelson, B. P. Berman, T. R. Laverty, R. A. George, L. Ciesiolka, M. Naeemuddin, A. D. Arenson, J. Durbin, R. G. David, P. E. Tabor, M. R. Bailey, D. R. DeShazo, J. Catanese, A. Mammoser, K. Osoegawa, P. J. de Jong, S. E. Celniker, R. A. Gibbs, G. M. Rubin, and S. E. Scherer.** 2000. A BAC-based physical map of the major autosomes of Drosophila melanogaster. *Science* **287:**2271–2274.
86. **Houck, M. A., J. B. Clark, K. R. Peterson, and M. G. Kidwell.** 1991. Possible horizontal transfer of Drosophila genes by the mite Proctolaelaps regalis. *Science* **253:**1125–1128.
87. **Jackson, M. S., D. M. Black, and G. A. Dover.** 1988. Amplification of KP elements associated with the repression of hybrid dysgenesis in Drosophila melanogaster. *Genetics* **120:** 1003–1013.
88. **Jacoby, D. B., and P. C. Wensink.** 1994. Yolk protein factor 1 is a Drosophila homolog of Ku, the DNA-binding subunit of a DNA-dependent protein kinase from humans. *J. Biol. Chem.* **269:**11484–11491.
89. **Jasin, M.** 1996. Genetic manipulation of genomes with rare-cutting endonucleases. *Trends Genet.* **12:**224–228.
90. **Johnson-Schlitz, D. M., and W. R. Engels.** 1993. P-element-induced interallelic gene conversion of insertions and deletions in Drosophila melanogaster. *Mol. Cell. Biol.* **13:** 7006–7018.
91. **Joyce, C. M., and T. A. Steitz.** 1995. Polymerase structures and function: variations on a theme? *J. Bacteriol.* **177:** 6321–6329.
92. **Kaiser, K., and S. F. Goodwin.** 1990. "Site-selected" transposon mutagenesis of Drosophila. *Proc. Natl. Acad. Sci. USA* **87:**1686–1690.
93. **Kanaar, R., J. H. Hoeijmakers, and D. C. van Gent.** 1998. Molecular mechanisms of DNA double strand break repair. *Trends Cell. Biol.* **8:**483–489.
94. **Karess, R. E., and G. M. Rubin.** 1984. Analysis of P transposable element function in Drosophila. *Cell* **38:**135–146.
95. **Karpen, G. H., and A. C. Spradling.** 1992. Analysis of subtelomeric heterochromatin in the Drosophila minichromosome Dp1187 by single P element insertional mutagenesis. *Genetics* **132:**737–753.
96. **Kassis, J. A., E. Noll, E. P. VanSickle, W. F. Odenwald, and N. Perrimon.** 1992. Altering the insertional specificity of a Drosophila transposable element. *Proc. Natl. Acad. Sci. USA* **89:**1919–1923.
97. **Kaufman, P. D., R. F. Doll, and D. C. Rio.** 1989. Drosophila P element transposase requires internal P element DNA sequences. *Cell* **38:**135–146.
98. **Kaufman, P. D., and D. C. Rio.** 1991. Drosophila P-element transposase is a transcriptional repressor in vitro. *Proc. Natl. Acad. Sci. USA* **88:**2613–2617.
99. **Kaufman, P. D., and D. C. Rio.** 1992. P element transposition in vitro proceeds by a cut-and-paste mechanism and uses GTP as a cofactor. *Cell* **69:**27–39.
100. **Keeler, K. J., T. Dray, J. E. Penney, and G. B. Gloor.** 1996. Gene targeting of a plasmid-borne sequence to a double-strand DNA break in Drosophila melanogaster. *Mol. Cell. Biol.* **16:**522–528.
101. **Keeler, K. J., and G. B. Gloor.** 1997. Efficient gap repair in Drosophila melanogaster requires a maximum of 31 nucleotides of homologous sequence at the searching ends. *Mol. Cell. Biol.* **17:**627–634.
102. **Ketting, R. F., T. H. Haverkamp, H. G. van Luenen, and R. H. Plasterk.** 1999. Mut-7 of C. elegans, required for transposon silencing and RNA interference, is a homolog of Werner syndrome helicase and RNaseD. *Cell* **99:**133–141.
103. **Kidwell, M. G.** 1981. Hybrid dysgenesis in Drosphila melanogaster: the genetics of cytotype determination in a neutral strain. *Genetics* **98:**275–290.
104. **Kidwell, M. G.** 1983. Evolution of hybrid dysgenesis determinants in Drosophila melanogaster. *Proc. Natl. Acad. Sci. USA* **80:**1655–1659.
105. **Kidwell, M. G.** 1985. Hybrid dysgenesis in Drosophila melanogaster: nature and inheritance of P element regulation. *Genetics* **111:**337–350.
106. **Kidwell, M. G.** 1992. Horizontal transfer of P elements and other short inverted repeat transposons. *Genetica* **86:** 275–286.
107. **Kidwell, M. G.** 1992. Horizontal transfer. *Curr. Opin. Genet. Dev.* **2:**868–873.
108. **Kidwell, M. G., and J. F. Kidwell.** 1975. Cytoplasm-chromosome interactions in Drosophila melanogaster. *Nature* **253:** 755–756.
109. **Kidwell, M. G., J. F. Kidwell, and M. Nei.** 1973. A case of high rate of spontaneous mutation affecting viability in Drosophila melanogaster. *Genetics* **75:**133–153.
110. **Kidwell, M. G., J. F. Kidwell, and J. A. Sved.** 1977. Hybrid dysgenesis in Drosophila melanogaster: a syndrome of aberrant traits including mutation, sterility and male recombination. *Genetics* **86:**813–833.
111. **Kidwell, M. G., and J. B. Novy.** 1979. Hybrid dysgenesis in Drosophila melanogaster: sterility resulting from gonadal dysgenesis in the P-M system. *Genetics* **92:**1127–1140.
112. **Kim, D. R., Y. Dai, C. L. Mundy, W. Yang, and M. A. Oettinger.** 1999. Mutations of acidic residues in RAG1 define the active site of the V(D)J recombinase. *Genes Dev.* **13:** 3070–3080.
113. **Kim, S. T., D. S. Lim, C. E. Canman, and M. B. Kastan.** 1999. Substrate specifications and identification of putative substrates of ATM kinase family members. *J. Biol. Chem.* **274:**37538–37543.
114. **Kooistra, R., A. Pastink, J. B. Zonneveld, P. H. Lohman, and J. C. Eeken.** 1999. The Drosophila melanogaster DmRAD54 gene plays a crucial role in double- strand break repair after P-element excision and acts synergistically with Ku70 in the repair of X-ray damage. *Mol. Cell. Biol.* **19:**6269–6275.
115. **Krasnow, M. A., S. Cumberledge, G. Manning, L. A. Herzenberg, and G. P. Nolan.** 1991. Whole animal cell sorting of Drosophila embryos. *Science* **251:**81–85.
116. **Kusano, K., D. M. Johnson-Schiltz, and W. R. Engels.** 2001. Sterility of Drosophila with mutations in the Bloom syndrome gene-complementation by Ku70. *Science* **291:**2600–2602.

117. **Labourier, E., M. D. Adams, and D. C. Rio.** 2001. Modulation of P element pre-mRNA splicing by a direct interaction between PSI and U1 snRNP 70K protein. *Mol. Cell* **8:** 363–373.
118. **Landree, M. A., J. A. Wibbenmeyer, and D. B. Roth.** 1999. Mutational analysis of RAG1 and RAG2 identifies three catalytic amino acids in RAG1 critical for both cleavage steps of V(D)J recombination. *Genes Dev.* **13:**3059–3069.
119. **Lankenau, D. H., V. G. Corces, and W. R. Engels.** 1996. Comparison of targeted-gene replacement frequencies in Drosophila melanogaster at the forked and white loci. *Mol. Cell. Biol.* **16:**3535–3544.
120. **Lankenau, D. H., and G. B. Gloor.** 1998. In vivo gap repair in Drosophila: a one-way street with many destinations. *Bioessays* **20:**317–327.
121. **Laski, F. A., D. C. Rio, and G. M. Rubin.** 1986. Tissue specificity of Drosophila P element transposition is regulated at the level of mRNA splicing. *Cell* **44:**7–19.
122. **Laski, F. A., and G. M. Rubin.** 1989. Analysis of the cis-acting requirements for germ-line-specific splicing of the P-element ORF2-ORF3 intron. *Genes Dev.* **3:**720–728.
123. **Lee, C. C., E. L. Beall, and D. C. Rio.** 1998. DNA binding by the KP repressor protein inhibits P element transposase activity in vitro. *EMBO J.* **17:**4166–4174.
124. **Lee, C. C., Y. M. Mul, and D. C. Rio.** 1996. The Drosophila P-element KP repressor protein dimerizes and interacts with multiple sites on P-element DNA. *Mol. Cell. Biol.* **16:** 5616–5622.
125. **Lemaitre, B., and D. Coen.** 1991. P regulatory products repress in vivo the P promoter activity in P-lacZ fusion genes. *Proc. Natl. Acad. Sci. USA* **88:**4419–4423.
126. **Lemaitre, B., S. Ronsseray, and D. Coen.** 1993. Maternal repression of the P element promoter in the germline of Drosophila melanogaster: a model for the P cytotype. *Genetics* **135:**149–160.
127. **Liao, G. C., E. J. Rehm, and G. M. Rubin.** 2000. Insertion site preferences of the P transposable element in Drosophila melanogaster. *Proc. Natl. Acad. Sci. USA* **97:**3347–3351.
128. **Lopez, A. J.** 1998. Alternative splicing of pre-mRNA: developmental consequences and mechanisms of regulation. *Annu. Rev. Genet.* **32:**279–305.
129. **Matunis, E. L., M. J. Matunis, and G. Dreyfuss.** 1992. Characterization of the major hnRNP proteins from Drosophila melanogaster. *J. Cell Biol.* **116:**257–269.
130. **Mayeda, A., and A. R. Krainer.** 1992. Regulation of alternative pre-mRNA splicing by hnRNP A1 and splicing factor SF2. *Cell* **68:**365–375.
131. **McClintock, B.** 1984. The significance of responses of the genome to challenge. *Science* **226:**792–801.
132. **Melek, M., and M. Gellert.** 2000. RAG1/2-mediated resolution of transposition intermediates: two pathways and possible consequences. *Cell* **101:**625–633.
133. **Miklos, G. L., and G. M. Rubin.** 1996. The role of the genome project in determining gene function: insights from model organisms. *Cell* **86:**521–529.
134. **Miller, W. J., N. Paricio, S. Hagemann, M. J. Martinez-Sebastian, W. Pinsker, and R. de Frutos.** 1995. Structure and expression of clustered P element homologues in Drosophila subobscura and Drosophila guanche. *Gene* **156:**167–174.
135. **Min, H., C. W. Turck, J. M. Nikolic, and D. L. Black.** 1997. A new regulatory protein, KSRP, mediates exon inclusion through an intronic splicing enhancer. *Genes Dev.* **11:** 1023–1036.
136. **Misra, S., R. M. Buratowski, T. Ohkawa, and D. C. Rio.** 1993. Cytotype control of Drosophila melanogaster P element transposition: genomic position determines maternal repression. *Genetics* **135:**785–800.
137. **Misra, S., and D. C. Rio.** 1990. Cytotype control of Drosophila P element transposition: the 66 kd protein is a repressor of transposase activity. *Cell* **62:**269–284.
138. **Moore, M. J., C. C. Query, and P. A. Sharp.** 1993. Splicing of precursors to messenger RNAs by the spliceosome. *In RNA World.* Cold Spring Harbor Laboratory Press, Cold Spring Harbor, N.Y.
139. **Mul, Y. M., and D. C. Rio.** 1997. Reprogramming the purine nucleotide cofactor requirement of Drosophila P element transposase in vivo. *EMBO J.* **16:**4441–4447.
140. **Mullins, M. C., D. C. Rio, and G. M. Rubin.** 1989. Cis-acting DNA sequence requirements for P-element transposition. *Genes Dev.* **3:**729–738.
141. **Myers, E. W., G. G. Sutton, A. L. Delcher, I. M. Dew, D. P. Fasulo, M. J. Flanigan, S. A. Kravitz, C. M. Mobarry, K. H. Reinert, K. A. Remington, E. L. Anson, R. A. Bolanos, H. H. Chou, C. M. Jordan, A. L. Halpern, S. Lonardi, E. M. Beasley, R. C. Brandon, L. Chen, P. J. Dunn, Z. Lai, Y. Liang, D. R. Nusskern, M. Zhan, Q. Zhang, X. Zheng, G. M. Rubin, M. D. Adams, and J. C. Venter.** 2000. A whole-genome assembly of Drosophila. *Science* **287:**2196–2204.
142. **Nassif, N., J. Penney, S. Pal, W. R. Engels, and G. B. Gloor.** 1994. Efficient copying of nonhomologous sequences from ectopic sites via P-element-induced gap repair. *Mol. Cell. Biol.* **14:**1613–1625.
143. **Nitasaka, E., T. Mukai, and T. Yamazaki.** 1987. Repressor of P elements in Drosophila melanogaster: Cytotype determination by a defective P element carrying only open reading frames 0 through 2. *Proc. Natl. Acad. Sci. USA* **84:** 7605–7608.
144. **O'Brochta, D. A., S. P. Gomez, and A. M. Handler.** 1991. P element excision in Drosophila melanogaster and related drosophilids. *Mol. Gen. Genet.* **225:**387–394.
145. **O'Hare, K., A. Driver, S. McGrath, and D. M. Johnson-Schiltz.** 1992. Distribution and structure of cloned P elements from the Drosophila melanogaster P strain pi 2. *Genet. Res.* **60:**33–41.
146. **O'Hare, K., and G. M. Rubin.** 1983. Structures of P transposable elements and their sites of insertion and excision in the Drosophila melanogaster genome. *Cell* **34:**25–35.
147. **O'Kane, C. J., and W. J. Gehring.** 1987. Detection in situ of genomic regulatory elements in Drosophila. *Proc. Natl. Acad. Sci. USA* **84:**9123–9127.
148. **Paques, F., and J. E. Haber.** 1999. Multiple pathways of recombination induced by double-strand breaks in *Saccharomyces cerevisiae. Microbiol. Mol. Biol. Rev.* **63:**349–404.
149. **Paterson, J., and K. O'Hare.** 1991. Structure and transcription of the singed locus of Drosophila melanogaster. *Genetics* **129:**1073–1084.
150. **Polard, P., and M. Chandler.** 1995. Bacterial transposases and retroviral integrases. *Mol. Microbiol.* **15:**13–23.
151. **Powers, T., and P. Walter.** 1995. Reciprocal stimulation of GTP hydrolysis by two directly interacting GTPases. *Science* **269:**1422–1424.
152. **Preston, C. R., and W. R. Engels.** 1996. P-element-induced male recombination and gene conversion in Drosophila. *Genetics* **144:**1611–1622.
153. **Preston, C. R., J. A. Sved, and W. R. Engels.** 1996. Flanking duplications and deletions associated with P-induced male recombination in Drosophila. *Genetics* **144:**1623–1638.
154. **Ramsden, D. A., D. C. van Gent, and M. Gellert.** 1997. Specificity in V(D)J recombination: new lessons from biochemistry and genetics. *Curr. Opin. Immunol.* **9:**114–120.
155. **Rasmusson, K. E., J. D. Raymond, and M. J. Simmons.** 1993.

Repression of hybrid dysgenesis in Drosophila melanogaster by individual naturally occurring P elements. *Genetics* **133:** 605–622.

156. **Rasmusson, K. E., M. J. Simmons, J. D. Raymond, and C. F. McLarnon.** 1990. Quantitative effects of P elements on hybrid dysgenesis in Drosophila melanogaster. *Genetics* **124:** 647–662.
157. **Reed, R.** 1996. Initial splice-site recognition and pairing during pre-mRNA splicing. *Curr. Opin. Genet. Dev.* **6:**215–220.
158. **Rice, P., R. Craigie, and D. R. Davies.** 1996. Retroviral integrases and their cousins. *Curr. Opin. Struct. Biol.* **6:**76–83.
159. **Rio, D. C.** 1990. Molecular mechanisms regulating Drosophila P element transposition. *Annu. Rev. Genet.* **24:** 543–578.
160. **Rio, D. C.** 1991. Regulation of Drosophila P element transposition. *Trends Genet.* **7:**282–287.
161. **Rio, D. C., F. A. Laski, and G. M. Rubin.** 1986. Identification and immunochemical analysis of biologically active Drosophila P element transposase. *Cell* **44:**21–32.
162. **Rio, D. C., and G. M. Rubin.** 1988. Identification and purification of a Drosophila protein that binds to the terminal 31-base-pair inverted repeats of the P transposable element. *Proc. Natl. Acad. Sci. USA* **85:**8929–8933.
163. **Robertson, H. M., and W. R. Engels.** 1989. Modified P elements that mimic the P cytotype in Drosophila melanogaster. *Genetics* **123:**815–824.
164. **Robertson, H. M., C. R. Preston, R. W. Phillis, D. M. Johnson-Schlitz, W. K. Benz, and W. R. Engels.** 1988. A stable genomic source of P element transposase in Drosophila melanogaster. *Genetics* **118:**461–470.
165. **Roche, S. E., and D. C. Rio.** 1998. Trans-silencing by P elements inserted in subtelomeric heterochromatin involves the Drosophila Polycomb group gene, Enhancer of zeste. *Genetics* **149:**1839–1855.
166. **Roche, S. E., M. Schiff, and D. C. Rio.** 1995. P-element repressor autoregulation involves germ-line transcriptional repression and reduction of third intron splicing. *Genes Dev.* **9:**1278–1288.
167. **Roiha, H., G. M. Rubin, and K. O'Hare.** 1988. P element insertions and rearrangements at the singed locus of Drosophila melanogaster. *Genetics* **119:**75–83.
168. **Rong, Y. S., and K. G. Golic.** 2000. Gene targeting by homologous recombination in drosophila. *Science* **228:**2013–2018.
169. **Ronsseray, S., M. Lehmann, and D. Anxolabehere.** 1991. The maternally inherited regulation of P elements in Drosophila melanogaster can be elicited by two P copies at cytological site 1A on the X chromosome. *Genetics* **129:**501–512.
170. **Ronsseray, S., M. Lehmann, D. Nouaud, and D. Anxolabehere.** 1996. The regulatory properties of autonomous subtelomeric P elements are sensitive to a suppressor of variegation in *Drosophila melanogaster. Genetics* **143:**1663–1674. (Erratum, **144:**1329.)
171. **Ronsseray, S., B. Lemaitre, and D. Coen.** 1993. Maternal inheritance of P cytotype in Drosophila melanogaster: a "pre-P cytotype" is strictly extra-chromosomally transmitted. *Mol. Gen. Genet.* **241:**115–123.
172. **Ronsseray, S., L. Marin, M. Lehmann, and D. Anxolabehere.** 1998. Repression of hybrid dysgenesis in Drosophila melanogaster by combinations of telomeric P-element reporters and naturally occurring P elements. *Genetics* **149:**1857–1866.
173. **Rørth, P.** 1996. A modular misexpression screen in Drosophila detecting tissue-specific phenotypes. *Proc. Natl. Acad. Sci. USA* **93:**12418–12422.
174. **Rørth, P.** 1998. Gal4 in the Drosophila female germline. *Mech. Dev.* **78:**113–118.
175. **Rørth, P., K. Szabo, A. Bailey, T. Laverty, J. Rehm, G. M. Rubin, K. Weigmann, M. Milan, V. Benes, and W. Ansorge.** 1998. Systematic gain-of-function genetics in Drosophila. *Development* **125:**1049–1057.
176. **Roth, D. B., and M. Gellert.** 2000. New guardians of the genome. *Nature* **404:**823–825.
177. **Rubin, G. M., L. Hong, P. Brokstein, M. Evans-Holm, E. Frise, M. Stapleton, and D. A. Harvey.** 2000. A Drosophila complementary DNA resource. *Science* **287:**2222–2224.
178. **Rubin, G. M., M. G. Kidwell, and P. M. Bingham.** 1982. The molecular basis of P-M hybrid dysgenesis: the nature of induced mutations. *Cell* **29:**987–994.
179. **Rubin, G. M., and E. B. Lewis.** 2000. A brief history of Drosophila's contributions to genome research. *Science* **287:** 2216–2218.
180. **Rubin, G. M., and A. C. Spradling.** 1982. Genetic transformation of Drosophila with transposable element vectors. *Science* **218:**348–353.
181. **Rubin, G. M., M. D. Yandell, J. R. Wortman, G. L. Gabor Miklos, C. R. Nelson, I. K. Hariharan, M. E. Fortini, P. W. Li, R. Apweiler, W. Fleischmann, J. M. Cherry, S. Henikoff, M. P. Skupski, S. Misra, M. Ashburner, E. Birney, M. S. Boguski, T. Brody, P. Brokstein, S. E. Celniker, S. A. Chervitz, D. Coates, A. Cravchik, A. Gabrielian, R. F. Galle, W. M. Gelbart, R. A. George, L. S. Goldstein, F. Gong, P. Guan, N. L. Harris, B. A. Hay, R. A. Hoskins, J. Li, Z. Li, R. O. Hynes, S. J. Jones, P. M. Kuehl, B. Lemaitre, J. T. Littleton, D. K. Morrison, C. Mungall, P. H. O'Farrell, O. K. Pickeral, C. Shue, L. B. Vosshall, J. Zhang, Q. Zhao, X. H. Zheng, F. Zhong, W. Zhong, R. Gibbs, J. C. Venter, M. D. Adams, and S. Lewis.** 2000. Comparative genomics of the eukaryotes. *Science* **287:**2204–2215.
182. **Savilahti, H., P. A. Rice, and K. Mizuuchi.** 1995. The phage Mu transpososome core: DNA requirements for assembly and function. *EMBO J.* **14:**4893–4903.
183. **Schmid, S. L., M. A. McNiven, and P. De Camilli.** 1998. Dynamin and its partners: a progress report. *Curr. Opin. Cell. Biol.* **10:**504–512.
184. **Scottoline, B. P., S. Chow, V. Ellison, and P. O. Brown.** 1997. Disruption of the terminal base pairs of retroviral DNA during integration. *Genes Dev.* **11:**371–382.
185. **Sentry, J. W., and K. Kaiser.** 1994. Application of inverse PCR to site-selected mutagenesis of Drosophila. *Nucleic Acids Res.* **22:**3429–3430.
186. **Sepp, K. J., and V. J. Auld.** 1999. Conversion of lacZ enhancer trap lines to GAL4 lines using targeted transposition in Drosophila melanogaster. *Genetics* **151:**1093–1101.
187. **Sharp, P. A.** 1999. RNAi and double-strand RNA. *Genes Dev.* **13:**139–141.
188. **Siebel, C. W., A. Admon, and D. C. Rio.** 1995. Soma-specific expression and cloning of PSI, a negative regulator of P element pre-mRNA splicing. *Genes Dev.* **9:**269–283.
189. **Siebel, C. W., L. D. Fresco, and D. C. Rio.** 1992. The mechanism of somatic inhibition of Drosophila P element pre-mRNA splicing: multiprotein complexes at an exon pseudo-5′ splice site control U1 snRNP binding. **6:**1386–1401.
190. **Siebel, C. W., R. Kanaar, and D. C. Rio.** 1994. Regulation of tissue-specific P-element pre-mRNA splicing requires the RNA-binding protein PSI. *Genes Dev.* **8:**1713–1725.
191. **Siebel, C. W., and D. C. Rio.** 1990. Regulated splicing of the Drosophila P transposable element third intron in vitro: somatic repression. *Science* **248:**1200–1208.
192. **Simmons, M. J., and L. M. Bucholz.** 1985. Transposase titration in Drosophila melanogaster: a model of cytotype in the P-M system of hybrid dysgenesis. *Proc. Natl. Acad. Sci. USA* **82:**8119–8123.
193. **Simmons, M. J., J. D. Raymond, C. D. Grimes, C. Belinco,**

B. C. Haake, M. Jordan, C. Lund, T. A. Ojala, and D. Papermaster. 1996. Repression of hybrid dysgenesis in Drosophila melanogaster by heat- shock-inducible sense and antisense P-element constructs. *Genetics* **144:**1529–1544.

194. **Simmons, M. J., J. D. Raymond, K. E. Rasmusson, L. M. Miller, C. F. McLarnon, and J. R. Zunt.** 1990. Repression of P element-mediated hybrid dysgenesis in Drosophila melanogaster. *Genetics* **124:**663–676.
195. **Smith, C. W., and J. Valcarcel.** 2000. Alternative pre-mRNA splicing: the logic of combinational control. *Trends Biochem. Sci.* **25:**381–388.
196. **Smith, G. C., R. B. Cary, N. D. Lakin, B. C. Hann, S. H. Teo, D. J. Chen, and S. P. Jackson.** 1999. Purification and DNA binding properties of the ataxia-telangiectasia gene product ATM. *Proc. Natl. Acad. Sci. USA* **96:**11134–11139.
197. **Smith, G. C., and S. P. Jackson.** 1999. The DNA-dependent protein kinase. *Genes Dev.* **13:**916–934.
198. **Spradling, A. C.** 1986. P element-mediated transformation, p. 175–197. *In Drosophila: A Practical Approach.* RL Press, Oxford, England.
199. **Spradling, A. C., and G. M. Rubin.** 1982. Transposition of cloned P elements into Drosophila germ line chromosomes. *Science* **218:**341–347.
200. **Spradling, A. C., D. Stern, E. J. Rhem, T. Laverty, N. Mozden, S. Misra, and G. M. Rubin.** 1999. The Berkeley *Drosophila* Genome Project gene disruption project: Single P-element insertions mutating 25% of vital Drosophila genes. *Genetics* **153:**135–177.
201. **Spradling, A. C., D. M. Stern, I. Kiss, J. Roote, T. Laverty, and G. M. Rubin.** 1995. Gene disruptions using P transposable elements: an integral component of the Drosophila genome project. *Proc. Natl. Acad. Sci. USA* **92:**10824–10830.
202. **Staveley, B. E., T. R. Heslip, R. B. Hodgetts, and J. B. Bell.** 1995. Protected P-element termini suggest a role for inverted-repeat-binding protein in transposase-induced gap repair in *Drosophila melanogaster. Genetics* **139:**1321–1329.
203. **Stellwagen, A. E., and N. L. Craig.** 1997. Gain-of-function mutations in TnsC, an ATP-dependent transposition protein that activates the bacterial transposon Tn7. *Genetics* **145:** 573–585.
204. **Stellwagen, A. E., and N. L. Craig.** 1998. Mobile DNA elements: controlling transposition with ATP-dependent molecular switches. *Trends Biochem. Sci.* **23:**486–490.
205. **Struhl, G., and K. Basler.** 1993. Organizing activity of wingless protein in Drosophila. *Cell* **72:**527–540.
206. **Sved, J. A., L. M. Blackman, A. S. Gilchrist, and W. R. Engels.** 1991. High levels of recombination induced by homologous P elements in Drosophila melanogaster. *Mol. Gen. Genet.* **225:**443–447.
207. **Sved, J. A., L. M. Blackman, Y. Svoboda, and R. Colless.** 1995. Male recombination with single and homologous P elements in Drosophila melanogaster. *Mol. Gen. Genet.* **246:** 381–386.
208. **Sved, J. A., W. B. Eggleston, and W. R. Engels.** 1990. Germline and somatic recombination induced by in vitro modified P elements in Drosophila melanogaster. *Genetics* **124:** 331–337.
209. **Svoboda, Y. H., M. K. Robson, and J. A. Sved.** 1995. P-element-induced male recombination can be produced in Drosophila melanogaster by combining end-deficient elements in trans. *Genetics* **139:**1601–1610.
210. **Tanaka, M. M., X. M. Liang, Y. H. Gray, and J. A. Sved.** 1997. The accumulation of P-element-induced recombinants in the germline of male Drosophila melanogaster. *Genetics* **147:**1769–1782.
211. **Theodosiou, N. A., and T. Xu.** 1998. Use of FLP/FRT system to study Drosophila development. *Methods* **14:**355–365.
212. **Thompson-Stewart, D., G. H. Karpen, and A. C. Spradling.** 1994. A transposable element can drive the concerted evolution of tandemly repetitious DNA. *Proc. Natl. Acad. Sci. USA* **91:**9042–9046.
213. **Tower, J., G. H. Karpen, N. Craig, and A. C. Spradling.** 1993. Preferential transposition of Drosophila P elements to nearby chromosomal sites. *Genetics* **133:**347–359.
214. **Tseng, J. C., S. Zollman, A. C. Chain, and F. A. Laski.** 1991. Splicing of the Drosophila P element ORF2-ORF3 intron is inhibited in a human cell extract. *Mech. Dev.* **35:**65–72.
215. **Vos, J. C., I. De Baere, and R. H. Plasterk.** 1996. Transposase is the only nematode protein required for in vitro transposition of Tc1. *Genes Dev.* **10:**755–761.
216. **Wang, J. C.** 1996. DNA topoisomerases. *Annu. Rev. Biochem.* **65:**635–692.
217. **Weaver, D. T.** 1995. What to do at an end: DNA double-strand-break repair. *Trends Genet.* **11:**388–392.
218. **Wilson, C., R. K. Pearson, H. J. Bellen, C. J. O'Kane, U. Grossniklaus, and W. J. Gehring.** 1989. P-element-mediated enhancer detection: an efficient method for isolating and characterizing developmentally regulated genes in Drosophila. *Genes Dev.* **3:**1301–1313.
219. **Wittinghofer, A., and E. F. Pai.** 1991. The structure of Ras protein: a model for a universal molecular switch. *Trends Biochem. Sci.* **16:**382–387.
220. **Xu, T., and S. D. Harrison.** 1994. Mosaic analysis using FLP recombinase. *Methods Cell Biol.* **44:**655–681.
221. **Xu, T., and G. M. Rubin.** 1993. Analysis of genetic mosaics in developing and adult Drosophila tissues. *Development* **117:** 1223–1237.
222. **Yamauchi, M., and T. A. Baker.** 1998. An ATP-ADP switch in MuB controls progression of the Mu transposition pathway. *EMBO J.* **17:**5509–5518.
223. **Zhang, P., and A. C. Spradling.** 1993. Efficient and dispersed local P element transposition from Drosophila females. *Genetics* **133:**361–373.

Mobile DNA II
Edited by N. L. Craig et al.

Chapter 22

The Tc1/mariner Family of Transposable Elements

RONALD H. A. PLASTERK AND HENRI G. A. M. VAN LUENEN

INTRODUCTION

The Tc1/mariner family is widespread (42, 92, 98). Members have been found in protozoa (19, 37), fungi (18, 61), nematodes (1, 39, 113), arthropods (4, 6, 8, 29, 36, 65, 97, 101, 105, 121), and chordates (42, 57) such as fish (32, 33, 41, 46, 56, 95) and humans (2, 83, 86, 99, 103, 104, 117). Actual transposition of these elements has in most cases not been detected. Although the human genome contains Tc1/mariner transposons, these all seem to contain multiple mutations (2, 83, 86, 99, 103, 104, 117) and are thus best viewed as fossils, relics of an ancient invasion by a once active element. This also seems to be the case with most homologs found in fish genomes (32, 45, 46, 95). Molecular reconstruction from multiple dead copies resulted in an active version, named Sleeping Beauty (44). In other species such as the much-studied nematode *Caenorhabditis elegans,* jumping of Tc1-type transposons is the main cause of spontaneous mutations (21, 24, 81).

In this chapter we will describe the family of Tc1/mariner elements, their mechanism of transposition, and the regulation of transposition. Finally, we will mention a few applications of this family of transposable element in forward and reverse genetics.

OVERVIEW OF THE FAMILY

Tc1/mariner elements are about 1.3 to 2.4 kb in length and contain a single gene encoding a similar transposase enzyme (approximately 15% amino acid identity between the different transposases) (15, 44, 48, 92, 106, 123). The transposase proteins contain the typical DDE or DDD motif (19, 44, 68, 97, 124, 128) found in most transposases and integrases (27, 54). The elements have terminal inverted repeats and always integrate at a TA sequence; the TA sequence is present on both sides of the integrated element (44, 48, 107, 125). The most terminal nucleotides of the inverted repeats of Tc1-like elements are conserved (5′-CAGT) (15, 92, 106). The structure of the elements most referred to in this chapter is depicted in Fig. 1.

Ronald H. A. Plasterk • Hubrecht Laboratory, Uppsalalaan 8, 3584 CT Utrecht, The Netherlands. **Henri G. A. M. van Luenen** • Division of Molecular Biology, The Netherlands Cancer Institute, Plesmanlaan 121, 1066 CX Amsterdam, The Netherlands.

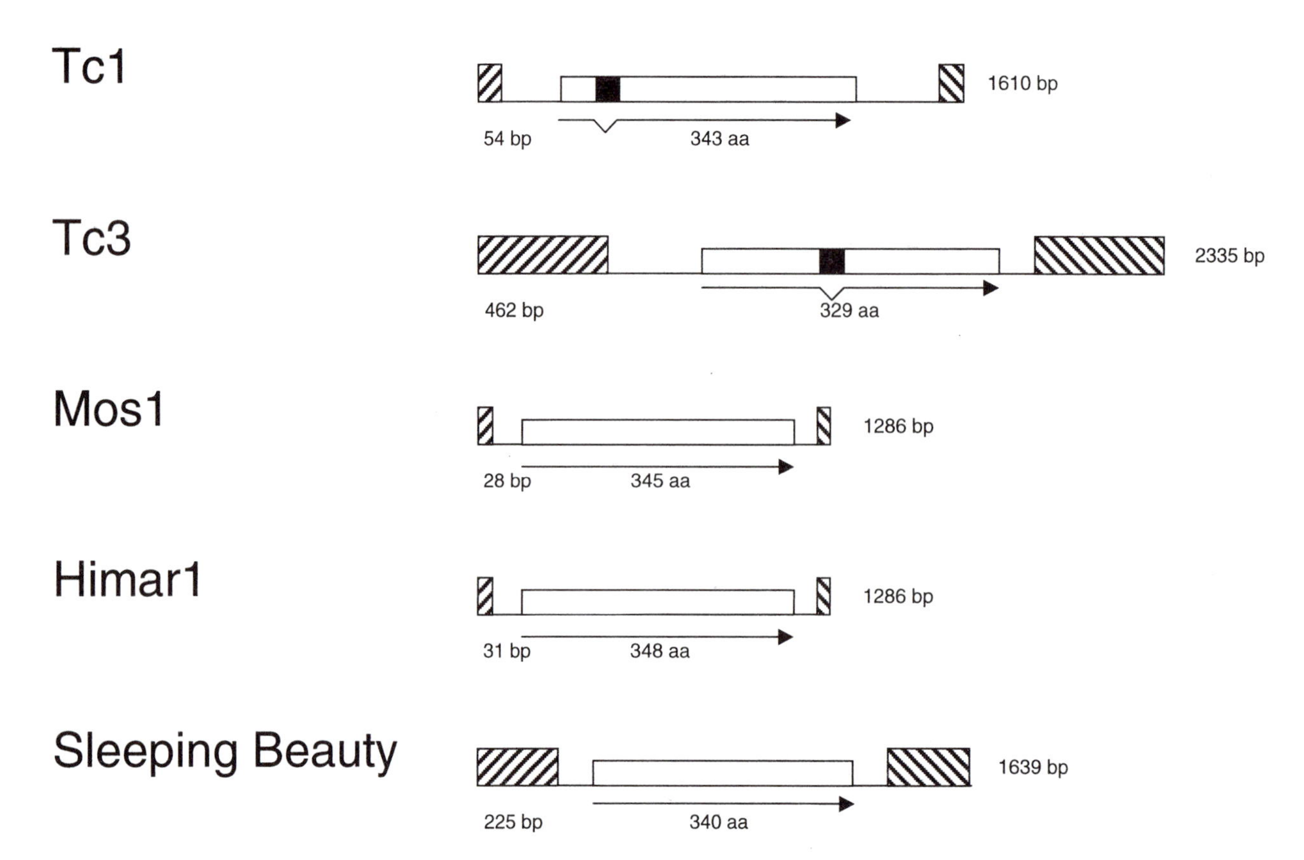

Figure 1. Structure of the most studied members of the Tc1/mariner family. The terminal inverted repeats (hatched boxes) and the open reading frames (open boxes) are indicated for Tc1, Tc3, Mos1, Himar1, and Sleeping Beauty. Black boxes indicate introns in the transposase gene.

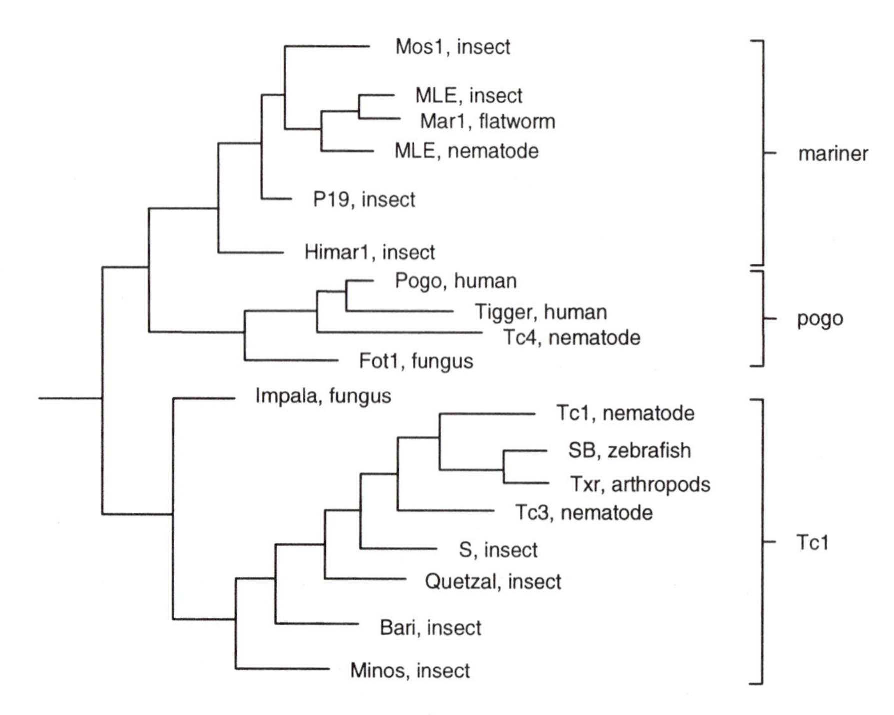

Figure 2. Phylogeny of the Tc1/mariner family (adapted from reference 92).

the ~40-bp RS1 direct repeats. At the 3′ ends of all groups of *pilS* sequences and just downstream of *pilE* are the 65-bp *SmaI/ClaI* direct repeats.

The lower part of Fig. 9 shows a few *pilS* sequences schematically and just below them a *pilE* gene at the start of an infection. The variation between the different possible *pilE* genes occurs in the terminal one-half to two-thirds of the gene. The diversity is provided by partial gene conversions by *pilS* sequences (60, 132). The first *pilE* gene shows a hatched variable region that has its duplicate among the *pilS* loci. The next event is a gene conversion with the gray *pilS* sequence. The cell remains piliated but now the pili have gray antigenic and adhesive characteristics. This recombination is unidirectional; there is not an exchange of the hatched *pilS* sequence for the gray *pilS* at the archival site. The next depicted event is a conversion by the dotted *pilS* sequence, but only part of the variable region is replaced. In this case the result is a mosaic that is part gray and part dotted. By serial partial gene conversions of different extents and locations a near infinite variety of pilin molecules is possible.

The third event, recombination, places the cross-hatched variable region cassette in *pilE,* but the reading frame is shifted such that a premature temination signal occurs. A truncated pilin, which is degraded or cannot be assembled into a pilus, is produced. The phenotype is Pil$^-$; a Pil$^+$ → Pil$^-$ phase variation has occurred. When another partial gene conversion with the black *pilS* sequence occurs, the reading frame extends to the usual stop codon. A Pil$^-$ → Pil$^+$ transition has occurred, and the cell has a pilin that is antigenically distinct from the last pilin it produced.

Figure 10 provides a more detailed view of representative products of the recombinations. An alignment of partial pilin sequences from two sets of data are shown. The sequences represent position 54 to position 160 of the prepilin of strain MSII (20, 92); processed pilin begins at position 8. Sequences from positions 1 to 54 and from positions 161 to 165 are highly conserved within and between strains. Sequences 307 to 320 are pilins of *N. gonorrhoeae* obtained from infected male volunteers who were inoculated with a clonal population of *N. gonorrhoeae* (117). Sequences 938 to 943 are from variants obtained in vitro from a clonal population (134). There is greater variation from position 54 to 99 among the in vivo isolates, but from position 100 onward there is comparable variation. The hypervariable region with insertions or deletions occurs between the two cysteines and in the beta-sheet part of the structure. This hypervariable region is exposed to solvent in the assembled pilus (96).

These events are RecA-dependent (84). In a *recA* knockout strain the frequency of these types of recombinations declined 1,000-fold. Another RecA-dependent recombination event, which is not shown in Fig. 9, is a deletion between an upstream *pilS* sequence and *pilE,* leading to loss of the 5′-terminal sequence and Pil$^-$ that cannot revert to Pil$^+$ (133). These cells no longer undergo phase variation of pili but are capable of phase variation of Opa. (Deletions between direct repeats within *pilE* can also produce nonpiliated cells, but in this case the *pilE* site is repaired by an archived *pilS* sequence [63].) Other, less frequent types of Pil$^-$ mutations at an unconverted *pilE* locus were not RecA dependent. These included single missense mutations that altered the proteolytic cleavage site of the leader peptide and single-base frame shifts

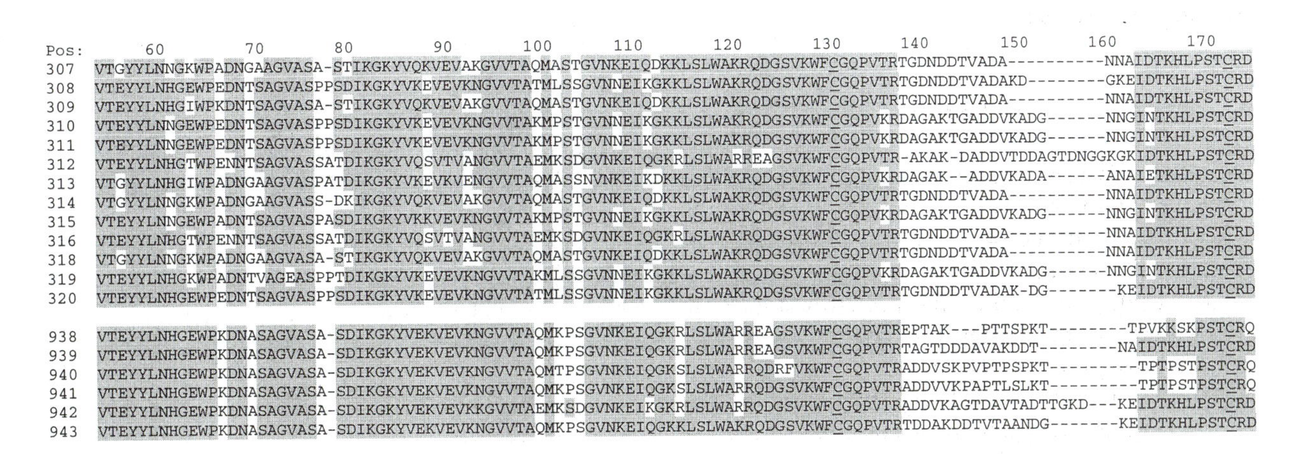

Figure 10. Alignment of variable regions of *N. gonorrhoeae* pilin amino acid sequences. Sequences 307 to 320 are from reference 117, and sequences 938 to 943 are from reference 134. Highlighted positions (Pos) indicate identical residues that were present in at least 10 of the 19 sequences. Conserved cysteines are underlined. The numbering refers to the sequence of the prepilin of strain MSII (20, 92).

in the open reading frame that produced premature terminal signals.

Although the *recA* mutation has had the greatest effect in decreasing pilin antigenic variation, other mutations have also been examined. Deletion of the *SmaI/ClaI* repeat at the pilE expression locus decreased the efficiency of pilin variation (139). Other constraints on pilin variation are less clear and may be strain dependent. For one strain, knockouts of the *recB* or *recC* genes had little effect on pilin variation, a finding which, together with mutations in other recombination genes, suggested that pilin switching proceeded by a RecF-like pathway (88). On the other hand, in another strain a *recD* mutation caused a 10-fold increase in pilin variation, a finding that suggested an alternative pathway (43). The role of DNA modification in antigenic and phase variation was investigated by mutating the gene for methyltransferase, but there was no discernible effect on pilin variation (123).

The natural competency and panmictic population structure of *N. gonorrheoae,* as well as the gonococcus' disposition for autolysis, suggested that the pilin gene conversion could be the result of high-frequency lateral gene transfer from other cells rather than strictly the result of intracellular events (116). Transformation-mediated gene conversion of pilin genes could account for the paradoxical rarity of reciprocal recombinations among these chromosomal genes. However, there was little difference in pilin variation in its frequency or diversity between mutants that are transformation defective and wild-type cells of the same background (53, 134, 144). When lateral, heteroallelic spread of pilin gene sequences does occur between otherwise isogenic isolates, it results in complete replacement of intact *pilE* alleles, not in the more limited conversion usually noted in vivo and in vitro (62).

C. FETUS

C. fetus is a spiral, gram-negative bacterium that is a colonzier and pathogen of domestic and wild animals. *C. fetus* subspecies *fetus,* the most commonly studied subspecies, has a host range that includes ungulates, birds, and reptiles and primarily infects the gastrointestinal tract. After oral ingestion of *C. fetus* in food or water, the organism invades the mucosa of the intestine and a transient bacteremia develops. This can lead to colonization of the biliary tree and chronic carriage with fecal shedding.

These organisms have a dense surface (S) layer over the outer membrane (24). The S layers are predominantly made up of single proteins in about 10^5 copies per cell and that are formed into either tetragonal or hexagonal regular arrays. These proteins are 97 to 147 kDa in size and are attached to the outer membrane through noncovalent interactions with lipopolysaccharides (LPSs). The S-layer proteins are grouped into either A or B serotypes, depending on the type of the LPS. Serotype A proteins bind to LPS of serotype A cells but not to LPS of serotype B cells. The S-layer proteins do not have signal peptides and instead are probably transported to the surface by the general secretory pathway.

The S-layer proteins provide resistance to the action of complement in serum and to ingestion by phagocytes. Mutants or protease-treated cells without the S layer have reduced infectivity and virulence. Although the S-layer proteins are highly immunogenic, infections with this organism are persistent. Earlier studies of infected cattle showed that antibody reactivity and size of the S-layer proteins from sequential isolates of *C. fetus* change over the course of the infection (46, 113). Similar heterogeneities in size and antibody reactivity have been noted among strains of *C. fetus* passaged in vitro (47). The frequency of variation in S-layer proteins is approximately 10^{-3} (25). These are indications of antigenic variation, but the function of the variation of these proteins has yet to be fully defined.

The studies of Blaser and his colleagues of the type A S-layer proteins of *C. fetus* subsp. *fetus* revealed the molecular basis for antigenic variation of these proteins (reviewed in reference 49). *C. fetus* subspecies *venerealis,* which is more restricted in its host range to the genital tract of cattle, appears to have a similar strategy for variation S-layer proteins (59).

The type A S-layer protein is encoded by *sapA,* of which there are eight to nine complete, different alleles that are clustered in a region of less than 93 kb on the chromosome. At the single expression site a σ^{70}-type promoter is 121 nucleotides upstream of the start codon. Each of the different cassettes starts with a highly conserved region beginning 74 bp upstream of the start codon and continues for 552 nucleotides into the open reading frame. The sequences of the alleles then become progressively more divergent in sequence over the rest of the lengths of the open reading frames, which vary from 2,760 to 3,327 bp in total length. Just beyond the 3′ ends of each of the *sapA* alleles are conserved sequences of 30 to 50 nucleotides.

Figure 11 shows a 6-kb region of the *C. fetus* genome with four *sapA* alleles, one of which is adjacent to the promoter and thus determines the antigenic type of S-layer protein. The three possible outcomes of a DNA inversion involving the expression locus and one of the three archived alleles are shown.

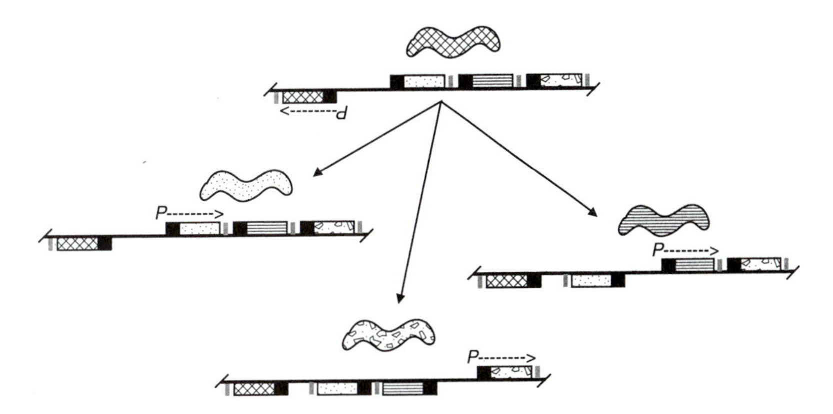

Figure 11. Schematic representation of the outcomes of inversions within the *sapA* locus of *C. fetus*. The phenotype of the bacterium is indicated by a pattern and corresponds to the gene downstream of the promoter (*P*). The arrow indicates the direction and extent of transcription. The black blocks at the 5′ ends of the *sapA* alleles are conserved sequences. The *sapA* alleles are followed by nontranscribed repeats, which are shown as short gray boxes.

These recombinations are RecA dependent and reciprocal (50).

OTHER BACTERIA

Borrelia burgdorferi and related species that cause Lyme disease have only one copy of the *vsp* orthologue, *ospC*, per cell, but they have several copies of sequences, called *vlsE* genes, that are homologous to the Vlp proteins of relapsing fever *Borrelia* spp. (142). During the course of infection of mice, but not detectably in vitro, there is variation in expressed VlsE proteins through partial gene conversions from an adjacent tandem array of cassettes containing different hypervariable regions of *vlsE* sequences. Variants appear in both immunodeficient and immunocompetent mice, but the rate of accumulation of amino acid changes in VlsE was higher in immunocompetent animals (143), an indication that the phenomenon is a form of antigenic variation. Although the role of RecA in these recombinations has not been established, the mechanism appears to be similar to pilin gene variation in *N. gonorrhoeae.*

Anaplasmosis is a persistent intraerythrocytic infection of cattle and goats and has a global distribution. Infected animals have a severe anemia and a higher rate of abortion. The infection is caused by members of the genus *Anaplasma,* an obligate intracelluar rickettsia-like bacterium. The infection is characterized by repetitive cycles of rickettsemia at intervals of 6 to 8 weeks. In cattle infected with *Anaplasma marginale* the number of organisms in the blood varies between a peak of 10^6 to 10^7 per ml to a low of 10^2 per ml. In each cycle the numbers of organisms in the blood rise over 10 to 14 days and then precipitously fall (79). Major surface protein 2 of *A. marginale* is an immunodominant outer membrane protein of about 40 kDa. Each strain has a large, polymorphic family of *msp2* genes; the variation occurs in the central region of the proteins. Each sequential cycle of rickettsemia is associated with a different transcript from at least 17 different *msp2* genes of the family (58). The genetic mechanisms for switches in *msp2* genes are not known but may involve partial gene conversions through homologous recombination.

Several pathogens, including *Streptococcus pyogenes* (74) and *Mycoplasma hyorhinis* (44), have surface proteins with highly repetitive domains (reviewed in reference 51). During the course of an infection or during in vitro cultivation, these microorganisms may vary in the number of repeats within the protein. This variability is associated with decreased binding of antibody and complement to the cells and with impaired access of antibody to other surface structures.

CONCLUSIONS

In their introduction to *Genome Rearrangement,* Simon and Herskowitz distinguished between stereotypic or "programmed" rearrangements and more stochastic or "unprogrammed" rearrangements (119). Clearly the bacterial examples of multiphasic antigenic variation of this chapter fall in the category of programmed rearrangements. But for relapsing fever spirochetes and gonococci at least, the process that varies expression among a repertoire of alleles is not as tidy and efficient as occurs in other programmed rearrangement systems, notably, yeast mating-type switches and the several examples of DNA inversions. In both *B. hermsii* and *N. gonorrhoeae* recombination products include premature termination and missense mutations in key structural regions. The cost to gonococci for having the capability of varying its pilin proteins is the occasional nonpiliated state, which may be of no benefit for initiating an infection. A *B. hermsii* cell that expresses a chimeric Vlp instead of a completely different Vlp remains susceptible to some of the circulating antibodies. In *B. turicatae* 10 kb or more of a plasmid are duplicated to achieve the switch of a single gene.

The observed sloppiness in the programming may be a stop along the way to a more efficient and precise mechanism in a more successful pathogen. But the process "as is" may also serve a less-considered role, that is, generation and maintenance of the antigen archives in the cell. The selection for diversity and retention of these genes in the archives is indirect, for after all, they are silent. The repertoire probably started with a duplication of a single copy gene like *ospC* of *B. burgdorferi*. But duplicated genes are at risk of homogenizing processes, deletions, and inversions. So diversification of the original set of duplicated genes was likely speeded by intragenic recombinations. For the panmictic *N. gonorrhoeae* the donors could be other strains as well as cells of the same strain (61, 116, 120). The more clonal *B. hermsii* may have horizontally acquired some genetic novelty, but much of the present polymorphism is the result of intragenomic recombination (64, 107). Duplication of an extensive region of a plasmid may be an inefficient way to activate a gene, but in another context it yields in a single step another set of allelic loci. These events may not be possible with a recombination mechanism that is more accurate and elegant.

Acknowledgments. I am grateful to Sven Bergström, Stuart Hill, and John Swanson for critical comments that improved the manuscript.

The work on *Borrelia* spp. in my laboratory was supported first by the intramural program of the National Institutes of Health and then by NIH grants AI24424 and AI37248, an Arthritis Foundation grant, and a National Science Foundation grant.

REFERENCES

1. **Apicella, M. A., M. Shero, G. A. Jarvis, J. M. Griffiss, R. E. Mandrell, and H. Schneider.** 1987. Phenotypic variation in epitope expression of the *Neisseria gonorrhoeae* lipooligosaccharide. *Infect. Immun.* **55:**1755–1761.
2. **Barbour, A. G.** 1989. Antigenic variation in relapsing fever *Borrelia* species: genetic aspects, p. 783–789. *In* D. E. Berg and M. M. Howe (ed.), *Mobile DNA*. American Society for Microbiology, Washington, D.C.
3. **Barbour, A. G.** 1987. Immunobiology of relapsing fever. *Contrib. Microbiol. Immunol.* **8:**125–137.
4. **Barbour, A. G.** 1999. Relapsing fever and other *Borrelia* infections, p. 535–546. *In* R. L. Guerrant, D. H. Walker, and P. F. Weller (ed.), *Tropical Infectious Diseases. Principles, Pathogens, & Practice*. Churchill Livingston, Philadelphia, Pa.
5. **Barbour, A. G., O. Barrera, and R. C. Judd.** 1983. Structural analysis of the variable major proteins of *Borrelia hermsii*. *J. Exp. Med.* **158:**2127–2140.
6. **Barbour, A. G., and V. Bundoc.** 2001. In vitro and in vivo neutralization of the relapsing fever agent *Borrelia hermsii* with serotype-specific immunoglobulin M antibodies. *Infect. Immun.* **69:**1009–1015.
7. **Barbour, A. G., N. Burman, C. J. Carter, T. Kitten, and S. Bergstrom.** 1991. Variable antigen genes of the relapsing fever agent *Borrelia hermsii* are activated by promoter addition. *Mol. Microbiol.* **5:**489–493.
8. **Barbour, A. G., C. J. Carter, V. Bundoc, and J. Hinnebusch.** 1996. The nucleotide sequence of a linear plasmid of *Borrelia burgdorferi* reveals similarities to those of circular plasmids of other prokaryotes. *J. Bacteriol.* **178:**6635–6639.
9. **Barbour, A. G., C. J. Carter, N. Burman, C. S. Freitag, C. F. Garon, and S. Bergstrom.** 1991. Tandem insertion sequence-like elements define the expression site for variable antigen genes of *Borrelia hermsii*. *Infect. Immun.* **59:**390–397.
10. **Barbour, A. G., C. J. Carter, and C. D. Sohaskey.** 2000. Surface protein variation by expression site switching in the relapsing fever agent *Borrelia hermsii*. *Infect. Immun.* **68:** 7114–7121.
11. **Barbour, A. G., and C. F. Garon.** 1987. Linear plasmids of the bacterium *Borrelia burgdorferi* have covalently closed ends. *Science* **237:**409–411.
12. **Barbour, A. G., and S. F. Hayes.** 1986. Biology of *Borrelia* species. *Microbiol. Rev.* **50:**381–400.
13. **Barbour, A. G., and B. I. Restrepo.** 2000. Antigenic variation of vector-borne pathogens. *Emerg. Infect. Dis.* **6:**449–457.
14. **Barbour, A. G., and H. G. Stoenner.** 1985. Antigenic variation of *Borrelia hermsii*, p. 123–135. *In* M. I. Simon and I. Herskowitz (ed.), *Genome Rearrangement*. Alan R. Liss, Inc., New York, N.Y.
15. **Barbour, A. G., S. L. Tessier, and H. G. Stoenner.** 1982. Variable major proteins of *Borrelia hermsii*. *J. Exp. Med.* **156:**1312–1324.
16. **Barbour, A. G., S. L. Tessier, and W. J. Todd.** 1983. Lyme disease spirochetes and ixodid tick spirochetes share a common surface antigenic determinant defined by a monoclonal antibody. *Infect. Immun.* **41:**795–804.
17. **Barstad, P. A., J. E. Coligan, M. G. Raum, and A. G. Barbour.** 1985. Variable major proteins of *Borrelia hermsii*. Epitope mapping and partial sequence analysis of CNBr peptides. *J. Exp. Med.* **161:**1302–1314.
18. **Belland, R. J., S. G. Morrison, J. H. Carlson, and D. M. Hogan.** 1997. Promoter strength influences phase variation of neisserial *opa* genes. *Mol. Microbiol.* **23:**123–135.
19. **Bergström, S., V. G. Bundoc, and A. G. Barbour.** 1989. Molecular analysis of linear plasmid-encoded major surface proteins, OspA and OspB, of the Lyme disease spirochaete *Borrelia burgdorferi*. *Mol. Microbiol.* **3:**479–486.
20. **Bergström, S., K. Robbins, J. M. Koomey, and J. Swanson.** 1986. Piliation control mechanisms in *Neisseria gonorrhoeae*. *Proc. Natl. Acad. Sci. USA* **83:**3890–3894.
21. **Bhat, K. S., C. P. Gibbs, O. Barrera, S. G. Morrison, F. Jahnig, A. Stern, E. M. Kupsch, T. F. Meyer, and J. Swanson.** 1991. The opacity proteins of *Neisseria gonorrhoeae* strain MS11 are encoded by a family of 11 complete genes. *Mol. Microbiol.* **5:**1889–1901.
22. **Biswas, G. D., T. Sox, E. Blackman, and P. F. Sparling.** 1977. Factors affecting genetic transformation of *Neisseria gonorrhoeae*. *J. Bacteriol.* **129:**983–992.
23. **Black, W. J., R. S. Schwalbe, I. Nachamkin, and J. G. Cannon.**

1984. Characterization of *Neisseria gonorrhoeae* protein II phase variation by use of monoclonal antibodies. *Infect. Immun.* **45**:453–457.
24. **Blaser, M. J., and Z. Pei.** 1993. Pathogenesis of *Campylobacter fetus* infections. Critical role of the high molecular weight S-layer proteins in virulence. *J. Infect. Dis.* **167**: 696–706.
25. **Blaser, M. J., E. Wang, M. K. Tummuru, R. Washburn, S. Fujimoto, and A. Labigne.** 1994. High-frequency S-layer protein variation in *Campylobacter fetus* revealed by *sapA* mutagenesis. *Mol. Microbiol.* **14**:453–462.
26. **Boslego, J. W., E. C. Tramont, R. C. Chung, D. G. McChesney, J. Ciak, J. C. Sadoff, M. V. Piziak, J. D. Brown, C. C. Brinton, Jr., S. W. Wood, et al.** 1991. Efficacy trial of a parenteral gonococcal pilus vaccine in men. *Vaccine* **9**:154–162.
27. **Brandt, M. E., B. S. Riley, J. D. Radolf, and M. V. Norgard.** 1990. Immunogenic integral membrane proteins of *Borrelia burgdorferi* are lipoproteins. *Infect. Immun.* **58**:983–991.
28. **Brinton, C. C., J. Bryan, J.-A. Dillion, N. Guerina, L. J. Jacobson, A. Labik, S. Lee, A. Levine, S. Lim, J. McMichael, S. Polen, K. Rodgers, A. C.-C. To, and S. C.-M. To.** 1978. Uses of pili in gonorrhea control: role of bacterial pili in disease, purification, and properties of gonococcal pili, and progress in the development of a gonococcal pilus vaccine for gonorrhea, p. 155–178. *In* G. F. Brooks, E. C. Gotschlich, K. K. Holmes, W. D. Sawyer, and F. E. Young (ed.), *Immunobiology of* Neisseria gonorrhoeae. American Society for Microbiology, Washington, D.C.
29. **Buchanan, T. M.** 1975. Antigenic heterogeneity of gonococcal pili. *J. Exp. Med.* **141**:1470–1475.
30. **Buchanan, T. M., and E. C. Gotschlich.** 1973. Studies on gonococcus infection. 3. Correlation of gonococcal colony morphology with infectivity for the chick embryo. *J. Exp. Med.* **137**:196–200.
31. **Bunikis, J., and A. G. Barbour.** 1999. Access of antibody or trypsin to an integral outer membrane protein (P66) of *Borrelia burgdorferi* is hindered by Osp lipoproteins. *Infect. Immun.* **67**:2874–2883.
32. **Burch, C. L., R. J. Danaher, and D. C. Stein.** 1997. Antigenic variation in *Neisseria gonorrhoeae:* production of multiple lipooligosaccharides. *J. Bacteriol.* **179**:982–986.
33. **Burman, N.** 1994. Antigenic variation in relapsing fever *Borrelia.* Thesis. Department of Microbiology, Umeå University, Umeå, Sweden.
34. **Burman, N., S. Bergström, B. I. Restrepo, and A. G. Barbour.** 1990. The variable antigens Vmp7 and Vmp21 of the relapsing fever bacterium *Borrelia hermsii* are structurally analogous to the VSG proteins of the African trypanosome. *Mol. Microbiol.* **4**:1715–1726.
35. **Burman, N., A. Shamaei-Tousi, and S. Bergström.** 1998. The spirochete *Borrelia crocidurae* causes erythrocyte rosetting during relapsing fever. *Infect. Immun.* **66**:815–819.
36. **Cadavid, D., and A. G. Barbour.** 1998. Neuroborreliosis during relapsing fever: review of the clinical manifestations, pathology, and treatment of infections in humans and experimental animals. *Clin. Infect. Dis.* **26**:151–164.
37. **Cadavid, D., V. Bundoc, and A. G. Barbour.** 1993. Experimental infection of the mouse brain by a relapsing fever *Borrelia* species: a molecular analysis. *J. Infect. Dis.* **168**: 143–151.
38. **Cadavid, D., P. M. Pennington, T. A. Kerentseva, S. Bergstrom, and A. G. Barbour.** 1997. Immunologic and genetic analyses of VmpA of a neurotropic strain of *Borrelia turicatae. Infect. Immun.* **65**:3352–3360.
39. **Cadavid, D., D. D. Thomas, R. Crawley, and A. G. Barbour.** 1994. Variability of a bacterial surface protein and disease expression in a possible mouse model of systemic Lyme borreliosis. *J. Exp. Med.* **179**:631–642.
40. **Carson, S. D., B. Stone, M. Beucher, J. Fu, and P. F. Sparling.** 2000. Phase variation of the gonococcal siderophore receptor FetA. *Mol. Microbiol.* **36**:585–593.
41. **Carter, C. J., S. Bergstrom, S. J. Norris, and A. G. Barbour.** 1994. A family of surface-exposed proteins of 20 kilodaltons in the genus *Borrelia. Infect. Immun.* **62**:2792–2799.
42. **Casjens, S., N. Palmer, R. Van Vugt, W. M. Huang, B. Stevenson, P. Rosa, R. Lathigra, G. Sutton, J. Peterson, R. J. Dodson, D. Haft, E. Hickey, M. Gwinn, O. White, and C. M. Fraser.** 2000. A bacterial genome in flux: the twelve linear and nine circular extrachromosomal DNAs in an infectious isolate of the lyme disease spirochete *Borrelia burgdorferi. Mol. Microbiol.* **35**:490–516.
43. **Chaussee, M. S., J. Wilson, and S. A. Hill.** 1999. Characterization of the *recD* gene of Neisseria gonorrhoeae MS11 and the effect of *recD* inactivation on pilin variation and DNA transformation. *Microbiology* **145**:389–400.
44. **Citti, C., M. F. Kim, and K. S. Wise.** 1997. Elongated versions of Vlp surface lipoproteins protect *Mycoplasma hyorhinis* escape variants from growth-inhibiting host antibodies. *Infect. Immun.* **65**:1773–1785.
45. **Coffey, E. M., and W. C. Eveland.** 1967. Experimental relapsing fever initiated by *Borellia hermsi.* I. Identification of major serotypes by immunofluorescence. *J. Infect. Dis.* **117**: 23–28.
46. **Corbeil, L. B., G. D. Schurig, P. J. Bier, and A. J. Winter.** 1975. Bovine venereal vibriosis: antigenic variation of the bacterium during infection. *Infect. Immun.* **11**:240–244.
47. **Dubreuil, J. D., M. Kostrzynska, J. W. Austin, and T. J. Trust.** 1990. Antigenic differences among *Campylobacter fetus* S-layer proteins. *J. Bacteriol.* **172**:5035–5043.
48. **Dunn, J. J., S. R. Buchstein, L. L. Butler, S. Fisenne, D. S. Polin, B. N. Lade, and B. J. Luft.** 1994. Complete nucleotide sequence of a circular plasmid from the Lyme disease spirochete, *Borrelia burgdorferi. J. Bacteriol.* **176**:2706–2717.
49. **Dworkin, J., and M. J. Blaser.** 1997. Molecular mechanisms of *Campylobacter fetus* surface layer protein expression. *Mol. Microbiol.* **26**:433–440.
50. **Dworkin, J., O. L. Shedd, and M. J. Blaser.** 1997. Nested DNA inversion of *Campylobacter fetus* S-layer genes is *recA* dependent. *J. Bacteriol.* **179**:7523–7529.
51. **Dybvig, K.** 1993. DNA rearrangements and phenotypic switching in prokaryotes. *Mol. Microbiol.* **10**:465–471.
52. **Eggers, C. H., and D. S. Samuels.** 1999. Molecular evidence for a new bacteriophage of *Borrelia burgdorferi. J. Bacteriol.* **181**:7308–7313.
53. **Facius, D., and T. F. Meyer.** 1993. A novel determinant (*comA*) essential for natural transformation competence in *Neisseria gonorrhoeae* and the effect of *comA*-defect on pilin variation. *Mol. Microbiol.* **10**:699–712.
54. **Felsenfeld, O.** 1971. *Borrelia. Strains, Vectors, Human and Animal Borreliosis.* Warren H. Greene, Inc., St. Louis, Mo.
55. **Ferdows, M. S., and A. G. Barbour.** 1989. Megabase-sized linear DNA in the bacterium *Borrelia burgdorferi*, the Lyme disease agent. *Proc. Natl. Acad. Sci. USA* **86**:5969–5973.
56. **Ferdows, M. S., P. Serwer, G. A. Griess, S. J. Norris, and A. G. Barbour.** 1996. Conversion of a linear to a circular plasmid in the relapsing fever agent *Borrelia hermsii. J. Bacteriol.* **178**: 793–800.
57. **Fraser, C. M., S. Casjens, W. M. Huang, G. G. Sutton, R. Clayton, R. Lathigra, O. White, K. A. Ketchum, R. Dodson, E. K. Hickey, M. Gwinn, B. Dougherty, J. F. Tomb, R. D. Fleischmann, D. Richardson, J. Peterson, A. R. Kerlavage, J. Quackenbush, S. Salzberg, M. Hanson, R. van Vugt, N. Pal-

mer, M. D. Adams, J. Gocayne, J. C. Venter, et al. 1997. Genomic sequence of a Lyme disease spirochaete, *Borrelia burgdorferi*. *Nature* **390:**580–586.

58. **French, D. M., W. C. Brown, and G. H. Palmer.** 1999. Emergence of *Anaplasma marginale* antigenic variants during persistent rickettsemia. *Infect. Immun.* **67:**5834–5840.
59. **Garcia, M. M., C. L. Lutze-Wallace, A. S. Denes, M. D. Eaglesome, E. Holst, and M. J. Blaser.** 1995. Protein shift and antigenic variation in the S-layer of *Campylobacter fetus* subsp. *venerealis* during bovine infection accompanied by genomic rearrangement of *sapA* homologs. *J. Bacteriol.* **177:** 1976–1980.
60. **Haas, R., and T. F. Meyer.** 1986. The repertoire of silent pilus genes in *Neisseria gonorrhoeae:* evidence for gene conversion. *Cell* **44:**107–115.
61. **Hill, S. A.** 1999. Cell to cell transmission of donor DNA overcomes differential incorporation of non-homologous and homologous markers in *Neisseria gonorrhoeae. Gene* **240:** 175–182.
62. **Hill, S. A.** 1996. Limited variation and maintenance of tight genetic linkage characterize heteroallelic pilE recombination following DNA transformation of *Neisseria gonorrhoeae. Mol. Microbiol.* **20:**507–518.
63. **Hill, S. A., S. G. Morrison, and J. Swanson.** 1990. The role of direct oligonucleotide repeats in gonococcal pilin gene variation. *Mol. Microbiol.* **4:**1341–1352.
64. **Hinnebusch, B. J., A. G. Barbour, B. I. Restrepo, and T. G. Schwan.** 1998. Population structure of the relapsing fever spirochete *Borrelia hermsii* as indicated by polymorphism of two multigene families that encode immunogenic outer surface lipoproteins. *Infect. Immun.* **66:**432–440.
65. **Hinnebusch, J., and A. G. Barbour.** 1991. Linear plasmids of *Borrelia burgdorferi* have a telomeric structure and sequence similar to those of a eukaryotic virus. *J. Bacteriol.* **173:** 7233–7239.
66. **Hinnebusch, J., and A. G. Barbour.** 1992. Linear- and circular-plasmid copy numbers in *Borrelia burgdorferi. J. Bacteriol.* **174:**5251–5257.
67. **Hinnebusch, J., S. Bergström, and A. G. Barbour.** 1990. Cloning and sequence analysis of linear plasmid telomeres of the bacterium *Borrelia burgdorferi. Mol. Microbiol.* **4:** 811–820.
68. **Hinnebusch, J., and K. Tilly.** 1993. Linear plasmids and chromosomes in bacteria. *Mol. Microbiol.* **10:**917–922.
69. **Holt, S. C.** 1978. Anatomy and chemistry of spirochetes. *Microbiol. Rev.* **38:**114–160.
70. **Hughes, C. A., and R. C. Johnson.** 1990. Methylated DNA in *Borrelia* species. *J. Bacteriol.* **172:**6602–6604.
71. **Hyde, F. W., and R. C. Johnson.** 1984. Genetic relationship of Lyme disease spirochetes to *Borrelia, Treponema,* and *Leptospira* spp. *J. Clin. Microbiol.* **20:**151–154.
72. **James, J. F., C. J. Lammel, D. L. Draper, and G. F. Brooks.** 1980. Attachment of *Neisseria gonorrhoeae* colony phenotype variants to eukaryotic cells and tissues, p. 213–216. *In* S. Normark and D. Danielsson (ed.), *Genetics and Immunobiology of Pathogenic Neisseria.* University of Umeå, Umeå, Sweden.
73. **Jerse, A. E., M. S. Cohen, P. M. Drown, L. G. Whicker, S. F. Isbey, H. S. Seifert, and J. G. Cannon.** 1994. Multiple gonococcal opacity proteins are expressed during experimental urethral infection in the male. *J. Exp. Med.* **179:**911–920.
74. **Jones, K. F., S. K. Hollingshead, J. R. Scott, and V. A. Fischetti.** 1988. Spontaneous M6 protein size mutants of group A steptococci display variation in antigenic and opsonogenic epitopes. *Proc. Natl. Acad. Sci. USA* **85:**8271–8275.
75. **Jonsson, A., G. Nyberg, and S. Normark.** 1991. Phase variation of gonococcal pili by frameshift mutation in pilC, a novel gene for pilus assembly. *EMBO J.* **10:**477–488.
76. **Jonsson, A. B., D. Ilver, P. Falk, J. Pepose, and S. Normark.** 1994. Sequence changes in the pilus subunit lead to tropism variation of *Neisseria gonorrhoeae* to human tissue. *Mol. Microbiol.* **13:**403–416.
77. **Kellogg, D. S., I. R. Cohen, L. C. Norins, A. L. Schroeter, and G. Reising.** 1968. *Neisseria gonorrhoeae.* II. Colonial variation and pathogenicity during 35 months in vitro. *J. Bacteriol.* **96:**596–605.
78. **Kellogg, D. S., Jr., W. L. Peacock, W. E. Deacon, L. Brown, and C. I. Pirkle.** 1963. *Neisseria gonorrhoeae.* I. Virulence genetically linked to colonial variation. *J. Bacteriol.* **85:** 1274–1279.
79. **Kieser, S. T., I. S. Eriks, and G. H. Palmer.** 1990. Cyclic rickettsemia during persistent *Anaplasma marginale* infection of cattle. *Infect. Immun.* **58:**1117–1119.
80. **Kitten, T., and A. G. Barbour.** 1990. Juxtaposition of expressed variable antigen genes with a conserved telomere in the bacterium *Borrelia hermsii. Proc. Natl. Acad. Sci. USA* **87:**6077–6081.
81. **Kitten, T., and A. G. Barbour.** 1992. The relapsing fever agent *Borrelia hermsii* has multiple copies of its chromosome and linear plasmids. *Genetics* **132:**311–324.
82. **Kitten, T., A. V. Barrera, and A. G. Barbour.** 1993. Intragenic recombination and a chimeric outer membrane protein in the relapsing fever agent *Borrelia hermsii. J. Bacteriol.* **175:** 2516–2522.
83. **Koomey, J. M.** 1994. Mechanisms of pilus antigenic variation in *Neisseria gonorrhoeae,* p. 113–126. *In* V. I. Miller, J. B. Kaper, D. A. Portnoy, and R. R. Isberg (ed.), *Molecular Genetics of Bacterial Pathogenesis.* American Society for Microbiology, Washington, D.C.
84. **Koomey, M., E. C. Gotschlich, K. Robbins, S. Bergström, and J. Swanson.** 1987. Effects of *recA* mutations on pilus antigenic variation and phase transitions in *Neisseria gonorrhoeae. Genetics* **117:**391–398.
85. **Kuzminov, A.** 1999. Recombinational repair of DNA damage in *Escherichia coli* and bacteriophage lambda. *Microbiol. Mol. Biol. Rev.* **63:**751–813.
86. **Lafay, B., A. T. Lloyd, M. J. McLean, K. M. Devine, P. M. Sharp, and K. H. Wolfe.** 1999. Proteome composition and codon usage in spirochaetes: species-specific and DNA strand-specific mutational biases. *Nucleic Acids Res.* **27:** 1642–1649.
87. **Magoun, L., W. R. Zuckert, D. Robbins, N. Parveen, K. R. Alugupalli, T. G. Schwan, A. G. Barbour, and J. M. Leong.** 2000. Variable small protein (Vsp)-dependent and Vsp-independent pathways for glycosaminoglycan recognition by relapsing fever spirochaetes. *Mol. Microbiol.* **36:**886–897.
88. **Mehr, I. J., and H. S. Seifert.** 1998. Differential roles of homologous recombination pathways in *Neisseria gonorrhoeae* pilin antigenic variation, DNA transformation and DNA repair. *Mol. Microbiol.* **30:**697–710.
89. **Meier, J. T., M. I. Simon, and A. G. Barbour.** 1985. Antigenic variation is associated with DNA rearrangements in a relapsing fever Borrelia. *Cell* **41:**403–409.
90. **Meleney, H. E.** 1928. Relapse phenomena of *Spironema recurrentis. J. Exp. Med.* **48:**65–82.
91. **Meyer, T. F.** 1999. Pathogenic neisseriae: complexity of pathogen-host cell interplay. *Clin. Infect. Dis.* **28:**433–441.
92. **Meyer, T. F., E. Billyard, R. Haas, S. Storzbach, and M. So.** 1984. Pilus genes of *Neisseria gonorrheae:* chromosomal organization and DNA sequence. *Proc. Natl. Acad. Sci. USA* **81:**6110–6114.
93. **Meyer, T. F., J. Pohlner, and J. P. van Putten.** 1994. Biology

of the pathogenic *Neisseriae*. *Curr. Top. Microbiol. Immunol.* **192:**283–317.

94. **Meyer, T. F., and J. P. van Putten.** 1989. Genetic mechanisms and biological implications of phase variation in pathogenic *Neisseriae*. *Clin. Microbiol. Rev.* **2:**S139–S145.
95. **Murphy, G. L., T. D. Connell, D. S. Barritt, M. Koomey, and J. G. Cannon.** 1989. Phase variation of gonococcal protein II: regulation of gene expression by slipped-strand mispairing of a repetitive DNA sequence. *Cell* **56:**539–547.
96. **Parge, H. E., K. T. Forest, M. J. Hickey, D. A. Christensen, E. D. Getzoff, and J. A. Tainer.** 1995. Structure of the fibre-forming protein pilin at 2.6 A resolution. *Nature* **378:**32–38.
97. **Paster, B., F. Dewhirst, W. Weisburg, L. Tordoff, G. Fraser, R. Hespell, T. Stanton, L. Zablen, L. Mandelco, and C. Woese.** 1991. Phylogenetic analysis of the spirochetes. *J. Bacteriol.* **173:**6101–6109.
98. **Pennington, P. M., C. D. Allred, C. S. West, R. Alvarez, and A. G. Barbour.** 1997. Arthritis severity and spirochete burden are determined by serotype in the *Borrelia turicatae*-mouse model of Lyme disease. *Infect. Immun.* **65:**285–292.
99. **Pennington, P. M., D. Cadavid, and A. G. Barbour.** 1999. Characterization of VspB of *Borrelia turicatae,* a major outer membrane protein expressed in blood and tissues of mice. *Infect. Immun.* **67:**4637–4645.
100. **Pennington, P. M., D. Cadavid, J. Bunikis, S. J. Norris, and A. G. Barbour.** 1999. Extensive interplasmidic duplications change the virulence phenotype of the relapsing fever agent *Borrelia turicatae. Mol. Microbiol.* **34:**1120–1132.
101. **Plasterk, R. H., M. I. Simon, and A. G. Barbour.** 1985. Transposition of structural genes to an expression sequence on a linear plasmid causes antigenic variation in the bacterium *Borrelia hermsii. Nature* **318:**257–263.
102. **Radolf, J. D., K. W. Bourell, D. R. Akins, J. S. Brusca, and M. V. Norgard.** 1994. Analysis of *Borrelia burgdorferi* membrane architecture by freeze-fracture electron microscopy. *J. Bacteriol.* **176:**21–31.
103. **Ras, N. M., B. Lascola, D. Postic, S. J. Cutler, F. Rodhain, G. Baranton, and D. Raoult.** 1996. Phylogenesis of relapsing fever *Borrelia* spp. *Int. J. Syst. Bacteriol.* **46:**859–865.
104. **Restrepo, B. I., and A. G. Barbour.** 1994. Antigen diversity in the bacterium *B. hermsii* through "somatic" mutations in rearranged vmp genes. *Cell* **78:**867–876.
105. **Restrepo, B. I., C. J. Carter, and A. G. Barbour.** 1994. Activation of a vmp pseudogene in *Borrelia hermsii:* an alternate mechanism of antigenic variation during relapsing fever. *Mol. Microbiol.* **13:**287–299.
106. **Restrepo, B. I., T. Kitten, C. J. Carter, D. Infante, and A. G. Barbour.** 1992. Subtelomeric expression regions of *Borrelia hermsii* linear plasmids are highly polymorphic. *Mol. Microbiol.* **6:**3299–3311.
107. **Rich, S. M., S. M. Sawyer, and A. G. Barbour.** Antigen polymorphism in *Borrelia hermsii,* a clonal pathogenic bacterium. *Proc. Natl. Acad. Sci. USA,* in press.
108. **Rudel, T., J. P. M. van Putten, C. P. Gibbs, R. Haas, and T. F. Meyer.** 1992. Interaction of two variable proteins (PilE and PilC) required for pilus-mediated adherence of *Neisseria gonorrhoeae* to human epithelial cells. *Mol. Microbiol.* **6:**3239–3450.
109. **Russell, H.** 1936. Observations on immunity in relapsing fever and trypanosomiasis. *Trans. R. Soc. Trop. Med. Hyg.* **30:**179–190.
110. **Sadziene, A., D. D. Thomas, and A. G. Barbour.** 1995. *Borrelia burgdorferi* mutant lacking Osp: biological and immunological characterization. *Infect. Immun.* **63:**1573–1580.
111. **Schoolnik, G. K., R. Fernandez, J. Y. Tai, J. Rothbard, and E. C. Gotschlich.** 1984. Gonococcal pili. Primary structure and receptor binding domain. *J. Exp. Med.* **159:**1351–1370.
112. **Schuhardt, V. T., and M. Wilkerson.** 1951. Relapse phenomena in rats infected with single spirochetes (*Borrelia recurrenis* var. turicatae). *J. Bacteriol.* **62:**215–219.
113. **Schurig, G. D., C. E. Hall, K. Burda, L. B. Corbeil, J. R. Duncan, and A. J. Winter.** 1973. Persistent genital tract infection with *Vibrio fetus intestinalis* associated with serotypic alteration of the infecting strain. *Am. J. Vet. Res.* **34:**1399–1403.
114. **Schwan, T. G., and B. J. Hinnebusch.** 1998. Bloodstream-versus tick-associated variants of a relapsing fever bacterium. *Science* **280:**1938–1940.
115. **Seifert, H. S.** 1996. Questions about gonococcal pilus phase- and antigenic variation. *Mol. Microbiol.* **21:**433–440.
116. **Seifert, H. S., R. S. Ajioka, C. Marchal, P. F. Sparling, and M. So.** 1988. DNA transformation leads to pilin antigenic variation in *Neisseria gonorrhoeae. Nature* **336:**392–395.
117. **Seifert, H. S., C. J. Wright, A. E. Jerse, M. S. Cohen, and J. G. Cannon.** 1994. Multiple gonococcal pilin antigenic variants are produced during experimental human infections. *J. Clin. Investig.* **93:**2744–2749.
118. **Shamaei-Tousi, A., P. Martin, A. Bergh, N. Burman, T. Brännström, and S. Bergström.** 1999. Erythrocyte-aggregating relapsing fever spirochete *Borrelia crocidurae* induces formation of microemboli. *J. Infect. Dis.* **180:**1929–1938.
119. **Simon, M. I., and I. Herskowitz.** 1985. Introduction, p. xv–xix. *In* M. I. Simon and I. Herskowitz (ed.), *Genome Rearrangement.* Alan R. Liss, Inc., New York, N.Y.
120. **Smith, J. M., N. H. Smith, M. O'Rourke, and B. G. Spratt.** 1993. How clonal are bacteria? *Proc. Natl. Acad. Sci. USA* **90:**4384–4388.
121. **Smith, N. H., E. C. Holmes, G. M. Donovan, G. A. Carpenter, and B. G. Spratt.** 1999. Networks and groups within the genus *Neisseria:* analysis of *argF, recA, rho,* and 16S rRNA sequences from human Neisseria species. *Mol. Biol. Evol.* **16:**773–783.
122. **Sohaskey, C. D., W. R. Zuckert, and A. G. Barbour.** 1999. The extended promoters for two outer membrane lipoprotein genes of *Borrelia* spp. uniquely include a T-rich region. *Mol. Microbiol.* **33:**41–51.
123. **Stein, D. C., R. Chien, and H. S. Seifert.** 1992. Construction of a *Neisseria gonorrhoeae* MS11 derivative deficient in NgoMI restriction and modification. *J. Bacteriol.* **174:**4899–4906.
124. **Stern, A., and T. F. Meyer.** 1987. Common mechanism controlling phase and antigenic variation in pathogenic neisseriae. *Mol. Microbiol.* **1:**5–12.
125. **Stevenson, B., S. F. Porcella, K. L. Oie, C. A. Fitzpatrick, S. J. Raffel, L. Lubke, M. E. Schrumpf, and T. G. Schwan.** 2000. The relapsing fever spirochete *Borrelia hermsii* contains multiple, antigen-encoding circular plasmids that are homologous to the cp32 plasmids of Lyme disease spirochetes. *Infect. Immun.* **68:**3900–3908.
126. **Stoenner, H. G., T. Dodd, and C. Larsen.** 1982. Antigenic variation of *Borrelia hermsii. J. Exp. Med.* **156:**1297–1311.
127. **Swanson, J.** 1982. Colony opacity and protein II composition of gonococci. *Infect. Immun.* **37:**359–368.
128. **Swanson, J.** 1973. Studies on gonococcus infection. IV. Pili: their role in attachment of gonococci to tissue culture cells. *J. Exp. Med.* **137:**571–589.
129. **Swanson, J., and O. Barrera.** 1983. Gonococcal size heterogeneity correlates with transitions in colony piliation phenotype, not with changes in colony opacity. *J. Exp. Med.* **158:**1459–1472.

130. **Swanson, J., O. Barrera, J. Sola, and J. Boslego.** 1988. Expression of outer membrane protein II by gonococci in experimental gonorrhea. *J. Exp. Med.* **168:**2121–2129.
131. **Swanson, J., R. J. Belland, and S. A. Hill.** 1992. Neisserial surface variation: how and why? *Curr. Opin. Genet. Dev.* **2:** 805–811.
132. **Swanson, J., S. Bergström, K. Robbins, O. Barrera, D. Corwin, and J. M. Koomey.** 1986. Gene conversion involving the pilin structural gene correlates with pilus$^+$ in equilibrium with pilus$^-$ changes in *Neisseria gonorrhoeae*. *Cell* **47:** 267–276.
133. **Swanson, J., and J. M. Koomey.** 1989. Mechanisms for variation of pili and outer membrane protein II in *Neisseria gonorrhoeae*, p. 743–761. *In* D. E. Berg and M. M. Howe (ed.), *Mobile DNA*. American Society for Microbiology, Washington, D.C.
134. **Swanson, J., S. Morrison, O. Barrera, and S. Hill.** 1990. Piliation changes in transformation-defective gonococci. *J. Exp. Med.* **171:**2131–2139.
135. **Swanson, J., K. Robbins, O. Barrera, D. Corwin, J. Boslego, J. Ciak, M. S. Blake, and J. M. Koomey.** 1987. Gonococcal pilin variants in experimental gonorrhea. *J. Exp. Med.* **165:** 1344–1357.
136. **Takahashi, N., K. Sakagami, K. Kusano, K. Yamamoto, H. Yoshikura, and I. Kobayashi.** 1997. Genetic recombination through double-strand break repair: shift from two-progeny to one-progeny mode by heterologous inserts. *Genetics* **146:** 9–26.
137. **Vidal, V., I. G. Scragg, S. J. Cutler, K. A. Rockett, D. Fekade, D. A. Warrell, D. J. Wright, and D. Kwiatkowski.** 1998. Variable major lipoprotein is a principal TNF-inducing factor of louse-borne relapsing fever. *Nat. Med.* **4:**1416–1420.
138. **Virji, M., J. E. Heckels, and P. J. Watt.** 1983. Monoclonal antibodies to gonococcal pili: studies on antigenic determinants on pili variants of strain P9. *J. Gen. Microbiol.* **129:** 1965–1973.
139. **Wainwright, L. A., K. H. Pritchard, and H. S. Seifert.** 1994. A conserved DNA sequence is required for efficient gonococcal pilin antigenic variation. *Mol. Microbiol.* **13:**75–87.
140. **Walker, E. M., L. A. Borenstein, D. R. Blanco, J. N. Miller, and M. A. Lovett.** 1991. Analysis of outer membrane ultrastructure of pathogenic *Treponema* and *Borrelia* species by freeze-fracture electron microscopy. *J. Bacteriol.* **173:** 5585–5588.
141. **Yang, Q. L., and E. C. Gotschlich.** 1996. Variation of gonococcal lipooligosaccharide structure is due to alterations in poly-G tracts in lgt genes encoding glycosyl transferases. *J. Exp. Med.* **183:**323–327.
142. **Zhang, J. R., J. M. Hardham, A. G. Barbour, and S. J. Norris.** 1997. Antigenic variation in Lyme disease borreliae by promiscuous recombination of VMP-like sequence cassettes. *Cell* **89:**275–285.
143. **Zhang, J. R., and S. J. Norris.** 1998. Kinetics and in vivo induction of genetic variation of *vlsE* in *Borrelia burgdorferi*. *Infect. Immun.* **66:**3689–3697.
144. **Zhang, Q. Y., D. DeRyckere, P. Lauer, and M. Koomey.** 1992. Gene conversion in *Neisseria gonorrhoeae:* evidence for its role in pilus antigenic variation. *Proc. Natl. Acad. Sci. USA* **89:**5366–5370.
145. **Zückert, W., T. A. Kerentseva, C. L. Lawson, and A. G. Barbour.** 2000. Structural conservation of neurotropism-associated VspA within the variable Borrelia Vsp-OspC lipoprotein family. *J. Biol. Chem.* **276:**457–463.
146. **Zückert, W. R., and J. Meyer.** 1996. Circular and linear plasmids of Lyme disease spirochetes have extensive homology: characterization of a repeated DNA element. *J. Bacteriol.* **178:**2287–2298.

IX. GENOMES AND MOBILE DNAs

Mobile DNA II
Edited by N. L. Craig et al.

Chapter 42

Control of Mobile DNA

NINA FEDOROFF

INTRODUCTION

McClintock commented in the introduction to her collected papers that the slim response to her first effort in 1950 to communicate her discovery of transposition in ". . . a journal with wide readership . . .," the *Proceedings of the National Academy of Sciences*, convinced her that ". . . the presented thesis, and evidence for it, could not be accepted by the majority of geneticists or by other biologists" (91). We now know that transposons constitute a large fraction—even a majority—of the DNA in many species of plants and animals, among them mice, humans, and such agriculturally important plants as corn and wheat. How is it that the discovery of transposition lagged so far behind the discovery of the basic laws of genetic transmission? Unlike Watson and Crick's brief paper on DNA (149), McClintock's early communications on transposition were not widely hailed for their explanatory power. Why has it taken so long since the discovery of transposition to recognize the ubiquity and generality of mobile DNA? Explanations sought in McClintock's communicational and personal style miss a fundamental and profound reason for the rather slow acceptance and understanding of the role of transposable elements in eukaryotic genome structure and evolution.

Transposons are almost invisible genetically. There are two reasons for this genetic invisibility. First, transposons and retrotransposons generally move at extremely low frequencies, and second, they participate in illegitimate recombination events at frequencies much lower than anticipated from homology. That is, by all experimental measures, genomic transposons and retrotransposons are capable of moving and recombining at much higher frequencies than they actually do in undisturbed genomes and when the systems that control them are intact. Some of the regulatory mechanisms are transposon-specific, such as the *Drosophila* P element repressor (118). However, what I will discuss here are more general mechanisms that regulate redundancy and recognize changes in gene expression levels. Although I discuss these topics under the rubric of "control of mobile DNA," I do not support the attribution of transposon specificity to such mechanisms, nor do I believe that they evolved in response to transposons. Indeed, I will develop the thesis that the mechanisms that minimize transposition and illegitimate recombination between transposons are likely to be a reflection of the more general capacity of eukaryotic organisms to monitor gene expression levels, as well as to detect, mark, and process duplicated DNA. Thus, I will invert current dogma, suggesting that transposons accumulate in

Nina Fedoroff • Biology Department and Life Sciences Consortium, 519 Wartik Laboratory, The Pennsylvania State University, University Park, PA 16802.

large numbers not despite, but precisely because of, the mechanisms that make large, redundant genomes possible.

THE DISCOVERY OF TRANSPOSITION

It is instructive to begin with a very brief synopsis of how transposition came to light. The study of the types of mutations, termed "unstable mutations," that are caused by transposons dates back to De Vries, who eventually came to the conclusion that unstable mutations did not obey Mendel's rules (22). The first person to make substantial sense of the inheritance patterns of unstable mutations was the maize geneticist Emerson. He carried out a detailed analysis of a variegating allele of the maize *P* locus during the first decades of this century (24–26). His first paper on the subject opens with the statement that variegation ". . . is distinguished from other color patterns by its incorrigible irregularity" (24). Nonetheless, Emerson was able to capture the behavior of unstable mutations in the Mendelian paradigm by postulating that variegation resulted from the temporary association of some type of inhibitor with a locus required for pigmentation.

Several prominent geneticists, among them Correns and Goldschmidt, dismissed unstable mutations as a special category of "diseased genes" (32, 40). It was their view that little could be learned from the study of such mutations that was relevant to the study of conventional genes. But the drosophilist Demerec and the maize geneticist Rhoades shared Emerson's view that there was no difference in principle between stable and unstable mutations. Rhoades' experiments had revealed that a standard recessive allele of the maize *A1* locus, isolated decades earlier and in wide use as a stable null allele, could become unstable in a different genetic background. The key ingredient of the destabilizing background was the presence of the *Dt* locus, which caused reversion of the *a1* allele to wild-type both somatically and germinally (115, 116).

In the late 1930s, McClintock had begun to work with broken chromosomes and by the early 1940s she had devised a method for producing deletion mutations commencing with parental plants, each of which contributed a broken chromosome 9 lacking a terminal segment and the telomere. Searching for mutants in the progeny of such crosses, she observed a high frequency of variegating mutants of all kinds (83). She noted that although reports of the appearance of new mutable genes were relatively rare in the maize literature, she had already isolated 14 new cases of such instability and observed more. She chose to follow the behavior of a locus, which she called *Dissociation* (*Ds*) for its propensity to cause the dissociation of the short arm of chromosome 9 at a position close to the centromere, although she soon appreciated that chromosome dissociation required the presence of another unlinked locus, which she designated the *Activator* (*Ac*) locus (83, 84).

By 1948, she had gained sufficient confidence that the *Ds* locus moves to report: "It is now known that the *Ds* locus may change its position in the chromosome . . ." (32, 85). The relationship between the chromosome-breaking *Ds* locus and variegation emerged as McClintock analyzed the progeny of a new variegating mutation of the *C* locus required for kernel pigmentation. She carried out a series of detailed cytological and genetic experiments on the new *c-m1* mutation, whose instability was conditional and depended on the presence of the *Ac* locus (32, 85, 86). She concluded that the origin of the unstable mutation coincided with the transposition of *Ds* from its original position near the centromere on chromosome 9 to a new site at the *C* locus and that *Ds* disappeared from the locus when the unstable mutant allele reverted to wild type. Having established that *Ds* could transpose into and out of the *C* locus germinally, she inferred that somatic variegation reflects the frequent transposition of *Ds* during development. Transposition provided an elegant explanation for both Emerson's and Rhoades' earlier observations.

The *Ac* and *Ds* elements are transposition-competent and transposition-defective members of a single transposon family. In the ensuing years, McClintock identified and studied a second transposon family, called *Suppressor-mutator* (*Spm*) (87, 88). Her studies on these element families were purely genetic, and she was able to make extraordinary progress in understanding the transposition mechanism because she studied the interactions between a single transposition-competent element and one or a small number of genes with insertions of cognate transposition-defective elements (29). Two points about this early history of transposition merit emphasis. First, the active elements were denumerable and manageable as genetic entities, despite their propensity to move. Second, the number of different transposon families and family members uncovered genetically was (and still is) small. McClintock recognized that the high frequency of new variegating mutations in her cultures was linked to the genetic perturbations associated with the presence of broken chromosomes (83, 90). Her inference, extraordinarily prescient, was that transposons are regular inhabitants of the genome but genetically silent.

MAIZE TRANSPOSONS IN THE AGE OF GENOMICS

With the cloning of the maize transposons, the picture changed radically (31, 81, 105). It became obvious immediately that the maize genome contains many more copies of a given transposon than there are genetically identifiable elements. Although most of these sequences are not complete transposons, there are nonetheless more complete transposons than can be perceived genetically (30). Another important realization was that a genetically active transposon can be distinguished by its methylation pattern from one that is intact but genetically silent (3, 30). Both of these observations bear on the genetic visibility of transposons. As maize genes and genome segments began to be cloned and sequenced, the discovery of new transposons accelerated. Although the transposons that McClintock identified and studied were DNA transposons, both *gypsy*-like and *copia*-like retrotransposons were soon identified in the maize genome and subsequently in many other plant genomes (36, 110, 132, 136, 137, 151). It became evident that non-long terminal repeat (LTR) retrotransposons are also abundant in maize, as well as in other plant genomes (98, 128). Many additional maize transposon families have been identified through their sequence organization and their presence in or near genes (10–13, 135, 150). We now know that transposons and retrotransposons comprise half or more of the maize genome (6).

TRANSPOSONS AND GENOME EVOLUTION

Given their abundance, one might expect that transposable elements would rapidly randomize genome order. Yet the results of a decade of comparative plant genome studies have revealed that gene order is surprisingly conserved. Close relationships among genomes have been demonstrated in crop plants belonging to the Solanaceae and the Gramineae, between *Brassica* crops and *Arabidopsis,* among several legumes, and others (38, 69, 140). The synteny among the genomes of economically important cereal grasses is so extensive that they are now represented by concentric circular maps (38). A relatively small number of major inversions and transpositions are required to harmonize the present-day maps, although additional rearrangements at a finer scale emerge as the level of resolution increases (60, 139–141).

A sequence comparison of the small region around the maize and sorghum *Adh1* loci revealed a surprising amount of change in a constant framework (141). The sorghum and maize genomes are 750 and 2,500 Mbp, respectively. The *Adh1* gene sequences are highly conserved, and complete sequencing of an adjacent region with homology in the two genomes revealed that there were 7 and 10 additional genes in the corresponding regions of maize and sorghum, respectively. The region of homology extends over about 65 kb of the sorghum genome but occupies more than 200 kb in the maize genome. The gene order and orientation are conserved, although three of the genes found in the sorghum *Adh1* region are not in the maize *Adh1* region. The genes are located elsewhere in the maize genome, suggesting that they transposed away from the *Adh1* region (141). Although homology is confined largely to genes, there are also homologous intergenic regions. There are simple sequence repeats and small transposons, called miniature inverted-repeat transposable elements (MITEs) as a group, scattered throughout this region in both sorghum and maize. In the *Adh1* region, MITEs are found neither in exons nor in retrotransposons but are located primarily between genes and occasionally in introns. There are, in addition, three non-LTR retrotransposons in the maize *Adh1* region and none in the sorghum *Adh1* region (141). The major difference between the maize and sorghum *Adh* regions is the presence of very large continuous blocks of retrotransposons in maize that are not present in sorghum.

Retrotransposon blocks occupy 74% of the 240-kb maize *Adh* region (141). These blocks contain 23 members of 11 different retrotransposon families, primarily as complete retrotransposons (141). Within these blocks, retrotransposons are commonly nested by insertion of retrotransposons into each other (123, 124). Solo LTRs are not common, which implies that illegitimate recombination between terminal LTRs is extremely infrequent. What is perhaps most surprising about the maize retrotransposon blocks that have been characterized is that they grow quite slowly. Using the sequence differences between nested retrotransposons, Bennetzen and his colleagues have estimated that the retrotransposon blocks have accumulated over the roughly 5 million years since the divergence of maize and sorghum (123). Importantly, no retrotransposons have been found in the corresponding *Adh1* flanking sequence in sorghum (123, 141).

New copies of transposons and retrotransposons provide new sites of homology for unequal crossing-over. Evidence that transposable elements are central to the evolutionary restructuring of genomes has accumulated in every organism for which sufficient sequence data exist. Exceptionally detailed examples of the role of transposition, retrotransposition, amplification, and transposon-mediated rearrangements in the evolution of a contemporary chromosome are provided by recent studies on the human Y chromosome (70, 71, 125, 127). Although the level of resolution is not yet sufficient in many cases to determine the molecular history of each duplication, it is evident that many, if not a majority of, plant genes belong to gene families ranging in size from a few members to hundreds (80, 92, 112, 117). *R* genes, for example, comprise a superfamily of similar *myc*-homologous, helix-loop-helix transcriptional activators of genes in anthocyanin biosynthesis (20, 74, 107). Detailed analysis of the *R-r* complex, a well-studied member of the *R* superfamily, reveals a history of transposon-catalyzed rearrangement and duplication (146).

CONTROLLING MOBILE (AND NONMOBILE) DNA

Despite our growing awareness of the abundance of transposable elements and the role they have played in shaping contemporary chromosome organization, it is a striking fact that they eluded discovery for the first half century of intensive genetic analysis. Thus, perhaps the most important property of mobile DNA is not its instability, but precisely the opposite: its stability. Not only are insertion mutations in genes infrequent, but retrotransposition events are so widely separated that the time interval between insertions in a particular region of the genome can be counted in hundreds of thousands to millions of years (123). Chromosomes containing many hundreds of thousands of transposable elements are as stable as chromosomes containing few. By what means are such sequences prevented from transposing, recombining, deleting, and rearranging?

The transposon problem is, in my view, a particular case of a general problem in genome structure and gene regulation: the retention and management of duplications. Duplications are a by-product of the properties of the DNA replication and recombination machinery. Short stretches of homology suffice to give rise to duplications by slippage during replication, homology-dependent unequal crossing-over, and double-strand breakage/repair (41, 72). But duplications are problematical. Once a duplication exists, the mechanisms that generated it also permit unequal crossing-over between identical repeats (1, 66, 106). Prokaryotes readily duplicate genetic material but do not retain duplications in the absence of selection (106, 119). This is simply because illegitimate recombination results in the rapid segregation of stable, single-copy progeny and progeny with increasing numbers of copies. Thus, the ability of genomes to expand by duplication is predicated on their ability to either sequester homologous sequences from the cell's recombination machinery or rapidly diverge duplicated sequences (68, 113).

Transcriptional control poses an additional problem. It has long been evident that genomes have compensatory mechanisms that adjust gene expression levels in response to differences in gene copy number, as evidenced by X inactivation and dosage compensation (46, 61, 76). Recent studies on imprinting, paramutation, and gene silencing have underscored the exquisite sensitivity of the genomic mechanisms that monitor gene copy number, irrespective of genomic position (33, 49, 79, 114).

Some years ago, Adrian Bird pointed out that there are two evolutionary discontinuities in the average number of genes per genome (9). The first is an increase between prokaryotes and eukaryotes and the second is between invertebrates and vertebrates. He suggested that, with a given cellular organization, there may be an upper limit on the tolerable gene numbers imposed by the imprecision of the biochemical mechanisms controlling gene expression. He suggested that the transcriptional "noise reduction" mechanisms that arose at the prokaryote/eukaryote boundary were the nuclear envelope, chromatin, and separation of the transcriptional and translational machinery, as well as RNA processing, capping, and polyadenylation to discriminate authentic from spurious transcripts. He also proposed that genome-wide DNA methylation is the novel "noise reduction" mechanism that has permitted the additional quantal leap in gene numbers characteristic of vertebrates.

Recent work in a wide variety of organisms has begun to reveal previously unsuspected, and often interconnected, mechanisms for monitoring both gene expression levels and gene redundancy. Moreover, it is becoming increasingly evident that such mechanisms are central to controlling the expression and transposition of mobile elements.

HOMOLOGY-DEPENDENT GENE SILENCING

Some lower eukaryotes, including *Neurospora crassa* (129, 131) and *Ascobolus immersus* (120, 121), have the capacity to recognize and mark duplicated sequences by methylating them. In *N. crassa*, the methylated sequences are mutated extremely rapidly, primarily through deamination of methylated C residues. Sequence methylation silences transcription, and has also been shown to inhibit recombination in *A. immersus* (78, 129). Thus, the suppression of recombination between duplicated sequences by methylation and (or) rapid divergence appears to be intimately linked with suppression of transcription.

The results of both classical and contemporary studies on the silencing of redundant gene copies in plants suggest that both methylation and other epigenetic mechanisms reflect a fundamental ability to recognize and regulate gene dosage (67). McClintock understood that transposable elements can exist in a genetically intact but cryptic form in the genome, and she carried out genetic analyses of *Spm* transposons undergoing epigenetic changes in their ability to transpose (89). It was later shown that the genetically inactive *Spm* transposons are methylated in critical regulatory sequences (3). It has also been reported that the large intergenic retrotransposon blocks in maize are extensively methylated (7). The results of recent studies on the classical epigenetic phenomenon of *R* locus paramutation in maize have revealed that local endoreduplication of a chromosomal segment both triggers silencing and can render the endoreduplicated locus capable of silencing an active allele of the gene on a homolog (63). Similar observations have been made with transgenes, as well as endogenous gene duplications at different chromosomal locations in tobacco and *Arabidopsis* sp. (75, 82).

The discovery that the introduction of a transgene into a plant can lead to the transcriptional silencing and methylation of both the introduced gene and its endogenous homolog brought gene silencing mechanisms under intense study (67, 104). This phenomenon has been termed "cosuppression" and has been described in both plants and animals (8, 39, 67, 100, 101, 104). Cosuppressed genes can be silenced either transcriptionally or posttranscriptionally (or both). Posttranscriptional gene silencing (PTGS) appears to be caused by RNA destabilization, whereas transcriptional gene silencing involves either DNA methylation or other stable repression mechanisms (39, 67, 100, 101, 144). Short 25-bp antisense oligonucleotides accumulate in plants undergoing PTGS (42). Perhaps the most extraordinary property of PTGS is its ability to spread systemically (102, 145). There is also evidence that posttranscriptional silencing triggers DNA methylation in plants (57, 58, 148). In a recent study, it was reported that methylation was confined to the promoter region of a transgene when silencing was induced by a promoter-containing construct, whereas methylation can spread within the transcribed region when silencing is induced by a segment of the gene's coding region (58).

Cosuppression has been observed both in animals that methylate genomic DNA and in animals that do not (8, 39, 65, 100, 101). In *Drosophila melanogaster,* there is an inverse correlation between transgene copy number and expression, and expression of both the transgene and the endogenous homolog is suppressed (100). Although there is essentially no genomic methylation in *Drosophila* sp., silencing is affected by the *Polycomb* and *Polycomblike* mutations. Recent analyses of cosuppression between two composite transgenes with no homology to each other, but each with homology to the endogenous *Adh* locus, revealed that the silencing of the transgenes was mediated by the regions of the endogenous gene shared with each transgene (101). The authors infer that cosuppression is attributable to transcriptional repression based on DNA homology. Cosuppression of β-galactosidase transgene expression was observed in multicopy mouse lines and accompanied by methylation of the transgenes (39). In situ reduction of the transgene copy number from 100 to 5 or 1 elevated the levels of reporter gene expression and decreased methylation. Thus, whether or not methylation is involved, it appears likely that the ability to detect and silence extra gene copies is a fundamental property of eukaryotic organisms.

QUELLING

A phenomenon termed "quelling," which resembles PTGS in plants, occurs in *Neurospora* sp. and has been subjected to genetic analysis (16, 17). Quelling is posttranscriptional and requires the introduction of a short segment of DNA homologous to the transcribed region of the test gene (16). Moreover, the signal mediating quelling is diffusible and is derived from the transcript. The isolation of quelling-defective mutants has resulted in the recent identification of two genes required for quelling, one with homology to the RecQ DNA helicase family and the other with homology to plant RNA-dependent RNA polymerase (18, 19, 126).

RNA INTERFERENCE

A phenomenon resembling both PTGS and quelling has emerged from investigations on the mechanism of gene suppression by sense and antisense transcripts in nematodes (50, 51, 95). Fire and his colleagues have provided extensive evidence that gene suppression is triggered by double-stranded RNA (dsRNA) homologous to the suppressed gene (34, 142). The ability of dsRNA to inhibit expression of the homologous gene has since been documented in a wide variety of organisms, including hydras (73), zebra fish (147), trypanosomes (97), planaria (122), *Drosophila* (15, 62), and mice (152). The dsRNA, now referred to as interfering RNA (RNAi), must correspond to exonic not intronic sequences (34, 35, 97). The RNAi causes a specific reduction in the levels of the target RNA (35, 62, 93, 122). RNAi exerts its effect at very low concentrations, suggesting that it either acts catalytically or is amplified (34, 35, 62). Recent genetic analyses have revealed the involvement of several genes, including *EGO-1,* which encodes a homolog of the plant RNA-dependent RNA polymerase (64, 134, 138).

The capacity of dsRNA to trigger the degradation of homologous RNAs has been reproduced in vitro using *Drosophila* embryonic extracts (43, 143, 154). Incubation of dsRNA with the extract enhances its ability to trigger degradation (143). The dsRNA is degraded to fragments of 21 to 23 nucleotides in an ATP-dependent reaction (154). Remarkably, mRNA degradation occurs in register and at intervals roughly corresponding to the lengths of the dsRNA degradation fragments produced, implying some type of processivity (154). The short dsRNA fragments appear to copurify with the nuclease (43). A model has been proposed in which the short dsRNA degradation products are strand-transferred to the mRNA by an RNA helicase (possibly a domain of the nuclease) to initiate mRNA degradation (4, 154).

TRANSPOSONS AND GENE SILENCING

Currently popular notions are that transposons are genomic "parasites" and that DNA methylation evolved to control them (153). The designation of transposons as parasitic, selfish DNA comes from two essays written 2 decades ago, one by Doolittle and Sapienza (23) and one by Orgel and Crick (99). These essays sought to free us from the then prevalent notion that genome structure is optimized by phenotypic selection. But the persistence of the moniker "selfish DNA" has become an impediment to further understanding of the origin, historical contribution, and contemporary role of transposons in chromosome structure.

DNA transposons may be an inevitable by-product of the evolution of sequence-specific endonucleases. Complete transposons have been shown to arise from a single cleavage site and an endonuclease gene (96). Although the successful constitution of a transposon from the recognition sequences used in immunoglobulin gene rearrangement and the RAG1 and RAG2 proteins was interpreted as evidence that the V(D)J recombination system evolved from an ancient mobile DNA element, the fact is that the critical components of a transposon and a site-specific rearrangement system are the same (47). Thus, questions about the origin of DNA transposons may devolve to questions about the association of sequence-specific DNA binding domains with endonuclease domains. It is a plausible hypothesis that the invention of nuclear mechanisms that monitor and tightly control gene expression levels and that prevent illegitimate recombination between duplicated sequences, both essential to the evolution of complex genomes by endoreduplication, has supported the extraordinary proliferation of transposons and retrotransposons.

Although the majority of methylated sequences in a genome can be transposable elements, the view that DNA methylation evolved to control transposons seems implausible in the light of evidence that duplications of any kind trigger methylation in organisms that methylate DNA (39, 130, 153). Moreover, organisms that do not methylate DNA also have mechanisms for detecting duplications and sequestering repeats (45, 109, 133). Genome expansion by duplication is predicated on preventing illegitimate recombination between duplicated sequences. Although different eukaryotic lineages appear to have invented slightly different mechanisms, what is common to repeat-induced silencing in all eukaryotes is the stable packaging of DNA into "repressive" chromatin. It may be that the evolution of mechanisms that recognize, mark, and sequester duplications into repressive chromatin structures, among which some involve DNA methylation, was prerequisite for expansion of genomes by endoreduplication at all scales. Since sequence duplication is inherent in transposition, the ability to recognize and repress duplications would serve to minimize both the activity and the adverse impacts of transposons, rendering them genetically invisible and favoring their gradual accumulation.

There is increasing, although still fragmentary, evidence that proteins involved in chromatin structure and recombination are important in controlling transposon mobility. Thus transposition of the LTR retrotransposon Tf1 in *Schizosaccharomyces pombe* is reduced in a mutant strain with a defect in the *pst1*$^{+}$ gene, a Sin3 homolog (21). Transposition of the *Saccharomyces cerevisiae* Ty1 element is elevated in a double mutant that is missing a component of the chromatin assembly factor I and has a mutation in a gene that regulates histone transcription (111). Both of these mutations appear to affect a step in retrotransposition after translation and reverse transcription. Mutations in *RAD6,* which encodes a ubiquitin-conjugating enzyme necessary for telomeric gene silencing, also enhance Ty transposition, possibly by a chromatin structural change that permits a higher level of transcription (59). *Arabidopsis* retrotransposons are demethylated and reactivated in a *ddml* mutant, which is missing a protein with homology to the SWI2/SNF2 chromatin remodeling protein (48, 53,

54). The DDM1 protein is necessary for maintaining genome-wide methylation in *Arabidopsis* sp., as well as for transcriptional gene silencing (37, 54).

The gene silencing mechanisms discussed in earlier sections are likely links between transposon endoreduplication and inactivation. It has been reported that the introduction of complete or partial I elements into *Drosophila* reduces the activity of endogenous elements (55, 56). Introduction of the tobacco retrotransposon *Tto1* into *Arabidopsis* results in a rapid increase in copy number, followed by its inactivation and methylation (48). As noted above, the retrotransposon is reactivated in a *ddm1* mutant (48). Transcriptionally silenced genes are reactivated in the same mutant background (37, 54, 94). There is growing evidence for a connection between posttranscriptional and transcriptional silencing in plants. Introduction of a homologous gene segment induces both transcriptional silencing and methylation in the endogenous copy of the gene (58, 148). Finally, certain *Caenorhabditis elegans* mutations, including the *mut-7* allele of a gene that encodes a protein with helicase and RNAse homologies, both reactivate transposons and interfere with RNAi-triggered gene inactivation (64, 138).

An important and as yet underappreciated property of compacted, inactive genomic regions is their ability to impose their organization on adjacent, as well as nonadjacent, active sequences in a homology-dependent manner. This is evidenced in position effect variegation in *Drosophila,* an organism that does not methylate its DNA, as well as in plant paramutation, which involves DNA methylation (45, 63). Increases in the copy number of Mu transposons resulting from inbreeding are associated with the concerted inactivation and methylation of transposons throughout the genome (14). Recent studies on transgene silencing and paramutation in plants suggest that it does not take many tandem duplications to trigger the formation of a compacted, silenced region. A silenced region may then become a "sink" for insertions within it, as well as a silencer for homologous sequences located adjacent to it or elsewhere in the genome (45, 52, 94).

An interesting recent observation suggests a potential link between retrotransposons and X inactivation. It is well known that X chromosome inactivation depends on transcription of the *Xist* locus, which encodes a large, untranslated RNA (44, 103). The RNA, in turn, coats the inactivating chromosome. On the basis of cytological data, Lyon has suggested that L1 elements are key components of the spread of inactivation along the chromosome. And indeed, it has recently been reported that long interspersed nuclear elements (LINE-1, or L1, elements) cluster preferentially on the X chromosome but are underrepresented in regions that are not inactivated (2, 77). It has been proposed that the responsible molecular mechanism resembles homology-dependent gene and transposon silencing.

What remains unclear is whether repeat- and homology-dependent silencing mechanisms, including such dosage-dependent mechanisms as imprinting and X chromosome inactivation, all require an RNA intermediate. And if so, are the RNA intermediates necessary just for initiation or for both initiation and maintenance? It is possible that the "spread" of inactivation along the X chromosome (44) and the systemic spread of gene silencing in plants (144) and animals (34) are all mediated by RNA. On the other hand, it is also possible that there are independent RNAi and DNA homology-dependent gene inactivation systems, as suggested by the ability of silenced genes to impose a heritable silent state on homologs, observed both in transgenic plants (45, 52, 94) and in the classical genetic phenomenon of paramutation (49).

SUMMARY

I have proposed that the explanation for the prevalence of transposons in contemporary genomes, as well as their genetic invisibility, lies in the fundamental ability of eukaryotic organisms to monitor gene expression levels and to recognize and process duplications. I have argued that such mechanisms, including the ability to mark, sequester, and transcriptionally inactivate duplicated DNA, are essential for the expansion of genomes by endoreduplication at all scales and are not specific to transposons. In contrast to the prevalent concept that gene silencing mechanisms evolved to control transposons and viruses, I suggest that transposons have accumulated in vast numbers because they are sheltered by such mechanisms. Whether transposons, retrotransposons, and other repetitive elements accumulate extensively in a given evolutionary lineage may depend on several factors, among them the efficiency of repressive mechanisms and the rate at which the sequences undergo mutational and deletional decay (108). There may be both selective pressure on overall genome size (5) and random "founder" effects that allow a new transposon to become established rapidly throughout a species (28). At the same time, the results of genome sequencing and analysis make it increasingly evident that the repressive mechanisms that control the expression and recombinogenicity of transposons are imperfect. However slowly, genomes are inexorably restructured by transposition and rearrangements arising from ectopic interactions between dispersed mobile elements (27).

REFERENCES

1. **Anderson, R. P., and J. R. Roth.** 1977. Tandem genetic duplications in phage and bacteria. *Annu. Rev. Microbiol.* **31:** 473–505.
2. **Bailey, J. A., L. Carrel, A. Chakravarti, and E. E. Eichler.** 2000. Molecular evidence for a relationship between LINE-1 elements and X chromosome inactivation: the Lyon repeat hypothesis. *Proc. Natl. Acad. Sci. USA* **97:**6634–6639.
3. **Banks, J. A., and N. Fedoroff.** 1989. Patterns of developmental and heritable change in methylation of the *Suppressor-mutator* transposable element. *Dev. Genet.* **10:**425–437.
4. **Bass, B. L.** 2000. Double-stranded RNA as a template for gene silencing. *Cell* **101:**235–238.
5. **Bennett, M. D., J. B. Smith, and J. S. Heslop-Harrison.** 1982. Nuclear DNA amounts in angiosperms. *Proc. R. Soc. Lond. Biol. Sci.* **216:**179–199.
6. **Bennetzen, J. L., P. San Miguel, M. Chen, A. Tikhonov, M. Francki, and Z. Avramova.** 1998. Grass genomes. *Proc. Natl. Acad. Sci. USA* **95:**1975–1978.
7. **Bennetzen, J. L., K. Schrick, P. S. Springer, W. E. Brown, and P. SanMiguel.** 1994. Active maize genes are unmodified and flanked by diverse classes of modified, highly repetitive DNA. *Genome* **37:**565–576.
8. **Bingham, P. M.** 1997. Cosuppression comes to the animals. *Cell* **90:**385–387.
9. **Bird, A. P.** 1995. Gene number, noise reduction and biological complexity. *Trends Genet.* **11:**94–100.
10. **Bureau, T. E., P. C. Ronald, and S. R. Wessler.** 1996. A computer-based systematic survey reveals the predominance of small inverted-repeat elements in wild-type rice genes. *Proc. Natl. Acad. Sci. USA* **93:**8524–8529.
11. **Bureau, T. E., and S. R. Wessler.** 1992. Tourist: a large family of small inverted repeat elements frequently associated with maize genes. *Plant Cell* **4:**1283–1294.
12. **Bureau, T. E., and S. R. Wessler.** 1994. Mobile inverted-repeat elements of the *Tourist* family are associated with the genes of many cereal grasses. *Proc. Natl. Acad. Sci. USA* **91:** 1411–1415.
13. **Bureau, T. E., and S. R. Wessler.** 1994. *Stowaway:* a new family of inverted repeat elements associated with the genes of both monocotyledonous and dicotyledonous plants. *Plant Cell* **6:**907–916.
14. **Chandler, V. L., and V. Walbot.** 1986. DNA modification of a maize transposable element correlates with loss of activity. *Proc. Natl. Acad. Sci. USA* **83:**1767–1771.
15. **Clemens, J. C., C. A. Worby, N. Simonson-Leff, M. Muda, T. Maehama, B. A. Hemmings, and J. E. Dixon.** 2000. Use of double-stranded RNA interference in *Drosophila* cell lines to dissect signal transduction pathways. *Proc. Natl. Acad. Sci. USA* **97:**6499–6503.
16. **Cogoni, C., J. T. Irelan, M. Schumacher, T. J. Schmidhauser, E. U. Selker, and G. Macino.** 1996. Transgene silencing of the *al-1* gene in vegetative cells of *Neurospora* is mediated by a cytoplasmic effector and does not depend on DNA-DNA interactions or DNA methylation. *EMBO J.* **15:**3153–3163.
17. **Cogoni, C., and G. Macino.** 1997. Isolation of quelling-defective (*qde*) mutants impaired in posttranscriptional transgene-induced gene silencing in *Neurospora crassa. Proc. Natl. Acad. Sci. USA* **94:**10233–10238.
18. **Cogoni, C., and G. Macino.** 1999. Gene silencing in *Neurospora crassa* requires a protein homologous to RNA-dependent RNA polymerase. *Nature* **399:**166–169.
19. **Cogoni, C., and G. Macino.** 1999. Posttranscriptional gene silencing in *Neurospora* by a RecQ DNA helicase. *Science* **286:**2342–2344.
20. **Consonni, G., F. Geuna, G. Gavazzi, and C. Tonelli.** 1993. Molecular homology among members of the *R* gene family in maize. *Plant J.* **3:**335–346.
21. **Dang, V. D., M. J. Benedik, K. Ekwall, J. Choi, R. C. Allshire, and H. L. Levin.** 1999. A new member of the Sin3 family of corepressors is essential for cell viability and required for retroelement propagation in fission yeast. *Mol. Cell. Biol.* **19:** 2351–2365.
22. **de Vries, H.** 1905. *Species and Varieties: Their Origin by Mutation.* Open Court Publishing Co., Chicago, Ill.
23. **Doolittle, W. F., and C. Sapienza.** 1980. Selfish genes, the phenotype paradigm and genome evolution. *Nature* **284:** 601–603.
24. **Emerson, R. A.** 1914. The inheritance of a recurring somatic variation in variegated ears of maize. *Am. Nat.* **48:**87–115.
25. **Emerson, R. A.** 1917. Genetical studies of variegated pericarp in maize. *Genetics* **2:**1–35.
26. **Emerson, R. A.** 1929. The frequency of somatic mutation in variegated pericarp of maize. *Genetics* **14:**488–511.
27. **Engels, W. R.** 1996. P elements in *Drosophila. Curr. Top. Microbiol. Immunol.* **204:**103–123.
28. **Engels, W. R.** 1997. Invasions of P elements. *Genetics* **145:** 11–15.
29. **Fedoroff, N.** 1989. Maize transposable elements, p. 375–411. *In* M. M. Howe and D. E. Berg (ed.), *Mobile DNA.* American Society for Microbiology, Washington, D.C.
30. **Fedoroff, N., D. Furtek, and O. Nelson.** 1984. Cloning of the *Bronze* locus in maize by a simple and generalizable procedure using the transposable controlling element *Ac. Proc. Natl. Acad. Sci. USA* **81:**3825–3829.
31. **Fedoroff, N., S. Wessler, and M. Shure.** 1983. Isolation of the transposable maize controlling elements *Ac* and *Ds. Cell* **35:**243–251.
32. **Fedoroff, N. V.** 1998. The discovery of transposable elements, p. 89–104. *In* S.-D. Kung and S.-F. Yang (ed.), *Discoveries in Plant Biology.* World Scientific Publishing Co., Singapore.
33. **Ferguson-Smith, A. C., H. Sasaki, B. M. Cattanach, and M. A. Surani.** 1993. Parental-origin-specific epigenetic modification of the mouse *H19* gene. *Nature* **362:**751–775.
34. **Fire, A.** 1999. RNA-triggered gene silencing. *Trends Genet.* **15:**358–363.
35. **Fire, A., S. Xu, M. K. Montgomery, S. A. Kostas, S. E. Driver, and C. C. Mello.** 1998. Potent and specific genetic interference by double-stranded RNA in *Caenorhabditis elegans. Nature* **391:**806–811.
36. **Flavell, A. J.** 1992. *Ty1-copia* group retrotransposons and the evolution of retroelements in the eukaryotes. *Genetica* **86:**203–214.
37. **Furner, I. J., M. A. Sheikh, and C. E. Collett.** 1998. Gene silencing and homology-dependent gene silencing in *Arabidopsis:* genetic modifiers and DNA methylation. *Genetics* **149:**651–662.
38. **Gale, M. D., and K. M. Devos.** 1998. Plant comparative genetics after 10 years. *Science* **282:**656–659.
39. **Garrick, D., S. Fiering, D. I. Martin, and E. Whitelaw.** 1998. Repeat-induced gene silencing in mammals. *Nat. Genet.* **18:** 56–59.
40. **Goldschmidt, R.** 1938. *Physiological Genetics.* McGraw-Hill, New York, N.Y.
41. **Gorbunova, V., and A. A. Levy.** 1997. Non-homologous DNA end joining in plant cells is associated with deletions and filler DNA insertions. *Nucleic Acids Res.* **25:**4650–4657.
42. **Hamilton, A. J., and D. C. Baulcombe.** 1999. A species of small antisense RNA in posttranscriptional gene silencing in plants. *Science* **286:**950–952.

43. **Hammond, S. M., E. Bernstein, D. Beach, and G. J. Hannon.** 2000. An RNA-directed nuclease mediates post-transcriptional gene silencing in *Drosophila* cells. *Nature* **404:** 293–296.
44. **Heard, E., P. Clerc, and P. Avner.** 1997. X-chromosome inactivation in mammals. *Annu. Rev. Genet.* **31:**571–610.
45. **Henikoff, S.** 1998. Conspiracy of silence among repeated transgenes. *Bioessays* **20:**532–535.
46. **Henikoff, S., and P. M. Meneely.** 1993. Unwinding dosage compensation *Cell* **72:**1–2.
47. **Hiom, K., M. Melek, and M. Gellert.** 1998. DNA transposition by the RAG1 and RAG2 proteins: a possible source of oncogenic translocations. *Cell* **94:**463–470.
48. **Hirochika, H., H. Okamoto, and T. Kakutani.** 2000. Silencing of retrotransposons in *Arabidopsis* and reactivation by the *ddm1* mutation. *Plant Cell* **12:**357–369.
49. **Hollick, J. B., J. E. Dorweiller, and V. L. Chandler.** 1997. Paramutation and related allelic interactions. *Trends Genet.* **13:**302–308.
50. **Hunter, C. P.** 1999. Genetics: a touch of elegance with RNAi. *Curr. Biol.* **9:**R440–R442.
51. **Hunter, C. P.** 2000. Gene silencing: shrinking the black box of RNAi. *Curr. Biol.* **10:**R137–R140.
52. **Jakowitsch, J., I. Papp, E. A. Moscone, J. van der Winden, M. Matzke, and A. J. Matzke.** 1999. Molecular and cytogenetic characterization of a transgene locus that induces silencing and methylation of homologous promoters in *trans*. *Plant J.* **17:**131–140.
53. **Jeddeloh, J. A., J. Bender, and E. J. Richards.** 1998. The DNA methylation locus *DDM1* is required for maintenance of gene silencing in *Arabidopsis*. *Genes Dev.* **12:**1714–1725.
54. **Jeddeloh, J. A., T. L. Stokes, and E. J. Richards.** 1999. Maintenance of genomic methylation requires a SWI2/SNF2-like protein. *Nat. Genet.* **22:**94–97.
55. **Jensen, S., L. Cavarec, M. P. Gassama, and T. Heidmann.** 1995. Defective *I* elements introduced into *Drosophila* as transgenes can regulate reactivity and prevent *I-R* hybrid dysgenesis. *Mol. Gen. Genet.* **248:**381–390.
56. **Jensen, S., M.-P. Gassama, and T. Heidmann.** 1999. Taming of transposable elements by homology-dependent gene silencing. *Nat. Genet.* **21:**209–215.
57. **Jones, A. L., C. L. Thomas, and A. J. Maule.** 1998. De novo methylation and co-suppression induced by a cytoplasmically replicating plant RNA virus. *EMBO J.* **17:**6385–6393.
58. **Jones, L., A. J. Hamilton, O. Voinnet, C. L. Thomas, A. J. Maule, and D. C. Baulcombe.** 1999. RNA-DNA interactions and DNA methylation in post-transcriptional gene silencing. *Plant Cell* **11:**2291–2302.
59. **Kang, X. L., F. Yadao, R. D. Gietz, and B. A. Kunz.** 1992. Elimination of the yeast RAD6 ubiquitin conjugase enhances base-pair transitions and G.C----T.A transversions as well as transposition of the Ty element: implications for the control of spontaneous mutation. *Genetics* **130:**285–294.
60. **Keller, B., and C. Feuillet.** 2000. Colinearity and gene density in grass genomes. *Trends Plant Sci.* **5:**246–251.
61. **Kelley, R. L., V. H. Meller, P. R. Gordadze, G. Roman, R. L. Davis, and M. I. Kuroda.** 1999. Epigenetic spreading of the *Drosophila* dosage compensation complex from *roX* RNA genes into flanking chromatin. *Cell* **98:**513–522.
62. **Kennerdell, J. R., and R. W. Carthew.** 1998. Use of dsRNA-mediated genetic interference to demonstrate that *frizzled* and *frizzled 2* act in the *wingless* pathway. *Cell* **95:**1017–1026.
63. **Kermicle, J. L., W. B. Eggleston, and M. Alleman.** 1995. Organization of paramutagenicity in *R-stippled* maize. *Genetics* **141:**361–372.
64. **Ketting, R. F., H. A. Haverkamp, H. G. A. M. van Luenen, and R. H. A. Plasterk.** 1999. *mut-7* of *C. elegans* required for transposon silencing and RNA interference, is a homolog of Werner syndrome helicase and RNaseD. *Cell* **99:**133–141.
65. **Ketting, R. F., and R. H. Plasterk.** 2000. A genetic link between co-suppression and RNA interference in *C. elegans*. *Nature* **404:**296–298.
66. **Koch, A. L.** 1979. Selection and recombination in populations containing tandem multiple genes. *J. Mol. Evol.* **14:** 273–285.
67. **Kooter, J. M., M. A. Matzke, and P. Meyer.** 1999. Listening to the silent genes: transgene silencing, gene regulation and pathogen control. *Trends Plant Sci.* **4:**340–347.
68. **Kricker, M. C., J. W. Drake, and M. Radman.** 1992. Duplication-targeted DNA methylation and mutagenesis in the evolution of eukaryotic chromosomes. *Proc. Natl. Acad. Sci. USA* **89:**1075–1079.
69. **Lagercrantz, U.** 1998. Comparative mapping between *Arabidopsis thaliana* and *Brassica nigra* indicates that *Brassica* genomes have evolved through extensive genome replication accompanied by chromosome fusions and frequent rearrangements. *Genetics* **150:**1217–1228.
70. **Lahn, B. T., and D. C. Page.** 1999. Four evolutionary strata on the human X chromosome. *Science* **286:**964–967.
71. **Lahn, B. T., and D. C. Page.** 1999. Retroposition of autosomal mRNA yielded testis-specific gene family on human Y choromosome. *Nat. Genet.* **21:**429–433.
72. **Liang, F., M. Han, P. J. Romanienko, and M. Jasin.** 1998. Homology-directed repair is a major double-strand break repair pathway in mammalian cells. *Proc. Natl. Acad. Sci. USA* **95:**5172–5177.
73. **Lohmann, J. U., I. Endl, and T. C. Bosch.** 1999. Silencing of developmental genes in *Hydra*. *Dev. Biol.* **214:**211–214.
74. **Ludwig, S. R., L. F. Habera, S. L. Dellaporta, and S. R. Wessler.** 1989. *Lc*, a member of the maize R gene family responsible for tissue-specific anthocyanin production, encodes a protein similar to transcriptional activators and contains the *myc*-homology region. *Proc. Natl. Acad. Sci. USA* **86:** 7092–7096.
75. **Luff, B., L. Pawlowski, and J. Bender.** 1999. An inverted repeat triggers cytosine methylation of identical sequences in *Arabidopsis*. *Mol. Cell* **3:**505–511.
76. **Lyon, M. F.** 1993. Epigenetic inheritance in mammals. *Trends Genet.* **9:**123–128.
77. **Lyon, M. F.** 1998. X-chromosome inactivation: a repeat hypothesis. *Cytogenet. Cell. Genet.* **80:**133–137.
78. **Maloisel, L., and J. L. Rossignol.** 1998. Suppression of crossing-over by DNA methylation in *Ascobolus*. *Genes Dev.* **12:** 1381–1389.
79. **Martienssen, R.** 1996. Epigenetic phenomena: paramutation and gene silencing in plants. *Curr. Biol.* **6:**810–813.
80. **Martienssen, R., and V. Irish.** 1999. Copying out our ABCs: the role of gene redundancy in interpreting genetic hierarchies. *Trends Genet.* **15:**435–437.
81. **Masson, P., R. Surosky, J. A. Kingsbury, and N. V. Fedoroff.** 1987. Genetic and molecular analysis of the *Spm*-dependent *a-m2* alleles of the maize *a* locus. *Genetics* **117:**117–137.
82. **Matzke, A. J., F. Neuhuber, Y. D. Park, P. F. Ambros, and M. A. Matzke.** 1994. Homology-dependent gene silencing in transgenic plants: epistatic silencing loci contain multiple copies of methylated transgenes. *Mol. Gen. Genet.* **244:** 219–229.
83. **McClintock, B.** 1946. Maize genetics. *Carnegie Inst. Wash. Year Book* **45:**176–186.
84. **McClintock, B.** 1947. Cytogenetic studie of maize and *Neurospora*. *Carnegie Inst. Wash. Year Book* **46:**146–152.

85. **McClintock, B.** 1948. Mutable loci in maize. *Carnegie Inst. Wash. Year Book* **47:**155–169.
86. **McClintock, B.** 1949. Mutable loci in maize. *Carnegie Inst. Wash Year Book* **48:**142–154.
87. **McClintock, B.** 1951. Mutable loci in maize. *Carnegie Inst. Wash. Year Book* **50:**174–181.
88. **McClintock, B.** 1954. Mutations in maize and chromosomal aberrations in *Neurospora. Carnegie Inst. Wash. Year Book* **53:**254–260.
89. **McClintock, B.** 1962. Topographical relations between elements of control systems in maize. *Carnegie Inst. Wash. Year Book* **61:**448–461.
90. **McClintock, B.** 1978. Mechanisms that rapidly reorganize the genome. *Stadler Genet. Symp.* **10:**25–48.
91. **McClintock, B.** 1987. *The Discovery and Characterization of Transposable Elements.* Garland Publishing, Inc., New York, N.Y.
92. **Michelmore, R. W., and B. C. Meyers.** 1998. Clusters of resistance genes in plants evolve by divergent selection and a birth-and-death process. *Genome Res.* **8:**1113–1130.
93. **Misquitta, L., and B. M. Paterson.** 1999. Targeted disruption of gene function in *Drosophila* by RNA interference (RNA-i): a role for *nautilus* in embryonic somatic muscle formation. *Proc. Natl. Acad. Sci. USA* **96:**1451–1456.
94. **Mittelsten Scheid, O., K. Afsar, and J. Paszkowski.** 1998. Release of epigenetic gene silencing by *trans*-acting mutations in *Arabidopsis. Proc. Natl. Acad. Sci. USA* **95:**632–637.
95. **Montgomery, M. K., S. Xu, and A. Fire.** 1998. RNA as a target of double-stranded RNA-mediated genetic interference in *Caenorhabditis elegans. Proc. Natl. Acad. Sci. USA* **95:** 15502–15507.
96. **Morita, M., A. Umemoto, H. Watanabe, N. Nakazono, and Y. Sugino.** 1999. Generation of new transposons in vivo: an evolutionary role for the "staggered" head-to-head dimer and one-ended transposition. *Mol. Gen. Genet.* **261:**953–957.
97. **Ngo, H., C. Tschudi, K. Gull, and E. Ullu.** 1998. Double-stranded RNA induces mRNA degradation in *Trypanosoma brucei. Proc. Natl. Acad. Sci. USA* **95:**14687–14692.
98. **Noma, K., E. Ohtsubo, and H. Ohtsubo.** 1999. Non-LTR retrotransposons (LINEs) as ubiquitous components of plant genomes. *Mol. Gen. Genet.* **261:**71–79.
99. **Orgel, L. E., and F. H. C. Crick.** 1980. Selfish DNA: the ultimate parasite. *Nature* **284:**604–607.
100. **Pal-Bhadra, M., U. Bhadra, and J. A. Birchler.** 1997. Cosuppression in *Drosophila:* gene silencing of alcohol dehydrogenase by *white-Adh* transgenes is *Polycomb* dependent. *Cell* **90:** 479–490.
101. **Pal-Bhadra, M., U. Bhadra, and J. A. Birchler.** 1999. Cosuppression of nonhomologous transgenes in *Drosophila* involves mutually related endogenous sequences. *Cell* **99:** 35–46.
102. **Palauqui, J. C., T. Elmayan, J. M. Pollien, and H. Vaucheret.** 1997. Systemic acquired silencing: transgene-specific post-transcriptional silencing is transmitted by grafting from silenced stocks to non-silenced scions. *EMBO J.* **16:** 4738–4745.
103. **Panning, B., and R. Jaenisch.** 1998. RNA and the epigenetic regulation of X chromosome inactivation. *Cell* **93:**305–308.
104. **Park, Y. D., I. Papp, E. A. Moscone, V. A. Iglesias, H. Vaucheret, A. J. Matzke, and M. A. Matzke.** 1996. Gene silencing mediated by promoter homology occurs at the level of transcription and results in meiotically heritable alterations in methylation and gene activity. *Plant J.* **9:**183–194.
105. **Pereira, A., Z. Schwarz-Sommer, A. Gierl, I. Bertram, P. A. Peterson, and H. Saedler.** 1985. Genetic and molecular analysis of the *Enhancer (En)* transposable element system of *Zea mays. EMBO J.* **4:**17–23.
106. **Perelson, A. S., and G. I. Bell.** 1977. Mathematical models for the evolution of multigene families by unequal crossing over. *Nature* **265:**304–310.
107. **Perrot, G. H., and K. C. Cone.** 1989. Nucleotide sequence of the maize *R-S* gene. *Nucleic Acids Res.* **17:**8003.
108. **Petrov, D. A., and D. L. Hartl.** 1998. High rate of DNA loss in the *Drosophila melanogaster* and *Drosophila virilis* species groups. *Mol. Biol. Evol.* **15:**293–302.
109. **Pirrotta, V.** 1997. Chromatin-silencing mechanisms in *Drosophila* maintain patterns of gene expression. *Trends Genet.* **13:**314–318.
110. **Purugganan, M. D., and S. R. Wessler.** 1994. Molecular evolution of *magellan,* a maize *Ty3/gypsy*-like retrotransposon. *Proc. Natl. Acad. Sci. USA* **91:**11674–11678.
111. **Qian, Z., H. Huang, J. Y. Hong, C. L. Burck, S. D. Johnston, J. Berman, A. Carol, and S. W. Liebman.** 1998. Yeast *Ty1* retrotransposition is stimulated by a synergistic interaction between mutations in chromatin assembly factor I and histone regulatory proteins. *Mol. Cell. Biol.* **18:**4783–4792.
112. **Rabinowicz, P. D., E. L. Braun, A. D. Wolfe, B. Bowen, and E. Grotewold.** 1999. Maize *R2R3 Myb* genes: sequence analysis reveals amplification in the higher plants. *Genetics* **153:** 427–444.
113. **Radman, M.** 1991. Avoidance of inter-repeat recombination by sequence divergence and a mechanism of neutral evolution. *Biochimie* **73:**357–361.
114. **Reik, W., and J. Walter.** 1998. Imprinting mechanisms in mammals *Curr. Opin. Genet. Dev.* **8:**154–164.
115. **Rhoades, M. M.** 1936. The effect of varying gene dosage on aleurone colour in maize. *J. Genet.* **33:**347–354.
116. **Rhoades, M. M.** 1938. Effect of the *Dt* gene on the mutability of the a_1 allele in maize. *Genetics* **23:**377–397.
117. **Riechmann, J. L., and E. M. Meyerowitz.** 1998. The AP2/EREBP family of plant transcription factors. *Biol. Chem.* **379:** 633–646.
118. **Rio, D. C.** 1991. Regulation of *Drosophila* P element transposition. *Trends Genet.* **7:**282–287.
119. **Romero, D., and R. Palacios.** 1997. Gene amplification and genome plasticity in prokaryotes. *Annu. Rev. Genet.* **31:** 91–111.
120. **Rossignol, J. L., and G. Faugeron.** 1994. Gene inactivation triggered by recognition between DNA repeats. *Experientia* **50:**307–317.
121. **Rossignol, J. L., and G. Faugeron.** 1995. MIP: an epigenetic gene silencing process in *Ascobolus immersus. Curr. Top. Microbiol. Immunol.* **197:**179–191.
122. **Sanchez Alvarado, A. and P. A. Newmark.** 1999. Double-stranded RNA specifically disrupts gene expression during planarian regeneration. *Proc. Natl. Acad. Sci. USA* **96:** 5049–5054.
123. **SanMiguel, P., B. S. Gaut, A. Tikhonov, Y. Nakajima, and J. L. Bennetzen.** 1998. The paleontology of intergene retrotransposons of maize. *Nat. Genet.* **20:**43–45.
124. **SanMiguel, P., A. Tikhonov, Y. K. Jin, N. Motchoulskaia, D. Zakharov, A. Melake-Berhan, P. S. Springer, K. J. Edwards, M. Lee, Z. Avramova, and J. L. Bennetzen.** 1996. Nested retrotransposons in the intergenic regions of the maize genome. *Science* **274:**765–768.
125. **Saxena, R., L. G. Brown, T. Hawkins, R. K. Alagappan, H. Skaletsky, M. P. Reeve, R. Reijo, S. Rozen, M. B. Dinulos, C. M. Disteche, and D. C. Page.** 1996. The *DAZ* gene cluster on the human Y chromosome arose from an autosomal gene that was transposed, repeatedly amplified and pruned. *Nat. Genet.* **14:**292–299.

126. **Schiebel, W., T. Pelissier, L. Riedel, S. Thalmeir, R. Schiebel, D. Kempe, F. Lottspeich, H. L. Sanger, and M. Wassenegger.** 1998. Isolation of an RNA-directed RNA polymerase-specific cDNA clone from tomato. *Plant Cell* **10:**2087–2101.
127. **Schwartz, A., D. C. Chan, L. G. Brown, R. Alagappan, D. Pettay, C. Disteche, B. McGillivray, A. de la Chapelle, and D. C. Page.** 1998. Reconstructing hominid Y evolution: X-homologous block, created by X-Y transposition, was disrupted by Yp inversion through LINE-LINE recombination. *Hum. Mol. Genet.* **7:**1–11.
128. **Schwartz-Sommer, Z., L. Leclercq, E. Gobel, and H. Saedler.** 1987. *CIN4*, an insert altering the structure of the *A1* gene in *Zea mays*, exhibits properties of nonviral retrotransposons. *EMBO J.* **6:**3873–3880.
129. **Selker, E. U.** 1997. Epigenetic phenomena in filamentous fungi: useful paradigms or repeat-induced confusion? *Trends Genet.* **13:**296–301.
130. **Selker, E. U.** 1999. Gene silencing: repeats that count. *Cell* **97:**157–160.
131. **Selker, E. U., and P. W. Garrett.** 1988. DNA sequence duplications trigger gene inactivation in *Neurospora crassa. Proc. Natl. Acad. Sci. USA* **85:**6870–6874.
132. **Shepherd, N. S., Z. Schwarz-Sommer, J. Blumberg vel Spalve, M. Gupta, U. Wienand, and H. Saedler.** 1984. Similarity of the *Cin1* repetitive family of *Zea mays* to eukaryotic transposable elements. *Nature* **307:**185–187.
133. **Sherman, J. M., and L. Pillus.** 1997. An uncertain silence. *Trends Genet.* **13:**308–313.
134. **Smardon, A., J. M. Spoerke, S. C. Stacey, M. E. Klein, N. Mackin, and E. M. Maine.** 2000. EGO-1 is related to RNA-directed RNA polymerase and functions in germ-line development and RNA interference in *C. elegans. Curr. Biol.* **10:** 169–178.
135. **Spell, M. L., G. Baran, and S. R. Wessler.** 1988. An RFLP adjacent to the maize *waxy* gene has the structure of a transposable element. *Mol. Gen. Genet.* **211:**364–366.
136. **Suoniemi, A., D. Schmidt, and A. H. Schulman.** 1997. *BARE-1* insertion site preferences and evolutionary conservation of RNA and cDNA processing sites. *Genetica* **100:**219–230.
137. **Suoniemi, A., J. Tanskanen, and A. H. Schulman.** 1998. Gypsy-like retrotransposons are widespread in the plant kingdom. *Plant J.* **13:**699–705.
138. **Tabara, H., M. Sarkissian, W. G. Kelly, J. Fleenor, A. Grishok, L. Timmons, A. Fire, and C. C. Mello.** 1999. The *rde-1* gene, RNA interference and transposon silencing in *C. elegans. Cell* **99:**123–132.
139. **Tanksley, S. D., R. Bernatzky, N. L. Lapitan, and J. P. Prince.** 1988. Conservation of gene repertoire but not gene order in pepper and tomato. *Proc. Natl. Acad. Sci. USA* **85:** 6419–6423.
140. **Tanksley, S. D., M. W. Ganal, J. P. Prince, M. C. de Vicente, M. W. Bonierbale, P. Broun, T. M. Fulton, J. J. Giovannoni, S. Grandillo, G. B. Martin, et al.** 1992. High density molecular linkage maps of the tomato and potato genomes. *Genetics* **132:**1141–1160.
141. **Tikhonov, A. P., P. J. SanMiguel, Y. Nakajima, N. M. Gorenstein, J. L. Bennetzen, and Z. Avramova.** 1999. Colinearity and its exceptions in orthologous I regions of maize and sorghum. *Proc. Natl. Acad. Sci. USA* **96:**7409–7414.
142. **Timmons, L., and A. Fire.** 1998. Specific interference by ingested dsRNA. *Nature* **395:**854.
143. **Tuschl, T., P. D. Zamore, R. Lehmann, D. P. Bartel, and P. A. Sharp.** 1999. Targeted mRNA degradation by double-stranded RNA in vitro. *Genes Dev.* **13:**3191–3197.
144. **Vaucheret, H., C. Beclin, T. Elmayan, F. Feuerbach, C. Godon, J. B. Morel, P. Mourrain, J. C. Palauqui, and S. Vernhettes.** 1998. Transgene-induced gene silencing in plants. *Plant J.* **16:**651–659.
145. **Voinnet, O., P. Vain, S. Angell, and D. C. Baulcombe.** 1998. Systemic spread of sequence-specific transgene RNA degradation in plants is initiated by localized introduction of ectopic promoterless DNA. *Cell* **95:**177–187.
146. **Walker, E. L., T. P. Robbins, T. E. Bureau, J. Kermicle, and S. L. Dellaporta.** 1995. Transposon-mediated chromosomal rearrangements and gene duplications in the formation of the maize *R-r* complex. *EMBO J.* **14:**2350–2363.
147. **Wargelius, A., S. Ellingsen, and A. Fjose.** 1999. Double-stranded RNA induces specific developmental defects in zebrafish embryos. *Biochem. Biophys. Res. Commun.* **263:** 156–161.
148. **Wassenegger, M., S. Heimes, L. Riedel, and H. L. Sanger.** 1994. RNA-directed de novo methylation of genomic sequences in plants. *Cell* **76:**567–576.
149. **Watson, J. D., and F. H. C. Crick.** 1953. Molecular structure of nucleic acids. *Nature* **171:**737–738.
150. **Wessler, S. R., T. E. Bureau, and S. E. White.** 1995. LTR-retrotransposons and MITEs: important players in the evolution of plant genomes. *Curr. Opin. Genet. Dev.* **5:**814–821.
151. **White, S. E., L. F. Habera, and S. R. Wessler.** 1994. Retrotransposons in the flanking regions of normal plant genes: a role for *copia*-like elements in the evolution of gene structure and expression. *Proc. Natl. Acad. Sci. USA* **91:** 11792–11796.
152. **Wianny, F., and M. Zernicka-Goetz.** 2000. Specific interference with gene function by double-stranded RNA in early mouse development. *Nat. Cell Biol.* **2:**70–75.
153. **Yoder, J. A., C. P. Walsh, and T. H. Bestor.** 1997. Cytosine methylation and the ecology of intragenomic parasites. *Trends Genet.* **13:**335–340.
154. **Zamore, P. D., T. Tuschl, P. A. Sharp, and D. P. Bartel.** 2000. RNAi: double-stranded RNA directs the ATP-dependent cleavage of mRNA at 21 to 23 nucleotide intervals. *Cell* **101:** 25–33.

Mobile DNA II
Edited by N. L. Craig et al.

Chapter 43

Interactions between Transposable Elements and the Host Genome

Mariano Labrador and Victor G. Corces

INTRODUCTION

The notion that transposable elements are universal components of both eukaryotic and prokaryotic genomes has been firmly established in biology. In our own genome, for example, the fraction of DNA that has been originated by transposition corresponds to more than 30%. This figure accounts for more than 30 times the amount of coding sequences and illustrates how significant transposable elements are in eukaryotic genomes from a structural and quantitative perspective. Their functional significance and whether transposable elements are DNA sequences beyond mere genome parasites have been a matter of debate among molecular and evolutionary biologists since the discovery of mobile DNA.

The observation that transposable elements are present in all characterized genomes, including prokaryotes, points to an ancient origin of mobile DNA. The phylogenetic relationships between transposable elements sustain this idea. Findings during the last 10 years have established the fact that integrases of retroviruses and retrotransposons share a common ancestor with the transposases from the widely distributed family of *Mariner/Tc1* elements and some prokaryotic transposons (14, 31). Likewise, the phylogeny of retroviral reverse transcriptase (RT) suggests that the reaction catalyzed by this enzyme was already in use by the prokaryotic ancestor of eukaryotes (56, 135). This reverse transcriptase probably already had the basic properties of this enzyme and was the origin of the present-day non-LTR (long terminal repeats) retrotransposon reverse transcriptase (36). These observations suggest that precursors of the two main classes of transposable elements (DNA elements and retrotransposons) might have been already present in the genome since very early in evolution and probably in the genome of early eukaryotes. The question of the nature of the relationship between transposable elements and their hosts would then have different answers depending on the time frame in which it is asked. Presumably, transposable elements were parasitic in nature at their origin. They probably appeared in the early genome as an inevitable product of the replicative machinery of the cell. As replicative sequences themselves, able to develop the genetic variability necessary to determine differences in the capability to exploit their host, mobile DNAs were also maintained in check by their environment, the genome. Natural selection is thus responsible for the existing diversity of transposable elements and for the many different ways they employ to interact with their

Mariano Labrador and Victor G. Corces • Department of Biology, The Johns Hopkins University, 3400 N. Charles St., Baltimore, MD 21218.

hosts. In the same manner, transposable elements benefit from their hosts and evolve, improving their replication efficiency; and because evolution is an opportunistic process, the host benefits also from the genetic variability offered by transposable elements. As a result of the selective process, transposable elements have become a natural component of modern genomes, and their endurance is due not only to their ability to replicate themselves inside the cell but also to the fact that the eukaryote genome found in these elements is an excellent tool that is constantly used to generate evolutionary novelties and to maintain its own integrity.

Here we explore the issue of how eukaryotic transposable elements interact with their hosts, trying to disclose the benefits they can bring to the genome by analyzing these interactions at the molecular level in every step of the transposition process. The invention of the telomerase and the function of telomeres, including the peculiar case of *Drosophila melanogaster*, the putative origin of introns from transposons, their possible contribution to the development of proteins involved in the control of gene expression at the level of chromatin structure, the healing of DNA double strand breaks, and the reshuffling of the genome through reverse transcription are among the more exciting phenomena that transposable elements may have triggered in eukaryotes. For the sake of brevity, we have deliberately omitted detailed descriptions of structure or molecular mechanisms of integration of transposable elements, considering that they are extensively discussed elsewhere in this volume.

TRANSPOSABLE ELEMENT EXPRESSION AND TRANSPOSITION RATE

The tendency of transposable elements to increase in copy number antagonizes the interest of the host genome to keep its genetic information free of the interference produced by new insertions. Because of this tradeoff, transposable elements and their hosts have developed mechanisms to control transposition, which is the most critical step in limiting their propensity to spread throughout the host genome (67). Although the mechanisms controlling transposition are largely unknown, it is evident that most eukaryotic transposable elements move only sporadically. Therefore, regulatory pathways controlled by the host and the element must be in place to regulate the transposition process. These controls could act at many different levels, as expected from the complexity of the transposition process, including transposon transcription and tissue-specific expression, processing encoded proteins, assembly of viral particles, reverse transcription, and integration. Such mechanisms are beginning to be understood and particular details depend on differences in the strategies used for transposition as well as differences in the host species. For example, invasive elements like *P* in *Drosophila melanogaster* have developed mechanisms to autoregulate their copy number (105), whereas structurally similar elements like *Tc1* in *Caenorhabditis elegans* are apparently controlled by the host genome through posttranscriptional mechanisms such as RNA interference (62, 122). Retrotransposons show a variety of mechanisms to control copy number, including very specific interactions with host factors (16, 21, 100), cosuppression (18, 58), and methylation (138).

Yeast Ty1: Transposition under Stress

The control of transposition at the level of expression is clearly evident in yeast. About 30 copies of the Ty1 retrotransposon are actively transcribed and account for 0.8% of the total yeast RNA, whereas the Ty1 pre-integration cDNA is present in less than 1 copy per haploid cell (21, 26). For that reason, significant levels of transposition are experimentally achieved only after overexpression of Ty1 under the control of a strong promoter like GAL1 (27). In a screen, searching for host factors controlling Ty1 transposition, Conte et al. (21) found that the mitogen-activated protein kinase Fus3 is a negative regulator of Ty1 transposition. They observed that the levels of Ty1 cDNA, Ty1-gag, RT, and integrase proteins increased significantly in mutants for the *fus3* gene, resulting in accumulation of cytoplasmic virus-like particles (VLPs). This increase in VLPs correlates with an increase in transposition levels between 16 and 80 times higher than in the presence of wild-type Fus3 proteins. The increment in transposition occurred without an apparent alteration in Ty1 RNA levels or Ty1 protein synthesis. Therefore, either Ty1 protein stability is increased in *fus3* mutants or VLP assembly is facilitated, slowing down protein degradation and causing an accumulation of VLPs. Recently, the same authors demonstrated that transcription of Ty1 also increases between two and four times in *fus3* mutants (22). This is expected because the internal control region of Ty1 contains consensus binding sites for transcription factors Ste12 and Tec1, which bind cooperatively to filamentous and invasive growth response elements (76), and the Fus3 kinase activity inhibits Tec1 and Ste12 function, resulting in the inactivation of the invasive growth pathway when functional Fus3 protein is present. Although the final mechanism responsible for the stability of Ty1-encoded proteins, and hence transposition, is not completely known, Conte and Curcio have shown that Fus3 inhibits Ty1 transposition directly by inhibiting the invasive/filamentous growth pathway (22). Interestingly, there are indications of low expression of *fus3* also during the normal yeast cell cycle. In normal conditions, this expression would be sufficient to prevent Ty1 transposition which, in turn, will only increase in the presence of stressing environmental factors such as starvation, capable of activating the invasive/filamentous growth pathway.

Transposable elements, not surprisingly, have developed specific interactions with their hosts, and these interactions could lead to apparently opposite strategies of transposition. The yeast Ty5 retrotransposon, for example, remains under transcriptional repression in silent chromatin associated with telomeres and the HM locus in diploid cells. Transcription and transposition are achieved only after the activation of the pheromone response pathway in haploid cells (61). This activation is also dependent on the binding of the Ste12 transcription factor, which is activated by the same Fus3 kinase that inhibits Ty1 transposition. Another yeast retrotransposon, Ty3, has been shown to be specifically repressed under chemical and temperature stress through protein degradation by a ubiquitin-dependent mechanism (79).

Activation of transposition following stress conditions is also frequently found in plant retrotransposons (66). Stresses like wounding, exposure to chemicals, or passing through tissue culture, elicit the activation of transcription and transposition of transposable elements that otherwise are transcriptionally dormant (55, 126). For example, the LTR promoter from the tobacco retrotransposon *Tto1* is responsible for the high level of expression of this element solely in cultured tissues and after wounding response. Interestingly, the responsiveness to tissue culture and wounding is conferred by a 13-bp *cis*-regulatory element located in the LTR that is also present in regulatory regions from plant stress response genes (125). Ty1 and plant retrotransposons are examples of how transposable elements and their hosts are coadapted to fine-tune transposition rates. Increasing transposition under stress conditions could be key to providing new genetic variability that can be useful to survival in certain environments. Accordingly, it has already been demonstrated in experimental populations that Ty1 transpositions could be directly responsible for the production of adaptive mutations in yeast (133).

gypsy and *copia:* with or without Envelope, They Are Here To Stay

As in yeast, *Drosophila* retrotransposons are normally quiescent and maintain transposition rates at very low levels, detected only after occurrence of spontaneous mutations induced by transposable element (TE) insertions. The rare and unpredictable TE movement also suggests a tight regulation of transposition, probably governed by the host genome. An example of this control is the demonstration that most retrotransposons have a developmentally regulated and tissue-specific expression pattern (10, 96). Host genes have been found to control the accumulation of the full-length RNA encoded by several *Drosophila* retrotransposons. Some of these genes have been detected because certain mutations have suppressive effects on the phenotype of mutations induced by particular retrotransposons at other loci. For example, it has been shown that the genes *Darkener of apricot* and *Weakener of white,* which modify the phenotype of *copia*-induced mutations in the *white* gene, also influence the accumulation of *copia* transcripts (6, 102, 139). These genes are often involved in the processing of the mRNA and do not regulate transposable element transcription directly (35, 86, 124).

Recently, however, a factor directly involved in *copia* expression has been characterized. Transcription of *copia* increases 30-fold when the retrotransposon is cotransfected into S2 *D. melanogaster* cells with a plasmid expressing a protein containing BTB domains and zinc fingers that binds an 18-bp sequence found in the 5′ transcribed untranslated region of the element (16, 17). The cDNA encoding this protein was originally cloned from *Drosophila hydei* and is the homolog of the *D. melanogaster* transcription factor *lola,* which is involved in axon growth and guidance in the central and peripheral nervous systems (46). This factor activates transcription of *copia* as well as of a reporter gene with a minimal promoter in *D. hydei* cells, whereas transcription is not induced in cells of *D. melanogaster* that are not expressing the *D. hydei* plasmid. In situ RNA hybridization and immunostaining using *lola* antibodies have shown that *copia* is highly expressed in the central nervous system of *D. melanogaster* embryos only in *lola* mutants. Wild-type embryos, however, do not show any *copia* RNA staining in the nervous system, where *lola* is highly expressed, but a strong expression is observed in the gonads. The authors suggest that two isoforms of *lola* have antagonistic regulatory properties and are differentially expressed, determining the observed expression pattern of *copia.* Considering the widespread distribution of *copia* in the *Drosophila* genus, it is interesting to point out that the genome of *D. hydei* is the only species devoid of *copia* retrotransposons. This observation suggests that the *lola*-*copia* system in *D. melanogaster,* where both transcription factor and transposable element coexist, has evolved in such a way that *copia* transcription is suppressed in somatic tissues to minimize the deleterious effects of transposition that is otherwise allowed in the germ line and also in somatic tissues of *D. hydei,* where *copia* ordinarily is absent.

The genomic distribution and the copy number of *copia* and *gypsy* retrotransposons are remarkably different. The *copia* retrotransposon is present at tens of euchromatic sites in all chromosomes, whereas most *D. melanogaster* strains have mostly heterochromatic copies of *gypsy,* with only a few or no insertions in the euchromatin. Since both elements seem to predate the genome of *D. melanogaster,* this unequal distribution probably reflects different regulatory strategies and/or an alternative strategy of interaction with the host. At least two main host genes are known to interact with the *gypsy* retrotransposon: *suppressor of Hairy-wing [su(Hw)]* and *flamenco (flam)* (67). Mutations in the *su(Hw)* gene suppress the phenotypes of *gypsy*-induced mutations (81, 113). The su(Hw) protein is a component of a chromatin insulator found in more than 500 sites in the genome in addition to *gypsy* (45, 118). The activity of the su(Hw) protein influences *gypsy* transcription but there is no evidence that this influence translates into transposition and, therefore, the role of the *su(Hw)-gypsy* interaction is not completely understood (114). It is important to note that contrary to Ty1 and *copia* (both high-copy-number retrotransposons), the number of active *gypsy* elements in the *D. melanogaster* genome could be very few. A possibility is that, in addition to the regulatory sequences located in the LTR, the insulator activity provided by su(Hw) is necessary to obtain the amount of transcription required to achieve transposition. The insulator may protect against position effects, increasing the chances that at least some of the few functional copies, often present in heterochromatin, will be transcribed. Because insulators might have a major role in establishing higher-order chromatin structure (45), it is also possible that a high number of *gypsy* euchromatic copies will interfere with the regular arrangement of the chromatin in the nucleus. If this hypothesis is true, there will be a limitation in the capacity of the genome to accommodate a certain number of *gypsy* sequences. Once this limiting point is reached, the copy number will be detrimental for the fitness of the fly.

As indicated previously, *gypsy* transpositions in wild-type stocks are rare and unpredictable. Interestingly, there are particular *Drosophila* stocks, containing multiple copies of *gypsy* in the euchromatin, in which *gypsy* insertions into the ovo^{D1} allele occur at a rate significantly higher than in wild-type stocks. Genetic analyses have shown that mutations in an X-linked recessive gene named *flamenco* (*flam*) dramatically change the transposition rate of *gypsy* and that the above-mentioned stocks are homozygous mutant for this gene (100). However, the high rate of *gypsy* mobilization (between 10 and 15% of *gypsy* insertions in the ovo^{D1} allele) is observed in the female progeny of homozygous *flam* females (97, 100). In fact, the females showing high rates of transposition do not necessarily have to be homozygous for the *flam* mutation. Therefore, any effect of the *flam* gene on *gypsy* transposition is possible only if the homozygous *flam* females maternally induce transposition in the germ line of their offspring. This inheritable transposition of *gypsy* would not be possible without the presence of *gypsy* RNA and proteins in the germ line cells of these females. Surprisingly, although *gypsy* is transcribed in various somatic tissues at different stages of development, its expression in the germ line is always very low (114) and yet *flam* mutations boost transposition by increasing levels of full-length *gypsy* RNA 10- to 20-fold in relation to wild-type females (97). Moreover, high levels of a new ORF3-specific spliced RNA, encoding a protein with the properties expected for a retroviral envelope protein not present in wild-type stocks, are also observed in Northern blots containing RNA from *flam* homozygous females (97). RNA in situ hybridization and immunostaining using antienvelope antibodies demonstrated that *gypsy* transcription and envelope expression occur only in the follicle cells surrounding the oocyte at stages 8–9 to 10 of oogenesis (97, 116). A hypothesis to explain these observations suggests that the assembly of infectious viral particles takes place in the follicle cells surrounding the oocyte. These particles should be able to infect the oocyte after budding out of the follicle cells, before the vitelline membrane is completely formed. Once in the oocyte, the virus particles might be transported to the posterior end of the embryo where they might be taken up by the pole cells to ensure that *gypsy* insertion takes place in the germ line (117). Several lines of evidence support this hypothesis. First, the model predicts the existence of infectious virus particles capable of infecting the oocyte. These particles have been isolated from *flam* females and demonstrated to be infectious after feeding *Drosophila* larvae lacking active *gypsy* elements (116). Second, retroviral particles have been detected by immunoelectron microscopy in follicle cells from *flam* females (69, 117) and found trapped in the vitelline membrane during late stages of oogenesis (117). This model, however, may be a simplification because *gypsy* elements defective in the envelope gene are also capable of reaching the oocyte and transpose in the offspring of *flam* females (19). Independently of the nature of the process, the somatic expression and subsequent invasion of the germ line seems a good strategy to ensure a stable relationship between the retrotransposon and the host. The addition of an envelope protein to the coding capacity of *gypsy* provides the opportunity of vertical transmission as well as horizontal transmission to other individuals of the same species and perhaps of different species. In fact, the phylogenetic analysis of *gypsy* sequences and that of their hosts suggest that *gypsy* has been transmitted horizontally (127). Infectious events will likely happen after *flam* homozygous mutants arise in the population; these mutants are perfectly viable in the laboratory with no visible phenotypes. In principle, this mechanism should be very efficient to avoid the elimination of the *gypsy* element from *Drosophila* populations in which the number of active copies could be already very low. As mentioned above, the transposition strategy of *gypsy* seems to consist of surviving with a very low copy number. The efficiency of a single element to expand in the population after reaching the germ line following infection could be greatly improved by the fact that the element is expressed in the follicle cells, where transcription is consequently amplified by hundreds of polyploid cells dumping *gypsy* particles into a single oocyte.

Transposable Elements, Repression by Chromatin Structure, and Posttranscriptional Gene Silencing: Who Is To Blame First?

Two major divisions can be established among all organisms in the phylogenetic scale: those in which the DNA is wrapped around histones and those lacking histones. All the major evolutionary lineages having complex genomes pack their DNA with nucleosomes. Histones have provided the eukaryotic genome with two features tightly related to each other: silencing and the possibility to grow bigger. The capacity to sustain a larger genome size opened the door to the expansion of repetitive DNA, including transposable elements, which in today's eukaryotic cells account for a very significant proportion of the total genomic DNA. This enrichment in repetitive sequences has been possible only because of the ability of histones to tightly package the chromatin in the nucleus. A result of such DNA packing is also the inaccessibility of the transcription and recombination machinery to these regions.

Since transposable elements predate both histones and the eukaryote genome, it is reasonable to think that the early eukaryotes would have had to deal with the problems generated by transposition before tackling the complex regulatory mechanisms involving chromatin structure. An interesting evolutionary possibility is that the proteins necessary to regulate the packing of nucleosomes into highly compacted transcriptionally inactive chromatin originally evolved to control the expansion and the undesirable effects of mobile and repetitive sequences. Besides transposition into genes, these effects also include ectopic recombination between repetitive sequences, which results in highly deleterious genome rearrangements. The same proteins would have evolved afterward to become major players in transcription regulation through alterations in chromatin structure. For this hypothesis to be true, the current genomes should still retain the ability to recognize repetitive sequences such as transposable elements and be able to silence them through changes in chromatin structure. The following examples illustrate how the same mechanisms determining the transcriptional state of homeotic genes in *Drosophila*, the epigenetic estate of several mammalian genes, or the complete inactivation of the X chromosome in mammals could also be involved in silencing transposable elements.

In mammals, DNA methylation in the 5′ cytosine of the dinucleotide CpG correlates with silencing and with inheritable inactivation of transcription, when found in promoter regions. Ninety percent of these methylated cytosine residues in human DNA are located on transposable elements and endogenous retroviruses, suggesting that perhaps the control of transposition by repressing transcription is the main function of methylation in mammals (138). The same control of transcription through methylation is responsible for X chromosome inactivation in females and for the maternal and paternal imprinting determination of the H19 and Igf2 genes in mice (138). In fact, transcription levels of intracisternal A particles (IAP) increase up to 50- to 100-fold in mice embryos deficient for DNA methyltransferase 1 (Dnmt1), the enzyme that is directly involved in cytosine methylation (130). Activation of IAP transcription occurs by demethylation of the sequences in the LTR, which is heavily methylated in normal mice. The same Dnmt1 mutants show ectopic activation of the inactivated X chromosome in females, whereas the H19 gene is expressed both from maternal and paternal alleles (129, 130). Consistent with these results, hybrids of the wallaby species *Macropus wugenii* and *Wallabia bicolor* display a genome-wide undermethylation that correlates with a major retroviral amplification (91).

The mechanisms by which methyltransferases recognize repetitive sequences in mammals are not completely understood. A possibility is that methylation is targeted to dispersed sequences sharing homology between them. For example, in plants, methylation of cytosine residues in promoter regions is also correlated with chromatin silencing of transposable elements as well as of other endogenous sequences (11, 57). In accordance with this idea, it has been shown that a gene arrangement in *Arabidopsis thaliana* consisting of an inverted repeat of the *PAI* gene triggers de novo cytosine methylation of three target *PAI* sequences located elsewhere in the genome (74). In addition, single *PAI* genes are not able to induce methylation on *PAI* sequences in other parts of the genome, which suggests that the involved methyltransferases recognize some kind of repetition of the target sequence (74). It is arguable whether the methylation of 5′ cytosines is a universal mechanism because it is not found in *Drosophila, C. elegans,* or yeast. In these organisms, as well as in mammals and plants, it is generally accepted that transcription correlates with histone acetylation, which will then induce changes in chromatin structure, making it more accessible to the transcription machinery (49). Several lines of evidence suggest that DNA methylation is also linked to histone acetylation (2, 60, 89). Nan et al. (89) and Jones et al. (60) demonstrated that MeCP2, a protein that specifically binds to methylated areas of the chromosomes to repress transcription, is associated with a corepressor complex containing the transcriptional repressor mSin3A together with histone deacetylases. Interestingly, mSin3A has been implicated in other deacetylation-related transcriptional gene silencing processes (104). In addition, Jones et al. (60) have shown that the deacetylase inhibitor trichostatin A lessens the repressor activity of methylation. Therefore, suppression of transcription in methylated DNA will be also associated with changes of chromatin structure through histone deacetylation.

Another hint of how transposable elements could be subject to silencing and heterochromatization by the host genomes comes from studies of the organization of the pericentromeric regions of the chromosomes and the heterochromatic sex chromosomes. The process by which pericentromeric regions become richer than other regions of the chromosomes in transposable elements and/or whether transposable elements are directly involved in the acquisition of heterochromatic properties in these regions are issues not well understood. In *Drosophila,* it has been suggested that because pericentromeric regions display very low recombination rates, the chances of ectopic recombination between homologous sequences is also reduced, allowing the accumulation of transposable elements over evolutionary time (68, 82, 115). One possibility is that the high concentration of transposable elements also determines the heterochromatic features of these regions. This seems to be the case for the *Drosophila* Y chromosome, which is completely heterochromatic and transcriptionally inactive, except in the male gonads during spermatogenesis. It has been postulated that in species where the sex is chromosomally determined, the pair of heteromorphic sex chromosomes has evolved from a pair of homologous chromosomes. Some X-autosome translocations in *D. miranda* are in this process of generating a neo-Y and a neo-X chromosome (78). Molecular analysis of the neo-Y chromosome suggests that the genetic degeneration is initiated by an enrichment of new transposable element insertions determining the progressive heterochromatization of the subsequent Y chromosome (120, 121). In these cases, transposable elements will contribute to the structure of the chromosomes and to the evolutionary processes leading to chromosome specialization. Interestingly, a recent publication by Bailey et al. (1) suggests that the *LINE-1* retrotransposon could also play a major role in X chromosome inactivation in mammals. The hypothesis suggests that *LINE-1* elements may act as nucleation centers along the X chromosome, facilitating the spreading of the inactivation mediated by the Xist RNA (1, 75). By analyzing data from the Human Genome Project, they demonstrated that 30% of the X chromosome corresponds to *LINE-1* sequences, whereas the same sequences in the autosomes account for only 13%. These differences are statistically significant and are bigger in the regions closer to Xq13, the locus where the X inactivation center is localized. Another intriguing result is that the *LINE-1* elements in the X chromosome are significantly enriched with insertions whose transposition dates correspond approximately to the period of separation between eutherian and metatherian mammals. The same comparison is not true for other repetitive sequences in the X chromosome. Interestingly, this time lapse corresponds to the development of the multistep eutherian X-inactivation pathway.

Silencing has also been observed in *Drosophila* when repeated transgenes form multiple tandem arrays (54) or when multiple copies of a transgene are dispersed in the genome (5, 7). In the first case, the expression of the *white* reporter gene, present in multiple tandem arrays of the *P* element transgene, displayed position-effect variegation (34). This effect is usually associated with partial inactivation of a gene by its proximity to heterochromatin. In this case, the variegated expression of *white* seems to be due to the pairing between the tandem repeats in the chromosome, generating heterochromatin-like structures able to interact in *trans* with endogenous chromosomal heterochromatin. To confirm this suggestion, it has been shown that the heterochromatin protein 1 (HP1) colocalizes with the tandem arrays of the transgenes using immunostaining of polytene chromosomes from larval salivary glands (41). In the second case (93, 94), it has been shown that multiple copies of a transgene, consisting of a *white-Adh* fusion dispersed in several locations in the genome, trigger a significant repression of transcription in both the transgenes and the endogenous *Adh* gene. This repression depends on the

Polycomb-group genes, which encode proteins involved in the maintenance of a repressed chromatin state in homeotic genes during *Drosophila* development (99). These are experimental examples of silencing triggered by the repetitive nature of otherwise nonheterochromatic sequences. Evidence of a similar mechanism acting on endogenous transposable elements comes from analysis of the repression of transposition of the *I* elements during hybrid dysgenesis in *Drosophila* (18). The activity of the promoter of the *I* element depends on the number of copies of the first 186 nucleotides of the *I* sequence, the activity of the endogenous *I* elements decreasing as the copy number of this sequence increases. However, although the authors rule out the possibility of posttranscriptional mechanisms of silencing, other evidence suggests that suppression of transposition of the *I* element could be also mediated by a mechanism involving RNA interference (RNAi) (58, 59).

RNAi is a posttranscriptional gene silencing mechanism, triggered by the presence of double-stranded RNA in the cell, that has been shown to be very effective in disrupting the expression of specific genes (43, 44). Although RNAi was only recently discovered in *C. elegans* (44), the universality of the mechanism has been already demonstrated by showing the same phenomena in a variety of organisms including fungi (106), *Drosophila* (52, 80), protozoa (90), and planaria (107). While the RNAi response in these organisms requires the direct application of a dsRNA to the cells, cosuppression in plants (also recently found in animals) has been shown to be related to RNAi. Cosuppression involves specific silencing of an endogenous gene triggered naturally, for example by transgenes, viral infections, or the introduction of gene fragments without a promoter (128). In both cosuppression and RNAi, the RNA seems to be an important initiator and also the target for the silencing mechanism (63). Although the molecular basis of RNAi is not completely understood, it seems to involve the degradation of the RNA target that is transcribed from the endogenous gene (43). Independently of the organism, this degradation is systematically carried out down to small 21 to 25 nucleotide pieces of the target RNA. These fragments probably function as an amplification of the silencing signal (51, 52). A putative RNase responsible for the degradation and an RNA-dependent RNA polymerase, probably involved in the generation of dsRNA in cases in which it is not directly available, have been implicated in this process (28, 52, 62).

It was initially suggested that cosuppression in plants could be a defense mechanism against virus infection or transposable elements spreading (103). In line with this suggestion, Ketting et al. (62) and Tabara et al. (122) have recently found that worm mutants for the *mut7* gene display a general mobilization of endogenous transposons such as *Tc1,* at the same time that the RNAi response was suppressed by the mutation. Since MUT7 has been identified as an RNase, the authors suggest that in a wild-type worm the general machinery of transposition is repressed by the degradation of transposon-specific messengers that might have a specific signal or might be produced as dsRNA (62). Interestingly, the above-mentioned mechanism of cosuppression of the *I* element by an increasing number of transgene copies in the genome has a counterpart in experiments suggesting that the suppression of transcription during hybrid dysgenesis could be also due to posttranscriptional gene silencing (58, 59). Transposition of endogenous *I* elements during hybrid dysgenesis can be repressed by progressively increasing the number of transgenes containing fragments of defective *I* elements, which can be transcribed but not translated. By obtaining a number of different transgenic flies, these authors show that in equivalent conditions, transgenes containing *I* sequences with an inducible promoter are able to repress endogenous *I* transposition, whereas the same transgene but with *I* transcription completely blocked by the absence of a promoter cannot repress endogenous *I* transposition. The same result is obtained when the experiments are done with transgenes expressing the sense or the antisense RNA. The degree of repression increases with the number of generations and is transmitted only maternally. Since the authors claim to have found antisense RNA in the lines containing only sense transgenes, it follows that posttranscriptional gene silencing of *I* elements is most probably triggered by a mechanism of RNA interference (58). Interestingly, in early work by Simmons et al. (111), an antisense *P* RNA was utilized to control *P* element transposition during hybrid dysgenesis induced by the *P* element in *Drosophila.* While the authors of the mentioned experiment suggested that the transgenic lines transcribing antisense *P* RNA repressed transposition of endogenous *P* elements by antisense inhibition of translation of the transposase mRNA, it is also possible that an RNAi response was activated instead. It would be worthwhile to go back to the original stocks and test this hypothesis by observing whether the expected products of RNAi degradation are present in flies expressing antisense RNA. In this context, it is interesting to note that both *I* and *P* elements are considered infectious transposable elements that have recently invaded the genome of *D. melanogaster* (39, 109). Since hybrid dysgenesis is a manifestation of such processes as invasion of a genome by alien transposable elements, this could be evidence for how RNAi might have evolved as a natural mechanism of defense to protect the cell against an assault by transposable elements.

SELECTION OF TRANSPOSABLE ELEMENT INSERTION SITES

Insertion of a transposable element in the coding or regulatory regions of a gene could reduce the fitness of the host. In the previous section, we have reviewed strategies by which transposable elements and their hosts minimize the chances of this transgression by suppressing transposable element expression and reducing the transposition rate. Since transposable elements continue to spread and populate the genome of their hosts, it is obvious that the mechanisms in operation only reduce the chances of transposition without completely inhibiting it. It has been argued that because transposition still occurs, transposable elements negotiate with the host a "not so bad" solution consisting of selecting the insertion site in places such that the fitness of the organism is not significantly affected (8, 25).

Apparently in contradiction with this idea, *P* elements insert preferentially in the 5′ regions of genes (119). The properties of the genomic DNA at these insertion sites, such as base composition and protein-induced DNA deformability, differ significantly from the average chromosomal DNA (72), indicating that the *P* transposase has evolved to preferentially insert the element at these sites. This preference could improve the chances for the new insertions to fall in a site not negatively regulated by chromatin structure, ensuring that the *P* element will remain transcriptionally active after the insertion. Interestingly, it has been demonstrated that *P* elements can develop the ability to insert in specific regions of the genome in a process that has been called homing. When a *P* element is structurally modified, introducing a sequence containing a binding site for a DNA binding protein, the resulting insertions are biased to chromosomal regions where the particular protein binds (50, 123). This capability of *P* elements brings up the question of why, evolutionarily, they have instead developed such aggressive strategy to insert in the 5′ region of genes. It is generally accepted that *P* elements are invasive in the genome of *D. melanogaster* (20, 39). The integration of *P* elements is not site specific, probably because they are able to regulate their copy number using the repressor protein encoded by defective *P* element copies, hence maximizing the invasiveness and minimizing the deleterious effects produced in the host population. This strategy has proven to be valid for the spreading of the *P* element through natural populations of *D. melanogaster,* without apparently reducing the fitness of the population. A case completely opposite to that of the *P* element is found in the *mariner*-like elements from hymenoptera, a DNA group of elements that spreads horizontally like *P* (53). Hymenopteran *mariner*-like elements are found in a specific insertion site in more than 20 species, suggesting that this site is conserved through evolution (4). This specificity probably reflects the choice of a selectively neutral insertion site to avoid the elimination of the element from hymenopteran populations in which the male is haploid. *Mariner*-like elements are present in a variety of groups of organisms and usually do not display such specificity of insertion (53).

Retroviral Integrases and Integration Specificity

Purified integrases in vitro have the ability of processing the 3′ ends of the LTRs and of carrying out the strand transfer reaction in only one of the ends of the viral DNA (38). These integrases are not specific for the site of integration, but show preference for integration in nucleosomal versus naked DNA, where they integrate in the more severely deformed major groove (101). This preference, perhaps, indicates an adaptation for insertion in eukaryotic DNA, which is always wrapped in nucleosomes, but demonstrates very little sequence specificity. In vivo, retroviral cDNA integration is mediated by a large nucleoprotein preintegration complex resulting from the reverse transcription of the RNA inside the viral particle. Preintegration complexes include integrase bound to several hundred base pairs at the ends of the retroviral cDNA, together with other protein factors (131). Among these factors, only a host protein called BAF (barrier against autointegration factor), involved in the protection against the integration of the ends of the retroviral cDNA into itself, has been identified (70). BAF does not share homologies with any other protein in the databases and its function in the host cell is unknown. In the absence of this factor, the integration-competent nucleoprotein complexes isolated from human immunodeficiency virus (HIV)- or Moloney murine leukemia virus-infected cells can only autointegrate in their own DNA (70). This result is an indication of the low specificity for integration sites shown in general by retroviruses. As in vitro, retroviral integrases, such as avian leukosis virus or HIV, show very little specificity for integration sites in vivo (15, 134). Only regions of alphoid satellite DNA seem to be refractory to integration, probably due to the inability of integrases to reach the DNA in highly compacted heterochromatin (15). Specificity for integration sites has been demonstrated, however, in experiments tethering DNA binding proteins such as phage lambda repressor or a zinc finger protein (zif268) to the HIV integrase (12, 13). In such experiments, both purified integrase-lambda repressor and preintegration complexes containing integrase-zif268 were able to direct integration near the respective binding sites in the DNA target in vitro.

As illustrated above for *P* and *mariner*-like elements, transposable elements show a surprising inherent ability to select specific insertion sites. The strategy by which some integrases acquire site specificity is exemplified by the HIV integrase tethering experiments. Several LTR retrotransposons, whose integrases are closely related to those of retroviruses, show site specificity for insertion in *S. cerevisiae*. Ty3 elements insert almost exclusively a few bases away from the transcription start site of genes transcribed by RNA polymerase III (Pol III) (65). Kirchner et al. (65) developed an in vitro assay where the insertionally active Ty3 element was extracted in the form of a preinitiation complex isolated from yeast. This complex promotes Ty3-specific integration in a target tRNA only when Pol III transcription extracts were added and the tRNA target DNA was transcriptionally active. Furthermore, fractions containing only TFIIIB and TFIIIC still specifically target Ty3 insertions to the tRNA promoter, even in the absence of transcription. Surprisingly, Ty1 also inserts upstream of the transcription start of Pol III genes, but utilizes a region between bp −75 and −700 (29). Like Ty3, Ty1 also requires actively transcribed Pol III genes, although in this case integration is not dependent on the presence of Pol III transcription complex. Ty1 is also able to insert with a high frequency in the mating and rDNA loci (112). The three DNA spots in which Ty1 is found have the common property of being actively silenced chromatin. The rDNA and HM loci are silenced by SIR proteins, and Pol II genes located in the upstream region of Pol III promoters are downregulated, suggesting the presence of silencing factors in this region.

The conclusion from these experiments is that one or several proteins of the integration complex, likely the integrase, is tethered by a protein or proteins binding in or near the region of specificity. In the case of Ty3, this protein is a component of the Pol III transcription complex that binds to the promoter of Pol III genes. For the Ty1 element, a chromatin protein required to confer the silencing properties of these three DNA regions is probably responsible for tethering the integration complex (8). In contrast to Ty1 and Ty3, Ty5 is found only near telomeres or the *HML* and *HMR* mating loci, suggesting that silent chromatin also targets Ty5 integration (143). Interestingly, the specificity in this case has been mapped to a single amino acid lying in the interface between integrase and reverse transcriptase (143). When mutated, this amino acid reduces the number of Ty5 transpositions by 20-fold, and the site specificity of insertion is lost. Moreover, the specificity of Ty5 depends on Sir4 protein (140). For example, in *sir3* or *sir4* mutants, insertions in the mentioned loci decreased almost 10-fold and most new insertions were found dispersed throughout the genome. But more interestingly, in the mutant Sir4–45, which redistributes SIR proteins to the rDNA genes, Ty5 insertions were also redirected to this region of the genome. This is the expected result if a protein from the SIR complex mediates the integration of Ty5.

The reason for insertional specificity into regions of silent chromatin by yeast retrotransposons has to be found in the small size of the yeast genome and the fact that most of the time yeast populations remain in the haploid phase. The small genome size results in very little intergenic space. Random insertions would cause mutations that would be eliminated from the population together with the element. The absence of non-site-specific elements in yeast has been attributed to this process (8). Since a boundary element has been described in the LTRs of Ty1 (32), an alternative possibility is that in some instances the insertion of these elements in silenced chromatin serves as a mechanism to stop the spreading of the silencing proteins to regions of active transcription. Bigger diploid genomes in organisms such as *Drosophila* or mammals are more permissive for the presence and movement of transposable elements and, therefore, elements showing non-site-specificity in their insertion are also found in these organisms.

Site-Specific versus Non-Site-Specific Endonucleases: How Do Genomes Grow Bigger?

Non-LTR retrotransposons are structurally less complex than LTR retrotransposons because they lack long terminal repeats. Thus, reverse transcription is not facilitated by LTRs and needs to be primed directly by the chromosomal DNA in the site of insertion. To complete transposition, non-LTR elements depend on an encoded endonuclease that directly nicks the target DNA, which is then used by the reverse transcriptase to prime the synthesis of the cDNA using the RNA from the element as a template (73). This process, known as target-primed reverse transcription, is completed by cellular repair mechanisms. Some endonucleases are highly specific in the selection of the integration site, whereas others can be completely sloppy with a total absence of insertion-site specificity and, on occasion, without regard for the nature of the RNA being reverse transcribed and integrated. This sloppiness can have a profound impact on the evolution of the genome.

Non-LTR retrotransposons are one of the more abundant classes of transposable elements in the genome and are widely distributed among all major taxonomic groups, including fungi, animals, plants, and protozoa. Recently, Malik et al. (77) have elaborated a phylogeny of non-LTR retrotransposons using all conserved reverse transcriptase motives. The 12 extant clades of non-LTR retrotransposons can be divided in two major groups. The first group includes four clades: *CRE* from trypanosomes, *R2* and *R4* from insects, and *NeSL-1* from *C. elegans*. This group is composed of elements encoding a restriction enzyme-like (REL) site-specific endonuclease (136). *CRE* and *NeSL* are site-specific elements that integrate in the 3′ end of the splice-leader sequences from trypanosomes and *C. elegans,* respectively, whereas *R1* and *R4* are also site-specific retrotransposons whose insertion sites are found at two different places in the 28S rDNA genes from arthropods. The second group includes most of the known non-LTR retrotransposons and contains the other eight clades (*LINE-1* from humans; *RTE* from *C. elegans; Tad1* from *Neurospora crassa*; *R1, LOA,* I, and *Jockey* from *Drosophila*; and *CR1* from chicken). This group differs from the former group by the loss of the REL-endonuclease and the acquisition of a non-site-specific apurinic/apyrimidinic (AP) nuclease, which results in a loss of the target site specificity and explains part of the sloppiness mentioned above (42). One of the more substantial contributions of this work is the conclusion that the phylogenetic distribution of elements in relation to their respective hosts can be explained without the necessity to assume horizontal transfer between species (77). This assumption, together with the fact that these elements date from at least early eukaryotes, provides a unique opportunity to evaluate the impact of these retrotransposons on the evolution of the genomes they inhabit, because it excludes the possibility that elements of recent origin would have invaded the full spectra of taxonomic groups by horizontal transmission.

To explain the relative success of non-site-specific elements compared with the more reduced group of site-specific elements, it has been suggested that non-LTR site-specific elements evolved from group II introns to suit the requirement of a compact genome, whereas the increasing size of eukaryotic nuclear genomes may have allowed the original site-specific non-LTR elements to exploit random insertions without the deleterious consequences otherwise found in small genomes (136). The trend for bigger genomes among eukaryotes explains why non-site-specific elements are much more abundant within the different taxa. This argument is in line with our previous discussion and with the claims made by several authors on the pressure that the genome size exerts on the specificity for transposable element insertion (8, 25). Notwithstanding, we suggest an alternative hypothesis that would explain the expansion of non-site-specific elements among eukaryotes and also the increase in genome size that accompanied the evolution of eukaryotes. This hypothesis upholds that non-site-specific non-LTR retrotransposons, among other factors, were the motors of eukaryotic evolution, determining the initial increase in genome size, starting from originally compacted and small genomes.

A strategy consisting of the addition of noncoding sequences to a compact genome to increase its size will not be selected favorably in a world in which fast replication means better chances for reproduction. If a bigger genome size were the product of a selective trend, it would probably consist of the addition of coding units providing useful new functions to the cells and still resulting in a compact genome. Such a trend will probably conduce to a genome equivalent to that of *S. cerevisiae,* with small intergenic distances, where insertion-site specificity will be still favored. Alternatively, the increase in genome size could be the consequence of a persistent nonselective trend originated by the expansion of repetitive elements such as non-site-specific non-LTR retrotransposons. Such an effect of transposition could determine the extinction of the particular lineage or eventually could become positively selected because of the acquired potential of the bigger genome to develop evolutionary novelties. The expansion in genome size will be due not only to the spreading of insertions but also to a series of side effects of retrotransposition associated with non-LTR retrotransposons that have come to light during the past few years (40, 83, 84). In an experiment designed to prove that *LINE-1* was actively moving in HeLa cells, results from the analysis of several *LINE-1* insertions indicated that the target sequence duplications induced upon insertion could reach more than 200 bp (84). These long duplications of the target site could contribute to the addition of new sequences to the genome if transpositions are frequent events. More interestingly, recently published results suggest that *LINE-1* can retrotranspose 3′-flanking genomic sequences to the new insertion site (83). Apparently, the endogenous *LINE-1* poly (A) signal is frequently read through and transcription proceeds until a stronger polyadenylation signal in the 3′ flanking DNA is found. This mechanism can actually be of great importance, not only because it contributes to increasing the amount of DNA in the genome but also because it provides a mechanism of gene shuffling when a coding sequence is transduced and therefore could generate new valuable functions for the cell (9, 83). It has been estimated that the amount of DNA originated in such a way in humans is about 0.6% of the total genome (47, 98). On the other hand, it has been always suspected that the presence of reverse transcriptase from *LINE-1* was the cause of retrotransposition of sequences such as SINES and pseudogenes that are devoid of their own retrotranscriptase (132). Esnault et al. (40) performed an experiment in which human cells were cotransformed with a *LINE-1* element driven by a strong promoter (as a source of reverse transcriptase), together with a reporter gene containing a selectable marker interrupted by an intron. If, after several rounds of cell replication, the reporter gene was expressed in a cell, it was considered the product of a retrotransposition, a process necessary to splice the intron and make possible the expression of the gene. The frequency of retrotransposition of the reporter gene was 10^{-4}, significantly higher than the 10^{-6} background frequency found when *LINE-1* reverse transcriptase is not present. This experiment demonstrates that the reverse transcriptase from *LINE-1* can retrotranspose a non-*LINE-1* mRNA, suggesting that it is also the source of the ubiquitous human pseudogenes and the source of the more than 300,000 copies of SINES, which account for more than 17% of the human genome (132). Certainly, non-LTR non-site-specific retrotransposons were not the only factors contributing to the changing size of the genome. Structures such as centromeres and telomeres and the proliferation of intervening sequences like introns played a major role in determining the size of the genome. Centromeres and telomeres permitted the presence of multiple linear chromosomes in a single cell, and introns added noncoding sequence to the genome, providing the opportunity for creating new functions by reshuffling preexisting ones. Recent data support the idea that telomeres and spliceosomal introns could also have evolved from retrotransposons.

Telomeres are composed of short tandem repeated DNA sequences that are replicated and maintained by a protein complex called telomerase, which uses RNA as a template for nucleotide polymerization (48). It has been recently shown that the catalytic unit responsible for the DNA polymerization of telomeres is a reverse transcriptase that contains all the conserved domains present also in reverse transcriptases from retrotransposons (88). Although it seems clear that telomerase RTs are closely related to non-LTR retrotransposons and group II introns from bacteria and yeast mtDNA, the precise origin of telomerase RTs is difficult to ascertain using phylogenetic analysis (37, 87). It is not evident, for example, whether telomerase RTs originated from a common ancestor of non-LTR retrotransposons or, on the contrary, telomerase RT also originated from non-LTR retrotransposons. Considering that non-LTR retrotransposons share the same target-primed reverse transcription mechanism as reverse transcriptases from telomerases and that non-LTR retrotransposons such as Het-A and TART can function in keeping the integrity of telomeres in *Drosophila* (telomerases are not described in *Drosophila*) (3, 71, 95), it seems reasonable to think that telomerase RTs were derived from an ancestor that was a non-LTR retrotransposon.

On the other hand, new evidence strongly suggests that group II introns are the evolutionary precursors of spliceosomal introns. Group II introns have been well characterized using yeast mitochondrial introns aI1 and aI2 from the *COXI* gene and in the intron L1.LtrB from the bacterium *Lactococcus lactis* (24, 141). They encode a protein with maturase, reverse transcriptase, and endonuclease activities. After splicing the intron via a lariat intermediate, the encoded protein remains bound to the excised intron lariat RNA to form a DNA endonuclease that mediates the site-specific insertion of the intron RNA by target-primed reverse transcription. This process is known as retrohoming and involves the insertion of the intron into a double-stranded DNA corresponding to an intronless allele of the same original gene (24, 142). It was suggested early that group II introns were the ancestors of spliceosomal introns because both use an RNA-catalyzed splicing reaction via lariat intermediates (110). A link between spliceosomal introns and group II introns came from the observation of ectopic insertions of mitochondrial group II introns into non-cognate genes of the mitocondrial chromosome (85, 108). However, due to the site specificity of the retrohoming reaction, the mechanism by which group II introns could transpose to ectopic sites was unclear. Two mechanisms have been suggested: a reverse splicing into RNA, followed by reverse transcription and recombination into the DNA of the ectopic gene, or, alternatively, illegitimate retrohoming (85, 108, 137). Recently, Cousineau et al. (23) reported experimental evidence indicating that the first mechanism is involved in a series of retrotranspositions of group II introns into ectopic genes. In these experiments, the intron lariat inserts into the RNA transcripts of the unrelated gene. After insertion, the RNA containing the group II intron is reverse transcribed and integrated back into the genome by a recombination-dependent mechanism. The insertions of group II introns in other genes are always in the sense strand and are properly spliced thereafter. Evidence for the second mechanism has also been obtained recently (30; see also chapter 31), suggesting that both mechanisms may have contributed to the spread of group II introns and spliceosomal introns in the eukaryote genome.

CONCLUSIONS

Without transposable elements we would not be here and the living world would probably look very different from the one we know. Twenty years after the controversy initiated by Doolittle and Sapienza (33) and Orgel and Crick (92), we suggest that transposable elements should be considered structural components of the genome, in the same category as exons, promoters, enhancers, telomeres, centromeres, introns, satellites, etc. Transposable elements, like other genome structures, are maintained by a variety of mechanisms that are conserved along the phylogenetic scale. Genomes are huge operation centers that orchestrate cell function by controlling gene expression and coordinating the activities of all its components. The recent development of a variety of genome projects allows, for the first time, the global comparison of the encoded protein sequences in different taxa. For example, 30% of the fly genes (about 4,000) have putative orthologs in the worm genome, as deduced by sequence similarity. These genes encode proteins that probably perform common functions in both organisms. The other 9,500 genes do not show enough sequence similarity to be identified as orthologs. Similar numbers are found after comparing flies with yeast or with humans. In the end, a large number of functions have diverged between both organisms. Because the total number of genes is also different (13,600 in flies and 18,400 in worms), new functions must have appeared, other functions might have been carried out by different proteins, and, finally, some functions were probably lost since the time the last common ancestor of worms and flies split to originate both lineages. Reverse transcriptases and integrases are part of the proteins whose functions and sequences are conserved between yeast, worms, humans, and flies. Transposable elements, by virtue of the properties discussed here, have facilitated the shuffling and reorganization of the genome necessary for the magnitude of protein evolution observed when distant genomes such as yeast, worms, humans, and flies are compared.

The short-term consequence of introducing transposable elements in the genome is probably a reduction in the relative fitness when competing with cells lacking them. Evolution, however, has dictated the presence of transposable elements in virtually all genomes. Certainly, any sequence with the ability to amplify itself moving from site to site in the genome is conceptually closer to a genomic parasite than to a functional component of the cell. Accordingly, the selfish DNA point of view sustains that mobile DNA can perpetuate itself in the genome without providing any benefit to the host. Facing this hypothesis we propose an alternative view suggesting that only the cells carrying transposable elements were able to promote the evolutionary change necessary to undergo the transformation from a simpler early eukaryote genome to the current, more complex, eukaryote genome. This evolutionary switch to complex eukaryote genomes carrying transposable elements early in evolution could also explain the wide distribution of mobile DNA among eukaryotes. In this context, we have already discussed evidence showing that since the early origin of eukaryotic cells, transposable elements have contributed at least to the generation of telomeres and possibly to the origin of introns. They could also have contributed to the evolution of proteins controlling gene expression at the level of chromatin structure and to the shaping of genomes into the final result that we see today. They could also be the original targets for a cellular defense mechanism that could have been later used to fend off viral infections. Due to space constraints, we have failed to discuss additional important roles of transposable elements such as the generation of new patterns of gene expression and how efficiently they can repair double-strand breaks in the absence of alternative mechanisms (64, 67). It should nevertheless be clear from the issues discussed above that transposable elements have been important contributors to the evolution of the genome and to the development of novel cellular functions.

REFERENCES

1. **Bailey, J. A., L. Carrel, A. Chakravarti, and E. E. Eichler.** 2000. From the cover: molecular evidence for a relationship between *LINE-1* elements and X chromosome inactivation: the Lyon repeat hypothesis. *Proc. Natl. Acad. Sci. USA* **97:** 6634–6639.
2. **Bestor, T. H.** 1998. Gene silencing. Methylation meets acetylation. *Nature* **393:**311–312.
3. **Biessmann, H., J. M. Mason, K. Ferry, M. d'Hulst, K. Valgeirsdottir, K. L. Traverse, and M. L. Pardue.** 1990. Addition of telomere-associated HeT DNA sequences "heals" broken chromosome ends in *Drosophila. Cell* **61:**663–673.
4. **Bigot, Y., M. H. Hamelin, P. Capy, and G. Periquet.** 1994. *Mariner*-like elements in hymenopteran species: insertion site and distribution. *Proc. Natl. Acad. Sci. USA* **91:**3408–3412.
5. **Bingham, P. M.** 1997. Cosuppression comes to the animals. *Cell* **90:**385–387.
6. **Birchler, J. A., U. Bhadra, L. Rainbow, R. Linsk, and A. T. Nguyen-Huynh.** 1994. *Weakener of white (Wow)*, a gene that modifies the expression of the white eye color locus and that suppresses position effect variegation in *Drosophila melanogaster. Genetics* **137:**1057–1070.

7. **Birchler, J. A., M. Pal-Bhadra, and U. Bhadra.** 1999. Less from more: cosuppression of transposable elements. *Nat. Genet.* **21:**148–149.
8. **Boeke, J. D., and S. E. Devine.** 1998. Yeast retrotransposons: finding a nice quiet neighborhood. *Cell* **93:**1087–1089.
9. **Boeke, J. D., and O. K. Pickeral.** 1999. Retroshuffling the genomic deck. *Nature* **398:**108–109, 111.
10. **Brookman, J. J., A. T. Toosy, L. S. Shashidhara, and R. A. White.** 1992. The *412* retrotransposon and the development of gonadal mesoderm in *Drosophila. Development* **116:** 1185–1192.
11. **Brutnell, T. P., and S. L. Dellaporta.** 1994. Somatic inactivation and reactivation of Ac associated with changes in cytosine methylation and transposase expression. *Genetics* **138:** 213–225.
12. **Bushman, F. D.** 1994. Tethering human immunodeficiency virus 1 integrase to a DNA site directs integration to nearby sequences. *Proc. Natl. Acad. Sci. USA* **91:**9233–9237.
13. **Bushman, F. D., and M. D. Miller.** 1997. Tethering human immunodeficiency virus type 1 preintegration complexes to target DNA promotes integration at nearby sites. *J. Virol.* **71:** 458–464.
14. **Capy, P., R. Vitalis, T. Langin, D. Higuet, and C. Bazin.** 1996. Relationships between transposable elements based upon the integrase-transposase domains: is there a common ancestor? *J. Mol. Evol.* **42:**359–368.
15. **Carteau, S., C. Hoffmann, and F. Bushman.** 1998. Chromosome structure and human immunodeficiency virus type 1 cDNA integration: centromeric alphoid repeats are a disfavored target. *J. Virol.* **72:**4005–4014.
16. **Cavarec, L., S. Jensen, J. F. Casella, S. A. Cristescu, and T. Heidmann.** 1997. Molecular cloning and characterization of a transcription factor for the *copia* retrotransposon with homology to the BTB-containing *lola* neurogenic factor. *Mol. Cell. Biol.* **17:**482–494.
17. **Cavarec, L., S. Jensen, and T. Heidmann.** 1994. Identification of a strong transcriptional activator for the *copia* retrotransposon responsible for its differential expression in *Drosophila hydei* and *melanogaster* cell lines. *Biochem. Biophys. Res. Commun.* **203:**392–399.
18. **Chaboissier, M. C., A. Bucheton, and D. J. Finnegan.** 1998. Copy number control of a transposable element, the *I* factor, a *LINE*-like element in *Drosophila. Proc. Natl. Acad. Sci. USA* **95:**11781–11785.
19. **Chalvet, F., L. Teysset, C. Terzian, N. Prud'homme, P. Santamaria, A. Bucheton, and A. Pelisson.** 1999. Proviral amplification of the *gypsy* endogenous retrovirus of *Drosophila melanogaster* involves env-independent invasion of the female germline. *EMBO J.* **18:**2659–2669.
20. **Clark, J. B., P. C. Kim, and M. G. Kidwell.** 1998. Molecular evolution of *P* transposable elements in the genus *Drosophila.* III. The *melanogaster* species group. *Mol. Biol. Evol.* **15:** 746–755.
21. **Conte, D., Jr., E. Barber, M. Banerjee, D. J. Garfinkel, and M. J. Curcio.** 1998. Posttranslational regulation of *Ty1* retrotransposition by mitogen-activated protein kinase Fus3. *Mol. Cell. Biol.* **18:**2502–2513.
22. **Conte, D., Jr., and M. J. Curcio.** 2000. *Fus3* controls Ty1 transpositional dormancy through the invasive growth MAPK pathway. *Mol. Microbiol.* **35:**415–427.
23. **Cousineau, B., S. Lawrence, D. Smith, and M. Belfort.** 2000. Retrotransposition of a bacterial group II intron. *Nature* **404:** 1018–1021.
24. **Cousineau, B., D. Smith, S. Lawrence-Cavanagh, J. E. Mueller, J. Yang, D. Mills, D. Manias, G. Dunny, A. M. Lambowitz, and M. Belfort.** 1998. Retrohoming of a bacterial group II intron: mobility via complete reverse splicing, independent of homologous DNA recombination. *Cell* **94:**451–462.
25. **Craigie, R.** 1992. Hotspots and warm spots: integration specificity of retroelements. *Trends Genet.* **8:**187–190.
26. **Curcio, M. J., A. M. Hedge, J. D. Boeke, and D. J. Garfinkel.** 1990. Ty RNA levels determine the spectrum of retrotransposition events that activate gene expression in *Saccharomyces cerevisiae. Mol. Gen. Genet.* **220:**213–221.
27. **Curcio, M. J., N. J. Sanders, and D. J. Garfinkel.** 1998. Transpositional competence and transcription of endogenous Ty elements in *Saccharomyces cerevisiae:* implications for regulation of transposition. *Mol. Cell. Biol.* **8:**3571–3581.
28. **Dalmay, T., A. Hamilton, S. Rudd, S. Angell, and D. C. Baulcombe.** 2000. An RNA-dependent RNA polymerase gene in *Arabidopsis* is required for posttranscriptional gene silencing mediated by a transgene but not by a virus. *Cell* **101:** 543–553.
29. **Devine, S. E., and J. D. Boeke.** 1996. Integration of the yeast retrotransposon Ty1 is targeted to regions upstream of genes transcribed by RNA polymerase III. *Genes Dev.* **10:**620–633.
30. **Dickson, L., L. Liu, M. Matsuura, A. M. Lambowitz, and P. S. Perlman.** 2001. Retrotransposition of a yeast group II intron occurs by reverse splicing directly into ectopic DNA sites. *Proc. Natl. Acad. Sci. USA* **98:**13207–13212.
31. **Doak, T. G., F. P. Doerder, C. L. Jahn, and G. Herrick.** 1994. A proposed superfamily of transposase genes: transposon-like elements in ciliated protozoa and a common "D35E" motif. *Proc. Natl. Acad. Sci. USA* **91:**942–946.
32. **Donze, D., C. R. Adams, J. Rine, and R. T. Kamakaka.** 1999. The boundaries of the silenced HMR domain in *Saccharomyces cerevisiae. Genes Dev.* **13:**698–708.
33. **Doolittle, W. F., and C. Sapienza.** 1980. Selfish genes, the phenotype paradigm and genome evolution. *Nature* **284:** 601–603.
34. **Dorer, D. R., and S. Henikoff.** 1997. Transgene repeat arrays interact with distant heterochromatin and cause silencing in *cis* and *trans. Genetics* **147:**1181–1190.
35. **Du, C., M. E. McGuffin, B. Dauwalder, L. Rabinow, and W. Mattox.** 1998. Protein phosphorylation plays an essential role in the regulation of alternative splicing and sex determination in *Drosophila. Mol. Cell* **2:**741–750.
36. **Eickbush, T. H.** 2000. Molecular biology. Introns gain ground. *Nature* **404:**940–941, 943.
37. **Eickbush, T. H.** 1997. Telomerase and retrotransposons: which came first? *Science* **277:**911–912.
38. **Engelman, A., K. Mizuuchi, and R. Craigie.** 1991. HIV-1 DNA integration: mechanism of viral DNA cleavage and DNA strand transfer. *Cell* **67:**1211–1221.
39. **Engels, W. R.** 1992. The origin of *P* elements in *Drosophila melanogaster. Bioessays* **14:**681–686.
40. **Esnault, C., J. Maestre, and T. Heidmann.** 2000. Human *LINE* retrotransposons generate processed pseudogenes. *Nat. Genet.* **24:**363–367.
41. **Fanti, L., D. R. Dorer, M. Berloco, S. Henikoff, and S. Pimpinelli.** 1998. Heterochromatin protein 1 binds transgene arrays. *Chromosoma* **107:**286–292.
42. **Feng, Q., J. V. Moran, H. H. Kazazian, Jr., and J. D. Boeke.** 1996. Human *L1* retrotransposon encodes a conserved endonuclease required for retrotransposition. *Cell* **87:**905–916.
43. **Fire, A.** 1999. RNA-triggered gene silencing. *Trends Genet.* **15:**358–363.
44. **Fire, A., S. Xu, M. K. Montgomery, S. A. Kostas, S. E. Driver, and C. C. Mello.** 1998. Potent and specific genetic interference by double-stranded RNA in *Caenorhabditis elegans. Nature* **391:**806–811.
45. **Gerasimova, T. I., and V. G. Corces.** 1998. Polycomb and

trithorax group proteins mediate the function of a chromatin insulator. *Cell* **92:**511–521.

46. **Giniger, E., K. Tietje, L. Y. Jan, and Y. N. Jan.** 1994. *lola* encodes a putative transcription factor required for axon growth and guidance in *Drosophila. Development* **120:** 1385–1398.
47. **Goodier, J. L., E. M. Ostertag, and H. H. Kazazian, Jr.** 2000. Transduction of 3′-flanking sequences is common in *L1* retrotransposition. *Hum. Mol. Genet.* **9:**653–657.
48. **Greider, C. W., and E. H. Blackburn.** 1987. The telomere terminal transferase of *Tetrahymena* is a ribonucleoprotein enzyme with two kinds of primer specificity. *Cell* **51:** 887–898.
49. **Grunstein, M.** 1997. Histone acetylation in chromatin structure and transcription. *Nature* **389:**349–352.
50. **Hama, C., Z. Ali, and T. B. Kornberg.** 1990. Region-specific recombination and expression are directed by portions of the *Drosophila* engrailed promoter. *Genes Dev.* **4:**1079–1093.
51. **Hamilton, A. J., and D. C. Baulcombe.** 1999. A species of small antisense RNA in posttranscriptional gene silencing in plants. *Science* **286:**950–952.
52. **Hammond, S. M., E. Bernstein, D. Beach, and G. J. Hannon.** 2000. An RNA-directed nuclease mediates post-transcriptional gene silencing in *Drosophila* cells. *Nature* **404:** 293–296.
53. **Hartl, D. L., A. R. Lohe, and E. R. Lozovskaya.** 1997. Modern thoughts on an ancyent marinere: function, evolution, regulation. *Annu. Rev. Genet.* **31:**337–358.
54. **Henikoff, S.** 1998. Conspiracy of silence among repeated transgenes. *Bioessays* **20:**532–535.
55. **Hirochika, H., K. Sugimoto, Y. Otsuki, H. Tsugawa, and M. Kanda.** 1996. Retrotransposons of rice involved in mutations induced by tissue culture. *Proc. Natl. Acad. Sci. USA* **93:** 7783–7788.
56. **Inouye, S., and M. Inouye.** 1993. The retron: a bacterial retroelement required for the synthesis of msDNA. *Curr. Opin. Genet. Dev.* **3:**713–718.
57. **Jacobsen, S. E., and E. M. Meyerowitz.** 1997. Hypermethylated SUPERMAN epigenetic alleles in *arabidopsis. Science* **277:**1100–1103.
58. **Jensen, S., M. P. Gassama, and T. Heidmann.** 1999. Cosuppression of *I* transposon activity in *Drosophila* by I-containing sense and antisense transgenes. *Genetics* **153:**1767–1774.
59. **Jensen, S., M. P. Gassama, and T. Heidmann.** 1999. Taming of transposable elements by homology-dependent gene silencing. *Nat. Genet.* **21:**209–212.
60. **Jones, P. L., G. J. Veenstra, P. A. Wade, D. Vermaak, S. U. Kass, N. Landsberger, J. Strouboulis, and A. P. Wolffe.** 1998. Methylated DNA and MeCP2 recruit histone deacetylase to repress transcription. *Nat. Genet.* **19:**187–191.
61. **Ke, N., P. A. Irwin, and D. F. Voytas.** 1997. The pheromone response pathway activates transcription of *Ty5* retrotransposons located within silent chromatin of *Saccharomyces cerevisiae. EMBO J.* **16:**6272–6280.
62. **Ketting, R. F., T. H. Haverkamp, H. G. van Luenen, and R. H. Plasterk.** 1999. *Mut-7* of *C. elegans,* required for transposon silencing and RNA interference, is a homolog of Werner syndrome helicase and RNaseD. *Cell* **99:**133–141.
63. **Ketting, R. F., and R. H. Plasterk.** 2000. A genetic link between co-suppression and RNA interference in *C. elegans. Nature* **404:**296–298.
64. **Kidwell, M. G., and D. Lisch.** 1997. Transposable elements as sources of variation in animals and plants. *Proc. Natl. Acad. Sci. USA* **94:**7704–7711.
65. **Kirchner, J., C. M. Connolly, and S. B. Sandmeyer.** 1995. Requirement of RNA polymerase III transcription factors for in vitro position-specific integration of a retroviruslike element. *Science* **267:**1488–1491.
66. **Kumar, A., and J. L. Bennetzen.** 1999. Plant retrotransposons. *Annu. Rev. Genet.* **33:**479–532.
67. **Labrador, M., and V. G. Corces.** 1997. Transposable element-host interactions: regulation of insertion and excision. *Annu. Rev. Genet.* **31:**381–404.
68. **Labrador, M., M. D. C. Seleme, and A. Fontdevila.** 1998. The evolutionary history of *Drosophila buzzatii.* XXXIV. The distribution of the retrotransposon *Osvaldo* in original and colonizing populations. *Mol. Biol. Evol.* **15:**1532–1547.
69. **Lecher, P., A. Bucheton, and A. Pelisson.** 1997. Expression of the *Drosophila* retrovirus *gypsy* as ultrastructurally detectable particles in the ovaries of flies carrying a permissive *flamenco* allele. *J. Gen. Virol.* **78:**2379–2388.
70. **Lee, M. S., and R. Craigie.** 1998. A previously unidentified host protein protects retroviral DNA from autointegration. *Proc. Natl. Acad. Sci. USA* **95:**1528–1533.
71. **Levis, R. W., R. Ganesan, K. Houtchens, L. A. Tolar, and F. M. Sheen.** 1993. Transposons in place of telomeric repeats at a *Drosophila* telomere. *Cell* **75:**1083–1093.
72. **Liao, G. C., E. J. Rehm, and G. M. Rubin.** 2000. Insertion site preferences of the *P* transposable element in *Drosophila melanogaster. Proc. Natl. Acad. Sci. USA* **97:**3347–3351.
73. **Luan, D. D., M. H. Korman, J. L. Jakubczak, and T. H. Eickbush.** 1993. Reverse transcription of *R2Bm* RNA is primed by a nick at the chromosomal target site: a mechanism for non-LTR retrotransposition. *Cell* **72:**595–605.
74. **Luff, B., L. Pawlowski, and J. Bender.** 1999. An inverted repeat triggers cytosine methylation of identical sequences in *Arabidopsis. Mol. Cell* **3:**505–511.
75. **Lyon, M. F.** 2000. *LINE-1* elements and X chromosome inactivation: a function for "junk" DNA? *Proc. Natl. Acad. Sci. USA* **97:**6248–6249.
76. **Madhani, H. D., C. A. Styles, and G. R. Fink.** 1997. MAP kinases with distinct inhibitory functions impart signaling specificity during yeast differentiation. *Cell* **91:**673–684.
77. **Malik, H. S., W. D. Burke, and T. H. Eickbush.** 1999. The age and evolution of non-LTR retrotransposable elements. *Mol. Biol. Evol.* **16:**793–805.
78. **Marin, I., A. Franke, G. J. Bashaw, and B. S. Baker.** 1996. The dosage compensation system of *Drosophila* is co-opted by newly evolved X chromosomes. *Nature* **383:**160–163.
79. **Menees, T. M., and S. B. Sandmeyer.** 1996. Cellular stress inhibits transposition of the yeast retrovirus-like element *Ty3* by a ubiquitin-dependent block of virus-like particle formation. *Proc. Natl. Acad. Sci. USA* **93:**5629–5634.
80. **Misquitta, L., and B. M. Paterson.** 1999. Targeted disruption of gene function in *Drosophila* by RNA interference (RNA-i): a role for nautilus in embryonic somatic muscle formation. *Proc. Natl. Acad. Sci. USA* **96:**1451–1456.
81. **Modolell, J., W. Bender, and M. Meselson.** 1983. *Drosophila melanogaster* mutations suppressible by the *suppressor of Hairy-wing* are insertions of a 7.3-kilobase mobile element. *Proc. Natl. Acad. Sci. USA* **80:**1678–1682.
82. **Montgomery, E. A., S. M. Huang, C. H. Langley, and B. H. Judd.** 1991. Chromosome rearrangement by ectopic recombination in *Drosophila melanogaster:* genome structure and evolution. *Genetics* **129:**1085–1098.
83. **Moran, J. V., R. J. DeBerardinis, and H. H. Kazazian, Jr.** 1999. Exon shuffling by *L1* retrotransposition. *Science* **283:** 1530–1534.
84. **Moran, J. V., S. E. Holmes, T. P. Naas, R. J. DeBerardinis, J. D. Boeke, and H. H. Kazazian, Jr.** 1996. High frequency retrotransposition in cultured mammalian cells. *Cell* **87:** 917–927.

85. **Mueller, M. W., M. Allmaier, R. Eskes, and R. J. Schweyen.** 1993. Transposition of group II intron aI1 in yeast and invasion of mitochondrial genes at new locations. *Nature* **366:** 174–176.
86. **Murray, M. V., M. A. Turnage, K. J. Williamson, W. R. Steinhauer, and L. L. Searles.** 1997. The *Drosophila* suppressor of sable protein binds to RNA and associates with a subset of polytene chromosome bands. *Mol. Cell. Biol.* **17:** 2291–2300.
87. **Nakamura, T. M., and T. R. Cech.** 1998. Reversing time: origin of telomerase. *Cell* **92:**587–590.
88. **Nakamura, T. M., G. B. Morin, K. B. Chapman, S. L. Weinrich, W. H. Andrews, J. Lingner, C. B. Harley, and T. R. Cech.** 1997. Telomerase catalytic subunit homologs from fission yeast and human. *Science* **277:**955–959.
89. **Nan, X., H. H. Ng, C. A. Johnson, C. D. Laherty, B. M. Turner, R. N. Eisenman, and A. Bird.** 1998. Transcriptional repression by the methyl-CpG-binding protein MeCP2 involves a histone deacetylase complex. *Nature* **393:**386–389.
90. **Ngo, H., C. Tschudi, K. Gull, and E. Ullu.** 1998. Double-stranded RNA induces mRNA degradation in Trypanosoma brucei. *Proc. Natl. Acad. Sci. USA* **95:**14687–14692.
91. **O'Neill, R. J., M. J. O'Neill, and J. A. Graves.** 1998. Undermethylation associated with retroelement activation and chromosome remodelling in an interspecific mammalian hybrid. *Nature* **393:**68–72.
92. **Orgel, L. E., and F. H. Crick.** 1980. Selfish DNA: the ultimate parasite. *Nature* **284:**604–607.
93. **Pal-Bhadra, M., U. Bhadra, and J. A. Birchler.** 1997. Cosuppression in *Drosophila:* gene silencing of *Alcohol dehydrogenase* by *white-Adh* transgenes is *Polycomb* dependent. *Cell* **90:**479–490.
94. **Pal-Bhadra, M., U. Bhadra, and J. A. Birchler.** 1999. Cosuppression of nonhomologous transgenes in *Drosophila* involves mutually related endogenous sequences. *Cell* **99:** 35–46.
95. **Pardue, M. L., O. N. Danilevskaya, K. Lowenhaupt, F. Slot, and K. L. Traverse.** 1996. *Drosophila* telomeres: new views on chromosome evolution. *Trends Genet.* **12:**48–52.
96. **Parkhurst, S. M., and V. G. Corces.** 1987. Developmental expression of *Drosophila melanogaster* retrovirus-like transposable elements. *EMBO J.* **6:**419–424.
97. **Pelisson, A., S. U. Song, N. Prud'homme, P. A. Smith, A. Bucheton, and V. G. Corces.** 1994. *Gypsy* transposition correlates with the production of a retroviral envelope-like protein under the tissue-specific control of the *Drosophila flamenco* gene. *EMBO J.* **13:**4401–4411.
98. **Pickeral, O. K., W. Makaowski, M. S. Boguski, and J. D. Boeke.** 2000. Frequent human genomic DNA transduction driven by *LINE*-1 retrotransposition. *Genome Res.* **10:** 411–415.
99. **Pirrotta, V.** 1997. Chromatin-silencing mechanisms in *Drosophila* maintain patterns of gene expression. *Trends Genet.* **13:**314–318.
100. **Prud'homme, N., M. Gans, M. Masson, C. Terzian, and A. Bucheton.** 1995. *Flamenco,* a gene controlling the *gypsy* retrovirus of *Drosophila melanogaster. Genetics* **139:**697–711.
101. **Pruss, D., R. Reeves, F. D. Bushman, and A. P. Wolffe.** 1994. The influence of DNA and nucleosome structure on integration events directed by HIV integrase. *J. Biol. Chem.* **269:** 25031–25041.
102. **Rabinow, L., S. L. Chiang, and J. A. Birchler.** 1993. Mutations at the Darkener of apricot *locus* modulate transcript levels of *copia* and *copia*-induced mutations in *Drosophila melanogaster. Genetics* **134:**1175–1185.
103. **Ratcliff, F., B. D. Harrison, and D. C. Baulcombe.** 1997. A similarity between viral defense and gene silencing in plants. *Science* **276:**1558–1560.
104. **Razin, A.** 1998. CpG methylation, chromatin structure and gene silencing—a three-way connection. *EMBO J.* **17:** 4905–4908.
105. **Rio, D. C.** 1990. Molecular mechanisms regulating *Drosophila P* element transposition. *Annu. Rev. Genet.* **24:** 543–578.
106. **Romano, N., and G. Macino.** 1992. Quelling: transient inactivation of gene expression in *Neurospora crassa* by transformation with homologous sequences. *Mol. Microbiol.* **6:** 3343–3353.
107. **Sanchez Alvarado, A., and P. A. Newmark.** 1999. Double-stranded RNA specifically disrupts gene expression during planarian regeneration. *Proc. Natl. Acad. Sci. USA* **96:** 5049–5054.
108. **Sellem, C. H., G. Lecellier, and L. Belcour.** 1993. Transposition of a group II intron. *Nature* **366:**176–178.
109. **Sezutsu, H., E. Nitasaka, and T. Yamazaki.** 1995. Evolution of the *LINE*-like *I* element in the *Drosophila melanogaster* species subgroup. *Mol. Gen. Genet.* **249:**168–178.
110. **Sharp, P. A.** 1991. "Five easy pieces." *Science* **254:**663.
111. **Simmons, M. J., J. D. Raymond, C. D. Grimes, C. Belinco, B. C. Haake, M. Jordan, C. Lund, T. A. Ojala, and D. Papermaster.** 1996. Repression of hybrid dysgenesis in *Drosophila melanogaster* by heat-shock-inducible sense and antisense *P*-element constructs. *Genetics* **144:**1529–1544.
112. **Smith, J. S., and J. D. Boeke.** 1997. An unusual form of transcriptional silencing in yeast ribosomal DNA. *Genes Dev.* **11:** 241–254.
113. **Smith, P. A., and V. G. Corces.** 1992. The suppressor of Hairy-wing binding region is required for *gypsy* mutagenesis. *Mol. Gen. Genet.* **233:**65–70.
114. **Smith, P. A., and V. G. Corces.** 1995. The suppressor of Hairy-wing protein regulates the tissue-specific expression of the *Drosophila gypsy* retrotransposon. *Genetics* **139:** 215–228.
115. **Sniegowski, P. D., and B. Charlesworth.** 1994. Transposable element numbers in cosmopolitan inversions from a natural population of *Drosophila melanogaster. Genetics* **137:** 815–827.
116. **Song, S. U., T. Gerasimova, M. Kurkulos, J. D. Boeke, and V. G. Corces.** 1994. An env-like protein encoded by a *Drosophila* retroelement: evidence that *gypsy* is an infectious retrovirus. *Genes Dev.* **8:**2046–2057.
117. **Song, S. U., M. Kurkulos, J. D. Boeke, and V. G. Corces.** 1997. Infection of the germ line by retroviral particles produced in the follicle cells: a possible mechanism for the mobilization of the *gypsy* retroelement of *Drosophila. Development* **124:**2789–2798.
118. **Spana, C., D. A. Harrison, and V. G. Corces.** 1988. The *Drosophila melanogaster* suppressor of Hairy-wing protein binds to specific sequences of the *gypsy* retrotransposon. *Genes Dev.* **2:**1414–1423.
119. **Spradling, A. C., D. M. Stern, I. Kiss, J. Roote, T. Laverty, and G. M. Rubin.** 1995. Gene disruptions using *P* transposable elements: an integral component of the *Drosophila* genome project. *Proc. Natl. Acad. Sci. USA* **92:**10824–10830.
120. **Steinemann, M., and S. Steinemann.** 1998. Enigma of Y chromosome degeneration: neo-Y and neo-X chromosomes of *Drosophila miranda* a model for sex chromosome evolution. *Genetica* **103:**409–420.
121. **Steinemann, M., and S. Steinemann.** 1997. The enigma of Y chromosome degeneration: *TRAM,* a novel retrotransposon is preferentially located on the Neo-Y chromosome of *Drosophila miranda. Genetics* **145:**261–266.

122. **Tabara, H., M. Sarkissian, W. G. Kelly, J. Fleenor, A. Grishok, L. Timmons, A. Fire, and C. C. Mello.** 1999. The *rde-1* gene, RNA interference, and transposon silencing in *C. elegans. Cell* **99:**123–132.
123. **Taillebourg, E., and J. M. Dura.** 1999. A novel mechanism for *P* element homing in *Drosophila. Proc. Natl. Acad. Sci. USA* **96:**6856–6861.
124. **Takagaki, Y., and J. L. Manley.** 1994. A polyadenylation factor subunit is the human homologue of the *Drosophila* suppressor of forked protein. *Nature* **372:**471–474.
125. **Takeda, S., K. Sugimoto, H. Otsuki, and H. Hirochika.** 1999. A 13-bp *cis*-regulatory element in the LTR promoter of the tobacco retrotransposon *Tto1* is involved in responsiveness to tissue culture, wounding, methyl jasmonate and fungal elicitors. *Plant J.* **18:**383–393.
126. **Takeda, S., K. Sugimoto, H. Otsuki, and H. Hirochika.** 1998. Transcriptional activation of the tobacco retrotransposon *Tto1* by wounding and methyl jasmonate. *Plant Mol. Biol.* **36:**365–376.
127. **Terzian, C., C. Ferraz, J. Demaille, and A. Bucheton.** 2000. Evolution of the gypsy endogenous retrovirus in the *Drosophila melanogaster* subgroup. *Mol. Biol. Evol.* **17:**908–914.
128. **Voinnet, O., Y. M. Pinto, and D. C. Baulcombe.** 1999. Suppression of gene silencing: a general strategy used by diverse DNA and RNA viruses of plants. *Proc. Natl. Acad. Sci. USA* **96:**14147–14152.
129. **Walsh, C. P., and T. H. Bestor.** 1999. Cytosine methylation and mammalian development. *Genes Dev.* **13:**26–34.
130. **Walsh, C. P., J. R. Chaillet, and T. H. Bestor.** 1998. Transcription of IAP endogenous retroviruses is constrained by cytosine methylation. *Nat. Genet.* **20:**116–117.
131. **Wei, S. Q., K. Mizuuchi, and R. Craigie.** 1998. Footprints on the viral DNA ends in moloney murine leukemia virus preintegration complexes reflect a specific association with integrase. *Proc. Natl. Acad. Sci. USA* **95:**10535–10540.
132. **Weiner, A. M.** 2000. Do all SINEs lead to LINEs? *Nat. Genet.* **24:**332–333.
133. **Wilke, C. M., and J. Adams.** 1992. Fitness effects of *Ty* transposition in *Saccharomyces cerevisiae. Genetics* **131:**31–42.
134. **Withers-Ward, E. S., Y. Kitamura, J. P. Barnes, and J. M. Coffin.** 1994. Distribution of targets for avian retrovirus DNA integration in vivo. *Genes Dev.* **8:**1473–1487.
135. **Xiong, Y., and T. H. Eickbush.** 1990. Origin and evolution of retroelements based upon their reverse transcriptase sequences. *EMBO J.* **9:**3353–3362.
136. **Yang, J., H. S. Malik, and T. H. Eickbush.** 1999. Identification of the endonuclease domain encoded by *R2* and other site-specific, non-long terminal repeat retrotransposable elements. *Proc. Natl. Acad. Sci. USA* **96:**7847–7852.
137. **Yang, J., G. Mohr, P. S. Perlman, and A. M. Lambowitz.** 1998. Group II intron mobility in yeast mitochondria: target DNA-primed reverse transcription activity of aI1 and reverse splicing into DNA transposition sites in vitro. *J. Mol. Biol.* **282:**505–523.
138. **Yoder, J. A., C. P. Walsh, and T. H. Bestor.** 1997. Cytosine methylation and the ecology of intragenomic parasites. *Trends Genet.* **13:**335–340.
139. **Yun, B., R. Farkas, K. Lee, and L. Rabinow.** 1994. The Doa *locus* encodes a member of a new protein kinase family and is essential for eye and embryonic development in *Drosophila melanogaster. Genes Dev.* **8:**1160–1173.
140. **Zhu, Y., S. Zou, D. A. Wright, and D. F. Voytas.** 1999. Tagging chromatin with retrotransposons: target specificity of the *Saccharomyces* Ty5 retrotransposon changes with the chromosomal localization of Sir3p and Sir4p. *Genes Dev.* **13:** 2738–2749.
141. **Zimmerly, S., H. Guo, R. Eskes, J. Yang, P. S. Perlman, and A. M. Lambowitz.** 1995. A group II intron RNA is a catalytic component of a DNA endonuclease involved in intron mobility. *Cell* **83:**529–538.
142. **Zimmerly, S., H. Guo, P. S. Perlman, and A. M. Lambowitz.** 1995. Group II intron mobility occurs by target DNA-primed reverse transcription. *Cell* **82:**545–554.
143. **Zou, S., N. Ke, J. M. Kim, and D. F. Voytas.** 1996. The *Saccharomyces* retrotransposon Ty5 integrates preferentially into regions of silent chromatin at the telomeres and mating *loci. Genes Dev.* **10:**634–645.

Mobile DNA II
Edited by N. L. Craig et al.

Chapter 44

Eubacterial Genomes

ALLAN CAMPBELL

INTRODUCTION

All DNA is potentially mobile. Most of this volume concerns mobile elements, which encode proteins specifically required for their own relocation. Other genes are, to various extents, mobilized by the proteins such elements encode. At the mechanistic level, this fact has long been appreciated through such examples as specialized transduction and gene incorporation into transposons or integrons.

The full extent of natural gene mobility has only begun to become apparent from comparative genomics, which indicates that genes have not only relocated within genomes but also have moved among genomes. The subject of horizontal gene transfer might seem beyond the scope of this volume. However, it is hard to ignore when some DNA segments (such as pathogenicity islands) have many of the hallmarks of known mobile elements but also show signs of having originated in hosts distant from those in which they presently reside. Eubacterial genomes are discussed first with respect to the presence of prophages and transposable elements, as well as their obvious derivatives, and finally with respect to horizontal gene transfer. Some of the phages and transposable elements are treated by other authors of this volume. This chapter covers only those aspects most relevant to their relationship with the rest of the genome.

PHAGES

General

Classified according to lifestyle, phages are either virulent (able to propagate only as infectious agents that kill and usually lyse the infected cell) or temperate

Allan Campbell • Department of Biological Sciences, Stanford University, Stanford, CA 94305.

Table 1. Types of temperate phages that insert into the chromosome

Mode of insertion	Integration enzyme	DNA in free phage particle	Preferred insertion site	Example
Conservation, site-specific recombination	Integrase family	ds, linear	tRNA	P22-
			Coding region	21
			Intergenic	λ
	Resolvase family	ds, linear	Intergenic	φC31
	Novel	ss, circular	Intergenic	φCTX
Transposition	Transposase	ds, linear	Nonspecific	M-1

(able to propagate also as part of the cell genome, transmitted vertically from mother to daughter at cell division). The latent phage genome thus transmitted is called a prophage. Prophages are either chromosomal or extrachromosomal. This chapter discusses temperate phages whose prophages are chromosomal.

Phages are found in virtually all bacterial species. Some of the variation relevant to their status as mobile elements is summarized in Table 1.

Members of the integrase and resolvase families both catalyze reactions with the same end result, although the two families are not detectably related in sequence and use somewhat different catalytic mechanisms. Temperate phages typically use integrases, whereas resolvases commonly cause specific DNA inversions or resolution of cointegrates. However, the definite exceptions suggest that this division of labor may be largely a historical accident. When enzymes of either family mediate integration, their natural substrates are a (usually circular) bacterial chromosome and a circular extrachromosomal molecule of phage DNA. Very few phages contain circular double-stranded (ds) DNA in their free infectious particles, so such DNA must be produced during intracellular development. In ds phages, this means ligation of ends (generally with complementary projecting single strands) by host ligase. Insertion of single-stranded (ss) phages requires replication to the ds form. In phage Mu-1 particles, the linear phage DNA is flanked by short stretches of host DNA, from which the phage DNA can transpose into the chromosome. Phages like P1 (not shown in Table 1 and not discussed further herein) lysogenize by establishing themselves as extrachromosomal plasmids able to replicate in harmony with the host cell.

The mechanism of conservative site-specific recombination generally entails DNA-DNA recognition of the two partners (phage and host), as well as protein recognition of the flanking sequences. As a result, each phage has a preferred insertion site on the chromosome, usually with a number of less favored sites that resemble the preferred one. At the other extreme, insertion of phage Mu-1, though not completely random, can occur almost anywhere on the chromosome.

Since most genes of an inserted prophage are silent, natural selection does little to retard their deterioration through genetic drift. Most prophages in natural strains have experienced at least one debilitating change and are therefore unable to undergo a complete lytic cycle following induction of phage development. Such prophages are called defective or cryptic.

I discuss the distribution of prophages first within one strain (*Escherichia coli* K-12), then within a species (*E. coli*), and last among species.

Distribution

Distribution within a strain

Most bacterial strains harbor one or more prophages or phage-related elements. Strain K-12 of *E. coli* seems fairly typical and has the benefit of in-depth characterization. *E. coli* K-12 was isolated from a Stanford hospital patient in 1922 and has been perpetuated in pure laboratory culture ever since. Therefore any of its descendants, such as MG1655, which was completely sequenced (5), may have lost some of the prophages K-12 contained when it was first isolated, but (unlike natural strains growing in the wild) it could not have acquired any foreign genetic elements during the 75 years between isolation and sequencing.

In 1951, E. Lederberg (26) discovered that K-12 is lysogenic for λ. The fact that it was baptized with a Greek letter, like cytoplasmic elements of *Paramecium* spp. or *Drosophila melanogaster,* reflects the state of knowledge at the time. Subsequent work, sharpened by genomic sequencing, revealed at least four other λ-related elements in the K-12 chromosomes, as well as several other segments that are homologous to other coliphages (Table 2). None of these elements constitutes a complete, rather than defective, prophage.

While other histories are possible, many of these sequences may have entered the *E. coli* genome as

Table 2. Phage-related elements of the *E. coli* K-12 chromosome[a]

Segment	Size (kb)	Coordinates[b]	Orientation[c]	Insertion site
CP4-6	34.3	262, 122–296, 489	A	*thrW* (anticodon)
DLP-12	21.4	563, 975–585, 326	A	*argU* (anticodon)
λ	48.5	806, 561–806, 575	A	intergenic
e14	15.2	1, 195, 432-1, 210, 646	A	*icd*
Rac	22.5	1, 409, 733–1432, 243	A	b1344/ b1875
Qin	≥15.5	~1, 632, 338-1, 647, 821	C	Unknown
P2 remnant	0.6	2, 165, 196-2, 165, 844	C	b2084
(*argW*)	?	~2, 464, 383-?	C	*argW*
(*eutB*)	6.8	2, 556, 711-2, 563, 508	C	*eutB*
CP4-57	22.0	2, 753, 956-2, 776, 007	C	*ssrA* (TΨC)
(*leuX*)	?	~4, 494, 037-?	C	*leuX* (TΨC)

[a]Blattner et al. (5) identify an additional element, CP4-44, because of its sequence similarity to CP4-6 and CP4-57. Since it has no integrase gene, no flanking direct repeats, and no homology with known phages, it is not included here.
[b]Coordinates are from the K-12 derivative MG-1655, which was cured of λ (5). Therefore, the coordinates of λ are given as the 15-bp identity between λ *attB* and λ *attP*. The endpoints of Rac are placed at an octamer repeat, although direct evidence is lacking. Qin has neither an integrase nor a recognized terminus, so coordinates are approximate.
[c]A, anticlockwise; C, clockwise. This is the direction of transcription of *int* on the circular *E. coli* map or (in the case of Qin and the P2 remnant) the direction *int* would have, deduced from the orientation of the phage homologs that are present.

complete prophages that subsequently suffered extensive deletions and disruptions by insertion of foreign elements, including transposons and insertion sequences. This is the expected fate of any chromosomal segment that imparts no selective value to the host. Depending on the proximity and importance of flanking genes, deletions may be predominantly either internal or terminal. Terminal deletions may remove the integrase gene and the attachment sites, rendering the margins of the element uncertain.

The sequences in K-12 are related to three separate families of phages: the λ family, the P4/φR73 family, and the P2 family. There are some characteristic differences among them that allow informed speculation about an element's origin. One of these concerns the integration system. All of these phages use members of the integrase family. Some λ-related phages insert in open reading frames (φ21), some in the anticodon loops of tRNA genes (P22), and some in intergenic DNA (λ itself). What all of these sites have in common is an imperfect, interrupted symmetry around the crossover site. When they insert into genes, the *int* gene is usually close to that end of the prophage closer to the 5′ end of the gene and is transcribed in the opposite direction from the gene, toward the crossover point. Disruption of the host gene is prevented by duplication in the phage genome of the 3′ end of the gene or operon (illustrated in Fig. 1). For insertion into an open reading frame, there are generally some synonymous codon changes that make the duplication imprecise; with tRNAs the sequence is frequently identical, so that the prophage is bracketed by a long direct repeat of more than 40 bp.

Members of the P4/φR73 family insert in or near the TψC loops of tRNA genes. No obvious symmetry is observable around the crossover site. As with the λ-related phages, the phage carries a duplication of DNA from the crossover point through the 3′ end of the tRNA gene. Since the TψC loop is closer to the 3′ end than the anticodon loop, the repeats flanking the prophage are shorter. The integrase genes of these phages are also near the prophage end closer to the complete tRNA gene, but they are transcribed in the same direction as the tRNA gene (away from the crossover point).

What follows may seem like a bill of particulars. No general interest attaches to what elements happen to be in K-12. The histories of individual elements are not subject to unique interpretation, but collectively their features focus attention on the possibilities and on the problems inherent in their study, as well as on the chromosomal localization of other insertions.

In CP4-6 (where "CP4" stands for "cryptic P4" and "6" stands for its location, in minutes, on the K-12 map), the flanking duplication extends from the anticodon loop through the 3′ end of the gene, a configuration shared by some of the λ family. Its integrase gene is transcribed in the opposite orientation from the tRNA gene *thrW*, again like a member of the λ family. It contains some genes of unknown function, with homologs in CP4-87 and CP4-44 (but not in P4), several insertion sequences, and a gene *argF* that duplicates a metabolic function encoded elsewhere on the chromosome. The G + C content of *argF* and several adjacent genes is much higher than that of *E. coli* in general; this and other sequence features had led to the characterization of these genes as alien DNA (discussed later in the chapter). One interpretation is

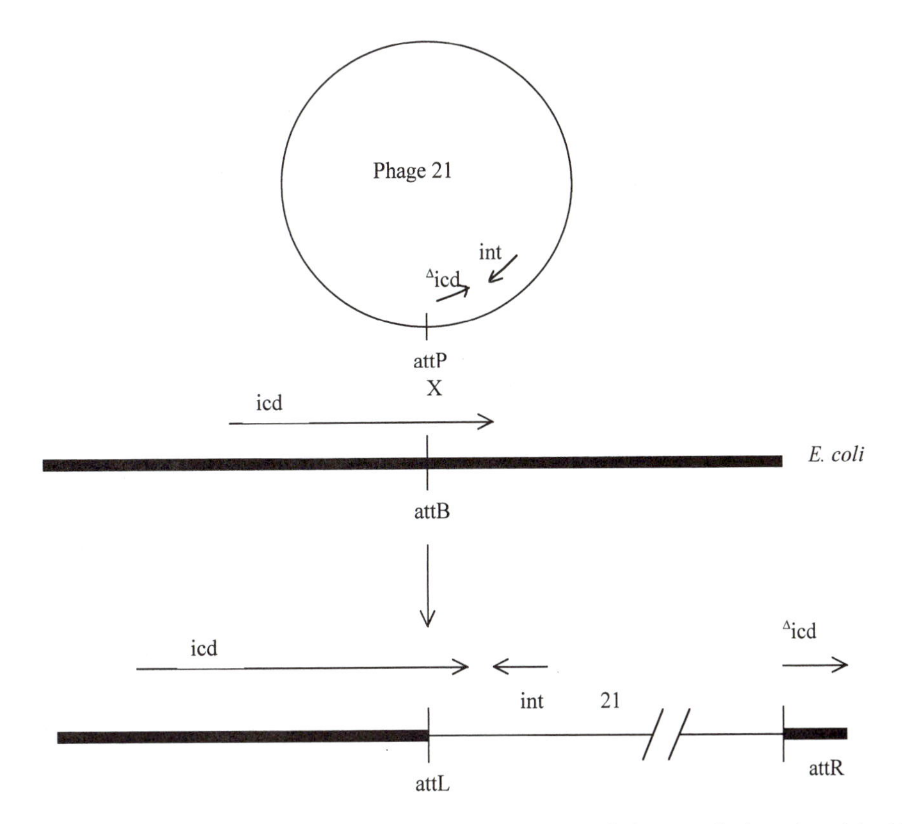

Figure 1. Insertion of phage 21 into the *icd* gene of *E. coli*. The phage DNA includes a 165-bp homolog of the 3′ end of *icd*.

that CP4-6 was once a λ relative, almost all of whose DNA has been deleted, and that extensive insertion has occurred between *int* and *attL*. CP4-6 has not been shown to excise, and its *int* gene is not known to be functional. Its insertion site is close to (perhaps identical to) that of the λ-related phage HK253.

DLP-12 ("defective lambda prophage at 12 min") is more clearly related to λ. It has a complete *int* gene, a truncated *xis* gene, analogs of λ *P, Q, S, R,* and *cos,* and *attL* and *attR* sequences, all in the same order as in λ. The *int* gene has not been shown to be functional, but the *QSR cos* segment is fully functional when rescued into λ by recombination. DLP-12 also contains insertion sequences and genes (*nmpC* and *appY*) that can serve bacterial functions and are not present in λ (27). The presence of such luxury genes in various λ-related phages will be covered later in the chapter.

e14 resembles CP4-6 in that few of its genes are obviously phage-related. It does have functional *int* and *xis* genes and a cell septation inhibitor (*sfiC*), controlled by a repressor that, like λ's, is inactivated by the SOS system in response to DNA damage. It can excise from the chromosome, and both cured chromosomes and excised circles have been observed (14). e14 is inserted at the same site within the *icd* gene as phage 21 (Fig. 1). Its *int* gene (like 21's) is transcribed toward the complete *icd* gene, but 3.5 kb of DNA separate *int* from *attL,* including a gene, *lit,* that interferes with the late development of phage T4. The rest of e14 DNA includes some genes with known activities (a restriction enzyme, *mcrA,* and an invertible *pin* region). e14 can be considered a λ-related phage that has lost much of its functions and acquired some extraneous DNA.

Rac, like DLP-12, has enough λ-homologous genes in the proper order to place it in the λ family. It was discovered as a source of recombination functions. The responsible gene (*recE*) is ordinarily repressed, but mutations or rearrangements with constitutive expression can be selected. Rac also has *int* and *xis* genes, which were demonstrated to be functional by recombinational rescue of a large segment of Rac (selecting for *recE*) into λ. The fact that the resulting phage, λ *rev,* has a substitution of λ DNA that includes the *att* site indicates that the two prophage ends must have been joined by an excision reaction and that λ *rev* itself can lysogenize and excise. Spontaneous deletion of the entire Rac prophage (precise at the level of restriction mapping and Southern hybridization) has been observed (13). Despite these facts, the exact termini of Rac are not known for certain. The coordinates on Table 1 mark an octanucleotide repeat, whose immediately flanking DNA shows little symmetry. The lysogenization frequency of λ *rev* is lower than that of λ and is further reduced in hosts deleted for the Rac prophage, as though the ancestral phage evolved to lysogenize some other *E. coli* strain and found the best available, but nonideal, site in K-12.

Like DLP-12, Qin was first recognized as a source of *Q*, *S*, and *R* analogs linked to a functional *cos*, flanked by enough λ homology to allow recombinational replacement of the corresponding λ genes. It also has homologs of *cI* and *cro* and genes (*dic*), whose products inhibit cell division. These genes have the same order and relative orientation as their λ counterparts, although the whole element is oriented oppositely from λ on the chromosome. No *int* gene or oligonucleotide repeats are known, so the exact limits are ill defined. It is what might be expected if one or both ends of a λ-like prophage had been deleted.

The P2 remnant is localized at a preferred *att* site for phage P2 and includes the P2 *ogr* gene and the 3′ end of P2 gene *D*. It is flanked by *att* sites similar to P2 *attL* and *attR* from *E. coli* C, where no such fragment is present. The b2084 open reading frame (ORF) (of unknown function) is an out-of-frame fusion with part of P2 gene *D*. The sequence at the junction point suggests that microhomologous recombination between *attL* and gene *D* caused a deletion of 98% of the prophage genome (3).

The element at *argW* is defined by an integrase gene and not much else. The integrase gene resembles the P4 family more than the λ family and, as in P4, is oriented away from, rather than toward, *argW*. The site is close or identical to the insertion site of the lambdoid phage PA-2.

A 6.8-kb element near *eutB* has an integrase similar to that of the P4 family. The boundaries have been taken as an octanucleotide repeat, and the DNA between these two repeats is missing from some *E. coli* strains. The integrase is oriented in the opposite direction from, and upstream from, *eutB*. Depending on which AUG or GUG initiates translation, the octanucleotide repeat is either upstream from *eutB* or within it, very close to the 5′ end.

CP4-57 has a P4-like integrase, with all its hallmarks: The integrase gene (originally named *slpA*, here called *int*) is close to the prophage *attL* and transcribed away from it, in the same direction as the tRNA-like gene *ssrA*, and the flanking oligonucleotide repeat extends from the TψC loop of *ssrA* through the 3′ end, with a one-nucleotide deletion in *attR* (35). Downstream of *int* is a gene (*alpA*) similar to ORF88 of P4 that activates *int*, so that *alpA* expression can induce loss of the entire element (39). When CP4-57 is lost, the one-nucleotide change in *attR* generates an altered *ssrA*, with reduced function. CP4-57 behaves in all respects like a P4-related prophage that has suffered extensive deletions and insertions of irrelevant DNA. The one base pair change in *attR* could have resulted from a mutation that happened after the element inserted. Insertion into a gene in the manner of Fig. 1 leaves the downstream repeat as a pseudogene no longer under selection.

Phage P4 inserts into the TψC region of *leuX*. Downstream of *leuX* in *E. coli* K-12 is an integrase gene very similar to that of P4, which is transcribed in the same direction as *leuX* (32). One interpretation is that a P4-like phage inserted at this site and its subsequent deletion removed most of the prophage, including *attR*.

In summary, the first six elements of Table 2 appear to be derived from phages similar to λ, although sometimes (as in CP4-6), the only remnant of λ is the integrase gene. The last four are related to P4. Since the elements are arranged in map order, it seems that (perhaps coincidentally) the phages of these two families are nonrandomly located along the chromosome.

That this nonrandom location is more than coincidence is strongly indicated by surveying not just the cryptic prophages in K-12 but the collection of λ-related plaque-forming phages that have been isolated from nature (generally on K-12 as plating host). The preferred insertion sites for all these phages lie within the half of the chromosome from 0.2 to 2.5 Mb (6, 8). Even more strikingly, 7/7 λ-related phages with insertion sites between 0.2 and 1.5 Mb are all oriented in the same direction as the first five prophages of Table 2. The one λ-related element known to be oppositely oriented (Qin) is within the terminus region, which has suggested a correlation with direction of chromosomal replication.

The reason for this nonrandomness is unknown. Three types of explanations can be considered.

1. Sampling bias can never be excluded completely. However, the phages under study were isolated at different times and places (Paris, Hong Kong, and Japan) and include *Salmonella* phage P22. In an early study (19), it was found that seven inducible prophages (which proved to be λ related) were in one part of the chromosome, whereas seven prophages that were noninducible (as is P4) mapped elsewhere.

2. History could have been critical. One (not very convincing) historical scenario is that the ancestral phage inserted at one site on the chromosome and evolved new specificities by some sort of directional walking (8). Historical scenarios are more plausible if the evolution of new site specificities happened long ago rather than being an ongoing process.

3. Selection may well be the major factor. Lysogens with prophages at secondary or translocated sites (7, 37) are stable and healthy in the laboratory, precluding strong selection. Weak selection, iterated over many generations, may be critical in the natural world. Until recently, an appeal to general principles of chromosome organization seemed too ill-defined to be constructive. However, some emerging facts suggest more specific hypotheses with heuristic value.

On the chromosomes of many bacteria (including *E. coli*), there is a consistent bias of the ratio (G-C/G+C) between leading and lagging strands (5, 29), the leading strand being more G-rich. This bias in λ reverses with the direction of transcription, but λ's orientation on the chromosome puts most of λ with the same strand bias as the chromosome it lysogenizes. The reason for the chromosomal strand bias itself is not fully understood.

The *dif* system for site-specific recombination that allows monomerization and segregation of dimeric chromosomes requires specific orientation of some as-yet-unknown sequences, most noticeably in the vicinity of the terminus (11). Moreover, a λ derivative inserted near *dif* by homologous recombination has an orientation-specific effect on its own excision, perhaps because of hyperrecombination attendant on inhibition of dimer resolution. The Qin prophage lies close to the *dif* site but beyond it, consistent with the notion that orientation toward *dif* could be important. Such effects underscore the possibility that, although geneticists may select mobile element transposition anywhere in the genome, the distribution of elements on natural chromosomes, refined through the sieve of long-term selection, may be more restricted.

The above discussion refers only to the λ-related phages. The statistics on P4 relatives are not extensive enough to discuss seriously. The four elements in Table 2, plus the seven noninducible phages mentioned earlier (19) and the known phages P4, φR-73, and Sf6, all have insertion sites in their half of the map, suggesting nonrandom location. As to orientation, all four *int* genes in Table 2 are transcribed clockwise; however, the *leuX* site is beyond the replication origin, so that its orientation with respect to direction of replication is opposite from the others.

One recurrent feature of the defective prophages of K-12 is the frequent presence of other insertions, especially IS elements. A combination of insertions and deletions can generate a very incomplete prophage. The less complete it is, the less compelling the argument becomes for the historical sequence assumed here, where a complete genome first inserted, then gradually deteriorated. In some of the elements to be discussed later (such as pathogenicity islands), the proximity of integrase genes, transposase genes, and possible insertion sites has led to alternative interpretations, which cannot be excluded.

Distribution among strains of a species

Bacteria in nature experience repeated phage infections. Thus the composition of natural strains represents a steady state determined by rates of infection, prophage decay, and prophage loss. Some sense of the dynamics is given by comparing strains of the same bacterial species. The *E. coli* collection most systemically studied, ECOR (31), includes isolates from various habitats and parts of the planet.

A word on methodology: how can we tell whether a new strain harbors complete or defective prophages? Historically, the primary criterion was the presence in culture supernatants of plaque-forming particles. Failing that, there is the induction of lysis, sometimes accompanied by the liberation of phage-like particles, using treatments, such as UV irradiation, that induce the SOS response. To observe plaque-forming particles requires not only a complete prophage in the tested strain but also the availability of a phage-sensitive tester strain. These conditions are not always met; *E. coli* K-12 was cultivated in laboratories for about 30 years before the existence of λ came to light, the first tester strains being generated as cured derivations of K-12 itself.

What about defective prophages? By definition, they do not determine complete, plaque-forming particles. Defective derivatives of laboratory lysogens frequently arise by deletion or mutation in the prophage. Depending on the nature of the event, many of these retain some detectable attributes of lysogeny, including repressor production, manifested as immunity to superinfection by the parent phage, or inducible lysis (sometimes with liberation of visible, though nonfunctional, phage particles or components thereof).

None of the defective λ-related prophages in Table 2 show such visible signs of lysogeny. As indicated in the previous section, each of them came to light in the course of other experiments. Most of them make repressor, but no tester phage with the same specificity is available for checking immunity. Although DLP-12 and Qin have functional lysis genes, *E. coli* derivatives cured of active λ do not lyse in response to SOS induction. Both e14 and Qin have genes (*sfiC* and *dicB*) that are induced by SOS and whose products interfere with cell septation. This contributes to the UV sensitivity of K-12 but is not specifically diagnostic for lysogeny.

An early study (1) examined numerous natural *E. coli* isolates for hybridization with a λ probe and found evidence that most contained one or more λ-related sequences. The absence of other obvious signs of lysogeny in this and other studies suggests that many, perhaps most, of these strains harbored defective rather than active prophages.

Are the defective prophages inserted at some of the same sites known in K-12? Available evidence shows that the same sites are frequently occupied. The *argU*-DLP-12 junction is apparently present in several *E. coli* strains and also in *Salmonella enterica* serovar Typhimurium and *Shigella boydii* (27). The λ *attB* site is unoccupied in most members of the ECOR collection that were tested, but sometimes it is occupied by an element that can be cured by superinfection with λ (22a), although these strains do not lyse after SOS induction or show other signs of lysogeny.

A survey of variability at the *icd* locus (where e14 inserts; see Fig. 1) showed that the site is occupied in many ECOR strains, but not by e14 (41). The e14 site is also used by phage 21, a λ relative. In 21, as in most known λ-related phages, the integrase gene is less than 1 kb from the *att* site; in e14, about 3.5 kb intervenes. Also, the *int* sequences, though similar, are only about 60% identical, and although both elements insert at the same site, their integrases are not functionally interchangeable (40). The prophages detected in strains other than K-12 are all of the 21 variety, with integrase genes that are more than 99% identical to those of 21 and are located next to the *att* site. Most of these are probably defective; in a PCR screen, the right end of the prophage was present in only two of five strains (H. Wang and A. Campbell, 2000, unpublished data). It remains to be seen whether elements similar to e14 can be detected by means other than the PCR screen for the e14/*icd* junction.

A survey of variation in the *trp tonB* region (38) showed that many ECOR strains harbor a λ-related prophage (dubbed Atlas) at a specific site. At least one strain produced active phage, whereas most of the others had shorter, presumably defective, elements. There was no correlation between strain phylogeny as judged from chromosomal markers and the presence of Atlas, 21, or λ prophages (active or defective) nor was the presence of any one prophage correlated with the others.

Altogether, these results suggest that prophages are transient and that most natural strains have experienced repeated rounds of lysogenization, decay, and loss. One implication is that phages with the insertion specificities of 21 and Atlas have been reasonably abundant and widespread in recent historical time, even though Atlas had not turned up in other screens for λ-related phages.

Distribution among species

Collectively, phages can infect almost all bacterial species; but most individual phage isolates have a rather narrow host range. Even within a species, many strains are frequently refractory to infection. Some phages, such as P1, can deliver DNA to many hosts, in most of which they cannot propagate. Although the same general types of phages recur throughout the biological world, phages that propagate on different hosts are typically distinct enough in sequence to appear almost unrelated.

Phages infecting a single host also fall into diverse, virtually unrelated, groups, e.g., coliphages λ, P2, Mu-1, and T4. Careful comparisons of the DNA sequences of such phages reveal occasional shared sequences, suggesting that the barriers to exchange are not absolute. This has been observed for genes encoding phage attachment proteins, which are presumably under strong selection for their ability to interact with the surface of whatever host cell type becomes prominent (16).

Similar evidence for horizontal transfer exists even among phages that infect very distant hosts, suggesting that phages and their genes get around in the natural world perhaps by processes (not necessarily restricted to phage infection) that involve multiple transfer steps from host A to host B, host B to host C, etc. Figure 2 shows some cross-relationships of individual genes of phages from highly divergent hosts. The defective ϕflu prophage is of special interest. In addition to a complete Mu-1 related prophage, the sequenced genome of *Haemophilus influenzae* Rd includes a 31.5-kb segment that contains (in addition to the *Hin*dIII restriction-modification genes) an integrase gene downstream from a tRNA gene, a holin gene resembling λ*S*, terminase genes resembling those of some *Bacillus* phages, tail fiber genes similar to those of *Haemophilus* phage HP1, and a DNA sequence related to *Mycobacterium* phage D29, among others (17).

The ϕflu genome was recognized only thanks to whole-genome sequencing and is not closely related to any known *Haemophilus* phages, although its tail fiber gene is similar to that of phage HP1 (which also integrates next to a tRNA gene). More examples of the type should surface as additional genomes are completed.

INSERTION SEQUENCES AND TRANSPOSONS

General

Soon after the discovery of transposing elements in bacteria, it became convenient to distinguish be-

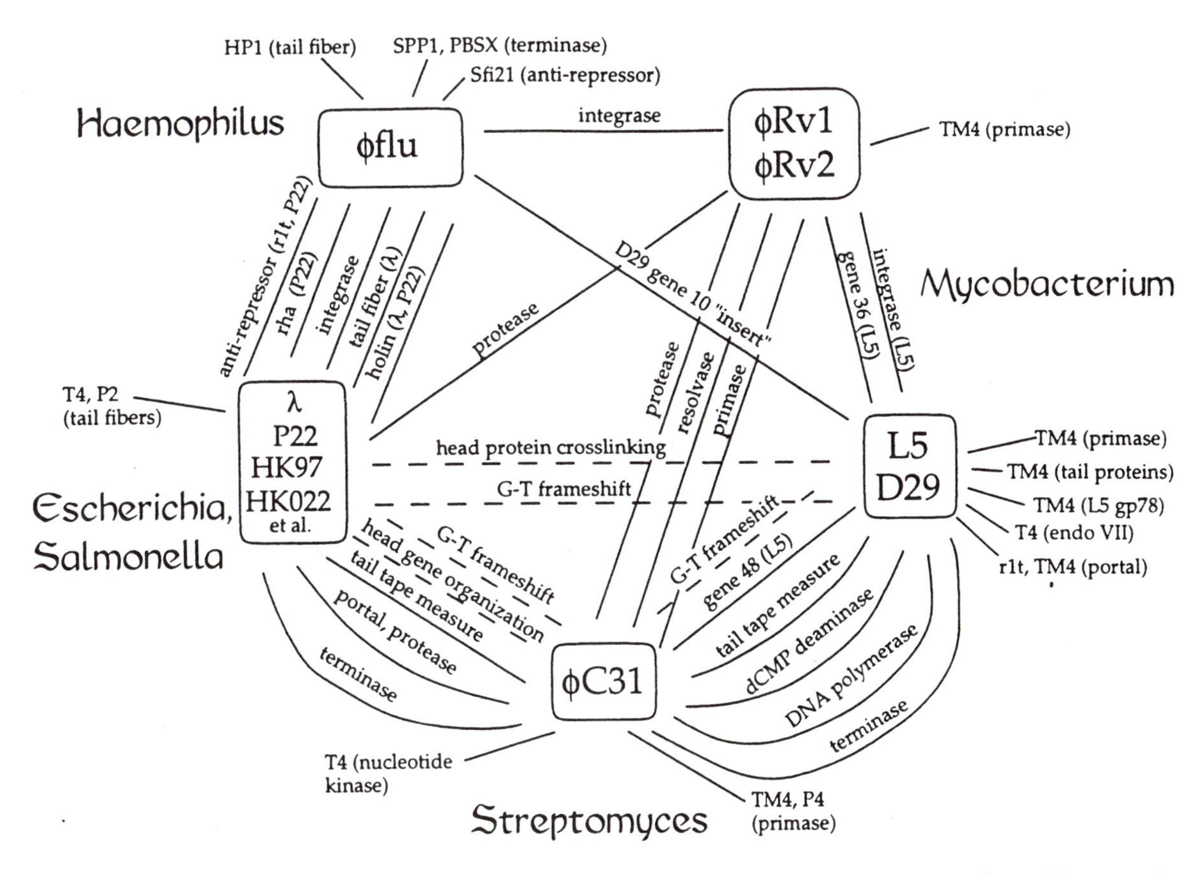

Figure 2. Individual gene homologies from phages of taxonomically distant hosts. Reprinted from *Proceedings of the National Academy of Sciences of the United States of America* (17) with permission from the publisher.

tween IS elements (simple insertion sequences that encode only transposition enzymes) and transposons (Tn), which encode other functions as well. Behind the distinction lay differences in how elements of the two groups were detected, studied, and exploited. Since IS elements confer no phenotype, they were usually discovered when they inserted into, thereby inactivating, known genes. Since they were necessarily indigenous to the strain in which they were discovered, the pedigrees connecting the multiple copies of a given IS element were unknowable. Accordingly, all elements that were indistinguishable from the original IS*1* by heteroduplex analysis were given the same name.

Transposons, on the other hand, were frequently discovered in plasmids or phages that could be transferred to a transposon-free recipient for analysis or use in engineering. Here the ancestry could be traced: all copies of Tn*10* in a strain were necessarily derived from the single Tn*10* transferred into their laboratory ancestor. Consequently, the name Tn*10* was restricted to the descendants of a transposon from a single source. This allowed investigators not directly involved in the work to be sure, without tracing pedigrees in the literature, of when transposons carrying the same antibiotic resistance gene had a recent common ancestor.

With time and sequencing technology, many of these consideration have lost force. A vast array of elements from various bacteria have been named, and independent isolates from different sources often prove to be identical or nearly so. Nomenclature more suited to the present state of knowledge and hierarchical classification of related elements is being developed (28).

Distribution within a Strain

Figure 3 shows the location of IS elements on the *E. coli* K-12 chromosome. These belong to several related families, grouped by sequence homology and arrangement of functional components: IS*1* family, IS*3* family (IS*2*, IS*3*, IS*600*, and IS*911*), IS*4* family (IS*4*, IS*186*, IS*5* family, IS*150* family, and IS*30* family). Wild-type K-12 also harbors the conjugative plasmid F, which has two copies of IS*3* and one of IS*2*.

Whereas the locations are fairly dispersed, some

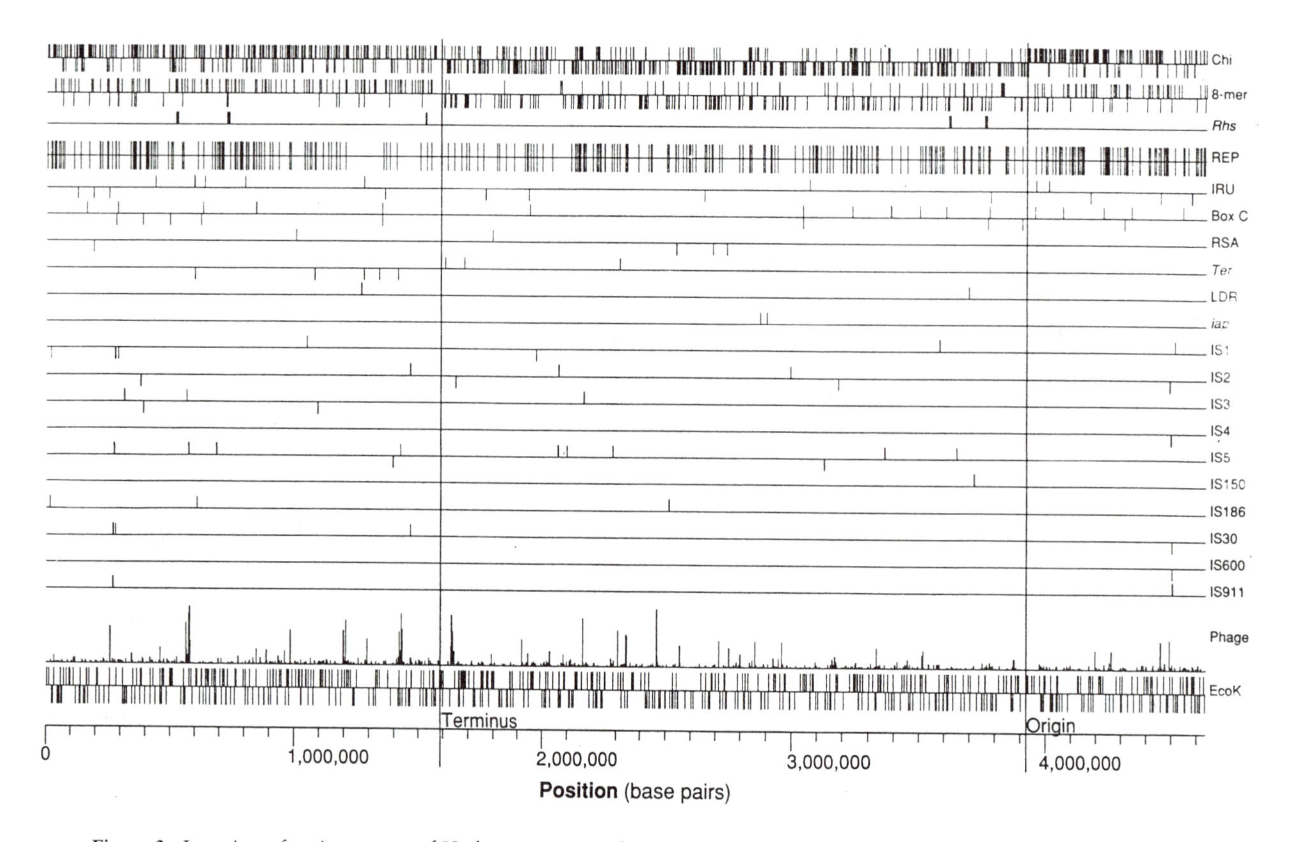

Figure 3. Location of major groups of IS elements in *E. coli* K-12. Tickmarks are either above or below the line, depending on orientation. Also shown are the locations of phage-like elements and *EcoK* restriction sites. Reprinted from *Science* (5) with permission of the publisher.

insertions (such as the two IS*1*s at about 0.3 Mb) are closer than expected on a random basis. In this particular case, the two IS*1*s lie within the defective prophage CP4-6. IS elements are overrepresented in defective prophages at least in part because insertions within transcriptionally silent DNA have little impact on cell growth. The 139 kb represented by the defective prophages of Table 2 (less than 3% of the genome) include 7 of the 41 IS insertions in Fig. 3 (or about 17%). The two directly oriented IS*1*s in CP4-6 also attracted attention because the intervening DNA is nonessential, so that the whole segment, which includes *argF,* can be lost by homologous recombination between the ISs.

The multiple copies of some ISs almost certainly resulted from transposition. Transposition is replicative for some elements and conservative ("cut and paste") for others, but even conservative transposition usually increases copy number because of donor suicide. Transposition should increase copy number, whereas transposon excision or selection against cells with high copy number should decrease it. Transposition frequencies are themselves frequently regulated by copy number. Figure 3 may be regarded as a snapshot in time of the dynamic interplay of the opposing factors in the particular strain K-12 (bearing in mind that this was a laboratory strain 75 years removed from its natural progenitor).

Relevant to within-strain dynamics is a survey of 118 separate subclones taken from a 30-year-old agar stab (30). Bacterial DNA cut with restriction enzymes was blotted with probes for the eight elements IS*1*, IS*2*, IS*3*, IS*4*, IS*5*, IS*30*, IS*150*, and IS*186*. No fewer than 68 different RFLP patterns were found, indicating frequent DNA rearrangement. The average copy number of individual ISs did not increase significantly (with the exception of IS*30*, which was sometimes present as a dimer that is known to transpose more rapidly); however, some variation in copy number among strains was noted. Under the conditions prevailing on the stab (low temperature, maintenance-level growth), IS*5* showed the greatest mobility, IS*1* never moved, and other elements were in between. On average, copy numbers neither increased nor decreased, as expected if a steady state had been reached.

Distribution among Strains

If the IS distribution in K-12 represents an equilibrium, one might expect the same distribution in other *E. coli* isolates. The expectation is firm only if the isolates are drawn from a panmictic population with no other genes on the chromosome that affect IS transfer and maintenance. Tests of genetic polymorphism within the ECOR collection show that *E. coli* is not panmictic, and the copy numbers of several common IS elements (IS*1* to IS*5* and IS*30*) vary widely among them (36). Each of these elements is absent from at least 11 of the 72 ECOR strains; when they are present, their copy numbers can exceed 20 for the most abundant elements (IS*1* and IS*5*). In two strains, none of the six IS elements was represented. There were correlations between the content of some ISs and strain origin (human versus other sources) and with strain genotype (assessed from electrophoretic polymorphisms). Moreover, the system is highly dynamic: each strain has its own pattern of IS-bearing restriction fragments, as though IS location had changed very frequently. The distribution could be fitted to the assumptions that IS*1* and IS*5* show no regulation of transposition and a strong decrease in fitness with increasing copy number, that IS*3* shows strong regulation and little effect on fitness, and that IS*2*, IS*4*, and IS*30* do not show strong regulation.

Sequencing of IS elements from different strains revealed much less sequence variation than in the average chromosomal gene, arguing that ISs are frequently transferred and that the current collection of elements may be of recent origin, hence probably transient (24). The frequent transfer of ISs among strains that this observation implies contrasts to the largely clonal (rather than panmictic) distribution of chromosomal genes, suggesting that ISs may frequently have been transferred on vehicles (phages, small plasmids) that do not readily transfer chromosomal DNA and/or that introduced ISs have often spread within genomes following entry.

The evolution of natural strains differs from the test tube evolution in the agar stab experiment (30) in that gene transfers from diverse, sometimes unknown, sources happen in the natural population. Most ECOR strains also harbor at least one plasmid, which may have played a role in determining their distribution. However, whatever that role may have been, plasmids account for only about 10% of the unique IS bands in the ECOR strains (36).

Distribution among Species

Because IS elements can hop onto and off of phages and conjugative plasmids, the whole prokaryotic world is in principle accessible to every IS element. Extensive data on IS distribution have been tabulated (28). Every such tabulation is necessarily biased because the known IS elements were characterized in well-studied species, such as *E. coli*, and recognized in other genomes by sequence homology. Thus IS elements restricted to archaea or mycoplasmas (for instance) may have been overlooked.

Nevertheless, the available data are informative. Some elements have a very narrow host range, e.g., IS*1* is restricted to *Escherichia* and *Shigella*. Others (like IS*5*) are broadly distributed among gram-negative bacteria, gram-positive bacteria, and archaebacteria. Sequence determinations of IS elements among enteric bacteria indicate that interspecific transfer is much rarer than intraspecific (24). Even though the species concept is notoriously difficult to define in prokaryotes, taxonomists seem to have identified the natural boundaries reasonably well.

Known IS elements are apparently absent from the laboratory strain of *Bacillus subtilis*, although present in other species of *Bacillus*. Remembering that two strains of the ECOR collection lacked all six common IS elements surveyed (36), such findings need not require any special explanation. And a strain that happens to be devoid of IS elements at this moment may well have possessed them and experienced their effects through most of its evolutionary history.

SPECIALIZATION ISLANDS

General

Specialization islands are blocks of contiguous genes (frequently in the range of 30 to 200 kb), most of which collaborate to perform some luxury functions. The best-known examples are the pathogenicity islands. Pathogenicity islands are typically present in some strains of a bacterial species and absent in others. They frequently differ from the rest of the genome in DNA composition. They are sometimes flanked by short direct repeats, like many phages and transposons. Some of these characteristics suggest that the islands have rather recently been transferred into the strain and species under study.

Some other DNA segments share with pathogenicity islands some (not always all) of their typical features: dedication to a specific function, erratic distribution either among strains of a species or among species of a taxon, atypical DNA composition, and terminal repeats. These include (for example) the cobalamin operon of *S. enterica* serovar Typhimurium (low in G + C content and not present in all salmonellae), the surface polysaccharide genes of enteric bacteria (alternative gene blocks at the same chromosomal locus, unrelated to one another and atypical in base composition), and the nitrogen-fixation gene of various species (erratically distributed among the major groups of bacteria).

Possible Evolution

A plausible scenario is that most specialization islands entered the species in which they now reside by lateral transfer (in one or more steps) from some distantly related species and integrated into the chromosome. Their subsequent evolution may have obscured their origin in several ways (arranged from fastest to slowest).

Deletion of a terminal duplication generated at the time of entry

As discussed above, many prophages seem to have undergone such deletions. The uropathogenic *E. coli* strain 536 contains a pathogenicity island (Pai I) flanked by an 18-bp direct repeat that includes the 3′ end of the *selC* gene. Some other pathogenic strains (such as E2348) have, in the same location with respect to *selC*, another island (Pai III), but here there is no direct repeat. It is as if the two elements had both integrated by a site-specific recombination (like λ or P4) but that the right end of Pai III had been deleted (15).

A cautionary word on the semantics of what constitute the "islands": the boundaries of an island are generally deduced from sequence comparisons with a natural strain from which the island is absent. However, as discussed earlier, defective prophages may accumulate insertions with time. Therefore, it is not at all excluded that the present "island" started out as a defective prophage and that the pathogenicity genes inserted into it later, by unknown mechanisms, a possibility that will be discussed further.

Amelioration of DNA composition

An introduced element is under mutational and/or selectional pressure to resemble the rest of the genomic DNA, although it is estimated that some compositional features (such as G+C content) require more than 100 million years for complete amelioration (23).

Species-wide establishment

Given occasional intraspecific recombination, an established island may, either through selection or drift, eventually become fixed in all strains of a species. This may be even slower than amelioration. For example, nitrogen-fixing genes, though distributed erratically among taxa, are frequently present throughout a species. They show no telltale terminal repetitions or atypical DNA composition. The same is true of some clusters of pathogenicity genes (15). It is of course only speculation that such genes ever had the properties of typical islands. It has been argued that the primary selective force underlying the organization of genes into operons is that, historically, the genes of each operon entered their current species through lateral transfer of a single block of clustered genes (25).

Possible Mechanisms of Insertion

In principle, the direct oligonucleotide repeats surrounding many pathogenicity islands might have at least two different origins. One possibility is that they were generated through transposition. However, the observed repeats (18 to 31 bp) are longer than those known to be produced in this manner.

The other possibility is that the element entered the cell as a DNA circle (perhaps phage or plasmid) that contains a short sequence homologous to some part of the recipient chromosome. Homology-dependent recombination might then insert the circle. The observed oligonucleotide repeats are generally too short to allow appreciable RecA-promoted recombination, but RecA-independent microhomologous recombination is a well-documented process. Alternatively, the recombination could be site-specific, promoted by a phage integrase.

Favoring site-specific recombination is the frequent location of insertions in places (especially tRNA genes) where phages are known to insert using an integrase mechanism. Furthermore, some islands contain integrase genes near the insertion site, although none has proven activity. A convincing example is the 12-kb (virulence-associated protein) island of *Dichelobacter nodosus*, which is inserted in the 3′ end of the *serV* gene. At the upstream end of the element (immediately 3′ to the complete *serV* gene) is an integrase gene, closely related to that of phage P4 and (as in the P4-related phages) oriented in the same direction as *serV* (10).

The known pathogenicity island insertions at tRNA genes all entail a duplication from the TψC loop through the 3′ end of the gene (as in P4). Duplications starting at the anticodon loop (as in λ-related phages) have not been reported. Pathogenicity islands have sometimes been described as inserted downstream from, rather than in, the tRNA genes. To use a Clintonesque phrase, it depends what is meant by "in." The problem is that the oligonucleotide repeat flanking an island frequently extends beyond the TψC loop through a few nucleotides of 3′ flanking DNA. Absent other information, all that can be said for certain is that the crossover took place somewhere within the repeat. A molecular evolutionist comparing the sequences of strains with and without the pathogenicity island places the boundary of the island at the first nucleotide where the two strains diverge, which is of course the 3′ end of the repeat (Fig. 1).

The only relevant information (for both pathogenicity islands and P4-like phages) concerns excision from sites where the homology is not quite perfect, so that the crossover site within the direct repeat can be localized more closely. These include the CP4-57 cryptic prophage of *E. coli* and a pathogenicity island (Cag Pai) of *Helicobacter pylori,* which inserts into a codon region rather than a tRNA gene (9). In both cases (as in the λ-related elements inserting at the anticodon loop) the actual crossover takes place within the gene, near the upstream end of the duplication.

Should pathogenicity islands then be regarded as elements that originally inserted by a phage-like mechanism? The questions again entails some semantic consideration concerning the boundaries of the elements, which may be appreciated by recalling the defective prophages of K-12. These contain DNA sequences (insertion sequences, alien genes) that may have survived simply by inserting into transcriptionally inert DNA. Whatever the origin of the pathogenicity genes, it seems equally possible that they reached their present locations by insertion into preexisting defective prophages. The situation is complicated by the frequent presence of insertion sequences in pathogenicity islands (as in defective prophages). The presence of these insertion sequences may or may not relate to the acquisition of the pathogenicity genes, or to the insertion of the entire island.

In summary, the linkage between pathogenicity islands and tRNA genes likely results from insertion of a P4-related element. The mechanism whereby pathogenicity genes came to reside within that element is open. They may have been part of the element before it inserted, or they may have inserted within it after it was already there; in either case, at some stage an event joining the pathogenicity genes to the rest of the element must be postulated. The same options hold for all defective prophages that include extraneous DNA.

Why pathogenicity genes should be associated with P4-type prophages rather than λ prophages is not clear. The number of observed cases is not large enough to be sure that it is more than a statistical accident.

It has been suggested (e.g., see reference 34) that tRNA genes might represent the primordial phage-insertion sites. Within either group of phages (such as the λ-related ones), the argument is not at all compelling. λ-related phages have evolved to prefer certain motifs (especially approximate symmetry around the center of the overlap region) that are provided by the anticodon loops as well as some sites elsewhere in the genome. From this phage group alone, it is hard to say which sites are primordial. However, the P4-like phages insert at the TΨC (loop, with no obvious symmetry around the insertion point). The fact that both phage types frequently use tRNA genes (apparently in rather different ways) invites speculation that both groups derived from an ancient association between tRNA genes and insertion systems. It has even been suggested (specifically for pathogenicity islands) that the 3′ end of the tRNA molecule might participate in the integration event itself (18), which would be unprecedented in integrase-promoted reactions.

ALIEN GENES

General Considerations

As just discussed, pathogenicity islands are distinguished from the rest of the genome on at least two counts: (i) their distribution among strains and their chromosomal positions (e.g., next to tRNA genes) suggest that they are inserted elements, and (ii) their DNA composition is atypical, as expected if they had been laterally transferred from some other species. This raises the question of how many other genes show similar signs of lateral transfer.

Cases where nearly identical genes are found in distantly related bacteria are rare and generally confined to genes typically located on transposable elements on plasmids. In those cases, lateral transfer seems a necessary postulate. For pathogenicity islands, lateral transfer is inferred from DNA composition, but good candidate donor strains are not at hand.

DNA Composition

Some features of DNA composition tend to be constant throughout a prokaryotic genomic. These include G + C content and signature (based on dinucleotide relative abundance). Codon usage patterns are also characteristic of a species, although each species has at least two characteristic patterns, depending on the rate of gene expression.

Long before DNA sequencing, it was recognized from physical properties of fragmented DNA (density, melting temperature) that the G + C content of each prokaryotic species tends to be uniform across the genome. However, some genome segments deviate significantly from the norm. Figure 4 (top panel) illustrates this feature for *Helicobacter pylori.* The average G + C content is 39%, but there are three segments (including one pathogenicity island) with a G + C content of $<36\%$.

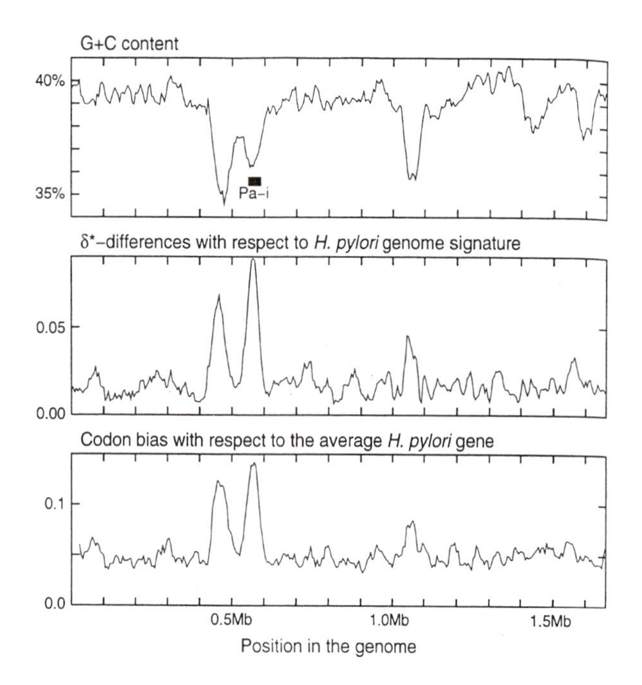

Figure 4. 50-kb sliding window analysis of the *Helicobacter pylori* genome for G+C content (top), dinucleotide relative abundance (middle), and codon bias (bottom). Reprinted from the *Annual Review of Genetics* (21) with permission of the publisher.

It turns out that these same three segments differ in other molecular statistics as well. The middle panel of Fig. 4 shows that the δ* differences between local 50-kb contigs and the entire genome peak at the same places. The δ* difference is given by

$$\delta^*(f,g) = \frac{1}{16}\Sigma_{xy}|\rho^*_{xy}(f) - \rho^*_{xy}(g)|$$

where f and g are two DNA sequences (in this case, the local segment versus the entire genome) and

$$\rho^*_{xy} = \frac{(XY)}{(X)(Y)}$$

where (XY) is the frequency of the dinucleotide XpY and (X) and (Y) are the mononucleotide frequencies. For random sequences, ρ^*_{xy} equals one for all X,Y. For real sequences, the signature (a vector composed of all 16 δ* values) is almost invariant throughout a genome but can be markedly different from one species to another.

The mutational and/or selectional pressures that push a genome toward uniformity are unknown for either G+C content or signature, but they are clearly pervasive. (For higher eukaryotes, the signature is uniform, but there is substantial variation in G+C content.) Both G+C content and signature indicate that the three aberrant segments of Fig. 4 entered the genome recently.

Figure 4 (bottom panel) shows that the codon bias of these segments is also high. Codon bias is defined as

$$B(F/C) = \Sigma\rho_a(F)\,\Sigma_{(x,y,z=a)}\,\rho_a(F)\,|f(x,y,z) - c(x,y,z)|$$

where ρ_a is the frequency of occurrence of amino acid a in the tested gene(s), $f(x,y,z)$ is the frequency of usage of codon x,y,z for amino acid a in the tested gene(s) (F), and $c(x,y,z)$ is the frequency in some comparison set C. Thus in Fig. 4, F is the coding regions within a 50-kb window and C is all the coding regions of the genome.

Those genes that are highly expressed in rapidly growing cells (such as those encoding ribosomal proteins, tRNA synthetases, and chaperones) have a high codon bias with respect to the average gene (21a). Codon usage in these genes is thought to be driven by relative tRNA abundance. When the bias of individual genes is tested against the set of all genes in the genome (as in Fig. 4, bottom panel) and also against the highly expressed genes, most genes show a low bias with respect to one set or the other. However, a significant number of genes are highly biased with respect to both these reference sets. These genes have been christened "alien," although the name should not prejudice interpretation of what these codon usage patterns may mean.

Alien Genes of *E. coli* K-12

The number of genes that score as alien depends completely on the cutoff values used. In an early study, about 2% of the genes of *E. coli* were classified as alien; later use of a more relaxed criterion increased the number to around 7%. From combined use of G+C content and codon bias, the frequency of genes that may have joined the *E. coli* genome by lateral transfer during the last 100 million years has been estimated to be as high as 15%.

Figure 5 shows the disposition of alien and highly expressed genes on the *E. coli* genome. The alien genes are distributed around the genome, with some small clusters. Most have a low G+C content (S3 < 45%) but a few have a high G+C content (S3 > 55%).

What are these genes? Of the 50 alien genes in Fig. 5, 17 lie within the defective prophages listed in Table 2. Another 9 function in the synthesis of surface antigens, which are possibly under high selection for variability. The function of the other 24 is unknown; some of them have no homologs in current databases.

The connection with defective prophages is thus a significant part of the story.

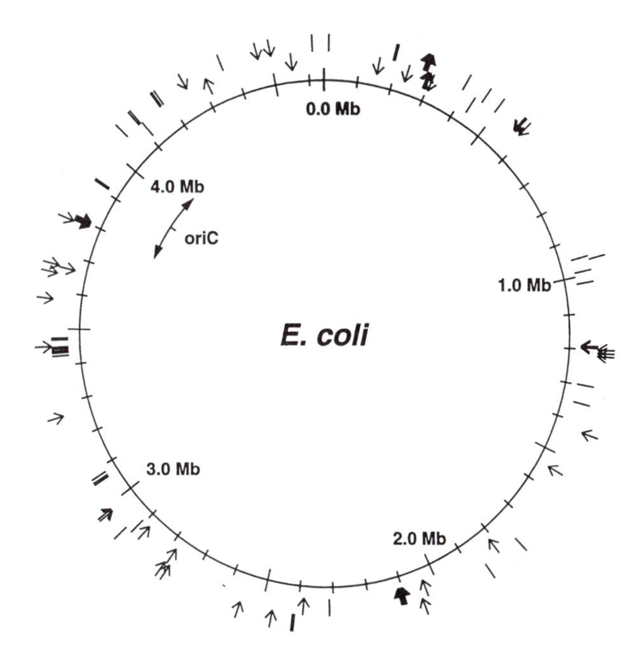

Figure 5. Genes of *E. coli* with codon bias relative to the average gene >0.520 and with S3 (= G+C content in third position) < 45% are shown as arrows pointing inwards; genes with S3 > 68% are shown as arrows pointing outwards. Marks without arrows indicate genes with S3 between 45% and 68%. The genes transcribed clockwise are in the outer circle; those transcribed inward are in the middle circle. Genes with arrows are considered alien. Those without arrows are predominantly highly expressed genes. Reprinted from *Molecular Microbiology* (22) with permission of the publisher.

Heterogeneity in Phage Genomes

As stated earlier, defective prophages might tend to accumulate insertions just because they are not under selection. Some alien genes may reside in prophage DNA for that reason. An alternative hypothesis is that these genes were already in the genome of a complete phage that lysogenized the ancestral strain and later deteriorated to become defective. The critical question is not just whether the gene entered the phage genome when that genome was replicating autonomously rather than integrated as a prophage, but whether it gets carried around by the phage in successive cycles of infection. Similar questions can be asked about the origins of pathogenicity islands.

More than one-third of the genes of bacteriophage λ can be classified as "accessory," i.e., essential for neither the lytic nor the lysogenic cycle (12). These are concentrated in several regions of the genome, perhaps positioned where their presence does not interrupt the major circuitry of the phage cycles. Some of these genes are expressed from an inserted prophage. These include *rexAB* (co-transcribed with the repressor gene) and also *lom* and *bor*, each of which is transcribed from its own promoter. Expression of the latter two genes renders bacterial growth resistant to serum, a correlate of virulence (2).

Unlike most genes of the major phage cycles, the accessory genes of λ are frequently absent from at least some λ-related phages. Conversely, λ-related phages sometimes contain accessory genes absent from λ. One such gene (*mmpC*) is present in the defective DLP-12 prophage and also in certain λ-related phages (such as PA-2) but absent from λ. Like *bor*, *mmpC* is located near the extreme right end of the linear phage genome and is transcribed leftward, oppositely from the major direction of transcription of the late operon in which it is embedded. In DLP-12, *mmpC* expression is inactivated by an adjacent IS insertion, but *mmpC* is expressed in PA-2 lysogens. Its product ("new membrane protein") can substitute for the bacterial outer membrane protein OmpC and is thought perhaps to reduce the ability of cellular debris to attach to and inactivate liberated phage particles.

Comparative genome analysis of λ-related phages has uncovered additional examples of genes that resemble *nmpC* in that (i) they each have their own promoter and terminator and (ii) they are present in one phage between genes that are contiguous in a second phage (20). Such genes have been dubbed "morons" by their discoverers and are regarded as independent modules that have inserted into the phage genome. Like alien genes of bacteria, morons frequently differ from flanking DNA in G+C content.

Neither for morons nor for the majority of bacterial alien genes are there any telltale signs to implicate specific systems (transposons or integrases) in their insertion. Probably such systems were used by some alien genes, but their traces, such as oligonucleotide repeats, have disappeared. It is also possible that nonspecific mechanisms, not currently understood, were frequently used, as in a reported case of nonspecific transposition (29).

CONCLUDING REMARKS

In a survey of bacterial genomes, what is impressive is not only the number of DNA segments that have been inserted but also the relationship between insertions of various types. Thus, insertion sequences and alien DNA are preferentially located within defective prophages, and alien DNA is inserted into active phage genomes as well. This clustering of insertion events frequently complicates historical interpretation of which mechanisms have been used. Despite this clustering, the majority of insertion events thought to have occurred during evolution have left no trace of past association with mobile elements.

Present inferences about lateral gene transfer depend primarily on aberrant DNA statistics (G+C content, dinucleotide relative abundance, or codon usage) in small segments of genomes. They would be enormously strengthened if more potential donor species had been identified. It is uncertain whether additional genome sequencing will eventually approach saturation, where every gene of a newly sequenced species has a homolog in the database.

REFERENCES

1. **Anilionis, A., and M. Riley.** 1980. Conservation and variation of nucleotide sequences within related bacterial genomes: *Escherichia coli* strains. *J. Bacteriol.* **143:**355–365.
2. **Barondess, J. J., and J. Beckwith.** 1990. A bacterial virulence determinant encoded by lysogenic coliphage lambda. *Nature* (London) **346:**871–872.
3. **Barreiro, V., and E. Haggård-Ljungquist.** 1992. Attachment sites for bacteriophage P2 on the *Escherichia coli* chromosome: DNA sequences, localization on the physical map, and detection of a P2-like remnant in *E. coli* K-12 derivatives. *J. Bacteriol.* **174:**4086–4093.
4. **Berlyn, M. K. B.** 1998. Linkage map of *Escherichia coli* K-12, edition 10: the traditional map. *Microbiol. Mol. Biol. Rev.* **62:** 814–984.
5. **Blattner, F. R., G. Plunkett III, C. A. Bloch, N. T. Perna, V. Burland, M. Riley, J. Collado-Vides, J. D. Glassner, C. K. Rode, G. F. Mayhew, J. Gregor, N. W. David, H. A. Kirkpatrick, M. A. Goeden, D. J. Rose, B. Mau, and Y. Shao.** 1997. The complete genome sequence of *Escherichia coli* K-12. *Science* **277:**1453–1468.
6. **Campbell, A.** 1992. Chromosomal insertion sites for phages and plasmids. *J. Bacteriol.* **174:**7495–7499.
7. **Campbell, A., J. Kim Ha, R. J. Limberger, and S. J. Schneider.** 1990. Bacteriophage evolution and population structure, p. 191–200. *In* M. Clegg and S. J. O'Brien (ed.), *Molecular Evolution.* Wiley-Liss, New York, N.Y.
8. **Campbell, A., S. J. Schneider, and B. Song.** 1992. Lambdoid phages as elements of bacterial genomes. *Genetica* **86:** 259–267.
9. **Censini, S., C. Lange, Z. Xiang, J. E. Crabtree, P. Ghiara, M. Borodovsky, R. Rappuoli, and A. Covacci.** 1996. *Cag,* a pathogenicity island of *Helicobacter pylori,* encodes I-specific and disease-associated virulence factors. *Proc. Natl. Acad. Sci. USA* **93:**14648–14652.
10. **Cheetham, B. F., D. B. Tattersoll, G. A. Bloomfield, J. I. Rood, and M. E. Katz.** 1995. Identification of a gene encoding a bacteriophage-related integrase in a *vap* region of the *Dichelobacter nodosus* genome. *Gene* **162:**53–58.
11. **Corre, J., J. Patte, and J.-M. Louarn.** 2000. Prophage λ induces terminal recombination in *Escherichia coli* by inhibiting chromosome dimer resolution: an orientation-dependent *cis* effect lending support to bipolarization of the terminus. *Genetics* **154:**39–48.
12. **Court, D., and A. Oppenheim.** 1983. Phage lambda's accessory genes, p. 251–278. *In* R. Hendrix, J. Roberts, F. Stahl, and R. Weisberg (ed.), *Lambda II.* Cold Spring Harbor Laboratory Press, Cold Spring Harbor, N.Y.
13. **Diaz, R., and K. Kaiser.** 1981. Rac *E. coli* K-12 strains carry a preferential attachment site for lambda rev. *Mol. Gen. Genet.* **183:**484–489.
14. **Greener, A., and C. W. Hill.** 1980. Identification of a novel genetic element in *Escherichia coli* K-12. *J. Bacteriol.* **144:** 312–321.
15. **Hacker, J., G. Blum-Ohler, L. Muhldorper, and H. Tschäpe.** 1997. Pathogenicity islands of virulent bacteria: structure, function, and impact on microbial evolution. *Mol. Microbiol.* **236:**1089–1097.
16. **Haggard-Ljungquist, E., C. Halling, and R. Calender.** 1992. DNA sequences of the tail fiber genes of bacteriophage P2: evidence for horizontal transfer of tail fiber genes among unrelated bacteriophages. *J. Bacteriol.* **174:**1462–1477.
17. **Hendrix, R. W., M. C. M. Smith, R. N. Burns, M. E. Ford, and G. F. Hatfull.** 1999. Evolutionary relationships among diverse bacteriophage and prophages: all the world's a phage. *Proc. Natl. Acad. Sci. USA* **96:**2192–2197.
18. **Hou, Y. M.** 1999. Transfer RNAs and pathogenicity islands. *Trends Biochem. Sci.* **24:**295–298.
19. **Jacob, F., and E. L. Wollman.** 1961. *Sexuality and the Genetics of Bacteria.* Academic Press, New York, N.Y.
20. **Juhala, R. J., M. E. Ford, R. L. Duda, A. Youlton, G. F. Hatfull, and R. W. Hendrix.** 2000. Genomic sequences of bacteriophages HK97 and HK022: pervasive genetic mosaicism in the lambdoid bacteriophages. *J. Mol. Biol.* **299:**27–52.
21. **Karlin, S., A. M. Campbell, and J. Mrázek.** 1998. Comparative DNA analysis across diverse genomes. *Annu. Rev. Genet.* **32:** 185–226.

21a. **Karlin, S., and J. Mrázek.** 2000. Predicted highly expressed genes of diverse prokaryotic genomes. *J. Bacteriol.* **182:** 5238–5250.

22. **Karlin, S., J. Mrázek, and A. M. Campbell.** 1998. Codon usages in different gene classes of the *Escherichia coli* genome. *Mol. Microbiol.* **29:**1341–1345.

22a. **Kuhn, J., and A. Campbell.** 2001. The bacteriophage λ attachment site in wild strains of *Escherichia coli. J. Mol. Evol.* **53:** 607–614.

23. **Lawrence, J. G., and H. Ochman.** 1997. Amelioration of bacterial genomes: rates of change and exchange. *J. Mol. Evol.* **44:** 383–397.
24. **Lawrence, J. G., H. Ochman, and D. L. Hartl.** 1992. The evolution of marker sequences within enteric bacteria. *Genetics* **131:** 9–20.
25. **Lawrence J. G., and J. R. Roth.** 1996. Selfish operons: horizontal transfer may drive the evolution of gene cluster. *Genetics* **143:**1843–1860.
26. **Lederberg, E. M.** 1951. Lysogenicity in *E. coli* K-12. *Genetics* **36:**560.
27. **Lindsey, D. R., D. A. Mullin, and J. P. Walker.** 1989. Characterization of the cryptic lambdoid prophage DLP12 of Escherichia coli and overlap of the DLP12 integrase gene with the tRNA gene *argU. J. Bacteriol.* **171:**6197–6205.
28. **Mahillon, J., and M. Chandler.** 1998. Insertion sequences. *Microbiol. Mol. Biol. Rev.* **62:**725–774.
29. **Mràzek, J., and S. Karlin.** 1998. Strand compositional symmetry in bacterial and large viral genomes. *Proc. Natl. Acad. Sci. USA* **95:**3720–3725.
30. **Noas, T., M. Blot, W. M. Fitch, and W. Arber.** 1994. Insertion sequence-related genetic variation in resting *Escherichia coli* K-12. *Genetics* **136:**721–730.
31. **Ochman, H., and R. K. Selander.** 1984. Evidence for clonal population structure in *Escherichia coli. Proc. Natl. Acad. Sci. USA* **81:**198–201.
32. **Pierson, L. S., and M. L. Kahn.** 1987. Integration of satellite bacteriophage P4 in *Escherichia coli* DNA sequence of the

phage and host regions involved in site-specific recombination. *J. Mol. Biol.* **196**:487–496.

33. **Rappleye, C. A., and J. R. Roth.** 1997. Transposition without a transposase: a spontaneous mutation in bacteria. *J. Bacteriol.* **179**:2047–2052.
34. **Reuter, W.-D., P. Palm, and S. Yeats.** 1989. Transfer RNA genes frequently serve as integration sites for prokaryotic genetic elements. *Nucleic Acids Res.* **17**:1907–1914.
35. **Retallack, D. M., L. L. Johnson, and D. I. Friedman.** 1994. Role of 10Sa RNA in the growth of λ-P22 hybrid phage. *J. Bacteriol.* **176**:2082–2089.
36. **Sawyer, S. A., D. E. Dykhuizen, R. F. DuBose, L. Green, T. Mutagandura-Mhlanga, D. F. Wolczyk, and D. L. Hartl.** 1987. Distribution and abundance of insertion sequences among natural isolates of *Escherichia coli*. *Genetics* **115**:51–63.
37. **Shimada, K., R. A. Weisberg, and M. E. Gottesman.** 1972. Prophage lambda at unusual chromosomal locations. I. Location of the secondary attachment sites and the properties of the lysogens. *J. Mol. Biol.* **63**:483–503.
38. **Stoltzfus, A. B.** 1991. A survey of natural variation in the *trp-tonB* region of the *E. coli* chromosome. Thesis. University of Iowa, Ames.
39. **Trempy, J. E. J. E. Kirby, and S. Gottesman.** 1994. Alp suppression of lon: dependence on the *slpA* gene. *J. Bacteriol.* **176**: 2061–2067.
40. **Wang, H., C.-H. Yang, G. Lee, F. Chang, H. Wilson, A. del Campillo-Campbell, and A. Campbell.** 1997. Integration specificities of two lambdoid phages (21 and el4) that insert at the same *attB* site. *J. Bacteriol.* **179**:5705–5711.
41. **Wang, F. S., T. S. Whittam, and R. K. Selander.** 1997. Evolutionary genetics of the isocitrate dehydrogenase gene (*icd*) in *Escherichia coli* and *Salmonella enterica*. *J. Bacteriol.* **179**: 6551–6559.

Mobile DNA II
Edited by N. L. Craig et al.

Chapter 45

Mobile Genetic Elements and Bacterial Pathogenesis

Brigid M. Davis and Matthew K. Waldor

INTRODUCTION

Bacterial pathogens have developed an impressive repertoire of strategies for exploitation of their eukaryotic hosts. Pathogenic species synthesize individualized tools that enable them to colonize inhospitable host niches, avoid host defenses, parasitize host resources, and generally enhance their own survival and dissemination. However, despite the remarkable diversity of their adaptations, most microbial pathogens arrived at their current specialized states by travelling along parallel evolutionary paths, in that most obtained critical virulence genes as components of horizontally transferred genetic elements. That is, bacterial virulence genes—by which we mean genes that contribute specifically to pathogenesis, rather than genes that enhance overall bacterial fitness—generally did not evolve independently and incrementally in each pathogen from ancestral loci. Instead, pathogens-to-be obtained genetic "building blocks," in the form of virulence gene cassettes, that enabled them to evolve new traits and abilities in "quantum leaps" rather than small steps (41). The impact of horizontal gene transfer was not limited to the evolution of pathogenicity; recent work, including analyses of the DNA sequences of several microbial genomes, has led to the conclusion that lateral gene transfer has had a profound influence on the evolution of a variety of prokaryotic capabilities.

Horizontally transferred genes were frequently disseminated among bacterial populations as components of mobile genetic elements, such as plasmids, phages, and transposons. It is likely that most virulence genes were merely passengers on mobile elements, because most have not been shown to encode functions that are necessary for the transfer of DNA. Today, many of the genes that encode critical bacterial virulence factors, including diverse toxins, adhesins, secretion systems, and other effectors of pathogenesis, are still contained within mobile or mobilizable genetic elements, and some continue to spread among bacterial populations. However, a significant subset is now found within pathogenicity islands (PAIs) instead. PAIs are chromosomal gene clusters linked to virulence that appear to be derived from mobile sequences, although most are no longer mobile. Chromosomal anchoring of PAIs appears to be one of the many ways in which virulence-linked gene clusters have been modified after acquisition by new hosts. In addition, horizontally transferred genes with common ancestors have gradually diverged because of selection of variants suited to the needs of each recipient species. Horizontally acquired virulence-linked genes have also been incorporated into regulatory loops, comprising both chromosomal loci and components of other mobile elements. The intricacy of the interactions that have developed between the products of mobile virulence genes and those of genet-

Brigid M. Davis and Matthew K. Waldor • Howard Hughes Medical Institute and Division of Geographic Medicine and Infectious Diseases, New England Medical Center and Tufts University School of Medicine, Boston, MA 02111.

ically unlinked loci clearly illustrates that acquisition of a mobile element is frequently only the first of many stages in the evolution of a specialized pathogen.

The previous chapters of this volume illustrate that the term mobile DNA applies to a plethora of genetic elements and that numerous distinct processes underlie mobilization of genetic material. In this chapter, we do not attempt to explore all of the ways in which DNA mobility contributes to pathogenesis. Instead, we focus on bacterial pathogens, and in particular, on the role of sequence-specific mechanisms for intercellular gene transfer in the evolution of such organisms. Thus, we do not explore the contributions to pathogenicity of natural competence or of generalized transducing phages. We also do not discuss phase variation or antigenic variation mediated by changes in DNA structure, although these clearly play a significant role in the course of bacterial infections. Finally, although antibiotic resistance is increasingly important in efforts to combat pathogenic bacteria, a review of R plasmids, R-linked transposons, and integrons is beyond the scope of this chapter. In summary, we concentrate on virulence factors encoded by genes on mobilizable or formerly mobilizable genetic elements, especially plasmids, bacteriophages, and PAIs.

MOBILE ELEMENTS LINKED TO VIRULENCE

There are many informative ways of grouping the horizontally transferred genetic elements that encode virulence factors: for example, by the bacterium in which they are found, by the type of virulence factor, and by the type of mobile element. Analyses that focus on the bacteria can emphasize the genetic complexity of individual pathogens and the steps by which they each evolved (e.g., see reference 49). In contrast, analyses centered on classes of virulence factors can demonstrate the diverse ways in which similar factors have been acquired and subsequently adapted by individual hosts (e.g., see references 67, 98, 117). Finally, analyses based on the different classes of mobile elements can highlight the processes by which virulence genes have been transported and illuminate the evolutionary relationships between types of mobile elements. In this chapter, we use the last-mentioned approach and sequentially discuss bacteriophages, plasmids, and PAIs. Thus, we proceed from elements that generally encode a solitary factor (bacteriophages, some plasmids) to elements that encode a multiprotein apparatus or numerous independent virulence-linked proteins (larger plasmids, most PAIs). These latter elements are frequently composite elements, assembled from smaller building blocks (e.g., insertion sequences, transposons, phage genomes); consequently, knowledge of the simpler elements should aid in understanding those that are more complex.

Bacteriophages

Phage-encoded virulence factors

The first discovery of a phage-encoded virulence factor was made in 1951, when it was found that diphtheria toxin is encoded within the genome of the β-phage of *Corynebacterium diphtheriae* (45). Since then, numerous additional virulence factors have been found to be phage encoded (Table 1). These virulence

Table 1. Virulence-linked bacteriophages[a]

Microorganism	Bacteriophage	Virulence gene(s)	Function	Reference(s)
C. diphtheriae	β-phage, ω-phage	*tox*	Diphtheria toxin	45, 64, 134
E. coli O157:H7 (EHEC)	933J, H19B	*stx-1*	Shiga toxin-1	22, 127
	933W	*stx-2*	Shiga toxin-2	8, 107, 127
		lom	Outer membrane protein	
		bor	Serum resistance	
P. aeruginosa	ϕCTX	*ctx*	Cytotoxin	59
S. flexneri	Sf6	*oac*	O-Antigen acetylase	30
	SfV, SfX, SfII	*gtr*	Glucosyltransferase	66, 86, 136
S. enterica	SopEϕ	*sopE*	SPI-1 effector	57, 93
	Gifsy-2	*sodC-1*	Superoxide dismutase	37, 40
V. cholerae	CTXϕ	*ctxAB*	Cholera toxin	9, 131, 139
		ace, *zot*	Enterotoxins	
S. aureus	ϕ13	*sak*	Staphylokinase	31
		entA	Enterotoxin A	
S. pyogenes	T12	*speA*	Erythrogenic toxin	140
	CS112	*speC*	Erythrogenic toxin	68

[a]This is not intended to be an exhaustive list.

factors include diverse toxins (e.g., Shiga toxin and cholera toxin), proteins that aid in evasion of host defenses (e.g., SodC), and proteins that alter bacterial cell surfaces and thereby interfere with host recognition of the bacterium (e.g., O antigen acetylase and glucosyltransferase). As mentioned above, many phage genomes encode only a single virulence factor (although some phages do contain multiple virulence genes). Still, these factors are frequently essential for bacterial pathogenicity. In fact, most disease symptoms induced by several pathogens, including *Vibrio cholerae* and *C. diphtheriae,* are caused by phage-encoded toxins.

Phage characteristics

Many types of phages have been found to encode virulence factors, including lambdoid phages (e.g., Shiga-toxin-encoding phages), P2-like phages (e.g., SopEϕ), and filamentous phages (e.g., CTXϕ). To our knowledge, all virulence-linked phages are temperate phages (i.e., phages whose genomes can be stably maintained as components of bacterial genomes, usually via integration of phage DNA and repression of most phage genes). This is not surprising, because temperate phages may have more opportunities to acquire virulence genes (see mechanism below) and they, unlike lytic phages, can be transmitted vertically as prophages integrated within the bacterial genome. (A few temperate phages, such as P1, are transmitted as plasmids instead, but we do not know of any such phages that have incorporated virulence genes into their genomes.) Phage DNA is generally integrated into bacterial chromosomes by site-specific recombination between phage and chromosomal attachment (*att*) sequences, which results in prophages flanked by short imperfect direct repeats (23). The chromosomal genes near the insertion sites are generally not disrupted, even by prophages that integrate within a gene, because replacement sequences are supplied by the phage genome. Prophages are frequently found near the 3′ ends of tRNA genes (23). Because tRNAs have been highly conserved among bacterial species and are usually found in multiple copies per genome, utilization of tRNA genes as insertion sites may have facilitated development of phages with relatively broad host ranges and thus aided in transfer of exogenous DNA (including virulence genes) between disparate organisms.

How were virulence genes acquired?

Although virulence genes are present within many phages, most are not required for replication of phage DNA or assembly of virions, and they appear to have evolved separately from the essential components of phage genomes. (One potential exception to this generalization, the VPIϕ, is discussed under "PAI mobility"; other possible exceptions are Ace and Zot, two essential CTXϕ proteins that may also function as toxins.) Toxin genes within phage genomes frequently have their own promoters and terminators; consequently, many can be expressed in lysogeny, a state in which most phage genes are repressed. Virulence genes are frequently found at one end of a prophage, adjacent to phage *att* sequences (e.g., in SFII, CTXϕ, and T12). Their terminal location suggests that they were acquired because of imprecise excision of prophages from chromosomal DNA, a process that can result in packaging both of phage genes and of DNA usually found adjacent to, rather than within, the prophage (specialized transduction [Fig. 1]). In support of this model, potential precursor phages, which lack toxin genes but are otherwise similar, have been identified for CTXϕ and T12 (89; E. F. Boyd, A. J. Heilpern, and M. K. Waldor, submitted for publication), and phages that contain tRNA genes, presumably obtained via the same mechanism, have been detected. Phages may also obtain virulence genes via integration of transposons or insertions sequences within the prophage and via recombination with other phage genomes. For example, the Shiga-toxin-encoding phage 933W, which also encodes several tRNAs, is thought to have a mosaic genome derived from a variety of phages (107).

Postacquisition adaptations

Most bacteria use bacteriophages solely as conveyors of virulence genes, and phage functions are not required once phage DNA has been integrated. In fact, from the perspective of the bacterial host, maintenance of functional prophages can be a liability, because such prophages can be induced to excise from the chromosome, resulting in loss of the virulence gene and frequently in lysis of the host. Not surprisingly, some lysogens have evolved in ways that circumvent these potential liabilities. For example, CTX prophages in *V. cholerae* yield CTXϕ via a process that is unlikely to depend on prophage excision, so toxin-encoding genes are retained within the chromosome (34). Furthermore, filamentous phages such as CTXϕ are secreted, rather than released via host cell lysis, so lysogens are not destroyed by production of CTXϕ virions. Other virulence-linked prophages have apparently been converted into cryptic phages (102) or incorporated into PAIs (see section on PAIs below), which are more permanent, and less detri-

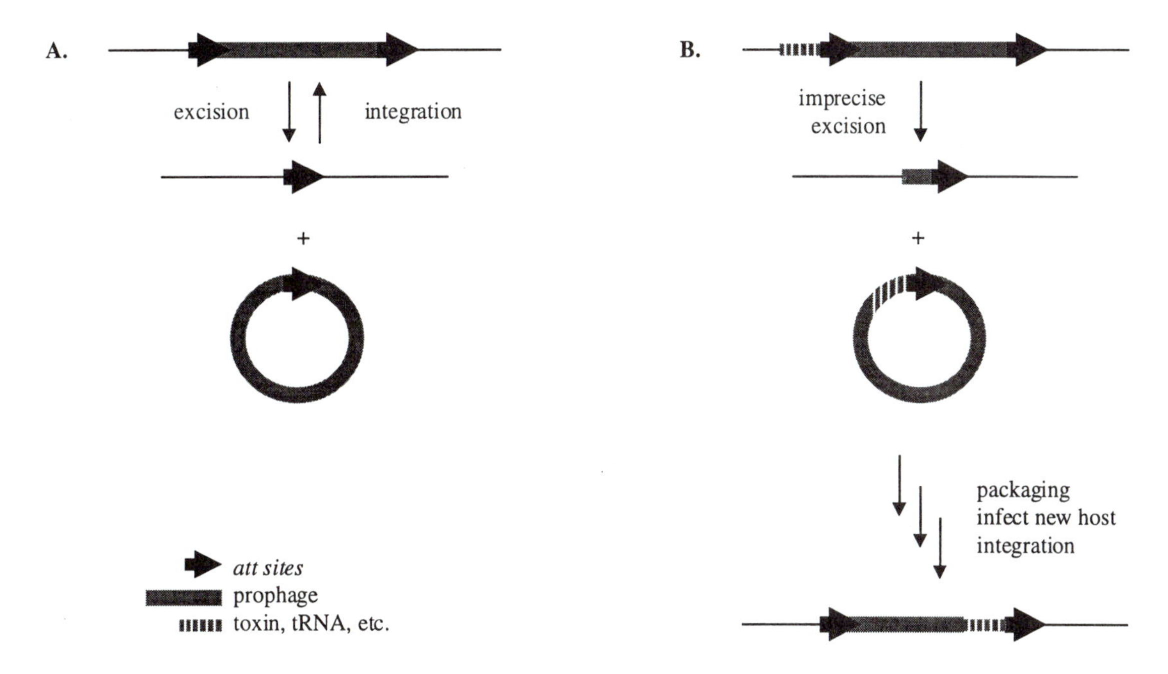

Figure 1. (A) Precise excision of a prophage from chromosomal flanking sequence. (B) Imprecise excision of a prophage, resulting in incorporation of adjacent chromosomal sequences into the phage genome and transfer of these sequences to a new host.

mental, chromosomal residents. Presumably, most prophages that retain both replicative and lytic capacities gradually develop means of balancing horizontal and vertical transmission that maximize perpetuation of the phage genome.

There is at least one instance in which a phage-encoded virulence factor only contributes effectively to pathogenesis when it is produced in conjunction with other phage proteins, so phage genes have to be maintained. Release of lambdoid phage-encoded Shiga toxin (Stx) from enterohemorrhagic *Escherichia coli* (EHEC) appears to depend on bacterial lysis induced by the phage proteins S and R (P. Wagner and M. K. Waldor, unpublished data). If the lytic process is blocked or delayed, Stx remains an intracellular, and thus effectively nontoxic, protein. *Shigella dysenteriae I* also can produce Stx; however, in contrast to EHEC, its toxin-encoding genes reside relatively stably on the chromosome, among numerous insertion sequences and some remnants of a lambdoid prophage (88). Although few intact phage genes are present, the genes encoding S and R have been maintained. Thus, although significant postacquisition adaptations have occurred in *S. dysenteriae,* they apparently have been constrained by a need to maintain the lysis-related genes. Further analyses may reveal that phage proteins produced in other pathogens make similar indirect contributions to virulence.

Plasmids

Virulence genes

Like phages, plasmids have been found to encode a diverse array of virulence factors, including toxins, pili, and type III secretion systems (Table 2). Many plasmids contain multiple virulence genes, some as elements of multigene clusters and/or operons and others as discrete, functionally independent entities. The plasmid-borne virulence genes of one species often appear to be related to virulence genes on mobile elements, plasmids or others, in other species. For example, the genes that encode heat-labile toxin in enterotoxigenic *E. coli* (ETEC), which are found in a variety of different plasmids, are very similar to the phage-linked cholera toxin genes of *V. cholerae* (32, 125). Similarly, the plasmid-encoded type III secretion systems of *Yersinia pestis* and of *Shigella* spp. are closely related to the PAI-encoded secretion systems of a variety of microorganisms (67). The similarities between the genes of each set strongly suggest that they evolved from ancestral virulence genes that were disseminated by horizontal transfer.

Plasmid characteristics

Most virulence plasmids are relatively large, especially in comparison with standard laboratory clon-

Table 2. Virulence-linked plasmids[a]

Microorganism (plasmid[b])	Virulence gene families	Function	Size (kb)	Reference(s)	Notes
Shigella spp.	*mxi*, *spa*, *ipa*	Type III secretion (invasion)	180–220	47, 105, 114	~31 kb are type III system
	icsA (*virG*)	Intracellular movement			
EPEC (EAF)	*bfp*	Type IV pili (BFP)	80	10, 61	
	perA	Transcriptional activator			
ETEC	*cfa*	Adherence pili	Various	36, 96, 121	Many types of adhesins and toxins, sometimes on same plasmids, frequently mobile or mobilizable
	etxAB	Heat-labile enterotoxin			
	estA, *estB*	(LT) heat-stabile enterotoxin (ST)			
EHEC (pO157)	*hly*	Hemolysin	92	21	Associated with virulence but no direct connection demonstrated
	esp	Protease			
		Cytotoxin?			
Salmonella spp. (pSLT, MP10, pQR28, pSTV)	*spv*	Intracellular proliferation	90	51, 81	Self-mobilizable (2); sometimes integrated (17)
	pef	Fimbriae			
Yersinia spp. (pYVe227, pIB1, pCD1)	*ysc*, *yop*, *syc*	Type III secretion (blocks macrophage phagocytosis, induces apoptosis)	66–72	67, 80	~20 kb conserved in all, core of type III secretion cluster
Y. pestis	*fra*	Capsular antigen fraction 1	100	28	Antiphagocytic
	pla	Plasminogen activator	9.5	123	Pla degrades many Yops (124)
B. anthracis (pXO1)	*cya*, *lef*, *pagA*	Anthrax toxin	174	103, 132	Nearly 70% of ORFs have no homologs in sequence databases
B. anthracis (pXO2)	*cap*, *dep*	Glutamic acid capsule	95	48, 133	
B. burgdorferi (many)	?		Many	44	Loss of plasmids correlates with loss of virulence. 58% of ORFs have no homologs in sequence databases
C. tetani	TeNT gene	Tetanus toxin	75	42	
C. perfringens	*cpb*, *iap*/*ibp*, *etx*	Toxins	Various	24, 111	All toxin genes are associated with insertion sequences
A. tumefaciens (Ti plasmids)	*vir*	Type IV secretion system	194–206	128, 142	
	ags, *iaa*, *mas*, *ocs*	Synthesis of plant hormones and opines (T-DNA genes)			

[a]This is not intended to be an exhaustive list.
[b]When no name is listed, plasmids are either referred to by a wide variety of names or simply denoted "virulence plasmids."

ing vectors. These plasmids comprise a variety of incompatibility groups (85). Many have not been fully sequenced, because analyses have focused on the virulence-linked genes; however, some similarities are apparent among those for which sequence data are available. First, virulence plasmids tend to have modular structures, with clusters of virulence genes, clusters of replication genes, and clusters (if present at all) of transfer-related genes. Individual clusters often differ in G + C content from their neighbors, which supports the theory that the clusters evolved and were acquired as independent units. Second, plasmids frequently contain remnants of small mobile elements, such as insertion sequences and transposons. These remnants can be identified by their characteristic repeat sequences or by the presence of genes such as integrases and transposases. Third, a high percentage of open reading frames (ORFs) on several virulence plasmids encode putative proteins with unknown functions; thus, these plasmids may be a repository for unusual or mutated sequences as well as for functional virulence genes. The average length of such plasmid ORFs is frequently shorter than the average length of chromosomal genes, whereas intergenic regions are larger, again suggesting that gene remnants, in addition to functional genes, are present (44, 103). Finally, at least a subset of virulence plasmids are relatively unstable and are readily lost by their hosts (e.g., plasmids in *Yersinia* and ETEC) (11, 36). However, this is not a universal characteristic; for example, pXO1 of *Bacillus anthracis* is well maintained even under apparently nonselective conditions (103), and the virulence plasmid of EHEC appears to encode a system for plasmid maintenance and partitioning (21).

How were virulence genes acquired?

Some of the characteristics of virulence plasmids—abundant insertion sequences and short ORFs—provide clues about how virulence genes came to be encoded on these plasmids, for they suggest that plasmids are sites of great genetic flux. It is impossible to be certain of the precise origins of the virulence plasmids; however, it is likely that insertion sequences and transposons facilitated their acquisition of virulence genes, either from chromosomal loci or from other, perhaps less versatile plasmids. Some transposons, such as Tn*1681* and Tn*4521*, routinely transport virulence genes to new loci (65, 122). Many virulence genes on contemporary plasmids are not directly adjacent to or flanked by insertion sequences; however, if such genes moved to plasmids in the distant past, it is likely that many would have become anchored there by loss or mutation of mobilizing sequences (which would generate the insertion sequence remnants noted above). It is also likely that some virulence genes were introduced into plasmids by formation of cointegrates between plasmids and lysogenic phages carrying virulence genes. Many plasmids have been found to contain phage-like genes, such as integrases, suggesting that such hybrids do form regularly (27). Finally, some lysogenic phages that lose their ability to integrate, and phages that infect a host without a functional integration site, can also be maintained as plasmids. Thus, there are a variety of mechanisms by which plasmids containing virulence genes can develop.

Intercellular transfer of plasmids

Some virulence plasmids contain genes that encode a DNA transfer apparatus (e.g., the *Salmonella* virulence plasmid) (2); however, many virulence plasmids lack such genes and are not capable of self-mobilization into a new host bacterium. Nonetheless, several lines of evidence generally suggest that these plasmids (or their precursors) were obtained by their current hosts through horizontal transfer of DNA. First, as mentioned above, the virulence plasmids of distinct pathogens are sometimes more closely related than are their chromosomal genes, which indicates that plasmid acquisition by at least one of the pathogens occurred after the two species diverged. Second, it is more likely that virulence plasmids were obtained by the few species or subspecies that use them in pathogenesis (e.g., ETEC and enteropathogenic *E. coli* [EPEC]) than that they were lost by all the closely related bacteria in which they are absent (e.g., nonpathogenic and uropathogenic *E. coli*, which may contain completely different plasmids). Finally, some plasmids, such as the large *Yersinia* virulence plasmid and the EHEC virulence plasmid, contain remnants of DNA transfer systems, suggesting that they were previously capable of self-mobilization. In addition, plasmids that cannot autonomously move to a new bacterial host can frequently do so with assistance from other mobile elements. For example, conjugative plasmids can mediate the transfer of other plasmids that do not encode a fully functional conjugation apparatus; the dependent plasmid only needs to contain an origin of transfer and a corresponding origin-nicking enzyme. Plasmids mobilizable in *trans* by other plasmids include the *Shigella* virulence plasmid and some of the virulence plasmids of ETEC (96, 115). Conjugative transposons and conjugal, self-transmissable, integrating elements (CONSTINS) can mediate plasmid transfer in *trans* as well (63, 113). In addition, nonconjugative virulence plasmids can be mobilized in *cis*, via formation of a cointegrate with a conjugative plasmid (e.g., in *Bacillus* spp.) (19) or a conjuga-

Table 3. Type IV secretion systems

Microorganism	Secretion substrate	Target	Biological effect	Reference(s)
H. pylori	CagA	Gastric epithelium (intracellular)	IL-8 induction, inflammation	26, 101
A. tumefaciens	T-DNA, VirE2	Plant cell cytoplasm	Crown gall tumor	29
B. pertussis	Pertussis toxin	Ciliated respiratory mucosa (extracellular)	Increased cAMP, cell death	141
L. pneumophila[a]	? (mobilizes exogenous plasmid DNA)	Alveolar macrophages	Prevents lysosomal fusion; enables growth within macrophages	118, 137

[a]The putative type IV secretion in *L. pneumophila* contains only two genes similar to those of other type IV secretion systems; however, it contains many genes homologous to components of bacterial conjugation systems, and it mediates transfer of plasmid DNA. Consequently, it is likely that it will be found to function in a manner analogous to that of type IV secretion systems, although it may use structurally distinct apparatus.

tive transposon (e.g., in *Bacteroides*) (113). Mobilization in *cis* allows for horizontal transfer even of plasmids that lack an origin of transfer. With such a mechanism available, basically all plasmids are potentially subject to horizontal transfer.

Related systems: type IV secretion

In addition to mediating intercellular transfer of DNA, both self-mobilization and mobilization of unlinked elements, some conjugative systems have become specialized for transport of proteins. Because conjugatively transferred DNA is typically part of a DNA-protein complex (29), these systems, known as type IV secretion systems, have not completely shifted substrates; however, it is clear that they exploit a conjugation-related apparatus for an unusual set of tasks (Table 3). Type IV secretion systems can mediate transport of proteins into animal and plant cells and into culture supernatants, and they make crucial contributions to pathogenesis. Several type IV systems are capable of transporting DNA as well as protein, but a physiological role for such DNA transfer has only been demonstrated for the Vir system of *Agrobacterium tumefaciens,* which transfers T-DNA into its plant host to prompt growth of crown gall tumors (29, 145). These type IV systems are probably no longer relevant for interbacterial DNA transfer (e.g., the Ti plasmid contains a separate gene cluster used for bacterial conjugation); however, they provide an interesting demonstration of an alternate means by which DNA transfer systems, frequently found as components of mobile elements, can contribute to pathogenesis.

PAIs

Virulence factors

The term pathogenicity island has been used with increasing frequency during the past decade to describe clusters of chromosome-encoded virulence genes (50, 52, 78). These clusters encode virulence factors such as toxins, type III secretion systems, pili, and iron acquisition systems (Table 4). Several dozen PAIs have now been described, and with the increasing availability of microbial genome sequences, it is likely that even more will soon be identified. There is no apparent limit on the number of PAIs per organism; some organisms, such as *Salmonella enterica,* have already been shown to contain four large islands and several smaller pathogen-specific chromosomal loci.

PAI definition

In a comprehensive review of PAIs in 1997, Hacker et al. (54) proposed that they are defined by the following characteristics:

1. presence in pathogenic strains and absence/sporadic presence in less pathogenic strains of the same or a related species
2. occupation of large chromosomal regions (often >30 kb) and carriage of virulence genes (often many)
3. tRNA genes, direct repeats, and/or insertion sequences at boundaries; (sometimes cryptic) mobility genes and sequences (e.g., insertion sequences, integrases, transposases, plasmid origins of replication) present within the element
4. different G+C content than the remainder of the chromosomal DNA of the host bacteria
5. found as compact, distinct units
6. instability (prone to deletion)

These characteristics can serve as a useful starting point for a discussion of PAIs; however, as Hacker et al. pointed out, many elements that are generally considered to be PAIs do not satisfy all these criteria. Overall, PAIs comprise a very diverse group of virulence loci, and the boundary lines of PAI territory are rather fuzzy. Unlike plasmids, which replicate as ex-

Table 4. Pathogenicity islands[a]

Microorganism	PAI	Virulence gene families	Function	G+C content (host/PAI)[b]	Size (kb)	Boundary (bp)[c]	tRNA[d]	Reference(s)
E. coli 536 (UPEC)	PAI I$_{536}$[e]	*hly*	Hemolysin	51/41	70	DR 16	*selC*	16
	PAI II$_{536}$[e]	*hly*	Hemolysin	51/41	190	DR 18	*leuX*	16
		prf	P fimbriae					
	PAI III$_{536}$	*sfa*	S fimbriae	51/46	25		*thrW*	53
	PAI IV$_{536}$	*fyuA, irp*	Yersiniabactin synthesis, uptake	51/~56	45		*asnT*	116
E. coli J96 (UPEC)	PAI I$_{J96}$	*hly*	Hemolysin	51/41	170		*pheV*	129
		pap	P fimbriae					
	PAI II$_{J96}$	*hly*	Hemolysin	51/41	110	DR 135	*pheR*	15, 129
		prs	P fimbriae					
		cnfl	Cytotoxic necrotizing factor I					
E. coli E2348/69 (EPEC)	LEE	*esc, esp, tir*	Type III secretion (attaching/effacing)	51/39	35		*selC*	87
		eaeA	Adhesin (intimin)					
E. coli EDL933 (EHEC)	LEE[f]	*esc, esp, tir*	Type III secretion (attaching/effacing)	51/41	43		*selC*	106
		eaeA	Adhesin (intimin)					
Y. pestis	pgm-HPI[g]	*hms*	Hemin storage	46–50/46–50 (*hms* = 47, *psn*/*ybt* = 56–59)	102	IS*100* (HPI has DR 16)	(HPI in *asnT*)	38
		psn	Yersiniabactin (Ybt) receptor					
		ybt	Ybt synthesis					
Y. enterocolitica	HPI	*fyuA*[h]	Yersiniabactin (Ybt) receptor	46–50/56	43		*asnT*	25
		irp	Ybt synthesis					
Y. pseudotuberculosis	HPI	*psn*	Yersiniabactin (Ybt) receptor	46–50/57 (Ybt region)	36	DR 16	*asnTUW*	20, 58
		ybt	Ybt synthesis					
S. enterica (all subspecies tested)	SPI-1	*inv, spa, sip*	Type III secretion (invasion, macrophage apoptosis, colonization)	52/40–47	40			92, 97
		hilA, invF	tx regulators					
S. enterica (most subspecies tested)	SPI-2	*ssa, sse, sse, spi*	Type III secretion[i] (intracellular proliferation, systemic disease)	52/40–47	31		*valV*	119, 135
		spiR	tx regulator					
S. enterica[j]	SPI-3	*mgt*	Survival in macrophages	52/47.5	17		*selC*	13, 14

Continued on following page

Table 4. *Continued*

Microorganism	PAI	Virulence gene families	Function	G+C content (host/PAI)[b]	Size (kb)	Boundary (bp)[c]	tRNA[d]	Reference(s)
S. enterica serovar Typhimurium	SPI-4	?	Survival in macrophages	52/37–54	25		tRNA-like	143
S. flexneri 2a	She (SHI-1)	*set1A, set1B* *she, sigA*	Enterotoxins Proteases	51/45–54	51	IS		4, 108
S. flexneri	SHI-2	*iuc*	Aerobactin synthesis/transport	52/34–56	24		*selC*	94, 138
V. cholerae	VPI	*tcp* *acf* *toxT, tcpP*	Type IV pilus Colonization factor Tx regulators	48/35	40	DR 13	*ssrA*[k]	71, 74
H. pylori	*cag* PAI	*cagA-T*	Type IV secretion IL-8 induction	38–45/35	40	DR 31		26
D. nodosus	*vap* region	*vapA-E*	Vap antigens	45/52	12	DR 19	*serV*	27
	vrl region	*vrl*	Vrl antigens	49/57	27		*ssrA*[k]	11
P. aeruginosa	Exo S regulon	*psc, pcr, exs, pop* *exsA*	Type III secretion[l] (ExoS secretion) Tx regulator	69/67.5	?			43
P. syringae	*hrp* cluster	*hrc, hrp* *hrpR, hrpS*	Type III secretion Tx regulator	60/46–58	~50		tRNA[Leu]	5
Erwinia spp.	*hrp* cluster	*hrp, hrc*	Type III secretion[l]		~25			
S. aureus	SaPI1	*tst*	Toxin (TSST-1)	35/35	15.2	DR 17	*tyrB*	82
L. monocytogenes	*prfA* cluster	*prfA, plc, hly*	Phospholipase, listeriolysin	No difference	9.5			75
C. difficile	PaLoc	*tcd*	Toxins	Different in PaLoc	19			18

[a]This table is not an exhaustive list of PAIs; rather, it contains a subset of them, composed primarily of large, well characterized PAIs with significant roles in pathogenesis. Much of the information in this table is also presented in reference 55 and in subsequent chapters in the same book.
[b]G+C contents for which a range is given contain regions with distinct compositions; some are clearly mosaic elements. Elements for which a single value is given may nonetheless contain similar variations.
[c]DR, direct repeat; IS, insertion sequence. If none is listed, no repeat sequences have been detected at the PAI boundaries.
[d]Site of insertion. PAIs for which no site is listed are not believed to be inserted in tRNA genes.
[e]UPEC PAI I and PAI II deletions are significant not only because of loss of virulence genes, but also because they inactivate the flanking tRNA genes (110).
[f]The EHEC LEE is very similar to the EPEC LEE; it is larger because it also contains a P4 family prophage at one end.
[g]*Y. pestis* pgm-HPI appears to be a compound locus, formed by sequential integration of independent virulence cassettes (58). The *psn* and *ybt* loci are contained within an ~36-kb region, known as the high pathogenicity island (HPI), which has also been found in other pathogenic *Yersinia* spp. These genes are adjacent to a second set of virulence genes—the pigmentation (pgm) locus—which was acquired as a separate unit. The pgm locus is important for survival of the bacteria within its insect host (63); it is also found in some strains of *Y. tuberculosis* and *E. coli* (not listed above).
[h]*fyuA* is an alternate name for *psn*; *irp* genes are orthologs of *ybt* genes.
[i]More related to the *E. coli* LEE than to genes in SPI-1; SPI-2 apparently did not develop via duplication and subsequent modification of SPI-1 (99).
[j]Some genes of SPI-3 are present in all subspecies, but others are not, suggesting that this PAI arose via a multistep process (14).
[k]Regulatory RNA, not a tRNA.
[l]It is thought that the genes encoding type III secretion systems arose initially from an ancestral flagellar gene cluster, then were distributed among a broad variety of bacterial species via horizontal gene transfer. Some type III secretion gene clusters are included in the table despite their lack of most features of PAIs, because they are clearly (i) virulence-linked, (ii) chromosomal, and (iii) derived from foreign sequences.

trachromosomal circles, and phages, which transduce DNA, PAIs have no obvious functional characteristic by which they can clearly be identified. Instead, PAIs are united by their apparently similar, although far from identical, evolutionary origins. All the PAI-defining characteristics listed above, except the requirement for virulence genes, are simply those indicative of horizontally transferred genetic loci. However, as we have already seen, a variety of molecular processes can mediate gene transfer, and these leave overlapping but distinct genetic fingerprints. Furthermore, acquisition of foreign sequences is only the first step in the evolution of PAIs, and subsequent steps can erase these signs of DNA transmission. Consequently, most PAIs possess a subset, rather than all, of the characteristics listed above. To faciliate understanding of PAIs, we discuss each of the definition points in turn, attempting to clarify both how widely each applies and what each reveals about the evolution of PAIs and pathogens.

Applicability of PAI criteria

Presence in pathogenic strains, absence in less/nonpathogenic strains. This is a fundamental characteristic of PAIs, and it describes most elements generally included under the PAI rubric. It is clear that if putative virulence genes are present in nonpathogens and in pathogens, then the genes' evolutionary history and their role in pathogenesis may have been misconstrued. Nonetheless, there can be difficulties in assessing whether this criterion has been met, because a pathogen may be evolutionarily isolated and lack close relatives, or all related organisms may be pathogens as well. Analyses can also be confounded by nonpathogens that contain but do not express virulence-linked loci, such as *Listeria seeligeri,* which contains an inactive form of the *Listeria monocytogenes* PAI (75). Gene disruption and gene transfer techniques frequently provide the most straightforward means of directly assessing a putative island's contribution to virulence.

Occupation of large chromosomal regions, carriage of (many) virulence genes. Like the first point, these are some of the most fundamental characteristics of PAIs, and as such are met by most elements generally included in the category. Still, there are instances in which it is not obvious whether these criteria have been satisfied. For example, it is not clear how large an element must be to be considered an island. Is the Vap region of *Dichelobacter nodosus,* which is <12 kb, truly an island, or is it perhaps an "islet", as some investigators have termed smaller regions? Similarly, how many virulence genes must actually be present within such regions; is one sufficient? Precise answers to such questions are probably unnecessary and would undoubtedly be arbitrary; however, posing the questions provides a useful reminder that even the most basic characteristics of PAIs are not sharply demarcated and can only serve as general guidelines.

In our minds, the chromosomal location of PAIs is one of their most definitive features. Consequently, although we recognize the similarities between some virulence gene clusters within plasmids and within PAIs, we think that referring to the plasmid genes as PAIs on plasmids, as has been done for the Yop virulon of the *Yersinia* spp. virulence plasmid, the T-DNA regions of the *A. tumefaciens* Ti plasmid, and the virulence genes of the *B. anthracis* plasmid pXO1, unnecessarily obscures some distinctions that are worth preserving. First, PAIs in general are not mobile or even mobilizable elements (see discussion below); they are elements tethered fairly firmly to the chromosome, whose days of horizontal transfer are past. In contrast, most plasmid encoded virulence genes could gain the capacity to be mobilized fairly readily, and several (including the Ti plasmid) are already mobile or mobilizable. Thus, PAIs far more than plasmids tend to be an evolutionary final destination for virulence genes. Second, it is not clear why only a portion of a virulence plasmid should count as a PAI, because if the plasmid were to integrate, the entire former plasmid would be considered a PAI, not just the virulence genes. In addition, once some plasmid-encoded virulence genes are counted as PAIs, it will be difficult to determine why all virulence genes on plasmids should not be called PAIs, and the PAI category will become so imprecisely bounded that it loses all meaning. The utility of the PAI designation is to distinguish chromosome-encoded virulence loci obtained via horizontal gene transfer from ancestral chromosomal genes that contribute to virulence. Plasmids are already recognized as agents of horizontal gene transfer; consequently, the term virulence plasmid should suffice to suggest that exogenously derived virulence sequences are present.

tRNAs, insertion sequences, and mobility sequences. tRNA genes are frequently adjacent to PAIs, direct repeats often mark their boundaries, and insertion sequences (IS) and mobility sequences are widespread within them (see Table 4). These elements provide strong clues about the nature of the processes by which PAIs evolved, even though it is not generally possible to determine the precise steps that gave rise to each PAI. Basically, PAIs appear to be chromosomal regions generated by integration of mobile or mobilizable DNA—e.g., from phages, plasmids, or conjugative transposons—that had incorporated virulence genes (by means described in preceding sections). In many cases it appears that hybrid elements (e.g., containing both phage and plasmid genes) were precursors to PAIs, although it is not clear whether the hybrids arose in the donors or the recipients. It appears that integration of precursor sequences was frequently mediated by phage or phagelike integrases, resulting in PAIs adjacent to tRNA genes and flanked by direct repeats. As in prophages, an integrase gene is often found next to one of the *att* sites (direct repeats; e.g., in the *V. cholerae* VPI, *Staphylococcus aureus* SaPI1 and SaPI2, and *Yersinia* HPIs). Some PAIs may have integrated instead as compound transposons flanked by ISs, although this mechanism was apparently less common, because only a few PAIs are known to be flanked by ISs. One of these, the pgm-HPI of *Y. pestis,* was not integrated via these loci, because it was acquired in two steps (see Table 4 and "Found as compact, distinct units" below). Plasmid integration, mediated by homologous recombination between plasmid and chromosomal IS, may also underlie formation of some PAIs.

Some putative PAIs lack many of the defining features of PAIs, and thus it is theoretically possible that they arose via different processes than those described above. Atypical PAIs are especially prevalent among gram-positive organisms (54). For example, several gram-positive PAIs lack mobility genes, flanking repeats, and nearby ISs (e.g., the *prfA* gene cluster of *L. monocytogenes* and the PaLoc of *Clostridium difficile*) (18, 75). (Note, in contrast, that *S. aureus* SaPI1 is marked by all of these features.) However, it is unlikely that these gram-positive PAIs were shaped by completely different means than other PAIs. Instead, their assembly may have been mediated by mobile elements that leave somewhat atypical molecular fingerprints, such as transposons like Tn*554* and conjugative transposons like Tn*916,* which do not duplicate target sequences and can lack terminal repeats (95). Alternatively, these PAIs may have been acquired in the distant past, and they may have since lost many defining features. Mobility sequences generally do not contribute to virulence; hence, there is no selective pressure promoting their maintenance (see also "Found as compact, distinct units"). PAIs obtained early within the evolution of a species (e.g., the *Salmonella* PAIs) often contain relatively few sequences associated with horizontal gene transfer.

Differences in G+C content. A difference in G+C content between the DNA of a putative PAI and surrounding chromosomal sequences provides unambiguous evidence that the PAI is in fact "foreign" DNA. Such differences were useful in identifying many of the PAIs listed in Table 4, and they provide some clues about the identities of the donor organisms. However, a lack of such differences does not rule out the possibility that a virulence-linked sequence is a PAI. For example, differences are not detectable when the DNA of the donor and recipient organisms has a similar G+C content, although it may still be possible under such circumstances to identify foreign sequences by differences in codon usage frequencies and other measures of nucleotide composition. In addition, differences may not be detected if the DNA transfer took place in the very distant past and/or if the recipient rapidly eliminates G+C imbalances. (If given enough time, the G+C content of "foreign" DNA eventually becomes indistinguishable from "ancestral" genes in all organisms [76].) The potential for confusion caused by such G+C normative processes is well demonstrated by comparisons of the PAI of *Pseudomonas aeruginosa* and the virulence plasmid of *Y. pestis,* which encode type III secretion systems more closely related to each other than to all other type III systems (67). These and other type III gene clusters are widely presumed to be derived from horizontally transferred sequences. However, in *Y. pestis,* these sequences have a G+C content of ~44%, similar to the chromosomal G+C content, whereas in *P. aeruginosa* both type III and chromosomal genes are ~68% G+C (67). Consequently, it is likely that at least one of the two type III gene clusters has undergone a dramatic shift in G+C content. *P. aeruginosa* is the most likely candidate for such change, because most other type III secretion systems are encoded by gene clusters with a G+C content of <50%. Thus, there is a strong possibility that a G+C imbalance once existed between the *P. aeruginosa* PAI and surrounding sequences, but this difference is no longer present as an indicator of horizontally transferred sequences.

Found as compact, distinct units. The description of PAIs as compact units stems from the fact that multiple virulence genes are frequently clustered within them, rather than distributed evenly along the host chromosome. This is especially apparent for loci that contain complex gene clusters, such as those encoding pili or type III secretion systems. PAIs are distinct units in that they (sometimes) differ in G+C content from flanking sequences (see "Differences in G+C content") and that they tend to be lost as discrete units if they are unstable (see "Instability"). Nonetheless, there are some PAIs, such as those of *S. aureus, L. monocytogenes,* and *P. aeruginosa,* that cannot be distinguished by either of these features. Furthermore, some loci that appear to be discrete units are actually hybrid elements, formed via several sequential integration events rather than acquisition of a single cluster of horizontally transferred genes. For example, the *hms, psn,* and *ybt* genes of *Y. pestis* (the pgm-HPI region) are lost together at a frequency of 10^{-4} to 10^{-5}, and thus were proposed to be a PAI (38, 39). However, comparisons of *Y. pestis* sequences with related sequences in *Y. enterocolitica, Y. pseudotuberculosis,* and some strains of *E. coli* indicate that the pgm-HPI region is actually composed of two subregions that were apparently acquired independently (58, 109). Each of these regions contains or is adjacent to a single IS*100,* and it appears that recombination between the two ISs mediates deletion of the intervening ~102 kb, which contains all the virulence genes associated with the two distinct PAIs. Thus, in this case, simultaneous loss of multiple virulence genes is coincidental; the genes are deleted as a distinct unit but they were not acquired as a single unit.

SPI-3 of *Salmonella* and the locus of enterocyte effacement (LEE) pathogenicity island of O157:H7 EHEC strains are also clearly composite elements. The EHEC LEE is basically equivalent to the LEE of enteropathogenic *E. coli,* except that it has acquired the genes of a P4 family prophage (106). Taking a larger view, all PAIs probably can be considered to be mosaic elements, because they were generated from mobile elements that had already become mosaics by acquiring virulence genes. However, it is not usually possible to determine whether a PAI was fully assembled before chromosomal integration (which is probably the case for many PAIs) or whether its gene clusters were integrated sequentially (as with the *Y. pestis* pgm-HPI, the *Salmonella* SPI-3, and the EHEC LEE).

Instability. Instability (i.e., an unusually high frequency of loss) is a feature of a subset of PAIs rather than of all PAIs (54). As a defining characteristic, it is like G + C content, useful if present, but uninformative if absent. It is likely that whether a PAI is stable primarily reflects the initial genetic content of the precursor and the extent of postacquisition modifications it has undergone. For example, PAIs that were integrated into the chromosome by a phagelike integration process should be unstable after acquisition as long as they still encode functional phage proteins needed for excision (integrase, sometimes excisionase). However, if the PAI progenitor did not encode an appropriate excisionase (e.g., probably SaPI1 and SaPI2) (82), or if the PAI's recombination genes or direct repeats were mutated after integration (e.g., in *Y. enterocolitica*) (109), then the PAI is likely to be stable. Because stabilizing mutations prevent loss of key virulence genes, from the bacterial perspective such mutations are probably beneficial rather than deleterious. It has been proposed that all PAIs gradually evolve to be fixed chromosomal elements (55).

PAI mobility

Although PAIs are derived from mobile elements, in general, they do not encode self-mobilization functions. Consequently, even those that can excise from the chromosome typically cannot transfer autonomously to new hosts. Thus, despite their origins, most PAIs are not considered to be mobile elements. There are a few PAIs that can be mobilized, however, at least under some laboratory conditions. For example, the SaPI1 island of *S. aureus,* which encodes the toxin responsible for toxic shock syndrome, can be transferred from donor strains that contain a specific helper phage, 80α (109). The helper phage apparently enables the island to excise and replicate, although the precise mechanism for this is not known. The island is then transferred to a recipient via transduction, where it integrates site-specifically, probably via an island-encoded integrase, into the site where it was found in the donor. Some toxigenic strains of *S. aureus* contain a variant island, SaPI2, which is inserted at a different chromosomal locus and can be mobilized by a different helper phage (109). Neither island tends to be excised or mobilized by natural isolates of *S. aureus,* because the helper phages are not widespread among these strains. It is not clear whether helper phages contributed to the initial distribution of SaPI1 and SaPI2; it is possible that the interactions between these PAIs and helper phages are primarily laboratory phenomena. It is also not clear whether these two elements truly fit under the rubric of PAIs. Perhaps they would be classified more accurately as satellite phages, akin to the *E. coli* bacteriophage P4, whose transmission depends on the phage P2. It would be worthwhile to investigate whether other PAIs can also be mobilized by helper phages, because such studies might yield greater understanding of how PAIs came to reside in their current hosts. Similar studies could explore whether any PAIs can be mobilized by conjugative transposons, because some PAIs might be analogous to the nonreplicating *Bacteriodes* units of *Bacteroides,* some of which also integrate into tRNA genes (112).

An additional PAI, the VPI of *V. cholerae,* was also recently transferred to a new host via transduction (72). In fact, the VPI was reported to correspond to the genome of an integrated filamentous phage, dubbed VPIϕ. However, the process underlying transmission of the VPI has not been extensively characterized. The VPI does not appear to contain a complete phage genome (72, 74), and the proposed major coat protein of VPIϕ already has a known function as the major subunit of the type IV pilus, TCP (130). The relationship between this VPI-encoded pilus and VPIϕ is ambiguous; they may be distinct entities, but no means for discriminating between them has been reported. TCP is already known to function as the receptor for the cholera-toxin-encoding phage, CTXϕ (139). The story is further complicated because very few *V. cholerae* strains were identified that could serve as either VPI donors or VPI recipients. One possible explanation for some of the puzzling aspects of VPI transmission is that VPI mobility, like that of SaPI1, may depend on a helper phage or some other accessory genetic element.

PAIs can also be transferred to new hosts by relatively nonspecific processes. For example, the SigE-encoding pathogenicity islet of *S. enterica* serovar Typhimurium can be mobilized at a low frequency by the SXT element, a chromosomally integrated conjugative element termed a CONSTIN, which can mediate transfer of chromosomal sequences in an Hfr-like fashion (63). At least in theory, other CONSTINS and integrated conjugal plasmids should be able to mobilize linked sequences in a similar manner. Such nonspecific mobilization of PAIs has not been thoroughly investigated. However, if homologous recombination mediates chromosomal integration of PAI DNA introduced via conjugal transfer, then most species could probably only acquire nonspecifically transferred sequences from a closely related strain. Alternatively, if PAI-encoded integrases are functional, then it is possible that conjugal transfer of PAI DNA could facilitate wider dispersal of PAIs.

REGULATION OF VIRULENCE GENES

Although the virulence genes acquired as components of mobile elements may not have been subject to host regulatory processes immediately after acquisition, they clearly have not remained autonomous agents, independent of host functions. Instead, they are woven into transcriptional regulatory networks that involve both genes of the ancestral bacterial chromosomes and genes on additional mobile elements. For example, production of the key virulence factors of *V. cholerae,* cholera toxin and the intestinal colonization factor TCP, is controlled both by the VPI-encoded regulator ToxT and by an ancestral chromosome-encoded regulator, ToxR (35, 120) (see Fig. 2). Thus, *toxT,* which is a resident of a PAI, regulates expression of additional horizontally acquired genes (*ctxAB*), which reside in a prophage. Similarly, in EPEC, operons of the LEE PAI are regulated both by LEE-encoded regulator (Ler) and by plasmid-encoded regulator (PerA) (90), whereas in *Salmonella* spp., the genes of SPI-1 are controlled by the SPI-1 products HilA and InvF and by the ancestral chromosomal regulators PhoP and SirA (7, 69, 70, 83). Thus, even virulence gene clusters acquired with their own control mechanisms, as were most type III secretion regulons, have become subject to exogenous control elements, such as global regulators that mediate responses to environmental conditions. In this way, the production of virulence factors has become precisely integrated in the cellular physiology of the pathogen. Stimuli that induce changes in virulence

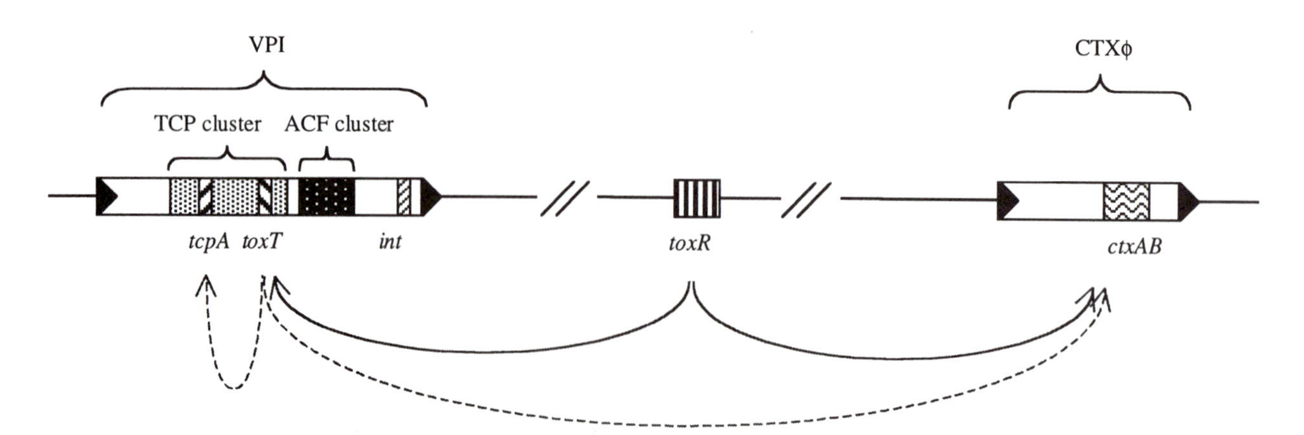

Figure 2. Regulation of virulence gene expression in *V. cholerae* by transcription factors encoded within the chromosome and within the VPI. An ancestral chromosomal gene, *toxR,* regulates expression both of *ctxAB* (encoding cholera toxin), which is part of the CTX prophage, and *toxT* (encoding another regulator of virulence gene expression), which is part of the vibrio pathogenicity island (VPI). ToxT is required for expression of *ctxAB* and other genes on the VPI, including the TCP and ACF gene clusters, which encode factors required for intestinal colonization (79). Additional factors encoded on the chromosome and on the VPI also participate in this regulatory network (not shown). Black triangles represent the direct repeats that flank the VPI and the CTX prophage, *int* encodes the putative VPI integrase, and *tcpA* encodes the major subunit of the toxin-coregulated pilus (TCP).

gene expression include temperature, the concentration of iron and other cations, osmolarity, pH, growth phase, and population density (41, 84).

Precise spatial and temporal regulation of virulence gene expression is essential to the process of pathogenesis. There is ordered, sequential expression of virulence factors during infection (79), and production of virulence factors at inappropriate times or levels can dramatically impair colonization by bacterial pathogens (e.g., as seen in *dam* or *phoP* mutants of *Salmonella* or *bvg* mutants of *Bordetella* [3, 60, 91]). The fact that a single host frequently contains genetically unlinked, (apparently) independently acquired virulence genes that are part of a single regulon (e.g., genes regulated by PerA, ToxR, and ToxT as mentioned above, and by PrfA in *Listeria* [75]) also highlights the importance of tight, coordinated regulation of virulence genes. These intricate means for controlling gene expression exemplify the remarkable coevolution of mobile elements and ancestral chromosomes that underlies development of bacterial pathogens (e.g., see Fig. 2).

Regulation of virulence genes within phage genomes may be more complex than that of other loci, because expression of phage-encoded genes can be influenced by several aspects of the phage life cycle and by traditional regulatory factors. First, phage induction generally results in extrachromosomal phage DNA replication, which can dramatically increase the copy number per cell of virulence genes. Consequently, the level of virulence gene transcripts may increase after phage induction, even if the activity of individual promoters is unchanged. (Whether comparable copy number increases ever modulate expression of plasmid-encoded virulence genes has not been investigated.) At least some phages have been produced during the course of experimental bacterial infections (e.g., CTXϕ by *V. cholerae* and Stx-encoding phages by EHEC [1, 73, 76]), and there is some indirect evidence for phage production during human infections as well (104). However, it is not yet known whether the frequency of prophage induction within a host is elevated relative to that occurring when bacteria are grown in experimental media.

In addition to its effect on the copy number of virulence genes, phage induction can also contribute directly to pathogenesis. As mentioned previously (see "Postacquisition Adaptations"), phage production by EHEC strains apparently enables release of Shiga toxin (Stx1 and Stx2). Furthermore, in lysogens of Stx2-encoding phages, toxin genes are primarily transcribed from the late-phage promoter, pR′, from which transcripts for the lysis genes *R* and *S* are also believed to originate (M. Neely, P. Wagner, M. K. Waldor, and D. Friedman, unpublished results). This means that the transcriptional regulatory cascade that governs phage development also controls *stx2* expression, so production and release of toxin are directly dependent on the phage life cycle. This coupling may have great clinical significance; some antibiotics used to treat EHEC infections trigger prophage induction and thereby also induce toxin synthesis and release (144). However, for most other prophage-encoded virulence factors it is not clear to what extent prophage induction contributes to the pathogenicity of the host bacterium. It should be remembered that bacterial gains due to increases in virulence factor expression or release after prophage induction may be offset by decreases in the bacterial population caused by phage-induced lysis. It is possible that prophage induction generally facilitates survival of the phage genome more than that of the bacterial host species.

WHERE DOES DNA TRANSFER OCCUR?

Little is known with certainty about the sites where past events of horizontal DNA transfer occurred. In most, if not all cases, the original donors of horizontally transferred sequences have not been identified, so the transfer sites cannot be delimited by knowledge of the sites that the donor species inhabit. Furthermore, some horizontal gene transfer can occur even between microorganisms that do not have a shared environment (e.g., transduction). It is likely that virulence genes have been transferred both outside and inside of host organisms. Some current pathogens probably acquired at least rudimentary virulence genes while outside their hosts, thereby gaining the ability to survive within them. However, there is indirect evidence suggesting that at least a subset of virulence genes may have been acquired while pathogens were within host environments. As mentioned above, some virulence-linked phages (e.g., CTXϕ, Stx-encoding phages) are produced during infections, and conditions that activate prophages (e.g., oxidative stress and DNA damage), which may be encountered because of host defenses, induce production of these virions and others (e.g., Gifsy-2) (1, 40, 73, 77). In addition, induction of CTXϕ's receptor, TCP, occurs during intestinal colonization by *V. cholerae* (79). These examples suggest that conditions within the host, and perhaps even specific signals produced by the host, may facilitate phage-mediated mobilization of virulence genes. Plasmid transfer can also be promoted during the course of an infection. Mobilization of the *A. tumifaciens* Ti plasmid is induced by signals produced by the plant host after transformation with plasmid-derived T-DNA (46, 145). Furthermore, it is likely that much of the recent spread of antibiotic resistance genes among bacteria, which was mediated by transposons, integrons, and plasmids, has occurred within eukaryotic hosts as a result of drug treatment, because such resistance was not widespread before the antibiotic era (33, 56, 100). Mobilization of many conjugative transposons that carry tetracycline resistance cassettes is induced by, or even dependent on, the presence of tetracycline; thus, the means as well as the need for resistance can be fostered by drug treatment (112). Because horizontal gene transfer clearly can occur within eukaryotic hosts, it seems likely that such sites have been used throughout the evolution of pathogens for transfer of virulence genes, although additional studies are warranted to identify the environmental conditions most conducive to the various forms of gene transfer.

WHY MOBILE DNA?

Mobile genetic elements have clearly distributed a diverse collection of virulence genes and thereby played essential roles in the evolution of bacterial pathogens. What may be less clear is why horizontally transferred genes have had such a profound influence. It appears that microbes, both pathogens and nonpathogens, frequently adopt foreign sequences, rather than simply adapting endogenous sequences, because doing so dramatically accelerates their rate of evolution. Point mutation-based evolution of endogenous sequences is a very slow and inefficient process, similar to a random walk that frequently arrives at a dead end. Such a process can eventually yield new genes; indeed, it must have given rise to all the sequences that were ultimately distributed by horizontal gene transfer, which includes all classes of genes, not just virulence genes. However, once a single species has evolved a new trait by this slow process, transfer of the corresponding DNA to additional microbes can immediately confer the new trait on them as well. Thus, horizontal transfer enables multiple species to, in effect, pool their resources, so each can obtain similar new traits without traversing lengthy parallel courses. The efficacy of this approach is well illustrated by the fact that, so far, none of the traits by which *E. coli* can be distinguished from the closely related species *S. enterica* are attributable to stepwise, point mutation-mediated development of new genes (76). Instead, horizontal DNA transfer and chromosomal deletion appear to account for all the phenotypic differences between these bacteria. Horizontal transfer was the source of more than 17% of the open reading frames in the genome of *E. coli* K12 (76). The evolutionary significance of horizontal gene flux is also hinted at by analysis of the *Chlamydia trachomatis* genome. This pathogen has obtained relatively few genes from other bacteria, probably because of early adoption of its obligate intracellular lifestyle, which isolated it from other microbes; however, its genome contains an unusually high number of genes with a eukaryotic origin instead (126).

Despite the capacity of many bacterial species to acquire DNA from exogenous sources, the sizes of their genomes are thought to have remained relatively constant throughout their evolution (76, 100). Increases in bacterial genome sizes caused by horizontal gene transfer have apparently been offset by the tendency of bacteria to lose, via deletion, nonbeneficial and nonessential genes. (This tendency is manifested even when their genomes are not augmented by foreign sequences, as in the intracellular pathogens *Chlamydia* and *Rickettsia,* whose genomes appear to be dwindling [6].) Ultimately, bacteria that retain useful new genes and lose nonessential sequences are thought to become predominant because of their enhanced fitness relative to competing strains. Many horizontally transferred virulence genes yield obvious benefits for their new hosts, so the selective advantages favoring their hosts are readily apparent. It is easy to see why genes that enable microorganisms to live as intracellular pathogens (e.g., the *mxi/spa* gene cluster of the *Shigella* virulence plasmid or SPI-2 of *S. enterica*) or to colonize new extracellular niches (e.g., the TCP and bundle-forming-pilus-encoding loci in *V. cholerae* and EPEC) would be selectively maintained; they can grant the bacteria access to abundant nutrients or uncontested space, both of which can facilitate proliferation. Bacteria of the gastrointestinal tract that induce diarrhea (e.g., cholera toxin-producing *V. cholerae* and attaching and effacing EPEC) and bacteria of the respiratory tract that provoke coughing (e.g., toxin-producing *B. pertussis*) also have clear advantages, for they promote their own dissemination. When pathogenicity and fitness are directly linked, as in all these examples, the factors favoring survival of the pathogen are obvious. However, a subset of virulence factors does not seem to confer any advantages on the bacterial hosts, so the forces underlying their continued production are a mystery. For example, it is not clear why *S. aureus, Clostridium botulinum,* and *Clostridium tetani* secrete their respective enterotoxins and neurotoxins, because these toxins do not obviously promote bacterial survival, proliferation, or dissemination. It is important to remember that pathogenicity in itself is not the bacterium's ultimate goal. Consequently, it is necessary when considering the evolution of a pathogen to adopt the perspective of the bacterium rather than that of the host, and to assess how virulence factors aid bacterial fitness, for it is the contributions of virulence genes to fitness, rather than to virulence, that ultimately account for their retention within the bacterial gene pool.

Acknowledgments. We are grateful to our colleagues C. Lee, A. Camilli, B. Hochhut, J. Solomon, G. Dumenil, and H. Kimsey for helpful suggestions and critical reading of this manuscript.

Our research is supported by National Institutes of Health grant AI-42347 to M.K.W. and grant GM20483-01 to B.M.D. and by the Howard Hughes Medical Institute. M.K.W. is a Pew Scholar in the Biomedical Sciences.

REFERENCES

1. **Acheson, D. W., J. Reidl, X. Zhang, G. T. Keusch, J. J. Mekalanos, and M. K. Waldor.** 1998. In vivo transduction with Shiga toxin 1-encoding phage. *Infect. Immun.* **66:** 4496–4498.
2. **Ahmer, B. M., M. Tran, and F. Heffron.** 1999. The virulence plasmid of *Salmonella typhimurium* is self-transmissible. *J. Bacteriol.* **181:**1364–1368.
3. **Akerley, B. J., P. A. Cotter, and J. F. Miller.** 1995. Ectopic expression of the flagellar regulon alters development of the *Bordetella*-host interaction. *Cell* **80:**611–620.
4. **Al-Hasani, K., I. R. Henderson, H. Sakellaris, K. Rajakumar, T. Grant, J. P. Nataro, R. Robins-Browne, and B. Adler.** 2000. The *sigA* gene which is borne on the She pathogenicity island of *Shigella flexneri 2a* encodes an exported cytopathic protease involved in intestinal fluid accumulation. *Infect. Immun.* **68:**2457–2463.
5. **Alfano, J. R., A. O. Charkowski, W. L. Deng, J. L. Badel, T. Petnicki-Ocwieja, K. van Dijk, and A. Collmer.** 2000. The *Pseudomonas syringae* Hrp pathogenicity island has a tripartite mosaic structure composed of a cluster of type III secretion genes bounded by exchangeable effector and conserved effector loci that contribute to parasitic fitness and pathogenicity in plants. *Proc. Natl. Acad. Sci. USA* **97:**4856–4861.
6. **Andersson, J. O., and S. G. Andersson.** 1999. Insights into the evolutionary process of genome degradation. *Curr. Opin. Genet. Dev.* **9:**664–671.
7. **Bajaj, V., R. L. Lucas, C. Hwang, and C. A. Lee.** 1996. Coordinate regulation of *Salmonella typhimurium* invasion genes by environmental and regulatory factors is mediated by control of *hilA* expression. *Mol. Microbiol.* **22:**703–714.
8. **Barondess, J. J., and J. Beckwith.** 1990. A bacterial virulence determinant encoded by lysogenic coliphage lambda. *Nature* **346:**871–874.
9. **Baudry, B., A. Fasano, J. Ketley, and J. B. Kaper.** 1992. Cloning of a gene *(zot)* encoding a new toxin produced by *Vibrio cholerae. Infect. Immun.* **60:**428–434.
10. **Bieber, D., S. W. Ramer, C. Y. Wu, W. J. Murray, T. Tobe, R. Fernandez, and G. K. Schoolnik.** 1998. Type IV pili, transient bacterial aggregates, and virulence of enteropathogenic *Escherichia coli. Science* **280:**2114–2118.
11. **Billington, S. J., J. L. Johnston, and J. I. Rood.** 1996. Virulence regions and virulence factors of the ovine footrot pathogen, *Dichelobacter nodosus. FEMS Microbiol. Lett.* **145:** 147–156.
12. **Biot, T., and G. R. Cornelis.** 1988. The replication, partition and yop regulation of the pYV plasmids are highly conserved in *Yersinia enterocolitica* and *Y. pseudotuberculosis. J. Gen. Microbiol.* **134:**1525–1534.
13. **Blanc-Potard, A. B., and E. A. Groisman.** 1997. The *Salmonella selC* locus contains a pathogenicity island mediating intramacrophage survival. *EMBO J.* **16:**5376–5385.
14. **Blanc-Potard, A. B., F. Solomon, J. Kayser, and E. A. Groisman.** 1999. The SPI-3 pathogenicity island of *Salmonella enterica. J. Bacteriol.* **181:**998–1004.
15. **Blum, G., V. Falbo, A. Caprioli, and J. Hacker.** 1995. Gene clusters encoding the cytotoxic necrotizing factor type 1, Prs-fimbriae and alpha-hemolysin form the pathogenicity island II of the uropathogenic *Escherichia coli* strain J96. *FEMS Microbiol. Lett.* **126:**189–195.
16. **Blum, G., M. Ott, A. Lischewski, A. Ritter, H. Imrich, H. Tschape, and J. Hacker.** 1994. Excision of large DNA regions termed pathogenicity islands from tRNA-specific loci in the chromosome of an *Escherichia coli* wild-type pathogen. *Infect. Immun.* **62:**606–614.
17. **Boyd, E. F., and D. L. Hartl.** 1998. *Salmonella* virulence plasmid. Modular acquisition of the spv virulence region by an F-plasmid in *Salmonella enterica* subspecies I and insertion into the chromosome of subspecies II, IIIa, IV and VII isolates. *Genetics* **149:**1183–1190.
18. **Braun, V., T. Hundsberger, P. Leukel, M. Sauerborn, and C. von Eichel-Streiber.** 1996. Definition of the single integration site of the pathogenicity locus in *Clostridium difficile. Gene* **181:**29–38.
19. **Braun, V., and C. von Eichel-Streiber.** 1999. Virulence-associated elements in bacilli and clostridia, p. 233–264. *In* J. B. Kaper and J. Hacker (ed.), *Pathogenicity Islands and Other Mobile Virulence Elements.* American Society for Microbiology, Washington, D.C.
20. **Buchrieser, C., R. Brosch, S. Bach, A. Guiyoule, and E. Carniel.** 1998. The high-pathogenicity island of *Yersinia pseudotuberculosis* can be inserted into any of the three chromosomal asn tRNA genes. *Mol. Microbiol.* **30:**965–978.
21. **Burland, V., Y. Shao, N. T. Perna, G. Plunkett, H. J. Sofia, and F. R. Blattner.** 1998. The complete DNA sequence and analysis of the large virulence plasmid of *Escherichia coli* O157:H7. *Nucleic Acids Res.* **26:**4196–4204.
22. **Calderwood, S. B., F. Auclair, A. Donohue-Rolfe, G. T. Keusch, and J. J. Mekalanos.** 1987. Nucleotide sequence of the Shiga-like toxin genes of *Escherichia coli. Proc. Natl. Acad. Sci. USA* **84:**4364–4368.
23. **Campbell, A. M.** 1992. Chromosomal insertion sites for phages and plasmids. *J. Bacteriol.* **174:**7495–7499.
24. **Canard, B., B. Saint-Joanis, and S. T. Cole.** 1992. Genomic diversity and organization of virulence genes in the pathogenic anaerobe *Clostridium perfringens. Mol. Microbiol.* **6:** 1421–1429.
25. **Carniel, E., I. Guilvout, and M. Prentice.** 1996. Characterization of a large chromosomal "high-pathogenicity island" in biotype 1B *Yersinia enterocolitica. J. Bacteriol.* **178:** 6743–6751.
26. **Censini, S., C. Lange, Z. Xiang, J. E. Crabtree, P. Ghiara, M. Borodovsky, R. Rappuoli, and A. Covacci.** 1996. cag, a pathogenicity island of *Helicobacter pylori,* encodes type I-specific and disease-associated virulence factors. *Proc. Natl. Acad. Sci. USA* **93:**14648–14653.
27. **Cheetham, B. F., and M. E. Katz.** 1995. A role for bacteriophages in the evolution and transfer of bacterial virulence determinants. *Mol. Microbiol.* **18:**201–208.
28. **Cherepanov, P. A., T. G. Mikhailova, G. A. Karimova, N. M. Zakharova, V. Ershov Iu, and K. I. Volkovoi.** 1991. Cloning and detailed mapping of the fra-ymt region of the *Yersinia pestis* pFra plasmid. *Mol. Genet. Microbiol. Virusol.* **1991:** 19–26.
29. **Christie, P. J.** 1997. *Agrobacterium tumefaciens* T-complex transport apparatus: a paradigm for a new family of multifunctional transporters in eubacteria. *J. Bacteriol.* **179:** 3085–3094.

30. **Clark, C. A., J. Beltrame, and P. A. Manning.** 1991. The *oac* gene encoding a lipopolysaccharide O-antigen acetylase maps adjacent to the integrase-encoding gene on the genome of *Shigella flexneri* bacteriophage Sf6. *Gene* **107:**43–52.
31. **Coleman, D. C., D. J. Sullivan, R. J. Russell, J. P. Arbuthnott, B. F. Carey, and H. M. Pomeroy.** 1989. *Staphylococcus aureus* bacteriophages mediating the simultaneous lysogenic conversion of beta-lysin, staphylokinase and enterotoxin A: molecular mechanism of triple conversion. *J. Gen. Microbiol.* **135:**1679–1697.
32. **Dallas, W. S., and S. Falkow.** 1980. Amino acid sequence homology between cholera toxin and *Escherichia coli* heat-labile toxin. *Nature* **288:**499–501.
33. **Davies, J.** 1996. Origins and evolution of antibiotic resistance. *Microbiologia* **12:**9–16.
34. **Davis, B. M., and M. K. Waldor.** CTXφ contains a hybrid genome derived from tandemly integrated elements. *Proc. Natl. Acad. Sci. USA,* in press.
35. **DiRita, V. J., C. Parsot, G. Jander, and J. J. Mekalanos.** 1991. Regulatory cascade controls virulence in *Vibrio cholerae. Proc. Natl. Acad. Sci. USA* **88:**5403–5407.
36. **Echeverria, P., J. Seriwatana, D. N. Taylor, S. Changchawalit, C. J. Smyth, J. Twohig, and B. Rowe.** 1986. Plasmids coding for colonization factor antigens I and II, heat-labile enterotoxin, and heat-stable enterotoxin A2 in *Escherichia coli. Infect. Immun.* **51:**626–630.
37. **Fang, F. C., M. A. DeGroote, J. W. Foster, A. J. Baumler, U. Ochsner, T. Testerman, S. Bearson, J. C. Giard, Y. Xu, G. Campbell, and T. Laessig.** 1999. Virulent Salmonella typhimurium has two periplasmic Cu, Zn-superoxide dismutases. *Proc. Natl. Acad. Sci. USA* **96:**7502–7507.
38. **Fetherston, J. D., and R. D. Perry.** 1994. The pigmentation locus of *Yersinia pestis* KIM6+ is flanked by an insertion sequence and includes the structural genes for pesticin sensitivity and HMWP2. *Mol. Microbiol.* **13:**697–708.
39. **Fetherston, J. D., P. Schuetze, and R. D. Perry.** 1992. Loss of the pigmentation phenotype in Yersinia pestis is due to the spontaneous deletion of 102 kb of chromosomal DNA which is flanked by a repetitive element. *Mol. Microbiol.* **6:** 2693–2704.
40. **Figueroa-Bossi, N., and L. Bossi.** 1999. Inducible prophages contribute to *Salmonella* virulence in mice. *Mol. Microbiol.* **33:**167–176.
41. **Finlay, B. B., and S. Falkow.** 1997. Common themes in microbial pathogenicity revisited. *Microbiol. Mol. Biol. Rev.* **61:** 136–169.
42. **Finn, C. W., Jr., R. P. Silver, W. H. Habig, M. C. Hardegree, G. Zon, and C. F. Garon.** 1984. The structural gene for tetanus neurotoxin is on a plasmid. *Science* **224:**881–884.
43. **Frank, D. W.** 1997. The exoenzyme S regulon of *Pseudomonas aeruginosa. Mol. Microbiol.* **26:**621–629.
44. **Fraser, C. M., S. Casjens, W. M. Huang, G. G. Sutton, R. Clayton, R. Lathigra, O. White, K. A. Ketchum, R. Dodson, E. K. Hickey, M. Gwinn, B. Dougherty, J. F. Tomb, R. D. Fleischmann, D. Richardson, J. Peterson, A. R. Kerlavage, J. Quackenbush, S. Salzberg, M. Hanson, R. van Vugt, N. Palmer, M. D. Adams, J. Gocayne, J. C. Venter, et al.** 1997. Genomic sequence of a Lyme disease spirochaete, *Borrelia burgdorferi. Nature* **390:**580–586.
45. **Freeman, V. J.** 1951. Studies of the virulence of bacteriophage-infected strains of *Corynebacterium diptheriae. J. Bacteriol.* **61:**675–688.
46. **Fuqua, W. C., and S. C. Winans.** 1994. A LuxR-LuxI type regulatory system activates *Agrobacterium* Ti plasmid conjugal transfer in the presence of a plant tumor metabolite. *J. Bacteriol.* **176:**2796–2806.
47. **Galan, J. E., and P. J. Sansonetti.** 1996. Molecular and cellular bases of *Salmonella* and *Shigella* interactions with host cells, p. 2757–2773. *In* F. C. Neidhardt et al. (ed.), Escherichia coli *and* Salmonella: *Cellular and Molecular Biology,* 2nd ed., vol. 2. ASM Press, Washington, D.C.
48. **Green, B. D., L. Battisti, T. M. Koehler, C. B. Thorne, and B. E. Ivins.** 1985. Demonstration of a capsule plasmid in *Bacillus anthracis. Infect. Immun.* **49:**291–297.
49. **Groisman, E. A., and H. Ochman.** 1997. How *Salmonella* became a pathogen. *Trends Microbiol.* **5:**343–349.
50. **Groisman, E. A., and H. Ochman.** 1996. Pathogenicity islands: bacterial evolution in quantum leaps. *Cell* **87:** 791–794.
51. **Gulig, P. A., H. Danbara, D. G. Guiney, A. J. Lax, F. Norel, and M. Rhen.** 1993. Molecular analysis of *spv* virulence genes of the *Salmonella* virulence plasmids. *Mol. Microbiol.* **7:** 825–830.
52. **Hacker, J., L. Bender, M. Ott, J. Wingender, B. Lund, R. Marre, and W. Goebel.** 1990. Deletions of chromosomal regions coding for fimbriae and hemolysins occur in vitro and in vivo in various extraintestinal *Escherichia coli* isolates. *Microb. Pathog.* **8:**213–225.
53. **Hacker, J., G. Blum-Oehler, B. Janke, G. Nagy, and W. Goebel.** 1999. Pathogenicity islands of extraintestinal *Escherichia coli,* p. 59–76. *In* J. B. Kaper and J. Hacker (ed.), *Pathogenicity Islands and Other Mobile Virulence Elements.* ASM Press, Washington, D.C.
54. **Hacker, J., G. Blum-Oehler, I. Muhldorfer, and H. Tschape.** 1997. Pathogenicity islands of virulent bacteria: structure, function and impact on microbial evolution. *Mol. Microbiol.* **23:**1089–1097.
55. **Hacker, J., and J. B. Kaper.** 1999. The concept of pathogenicity islands, p. 1–11. *In* J. B. Kaper and J. Hacker (ed.), *Pathogenicity Islands and Other Mobile Virulence Elements.* ASM Press, Washington, D.C.
56. **Hall, R. M.** 1997. Mobile gene cassettes and integrons: moving antibiotic resistance genes in Gram-negative bacteria. *Ciba Found. Symp.* **207:**192–205.
57. **Hardt, W. D., H. Urlaub, and J. E. Galan.** 1998. A substrate of the centisome 63 type III protein secretion system of *Salmonella typhimurium* is encoded by a cryptic bacteriophage. *Proc. Natl. Acad. Sci. USA* **95:**2574–2579.
58. **Hare, J. M., A. K. Wagner, and K. A. McDonough.** 1999. Independent acquisition and insertion into different chromosomal locations of the same pathogenicity island in *Yersinia pestis* and *Yersinia pseudotuberculosis. Mol. Microbiol.* **31:** 291–303.
59. **Hayashi, T., H. Matsumoto, M. Ohnishi, and Y. Terawaki.** 1993. Molecular analysis of a cytotoxin-converting phage, phi CTX, of *Pseudomonas aeruginosa:* structure of the attP-cos-ctx region and integration into the serine tRNA gene. *Mol. Microbiol.* **7:**657–667.
60. **Heithoff, D. M., R. L. Sinsheimer, D. A. Low, and M. J. Mahan.** 1999. An essential role for DNA adenine methylation in bacterial virulence. *Science* **284:**967–970.
61. **Hicks, S., G. Frankel, J. B. Kaper, G. Dougan, and A. D. Phillips.** 1998. Role of intimin and bundle-forming pili in enteropathogenic *Escherichia coli* adhesion to pediatric intestinal tissue in vitro. *Infect. Immun.* **66:**1570–1578.
62. **Hinnebusch, B. J., R. D. Perry, and T. G. Schwan.** 1996. Role of the *Yersinia pestis* hemin storage (hms) locus in the transmission of plague by fleas. *Science* **273:**367–370.
63. **Hochhut, B., J. Marrero, and M. K. Waldor.** 2000. Mobilization of plasmids and chromosomal DNA mediated by the SXT element, a constin found in *Vibrio cholerae* O139. *J. Bacteriol.* **182:**2043–2047.

64. **Holmes, R. K., and L. Barksdale.** 1969. Genetic analysis of tox+ and tox− bacteriophages of *Corynebacterium diphtheriae*. *J. Virol.* **3:**586–598.
65. **Hu, S. T., and C. H. Lee.** 1988. Characterization of the transposon carrying the STII gene of enterotoxigenic *Escherichia coli*. *Mol. Gen. Genet.* **214:**490–495.
66. **Huan, P. T., B. L. Whittle, D. A. Bastin, A. A. Lindberg, and N. K. Verma.** 1997. *Shigella flexneri* type-specific antigen V: cloning, sequencing and characterization of the glucosyl transferase gene of temperate bacteriophage SfV. *Gene* **195:** 207–216.
67. **Hueck, C. J.** 1998. Type III protein secretion systems in bacterial pathogens of animals and plants. *Microbiol. Mol. Biol. Rev.* **62:**379–433.
68. **Johnson, L. P., P. M. Schlievert, and D. W. Watson.** 1980. Transfer of group A streptococcal pyrogenic exotoxin production to nontoxigenic strains of lysogenic conversion. *Infect. Immun.* **28:**254–257.
69. **Johnston, C., D. A. Pegues, C. J. Hueck, A. Lee, and S. I. Miller.** 1996. Transcriptional activation of *Salmonella typhimurium* invasion genes by a member of the phosphorylated response-regulator superfamily. *Mol. Microbiol.* **22:** 715–727.
70. **Kaniga, K., J. C. Bossio, and J. E. Galan.** 1994. The *Salmonella typhimurium* invasion genes *invF* and *invG* encode homologues of the AraC and PulD family of proteins. *Mol. Microbiol.* **13:**555–568.
71. **Karaolis, D. K., J. A. Johnson, C. C. Bailey, E. C. Boedeker, J. B. Kaper, and P. R. Reeves.** 1998. A *Vibrio cholerae* pathogenicity island associated with epidemic and pandemic strains. *Proc. Natl. Acad. Sci. USA* **95:**3134–3139.
72. **Karaolis, D. K., S. Somara, D. R. Maneval, Jr., J. A. Johnson, and J. B. Kaper.** 1999. A bacteriophage encoding a pathogenicity island, a type-IV pilus and a phage receptor in cholera bacteria. *Nature* **399:**375–379.
73. **Kimsey, H. H., and M. K. Waldor.** 1998. CTXϕ immunity: application in the development of cholera vaccines. *Proc. Natl. Acad. Sci. USA* **95:**7035–7039.
74. **Kovach, M. E., M. D. Shaffer, and K. M. Peterson.** 1996. A putative integrase gene defines the distal end of a large cluster of ToxR-regulated colonization genes in *Vibrio cholerae*. *Microbiology* **142:**2165–2174.
75. **Kreft, J., J.-A. Vasquez-Boland, E. Ng, and W. Goebel.** 1999. Virulence gene clusters and putative pathogenicity islands in Listeriae, p. 219–232. *In* J. B. Kaper and J. Hacker (ed.), *Pathogenicity Islands and Other Mobile Virulence Elements*. ASM Press, Washington, D.C.
76. **Lawrence, J. G., and H. Ochman.** 1998. Molecular archaeology of the *Escherichia coli* genome. *Proc. Natl. Acad. Sci. USA* **95:**9413–9417.
77. **Lazar, S., and M. K. Waldor.** 1998. ToxR-independent expression of cholera toxin from the replicative form of CTXϕ. *Infect. Immun.* **66:**394–397.
78. **Lee, C. A.** 1996. Pathogenicity islands and the evolution of bacterial pathogens. *Infect. Agents Dis.* **5:**1–7.
79. **Lee, S. H., D. L. Hava, M. K. Waldor, and A. Camilli.** 1999. Regulation and temporal expression patterns of *Vibrio cholerae* virulence genes during infection. *Cell* **99:**625–634.
80. **Lee, V. T., and O. Schneewind.** 1999. Type III machines of pathogenic yersiniae secrete virulence factors into the extracellular milieu. *Mol. Microbiol.* **31:**1619–1629.
81. **Libby, S. J., L. G. Adams, T. A. Ficht, C. Allen, H. A. Whitford, N. A. Buchmeier, S. Bossie, and D. G. Guiney.** 1997. The *spv* genes on the *Salmonella dublin* virulence plasmid are required for severe enteritis and systemic infection in the natural host. *Infect. Immun.* **65:**1786–1792.
82. **Lindsay, J. A., A. Ruzin, H. F. Ross, N. Kurepina, and R. P. Novick.** 1998. The gene for toxic shock toxin is carried by a family of mobile pathogenicity islands in *Staphylococcus aureus*. *Mol. Microbiol.* **29:**527–543.
83. **Lucas, R. L., and C. A. Lee.** 2000. Unravelling the mysteries of virulence gene regulation in *Salmonella typhimurium*. *Mol. Microbiol.* **36:**1024–1033.
84. **Mahan, M. J., J. M. Slaugh, and J. J. Mekalanos.** 1996. Environmental regulation of virulence gene expression in *Escherichia*, *Salmonella*, and *Shigella* spp., p. 2803–2815. *In* F. C. Neidhardt et al. (ed.), Escherichia coli *and* Salmonella: *Cellular and Molecular Biology*, 2nd ed., vol. 2. ASM Press, Washington, D.C.
85. **Mainil, J. G., F. Bex, P. Dreze, A. Kaeckenbeeck, and M. Couturier.** 1992. Replicon typing of virulence plasmids of enterotoxigenic *Escherichia coli* isolates from cattle. *Infect. Immun.* **60:**3376–3380.
86. **Mavris, M., P. A. Manning, and R. Morona.** 1997. Mechanism of bacteriophage SfII-mediated serotype conversion in *Shigella flexneri*. *Mol. Microbiol.* **26:**939–950.
87. **McDaniel, T. K., K. G. Jarvis, M. S. Donnenberg, and J. B. Kaper.** 1995. A genetic locus of enterocyte effacement conserved among diverse enterobacterial pathogens. *Proc. Natl. Acad. Sci. USA* **92:**1664–1668.
88. **McDonough, M. A., and J. R. Butterton.** 1999. Spontaneous tandem amplification and deletion of the Shiga toxin operon in *Shigella dysenteriae 1*. *Mol. Microbiol.* **34:**1058–1069.
89. **McShan, W. M., and J. J. Ferretti.** 1997. Genetic diversity in temperate bacteriophages of *Streptococcus pyogenes:* identification of a second attachment site for phages carrying the erythrogenic toxin A gene. *J. Bacteriol.* **179:**6509–6511.
90. **Mellies, J. L., S. J. Elliott, V. Sperandio, M. S. Donnenberg, and J. B. Kaper.** 1999. The Per regulon of enteropathogenic *Escherichia coli:* identification of a regulatory cascade and a novel transcriptional activator, the locus of enterocyte effacement (LEE)-encoded regulator (Ler). *Mol. Microbiol.* **33:** 296–306.
91. **Miller, S. I., and J. J. Mekalanos.** 1990. Constitutive expression of the phoP regulon attenuates *Salmonella* virulence and survival within macrophages. *J. Bacteriol.* **172:**2485–2490.
92. **Mills, D. M., V. Bajaj, and C. A. Lee.** 1995. A 40 kb chromosomal fragment encoding *Salmonella typhimurium* invasion genes is absent from the corresponding region of the *Escherichia coli* K-12 chromosome. *Mol. Microbiol.* **15:**749–759.
93. **Mirold, S., W. Rabsch, M. Rohde, S. Stender, H. Tschape, H. Russmann, E. Igwe, and W. D. Hardt.** 1999. Isolation of a temperate bacteriophage encoding the type III effector protein SopE from an epidemic *Salmonella typhimurium* strain. *Proc. Natl. Acad. Sci. USA* **96:**9845–9850.
94. **Moss, J. E., T. J. Cardozo, A. Zychlinsky, and E. A. Groisman.** 1999. The *sel*C-associated SHI-2 pathogenicity island of *Shigella flexneri*. *Mol. Microbiol.* **33:**74–83.
95. **Murphy, E.** 1989. Transposable elements in gram-positive bacteria, p. 269–288. *In* D. E. Berg and M. M. Howe (ed.), *Mobile DNA*. American Society for Microbiology, Washington, D.C.
96. **Murray, B. E., D. J. Evans, Jr., M. E. Penaranda, and D. G. Evans.** 1983. CFA/I-ST plasmids: comparison of enterotoxigenic *Escherichia coli* (ETEC) of serogroups O25, O63, O78, and O128 and mobilization from an R factor-containing epidemic ETEC isolate. *J. Bacteriol.* **153:**566–570.
97. **Murray, R. A., and C. A. Lee.** Invasion genes are not required for *Salmonella enterica* serovar Typhimurium to breach the intestinal epithelium: evidence that *Salmonella* pathogenicity island 1 has alternative functions during infection. *Infect. Immun.*, in press.

98. **O'Brien, A. D., and R. K. Holmes.** 1996. Protein toxins of *Escherichia coli* and *Salmonella,* p. 2788–2802. *In* F. C. Neidhardt et al. (ed.), Escherichia coli *and* Salmonella: *Cellular and Molecular Biology,* 2nd ed., vol. 2. ASM Press, Washington, D.C.
99. **Ochman, H., and E. A. Groisman.** 1996. Distribution of pathogenicity islands in *Salmonella* spp. *Infect. Immun.* **64:** 5410–5412.
100. **Ochman, H., J. G. Lawrence, and E. A. Groisman.** 2000. Lateral gene transfer and the nature of bacterial innovation. *Nature* **405:**299–304.
101. **Odenbreit, S., J. Puls, B. Sedlmaier, E. Gerland, W. Fischer, and R. Haas.** 2000. Translocation of *Helicobacter pylori* CagA into gastric epithelial cells by type IV secretion. *Science* **287:**1497–1500.
102. **Ohlsen, K., W. Ziebuhr, W. Reichardt, W. Witte, F. Götz, and J. Hacker.** 1999. Mobile elements, phages, and genomic islands of staphylococci and streptococci, p. 265–287. *In* J. B. Kaper and J. Hacker (ed.), *Pathogenicity Islands and Other Mobile Virulence Elements.* ASM Press, Washington, D.C.
103. **Okinaka, R. T., K. Cloud, O. Hampton, A. R. Hoffmaster, K. K. Hill, P. Keim, T. M. Koehler, G. Lamke, S. Kumano, J. Mahillon, D. Manter, Y. Martinez, D. Ricke, R. Svensson, and P. J. Jackson.** 1999. Sequence and organization of pXO1, the large *Bacillus anthracis* plasmid harboring the anthrax toxin genes. *J. Bacteriol.* **181:**6509–6515.
104. **Pappenheimer, A. M., Jr., and J. R. Murphy.** 1983. Studies on the molecular epidemiology of diphtheria. *Lancet* **11:** 923–926.
105. **Parsot, C., and P. J. Sansonetti.** 1999. The virulence plasmid of shigellae: an archipelago of pathogenicity islands?, p. 151–165. *In* J. B. Kaper and J. Hacker (ed.), *Pathogenicity Islands and Other Mobile Virulence Elements.* ASM Press, Washington, D.C.
106. **Perna, N. T., G. F. Mayhew, G. Posfai, S. Elliott, M. S. Donnenberg, J. B. Kaper, and F. R. Blattner.** 1998. Molecular evolution of a pathogenicity island from enterohemorrhagic *Escherichia coli* O157:H7. *Infect. Immun.* **66:**3810–3817.
107. **Plunkett, G., 3rd, D. J. Rose, T. J. Durfee, and F. R. Blattner.** 1999. Sequence of Shiga toxin 2 phage 933W from *Escherichia coli* O157:H7: Shiga toxin as a phage late-gene product. *J. Bacteriol.* **18:**1767–1778.
108. **Rajakumar, K., C. Sasakawa, and B. Adler.** 1997. Use of a novel approach, termed island probing, identifies the *Shigella flexneri* she pathogenicity island which encodes a homolog of the immunoglobulin A protease-like family of proteins. *Infect. Immun.* **65:**4606–4614.
109. **Rakin, A., S. Schubert, C. Pelludat, D. Brem, and J. Heesemann.** 1999. The high-pathogenicity island of yersiniae, p. 77–90. *In* J. B. Kaper and J. Hacker (ed.), *Pathogenicity Islands and Other Mobile Virulence Elements.* ASM Press, Washington, D.C.
110. **Ritter, A., G. Blum, L. Emody, M. Kerenyi, A. Bock, B. Neuhierl, W. Rabsch, F. Scheutz, and J. Hacker.** 1995. tRNA genes and pathogenicity islands: influence on virulence and metabolic properties of uropathogenic *Escherichia coli. Mol. Microbiol.* **17:**109–121.
111. **Rood, J. I.** 1998. Virulence genes of *Clostridium perfringens. Annu. Rev. Microbiol.* **52:**333–360.
112. **Salyers, A., N. Shoemaker, G. Bonheyo, and J. Frias.** 1999. Conjugative transposons: transmissible resistance islands, p. 331–346. *In* J. B. Kaper and J. Hacker (ed.), *Pathogenicity Islands and Other Mobile Virulence Elements.* ASM Press, Washington, D.C.
113. **Salyers, A. A., N. B. Shoemaker, A. M. Stevens, and L. Y. Li.** 1995. Conjugative transposons: an unusual and diverse set of integrated gene transfer elements. *Microbiol. Rev.* **59:** 579–590.
114. **Sansonetti, P. J., T. L. Hale, G. J. Dammin, C. Kapfer, H. H. Collins, Jr., and S. B. Formal.** 1983. Alterations in the pathogenicity of *Escherichia coli* K-12 after transfer of plasmid and chromosomal genes from *Shigella flexneri. Infect. Immun.* **39:**1392–1402.
115. **Sansonetti, P. J., D. J. Kopecko, and S. B. Formal.** 1982. Involvement of a plasmid in the invasive ability of *Shigella flexneri. Infect. Immun.* **35:**852–860.
116. **Schubert, S., A. Rakin, D. Fischer, J. Sorsa, and J. Heesemann.** 1999. Characterization of the integration site of Yersinia high-pathogenicity island in *Escherichia coli. FEMS Microbiol. Lett.* **179:**409–414.
117. **Sears, C. L., and J. B. Kaper.** 1996. Enteric bacterial toxins: mechanisms of action and linkage to intestinal secretion. *Microbiol. Rev.* **60:**167–215.
118. **Segal, G., and H. A. Shuman.** 1997. Characterization of a new region required for macrophage killing by *Legionella pneumophila. Infect. Immun.* **65:**5057–5066.
119. **Shea, J. E., M. Hensel, C. Gleeson, and D. W. Holden.** 1996. Identification of a virulence locus encoding a second type III secretion system in *Salmonella typhimurium. Proc. Natl. Acad. Sci. USA* **93:**2593–2597.
120. **Skorupski, K., and R. K. Taylor.** 1997. Control of the ToxR virulence regulon in *Vibrio cholerae* by environmental stimuli. *Mol. Microbiol.* **25:**1003–1009.
121. **Smith, H. W., and M. A. Linggood.** 1971. The transmissible nature of enterotoxin production in a human enteropathogenic strain of *Escherichia coli. J. Med. Microbiol.* **4:** 301–305.
122. **So, M., and B. J. McCarthy.** 1980. Nucleotide sequence of the bacterial transposon Tn*1681* encoding a heat-stable (ST) toxin and its identification in enterotoxigenic *Escherichia coli* strains. *Proc. Natl. Acad. Sci. USA* **77:**4011–4015.
123. **Sodeinde, O. A., and J. D. Goguen.** 1988. Genetic analysis of the 9.5-kilobase virulence plasmid of *Yersinia pestis. Infect. Immun.* **56:**2743–2748.
124. **Sodeinde, O. A., A. K. Sample, R. R. Brubaker, and J. D. Goguen.** 1988. Plasminogen activator/coagulase gene of *Yersinia pestis* is responsible for degradation of plasmid-encoded outer membrane proteins. *Infect. Immun.* **56:**2749–2752.
125. **Spicer, E. K., W. M. Kavanaugh, W. S. Dallas, S. Falkow, W. H. Konigsberg, and D. E. Schafer.** 1981. Sequence homologies between A subunits of *Escherichia coli* and *Vibrio cholerae* enterotoxins. *Proc. Natl. Acad. Sci. USA* **78:**50–54.
126. **Stephens, R. S., S. Kalman, C. Lammel, J. Fan, R. Marathe, L. Aravind, W. Mitchell, L. Olinger, R. L. Tatusov, Q. Zhao, E. V. Koonin, and R. W. Davis.** 1998. Genome sequence of an obligate intracellular pathogen of humans: *Chlamydia trachomatis. Science* **282:**754–759.
127. **Strockbine, N. A., L. R. Marques, J. W. Newland, H. W. Smith, R. K. Holmes, and A. D. O'Brien.** 1986. Two toxin-converting phages from *Escherichia coli* O157:H7 strain 933 encode antigenically distinct toxins with similar biologic activities. *Infect. Immun.* **53:**135–140.
128. **Suzuki, K., Y. Hattori, M. Uraji, N. Ohta, K. Iwata, K. Murata, A. Kato, and K. Yoshida.** 2000. Complete nucleotide sequence of a plant tumor-inducing Ti plasmid. *Gene* **242:** 331–336.
129. **Swenson, D. L., N. O. Bukanov, D. E. Berg, and R. A. Welch.** 1996. Two pathogenicity islands in uropathogenic *Escherichia coli* J96: cosmid cloning and sample sequencing. *Infect. Immun.* **64:**3736–3743.
130. **Taylor, R. K., V. L. Miller, D. B. Furlong, and J. J. Mekalanos.** 1987. Use of phoA gene fusions to identify a pilus

colonization factor coordinately regulated with cholera toxin. *Proc. Natl. Acad. Sci. USA* **84:**2833–2837.

131. **Trucksis, M., J. E. Galen, J. Michalski, A. Fasano, and J. B. Kaper.** 1993. Accessory cholera enterotoxin (Ace), the third toxin of a *Vibrio cholerae* virulence cassette. *Proc. Natl. Acad. Sci. USA* **90:**5267–5271.
132. **Uchida, I., K. Hashimoto, and N. Terakado.** 1986. Virulence and immunogenicity in experimental animals of *Bacillus anthracis* strains harbouring or lacking 110 MDa and 60 MDa plasmids. *J. Gen. Microbiol.* **132:**557–559.
133. **Uchida, I., S. Makino, C. Sasakawa, M. Yoshikawa, C. Sugimoto, and N. Terakado.** 1993. Identification of a novel gene, *dep*, associated with depolymerization of the capsular polymer in *Bacillus anthracis*. *Mol. Microbiol.* **9:**487–496.
134. **Uchida, T., D. M. Gill, and A. M. Pappenheimer, Jr.** 1971. Mutation in the structural gene for diphtheria toxin carried by temperate phage. *Nat. New Biol.* **233:**8–11.
135. **Vazquez-Torres, A., Y. Xu, J. Jones-Carson, D. W. Holden, S. M. Lucia, M. C. Dinauer, P. Mastroeni, and F. C. Fang.** 2000. Salmonella pathogenicity island 2-dependent evasion of the phagocyte NADPH oxidase. *Science* **287:**1655–1658.
136. **Verma, N. K., D. J. Verma, P. T. Huan, and A. A. Lindberg.** 1993. Cloning and sequencing of the glucosyl transferase-encoding gene from converting bacteriophage X (SFX) of *Shigella flexneri*. *Gene* **129:**99–101.
137. **Vogel, J. P., H. L. Andrews, S. K. Wong, and R. R. Isberg.** 1998. Conjugative transfer by the virulence system of *Legionella pneumophila*. *Science* **279:**873–876.
138. **Vokes, S. A., S. A. Reeves, A. G. Torres, and S. M. Payne.** 1999. The aerobactin iron transport system genes in Shigella flexneri are present within a pathogenicity island. *Mol. Microbiol.* **33:**63–73.
139. **Waldor, M. K., and J. J. Mekalanos.** 1996. Lysogenic conversion by a filamentous phage encoding cholera toxin. *Science* **272:**1910–1914.
140. **Weeks, C. R., and J. J. Ferretti.** 1984. The gene for type A streptococcal exotoxin (erythrogenic toxin) is located in bacteriophage T12. *Infect. Immun.* **46:**531–536.
141. **Weiss, A. A., F. D. Johnson, and D. L. Burns.** 1993. Molecular characterization of an operon required for pertussis toxin secretion. *Proc. Natl. Acad. Sci. USA* **90:**2970–2974.
142. **Winans, S. C., V. Kalogeraki, S. Jafri, R. Akakura, and Q. Xia.** 1999. Diverse roles of *Agrobacterium* Ti plasmid-borne genes in the formation and colonization of plant tumors, p. 289–307. *In* J. B. Kaper and J. Hacker (ed.), *Pathogenicity Islands and Other Mobile Virulence Elements*. ASM Press, Washington, D.C.
143. **Wong, K. K., M. McClelland, L. C. Stillwell, E. C. Sisk, S. J. Thurston, and J. D. Saffer.** 1998. Identification and sequence analysis of a 27-kilobase chromosomal fragment containing a Salmonella pathogenicity island located at 92 minutes on the chromosome map of *Salmonella enterica* serovar Typhimurium LT2. *Infect. Immun.* **66:**3365–3371.
144. **Zhang, X., A. D. McDaniel, L. E. Wolf, G. T. Keusch, M. K. Waldor, and D. W. Acheson.** 2000. Quinolone antibiotics induce Shiga toxin-encoding bacteriophages, toxin production, and death in mice. *J. Infect. Dis.* **181:**664–670.
145. **Zhu, J., P. M. Oger, B. Schrammeijer, P. J. J. Hooykaas, S. K. Farrand, and S. C. Winans.** 2000. The bases of crown gall tumorigenesis. *J. Bacteriol.* **182:**3885–3895.

Mobile DNA II
Edited by N. L. Craig et al.

Chapter 46

Archaeal Mobile DNA

PETER REDDER, DAVID FAGUY, KIM BRÜGGER, QUNXIN SHE, AND ROGER A. GARRETT

INTRODUCTION

The discovery of the archaea in 1977 was a major biological milestone. They were first identified on the basis of their rRNA sequences where, in particular, the nucleotide modification patterns differed strongly from those of bacteria (87). Subsequent studies of their physiology, biochemistry, molecular biology, and phylogeny using molecular chronometers other than rRNA have demonstrated unequivocally that they are a distinct group of organisms (88). The first archaea to be characterized were all from extreme environments (strictly anaerobic, extremes of pH, and high temperatures), but they are now known to thrive in most of the earth's major environments (4, 18) and they constitute a substantial fraction of the biosphere (51). They fall into two major kingdoms, the euryarchaeotes and crenarchaeotes, and analyses of environmental rDNA samples indicate the possible existence of a third kingdom, the korarchaeotes (9, 88). The euryarchaeotes exhibit a wide range of metabolic properties extending from strictly anaerobic methanogens to aerobic extreme halophiles and from psychrophiles to hyperthermophiles. Crenarchaeotes are also very diverse metabolically, although most of those that have been cultured, to date, are extreme thermophiles or hyperthermophiles (19, 82).

Recently, chromosomes of several archaea have been sequenced completely (http://www.tigr.org). Sequenced euryarchaeotes include *Archaeoglobus fulgidus*, *Halobacterium* sp. NRC1, *Methanococcus jannaschii*, *Methanobacterium thermoautotrophicum*, *Methanosarcina mazei*, *Pyrococcus abyssi*, *Pyrococcus horikoshii*, *Thermoplasma acidophilum*, and *Thermoplasma volcanium*. The chromosomes of the crenarchaeotes *Aeropyrum pernix*, *Pyrobaculum aerophilum*, *Sulfolobus solfataricus*, and *Sulfolobus tokodaii* are completely sequenced, and several others

Peter Redder, Kim Brügger, Qunxin She, and Roger A. Garrett • Institute of Molecular Biology, University of Copenhagen, Sølvgade 83H, DK-1307 Copenhagen K, Denmark. **David Faguy** • Department of Biology, University of New Mexico, Albuquerque, NM 87131-1091.

are underway. These sequences have provided a basis for establishing the extent to which the archaeal chromosome is bacterium-like and eukaryote-like (36), and they have produced a preliminary classification of archaea-specific or "signature" genes (23).

A major effort is also underway to characterize and sequence genetic elements, especially those from crenarchaeotes. They include several different classes of virus, including circular and linear viruses, lysogenic and lytic viruses, and many plasmids, including high- and low-copy-number cryptic plasmids, a plasmid-virus hybrid, and conjugative plasmids. Moreover, insertion elements (IS elements), nonautonomous transposable elements, archaeal introns, and inteins are common to archaeal genomes. Studies of all these have revealed many novel biological properties, which include mechanisms for conjugation and intron splicing that differ radically from those found in bacteria and/or eukaryotes (42, 77, 81). There is also evidence for a general mechanism of integration into chromosomes for some viruses and, possibly, plasmids that requires an archaeal integrase for insertion (53, 64). Here, we summarize the different classes of mobile DNA elements of archaea and consider their impact on archaeal evolution.

VIRUSES

Several viruses and virus-like particles have been characterized for the archaea, although, with the exception of *Sulfolobus*, no comprehensive studies have been performed to examine the diversity of viruses that are present (6). Those isolated, to date, from euryarchaeotes mainly resemble coliphages with heads and tails, while for the crenarchaeotes, and *Sulfolobus* in particular, a wide range of virus types have been characterized, which have been classified into four main families on the basis of their unusual features (91).

Euryarchaeal Viruses

The few viruses that have been investigated are listed in Table 1. They carry double-stranded DNA genomes in their heads, mainly in the size range 30 to 60 kb, and their tails are variable in size and often contractile and flexible. The *Halobacterium* virus ΦH, which has a linear DNA genome, is the best characterized of the haloviruses. It generally has a lytic life cycle, but at low frequency it has a lysogenic one. The lysogenic strain carries the virus (57 kb) as an autonomously replicating plasmid and is resistant to infection by wild-type virus. In some respects, its gene organization and expression are similar to bacterial phages, albeit at a very large evolutionary distance (6). The virulent virus ΨM1, which infects the *M. thermoautotrophicum* strain Marburg, is the best characterized of the methanogenic viruses. It has a linear double-stranded DNA genome (30.4 kb) which may recombine and circularize at its 3-kb terminal

Table 1. Euryarchaeal viruses

Virus	Host	Type[a]	Size (kb)	Family[a]	Reference(s)
Halophile					
ΦH	*Halobacterium halobium*	Linear	59	*Myoviridae*	93
ΦN	*Halobacterium halobium*	Linear	56	*Siphoviridae*	93
Hh1	*Halobacterium halobium*	ND	37.2	ND	93
Hh3	*Halobacterium halobium*	ND	29.4	ND	93
S45	*Halobacterium halobium*	ND		ND	93
Ja1	*Halobacterium halobium*	ND	230	ND	93
Hs1	*Halobacterium salinarium*	ND		*Myoviridae*	93
HF1	*Haloferax* strain Aa2.2	Linear	79	*Myoviridae*	6, 60
HF2	*Halobacterium* strain Ch2	Linear	79.7	*Myoviridae*	6, 60
ΦCh1	*Natronobacterium magadii*	Linear	55	*Myoviridae*	86
Methanogen					
ψM1	*Methanobacterium thermoautotrophicum*	Linear	30.4	*Siphoviridae*	6, 40
ΦF1	*Methanobacterium thermoformicicum*	Linear	85	ND	57
ΦF3	*Methanobacterium thermoformicicum*	Linear	36	ND	57
PG	*Methanobrevibacter smithii*	Linear	50	ND	93
PMS1	*Methanobrevibacter smithii*	ND	35	ND	57
VLP	*Methanococcus voltae*	Circular	23	ND	6
Thermophile					
CWP	*Pyrococcus woesei*	ND		ND	93

[a]ND, not determined.

repeat. The genome encodes a pseudomurein endopeptidase that facilitates host cell lysis by cleaving the isopeptide bond of the cell wall polymer pseudomurein (40). A related enzyme is also produced by *Methanobacterium wolfei* under certain growth conditions, but since no extrachromosomal elements have been detected in this organism, the genetic element may have been integrated into the chromosome (66). Lemon-shaped particles resembling the crenarchaeal SSV1 virus were found in the supernatant of a strain of *Methanococcus voltae,* but no infectivity or inducibility of virus-like particles was demonstrated (89). To date, although there is strong evidence for exchange of IS elements between viruses and chromosomes, there is no direct evidence for chromosome integration of any euryarchaeal viruses.

Crenarchaeal Viruses

All of the crenarchaeal viruses that have been characterized so far carry double-stranded DNA genomes and differ in morphology from their euryarchaeal counterparts. So far, they have been isolated from *Thermoproteus tenax* of the order *Thermoproteales* or from *Sulfolobus* strains of the order *Sulfolobales* and are listed in Table 2.

T. tenax is a strict anaerobe and hyperthermophile and is facultatively chemolithotrophic. Four viruses of the family *Lithothrixviridae* have been characterized: TTV1, a flexible rod; TTV2 and TTV3, flexible filaments; and TTV4, a stiff rod (28, 91). They differ in their morphology and protein composition, and hybridization studies reveal that they do not show high sequence similarity. TTV1 is the best-characterized of the four and the only one to have its genome sequenced. It is temperate and causes lysis when sulfur in the host culture is consumed.

Sulfolobus is one of the best-characterized archaeal genera. It grows aerobially and either facultatively or obligately heterotrophically. Several *Sulfolobus* viruses have been characterized, none of which are lytic. They include SIRV1 and SIRV2, which are closely related rod-shaped rudiviruses from *Sulfolobus islandicus* strains. They consist of a superhelix of DNA and a single DNA-binding protein, which produce a hollow tube that is plugged terminally and carries tail fibers at both ends. The ends of the linear genomes are covalently linked, and the smaller one, SIRV1, mutates at a much higher rate than SIRV2 (68). Both viral genomes have been sequenced and some of their open reading frames (ORFs) show sequence similarity to eukaryotic poxviruses (64a). The virus SNDV from *Sulfolobus neozealandicus* resembles a droplet and carries a beard with densely packed thin fibers protruding from the pointed end. Its circular genome is exceptional in being methylated, as also occurs for the haloviruses ΦCh1 and Hs1 (Table 1). It is infectious for only a few *Sulfolobus* strains isolated in New Zealand (7). The lithothrixvirus SIFV is filamentous and flexible with a phospholipid-containing coat and is likely to be related to TTV1. Although its genome has been sequenced, the termini of the linear DNA genome carry unknown modifications (8). The lipothrixvirus DAFV from *Acidianus ambivalens,* an obligately chemolithotrophic relative of *Sulfolobus,* is filamentous, flexible, and morphologically similar to viruses TTV2 and TTV3 from *T. tenax* (92). To date, there is no evidence for any of these viruses integrating into crenarchaeal genomes.

The best-characterized *Sulfolobus* virus is SSV1 from *Sulfolobus shibatae,* which infects some other *Sulfolobus* species. It is a spindle-shaped virus with a very short tail and tail fibers that can attach to a host membrane. The virus is temperate, and after infection, the viral genome (15.5 kb) integrates into the host chromosome at a specific attachment site. Insertion is mediated by an integrase encoded in the viral genome and recombination occurs at a 44-bp *att* site within

Table 2. Crenarchaeal viruses

Virus	Host	Type[a]	Size (kb)	Family	Reference(s)
SSV1	*S. shibatae*	Circular	15.5	*Fuselloviridae*	91
SSV2	*S. islandicus* REY15/4	Circular	15	*Fuselloviridae*	Q. She, unpublished data
SSV3	*S. islandicus*	Circular	15	*Fuselloviridae*	Q. She, unpublished data
SIRV1	*S. islandicus*	Linear	33	*Rudiviridae*	64a, 68
SIRV2	*S. islandicus*	Linear	36	*Rudiviridae*	64a, 68
SNDV	*Sulfolobus neozealandicus*	Circular	20	*Guttaviridae*	7
SIFV	*S. islandicus*	Linear	42	*Lipothrixviridae*	8
DAFV	*A. ambivalens*	Linear	40	*Lipothrixviridae*	91
TTV1	*T. tenax*	Linear	16	*Lipothrixviridae*	28
TTV2	*T. tenax*	Linear	16	*Lipothrixviridae*	28
TTV3	*T. tenax*	Linear	27	*Lipothrixviridae*	28
TTV4	*T. tenax*	Linear	17	*Lipothrixviridae*	91

the downstream half of a $tRNA^{Arg}$ gene. Integration produces a partitioned *int* gene in the chromosome and does not disrupt the tRNA gene (53). The same integration mechanism also appears to operate for other members of this potentially large viral family, including SSV2 and SSV3, isolated from *S. islandicus* strains (Q. She, unpublished data; 91, 93).

PLASMIDS

Several plasmids have been isolated and characterized, at least partially, from both archaeal kingdoms. Several of them are small, cryptic plasmids, which have not yet been sequenced and, with a few exceptions, we still know very little about their potential for mobility (40, 91, 93).

Euryarchaeal Plasmids

Some haloarchaea contain very large megaplasmids (up to 700 kb) that comprise a major fraction of the genome. *Halobacterium* sp. NRC1 contains two such plasmids, of which one, pNRC100 (191 kb), encodes (i) proteins essential for replication and partitioning; (ii) proteins responsible for other cellular functions, including cytochrome d oxidase, which participates in the electron transport chain; and (iii) proteins involved in gas vesicle production. pNRC100 and pGT5 from *Pyrococcus abyssi* have been shown to replicate by rolling circle mechanisms (21, 55). pNRC100 may integrate, reversibly, into the host chromosome in a manner similar to that of the F′ factor of *Escherichia coli* (54).

The pRN Family

The pRN plasmid family of the *Sulfolobales* includes plasmids pRN1, pRN2, pSSVx, pDL10, and pHEN7 (Table 3), which have all been sequenced and shown to share a large region that is conserved in sequence (35, 64). Plasmids pDL10 and pHEN7 also carry a large variant region between a direct sequence repeat TTAGAATGGGGATTC that is also present in the other plasmids. It is inferred that recombination at these sites produces the main genetic variability in this plasmid family (64).

Table 3. Crenarchaeal pRN plasmids

Plasmid	Host	Size (kb)	Copy number	Reference
pDL10	*A. ambivalens*	7	High	35
pRN1	*S. islandicus* REN1H1	5.5	~20	31
pRN2	*S. islandicus* REN1H1	6.9	~35	31
pSSVx	*S. islandicus* REY15/4	5.7	High	5
pHEN7	*S. islandicus* HEN7H2	7.5	~15	64
pXQ1	*S. solfataricus* P2	7.5	1	64

The chromosome sequence of *S. solfataricus* P2 also carries a pRN-like plasmid. It contains the characteristic conserved region bordered by a direct repeat and differs only in that its putative *repA* gene is interrupted by the IS element ISC1439 (Fig. 1). The linear sequence is bordered by two ORFs, which show high sequence similarity and identity (67 and 87%, respectively) to the N-terminal (124-aa) and C-terminal (290-aa) regions, respectively, of the integrase of the SSV1 virus (Int). Moreover, perfect repeats of 44-bp sequences occur within these two ORFs, which correspond closely (71% identity) to the *att* sequences that flank the SSV1 virus when it is integrated into the *S. shibatae* chromosome in the downstream half of a $tRNA^{Arg}$ gene (53, 64). In the *S. solfataricus* genome, the downstream *att* sequence lies in the downstream half of a $tRNA^{Val}$ gene. Excision from the chromosome by recombination at these direct repeats, and circularization, would produce a plasmid pXQ1 (Fig. 1) which, on the basis of both sequence similarity and the organization of its ORFs and sequence elements, would constitute a member of the pRN plasmid family (Table 3) and lie equidistant from the other members (64). This observation strongly suggests that the same mechanism of integrase-mediated insertion at *att* sequences within tRNA genes is used for both viruses and plasmids, at least among the *Sulfolobales*.

Plasmid-Virus Hybrid pSSVx

One strain of *S. islandicus* harbors both the fusellovirus SSV2 and a small plasmid, pSSVx. The plasmid spreads in *S. solfataricus* P1, together with the virus, after infection with the supernatant or the *S. islandicus* culture. The purified virus always constitutes a mixture of two particle sizes, one the size of a normal fusellovirus and the other smaller, and it was inferred that the SSV2 occupied the larger particle and pSSVx the smaller (5). Apart from containing the large conserved region of the pRN family, pSSVx contains three additional ORFs, two of which are juxtapositioned and show high similarity to tandem ORFs of the fusellovirus genomes (Table 3). Since the other pRN plasmids cannot spread in the presence of SSV2, it was concluded that the two gene products shared by pSSVx enable pSSVx to use the same packaging system as the SSV2 virus in order to spread (5).

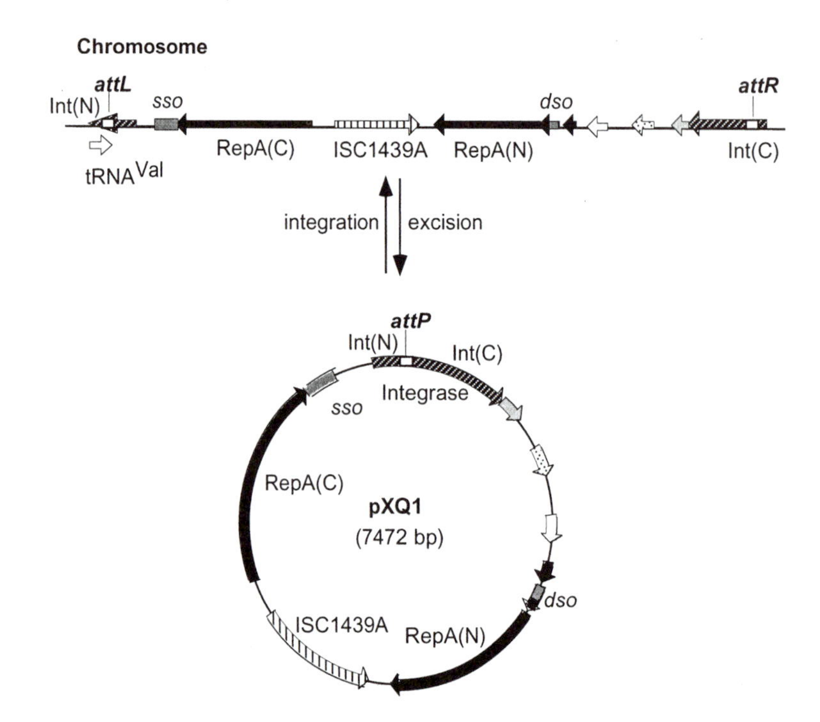

Figure 1. A model for the excision/integration of the putative plasmid pXQ1 in the chromosome of *S. solfataricus* P2. The integrase target sites in the plasmid and chromosome are the 45-bp direct repeats, *att,* where L and R denote left and right, respectively. Excision and circularization produce an intact integrase gene (*int*). The IS element ISC1439A interrupts *repA*. Four ORFs from pXQ1 show high sequence similarity to genes of other members of the pRN family. *dso* and *sso* represent putative double-strand and single-strand origins of replication, respectively (64).

Conjugative Plasmids

Conjugative plasmids in Archaea have so far been found only in closely related strains of *Sulfolobus,* where they occur frequently (67). Members of three subfamilies of these conjugative plasmids, pNOB8 (77), pING (81), and pSOG2/4 (G. Erauso and J. Van Der Oost, unpublished data), have now been sequenced (Table 4). They lie in the size range of 28 to 43 kb and carry up to 50 ORFs, and they are providing important insights into novel archaeal mechanisms including conjugation, maintenance, and copy number control. Since vector-host systems for investigating archaeal gene expression are still in a developmental stage (3, 12, 80), insight into function has been limited to studying the properties of sequenced variants of plasmids pNOB8 and pING (77, 81).

Table 4. Crenarchaeal conjugative plasmids

Plasmid	Host	Size (kb)	Derived plasmid(s)	Reference
pNOB8	*Sulfolobus* sp. strain NOB8H2	42.6	pNOB8-33	77
pING1	*S. islandicus* HEN2P2	24.6	pING3, pING4, pING5, pING6	81
pSOG2/4	*S. islandicus*	28.4	pSOG2/4 clone 1	91
pARN3/2	*S. islandicus*	26.1		91
pARN4/2	*S. islandicus*	26.5		91
pKEF9/1	*S. islandicus*	29.8		91
pHVE12/4	*S. islandicus*	27.3		91
pHVE14/5	*S. islandicus*	36.5		91

In bacteria, conjugative systems share the following features: (i) adhesion of donor and recipient cells via an extracellular filamentous structure for some proteobacteria or by a fibrillar "adhesion substance" for some gram-positive bacteria, (ii) covalent association of a relaxase protein to the leading 5′ end of the DNA that initiates DNA replication and transfer via a strand- and site-specific cleavage event, and (iii) transfer of single-stranded DNA that is generated by rolling-circle-type replication (63). Transmembrane or membrane binding segments are especially important for facilitating cell-cell contacts and forming membrane pores during conjugation. For pNOB8, 16 ORFs were predicted to contain such transmembrane or membrane binding segments, but only two putative homologs of bacterial conjugative proteins were detected. These show significant sequence similarity in their C-terminal halves with the TrbE and TraG families of ATPases (77) that are involved in bacterial conjugation and virulence, although their detailed functions remain obscure (37). Both ORFs contain ATP-binding motifs and may therefore generate energy for the conjugation process. The two archaeal ORFs also show a significant level of similarity to one another, suggesting that the TrbE and TraG families may also have a common evolutionary origin (77). Therefore, although the conjugation proteins of *Sulfolobus* and bacteria show very little similarity, since the TrbE and TraG proteins are the most conserved of the bacterial conjugative proteins (38, 39), it is likely that the bacterial and archaeal conjugation systems have evolved, at least in part, from a common ancestral system (77).

INTEGRON-LIKE ELEMENTS

In bacteria, integrons are genetic elements that acquire ORFs and convert them into functional units (26). They are mobilized by coupling to transposons or by incorporation into viruses or conjugative plasmids. Integrons can readily be enlarged by the insertion of gene cassettes at a complementary *att* site in the integron and gene cassette. They can confer antibiotic resistance or pathogenicity, but the discovery of megaintegrons, first in *Vibrio cholerae* (52) and then in several bacteria belonging to distinct genera of the β- and γ-proteobacteria (72), suggests that their occurrence is widespread and their functions are diverse. The megaintegron of *V. cholerae* contains hundreds of gene cassettes representing up to 10% of the chromosome. Moreover, integrons of different *Vibrio* species carry gene cassettes that are species-specific, which creates an enormous level of genetic diversity. Integrated cassettes are flanked by the recombination sites GTTRRRY, and they contain a gene with an imperfect inverted repeat located at the downstream end, called a 59-bp element, which can vary in size from 57 to 141 bp. These elements are partially conserved in sequence and constitute recognition sites for the site-specific integrase (15).

Putative large inserts, which resemble bacterial integrons, have been detected in archaeal chromosomes (76). They carry gene cassettes, which may act as functional units, although they lack the 59-bp motif common to the bacterial integrons. Instead, they are bordered by partitioned *int* genes and the *int*(N) overlaps with downstream halves of tRNA genes containing *att* sites, as was found for the virus SSV1 and putative plasmid pXQ1 inserted within *Sulfolobus* chromosomes (53, 64). For example, in the chromosome of *S. solfataricus,* a partitioned *int* gene borders a segment of 67.7 kb, which encodes enzymes involved in central metabolism: dTDP-glucose 4,6-dehydratase, glucose-1-phosphate-thymidyl-transferase, dTDP-4-dehydrorhamnose reductase, and dTDP-4-dehydrorhamnose 3,5-epimerase, three of which are generally clustered in one operon. Since additional copies of all these enzymes are encoded in the *S. solfataricus* chromosome and show high sequence identity and similarity (76 to 84% and 86 to 96%, respectively), it is very likely that the DNA segment entered the chromosome aided by its own integrase (76).

Analysis of the other archaeal genome sequences has revealed partitioned integrase homologs in the euryarchaeote *P. horikoshii* and the crenarchaeote *A. pernix* containing *att* sites and bordering regions of 21.5 and 4 kb in *P. horikoshii* and 17.5 kb in *A. pernix* (76). Two similar regions have also been reported for the unpublished genome of *Pyrococcus* OT3 (49). Since none of their ORFs contained in these segments show any hits in the GenBank/EMBL sequence databases, it is likely that they are integrons, of unknown function, deriving from unknown organisms.

Mechanisms of Integrase-Mediated Insertion into Archaeal and Bacterial Chromosomes

Many bacterial integrases belong to a superfamily of tyrosine recombinases, and they have been strongly implicated in site-specific DNA recombination between extrachromosomal elements and chromosomes. In bacteria and eukaryotes, chromosome integration depends on two key elements, a site-specific recombinase, Int, and a recombination site, *att*. Generally, the *int* gene is adjacent to one of the recombination sites in the genetic element. More than 140 members of the integrase family have been identified and are classified into four subgroups on the basis of sequence similarity (59). The crystal structures of representatives from each group demonstrate that despite wide differences in sequence, the structures of the active sites are highly conserved (24).

Although chromosomal integration in bacteria also frequently occurs within tRNA genes (70), the archaeal integration process differs from that found in bacteria in the following respects: (i) the archaeal integrase lacks a conserved motif common to the bacterial integrases, (ii) the coding region of the archaeal integrase contains the *att* sites, and (iii) the archaeal integrase gene is partitioned on insertion into the chromosome. The last may have implications for chromosomal evolution. Since the intact integrase recognizes the *att* site and is therefore implicated in chromosomal insertion and excision (53), loss of the integrase function will lead to chromosomal gene capture (76).

IS ELEMENTS

Autonomous transposable elements are common in the three domains of life. They encode enzymes that

Table 5. Summary of IS elements and MITEs found in archaeal genomes

Organism	No. of types of IS elements[a]	Total no.[a]	IS families	No. of MITE-like elements	Reference(s)
Halobacterium sp. strain NRC-1	14 (3)	23 (3)	IS*4*, IS*5*, IS*605*, IS*L3*, unclassified	0	56
Haloferax volcanii	1	ND[b]	IS*4*	ND	48
Methanococcus jannaschii	5 (1)	7 (2)	IS*6*, IS*605*, unclassified	9 (265, 360 bp)	10, 69
Methanobrevibacter smithii	1	ND	IS*4*	ND	48
Methanobacterium thermoautotrophicum	0	0	ND	0	79
Pyrococcus abyssi	1	1	IS*L3*	0	Unpublished data
Pyrococcus furiosus	3	ND	IS*6*, IS*605*, IS*982*	0	Unpublished data
Pyrococcus horikoshii	2	3	IS*605*, IS*L3*	0	30
Archaeoglobus fulgidus	4	18	IS*3*, IS*6*, IS*630*/Tc1, IS*L3*	0	34
Aeropyrum pernix	1 (1)	2 (2)	Unclassified	9 (306 bp)	29
Sulfolobus solfataricus P2	25 (3)	201 (18)	IS*4*, IS*5*, IS*110*, IS*256*, IS*605*, IS*630*/Tc1, IS*L3*, unclassified	143 (79, 131, 163, 184 bp)	69, 75
Sulfolobus tokodaii	12	34	IS*4*, IS*110*, IS*605*, unclassified	46 (79, 184, 352 bp)	30a
Thermoplasma acidophilum	4	4	IS*256*, IS*605*	0	73
Thermoplasma volcanium	24 (3)	27 (3)	IS*3*, IS*4*, IS*5*, IS*256*, IS*605*, IS*630*/Tc1, IS*982*, IS*L3*, unclassified	2 (598 bp)	30b

[a]Numbers in parentheses are numbers of unclassified IS elements.
[b]ND, not determined.

facilitate their excision from, and integration into, genomes and they can cross barriers between species, kingdoms, and even domains (48). While bacterial IS elements migrate via DNA intermediates, in eukaryotes, transposons can migrate via either DNA or RNA (retrotransposon) intermediates.

Some archaeal genomes, and in particular those from some *Halobacterium* and *Sulfolobus* species, are rich in IS elements, both in their chromosomes and in their extrachromosomal genetic elements, while others appear to contain very few (Tables 5 and 6). They also produce phenotypic variants at high frequencies as a result of transposition events (14, 50). Most archaeal IS elements carry terminal inverted repeats, flanked by short direct repeats, and encode a transposase. Many are members of families of IS elements found in bacteria and eukaryotes, but some show no similarities to known families of IS elements (Tables 5 and 6).

In the genome of *S. solfataricus* P2, 201 copies of IS elements were identified and they constitute about 10.5% of the genome (75). Those that could be classified belong to seven known bacterial/eukaryotic families, while the 18 that were not classified may represent new families, some of which could be archaea-specific (Table 6). Those that have been shown to be active experimentally are ISC1058, ISC1217, ISC1359, and ISC1439 (50), but several others are thought to have moved recently because two or more copies are 100% identical. Five classes of the IS elements contain two ORFs (Table 6). For ISC1904 and ISC1913, one of them shows sequence similarity to a resolvase, which is common to some transposons, whereas for the shorter IS elements ISC774, ISC1212, and ISC1229, the ORFs overlap, which may indicate that they undergo frameshifting (13).

Several IS elements have also been detected in archaeal genetic elements. These include two copies of ISH1800 in each of the haloarchaeal viruses ΦH2 and ΦH5 (Table 1). Moreover, 9 IS elements occur in the haloarchaeal plasmid pHH1 (65), and another 54 copies occur in pNRC100 and pNRC200 (54). For the conjugative plasmids of *Sulfolobus*, pNOB8 contains ISC1316 and ISC1332, while pING contains ISC1913, all of which occur in the *S. solfataricus* P2 chromosome (Table 6).

Table 6. Summary of the IS elements in the *S. solfataricus* genome[a]

Name	Consensus size (bp)	ORF-encoded sequence length (aa)[b]	Inverted repeat (bp)	Direct repeat (bp)	Family	No. of copies
SSO0008	>846	282	ND[c]	ND	IS*605*	1
ISC774	782	142 + 84	19	None	ND	2
ISC1043	1,043	220	14	0	IS*L3*	5
ISC1048	1,048	328	23	2	IS*630*/Tc1	11
ISC1058	1,058	300	19	9	IS*5*	14
ISC1078	1,078	337	19	2	IS*630*/Tc1	8
ISC1160	1,160	34 + 90 + 24	12	6	IS*4*	3
ISC1173	1,173	321	46	8	ND	5
ISC1190	1,190	316	None	8	IS*110*	13
ISC1212	1,212	84 + 138	27	0, 4 or 7	IS*5*	9
ISC1217	1,148	355	13	6	ND	11
ISC1225	1,225	365	17	4 or 5	IS*4*	11
ISC1229	1,229	126 + 251	None	None	IS*110*	7
ISC1234	1,234	319	19	4	IS*5*	16
ISC1250	1,250	342	9	0	IS*256*	3
ISC1290	1,290	307	40	0 or 9	IS*5*	3
ISC1316	1,316	403	None	None	IS*605*	13
ISC1332	1,332	414	22	9	IS*256*	1
ISC1359	1,359	333	25	4	IS*4*	10
ISC1395	1,395	127 + 267	68	0	IS*630*/Tc1	4
ISC1439	1,439	322	20	9	IS*4*	32
ISC1476	1,476	460	20	0	IS*605*	2
ISC1491	1,491	245	None	None	IS*110*	5
ISC1904	1,904	192 + 423	None	None	IS*605*	10
ISC1913	1,913	211 + 410	None	None	IS*650*	2

[a]The numbers given are averaged for each class of IS element.
[b]A plus sign indicates additional ORFs, possibly frameshifts. aa, amino acids.
[c]ND, not determined.

NONAUTONOMOUS ELEMENTS

MITEs and SINEs are small nonautonomous elements present in eukaryal genomes (61, 90). They do not encode proteins and their mobility is facilitated in *trans* by transposases or reverse transcriptases, respectively, which are encoded by transposons (84, 85, 90). The genome of the archaeon *S. solfataricus* P2 contains at least four types of short nonautonomous sequence elements lacking ORFs (69, 75). They are similar to a subclass of eukaryotic MITEs that includes *Ds1* from maize and *Cele11, Cele12,* and *CeleTc2* from *C. elegans,* since they share terminal inverted repeats with their partner IS element/transposon (22, 62). A total of 143 SM elements (for *Sulfolobus* MITEs) have been identified, of which about 30 occur in ORFs where they introduce stop codons. The remainder lie between ORFs or within IS elements. They have been classified into four main groups on the basis of the degree of similarity of their sequences (69). Their general properties are summarized in Table 7.

The SM elements appear to be mobile because they are both highly conserved and widely distributed in the genome. Their terminal inverted repeats show sequence similarities to the terminal repeats of putative partner IS elements found in the chromosome, which are SM1 and ISC1048 (87% identity) (Fig. 2a); SM2 and ISC1217 (87% identity), SM3 and ISC1058 (81% identity), and, at a lower level, SM4 and ISC1173 (49% identity). These pairs of elements also exhibit similar terminal direct repeats. However, no sequence similarity was detected between the central regions of the SM elements and their corresponding

Table 7. SM elements found in the *S. solfataricus* genome

Element	Length (bp)[a]	Inverted repeat (bp)	Direct repeat (bp)[a]	Total no.	Partner IS element
SM1	79	23	2	40	ISC1048
SM2	184	16	6	25	ISC1217
SM3	131	24	9	44	ISC1058
SM4	164	27	8	34	ISC1173

[a]Averaged for each class of SM element.

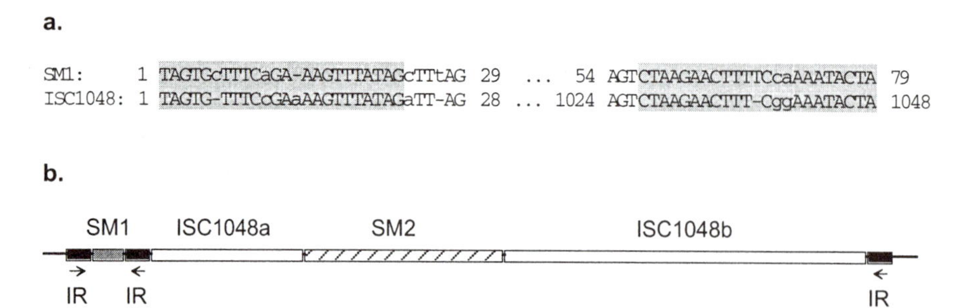

Figure 2. (a) Consensus sequences of the aligned terminal inverted repeats for the *Sulfolobus* MITE SM1 and its partner IS element ISC1048. The terminal inverted repeats are shaded, and capital letters indicate identical nucleotides. (b) Potentially mobile unit (800 bp) containing a copy of ISC1048 where the upstream half has been replaced by SM1 and the remainder contains a copy of SM2 (69). The left inverted repeat (IR) of SM1 and the right IR of ISC1048 are complementary, as shown in panel a.

IS elements, indicating that they may not share a common evolutionary origin (69). The transposases encoded by the IS elements, which are considered responsible for mobility of the SM elements, are classified in Table 7. The SM1/ISC1048 transposase belongs to the IS630/Tc1 family and has both bacterial and eukaryotic homologs, while the transposase of SM3/ISC1058 belongs to the bacterial IS5 family. The transposases of SM2/ISC1217 and SM4/ISC1173 remain unclassified.

For the chromosome of *S. solfataricus*, the SM elements constitute about 0.6% of the sequence. They are partly concentrated in four regions, denoted IS-1 to IS-4, that are also rich in IS elements (Fig. 3). Elements SM1 and SM2 are more conserved in sequence than SM4 (Table 7), which suggests that SM4 entered the genome, or arose within it, first.

Some of the SM and IS elements are interwoven and can generate potentially mobile units. For example, a copy of ISC1048 contains an SM2 element; moreover, its upstream end (550 bp), which includes part of the transposase gene (169 aa), has been replaced by a copy of SM1 (Fig. 2b). This creates a potentially novel mobile unit because the upstream inverted repeat of the SM1 and the downstream inverted repeat of the ISC1048 are sufficiently similar (Fig. 2a) to create a nonautonomous transposon. The latter also reveals a mechanism whereby SM elements can transfer between archaea, and possibly the other domains, together with their partner IS element.

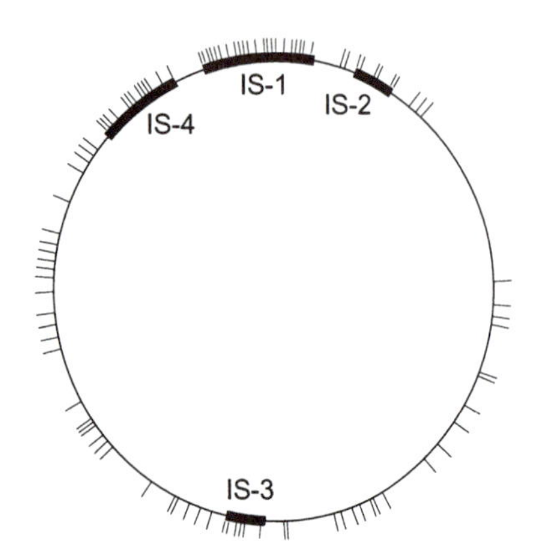

Figure 3. Distribution of SM elements in the chromosome of *S. solfataricus* P2. The areas marked IS-1 to IS-4 represent high concentrations of IS and SM elements (69).

MITE-like elements are also present in other archaeal genomes, and corresponding IS elements have been identified there on the basis of sequence similarity (69) (see Table 5). The two MITEs from *M. jannaschii* and the single element from *A. pernix* (which lacks terminal inverted repeats) show sequence similarity to IS elements ISE703 and ISC1313, respectively. The latter MITEs, in contrast to those of *Sulfolobus* (Table 5), clearly result from deletions in their corresponding IS element and thus belong to a different subclass of MITEs (22, 62). Short repeating elements (55 bp) in the *A. fulgidus* chromosome, which carry 5-bp inverted repeats, could not be linked to any IS element.

INTRONS

Archaea carry introns in their tRNA, rRNA, and mRNA genes that are archaea-specific with respect to both the structures of their RNA intron-exon junctions and their splicing mechanism. To date, no other intron types have been detected in archaea. Introns in tRNAs occur most frequently at one nucleotide 5′ to the anticodon where they occur exclusively in eukaryotic tRNAs, and between 2 and 18 copies have been

Table 8. Archaeal rRNA introns

rRNA and organism	No.	Size(s) (bp)	rRNA position(s)[b]	Reference
16S rRNA				
Aeropyrum pernix	1	699	908	58
Desulfurococcus sp. strain OC11	1	182	548	M. Bjerregaard and R. A. Garrett, unpublished data
Pyrobaculum aerophilum[a]	1	713	373	11
Staphylothermus marinus	1	64	548	M. Bjerregaard and R. A. Garrett, unpublished data
Thermoplasma neutrophilus JCM 9278	2		1205, 1213	27
Thermoplasma sp. strain IC-061	3	764, 32, 688	781, 1205, 1213	27
Thermoplasma sp. strain IC-033	5	627, 762, 636, 33, 682	548, 781, 1092, 1205, 1213	27
23S rRNA				
Aeropyrum pernix	2	202, 575	1085, 1927	58
Desulfurococcus mobilis	1	622	1952	32
Pyrobaculum organotrophum[a]	2	607, 598	1927, 2552	16
Staphylothermus marinus	2	56, 54	1945, 2449	33

[a]Environmental samples closely related to the genus *Pyrobaculum* have been shown to contain multiple rRNA introns (83).
[b]*E. coli* numbering is used for the rRNA positions.

detected in the sequenced archaeal genomes. Introns in rRNAs are also located at important functional centers in the 16S and 23S rRNAs but so far have been located only in the orders *Thermoproteales* and *Sulfolobales* of the crenarchaeotes, where some of the rRNA genes contain multiple introns (Table 8). Splicing is a prerequisite for tRNA and rRNA function and cleavage occurs via a homodimeric, or homotetrameric, archaeal enzyme at a "bulge-helix-bulge" motif at the intron-exon junction. The enzyme cuts, at symmetrical positions, in each of the 3-nucleotide bulges, and exon ligation and intron circularization ensue (20, 43). The archaeal splicing enzyme is a more primitive form of the hetero-pentameric enzyme complex that exclusively directs splicing of tRNA introns in eukaryotes; the archaeal monomer shows sequence similarity to two subunits of this enzyme complex (41, 44).

Small archaeal introns generally constitute a long irregular "hairpin" core structure where the terminal loop is relatively large and unstructured (43, 84a). In some of the rRNA introns, the terminal loop contains ORFs that correspond to LAGLIDADG proteins (Table 8). Two of these, I-*Dmo*I from *Desulfurococcus mobilis* and I-*Por*I from *Pyrobaculum organotrophum,* have been shown to exhibit DNA homing activity (17, 43), and the crystal structure of I-*Dmo*I partially resembles those of eukaryotic homing enzymes I-*Cre*I, from chloroplasts of *Chlamydomonas reinhardtii,* and PI-*Sce*I from yeast (78).

Like their eukaryal counterparts, both archaeal enzymes recognize large DNA sites (15 to 20 bp). They bind as monomers, bridging the minor groove, where they cut the sugar-phosphate backbone on either side and bind within the adjoining major grooves inducing some distortion or bending of the DNA (1, 45). The DNA binding regions of the homing enzymes carry some archaea-specific motifs. Cleavage activity requires the presence of divalent metal ions, which are positioned at the catalytic site, and the LAGLIDADG motifs have been implicated in their binding (46).

The presence of ORFs encoding homing enzymes in some rRNA introns and their absence from other rRNA introns and all tRNA introns of the archaea suggest that the ORFs themselves may enter and leave the RNA intron core at the accessible terminal loop, although no mechanism for this has been established (16). The same argument applies to the inteins, which are also found in some ORFs in all the sequenced archaeal genomes at between 1 and 18 copies. They also carry LAGLIDADG motifs and constitute putative homing endonucleases.

Intercellular Intron Mobility

Most crenarchaeotes investigated contain one set of rRNA genes. Thus, if an rRNA gene carries an intron that encodes a homing endonuclease, the latter will have no homologous site to which to migrate within the cell unless similar homing sequences exist within other types of genes (16). This led to the hypothesis that the introns are transmittable, intercellularly, between rRNA genes, and this concept receives support from the fact that the rRNA sequences flanking intron locations are generally highly conserved because they constitute important functional sites (2, 16, 32). To test for intercellular mobility, the intron from *D. mobilis,* which encodes I-*Dmo*I, was electroporated, together with the free I-*Dmo*I protein, into *Sulfolobus acidocaldarius* cells on a bacterial plasmid that could not replicate in the archaeon. The intron was shown not only to recombine into the *Sulfolobus* chromosome at the corresponding homing site of the 23S rRNA, but also to spread through the *Sulfolobus* population (2). To demonstrate that the intron was mobile intercellularly, a double antibiotic-resistant mutant was used; it was demonstrated that the intron moved from the wild-type cells to the mutant cells (2). However, attempts to repeat these experiments with other *Sulfolobus* species, including *S. shibatae* and *S. solfataricus* P2, failed to demonstrate intercellular intron mobility, and it was therefore inferred that *S. acidocaldarius* is in some way special (R. N. Aravalli and R. A. Garrett, unpublished data).

The mechanism of intron transfer between the *S. acidocaldarius* cells is still unknown. The high degree of structuring and stability of the RNA intron from *D. mobilis* raises the possibility of transmission via an RNA intermediate (43). However, a DNA transfer mechanism is also suggested by the observation that *S. acidocaldarius* can exchange and recombine a variety of chromosomal markers, presumably as a result of cell-cell contact and DNA conjugation (25, 74). Evidence for such a conjugative mechanism has also been provided by the detection of a two-way exchange of plasmids between cells of the euryarchaeote *Haloferax volcanii* (71). Thus, DNA exchange between cells may be a general phenomenon, at least for some archaea.

CONCLUSIONS

The study of DNA mobility events in archaea is still at an early stage. A general mechanism of chromosomal integration appears to operate in archaea for some viruses, plasmids, and possibly integrons. It involves integrase-mediated insertion into the downstream halves of tRNA genes that produces a partitioned integrase gene in the chromosome. Furthermore, mobile elements, including IS elements and nonautonomous elements, which are probably all mobilized by transposases, are widespread throughout the archaeal domain. Archaeal rRNA introns and tRNA introns are spliced by a mechanism specific to the archaeal domain, and the former, at least, can exchange between archaeal cells with the help of their encoded homing enzymes. Potentially mobile inteins are also common in the archaeal domain. Comparing and contrasting the detailed mechanisms of DNA mobility that occur among archaea with those operating in the bacterial and eukaryal domains will lead to a deeper understanding of how these complex processes have evolved.

REFERENCES

1. **Aagaard, C., M. J. Awayez, and R. A. Garrett.** 1997. Profile of the DNA recognition site of the archaeal homing endonuclease I-*Dmo*I. *Nucleic Acids Res.* **25:**1523–1530.
2. **Aagaard, C., J. Dalgaard, and R. A. Garrett.** 1995. Inter-cellular mobility and homing of an archaeal rDNA intron confers selective advantage over intron-cells of *Sulfolobus acidocaldarius. Proc. Natl. Acad. Sci. USA* **92:**12285–12289.
3. **Aravalli, R. N., and R. A. Garrett.** 1997. Shuttle vectors for hyperthermophilic archaea. *Extremophiles* **1:**183–191.
4. **Aravalli, R. N., Q. She, and R. A. Garrett.** 1998. Archaea and the new age of microorganisms. *Trends Ecol. Evol.* **13:** 190–194.
5. **Arnold, H. P., Q. She, H. Phan, K. Stedman, D. Prangishvili, I. Holz, J. K. Kristjansson, R. Garrett, and W. Zillig.** 1999. The genetic element pSSVx of the extremely thermophilic crenarchaeon *Sulfolobus* is a hybrid between a plasmid and a virus. *Mol. Microbiol.* **34:**217–226.
6. **Arnold, H. P., K. M. Stedman, and W. Zillig.** 1999. Archaeal phages, p. 76–89. *In* A. Granoff and R. G. Webster (ed.), *The Encyclopedia of Virology.* Academic Press, San Diego, Calif.
7. **Arnold, H. P., U. Ziese, and W. Zillig.** 2000. SNDV, a novel virus of the extremely thermophilic and acidophilic archaeon *Sulfolobus. Virology* **272:**409–416.
8. **Arnold, H. P., W. Zillig, U. Ziese, I. Holz, M. Crosby, T. Utterback, J. F. Weidman, J. K. Kristjansson, H. P. Klenk, K. E. Nelson, and C. M. Fraser.** 2000. A novel lipothrixvirus, SIFV, of the extremely thermophilic crenarchaeon *Sulfolobus. Virology* **267:**252–266.
9. **Barns, S. A., C. F. Delwiche, J. D., Palmer, and N. R. Pace.** 1996. Perspectives on archaeal diversity, thermophily and monophyly from environmental rRNA sequences. *Proc. Natl. Acad. Sci. USA* **93:**9188–9193.

9a. **Blum, H., W. Zillig, S. Mallok, H. Domdey, and D. Prangishvili.** 2001. The genome of the archaeal virus SIRV1 has features in common with genomes of eukaryal viruses. *Virology* **281:** 6–9.

10. **Bult, C. J., O. White, G. J. Olsen, L. Zhou, R. D. Fleischmann, G. G. Sutton, J. A. Blake, L. M. FitzGerald, R. A. Clayton, J. D. Gocayne, A. R. Kerlavage, B. A. Dougherty, J. F. Tomb, M. D. Adams, C. I. Reich, R. Overbeek, E. F. Kirkness, K. G. Weinstock, J. M. Merrick, A. Glodek, J. L. Scott, N. S. M. Geoghagen, and J. C. Venter.** 1996. Complete genome sequence of the methanogenic archaeon, *Methanococcus jannaschii. Science* **273:**1058–1073.
11. **Burggraf, S., N. Larsen, C. R. Woese, and K. O. Stetter.** 1993. An intron within the 16S rRNA gene of the archaeon *Pyrobaculum aerophilum. Proc. Natl. Acad. Sci. USA* **90:**2547–2550.
12. **Cannio, R. P., P. Contursi, M. Rossi, and S. Bartolucci.** 1998. An autonomously replicating transforming vector for *Sulfolobus solfataricus. J. Bacteriol.* **180:**3237–3240.
13. **Chandler, M., and O. Fayet.** 1993. Translational frameshifting in the control of transposition in bacteria. *Mol. Microbiol.* **7:** 497–503.
14. **Charlebois, R. L., and W. F. Doolittle.** 1989. Transposable elements and genome structure in *Halobacteria,* p. 297–308. In D. E. Berg and M. M. Howe (ed.), *Mobile DNA.* American Society for Microbiology, Washington, D.C.

15. **Collis, C. M., M. J. Kim, H. W. Stokes, and R. M. Hall.** 1998. Nucleotide binding of the purified integron DNA integrase Intl1 to integron- and cassette-associated recombination sites. *Mol. Microbiol.* **29:**477–490.
16. **Dalgaard, J. Z., and R. A. Garrett.** 1992. Protein-coding introns from the 23S rRNA-encoding gene from stable circles in the hyperthermophilic archaeon *Pyrobaculum organotrophum. Gene* **121:**103–110.
17. **Dalgaard, J. Z., R. A. Garrett, and M. Belfort.** 1993. A site-specific endonuclease encoded by a typical archaeal intron. *Proc. Natl. Acad. Sci. USA* **90:**5414–5417.
18. **DeLong, E. F.** 1998. Everything in moderation: archaea as 'non-extremophiles.' *Curr. Opin. Genet. Dev.* **8:**649–654.
19. **DeLong, E. F., K. Y. Wu, B. B. Prezelin, and R. V. Jovine.** 1994. High abundance of Archaea in Antarctic marine picoplankton. *Nature* **371:**695–697.
20. **Diener, J. L., and P. B. Moore.** 1998. Solution structure for the archaeal pre-tRNA splicing endonucleases: the bulge-helix-bulge motif. *Mol. Cell* **1:**883–894.
21. **Erauso, G., S. Marsin, N. Benbouzid-Rollet, M. F. Baucher, T. Barbeyron, Y. Zivanovic, D. Prieur, and P. Forterre.** 1996. Sequence of plasmid pGT5 from the archaeon *Pyrococcus abyssi:* evidence for rolling-circle replication in a hyperthermophile. *J. Bacteriol.* **178:**3232–3237.
22. **Fedoroff, N. V.** 1989. Maize transposable elements, p. 375–413. *In* D. E. Berg and M. M. Howe (ed.), *Mobile DNA.* American Society for Microbiology, Washington, D.C.
23. **Graham, D. E., R. Overbeek, G. J. Olsen, and C. Woese.** 2000. An archaeal genomic signature. *Proc. Natl. Acad. Sci. USA* **97:**3304–3308.
24. **Grainge, I., and M. Jayaram.** 1999. The integrase family of recombinase: organization and function of the active site. *Mol. Microbiol.* **33:**449–456.
25. **Grogan, D. W.** 1996. Exchange of genetic markers at extremely high temperatures in the archaeon *Sulfolobus acidocaldarius. J. Bacteriol.* **178:**3207–3211.
26. **Hall, R. M., C. M. Collis, M. J. Kim, S. R. Partridge, G. D. Recchia, and H. W. Stokes.** 1998. Mobile gene cassettes and integrons in evolution. *Ann. N. Y. Acad. Sci.* **870:**68–80.
27. **Itoh, T., K. Suzuki, and T. Nakase.** 1998. Occurrence of introns in the 16S rRNA genes of members of the genus *Thermoproteus. Arch. Microbiol.* **170:**151–166.
28. **Janekovic, D., S. Wunderl, I. Holz, W. Zillig, A. Gierl, and H. Neumann.** 1983. TTV1, TTV2 and TTV3, a family of viruses of the extremely thermophilic, anaerobic sulfur reducing archaebacterium *Thermoproteus tenax. Mol. Gen. Genet.* **192:** 39–45.
29. **Kawarabayasi, Y., Y. Hino, H. Horikawa, S. Yamazaki, Y. Haikawa, K. Jin-no, M. Takahashi, M. Sekine, S. Baba, A. Ankai, H. Kosugi, A. Hosoyama, S. Fukui, Y. Nagai, K. Nishijima, H. Nakazawa, M. Takamiya, S. Masuda, T. Funahashi, T. Tanaka, Y. Kudoh, J. Yamazaki, N. Kushida, A. Oguchi, K. Aoki, K. Kubota, Y. Nakamura, N. Nomura, Y. Sako, and H. Kikuchi.** 1999. Complete genome sequence of an aerobic hyperthermophilic crenarchaeon, *Aeropyrum pernix* K1. *DNA Res.* **6:**83–101.
30. **Kawarabayasi, Y., M. Sawada, H. Horikawa, Y. Haikawa, Y. Hino, S. Yamamoto, M. Sekine, S. Baba, H. Kosugi, A. Hosoyama, Y. Nagai, M. Sakai, K. Ogura, R. Otsuka, H. Nakazawa, M. Takamiya, Y. Ohfuku, T. Funahashi, T. Tanaka, Y. Kudoh, J. Yamazaki, N. Kushida, A. Oguchi, K. Aoki, and H. Kikuchi.** 1998. Complete sequence and gene organization of the genome of a hyperthermophilic archaebacterium, *Pyrococcus horikoshii* OT3. *DNA Res.* **5:**55–76.

30a. **Kawarabayasi, Y., Y. Hino, H. Horikawa, K. Jin-no, M. Takahashi, M. Sekine, S. Baba, A. Ankai, H. Kosugi, A. Hosoyama, S. Fukui, Y. Nagai, K. Nishijima, R. Otsuka, H. Nakazawa, M. Takamiya, Y. Kato, T. Yoshizawa, T. Tanaka, Y. Kudoh, J. Yamazaki, N. Kushida, A. Oguchi, K. Aoki, S. Masuda, M. Yanagii, M. Nishimura, A. Yamagishi, T. Oshima, and H. Kikuchi.** 2001. Complete genome sequence of an aerobic thermoacidophilic crenarchaeon, *Sulfolobus tokodaii* strain 7. *DNA Res.* **8:**123–140.

30b. **Kawashima, T., N. Aman, H. Koike, S. Makino, S. Higuchi, Y. Kawashima-Ohya, K. Watanabe, M. Yamazaki, K. Kanehori, T. Kawamoto, T. Nunoshiba, Y. Yamamoto, H. Aramaki, K. Makino, and M. Suzuki.** 2000. Archaeal adaptation to higher temperatures revealed by genomic sequence of *Thermoplasma volcanium. Proc. Natl. Acad. Sci. USA* **97:** 14257–14262.

31. **Keeling, P. J., H. P. Klenk, R. K. Singh, M. E. Schenk, C. W. Sensen, W. Zillig, and W. F. Doolittle.** 1998. *Sulfolobus islandicus* plasmid pRN1 and pRN2 share distant but common evolutionary distance. *Extremophiles* **2:**391–393.
32. **Kjems, J., and R. A. Garrett.** 1985. An intron in the 23S rRNA gene of the archaebacterium *Desulfurococcus mobilis. Nature* **318:**675–677.
33. **Kjems, J., and R. A. Garrett.** 1991. Ribosomal RNA introns in archaea and evidence for RNA conformational changes associated with splicing. *Proc. Natl. Acad. Sci. USA* **88:**439–443.
34. **Klenk, H. P., R. A. Clayton, J. F. Tomb, O. White, K. E. Nelson, K. A. Ketchum, R. J. Dodson, M. Gwinn, E. K. Hickey, J. D. Peterson, D. L. Richardson, A. R. Kerlavage, D. E. Graham, N. C. Kyrpides, R. D. Fleischmann, J. Quackenbush, N. H. Lee, G. G. Sutton, S. Gill, E. F. Kirkness, B. A. Dougherty, K. McKenney, M. D. Adams, B. Loftus, and J. C. Venter.** 1997. The complete genome sequence of the hyperthermophilic, sulphate-reducing archaeon *Archaeoglobus fulgidus. Nature* **390:** 364–370.
35. **Kletzin, A., A. Lieke, T. Urich, R. L. Charlebois, and C. W. Sensen.** 1999. Molecular analysis of pDL10 from *Acidianus ambivalens* reveals a family of related plasmids from extremely thermophilic and acidophilic archaea. *Genetics* **152:**1307–1314.
36. **Koonin, E. V., A. R. Mushegian, M. Y. Galperin, and D. R. Walker.** 1997. Comparison of archaeal and bacterial genomes: computer analysis of protein sequences predicts novel functions and suggests a chimeric origin for the archaea. *Mol. Microbiol.* **25:**619–637.
37. **Lanka, E., and B. Wilkins.** 1995. DNA processing reactions in bacterial conjugation. *Annu. Rev. Biochem.* **64:**141–169.
38. **Lessl, M., D. Balzer, W. Pansegrau, and E. Lanka.** 1992. Sequence similarities between the RP4 Tra2 and the Ti VirB region strongly support the conjugation model for T-DNA transfer. *J. Biol. Chem.* **267:**20471–20480.
39. **Lessl, M., W. Pansegrau, and E. Lanka.** 1992. Relationship of DNA-transfer-systems: essential transfer factors of plasmids RP4, Ti and F share common sequences. *Nucleic Acids Res.* **20:**6099–6100.
40. **Leisinger, T., and L. Meile.** 1992. Plasmids, phages, and gene transfer in methanogenic bacteria. *In* M. Sebald (ed.), *Genetics and Molecular Biology of Anaerobic Bacteria.* Springer-Verlag, New York, N.Y.
41. **Li, H., C. R. Trotta, and J. Abelson.** 1998. Crystal structure and evolution of a tRNA splicing enzyme. *Science* **280:** 279–284.
42. **Lykke-Andersen, J., C. Aagaard, M. Semionenkov, and R. A. Garrett.** 1997. Archaeal introns: splicing, intercellular mobility and evolution. *Trends Biochem. Sci.* **22:**326–331.
43. **Lykke-Andersen, J., and R. A. Garrett.** 1994. Structural characteristics of the stable RNA introns of archaeal hyperther-

mophiles and their splicing junctions. *J. Mol. Biol.* **243:** 846–855.
44. **Lykke-Andersen, J., and R. A. Garrett.** 1997. RNA-protein interactions of an archaeal homotetrameric splicing endonuclease with an exceptional evolutionary history. *EMBO J.* **16:** 6290–6300.
45. **Lykke-Andersen, J., R. A. Garrett, and J. Kjems.** 1996. Protein footprinting approach to mapping DNA binding sites of two archaeal homing endonucleases: evidence for a two domain protein structure. *Nucleic Acids Res.* **24:**3982–3989.
46. **Lykke-Andersen, J., R. A. Garrett, and J. Kjems.** 1997. Mapping metal ions at the catalytic centres of two intron-encoded endonucleases. *EMBO J.* **16:**3272–3281.
47. **Lykke-Andersen, J., H. Phan Thi-Ngoc, and R. A. Garrett.** 1994. DNA substrate specificity and cleavage kinetics of an archaeal homing-type endonuclease from *Pyrobaculum organotrophum. Nucleic Acids Res.* **22:**4583–4590.
48. **Mahillon, J., and M. Chandler.** 1998. Insertion sequences. *Microbiol. Mol. Biol. Rev.* **62:**725–774.
49. **Makino, S., N. Amano, H. Koike, and M. Suzuki.** 1999. Prophages inserted in archaebacterial genomes. *Proc. Jpn. Acad.* **75:**166–171.
50. **Martusewitsch, E., C. W. Sensen, and C. Schleper.** 2000. High spontaneous mutation rate in the hyperthermophilic archaeon *Sulfolobus solfataricus* is mediated by transposable elements. *J. Bacteriol.* **182:**2574–2581.
51. **Massana R., A. E. Murray, C. M. Preston, and E. F. DeLong.** 1997. Vertical distribution and phylogenetic characterization of marine planktonic Archaea in the Santa Barbara Channel. *Appl. Environ. Microbiol.* **63:**50–56.
52. **Mazel D., B. Dychinco, V. A. Webb, and J. Davies.** 1998. A distinctive class of integron in the *Vibrio cholerae* genome. *Science* **280:**605–608.
53. **Muskhelishvili, G., P. Palm, and W. Zillig.** 1993. SSV1-encoded site-specific recombination system in *Sulfolobus shibatae. Mol. Gen. Genet.* **237:**334–342.
54. **Ng, W. V., S. A. Ciufo, T. M. Smith, R. E. Bumgarner, D. Baskin, J. Faust, B. Hall, C. Loretz, J. Seto, J. Slagel, L. Hood, and S. DasSarma.** 1998. Snapshot of a large dynamic replicon in a halophilic archaeon: megaplasmid or minichromosome? *Genome Res.* **8:**1131–1141.
55. **Ng, W. L., and S. DasSarma.** 1993. Minimal replication origin of the 200-kilobase *Halobacterium* plasmid pNRC100. *J. Bacteriol.* **175:**4584–4596.
56. **Ng, W. V., S. P. Kennedy, G. G. Mahairas, B. Berquist, M. Pan, H. D. Shukla, S. R. Lasky, N. Baliga, V. Thorsson, J. Sbrogna, S. Swartzell, D. Weir, J. Hall, T. A. Dahl, R. Welti, Y. A. Goo, B. Leithauser, K. Keller, R. Cruz, M. J. Danson, D. W. Hough, D. G. Maddocks, P. E. Jablonski, M. P. Krebs, C. M. Angevine, H. Dale, T. A. Isenbarger, R. F. Peck, M. Pohlschrod, J. L. Spudich, K.-H. Jung, M. Alam, T. Freitas, S. Hou, C. J. Daniels, P. P. Dennis, A. D. Omer, H. Ebhardt, T. M. Lowe, P. Liang, M. Riley, L. Hood, and S. DasSarma.** 2000. Genome sequence of *Halobacterium* species NRC-1. *Proc. Natl. Acad. Sci. USA* **97:**12176–12181.
57. **Nölling, J., A. Groffen, and W. M. De Vos.** 1993. ΦF1 and ΦF3, two novel virulent phages infecting thermophilic strains of the genus *M. thermoautotrophicum. J. Gen. Microbiol.* **139:** 2511–2516.
58. **Nomura, N., Y. Sako, and A. Uchida.** 1998. Molecular characterization and postsplicing fate of three introns within the single rRNA operon of the hyperthermophilic archaeon *Aeropyrum pernix* K1. *J. Bacteriol.* **180:**3635–3643.
59. **Nunes-Duby, S. E., H. J. Kwon, R. S. Tirumalai, T. Ellenberger, and A. Landy.** 1998. Similarities and differences among 105 members of the Int family of site-specific recombinases. *Nucleic Acids Res.* **26:**391–406.
60. **Nutall, S. D., and M. L. Dyall-Smith.** 1993. HF1 and HF2: novel bacteriophages of halophilic archaea. *Virology* **197:** 678–684.
61. **Okada, N., M. Hamada, I. Ogiwara, and K. Ohshima.** 1997. SINEs and LINEs share common 3′ sequences: a review. *Gene* **205:**229–243.
62. **Oosumi, T., B. Garlick, and W. R. Belknap.** 1996. Identification of putative nonautonomous transposable elements associated with several transposon families in *Caenorhabditis elegans. J. Mol. Evol.* **43:**11–18.
63. **Pansegrau, W., and E. Lanka.** 1996. Enzymology of DNA transfer by conjugative mechanisms. *Prog. Nucleic Acids Res. Mol. Biol.* **54:**197–251.
64. **Peng, X., I. Holz, W. Zillig, R. A. Garrett, and Q. She.** 2000. Evolution of the family of pRN plasmids and their integrase-mediated insertion into the chromosome of the crenarchaeon *Sulfolobus solfataricus. J. Mol. Biol.* **303:**449–454.
64a. **Peng, X., H. Blum, Q. She, H. Domdey, S. Mallok, K. Brügger, R. A. Garrett, W. Zillig, and D. Prangishvili.** Sequences and replication of the archaeal rudiviruses SIRV1 and SIRV2: relationships to the archaeal lipothrixvirus SIFV and some eukaryal viruses: similarities between rudiviruses SIRV1 and SIRV2 which infect the extremely thermophilic crenarchaeon *Sulfolobus* and eukaryal pox viruses. *Virology*, in press.
65. **Pfeifer, F., and P. Ghahraman.** 1993. Plasmid pHH1 of *Halobacterium salinarium:* characterization of the replicon region, the gas vesicle gene cluster and insertion elements. *Mol. Gen. Genet.* **238:**193–200.
66. **Pfister, P., A. Wasserfallen, R. Stettler, and T. Leisinger.** 1998. Molecular analysis of *Methanobacterium* phage ΨM2. *Mol. Microbiol.* **30:**233–244.
67. **Prangishvili, D., S.-V. Albers, I. Holz, H. P. Arnold, K. Stedman, T. Klein, H. Singh, J. Hiort, A. Schweier, J. K. Kristjansson, and W. Zillig.** 1998. Conjugation in archaea: frequent occurrence of conjugative plasmids in *Sulfolobus. Plasmid* **40:** 190–202.
68. **Prangishvili, D., H. P. Arnold, D. Götz, U. Ziese, I. Holz, J. K. Kristjansson, and W. Zillig.** 1999. A novel virus family, the Rudiviridae: structure, virus-host interactions and genome variability of the *Sulfolobus* viruses SIRV1 and SIRV2. *Genetics* **152:**1387–1396.
69. **Redder, P., Q. She, and R. A. Garrett.** 2001. Non-autonomous mobile elements in the crenarchaeon *Sulfolobus solfataricus. J. Mol. Biol.* **306:**1–6.
70. **Reiter, W.-D., P. Palm, and S. Yeats.** 1989. tRNA genes frequently serve as integration sites for prokaryotic genetic elements. *Nucleic Acids Res.* **17:**1907–1914.
71. **Rosenshine, I., R. Tchelet, and M. Mevarech.** 1989. The mechanism of DNA transfer in the mating system of an archaebacterium. *Science* **245:**1387–1389.
72. **Rowe-Magnus, D. A., A. M. Guerout, and D. Mazel.** 1999. Super-integrons. *Res. Microbiol.* **150:**641–651.
73. **Ruepp, A., W. Graml, M.-L. Santos-Martinez, K. K. Koretke, C. Volker, H. W. Mewes, D. Frishman, S. Stocker, A. N. Lupas, and W. Baumeister.** 2000. The genome sequence of the thermacidophilic scavenger *Thermoplasma acidophilum. Nature* **407:** 508–513.
74. **Schmidt, K. J., K. E. Beck, and D. W. Grogan.** 1999. UV stimulation of chromosomal marker exchange in *Sulfolobus acidocaldarius:* implications for DNA repair, conjugation and homologous recombination at extremely high temperatures. *Genetics* **152:**1407–1415.
75. **She, Q., R. K. Singh, F. Confalonieri, Y. Zivanovic, P. Gordon, G. Allard, M. J. Awayez, C.-Y. Chan-Weiher, I. G. Clausen,**

B. Curtis, A. De Moors, G. Erauso, C. Fletcher, P. M. K. Gordon, I. Heidekamp de Jong, A. Jeffries, C. J. Kozera, N. Medina, X. Peng, H. Phan Thi-Ngoc, P. Redder, M. E. Schenk, C. Theriault, N. Tolstrup, R. L. M. Charlebois, W. F. Doolittle, M. Duguet, T. Gaasterland, R. A. Garrett, M. Ragan, C. W. Sensen, and J. Van der Oost. 2001. The complete genome of the crenarchaeon *Sulfolobus solfataricus* P2. *Proc. Natl. Acad. Sci. USA* **98**:7835–7840.

76. **She, Q., X. Peng, W. Zillig, and R. A. Garrett.** 2001. Gene capture events in archaeal chromosomes. *Nature* **409**:478.
77. **She, Q., H. Phan, R. A. Garrett, S.-V. Albers, K. M. Stedman, and W. Zillig.** 1998. Genetic profile of pNOB8 from *Sulfolobus:* the first conjugative plasmid from an archaeon. *Extremophiles* **2**:417–425.
78. **Silva, G. H., J. Z. Dalgaard, M. Belfort, and P. Van Roey.** 1999. Crystal structure of the thermostable archaeal intron-encoded endonuclease I-*Dmo*I. *J. Mol. Biol.* **286**:1123–1136.
79. **Smith, D. R., L. A. Doucette-Stamm, C. Deloughery, H. Lee, J. Dubois, T. Aldredge, R. Bashirzadeh, D. Blakely, R. Cook, K. Gilbert, D. Harrison, L. Hoang, P. Keagle, W. Lumm, B. Pothier, D. Qiu, R. Spadafora, R. Vicaire, Y. Wang, J. Wierzbowski, R. Gibson, N. Jiwani, A. Caruso, D. Bush, H. Safer, D. Patwell, S. Prabhakar, S. McDougall, G. Shimer, A. Goyal, S. Pietrokovski, G. M. Church, C. J. Daniels, J.-I. Mao, P. Rice, J. Nölling, and J. N. Reeve.** 1997. Complete genome sequence of *Methanobacterium thermoautotrophicum* ΔH: functional analysis and comparative genomics. *J. Bacteriol.* **179**:7135–7155.
80. **Stedman, K. M., C. Schleper, E. Rumpf, and W. Zillig.** 1999. Genetic requirements for viral function in the extremely thermophilic archaeon *Sulfolobus solfataricus:* construction and testing of viral shuttle vectors. *Genetics* **152**:1397–1405.
81. **Stedman, K. M., Q. She, H. Phan, I. Holz, H. Singh, D. Prangishvili, R. Garrett, and W. Zillig.** 2000. pING family of conjugative plasmids from the extremely thermophilic archaeon *Sulfolobus islandicus:* insights into recombination and conjugation in crenarchaeota. *J. Bacteriol.* **182**:7014–7020.
82. **Stetter, K. O., G. Fiala, G. Huber, R. Huber, and A. Segerer.** 1990. Hyperthermophilic microorganisms. *FEMS Microbiol. Rev.* **75**:117–124.
83. **Takai, K., and K. Horikoshi.** 1999. Molecular phylogenetic analysis of archaeal intron-containing genes coding for rRNA obtained from a deep-subsurface geothermal water pool. *Appl. Environ. Microbiol.* **65**:5586–5589.
84. **Varmus, H., and P. Brown.** 1989. Retroviruses, p. 53–109 *In* D. E. Berg and M. M. Howe (ed.), *Mobile DNA*. American Society for Microbiology, Washington, D.C.

84a. **Watanabe, Y., S. Yokobori, T. Inaba, A. Yamagishi, T. Oshima, Y. Kawarabayasi, H. Kikuchi, and K. Kita.** Introns in protein-coding genes in Archaea. *FEBS Lett.*, in press.

85. **Wessler, S. R., T. E. Bureau, and S. E. White.** 1995. LTR-retrotransposons and MITEs: important players in the evolution of plant genomes. *Curr. Opin. Genet. Dev.* **5**:814–821.
86. **Witte, A., U. Baranyi, R. Klein, M. Sulzner, C. Luo, G. Wanner, D. H. Kruger, and W. Lubitz.** 1997. Characterization of *Natronobacterium magadii* phage ΦCh1, a unique phage containing DNA and RNA. *Mol. Microbiol.* **23**:603–616.
87. **Woese, C. R., and G. E. Fox.** 1977. Phylogenetic structure of the prokaryotic domain: the primary kingdoms. *Proc. Natl. Acad. Sci. USA* **74**:5088–5090.
88. **Woese, C. R., D. Kandler, and M. L. Wheelis.** 1990. Towards a natural system of organisms: proposal for the domains Archaea, Bacteria, and Eucarya. *Proc. Natl. Acad. Sci. USA* **87**: 4576–4579.
89. **Wood, A. G., W. B. Whitman, and J. Koniski.** 1989. Isolation and characterization of an archaebacterial virus-like particle from *Methanococcus voltae* A3. *J. Bacteriol.* **171**:93–98.
90. **Xiong, Y., and T. H. Eickbush.** 1990. Origin and evolution of retroelements based upon their reverse transcriptase sequences. *EMBO J.* **9**:3353–3362.
91. **Zillig, W., H. P. Arnold, I. Holz, D. Prangishvili, A. Schweier, K. Stedman, Q. She, H. Phan, R. Garrett, and J. K. Kristjansson.** 1998. Genetic elements in the extremely thermophilic archaeon *Sulfolobus*. *Extremophiles* **2**:131–140.
92. **Zillig, W., A. Kletzin, C. Schleper, I. Holz, D. Janekovic, J. Hain, M. Lanzendorfer, and J. K. Kristjansson.** 1994. Screening for Sulfolobales, their plasmids and their viruses in Icelandic solfataras. *Syst. Appl. Microbiol.* **16**:609–628.
93. **Zillig, W., W. D. Reiter, P. Palm, F. Gropp, H. Neumann, and M. Rettenberger.** 1988. Viruses of archaebacteria, p. 517–568. *In* R. Calendar (ed.), *The Bacteriophages,* vol. 1. Plenum Press, New York, N.Y.

Mobile DNA II
Edited by N. L. Craig et al.

Chapter 47

Mobile Elements in Animal and Plant Genomes

PRESCOTT L. DEININGER AND ASTRID M. ROY-ENGEL

INTRODUCTION

The genomes of all higher eukaryotes are littered with copies of mobile elements, some active, but most representing old, defective copies. One view is that the animal and plant genomes are at war with their mobile elements and, to a lesser extent, mobile elements war with one another. One of the most obvious analogies would be to consider the retroviruses as mobile elements that are simply extremely efficient at horizontal transmission. Genomes and organisms often fight viral epidemics, and it would be reasonable to consider that they might be in conflict with their endogenous elements in the same manner. Mobile elements contribute to insertional mutagenesis, as well as to postinsertional genomic changes that lead to altered gene expression and genetic recombination. The genomes want to suppress the damage caused by mo-

Prescott L. Deininger and Astrid M. Roy-Engel • Tulane Cancer Center and Department of Environmental Health Sciences, Tulane University Medical Center, 1430 Tulane Ave., New Orleans, LA 70112.

bile element amplification, whereas mobile elements continue to evolve new approaches for efficient amplification. It is equally possible that different families of mobile elements are at war with one another. They may be competing for limited cellular resources, as well as for the highest possible share of the genetic load caused by their mobility. Others believe that this selfish view is too limited and that purely hostile mobile elements would not be successful if they did not provide some advantage to the host (11, 51) and adapt to the genomes in which they survive. Whether one accepts the selfish view, or believes that the elements are positive forces allowing adaptation to new functions, it is true that there are incredible variations both in the types and copy number of mobile elements within different genomes. Thus, mobile elements have significantly influenced the genomes of all the higher eukaryotes, but with more profound effects on some than others.

We briefly discuss the different classes of mobile elements seen in some of the best-characterized genomes and then discuss some of the general concepts behind the colonization of genomes by mobile elements.

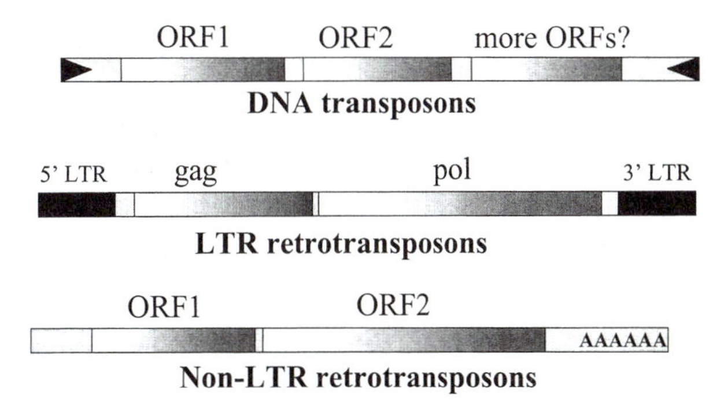

Figure 1. Major classes of autonomous mobile elements in animals and plants. Schematic diagrams for the three major classes of mobile elements are shown. The DNA transposons have two or three open reading frames coding for enzymes involved in the transposition process. They are also flanked by inverted repeats (arrowheads) at the ends that act as *cis*-acting elements in the integration process. The transposition process involves direct integration through DNA intermediates. Both the LTR retrotransposons and the non-LTR retrotransposons use an RNA intermediate in the amplification process. They generally encode two open reading frames coding for RNA-binding proteins and enzymes involved in reverse transcription and endonuclease cleavage at the site of integration. The LTR retrotransposons have LTRs that contain signals for transcription of the elements and are involved in the integration process. The non-LTR retrotransposons use a promoter region found in the 5′ noncoding region of the RNA and terminate in a poly(A) tract like a typical mRNA.

CLASSES OF MOBILE ELEMENTS IN EUKARYOTIC GENOMES

There are basically three classes of autonomous mobile elements (Fig. 1), all three of which can be found to various extents in different genomes of all animals and plants. These are the DNA transposons, the long terminal repeat (LTR) retrotransposons, and the non-LTR retrotransposons. They are classified on the basis of their structure and transposition mechanism, and examples of each of these classes are discussed in detail elsewhere in this book.

In addition to the autonomous mobile elements, there are nonautonomous mobile elements that rely on the autonomous elements for at least some of their transposition apparatus. Some of the nonautonomous elements have evolved separately, but in parallel with the autonomous elements (such as the mammalian short, interspersed elements [SINEs] [29]). However, many of the nonautonomous elements are derived from the autonomous elements through deletions of portions of the genome that leave critical *cis*-acting elements in place, but delete portions that can act in *trans* (e.g., miniature inverted-repeat transposable elements, MITEs [36]). Thus, elements that delete critical transposition proteins may be able to amplify as long as there is a functional copy somewhere in the genome to supply the activity in *trans*. Having a smaller genetic unit may allow some of the nonautonomous elements to amplify much more effectively, amplifying to much higher copy numbers than those found for the autonomous counterpart. It even seems likely that some of the nonautonomous elements compete for limiting transposition machinery and therefore result in less amplification of their autonomous elements on which they are dependent.

THE HUMAN GENOME AND OTHER MAMMALIAN GENOMES

The sequences of human chromosomes 21 and 22 are almost complete (30, 43), as well as an analysis of mobile elements on the X chromosome (2, 43). These analyses allow for a very detailed picture of the genomic organization of mobile elements and demonstrate that approximately 40% of the human genome is made up of mobile elements. In fact, this represents only those elements that are recognized as repetitive. It is quite possible that significant portions of the rest of the genome could be made up of ancient elements that have diverged to the point that they essentially represent unique sequences. The classification and abundance of the recognizable elements are described in Table 1. Both DNA elements and LTR elements make up a significant portion of the chromosomes.

Table 1. Abundance of human mobile elements on chromosomes 21, 22, and X

Element	Copy no. on chromosome 22[a]	% of chromosome 22 (34 Mb)[a]	Copy no. on chromosome 21	% of chromosome 21 (33.8 Mb)	Copy no. on X chromosome	% of X chromosome (56.7 Mb)
DNA elements			3,572	2.20	5,483	2.49
MER1&2 type			3,404	2.12	4,684	2.22
Mariners			168	0.08	190	0.09
LTR elements		3.44	9,590	9.22	8,941	8.98
MaLRs			5,379	4.87	4,315	3.97
Retroviral	255	0.48	2,115	2.25	2,590	2.98
Mer4			1,396	1.42	1,692	1.69
Other LTR	3,635	2.86	708	0.68		
LINEs	14,424	13.54	12,723	15.51	19,682	30.16
L1	8,043	9.73	8,982	12.93	13,018	26.50
L2	6,381	3.81	3,741	2.58	6,394	3.56
SINEs	28,614	19.98	15,748	10.84	23,405	9.56
Alu	20,188	16.80	12,341	9.48	15,859	7.51
MIRs	8,426	3.18	3,407	1.36	7,546	2.06

[a]Blank entries indicate that these data were not presented or analyzed when published.

However, the vast majority of recent mobile element activity has been due to the non-LTR class of elements and their nonautonomous relatives.

DNA Transposons

DNA transposons make up only about 2 to 3% of the human genome (43, 98). This is primarily composed of MER1 and MER2 elements, although there are about 100 to 300 copies of a mariner/Tc1 family of transposons. There are also several thousand related, nonautonomous elements that appear to be a family of MITEs created by deletions from these mariner elements. The DNA elements in general show high levels of sequence heterogeneity, suggesting that they have been in the genome for quite a long time (generally 10 to 35% mismatch [98]). There is very little evidence for recent activity from the mammalian DNA transposons and the presence of most of them probably predates the mammalian radiation. However, many human mariner elements are not present in orthologous positions in the bovine genome, suggesting at least some activity in the mammals (77).

LTR Retrotransposons

The LTR retrotransposons make up 5 to 10% of the human genome (30, 43). This includes abundant families of MaLR, Mer4, and abundant members of several families of endogenous retroviruses. Almost all of the copies are not only defective, but appear to have deletions typical of nonautonomous families. The MaLR elements appear to be related to the ERV-L class of elements (96). Many of the LTR elements are quite old. However, several of the retroviral elements show extensive polymorphisms between primate genomes (48) and therefore have only been present for the past few tens of millions of years. There is little evidence for modern activity of these LTR elements, although there may have been several hundred HERV-K elements that have inserted since the divergence of chimpanzee and human (74).

Non-LTR Retrotransposons

There are two dominant forms of non-LTR retrotransposons in the human genome. The L2 family is abundant, making up 3 to 4% of the human genome. However, these elements are quite battered, almost all showing greater than 20% mismatch (97). Thus, they are likely to have been active early in the mammalian radiation, but have had little or no activity in recent evolutionary times, at least in the primate genome.

L1 elements are the major form of currently active, autonomous non-LTR element in the mammalian genome. They make up 10 to 13% of chromosomes 21 and 22 (Table 1 [30, 43]) and more than 25% of the X chromosome (2), and show evidence of activity throughout mammalian evolution. This represents a copy number that may approach one million elements in the human genome. Although L1 elements appear to have predated the mammalian radiation, most have amplified in the past 30 million years (97). Recent activity has led to approximately 0.1% of human genetic disease caused by direct insertion of these elements in the human genome (55).

The human genome also has a dominant form of

nonautonomous, non-LTR retrotransposon, referred to as *Alu* elements. Like the MIR elements, these elements fall into the general class termed SINEs (112). They retropose via an RNA intermediate that is synthesized by RNA polymerase III. They share several features in common with L1 elements leading to the hypothesis that they specifically use the L1 retrotransposition machinery for their amplification (8, 50, 55). One of the major similarities is a 3′ A tract which resembles the poly (A) stretch at the 3′ end of L1 elements thought to be involved in the priming of the reverse transcription (Fig. 1). The other similarity is that although the direct repeats formed at the site of integration are quite heterogeneous in length and sequence, they share significant sequence similarity at their ends with the preferred cleavage site of the L1 endonuclease (34, 50).

Alu elements are an extremely successful form of nonautonomous element, with more than 10^6 copies making up 12 to 15% of the genome (25, 30, 43, 97). This represents the single highest copy number of mobile elements identified to date. MIRs represent a second major class of SINEs. However, as discussed below, the MIRs and many of the other mobile element families in humans appear to be quiescent and represent only fossils of previously active elements that may have been dependent on the L2 elements for their activity based on their similar periods of amplification (98).

Processed pseudogenes are also a significant factor in the mammalian genome and may be classified as nonautonomous, non-LTR elements. Approximately 16% of all human genes identified on chromosome 22 appear to be processed pseudogenes (30). This would extrapolate to more than 10,000 processed pseudogenes in the human genome. The processed pseudogenes have the two features discussed above for *Alu,* a 3′ A tail and flanking direct repeats similar to those of L1. Thus, processed pseudogenes are also likely to be dependent on L1 amplification. Although some genes have generated multiple processed pseudogenes, most pseudogenes are present in only one or two copies. Thus, the process is extremely inefficient for any given gene, in contrast to the remarkable efficiency of *Alu* elements and other SINEs.

Other Mammals

Although the same volume of large contiguous sequence assemblies is not available for other mammals, extensive studies show that they have some properties similar to those in the human genome. For instance, the older families of elements, like the DNA transposons and LTR retrotransposons, were likely to have been active before the separation of most of the mammalian orders (99). Thus, a very similar pattern of these sequences should be present in all of the mammalian species, many in orthologous loci.

We cannot be sure that all of these families lost amplification capability to the same extent in all orders; therefore, the biggest differences are likely to be in the modern amplification pattern. For instance, in the mouse genome there are approximately 2,000 endogenous intracisternal A particle elements, representing LTR elements, which show tremendous levels of strain differences consistent with high levels of recent activity (68). Similarly, most of the activity generating endogenous retroviral elements in genomes appears to have occurred in the past 20 million to 30 million years. Thus, a high proportion of mutations in the mouse (54) can be attributed to insertions of mobile elements, compared with a much smaller proportion in humans (approximately 0.2% of human disease [25, 54]).

Levels of L1 and SINE activity have varied significantly between mammalian orders and even in some fairly closely related species. Older subfamilies of L1 are present in all orders (99). However, many of the L1 elements represent order-specific variants that have shown different patterns of amplification rate during recent evolutionary time in different species. Recently, strong evidence has been reported to show that L1 elements have largely become inactive in South American rice rats (17), whereas the activity remains strong in most other rodent species. SINEs are even more variable between species. MIRs represent very old SINEs, but most of the other major SINE families are order- and sometimes species-specific (26) and therefore have undergone massive and recent expansions that vary between species. For instance, almost all of the greater than 10^6 *Alu* elements have inserted in the primate genome since the divergence of higher primates from the prosimian primates. Similarly, there are three different SINE families that are specific for rodents. However, even amplification of these SINEs has been sporadic and varied in the different rodent species (52). Thus, although the SINEs all apparently initiated early in rodent evolution, they either died out at various rates or were sporadically activated to extremely high levels in different species. Equally important as the sporadic nature of this expansion is that different SINEs show different expansion patterns within the same species. For instance, B2 elements are relatively well amplified in all rodent species tested, whereas the ID elements are sporadic (52). Thus, if SINEs are all dependent on L1 elements, it suggests that there are regulatory influences other than just L1 activity. This will be discussed more in the section on amplification dynamics below.

THE *CAENORHABDITIS ELEGANS* GENOME

Almost complete analysis of the *C. elegans* genome has led to the identification of 38 different families of mobile elements (18), although a complete analysis of the transposons based on copy number and type is not available. Most of the elements have mutations that make them clearly defective. Among the elements are the LTR retrotransposons of the Ty3/-gypsy family and several non-LTR retrotransposons, NeSL-1 and RTE-1 (69). There is indirect evidence that at least one Ty3/gypsy LTR retrotransposon has been active recently in the genome (9). NeSL-1 represents an ancient branch from most of the other non-LTR retrotransposons. DNA transposons are very common in the *C. elegans* genome, as well as several relatively high-copy-number families of their nonautonomous relatives, MITEs (100). This includes the Tc7 family of MITEs, which have two large inverted repeats with strong homology to the ends of the Tc1 element, separated by only a short unique sequence (87).

INSECT GENOMES

The *Drosophila melanogaster* genome sequence is now available (1), although a complete analysis of the data in terms of their mobile elements has not been presented. There are more than 50 classes of transposons identified, almost all of which have been found in previous studies (111). The *Drosophila* genome appears to be dominated by DNA transposons and LTR retrotransposons, with relatively little data to suggest high-copy numbers of nonautonomous elements. Some of these elements are among the best characterized of the mobile elements and are considered in detail in chapters 21 and 33.

In *Drosophila* species, it is rare for a mobile element to exceed 100 copies (Table 2) with tremendous diversity in copy numbers for almost all families of elements between species, and even between different populations of the same species (111). This tremendous diversity suggests a great deal of recent mobile element activity in *Drosophila,* which is reinforced by the high proportion of mutations in *Drosophila* caused by mobile element insertions. The apparent lack of many older and high-copy-number families of elements suggests either that many of the elements are new to the *Drosophila* genome or that there are forces that cleanse the genome of such elements. These potential controlling influences will be discussed in the section on amplification dynamics below.

Table 2. Species and population variation in *Drosophila* mobile elements

Mobile element	Copy no.	
	D. melanogaster[a]	*D. simulans*[a]
DNA elements		
bari-1	4.37	4.88
hobo	49.90	66.23
pogo	13.25	0
P[c]	14–33[c]	0
LTR elements		
412	28.45 (26–35,[b] 2–54[d])	13.88 (1–73)[b]
Blood	17.45	2.50
Copia	24.05	3.88
Tirant	11.45	1.62
Non-LTR elements		
Helena	0.25	10.23
Jockey	31.60	3.27

[a]Data from reference 111.
[b]Data from reference 71.
[c]Reviewed in reference 35.
[d]From reference 5.

Site-Specific Mobile Elements

There are several well-characterized examples of mobile elements with extremely high levels of site specificity for their insertion in the insects. Among these are the R1 and R2 elements that were originally characterized in *Bombyx mori* as having highly specific insertion sites in the 28S rRNA genes. These elements have now been found throughout the arthropods, consistent with their vertical transmission during the past 500 million years (14).

Other site-specific mobile elements are the HetA and TART elements that make up the telomeres in *Drosophila* (6, 24, 63). All telomeres may be considered nonautonomous site-specific mobile elements that simply insert at the ends of chromosomes using reverse transcriptase supplied in *trans.* However, the *Drosophila* elements are autonomous elements and show that this important cellular function has changed significantly within this species.

REPTILES, AMPHIBIANS, AND FISH

Only sporadic data are available on mobile elements in reptiles, amphibians, and fish. Most of the data consist of a large variety of individual reports of specific autonomous and nonautonomous mobile elements, some of which will be covered in this section as representative of the existing knowledge.

One of the better characterized is the pufferfish genome. The genome of the pufferfish, *Fugu rubripes,* is considered as a model vertebrate genome, because it contains a gene set similar to humans in a genome eight times smaller (44 Mb). Densely packed genes with short intergenic and intronic sequences devoid of repetitive elements characterize the *F. rubripes* genome. Initial evaluation of a pufferfish genomic library covered 25 Mb by sequence scanning (31). The analysis detected several mobile element families that represented a total of only 2.19% of the genome, confirming the relatively low number of mobile elements. The most abundant of these, the CR1-like repeat (non-LTR retrotransposon), contributes about half (1.2% of the genome) of existing elements. An earlier study previously detected and characterized the CR1-like repeat in *Fugu,* naming it Maui (86).

On the opposite end of genome sizes, the salamander *Hydromantes* possesses one of the largest genomes among terrestrial vertebrates. Among the mobile elements in the *Hydromantes* genome, about a million copies of the LTR retrotransposon Hsr1 (Ty3/-gypsy) can be detected (70). However, most appear to be transcriptionally silent, as indicated by methylation analysis and detection of truncated copies.

DNA Transposons

Among the DNA transposons, Tc1-like families can be found widespread throughout Vertebrata, in invertebrates such as insects, and even among fungi. Tc-1 like transposons (Tzf and Tdr1) have been reported to be active in zebrafish (61). Some of the Tc-1-like members can be present in high numbers, as in the Atlantic salmon with 15,000 SALT1 copies/haploid genome (40). Another transposon with similarity to the hAt family, Tol2, is active in fish (53). Recent activity of this DNA transposon is associated with an albino phenotype in *medaka* fish (56), even though there are 10 copies/diploid genome in this species.

Tx1 and Tx2 families of DNA transposons detected in *Xenopus laevis* represent examples of DNA transposons and their derived families of MITEs. The autonomous elements span 6.9 kb with two open reading frames (ORFs) (Tx1c and Tx2c), and the related MITE spans only 400 bp with a tandem internal repeat flanked by inverted repeats (Tx1d and Tx2d) (36). An additional example, the Xbr elements isolated from *Xenopus* genomes are also short sequences framed by a 46-bp inverted repeat that contains a consensus sequence termed the T2 motif. Although the small elements are referred to as SINEs in all three examples, they are better described as MITEs, because there is clear evidence they spread via DNA-mediated transposition.

LTR Retrotransposons

The LTR retrotransposons can be divided into two major families, Ty1/copia and Ty3/gypsy elements, based on sequence similarity and order of the protease, reverse transcriptase/RNase H, and integrase domains (115). Examples from both families can be found in the genomes of fish, reptiles, and amphibians. The Ty3/gypsy class is present in herring (Cpr1), salmonids (easel), brook trout (B2), tunicate (Cir2), and echinoderms. A member of this class (sushi-ischi) was detected even within the small genome of *Fugu* (85). As mentioned above, one highly efficient LTR retrotransposon from this family is the Hsr1 in the salamanders with $\sim 10^6$ copies, possibly the highest number of copies for an LTR retrotransposon in eukaryotes (70). Multiple members of the Ty-1/copia group have been reported in Amphibia and Reptilia. The Tcs1 in iguanas is suggested to be inactive because of the lack of polymorphisms (35). Another member of the Ty1/copia group, the reptilian Tpm1, has a discontinuous distribution in snakes, suggesting loss in some species or horizontal transfer.

Non-LTR Retrotransposons

Non-LTR retrotransposons can be found throughout most, if not all, eukaryotic genomes. Three currently active groups have been described: L1-like elements found in humans and rodents; CR-1-like elements observed in avian, fish, and reptile genomes; and the RTE group found in ungulates, *C. elegans,* silkmoth, mosquito, and fish.

Some of the nonautonomous, tRNA-derived SINE elements are suggested to have been created by a fusion with LINE 3′ sequences (82). This was proposed to be created by a fusion event in which a strong-stop DNA with a primer tRNA was integrated at the 3′ end of a LINE and began to retrotranspose along with appended L1 sequences. The L1 sequences incorporated in this SINE then supply necessary *cis*-acting sequences to use LINE retrotransposition apparatus to amplify the SINE. The turtle CR-1-like LINE proved to share 3′-end sequence identity with the 3′ end of the tortoise SINEs (82). Other examples of this tight non-LTR retrotransposon-SINE relationship are observed in the tRNA-derived SINEs (*Hpa*I and *Sma*I) from the salmonids family and the tobacco SINE and their respective LINEs (82). However, this relationship has not been observed for all the known SINEs or LINEs, because in many cases one of the corresponding partners has not yet been found. Many tRNA-derived SINEs did not originate from fusion with a LINE and developed by a yet unknown mechanism. Some SINEs appear to be composites of different insertions. The DANA element in zebrafish appears to have been assembled by insertions of short sequences into a progenitor tRNA-derived element. Once associated, they amplified as one unit (46). In the Rana/PolIII SINE in Palearctic green water frogs, it also appears that a transposition event inserted a segment in a tRNA element making it a composite SINE (12).

PLANT GENOMES

Almost the entire sequence of chromosome 2 of *Arabidopsis thaliana* has been published (64). This includes almost 20 Mb of sequence with 563 recognizable, individual mobile elements. These include 159 DNA transposable elements, 251 LTR retrotransposons (with an additional 35 retroviral-related insertions), and 153 non-LTR retrotransposons. Thus, there is a relatively even distribution of classes of elements, with LTR retrotransposons the most prevalent. There are five different classes of DNA elements, with "Mutator" and "En-Spm/Tam1/Psl" dominating. All but one of the non-LTR retrotransposons are from the Tal1-like class, and there are at least three common families of LTR retroelements. At least 50% of these elements are clearly pseudogenes. Thus, *Arabidopsis* has representatives of most of the major families of plant mobile elements, although they represent a relatively modest proportion of the genome.

DNA Transposons

In general, the autonomous DNA transposons have not amplified to extremely high-copy numbers in plants, although there is a tremendous diversity of different families of DNA transposons in plant genomes (3, 59). However, the nonautonomous MITEs have amplified to high copies in a broad range of plant genomes.

LTR Retrotransposons

LTR retrotransposons are the most abundant form of mobile elements in plants. As mentioned previously, these elements can be broken down into two broad classes, the Ty1/copia and the Ty3/gypsy groups (115), which can then be subdivided into many families and subfamilies of elements.

In contrast to the relative paucity of mobile elements in the *Arabidopsis* genome, approximately 70 to 85% of the maize genome is composed of mobile elements, with the majority being LTR retrotransposons. Apparently a great deal of this amplification in maize has occurred during the past few million years. The maize genome is 3 to 4 times the size of the sorghum genome, which diverged only 15 million to 20 million years ago (37). Molecular analysis of the maize elements suggests that most of them have arisen and amplified within the past 2 million to 6 million years (90). Analysis of the adh1 locus in maize (105) and several other studies show that these elements are arranged in clusters in which elements have inserted into other elements. Although some of this pattern represents preferential insertion into elements (91), a great deal of it is likely to result from fairly random insertions into a genome that was becoming crowded with elements. Some of the maize element families have amplified into the tens of thousands of copies, such as Cinful-1 (90), Opie-1 (91), and PREM-2 (107), whereas many others are present at more modest copy numbers.

Non-LTR Retrotransposons

Non-LTR retrotransposons are also present in most, if not all, plant genomes. These include Del2, which has reached 250,000 copies in the Lilium genome (4% of the genome, [62]) and other low-copy-number elements in *Arabidopsis* (114) and maize and other plants (80). Nonautonomous SINE elements are also found in numerous plant genomes (59), although generally in relatively low-copy-number families. However, TS elements in tobacco are present at greater than 50,000 copies (117), which demonstrates sporadic potential for extreme amplification.

SITE PREFERENCES AND GENOMIC LOCATIONS

Patterns of Specificity

Almost all the classes of mobile elements are present to some degree in almost all animal and plant genomes. The differences seem to be in the quantity and timing of amplification of the different types of elements (discussed under "Amplification Dynamics" below) and in their genomic locations. Fairly complete maps of mobile elements on several chromosomes and whole genomes that illustrate the diversity well now exist. Figure 2 shows a schematic of some of the general patterns of genomic distribution of different elements.

The highest degree of specific localization is observed in the site-specific elements. These are best illustrated by the R1 and R2 retrotransposons in insects (67). These elements insert in a specific sequence found only within the rRNA gene cluster. Similarly, the *Drosophila* HetA and TART retrotransposons insert almost exclusively at the telomeres and serve as the telomeric repeat units in that species (6, 24, 63). These represent examples whereby the mechanism of insertion limits the presence of these mobile elements

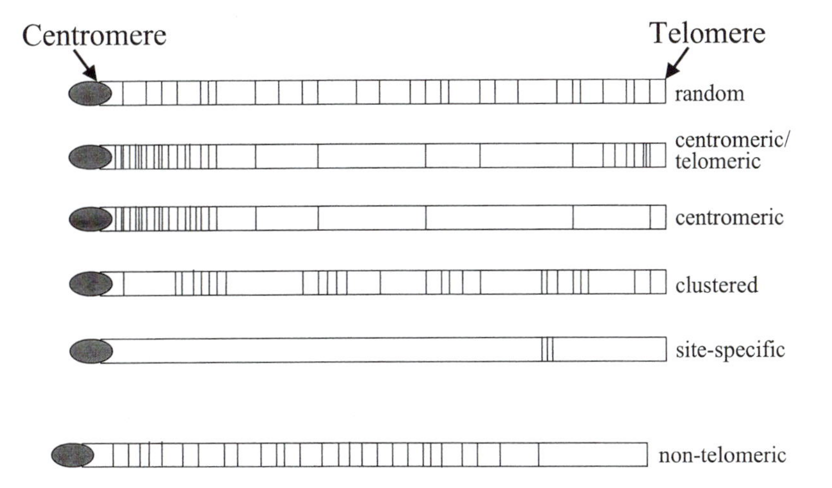

Figure 2. Genomic patterns for mobile elements. The patterns show schematically some of the basic density patterns found for mobile elements in animal and plant genomes. These include elements that are distributed fairly randomly, some showing either a combined centromeric/telomeric preference, and many with a preference for the juxtacentromeric region. In more complex genomes, it is typical to see a complex series of clusters spread throughout the genome. There are also classes of mobile elements with a very high degree of specificity for very specific gene or chromosomal regions.

to very specific chromosomal sites and places limits on copy numbers as well.

Other mobile elements show some site preference, but are clearly much more promiscuous than the site-specific elements. Mammalian L1 and SINE elements are an excellent example of this lesser level of preference. The sequences of human chromosomes 21 (43) and 22 (30) confirm previous cytogenetic analysis (57) and illustrate that although L1 elements are dispersed throughout the chromosomes, strong preferences for broad chromosomal regions are observed. Several factors may cause this weaker form of clustering. First, the endonuclease associated with L1 elements has been shown to have some sequence preference (20). The endonuclease consensus sequence, TTAAAA, would be likely to be widely dispersed, but may be found preferentially in some chromosomal regions. DNA elements, like the *Xenopus* Xbr elements, also often show strong preference for short consensus sequences at their integration site (108). Second, there may be some chromatin structure or other more general feature present that influences insertion preferences. Third, there may be strong selective pressures to eliminate elements that insert in some chromosomal regions.

It seems likely that *Alu* elements and other mammalian SINEs use the L1 endonuclease for their own insertion and therefore might be expected to show a similar clustering (8, 50). *Alu* elements on chromosomes 21 and 22 show an overlapping, but distinct clustering relative to the L1. This is in reasonable agreement with previous cytogenetic studies showing distinctly different banding patterns for L1 and Alu (57). Several possible, broader influences on mobile element location are discussed below.

Rather than a series of clusters, some genomes have mobile elements preferentially located in very specific chromosomal domains. In Fig. 2, these might correspond to the juxtacentromeric or centromeric/telomeric types of organization. In *C. elegans,* a series of MITEs have very strong biases for different chromosomal regions, but different MITEs show different biases (100). For instance, Cele14 and 42 show strong biases toward the centromeric and telomeric quarters of each chromosome, whereas the center half of the chromosome arm is relatively devoid of elements. Cele2 shows no particular clustering, except for a bias against the telomeric quarter of the chromosome. In contrast, Cele1 is spread throughout, but with a pretty strong enrichment toward the centromere. Thus, although there appear to be specific influences of particular chromosomal regions that are consistent for specific elements, they are not a universal chromosomal property in *C. elegans.* Perhaps more striking is the observation that the nonautonomous MITE, Tc7, shows a very strong clustering to the X chromosome (20 of 33 elements), whereas its apparent progenitor, Tc1, is distributed broadly throughout the chromosomes (87). This is similar to the mammalian L1/*Alu* relationship in which the apparent autonomous element does not integrate with the same pattern of specificity as the nonautonomous elements that depend on it.

Mobile elements are spread throughout the *Drosophila* genome, but with a significant level of clustering in the juxtacentromeric region, as well as a noted tendency to cluster in gene-poor regions (1) and heterochromatin (49, 103). *Arabidopsis* chromosome 2 shows an even more striking concentration of mobile elements in the juxtacentromeric region, which also corresponds to the most gene-poor region (64). The density of mobile elements has also been shown to extend into the centromeric region in *Arabidopsis* (44, 58). This contrasts somewhat with the relative paucity of L1 and *Alu* elements in human centromeric regions. This difference may reflect non-LTR retrotransposons having difficulty inserting in the centromeres, relative to the more abundant LTR and DNA mobile elements in *Drosophila* favoring centromeric regions. Alternatively, it may reflect the more complex chromosomal structure in mammals and different selective pressures.

Site Preference versus Selection

There are many arguments why it would be advantageous for mobile elements to develop a high level of site-insertion preference. The most obvious is that random insertion of mobile elements would result in random mutation. Initially, this might result in a paucity of elements in the gene-rich regions of an organism purely from selection. However, given time, this selection would lead to an advantage for those elements that minimized the deleterious effects on their host. Some of the different strategies that mobile elements may have developed for site preference are discussed below.

There is at least one additional reason why site specificity may be a useful strategy for elements. Some elements apparently show insertion preference within other elements (42, 91). This serves a 2-fold advantage for those elements. First, they will be inserting into a chromosomal region that has already tolerated a mobile element insertion and therefore there will be little likelihood for negative selection. This may also be why some elements, like *Alu* elements, like to insert adjacent to one another. In addition, the insertion within another element also serves to inactivate that element. This may prove a selective advantage for the secondary element. By damaging the first element it eliminates an element that could otherwise damage the host, allowing the first element to amplify more before some destructive level of element insertions is reached in that genome. In addition, it is possible that different elements may be competing for other host factors in their amplification process. Thus, eleminating competing elements could allow the new element better access to host factors for the amplification process.

Site preference for insertion may occur at either a short- or long-range level. The highest levels of specificity come from the short-range specificities. The most extreme cases would be the almost total site specificity of the R1 and R2 elements, and the *Drosophila* telomeric elements (see above). These would be the ultimate strategies for avoiding damage to the host. In the telomeric elements, there would be a strong positive selection for the elements. Lesser degrees of specificity may arise from mobile element endonucleases that have only a partial specificity. Sequence preferences for insertion would only be expected to result in major chromosomal biases if the preferred sites were distributed nonrandomly. In the L1 elements, the A + T-rich recognition site would be expected to be present in A + T-rich domains at higher levels. This could result in some bias to the extent that large domains of this nature exist in the chromosomes.

Some long-range factors must also contribute to nonrandom integration in chromosomal regions. These could be regional A + T richness, as described above. However, although *Alu* elements in humans prefer to insert in locally A + T-rich regions, they show a strong preference for G + C-rich chromosomal domains. Thus, alternative factors relating to chromatin structure, heterochromatin versus the more open euchromatin structure for instance, are likely to have an influence. In human *Alu* elements, it has been reported that their abundance in euchromatic chromosome bands (57) and housekeeping genes (95) suggests that they may insert preferentially in the more open chromatin. This could simply be the result of accessibility of the DNA, or it could be related to factors such as topoisomerase sites influencing supercoiling or transcription influencing the openness of gene regions. Alternatively, the correlation with more active genes may be indirect, reflecting a factor such as replication timing (22, 39). Gene-rich regions tend to replicate earlier in S phase than the heterochromatic regions. If *Alu* elements were more prone to insertion early in S phase for some reason unrelated to gene expression, they might still have the tendency to insert in the gene-rich regions. On the other hand, several mobile elements show preference for location in heterochromatin. This includes those in the simpler eukaryotic chromosomes, like *Arabidopsis* and *C. elegans,* that show preferences for centromeric and juxtacentromeric heterochromatin. It also includes L1 elements in mammals, which show preferences for the more heterochromatic R bands (57).

In some cases the elements may have developed chromosomal or site preferences that allow them to target heterochromatin. Alternatively, it is likely that this is still at least partially the result of selection. The *Arabidopsis* (64, 72) and *C. elegans* (18) chromosomes have approximately half of the sequences on the chromosome arms actually occupied by coding sequences. Thus, there may be relatively few sites in the gene-rich regions that can accept a mobile element insertion without some negative selective pressure. Although these pressures may not work quickly, in the long run, one would expect even a mild selective pressure to result in loss of that insertion from the population. In L1 and *Alu* elements, where *Alu* is proposed to be dependent on L1, it is still possible to explain some of the difference with preferential amplification of each type of element in different cell types or times of the cell cycle. However, it seems likely that some of the differences are because the longer L1 elements are more likely to have a greater negative impact than *Alu* and therefore can be tolerated less well within genes.

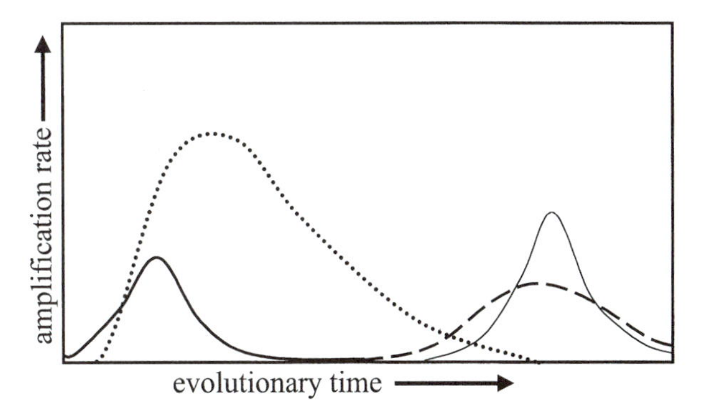

Figure 3. Schematic of evolutionary peaks of mobile element amplification. Each of the different line types represents a theoretical amplification rate versus evolutionary time for a different mobile element. These simply show the concept that when a mobile element enters or is activated in a naïve genome, they often amplify efficiently and reach a peak amplification rate. Processes such as mobile elements regulating their own amplification, development of genomic suppression mechanisms, or simply negative selection due to the damage caused by the elements, eventually lead most elements to greatly decrease or lose all amplification capability. Because very few elements are actively removed from the genome, the modern genome will be littered with pseudogene copies of the older elements, as well as members of any active families of elements. Also, some of the highest-copy-number elements are nonautonomous and therefore only amplify in conjunction with an active, autonomous family of elements. The modern genome is littered with both the elements that amplified earlier in evolution and families of currently active elements.

AMPLIFICATION DYNAMICS

Looking at the proportions of the different classes of mobile elements in a genome provides only a static summary of elements and their past activity. It does not provide an accurate picture of the current impact of those elements. For instance, the vast majority of the recent human mobile element amplification has been from the L1 and *Alu* elements. Both L1 and *Alu* are amplifying at a rate consistent with about one insert in every 200 new births in the human population (25, 54). This has led to each of them being responsible for about 0.1% of all human genetic defects. This rate, however, appears to represent a slowing down in amplification rate of almost 100-fold over the peak amplification time (94), at least for *Alu*. Almost all of the other major classes of mobile elements appear to be either fossils of extinct classes of elements or families that are extremely inefficient and therefore have had only modest recent impact on the human genome. Thus, these elements appear to amplify rapidly for only limited periods of evolutionary time.

There must be a first time in which a genome was exposed to a particular type of element. When an element is first introduced to a genome (either through horizontal transmission or de novo formation) it may begin to expand. The initial amplification may be relatively slow, because the element may need to evolve to a more active form within that organism. More active copies that evolve will then spread more effectively through the species and the rate will increase. The rate is likely to continue to increase until the regulatory, or selective, pressures discussed below lead to a decrease, or complete loss, of amplification. Figure 3 shows a simple general concept for the population of genomes by mobile elements. Although some of the lower-copy-number elements that are more highly regulated may maintain low, steady levels of amplification for very long periods, most of the elements that have contributed higher copy numbers to genomes appear to accumulate in evolutionary bursts. Because very few elements are removed from genomes, the modern genome contains a collection that includes both the active, modern elements, and a vast array of elements from families that are no longer active. For instance, in humans, almost all of the DNA elements and the LTR elements currently appear to be inactive (97). Two types of data allow estimates of the time of amplification of families of elements. The first, and most definitive, is the analysis of orthologous genomes and the ability to time the insertion of a specific element by its presence or absence in different species (78, 94). The second estimates amplification time by determining the average divergence between elements. Higher levels of divergence imply presence in the genome for a longer period accumulating mutations. Many of the mammalian MIR and MER families of repeats appear to represent quite old and battered elements (98). It is possible that other members of these families or entirely different families that are old enough that they are no longer recognizable as members of the family still exist. They may have been degraded to the point that they now represent unique sequence.

Overall the findings support the presence of almost all classes of mobile elements in almost all animal and plant genomes. Most of these types of elements must have extremely deep evolutionary roots and have coevolved with the various animals and plants. Although the DNA elements have some capability to excise themselves so that part of the time they may move rather than amplify, most of the time they appear to undergo an amplification (32, 84). LTR-based elements are almost always replicative, although their LTRs have a tendency to recombine, which results in deletion of most of the element. Non-LTR elements appear to always be replicative and therefore always result in higher copy number.

The biggest differences in mobile elements between genomes are the result of which types of elements have amplified most efficiently at different stages in that organism's evolution. Perhaps the biggest question is why so many genomes have apparently been subjected to cycles of major amplification. Although an element remains at low-copy number in a genome, it is possible that random mutations may silence the few copies available, resulting in a decrease. However, once a critical level of copy number is established, one would think that it would be self-sustaining, unless there was some specific limiting mechanism. On the other hand, if amplification continues to accelerate for a family of elements, it would eventually have to reach a point where it would be so deleterious that it could lead to extinction of itself (except for old "pseudogene" fossils left in the genome), along with its host. This would be closely analogous to the viral plagues that periodically sweep through populations but become self-limiting when the host population decreases sufficiently.

Regulatory Mechanisms

Some mobile elements have evolved to the point that they include mechanisms to limit their amplification and therefore result in minimal damage to their host. P elements, in *Drosophila,* illustrate several levels of regulation. One, they produce an inhibitory protein that suppresses element expression when the copy number reaches a certain level (66). As those elements mutate and lose capability of expression, or if the host mates with a fly that does not have P elements, reactivation occurs to maintain a reasonable steady state of functional elements. In addition, P elements have developed a germ-line-specific splicing pattern that allows them to limit their amplification to the germ line, where the amplification events can be passed on. This eliminates damage to the somatic cells that might lessen host viability. These types of regulatory processes allow P elements to minimize their negative impact on the genome, while allowing them to maintain activity in the appropriate tissues and other conditions necessary to maintain the overall activity of P elements in the population.

Elements that have not developed methods of control have tremendous potential for damaging their host. This may result in selective pressure for the host to develop mechanisms to suppress the mobile elements. Host suppression would be a likely mechanism to cause cyclic mobile element amplification. It has been suggested that one of the primary reasons for methylation in most eukaryotic genomes is the suppression of mobile elements (116). If it were not to their advantage, however, it would seem relatively easy for a mobile element to evolve to the point where it is not subject to suppression by methylation because of selection for loss of CpG dinucleotides. Mammalian L1 elements, for instance, have a major CpG island at their 5′ end around the promoter, and there are data suggesting that methylation limits their mobility. There is also a general mechanism involving cosuppression of elements based on homology-dependent gene silencing that has been observed for elements in several different genomes (7, 47, 89). For instance, the non-LTR I element can be introduced to *Drosophila* transgenically where they increase in copy number for several generations before they somehow become suppressed (47). This type of suppression appears to be homology dependent, but it does not necessarily depend on full copies or gene products (reviewed in reference 7). There is conflicting evidence for whether this acts through an RNA-dependent mechanism or through genomic homology-based suppression. This is a general method that may have developed to regulate a broad range of elements. However, it does not seem to be equally effective in all genomes or on all elements, as evidenced by some of the extremely high-copy-number elements found in some genomes. Elements that lose this level of control may be too deleterious and result in strong negative selection for that variant.

Alternatively, to the extent that mobile elements rely on cellular factors during their amplification, those factors may change in a manner that makes them less effective for the mobile element. For instance, transcription of copia elements in *Drosophila* may vary more than 100-fold because of host-specific differences (21). Several other host-specific factors have been reported in *Drosophila* elements (66). Although less clear in mammals, similar possibilities have been proposed. There are data demonstrating that the Sry family of proteins is critical for L1 expression in mammals (102), and it is hypothesized that SRP9/14 protein and *Alu* elements have coevolved with one another and that this coevolution may have helped the damping of *Alu* amplification rates (19). In a similar manner, elements that find themselves competing with, or under attack by, other elements may select for forms that evolve to avoid these competitions.

The massive levels of evolutionary amplification seem more likely to arise from elements with fewer controls on their amplification. In fact, some of the more controlled elements present today are probably evolved from less controlled ancestors whose activity may have been eliminated through negative selection. Alternatively, massive levels of amplification may be more acceptable with elements whose insertion is less deleterious. Site preferences (discussed above) would be one selective strategy to allow high levels of amplification with minimal impact. Many of the highly successful nonautonomous elements may also be effective because their small size and minimal functional sequence content may result in lower levels of damage to the genome and therefore less negative selection.

In many cases, however, an active cellular process may not silence the really high copy number elements, but mutation may simply silence the most active copies. Sequencing within *C. elegans* (18) has demonstrated that most mobile elements have defects that would keep them from amplifying. In some types of elements, only a small portion of new copies has any activity. The non-LTR retrotransposons are the best example, because most of the copies are severely 5′ truncated upon insertion, resulting in at least 90% of the copies being defective immediately. Of the approximately 500,000 L1 elements in the human genome, it has been estimated that only about 60 copies are not defective (92). Even these 60 copies are relatively repressed, and it is unclear how many are truly functional in the organism. When an apparently large mobile element family has only a few functional elements, it allows stochastic processes, such as random mutation, to potentially damage elements faster than new master elements can be created (27, 28). This leads to an evolutionary process where distinct sequence variants, or subfamilies, are made at different times in an organism's evolution. It can also result in the relative extinction of a family of mobile elements.

Induction of Amplification Cycles

Extremely active amplification cycles, like those schematized in Fig. 3, may be initiated either when an element is first introduced to a naïve genome, or through reactivation of a relatively quiescent class of elements. Reactivation may occur through either the accumulation of mutations or recombination between different elements, which alter the element in such a way to recover its activity, possibly by changing in a way that avoids any suppressive strategies used by the organism. Alternatively, there may be times in evolution when the host can accept a higher insertional load without too much adverse selection. This may lessen negative selective pressure, resulting in more rapid amplification within the population. For instance, there have been periods in evolution where new opportunities for speciation have been abundant and genetic diversity may have been more favorable. During these speciation bursts, one might expect that all types of genetic diversity would be tolerated better and that mobile elements in particular might help create the necessary diversity. As any new biological niches became filled and competition became more intense, the high levels of damage associated with genetic diversity might have been less advantageous and therefore created more negative selection.

It seems that stochastic processes must play a role in the initiation of mobile element amplification peaks. For instance, an older and relatively ineffective element may accumulate chance mutations that allow the element to overcome any cellular regulations or weaknesses in its own amplification. Alternatively, new active elements may be created by chance recombination events that rebuild a functional element from two previously defective elements. This could even occur between members of different families of the same type of element, resulting in an entirely new family. Last, new families may be created by horizontal transmission of an element from another species. The mariner family of elements probably represents the most widespread example of an element that has spread horizontally throughout the plant and animal kingdoms (15). This type of transmission would open a virgin genome to an element that would then avoid any regulations that previously evolved to control that element. Evolutionary analyses of the various mobile elements suggest very deep common origins for the integrase and transposase domains (16) and similarities among reverse transcriptases of all types (115). In very few cases, new elements may be created from genes that were not previously mobile elements. The first mobile elements certainly had to come from somewhere. Clear examples of this type of evolution have occurred with the nonautonomous SINE elements. These elements all appear to be derived from either tRNA genes or the 7SL RNA gene (23), all of which are transcribed by RNA polymerase III (29). All the different families of SINEs appear to be specific for individual orders, and some are limited to a relatively small clade of closely related species. All mammals studied to date appear to have one to three relatively abundant (>100,000 copies) SINEs (26). SINE families can vary widely in copy number between relatively closely related species (52). They also occur, although more sporadically and often in lower-copy number, in other phylla, such as fish (101, 117), amphibia (82), and plants (59, 75). The relatively few families of these elements and their sporadic occurrence in other species suggest that their formation and high level of amplification is subject to stochastic processes. Because they are nonautonomous, they are certainly dependent on the presence of autonomous elements. In addition, however, a tRNA pseudogene must evolve in just the right manner to allow initiation of a SINE family (28), and other changes must influence how the SINE interacts with the autonomous element or cellular factors for amplification efficiency.

Population Influences

When we study the evolution of mobile elements and their amplification, we are generally seeing a static picture of those elements that have fixed in the population during speciation. However, there are striking examples of studies that demonstrate that mobile elements can spread through a population with astonishing rapidity. *Drosophila* represents one of the best-studied models for modern population dynamics of mobile elements. Table 2 shows a selection of data regarding both copy-number differences between two *Drosophila* species, *melanogaster* and *simulans,* as well as some indications of the copy-number variation within a species. In the P elements, it is thought that they entered the *D. melanogaster* species quite recently from another *Drosophila* species (33). The possible mechanisms of horizontal transmission are covered elsewhere in this text. Thus, they are spreading through a relatively naive population and have not reached an equilibrium value, where processes for elimination of copies from the population equal rates of formation. In fact, the variability of many of these elements suggests a high level of activity of many elements within the *Drosophila* genus, and that activity has been particularly high, in general, in *melanogaster* relative to *simulans.*

There is some evidence for clines in mobile element copies. This has been observed as a general latitudinal cline for the 412 elements in *Drosophila* populations (5). This might be expected after introduction of a mobile element into a population through either horizontal transmission or a process which de novo forms or activates a new element. The element would immediately propagate in the population in which it is first introduced and then spread out from that site to form a gradient as that population gradually mixes with neighboring populations. In the laboratory, the mdg-1 elements equilibrated to the same average copy numbers in small, isolated populations, within 20 generations (104). In larger, natural populations this equilibration will be much larger, and it will be influenced by migration patterns and mixing with other population groups of that species.

Besides the dynamics of spread of elements, populations are the units in which selection occurs. Several lines of evidence show that several transposable elements are active in *Drosophila,* but that there is very little evidence for excision, suggesting that the primary mode of elimination of elements is through selection (4, 5, 71). This selection could be caused by direct deleterious effects of insertions or could occur through an increased potential for unequal recombination as discussed below. This selection could then be a major cause of the chromosomal distribution patterns seen (as discussed above). It could also suggest that the level of selective pressure on a particular species might influence the equilibrium copy number of elements. Thus, during periods of population expansion into new ecological niches, variation within a species could be more beneficial and selection lower, resulting in a net increase in mobile element copy number. Under intense competition, individuals with lower copy numbers of mobile elements may have sufficient selective advantage to allow selection for lower copy numbers in the population. The high levels of recent mobility in plant species, such as maize, might be explained in this way. Of particular interest in maize and other plants is whether the high levels of somatic mobility are simply caused by recent amplification bursts which selection has not had time to control or whether somatic mobility is less negative in plants than animals.

Interspecific hybridization in *Drosophila* can result in a greatly increased transposition rate (60). The increased transposition rate in these hybrids would be likely to contribute a strong negative selection against the viability of such hybrids and could provide an effective breeding barrier to help in speciation. One might hypothesize that a specific population could have genetically adapted to minimize transposition of a family of elements, whereas a different population of the species was naïve for the element. Were these populations to interbreed, controls on transposition could be eliminated, causing negative selection and poor viability of the progeny. Thus, the two populations would already have some genetic barrier to breeding and be partly along the path to speciation.

SECONDARY IMPACT OF MOBILE ELEMENTS

Not only is the impact of mobile elements on their genome a direct result of insertional mutagenesis, but there are also a series of secondary impacts on the genome including recombination and many more subtle changes that may alter the stability or evolution of an organism's genome.

Recombination

For some elements and genomes, the most damaging aspect of mobile elements may not be their insertion into a genomic site, but instead the possibility of secondary rearrangements caused by the dispersal of homologous sequences to different chromosomal regions. There are several reasons why mobile elements are potentially very damaging through recombination. First, any element that inserts and is tolerated by the genome (i.e., does not provide a major negative selection) will continue to provide a potential site for unequal homologous recombination with other elements for millions of years. This threat is lost only when the element is lost from the population through fixation or evolves to the point that it has sufficient levels of sequence mismatch relative to other elements to lessen its recombination potential. Second, recombination between two elements will result in either deletion or duplication of the sequences between them, or alternatively, in a chromosomal translocation. These types of events are likely to be extremely deleterious to any gene involved and may influence multiple genes.

The impact of recombination between homologous mobile elements may vary depending on several conditions. First, the copy number of the elements should have a strong effect on the probability of recombination. Second, the spacing of the elements (which will be somewhat influenced by copy number) may also be important. Third, the site preferences of elements may have a major influence. Those elements that are located in relatively gene-poor chromosomal regions will have less negative consequences upon recombination. Fourth, the length of the elements is likely to play an important role because every study of recombination to date suggests that longer nucleation regions favor recombination. Fifth, the sequence identity between elements is likely to be important for similar reasons. Sequence identity is influenced by the age of elements in the genome, but it might also be influenced by the presence of different evolutionary subsets of active mobile elements in some cases. Sixth, some organisms have much higher homologous recombination levels, such as that of yeast relative to that of humans. This might make the organism much less tolerant of high copy number repeats and may be a major selective factor of whether a genome can tolerate high proportions of high copy number elements. Thus, if an element begins to spread rapidly, there may be selection for lessening of homologous recombination in the organism. If this does not happen, the element may be overly destructive to the genome.

In humans, a significant portion of genomic deletions and duplications are caused by unequal recombination between *Alu* elements (25). This represents about 0.3% of all human genetic diseases, or nearly three times the level of disease currently caused by *Alu* insertions. There is indirect evidence that recombination greatly increases in tumors (38). Furthermore, a reporter gene system has recently been used to demonstrate that *Alu* elements in an inverted orientation are even more prone than those in the direct repeat orientation to cause deletions in cells carrying a mutant p53 gene (38). However, these deletions are nonhomologous and may be the result of some local structural change in the DNA. This would be similar to the observation that inverted repeats are unstable in the bacterial chromosome. This was further extended with human *Alu* elements in a yeast model system to show that inverted-repeated *Alu* elements stimulated the recombination of other nearby homologous sequences (65).

It is surprising that L1 elements in humans contribute to many fewer recombination events than the *Alu* elements (25). The generally longer length of L1 elements would be likely to aid in their recombination. It has been proposed that *Alu* elements have a Chi-like sequence that may make them specifically prone to recombination (88). Our analysis of large numbers of these recombination events does not strongly support the Chi sequence model (M. El-Sawi and P. Deininger, unpublished data), but no direct test has been performed. The lower level of observed L1 recombination may be caused by their somewhat lower copy number and larger spacing. Alternatively, L1 elements may be causing high levels of recombination, but ascertainment may be ineffective. One might expect L1 elements to be more destructive when they do recombine because their spacing is larger (particularly between the longer, less-truncated copies). This larger spacing would result in larger, more destructive deletions and therefore become more lethal. *Alu*, on the other hand, may often cause more modest damage, resulting in the types of intermediate levels of damage that can cause disease without immediate lethality.

This type of genomic instability is clearly not just a human or mammalian phenomenon. Other genomes have been reported subject to high levels of rearrangement triggered by mobile elements as well (73). Reciprocal translocations and other rearrangements apparently associated with mobile elements have been documented in yeast and *Drosophila* (113, 118) and almost certainly occur in all genomes carrying mobile elements. It is possible, however, that all rearrangements ascribed to mobile elements are not caused by the elements themselves, but to the possibility that the site of insertion of the mobile element may itself be a hotspot for genomic instability. In this hypothesis, some of the mobile elements may serve only as markers of genomic instability and they inserted in those regions because the same properties that contribute to recombination may also contribute to some of the mobile element insertion preferences.

Besides their potential for recombination, mobile elements also have a documented potential for transduction of other genomic sequences through their mobility. Several recent studies show that L1 elements (non-LTR retrotransposons) not only have the strong potential (76), but also have consistently been able to retrotranspose small segments (41, 83) downstream from the element to new genomic locations. This occurs because the elements have only a weak polyadenylation signal that can allow a stronger signal downstream to serve as termination for the RNA. Thus, this type of nonstringent, non-LTR retrotransposon may move extra sequences with up to 20% of its amplification events. Similarly, the formation of processed pseudogenes leads to duplication of whole-gene coding regions, and occasionally leads to additional functional genes (11) that are clearly of evolutionary benefit to the organism. The Bs1 element (an LTR retrotransposon) has acquired part of a cellular gene and transduced it throughout the genome (13). Similarly, retroviruses have been known to transduce genomic sequences as illustrated by their ability to incorporate oncogenes into their structure (110). These types of transductions have the potential to duplicate exons, control regions, and other key functional elements that may play a role in evolution of gene structure and regulation.

Benefits of Mobile Element-Mediated Genetic Rearrangements

Although the vast majority of mobile element-mediated recombination events will be deleterious, a few will be beneficial to an organism. These few events may play an important role in the evolution of an organism. Such events could lead to duplications of exons or entire genes. This could lead to proteins with duplicated domains that allow them to evolve to function better, or allow for a duplicated copy to evolve to an entirely different function. Mobile element-mediated chromosomal rearrangements could also lead to chromosomal changes that ultimately play a role in speciation by creating incompatabilities between chromosomes.

Recombination may also play an important role in controlling mobile elements. For instance, recombination between LTRs of the LTR-retrotransposons often results in the deletion of the bulk of the element. This is a rapid and effective manner for a genome to help silence LTR-retrotransposons.

Last, normal portions of mobile elements have also contributed positively to genome function. There is extensive evidence that mobile element insertions have contributed genetic elements that regulate gene expression. This could include single regulatory elements, such as the hormone receptor elements sometimes found in *Alu* elements (81, 109), or it could include whole promoters as seen in some LTR-driven carcinogenesis by retroviruses (93). There are also numerous cases in which it has been proposed that organisms have evolved genes from the original mobile element genes to serve a new function for the organism (98). Examples of these types of events include telomerase (79), the immunoglobin recombination genes RAG 1 and 2 (45), and the centromere-binding protein CENP-B (106). Retrotransposition has also given rise to gene duplications or other useful gene modifications that allowed evolution to new function (10, 11). Thus, although the vast majority of examples of element mobility must contribute overall to genome damage, there have been a limited number of element insertions that have served ultimately as positive changes and been incorporated through selection into the genome of the species.

REFERENCES

1. Adams, M. D., S. E. Celniker, R. A. Holt, C. A. Evans, J. D. Gocayne, P. G. Amanatides, S. E. Scherer, P. W. Li, R. A. Hoskins, R. F. Galle, R. A. George, S. E. Lewis, S. Richards, M. Ashburner, S. N. Henderson, G. G. Sutton, J. R. Wortman, M. D. Yandell, Q. Zhang, L. X. Chen, R. C. Brandon, Y. H. Rogers, R. G. Blazej, M. Champe, B. D. Pfeiffer, K. H. Wan, C. Doyle, E. G. Baxter, G. Helt, C. R. Nelson, G. L. Gabor Miklos, J. F. Abril, A. Agbayani, H. J. An, C. Andrews-Pfannkoch, D. Baldwin, R. M. Ballew, A. Basu, J. Baxendale, L. Bayraktaroglu, E. M. Beasley, K. Y. Beeson, P. V. Benos, B. P. Berman, D. Bhandari, S. Bolshakov, D. Borkova, M. R. Botchan, J. Bouck, P. Brokstein, P. Brottier, K. C. Burtis, D. A. Busam, H. Butler, E. Cadieu, A. Center, I. Chandra, J. M. Cherry, S. Cawley, C. Dahlke, L. B. Davenport, P. Davies, B. de Pablos, A. Delcher, Z. Deng, A. D. Mays, I. Dew, S. M. Dietz, K. Dodson, L. E. Doup, M. Downes, S. Dugan-Rocha, B. C. Dunkov, P. Dunn, K. J. Durbin, C. C. Evangelista, C. Ferraz, S. Ferriera, W. Fleischmann, C. Fosler, A. E. Gabrielian, N. S. Garg, W. M. Gelbart, K. Glasser, A. Glodek, F. Gong, J. H. Gorrell, Z. Gu, P. Guan, M. Harris, N. L.

Harris, D. Harvey, T. J. Heiman, J. R. Hernandez, J. Houck, D. Hostin, K. A. Houston, T. J. Howland, M. H. Wei, C. Ibegwam, M. Jalali, F. Kalush, G. H. Karpen, Z. Ke, J. A. Kennison, K. A. Ketchum, B. E. Kimmel, C. D. Kodira, C. Kraft, S. Kravitz, D. Kulp, Z. Lai, P. Lasko, Y. Lei, A. A. Levitsky, J. Li, Z. Li, Y. Liang, X. Lin, X. Liu, B. Mattei, T. C. McIntosh, M. P. McLeod, D. McPherson, G. Merkulov, N. V. Milshina, C. Mobarry, J. Morris, A. Moshrefi, S. M. Mount, M. Moy, B. Murphy, L. Murphy, D. M. Muzny, D. L. Nelson, D. R. Nelson, K. A. Nelson, K. Nixon, D. R. Nusskern, J. M. Pacleb, M. Palazzolo, G. S. Pittman, S. Pan, J. Pollard, V. Puri, M. G. Reese, K. Reinert, K. Remington, R. D. Saunders, F. Scheeler, H. Shen, B. C. Shue, I. Siden-Kiamos, M. Simpson, M. P. Skupski, T. Smith, E. Spier, A. C. Spradling, M. Stapleton, R. Strong, E. Sun, R. Svirskas, C. Tector, R. Turner, E. Venter, A. H. Wang, X. Wang, Z. Y. Wang, D. A. Wassarman, G. M. Weinstock, J. Weissenbach, S. M. Williams, T. Woodage, K. C. Worley, D. Wu, S. Yang, Q. A. Yao, J. Ye, R. F. Yeh, J. S. Zaveri, M. Zhan, G. Zhang, Q. Zhao, L. Zheng, X. H. Zheng, F. N. Zhong, W. Zhong, X. Zhou, S. Zhu, X. Zhu, H. O. Smith, R. A. Gibbs, E. W. Myers, G. M. Rubin, and J. C. Venter.** 2000. The genome sequence of *Drosophila melanogaster. Science* **287:** 2185–2195.

2. **Bailey, J. A., L. Carrel, A. Chakravarti, and E. E. Eichler.** 2000. From the cover: molecular evidence for a relationship between LINE-1 elements and X chromosome inactivation: the Lyon repeat hypothesis. *Proc. Natl. Acad. Sci. USA* **97:** 6634–6639.
3. **Bennetzen, J. L.** 2000. Transposable element contributions to plant gene and genome evolution. *Plant Mol. Biol.* **42:** 251–269.
4. **Biemont, C., and C. Terzian.** 1988. Mdg-1 mobile element polymorphism in selected *Drosophila melanogaster* populations. *Genetica* **76:**7–14.
5. **Biemont, C., C. Vieira, C. Hoogland, G. Cizeron, C. Loevenbruck, C. Arnault, and J. P. Carante.** 1997. Maintenance of transposable element copy number in natural populations of *Drosophila melanogaster* and *D. simulans. Genetica* **100:** 161–166.
6. **Biessmann, H., and J. M. Mason.** 1997. Telomere maintenance without telomerase. *Chromosoma* **106:**63–69.
7. **Birchler, J. A., M. P. Bhadra, and U. Bhadra.** 2000. Making noise about silence: repression of repeated genes in animals. *Curr. Opin. Genet. Dev.* **10:**211–216.
8. **Boeke, J. D.** 1997. LINEs and Alus—the polyA connection. *Nat. Genet.* **16:**6–7.
9. **Britten, R. J.** 1995. Active gypsy/Ty3 retrotransposons or retroviruses in *Caenorhabditis elegans. Proc. Natl. Acad. Sci. USA* **92:**599–601.
10. **Britten, R. J.** 1997. Mobile elements inserted in the distant past have taken on important functions. *Gene* **205:**177–182.
11. **Brosius, J., and H. Tiedge.** 1995. Reverse transcriptase: mediator of genomic plasticity. *Virus Genes* **11:**163–179.
12. **Bucci, S., M. Ragghianti, G. Mancino, G. Petroni, F. Guerrini, and S. Giampaoli.** 1999. Rana/Pol III: a family of SINE-like sequences in the genomes of western Palearctic water frogs. *Genome* **42:**504–511.
13. **Bureau, T. E., S. E. White, and S. R. Wessler.** 1994. Transduction of a cellular gene by a plant retroelement. *Cell* **77:** 479–480.
14. **Burke, W. D., H. S. Malik, J. P. Jones, and T. H. Eickbush.** 1999. The domain structure and retrotransposition mechanism of R2 elements are conserved throughout arthropods. *Mol. Biol. Evol.* **16:**502–511.
15. **Capy, P., T. Langin, Y. Bigot, F. Brunet, M. J. Daboussi, G. Periquet, J. R. David, and D. L. Hartl.** 1994. Horizontal transmission versus ancient origin: mariner in the witness box. *Genetica* **93:**161–170.
16. **Capy, P., R. Vitalis, T. Langin, D. Higuet, and C. Bazin.** 1996. Relationships between transposable elements based upon the integrase-transposase domains: is there a common ancestor? *J. Mol. Evol.* **42:**359–368.
17. **Casavant, N. C., L. Scott, M. A. Cantrell, L. E. Wiggins, R. J. Baker, and H. A. Wichman.** 2000. The end of the LINE? Lack of recent 11 activity in a group of South American rodents. *Genetics* **154:**1809–1817.
18. **The *C. elegans* Sequencing Consortium.** 1998. Genome sequence of the nematode C. elegans: a platform for investigating biology. *Science* **282:**2012–2018. (Errata, **283:**35, 1999; **283:**2103, 1999; **285:**1493, 1999.)
19. **Chang, D. Y., N. Sasaki-Tozawa, L. K. Green, and R. J. Maraia.** 1995. A trinucleotide repeat-associated increase in the level of Alu RNA-binding protein occurred during the same period as the major Alu amplification that accompanied anthropoid evolution. *Mol. Cell. Biol.* **15:**2109–2116.
20. **Cost, G. J., and J. D. Boeke.** 1998. Targeting of human retrotransposon integration is directed by the specificity of the L1 endonuclease for regions of unusual DNA structure. *Biochemistry* **37:**18081–18093.
21. **Csink, A. K., and J. F. McDonald.** 1990. copia expression is variable among natural populations of *Drosophila. Genetics* **126:**375–385.
22. **D'Andrea, A. D., U. Tantravahi, M. Lalande, M. A. Perle, and S. A. Latt.** 1983. High resolution analysis of the timing of replication of specific DNA sequences during S phase of mammalian cells. *Nucleic Acids Res.* **11:**4753–4774.
23. **Daniels, G., and P. Deininger.** 1985. Repeat sequence families derived from mammalian tRNA genes. *Nature* **317:**819–822.
24. **Danilevskaya, O. N., K. Lowenhaupt, and M. L. Pardue.** 1998. Conserved subfamilies of the *Drosophila* HeT-A telomere-specific retrotransposon. *Genetics* **148:**233–242.
25. **Deininger, P. L., and M. A. Batzer.** 1999. Alu repeats and human disease. *Mol. Genet. Metab.* **67:**183–193.
26. **Deininger, P., and M. A. Batzer.** 1993. Evolution of retroposons. *Evol. Biol.* **27:**157–196.
27. **Deininger, P., and M. Batzer.** 1995. SINE master genes and population biology, p. 43–60. *In* R. Maraia (ed.), *The Impact of Short, Interspersed Elements (SINEs) on the Host Genome.* R. G. Landes, Georgetown, Tex.
28. **Deininger, P., M. Batzer, I. C. Hutchison, and M. Edgell.** 1992. Master genes in mammalian repetitive DNA amplification. *Trends Genet.* **8:**307–312.
29. **Deininger, P., and G. Daniels.** 1986. The recent evolution of mammalian repetitive DNA elements. *Trends Genet.* **2:** 76–80.
30. **Dunham, I., N. Shimizu, B. A. Roe, S. Chissoe, A. R. Hunt, J. E. Collins, R. Bruskiewich, D. M. Beare, M. Clamp, L. J. Smink, R. Ainscough, J. P. Almeida, A. Babbage, C. Bagguley, J. Bailey, K. Barlow, K. N. Bates, O. Beasley, C. P. Bird, S. Blakey, A. M. Bridgeman, D. Buck, J. Burgess, W. D. Burrill, and K. P. O'Brien.** 1999. The DNA sequence of human chromosome 22. *Nature* **402:**489–495. (Erratum, **404:**904, 2000.)
31. **Elgar, G., M. S. Clark, S. Meek, S. Smith, S. Warner, Y. J. Edwards, N. Bouchireb, A. Cottage, G. S. Yeo, Y. Umrania, G. Williams, and S. Brenner.** 1999. Generation and analysis of 25 Mb of genomic DNA from the pufferfish *Fugu rubripes* by sequence scanning. *Genome Res.* **9:**960–971.
32. **Emmons, S. W., S. Roberts, and K. S. Ruan.** 1986. Evidence in a nematode for regulation of transposon excision by tissuespecific factors. *Mol. Gen. Genet.* **202:**410–415.

33. **Engels, W. R.** 1992. The origin of P elements in *Drosophila melanogaster*. *Bioessays* **14:**681–686.
34. **Feng, Q., J. V. Moran, H. H. Kazazian, Jr., and J. D. Boeke.** 1996. Human L1 retrotransposon encodes a conserved endonuclease required for retrotransposition. *Cell* **87:**905–916.
35. **Flavell, A. J., V. Jackson, M. P. Iqbal, I. Riach, and S. Waddell.** 1995. Ty1-copia group retrotransposon sequences in amphibia and reptilia. *Mol. Gen. Genet.* **246:**65–71.
36. **Garrett, J. E., D. S. Knutzon, and D. Carroll.** 1989. Composite transposable elements in the *Xenopus laevis* genome. *Mol. Cell. Biol.* **9:**3018–3027.
37. **Gaut, B. S., D. E. Le Thierry, A. S. Peek, and M. C. Sawkins.** 2000. Maize as a model for the evolution of plant nuclear genomes. *Proc. Natl. Acad. Sci. USA* **97:**7008–7015.
38. **Gebow, D., N. Miselis, and H. L. Liber.** 2000. Homologous and nonhomologous recombination resulting in deletion: effects of p53 status, microhomology, and repetitive DNA length and orientation. *Mol. Cell. Biol.* **20:**4028–4035.
39. **Goldman, M. A., G. P. Holmquist, M. C. Gray, L. A. Caston, and A. Nag.** 1984. Replication timing of genes and middle repetitive sequences. *Science* **224:**686–692.
40. **Goodier, J. L., and W. S. Davidson.** 1994. Tc1 transposon-like sequences are widely distributed in salmonids. *J. Mol. Biol.* **241:**26–34.
41. **Goodier, J. L., E. M. Ostertag, and H. H. Kazazian, Jr.** 2000. Transduction of 3′-flanking sequences is common in L1 retrotransposition. *Hum. Mol. Genet.* **9:**653–657.
42. **Hagemann, S., W. J. Miller, E. Haring, and W. Pinsker.** 1998. Nested insertions of short mobile sequences in Drosophila P elements. *Chromosoma* **107:**6–16.
43. **Hattori, M., A. Fujiyama, T. D. Taylor, H. Watanabe, T. Yada, H. S. Park, A. Toyoda, K. Ishii, Y. Totoki, D. K. Choi, E. Soeda, M. Ohki, T. Takagi, Y. Sakaki, S. Taudien, K. Blechschmidt, A. Polley, U. Menzel, J. Delabar, K. Kumpf, R. Lehmann, D. Patterson, K. Reichwald, A. Rump, M. Schillhabel, and A. Schudy.** 2000. The DNA sequence of human chromosome 21. The chromosome 21 mapping and sequencing consortium. *Nature* **405:**311–319.
44. **Heslop-Harrison, J. S., M. Murata, Y. Ogura, T. Schwarzacher, and F. Motoyoshi.** 1999. Polymorphisms and genomic organization of repetitive DNA from centromeric regions of Arabidopsis chromosomes. *Plant Cell* **11:**31–42.
45. **Hiom, K., M. Melek, and M. Gellert.** 1998. DNA transposition by the RAG1 and RAG2 proteins: a possible source of oncogenic translocations. *Cell* **94:**463–470.
46. **Izsvak, Z., Z. Ivics, D. Garcia-Estefania, S. C. Fahrenkrug, and P. B. Hackett.** 1996. DANA elements: a family of composite, tRNA-derived short interspersed DNA elements associated with mutational activities in zebrafish (Danio rerio). *Proc. Natl. Acad. Sci. USA* **93:**1077–1081.
47. **Jensen, S., M. P. Gassama, and T. Heidmann.** 1999. Taming of transposable elements by homology-dependent gene silencing. *Nat. Genet.* **21:**209–212.
48. **Johnson, W. E., and J. M. Coffin.** 1999. Constructing primate phylogenies from ancient retrovirus sequences. *Proc. Natl. Acad. Sci. USA* **96:**10254–10260.
49. **Junakovic, N., A. Terrinoni, C. Di Franco, C. Vieira, and C. Loevenbruck.** 1998. Accumulation of transposable elements in the heterochromatin and on the Y chromosome of *Drosophila simulans* and *Drosophila melanogaster*. *J. Mol. Evol.* **46:**661–668.
50. **Jurka, J.** 1997. Sequence patterns indicate an enzymatic involvement in integration of mammalian retroposons. *Proc. Natl. Acad. Sci. USA* **94:**1872–1877.
51. **Jurka, J.** 1998. Repeats in genomic DNA: mining and meaning. *Curr. Opin. Struct. Biol.* **8:**333–337.
52. **Kass, D. H., J. Kim, and P. L. Deininger.** 1996. Sporadic amplification of ID elements in rodents. *J. Mol. Evol.* **42:** 7–14.
53. **Kawakami, K., and A. Shima.** 1999. Identification of the Tol2 transposase of the medaka fish Oryzias latipes that catalyzes excision of a nonautonomous Tol2 element in zebrafish *Danio rerio*. *Gene* **240:**239–244.
54. **Kazazian, H. H. J.** 1998. Mobile elements and disease. *Curr. Opin. Genet. Dev.* **8:**343–350.
55. **Kazazian, H. H. J., and J. V. Moran.** 1998. The impact of L1 retrotransposons on the human genome. *Nat. Genet.* **19:** 19–24.
56. **Koga, A., M. Suzuki, H. Inagaki, Y. Bessho, and H. Hori.** 1996. Transposable element in fish. *Nature* **383:**30.
57. **Korenberg, J. R., and M. C. Rykowski.** 1988. Human genome organization: Alu, lines, and the molecular structure of metaphase chromosome bands. *Cell* **53:**391–400.
58. **Kotani, H., T. Hosouchi, and H. Tsuruoka.** 1999. Structural analysis and complete physical map of Arabidopsis thaliana chromosome 5 including centromeric and telomeric regions. *DNA Res.* **6:**381–386.
59. **Kumar, A., and J. L. Bennetzen.** 1999. Plant retrotransposons. *Annu. Rev. Genet.* **33:**479–532.
60. **Labrador, M., M. Farre, F. Utzet, and A. Fontdevila.** 1999. Interspecific hybridization increases transposition rates of Osvaldo. *Mol. Biol. Evol.* **16:**931–937.
61. **Lam, W. L., T. S. Lee, and W. Gilbert.** 1996. Active transposition in zebrafish. *Proc. Natl. Acad. Sci. USA* **93:** 10870–10875.
62. **Leeton, P. R., and D. R. Smyth.** 1993. An abundant LINE-like element amplified in the genome of Lilium speciosum. *Mol. Gen. Genet.* **237:**97–104.
63. **Levis, R. W., R. Ganesan, K. Houtchens, L. A. Tolar, and F. M. Sheen.** 1993. Transposons in place of telomeric repeats at a Drosophila telomere. *Cell* **75:**1083–1093.
64. **Lin, X., S. Kaul, S. Rounsley, T. P. Shea, M. I. Benito, C. D. Town, C. Y. Fujii, T. Mason, C. L. Bowman, M. Barnstead, T. V. Feldblyum, C. R. Buell, K. A. Ketchum, J. Lee, C. M. Ronning, H. L. Koo, K. S. Moffat, L. A. Cronin, M. Shen, G. Pai, S. Van Aken, L. Umayam, L. J. Tallon, J. E. Gill, and J. C. Venter.** 1999. Sequence and analysis of chromosome 2 of the plant *Arabidopsis thaliana*. *Nature* **402:**761–768.
65. **Lobachev, K. S., J. E. Stenger, O. G. Kozyreva, J. Jurka, D. A. Gordenin, and M. A. Resnick.** 2000. Inverted Alu repeats unstable in yeast are excluded from the human genome. *EMBO J.* **19:**3822–3830.
66. **Lozovskaya, E. R., D. L. Hartl, and D. A. Petrov.** 1995. Genomic regulation of transposable elements in *Drosophila*. *Curr. Opin. Genet. Dev.* **5:**768–773.
67. **Luan, D. D., M. H. Korman, J. L. Jakubczak, and T. H. Eickbush.** 1993. Reverse transcription of R2Bm RNA is primed by a nick at the chromosomal target site: a mechanism for non-LTR retrotransposition. *Cell* **72:**595–605.
68. **Lueders, K. K., and W. N. Frankel.** 1994. Mapping of mouse intracisternal A-particle proviral markers in an interspecific backcross. *Mamm. Genome* **5:**473–478.
69. **Malik, H. S., and T. H. Eickbush.** 1998. The RTE class of non-LTR retrotransposons is widely distributed in animals and is the origin of many SINEs. *Mol. Biol. Evol.* **15:** 1123–1134.
70. **Marracci, S., R. Batistoni, G. Pesole, L. Citti, and I. Nardi.** 1996. Gypsy/Ty3-like elements in the genome of the terrestrial Salamander hydromantes (Amphibia, Urodela). *J. Mol. Evol.* **43:**584–593.
71. **Maside, X., S. Assimacopoulos, and B. Charlesworth.** 2000. Rates of movement of transposable elements on the second

chromosome of *Drosophila melanogaster*. *Genet. Res.* **75**: 275–284.

72. **Mayer, K., C. Schuller, R. Wambutt, G. Murphy, G. Volckaert, T. Pohl, A. Dusterhoft, W. Stiekema, K. D. Entian, N. Terryn, B. Harris, W. Ansorge, P. Brandt, L. Grivell, M. Rieger, M. Weichselgartner, V. de Simone, B. Obermaier, R. Mache, M. Muller, M. Kreis, M. Delseny, P. Puigdomenech, M. Watson, W. R. McCombie, et al.** 1999. Sequence and analysis of chromosome 4 of the plant *Arabidopsis thaliana*. *Nature* **402**:769–777.
73. **McCarron, M., A. Duttaroy, G. Doughty, and A. Chovnick.** 1994. *Drosophila* P element transposase induces male recombination additively and without a requirement for P element excision or insertion. *Genetics* **136**:1013–1023.
74. **Medstrand, P., and D. L. Mager.** 1998. Human-specific integrations of the HERV-K endogenous retrovirus family. *J. Virol.* **72**:9782–9787.
75. **Mochizuki, K., M. Umeda, H. Ohtsubo, and E. Ohtsubo.** 1992. Characterization of a plant SINE, p-SINE1, in rice genomes. *Jpn. J. Genet.* **67**:155–166.
76. **Moran, J. V., R. J. DeBerardinis, and H. H. Kazazian, Jr.** 1999. Exon shuffling by L1 retrotransposition. *Science* **283**: 1530–1534.
77. **Morgan, G. T.** 1995. Identification in the human genome of mobile elements spread by DNA-mediated transposition. *J. Mol. Biol.* **254**:1–5.
78. **Murata, S., N. Takasaki, M. Saitoh, and N. Okada.** 1993. Determination of the phylogenetic relationships among Pacific salmonids by using short interspersed elements (SINEs) as temporal landmarks of evolution. *Proc. Natl. Acad. Sci. USA* **90**:6995–6999.
79. **Nakamura, T. M., and T. R. Cech.** 1998. Reversing time: origin of telomerase. *Cell* **92**:587–590.
80. **Noma, K., E. Ohtsubo, and H. Ohtsubo.** 1999. Non-LTR retrotransposons (LINEs) as ubiquitous components of plant genomes. *Mol. Gen. Genet.* **261**:71–79.
81. **Norris, J., D. Fan, C. Aleman, J. R. Marks, P. A. Futreal, R. W. Wiseman, J. D. Iglehart, P. L. Deininger, and D. P. McDonnell.** 1995. Identification of a new subclass of Alu DNA repeats which can function as estrogen receptor-dependent transcriptional enhancers. *J. Biol. Chem.* **270**: 22777–22782.
82. **Ohshima, K., M. Hamada, Y. Terai, and N. Okada.** 1996. The 3′ ends of tRNA-derived short interspersed repetitive elements are derived from the 3′ ends of long interpersed repetitive elements. *Mol. Cell. Biol.* **16**:3756–3764.
83. **Pickeral, O. K., W. Makaowski, M. S. Boguski, and J. D. Boeke.** 2000. Frequent human genomic DNA transduction driven by LINE-1 retrotransposition. *Genome Res.* **10**: 411–415.
84. **Plasterk, R. H.** 1991. The origin of footprints of the Tc1 transposon of *Caenorhabditis elegans*. *EMBO J.* **10**: 1919–1925.
85. **Poulter, R., and M. Butler.** 1998. A retrotransposon family from the pufferfish (fugu) *Fugu rubripes*. *Gene* **215**:241–249.
86. **Poulter, R., M. Butler, and J. Ormandy.** 1999. A LINE element from the pufferfish (fugu) *Fugu rubripes* which shows similarity to the CR1 family of non-LTR retrotransposons. *Gene* **227**:169–179.
87. **Rezsohazy, R., H. G. van Luenen, R. M. Durbin, and R. H. Plasterk.** 1997. Tc7, a Tc1-hitch hiking transposon in *Caenorhabditis elegans*. *Nucleic Acids Res.* **25**:4048–4054.
88. **Rudiger, N. S., N. Gregersen, and M. C. Kielland-Brandt.** 1995. One short well conserved region of Alu-sequences is involved in human gene rearrangements and has homology with prokaryotic chi. *Nucleic Acids Res.* **23**:256–260.
89. **Ruiz, F., L. Vayssie, C. Klotz, L. Sperling, and L. Madeddu.** 1998. Homology-dependent gene silencing in Paramecium. *Mol. Biol. Cell* **9**:931–943.
90. **SanMiguel, P., B. S. Gaut, A. Tikhonov, Y. Nakajima, and J. L. Bennetzen.** 1998. The paleontology of intergene retrotransposons of maize. *Nat. Genet.* **20**:43–45.
91. **SanMiguel, P., A. Tikhonov, Y. K. Jin, N. Motchoulskaia, D. Zakharov, A. Melake-Berhan, P. S. Springer, K. J. Edwards, M. Lee, Z. Avramova, and J. L. Bennetzen.** 1996. Nested retrotransposons in the intergenic regions of the maize genome. *Science* **274**:765–768.
92. **Sassaman, D. M., B. A. Dombroski, J. V. Moran, M. L. Kimberland, T. P. Naas, R. J. DeBerardinis, A. Gabriel, G. D. Swergold, and H. H. Kazazian, Jr.** 1997. Many human L1 elements are capable of retrotransposition. *Nat. Genet.* **16**: 37–43.
93. **Shackleford, G. M., C. A. MacArthur, H. C. Kwan, and H. E. Varmus.** 1993. Mouse mammary tumor virus infection accelerates mammary carcinogenesis in Wnt-1 transgenic mice by insertional activation of int-2/Fgf-3 and hst/Fgf-4. *Proc. Natl. Acad. Sci. USA* **90**:740–744.
94. **Shen, M., M. Batzer, and P. Deininger.** 1991. Evolution of the Master Alu Gene(s). *J. Mol. Evol.* **33**:311–320.
95. **Slagel, V., E. Flemington, V. Traina-Dorge, H. Bradshaw, and P. Deininger.** 1987. Clustering and sub-family relationships of the Alu family in the human genome. *Mol. Biol. Evol.* **4**:19–29.
96. **Smit, A. F.** 1993. Identification of a new, abundant superfamily of mammalian LTR-transposons. *Nucleic Acids Res.* **21**: 1863–1872.
97. **Smit, A. F.** 1996. The origin of interspersed repeats in the human genome. *Curr. Opin. Genet. Dev.* **6**:743–748.
98. **Smit, A. F.** 1999. Interspersed repeats and other mementos of transposable elements in mammalian genomes. *Curr. Opin. Genet. Dev.* **9**:657–663.
99. **Smit, A. F., G. Toth, A. D. Riggs, and J. Jurka.** 1995. Ancestral, mammalian-wide subfamilies of LINE-1 repetitive sequences. *J. Mol. Biol.* **246**:401–417.
100. **Surzycki, S. A., and W. R. Belknap.** 2000. Repetitive-DNA elements are similarly distributed on *Caenorhabditis elegans* autosomes. *Proc. Natl. Acad. Sci. USA* **97**:245–249.
101. **Takasaki, N., L. Park, M. Kaeriyama, A. J. Gharrett, and N. Okada.** 1996. Characterization of species-specifically amplified SINEs in three salmonid species—chum salmon, pink salmon, and kokanee: the local environment of the genome may be important for the generation of a dominant source gene at a newly retroposed locus. *J. Mol. Evol.* **42**:103–116.
102. **Tchenio, T., J. F. Casella, and T. Heidmann.** 2000. Members of the SRY family regulate the human LINE retrotransposons. *Nucleic Acids Res.* **28**:411–415.
103. **Terrinoni, A., C. D. Franco, P. Dimitri, and N. Junakovic.** 1997. Intragenomic distribution and stability of transposable elements in euchromatin and heterochromatin of Drosophila melanogaster: non-LTR retrotransposon. *J. Mol. Evol.* **45**: 145–153.
104. **Terzian, C., and C. Biemont.** 1988. The founder effect theory: quantitative variation and mdg-1 mobile element polymorphism in experimental populations of *Drosophila melanogaster*. *Genetica* **76**:53–63.
105. **Tikhonov, A. P., P. J. SanMiguel, Y. Nakajima, N. M. Gorenstein, J. L. Bennetzen, and Z. Avramova.** 1999. Colinearity and its exceptions in orthologous adh regions of maize and sorghum. *Proc. Natl. Acad. Sci. USA* **96**:7409–7414.
106. **Tudor, M., M. Lobocka, M. Goodell, J. Pettitt, and K. O'Hare.** 1992. The pogo transposable element family of *Drosophila melanogaster*. *Mol. Gen. Genet.* **232**:126–134.

107. **Turcich, M. P., A. Bokharririza, D. A. Hamilton, C. P. He, W. Messier, et al.** 1996. PREM-2, a copia-type retroelement in maize is expressed preferentially in early microspores. *Sex. Plant Reprod.* **9:**65–74.
108. **Unsal, K., and G. T. Morgan.** 1995. A novel group of families of short interspersed repetitive elements (SINEs) in Xenopus: evidence of a specific target site for DNA-mediated transposition of inverted-repeat SINEs. *J. Mol. Biol.* **248:**812–823.
109. **Vansant, G., and W. F. Reynolds.** 1995. The consensus sequence of a major Alu subfamily contains a functional retinoic acid response element. *Proc. Natl. Acad. Sci. USA* **92:**8229–8233.
110. **Varmus, H., and J. M. Bishop.** 1986. Biochemical mechanisms of oncogene activity: proteins encoded by oncogenes. Introduction. *Cancer Surv.* **5:**153–158.
111. **Vieira, C., D. Lepetit, S. Dumont, and C. Biemont.** 1999. Wake up of transposable elements following *Drosophila simulans* worldwide colonization. *Mol. Biol. Evol.* **16:**1251–1255.
112. **Weiner, A., P. Deininger, and A. Efstratiadis.** 1986. The reverse flow of genetic information: pseudogenes and transposable elements derived from nonviral cellular RNA. *Annu. Rev. Biochem.* **55:**631–661.
113. **Williamson, V. M., D. Cox, E. T. Young, D. W. Russell, and M. Smith.** 1983. Characterization of transposable element-associated mutations that alter yeast alcohol dehydrogenase II expression. *Mol. Cell. Biol.* **3:**20–31.
114. **Wright, D. A., N. Ke, J. Smalle, B. M. Hauge, H. M. Goodman, and D. F. Voytas.** 1996. Multiple non-LTR retrotransposons in the genome of *Arabidopsis thaliana. Genetics* **142:**569–578.
115. **Xiong, Y., and T. H. Eickbush.** 1990. Origin and evolution of retroelements based upon their reverse transcriptase sequences. *EMBO J.* **9:**3353–3362.
116. **Yoder, J. A., C. P. Walsh, and T. H. Bestor.** 1997. Cytosine methylation and the ecology of intragenomic parasites. *Trends Genet.* **13:**335–340.
117. **Yoshioka, Y., S. Matsumoto, S. Kojima, K. Ohshima, N. Okada, and Y. Machida.** 1993. Molecular characterization of a short interspersed repetitive element from tobacco that exhibits sequence homology to specific tRNAs. *Proc. Natl. Acad. Sci. USA* **90:**6562–6566.
118. **Zelentsova, H., H. Poluectova, L. Mnjoian, G. Lyozin, V. Veleikodvorskaja, L. Zhivotovsky, M. G. Kidwell, and M. B. Evgen'ev.** 1999. Distribution and evolution of mobile elements in the virilis species group of Drosophila. *Chromosoma* **108:**443–456.

Mobile DNA II
Edited by N. L. Craig et al.

Chapter 48

Evolution of DNA Transposons in Eukaryotes

HUGH M. ROBERTSON

INTRODUCTION

The evolution of DNA transposons can be divided into two major topics by the time frame being considered. First, we can consider how various DNA transposons are related to each other and the consequent inferences about their evolution over very long periods. Second, we can ask about the short-term evolutionary dynamics that reflect the ability of these transposons to persist and spread in host genomes. An enormous increase in our understanding of long- and short-term DNA transposon evolution has occurred during the past decade. I will overview both, which I will term the macroevolutionary and microevolutionary levels. The availability of genomic sequences from all kingdoms of life and sophisticated searching algorithms has revolutionized our ability to detect distantly related transposons purely on the basis of sequence similarity, thus improving our understanding of their macroevolutionary trends. These genomic data also for the first time provide complete data sets of the complements of transposons in particular hosts and, hence, improved understanding of their microevolutionary dynamics. I include analyses at both levels to illustrate the utility of genomic sequences in understanding DNA transposons. This review focuses on these transposons in eukaryotic genomes and particularly the four available genomes from plants and animals.

DNA TRANSPOSONS AT THE MACROEVOLUTIONARY LEVEL

The D,D35E Transposase/Integrase Megafamily

In the 1980s, DNA transposons from many diverse organisms and with an enormous variety of DNA structures and encoded transposases and other proteins were described. A number were reported in the first edition of *Mobile DNA* with little indication that they might be evolutionarily and functionally related to each other. Some researchers realized that there were certain similarities among various groups of these transposons, but Fayet et al. (31) provided one of the first major leaps by proposing that the transposases of the IS*3* family of bacterial transposons

Hugh M. Robertson • Department of Entomology, University of Illinois, 320 Morrill Hall, MC118, 505 S. Goodwin, Urbana, IL 61801.

were functionally and evolutionarily related to the integrase proteins of retroviruses and long-terminal-repeat retrotransposons. Specifically, they identified the D,D35E motif of two aspartates and a glutamate spaced in a characteristic fashion as being the core catalytic domain of these diverse and highly divergent proteins. As described in detail in several other chapters in this book, this insight has been substantiated by a large amount of functional and structural work.

Identification of this core catalytic motif led to an improved understanding of a large fraction of all DNA transposons because their transposases share this motif. Using improved algorithms for detecting amino acid sequence similarity, Henikoff and Henikoff (49) were able to find similarities between a number of animal transposons and the *Tc1* transposon of the nematode *Caenorhabditis elegans*. Two years later, Doak et al. (26) showed that this grouping, together with other diverse transposons from animals and ciliates, shared the D,D35E domain. This observation was made by sequence gazing but confirmed, at least for the D35E part of the domain, by rigorous statistical methods. Several additional transposon families are now members of the megafamily and include the *pogo* superfamily in organisms ranging from fungi to humans (92, 104).

I am calling this a megafamily of proteins because the level of sequence similarity is so low (commonly reducing to just the three conserved acidic amino acids and a few others within the D,D35E domain) that their evolutionary relationships are only fully confirmed by their shared tertiary structures. Phylogenetic analysis of such a megafamily is fraught with difficulties because only this shared D,D35E domain can be employed, and even then the alignment of many of the amino acids in-between is uncertain without structural information, which is only available for several integrases and the Tn*5* transposase (see comparisons in reference 23; also see chapters 2, 14, 18, 19, and 22). Capy et al. (11, 12) attempted such a phylogenetic analysis; however, the resultant trees have poor statistical support, so I will simply discuss the major differentiable groups in a roughly phylogenetic order.

The closest link to the transposases of eukaryotic members of the megafamily is provided by the bacterial IS*630* family (26), whose transposases can be fully aligned with those of the *Ant1*/*Tc1*/*mariner* groups (Fig. 1). This family includes the IS*630* transposase from *Shigella sonnei*, the M5 protein of *Salmonella enterica* serovar Typhimurium, the IS*870* transposase from *Agrobacterium tumefaciens*, TnpA from *Pseudomonas putida*, and two transposases from *Pseudomonas syringae*. These proteins have slightly longer N termini than do those of the *Tc1*/*mariner* group. They share as little as 15% amino acid identity with each other and 10 to 15% amino acid identity with the remainder of the megafamily given below.

Another differentiable group is related to the transposases from IS*Rm10*, IS*Rm10-1*, and IS*Rm2011-2* from *Sinorhizobium meliloti* and transposons from *Anabaena* sp. strain PCC7120 and *Sphingomonas aromaticivorans*, *Ant1* from *Aspergillus niger* (35), and *hupfer* from *Beauveria bassiana* (77). The transposases from the *Ant1* and *hupfer* elements in fungi seem to be the most basal members of the transposase megafamily in eukaryotes. These are some of the shortest transposases in the megafamily, being around 300 amino acids long. Again, the level of amino acid identity within this grouping is as low as 15%, with 10 to 15% identity to the remainder of the eukaryotic transposases. IS*1471* from *Burkholderia cepacia* and transposons from a *Synechocystis* sp. and *Sulfolobus solfataricus* have more divergent transposases that cluster phylogenetically with these two groups, but without bootstrap support, as shown in Fig. 1.

A highly divergent group of eukaryotic elements are the apparent transposases of several ciliate elements; their sequences spurred linking of the megafamily (26). These are the *TBE1*, *tec1*, and *tec2* elements from *Oxytricha fallax* and *Euplotes crassus* (see references 26, 64, and 116 and chapter 30).

Another divergent eukaryotic lineage is the *pogo* superfamily first recognized in animals and fungi (92, 104, 111) but also recently found in plants (32, 59, 69). These are generally much longer transposons than most of the eukaryotic megafamily members, and their transposases have a longer unalignable C-terminal region. Related transposons include the *Fot1* group from various fungi (see reference 61) and the *Tc2*/*4*/*5* transposons from the nematode *C. elegans* (see references 92 and 104). Although clearly members of the D,D35E megafamily, only this domain can be aligned with the other members, but even then it is quite divergent, for example, because the final E residue has changed to a D and the spacing ranges from D,D30D in *pogo* to D,D37D in *Tc4*/*5*. Capy et al. (12) allied the *pogo* superfamily directly with the *mariner* family, and indeed they share substitution of the final E residue for a D; however, the *pogo* superfamily appears to be more divergent based on full-length comparisons, so this is probably an independent substitution.

Finally, the *Tc1*/*mariner* superfamily has been recognized for some time and is the largest and most studied grouping among the eukaryotic members (26, 47, 49, 87, 91; chapter 22). Figure 1 provides a phylogeny for the superfamily, rooted with the IS*630*/*Ant1* groups. The *Tc1* family is fairly well defined, although

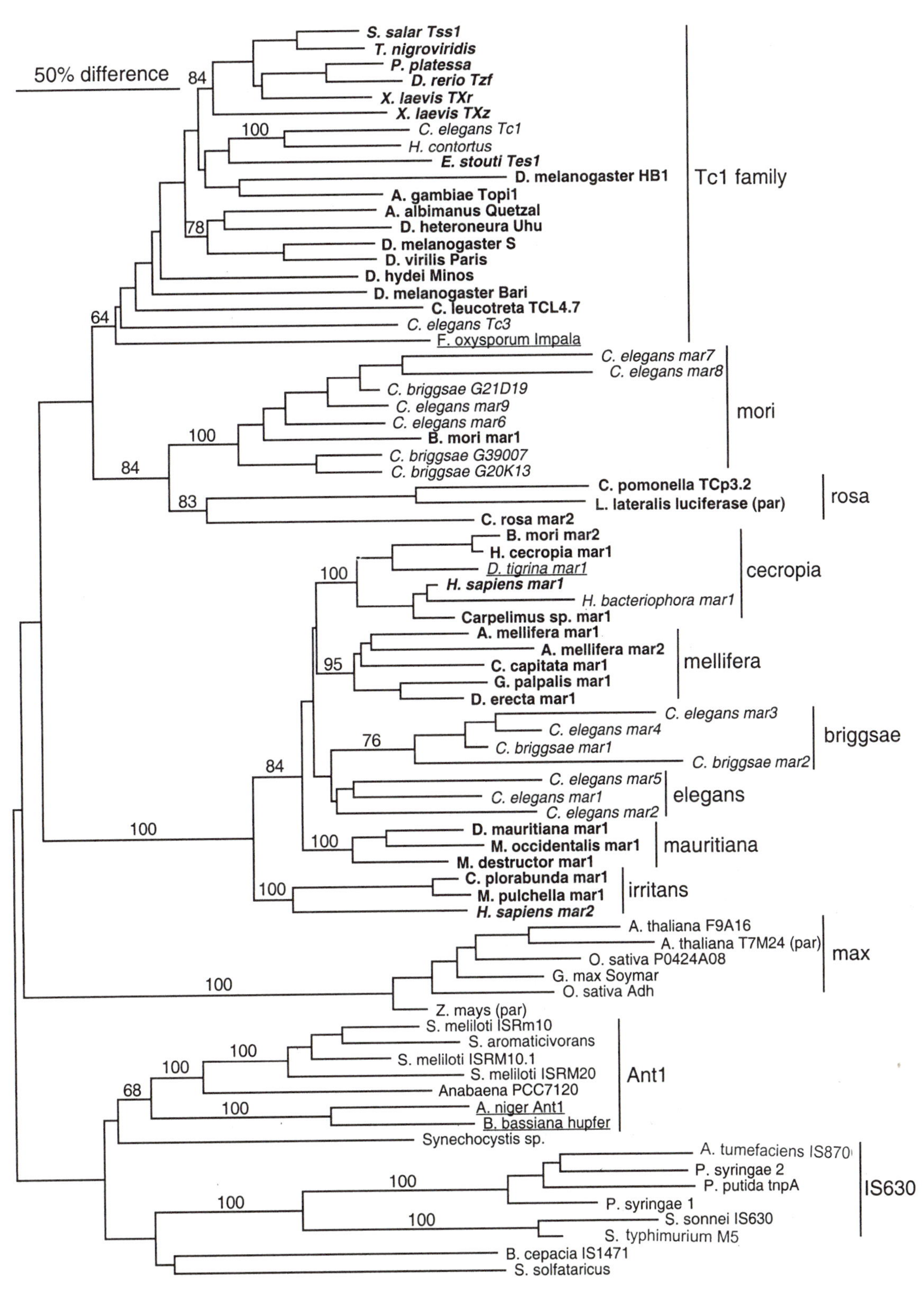

Figure 1. Relationships within the *Tc1/mariner* superfamily rooted with the IS*630-Ant1* grouping of bacterial and fungal transposons. Subfamilies of *mariner* elements are indicated. Bacterial and plant sequence names are in plain text, fungal names are underlined, nematode names are in italic, insect names are in boldface, and vertebrate names are in bold italic. Numbers above the major branches indicate the percentage of 1,000 bootstrap replications in which that branch was present. Transposase sequences were aligned using CLUSTAL X (55), and the tree was constructed using neighbor joining in PAUP* version 4.0b4(PPC) (107) with distances corrected for multiple replacements by TREE-PUZZLE v5.0 (106) using the BLOSUM62 matrix and maximum likelihood with uniform rates.

relationships within it remain uncertain, except for lineages of elements in vertebrates, insects, and nematodes. A major vertebrate lineage in particular has been documented in diverse hosts (e.g., references 53, 66, and 67) and includes representatives from the pufferfish genomes; an example is the consensus sequence of *Tetraodon nigroviridis* in Fig. 1 built from hundreds of partial bacterial artificial chromosome clone end sequences (21). Several additional members from mosquitoes have also been described (38).

Relationships within the *mariner* family, as originally defined (91), are fairly well resolved and mostly supported by bootstrapping; however, the discovery of several highly divergent lineages described below has brought the definition of this family into question. When we described the basal *Bmmar1* element from the silkmoth *Bombyx mori,* we assigned it to the family based on careful comparisons (95); however, there are now three such basal divergent lineages known, and the phylogenetic analysis in Fig. 1 suggests that the family will eventually require redefinition. We also proposed a naming convention for *mariner* family elements that will be used here (95).

Genomic sequences have revealed several additional lineages of interest within and related to the *mariner* family. First, the completed *C. elegans* and partial *Caenorhabditis briggsae* sequences reveal a large number of different *mariner* family transposons of various ages (D. J. Witherspoon and H. M. Robertson, unpublished results). These range from the very young *Cemar1* element, having 66 copies with less than 1% DNA divergence from each other (101), to the ancient *Cemar6* to *Cemar9* elements with only 1 to 5 copies each and for which consensus sequences are at best tentative. *C. briggsae* contains several *mariner* transposons, too, including some related to *Cemar6* to *Cemar9* represented by single copies to date. These nematode *mariner* transposons cluster into three subfamilies, as shown by phylogenetic analyses: the elegans, briggsae, and mori subfamilies (Fig. 1). The latter forms a divergent basal lineage of the *mariner* family, with clear affinities to the *Tc1* family (95). Another basal *mariner* subfamily called the rosa subfamily consists of elements known from several tephritid flies such as *Ceratitis rosa* (L. M. Gomulski and R. A. Malacrida, personal communication), the *TCp3.2* element from the moth *Cydia pomonella,* which was described as a *Tc1* family element (56), and a partial *mariner* sequence in the 5′ upstream region of the firefly *Luciola lateralis* luciferase gene (17). As noted above, these two divergent basal lineages might eventually best be treated as distinct families.

In addition, Jarvik and Lark (54) described the first member of this megafamily of DNA transposons in a plant; it is a copy of a highly divergent *mariner* family transposon in the genome of soybean (*Glycine max*). This sequence represents a highly divergent group of *mariner*-like elements unique to plant genomes, and it has also been found in genome sequences from *Arabidopsis thaliana* (72) and rice (76, 109). Searches of all public databases, including the high throughput genomic sequence, genomic survey sequence, and expressed sequence tag (EST) databases at the National Center for Biotechnology, Bethesda, Md., revealed that related sequences are also present in maize. Like the original sequence from soybean, all of these sequences are from highly defective copies or molecular fossils. Figure 1 shows that these all form a discrete divergent subfamily of *mariner* elements that are highly divergent in length as well as in sequence, with numerous insertions of amino acid segments and a longer C terminus (54). It would appear that a particular *mariner* transposon invaded a plant genome long ago, successfully adapted to its peculiarities, and led to diversification of an entire subfamily of *mariner* transposons within plants. The presence of these transposons in both monocot and dicot plants, plus the presence of two divergent *mariner* transposons in the rice genome, suggests that they have undergone considerable movement among plants, presumably by horizontal transfer as occurs in animals (see below). No other members of the *mariner* family or the *Tc1* family have been found in the available plant genome sequences, suggesting that only a single invasion of a plant genome by the *Tc1/mariner* superfamily was ever successful, although divergent DNA transposons with D,D35E megafamily transposases, like the *Lemi* element in the *pogo* superfamily (32), may remain undetected. Furthermore, despite the abundant opportunities for horizontal transfer to diverse animals that eat or in other ways interact closely with plants, no horizontal transfers back to animal genomes have been detected to date, so this transposon lineage may now be adapted to plants in some host-specific fashion.

An interesting feature of this superfamily is that the transposase genes of members of the *pogo* family have occasionally been recruited to perform host functions. Tudor et al. (111) noted that the only relative of the *pogo* transposase available at the time was the CENP-B protein of mammals, which is a major constituent of mammalian centromeres but which is clearly derived from a *pogo* superfamily transposase (62, 104). Remarkably, it is not an essential protein for mitosis, meiosis, or viability in mice (52), although null mice are smaller. Homologs of CENP-B can be found in nematode and *Drosophila melanogaster* genomes and are probably present in all multicellular eukaryotes given their presence in yeasts (41, 70) and plants (5, 72). The *jerky* gene of mice and humans is derived from a *pogo* superfamily transposase and is specifically related to the *Tigger* transposons of mammals (103, 110). Its involvement is implicated in epilepsy in both mice and humans (81). Another related human gene called JRKL has also been recognized (118). On the basis of phylogenetic relationships, Jurka and Kapitonov (58) suggested that *jerky* is derived from a copy of the *Tigger4/ZOMBI* transposon and noted yet another open reading frame (ORF) in the human genome sequences encoding a protein belonging to this group, which they called *Sancho*. A search of the human and mouse genomic and EST sequences indicated that there are at least three more such proteins, including a hypothetical protein defined by cDNA DKFZp761E2110 (GenBank accession no. AL136539). An additional two proteins are also encoded by single long ORFs, New1 (at 4q22.1, GenBank accession no. AC055833) and New2 (at 11q13.2 immediately downstream of the *REQ* gene). All of these host genes are single copy, indicating that in fact they do not derive directly from any of the copies of the extant *Tigger* transposons in the human genome but are instead much older; there are clear mouse orthologs of JRKL, New1, and New2, in addition to *jerky* and CENP-B, and orthologs of the others can be expected in the soon-to-be-completed mouse genome (Fig. 2).

I have recently detected, among the approximately 3,000 fossil copies of the *pogo* family *Tigger1* transposon in our genome, 1 copy at 2q37.1, provisionally named *Eeyore*, with an intact and conserved full-length ORF. Only the long colinear 1,773-bp ORF is conserved, with just 5.2% divergence from the *Tigger1* consensus, while the flanking 5′ and 3′ regions have diverged by 12.2% and nine indels in 645 bp, which is divergence typical of the other thousands of copies having been in our genome lineage for 80 to 90 million years (Myr) (92). Evidence from human ESTs suggests that this copy is transcribed, but again the function of the encoded protein is unclear. The most conserved region relative to the consensus is the N-terminal DNA-binding domain, but the D,D32D catalytic domain appears to be intact as well (as is true for jerky, JRKL, and New1, although Sancho, cDNA DKFZ, New2, and CENP-B have various changes in the catalytic domain that might inactivate it). A comparison of the coding region with that of the consensus sequence reveals a K_s/K_a ratio of silent-to-replacement changes of 4.2, clearly indicative of purifying selection at the level of the host. *Eeyore* therefore represents a most recent example, albeit 80 to 90 Myr old, of a *Tigger* transposase gene recruited for host function. Phylogenetic analysis for all of these proteins and the available human *pogo* family of transposases (all consensuses of *Tigger* transposons) shows how recently *Eeyore* was formed (Fig. 2). As Jurka and Kapitonov (58) noted, *jerky* and *Sancho* (and JRKL and New1) are most closely related to *Tigger4/Zombie*; however, they are not derived from particular copies of it but are much older, probably

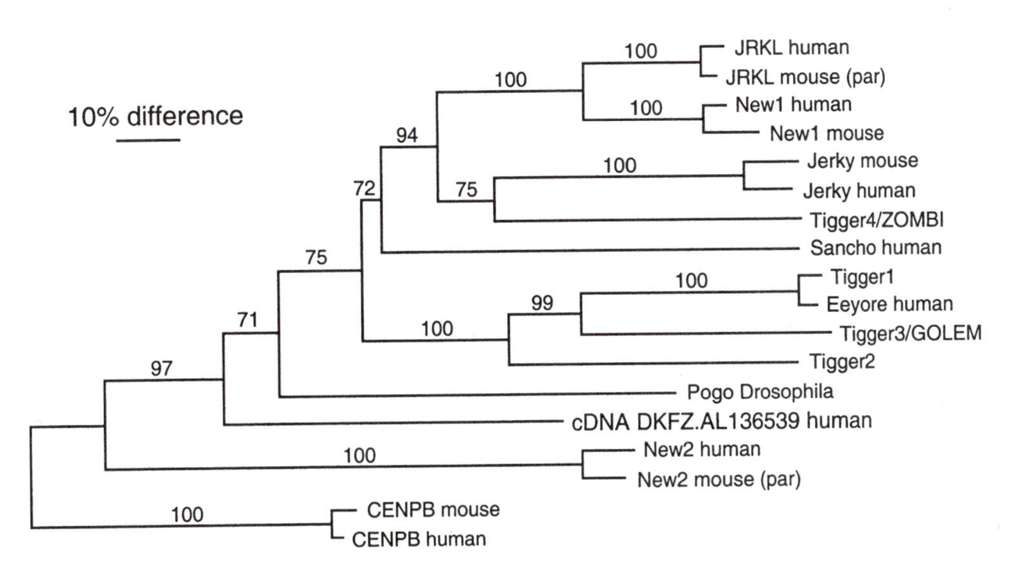

Figure 2. Host genes derived from the mammalian *Tigger* transposons. The tree includes the available transposase translations of the *Tigger1* to *Tigger4* consensus sequences (from the humrep.ref file of RepBase 6.1 at RepBase Update; http://charon.girinst.org/server/RepBase/) and eight genes in mammalian genomes apparently derived from the *Tigger/pogo* transposons. The mammalian CENP-B proteins were used to root the tree, based on their apparent antiquity and homologs being present in all multicellular eukaryotes. The C-terminal regions, which can be aligned among most of these but not with the CENP-B protein, were excluded; otherwise, see the Fig. 1 legend for phylogenetic methods.

originating well before the mammalian radiation (note that the divergence of the human and mouse comparisons is far smaller than is the divergence of these three proteins and the *Tigger4* transposase). Finally, cDNA DKFZ and the New2 gene are much older, perhaps originating in early vertebrates, while the CENP-B gene is highly divergent, befitting the presence of apparent homologs in all multicellular eukaryotes.

The *mariner* family has also contributed to a functional host gene, the SETMAR gene of humans and other anthropoid primates; it is a ±50-Myr-old chimeric gene at 3p26.1 in the human genome derived from the fusion of an N-terminal SET domain and a C-terminal *Hsmar1 mariner* transposase gene domain, which are spliced together from three exons in the human genome (100). SETMAR is expressed widely in all human tissues examined (K. K. O. Walden and H. M. Robertson, unpublished results), but its function remains unknown; however, some SET domains have been shown to have histone H3-specific methyltransferase activity (88), and the MAR domain shows evidence of conservative evolution, indicative of serving some function in the protein, presumably in part DNA binding (100). Other examples of such recruitment of transposases to host functions are coming to light with completion of the sequencing of various genomes (e.g., reference 103 and below). They represent a subset of the many ways in which transposons sometimes contribute positively to host genomes (e.g., reference 79).

The hAT Superfamily

The first evidence of transposons was described in plants by Barbara McClintock, and among the first characterized molecularly were her *Activator* (*Ac*)/*Dissociation* family in *Zea mays* (chapter 24). By the early 1990s, a superfamily of transposons related to *Ac* in plants and animals was recognized with the sequencing of the *hobo* element from *D. melanogaster* (10) and named the hAT superfamily for *hobo, Ac,* and *Tam3* of snapdragon. An increase in member diversity and range of hosts marks this superfamily, although not quite to the same extent as the D,D35E megafamily. I have undertaken a preliminary phylogenetic analysis of the superfamily by using many of the genomic sequences now available. PSI-BLASTP searches (2) that began with transposases of several of the divergent members of the superfamily were used to identify many annotated proteins with amino acid similarity. TBLASTN searches identified several additional genomic sequences encoding proteins of this superfamily, but only a few of these were used in the analysis. Figure 3 shows the relationships of this superfamily.

Several additional hAT superfamily elements from various plants have been described (see also chapter 24). Pearl millet (*Pennisetum glaucum*) has a close relative of *Ac* (75). Next most similar is the *Tam3* element from snapdragon (*Antirrhinum majus*) (48). The *Tag1* element of *A. thaliana* (73) has a relative from rice (*Oryza sativa*) (GenBank accession no. AP001800) but is highly divergent from the *Ac*/*Tam3* grouping. The same is true for the *Tip100* element from *Ipomoea purpurea* (40), which is related to the *IR* element from maize (W. Eggleston, personal communication), while the *slide* (37) and AJ133502 elements of tobacco and gene *F5K24.11* of *Arabidopsis* are all highly divergent lineages with no close relatives found in the databases. The *Bergamo* element from maize was not included in the analysis because its coding capacity is unclear (46). The genomic sequences from *Arabidopsis* and *Oryza* have also revealed many sequences encoding related proteins (e.g., 72, 109). For example, in Fig. 3, 12 sequences from *Arabidopsis* are included in the immediate *Ac*-like cluster, while another 4 that are related to *Tag1* and 2 are related to *Tip100.* Several more unannotated sequences like these are also present in these genomes.

Four hAT superfamily elements have been described from various fungi: *restless* from *Tolypocladium inflatum* (60), *Tfo1* and *Folyt1* from *Fusarium oxysporum* (36, 82), and *Tascot* from *Ascobolus immersus* (GenBank accession no. Y07695). Three of these form a cluster in the tree, while *Folyt1* is a divergent lineage.

For insects, a collaboration of David O'Brochta and Peter Atkinson has led to the characterization of several *hobo*-like elements of various flies. These are *Hermes* from the housefly *Musca domestica* (113), *Hermit* from the blowfly *Lucilia cuprina* (20), and *Homer* and a nonautonomous *Homer*-like element from the tephritid fly *Bactrocera tryoni* (86). They form a clearly related grouping of *hobo*-like elements in the phylogenetic tree (Fig. 3). Others have reported more distantly related hAT superfamily elements from tephritid flies (42) and moths (25), but full-length sequences are only available for a highly divergent nonautonomous copy of the *Bactrocera dorsalis hopper* element (43), which may have its closest allies in sequences from *D. melanogaster, C. elegans,* and *Homo sapiens* (Fig. 3). The PSI-BLASTP search revealed several *D. melanogaster* genes that are members of the superfamily, including the DREF transcription factor (108) and annotated genes *BG.DS00797.6* (7) and *CG13775* (1).

Bigot et al. (8) described a member of the hAT superfamily from the nematode *C. elegans* genomic

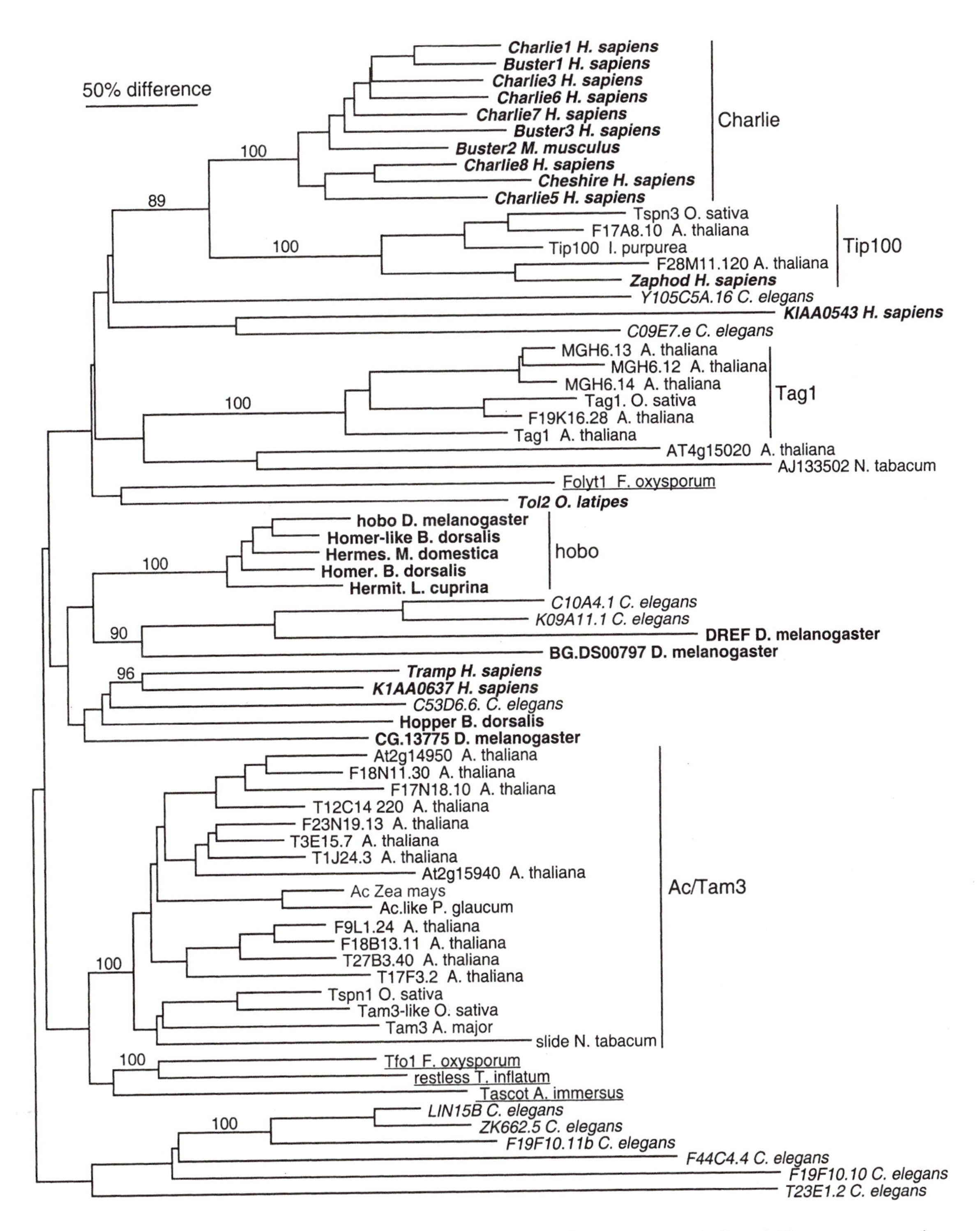

Figure 3. Relationships within the hAT superfamily. Clear groupings of transposons are indicated. The tree was rooted at the midpoint in the absence of a convincing outgroup; otherwise, see the Fig. 1 legend for phylogenetic methods and typeface treatment. The published translation for *slide* was extended to full length. *Charlie4* is not present because no consensus sequence capable of encoding a transposase is available; the other *Charlie* consensus sequences generally still have many ambiguous positions (RepBase 6.1).

sequences. Several other annotated *C. elegans* genes also belong to this superfamily (Fig. 3). They also attempted to show that the hAT superfamily transposases contain a D,D35E motif; however, this assertion is not supported by my analysis. The hAT superfamily transposases share several highly conserved aspartate and glutamate residues that might well be involved in the chelation of magnesium ions at an active site, but the middle residue highlighted by Bigot et al. (8) as corresponding to the D,D35E domain is not highly conserved among hAT transposases. Furthermore, no other conserved residues within the putative D,D35E domain align between these transposase superfamilies, putting their suggestion of homology into question. Eventually structural data for a hAT superfamily transposase will provide an answer to this question.

Among vertebrates, a member of the hAT superfamily, *Tol2*, has been characterized from the medaka fish (65). Smit (103) has also described several ancient *Charlie* elements in this superfamily, identified as consensus sequences of numerous ancient highly diverged copies (available from the humrep.ref file of RepBase 6.1 at RepBase Update; http://charon.girinst.org/server/RepBase/). In addition, he has identified three human genes potentially derived from hAT superfamily transposases, named *Buster1* to *Buster3*. The *Tramp* gene identified by Esposito et al. (30) appears to have had a similar origin. In addition, the proteins encoded by cDNAs K1AA0637 and K1AA0543 are fusions of zinc finger N-terminal domains to hAT superfamily transposase C-terminal domains (103). With genomic sequences becoming available, it is possible to search for similar transposon-derived genes in other organisms, and a preliminary effort was undertaken using PSI-BLASTP. The limitations of this approach are that only those sequences annotated as encoding proteins will be recovered, and sometimes these are partial sequences that might be extended upon further study (and others have insertions that might be removed from the annotations upon review). Nevertheless, this analysis provides considerable additional insight into the evolution of this superfamily.

First, the Buster1 to Buster3 proteins are indeed most closely related to the transposases of the *Charlie* transposons in the human genome, although not originating from any particular known *Charlie*; however, the other three proteins (Tramp, K1AA0637, and K1AA0543) appear to have had separate origins. Second, multiple host genes appear to have been derived similarly from hAT superfamily transposases in all of the other genomes available. These include the DREF transcription factor and the genes *BG.DS00797.6* and *CG13775* of *D. melanogaster,* at least 11 genes from *C. elegans,* including the gene encoding LIN15B, and at least 13 genes of *A. thaliana,* most of which are apparently derived from *Ac*-type transposons. Each of these is a unique gene, with no highly similar sequences among the available genomic sequences, strongly suggesting that they are transposase genes recruited to host function. Thus, hAT superfamily transposons have made remarkable contributions to their host genomes in the form of proteins that are involved in a variety of functions, probably all involving DNA binding and hence some kind of gene regulation.

Finally, there is little resolution to the basal lineages of this tree and no bootstrap support; hence the overall relationships of hAT superfamily elements remain unclear, except the similarity of the human *Charlie* transposons to those of the plant *Tip100* family, to which the unusual human *Zaphod* consensus already belongs. In addition, there are no known protist or prokaryotic members of the superfamily, so it is possible that this transposase evolved within the eukaryote lineage before the plant/fungi/animal split approximately 1,000 Myr ago. The presence of active elements or their remnants, plus genomic acquisitions, in each of these major kingdoms, and each of the genomes sequenced to date, shows that they will be found widely within the higher eukaryotes.

The P-Element Family

P elements were among the first DNA transposons described; however, their evolutionary relationships remain rather mysterious. Many members of the family have now been identified from diverse drosophilid flies and their close relatives (see chapter 21 and below), but they are all very similar to each other. The discovery of two distantly related, albeit defective, copies of a family member in the blowfly *L. cuprina* (83) raised hopes of expanding the family; however, considerable effort has revealed that related sequences can only be found in other muscamorphan or higher flies (16), and one such sequence has been published (71). No related sequences have been recognized among the available genomic sequences, so this family may be insect specific.

The *piggyBac* Family

The first insect members of the *piggyBac* family were recognized as insertions within the *Few Polyhedra* locus of baculoviruses passaged through various moth cell lines (6, 14, 112). The novel transposons revealed by sequencing these insertions shared a few features, primarily that they duplicated a TTAA target site when inserted and their termini are short inverted repeats ending in 5′-CC . . . GG-3′. Therefore, they resemble the *Tx* elements of amphibians

(13). Subsequently, a full-length active version of one of these was obtained, named *piggyBac* for its original detection, and shown to encode an entirely novel transposase unrelated to any member of the major superfamilies (28). Remarkably, a related transposase is encoded by the consensus sequence of numerous ancient copies of a transposon in the human genome, called *looper* for its relationship to *piggyBac* from the cabbage looper, *Trichoplusia ni* (information found at the RepBase Update website, http://charon.girinst.org/server/RepBase/). These ±280-amino-acid transposases share only 27% identity. Recently, a very closely related transposon was serendipitously discovered in a tephritid fly genome, indicating both that *piggyBac* relatives will be found widely in insects and that they undergo horizontal transfer between distantly related hosts (44). Nevertheless, these cannot be extremely common transposons because none of the other completed animal or plant genomes reveal relatives.

The *Mutator* Family

The *Mutator* element of maize can be its most active transposon (chapter 23). Related transposons have been identified in the genomic sequences of several plants, including rice and sorghum, by their similarity to the MURA transposase (50). The *A. thaliana* genome has hundreds of these genes, with the closest transposase having 25% amino acid identity, not all of which are likely to be transposases of active transposons (59, 69, 72, 117). Plants also have distantly related transposons, such as *jittery* (GenBank accession no. AAF66982) which encodes a transposase with 15% identity to MURA, and even more distant relationships have been detected with a group of bacterial transposases (27), so this is another ancient family but one that has apparently not invaded fungal and animal genomes.

The *En/Spm* Family

Another of Barbara McClintock's original discoveries, *En/Spm,* remains a plant-specific lineage of transposons, with relatives like *Tgm* in soybeans (89) and *Psl* in petunia (105) and other plants (34). Kapitonov and Jurka (59) identified four highly divergent members of the family in the *A. thaliana* genome, but relatives have yet to be detected in species other than plants.

Other DNA Transposons

There are many other transposons with characteristics of DNA-mediated or class II transposons known from various eukaryotes; however, most are nonautonomous copies with their transposase genes deleted, for example, the numerous miniature inverted repeat transposable elements (MITEs) known from plants, nematodes, humans, and others. When the corresponding full-length elements have been characterized, they have generally fallen into one of the known groups; it will be interesting to see whether they all belong to these six major evolutionarily independent groups of transposases or whether they represent other novel but rarer transposases. For example, Kapitonov and Jurka (59) describe two novel DNA transposons from *A. thaliana* that they call *Harbinger* and *Arnold,* although the former is related to some bacterial transposons while the latter is distantly related to the *Mutator* family. These have apparently been responsible for generating large numbers of MITEs. Similarly, Le et al. (69) identify a new group of transposons called *Basho* in this genome, with features that do not even allow immediate classification as DNA or RNA transposons.

DNA TRANSPOSONS AT THE MICROEVOLUTIONARY LEVEL

The evolutionary dynamics of DNA transposons within particular host species, and over the relatively short term, say 1 Myr, are expected to be extremely diverse. Bacterial transposons experience a very different host environment from those in eukaryotes, and even among eukaryotes, there is huge variation in these dynamics due to vagaries of the transposon-host interaction. So many factors affect the evolution of transposons at this level that it is not feasible to review them all here. I will focus instead on what recent eukaryotic genomic sequences tell us about the history of DNA transposons.

In eukaryotes, the link between transcription and translation is broken by the nuclear membrane, which means that the transposase from any particular copy of a DNA transposon can mobilize any other copy, no matter how defective, as long as it has the requisite *cis* sequences, usually, but not always, the inverted terminal repeats (ITRs). Indeed, the proliferation of MITEs within most plant and animal genomes clearly results from this process. Foldback and other exotic constructs flanked by the appropriate repeats presumably proliferate in the same fashion by parasitizing the transposase of cognate active transposons. The result is that the link between the coding capacity of a DNA transposon and its success in making copies of itself is effectively broken. A simple prediction is that all copies of a DNA transposon within a genome should acquire incapacitating mutations at a neutral rate and eventually become completely inactive. Indeed for several transposons, particularly of the *Tc1/mariner* superfamily, this appears to be the case because commonly all copies discovered within a particular genome are defective. Given that the copies of a transposon are likely to be active for only a relatively short period followed by a very long period of slow mutation decay, the prediction is that the vast majority of copies found by nonselective methods (e.g., genomic sequencing or homology determined by PCR rather than insertional traps or phenotypic evidence of dysgenesis) should be defective. This is the case for the *Tc1/mariner* superfamily where the majority of copies found in genomic sequences, primarily human and nematode, or by PCR surveys of diverse animals are defective. Only a few active transposon candidates have been found nonselectively. The evidence for these processes is nevertheless tainted, and eventually completely removed, in genomes such as those from *Drosophila* flies and *Caenorhabditis* nematodes and probably *A. thaliana* because their small genomes are maintained that way by deletion of junk DNA, including transposons (84, 85, 94). Thus, the human genome has provided some of the most remarkably complete evidence, simply because the rate at which genomic DNA is removed from our genome lineage has been extremely low (39, 98).

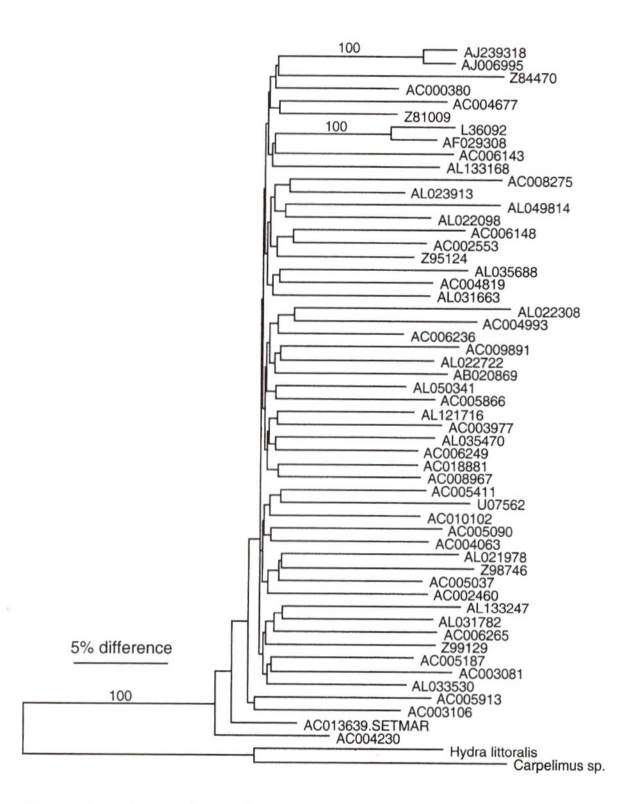

Figure 4. Relationships of 54 *Hsmar1* copies in the human genome. The various copies are indicated by the accession number of the sequenced clone containing them. The tree is based on full-length DNA sequences rooted by the hydra and beetle consensus sequences. Phylogenetic methods are as given in the Fig. 1 legend, except that the HKY model was used to correct DNA distances.

The Human Genome

There are two major kinds of *mariner* transposons in the human genome. I have called these *Hsmar1* in the cecropia subfamily (100) and *Hsmar2* in the irritans subfamily (98), in keeping with a convention that we proposed for naming *mariner* family transposons (95), although others had first published and sometimes named them (references in 98 and 100) (see Fig. 1 for their relationships with other *mariner* transposons). *Hsmar1* invaded our genome lineage approximately 50 Myr ago, and we retain approximately 200 copies of it. All but one in the nearly complete genomic sequences are defective, with multiple stop codons or frameshifting indels incapacitating their coding regions; this sequence is the copy recruited into the SETMAR gene mentioned earlier. Even this copy cannot be mobilized because its 5′ ITR is removed by an Alu insertion. *Hsmar1* was responsible for generating many thousand copies of a related MITE that shares the 30-bp ITRs and an additional 20 bp in-between them (80, 100, 104). A simple consensus sequence from which individual copies, except for SETMAR, differ by an average of 9% encodes a respectable cecropia subfamily of *mariner* transposases. In turn, this *Hsmar1* transposase is most closely related to those from a hydra (*Hydra littoralis*) and a staphylinid beetle (*Carpelimus* sp.), for which we have obtained multiple full-length sequences from genomic libraries to derive consensus sequences (the individual copies in the hydra and beetle are very similar to each other and have full-length ORFs). A phylogenetic analysis of the first 54 full-length *Hsmar1* copies in the public databases, rooted with these hydra and beetle sequences, is shown in Fig. 4. It reveals the expected pattern of a rapid radiation of the different copies, followed by a long period of idiosyncratic evolution during which each copy acquired private mutations. The only exceptions are two pairs of copies that are far more closely related to each other, and in both cases, these involve more recent large duplications, for example, of the T-cell receptor region. The shortest branch leads to the SETMAR copy. Searches of the public databases in late 2000, when 93% of the human genome was available in at least draft form, revealed no additional intact copies of *Hsmar1*. Thus, *Hsmar1* appears to epitomize the pattern of neutral evolution expected of DNA transposons leading to inactivation of all copies in the genome. The approximately 1,000 copies of *Hsmar2* reveal the same pat-

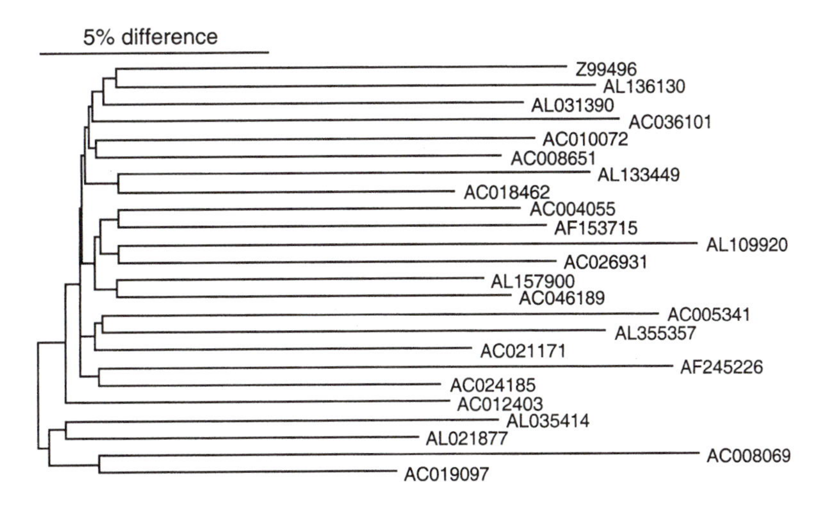

Figure 5. Relationships of 24 *Charlie3* copies in the human genome. In the absence of a close relative, the tree was rooted at the midpoint. See the Fig. 4 legend for details.

tern, except that they are older, having entered our genome lineage approximately 80 Myr ago (98). There are no copies with intact coding capacity, and no MITEs were generated by *Hsmar2*.

Essentially, the same pattern appears to be true for all of the other DNA-mediated transposons in the human genome, including the ±3,000 copies of *Tigger1,* which is similar in age to *Hsmar2,* the other three *Tigger* transposons for which full-length consensus sequences are available in RepBase Update (57), and all of the *Charlie* elements (103). The youngest of the latter is *Charlie3,* of which there are about 100 full-length copies averaging about 13% divergence from their consensus, and again their relationships show the expected pattern of largely private mutations (Fig. 5). Thus, in the one available genome where junk DNA is only slowly discarded, the pattern of DNA transposon evolution is as expected. Unfortunately, the youngest of these transposons in our genome is approximately 50 Myr old, so they do not reveal much about evolutionary dynamics on a shorter timescale.

The *Caenorhabditis* Nematode Genomes

For shorter timescales, we must turn to other genomes; however, the additional complexities of these often make analysis more difficult. The most obvious starting place is the *C. elegans* genome (15), which has nine different kinds of *mariner* transposons (Witherspoon and Robertson, unpublished), the relationships of which are shown in Fig. 1. *Cemar1* has the most copies, 66, and is the most recently amplified in the genome, with most of the copies differing from the consensus sequence by just a few base pairs. Most of the copies encode a potentially active transposase. The phylogenetic relationships of these copies shown in Fig. 6 reveal that the vast majority of mutations (92 of 103) are private, again supporting the pattern of independent evolution of each copy. There is nevertheless some structure to the tree, with certain copies being derived from others after acquiring mutations, but this is to be expected based on chance alone. *Cemar6* exhibits a similar pattern, but with just 11 copies. *Cemar2, -3,* and *-4* reveal a very different history. For example, the 46 copies of *Cemar2* range in sequence divergence from each other from 0 to 20%. A phylogenetic tree of their relationships shows a very different pattern in which there are several old divergent lineages and then some recently amplified lineages with very similar copies (Fig. 7). None of these copies can encode a functional transposase; indeed their consensus sequences cannot, because they contain two frameshifts and an in-frame stop codon. Thus, they and the *Cemar3* and *Cemar4* copies appear to have been generated by another *mariner* transposase acting in *trans.* Enigmatically, this cannot be the *Cemar1* transposase because that *mariner* has only been in this genome for a very short time and in any case is probably too divergent to mobilize the *Cemar2* to *Cemar4* copies, because only closely related *mariner* elements are likely to be able to cross-mobilize each other (24, 68). *Cemar5, -7, -8,* and *-9* are represented by only one to five extremely old defective copies and, presumably, are the last remnants of once common transposons in this genome, most copies having been removed by deletions. *C. briggsae* similarly has a complicated set of recent (*Cbmar1*) and ancient (*Cbmar2* and several single copies of other types) *mariner* transposons in its genome (see Fig. 1).

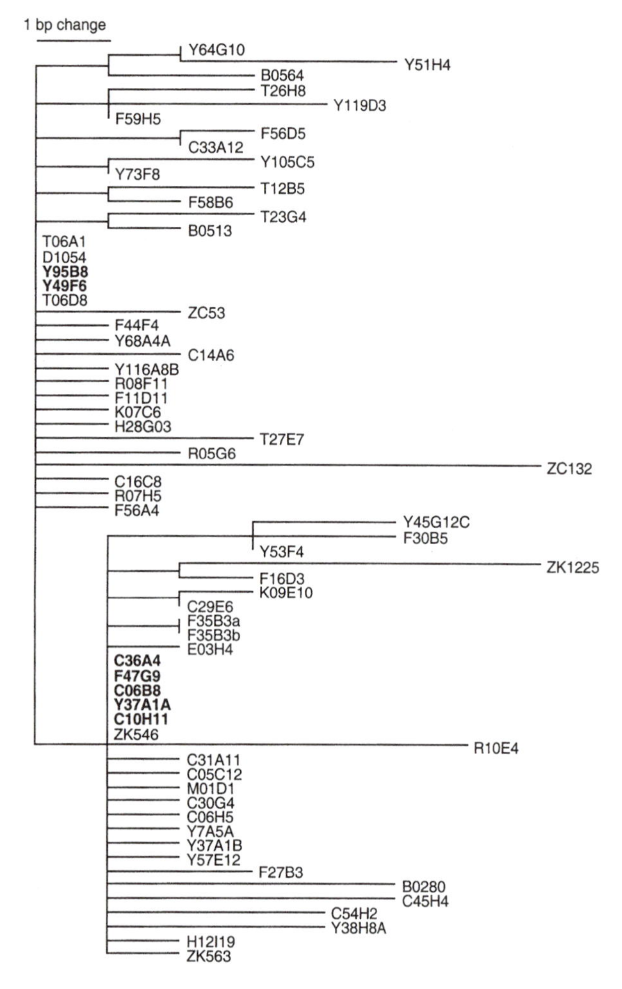

Figure 6. Relationships of the 66 *Cemar1* copies in the nematode *C. elegans* genome. Copies are named for the cosmid or yeast artificial chromosome clone in which they occur. Those in boldface are identical to one of the two consensus sequences which differ by a nonsynonymous transition; the others showing no sequence differences from the consensus sequences have small indels. Maximum parsimony was employed using PAUP* version 4.0b4a to build the tree because so few changes have occurred that correction for multiple changes is unnecessary; the tree is rooted at the midpoint.

Thus, evidence for *mariner* transposons from nematode genomes supports the model, except that sometimes the evolutionary history is complicated by interactions between *mariner* elements and the long-term history is removed by deletions. Preliminary examination of the other DNA transposons in these two genomes suggests that they exhibit a similar range of evolutionary histories.

The *D. melanogaster* Genome

The *D. melanogaster* genome before the arrival of *hobo* and the P elements was relatively depauperate in DNA transposons, with only the *S* and *Bari* elements in the *Tc1* family being fairly common, along with *pogo* and relics of *hobo*-like elements. *S* elements present a complex picture that calls the standard model into question. Merriman et al. (78) described nine copies from genomic clones, all but one of which is defective in their coding regions. Searches of the available euchromatic portion of the *D. melanogaster* genome reveal 18 reasonably full-length copies of this transposon, as well as at least 60 partial sequences or copies with major internal deletions. The latter arise in two ways. Those with two ends and large internal deletions probably result from interrupted double-strand gap repair (29). Most of the remainder are ends truncated by the end of a contig, presumably because they are in regions of heterochromatin that the automated assembly of Celera was unable to master (caution is needed in evaluating even these copies, because during the automated assembly of the Celera *Drosophila* sequences, some transposons were represented by "consensus" sequences). Indeed, it is remarkable that none of the nine copies cloned and sequenced by Merriman et al. (78) are represented in the Celera euchromatic sequence, indicating that the copies mostly came from heterochromatic sites, or are in gaps within scaffolds, and that many additional copies must exist in the heterochromatin. Phylogenetic analysis of 18 fairly full-length copies from the genomic sequences along with the nine from Merriman et al. (78) reveals that there are two distinct lineages of *S* elements (Fig. 8). The copy with an ORF is pS3, part of the smaller lineage. All other copies are highly defective with multiple incapacitating mutations. Nevertheless, Merriman et al. (78) detected the *S* element by observing a transposition of the pS2 copy into the *suppressor of sable* gene, indicating that there is at least one copy capable of encoding a functional transposase, perhaps the pS3 copy. How this one or a few intact autonomous copies have persisted in the genome remains to be determined.

Horizontal Transfer versus Vertical Inheritance

If DNA transposons are doomed to extinction within any particular eukaryotic host genome, then the only way that they could have perpetuated and spread as widely as they have would be by regular horizontal transfer to new hosts. The original canonical example of horizontal transfer of a transposon is that of the P element from *Drosophila willistoni*, or a closely related species, to *D. melanogaster*, an event that might have occurred last century (22). Was this simply an exceptional event or instead the norm for these and other DNA transposons? Clark et al. (19) analyzed the available full-length P-element sequences from diverse drosophilid flies and inferred that at least

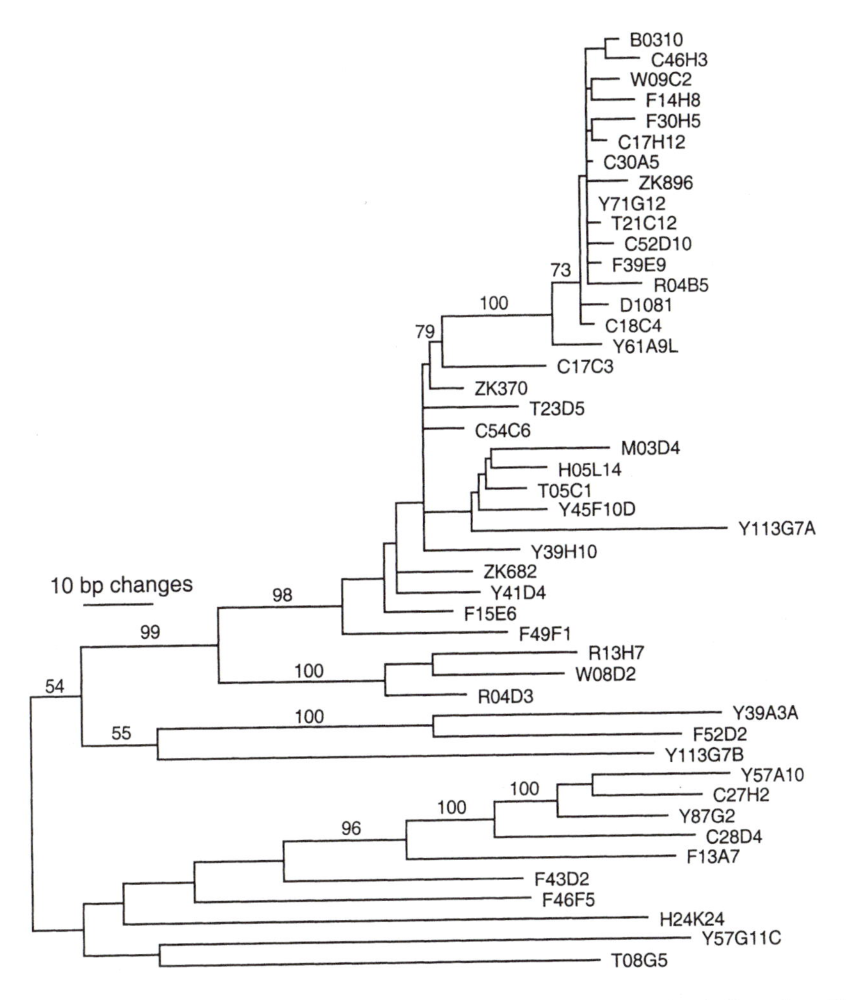

Figure 7. Relationships of the 46 *Cemar2* copies in the nematode *C. elegans* genome. See the legends to Figs. 1 and 6 for details.

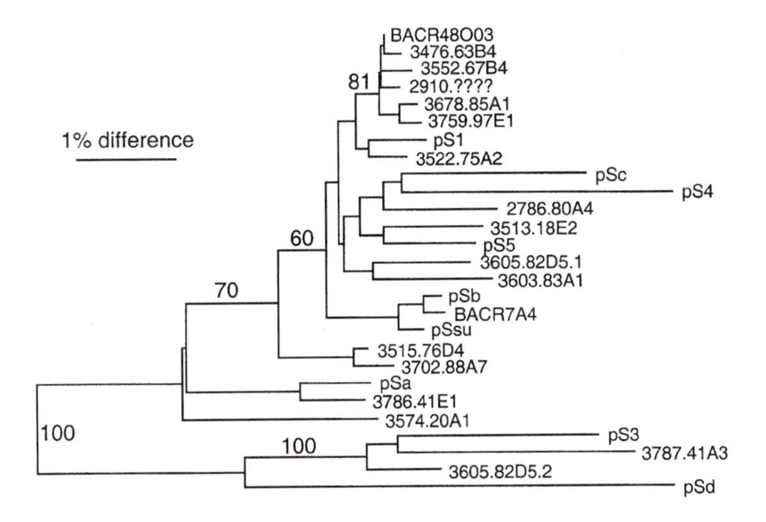

Figure 8. Relationships of *S* elements in the *D. melanogaster* genome. The pSX copies are from Merriman et al. (78), while the remainder are indicated by the GenBank accession (preceded with AE00) in which they are found (scaffolds are from the Celera data), with their chromosomal locations, if known, shown after a period. Phylogenetic methods are as given in the Fig. 6 legend; values are as given in the Fig. 1 legend.

four such horizontal transfers had occurred within this genus of flies. Subsequent generation of large numbers of partial P sequences from many additional species have allowed additional analyses showing that many horizontal transfers must have occurred to yield the current distribution of P elements in this one genus (18); indeed, Silva and Kidwell (102) argue persuasively that horizontal transfers alone might provide all of the selection pressure maintaining the integrity and activity of P elements. Witherspoon (115) came to the same conclusion from careful analysis of the evidence for selection on P sequences, while Haring et al. (45) reached similar conclusions from their analysis of a separate lineage of P elements within drosophilids.

I have come to the same conclusion from our work with the *mariner* family of transposons from a far greater diversity of animal hosts (90, 93, 96, 97, 99). It seems inescapable that even the most closely related *mariner* transposons that we have discovered in different hosts (with greater than 90% DNA identity) represent a tiny fraction of the hosts that have been invaded by these particular *mariner* elements; indeed, many hundreds if not thousands of intermediate horizontal transfers have occurred, each involving few if any nucleotide changes before the next horizontal transfer. Similar evidence from other laboratories supports this view of *mariner* evolution (33, 74), and this model has been summarized several times (47). This conclusion gains additional support from a recent study of the anciently asexual bdelloid rotifers whose genomes appear to be devoid of vertically inherited RNA transposons but who have acquired at least four lineages of *mariner* transposons, presumably by horizontal transfer (3). It seems reasonable to expect that the same patterns of evolution through horizontal transfer will apply to the *Tc1* family (4), but there are hints of limits to this process, for example, in the apparent restriction of the *mariner*-like elements in plants to that kingdom discussed earlier, and of course, the absence of *Tc1/mariner* lineages from bacterial and protist genomes indicates that they cannot readily establish themselves there.

Similar observations of invasions of new genomes have been made in less systematic studies of elements such as the hAT superfamily elements *hobo* of *D. melanogaster* (9) and *Tol2* of the medaka fish (65) and the *piggyBac* element of moths and flies (44). While the mechanisms of these horizontal transfers remain as obscure as ever, they do appear to be the primary way in which DNA transposons of most eukaryotes perpetuate themselves.

There are, nevertheless, examples that are less easily reconciled with the perpetuation of DNA transposons via horizontal transfer. The ciliate transposons of the D,D35E megafamily are an extreme example; in this case, the transposase has become essential for host existence in the context of precise removal of thousands of copies from inside genes during the maturation of the macronucleus from the micronucleus (64, 116; chapter 30). But there also appear to be examples involving more standard DNA transposons. Witherspoon (114) has reanalyzed the multiple sequences of the hAT superfamily *Tam3* element from *A. majus* generated by Kishima et al. (63) and found evidence of selection on these transposons within their host genome. He proposes a model analogous to trait group selection that would enable perpetuation of a DNA transposon family within a host but which is mainly applicable to low-copy-number DNA transposons. It may therefore be applicable to situations involving the *pogo* element of *D. melanogaster* (111), where the genomic sequences reveal the predicted presence of only a few very similar intact copies, the *S* element discussed above, the low-copy-number but active *Psl* element of petunia (105), and the *impala* element of *F. oxysporum* (51), where there appear to be two lineages of active elements despite a seemingly long history in this genome. Many other DNA transposons exist at relatively low copy numbers, for example, in filamentous fungi (61), and have multiple lineages in particular hosts, for example, the *Mutator*-like elements of plants (117); however, these situations require extensive analysis of large sequence sets within species and consideration of the evolution of the transposon family across species. Generating additional genomic sequences from representatives of most major protist, fungal, plant, and animal lineages during the next decade will provide ample opportunities for analyses of this kind, which should help to unravel the contributions of vertical inheritance within hosts and horizontal transfer between hosts to the remarkable ability of DNA transposons to perpetuate and spread over periods of hundreds of millions of years.

Acknowledgments. I thank the various public genome projects for making genomic sequences available before publication and David Witherspoon, Bill Eggleston, Ludvik Gomulski, and Anna Malacrida for communicating results and manuscripts before publication.

My work with transposons has been supported by grants from the NSF and NIH.

ADDENDUM IN PROOF

Since preparation of this chapter, the completed *Arabidopsis thaliana* genome sequence has been published (The Arabidopsis Genome Initiative, *Nature* 408:796–815, 2000). The features of some of the novel transposons in this genome mentioned above have been clarified. The *Harbinger* elements are allied in a new superfamily with the PIF elements from maize that were responsible for generation of numerous Tourist-like MITEs (X. Zhang, C. Feschotte, Q. Zhang, N. Jiang, W. B. Eggleston, and S. R. Wessler, *Proc. Natl. Acad. Sci. USA* 98:12572–12577, 2001; J. Jurka and V. V. Kapitonov, *Proc. Natl. Acad. Sci. USA* **98**:12315–12316, 2001). The *Basho* elements belong with others from *Arabidopsis,* rice, and the *Caenorhabditis* nematode genomes in a third entire class of eukaryotic transposons called Helitrons that replicate via a rolling-circle mechanism shared with a small group of prokaryotic elements (V. V. Kapitonov and J. Jurka, *Proc. Natl. Acad. Sci. USA* **98**: 8714–8719, 2001; C. Feschotte and S. R. Wessler, *Proc. Natl. Acad. Sci. USA* **98**:8923–8924, 2001).

The draft human genome sequences have also been published (International Human Genome Sequencing Consortium, *Nature* **409**:860–921, 2001; J. C. Venter et al., *Science* **291**:1304–1351, 2001). The public human genome consortium included a table of 40 human genes derived from transposons, primarily DNA-mediated transposons, including 6 of the 8 that I include in Fig. 2. However, they do not include the *Eeyore* gene recently derived from a particular copy of *Tigger1* described here.

There have also been significant additions to the *Tc1/mariner* superfamily described in the first part of the chapter and Fig. 1. Shao and Tu (*Genetics* **159**:1103–1115, 2001) described an additional lineage of transposons from *Aedes* mosquitoes within the *Tc1/mariner* superfamily. The transposases of these elements have a distinctive D,D37E catalytic domain motif, and the authors propose that these, the basal *mori* subfamily of *mariner* elements, and the highly divergent *max* subfamily of plant *mariner* elements be raised to family status and named the *D37E, D37D,* and *D39D* families, respectively, after their distinctive catalytic domain motifs. I noted that these subfamilies might warrant family status and add that in this classification, the *rosa* subfamily represented by *Cemar2* from *Ceratitis rosa* (L. M. Gomulski, C. Torti, M. Bonizzoni, D. Moralli, E. Raimondi, P. Capy, G. Gasperi, and A. R. Malacrida, *J. Mol. Evol.* **53**:597–606, 2001) would become the *D41D* family (Fig. 1). We have identified an additional member of the *mori* subfamily (*D37D* family) from *Bombyx mori,* named *Bmmar3* (GenBank accession no. AF461149). Feschotte and Wessler (*Proc. Natl. Acad. Sci. USA,* in press) surveyed plants for members of the *max* subfamily (*D39D* family) and report their occurrence in numerous monocots and a few dicots. Additional full-length *mariner* transposons have been described, including *B. tryoni mar1* and *mar2* from the tephritid fruit fly *Bactrocera tryoni* (C. L. Green and M. Frommer, *Insect Mol. Biol.* **10**:371–386, 2001) and a case of relatively recent horizontal transfer involving *mariners* in a parasitoid wasp and its moth host (M. Yoshiyama, Z. Tu, Y. Kainoh, H. Honda, T. Shono, and K. Kimura, *Mol. Biol. Evol.* **18**:1952–1958, 2001). Finally, additional members of the *Tc1* family have been described, including *BmTc1* from *B. mori* (K. Mikitani, T. Sugasaki, T. Shimada, M. Kobayashi, and J.-A. Gustafsson, *J. Biol. Chem.* **275**:37725–37732, 2000), *MsqTc3* from *Aedes aegypti* (H. Shao, Y. Qi, and Z. Tu, *Insect Mol. Biol.,* in press), and several additional transposons from fish and frogs that show the expected pattern of horizontal transfer across diverse animal hosts (M. J. Leaver, *Gene* **271**:203–214, 2001).

REFERENCES

1. **Adams, M. D., et al.** 2000. The genome sequence of *Drosophila melanogaster. Science* **287**:2185–2195.
2. **Altschul, S. F., T. L. Madden, A. A. Schäffer, J. Zhang, Z. Zhang, W. Miller, and D. J. Lipman.** 1997. Gapped BLAST and PSI-BLAST: a new generation of protein database search programs. *Nucleic Acids Res.* **25**:3389–3402.
3. **Arkhipova, I., and M. Meselson.** 2000. Transposable elements in sexual and ancient asexual taxa. *Proc. Natl. Acad. Sci. USA* **97**:14473–14477.
4. **Avancini, R. M. P., K. K. O. Walden, and H. M. Robertson.** 1996. The genomes of most animals have multiple members of the *Tcl* family of transposable elements. *Genetica* **98**: 131–140.
5. **Barbosa-Cisneros, O., S. Fraire-Velazquez, J. Moreno, and R. Herrera-Esparza.** 1997. CENP-B autoantigen is a conserved protein from humans to high plants: identification of the aminoterminal domain in *Phaseolus vulgaris. Rev. Rheumatol.* **64**:368–374.
6. **Beames, B., and M. D. Summers.** 1990. Sequence comparison of cellular and viral copies of host cell DNA insertions found in *Autographa californica* nuclear polyhedrosis virus. *Virology* **174**:354–363.
7. **Benos, P. V., et al.** 2000. From sequence to chromosome: the tip of the X chromosome of *D. melanogaster. Science* **287**: 2220–2222.
8. **Bigot, Y., C. Augé-Gouillou, and G. Periquet.** 1996. Computer analyses reveal a *hobo*-like element in the nematode *Caenorhabditis elegans,* which presents a conserved transposase domain common with the Tc1-*mariner* transposon family. *Gene* **174**:265–271.
9. **Bonnivard, E., C. Bazin, B. Denis, and D. Higuet.** 2000. A scenario for the *hobo* transposable element invasion, deduced from the structure of natural populations of *Drosophila melanogaster* using tandem TPE repeats. *Genet. Res.* **75**:13–23.
10. **Calvi, B. R., T. J. Hong, S. D. Findley, and W. M. Gelbart.** 1991. Evidence for a common evolutionary origin of inverted repeat transposons in Drosophila and plants: hobo, Activator, and Tam3. *Cell* **66**:465–471.
11. **Capy, P., R. Vitalis, T. Langin, D. Higuet, and C. Bazin.** 1996. Relationships between transposable elements based upon the integrase-transposase domains: is there a common ancestor? *J. Mol. Evol.* **42**:359–368.
12. **Capy, P., T. Langin, D. Higuet, P. Maurer, and C. Bazin.** 1997. Do the integrases of LTR-retrotransposons and class II element transposases have a common ancestor? *Genetica* **100**:63–72.
13. **Carroll, D., D. S. Knutzon, and J. E. Garrett.** 1989. Transposable elements in *Xenopus* species, p. 567–574. *In* D. E. Berg and M. M. Howe (ed.), *Mobile DNA.* American Society for Microbiology, Washington, D.C.
14. **Cary, L. C., M. Goebel, B. G. Corsaro, H. Wang, E. Rosen, and M. J. Fraser.** 1989. Transposon mutagenesis of baculoviruses: analysis of *Trichoplusia ni* transposon IFP2 insertions within the FP-locus of nuclear polyhedrosis viruses. *Virology* **172**:156–169.
15. ***C. elegans* Sequencing Consortium.** 1998. Genome sequence of the nematode. *C. elegans:* a platform for investigating biology. *Science* **282**:2012–2018.
16. **Chen, C.-L.** 1998. Distribution of P elements in Muscamorpha. Ph.D. thesis. University of Illinois, Urbana-Champaign.
17. **Cho, K. H., J. S. Lee, Y. D. Choi, and K. S. Boo.** 1999. Structural polymorphism of the luciferase gene in the firefly, *Luciola lateralis. Insect Mol. Biol.* **8**:193–200.
18. **Clark, J. B., and M. G. Kidwell.** 1997. A phylogenetic perspective on P transposable element evolution in *Drosophila. Proc. Natl. Acad. Sci. USA* **94**:11428–11433.
19. **Clark, J. B., W. P. Maddison, and M. G. Kidwell.** 1994. Phylogenetic analysis supports horizontal transfer of *P* transposable elements. *Mol. Biol. Evol.* **11**:40–50.
20. **Coates, C. J., K. M. Johnson, H. D. Perkins, A. J. Howells, D. A. O'Brochta, and P. W. Atkinson.** 1996. The *hermit* transposable element of the Australian sheep blowfly, *Lucilia cuprina,* belongs to the *hAT* family of transposable elements. *Genetica* **97**:23–31.
21. **Crollius, H. R., O. Jaillon, C. Dasilva, C. Ozouf-Costaz, C. Fizames, C. Fischer, L. Bouneau, A. Billault, F. Quetier, W. Saurin, A. Bernot, and J. Weissenbach.** 2000. Characteriza-

tion and repeat analysis of the compact genome of the freshwater pufferfish *Tetraodon nigroviridis*. *Genome Res.* **10:** 939–949.

22. **Daniels, S. B., K. R. Petersen, L. D. Strausbaugh, M. G. Kidwell, and A. Chovnick.** 1990. Evidence for horizontal transmission of the P transposable element between *Drosophila* species. *Genetics* **124:**339–355.
23. **Davies, D. R., L. M. Braam, W. S. Reznikoff, and I. Rayment.** 1999. The three-dimensional structure of a *Tn5* transposase-related protein determined to 2.9-A resolution. *J. Biol. Chem.* **274:**11904–11913.
24. **De Aguiar, D., and D. L. Hartl.** 1999. Regulatory potential of nonautonomous *mariner* elements and subfamily crosstalk. *Genetica* **107:**79–85.
25. **DeVault, J. D., and S. K. Narang.** 1994. Transposable element in Lepidoptera: hobo-like transposons in *Heliothis virescens* and *Helicoverpa zea*. *Biochem. Biophys. Res. Commun.* **203:**169–175.
26. **Doak, T. G., F. P. Doerder, C. L. Jahn, and G. Herrick.** 1994. A proposed superfamily of transposase-related genes: new members in transposon-like elements of ciliated protozoa and a common "D35E" motif. *Proc. Natl. Acad. Sci. USA* **91:** 942–946.
27. **Eisen, J. A., M.-I. Benito, and W. Walbot.** 1994. Sequence similarity of putative transposases links the maize *Mutator* autonomous element and a group of bacterial insertion sequences. *Nucleic Acids Res.* **22:**2634–2636.
28. **Elick, T. A., C. A. Bauser, and M. J. Fraser.** 1996. Excision of the *piggyBac* transposable element in vitro is a precise event that is enhanced by the expression of its encoded transposase. *Genetica* **98:**33–41.
29. **Engels, W. R., D. M. Johnson-Schlitz, W. B. Eggleston, and J. Sved.** 1990. High-frequency P element loss in Drosophila is homolog dependent. *Cell* **62:**515–525.
30. **Esposito, T., F. Gianfrancesco, A. Ciccodicola, L. Montanini, S. Mumm, M. D'Urso, and A. Forabosco.** 1999. A novel pseudoautosomal human gene encodes a putative protein similar to Ac-like transposases. *Hum. Mol. Genet.* **8:**61–67.
31. **Fayet, O., P. Ramond, P. Poland, M. F. Prère, and M. Chandler.** 1990. Functional similarities between retroviruses and the IS3 family of bacterial insertion sequences? *Mol. Microbiol.* **4:**1771–1777.
32. **Feschotte, C., and C. Mouches.** 2000. Evidence that a family of miniature inverted-repeat transposable elements (MITEs) from the *Arabidopsis thaliana* genome has arisen from a *pogo*-like DNA transposon. *Mol. Biol. Evol.* **17:**730–737.
33. **Garcia-Fernàndez, J., J. R. Bayascas-Ramírez, G. Marfany, A. M. Muñoz-Mármol, A. Casali, J. Baguñà, and E. Saló.** 1995. High copy number of highly similar *mariner*-like transposons in planarian (Platyhelminthes): evidence for a transphyla horizontal transfer. *Mol. Biol. Evol.* **12:**421–431.
34. **Gierl, A.** 1996. The *En/Spm* transposable element of maize. *Curr. Top. Microbiol. Immunol.* **1996:**145–159.
35. **Glayzer, D. C., I. N. Roberts, D. B. Archer, and R. P. Oliver.** 1995. The isolation of *Ant1*, a transposable element from *Aspergillus niger*. *Mol. Gen. Genet.* **249:**432–438.
36. **Gomez-Gomez, E., N. Anaya, M. I. Roncero, and C. Hera.** 1999. *Folyt1*, a new member of the hAT family, is active in the genome of the plant pathogen *Fusarium oxysporum*. *Fungal Genet. Biol.* **27:**67–76.
37. **Grappin, P., C. Audeon, M. C. Chupeau, and M. A. Grandbastien.** 1996. Molecular and functional characterization of *Slide*, an *Ac*-like autonomous transposable element from tobacco. *Mol. Gen. Genet.* **252:**386–397.
38. **Grossman, G. L., A. J. Cornel, C. S. Rafferty, H. M. Robertson, and F. H. Collins.** 1999. *Tsessebe*, *Topi*, and *Tiang:* three distinct *Tc1*-like transposable elements in the malaria vector, *Anopheles gambiae*. *Genetica* **105:**69–80.
39. **Gu, X., and W.-H. Li.** 1995. The size distribution of insertions and deletions in human and rodent pseudogenes suggests the logarithmic gap penalty for sequence alignment. *J. Mol. Evol.* **40:**464–473.
40. **Habu, Y., Y. Hisatomi, and S. Iida.** 1998. Molecular characterization of the mutable flaked allele for flower variegation in the common morning glory. *Plant J.* **16:**371–376.
41. **Halverson, D., M. Baum, J. Stryker, J. Carbon, and L. Clarke.** 1997. A centromere DNA-binding protein from fission yeast affects chromosome segregation and has homology to human CENP-B. *J. Cell Biol.* **136:**487–500.
42. **Handler, A. M., and S. P. Gomez.** 1996. The *hobo* transposable element excises and has related elements in tephritid species. *Genetics* **143:**1339–1347.
43. **Handler, A. M., and S. P. Gomez.** 1997. A new *hobo, Ac, Tam3* transposable element, *hopper,* from *Bactrocera dorsalis* is distantly related to *hobo* and *Ac*. *Gene* **185:**133–135.
44. **Handler, A. M., and S. D. McCombs.** 2000. The *piggyBac* transposon mediates germ-line transformation in the Oriental fruit fly and closely related elements exist in its genome. *Insect Mol. Biol.* **9:**605–612.
45. **Haring, E., S. Hagemann, and W. Pinsker.** 2000. Ancient and recent horizontal invasion of drosophilids by P elements. *J. Mol. Evol.* **51:**577–586.
46. **Hartings, H., C. Spilmont, N. Lazzaroni, V. Rossi, F. Salamini, R. D. Thompson, and M. Motto.** 1991. Molecular analysis of the *Bg-rbg* transposable element system of *Zea mays* L. *Mol. Gen. Genet.* **227:**91–96.
47. **Hartl, D. L., A. R. Lohe, and E. R. Lozovskaya.** 1997. Modern thoughts on an ancyent marinere: function, evolution, regulation. *Annu. Rev. Genet.* **31:**337–358.
48. **Hehl, R., W. K. Nacken, A. Krause, H. Saedler, and H. Sommer.** 1991. Structural analysis of *Tam3*, a transposable element from *Antirrhinum majus,* reveals homologies to the *Ac* element from maize. *Plant Mol. Biol.* **16:**369–371.
49. **Henikoff, A., and J. G. Henikoff.** 1992. Amino acid substitution matrices from protein blocks. *Proc. Natl. Acad. Sci. USA* **89:**10915–10919.
50. **Hershberger, R. J., C. A. Warren, and V. Walbot.** 1991. Mutator activity in maize correlates with the presence and expression of the Mu transposable element Mu9. *Proc. Natl. Acad. Sci. USA* **88:**10198–10202.
51. **Hua-Van, A., F. Héricourt, P. Capy, M.-J. Daboussi, and T. Langin.** 1998. Three highly divergent subfamilies of the *impala* transposable element coexist in the genome of the fungus *Fusarium oxysporum*. *Mol. Gen. Genet.* **259:**354–362.
52. **Hudson, D. F., K. J. Fowler, E. Earle, R. Saffery, P. Kalitsis, H. Torwell, J. Hill, N. G. Wreford, D. M. de Kretzer, M. R. Cancilla, E. Howman, L. Hii, S. M. Cutts, D. V. Irvine, and K. H. Choo.** 1998. Centromere protein B null mice are mitotically and meiotically normal but have lower body and testis weight. *J. Cell Biol.* **141:**309–319.
53. **Izsvák, Z., Z. Ivics, and R. H. Plasterk.** 2000. *Sleeping Beauty,* a wide host-range transposon vector for genetic transformation in vertebrates. *J. Mol. Biol.* **302:**93–102.
54. **Jarvik, T., and K. G. Lark.** 1998. Characterization of *Soymar1*, a *mariner* element in soybean. *Genetics* **149:** 1569–1574.
55. **Jeanmougin, F., J. D. Thompson, M. Gouy, D. G. Higgins, and T. J. Gibson.** 1998. Multiple sequence alignment with Clustal X. *Trends Biochem. Sci.* **23:**403–405.
56. **Jehle, J. A., A. Nickel, J. M. Vlak, and H. Backhaus.** 1998. Horizontal escape of the novel *Tc1*-like lepidopteran transpo-

son *TCp3.2* into *Cydia pomonella* granulovirus. *J. Mol. Evol.* 46:215–224.

57. **Jurka, J.** 2000. Repbase Update: a database and an electronic journal of repetitive elements. *Trends Genet.* **16**:418–420.
58. **Jurka, J., and V. V. Kapitonov.** 1999. Sectorial mutagenesis by transposable elements. *Genetica* **107**:239–248.
59. **Kapitonov, V. V., and J. Jurka.** 1999. Molecular paleontology of transposable elements from *Arabidopsis thaliana. Genetica* **107**:27–37.
60. **Kempken, F., and U. Kück.** 1996. *restless,* an active *Ac*-like transposon from the fungus *Tolypocladium inflatum:* structure, expression, and alternative RNA splicing. *Mol. Cell. Biol.* **16**:6563–6572.
61. **Kempken, F., and U. Kück.** 1998. Transposons in filamentous fungi—facts and perspectives. *BioEssays* **20**:652–659.
62. **Kipling, D., and P. E. Wharburton.** 1997. Centromeres, CENP-B and *Tigger* too. *Trends Genet.* **13**:141–145.
63. **Kishima, Y., S. Yamashita, C. Martin, and T. Mikami.** 1999. Structural conservation of the transposon Tam3 family in *Antirrhinum majus* and estimation of the number of copies able to transpose. *Plant Mol. Biol.* **39**:299–308.
64. **Klobutcher, L. A., and G. Herrick.** 1997. Developmental genome reorganization in ciliated protozoa: the transposon link. *Prog. Nucleic Acid Res. Mol. Biol.* **56**:1–62.
65. **Koga, A., A. Shimada, A. Shima, M. Sakaizumi, H. Tachida, and H. Hori.** 2000. Evidence for recent invasion of the medaka fish genome by the *Tol2* transposable element. *Genetics* **155**:273–281.
66. **Lam, W. L., P. Seo, K. Robison, S. Virk, and W. Gilbert.** 1996. Discovery of amphibian Tc1-like transposon families. *J. Mol. Biol.* **257**:359–366.
67. **Lam, W. L., T.-S. Lee, and W. Gilbert.** 1997. Active transposition in zebrafish. *Proc. Natl. Acad. Sci. USA* **93**: 10870–10875.
68. **Lampe, D. J., K. K. O. Walden, and H. M. Robertson.** 2001. Loss of transposase-DNA interaction may underlie the divergence of *mariner* family transposable elements and the ability of more than one *mariner* to occupy the same genome. *Mol. Biol. Evol.* **18**:954–961.
69. **Le, Q. H., S. Wright, Z. Yu, and T. Bureau.** 2000. Transposon diversity in *Arabidopsis thaliana. Proc. Natl. Acad. Sci. USA* **97**:7376–7381.
70. **Lee, J. K., J. A. Huberman, and J. Hurwitz.** 1997. Purification and characterization of a CENP-B homologue protein that binds to the centromeric K-type repeat DNA of *Schizosaccharomyces pombe. Proc. Natl. Acad. Sci. USA* **94**: 8427–8432.
71. **Lee, S. H., J. B. Clark, and M. G. Kidwell.** 1999. A P element-homologous sequence in the house fly, *Musca domestica. Insect Mol. Biol.* **8**:491–500.
72. **Lin, X., et al.** 1999. Sequence and analysis of chromosome 2 of the plant *Arabidopsis thaliana. Nature* **402**:761–768.
73. **Liu, D., and N. M. Crawford.** 1998. Characterization of the putative transposase mRNA of *Tag1,* which is ubiquitously expressed in *Arabidopsis* and can be induced by *Agrobacterium*-mediated transformation with dTag1 DNA. *Genetics* **149**:693–701.
74. **Lohe, A. R., E. N. Moriyama, D.-A. Lidholm, and D. L. Hartl.** 1995. Horizontal transmission, vertical inactivation, and stochastic loss of *mariner*-like transposable elements. *Mol. Biol. Evol.* **12**:62–72.
75. **MacRae, A. F., G. A. Huttley, and M. T. Clegg.** 1994. Molecular evolutionary characterization of an *Activator* (*Ac*)-like transposable element sequence from pearl millet (*Pennisetum glaucum*) (*Poaceae*). *Genetica* **92**:77–89.
76. **Mao, L., T. C. Wood, Y. Yu, M. A. Budiman, J. Tomkins, S. Woo, M. Sasinowski, G. Presting, D. Frisch, S. Goff, R. A. Dean, and R. A. Wing.** 2000. Rice transposable elements: a survey of 73,000 sequence-tagged-connectors. *Genome Res.* **10**:982–990.
77. **Maurer, P., A. Réjasse, P. Capy, T. Langin, and G. Riba.** 1997. Isolation of the transposable element *hupfer* from the entomopathogenic fungus *Beauveria bassiana* by insertion mutagenesis of the nitrate reductase structural gene. *Mol. Gen. Genet.* **256**:195–202.
78. **Merriman, P. J., C. D. Grimes, J. Ambroziak, D. A. Hackett, P. Skinner, and M. J. Simmons.** 1995. *S* elements: a family of *Tc1*-like transposons in the genome of *Drosophila melanogaster. Genetics* **141**:1425–1438.
79. **Miller, W. J., J. F. McDonald, D. Nouaud, and D. Anxolabéhère.** 1999. Molecular domestication—more than a sporadic episode in evolution. *Genetica* **107**:197–207.
80. **Morgan, G. T.** 1995. Identification in the human genome of mobile elements spread by DNA-mediated transposition. *J. Mol. Biol.* **254**:1–5.
81. **Morita, R., E. Miyazaki, C. Y. Fong, X. N. Chen, J. R. Korenberg, A. V. Delgado-Escueta, and K. Yamakawa.** 1998. JH8, a gene highly homologous to the mouse jerky gene, maps to the region for childhood absence epilepsy on 8q24. *Biochem. Biophys. Res. Communun.* **248**:307–314.
82. **Okuda, M., K. Ikeda, F. Namiki, K. Nishi, and T. Tsuge.** 1998. *Tfo1:* an *Ac*-like transposon from the plant pathogenic fungus *Fusarium oxysporum. Mol. Gen. Genet.* **258**: 599–607.
83. **Perkins, H. D., and A. J. Howells.** 1992. Genomic sequences with homology to the P element of *Drosophila melanogaster* occur in the blowfly *Lucilia cuprina. Proc. Natl. Acad. Sci. USA* **89**:10753–10757.
84. **Petrov, D. A., E. R. Lozovskaya, and D. L. Hartl.** 1996. High intrinsic rate of DNA loss in *Drosophila. Nature* **384**: 346–349.
85. **Petrov, D. A., T. A. Sangster, J. S. Johnston, D. L. Hartl, and K. L. Shaw.** 2000. Evidence for DNA loss as a determinant of genome size. *Science* **287**:1060–1062.
86. **Pinkerton, A. C., S. Whyard, H. A. Mende, C. J. Coates, D. A. O'Brochta, and P. W. Atkinson.** 1999. The Queensland fruit fly, *Bactrocera tryoni,* contains multiple members of the *hAT* family of transposable elements. *Insect Mol. Biol.* **8**: 423–434.
87. **Plasterk, R. H., Z. Izsvák, and Z. Ivics.** 1999. Resident aliens: the Tc1/mariner superfamily of transposable elements. *Trends Genet.* **15**:326–332.
88. **Rea, S., F. Eisenhaber, D. O'Carroll, B. D. Strahl, Z.-W. Sun, M. Schmid, S. Opravil, K. Mechtler, C. P. Ponting, C. D. Allis, and T. Jenuwein.** 2000. Regulation of chromatin structure by site-specific histone H3 methyltransferases. *Nature* **406**: 593–599.
89. **Rhodes, P. R., and L. O. Vodkin.** 1988. Organization of the *Tgm* family of transposable elements in soybean. *Genetics* **120**:597–604.
90. **Robertson, H. M.** 1993. The *mariner* transposable element is widespread in insects. *Nature* **362**:241–245.
91. **Robertson, H. M.** 1995. The *Tc1-mariner* superfamily of transposons in animals. *J. Insect Physiol.* **41**:99–105.
92. **Robertson, H. M.** 1996. Members of the pogo superfamily of DNA-mediated transposons in the human genome. *Mol. Gen. Genet.* **252**:761–766.
93. **Robertson, H. M.** 1997. Multiple *mariner* transposons in flatworms and hydras are related to those of insects. *J. Hered.* **88**:195–201.
94. **Robertson, H. M.** 2000. The large *srh* family of chemoreceptor genes in *Caenorhabditis* nematodes reveals processes of

genome evolution involving large duplications and deletions and intron gains and losses. *Genome Res.* **10**:192–203.

95. **Robertson, H. M., and M. L. Asplund.** 1996. *Bmmar1:* a basal lineage of the *mariner* family of transposable elements in the silkworm moth, *Bombyx mori. Insect Biochem. Mol. Biol.* **26**:945–954.
96. **Robertson, H. M., and D. J. Lampe.** 1995. Recent horizontal transfer of a *mariner* element between Diptera and Neuroptera. *Mol. Biol. Evol.* **12**:850–862.
97. **Robertson, H. M., and E. G. MacLeod.** 1993. Five major subfamilies of *mariner* transposable elements in insects, including the Mediterranean fruit fly, and related arthropods. *Insect Mol. Biol.* **2**:125–139.
98. **Robertson, H. M., and R. Martos.** 1997. Molecular evolution of the second ancient human *mariner* transposon, *Hsmar2,* illustrates patterns of neutral evolution in the human genome lineage. *Gene* **205**:219–228.
99. **Robertson, H. M., F. N. Soto-Adames, K. K. O. Walden, R. M. P. Avancini, and D. J. Lampe.** 1998. The *mariner* transposons of animals: horizontally jumping genes, p. 268–284. *In* M. Syvanen and C. Kado (ed.), *Horizontal Gene Transfer.* Chapman & Hall, London, United Kingdom.
100. **Robertson, H. M., and K. L. Zumpano.** 1997. Molecular evolution of an ancient *mariner* transposon, *Hsmar1,* in the human genome. *Gene* **205**:203–217.
101. **Sedensky, M. M., S. J. Hudson., B. Everson, and P. G. Morgan.** 1994. Identification of a *mariner*-like repetitive sequence in *C. elegans. Nucleic Acids Res.* **22**:1719–1723.
102. **Silva, J. C., and M. G. Kidwell.** 2000. Horizontal transfer and selection in the evolution of P elements. *Mol. Biol. Evol.* **17**:1542–1557.
103. **Smit, A. F. A.** 1999. Interspersed repeats and other mementos of transposable elements in mammalian genomes. *Curr. Opin. Genet. Dev.* **9**:657–663.
104. **Smit, A. F. A., and A. D. Riggs.** 1996. *Tiggers* and other DNA transposon fossils in the human genome. *Proc. Natl. Acad. Sci. USA* **93**:1443–1448.
105. **Snowden, K. C., and C. A. Napoli.** 1998. *Psl:* a novel *Spm*-like transposable element from *Petunia hybrida. Plant J.* **14**: 43–54.
106. **Strimmer, K., and A. von Haeseler.** 1996. Quartet puzzling: a quartet maximum likelihood method for reconstructing tree topologies. *Mol. Biol. Evol.* **13**:964–969.
107. **Swofford, D. L.** 1998. *PAUP*: Phylogenetic Analysis Using Parsimony and Other Methods,* version 4. Sinauer Press, New York, N.Y.
108. **Takahashi, Y., F. Hirose, A. Matsukage, and M. Yamaguchi.** 1999. Identification of three conserved regions in the DREF transcription factors from *Drosophila melanogaster* and *Drosophila virilis. Nucleic Acids Res.* **27**:510–516.
109. **Tarchini, R., P. Biddle, R. Wineland, S. Tingey, and A. Rafalski.** 2000. The complete sequence of 340 kb of DNA around the rice *Adh1-Adh2* region reveals interrupted colinearity with maize chromosome 4. *Plant Cell* **12**:381–391.
110. **Toth, M., J. Grimsby, G. Buzsaki, and G. P. Donovan.** 1995. Epileptic seizures caused by inactivation of a novel gene, *jerky,* related to centromere binding protein-B in transgenic mice. *Nature Genet.* **11**:71–75.
111. **Tudor, M., M. Lobocka, M. Goodell, J. Pettitt, and K. O'Hare.** 1992. The *pogo* transposable element family of *Drosophila melanogaster. Mol. Gen. Genet.* **232**:126–134.
112. **Wang, H. H., M. J. Fraser, and L. C. Cary.** 1989. Transposon mutagenesis of baculoviruses: analysis of TFP3 lepidopteran transposon insertions at the *FP* locus of nuclear polyhedrosis viruses. *Gene* **81**:97–108.
113. **Warren, W. D., P. W. Atkinson, and D. A. O'Brochta.** 1994. The *Hermes* transposable element from the house fly, *Musca domestica,* is a short inverted repeat-type element of the *hobo, Ac,* and *Tam3* (*hAT*) element family. *Genet. Res.* **64**:87–97.
114. **Witherspoon, D. J.** 2000. Natural selection on transposable elements of eukaryotes. Ph.D. thesis. University of Utah, Salt Lake City.
115. **Witherspoon, D. J.** 1999. Selective constraints on P-element evolution. *Mol. Biol. Evol.* **16**:472–478.
116. **Witherspoon, D. J., T. G. Doak, K. R. Williams, A. Seegmiller, J. Seger, and G. Herrick.** 1997. Selection on the protein-coding genes of the TBE1 family of transposable elements in the ciliates *Oxytricha fallax* and *O. trifallax. Mol. Biol. Evol.* **14**:696–706.
117. **Yu, Z., S. I. Wright, and T. E. Bureau.** 2000. *Mutator*-like elements in *Arabidopsis thaliana:* structure, diversity and evolution. *Genetics* **156**:2019–2031.
118. **Zeng, Z., H. Kyaw, K. R. Gakenheimer, M. Augustus, P. Fan, X. Zhang, K. Su, K. C. Carter, and Y. Li.** 1997. Cloning, mapping, and tissue distribution of a human homologue of the mouse *jerky* gene product. *Biochem. Biophys. Res. Commun.* **236**:389–395.

Mobile DNA II
Edited by N. L. Craig et al.

Chapter 49

Origins and Evolution of Retrotransposons

THOMAS H. EICKBUSH AND HARMIT S. MALIK

INTRODUCTION

The analysis of eukaryotic genomes reveals a surprising abundance of repeated sequences formed by the reverse transcription of RNA. Originally attributed to retroviruses, it is now clear that most of these reverse transcribed sequences are the products of endogenous mobile elements. Mobile elements that utilize reverse transcription of RNA templates to make additional copies of themselves are generally referred to as retroelements. The abundance and variety of these retroelements differ strikingly between species. For example, sequence analysis of the yeast genome reveals five families of retroelements that together represent 3% of the genome (75). These yeast elements are located next to tRNA genes or within silenced heterochromatin where they have little effect on the expression of host genes. In contrast, nearly 50 retroelement families occupy almost 20% of the *Drosophila melanogaster* genome and cause more than half of all spontaneous mutations (23). In spite of this apparent high activity, element numbers have remained relatively constant in *D. melanogaster,* due to mechanisms that gradually remove insertions (23, 120). Astonishingly, nearly 35% of our own genome is composed of a single retroelement family and its byproducts (134). This family of elements is relatively inactive; thus, few spontaneous mutations in humans arise from these insertions (73). However, unlike *D. melanogaster,* the human genome does not have the means to efficiently rid itself of insertions, which has resulted in the accumulation of hundreds of thousands of copies in the past 80 million years. This accumulation of mobile elements in humans pales in comparison to what has happened in some plants. More than two-thirds of the maize genome has been estimated to be composed of approximately 20 families of retroelements. Most copies of these elements have accumulated in the last six million years, tripling the size of the genome in this short period (125).

Thomas H. Eickbush • Department of Biology, University of Rochester, Rochester, NY 14627. **Harmit S. Malik** • Fred Hutchinson Cancer Research Center, Seattle, WA 98109.

To date, all retroelements identified in eukaryotic nuclear genomes can be divided into two major classes (4, 157). Elements of the first-described class are flanked by long terminal repeats (LTRs), which play a key role in all aspects of their life cycle. These retroelements synthesize a double-stranded DNA intermediate from their RNA template that is then integrated into chromosomes by a mechanism similar to that used by DNA-mediated mobile elements (transposons). Retroelements of the second-described class are phylogenetically distinct from the first class, lack terminal repeats, and simply reverse transcribe a cDNA copy of their RNA transcript directly onto the chromosomal target site.

One of the confusing aspects faced by those individuals attempting to follow the study of eukaryotic mobile elements is that there is no clear agreement as to what to call these two classes of retroelements. The first retroelements identified were simply called retrotransposons or retroposons to differentiate them from the DNA transposons. However with the discovery of two classes of retroelements that have entirely different integration mechanisms, further descriptors became necessary. Because of the similarity in structures and mechanisms of one class of retroelements to retroviruses, the two classes were originally identified as the viral-like and nonviral retroposons (157). Given the importance of the LTRs in one class and the ease with which these repeats can be identified, the two classes of retroelements have also been called the LTR and the non-LTR retrotransposons (159, 160). Further complicating the nomenclature, the non-LTR (or nonviral) class have also been called LINE-like elements, after the abundant mobile element family in humans (132); the poly(A) retrotransposons, after their tendency to end in poly(A) tails (6, 11); or simply retroposons (142). The LTR and non-LTR retrotransposons nomenclature is most frequently used today and will be used throughout this chapter.

The rate at which new LTR and non-LTR retrotransposons are being discovered has increased dramatically in recent years as a result of the various genome sequencing projects. Indeed, in many of these genome projects, more copies of retrotransposons are being sequenced than all the host genes combined. In this chapter, we summarize their similarities and differences, as well as the phylogenetic relationships of the various elements that have been characterized from each of the two classes of retrotransposons. Using their phylogeny, we then attempt to summarize their evolution, emphasizing recurring themes that provide clues to the evolutionary pressures they encounter. Finally, we discuss what can be inferred about the ages and origins of retroelements and their relationship to other cellular components.

LTR RETROTRANSPOSONS

The first class of retroelements to be discovered in eukaryotes, the LTR retrotransposons, were similar in structure and coding capacity to retroviruses. However, unlike retroviruses, which are limited to vertebrate genomes, LTR retrotransposons proved to be widely distributed in eukaryotes. For example, the very first elements to be studied were the Ty1 elements of *Saccharomyces cerevisiae* (10) and the copia elements of *D. melanogaster* (110). Similar elements were soon found in many taxa from slime molds to plants (20, 152). Given their remarkable similarity in structure to retroviruses, it was immediately assumed that these elements also transposed via an RNA intermediate. Direct demonstration that these elements utilize an RNA intermediate in the formation of a DNA copy for insertion was first shown for the Ty1 element of yeast (10). A nuclear spliceosomal intron was placed within a Ty1 element and the element reintroduced to yeast. When newly transposed copies were characterized, most had precisely lost this intron sequence. Loss of the intron sequence would not be expected via a DNA-mediated reaction; thus, an RNA transcript that had first been spliced must have been used as a template for formation of the new copies. This retrotransposition mechanism was subsequently shown to be similar in most respects to that of retroviruses. The RNA of these retrotransposons could even be found packaged into virus-like particles within cells (11, 58, 131).

Mechanism of LTR Retrotransposition

The mechanism used by the LTR retrotransposons has been extensively reviewed (6, 11, 147; see also chapters 26 and 27) and will only be briefly summarized here to compare it with that used by the non-LTR retrotransposons. The basic steps of retrotransposition are diagrammed in Fig. 1. Under the control of a promoter located within the left LTR, RNA synthesis initiates within the left LTR, and this transcript is terminated with polyadenylation a short distance downstream of this same site in the right LTR of the element. Thus the RNA templates of all LTR retrotransposons contain repeated sequences at their 5′ and 3′ ends. This terminally redundant RNA is used for both translation and as a template for retrotransposition.

The initial step of retrotransposition, formation of a double-stranded DNA intermediate, occurs

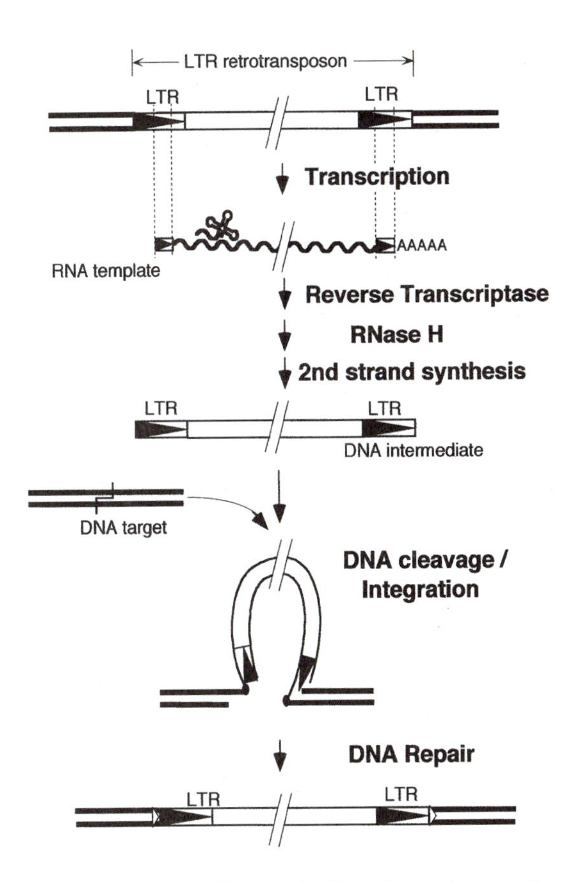

Figure 1. Retrotransposition mechanism of retroviruses and most LTR retrotransposons. Only the basic steps of this reaction are shown for comparison to that of the non-LTR retrotransposition mechanism (Fig. 7). Thick lines, flanking chromosomal sequences; thin lines, element sequences with long terminal repeats (LTRs) indicated as boxed triangles. The RNA template (wavy line) used for protein translation and reverse transcription begins and ends within the LTR sequences. The sequences within these terminal repeats that are used by the reverse transcriptase to template jump between ends are indicated with smaller boxed triangles. First-strand DNA synthesis is primed by a tRNA annealed near the 5′ terminal repeat of the RNA (clover leaf), while second-strand synthesis is primed by an RNase H-resistant RNA near the 3′ terminal repeat (not shown). After formation in the cytoplasm of a linear DNA intermediate, the element-encoded integrase binds to termini of the intermediate, migrates to the nucleus, cleaves a target site on chromosomal DNA, and inserts the DNA intermediate in a reaction similar to that of DNA transposons. Finally, DNA repair fills in the gaps that had been generated by the staggered cut in the chromosome, generating a target-site duplication of uniform length for each element (small open triangles).

within cytoplasmic particles (the virus-like particles). Reverse transcription is primed by a specific host tRNA (often a methionine tRNA) annealed to the RNA template. Typically 15 to 20 nucleotides at the 3′ end of the tRNA base pair to the RNA template immediately downstream of the 5′ LTR sequences. When the reverse transcriptase reaches the 5′ end of the RNA, the presence of identical sequences at the two ends of the RNA template enables the reverse transcriptase to undergo strand transfer (also called a template jump) to the 3′ end of the same or a second RNA template to complete first-strand synthesis. During reverse transcription, the original RNA template, now part of the newly formed RNA:DNA heteroduplex, is removed by an RNase H encoded by the element. Second-strand synthesis (utilizing, in this case, the DNA-dependent DNA polymerase activity of the reverse transcriptase) is primed by resistant oligoribonucleotides that remain annealed to a polypurine stretch of nucleotides immediately upstream of the 3′ LTR sequences. Again, the presence of identical sequences at the 5′ and 3′ ends of the template molecule allows strand transfer and completion of second-strand synthesis. Both first- and second-strand synthesis displace the DNA strand already synthesized in the region of the terminal repeat, resulting in a double-stranded linear DNA intermediate that has a complete LTR sequence at both ends.

In the final steps of retrotransposition, integrase binds to terminal sequences at each end of this DNA intermediate and the large DNA:protein complex, possibly involving a number of cellular proteins, either migrates to the nucleus or waits for the nuclear breakdown during cell division. At the chromosome, the integrase makes a staggered cut in a target site and covalently joins the ends of the linear DNA intermediate. Because of the staggered cut in the target site, DNA repair of the integration product generates a target-site duplication that is the same length for all copies of an element.

Considerable variation has been found in this basic integration mechanism used by LTR retrotransposons. Minor variations include the choice of tRNA primer, the distance this primer anneals downstream of the LTR sequence, and the length of the stagger generated by cleavage of the target site. In addition, while most elements use an intact tRNA, a few LTR retrotransposons use as primer an initiator methionine tRNA half-molecule, generated by cleaving the tRNA in the anticodon stem (8). In a more radical change, some LTR retrotransposons undergo a mechanism of self-priming in which a short stretch of nucleotides at the extreme 5′ end of the template RNA folds back and anneals downstream of the LTR in the region where the tRNA primer binds in other retrotransposons. RNase cleavage releases a 3′ end of the RNA, which is then used to prime reverse transcription (see chapter 28). Finally, in the most unusual group of LTR retrotransposons (the DIRS group described below), complete elements have unusual inverted or split terminal repeats and do not encode an integrase. The mechanism of integration by this rare group of elements is largely unknown (20).

Phylogeny

An updated phylogram of the LTR retrotransposons, based on the amino acid sequence of their common reverse transcriptase (RT) domain, is shown in Fig. 2. The 250- to 275-residue segment used for this phylogeny includes seven blocks of conserved sequences previously found in the fingers and palm region of reverse transcriptases from all retroelements (160), as well as an adjacent C-terminal eighth block corresponding to part of the thumb region (77). The phylogram is based on neighbor-joining distance algorithms and is rooted using non-LTR retrotransposon sequences. The elements used in Fig. 2 are a subset of the published sequences that were selected to represent the diversity of the various groups of LTR retrotransposons, as well as three classes of viruses whose origins appear closely tied to these elements: retroviruses, caulimoviruses, and hepadnaviruses. In a few instances, previously unreported LTR retrotransposon sequences that are available in public databases have been included in the analysis because these sequences expanded the distribution of specific groups of elements beyond what could be concluded based only on published reports.

The RT phylogeny reveals four distinct lineages of LTR retrotransposons. Two of these lineages are the abundant and extensively described Ty1/copia and Ty3/gypsy groups. The remaining two lineages, the DIRS and BEL groups, are not as abundant and have only more recently been documented. The phylogeny in Fig. 2 is essentially the same as that originally reported by us (41, 160), and essentially the same as the more updated phylogenies reported by Bowen and McDonald (12) and others (27, 98, 99, 119). To aid the following discussion, in which the structure and distribution of each group is compared with the vertebrate retroviruses, Fig. 3 shows the open reading frames (ORFs) of representative elements from each of the LTR retrotransposon groups and from the three viruses. (For a complete description of retroviral genomes, see chapter 25.) Two elements are shown for three of the LTR retrotransposon groups, one with and one without an ORF that shares sequence features with retroviral *env* genes. The possible origins of these *env*-like ORFs will be discussed later in this chapter.

Ty1/copia group

This group is named after the first retrotransposons to be identified in yeast and *Drosophila* (10, 58, 110; see also chapter 26). On the basis of the divergence of their reverse transcriptase sequences, this group represents the most ancient lineage of elements. As shown in Fig. 3, Ty1/copia elements contain ORFs similar to the *gag* and *pol* genes of retroviruses. These ORFs are typically in different frames and require frame shifting during translation; however, a few elements (e.g., copia) contain a fusion of the *gag* and *pol* ORFs. The *gag* protein is required for assembly of the RNA transcript into the cytoplasmic particles; the protease (PR) is required for processing of the *gag* and *pol* polyproteins; and the reverse transcriptase (RT), RNase H (RH), and integrase (IN) domains are responsible for formation of the DNA intermediate and its insertion into the genome. A major difference between the organization of these protein domains in the Ty1/copia group and the retroviruses is in the location of the IN domain. The IN domain is immediately upstream of the RT domain in the Ty1/copia group and downstream of the RT/RH domain in retroviruses and all other LTR retrotransposons. A number of highly divergent and poorly resolved Ty1/copia lineages have been found in fungi, insects, and higher plants. There have also been reports identifying Ty1/copia elements in vertebrates, however, no complete reverse transcriptase sequence from a vertebrate element is currently available (54). The Ty1/copia group of elements has recently been termed the *Pseudoviridae,* reflecting both their structural similarity to the *Retroviridae* (retroviruses) and the uncertainty as to whether these elements are true viruses (8).

Ty3/gypsy group

This second major group of retrotransposons is again named after initially characterized elements from yeast and *Drosophila* (62, 101; see also chapters 27 and 28). On the basis of their reverse transcriptase sequences, elements of this group have a significantly closer relationship to the retroviruses and caulimoviruses than they do to elements of the Ty1/copia group. All elements of the Ty3/gypsy group have the same organization of their *gag* and *pol* genes as the retroviruses, including the location of the IN domain. The Ty3/gypsy group has been called the *Metaviridae,* implying their close phylogenetic relationship to the retroviruses while recognizing the uncertainty as to whether they should be considered true viruses (9). Based on all eukaryotic genomic sequences available to date, the Ty3/gypsy elements are the most abundant group of LTR retrotransposons. Like the Ty1/copia group, many lineages have been found in fungi, insect, and plant species. They have also been extensively described in slime molds, sea urchins, nematodes, and vertebrates. Recent comprehensive analyses of elements from this group now include more than 85 distinct lineages from which an entire reverse transcriptase domain has been sequenced (12, 96, 99, 119). Most higher organisms contain multiple lineages of the Ty3/gypsy group. For example, there are

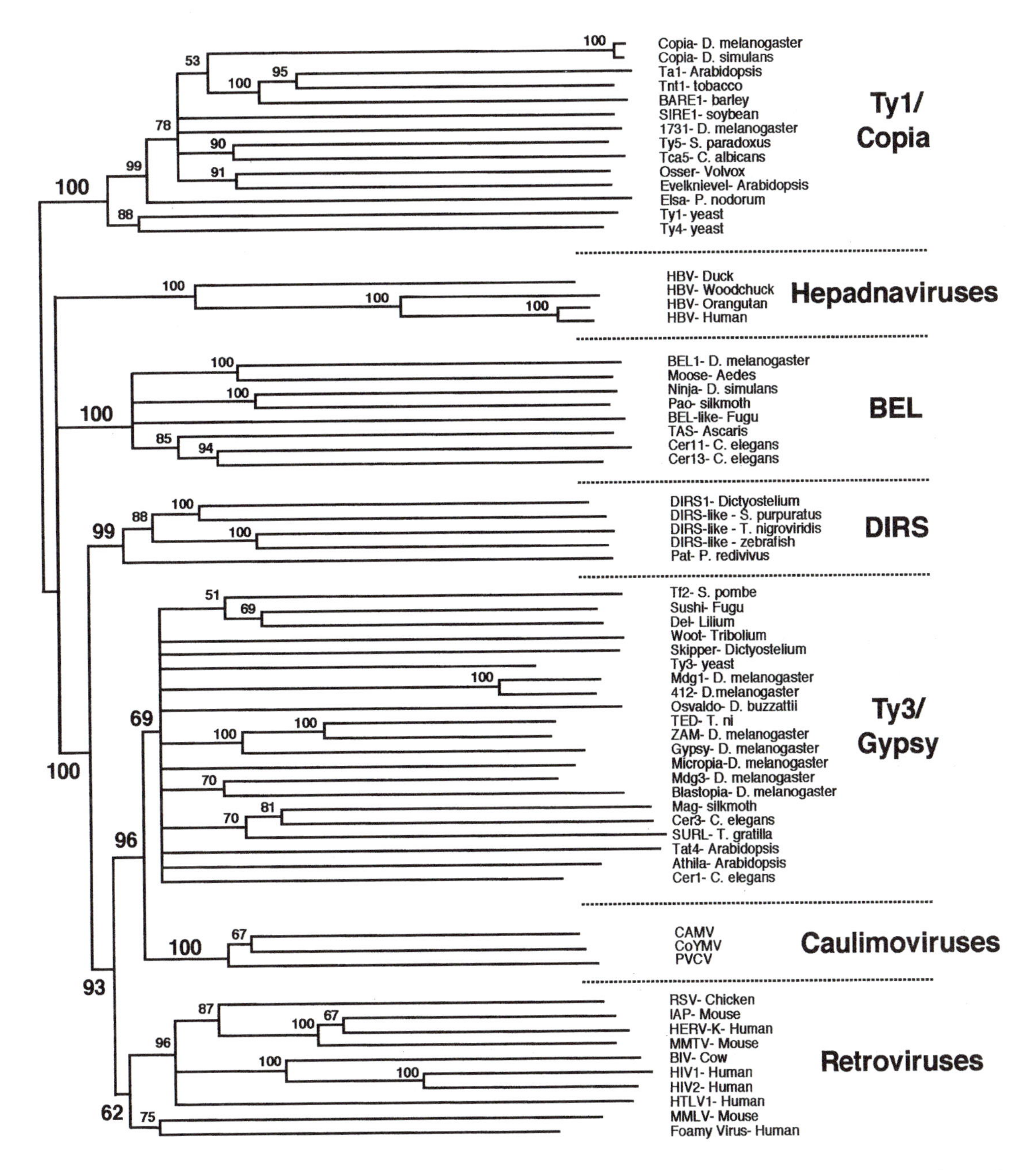

Figure 2. Phylogeny of the LTR retrotransposons and related viruses, based on their reverse transcriptase domain. The portion of the reverse transcriptase domain of each element used in this analysis (approximately 250 residues) includes the seven blocks of conserved residues found in the finger and palm regions of all retroelements (152), as well as an eighth block corresponding to part of the thumb region (93). The phylogram is a 50% consensus tree of the elements based on the neighbor-joining distance algorithms (126) and is rooted using non-LTR retrotransposon sequences (not shown). Numbers adjacent to branch points indicate bootstrap values (percentage from 1,000 trials). The name of each published element is shown at the right of each branch along with either the common name, well-known genus name, or complete species identification of the host in which the element was found. The names of the four major groups of LTR retrotransposons are shown to the right of the figure and correspond to one or more of the first elements identified in the group (see text). Several elements of the DIRS and BEL clade have been obtained directly from sequence databases and have not been named. The phylogeny does not include all available elements in each group; rather, it represents only the diversity of elements known to date. Accession numbers for each element sequence, as well as analyses involving more comprehensive sets of elements, can be found in several previous reports (11, 93, 96, 152).

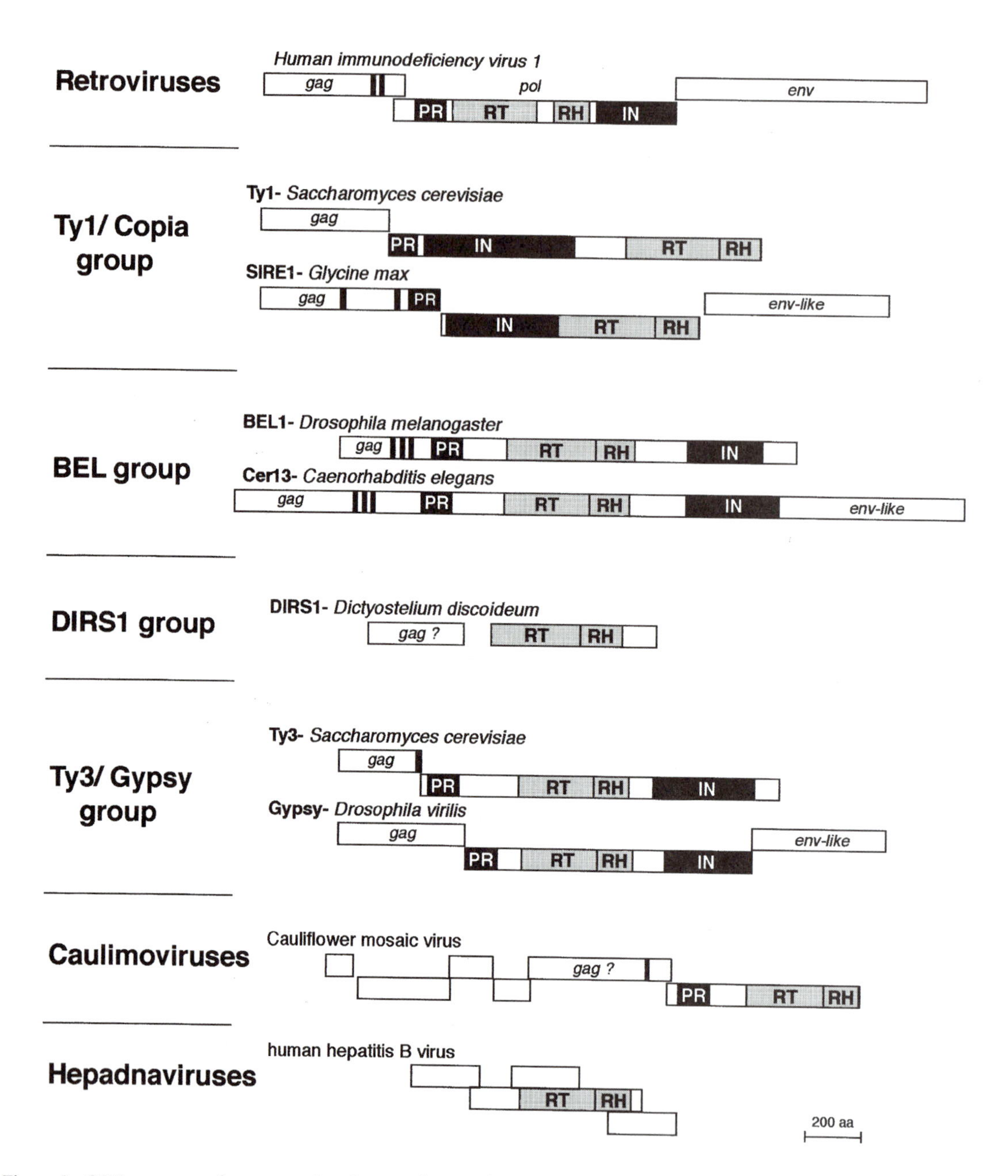

Figure 3. ORF structure of representative elements from each major group of LTR retrotransposons and of related viruses. The multiple ORFs of each element are indicated by the rectangular boxes, with boxes on different levels indicating ORFs in different reading frames or separated by termination codons. The three consensus ORFs found in all retroviruses (*gag, pol, env*) are shown, but additional small ORFs encoded by HIV are not indicated. Shaded regions indicate the enzymatic domains associated with the *pol* gene: RT, reverse transcriptase; RH, RNase H; IN, integrase; PR, protease. Vertical black bars within some *gag* or *gag*-like ORFs represent cysteine-histidine motifs believed to be associated with nucleic-acid binding domains. For the Ty3/copia, BEL, and Ty3/gypsy groups, two elements are shown, one with an *env*-like ORF and one without an *env*-like ORF.

at least 6 distinct families in *Caenorhabditis elegans*, 9 in *Arabidopsis thaliana*, and 20 in *D. melanogaster*. Unfortunately the phylogeny of the Ty3/gypsy group remains little more than a polytomy. At least 20 poorly resolved lineages have been found in a comprehensive survey of these elements, of which only 13 are shown in Fig. 2 (99).

BEL group

The original members of this group, Pao of *Bombyx mori* (161) and TAS of *Ascaris lumbricoides* (50), are both defective and remain poorly described. Therefore this group has been named after an abundant, active element from *D. melanogaster* (33). Elements of the Bel group are abundant in *C. elegans* (12), and a BEL-like element can be identified in the *Fugu rubripes* genome project database (accession no. AF108423). Partial BEL-like elements can be found in the ascidian urochordate *Ciona intestinalis* (accession no. AJ226777), and the blood fluke *Schistosoma mansoni* (accession no. AJ073334), suggesting that the BEL group is widespread in metazoans. However, unlike the Ty1/copia and Ty3/gypsy groups, no element from the BEL group has been identified to date in fungi or plants. On the basis of their reverse transcriptase domain, the BEL lineage is second only to the Ty1/copia lineage in terms of age. Indeed, the bootstrap support for the somewhat closer relationship of the BEL group to the Ty3/gypsy group is not significant. However, the BEL group is the more likely progenitor of the Ty3/gypsy and retroviral groups, on the basis of the location of the BEL IN domain downstream of the RT domain (Fig. 3).

DIRS group

This group has been named after an unusual retrotransposon discovered in *Dictyostelium discoideum* (20). The only other complete element in this group is the PAT element from the nematode *Panagrellus redivivus* (34). Partial elements can be identified in the genomic sequences of one species of sea urchin and two species of fish (accession no. AZ181274, AL305423, and AF112374, respectively). Elements of the DIRS group are unusual in that they do not contain an IN domain (or PR domain), and while they have terminal repeats, these repeats are not the typical LTR arrangement found in the other three groups. For example, DIRS elements contain inverted terminal repeats and an internal sequence that precisely complements the terminal nucleotides (20). PAT elements contain split terminal repeats that are again duplicated at an internal site (34). Priming of reverse transcription in either element appears not to utilize a tRNA molecule. Using features of the retroviral mechanism, Cappello and co-workers (20) have suggested an ingenious model for the reverse transcription of the DIRS elements in which the internal repeat site plays part of the role of LTR. The lack of an integrase domain suggests that once the DNA intermediate is formed, it "recombines" into the host genome. This recombination model for insertion is supported by the finding that five of six DIRS1 elements have inserted into preexisting DIRS1 sequences (19). It will be interesting to see if the DIRS-like elements of sea urchins and fish support this model, or provide insights into possible alternate mechanisms of insertion.

Caulimoviruses

On the basis of their reverse transcriptase sequences, these exclusively plant viruses are closely related to the Ty3/gypsy group of LTR retrotransposons. The predominant extracellular form of the caulimoviruses is DNA, which does not require integration into the host genome for completion of the viral life cycle. Of the variety of genes encoded by these viruses, two appear similar to the *gag* and *pol* genes of LTR retrotransposons. The *pol* gene, however, does not contain an IN domain. Formation of the DNA genome from a terminally redundant RNA template occurs by a mechanism that is similar to that of LTR retrotransposons (121). A specific host tRNA primes first-strand synthesis near the 5′ end of the RNA template and strand-transfers to the 3′ end of the same RNA template. Second-strand synthesis is primed by means of a polypurine tract upstream of the 3′ end of the RNA. However, unlike LTR retrotransposons, neither first- nor second-strand synthesis of the CaMV genome displaces the DNA strand already synthesized in the region of the terminal repeat. The result is a double-stranded circular DNA molecule with one repeat sequence rather than the linear DNA molecule with complete terminal repeats on each end as generated in retroviruses and LTR retrotransposons. Recently it has been noted that some viruses, e.g., the Petunia vein clearing virus, might contain an integrase domain (122). While integrated copies of caulimoviruses have been detected in some species, it is unclear whether these integrations are specific events or simply reflect the ability of plant host genomes to incorporate exogenous DNA (88).

Hepadnaviruses

While their reverse transcriptase domains suggest that the hepadnaviruses evolved early in the evolution of the LTR retrotransposons, their phylogenetic range is presently limited to vertebrates. Like caulimoviruses, the extracellular form of the hepadnaviruses is DNA and integration is not a required component of the life cycle. The only hepadnavirus encoded proteins with similarity to retrotransposons are the RT and RH domains. Several aspects of hepadnavirus replication are similar to the first steps of LTR retrotransposition (57). The RNA template is terminally redundant, and both first- and second-strand synthesis require strand transfers made possible by these terminal repeats. However, priming of reverse transcription is an entirely different mechanism, which involves a hydroxyl group of the RT protein itself (153). The primer for second-strand synthesis is a purine-rich oligomer of template RNA sequence located near the 3′ repeat sequences.

Rearrangement of Structural Domains

The ORF structures of the LTR retrotransposons and viruses diagrammed in Fig. 3 show considerable variation in coding capacity and in organization. This variation is most extreme in the gain and loss of domains associated with the viruses, but flexibility is also evident between retrotransposon groups. The IN domain is located upstream of the RT domain in the Ty1/copia group but downstream in the other groups. Assuming that the DIRS group evolved from a BEL-like retrotransposon, then these elements have lost both their IN and PR domains.

Given the ability of these elements to gain, lose, or shuffle the position of these enzymatic domains, it is possible that protein domains may also have been exchanged between lineages. If this were true, the phylogram in Fig. 2, which only reflects the history of the RT domains, would not accurately reflect the phylogeny of the elements themselves. Fortunately all LTR retrotransposons and the three related viruses contain an ~140-residue RNase H (RH) domain, which can also be used to determine the phylogeny of these elements (97). A phylogram of the RH sequences rooted using cellular RNase H sequences is shown in Fig. 4. Comparison of Fig. 2 and 4 reveals a striking difference. Retroviruses are located distal to the four retrotransposon groups, as well as to caulimoviruses and hepadnaviruses in the RH phylogeny, while retroviruses are a sister group to the Ty3/gypsy elements and caulimoviruses in the RT phylogeny (Fig. 2). The RH domains of retroviruses and these other elements are believed to have identical functions. It is therefore unlikely that the RH domain is under different selective constraints in retroviruses, which would result in its rapid sequence change and "misplacement" on the tree. It seems more likely that the difference between the two phylograms is due to the RH domain in the lineage leading to the retroviruses having been replaced by an RH from an outside source. Consistent with this hypothesis, retroviruses are the only elements that contain an additional "spacer or hinge" region between the RT and RH catalytic domains (Fig. 3). This extra segment could represent remnants of the original RH domain or flanking segments of the introduced sequences. The possible origin of the RH domains of LTR retrotransposons and related viruses will be discussed at the end of this chapter.

Other than the location of the retroviral branch and minor differences in the poorly resolved Ty1/copia, BEL, and hepadnavirus branches, the RH phylogram is consistent with that of the RT phylogram. Most importantly, all elements identified within a retrotransposon group on the basis of their RT domain are also located in the same group on the basis of their RH domain. This suggests that there has been no independent exchange of either the RT or RH domains between groups. The similarity of the two phylogenies within each group further suggests that there has been no exchange within a group of retrotransposons. Unfortunately, the lower resolution of the phylogenies within each group means that for this latter conclusion, we can only exclude recent exchanges.

Other protein sequences that might help define the phylogenetic relationships of the LTR retrotransposons and related viruses, namely, the *gag* gene and the PR and IN domains, are too poorly conserved to enable one to build a complete phylogeny (21). This inability to trace the phylogeny of the IN domain means we are unable to distinguish whether the different location of the IN domain in the Ty1/copia group represents a shift in the position of this domain or represents the independent acquisition of an IN domain (see reference 21 for a more extensive discussion of attempts to determine the phylogeny of IN domains). In spite of our inability to build a complete phylogeny, a comparison of the IN sequences as shown in Fig. 5 reveals common features that are limited to the members of a particular group of retrotransposons or common to groups of elements, which greatly supports the phylogeny of these elements on the basis of their RT domains.

The 150- to 200-residue N-terminal region of all IN domains contains a series of conserved residues (Fig. 5). These include H-X_5-H and C-X_4-C motifs, believed to be involved in DNA binding (74), and a set of three acidic residues (D or E) that are part of the catalytic domain (the D..D35E found in many transposases) (36, 49, 80). The spacing of these conserved motifs is similar in the Ty3/gypsy group and retroviruses, while quite different in the BEL and Ty1/copia groups. C-terminal to these conserved motifs, each retrotransposon group contains an additional segment shared by members of the same group but not by other groups. For example, in the Ty1/copia group, this C-terminal extension is from 250 to 500 residues in length, due to a variable internal segment. In the case of the BEL group, all elements contain a similar C-terminal domain of approximately 200 residues. The C-terminal BEL domain has no significant sequence similarity with the IN domains of either the Ty1/copia or Ty3/gypsy groups.

Elements of the Ty3/gypsy group, on the other hand, share a 75-residue extension with the retroviruses that is not found in either of the other two retrotransposon groups. This common segment, which is also believed to be involved in DNA binding (45), strongly supports the close relationship of the Ty3/gypsy group and retroviruses that was suggested by the RT phylogeny (Fig. 2). In addition to this 75-resi-

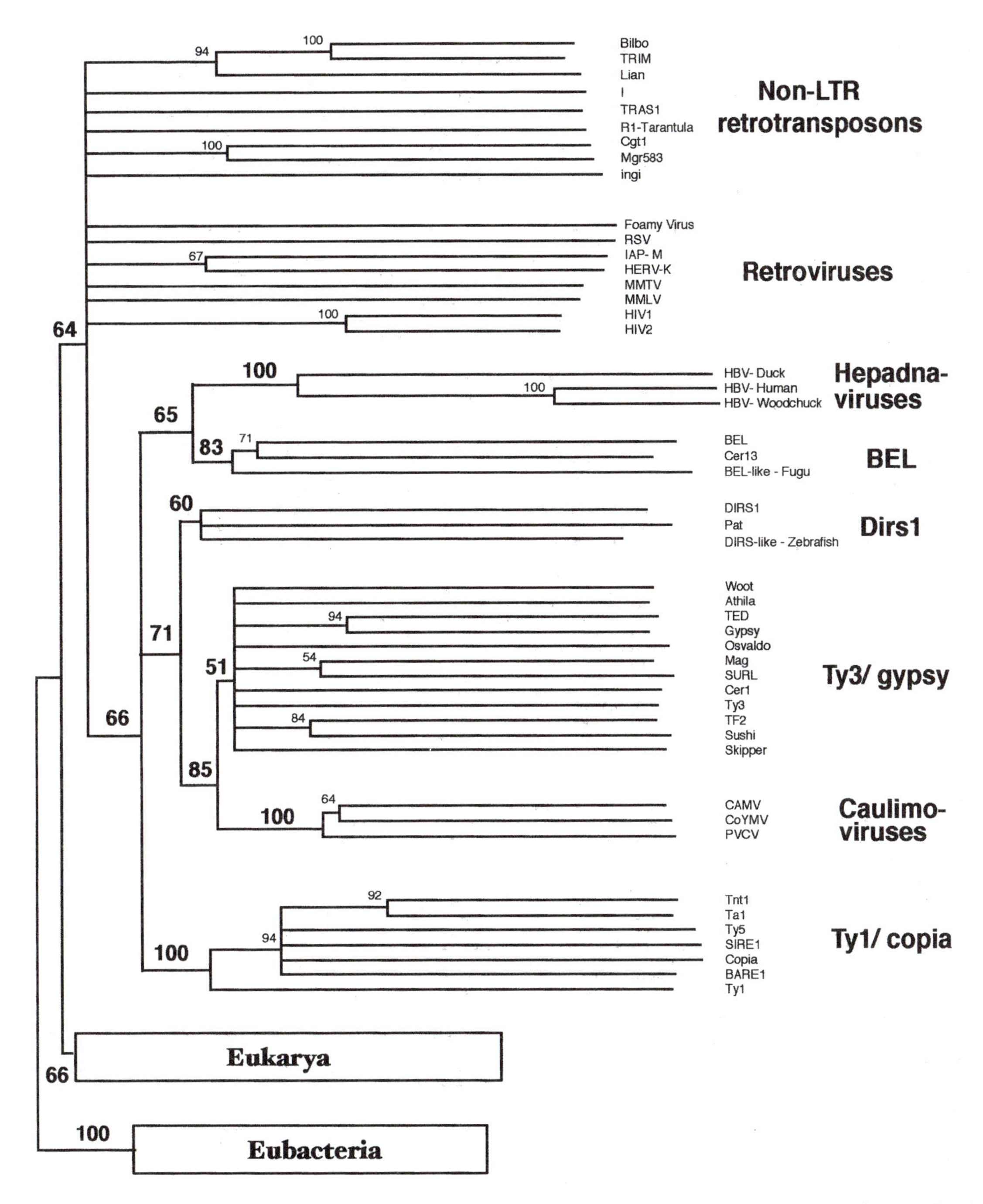

Figure 4. Phylogeny of the LTR retrotransposons and related viruses based on their RH domain. The RH domain of each element is approximately 140 residues in length (90, 93). The phylogram is based on the neighbor-joining distance algorithms and is rooted using cellular RH sequences of eubacteria. Seven eukaryotic and seven eubacterial sequences selected to represent the diversity in each kingdom were used in the analysis, but individual branches have not been shown to conserve space. Also to conserve space, only a subset of the LTR retrotransposons used in Fig. 2 is presented. The species used and accession numbers can be found in reference 91. Numbers adjacent to branch points indicate bootstrap values (percentage from 1,000 trials). Also included in the phylogram are the RH domains found in non-LTR retrotransposons of the I group (Fig. 8 and 9). Only the name of each published element is shown at the right of each branch, since the host species can be found in either Fig. 2 or 9.

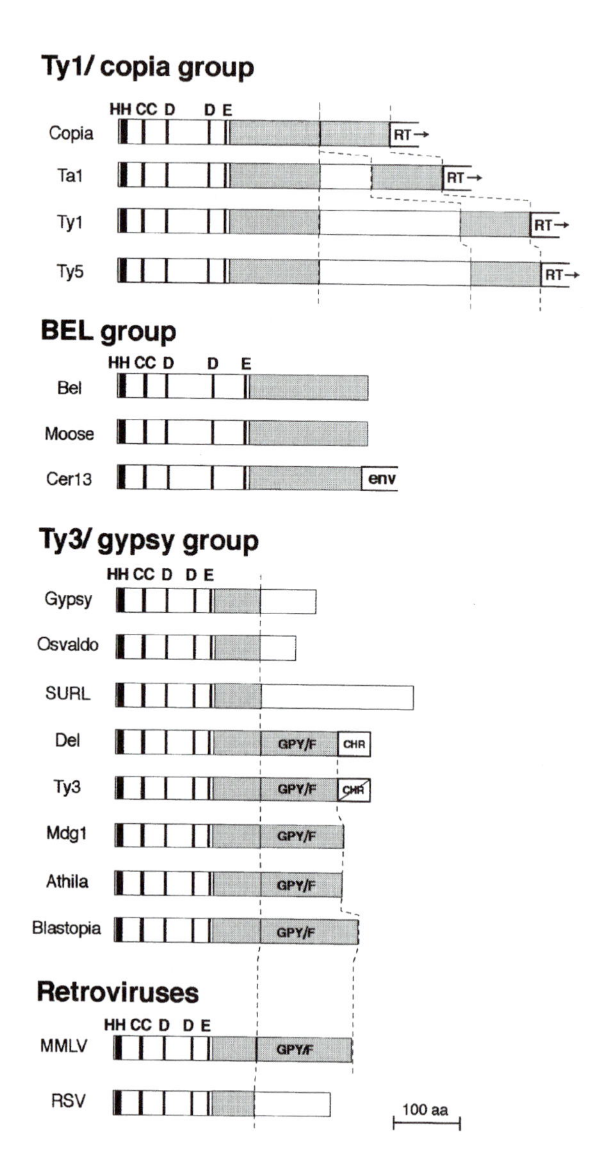

Figure 5. Comparison of the integrase (IN) domain of LTR retrotransposons. Representative elements were selected to show the diversity of structure in each group. Members of the DIRS, hepadnavirus, and caulimovirus groups of elements shown in Fig. 2 do not appear to contain an IN domain (Fig. 3). The IN domain represents the end of the *pol* ORF in most elements; however, it is followed by the RT domain in the Ty1/copia group and by an *env*-like domain in Cer13. All IN domains contain a cysteine-histidine motif (HH-CC) at the N-terminal end and three conserved acidic residues (aspartic [D] or glutamic [E]), which form the active site of the catalytic domain. Spacing between these acidic residues is similar in retrovirus and gypsy/Ty3 groups but is different in the BEL and Ty1/copia groups. In the C-terminal direction from these common IN regions, the proteins differ between groups and even among elements of the same group. Regions with sequence identity between members of the same group are shaded gray. C-terminal protein regions that are poorly conserved within a group are not shaded. In the case of the Ty3/gypsy and retrovirus groups, sequence similarity between groups can also be detected. This similarity includes a short region found in all elements and a longer region (identified as GPY/F) found in only some elements of each group (93). The C-terminal end of some members of the Ty3/gypsy group contains a chromodomain or, in the case of Ty3, a modified chromodomain (93).

due segment, approximately half of all Ty3/gypsy elements and the MMLV lineage of retroviruses share another segment extending an additional 120 residues. We have called this segment the GPY/F domain (95). The GPY/F domain also appears to have been present in the common ancestor of the Ty3/gypsy group and retroviruses, but it has subsequently been lost or has undergone extensive sequence change in many of the individual lineages. Finally, a few members of the Ty3/gypsy group contain yet another C-terminal extension. This domain contains sequence similarity to a chromodomain (95), and thus may be involved in the association of the IN protein with chromosomal proteins and the selection of potential target sites.

In summary, LTR retrotransposons have a flexible structure that has permitted the gain, loss, and possible rearrangement of their protein domains. Perhaps the most unexpected finding is the acquisition of a new RH domain by the retroviral branch. However, in spite of this flexibility, there are no data to suggest the exchange of protein domains between groups of elements. As a result, the RT phylogram in Fig. 2 continues to represent our best estimate of the relative age and phylogenetic relationships of the LTR retrotransposons.

The Propensity To Form Viruses

One of the most interesting aspects of the evolution of the LTR retrotransposons is their close relationship to retroviruses, caulimoviruses, and hepadnaviruses. How did these viruses evolve? The origin of caulimoviruses and hepadnaviruses involved the acquisition of multiple new domains. Koonin and coworkers (78) have suggested that caulimoviruses arose from the fusion of an LTR retrotransposon with a single-stranded RNA virus. This hypothesis is based on the finding that the cell-to-cell movement proteins of the caulimoviruses are both functionally and phylogenetically related to those from a number of other plant viruses. We are unaware of any similar evidence pertaining to the origin of the hepadnaviruses. Indeed, the antiquity of the hepadnaviruses and the long branch length leading to these elements in the RT and RH phylograms (indicating extensive sequence change) may make it impossible to ever trace their origin.

An origin of the vertebrate retroviruses is much easier to propose, since one need only suggest that a member of the Ty3/gypsy group acquired a third ORF, the present-day *env* gene. The *env* gene encodes a transmembrane host receptor-binding protein that is responsible for transmission of the virus (26). What is the origin of this third ORF? Given that current retroviruses have readily incorporated host genes during their evolution (e.g., see reference 141), one possibility is that the *env* gene was acquired from a host gene encoding a normal receptor-binding or membrane fusion protein. A second possibility is that a retrotransposition event resulted in the fusion of two protein domains and the de novo formation of the *env* gene. Finally an LTR retrotransposon may have acquired an *env*-like gene from a virus and usurped the already evolved machinery of the latter for its own ends. Unfortunately, little sequence similarity is found between retroviral *env* genes and any other protein or even between *env* genes from different retroviral lineages. This inability to trace the origin of the *env* gene of retroviruses is likely the result of the strong selective pressure exerted on the *env* proteins to diverge because they represent antigenic sites that elicit vertebrate immune responses.

Surprisingly a survey of LTR retrotransposons reveals at least six additional lineages within either the Ty3/gypsy, BEL, or Ty1/copia groups that have also acquired *env*-like ORFs (see Fig. 6, as well as the structure of three examples in Fig. 3). The various structural features reminiscent of retroviral *env* genes include leader peptides, N-glycosylation sites, and transmembrane regions. The best-studied example of an LTR retrotransposon with an expressed *env*-like domain is the gypsy element of *D. melanogaster* (101). A large number of insect retrotransposons closely related to gypsy have been discovered that also contain a third *env*-like ORF (17.6, 297, Tom, Tv1, Indefix, TED, ZAM, Tirant, and Yoyo) (35). The phylogeny of the *env* genes encoded by these many gypsy-like elements is similar to that of their RT domains, suggesting this acquisition was a single event in their common ancestor (98).

In addition to this gypsy lineage, two other lineages within the Ty3/gypsy group appear to have independently acquired an *env*-like domain. These include Osvaldo from *Drosophila buzzatii* (117) and a related group of elements from higher plants, including Athila, Cyclops-2, and Calypso (158). A fourth member of the Ty3/gypsy group, Cer1 of *C. elegans*, contains a large extension downstream of its IN domain (14); however, this Cer1 domain does not have the predicted properties of an *env* gene. In the case of the BEL group, two lineages of elements contain an *env*-like ORF: Cer7 (and related CER13 and CER14) from *C. elegans* (12) and Tas from *A. lumbricoides* (50). Finally, within the Ty1/copia group, SIRE-1 elements from soybeans, as well as the related Endovir1, ToRTL1, and Opie-2 elements of other higher plants, have been shown to encode a third *env*-like ORF (83, 119).

Despite the many examples of an *env*-like ORF encoded by LTR retrotransposons, only in the case of gypsy have virus-like particles generated by the elements been shown to be infective (135). Consistent with the model that gypsy is an authentic virus that cannot only travel from organism to organism but also from species to species, the sequence relationship of gypsy elements isolated from the eight members of the melanogaster species subgroup bears no relation-

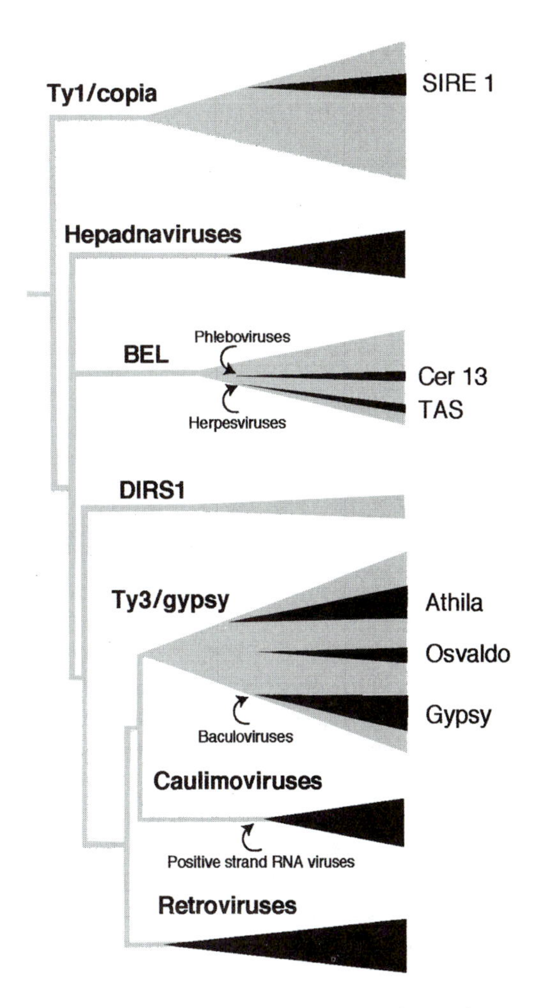

Figure 6. Schematic of LTR retrotransposon evolution showing the nine instances in which the lineage has given rise to viruses or potential viruses. The figure shows only a summary of the phylogenetic relationships derived from the reverse transcriptase sequences of the seven groups of elements (the detailed phylogeny is shown in Fig. 2). The length and height of each triangle represent, respectively, the presumed age (based on the diversity) and abundance of each group. The nine instances where an LTR-retrotransposon lineage (gray lines and triangles) can be proposed to have assumed viral-like properties are indicated in black. In only four instances (hepadnaviruses, caulimoviruses, retroviruses, and gypsy elements) have the elements directly been shown to be viruses. In the five other instances, the lineage is assumed to have viral-like properties because of the acquisition of an *env*-like ORF. In four cases, the evolutionary origin of the protein that enables (or presumably enables) the element to leave a cell has been traced to the indicated viral source. Adapted from reference 95.

ship to the species phylogeny, and nearly identical elements were found in species separated by 10 to 12 million years (143). In a similar manner, gypsy elements from the *obscura, hydei,* and *virilis* species group of *Drosophila* are nearly identical in sequence, even though these species groups have been separated by more than 40 million years (148).

The *env*-like ORFs found in the six lineages within the Ty3/gypsy, BEL, and Ty1/copia groups represent acquisition events that have occurred more recently than the origin of hepadnaviruses, retroviruses, or even caulimoviruses. It should therefore be easier to identify the possible origins of these *env*-like sequences, and indeed the probable origins for three of these ORFs have recently been suggested (98).

In the case of the gypsy-like elements, alignment of the *env* ORF from 10 elements has revealed a highly conserved 26-residue protein segment (35, 86). When this segment was used to screen the databases of protein sequences, the highest matches were found to an ORF of a related group of baculoviruses. When the entire sequence of this baculoviral protein was compared with the gypsy *env* proteins, sequence similarity was identified throughout. Equally important, the baculoviral protein contained N-terminal signal peptide and C-terminal transmembrane domains and had previously been identified as the protein likely responsible for the infectious ability of these baculoviruses (81). The similarity of sequence, structure, and probable function of these proteins suggested that either the baculoviruses acquired their protein from a gypsy-like retrotransposon or the gypsy elements acquired their *env* from a baculovirus. While there is insufficient phylogenetic data to resolve between these two possibilities, we prefer the latter, since baculoviruses were presumably always infectious agents. It is satisfying to trace the *env* gene of gypsy to a baculovirus because the gypsy elements and baculoviruses have the same host range (insects). In addition, one of these gypsy-like elements, TED, has been found inserted within a baculoviral genome, suggesting a likely scenario for the acquisition of the *env* gene (55).

The origin of the *env*-like ORF of the Cer7 and 13 elements of *C. elegans* may also have been traced (12). Screening protein databases with the Cer *env*-like protein revealed strong sequence similarity to the G2 glycoprotein of the phleboviruses, a group of viruses belonging to the bunyaviral genus of single-stranded RNA viruses (3). Again, sequence similarity between the *env*-like gene of the LTR retrotransposon and the viral protein spanned the entire protein, including the conservation of 19 cysteine residues and a predicted transmembrane domain near the C-terminal end. Remarkably, even the manner in which this protein is expressed may be the same in these two elements. In the phleboviruses, the G2 protein is processed from a single polypeptide that also encodes the nonstructural protein G1. Likewise the *env* gene in Cer13, the most intact member of the *C. elegans* elements, is in the same frame as the rest of the *pol* domains. The predicted proteolytic cleavage sites of the G2 and Cer13 *env* proteins coincide.

Finally, in the case of the TAS element of *A. lumbricoides* (50), a search of the protein databases with its *env*-like ORF revealed significant matches to the gB glycoproteins from the *Herpesviridae,* a class of double-stranded DNA viruses. The gB glycoprotein represents nearly 50% of the protein mass of the envelope of these viruses and has been implicated in viral attachment and cell-membrane fusion (13). The region of similarity between the TAS *env* and gB protein includes only part of the protein, but it is highly significant that the regions of sequence similarity correspond to segments of the gB protein that are believed to be responsible for the viral attachment to the cell surface (13). Once again, transmembrane domains reside at the C-terminal end of both proteins.

Given that the reverse transcriptase domain most accurately reflects the history of the LTR retrotransposons, then as summarized in Fig. 6, these elements have acquired the proteins necessary to become viruses on at least nine separate occasions. In two instances—hepadnaviruses and caulimoviruses—there was a major change in the element lifestyle (elimination of the requirement for integration). In the remaining seven cases there was a simple acquisition of a third ORF that has presumably given the element the ability to leave a cell. The formation of these viruses (or putative viruses in those cases where viral transmission has not been documented) has occurred twice each in vertebrates, insects, and nematodes, and three times in plants. The acquisition of an *env*-like ORF to become a virus clearly represents a general paradigm under which the LTR retrotransposons evolve. In all instances, the positioning of the new *env*-like genes downstream of the *gag* and *pol* genes would not disrupt the expression of the critical structural and enzymatic components for retrotransposition. The mechanism for the acquisition of an *env* gene is unclear, but the viral infection of a host cell that is also undergoing retrotransposition of an active mobile element could presumably lead to illegitimate recombination events. The immediate selective advantage to a mobile element of being able to occasionally leave its host's genome and spread to other species suggests that many more examples will be found as the genomes of more diverse organisms are studied.

Mode of Evolution

What is the mode of evolution of the retrotransposons that have not gained an *env* gene? Are they restricted to purely vertical descent through the germ line of the host? One fascinating aspect of the phylogeny of the true retrotransposons (i.e., those without an *env*-like gene) that can be seen in Fig. 2 is that elements from the same organism or group of organisms frequently do not form distinct lineages. Indeed, elements present in the same species are frequently no more related to each other than elements from highly divergent species. For example, within the Ty1/copia group, Ta1 of *A. thaliana* is in the same lineage as copia from *D. melanogaster,* while Evelkneivel elements of *A. thaliana* are most closely related to the Volvox element Osser. Both lineages are distinct from many other elements of higher plants (e.g., SIRE-1 of soybean). In the Ty3/gypsy group, Tf2, Sushi, and Del (from a fungus, vertebrate, and plant, respectively) represent one lineage, while Mag, Cer3, and SURL (from an insect, nematode, and sea urchin) are a second lineage. On the basis of the data available to date, there are at least 9 such deep lineages of Ty1/copia elements, while the more abundant Ty3/gypsy group has more than 20 lineages.

There are two alternative interpretations for these many lineages. First, if the elements are only transmitted vertically, then these groups of retrotransposons have propagated multiple distinct lineages since the divergence of plants and animals. Alternatively, if LTR retrotransposons are capable of horizontal transfers between highly divergent species, then the presence of deep lineages within the same species could simply have arisen by more recent horizontal events. Direct evidence to support the horizontal transfer model for the transmission of mobile elements is obtained when elements in two species have a level of sequence divergence that is dramatically less than that expected by vertical descent (31, 65, 123).

Strikingly clear examples of the horizontal transfer of the DNA-mediated transposon, mariner, have been obtained (123; see also chapter 48). While the data are not as spectacular as those for mariner, evidence is accumulating that LTR retrotransposons have undergone horizontal transfers. The most extensive attempt to trace the evolutionary history of an LTR retrotransposon has been conducted with the SURL elements (Ty3/gypsy group) of echinoid species (sea urchin, sand dollar, heart urchin) (60, 136). The 260 million-year-old echinoid phylogeny has been extensively characterized with most dates for the separation of the major lineages supported by the fossil record. Analysis of SURL elements from 40 species has indicated that the predominant means of evolution of SURL is vertical, with the elements evolving at a rate that averages only 50% faster than that of nuclear genes (60). In six cases, however, strong evidence for the horizontal transmission of SURL elements was obtained. In these instances, nucleotide divergence by vertical descent would require that the elements evolved at rates more than 10-fold slower than nuclear genes. In three additional cases, identical SURL elements were discovered in species separated by 3 to 9 million years, indicating recent horizontal transfers.

A clear example of the horizontal transmission of an LTR retrotransposon has also been found in the *Drosophila* genus. Jordan and coworkers (71) reported the sequences of copia elements that exhibited less than 1% nucleotide divergence in two species, *D. melanogaster* and *Drosophila willistoni,* which diverged more than 40 million years ago. Because copia is abundant in all species closely related to *D. melanogaster* but is rare in species closely related to *D. willistoni,* the direction of the transfer from *melanogaster* to *willistoni* seems clear. Thus, while LTR retrotransposons may not be as prolific at horizontal transfers as the DNA transposons, they clearly do have the ability to jump between species. Therefore, the presence of a single lineage of LTR retrotransposon in two species does not mean that this lineage was present in the common ancestor of both species. Consequently, it is difficult to estimate the age of this group of mobile elements, but we will make such an estimate at the end of this chapter.

NON-LTR RETROTRANSPOSONS

The second major class of eukaryotic mobile elements that utilize reverse transcriptase is the non-LTR retrotransposons [also called LINE-like elements, poly(A) retrotransposons, and retroposons]. While these elements are abundant in all lineages of eukaryotes, for many years they were not recognized as a distinct class of self-propagating mobile elements. Their delayed recognition was a result of their sloppy means of integration and their absence in *S. cerevisiae* (one of the very few eukaryotic species without these elements). The presence of a poly(A) tail at the 3′ end of many copies clearly suggested that these elements were derived by reverse transcription of an RNA template. However, the initial sequences of these elements from the human genome did not reveal intact ORFs with similarity to known proteins. The absence of LTRs or inverted terminal repeats associated with previously known retrotransposons and transposons also appeared to confirm that these insertions were probably not autonomous elements. Thus these sequences were originally grouped with processed pseudogenes and short interspersed nucleotide elements (SINEs) and called retroposons (124) or long interspersed nucleotide elements (LINEs) (132).

The integration of these non-LTR retrotransposons was interpreted as a product of the enzymatic activity supplied in *trans* by the LTR retrotransposons (124). It was not until the LINE sequences from primates and mouse revealed ORFs with similarity to reverse transcriptase (66, 89) that these elements were postulated to be a new type of autonomous mobile element. Sequences of similar elements from insects, plants, and trypanosomes soon followed, revealing that these elements were widespread (16, 48, 76, 127). Employing the same approach that had originally been used to demonstrate retrotransposition of the Ty1 element of yeast (10), it was shown that insertion of new copies of both the I element of *D. melanogaster* and the LINE1 (shortened to L1) element of mammals resulted in the accurate removal of an inserted intron (70, 118). These findings left little doubt that these non-LTR retrotransposons were autonomous and used an RNA intermediate.

Target-Primed Reverse Transcription (TPRT)

Early models of non-LTR retrotransposon integration needed to explain three unusual features of these elements. First, these elements did not encode an integrase (IN) domain like that of the LTR retrotransposons or the DNA transposons. Second, different copies of a non-LTR retrotransposon had the same 3′ ends but were frequently truncated at their 5′ ends, suggesting the 5′ end of each element had no role in the insertion process. Third, the target-site duplications generated by different copies of the same element were frequently of different lengths, suggesting either that the elements did not encode their own endonuclease or that their encoded endonuclease was imprecise in its cleavage of the two DNA strands.

On the basis of these insertion features, the original models of non-LTR retrotransposition suggested that the reverse transcriptase encoded by the element used preexisting breaks on the chromosome as primers to polymerize a cDNA copy of the RNA transcript directly onto the DNA break (15, 53, 127, 151). Our current model for non-LTR retrotransposition is shown in Fig. 7 and retains many aspects of these original models. Direct support for this model is based mainly on in vitro studies of R2 retrotransposition (90–92, 163), but its general features are supported by in vivo assays of L1 and I retrotransposition (24, 47, 109; also see chapters 33 to 35 for more detailed descriptions of these models).

RNA synthesis of most non-LTR retrotransposons is probably controlled by an internal promoter that initiates upstream at the 5′ end of the element (106, 137). Other elements may depend upon their chance cotranscription by an upstream cellular pro-

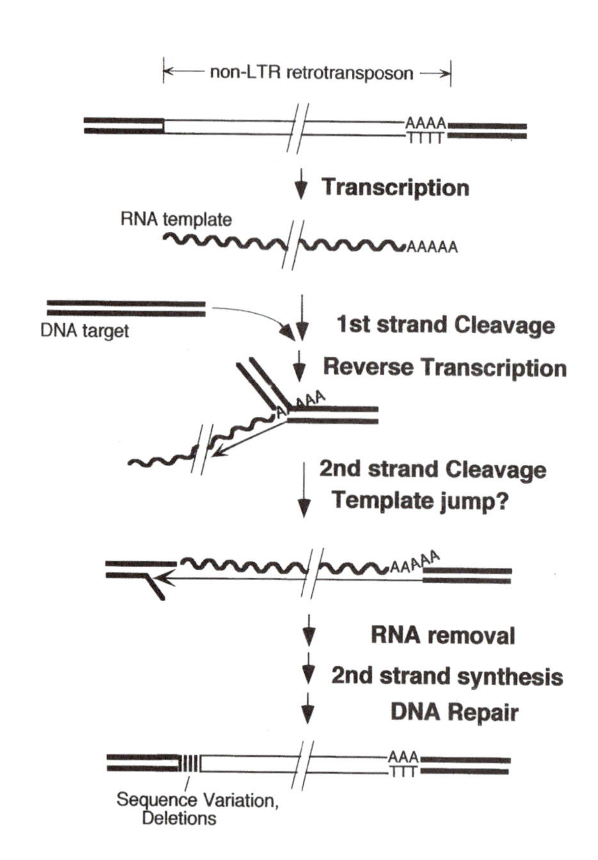

Figure 7. Target-primed reverse transcription (TPRT) model for the integration of non-LTR retrotransposons. Most elements end in an A-rich or poly(A) 3′ end. Transcription is usually mediated by an internal promoter, which initiates RNA synthesis upstream of itself. However, some elements appear to be expressed as cotranscripts with host genes. Most of the information concerning the TPRT reaction is based on in vitro studies with the protein encoded by the R2 element (87). This TPRT accounts for many of the integration properties of other non-LTR retrotransposons, but the degree to which this mechanism reflects that used by these different elements is not known. In the initial step of the integration reaction, an element encoded endonuclease cleaves the first (primer) strand of the target site and uses the released 3′ hydroxyl of the terminal nucleotide to prime reverse transcription starting within the poly(A) tail. Cleavage of the second (non-primer) strand occurs after reverse transcription. Thick lines, DNA target sequences; wavy lines, element RNA sequences; thin lines, element DNA sequences. The mechanism by which the 5′ end of the newly made cDNA is attached to the upstream target sequences is unclear, but it appears to be functionally equivalent to the reverse transcriptase simply jumping from the RNA template onto the DNA target. If this jump occurs before the reverse transcriptase reaches the end of the RNA template, a 5′ truncated copy is generated. The process of 5′ attachment generates considerable sequence variation at the junction of the element with the target site, even in the absence of 5′ truncations. The means by which the RNA is removed and the second strand of the element synthesized is not known. Non-LTR retrotransposons differ in whether they generate a target-site duplication and in whether that duplication is of a defined length.

moter, while some elements specifically insert into genes and may be expressed as precise cotranscripts (40). Integration starts with the element-encoded endonuclease generating a break in the target DNA. While the integration reaction could presumably occur at a preexisting nick or break on the chromosome, as was proposed in the original models, all non-LTR retrotransposons encode their own endonuclease domain. The cleavage activity of all non-LTR retrotransposon endonucleases studied to date is also consistent with the generation of an initial nick rather than a double-stranded break in the DNA (2, 25, 51, 52, 92).

The critical step in the retrotransposition mechanism is the utilization of the 3′ hydroxyl group released by first-strand cleavage (lower strand in Fig. 7) to prime reverse transcription of the element's RNA template. This reaction has been called target-primed reverse transcription (TPRT). Priming of reverse transcription in the case of the R2 element has been shown to occur in the absence of the RNA template annealing to the DNA target (91). The R2 reverse transcriptase specifically binds the 3′ untranslated region of the RNA template and is able to position this sequence next to the target site for reverse transcription. Because the RNA template is not annealed to the DNA target, the reverse transcriptase sometimes will make several attempts before successfully engaging the RNA template. Nucleotides are added to the target DNA during these aborted attempts, which generate simple repeat sequences at the 3′ end of the newly integrated elements. If the RNA template used in the R2 TPRT reaction ends in A residues, the extra nucleotides added to the target site during these abortive attempts are predominantly T residues (90). Thus the TPRT reaction can itself generate A-rich tails at the 3′ end of the element. This potentially explains why some non-LTR retrotransposons have poly(A) tails at their 3′ end even though they do not encode a polyadenylation signal and why other elements have simple A-rich sequences or tandem copies of simple repeats (e.g., TAA) (6, 40).

After engaging the 3′ end of the RNA template, the reverse transcriptase can proceed the entire length of the template to generate a full-length cDNA product. However, as postulated in the original models, if the RNA is degraded or if the reverse transcriptase falls off the RNA before reaching the 5′ end of the template, then a 5′ truncated insertion is generated. Once reverse transcription is initiated, the second strand of the DNA target is cleaved (upper strand in Fig. 7). It should be noted that cleavage of the second DNA strand after reverse transcription has only been shown for the R2 element, but can be inferred from the delayed kinetics or reduced cleavage of the second DNA strand by the endonuclease of several other non-LTR retrotransposons (2, 25, 51, 52). The manner in which the cDNA is attached to the upstream target site is not known for any non-LTR retrotransposons, but considerable sequence variation is often generated at the 5′ end even of full-length copies. This variation generated during 5′ attachment could be explained by a mechanism similar to the template jumping detected with the reverse transcriptases of the LTR retrotransposons. In the case of the non-LTR retrotransposons, however, the reverse transcriptase would typically jump from the RNA template onto the upstream DNA target site in a reaction that would not involve homologous sequences. The R2 endonuclease cleaves the second DNA strand two nucleotides upstream of the first DNA strand (92); thus this postulated jump readily generates the two base pair deletion of the target site seen in most integrated R2 copies. Because most non-LTR retrotransposons generate target-site duplications, cleavage of the second strand is likely to be downstream of the first-strand cleavage site. For many elements the length of the target-site duplication is variable, suggesting that either the location of the second-strand cleavage with respect to the first-strand cleavage or the jump may not be precise.

Finally, complete integration of the newly synthesized non-LTR retrotransposons requires removal of the RNA template and second-strand DNA synthesis. While a few elements encode an RNase H domain, the vast majority do not, and presumably rely on a cellular enzymatic activity. It also remains uncertain whether non-LTR retrotransposons are responsible for the synthesis of the second DNA strand or whether synthesis is accomplished via the DNA repair machinery.

An alternative model to that shown in Fig. 7 postulates that instead of a template jump by the reverse transcriptase, cleavage of the second DNA strand generates a new primer for the reverse transcriptase to synthesize the second DNA strand in a DNA-dependent manner. Second-strand cleavage at a site either upstream or downstream of the first strand would again generate target-site deletions or duplications. While this model is attractive because of its symmetry, its greater similarity to the LTR retrotransposition mechanisms, and its reduced dependency on the DNA repair machinery, there is no direct evidence, to date, to support this model. Clearly our understanding of the TPRT process lags far behind that of the LTR retrotransposition mechanism, and additional biochemical analysis of the reverse transcriptases from several retroposons is needed. As in the case of the LTR retrotransposons, many idiosyncratic differences in the basic TPRT mechanism used by non-LTR retrotransposons will probably be encountered.

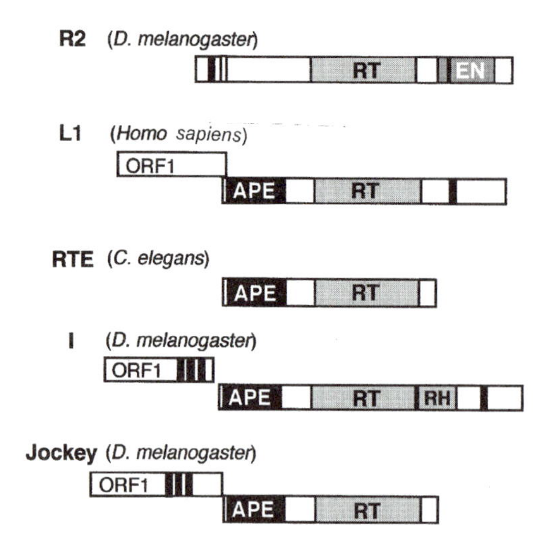

Figure 8. Diagram of the ORF structure of five non-LTR retrotransposons selected to represent the diversity of structures found in this class of elements. ORFs are represented by rectangular boxes, with boxes at different levels indicating ORFs in different reading frames or separated by termination codons. Shaded regions indicate identified enzymatic domains: RT, reverse transcriptase; RH, RNase H; APE, apurinic-apyrimidinic endonuclease; EN, endonuclease. Vertical black bars represent cysteine-histidine motifs believed to be associated with nucleic-acid binding domains. These elements represent the first element to be characterized from each of the five major groups of non-LTR retrotransposons identified in Fig. 9 and Table 1.

TPRT Allows Structural Flexibility

The simplicity of the TPRT mechanism for inserting new copies appears to permit considerable structural flexibility in the evolution of different non-LTR retrotransposons. As shown in Fig. 8, these elements contain ORFs with very different coding capacities. Indeed, the only features common to all non-LTR retrotransposons are the two basic enzymatic activities needed for the TPRT reaction: an endonuclease and a reverse transcriptase. The RTE element of *C. elegans* encodes only these two domains, thus representing the minimal non-LTR retrotransposon. The reverse transcriptase (RT) domain is highly conserved in all elements (93). The endonuclease domains, on the other hand, are of two types (APE and EN). Some of these endonucleases make highly precise cleavages at specific target sites, while others make nonspecific nicks at generic target sites.

In non-LTR retrotransposons with two ORFs, the first ORF encodes a protein that binds nucleic acids (103). In some elements (e.g., I and jockey), this ORF contains three cysteine-histidine motifs of the type CCHC, which is similar to those found in the *gag* protein of many LTR retrotransposons. In other non-LTR retrotransposons (e.g., L1), no such CCHC motifs are present. ORF1 has been shown to be required for L1 retrotransposition (109); however, its precise function remains a mystery. As described earlier, the *gag* proteins of LTR retrotransposons package the RNA template into virus-like particles where the double-stranded DNA intermediate is synthesized. Presumably in the case of the non-LTR elements, reverse transcription of the RNA template directly onto the DNA target site would not require such particles. In mice, the ORF1 protein of L1 cofractionates with full-length L1 RNA (68), and its expression is developmentally regulated in both spermatogenesis and oogenesis (145). In *D. melanogaster,* the ORF1 protein of I is present during oogenesis, but only when I elements are active (129). One possible role of the ORF1 protein is to associate with the RNA template after translation and import the template back into the nucleus for reverse transcription (68, 129). Other reports have suggested that the ORF1 protein may play a role in the binding of the ORF2 protein to the same RNA that was used to translate the ORF (*cis*-binding) (see chapter 35).

In non-LTR retrotransposons with two ORFs, expression of the second ORF is poorly understood but presumably involves either translational frameshifting, bypassing of termination codons, or the utilization of internal ribosomal entry sites. Protein expression in those elements that contain a single ORF is also poorly understood. The conservation of termination codons but not initiation codons at the start of the single ORFs in these elements suggests expression may involve translational regulatory mechanisms similar to those in elements with two ORFs (16, 94).

Other than the RT domain, the only other domain of an LTR retrotransposon *pol* gene that can be occasionally found in non-LTR retrotransposons is an RNase H domain (RH domain) (e.g., the *I* element in Fig. 8) (48). However, most non-LTR retrotransposons do not include this domain. Because reverse transcription occurs within the nucleus, rather than within a virus-like particle in the cytoplasm, endogenous RNase H activity is presumably sufficient to remove the template RNA in preparation for second-strand synthesis. The apparent ease with which cellular RHase H can substitute for an encoded activity is demonstrated in many lineages where only a subset of elements encode an RH domain. For example, R1 elements in some species of arthropods contain an RH domain while R1 elements in other species do not (18, 93).

Upstream of the RT domain, most non-LTR retrotransposons encode an endonuclease domain (APE) with sequence similarity to the apurinic-apyrimidinic endonucleases involved in DNA repair (51, 102). The

APE domains from L1, R1, Tx1, and TRAS (2, 25, 51, 52) have been expressed and shown to encode an endonuclease capable of cleaving DNA at preferred sites. The APE domains of all non-LTR retrotransposons are more similar in sequence to each other than to eukaryotic or prokaryotic cellular APE sequences. This suggests that the acquisition of this endonuclease domain from the cellular repair machinery was a monophyletic event (93). The non-LTR retrotransposon APE sequences are equally similar to both eukaryotic and prokaryotic sequences, further suggesting this acquisition occurred very early in the evolution of eukaryotes. The remaining non-LTR retrotransposons (e.g., R2) encode a different endonuclease (EN) C-terminal of the RT domain. All elements that encode this C-terminal endonuclease have sequence specificity for unique target sites in the genome (164). The active site of this EN domain is similar in sequence to that of various type II and type IIs restriction enzymes, e.g., *Eco*RI, *Eco*RV, and *Fok*I. It is not clear whether these different restriction enzymes and the non-LTR retrotransposons have independently evolved this active site or whether this similarity represents common ancestry.

Finally, there are two other protein domains encoded by some non-LTR retrotransposons. First, all elements like R2 that have a C-terminal EN domain also possess an N-terminal domain that frequently contains conserved C_2H_2 zinc-finger and/or c-myb DNA binding motifs. These N-terminal domains are presumably involved in DNA binding. Second, elements such as L1 and I that encode an N-terminal APE possess a C-terminal domain that contains a cysteine-histidine nucleic acid binding motif similar to that found in the EN domain of elements like R2. Mutation of this C-terminal motif eliminates retrotransposition in L1 (109). Therefore, for many non-LTR retrotransposons, DNA binding may be mediated by both the N- and C-terminal domains of the protein. In the case of non-LTR retrotransposons with no C-terminal domain (e.g., RTE and Jockey), DNA binding is presumably controlled entirely by the APE domain.

Phylogeny

Phylogenetic analysis of the non-LTR retrotransposons is restricted to the only sequences found in all elements, the RT domain. Recent analysis of these sequences has suggested that their RT domains can be folded into a "right hand" structure similar to that determined for retroviral reverse transcriptases (17, 93). The non-LTR retrotransposon RT domains also contain two additional segments within the "fingers" and "palm" regions that are not found in the LTR retrotransposons, as well as a 100-residue region corresponding to the "thumb" (93). The total length of these regions that can be used in the phylogenetic analysis is approximately 440 residues. This is more than twice the length of sequence used in earlier phylogenetic analyses of retroelements (37, 41, 160) and has greatly increased resolution of the sequence relationships of the elements.

The phylogram in Fig. 9 is an updated analysis of the non-LTR retrotransposons that have been characterized to date (June 2000). To the best of our knowledge it includes all published elements except for those that have incomplete RT domains and those that represent highly degenerate elements requiring repeated shifts of frame or by-passing of termination codons. To conserve space, sequences that represent different lineages of the same element in the same or related species (100, 138, 150) have often been omitted. The phylogram shown in Fig. 9 is a 50% consensus tree based on the neighbor-joining distance algorithms (125). The phylogeny was rooted using the closely related reverse transcriptase sequences of group II introns (105).

In our recent report (93), we proposed the use of the term "clade" to represent those non-LTR retrotransposons that share the same structural features, are grouped together with ample phylogenetic support, and likely date back to the pre-Cambrian era (see discussion in the next section). Each clade was named after one of the best-characterized elements within that clade. On the basis of these criteria there were 11 clades of non-LTR retrotransposons in 1999. While most newly published elements fall within these clades, 3 of 13 new elements have given rise to new clades (Fig. 9): NeSL1 (96), Rex1 (150), and the I clade, which has been separated into the I and Ingi clades. Partial or unpublished data of other elements suggest that at least two additional clades will soon need to be added (17). Unfortunately the addition of these new sequences has not helped resolve the deeper relationships of the many non-LTR retrotransposon clades. Clearly it would be helpful to divide these many clades into a limited number of related groups. Therefore, we propose here an additional classification scheme in which the various clades are grouped on the basis of both the phylogenetic relationship of their RT sequences and the nature and arrangement of their protein domains. On the basis of these two criteria, all non-LTR retrotransposons can be divided into five groups. The structure and distribution of these five groups are summarized in Table 1. Each group has been named after the original element to be characterized in that group and corresponds to the five elements shown in Fig. 8.

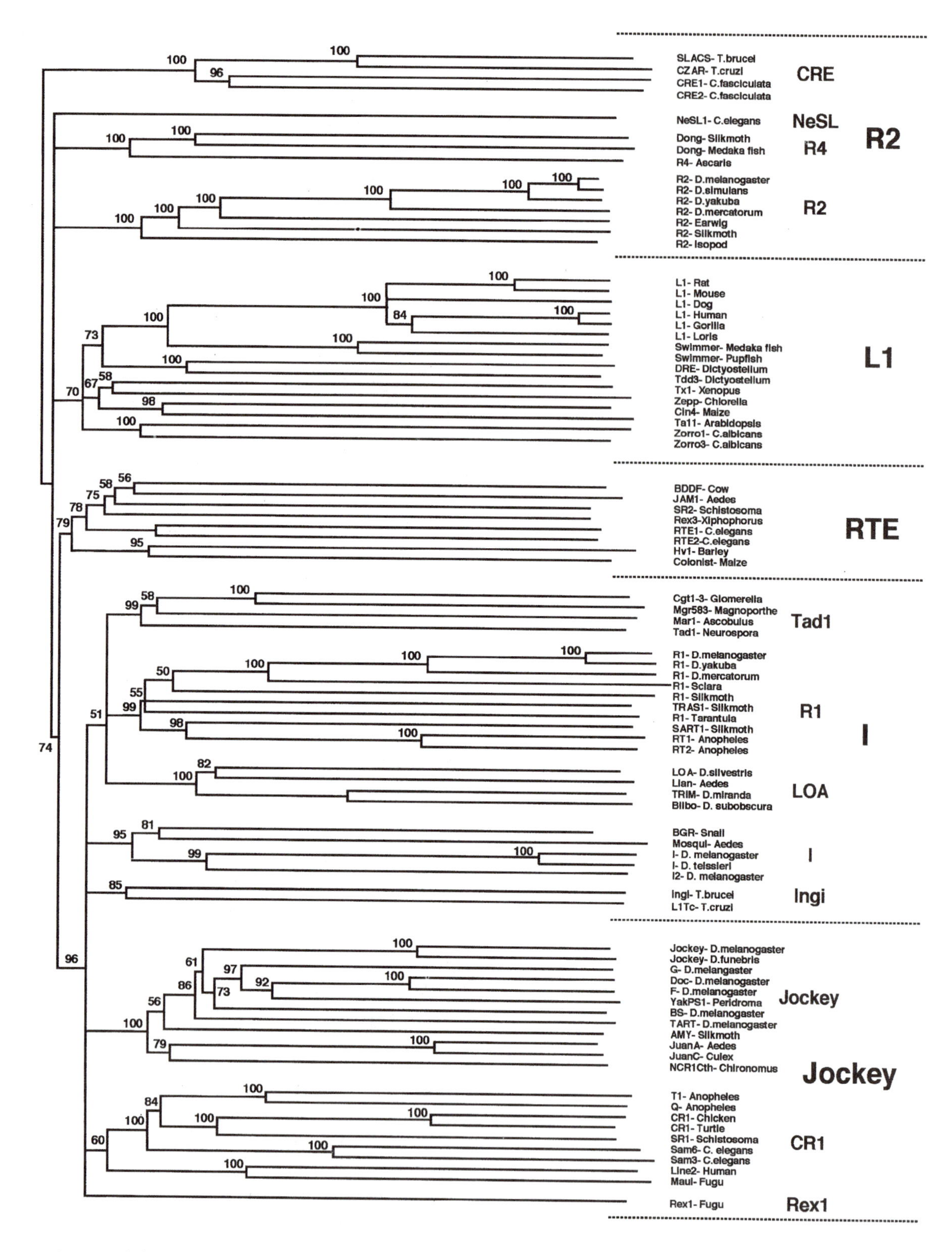

Figure 9. Phylogeny of the non-LTR retrotransposons based on their reverse transcriptase domain. This domain corresponds to approximately 440 amino acid residues as defined by Malik et al. (91). The phylogeny is a 50% consensus tree of the elements using the neighbor-joining algorithms and is rooted with the reverse transcriptase sequences of the group II introns (not shown). Numbers at each node indicate bootstrap values (percentage of 1,000 trials). The name of each element and the host organism from which it was isolated are given to the right. The elements can be divided into distinct clades, with each clade named after one of the earliest elements described from that clade (intermediate size names to the right). The tree is similar to that previously reported (91) except the addition of 13 new elements has defined three new clades. Individual, well-resolved clades or several clades containing elements with the same ORF structure have been placed into five groups (large names to the right). The elements within each group are delineated by the dotted horizontal lines. The basic structure of the elements in each group is shown in Fig. 8.

Table 1. Classification of non-LTR retrotransposons

Group	Clade and subgroup	ORF structures						Distribution	Site specific
		ORF1	APE	RT	RH	EN	CCHC		
R2	CRE	–	–	+	–	+	+	Trypanosomes	+
	NeSL	–	–	+	–	+	+	Nematodes	+
	R4								
	R4	–	–	+	–	+	+	Nematodes	+
	Dong	–	–	+	–	+	+	Insects and vertebrates	+
	R2	–	–	+	–	+	+	Arthropods	+
L1	L1								
	L1	+	+	+	–	–	+	Vertebrates	+/–
	DRE	+	+	+	–	–	+	Slime molds	+
	Cin4	+	+	+	–	–	+	Plants	–
	Tx1	+	+	+	–	–	+	Vertebrates	+
	Zepp	+	+	+	–	–	+	Algae and vertebrates	+
	Zorro	+	+	+	–	–	+	Fungi	–
RTE	RTE								
	RTE	–	+	+	–	–	–	Nematodes	–
	SR2	?	+	+	–	–	–	Flatworms, insects, and vertebrates	–
	Rex3	?	+	+	–	–	–	Vertebrates	–
	Colonist	?	+	+	–	–	–	Plants	–
I	Tad	+	+	+	+/–	–	+	Fungi	–
	R1	+	+	+	+/–	–	+	Arthropods	+
	LOA	+	+	+	+	–	+	Insects	–
	I	+	+	+	+	–	+	Insects and molluscs	–
	Ingi	+	+	+	+	–	+	Trypanosomes	–
Jockey	Jockey	+	+	+	–	–	–	Insects	+/–
	CR1								
	CR1	+	+	+	–	–	–	Vertebrates and flatworms	–
	Sam	+	+	+	–	–	–	Nematodes	–
	T1	+	+	+	–	–	–	Insects	–
	L2	+	+	+	–	–	–	Vertebrates	–
	Rex1	+	+	+	–	–	–	Vertebrates	–

The R2 group

The R2 group of elements is composed of the oldest clades of non-LTR retrotransposons. The phylogenetic relationships of these clades are poorly defined. All elements within this group insert into specific locations in their host's genome. In the case of the CRE and NeSL clades, the target sites are the tandemly repeated spliced leader (SL) genes of trypanosomes and nematodes, respectively (1, 56, 96). The spliced leader genes in these organisms are individually transcribed and trans-spliced onto the protein-encoding RNA sequences (5). Elements of the R2 and R4 clades insert into the 28S genes of arthropods and nematodes (see chapter 34 for a more detailed discussion of these elements). The insertion of both the SL and 28S gene-specific elements into host genes means these elements do not need their own promoter. While these elements appear to abolish function of the inserted gene, the host is not too adversely affected because their genomes contain an excess in the number of copies of these genes. Finally, a subset of the R4 clade, represented by Dong elements, inserts into tandem repeats of the simple sequence, TAA (162). These simple repeats are unlikely to play a role in the structure or expression of a gene and thus also serve as a safe haven for mobile element insertions.

The major distinguishing feature of the R2 group of elements is a single ORF and the presence of a C-terminal endonuclease domain (9, 164). All elements have an extensive N-terminal region that usually contains C_2H_2 zinc-finger and/or c-myb DNA binding domains. Some elements also have longer N-terminal domains that extend beyond these DNA binding motifs, but the potential function of these additional domains is unknown (1, 56). More members of these sequence-specific clades are likely to be discovered, since spliced leaders have been identified in other organisms (144), and of course rRNA genes and TAA repeats are present throughout eukaryotes. As described below, a number of additional targets in the eukaryotic genome have become home to site-specific elements from the other major groups of non-LTR retrotransposons. On the basis of those already identified, the only properties required of a "successful niche" are sequence conservation and multiple copies.

The L1 group

All elements of this group belong to a single lineage that is well resolved from all other non-LTR retrotransposon lineages. These elements are present in diverse taxa of eukaryotes including plants, slime molds, fungi, green algae, and vertebrates. As summarized in Table 1, the L1 group can be subdivided into six deep lineages corresponding to these taxonomic groups. All elements of the L1 group have two ORFs, with the second containing an N-terminal APE domain and a probable C-terminal DNA binding domain. It has been suggested that some members of this group also encode an RNase H domain (104), but this has not been confirmed.

Most members of the L1 group have little site specificity and are thus distributed throughout their host's genome. In three separate cases however, members of this group have evolved target specificity for unique locations in the genome that would have minimal deleterious effects on the host. Tx1 elements of *Xenopus laevis* insert exclusively into an abundant DNA-mediated mobile element in the genome (Tx1d) (59). This site-specificity is at least in part a result of the sequence specificity of its APE endonuclease (25). DRE, Tdd-3, and several related lineages of elements in *D. discoideum* insert adjacent to tRNA genes (138). This family of elements does not insert into a specific DNA sequence; thus, their encoded endonuclease may be directed to these genes by interaction with the transcriptional apparatus of the tRNA gene in a manner similar to that of the Ty1 and Ty3 elements of yeast (75). Finally, Zepp elements of *Chlorella vulgaris* are found near telomeric regions of chromosomes, where they have been suggested to supply part of the telomere function for the cell (67). The Zepp endonuclease is also unlikely to be sequence-specific and may direct insertions to the chromatin structure of telomeric regions in a manner similar to the Ty5 element of yeast (75).

The RTE group

This group corresponds to another widely distributed collection of elements that have a common lineage well resolved from all other non-LTR retrotransposons. Like L1, the RTE group can be subdivided into a number of deep lineages (Table 1). Many of the RTE members are still incompletely characterized, but all data suggest that the elements in this group encode an ORF with RT and APE domains but lack both an upstream ORF1 and a downstream C-terminal domain (94). RTE elements are thus the simplest non-LTR retrotransposons yet to be characterized. While no ORF1 has been identified in any member of this group, 5′ truncations are so frequent for many lineages that it is not yet clear whether a full-length element has been characterized (thus the uncertainty shown in Table 1 as to whether some have an ORF1). Indeed, the 5′ truncations are so large in size for certain of the RTE group that some of the members were originally characterized as SINEs (113, 139). In addition to their short ORF, all elements of the RTE group have extremely short 3′ untranslated regions ending in tandem repeats of a 3- to 5-bp sequence. In many elements these tandem repeats constitute the entire 3′ untranslated region. Thus, economy of length appears to be a major factor driving the evolution of this group of non-LTR retrotransposons. No site specificity has been documented for any member of this group.

The I group

This group is composed of five clades of elements that are well resolved phylogenetically from the R2, L1, and RTE groups but not from the Jockey group (described below). These five clades have been grouped together because they possess both an RNase H and a C-terminal domain. The phylogeny of the RNase H domain from these elements is consistent with that of the RT domain, suggesting a monophyletic origin of this domain in the common ancestor of this group (93) (Fig. 4). Individual elements of the R1 and Tad clades do not contain this RNase H domain, suggesting that it has been lost on several occasions. Elements of the R1 clade are highly site-specific in their target-site selection. R1 elements themselves insert within the 28S rRNA genes at a site that is only 74 bp downstream of the R2 elements. RT1 and RT2 elements of mosquitoes also insert in the 28S gene at a site that is ~700 bp downstream of the R2 site (4). Finally, TRAS and SART elements insert into the telomeric TTAGG repeats of silkmoths (115, 140). The APE domains of R1 and TRAS have been shown to have sequence specificity for these target sites (2, 52).

The Jockey group

This group includes the Jockey, CR1, and REX1 clades. Members of this group encode a first ORF and lack the C-terminal domain. The Jockey clade itself is extremely abundant in insects with at least six distinct lineages present in *D. melanogaster* alone. The REX1 clade is abundant in teleosts, while the CR1 clade is more broadly distributed with lineages in flatworms, nematodes, insects, and two lineages in vertebrates. One of these vertebrate CR1 lineages, L2, is only the second non-LTR retrotransposons to be identified in the human genome (133). CR1 elements in vertebrates have been suggested to generate a number of SINE elements by inserting highly 5′ truncated copies downstream of tRNA genes (114). Only one member of the Jockey group is site-specific. TART elements of *D. melanogaster* specifically insert at the telomeres of chromosomes (130). Because *D. melanogaster* chromosomes do not contain telomere repeats and the genome sequence has not revealed an active telomerase (see chapter 36), TART, and a related element HeT-A, appear to supply the telomere function in this organism.

Mode of Evolution

Having defined the phylogenetic relationship of the non-LTR retrotransposons, the next question to be addressed is whether they undergo horizontal transfers. As described earlier for the LTR retrotransposons, the existence of nearly identical elements in species that have diverged for many millions of years has served as strong support for the horizontal transfer of both transposons and LTR retrotransposons (see chapter 48). Such clear examples of horizontal transfer have not been found for the non-LTR retrotransposons. On the other hand, numerous examples have been found in which the phylogeny of the non-LTR retrotransposons in various species does not agree with the phylogeny of the host organisms. In some cases, these violations have been associated with higher than expected levels of sequence similarity, and therefore horizontal transfers have been proposed. However, tracing the evolutionary history of a mobile element by simple phylogenetic analysis is often complicated by the existence of multiple lineages of these elements. Multiple lineages of a mobile element arise because there is typically no means to maintain uniformity among the different copies of a mobile element family within a genome. Because there is also no selective pressure to maintain these lineages, such multiple lineages will be co-propagated with various degrees of success in a species phylogeny. The result is that sequence comparisons are frequently made between paralogous lineages of mobile elements in different species rather than within an orthologous lineage. For example, in the hypothetical species phylogeny in Fig. 10A, two recently diverged lineages of a mobile element (solid and broken lines) are differentially lost in the four species analyzed (species A through D). As a result of these losses, the elements obtained from the divergent species A and C are more closely related in sequence than the elements from the more closely related species C and D. This is because the A to D comparisons represent an orthologous lineage, while the C to D comparison represents paralogous lineages. Because species B has not retained either lineage of the element, it can be mistakenly assumed that the distribution of the mobile element originally included only species C and D, and the presence of a more closely related element in species A has resulted from the horizontal transfer of an element from C to A (interpretation in Fig. 10B).

To differentiate between these two explanations, elements from additional intermediate species could be sequenced. However, this approach is not an option if new lineages are frequently formed and lost or if additional species are not available. The alternative approach would be to estimate the rate at which mobile element sequences evolved in the species lineage. Horizontal transfer is suggested if the C versus D comparison in Fig. 10 is consistent with this rate of mobile element evolution, while vertical descent is suggested if the A versus C comparison is more consistent with the rate of mobile element evolution. Unfortunately, accurate rate estimates are not easy to obtain for a mobile element because of the very presence of these paralogous lineages. Comparison between paralogous lineages will inflate the apparent divergence rate, and because reverse transcription is error-prone, high rates of sequence evolution are easy to accept for retrotransposons. However, accurate estimates of the rates of sequence evolution have been made for three separate lineages of non-LTR retrotransposons: R1 and R2 elements of arthropods and L1 elements of vertebrates (18, 43, 44, 64, 84, 85, 146). In each case, the rates of evolution for these element lineages were only slightly faster than those of nuclear genes. These findings are similar to those described above for the evolution of the LTR retrotransposon SURL in echinoids (60). However, unlike the SURL data, no evidence of horizontal transfers was found for these non-LTR retrotransposons. Evidence for the lack of horizontal transfers is sufficiently clear in the case of L1 that these elements have been used to estimate rodent speciation events (149).

The divergence rates of the three non-LTR retrotransposon lineages are graphically presented in Fig. 11. The x axis represents estimated time of host species divergence and the y axis represents amino-acid sequence divergence between the RT domains of the elements obtained from those species. Only the com-

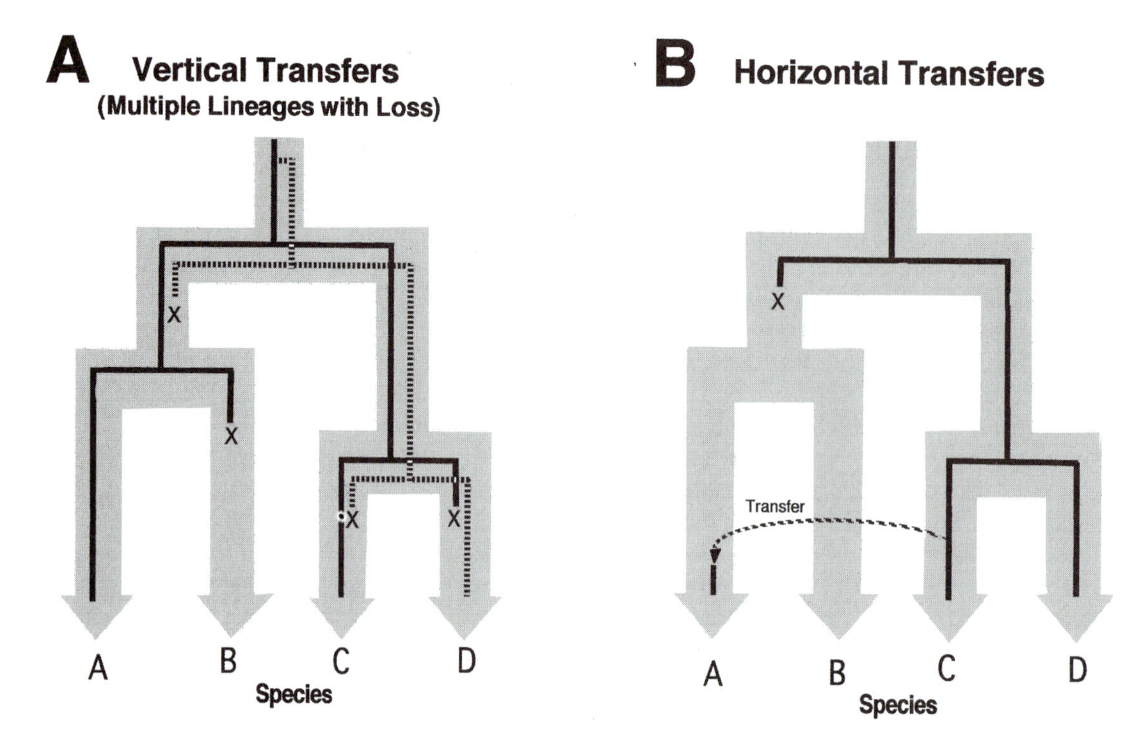

Figure 10. Alternative explanations as to why the phylogeny of a mobile element can violate the phylogeny of the host species. In scenario A, the mobile element lineage separates into two independent lineages (solid and broken lines) that are differentially maintained in the four extant species. Species A, C, and D each retain one lineage, while both lineages are lost in species B. The elements in species A and C represent an orthologous lineage, while the elements in species C and D represent paralogous lineages. In scenario B, a single lineage of the mobile element is lost from the ancestor of species A and B and retained in species C and D. A horizontal transfer of the element from species C introduces the elements into species A. Either series of events (panels A and B) can explain why the elements isolated from species C are more closely related to the elements from the distant species A, rather than to the elements from the related species D. The best means to differentiate between these two explanations is to estimate the rate of sequence evolution of the element and then to determine whether it is the divergence of elements in species A and D or the divergence of elements in species D and C that is most consistent with this rate of evolution.

parisons between the most closely related lineages of elements for each pair of species were plotted since they most likely represented true orthologous comparisons. The upper curve (solid line) is based on R1 and R2 comparisons in arthropods while the lower curve (dotted line) is based on L1 comparisons in vertebrates. The R1 and R2 lineages show increasing levels of divergence with time that approach saturation near the origin of arthropods (~600 million years). The L1 elements in vertebrates are evolving at a rate that is clearly slower than that of the arthropod elements. The slower rate of evolution of elements in vertebrates compared with arthropods is consistent with the slower rate of sequence change in nuclear genes of vertebrates compared with arthropods (reviewed in reference 87).

Thirteen other comparisons of non-LTR retrotransposon sequence divergence versus host divergence time are also plotted in Fig. 11. Seven of these points are from arthropod lineages (Jockey, I, LOA, and R4 clades) and six are from vertebrates (CR1, RTF, and REX clades). Significantly, all 13 comparisons fall close to the appropriate curve in Fig. 11. The many other comparisons that are possible between the elements used in the phylogeny in Fig. 9 have not been included in Fig. 11 because they fall well above these two curves, suggesting that they are paralogous comparisons.

It should be noted that these 13 comparisons include previous reports of the possible horizontal transmission of retroposons. These are (i) Jockey elements between *D. melanogaster* and *Drosophila funebris* (107) (circle labeled J at 40 million years), (ii) CR1 elements between a turtle and chicken (72) (square labeled CR1 at 225 million years), (iii) CR1 elements between a schistosome and chicken (39) (square labeled CR1 at 900 million years), (iv) RTE elements between cow and viper genomes (79) (square labeled RTE at 275 million years), and finally (v) Rex1 elements between a fish and an eel (150) (square

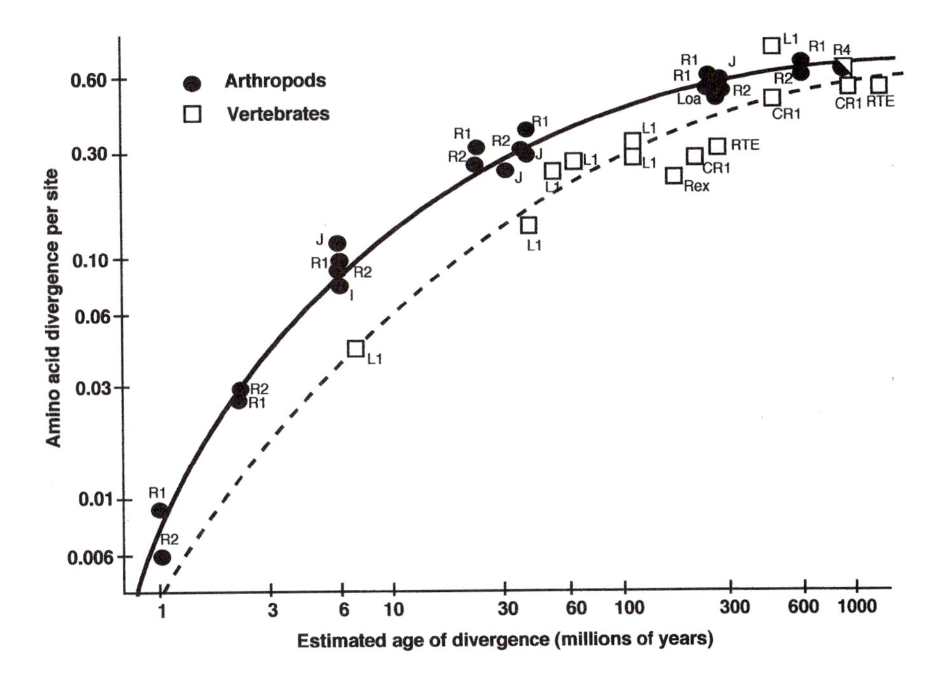

Figure 11. Plot of sequence divergence versus age estimates of non-LTR retrotransposons. Divergence calculations are based on the amino acid sequence of the RT domain. Circles, comparisons between arthropod elements; squares, comparisons between vertebrate elements. The clade from which each comparison is made (Fig. 9 and Table 1) is shown immediately adjacent to each data point (J, Jockey clade). The curves are average rates of divergence in arthropods (solid line) based on data from the R1 and R2 clades and in vertebrates (dotted lines) based on data from the L1 clade (91). Most divergences are based on the entire RT domain; however, in some cases the divergences are based on ~50% of this domain corrected to the complete RT domain using several outgroup comparisons. Divergence time estimates are described in reference 91.

labeled Rex at 180 million years). While three of the comparisons of the vertebrate elements fall somewhat below the vertebrate curve determined for L1 elements and therefore could be argued to represent horizontal events, more accurate determinations of the rates of non-LTR retrotransposon sequence evolution in the various lineages of vertebrates are necessary before these comparisons can be considered as clear evidence for horizontal transfers.

To summarize, there is no convincing evidence to date for the horizontal transfer of non-LTR retrotransposons during the past 600 million years. If the assumption of vertical descent is projected through the entire history of these elements, then the presence of highly divergent eukaryotic lineages in all five groups of non-LTR retrotransposons suggests that all groups date back to the origin of the eukaryote crown group. Consistent with this vertical evolution model, many very deep lineages of the non-LTR retrotransposons (those originally identified as clades in Fig. 9 and Table 1) contain elements from single taxonomic groups of eukaryotes (e.g., R2, Tad, LOA, Jockey). Even within some of the more diverse clades (e.g., L1 and RTE), elements from the same taxonomic groups are in the same subgroup (e.g., the plant Cin4/Ta11 lineage in the L1 group, the plant Hv1/colonist lineage in the RTE group). This clustering of elements from similar taxonomic groups of organisms is different from the interspersion of plants, fungi, and insect elements in the Ty3/gypsy and Ty1/copia groups of LTR retrotransposons (Fig. 2). Thus, even if isolated instances of the horizontal transfer of a non-LTR retrotransposons were to be discovered, this clustering of elements from the same taxonomic groups suggests such events have not crossed large taxonomic groups.

WHY DO MOBILE ELEMENTS DIFFER IN THEIR MODES OF EVOLUTION?

Why are non-LTR retrotransposons stably maintained by vertical descent in eukaryotic genomes, while transposons undergo frequent horizontal transfers? Why do LTR retrotransposons exhibit intermediate frequencies of horizontal transmission? These three classes of eukaryotic mobile elements are found in the same taxonomic groups of organisms; therefore, the answer to these questions must reside in their mechanism of movement. One critical difference resulting from the integration mechanisms used by these classes of elements is the ability of these elements to maintain a population of active copies within a species. The integrases encoded by transposons, once synthesized in the cytoplasm, must migrate into the nucleus where they bind the short, terminal DNA repeats of the first chromosomal element encountered. As a result, all defective copies of the transposon that still retain the correct short terminal repeats are just as likely to transpose as active copies. Because mutations are inevitable, this ever-expanding population of inactive copies will eventually overwhelm the active copies (65). A new cycle of increased numbers of active copies can be obtained only by horizontal transmission of an active element to a new species.

The reverse transcriptases of LTR and non-LTR retrotransposons, on the other hand, associate with their mRNAs in the cytoplasm. In the case of the LTR retrotransposons, the only copies of elements that can be duplicated are those expressed at the same time and whose RNA transcripts contain the correct 5′ and 3′ ends as well as internal regions needed for formation of the double-stranded DNA intermediate (Fig. 1). Thus, it is less likely that defective copies of LTR retrotransposons will overwhelm the system. In the case of the non-LTR retrotransposons, transcriptional activity is again required of the element but there are few sequence requirements for the RNA to be used as a template for reverse transcription (Fig. 7). However, evidence has recently been obtained indicating that the newly translated proteins of a non-LTR retrotransposon will predominately bind to the RNA from which they were translated (47, 109, 156). This "*cis* preference" provides an efficient mechanism to enable active elements to be preferentially duplicated. Thus, for both the LTR and non-LTR retrotransposons, the utilization of RNA templates can provide longer-term stability in a genome. *cis* preference by non-LTR retrotransposons would provide the most efficient means of stability.

While these arguments would suggest that transposons should be most dependent and non-LTR retrotransposons least dependent upon horizontal transfers for their long-term survival, the next question becomes how do these horizontal events occur? One possible means by which mobile elements can undergo horizontal transfer is the direct transfer of the nonchromosomal intermediate between species. The intermediate of a transposon is the cleaved DNA intermediate with bound transposase. Because of their dependence on horizontal events for long-term survival, these intermediates may have been selected through time for high stability outside a cell. LTR retrotransposons also generate double-stranded DNA intermediates, suggesting they would also be subject to this horizontal transfer mechanism. However, the increased stability of LTR retrotransposons within a lineage of organisms would suggest that selection for these horizontal transfers may be lower. Non-LTR retrotransposons, on the other hand, utilize TPRT. The only extrachromosomal copy of a non-LTR element that could be transferred between species is an unstable RNA intermediate. Thus the structure and stability of the nonchromosomal intermediate could explain, at least in part, why both transposons and LTR retrotransposons can undergo horizontal transfers while non-LTR retrotransposons do not.

A second frequently suggested mechanism for the horizontal transfer of mobile elements is hitchhiking via a viral genome. Insertion of the DNA intermediate of a transposon or LTR retrotransposon into the genome of a DNA virus would presumably be similar to their insertion into the host chromosomes. Such integrations have been observed (e.g., reference 55). Integration of a non-LTR retrotransposon, on the other hand, requires a more extensive participation of the host cellular DNA repair machinery (Fig. 7). It seems unlikely that the DNA repair/replication machinery required for non-LTR retrotransposons integration would function efficiently in the production of extra-chromosomal viral genomes. Thus, once again, non-LTR elements are deprived of a potential means used by the other mobile element classes for their horizontal transfers.

In summary, DNA-mediated transposons appear to need horizontal transfers for their long-term evolution because in any particular species their integration machinery is eventually overwhelmed by inactive copies. LTR retrotransposons are more adept at maintaining a pool of active elements within a species. While they do not need to undergo frequent horizontal transfers, their similar means of integration suggests that LTR retrotransposons have access to the same mechanisms of horizontal transfer as the transposons. Only in the case of the non-LTR retrotransposons are the likely avenues of horizontal transfer closed, and thus these elements are confined to their current host genome.

If non-LTR retrotransposons are not capable of being transferred between species, those elements observed today must be highly adept at maintaining themselves in their host lineage. This greater stability may explain one final difference between non-LTR retrotransposons and both transposons and LTR retrotransposons. A significant number of the non-LTR retrotransposons have evolved specificity in the selection of their target sites (Table 1). By eliminating the uncertainty associated with random insertions in the genome, site-specific elements can more easily regulate their own activity and their effect on the host's fitness. As a consequence, they should have a more stable future. This propensity to become site specific is not found for either eukaryotic transposons or LTR retrotransposons. In fact, no eukaryotic transposon has been found to be site specific, and the only known

LTR retrotransposons that have evolved site specificity are the five families of elements in *S. cerevisiae* (75). The evolution of site specificity for all yeast retrotransposons has been suggested to be a direct consequence of a shrinking yeast genome (7). Indeed, conditions in the small yeast genome appear to have proven too ruthless for most mobile elements: *S. cerevisiae* is the only characterized eukaryote that does not contain either transposons or non-LTR retrotransposons.

ORIGIN OF LTR AND NON-LTR RETROTRANSPOSONS

What is the evolutionary relationship between the LTR and non-LTR retrotransposons? Do they have a common ancestry or have they independently evolved? Several phylogenetic analyses have been conducted to determine the relationship of these two classes of retrotransposons to each other and to reverse transcriptase-containing elements in general (37, 41, 112, 159, 160). These other retroelements include the group II introns of mitochondria and eubacteria (29, 105; see also chapter 31), retrons of eubacteria (69; see also chapter 32), the mitochondrial plasmids of *Neurospora crassa* (154), and finally eukaryotic telomerases (112). A summary phylogram of one such recent phylogenetic analysis is shown in Fig. 12. (97). This phylogram is shown in an unrooted form because the rooting of this tree is highly problematic. The nucleotide polymerases with the greatest similarity in sequence and structure to that of reverse

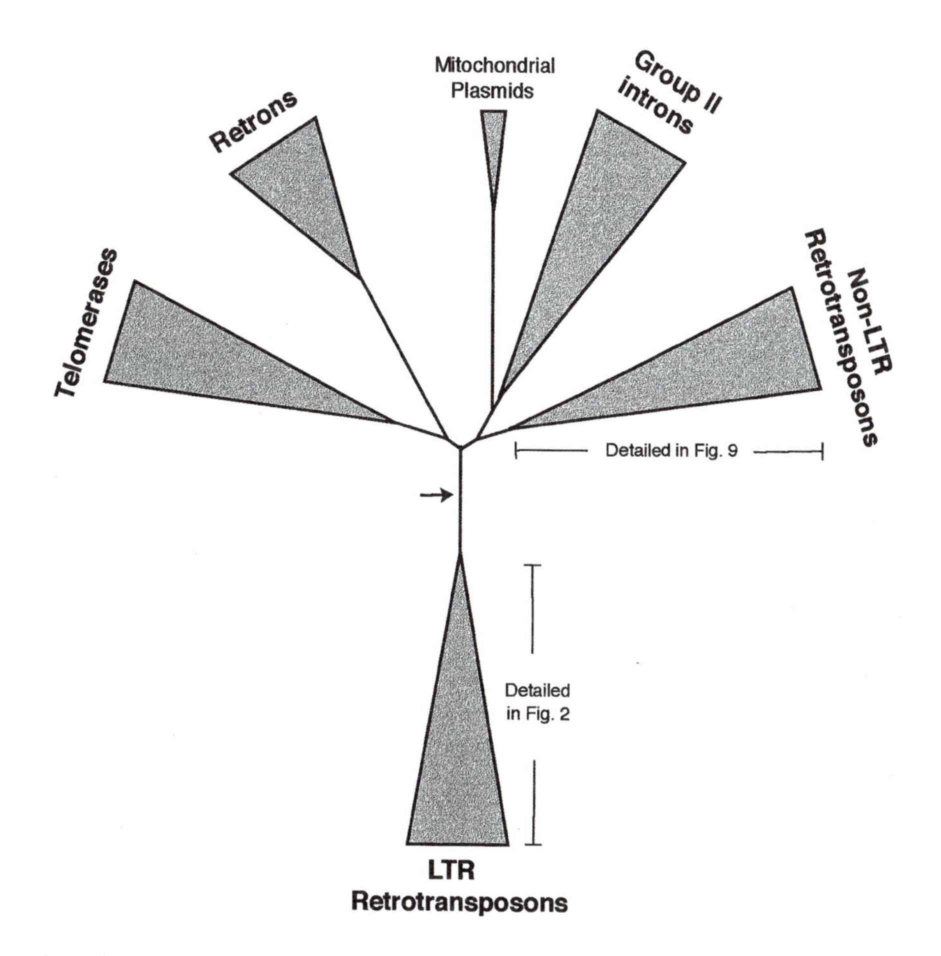

Figure 12. Schematic diagram of the unrooted phylogenetic tree of eukaryotic and prokaryotic retroelements. The phylogram is derived from the amino acid sequence of their encoded reverse transcriptase domain. Each group of retroelements is represented by a triangle, with the length of that triangle related to the sequence diversity with the group. The phylogenetic analysis is similar to those used in previous reports (109, 152), except that the regions of conserved sequences were not forced into blocks and an additional conserved region (segment 2A, reference 152) was included (90). The extensive sequence changes have resulted in a long branch length leading to the LTR retrotransposon lineage, suggesting that its position on the phylogeny is in question (see text). The arrow indicates where the branch leading to the RNA-directed RNA polymerases would be located on this tree of retroelement sequences. As described in the text, utilization of these RNA polymerases only identifies the most divergent branch of the phylogeny (mid-point roots) and thus is of limited value.

transcriptases are the RNA-directed RNA polymerases of RNA viruses (63). While these RNA polymerases have been previously used to root the retroelement tree (112, 160), the regions of similarity between these RNA and DNA polymerases are short and highly divergent; thus, they serve only to midpoint root the retroelement tree (i.e., place the root between the most divergent sequences; arrow in Fig. 12). Assuming that the reverse transcriptases of different retroelements have been under different selective constraints and quite likely different rates of evolution, there is no reason to believe that the most divergent reverse transcriptases are the oldest (41, 42). This view has also been recently supported by Nakamura and Cech (111). Thus, at present it seems unreliable to root reverse transcriptase trees like those in Fig. 12 using other polymerase sequences.

Without this standard means of rooting the phylogeny of a protein, the only alternative option is to assume that reverse transcriptases are ancient and to superimpose their phylogeny upon the tree of life. Extensive analyses of numerous genes have suggested that eukaryote genomes are descended from a fusion of eubacterial and archaebacterial genomes (for review, see reference 38). Because retroelements have been detected in *Eubacteria* but not *Archaea,* eukaryotic retroelements would appear to have evolved from eubacterial sequences. This eubacterial origin of eukaryotic reverse transcriptases has been previously suggested in various contexts (69, 82, 105, 142, 155). Therefore, even though the alternative model that reverse transcriptase originated in eukaryotes (or was only retained in eukaryotes) and subsequently spread into bacteria cannot be formally excluded, the following discussion of the origin of LTR and non-LTR retrotransposons assumes that the original reverse transcriptases were of eubacterial origin.

Origin of the Non-LTR Retrotransposons

On the basis of the reverse transcriptase phylogeny (Fig. 12), non-LTR retrotransposons are most closely related to the group II introns of mitochondria and eubacteria. Consistent with this phylogenetic relationship, group II intron "retrohoming" to unoccupied target sites is based on a TPRT mechanism similar to that used by non-LTR retrotransposons (29, 166, 167). Reverse transcription of the intron RNA is primed by a 3′ hydroxyl group generated by DNA cleavage. Cleavage of this DNA is catalyzed by a protein endonuclease domain located downstream of the group II intron's reverse transcriptase domain. This arrangement of catalytic domains is similar to the organization of protein domains in the oldest lineages of non-LTR retrotransposons, members of the R2 group (Fig. 8). The endonucleases encoded by group II introns and these R2-like elements are not homologous however. The active sites of the group II intron proteins are similar to endonucleases with conserved H-N-H motifs found in some restriction enzymes (61, 128). Elements of the R2 group also have active sites similar to restriction enzymes, but to those with D-D-K motifs (96, 164). Group II introns and the earliest lineages of non-LTR retrotransposons also differ in two additional aspects. First, group II elements use reverse splicing of the intron RNA for cleavage of the upper DNA strand; non-LTR retrotransposons cleave both DNA strands via the protein endonuclease domain. Second, DNA cleavage specificity by group II introns is predominately accomplished by their RNA sequences annealing to the target DNA in the reverse splicing reaction and is only aided by protein recognition (46); DNA recognition for the non-LTR retrotransposons is based entirely on protein:DNA recognition.

Current phylogenetic analysis cannot distinguish whether the non-LTR retrotransposons evolved from the group II introns after the origin of eukaryotes or whether the TPRT machinery of these two types of elements simply shares a common ancestor. Most current models (22, 82, 105, 116, 155, 164, 166) suggest that the mobile group II introns of eubacteria were transmitted to eukaryotes during the initial fusion of the eubacterial and archaebacterial genomes or during the symbiosis that gave rise to the mitochondria. These mobile introns subsequently evolved into the modern-day spliceosomal introns. Recent studies of group II introns have suggested that in addition to the efficient means they have to insert at a specific (homing) site, these introns can also insert at low levels into "ectopic sites." Ectopic insertions result from reverse splicing either directly into additional DNA sites (165) or indirectly into sites on cellular RNAs (28). In the latter mechanism, reverse transcription of this reverse-spliced RNA, followed by recombination of this DNA with the host genes results in the occasional establishment of copies of the intron at new locations in the host genome. Presumably an increasing number of these self-splicing introns accumulated in early eukaryotes and drove the host cell to take over the responsibility for RNA splicing by evolving the spliceosome. This adaptation by the host cell may have been in response to the need to improve the efficiency of the RNA splicing reaction or, alternatively, as a defense mechanism to prevent reverse splicing and the further spread of group II introns.

Once the group II introns were no longer able to reverse-splice into new locations, regaining the ability to spread by the TPRT reaction required an alternative means to cleave the nonprimer strand of a DNA target. The acquisition of an autonomous endonuclease by these introns would thus represent the first step in the evolution of non-LTR retrotransposons. Consistent with this model, phylogenetic analysis of the non-LTR retrotransposons (Fig. 9) suggests the oldest lineages of elements were mostly site specific and, like the group II introns, relied on cotranscription from specific genes in the genome. Thus, these site-specific non-LTR retrotransposons are still undergoing a reaction functionally similar to retrohoming. However, because these retrotransposons cannot splice themselves out of a cotranscript, they can only insert into a fraction of the genes in a multigene family. A second feature of eukaryotic genomes is that their large size allowed great opportunity for random insertions of a mobile element without deleterious consequences. As a result, while a few non-LTR retrotransposons have retained site specificity for various locations in the genome, in the next major step of non-LTR retrotransposon evolution, a lineage of elements found advantage in acquiring a less specific endonuclease, the APE endonuclease. Combined with the acquisition of their own promoter, these new non-LTR retrotransposons now had free rein to insert into many locations throughout the genome.

Origin of the LTR Retrotransposons

The reverse transcriptases encoded by LTR retrotransposons and related viruses differ substantially from the reverse transcriptases of all other retroelements. Compared with these other retroelements, the reverse transcriptases encoded by LTR retrotransposons are missing a conserved motif (segment 2A) and have duplicated another motif (segment 7) (see references 41, 112, and 160 for a description of the common segments found in all reverse transcriptases). LTR retrotransposons also encode the only reverse transcriptases that do not specifically bind their own RNA template. LTR retrotransposons use the 3′ end of an "annealed" RNA molecule to prime reverse transcription. Non-LTR retrotransposons, group II introns, and telomerases all use the 3′ end of DNA to prime reverse transcription, while retrons use the 2′ hydroxyl of an internal G base of the RNA template itself to prime reverse transcription (69).

Consistent with these major structural differences, phylogenetic analysis of the reverse transcriptase sequences places the LTR retrotransposons as the most distant branch of retroelements (Fig. 12). However, as described above, just because the reverse transcriptase of the LTR retrotransposons are the most divergent in sequence does not mean that these elements represent the oldest lineage of retroelements. Indeed, no prokaryotic member of the LTR retrotransposon lineage has been identified to date. We therefore suggest that it is more likely that the LTR retrotransposons evolved from the same lineages of eubacterial sequences as all other retroelements. Their reverse transcriptase domains have been under different, possibly relaxed, constraints, which has enabled their sequences to change dramatically. As a result of these extensive sequence changes, the algorithms used to derive the RT protein phylogeny are unable to properly position the LTR retrotransposons on the retroelement phylogeny.

What is the origin of the LTR retrotransposons? The retrotransposon mechanism of integration is a hybrid of two methods: reverse transcription of an RNA template to form a double-stranded DNA intermediate, and integration of this DNA intermediate by way of a transposase (Fig. 1). The integrases of LTR retrotransposons have been shown to have homologous structures and similar modes of action to those of bacterial and eukaryotic transposons (30, 32, 108), but similar to the situation with the RT phylogeny sequence analysis is unable to resolve the phylogeny of these transposases (21, 36, 80). Therefore, while there is no clear phylogenetic support based on the RT or transposase domains, the model that seems most consistent with the hybrid mechanisms of LTR retrotransposon mobility is one in which these retrotransposons arose as a fusion of a non-LTR retrotransposon and a transposon (Fig. 13). This model differs dramatically from previous models for the origin of LTR retrotransposons that were based on the assumption that their highly divergent reverse transcriptases accurately reflected their age (41, 42, 112, 160), but does agree with models for the origin of LTR retrotransposons suggested by Capy and co-workers (21).

What non-LTR retrotransposon lineage gave rise to the LTR retrotransposons? Elements of the I group contain both a *gag*-like first ORF and an RH domain (Fig. 8). Thus the most parsimonious model is that the initial step in the formation of an LTR retrotransposon involved the fusion of a DNA transposon with an I-like non-LTR retrotransposon. Integration of this hybrid element would have required the evolution of a number of new processes. For example, instead of requiring a DNA target to initiate reverse transcription, the hybrid element used a small, ubiquitous host RNA (tRNA). The acquisition of terminal repeats on the RNA template promoted end-to-end jumps, thus resolving the "replication of ends" problem generated by not using the target DNA as primer. The only enzymatic activity not supplied by either the transposon or the non-LTR retrotransposon was a protease needed to separate the integrase from the reverse transcriptase/RNase H activities required at different times and places in the cell. This protease

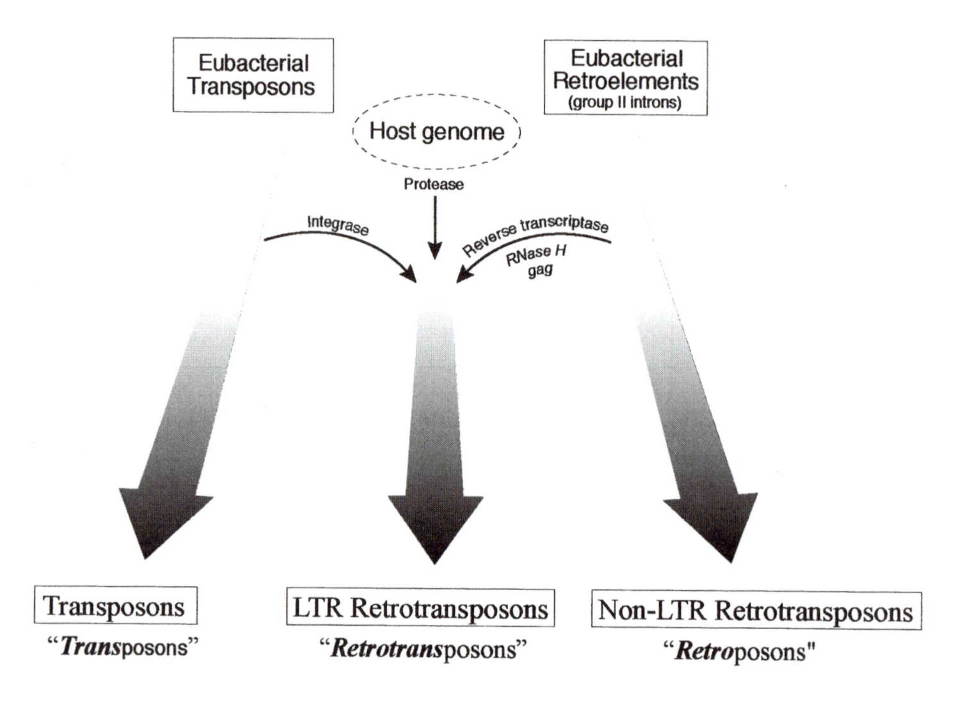

Figure 13. Mosaic origin of the LTR retrotransposons. All mobile elements in eukaryotes are assumed to have descended from bacterial elements. The structure and function of eukaryotic DNA transposons are virtually identical to those in bacteria (28, 30, 105). Non-LTR retrotransposons are suggested to have descended from a bacterial retroelement, probably a mobile group II intron. The origin of the LTR retrotransposons is postulated to have resulted from the fusion of these two original classes of mobile elements. The reverse transcriptase, RNase H, and *gag*-like proteins were derived from a non-LTR retrotransposon, and the integrase was derived from a transposon. The only other protein component of LTR retrotransposons is a protease, presumably derived from the host genome. This model for the origin of the LTR retrotransposons suggests a change in the nomenclature. The originally termed non-LTR retrotransposons (or LINEs) are referred to simply as retroposons. This term faithfully reflects the original use of the term (120) as elements which are simple reverse transcription products of an RNA. We suggest the originally termed LTR-retrotransposons continue to be referred to as *retrotrans*posons. This term accurately reflects the mosaic nature of their structure and integration mechanism as part *retro*poson and part *trans*poson.

activity was presumably acquired from the eukaryotic host.

This highly speculative model makes two predictions: first, that the non-LTR retrotransposons are older than the LTR retrotransposon lineage, and second that the RH domain and *gag*-like ORF were also derived from the non-LTR retrotransposons. In support of the greater age of the non-LTR retrotransposons, the earliest branching eukaryotes such as *Giardia* and *Trypanosoma* appear to lack LTR retrotransposons but having multiple lineages of non-LTR retrotransposons (1, 17, 56, 76). In the case of the origin of the *gag* domain, sequence conservation is too low to allow phylogenetic analysis. However, at least one feature of the LTR retrotransposon *gag* genes and the I group ORF1 are conserved. Protein of both types of elements shares one or more CCHC motifs with a unique spacing of cysteine and histidine residues not found in other cellular proteins. More direct support for the prediction that the LTR retrotransposons are derived from the I group of non-LTR retrotransposons comes from phylogenetic analysis of the RH domains. RNase H is the only protein domain present in both classes of retrotransposons, as well as in eukaryotic and prokaryotic genomes. Referring back to Fig. 4, and ignoring for the present the unusual location of the retroviral sequences, the RNase H phylogeny suggests that the same lineage of RNase H is found in both LTR and non-LTR retrotransposons. On the basis of the low bootstrap values separating the eukaryotic sequences and the retroelement sequences, the origin of this lineage is most likely that of a eukaryotic gene (see also reference 95). Consistent with the proposed greater age of the non-LTR retrotransposons, the poorly resolved branch containing the RH domains from the I group of elements predates that of the entire LTR retrotransposon lineage. As was discussed earlier in the analysis of LTR retrotransposons, the RH domain of retroviruses appears to have had an independent origin from that of the other lineages of LTR retrotransposons. This origin was more recent and again appears to have been derived from the I group of elements. This unusual similarity of the I element RNase H and that of retroviruses was originally noted by Doolittle and coworkers (37).

In summary, we can now present a highly speculative scenario for the origin of eukaryotic retrotransposons, with the non-LTR retrotransposons derived from group II introns and the LTR retrotransposons derived in turn from the non-LTR retrotransposons. As more eukaryotic and prokaryotic genomes are sequenced, evidence of new types of retroelements may be discovered that will necessitate a reevaluation of this proposal. Most significant would be the discovery of any type of retroelement in *Archaea,* or LTR or non-LTR retrotransposons in a eubacterium.

SUMMARY AND CONCLUSIONS

We have attempted in this chapter to provide an overview of the diversity and mode of evolution of eukaryotic retrotransposable elements and to trace their origins back to prokaryotic sequences. While many of the original elements were discovered in a few model genetic organisms because they produce insertional mutations, new elements today are more frequently discovered in a wider range of organisms as a result of genome sequencing projects. These genomic studies have revealed that mobile elements are widespread in all eukaryotes. These studies have also shown that while eukaryotic genomes differ dramatically in size and apparent complexity, these differences are due not to large variations in the number of genes, but rather to a difference in the abundance and diversity of their mobile elements. We currently have a reasonable understanding of the conserved structures and mechanisms these elements use to make new copies. Less well known are the mechanisms the elements use to spread within or between genomes and what means eukaryotes use to control this spread. Clearly some genomes, including our own, do not appear to be effective in this regard and thus are on a path to ever-increasing size.

As we have repeatedly emphasized in this chapter, two distinct classes of mobile elements utilize a reverse transcription step in their integration cycle. These two classes of elements differ substantially in their mechanism of insertion and in their mode of evolution. During the preparation of this chapter, it became increasingly apparent to us that a new nomenclature for these mobile elements was in order. The nomenclature most frequently used today, LTR and non-LTR retrotransposons, does not taken into consideration that these classes of elements utilize different mechanisms. Indeed, non-LTR retrotransposons are not named after their unique features or means of insertion but rather after what they do not possess. We feel the nomenclature should reflect the fundamental differences and similarities of the three major types of eukaryotic mobile elements as well as their probable origins (Fig. 13).

We suggest that the non-LTR retrotransposons be referred to simply as retroposons. As coined by Rogers (124) and Weiner et al. (157), this term implies that new copies of these elements are reverse transcripts of an RNA molecule. Unfortunately, this term was originally used to include both the LTR and non-LTR elements and more recently has been used to refer to those reverse transcripts that do not encode their own machinery for replication (e.g., SINEs and processed pseudogenes). Because it is now generally assumed that the replication machinery of the non-LTR retrotransposons (retroposons) is used for the integration of new copies of SINEs and processed pseudogenes (see chapter 35), SINEs can be referred to as the nonautonomous retroposons. We suggest that LTR retrotransposons be shortened to simply retrotransposon. This term, however, now reflects the mosaic nature of their integration as part *retro*poson, in that they use reverse transcription, and part *trans*poson, in that they use a transposase (or integrase). It should be noted that this terminology was originally suggested by one of the founding fathers in the study of reverse transcription, Howard Temin, more than 10 years ago (142).

Acknowledgments. We gratefully acknowledge the many stimulating discussions and the generous communication of unpublished manuscripts and data by our colleagues. We thank W. Burke and C. Perez-Gonzalez for their comments and assistance on this manuscript. We are especially indebted to D. Eickbush for her tireless attempts to make certain this chapter actually said what we had intended.

Research in the laboratory of T. H. Eickbush was supported by grants from the National Institutes of Health, the National Science Foundation, and the American Cancer Society. H. Malik is currently supported by a Helen Hay Whitney Fellowship.

REFERENCES

1. **Aksoy, S., S. Williams, S. Chang, and F. F. Richards.** 1990. SLACS retrotransposon from *Trypanosoma brucei gambiense* is similar to mammalian LINEs. *Nucleic Acids Res.* **18:**785–792.
2. **Anzai, T., H. Takahashi, and H. Fujiwara.** 2001. Sequence-specific recognition and cleavage of Telomeric Repeat $(TTAGG)_n$ by Endonuclease of Non-Long Terminal Repeat Retrotransposon TRAS1. *Mol. Cell. Biol.* **21:**100–108.
3. **Bateman, A., E. Birney, R. Durbin, S. R. Eddy, R. D. Finn,**

and **E. L. L. Sonnhammer.** 1999. Pfam 3.1: 1313 multiple alignments match the majority of proteins. *Nucleic Acids Res.* **27:**260–262.
4. **Besansky, N. J.** 1990. A retrotransposable element from the mosquito *Anopheles gambiae. Mol. Cell. Biol.* **10:**863–871.
5. **Blumenthal, T.** 1998. Gene clusters and polycistronic transcription in eukaryotes. *Bioessays* **20:**480–487.
6. **Boeke, J. D., and V. G. Corces.** 1989. Transcription and reverse transcription of retrotransposons. *Annu. Rev. Microbiol.* **43:**403–434.
7. **Boeke, J. D., and S. E. Devine.** 1998. Yeast retrotransposons: finding a nice quiet neighborhood. *Cell* **93:**1087–1089.
8. **Boeke, J. D., T. H. Eickbush, S. B. Sandmeyer, and D. F. Voytas.** 2000. Pseudoviridae, p. 349–357. *In* M. H. V. van Regenmortel et al. (ed.), *Virus Taxonomy: VIIth Report of the ICTV.* Springer-Verlag, New York, N.Y.
9. **Boeke, J. D., T. H. Eickbush, S. B. Sandmeyer, and D. F. Voytas.** 1999. Metaviridae, p. 359–367. *In* M. H. V. van Regenmortel et al. (ed.), *Virus Taxonomy: VIIth Report of the ICTV.* Springer-Verlag, New York, N.Y.
10. **Boeke, J. D., C. A. Garfinkel, A. Styles, and G. R. Fink.** 1985. Ty elements transpose through an RNA intermediate. *Cell* **40:**491–500.
11. **Boeke, J. D., and J. P. Stoye.** 1997. Retrotransposons, endogenous retroviruses, and the evolution of retroelements, p. 343–435. In J. M. Coffin, S. H. Hughes, and H. E. Varmus (ed.), *Retroviruses.* Cold Spring Harbor Laboratory Press, Cold Spring Harbor, N.Y.
12. **Bowen, N. J., and J. F. McDonald.** 1999. Genomic analysis of *Caenorhabditis elegans* reveals ancient families of retroviral-like elements. *Genome Res.* **9:**924–935.
13. **Britt, W. J., and M. Mach.** 1996. Human cytomegalovirus glycoproteins. *Intervirology* **39:**401–412.
14. **Britten, R. J.** 1995. Active gypsy/Ty3 retrotransposons or retroviruses in *Caenorhabditis elegans. Proc. Natl. Acad. Sci. USA* **92:**599–601.
15. **Bucheton, A.** 1990. I transposable elements and I-R hybrid dysgenesis in Drosophila. *Trends Genet.* **6:**16–21.
16. **Burke, W. D., C. C. Calalang, and T. H. Eickbush.** 1987. The site-specific ribosomal insertion element type II of *Bombyx mori* (R2Bm) contains the coding sequence for a reverse transcriptase-like enzyme. *Mol. Cell. Biol.* **7:**2221–2230.
17. **Burke, W. D., H. S. Malik, J. P. Jones, and T. H. Eickbush.** 1999. The domain structure and retrotransposition mechanism of R2 elements are conserved throughout arthropods. *Mol. Biol. Evol.* **16:**502–511.
18. **Burke, W. D., H. S. Malik, W. C. Lathe, and T. H. Eickbush.** 1998. Are retrotransposons long term hitchhikers? *Nature* **239:**141–142.
19. Reference deleted.
20. **Cappello, J., K. Handelsman, and H. Lodish.** 1985. Sequence of Dictyostelium DIRS-1: an apparent retrotransposon with inverted terminal repeats and an internal circle junction sequence. *Cell* **43:**105–115.
21. **Capy, P., C. Basin, D. Higuet, and T. Langin.** 1998. *Dynamics and Evolution of Transposable Elements.* Springer-Verlag, New York, N.Y.
22. **Cavalier-Smith, T.** 1991. Intron phylogeny: a new hypothesis. *Trends Genet.* **7:**145–148.
23. **Charlesworth, B., and C. H. Langley.** 1989. The population genetics of Drosophila transposable elements. *Annu. Rev. Genet.* **23:**251–287.
24. **Chaboissier, M. C., D. Finnegan, and A. Bucheton.** 2000. Retrotransposition of the I factor, a non-long terminal repeat retrotransposon of Drosophila, generates tandem repeats at the 3′ end. *Nucleic Acids Res.* **28:**2467–2472.
25. **Christensen, S., G. Pont-Kingdom, and D. Carroll.** 2000. Target specificity of the endonuclease from the *Xenopus laevis* non-long terminal repeat retrotransposon, Tx1L. *Mol. Cell. Biol.* **20:**1219–1226.
26. **Coffin, J. M., S. H. Hughes, and H. E. Varmus.** 1997. *Retroviruses.* Cold Spring Harbor Laboratory Press, New York, N.Y.
27. **Cook, J. M., J. Martin, A. Lewin, R. E. Sinden, and M. Tristem.** 2000. Systematic screening of Anopheles mosquito genomes yields evidence for a major clade of Pao-like retrotransposons. *Insect. Mol. Biol.* **9:**109–117.
28. **Cousineau, B., S. Lawrence, D. Smith, and M. Belfort.** 2000. Retrotransposition of a bacterial group II intron. *Nature* **404:** 1018–1021.
29. **Cousineau, B., D. Smith, S. Lawrence-Cavanagh, J. E. Mueller, J. Yang, D. Mills, D. Manias, G. Dunny, A. M. Lambowitz, and M. Belfort.** 1998. Retrohoming of a bacterial group II intron: mobility via complete reverse splicing, independent of homologous DNA recombination. *Cell* **94:**451–462.
30. **Craig, N. L.** 1995. Unity in transposition reactions. *Science* **270:**253–254.
31. **Cummings, M.** 1994. Transmission patterns of eukaryotic transposable elements: arguments for and against horizontal transfer. *Trends Ecol. Evol.* **9:**141–145.
32. **Davies, D. R., I. Y. Goryshin, W. S. Reznikoff, and I. Rayment.** 2000. Three-dimensional structure of the Tn5 synaptic complex transposition intermediate. *Science* **289:**77–85.
33. **Davis, P. S., and B. H. Judd.** 1995. Nucleotide sequence of the transposable element BEL of *Drosophila melanogaster. Drosoph. Inf. Ser.* **76:**134–136.
34. **de Chastonay, Y., H. Felder, C. Link, P. Aeby, H. Tobler, and F. Müller.** 1992. Unusual features of the retroid element PAT from the nematode *Panagrellus redivivus. Nucleic Acids Res.* **20:**1623–1628.
35. **Desset, S., C. Conte, P. Dimitri, V. Calco, B. Dastugue, and C. Vaury.** 1999. Mobilization of two retroelements, ZAM and Idefix, in a novel unstable line of *Drosophila melanogaster. Mol. Biol. Evol.* **16:**54–66.
36. **Doak, T. G., F. P. Doerder, C. L. Jahn, and G. Herrick.** 1994. A proposed superfamily of transposase genes: transposon-like elements in ciliated protozoa and a common "D35E" motif. *Proc. Natl. Acad. Sci. USA* **91:**942–946.
37. **Doolittle, R. F., D. F. Feng, M. S. Johnson, and M. A. McClure.** 1989. Origins and evolutionary relationships of retroviruses. *Q. Rev. Biol.* **64:**1–30.
38. **Doolittle, W. F.** 1999. Phylogenetic classification and the universal tree. *Science* **284:**2124–2128.
39. **Drew, A. C., and P. J. Brindley.** 1997. A retrotransposon of the non-long terminal repeat class from the human blood fluke *Schistosoma mansoni.* Similarities with the chicken repeat 1-like elements from vertebrates. *Mol. Biol. Evol.* **14:** 602–610.
40. **Eickbush, T. H.** 1992. Transposing without ends: the non-LTR retrotransposable elements. *New Biol.* **4:**430–440.
41. **Eickbush, T. H.** 1994. Origin and evolutionary relationships of retroelements, p. 121–157. *In* S. S. Morse (ed.), *The Evolutionary Biology of Viruses.* Raven Press, New York, N.Y.
42. **Eickbush, T. H.** 1997. Telomerase and retrotransposons: which came first? *Science* **277:**911–912.
43. **Eickbush, D. G., and T. H. Eickbush.** 1995. Vertical transmission of the retrotransposable elements R1 and R2 during the evolution of the *Drosophila melanogaster* species subgroup. *Genetics* **139:**671–684.
44. **Eickbush, D. G., W. D. Lathe, M. P. Francino, and T. H. Eickbush.** 1995. R1 and R2 retrotransposable elements of *Drosophila* evolve at rates similar to that of nuclear genes. *Genetics* **139:**685–695.

45. **Engelman, A., A. B. Hickman, and R. Craigie.** 1994. The core and carboxyl-terminal domains of the integrase protein of human immunodeficiency virus type 1 each contribute to nonspecific DNA binding. *J. Virol.* **68:**5911–5917.
46. **Eskes R., J. Yang, A. M. Lambowitz, and P. S. Perlman.** 1997. Mobility of yeast mitochondrial group II introns: engineering a new site specificity and retrohoming via full reverse splicing. *Cell* **88:**865–874.
47. **Esnault, C., J. Maestre, and T. Heidmann.** 2000. Human LINE retrotransposons generate processed pseudogenes. *Nat. Genet.* **24:**363–367.
48. **Fawcett, D. H., C. K. Lister, E. Kellett, and D. J. Finnegan.** 1986. Transposable elements controlling I-R hybrid dysgenesis in *D. melanogaster* are similar to mammalian LINEs. *Cell* **47:**1007–1015.
49. **Fayet, O., P. Raymond, P. Polard, M. F. Prere, and M. Chandler.** 1990. Functional similarities between retroviruses and the IS3 family of bacterial insertion sequences? *Mol. Microbiol.* **4:**1771–1777.
50. **Felder, H., A. Herzceg, Y. de Chastonay, P. Aeby, H. Tobler, and F. Müller.** 1994. Tas, a retrotransposon from the parasitic nematode *Ascaris lumbricoides. Gene* **149:**219–225.
51. **Feng, Q., J. V. Moran, H. H. Kazazian, and J. D. Boeke.** 1996. Human L1 retrotransposon encodes a conserved endonuclease required for retrotransposition. *Cell* **87:**905–916.
52. **Feng, Q., G. Schumann, and J. D. Boeke.** 1998. Retrotransposon R1Bm endonuclease cleaves the target sequence. *Proc. Natl. Acad. Sci. USA* **95:**2083–2088.
53. **Finnegan, D. J.** 1989. The I factor and I-R hybrid dysgenesis in *Drosophila melanogaster,* p. 503–517. *In* D. E. Berg and M. M. Howe (ed.), *Mobile DNA.* American Society for Microbiology, Washington, D.C.
54. **Flavell, A. J., V. Jackson, M. P. Iqbal, I. Riach, and S. Waddell.** 1995. Ty1-copia group retrotransposon sequences in amphibia and reptilia. *Mol. Gen. Genet.* **246:**65–71.
55. **Friesen, P. D., and M. S. Nissen.** 1990. Gene organization and transcription of TED, a lepidopteran retrotransposon integrated within the baculovirus genome. *Mol. Cell. Biol.* **10:** 3067–3077.
56. **Gabriel, A., T. J. Yen, D. C. Schwartz, C. L. Smith, J. D. Boeke, B. Sollner-Webb, and D. W. Cleveland.** 1990. A rapidly rearranging retrotransposon within the miniexon gene locus of *Crithidia fasciculata. Mol. Cell. Biol.* **10:**615–624.
57. **Ganem, D., and H. E. Varmus.** 1987. The molecular biology of the hepatitis B viruses. *Annu. Rev. Biochem.* **56:**651–693.
58. **Garfinkel, D. J., J. D. Boeke, and G. R. Fink.** 1985. Ty element transposition: reverse transcriptase and virus-like particles. *Cell* **42:**507–517.
59. **Garrett, J., D. S. Knutzon, and D. Carroll.** 1989. Composite transposable elements in the *Xenopus laevis* genome. *Mol. Cell. Biol.* **9:**3018–3027.
60. **Gonzalez, P., and H. A. Lessios.** 1999. Evolution of sea urchin retroviral-like (SURL) elements: evidence from 40 echinoid species. *Mol. Biol. Evol.* **16:**938–952.
61. **Gorbalenya, A. E.** 1994. Self-splicing group I and group II introns encode homologous (putative) DNA endonucleases of a new family. *Protein Sci.* **3:**1117–1120.
62. **Hansen, L. J., D. L. Chalker, and S. B. Sandmeyer.** 1988. Ty3, a yeast retrotransposon associated with tRNA genes, has homology to animal retroviruses. *Mol. Cell. Biol.* **8:** 5245–5256.
63. **Hansen, J. L., A. M. Long, and S. C. Schultz.** 1997. Structure of the RNA-dependent RNA polymerase of poliovirus. *Structure* **5:**1109–1122.
64. **Hardies, S. C., C. H. Martin, C. F. Voliva, C. A. Hutchinson, and M. H. Edgell.** 1986. An analysis of replacement and synonymous changes in the rodent L1 repeat family. *Mol. Biol. Evol.* **3:**109–125.
65. **Hartl, D. L., A. R. Lohe, and E. R. Lozovskya.** 1997. Modern thoughts on an ancyent marinere: function, evolution, regulation. *Annu. Rev. Genet.* **31:**337–358.
66. **Hattori, M., S. Kuhara, O. Takenaka, and Y. Sakaki.** 1986. L1 family of repetitive DNA sequences in primates may be derived from a sequence encoding a reverse transcriptase-related protein. *Nature* **321:**625–628.
67. **Higashiyama, T., Y. Noutoshi, M. Fujie, and T. Yamada.** 1997. Zepp, a LINE-like retrotransposon accumulated in the Chorella telomeric region. *EMBO J.* **16:**3715–3723.
68. **Hohjoh, H., and M. F. Singer.** 1997. Sequence-specific single-stranded RNA binding protein encoded by the human LINE-1 protein and RNA. *EMBO J.* **15:**630–639.
69. **Inouye, S., and M. Inouye.** 1993. The retron: a bacterial retroelement required for the synthesis of msDNA. *Curr. Opin. Genet. Dev.* **3:**713–718.
70. **Jensen, S., and T. Heidmann.** 1991. An indicator gene for detection of germline retrotransposition in transgenic *Drosophila* demonstrates RNA-mediated transposition of the LINE I element. *EMBO J.* **10:**1927–1937.
71. **Jordan, I. K., L. V. Matyunia, and J. F. McDonald.** 1999. Evidence for the recent horizontal transfer of long terminal repeat retrotransposon. *Proc. Natl. Acad. Sci. USA* **96:** 12621–12625.
72. **Kajikawa, M., K. Ohshima, and N. Okada.** 1997. Determination of the entire sequence of turtle CR1: the first open reading frame of the turtle CR1 element encodes a protein with a novel zinc finger motif. *Mol. Biol. Evol.* **14:**1206–1217.
73. **Kazazian, H. H., and J. V. Moran.** 1998. The impact of L1 retrotransposition on the human genome. *Nat. Genet.* **19:** 19–24.
74. **Khan, E., J. P. Mack, R. A. Katz, J. Kulkosky, and A. M. Skalka.** 1991. Retroviral integrase domains: DNA binding and the recognition of LTR sequences. *Nucleic Acids Res.* **19:**851–860.
75. **Kim, J. M., S. Vanguri, J. D. Boeke, A. Gabriel, and D. F. Voytas.** 1998. Transposable elements and genome organization: a comprehensive survey of retrotransposons revealed by the complete *Saccharomyces cerevisiae* genome sequence. *Genome Res.* **8:**464–478.
76. **Kimmel, B. E., O. K. Ole-Moiyoi, and J. R. Young.** 1987. Ingi, a 5.2-kilobase dispersed sequence element from *Trypanosoma brucei* that carries half a smaller mobile element at their end and has homology with mammalian LINEs. *Mol. Cell. Biol.* **7:**1465–1475.
77. **Kohlstaedt, L. A., J. Wang, J. M. Friedman, P. A. Rice, and T. A. Steitz.** 1992. Crystal structure at 3.5 angstrom resolution of HIV-1 reverse transcriptase complexed with an inhibitor. *Science* **256:**1783–1790.
78. **Koonin, E. V., A. R. Mushegian, E. V. Ryabov, and V. V. Dolja.** 1991. Diverse groups of plant RNA and DNA viruses share related movement proteins that may possess chaperone-like activity. *J. Gen. Virol.* **72:**2895–2903.
79. **Kordis, D., and F. Gusenbek.** 1998. Unusual horizontal transfer of a long interspersed nuclear element between distant vertebrate classes. *Proc. Natl. Acad. Sci. USA* **95:** 10704–10709.
80. **Kulkosky, J., K. S. Jones, R. A. Katz, J. P. G. Mack, and A. M. Skalka.** 1992. Residues critical for retroviral integrative recombination in a region that is highly conserved among retroviral/retrotransposon integrases and the bacterial insertion sequence transposases. *Mol. Cell. Biol.* **12:**2331–2338.
81. **Kuzio, J., M. N. Pearson, S. H. Harwood, C. J. Funk, J. T. Evans, J. M. Slavicek, and G. F. Rohrmann.** 1999. Sequence

and analysis of the genome of a baculovirus pathogenic for *Lymantria dispar*. *Virology* **253**:17–34.

82. **Lambowitz, A. M., M. G. Caprara, S. Zimmerly, and P. S. Perlman.** 1999. Group I and group II ribozymes as RNPs: clues to the past and guides to the future, p. 451–485. *The RNA World II*. Cold Spring Harbor Laboratory, Cold Spring Harbor, N.Y.
83. **Laten, H. M., A. Majumdar, and E. A. Gaucher.** 1998. SIRE-1, a copia/Ty1-like retroelement from soybean, encodes a retroviral envelope-like protein. *Proc. Natl. Acad. Sci. USA* **95**: 6897–6902.
84. **Lathe, W. C., W. D. Burke, and T. H. Eickbush.** 1995. Evolutionary stability of the R1 retrotransposable element in the genus *Drosophila*. *Mol. Biol. Evol.* **12**:1094–1105.
85. **Lathe, W. C., III, and T. H. Eickbush.** 1997. A single lineage of R2 retrotransposable elements is an active, evolutionarily stable component of the *Drosophila* rDNA locus. *Mol. Biol. Evol.* **14**:1232–1241.
86. **Lerat, E., and P. Capy.** 1999. Retrotransposons and retroviruses: analysis of the envelope gene. *Mol. Biol. Evol.* **16**: 1198–1207.
87. **Li, W. H.** 1993. So, what about the molecular clock hypothesis? *Curr. Opin. Genet. Dev.* **3**:896–901.
88. **Lockhart, B. E., J. Menke, G. Dahal, and N. E. Olszewski.** 2000. Characterization and genomic analysis of tobacco vein clearing virus, a plant pararetrovirus that is transmitted vertically and related to sequences integrated in the host genome. *J. Gen. Virol.* **81**:1579–1585.
89. **Loeb, D. D., R. W. Padgett, S. C. Hardies, W. R. Shehee, M. B. Comer, M. H. Edgell, and C. A. Hutchison III.** 1986. The sequence of a large L1Md element reveals a tandemly repeated 5′ end and several features found in retrotransposons. *Mol. Cell. Biol.* **6**:168–182.
90. **Luan, D. D., and T. H. Eickbush.** 1995. RNA template requirements for target DNA-primed reverse transcription by the R2 retrotransposable element. *Mol. Cell. Biol.* **15**: 3882–3891.
91. **Luan, D. D., and T. H. Eickbush.** 1996. Downstream 28S gene sequences on the RNA template affect the choice of primer and the accuracy of initiation by the R2 reverse transcriptase. *Mol. Cell. Biol.* **16**:4726–4734.
92. **Luan, D. D., M. H. Korman, J. L. Jakubczak, and T. H. Eickbush.** 1993. Reverse transcription of R2Bm RNA is primed by a nick at the chromosomal target site: a mechanism for non-LTR retrotransposition. *Cell* **72**:595–605.
93. **Malik, H. S., W. D. Burke, and T. H. Eickbush.** 1999. The age and evolution of non-LTR retrotransposable elements. *Mol. Biol. Evol.* **16**:793–805.
94. **Malik, H. S., and T. H. Eickbush.** 1999. The RTE class of non-LTR retrotransposons is widely distributed in animals and is the origin of many SINEs. *Mol. Biol. Evol.* **15**: 1123–1134.
95. **Malik, H. S., and T. H. Eickbush.** 1999. Modular evolution of the integrase domain in the Ty3/Gypsy class of LTR retrotransposons. *J. Virol.* **73**:5186–5190.
96. **Malik, H. S., and T. H. Eickbush.** 2000. NeSL-1, an ancient lineage of site-specific non-LTR retrotransposons from *Caenorhabditis elegans*. *Genetics* **154**:193–203.
97. **Malik, H. S., and T. H. Eickbush.** 2001. Phylogenetic analysis of ribonuclease H domains suggests a late, chimeric origin of LTR retrotransposable elements and retroviruses. *Genome Res.* **11**:1187–1197.
98. **Malik, H. S., S. Henikoff, and T. H. Eickbush.** 2000. Poised for contagion: evolutionary origins of the infectious abilities of invertebrate retroviruses. *Genome Res.* **10**:1307–1318.
99. **Marin, I., and C. Llorens.** 2000. Ty3/Gypsy retrotransposons: description of new *Arabidopsis thaliana* elements and evolutionary perspectives derived from comparative genomic data. *Mol. Biol. Evol.* **17**:1040–1049.
100. **Marin, I., P. Plata-Rengifo, M. Labrador, and A. Fontdevila.** 1998. Evolutionary relationships among the members of an ancient class on non-LTR retrotransposons found in the nematode *Caenorhabditis elegans*. *Mol. Biol. Evol.* **15**: 1390–1402.
101. **Marlor, R., S. Parkhurst, and V. Corces.** 1986. The *Drosophila melanogaster* gypsy transposable element encodes putative gene products homologous to retroviral proteins. *Mol. Cell. Biol.* **6**:1129–1134.
102. **Martin, F., C. Maranon, M. Olivares, C. Alonso, and M. C. Lopez.** 1995. Characterization of a non-long terminal repeat retrotransposon cDNA (L1Tc) from *Trypanosoma cruzi*: homology of the first ORF with the Ape family of DNA repair enzymes. *J. Mol. Biol.* **247**:49–59.
103. **Martin, S. L.** 1991. LINEs. *Curr. Opin. Genet. Dev.* **1**: 505–508.
104. **McClure, M. A.** 1991. Evolution of retroposons by acquisition or deletion of retrovirus-like genes. *Mol. Biol. Evol.* **8**: 835–856.
105. **Michel, F., and J.-L. Ferat.** 1995. Structure and activities of group II introns. *Annu. Rev. Biochem.* **64**:435–461.
106. **Mizrokhi, L. J., S. G. Georgieva, and Y. V. Ilyin.** 1988. Jockey, a mobile *Drosophila* element similar to mammalian LINEs, is transcribed from the internal promoter by RNA polymerase II. *Cell* **54**:685–691.
107. **Mizrokhi, L. J., and A. M. Mazo.** 1990. Evidence for horizontal transmission of the mobile element jockey between distant Drosophila species. *Proc. Natl. Acad. Sci. USA* **86**: 9216–9220.
108. **Mizuuchi, K.** 1992. Transpositional recombination: mechanistic insights from studies of mu and other elements. *Annu. Rev. Biochem.* **61**:1011–1051.
109. **Moran, J. V., S. E. Holmes, T. P. Naas, R. J. DeBerardinis, J. D. Boeke, and H. H. Kazazian.** 1996. High frequency retrotransposition in cultured mammalian cells. *Cell* **87**:917–927.
110. **Mount, S. M., and G. M. Rubin.** 1985. Complete nucleotide sequence of the *Drosophila* transposable element copia: homology between copia and retroviral proteins. *Mol. Cell. Biol.* **5**:1630–1638.
111. **Nakamura, T. M., and T. R. Cech.** 1998. Reversing time: origin of telomerase. *Cell* **92**:587–600.
112. **Nakamura, T. M., G. B. Morin, K. B. Chapman, S. L. Weinrich, W. H. Andrews, L. Lingner, C. B. Harley, and T. R. Cech.** 1997. Telomerase catalytic subunit homologs from fission yeast and human. *Science* **277**:955–959.
113. **Okada, N., and M. Hamada.** 1997. The 3′ ends of tRNA-derived SINEs originated from the 3′ ends of LINEs: a example from the bovine genome. *J. Mol. Evol.* **44**:S52–S56.
114. **Okada, N., M. Hamada, I. Ogiwara, and K. Ohshima.** 1997. SINEs and LINEs share common 3′ sequences: a review. *Gene* **205**:229–243.
115. **Okazaki, S., H. Ishikawa, and H. Fujiwara.** 1995. Structural analysis of TRAS1, a novel family of telomeric repeat-associated retrotransposons in the silkworm, *Bombyx mori*. *Mol. Cell. Biol.* **15**:4545–4552.
116. **Palmer, J. D., and J. M. Logsdon, Jr.** 1991. The recent origins of introns. *Curr. Opin. Genet. Dev.* **1**:470–477.
117. **Pantazidis, A., M. Labrador, and A. Fontdevila.** 1999. The retrotransposon Osvaldo from *Drosophila buzzatii* displays all structural features of a functional retrovirus. *Mol. Biol. Evol.* **16**:909–921.
118. **Pelisson, A., D. J. Finnegan, and A. Bucheton.** 1991. Evidence for retrotransposition of the I factor, a LINE element of *Dro-*

sophila melanogaster. *Proc. Natl. Acad. Sci. USA* **88:** 4907–4910.

119. **Peterson-Burch, B. D., D. A. Wright, H. M. Laten, and D. F. Voytas.** 2000. Retroviruses in plants? *Trends Genet.* **14:** 151–152.
120. **Petrov, D. A., E. L. Lozovskaya, and D. L. Hartl.** 1996. High intrinsic rate of DNA loss in Drosophila. *Nature* **384:** 346–349.
121. **Pfeiffer, P., and T. Hohn.** 1983. Involvement of reverse transcription in the replication of cauliflower mosaic virus: a detailed model and test of some aspects. *Cell* **33:**781–789.
122. **Richert-Poggeler, K. R., and R. J. Shepherd.** 1997. Petunia vein-clearing virus: a plant pararetrovirus with the core sequences for an integrase function. *Virology* **236:**137–146.
123. **Robertson, H. M.** 1993. The mariner transposable element is widespread in insects. *Nature* **362:**241–245.
124. **Rogers, J.** 1985. The origin and evolution of retroposons. *Int. Rev. Cytol.* **93:**187–279.
125. **SanMiguel, P., B. S. Gaut, A. Tiknonov, Y. Nakajima, and J. L. Bennetzen.** 1998. The paleontology of intergene retrotransposons of maize. *Nat. Genet.* **20:**43–45.
126. **Saitou, N., and M. Nei.** 1987. The neighbor-joining method: a new method for reconstructing phylogenetic trees. *Mol. Biol. Evol.* **4:**406–425.
127. **Schwarz-Sommer, Z., L. Leclercq, E. Gobel, and H. Saedler.** 1987. Cin4, an insert altering the structure of the A1 gene in *Zea mays*, exhibits properties of nonviral retrotransposons. *EMBO J.* **6:**3873–3880.
128. **Shub, D. A., H. Goodrich-Blair, and S. R. Eddy.** 1994. Amino acid sequence motif of group I intron endonucleases is conserved in open reading frames of group II introns. *Trends Biochem. Sci.* **19:**402–406.
129. **Seleme, M.-D. C., I. Busseau, S. Malinsky, A. Bucheton, and D. Teninges.** 1999. High-frequency retrotransposition of a marked I factor in *Drosophila melanogaster* correlates with a dynamic expression pattern of the ORF1 protein in the cytoplasm of oocytes. *Genetics* **151:**761–771.
130. **Sheen, F. M., and R. W. Levis.** 1994. Transposition of the LINE-like retrotransposon TART to Drosophila chromosome termini. *Proc. Natl. Acad. Sci. USA* **91:**12510–12514.
131. **Shiba, T., and K. Saigo.** 1983. Retrovirus-like particles containing RNA homologous to the transposable element copia in *Drosophila melanogaster*. *Nature* **302:**119–124.
132. **Singer, M. F., and J. Skowronski.** 1985. Making sense out of LINEs: long interspersed repeat sequences in mammalian genomes. *Trends Biochem. Sci.* **10:**119–122.
133. **Smit, A. F.** 1996. The origin of interspersed repeats in the human genome. *Curr. Opin. Genet. Dev.* **6:**743–748.
134. **Smit, A. F.** 1999. Interspersed repeats and other mementos of transposable elements in mammalian genomes. *Curr. Opin. Genet. Dev.* **9:**657–663.
135. **Song, S. U., T. Gerasimova, M. Kurkulos, J. D. Boeke, and V. G. Corces.** 1994. An env-like protein encoded by a Drosophila retroelement: evidence that gypsy is an infectious retrovirus. *Genes Dev.* **8:**2046–2057.
136. **Springer, M. S., N. A. Tusneem, E. H. Davidson, and R. J. Britten.** 1995. Phylogeny, rates of evolution, and patterns of codon usage among sea urchin retroviral-like elements, with implications for the recognition of horizontal transfer. *Mol. Biol. Evol.* **12:**219–230.
137. **Swergold, G. D.** 1990. Identification, characterization, and cell specificity of a human LINE-1 promoter. *Mol. Cell. Biol.* **10:**6718–6729.
138. **Szafranski, K., G. Glockner, T. Dingermann, K. Dannat, A. A. Noegel, L. Eichinger, A. Rosenthal, and T. Winckler.** 1999. Non-LTR retrotransposons with unique integration preferences downstream of *Dictyostelium discoideum* tRNA genes. *Mol. Gen. Genet.* **262:**772–780.
139. **Szemraj, J., G. Plucienniczak, J. Jaworski, and A. Plucienniczak.** 1995. Bovine *Alu*-like sequences mediate transposition of a new site-specific retroelement. *Gene* **152:**261–264.
140. **Takahashi, H., S. Okazaki, and H. Fujiwara.** 1997. A new family of site-specific retrotransposons, SART1, is inserted into telomeric repeats of the silkworm, *Bombyx mori*. *Nucleic Acids Res.* **25:**1578–1584.
141. **Takeya, T., H. Hanafusa, R. P. Junghans, G. Ju, and A. M. Skalka.** 1981. Comparison between the viral transforming gene (src) of recovered avian sarcoma virus and its cellular homolog. *Mol. Cell. Biol.* **1:**1024–1037.
142. **Temin, H. M.** 1989. Retrons in bacteria. *Nature* **339:** 254–255.
143. **Terzian, C., C. Ferraz, J. Demaille, and A. Bucheton.** 2000. Evolution of the Gypsy endogenous retrovirus in the Drosophila melanogaster subgroup. *Mol. Biol. Evol.* **17:**908–914.
144. **Tessier L. H., M. Keller, R. L. Chan, R. Fournier, J. H. Weil, and P. Imbault.** 1991. Short leader sequences may be transferred from small RNAs to pre-mature mRNAs by trans-splicing in Euglena. *EMBO J.* **10:**2621–2625.
145. **Trelogan, S. A., and S. L. Martin.** 1995. Tightly regulated, developmentally specific expression of the first open reading frame from LINE-1 during mouse embryogenesis. *Proc. Natl. Acad. Sci. USA* **92:**1520–1524.
146. **Usdin, K., P. Chevret, F. M. Catzeflis, R. Verona, and A. V. Furano.** 1995. L1 (LINE-1) retrotransposable elements provide a "fossil" record of the phylogenetic history of murid rodents. *Mol. Biol. Evol.* **12:**73–82.
147. **Varmus, H., and P. Brown.** 1989. Retroviruses, p. 53–108. *In* D. E. Berg and M. M. Howe (ed.), *Mobile DNA*. American Society for Microbiology, Washington, D.C.
148. **Vazquez-Manrique, R. P., M. Hernandez, M. J. Martinez-Sebastian, and R. de Frutos.** 2000. Evolution of gypsy endogenous retroviruses in the *Drosophila obscura* species group. *Mol. Biol. Evol.* **17:**1185–1193.
149. **Verneau, O., F. Catzeflis, and A. V. Furano.** 1998. Determining and dating recent rodent speciation events by using L1 (LINE-1) retrotransposons. *Proc. Natl. Acad. Sci. USA* **95:** 11284–11289.
150. **Volff, J.-N., C. Korting, and M. Schartl.** 2000. Multiple lineages of the non-LTR retrotransposon Rex1 with various success in invading fish genomes. *Mol. Biol. Evol.* **17:** 1673–1684.
151. **Voliva, C. F., S. L. Martin, C. A. Hutchison III, and M. H. Edgell.** 1984. Dispersal process associated with the L1 family of interspersed repetitive DNA sequences. *J. Mol. Biol.* **178:** 795–813.
152. **Voytas, D. F., and F. M. Ausubel.** 1988. A copia-like transposable element family in *Arabidopsis thaliana*. *Nature* **336:** 242–244.
153. **Wang, G.-H., and C. Seeger.** 1992. The reverse transcriptase of hepatitis B virus acts as a primer for viral DNA synthesis. *Cell* **71:**663–670.
154. **Wang, H., and A. M. Lambowitz.** 1993. The Mauriceville plasmid reverse transcriptase can initiate cDNA synthesis de novo and may be related to reverse transcriptase and DNA polymerase progenitor. *Cell* **75:**1071–1081.
155. **Wank, H., J. SanFilippo, R. N. Singh, M. Matsuura, and A. M. Lambowitz.** 1999. A reverse-transcriptase/maturase promotes splicing by binding at its own coding segment in a group II intron RNA. *Mol. Cell.* **4:**239–250.
156. **Wei, W., N. Gilbert, S. L. Ooi, J. F. Lawler, E. M. Ostertag, H. H. Kazazian, J. D. Boeke, and J. V. Moran.** 2001. Human

L1 retrotransposition: *cis* preference versus *trans* complementation. *Mol. Cell. Biol.* **21**:1429–1439.

157. **Weiner, A. M., P. L. Deininger, and A. Efstratiadis.** 1986. Nonviral retroposons: genes, pseudogenes, and transposable elements generated by the reverse flow of genetic information. *Annu. Rev. Biochem.* **55**:631–661.
158. **Wright, D. A., and D. F. Voytas.** 1998. Potential retroviruses in plants: Tat1 is related to a group of *Arabidopsis thaliana* Ty3/gypsy retrotransposons that encode envelope-like proteins. *Genetics* **149**:703–715.
159. **Xiong, Y., and T. H. Eickbush.** 1988. Similarity of reverse transcriptase-like sequences of viruses, transposable elements, and mitochondrial introns. *Mol. Biol. Evol.* **5**: 675–690.
160. **Xiong, Y., and T. H. Eickbush.** 1990. Origin and evolution of retroelements based upon their reverse transcriptase sequences. *EMBO J.* **9**:3353–3362.
161. **Xiong, Y., W. D. Burke, and T. H. Eickbush.** 1993. Pao, a highly divergent retrotransposable element from *Bombyx mori* containing long terminal repeats with tandem copies of the putative R region. *Nucleic Acids Res.* **21**:2117–2123.
162. **Xiong, Y., and T. H. Eickbush.** 1993. *Dong*, a new non-long terminal repeat (non-LTR) retrotransposable element from *Bombyx mori. Nucleic Acids Res.* **21**:1318.
163. **Yang, J., and T. H. Eickbush.** 1998. RNA-induced changes in the activity of the endonuclease encoded by the R2 retrotransposable element. *Mol. Cell. Biol.* **18**:3455–3465.
164. **Yang, J., H. S. Malik, and T. H. Eickbush.** 1999. Identification of the endonuclease domain encoded by R2 and other site-specific, non-long terminal repeat retrotransposable elements. *Proc. Natl. Acad. Sci. USA* **96**:7847–7852.
165. **Yang, J., G. Mohr, P. S. Perlman, and A. M. Lambowitz.** 1998. Group II intron mobility in yeast mitochondria: target DNA-primed reverse transcription activity in aI1 and reverse splicing into DNA transposition sites in vitro. *J. Mol. Biol.* **282**:505–523.
166. **Zimmerly, S., H. Guo, P. S. Perlman, and A. M. Lambowitz.** 1995. Group II intron mobility occurs by target DNA-primed reverse transcription. *Cell* **82**:1–10.
167. **Zimmerly, S., H. Guo, R. Eskes, J. Yang, P. S. Perlman, and A. M. Lambowitz.** 1995. A group II intron RNA is a catalytic component of a DNA endonuclease involved in intron mobility. *Cell* **83**:529–538.

X. NEWLY DISCOVERED TRANSPOSON FAMILIES

Mobile DNA II
Edited by N. L. Craig et al.

Chapter 50

Miniature Inverted-Repeat Transposable Elements and Their Relationship to Established DNA Transposons

Cédric Feschotte, Xiaoyu Zhang, and Susan R. Wessler

THE DISCOVERY OF MITEs IN PLANT AND ANIMAL GENOMES

A 128-bp insertion in a mutant maize *waxy* gene (the *wxB2* allele) led to the identification of a group of related elements called *Tourist* in the untranslated regions of genes from diverse grass species (5, 7). A 257-bp insertion in a sorghum *Tourist* element led to the discovery of a second family of elements, called *Stowaway*, in the genes of a diversity of flowering plants (6). *Tourist* and *Stowaway* elements share structural but not sequence similarity. Both are short (~100 to 500 bp), have conserved terminal repeats, have target site preference (*Tourist*, TAA; *Stowaway*, TA), and, of most significance for this chapter, have no coding potential.

All of these features are reminiscent of the nonautonomous members of some DNA transposon families. However, it was the high copy number of *Tourist* and *Stowaway* elements and the uniformity of related elements that served to set them apart from previously described nonautonomous elements. For this reason, and because they were being mistakenly classified as SINEs, it was decided to name these and other structurally related elements miniature inverted-repeat transposable elements (MITEs) (4, 64). Although first discovered in plants, MITEs were soon found in several animal genomes, including those of *Caenorhabditis elegans* (41, 42), mosquitoes (16, 58), fish (25), *Xenopus* spp. (62), and humans (38, 51).

With the advent of genome sequencing projects, vast amounts of DNA sequence, from a wide variety of plant and animal species, have become available for analysis. MITEs, with their high copy number, distinct structural features (target site duplications [TSDs] and terminal inverted repeats [TIRs]), and compact stature, are relatively easy to mine from DNA sequence databases. As such, the number of MITEs and MITE families has proliferated in the literature much as the MITEs themselves have proliferated

Cédric Feschotte, Xiaoyu Zhang, and Susan R. Wessler • Departments of Botany and Genetics, University of Georgia, Athens, GA 30602.

in the genome. Initially this led to a confusing jumble of names and hypothesized relationships. This confusion reflected two important facts about MITEs. First, as very short, nonautonomous elements, none of the available MITE sequences revealed clear-cut relationships with known transposases. Second, and perhaps most significantly, no MITE family to date has been shown to be actively transposing. In the absence of coding sequences and activity, it has been difficult to determine how MITEs originate and how they attain their high copy numbers.

Fortunately, this situation has changed dramatically in the past few years as two approaches have been successfully employed to establish relationships between MITEs and existing DNA transposon families. We call these approaches "bottom up" and "top down," and they are illustrated below in the descriptions of Tc1/*mariner*-like MITEs and *PIF*/*Harbinger*-like MITEs, respectively. With a bottom up approach, the sequences of nonautonomous family members are used as queries to identify potentially autonomous family members through similarities in their TSDs and TIRs. This approach has revealed complex relationships between hundreds of MITE families and the well-characterized Tc1/*mariner* superfamily of transposases. In contrast, the top down approach began with the discovery that a genetically active DNA transposon system in maize (called *P Instability Factor* [*PIF*]) had MITE members (called *miniature PIF* [*mPIF*]) and led to the identification of a new superfamily of transposases that are responsible for the transposition of *Tourist*-like MITEs.

ORGANIZING THE DIVERSITY OF MITEs

The data summarized in Table 1 are the first attempt to classify most or all of the previously published MITEs as well as those recently generated by the systematic mining of the complete genome sequences of *Arabidopsis thaliana* and *C. elegans* (consensus sequences are available through Repbase at http://www.girinst.org [28]). As mentioned above, this task is complicated by the fact that MITEs lack coding capacity and different families generally have common structural characteristics but little if any sequence similarity. In Table 1, MITEs are grouped into superfamilies based on their association with established superfamilies of transposases. The link between a given MITE family and a source of transposase is first based on the length and sequence of the TSD, as this feature is a function of the transposase (chapter 1). This relationship is strengthened when significant sequence similarity, in particular in the TIRs, is shared between MITEs and a transposon(s) with coding capacity for the transposase. We have attempted to quantify these various degrees of association in Table 1 by assigning each match between a MITE family and a potential partner with one of four levels of sequence similarity. These are as follows.

At level 1, the MITE and the DNA transposon share significant sequence similarity over the entire MITE sequence; the MITE is likely to be derived from the larger element by an internal deletion. At level 2, the MITE and DNA transposon share sequence similarity in their terminal and subterminal regions; only an internal segment of the MITE appears unrelated to the partner element. At level 3, the MITE only has the TIRs identical or almost identical to those of a DNA transposon from the same genome. At level 4, the MITE only has the TIRs identical or almost identical to those of a DNA transposon from another species.

A level 1 or 2 designation indicates that there is strong evidence for the involvement of the larger element in both MITE origin and amplification. A level 3 designation indicates strong evidence that the MITE family was mobilized by the transposase encoded by the larger transposon (or by a close relative) but does not provide information as to the origin of the MITE. Finally, a level 4 designation provides only indirect evidence of a possible relationship between a MITE family and an existing superfamily of transposase and thus reflects the most tentative assignment. Evidence supporting the data presented in Table 1 is presented in sections that follow.

THE Tc1/*mariner* SUPERFAMILY AS A SOURCE OF MITEs

As discussed above, *Stowaway* was one of the first described MITE families. Fifty *Stowaway* elements were initially discovered in close association with the genes of diverse flowering plants, including six grass species and 12 species of dicotyledonous plants (6). All *Stowaway* elements described to date have conserved 11-bp termini (5′-CTCCCTCCGTT-3′), are short (70 to 350 bp) and homogeneous in length within subfamilies, and have a preference for insertion into the TA dinucleotide (the TSD) (6, 32, 64). From the initial discovery of *Stowaway* elements, a connection was noted between *Stowaway*'s preference for TA dinucleotide insertion sites and the fact that this feature is also a hallmark of the Tc1/*mariner* superfamily of DNA transposons (6).

MITEs Related to Tc1/*mariner* Transposons in *C. elegans*

More-extensive sequence relationships between MITEs and Tc1/*mariner* elements were first established with *C. elegans,* where MITEs make up more than 2% of the genome (42, 54). Using computer-assisted searches, Oosumi et al. (41, 42) identified several MITE families with copy numbers that range from a few dozen (*Cele5*) to several hundred (*Cele2*). Most of the *C. elegans* MITE families share their termini (~20 to 150 bp) and TSD sequence with one of the many Tc1/*mariner* transposons described in this species (including Tc1, Tc2, Tc5, and *mariner*-like elements [MLEs] [see references 41 and 42, Table 1, and chapter 22).

Comparison of the transposase-encoding Tc elements and the numerous MITE families suggested possible scenarios for the origin of MITEs in *C. elegans.* For example, *CeleTc2, Cele11,* and *Cele12* all have Tc2-like TIRs, while their internal sequences

Table 1. Classification of MITEs

Possible transposase superfamily	Levels of similarity[a]	Species	Family	Approx. copy no.	TSD[b]	No. of TIRs	Approx. size (bp)	Reference(s)[c]
Tc1/*mariner*								
D39D mariner-like	3	Various flowering plants	*Stowaway*[d]	ND[e]	TA	>10	70–350	6, 11
	2, 3	*Oryza sativa*	Various *Stowaway*[d]	40,000	TA	20–150	100–350	61, b
	4	*Arabidopsis thaliana*	Various *Stowaway*[d]	300	TA	25–100	200–300	32, c
D34D	3	*Caenorhabditis elegans*	*Cele1*	1,000	TA	120	330	41
mariner-like	3	*C. elegans*	*Cele2*	1,500	TA	90	325	41
	3	*C. elegans*	*Cele4*	300	TA	11	470	41
	3	*C. elegans*	*Cele6*	100	TA	50	160	41
	3	*C. elegans*	*PALTA1_CE*	300	TA	580	1,466	a
	3	*C. elegans*	*PALTA2_CE*	50	TA	600	1,534	a
	3	*C. elegans*	*PALTA4_CE*	20	TA	87	198	a
	3	*C. elegans*	*TIR54TA1_CE*	100	TA	54	200	a
mariner-like	4	*Culex pipiens*	*Milord*	3,000	TA	118	521	c
D34D mariner-like	1	*Homo sapiens*	*Mrs/Madel*	2,400	TA	37	80	38
pogo-like	1	*A. thaliana*	*Emigrant*	250	TA	24	550	9
	4	*Aedes aegypti*	*Wujin*	3	TA	23	185	58
	4	*C. pipiens*	*Mimo*	1,000	TA	23	350	16
	4	*C. pipiens*	*Nemo*	ND	TA	25	324	16
	4	*Anopheles gambiae*	*TA-I*	1,840	TA	24	355	56
	4	*A. gambiae*	*TA-II*	1,080	TA	23	368	56
	4	*A. gambiae*	*TA-IV*	130	TA	20	363	56
	4	*A. gambiae*	*TA-V*	300	TA	22	348	56
	1	*Drosophila melanogaster*	*Dm-mPogo*	30–50	TA	25	180	56
	1, 3	*H. sapiens*	MER2 group[d]	50,000	TA	23–25	250–700	51
impala-like	3	*Fusarium oxysporum*	*mimp*	ND	TA	27	~220	22
Tc1-like	3	*C. elegans*	*Cele*Tc1/Tc7	30	TA	60–350	920	a
	3	*C. elegans*	Tc6	30	TA	765	1,600	12
	3	*C. elegans*	*NPALTA1_CE*	50–100	TA	49	173	a
	4	*A. gambiae*	*TA-III*	970	TA	54	245	56
	4	*C. pipiens*	*Mikado*	~1,500	TA	30	830	c
	4	*C. pipiens*	*Mirza*	ND	TA	72	170	c
	4	*Tenebrio molitor*	*DEC*	3,500	TA	26	475	3
Tc2-like	3	*C. elegans*	*Cele11*	25	TA	52	220	42
	3	*C. elegans*	*Cele12*	50	TA	39	370	42
	3	*C. elegans*	*Cele*Tc2	150	TA	111	210	41
	4	*A. aegypti*	*Pony*	15,000	TA	24	500	57
Tc5-like	3	*C. elegans*	*Cele*Tc5	15	TA	280–680	500–1,400	41

Continued on following page

Table 1. *Continued*

Possible transposase superfamily	Levels of similarity[a]	Species	Family	Approx. copy no.	TSD[b]	No. of TIRs	Approx. size (bp)	Reference(s)[c]
PIF/Harbinger	2, 3	Grasses	Various *Tourist*[d]	>5,000	TWA	13–100	100–400	5, 7, 45
	3	*Zea mays*	*Hbr*	4,000	TWW	14	310	67
	3	*Z. mays*	*Zm-mPIF*	6,000	TWA	13	350	68
	3	*Z. mays*	*B2*	1,000	TWA	14	130	7, 27
	3	*O. sativa*	*Castaway*	1,000	TWA	13	350	4
	3	*O. sativa*	*Ditto*	2,000	TWA	15	300	4
	3	*O. sativa*	*Wanderer*	4,000	TWA	10	300	4
	3	*O. sativa*	*Gaijin*	3,000	WWW	17	180	4
	3	*O. sativa*	*Explorer*	2,000	TWA	13	240	4
	2	*O. sativa*	*Os-mPIF2*	150	WWW	14	260	68
	3	*A. thaliana*	*MathE1*	50	TWA	25	400	53
	3	*A. thaliana*	*ATTIRX1*	70	TWA	16–40	350–400	29
	3	*A. thaliana*	*ATTIR16T3A*	100	TWA	16	500	29
	3	*A. thaliana*	Various *Tourist*[d]	>300	TWA	>14	300–500	32
	2	*A. thaliana*	*At-mPIF2*	20	TWA	14	400	68
	4	Bell pepper and Solanaceae	*Alien*	2,400	TWA	25	400	43
	3	*C. elegans*	*Cele7*	300	TWA	170	360	40
	1	*C. elegans*	*PAL3A_CE*	100	TWA	28	150	61, a
	2	*Caenorhabditis briggsae*	*Cb-mPIF1a*	50	TWA	60	244	68
	1	*C. briggsae*	*Cb-mPIF1b*	30	TWA	27	60	58
	4	*A. gambiae*	*Joey*	1,120	TNA	70	350	2, 56
	4	*Anopheles stephensi*	*NOS*	ND	TAA	29	492	34
	4	*A. gambiae*	*TAA-I*	40	TAA	38	184	56
	4	*A. gambiae*	*TAA-II*	320	TAA	26	142	56
	4	*Xenopus laevis*	*Vi*	7,500	TWA	16	100–470	48
piggyBac/TTAA	4	*C. elegans*	*PALTTAA1_CE*	100	TTAA	285	592	a
	4	*C. elegans*	*PALTTAA2_CE*	100	TTAA	65	174	a
	4	*C. elegans*	*PALTTAA3_CE*	50	TTAA	285	594	a
	4	*A. aegypti*	*Wuneng*	2,700	TTAA	19	256	58
	4	*X. laevis*	*REM1*	25,000	TTAA	15	500	19
	4	*Xenopus* spp.	*Xbr*	5×10^3–2×10^4	TTAA	42	475	62
	4	*Xenopus* spp.	*XR*	ND	TTAA	12	500	62
	4	*Xenopus* spp.	*Ub3*	ND	TTAA	44	450	62
	4	*Xenopus* spp.	*Ub7*	20,000	TTAA	22	500	62
	4	*Xenopus* spp.	*Xfb*	ND	TTAA	127	500	62
	4	*Xenopus* spp.	*Pir*	4,000	TTAA	15	500	23
	4	*Danio rerio* and other fish	*Angel*	10^3–10^4	TTAA	26	315	25
	3	*H. sapiens*	MER75	ND	TTAA	14	242/540	d
	3	*H. sapiens*	MER85	2,000	TTAA	13	140	d
hAT	1, 3	*O. sativa*	Various *hAT*-like[d]	2,000	8 bp	12–15	100–600	b
	3	*A. thaliana*	*MathE3*	<100	8 bp	70–135	300–1,200	53
	1, 3	*A. thaliana*	Various *hAT*-like[d]	20–200	8 bp	12–20	250–800	c, e
	4	*A. gambiae*	*Pegasus*	90	8 bp	8	534	2, 56
	4	*A. gambiae*	*8bp-I*	725	8 bp	8	534	56
	3	*X. laevis*	*Ocr*	1×10^3–5×10^3	8 bp	19	300–400	39
	3	*X. laevis*	*Vision*	300	7 bp	14	284	33
	1, 3	*H. sapiens*	MER1 group[d]	100,000	8 bp	14–15	200–500	51

Continued on following page

Table 1. *Continued*

Possible transposase superfamily	Levels of similarity[a]	Species	Family	Approx. copy no.	TSD[b]	No. of TIRs	Approx. size (bp)	Reference(s)[c]
Mutator-like	2, 3	*O. sativa*	Various *Os-mMu*[d]	3,000	9 bp	10–300	100–700	b
	2, 3	*A. thaliana*	Various *At-mMu*[d]	<100	9 bp	~300	1 kb	e
	4	*Medicago truncatulata*	*Bigfoot*	10^3–10^4	9 bp	38	175	10
Mirage	3	*C. elegans*	*PALNN1_CE*	20	2 bp	70	210	a
	3	*C. elegans*	*NPAL0A_CE*	20–30	2 bp	118	286	a
Merlin/IS1016	4	*C. elegans*	*PAL8C_CE*	>100	8 bp	50	200–350	a, c
Unclassified (unknown transposase)		*Ipomoea nil*	*MELS3*	ND	3 bp?	50	185	24
		Ipomoea nil	*MELS4*	ND	3 bp?	49	215	24
		Ipomoea nil	*MELS5*	ND	3 bp?	13	251	24
		Ipomoea nil, *Ipomoea purpurea*	*MELS8*	ND	ND	10	355	13
		Neurospora crassa	*Guest*	ND	GTA	15	98	66
		O. sativa	*Snabo-2*	150	4 bp?	107	383	11
		A. aegypti	*Wukong*	2,200–3,000	TAYA	17	430	58
		Ciona intestinalis	*Cimi-1*	17,000	T(A/T)ATA	30	193	49
		A. thaliana	*Hairpin*	10	CTWAR	114	238	1
		X. laevis	*Glider*	20,000	6–11 bp	14	150	33
		C. elegans	*Cele42*	600	6 bp	23	240	54
		C. elegans	*Cele14*	2,000	6 bp	58	180	42
		C. elegans	*PALTTTAAA1_CE*	50	TTTAAA	320	680	a
		C. elegans	*PALTTTAAA2_CE*	50	TTTAAA	320	680	a
		O. sativa	*Snap*	100	7 bp	SIRs[f]	170	52
		O. sativa	*Crackle*	500	8 bp	SIRs	385	52
		O. sativa	*Pop*	50	8 bp	SIRs	125	52
		A. aegypti	*Microuli*	3,000	TTAA	SIRs	209	56
		C. pipiens	*Mint*	ND	CA	SIRs	141	16
		Drosophila obscura group	*SGM-IS*	>100?	ND	SIRs	600–1,200	37

[a]Between MITEs and potential partner DNA transposon (see text for description of each level).
[b]N = any nucleotide; W = A or T; Y = C or T; R = A or G.
[c]a, V. Kapitonov and J. Jurka, Repbase update, *C. elegans* section (http://www.girinst.org); b, N. Jiang, unpublished data; c, C. Feschotte, unpublished data; d, J. Jurka and A. Smit, Repbase update, *H. sapiens* section; e, V. Kapitonov and J. Jurka, Repbase update, *A. thaliana* section.
[d]Refers to several MITE families grouped into the same superfamily based on TIR and TSD sequence similarities.
[e]ND, not determined.
[f]These elements have no TIRs but have subterminal inverted repeats (SIRs).

have little similarity to each other or to other Tc2 sequences (42). To explain this and related instances, the authors hypothesized that a single Tc element type can mobilize a variety of highly divergent sequences. Supporting this notion was the finding that Tc7 MITEs could be mobilized in vivo and in vitro by an autonomous Tc1 element, even though Tc7 and Tc1 share only their 36 terminal nucleotides (44). It is possible that Tc7 was initially derived from a complete Tc1-like element that was subsequently lost from the genome of most *C. elegans* strains (44). Alternatively, new MITEs may arise de novo from the fortuitous association of TIRs flanking unrelated segments of DNA.

MITEs Related to Tc1/*mariner* Transposons in Humans

Examples of Tc1/*mariner*-related MITEs have also been found in the human genome. Several groups independently discovered an abundant family of short (80-bp) palindromic elements, called *Mrs* or *Made1* (38, 40, 51). These elements are extremely homogeneous in length, consisting of two 37-bp TIRs separated by a 6-bp sequence. The TIRs are 80 to 100% similar to those of MLEs that are dispersed in the genome (38, 40, 51). These were the first MLEs identified in mammals and were subsequently grouped into the *Hsmar1* family (46; also see chapter 48). There are ~200 *Hsmar1* copies in the human genome, while the copy number of the 80-bp *Mrs* MITE is estimated to be ~2,400 (46). Based on their sequence and size homogeneity, it was proposed that *Mrs* MITEs originated from a single *Hsmar1* deletion derivative (38, 46).

The human genome also harbors another large family of MITEs called the MER2 group (51). MER2 elements are short (200 to 800 bp) and form numerous families that are distinguished by their homogeneity of length and sequence, once again hinting that a family originated from a single element (Repbase update, *H. sapiens* section [http://www.girinst.org]). Overall it is estimated that our genome harbors more than 30,000 MER2 MITEs. Based on similar TIR sequences and TA target site duplication, MER2 is associated with larger transposons, called *Tiggers,* that contain large open reading frames (ORFs) with similarity to the *pogo* subgroup of Tc1/*mariner* transposases (51; also see chapter 48). *Tigger* subfamilies can be directly connected with MER2 subfamilies by sequence similarity that extends into internal regions. For example, MER28 MITEs (~435 bp) resemble internal deletion derivatives of *Tigger2,* but they are five times more abundant than *Tigger2* (~5,000 versus 1,000 copies [51]).

MITEs Related to Tc1/*mariner* Transposons in Insects

MITEs with TIRs that are strikingly similar to those of *pogo*-like elements have been described in three distantly related species of mosquitoes (16, 56, 58) and fragments of *pogo*-like transposases have been detected in the mosquito *Anopheles gambiae* (http://bioweb.pasteur.fr/BBMI/trans.html; C. Feschotte, unpublished data). It is thus likely that *pogo*-like transposases are also responsible for the proliferation of multiple MITE families in mosquitoes. The original *pogo* element from *Drosophila melanogaster* has also given rise to a homogeneous group of deletion derivatives (180 bp [60]) that could be viewed as one of the rare MITE families in this species (with ~40 copies in the available genomic sequence). Finally, another abundant MITE family, *Pony* (~18,000 copies), from the genome of the mosquito *Aedes aegypti,* displays TIRs with striking similarity to those of the Tc2 transposon from *C. elegans* and also has the TA target site duplication (57). Thus, *Pony* MITEs may have proliferated by using endogenous Tc2-like transposases. Taken together, these data suggest that Tc1/*mariner* transposons have been a common source of transposase for the origin and/or amplification of MITEs in animals.

MITEs Related to Tc1/*mariner* Transposons in Plants

Tc1/*mariner* transposons were long believed to be absent or rare in the plant kingdom (8, 20). However, recent studies indicate that Tc1/*mariner* transposons are actually widespread in plant genomes and have probably given rise to a large fraction of plant MITEs (15, 17). Evidence connecting a plant MITE family with a Tc1/*mariner* transposon was first obtained by analyzing the genome sequence of *Arabidopsis.* Homology-based searches revealed that *Emigrant,* the first MITE family identified in this species (9), originated from the larger *Lemi1,* which has coding capacity for a *pogo*-like transposase (15, 29). *Lemi1* is present as a single copy in the Columbia ecotype, where there are ~250 copies of *Emigrant.* Sequence similarity between *Emigrant* and *Lemi1* is moderate (~70%) but encompasses the entire MITE consensus sequence (15). Therefore, *Emigrant* MITEs probably originated by internal deletion of *Lemi1* or from a closely related element.

To date, plant MITEs related to *pogo*-like transposons have only been identified in *Arabidopsis* (15, 29). However, there are now several lines of evidence that the widespread *Stowaway* MITEs are related to a new group of Tc1/*mariner* transposons. That *Stowaway* MITEs display a strong preference for TA targets was the first indication that this heterogeneous group might be related to the Tc1/*mariner* superfamily (6). Recently it was shown that the 10 terminal nucleotides characteristic of *Stowaway* MITEs match those of the two elements identified in soybean and rice that possess long ORFs with similarity to animal *mariner* transposases (26, 55, 61). This provides additional evidence that *Stowaway* MITEs were mobilized by transposases encoded in *trans* by MLEs.

Given the wide distribution of *Stowaway* in plants, it follows that MLEs should also be widespread in their genomes. Database searches and a PCR approach exploiting newly designed plant-specific primers were recently combined to demonstrate that MLEs are present in a wide range of flowering plants (17). Phylogenetic analyses of over 100 plant MLE transposase sequences revealed the existence of multiple and divergent lineages of MLE transposases (17). Together these results provide an explanation for the proliferation, diversity, and success of *Stowaway* MITEs in plant genomes.

Tourist-LIKE MITEs ARE RELATED TO MEMBERS OF THE *PIF/Harbinger* SUPERFAMILY

Connections between several families of MITEs and the Tc1/*mariner* superfamily of transposases are numerous and widespread in animal genomes and now also in plant genomes. In contrast, connections between the large numbers of *Tourist*-like MITEs and possible sources of transposase were much more elusive because *Tourist* TSDs and TIRs were not related to any well-characterized transposon family (7).

This situation began to change with the discovery of *Harbinger*, a 5.4-kb element that was mined from the *Arabidopsis* genome sequence (29). *Harbinger* contains an ORF that could potentially encode a transposase related to transposases from bacterial insertion elements of the IS5 group (29, 32; also see chapter 15). Similarities between *Harbinger* and two diverse nonautonomous transposons in *C. elegans*, *Turmoil1* and *Turmoil2*, were also noted (29). These elements all have similar 3-bp TSDs and TIRs. Further analysis of *Turmoil* family members revealed their relationship with *Tourist*-like MITEs (31).

Maize *PIF* and *mPIF* MITEs

The relationships between *Harbinger*, *Turmoil*, and *Tourist* MITEs reflect the association of a putative transposase from one kingdom with MITEs in another kingdom. Based on the four levels of confidence used in Table 1, this evidence would be classified as level 4. Thus, additional evidence was needed to bolster the ties between transposase source and MITEs. This could be the identification of transposase-encoding elements and related MITEs in the same genome or the identification of additional and possibly active elements related to the *Harbinger* family. Fortunately, the *PIF* transposon system of maize provided the additional evidence needed to unequivocally associate a transposase source with *Tourist*-like MITEs.

PIF is an active DNA element family first discovered as multiple mutagenic insertions into the maize *R* gene (63). Additional *PIF* elements were later isolated and characterized, including a putative autonomous element, *PIFa*, which has coding sequences related to *Harbinger*, *Turmoil*, and distantly to bacterial IS5 elements (68). The ~25 *PIF* elements in the maize genome have 14-bp TIRs and are flanked by the 3-bp TTA TSDs. Of particular interest was the finding that *PIF* is associated with a maize *Tourist*-like MITE named *mPIF*. There are many similarities between *PIF* and *mPIF* (Fig. 1) (68). First, they share identical 14-bp TIRs and similar subterminal sequences (~70% over ~100 bp at each end). In fact, the discovery of the large *mPIF* family was due to its sequence similarities to *PIF*. In addition, they both insert preferentially into the 9-bp imperfect palindrome CWCTTAGWG (W stands for either A or T), and insertion leads to duplication of the central TTA. While the extent of sequence similarity alone indicates that *mPIF* was probably derived from *PIF* or from a closely related element, their identical, extended target sites provide the strongest evidence that both elements were mobilized by the same or a related transposase (68).

PIF-Like Elements and Their MITEs in Other Organisms

The discovery of *PIF* led to the recognition of a new superfamily, *PIF/Harbinger*, with members identified thus far in plants (maize, rice, and *Arabidopsis*), nematodes (*C. elegans* and *Caenorhabditis briggsae*), and a fungus (*Fusarium neoformans*). All elements encode a putative transposase with 45 to 65% amino acid identity that is also distantly related to bacterial IS5 transposases (30, 68). Like the maize *PIF*, these *PIF*-like elements also have TIRs and TSDs that are similar to those of *Tourist*-like MITEs. In fact, once *PIF*-like elements were uncovered in the genomic sequences of rice, *Arabidopsis*, and *C. briggsae*, it was not difficult to identify their associated *Tourist*-like MITEs (Fig. 1). For example, the rice *Os-PIF2* element is associated with a MITE family called *Os-mPIF2*. Sequence similarity between *Os-PIF2* and *Os-mPIF2* encompasses the entire *Os-mPIF2* length (70 to 90% overall), and a deletion breakpoint can be clearly defined (Fig. 1). Another example is the association between the *Arabidopsis At-PIF2* element and *At-mPIF2*. These elements even share an identical mismatch in their imperfect TIRs. Finally, the *PIF*-like element in *C. briggsae*, *Cb-PIF1*, is associated with two MITE families, the longer *Cb-mPIF1a* and the shorter *Cb-mPIF1b*. As with *Os-PIF2* and *Os-mPIF2*, a clear deletion breakpoint can be defined both within *Cb-PIF1* and *Cb-mPIF1a* (Fig. 1).

OTHER MITEs AND NONAUTONOMOUS DNA TRANSPOSONS

Other MITEs

Table 1 summarizes the evidence that most of the MITE families described to date can be assigned to one of two superfamilies, Tc1/*mariner* and *PIF/Har-*

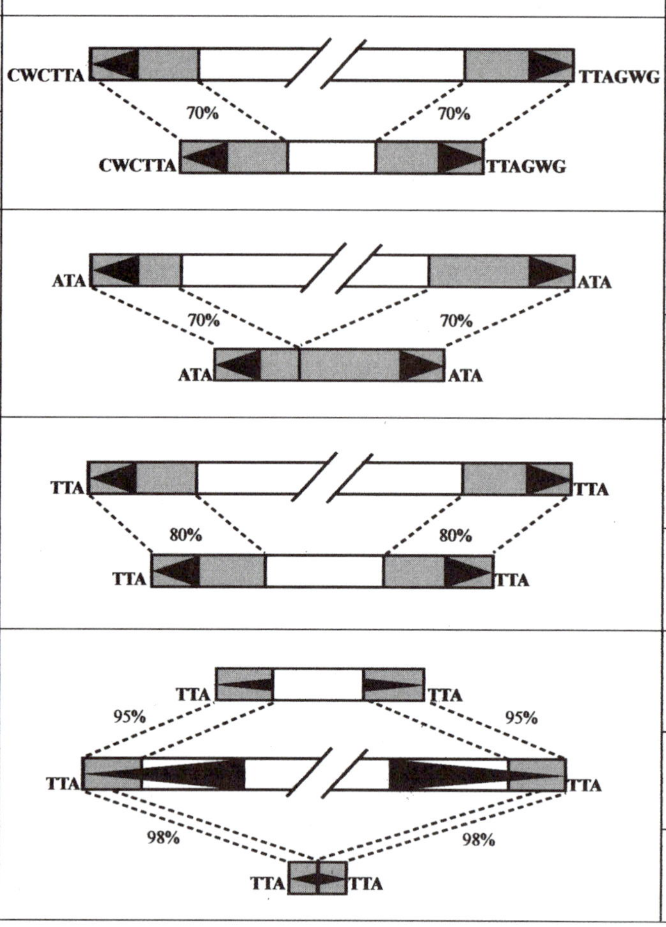

Name	Comparison between *PIFs* and *mPIFs*	Size (bp)	Copy No.	*mPIF* Similarity
Zm-PIFa	CWCTTA ... TTAGWG; 70% ... 70%	3,728	1	—
Zm-mPIF	CWCTTA ... TTAGWG	358	~6,000	90-95%
Os-PIF2	ATA ... ATA; 70% ... 70%	3,770	ND	—
Os-mPIF2	ATA ... ATA	270	~150	70-90%
At-PIF2	TTA ... TTA; 80% ... 80%	4,229	1	—
At-mPIF2	TTA ... TTA	410	~20	91-98%
Cb-PIF1	TTA ... TTA; 95% ... 95%	~2,000	<10	—
Cb-mPIF1a	TTA ... TTA; 98% ... 98%	244	~50	85-91%
Cb-mPIF1b	TTA ... TTA	60	~30	99-100%

Figure 1. Similarities between *PIF*-like elements and *Tourist*-like MITEs. Grey rectangles represent regions conserved between *PIF* elements and related *mPIF* MITES (nucleotide homology shown in percentage). Black triangles represent element TIRs.

binger. Analysis of large data sets of genome sequences harboring a vast number of MITEs, such as those of *C. elegans, Arabidopsis,* or *Oryza sativa,* has confirmed that most MITE families are related to these two superfamilies of transposases. However, in addition to these associations, a survey of mined MITEs from the rapidly expanding rice database indicates that the remaining rice MITEs are most likely to be derived from other DNA transposon superfamilies such as *hAT* (*Ac*-like) or *Mutator* (Table 1 and chapters 23 and 24). Although poorly represented among the MITEs identified so far in *Arabidopsis,* rice, and other grasses, *hAT*- and *Mutator*-related MITEs might be abundant in other plant species. For example, *Bigfoot* MITEs in the *Medicago* genus (alfalfa) are present at 10^3 to 10^4 copies per genome, and they share several structural features reminiscent of *Mutator*-like transposons such as a 9-bp TSD (10).

The large vertebrate genomes, such as those of *Xenopus,* fish, and humans, also harbor several MITE families that are probably unrelated to the two superfamilies. One group is referred to as the TTAA (or T2) superfamily and is characterized by TTAA target site duplications and a particular sequence motif in the TIRs (62). Several TTAA MITE families were identified in *Xenopus* (62), fish (25), and more recently *C. briggsae* (Feschotte, unpublished data), *C. elegans,* and humans (Repbase update [http://www.girinst.org]). Although no related elements with coding capacity for a transposase have as yet been identified in these species (except perhaps in humans; see below), the TTAA target preference may suggest a link with a newly recognized superfamily of DNA transposons called *piggyBac* (chapter 48). The founding member of this superfamily is an autonomous transposon, *piggyBac,* from the lepidopteran *Tri-*

choplusia ni (18). The *piggyBac* transposase is responsible for the specific integration into TTAA targets (18). Elements with the same target site preference and coding capacity for similar putative transposases have recently been identified in other insects and humans (50; also see chapter 48). Thus, it is tempting to connect the TTAA MITEs from *Xenopus*, fish, and nematodes with the *piggyBac*-like transposases that may reside in their respective genomes.

Finally, *hAT*-related transposases appear to be involved in the propagation of a number of animal MITEs. In humans, the MER1 group of MITEs are flanked by an 8-bp TSD and they possess TIRs similar (or identical) to those of *hAT* transposon fossils present in the same genome (50, 51). Based on similarities in TIRs and TSDs, a relationship to *hAT*-like transposons was also proposed for several *Xenopus* and mosquito MITE families (56, 62).

Several MITE families described from plants, animals, and fungi do not share any structural or sequence features with known DNA transposon families. Therefore, their classification and source of transposase remain elusive (Table 1). An intriguing example is *Microuli,* a homogeneous family of elements from the mosquito *A. aegypti* that lack TIRs but have subterminal inverted repeats (59). Elements similar in structure but not in sequence have been found in rice and in other dipteran species (4, 16, 37) (Table 1). Along with their subterminal inverted repeats, the fact that they exhibit a preference for insertion into targets of conserved length (and sequence in the case of *Microuli*) suggests that they are nonautonomous DNA elements mobilized by transposases encoded in *trans* (59).

Helitrons: Rolling-Circle Transposons in Eukaryotes

An even more puzzling group of nonautonomous transposons, designated *Helitron* transposons, was recently identified in the *A. thaliana, C. elegans,* and rice genomes (29, 32, 53, 61). Unlike MITEs, these elements have no inverted repeats and do not generate TSDs but have conserved ends and form homogeneous subfamilies with relatively high copy numbers (~2% of their genomes). It was only when the complete genome sequence of *Arabidopsis* became available that the autonomous partners could be identified, leading to the discovery of a new type of eukaryotic DNA transposons (30). Autonomous *Helitron* transposons were subsequently identified in *C. elegans,* and both autonomous and nonautonomous forms were found in the rice genome (30). The autonomous *Helitron* transposons are large elements (5.5 to 15 kb) with coding capacity for a product sharing similarities to DNA helicases and to the replicator initiator proteins of rolling-circle plasmids and certain single-stranded DNA viruses (see chapter 37). Along with other structural characteristics, these features suggest that *Helitron* transposons define a new type of transposable elements employing a rolling-circle mode of transposition (30; also see chapter 37).

A MODEL FOR THE ORIGIN OF MITEs

MITEs were discovered only 10 years ago. Until very recently, investigators wondered whether they were class 1 or class 2 elements and, if they were class 2 elements, how they were able to attain such high copy numbers. From the summary presented in this chapter, it is now evident that MITEs are nonautonomous DNA elements that originated from a subset of the existing DNA transposons. One hallmark of these transposons appears to be target site preference. Whether all DNA transposons are able to give rise to MITEs remains an open question.

The issue of MITE copy number has become more complex. The high copy numbers attributed to many MITE families may, in the majority of instances, result from independent amplifications of subfamilies in the same genome. This is illustrated best in rice, where *Stowaway* MITEs account for over 2% of genomic DNA (35). However, upon closer inspection it can be seen that there are over 30 subfamilies of *Stowaway* MITEs and none of these have attained copy numbers significantly greater than 1,000 (N. Jiang, C. Feschotte, and S. R. Wessler, unpublished data). In contrast, larger genomes (such as maize, human, and *Xenopus*) harbor very-high-copy-number MITE families. For example, there are over 6,000 copies of *mPIF* in maize that appear to have arisen from a single ancestral element (Fig. 1) (68).

A model for the origin and amplification of MITEs, based largely on the data summarized in this review, is shown in Fig. 2. According to this model, a MITE family is composed of MITE subfamilies that have arisen from related autonomous elements in a single genome. A single type of autonomous element can give rise to one or multiple MITE families or can activate nonautonomous elements derived from a related autonomous element (if, for example, that autonomous element has become inactive or is no longer in the genome). Another aspect of this model is that MITEs originate from autonomous elements like previously described (conventional) nonautonomous elements. This may result from an abortive gap repair mechanism following transposition or another transposase-dependent deletion event, such as those described for the *Drosophila P* element or the *Ac/Ds*,

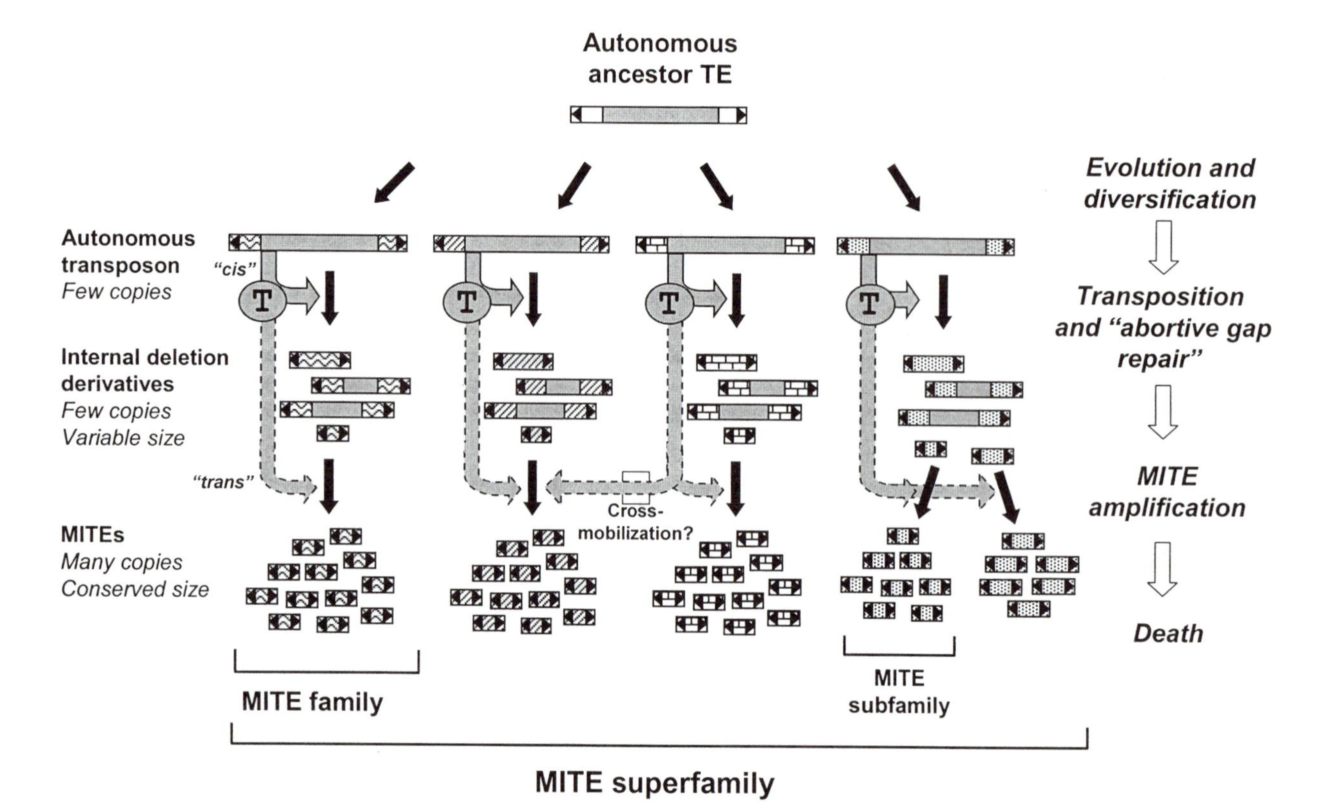

Figure 2. Model for the origin and amplification of MITEs. See text for discussion. The circled T stands for the transposase. Transposase is known to mediate the formation of nonautonomous derivatives through mechanisms such as abortive gap repair (grey arrows) (14, 21, 47, 65). The subsequent amplification of one or a few deletion derivatives (i.e., MITE amplification [dashed grey arrows]) is likely to be mediated by the same transposase or those produced by a close relative (trans- or cross-mobilization, respectively). The different patterns at the ends of the autonomous elements represent different subterminal sequences with identical or near-identical TIRs (black triangles).

Mutator, and *En/Spm* systems of maize (14, 21, 36, 47, 65; also see chapters 21 and 24). However, MITE derivatives are proposed to possess some feature(s) that allows them to be subsequently amplified to higher copy numbers than their sibling conventional nonautonomous elements (Fig. 2). Testing this aspect of the model will require in vivo and in vitro systems in which the requirements for MITE transposition can be assessed. This should now be possible since active transposases that are related to those involved in the amplification of MITEs are now available.

The final aspect of the model involves the possible impact of MITE amplification on the evolution of autonomous elements. The proliferation of nonautonomous elements has been hypothesized to lead to the extinction of the cognate autonomous element through titration of active transposase (20). In this regard, the birth and explosive amplification of MITEs could paradoxically be a death sentence for the transposase and consequently for the whole subfamily. However, selection would then lead to the diversification of the transposase by favoring variants with altered binding sites, thus ushering in a new cycle of birth and death.

Acknowledgments. We thank members of the *Zea* group at the University of Georgia for comments and insightful discussions. We are grateful to Jerzy Jurka, Vladimir Kapitonov, and Ning Jiang for sharing information and unpublished data on *C. elegans* and rice MITEs.

This work was supported by grants from the NSF Plant Genome Initiative and the NIH to S.R.W.

REFERENCES

1. **Adé, J., and F. J. Belzile.** 1999. *Hairpin* elements, the first family of foldback transposons (FTs) in *Arabidopsis thaliana. Plant J.* **19:**591–597.
2. **Besansky, N. J., O. Mukabayire, J. A. Bedell, and H. Lusz.** 1996. *Pegasus,* a small terminal inverted repeat transposable element found in the white gene of *Anopheles gambiae. Genetica* **98:**119–129.
3. **Braquart, C., V. Royer, and H. Bouhin.** 1999. DEC: a new miniature inverted-repeat transposable element from the genome of the beetle *Tenebrio molitor. Insect Mol. Biol.* **8:** 571–574.
4. **Bureau, T. E., P. C. Ronald, and S. R. Wessler.** 1996. A com-

puter-based systematic survey reveals the predominance of small inverted-repeat elements in wild-type rice genes. *Proc. Natl. Acad. Sci. USA* **93:**8524–8529.
5. **Bureau, T. E., and S. R. Wessler.** 1994. Mobile inverted-repeat elements of the Tourist family are associated with the genes of many cereal grasses. *Proc. Natl. Acad. Sci. USA* **91:** 1411–1415.
6. **Bureau, T. E., and S. R. Wessler.** 1994. *Stowaway:* a new family of inverted-repeat elements associated with genes of both monocotyledonous and dicotyledonous plants. *Plant Cell* **6:**907–916.
7. **Bureau, T. E., and S. R. Wessler.** 1992. *Tourist:* a large family of inverted-repeat element frequently associated with maize genes. *Plant Cell* **4:**1283–1294.
8. **Capy, P., C. Bazin, D. Higuet, and T. Langin.** 1998. *Dynamics and Evolution of Transposable Elements.* Springer-Verlag, Austin, Tex.
9. **Casacuberta, E., J. M. Casacuberta, P. Puigdomenech, and A. Monfort.** 1998. Presence of miniature inverted-repeat transposable elements (MITEs) in the genome of *Arabidopsis thaliana:* characterisation of the *Emigrant* family of elements. *Plant J.* **16:**79–85.
10. **Charrier, B., F. Foucher, E. Kondorsi, Y. d'Aubenton-Carafa, C. Thermes, and P. Ratet.** 1999. *Bigfoot,* a new family of MITE elements characterized from the *Medicago* genus. *Plant J.* **18:** 431–441.
11. **Chen, M. P., P. SanMiguel, A. C. de Oliveira, S. S. Woo, H. Zhang, R. A. Wing, and J. L. Bennetzen.** 1997. Microcolinearity in *sh2*-homologous regions of the maize, rice and sorghum genomes. *Proc. Natl. Acad. Sci. USA* **94:**3431–3455.
12. **Dreyfus, D. H., and S. W. Emmons.** 1991. A transposon related palindromic repetitive sequence from *C. elegans. Nucleic Acids Res.* **19:**1871–1877.
13. **Durbin, M. L., A. L. Denton, and M. T. Clegg.** 2001. Dynamics of mobile element activity in chalcone synthase loci in the common morning glory (*Ipomoea purpurea*). *Proc. Natl. Acad. Sci. USA* **98:**5084–5089.
14. **Engels, W. R., D. M. Johnson-Schlitz, W. B. Eggleston, and J. Sved.** 1990. High-frequency P element loss in *Drosophila* is homolog dependent. *Cell* **62:**515–525.
15. **Feschotte, C., and C. Mouchès.** 2000. Evidence that a family of miniature inverted-repeat transposable elements (MITEs) from the *Arabidopsis thaliana* genome has arisen from a *pogo*-like DNA transposon. *Mol. Biol. Evol.* **17:**730–737.
16. **Feschotte, C., and C. Mouchès.** 2000. Recent amplification of miniature inverted-repeat transposable elements in the vector mosquito *Culex pipiens:* characterization of the *Mimo* family. *Gene* **250:**109–116.
17. **Feschotte, C., and S. R. Wessler.** *Mariner*-like transposases are widespread and diverse in flowering plants. *Proc. Natl. Acad. Sci. USA,* in press.
18. **Fraser, M. J., T. Ciszczon, T. Elick, and C. Bauser.** 1996. Precise excision of TTAA-specific lepidopteran transposons *piggyBac* (IFP2) and *tagalong* (TFP3) from the baculovirus genome in cell lines from two species of Lepidoptera. *Insect Mol. Biol.* **5:**141–151.
19. **Gerber-Huber, S., D. Nardelli, J. A. Haefliger, D. N. Cooper, F. Givel, J. E. Germond, J. Engel, N. M. Green, and W. Wahli.** 1987. Precursor-product relationship between vitellogenin and the yolk proteins as derived from the complete sequence of a Xenopus vitellogenin gene. *Nucleic Acids Res.* **15:**4737–4760.
20. **Hartl, D. L., A. R. Lohe, and E. R. Lozovskaya.** 1997. Modern thoughts on an ancyent *marinere:* function, evolution, regulation. *Annu. Rev. Genet.* **31:**337–358.
21. **Hsia, A. P., and P. S. Schnable.** 1996. DNA sequence analyses support the role of interrupted gap repair in the origin of internal deletions of the maize transposon, *MuDR. Genetics* **142:** 603–618.
22. **Hua-Van, A., J. M. Daviere, F. Kaper, T. Langin, and M. J. Daboussi.** 2000. Genome organization in *Fusarium oxysporum:* clusters of class II transposons. *Curr. Genet.* **37:** 339–347.
23. **Hummel, S., W. Meyerhof, E. Korge, and W. Knochel.** 1984. Characterization of highly and moderately repetitive 500 bp EcoRI fragments from *Xenopus laevis. Nucleic Acids Res.* **12:** 4921–4938.
24. **Inagaki, Y., Y. Johzuka-Hisatomi, T. Mori, S. Takahashi, Y. Hayakawa, S. Peyachoknagul, Y. Ozeki, and S. Iida.** 1999. Genomic organization of the genes encoding dihydroflavonol 4-reductase for flower pigmentation in the Japanese and common morning glories. *Gene* **226:**181–188.
25. **Izsvák, Z., Z. Ivics, N. Shimoda, D. Mohn, H. Okamoto, and P. B. Hackett.** 1999. Short inverted-repeat transposable elements in teleost fish and implications for a mechanism of their amplification. *J. Mol. Evol.* **48:**13–21.
26. **Jarvik, T., and K. G. Lark.** 1998. Characterization of *Soymar1*, a *mariner* element in soybean. *Genetics* **149:**1569–1574.
27. **Jiang, N., and S. R. Wessler.** 2001. Insertion preference of maize and rice miniature inverted repeat transposable elements as revealed by the analysis of nested elements. *Plant Cell* **13:** 2553–2564.
28. **Jurka, J.** 2000. Repbase update: a database and an electronic journal of repetitive elements. *Trends Genet.* **16:**418–420.
29. **Kapitonov, V. V., and J. Jurka.** 1999. Molecular paleontology of transposable elements from Arabidopsis thaliana. *Genetica* **107:**27–37.
30. **Kapitonov, V. V., and J. Jurka.** 2001. Rolling-circle transposons in eukaryotes. *Proc. Natl. Acad. Sci. USA* **98:**8714–8719.
31. **Le, Q. H., K. Turcotte, and T. Bureau.** 2001. *Tc8,* a *Tourist*-like transposon in *Caenorhabditis elegans. Genetics* **158:** 1081–1088.
32. **Le, Q. H., S. Wright, Z. Yu, and T. Bureau.** 2000. Transposon diversity in *Arabidopsis thaliana. Proc. Natl. Acad. Sci. USA* **97:**7376–7381.
33. **Lepetit, D., S. Pasquet, M. Olive, N. Theze, and P. Thiebaud.** 2000. *Glider* and *Vision:* two new families of miniature inverted-repeat transposable elements in *Xenopus laevis* genome. *Genetica* **108:**163–169.
34. **Luckhart, S., and R. Rosenberg.** 1999. Gene structure and polymorphism of an invertebrate nitric oxide synthase gene. *Gene* **232:**25–34.
35. **Mao, L., T. C. Wood, Y. Yu, M. A. Budiman, J. Tomkins, S. Woo, M. Sasinowski, G. Presting, D. Frisch, S. Goff, R. A. Dean, and R. A. Wing.** 2000. Rice transposable elements: a survey of 73,000 sequence-tagged-connectors. *Genome Res.* **10:**982–990.
36. **Masson, P., R. Surosky, J. A. Kingsbury, and N. V. Fedoroff.** 1987. Genetic and molecular analysis of the *Spm*-dependent *a-m2* alleles of the maize a locus. *Genetics* **117:**117–137.
37. **Miller, W. J., A. Nagel, J. Bachmann, and L. Bachmann.** 2000. Evolutionary dynamics of the *SGM* transposon family in the *Drosophila obscura* species group. *Mol. Biol. Evol.* **17:** 1597–1609.
38. **Morgan, G. T.** 1995. Identification in the human genome of mobile elements spread by DNA-mediated transposition. *J. Mol. Biol.* **254:**1–5.
39. **Morgan, G. T., and K. M. Middleton.** 1990. Short interspersed repeats from Xenopus that contain multiple octamer motifs are related to known transposable elements. *Nucleic Acids Res.* **18:**5781–5786.
40. **Oosumi, T., W. R. Belknap, and B. Garlick.** 1995. *Mariner* transposons in humans. *Nature* **378:**672.

41. **Oosumi, T., B. Garlick, and W. R. Belknap.** 1995. Identification of putative nonautonomous transposable elements associated with several transposon families in *Caenorhabditis elegans*. *Proc. Natl. Acad. Sci. USA* **92:**8886–8890.
42. **Oosumi, T., B. Garlick, and W. R. Belknap.** 1996. Identification of putative nonautonomous transposable elements associated with several transposon families in *Caenorhabditis elegans*. *J. Mol. Evol.* **43:**11–18.
43. **Pozueta-Romero, J., G. Houlné, and R. Schantz.** 1996. Nonautonomous inverted repeat *Alien* transposable elements are associated with genes of both monocotyledonous and dicotyledonous plants. *Gene* **171:**147–153.
44. **Rezsohazy, R., H. G. A. M. van Luenen, R. M. Durbin, and R. H. A. Plasterk.** 1997. Tc7, a Tc1-hitch hiking transposon in *Caenorhabditis elegans*. *Nucleic Acids Res.* **25:**4048–4054.
45. **Rio, A., P. Puigdomenech, and J. M. Casacuberta.** 1996. *Mrs*, a new subfamily of *Tourist* transposable elements. *Plant Mol. Biol.* **32:**1221–1226.
46. **Robertson, H. M., and K. L. Zumpano.** 1997. Molecular evolution of an ancient *mariner* transposon, *Hsmar1*, in the human genome. *Gene* **205:**203–217.
47. **Rubin, E., and A. A. Levy.** 1997. Abortive gap repair: underlying mechanism for *Ds* element formation. *Mol. Cell. Biol.* **17:** 6294–6302.
48. **Schlubiger, J.-L., J.-E. Germond, B. ten Heggler, and W. Wahli.** 1985. The Vi element. A transposon-like repeated DNA sequence interspersed in the vitellogenin locus of Xenopus laevis. *J. Mol. Biol.* **186:**491–503.
49. **Simmen, M. W., and A. Bird.** 2000. Sequence analysis of transposable elements in the sea squirt, *Ciona intestinalis*. *Mol. Biol. Evol.* **17:**1685–1694.
50. **Smit, A. F.** 1999. Interspersed repeats and other mementos of transposable elements in mammalian genomes. *Curr. Opin. Genet. Dev.* **9:**657–663.
51. **Smit, A. F. A., and A. D. Riggs.** 1996. *Tiggers* and DNA transposon fossils in the human genome. *Proc. Natl. Acad. Sci. USA* **93:**1443–1448.
52. **Song, W.-Y., L.-Y. Pi, T. E. Bureau, and P. C. Ronald.** 1998. Identification and characterization of 14 transposon-like elements in the noncoding regions of members of the *Xa21* family of disease resistance genes in rice. *Mol. Gen. Genet.* **258:** 449–456.
53. **Surzycki, S. A., and W. R. Belknap.** 1999. Characterization of repetitive DNA elements in *Arabidopsis*. *J. Mol. Evol.* **48:** 684–691.
54. **Surzycki, S. A., and W. R. Belknap.** 2000. Repetitive-DNA elements are similarly distributed on *Caenorhabditis elegans* autosomes. *Proc. Natl. Acad. Sci. USA* **97:**245–249.
55. **Tarchini, R., P. Biddle, R. Wineland, S. Tingey, and A. Rafalski.** 2000. The complete sequence of 340 kb of DNA around the rice Adh1-adh2 region reveals interrupted colinearity with maize chromosome 4. *Plant Cell* **12:**381–391.
56. **Tu, Z.** 2001. Eight novel families of miniature inverted repeat transposable elements in the African malaria mosquito, *Anopheles gambiae*. *Proc. Natl. Acad. Sci. USA* **98:** 1699–1704.
57. **Tu, Z.** 2000. Molecular and evolutionary analysis of two divergent subfamilies of a novel miniature inverted repeat transposable element in the yellow fever mosquito, *Aedes aegypti*. *Mol. Biol. Evol.* **17:**1313–1325.
58. **Tu, Z.** 1997. Three novel families of miniature inverted-repeat transposable elements are associated with genes of the yellow fever mosquito, *Aedes aegypti*. *Proc. Natl. Acad. Sci. USA* **94:** 7475–7480.
59. **Tu, Z., and S. P. Orphanidis.** 2001. *Microuli*, a family of miniature subterminal inverted-repeat transposable elements (MSITEs): transposition without terminal inverted repeats. *Mol. Biol. Evol.* **18:**893–895.
60. **Tudor, M., M. Lobocka, M. Goodwell, J. Pettitt, and K. O'Hare.** 1992. The *pogo* transposable element family of *Drosophila melanogaster*. *Mol. Gen. Genet.* **232:**126–134.
61. **Turcotte, K., S. Srinivasan, and T. Bureau.** 2001. Survey of transposable elements from rice genomic sequences. *Plant J.* **25:**169–179.
62. **Unsal, K., and G. T. Morgan.** 1995. A novel group of families of short interspersed repetitive elements (SINEs) in Xenopus: evidence of a specific target site for DNA-mediated transposition of inverted-repeat SINEs. *J. Mol. Biol.* **248:**812–823.
63. **Walker, E. L., W. B. Eggleston, D. Demopulos, J. Kermicle, and S. L. Dellaporta.** 1997. Insertions of a novel class of transposable elements with a strong target site preference at the *r* locus of maize. *Genetics* **146:**681–693.
64. **Wessler, S. R., T. E. Bureau, and S. E. White.** 1995. LTR-retrotransposons and MITEs: important players in the evolution of plant genomes. *Curr. Opin. Genet. Dev.* **5:**814–821.
65. **Yan, X., I. M. Martinez-Ferez, S. Kavchok, and H. K. Dooner.** 1999. Origination of Ds elements from Ac elements in maize: evidence for rare repair synthesis at the site of Ac excision. *Genetics* **152:**1733–1740.
66. **Yeadon, P. J., and D. E. Catcheside.** 1995. *Guest*: a 98 bp inverted repeat transposable element in *Neurospora crassa*. *Mol. Gen. Genet.* **247:**105–109.
67. **Zhang, Q., J. Arbuckle, and S. R. Wessler.** 2000. Recent, extensive and preferential insertion of members of the miniature inverted-repeat transposable element family *Heartbreaker* (*Hbr*) into genic regions of maize. *Proc. Natl. Acad. Sci. USA* **97:**1160–1165.
68. **Zhang, X., C. Feschotte, Q. Zhang, N. Jiang, W. B. Eggleston, and S. R. Wessler.** 2001. *P Instability Factor*: an active maize transposon system associated with the amplification of *Tourist*-like MITEs and a new superfamily of transposases. *Proc. Natl. Acad. Sci. USA* **98:**12572–12577.

INDEX

Note: Page numbers followed by "f" indicate figures; page numbers followed by "t" indicate tables.